DESK
REFERENCE
FOR

HEMATOLOGY

SECOND EDITION

DESK REFERENCE FOR

HEMATOLOGY

SECOND EDITION

EDITED BY
N.K. SHINTON

CRC Press
Taylor & Francis Group
Boca Raton   London   New York

CRC Press is an imprint of the
Taylor & Francis Group, an **informa** business

CRC Press
Taylor & Francis Group
6000 Broken Sound Parkway NW, Suite 300
Boca Raton, FL 33487-2742

© 2008 by Taylor & Francis Group, LLC
CRC Press is an imprint of Taylor & Francis Group, an Informa business

International Standard Book Number-13: 978-0-8493-3393-4 (Hardcover)

Library of Congress Cataloging-in-Publication Data

Desk reference for hematology / edited by N.K. Shinton. -- 2nd ed.
    p. ; cm.
  Rev. ed. of: CRC desk reference for hematology. c1998.
  Includes bibliographical references and index.
  ISBN 0-8493-3393-8 (alk. paper)
  1. Hematology--Handbooks, manuals, etc. 2. Blood--Diseases--Handbooks, manuals, etc. 3.
Blood--Handbooks, manuals, etc. I. Shinton, N. K. (N. Keith) II. CRC desk reference for hematology.
    [DNLM: 1. Hematologic Diseases--Handbooks. 2. Hematology--Handbooks. WH 39 D459 2006]

RB145.C69 2006
616.1'5--dc22
                                                   2006045568

Visit the Taylor & Francis Web site at
http://www.taylorandfrancis.com

and the CRC Press Web site at
http://www.crcpress.com

Notice

The indications and dosage of all drugs in this book are those recommended in the medical literature and in conformance with general medical practice. The drugs do not necessarily have specific approval by the U.S. Food and Drug Administration or of the European Pharmaceutical Commission, either for use in the disorders or in the dosages recommended. The package insert for each drug should be consulted for indications, contraindications, dosage, and adverse drug reactions, particularly in respect to weight and age in children. It is advisable for those using a specific drug to be aware of any revised recommendations, especially when prescribing drugs that have recently become available.

Preface

The intention of the publishers, editor, and authors of this book is to provide a major source of easily obtained, reliable information on hematology in one book, with specific topics arranged in alphabetical order. As this is mainly based on well-established knowledge, it has not been closely referenced. The references provided here have been confined to guidelines, reviews, and recently published articles on topics that are either controversial or for which opinion has yet to be consolidated. In addition, there is a bibliography of recommended further reading.

The information has been made concise by the use of extensive cross-referencing within the book, as indicated by words printed in boldface. Words thus highlighted within an article indicate that further information on the subject is available in a separate article identified by the emboldened words.

Common abbreviations are listed in a separate table at the front of the book and are widely identified throughout the text. Acronyms for combination therapies are listed under cytotoxic agents. Manuscript references are superscript numbers; in-text references are numbers in brackets [1].

The nomenclature regarding hematopoietic and lymphoid tissue is that used in the World Health Organization (WHO) Classification of Tumors of Hematopoietic and Lymphoid Tissues, a summary of which is included as an Appendix. Color photographs of cells or tissues have not been included, as these are well presented in many easily obtainable publications. Likewise, details of laboratory technical procedures are not included.

It is my great pleasure to acknowledge the help that I have received from the authors and from colleagues, in particular Dr. R. A. Shinton, senior lecturer in medicine, University of Birmingham, Birmingham, U.K., for reading the manuscript and Mrs. Anne Caine, administrator, Division of Clinical Sciences, Warwick Medical School, University of Warwick, Coventry, U.K., for help in data processing.

N. K. Shinton
Editor

Contributors

James Barton, M.D.
Southern Hematology and Iron Disorders
  Center
Birmingham, Alabama

**Rajesh Chopra, Ph.D., F.R.C.P., F.R.C.
Path.**
University College Hospital Medical
  School
London, U.K.

**Christopher D. Fegan, M.D., F.R.C.P.,
  F.R.C.Path.**
Llandaff Hospital
Cardiff, Wales, U.K.

**Edward C. Gordon-Smith, M.A., M.Sc.,
  M.B., F.R.C.P., F.R.C.Path.**
Department of Haematology
St. George's Hospital Medical School
London, U.K.

Thomas M. Habermann, M.D.
Department of Hematology and Internal
  Medicine
Mayo Clinic
Rochester, Minnesota

Robert A. Kyle, M.D.
Department of Hematology and Internal
  Medicine
Mayo Clinic
Rochester, Minnesota

**Edwin Massey, M.B., F.R.C.P.,
  F.R.C.Path.**
National Blood Service
Bristol, U.K.

Alan Morris, B.A., D.Phil.
Department of Biological Sciences
University of Warwick
Coventry, U.K.

Miguel Ortin, M.D., Ph.D., M.R.C. Path.
St. George's Hospital
London, U.K.

**K. John Pasi, M.B., Ph.D., F.R.C.P.,
  F.R.C.Path., F.R.C.P.C.H.**
Department of Haematology
The Royal London Hospital
London, U.K.

Geoffrey D. Poole, M.Sc., F.I.B.M.S
National Blood Service
Bristol, U.K.

**R. Martin Rowan (deceased), F.R.C.P.(G),
  F.R.C.P.(E), F.R.C.Path.**
TOA Medical Electronics (Europe)
and
University of Glasgow
Glasgow, Scotland, U.K.

**N. Keith Shinton, M.D., F.R.C.P.,
  F.R.C.Path.**
Clinical Sciences, Research Institute
Warwick Medical School
University of Warwick
Coventry, U.K.

Jerry Spivak, M.D.
Division of Haematology
Johns Hopkins University Medical School
Baltimore Maryland

List of Comprehensive Entries

List of Illustrations

List of Tables

Reference Range Tables

Abbreviations

AA	aplastic anemia
ABMT	autologous bone marrow transplantation
ABT	autologous blood transfusion
AChE	acetylcholinesterase
aCL	anticardiolipin antibody
ACLE	acetylcholinesterase
aCML	atypical chronic myeloid leukemia
ADA	adenosine deaminase
ADCC	antigen-dependent cellular cytotoxicity
ADP	adenosine diphosphate
ADR	adverse drug reaction
AERA	acute event-related anemia
AIDS	acquired immune deficiency syndrome
AIHA	autoimmune hemolytic anemia
AILD	angioimmunoblastic lymphadenopathy with dysproteinemia
AIP	acute intermittent porphyria
AK	adenylate kinase
ALA	δ-aminolevulinic acid
ALCL	anaplastic large-cell lymphoma
ALG	antilymphocyte globulin
ALL	acute lymphoblastic leukemia
ALPS	autoimmune lymphoproliferative syndrome
ALT	alanine aminotransferase (transaminase)
AML	acute myeloid leukemia; acute myeloblastic leukemia
AMML Eo	abnormal eosinophil component
AMP	adenosine monophosphate
ANA/ANF	antinuclear factor
ANCA	antineutrophilic cytoplasmic antibodies
α_2-AP	antiplasmin
APC	antigen-presenting cell; activated protein C
APCr	activated protein C resistance
aPL	antiphospholipid antibody
APL	acute promyelocytic leukemia
APRT	adenine phosphoribosyltransferase (transaminase)
APS	antiphospholipid antibody syndrome
APSAC	acylated plasma streptokinase activator complex
APTT	activated partial thromboplastin time
ARAM	antigen-recognizing activation motif
ARC	AIDS-related complex
ASO	antistreptolysin titer
AST	aspartate aminotransferase (transaminase)

AT/AT-III	antithrombin-III
ATG	antithymocyte globulin
ATLL	adult T-cell lymphoma/leukemia
ATP	adenosine triphosphate
B-CLL	B-cell chronic lymphatic lymphoma/leukemia
B-CR	B-cell receptor
BCR	break-point cluster region (genetics)
BDNF	brain-derived neurotropic factor
BFU	burst-forming cell
BFU-E	burst-forming unit-erythroid
B-LBL	B-lymphoblastic lymphoma/leukemia
BM	bone marrow
BMN	bone marrow necrosis
BMP	bone morphogenetic protein
BMSC	bone marrow stem cells
BMT	bone marrow transplantation
BPA	burst-promoting activity
B-PLL	B-prolymphocytic lymphoma/leukemia
BSE	bovine spongiform encephalopathy
B-SLL	B-small lymphocytic lymphoma/leukemia
c	cyclic
C	complement
C-ALL	common acute lymphoblastic leukemia
CAM	cellular adhesion molecules
cAMP	cyclic adenosine monophosphate
CARS	compensating anti-inflammatory response syndrome
CBG	cortisol bonding globulin
CCI	corrected count increment
CDA	congenital dyserythropoietic anemia; 2′-chlorodeoxyadenosine
CDK	cyclin-dependent kinases
cDNA	complementary DNA
CEL	chronic eosinophilic leukemia
CFC	colony forming cell
CFI	complement factor inhibitor
CFU	colony forming unit
CFU-A	colony forming unit-type A
CFU-Bl	colony forming unit-blasts
CFU-C	colony forming unit-culture
CFU-E	colony forming unit-erythrocytes
CFU-GEMM	colony forming unit-granulocytes/erythroid/macrophages/ megakaryocytes
CFU-GM	colony forming unit-granulocytes/macrophages
CFU-L	colony forming unit-lymphocytes
CFU-Meg	colony-forming unit-megakaryocyte
CFU-S	colony forming unit-spleen
CGH	comparative genomic hybridization
CGL	chronic granulocytic leukemia

CHAD	cold hemolytic anemia disease
CHBHA	congenital Heinz body hemolytic anemia
CHOP	cyclophosphamide, adriamycin, vincristine, and prednisone or prednisolone
CIMF	chronic idiopathic myelofibrosis
CJD	Creutzfeldt-Jacob disease
CKI	cyclin kinase inhibitor
CKR	chemokine receptor
CLA	cutaneous lymphocyte antigen
CLEVER	common lymphatic and endothelial vascular endothelial receptor
CLL	chronic lymphocytic leukemia
CLN	cyclins
CLP	common lymphoid progenitor
CMI	cell-mediated immunity
CML	chronic myeloid leukemia; chronic myelogenous leukemia
CMML	chronic myelomonocytic leukemia
cMP	cyclic guanosine monophosphate
CMV	cytomegalovirus
CNL	chronic neutrophilic leukemia
CNS	central nervous system
CNSHA	congenital nonspherocytic hemolytic anemia
CNTF	ciliary neurotrophic factor
CNTFR	ciliary neurotrophic factor receptor
conA	concanavalin
CR	complete remission
CRM	cross-reactive material
CRP	C-reactive protein
CSA	colony stimulating activity
CSF	colony stimulating factor; cerebro-spinal fluid
CSHH	congenital self-healing histiocytosis
CT	computed tomography
CT-1	cardiotrophin-1
CTL	cytotoxic T-lymphocyte
CTLP	cytotoxic T-lymphocyte precursor
CV	coefficient of variation
CVID	common variable immunodeficiency
CYS	cysteine rich
DAF	decay-accelerating factor
DAG	diacyl glycerol
DAT	direct antiglobulin (Coombs) test
DAVP	synthetic analogue of vasopressin
DCF	2'deoxycoformycin
DDAVP	1-desamino-8-d-arginine vasopressin
DDP	double donor product
DHF	dengue hemorrhagic fever
DIC	disseminated intravascular coagulation
DILS	diffuse infiltrative lymphocytosis syndrome

DLBCL	diffuse large B-cell lymphoma
DNA	deoxyribonucleic acid
DPG	diphosphoglycerate
2,3-DPG	2,3-diphosphoglycerate
DPGM	diphosphoglyceromutase
DPGP	diphosphoglycerophosphatase
DRVVT	diluted Russell viper venom test
DS	Down syndrome
DTH	delayed type hypersensitivity
dU	deoxyuridine
DVT	deep-vein thrombosis
EACA/EAPA	epsilon-aminocaproic acid
EBV	Epstein-Barr virus
ECLT	euglobulin clot lysis time
ECM	extracellular matrix
EDRF	endothelial-derived relaxing factor
EDTA	ethylenediamine tetraacetic acid
ELISA	enzyme-linked immunosorbent assay
EM	electron microscopy
EPO	erythropoietin
EPOR	erythropoietin receptor
ER	endoplasmic reticulum
ESR	erythrocyte sedimentation rate
ET	essential thrombocythemia
ETP	endogenous thrombin potential
FAB	French-American-British (classification of leukemia)
Fc	constant portion of heavy chain of immunoglobulin
FCAS	familial cold-associated autoinflammatory disorder
FCL	follicle center lymphoma
FcR	Fc receptor
FDC	follicle dendritic cells
FDP	fibrin/fibrinogen degradation products
FEP	free erythrocyte protoporphyrin
FFP	fresh-frozen plasma
FGF	fibroblast growth factor
FHH	familial hemophagocytic histiocytosis
FIGLU	formiminoglutamic acid
FISH	fluorescence *in situ* hybridization
FITC	fluorescein isothiocyanate
FMAIT	fetomaternal alloimmune thrombocytopenic purpura
FMLP	formyl-methionyl-leucyl-phenylalanine
FpA/FpB	fibrinopeptides A and B
FT	farnesyl transferase
FTI	farnesyl transferase inhibitor
rFVIII;c	recombinant factor VIII coagulation
G6PD	glucose-6-phosphate dehydrogenase
G-CSF	granulocyte colony stimulating factor

G-CSFR	granulocyte colony stimulating factor receptor
GDP	guanosine 5'-diphosphate
GFR	growth factor receptor
GHR	growth hormone receptor
GI	gastrointestinal tract
GM-CSF	granulocyte/macrophage colony stimulating factor
GP/Gp	glycoprotein
GPI	glycosylphosphotidylinositol
GSH	glutathione reduced form
GSSG	glutathione, oxidized form
GTP	guanosine 5'-triphosphate
GTPase	guanosine triphosphatase
GVHD	graft-versus-host disease
GVL	graft-versus-leukemia
Gy	Gray = 100 rads
HAART	highly active antiretroviral therapy
HAE	hereditary angioedema
HAS	human albumin solution
HAT/HIT	heparin-associated/induced thrombocytopenia
Hb	hemoglobin
HbA	adult hemoglobin
HbCO	carboxyhemoglobin
HbF	fetal hemoglobin
HbO_2	oxyhemoglobin
HBV	hepatitis B virus
HCD	heavy-chain disease
HCII	heparin cofactor II
HCL	hairy cell lymphoma/leukemia
Hct	hematocrit
HCV	hepatitis C virus
HD	Hodgkin disease
HDN	hemolytic disease of the newborn
HE	hereditary elliptocytosis
HELLP	hemolysis, elevated liver enzymes, low platelets, and pregnancy
HEMPAS	hereditary erythroblastic multinuclearity associated with a positive acidified serum test
HES	hypereosinophilic syndrome
HGF	hematopoietic growth factor
HGFR	hematopoietic growth factor receptor
HGPRT	hypoxanthine-guanine phosphoribosyltransferase
HHT	hereditary hemorrhagic telangiectasia
HHV	human herpes virus
HIT	heparin-induced thrombocytopenia
HIV	human immunodeficiency virus
HK	hexokinase
HLA	human leukocyte antigen
HMWK	high-molecular-weight kininogen

HPA	human platelet antigen
HPC	hematopoietic progenitor cell
HPFH	hereditary persistent fetal hemoglobin
HPLC	high-pressure liquid chromatography
HPP	hereditary infantile pyropoikilocytosis
HPP-CFC	high-proliferative-potential colony forming cell
HPRT	hypoxanthine ribosyltransferase
HPS	hematopoietic stem cell
HRG	histidine-rich glycoprotein
HS	hereditary spherocytosis; hemoglobin solution
HSA	hereditary sideroblastic anemia
HSC	hematopoietic stem cell
HSCT	hematopoietic stem cell transplantation
HSV	herpes simplex virus
5-HT	5-hydroxytryptamine
HTLV	human T-cell leukemia virus
HUMARA	human androgen receptor
HUS	hemolytic uremic syndrome
HVG	host versus graft
IBS	intraoperative blood salvage
ICAM	immunoglobulin cell adhesion molecule
ICSH	International Council for Standardization in Hematology
IDAT	indirect antiglobulin (Coombs) test
IE	intestinal enteropathy
IF	intrinsic factor
IFN	interferon
IgG, IgA, etc.	immunoglobulins
IL	interleukins
IMF	idiopathic myelofibrosis
IMIg	intramuscular immunoglobulin
INN	international nonproprietary names
INR	international normalized ratio (for prothrombin time)
IP	inositol phosphate
IPG	impedance plethysmography
IR	ionizing radiation
IRE	iron-responsive elements
IRP	iron-regulating protein
IPSID	immunoproliferative small intestinal disease
IS	immunosuppression
ISBT	International Society of Blood Transfusion
ISI	international sensitivity index
ISTH	International Society of Thrombosis and Hemostasis
ITP	immune thrombocytopenic purpura
IUT	intrauterine (blood) transfusion
IVC	inferior vena cava
IVH	intravascular hemolysis
IVIg	intravascular immunoglobulin

IVT	intravascular fetal transfusion
J-CML	juvenile chronic myelocytic leukemia
KGF	keratinocyte growth factor
KS	Kaposi's sarcoma
LA	lupus anticoagulant
LAD	leukocyte adhesion deficiency
LAK	lymphokine-activated killer (cell)
LAP	leukocyte alkaline phosphatase
LCAT	lecithin cholesterol acyltransferase
LDH	lactate dehydrogenase
LDL	low-density lipoprotein
LE	lupus erythematosus (cell)
Lf	lactoferrin
LGL	large granular lymphocytes
LIF	leukocyte migration inhibiting factor
LIFR	leukocyte migration inhibiting factor receptor
LISS	low-ionic-strength saline
LMP	large multimeric proteases
LMWH	low-molecular-weight heparin
LPS	lipopolysaccharide
LRS	lymphoreticular system
LT	lymphotoxin
LTBMC	long-term bone marrow culture
LU/Lu	Lutheran (blood group)
MAB	monoclonal antibodies
MAC	membrane attack complex
MAHA	microangiopathic hemolytic anemia
MALT	mucosa-associated lymphocytic tumor
MCF	mean corpuscular fragility
MCH	mean corpuscular hemoglobin
MCHC	mean corpuscular hemoglobin concentration
M-CSF	macrophage colony stimulating factor
MCV	mean cell volume
MDS	myelodysplasia; myelodysplastic syndrome
MF	mycosis fungoides
MGUS	monoclonal gammopathy of undetermined significance
MHC	major histocompatibility complex
MI	myocardial infarction
MIF	migration inhibition factor
mIg	membrane immunoglobulin
MIP	macrophage inflammatory protein
MLA	mucosa-specific lymphocyte antigen
MLC	mixed lymphocyte culture
MM	multiple myeloma
MNC	mononuclear cell
MPO	myeloperoxide
MPV	mean platelet volume

MRI	magnetic resonance irradiation
mRNA	messenger ribonucleic acid
MTP	microsomal triglyceride transfer protein
MUD	matched unrelated donor
MW	molecular weight
NADH	nicotinamide adenine dinucleotide
NADP	nicotinamide adenine dinucleotide phosphate
NADPH	nicotinamide adenine dinucleotide phosphate reduced
NAIT	neonatal alloimmune thrombocytopenia
NAP	neutrophil alkaline phosphatase
NAPTT	nonactivated partial thromboplastin time
NBT	nitroblue tetrazolium test
NCI	nuclear contour index
NGF	nerve growth factor
NHFTR	nonhemolytic febrile transfusion reaction
NHL	non-Hodgkin lymphoma
NK	natural killer (cells)
NOD	nucleotide-binding oligomerization domain protein
NRBC	nucleated red blood cell
NRTi	nonnucleoside reverse transcriptase inhibitor
NSAID	nonsteroidal anti-inflammatory drugs
NSE	nonspecific esterase
NT	neurotrophin
OPSI	overwhelming postsplenectomy infection
OSM	oncostatin-M
PAI	plasminogen activator inhibitor
PAS	para-aminosalicylic acid; periodic acid-Schiff
PB	peripheral blood
PBSC	peripheral blood stem cell
PBSCT	peripheral blood stem cell transplantation
PCC	prothrombin complex concentration
PCH	paroxysmal cold hemoglobinuria
PCI	protein C inhibitor; percutaneous coronary artery intervention
PCR	polymerase chain reaction
PCR-SSCP	polymerase chain reaction-single stranded conformational polymorphisms
PCT	plateletcrit
PCV	packed-cell volume (hematocrit)
PDGF	platelet-derived growth factor
PDW	platelet distribution width
PE	pulmonary embolism
PET	positive emission tomography
PEX	plasma exchange
PF3	platelet factor 3
PF4	platelet factor 4
PFC	perfluorochemicals
PFGE	pulse-field gel electrophoresis

PG	prostaglandin (e.g., PGE_1)
PGI2	prostacyclin
PGK	phosphoglycerate kinase
Ph'	Philadelphia chromosome
pH	Reciprocal of logarithm$_{10}$ of hydrogen ion concentration
PHA	phytohemagglutinin
PI	protease inhibitor
PIE	pulmonary infiltrates with eosinophilia
PIP	phosphotidyl inositol
PIT	plasma iron turnover
PIVKA	protein-induced vitamin K absence (or antagonist)
PK	pyruvate kinase
PKC	protein kinase-C
PlGF	placenta growth factor
PLC	phospholipase C
PLT	platelet
PNH	paroxysmal nocturnal hemoglobinuria
PNP	purine nucleoside phosphorylase
POEMS	polyneuropathy, organomegaly, endocrinopathy, monoclonal gammopathy, and skin changes
PPP	platelet-poor plasma
PRCA	pure red cell aplasia
PRM	pathogen recognition molecules
PRP	platelet-rich plasma
PRV	polycythemia rubra vera
P/S	plastic/stroma
PT	prothrombin time
PTLD	posttransplant lymphoproliferative disorder
PTP	posttransfusion purpura
PTTK	partial thromboplastin time with kaolin
PUVA	psoralen and ultraviolet-A radiation therapy
PVP	polyvinylpyrolidine
PWM	pokeweed mitogen
RA	refractory anemia
RAEB	refractory anemia with excess blasts
RBC	red blood cell
RCA	regulator of complement activation
RCM	red blood cell mass
RCV	red blood cell volume
RDW	red blood cell distribution width
RF	rheumatoid factor
RFLP	restriction fragment length polymorphism
rFVIII;c	recombinant factor VIII coagulation
RhD	Rh blood group D
RIA	radioimmunoassay
RNA	ribonucleic acid
mRNA	messenger ribonucleic acid

RNI	reactive nitrogen intermediate
RSC	Reed-Sternberg cell
RT	reverse transcriptase
RTK	receptor tyrosine kinase
RVV	Russell viper venom
SAG-M	saline, adenine, glucose, and mannitol
SC	stem cell
SCD	sickle cell disorder/disease
SCF	stem cell factor
SCID	severe combined immunodeficiency
SCN	severe congenital neutropenia
SCT	stem cell transplantation
SDS	Schwachman-Diamond syndrome
SERPINS	serine protease inhibitors
SH-	sulfhydryl groups
SIRS	systemic inflammatory response syndrome
SK	streptokinase
SKY	spectral karyotyping
SLE	systemic lupus erythematosus
SLT	sucrose lysis test
SPD	storage pool disease
S-phase	cells in DNA-synthesis phase of cycle
SS	homozygote sickle cell disease
S/T	serine/threonine
STATS	signal transducers and activators of transcription
SVC	superior vena cava
TAA	tumor-associated antigen
TAFI	thrombin-activatable fibrinolysis inhibitor
TA-GVHD	transfusion-associated graft-versus-host disease
TAP	transporters in antigen presentation; T-cell antigen presentation
TAR	thrombocytopenia with absent radii
TBG	thyroxin-binding globulin
TBI	total body irradiation
Tc	T-lymphocyte-cytotoxic
TC	transcobalamin
TC-I, TC-II, TC-III	transcobalamin I, II, III
TCLL	T-cell chronic lymphatic lymphoma/leukemia
TCR	T-cell receptor
TdT	terminal deoxynucleotidyl transferase
TEC	transient erythroblastopenia of childhood
TEG	thromboelastograph
Tf	transferrin
TFPI	tissue factor pathway inhibitor
TGF	transforming growth factor
Th/TH/T_H	T-lymphocyte helper cell
THFA	tetrahydrofolic acid
TIBC	total iron-binding capacity

TIL	tumor-infiltrating lymphocytes
TK	tyrosine kinase
T-LBL	T-lymphoblastic lymphoma/leukemia
Tm	thrombomodulin
TMA	thrombotic microangiopathy
TMD	transient myeloproliferative disorder
tMDS	therapy-related myelodysplasia
TNF	tumor-necrosis factor
TNFR	tumor-necrosis-factor receptor
t-PA	tissue plasminogen activator
TPH	transplacental hemorrhage
TPO/Tpo	thrombopoietin
TRALI	transfusion-related acute lung injury
TRAP	tartrate-resistant acid phosphatase
TSA	tumor-specific antigen
TSP	thrombospondin
TT	thrombin time
TTP	thrombotic thrombocytopenic purpura
T_xA_2	thromboxane A_2
UFH	unfractionated heparin
UIBC	unsaturated iron-binding capacity
UK	urokinase
U/S	ultrasonic
VAHS	virus-associated hemangiophagocytic syndrome
VCAM	vascular cell adhesion molecule
VEGF	vascular endothelial growth factor
VTE	venous thromboembolism
VUD	volunteer unrelated donor
VWD	Von Willebrand Disease
VWF	Von Willebrand's factor
WAS	Wiscott-Aldrich syndrome
WBC	white blood cells
WHO	World Health Organization
WNV	West Nile virus
XIC	X-inactivation center
XIST	X-inactive specific transcript

Contents

A

ABC7

A member of a large protein family of **adenosine triphosphate** (ATP)-binding cassette (ABC) transporters probably involved in metal homeostasis that is extremely conserved across evolution from bacteria to humans. ABC7 is a half-transporter involved in the transport of **heme** from the mitochondria to the cytosol. The ABC7 gene is located on Xq12-q13. Allelic variants have been detected in hemizygous males with **sideroblastic anemia** and spinocerebellar ataxia.

ABCIXIMAB

A **monoclonal antibody** that binds to **glycoprotein IIb/IIIa** (GpIIb/IIIa) on **platelets** and consequently inhibits platelet function. It is a Fab fragment of a human–mouse chimeric monoclonal antibody. It has a very high affinity for (but a slow dissociation rate from) the GpIIb/IIIa. Although it has a short half-life of 10 to 30 min, it has a long biological half-life due to its strong affinity to the GpIIb/IIIa receptor, which remains bound in circulation for up to 15 days. However, this prolonged presence has minimal residual activity, and platelet function returns to baseline 12 to 36 h after therapy. More than 80% of GpIIb/IIIa must be blocked to inhibit platelet function. It has provided robust, consistent, and highly significant reductions in death or myocardial infarction in several large protein C inhibitor (PCI) trials; it is a preferred drug used during coronary interventions such as percutaneous transluminal stenting. **Adverse drug reactions** are those associated with **hemorrhage** and **hypersensitivity**. It should not be given to patients receiving drugs for activation of **fibrinolysis**.

ABETALIPOPROTEINEMIA

(Bassen-Kornweig syndrome) An autosomally recessive disorder caused by mutations of the microsomal triglyceride transfer protein (MTP), resulting in defective synthesis of apoprotein B-100 in the liver and apoprotein B-48 in the interstitial cells. Lipids, principally triglyceride, cholesterol, and cholesterol esters, are normally surrounded by a stabilizing coat of phospholipid. Apoproteins are embedded in the surface of the complex of lipids and phospholipids to form lipoproteins, which in this condition cannot be synthesized. Low-density lipoproteins, such as chylomicrons, are absent from serum, leading to defective mobilization of triglycerides from the intestines and liver. Plasma cholesterol and phospholipids are reduced, and the enzyme lecithin cholesterol acyltransferase (LCAT) is greatly reduced. Triglycerides accumulate in the absorptive cells of the small intestine, resulting in malabsorption in infancy with failure to develop. Fat intolerance and fat-soluble vitamin deficiency result in progressive cerebellar ataxia with peripheral neuropathy and retinitis pigmentosa. **Acanthocytes** are seen in the peripheral blood, their shortened red blood cell survival giving rise to a mild chronic **hemolytic anemia**. No specific treatment is available.

ABO(H) BLOOD GROUPS

Blood groups arising from two specific antigen–antibody systems located on **red blood cells** (RBCs) and **platelets**.

Biochemistry

The ABH antigens are carbohydrates, the A and B determinants being found at the termini of the oligosaccharide chains of glycoproteins and glycolipids.

Most ABH antigens on RBCs are glycoproteins, the carbohydrate portion usually being linked to protein, e.g., the anion transport protein (AE1 or Band 3). Because the ABH blood group antigens are not localized to a particular protein, the antigens are also found widely distributed in body tissues and occur in soluble form in secretions. The presence or absence of ABH substances in saliva determines the "secretor" status of the person (see **Lewis blood groups**).

Genetics and Phenotypes

The ABH antigens of RBCs are determined by genes belonging to the ABO and Hh blood group systems. Blood group genes in these systems specify glycosyl transferases that catalyze the transfer of monosaccharides from nucleotide sugars onto carbohydrate precursors. The product of the *H* gene is an alpha 1,2-L-fucosyl transferase. The H antigen that is produced is the biosynthetic precursor of the A and B antigens: the *A* gene specifies an alpha 1,3-N-acetyl-D-galactosaminyltransferase, and the *B* gene specifies an (alpha)1,3-D-galactosyltransferase (see Figure 1).

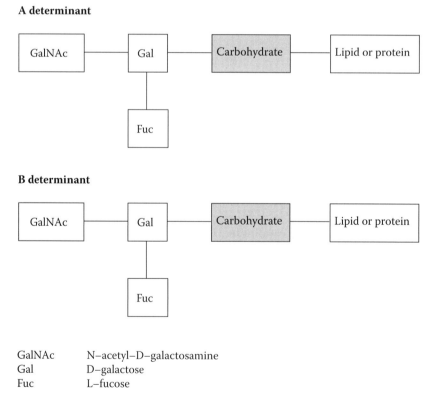

GalNAc	N–acetyl–D–galactosamine
Gal	D–galactose
Fuc	L–fucose

FIGURE 1
Diagrammatic representation of ABH blood group biochemistry.

TABLE 1

Frequencies of the Major ABO Blood Groups

Phenotype	Genotype	Frequency in Caucasians (%)	ABO Antibody Regularly Present in Plasma
O	OO	47	anti-A and anti-B
A	AA or AO	42	anti-B
B	BB or BO	9	anti-A
AB	AB	3	none

The *O* gene is a common allele at the ABO locus on chromosome 9 and does not result in the production of A or B antigens. A number of alleles at the ABO locus result in a weakened or altered expression of A and B antigens. For example, RBCs of the A3 and Ax phenotypes react with some anti-A antibodies, but not with others. Weak **phenotypes** may also have a nongenetic cause, e.g., old age and disease (particularly leukemia). Table 1 shows the incidence of the major ABO phenotypes. These frequencies vary significantly between different ethnic groups; for example, the incidence of blood group B is much higher in Africans and in Vietnamese. ABO polymorphisms have been claimed to have putative relationships with disease.[1]

A very rare allele at the H locus on chromosome 19 is the *h* gene, which results in no expression of H, and since H is the precursor for A and B, people with the genotype hh have no A, B, or H antigens on their RBCs. These people are said to have the "Bombay" (Oh) null phenotype, because of its relatively high frequency in Bombay, India.

Antibodies and Their Clinical Significance

Alloantibodies to the A, B, and H antigens are naturally occurring. They are found in the plasma of people who lack the corresponding antigen and who have had no known RBC stimulus, e.g., transfusion or pregnancy (see Table 1). Exposure to widely distributed bacteria carrying ABH-determinant structures accounts for the production of anti-A and anti-B in infants. Maternal ABO antibodies are often found in the plasma of newborns.

Anti-A and anti-B (and anti-H found in Bombay persons) are often active at 37°C and bind complement, and may give rise to severe hemolytic transfusion reactions if incompatible RBCs are transfused (see **Red blood transfusion**). ABO antibodies may cause **hemolytic disease of the newborn**, but this is rarely severe, because:

ABO antigens are not fully developed on fetal RBCs.

ABO antigens are present on tissues other than RBCs, allowing neutralization of the antibody to occur.

Maternal ABO antibodies may be partly or wholly IgM.

Weak examples of anti-A, anti-H, and, occasionally, anti-B are sometimes found in the plasma of persons who have weak or altered expressions of the relevant antigens. For example, anti-A1 may be found in people with the phenotypes A2B and Ax. These antibodies are rarely of clinical significance, but may cause problems in **pretransfusion testing**.

ABO Blood Grouping

Determination of ABO groups is an essential requirement of pretransfusion testing. Anti-A and anti-B reagents are used to define the antigens on the patient's RBCs, and the

TABLE 2

Laboratory Determination of ABO Blood Groups

Reagent	Patient's Blood Group				Description
	A	B	O	AB	
anti-A	+	–	–	+	Cell or forward group
anti-B	–	+	–	+	(reagent antibody tested against patient's RBCs)
A cells	–	+	+	–	Serum or reverse group
B-cells	+	–	+	–	(patient's serum or plasma tested against reagent RBCs)
O cells	–	–	–	–	—

presence or absence of naturally occurring anti-A and anti-B is determined by testing the patient's serum or plasma against reagent RBCs (see Table 2).

Ideally, donor blood that is of the identical ABO group as the recipient should be selected for transfusion (see **Pretransfusion testing**). It is usually acceptable to select blood that is compatible with actual or potential ABO antibodies in the recipient, e.g., group B RBCs can be selected for a group AB recipient. Group O blood is used in emergency situations when the recipient's group is not known, but it should be recognized that alloantibodies other than anti-A and anti-B may be present. If group O blood is selected for patients of other ABO blood groups (particularly if the recipient is less than 5 years old and is receiving repeated transfusions), then the blood should be plasma depleted or should be screened to ensure that high-titer anti-A/anti-B is not present.

ACANTHOCYTE

(Spur cells) Abnormal **red blood cells** (RBCs) characterized by having 2 to 20 irregularly placed spicules or thornlike projections of unequal length. These spicules are caused by changes in the lipid content of the red blood cell membrane, possibly from reduction of glycophospholipids with relative increase of sphingomyelin. Once produced, the change is usually irreversible. Acanthocytes are associated with other disorders in some patients.

Hereditary
- **Abetalipoproteinemia**
- Surface antigen polymorphisms:
    McLeod phenotype of the **Kell blood group** – X-linked anomaly
    Inhibition of **Lutheran blood groups** Lua and Lub – somatic dominant anomaly
- **Red blood cell membrane disorders**
    High RBC membrane phosphatidylcholine
    Abnormal RBC membrane Band 3 protein
- Chorea-acanthocytosis syndromes
    Recessive acanthocytosis: parkinsonism with occasional motor neuron disease
    Mitochondrial myopathy: encephalopathy, lactic acidosis
    Hallervorden-Spatz disease: progressive dementia, spasticity, pallidal and retinal degeneration
    HARP: hypoprebirth lipoproteinemia, acanthocytosis, retinitis pigmentosa, pallidal degeneration with iron deposition

Acquired

- **Liver disorders**, particularly alcoholic cirrhosis
- Malnutrition

    Hypobetalipoproteinemia

    Anorexia nervosa, cystic fibrosis
- Hypothyroidism and panhypopituitarism
- **Infantile** pyknocytosis
- **Vitamin E** deficiency in premature newborns
- **Splenic hypofunction** and **splenectomy**

Acanthocytes have a reduced red blood cell survival. This may give rise to a mild chronic **hemolytic anemia** (spur cell hemolytic anemia, Zieve's syndrome).

ACETYLCHOLINESTERASE

(AChE) An enzyme of the erythrocyte (**red blood cell)** membrane that has greater activity in young erythrocytes and decreases progressively with age. This decrease in membrane AChE activity is a consistent erythrocyte abnormality in **paroxysmal nocturnal hemoglobinuria** (PNH), reflecting defects in the phosphotidyl-inositol anchor.

ACIDIFIED GLYCEROL LYSIS TEST

A screening test for **hereditary spherocytosis**. In symptomless relatives of known cases, it has a higher detection rate than the conventional **osmotic fragility test** of **red blood cells**, although the same principles apply, but it is also positive in other causes of spherocytosis. It requires less blood than the osmotic fragility test and can be performed more rapidly. Measurement is carried out in a recording spectrophotometer with the wavelength set at 625 nm.

ACIDIFIED SERUM LYSIS TEST

(Ham test) A test used for the diagnosis of **paroxysmal nocturnal hemoglobinuria** (PNH).[2] **Red blood cells** from defibrinated, oxalated, citrated, or ethylenediamine tetraacetate (EDTA) anticoagulated patient's blood are treated at 37°C with normal or patient's serum acidified to pH 6.5 to 7.0. Normal serum known to be strongly lytic to PNH cells should be used, but the test should be repeated with the patient's own serum to exclude a form of **congenital dyserythropoietic anemia** — HEMPAS (hereditary erythroblastic multinuclearity associated with a positive acidified serum test) — in which lysis occurs in only about 30% of cells with normal serum and not at all with the patient's own serum. HEMPAS patients have a negative sucrose lysis test, but this test is positive in those with PNH. The lytic potency of normal serum varies and is destroyed by mild heat, even at low temperature within a few days, so the serum should be used fresh. Lysis is measured as liberated hemoglobin in a spectrophotometer at the wavelength of 540 nm. In PNH, 10 to 15% lysis is usually obtained, with a range of 5 to 80%. If the patient has been recently transfused, the degree of hemolysis will be reduced. Confirmation of the diagnosis of PNH requires the demonstration of GPI-linked molecules within the red blood cell membrane by **flow cytometry**.

TABLE 3

Drugs Associated with Acquired Aplastic Anemia

Class	Examples
Antibiotic	chloramphenicol
	sulfonamides
	cotrimoxazole
Antimalarial	chloroquine
	quinacrine
Nonsteroidal anti-inflammatory drugs	phenylbutazone
	indomethacin
	naproxen
	diclofenac
	ibuprofen
	piroxicam
Antirheumatic	gold salts
	D-penicillamine
Antithyroid	carbimazole
	methylthiouracil
	propylthiouracil
Psychotropic/antidepressants	phenothiazines
	mianserin
	dothiepin
Anticonvulsants	carbamazepine
	phenytoin
Antidiabetics	chlorpropamide
Carbonic anhydrase inhibitors	acetazolamide

ACQUIRED APLASTIC ANEMIA

(Idiopathic aplastic anemia) An uncommon disorder, with an incidence of 1 to 2 per million per year in Western countries and perhaps twice or three times as common in China, Southeast Asia, and Japan.[3] All ages are affected, with a peak incidence in young adults and again in older patients. In children and young adults, acquired aplastic anemia must be distinguished from congenital types; in the elderly, it must be distinguished from myelodysplastic syndrome. The majority of cases (70%) are idiopathic; 15% follow exposure to drugs, which only rarely cause blood dyscrasias, and about 10% follow a hepatic illness, presumed to be viral, although no specific agent has yet been identified. The remaining cases follow other known viral infections or exposure to industrial or domestic agents. Drugs associated with aplastic anemia are shown in Table 3.

Pathogenesis

The disease is the result of dysfunction in the hematopoietic stem cell population. There is reduction in the number of CD34+ hematopoietic precursors. Remaining cells have a proliferative defect. The most convincing evidence suggests that the damage is caused by autoimmune cytotoxic T-cell attack. *In vitro* evidence to support the autoimmune hypothesis has been summarized[3–5a]:

60 to 80% respond to immunosuppressive therapy using ATG and ciclosporin

Increased levels of cytokines that inhibit hematopoiesis (IFN-γ, TNF-α)

Increased Fas-antigen expression on bone marrow CD34+ cells

Activated cytotoxic CD8+ T-cells present in blood and bone marrow

Oligoclonal expansion of CD8[+] T-cells

Upregulation of **apoptosis** and immune response genes

Human leukocyte antigen (HLA) restriction with overrepresentation of HLA-DR 15

In acquired aplastic anemia, activated cytotoxic T-cells secrete cytokines such as tumor necrosis factor (TNF)-α and interferon (IFN)-γ, which inhibit hematopoietic progenitor cells (HPC). TNF-α and IFN-γ upregulate the expression of the Fas receptor on HPCs, triggering apoptosis. Increased production of interleukin-2 results in expansion of T-cells. TNF-α and IFN-γ also increase nitric oxide (NO) production by marrow cells, which may contribute to immune-mediated cytotoxicity and elimination of HPCs. This is summarized in Figure 2.

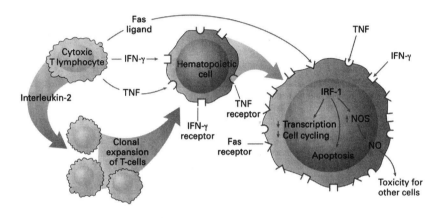

FIGURE 2
Schematic diagram of proximal events in immune-mediated marrow failure: IFN-γ = interferon-γ; IRF-1 = IFN regulatory factor-1; NOS = nitric oxide synthase; TNF = tumor necrosis factor. (From Young, N.S., *Rev. Clin. Exp. Haematol.*, 4, 426–459, 2000. With permission.)

Laboratory Features

The laboratory diagnosis depends upon the peripheral-blood cell levels, peripheral-blood-film examination, and bone marrow aspirate and trephine. Other conditions that lead to pancytopenia, particularly acute leukemias, must be excluded. In the peripheral blood there is anemia, granulocytopenia, and a low platelet count. Lymphocytes are normal in number or reduced. Typically there is some red blood cell macrocytosis and the reticulocyte count is low, either absolutely or in proportion to the degree of anemia. Fetal hemoglobin levels may be normal or slightly elevated, but these are not helpful in the differential diagnosis. There may be some variation in size and shape of red cells, although this is not usually marked. The presence of nucleated red cells raises the possibility of paroxysmal nocturnal hemoglobinuria (PNH) or myelodysplastic syndrome (MDS). Neutrophils are reduced, but their morphology is normal apart from some increase in granulation, so-called toxic granulation. There is typically no shift to the left. Platelet numbers are reduced, and the platelets are mostly small in size. Abnormal cells including blasts or hairy cells are significantly absent. The diagnosis ultimately rests on bone marrow examination. Aspiration in typical cases is "easy." The marrow films contain many fragments that are of hypocellular, "lacy" appearance, and the cell trails are hypocellular, with mainly lymphocytes and nonhematopoietic cells remaining. The bone marrow trephine is the key investigation. There is usually overall hypocellularity, with hematopoietic marrow replaced by fat cells, although cellular, or even hypercellular islands ("hot spots"), may be found. Reticulin is not increased, and no abnormal cells are seen. Cytogenetic analysis of the marrow is difficult, but it is usually normal, although abnormal clones, particularly of trisomy 8 or monosomy 7, are occasionally found. The significance is uncertain, sometimes indicating progression to MDS or acute leukemia. There is no increased chromosomal instability in response to clastogens, as is seen in Fanconi anemia. Mutations in telomeres, with alterations in the ribonucleic protein complex responsible for synthesis and maintenance of telomere repeats, has been reported.[6]

Clinical Features

The pancytopenia develops slowly, such that symptoms, which are the consequence of the cytopenias, may be delayed.[7] Purpura, easy bruising, or repeated epistaxis that is difficult to control are common presentations. Persistent sore throat with or without fever and pallor, tiredness, or other effects of anemia may also be presenting features. The course of the untreated disease depends upon the severity of the bone marrow failure, as reflected in the peripheral blood. The degree of hypoplasia may remain stable over the ensuing months or years or become more severe, necessitating therapeutic intervention. **Paroxysmal nocturnal hemoglobinuria** and **myelodysplasia,** which may further progress to **acute myeloid leukemia**, develop in 10 to 20% of patients. Spontaneous recovery rarely occurs.

Treatment

General Supportive Measures

Before specific treatment became available, only about 10% of patients with SAA or VSAA (see Table 4) survived more than 1 year. Current management depends on **red blood cell transfusion** and **platelet transfusion** to correct the cytopenias and on prevention and treatment of infection consequent upon any **neutropenia.** Definitive treatments to restore stem cell function rely either on **immunosuppression** to allow partial or complete recovery of peripheral blood pancytopenia, or **stem cell transplantation** to replace stem cells.[8]

TABLE 4

Definition of Disease Severity of Aplastic Anemia (AA)

Severe (SAA)	bone-marrow cellularity <25%, or 25–50% with <30% residual hematopoietic cells; with 2/3 of the following: neutrophils <0.5 × 10⁹/l platelets <20 × 10⁹/l reticulocytes <20 × 10⁹/l
Very severe (VSAA)	as for severe AA but with neutrophils <0.2 × 10⁹/l
Nonsevere (NSAA)	patients not fulfilling the criteria for severe or very severe AA; with a hypocellular marrow with 2/3 of the following: neutrophils <1.5 × 10⁹/l platelets <100 × 10⁹/l hemoglobin <10 g/dl

Immunosuppression (IS)

The standard therapy is with intravenous **immunoglobulin**, antilymphocyte globulin (ALG), given daily for 5 days. (Some preparations are labeled antithymocyte globulin, or ATG.) **Ciclosporin**, started after the ALG, may increase the rate and incidence of remission, and some remissions are ciclosporin dependent. About 65% of patients will achieve remission as defined as freedom from transfusion dependence. Response is similar at all ages, and IS remains the treatment preferred for patients over the age of 40 years with severe or very severe aplastic anemia, even if a suitable donor is available. Improvement is slow, counts rarely improving before 6 weeks following ALG and may be delayed for 6 months or more. Second or third courses of IS may be effective in inducing remission in a few patients. Relapse, defined as a return to transfusion dependence, occurs in up to 40% of responders over 10 years.

High-dose cyclophosphamide, with or without ATG, apart from its use as a conditioning therapy prior to transplantation, can be particularly useful when a suitable donor is not available. Monoclonal antibody therapy with anti-CD20 (**rituximab**) and anti-CD52 (**alemtuximab**) is being evaluated for refractory cases.

Allogeneic Hematopoietic Stem Cell Transplantation

(HSCT) See also **Allogeneic stem cell transplantation**; **Hematopoietic stem cell assays**. When a **human leukocyte antigen** (HLA)-compatible sibling donor is available, HSCT is the treatment of choice for children with SAA or VSAA and for adults under 40 years of age with VSAA. HSCT from unrelated, HLA-matched volunteer donors has become more successful with the development of less-toxic conditioning regimens and is offered following failure of IS treatment for younger patients. Event-free survival is about 70% and is age dependent, as with all HSCT. Survivors do not have a higher incidence of abnormal clones of PNH or MDS cells arising. Children have normal growth and fertility so long as irradiation is avoided in conditioning for transplant.

ACQUIRED HEMOLYTIC ANEMIA

See **Hemolytic anemias**.

ACQUIRED IMMUNODEFICIENCY SYNDROME

(AIDS) A clinical disorder arising from **immunodeficiency** following infection, usually by the **human immunodeficiency virus**-1 (HIV-1), less commonly by HIV-2, and rarely by an undetermined cause. The first patients to be infected were identified in 1981. Since then, the incidence has rapidly increased throughout the world; by 2003, an estimated 20 million had died from the disease. Early diagnosis by HIV screening is recommended.[20]

Clinical Features

These occur in stages over a number of years. Due to differences in immunological response, they must be considered separately for adults and infants.

Adults

The best indicator of progression is the level of CD4 **lymphocytes**, as measured by **flow cytometry** (see Figure 3).

Acute Seroconversion Syndrome

During the prodromal stage (10 to 40 days), evidence of viremia can be detected. At the end of this time, approximately 50% of individuals develop an inflammatory illness similar to **infectious mononucleosis** with transient **lymphadenopathy**, fever, pharyngitis, myalgia, arthralgia, maculopapular skin eruptions, aphthous ulcerations, diarrhea, nausea, vomiting, and headache. Neuropathic symptoms such as peripheral neuropathy or asymptomatic

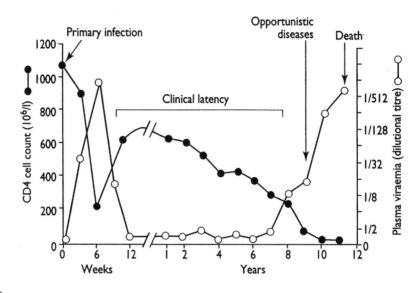

FIGURE 3
Changes in CD4 cell counts (●) and plasma viremia (○) during HIV infection. (Modified from Fauci, A.S., *Science*, 262, 1011–1018, 1993; and from Jolles, S. et al., *Br. Med. J.*, 312, 1243–1244, 1996. With permission.)

meningitis occur in about 15% of those infected. The illness usually resolves in 1 to 2 weeks, but it can be more prolonged. A marked **lymphocytopenia** with depletion of CD4 and CD8 lymphocytes occurs. With resolution of symptoms, there is a rise in the CD4 lymphocyte count, but in some patients it may remain depressed.

Early HIV-1 Disease

Most affected individuals remain asymptomatic for 5 to 10 years. The most common early symptoms are mild generalized lymphadenopathy and dermatological manifestations, such as seborrheic dermatitis, psoriasis, and eosinophilic folliculitis. At this stage, the CD4 lymphocyte count remains above $0.5 \times 10^9/l$.

Intermediate Stage of HIV Disease

This was previously termed AIDs-related complex and occurs from 2 to 5 years after infection. Mild skin and oral lesions become more common, particularly recurrent infection by herpes simplex, varicella zoster, and candidiasis. This is often associated with weight loss. The CD4 lymphocyte count is 0.2 to $0.5 \times 10^9/l$, and progression is indicated by a fall in this level.

Late-Stage Disease

Progression is indicated by a continued fall of CD4 lymphocytes to 0.05 to $0.2 \times 10^9/l$. The consequences of T-cell depletion with impaired cell-mediated immunity and of B-cell proliferation are:

   Infections. Bacterial and viral infections by organisms commonly present in the respiratory, gastrointestinal, and renal tracts; opportunistic infection by *Pneumocystis carinii*, *Toxoplasma gondii*, *Mycobacterium tuberculosis*, *Cryptococcus neoformans*, *Cryptosporidium* spp., *Cyclospora* spp., *Histoplasma capsulatum*, *Coccidiosis immitis*, or *Aspergillus* spp.; these infections are usually associated with a febrile illness

Anemia

- **Anemia of chronic disorders**
- **Nutritional deficiency disorders** following intestinal malabsorption
- Failure of **erythropoiesis** with inappropriately low levels of erythropoietin (<500 mU/mm^3)
- **Immune hemolytic anemia** (rare)

Neutropenia

Thrombocytopenia

- **Bone marrow hypoplasia**
- **Immune thrombocytopenic purpura**

Malignancy

- Human immunodeficiency virus-related lymphomas:
  - **Diffuse large B-cell lymphoma** (60 to 70%)
  - **Burkitt lymphoma** (30 to 40%): 75% are extranodal
  - **Hodgkin disease**: risk increased fivefold and is particularly associated with intravenous drug abuse; mixed cellularity and lymphocyte-depleted histology seen at an advanced stage of AIDS
  - **Primary effusion lymphoma** (rare)
  - **Polymorphic lymphoid proliferations** (rare)
  - **Plasmablast**oma (localized to the oral mucosa)
  - **Kaposi's sarcoma**: the first reported malignant association in homosexual men
- Cervical carcinoma in women and rectal carcinoma in men, together with other disorders associated with human papilloma virus infection
- Bronchiogenic carcinoma

Thromboembolic disorders due to **antiphospholipid antibody syndrome** and reduced active **Protein S**

From this phase, without treatment, the disease always progresses, but at differing individual time scales.

Advanced HIV-1 Disease

The rate of progression to clinical AIDS probably depends upon the plasma load of HIV-1 antigen, but rapidity also increases with the age of the patient. At this stage, the CD4 lymphocyte count has fallen to less than $0.05 \times 10^9/l$. Opportunistic infections become more resistant to treatment, and neurological disorders such as peripheral neuropathy, dementia, and progressive multifocal leukoencephalopathy become common.

Terminal HIV-1 Disease

This is inevitable in all patients and occurs when treatment of the complications is no longer effective.

Infants

Diagnosis up to 18 months (due to unreliability of HIV antibody testing from persistence of placental transfer of IgG) depends upon evidence of HIV culture, HIV polymerase chain reaction (PCR), or HIV antigen (p24).

During the first 6 months of life, prodromal lymphadenopathy, hepatomegaly, diarrhea, fever for over 1 month, and failure to thrive are features suggestive of infection.[11] Clinical AIDS in infancy presents as:

Opportunistic infection: oral candidiasis, lymphoid interstitial pneumonitis; *Pneumocystis carinii* and other infections as in adults (see above)

Encephalopathy, peripheral neuropathy

Malignancy (rare)

Treatment

Antiretroviral Therapy[12-14]

The goal of therapy is long-term suppression of HIV-1, indicated by undetectable plasma HIV-1 RNA, but it is unlikely to totally eliminate HIV reservoirs of infection from the body. Treatment should be offered before substantial immunodeficiency ensues, such as when the CD4 count falls below 500 cells/mm^3, or the plasma viral load exceeds 10,000 to 50,000 HIV RNA copies/ml, with or without symptoms. Initial treatment (highly active antiretroviral therapy, HAART) should include a combination of at least two nucleoside reverse-transcriptase inhibitors (RTi) — abacavir, didanosine, lamivudine, stavudine, zalcitabine, zidovudine — thereby impairing the production of DNA replication of the virus. To these should be added a nonnucleoside reverse-transcriptase inhibitor (nRTi) — efavirenz, nevirapine, or one or two protease inhibitors (PI) such as amprenavir, indinavir, lopinavir, nelfinavir, ritonavir, or saquinavir — to block virus replication. More recent regimes have included the addition of **rituximab**. For assessment of response viral load and CD4, measurements are essential. Reduction in viral load to below the detection level of a sensitive assay represents the optimal treatment response, and failure to achieve or sustain this control should prompt consideration of therapy modification. Switches in therapy should involve substitution or addition of at least two agents. The emergence of drug-resistant strains has been encountered. Maternal therapy, regardless of HIV RNA load or CD4 cell count, reduces transmission to the infant, particularly when administered during the last 3 to 4 weeks of pregnancy. Fortunately, combination therapy does not increase the risk of premature delivery or other adverse effects on the fetus.

While HAART has greatly reduced the level of HIV-1 infection in patients, chronic low levels of virus replication occur, and HIV genomes can persist in some infected cells. It has recently been claimed that this state can be reduced by the administration of valproic acid.[14] To assess this claim, a randomized trial is necessary.

Adverse drug reactions are common. Zidovudine often gives rise to varying degrees of bone marrow hypoplasia, less frequently to neuropathy and myopathy. Reactions to zalcitabine, didanosine, stavudine, and lamivudine include peripheral neuropathy and pancreatitis. Protease inhibitors may induce hemorrhage in patients with hemophilia and can cause a hemolytic anemia. After some years of treatment with retroviral drugs, there arises the possibility of premature atherosclerosis. Whatever drug regime is chosen, patient compliance is essential on the understanding that it may well continue for many years.

Opportunistic Infection Treatment

This depends upon the infecting organism[15] (see **Immunodeficiency**). Prophylactic therapy for pneumocystis infection can be given with pentamidine by inhalation. Oral cotrimoxazole has also been advocated to prevent secondary infection in the severely immunocompromised patients, but its use is limited by adverse drug reactions.

Lymphoma Treatment

Aggressive chemotherapy can be administered with CD4 counts greater than 200, depending upon the type of lymphoma. Approaches have included dose-adjusted CDE, R-CDE, and R-CHOP in diffuse large B-cell lymphoma (see **Cytotoxic agents**).

Vaccine

Considerable research to find a suitable vaccine continues, but none of proven value has yet been developed. To achieve a cure, innovative strategies such as gene therapy, immune reconstruction, and nucleic acid therapeutic vaccines may be necessary to target viral, immunological, and cellular components.[16]

Prognosis

Progression from initial infection to an AIDS-defining illness without retroviral therapy is 10 to 12 years, followed by 1 to 2 years postclinical AIDS before death. Longer delays in progression sometimes occur in cases where transmission has been by plasma, such as patients with hemophilia A. Following perinatal infection, some children become clear of viral DNA/RNA and HIV antibody and are symptom free after periods of 1 to 5 years. Rapid progression is indicated by a fall of CD4 lymphocytes below 200 cells/μl, supplemented by a high concentration of HIV-1 RNA load (qRNA)[17] and associated with low amounts of antibody-to-HIV-1 gag proteins p17 and p24. Survival can be prolonged with retroviral therapy, but this is limited by drug resistance or toxicity. The use of HAART has reduced the incidence of primary central nervous system (CNS) lymphoma.

Prevention

A wide variety of methods are available, depending much on personal attitudes and circumstances. They include:

Monogamy

Use of condoms

Screening of blood donors for HIV antibodies: direct HIV antigen testing by polymerase chain reaction (PCR) further reduces risk, but at markedly increased cost; these procedures should be accompanied by limitation of transfusion to essential requirements and the use of **autologous blood transfusions**

Heating of plasma during **blood components for transfusion**, where possible

Single use of needles, particularly for those of drug abusers

Care in specimen collection and transport: laboratory safety measures; limitation of autopsies on known HIV-positive bodies

Antenatal testing with treatment by antiretroviral drugs of HIV-positive pregnant mothers

Avoidance of breast feeding for infected mothers, provided that feeding with artificial milk is available and safe

Early treatment of genital herpes and other sexually transmitted infections

Chemoprophylaxis with tenofovir

Advocation of male circumcision

ACQUIRED REFRACTORY ANEMIA
See **Myelodysplasia**.

ACQUIRED VON WILLEBRAND DISEASE
An acquired syndrome similar to hereditary **Von Willebrand Disease** (VWD) but with a later onset. It is due to the development of an **autoantibody** to **Von Willebrand Factor** (VWF), which can occur in previously healthy people, or in association with **systemic lupus erythematosus, lymphoproliferative disorders, myeloproliferative disorders**, cardiovascular structural disorders, or **angiodysplasia.** The clinical course of these patients can range from a mild bleeding disorder to severe and life-threatening bleeding. Patients show low levels of VWF antigen activity, factor VIII, and prolonged bleeding times. Inhibitory activity to VWF may be detected in plasma from affected patients, although this is not always the case, as a range of different pathogenetic mechanisms, apart from autoantibody activity, can be involved. These pathological mechanisms include decreased synthesis of VWF and increased degradation of VWF, either by proteolysis, by vascular high-shear stress, or by its cellular absorption. Treatment is aimed at the underlying disorder and VWF replacement for bleeding, as in hereditary VWD, although because of the presence of an inhibitor, VWF replacement may not be as effective. The administration of intravenous **immunoglobulin** has been advocated.[18]

ACTIN
See **Cell locomotion; Red blood cell** — membrane (Band 5).

ACTIVATED PARTIAL THROMBOPLASTIN TIME
(APTT) The time taken for citrated **plasma** to clot at 37°C after **calcium** has been added to plasma that has had a contact activator previously added. A standard amount of phospholipid is also added. There is no addition of thromboplastin (tissue factor). Contact activators are insoluble compounds that present a large surface area and include kaolin, ellagic acid, or micronized silica. The APTT is also known as the partial thromboplastin time with kaolin (PTTK). The APTT is a measure of the overall efficiency of the intrinsic pathway of blood coagulation. It is dependent upon factors **II, V, VIII, IX, X, XI**, and **XII, fibrinogen, prekallikrein**, and high-molecular-weight **kininogen**. The degree of activation of contact factors in the APTT is standardized by fixing the time of incubation prior to the final addition of calcium. The actual time of the APTT is dependent upon the reagents used, length of the preincubation, and the quantity and quality of added phospholipid. The APTT is prolonged when there is a deficiency in any one of the coagulation proteins involved or when there is a coagulation inhibitor (or heparin) present. The APTT should be particularly sensitive to factor VIII, IX, X, or XI deficiencies and to phospholipid-dependent inhibitors (**lupus anticoagulants**). Sensitivity to contact-factor activation (factor XII, prekallikrein, and high-molecular-weight kininogen) is dependent upon incubation time employed: the longer the incubation time, the less sensitive is the APTT to deficiency of these components.

In cases where the APTT is prolonged, it is important to determine if the prolongation is due to a coagulation-factor deficiency or to the presence of a circulating inhibitor. In such cases, a mixture of patient plasma and normal pool plasma (50:50 or 80:20) should be made, and the APTT of the mixture then measured. If the APTT of the mixture is decreased by more than 50% of the difference between the APTTs of the two plasmas tested separately, the patient probably has a deficiency of one or more coagulation factors.

If the addition of normal plasma fails to correct the prolonged APTT of the patient plasma, an anticoagulant may be present. The presence of heparin can be determined by use of the **thrombin time** and **reptilase time**. Not all inhibitors will be seen to prolong the APTT, as not all act immediately. The mixing-test APTT forms the basis of an assay for screening plasma for the presence of time-dependent inhibitors. In this assay, correction of the APTT is tested at time zero and after 1 h. The presence of a time-dependent inhibitor will be seen by the failure of the plasma mixture to be corrected at 1 h compared with correction at time zero. Inhibitors seen in **hemophilia A** are usually time dependent.

The **reference range** for adults is 30 to 40 sec. Ranges for full-term and premature infants over the initial 6 months of life are given in **Reference Range Tables XV** and **XVI**.

ACTIVATED PROTEIN C
(APC) See **Protein C**.

ACUTE EVENT-RELATED ANEMIA
(AERA) Acute onset of **anemia** with acute **inflammation** or surgical trauma without blood loss. It is due to functional **iron deficiency** as a consequence of iron entrapment by **histiocytes** (macrophages) and decreased responsiveness to erythropoietin.[20] It is related to **anemia of chronic disorders**.

ACUTE FEBRILE NEUTROPHILIC DERMATOSIS
(Sweet's syndrome) The association of fever, **neutrophilia**, and painful erythematous cutaneous plaques due to dense, diffuse clustering of granulocytes. **Leukocytoclastic vasculitis** in the skin with dermal edema is prominent. It particularly occurs with **lymphedema**, typically postmastectomy, and is more severe if granulocytes have been stimulated by granulocyte **colony stimulating factor**.[187a] The cause and pathogenesis are unknown, but the syndrome is regarded as a **hypersensitivity** reaction to a hematological malignancy, usually **acute myeloid leukemia** or **chronic myelogenous leukemia**.

ACUTE INTERMITTENT PORPHYRIA
See also **Porphyrias**.
(AIP) An autosomally dominant disorder due to a partial deficiency in porphobilinogen (PBG) deaminase (see **Heme**). This enzyme is encoded on Ch11q; numerous mutations have been described and account in part for clinical and laboratory variants of AIP.

Clinical Features
AIP is relatively common, especially in some Western European subpopulations. Onset of attacks is rare before puberty; females predominate in many clinical series. The clinical course is marked by recurrent attacks lasting several days to several months, interspersed with long but variable periods without symptoms. The frequency of attacks is variable, from as few as two or three in a lifetime to two or three each year. Acute attacks may be precipitated by exposure to many drugs, of which barbiturates, sulfonamides, antibiotics, anticonvulsants, and alcohol are the most common. Attacks may also be precipitated by the onset of menses, high-progesterone-containing contraceptives, pregnancy, stress, or a sudden decrease in calorie intake.

Typical attacks are characterized by diffuse abdominal pain of varying severity accompanied by nausea, vomiting, constipation, diarrhea, and ileus; there is usually no sign of peritoneal irritation. Some previously undiagnosed patients have undergone unnecessary exploratory laparotomy for investigation of abdominal symptoms. Mild or moderate motor weakness in one or more extremities is common, and this may progress to flaccid quadriparesis. Respiratory paralysis is uncommon. Deep-tendon reflexes are commonly absent. Autonomic neuropathic features include bladder distension, inappropriate sweating, tachycardia, hypertension, and postural hypotension. Pain or paresthesias in the extremities sometimes occur. Acute psychosis with manic-depressive or schizophrenic behavior may develop during an attack, and depression with suicidal tendency may follow. Generalized seizures sometimes occur, especially in patients with hyponatremia.

Laboratory Features

There are three molecular types of AIP:

Type I, intermediately reduced PBG deaminase activity and protein content.

Type II, partially decreased PBG activity in nonerythroid cells and normal erythrocyte PBG deaminase activity.

Type III, decreased PBG deaminase activity with a structurally abnormal enzyme. During an attack, the urinary excretion of both δ-aminolevulinic acid (δ-ALA) and PBG are increased. Between attacks, the levels of both substances are somewhat elevated, although diagnostic testing is often unproductive. Population testing using PBG deaminase molecular and genetic techniques reveals that there are many persons with asymptomatic AIP.

Treatment

Heme suppresses δ-ALA and PBG synthesis. Thus, **hematin**, administered intravenously 1 to 4 mg/kg once or twice daily, is the treatment of choice. The onset of action is slower than that of glucose loading, and the full response is not observed for 24 to 48 h. The prime indications for hematin therapy are severe abdominal pain during an attack or when serious neurological manifestations occur. Excessive doses of hematin can cause renal tubular damage, and hematin degradation products possess anticoagulant properties. Narcotic analgesics should be used for pain control; chlorpromazine controls nausea and vomiting. Diminished calorie intake worsens attacks, and a high-carbohydrate regimen depresses hepatic porphyrin biosynthesis. Thus, a high carbohydrate intake (at least 300 g daily) should be ensured either by mouth or parenterally. Prevention of attacks involves identification and avoidance of precipitating factors. Gonadotrophin-releasing hormone antagonists such as Gonadarelin (LH-RH) inhibit ovulation and reduce the incidence of perimenstrual attacks in women.

ACUTE LEUKEMIA

Hematological malignancies characterized by sudden onset of febrile illness with pallor and often **purpura**. Their causes remain unknown, but certain major factors have been implicated in some cases.[21] These include **ionizing radiation**, certain **chemical toxic disorders**, especially those due to organic solvents (benzene), and at least one **viral infection disorder**[12] (human T-cell leukemia virus, HTLV-1). Knowledge of their biology,[22] molecular genetics,[23] and treatment[24] has drastically changed the outlook over the past 30 years from an inevitably fatal illness to one that is less predictable but carries the hope, especially in children, of long survival following a period of intensive treatment. The underlying abnormality is excessive proliferation of a clone of hematopoietic cells arrested in maturation and characterized by a failure to undergo **apoptosis**. The arrest of the cell in the bone marrow occurs at either the stem cell or progenitor stage of maturation (see **Cytogenetic analysis**; **Oncogenesis**). There is usually a marked increase of proliferating cells in the peripheral blood. Their classification, clinical features, laboratory features, treatment, and prognosis are considered under the following groups:

Acute myeloid leukemia (AML) including
- Acute promyelocytic leukemia (APL)
- Acute myelomonocytic leukemia
- Acute monoblastic and monocytic leukemia
- Acute erythroid leukemia
- Acute megakaryoblastic leukemia
- Acute basophilic leukemia
- Acute eosinophilic leukemia

Acute lymphoblastic leukemia (ALL)

Adult T-cell leukemia/lymphoma

Acute leukemia of ambiguous lineage

Acute myelosis and myelofibrosis

Myeloid sarcoma

ACUTE LEUKEMIA OF AMBIGUOUS LINEAGE

(Hybrid acute leukemias; mixed-lineage acute leukemia; mixed-phenotype acute leukemia) These are biphenotypic, where >10% of blasts show lymphoid and myeloid markers, and bilineal, where a mixture of blasts displays either lymphoid or myeloid characteristics. They have generated debate about the lineage commitment of progenitor cells, which may either express inappropriate phenotypic features or, as a consequence of leukemic transformation, express aberrant phenotypes. Their presenting features are those of **acute myeloid leukemia** related to their degree of **anemia**, **neutropenia**, and **thrombocytopenia**. They have a high percentage of cytogenetic abnormalities. About one-third have a Philadelphia chromosome with a CD10+ precursor lymphoid component; others are associated with t(4;11)(q21;q23) or 11q23 abnormalities.

Gene *MLL1*, a histone methyl transferase, rearranged with binding to *HOXa7* and *HOXa9*, has been reported. The prognosis is poor for all types. Treatment is by aggressive chemotherapy and **allogeneic stem cell transplantation** or by **autologous stem cell transplantation**.

ACUTE LYMPHOBLASTIC LEUKEMIA

See also **Fetal disorders**; **Lymphoproliferative disorders**; **Oncogenesis**.
(ALL) Malignant transformation of **lymphocyte** precursors of B-cell and T-cell lineage. Invasion of the **bone marrow** by lymphoblasts interferes with normal **hematopoiesis** and in other tissues creates space compromise or alteration of function.

Incidence

ALL is the most common form of malignancy of childhood, about one child in every 2000, probably due to a combination of genetic susceptibility and environmental factors.[25] It has been proposed that the major subtype (common ALL) is caused by at least two events, one inducing a mutation before birth, followed by a second event in early postnatal life that triggers the disease. Early exposure to infection, as occurs in babies attending day-care centers, reduces the incidence. There is an unrelated increase of incidence in the elderly.

Classification

Morphological (FAB Classification)

This is summarized in Table 5.

TABLE 5

Morphological Features of Acute Lymphoblastic Leukemia

	L1	L2	L3
Cell size	small to intermediate	large, heterogeneous	large, homogeneous
Nuclear shape	uniform	pleiomorphic indentations common	uniform
Nucleoli	small or absent	large, prominent	often single
Cytoplasm	scanty	moderate to abundant	moderate (basophilic)
Cytoplasmic vacuoles	none to few	none to few	prominent

FAB L1

The cells are relatively homogeneous small blasts up to twice the size of small lymphocytes. Cytoplasm is scanty, and the nucleoli are absent or poorly visualized. Occasionally, vacuoles with a few azurophilic granules are seen.

FAB L2

The cells are heterogeneous and the nuclei are irregularly shaped, often folded or indented with prominent nucleoli. The cytoplasm varies in quantity and may contain occasional azurophilic granules.

FAB L3 (Burkitt Tumor Cells)

The cells are large, homogeneous blasts with deeply basophilic cytoplasm containing prominent, well-defined vacuoles. The nucleus is regular, with prominent nucleoli.

Immunological[26,27]

See also Appendix to this book, "WHO Classification of Tumors of Hematopoietic and Lymphoid Tissues."

Lineages are based on patterns of expression rather than on the presence or absence of any single antigen, but several subgroups can be separated by further **immunophenotype** studies supplemented by studies of Ig and T-cell receptor gene patterns (Table 6). **Acute myeloid leukemia** can be separated by pan-myeloid surface antigens (including CD13, CD33, CD15, MPO, and CD117).

Precursor B-Cell Lymphoblastic Leukemia/Lymphoblastic Lymphoma

(B-LBL; Rappaport: diffuse poorly differentiated lymphocytic lymphoma) Characteristically composed of small- to medium-sized blast cells with scant cytoplasm, with moderately

TABLE 6

Classification and Incidence of Acute Lymphoblastic Leukemias by Immunophenotyping

Classification	Immunophenotype	Frequency (%) Child	Adult
B-lineage ALL			
Pre pre-B ALL	Tdt, HLA-DR, CD19, cyCD22	5	11
Common ALL	TdT, HLA-DR, CD10, CD19, cyCD22, CD24	63	52
Pre-B ALL	TdT, HLA-DR, cyIgM, CD19, cyCD22, CD10, CD24	16	19
B-cell ALL	HLA-DR, SIgM, CyIgM, CD19, CD22, CD20, CD24	3	3
T-lineage ALL			
Pre-T ALL	Tdt, CD7, cyCD3	1	6
T-cell ALL	Tdt, CD7, cyCD3, CD2, CD5, CD4, CD8	12	18

Note: The antigens underlined are usually needed for the diagnosis of each of the individual subtypes.

condensed to dispersed chromatin and inconspicuous nucleoli, involving bone marrow and blood (B-lymphoblastic leukemia), and occasionally presenting with primary involvement of nodal or extranodal sites (B-lymphoblastic lymphoma). They are morphologically identical to precursor T-lymphoblasts. The probable origin is a bone marrow-derived precursor B-cell. Their immunophenotype is TdT+, HLA-DR+, CD19+, cyCD22+; other antigens vary with the subgroups (Table 6).

Precursor T-Lymphoblastic Leukemia/Lymphoblastic Lymphoma

(T-LBL; Rappaport: poorly differentiated lymphocytic lymphoma) The cells are morphologically indistinguishable from those of precursor B-lymphoblasts. Their probable origin is a precursor T-cell, either prothymocyte, early thymocyte, or common thymocyte. Their immunophenotype is usually TdT+, CD7+, CD3+; other T-cell-associated antigens are variable, including T-cell receptor (TCR) ab or gd+ or no TCR; CD1a+/−; often CD4, 8 double positive or double negative; Ig−; and B-cell-associated antigens−. Occasional cases express natural killer (NK) antigens. Subgroups are shown in Table 6.

Cytogenetic Classification

See also **Cytogenetic analysis**; **Leukemogenesis**.
These are in groups, depending upon the chromosome number: hypoploidy >50, hypoploidy <50, translocations, and pseudodiploid {974, 1060, 1075}. Their classification and frequency are shown in Table 7.

TABLE 7

Chromosomal Translocations and Genetic Alterations
in Acute Lymphoblastic Leukemia

Cytogenetics	Genetic Alteration	Frequency (%)	Prognosis
t(9;22)(q34;q11.2)	*BCR/ABL*	3–4	poor
t(4;11)(q21;q23)	*AF4/MLL*	2–3	poor
t(1;19)(q21;p13.3)	*PBX/E2A*	6	variable
t(12;21)(p13;q22)	*TEL/AML1*	16–29	good
t(1.14)(p32;q11)	—	20	good
Hyperploidy	—	20–25	good
Hypoploidy	—	5	poor

Clinical and Laboratory Features

Most patients present with a sudden febrile illness with pallor and sometimes **purpura**. Children are more commonly of B-cell lineage (80% cases), with peripheral blood and bone marrow involvement. A small proportion present as solid tumors, most often in skin, bone, and lymph nodes, with or without leukemia. Adolescents and young adults are predominantly of T-cell lineage, presenting with rapidly enlarging mediastinal (thymic) masses and/or peripheral **lymphadenopath**y. Superior vena caval obstruction, occasionally with pleural effusion, is a less common form of presentation. **Splenomegaly** and **hepatomegaly** are common, especially in adults but, usually, less frequent and less prominent than in acute myeloid leukemia (AML). Generalized lymphadenopathy is the most frequent physical finding, especially in children, and is overall more frequent than in AML. The symptoms and signs of marked **anemia** are uncommon, although pallor is frequent. Marked **thrombocytopenia** is infrequent, and **disseminated intravascular coagulation** (DIC) is exceptional, so that bleeding is not common. Although fever is a frequent feature, particularly in children, infection at presentation is not common because severe **neutropenia** is also exceptional. The presenting features of both lineages are of short duration and generally less dramatic than in AML. Symptoms and signs can resemble those of chronic infection, which can delay diagnosis for weeks or even months.

Because ALL usually presents with high blast cell counts in peripheral blood, features of **tumor lysis syndrome** — raised potassium, calcium, and creatinine or frank renal failure — are not uncommon. Central nervous system (CNS) infiltration is an unusual form of presentation, although it is, by far, more common than in AML. In up to 5% of patients, routine examination of the colony stimulating factor (CSF) at diagnosis reveals CNS infiltration (defined as counts above five cells per high-power field with recognizable blasts in a CSF cytocentrifuge preparation). This occasionally leads to symptoms that may include cranial nerve palsies and signs of raised intracranial pressure.

Exceptional features of ALL are hypercalcemia due to tumor lysis syndrome or bone consumption in localized areas of tumor burden, inadequate secretion of ADH with subsequent abnormal weight gain (suggesting CNS infiltration), and **autoimmune disorders**, usually with accompanying antinuclear antibodies. All of these are more frequent in precursor T-cell ALL.

Treatment

General Considerations

Modern treatment regimens are increasingly achieving disease-free long-term survival in ALL, especially in childhood. Treatments are intensive, complex schedules that often render patients repeatedly severely neutropenic and thrombocytopenic. Good supportive care is essential, and the best results can only be expected in centers where expertise is concentrated. Treatment combines various strategies based on the natural biology of ALL, where a significant proportion of blasts are located in lymph nodes, in G0 phase ("dormant cells"), and do not take up **cytotoxic agents**. After achieving a complete remission in peripheral blood and bone marrow by using high doses of chemotherapy ("induction" plus "consolidation" courses), patients receive "maintenance treatment" for several months with monthly courses of **vincristine** and **corticosteroids**, followed by 4 weeks of oral **mercaptopurine** and **methotrexate** to progressively deplete the pool of inactive cells.[28,29]

Treatments usually combine a large number of drugs (up to 12) to cover all possible sensitivities, and in recent years, progressively higher doses of certain agents (**cytarabine**) have proven to be effective to clear high-risk leukemia blasts in the so-called BFM approach (Berlin-Frankfurt-Munster). Most treatment regimens follow a risk-based stratification approach, so that patients with high-risk disease (according to their age, blast count at diagnosis, and cytogenetic findings) receive a more aggressive treatment than those without such factors. **Autologous stem cell transplantation** or **allogeneic stem cell transplantation** should only be considered following poor outcome from chemotherapy.

Remission-Induction Chemotherapy

Complete remission rates of up to 90% for children and 80% for adults are possible using vincristine intravenously (IV) once weekly for 4 weeks, prednisolone or prednisone orally each day for 3 to 4 weeks, and IV L-**asparaginase** each week. The use of other corticosteroids (such as dexamethasone) has been reported, and, overall, a higher percentage of complete remission (CR) and a lower incidence of CNS relapse has been observed. However, added toxicity (aseptic necrosis and benign intracranial hypertension being the most relevant) has also been found. Because L-asparaginase is associated with allergic reactions in a few patients, particular care should be taken when using the IV route. If allergic reactions occur, L-asparaginase from a different source should be used. The addition of an anthracycline, usually daunorubicin, 45 mg/m^2 on the first two days of each week, improves CR rates in adults and in children with high-risk disease, and is a standard practice in such patients.

Rituximab may be effective against precursor B-cell CD20$^+$ ALL, especially if there is Ph$^+$ *BCR/ABL* translocation. Ph$^+$ adults respond well to therapy with **imatinib**. The use of Campath® is under evaluation.[30]

Consolidation Treatment

Drug-resistant cells in the tumor population probably increase in number with high mutation rates. It is important to decrease the chance of drug resistance by reducing the blast cell population quickly. Thus, following the initial induction treatment, it is probably important to give early intensification treatment. A variety of drugs has been used for this purpose, and these include cyclophosphamide, cytosine arabinoside, high-dose IV methotrexate, 6-thioguanine, or the anthracyclines if these have not been used in the induction phase. In the specific case of children with high-risk disease, consolidation is based on high doses of cytarabine ("BFM augmented consolidation") rather than on the addition of multiple drugs. Consolidation treatment is standard practice for all patients, and subsequent intensification courses ("delayed intensifications") are added to the treatment of children with high-risk disease, delaying the beginning of maintenance treatment.

Maintenance Treatment

Further chemotherapy using oral agents is essential to maintain remission. The most commonly used drugs are 6-mercaptopurine and methotrexate. Other drugs, such as **thioguanine**, have proven to be more effective in reducing the rate of relapse but add to the risk of very severe long-term liver toxicity. The maintenance phase of treatment should last for about 1∫ years, so that the complete treatment lasts for 2 years in girls and 3 years in boys. This stratification abrogates the higher rate of relapse seen in boys in several studies, and no added toxicity has been evidenced.

Central Nervous System Treatment

Prophylaxis

Without some form of CNS treatment, up to 70% of patients will relapse with CNS disease. Using CNS prophylaxis, this figure falls to less than 5% in children and to about 10% in adult ALL. Current practice includes the administration of intrathecal methotrexate every

4 weeks during the first 6 to 8 months and then every 3 months. Total number of doses is usually between 12 and 14. No definite added benefit of combining hydrocortisone and cytarabine with methotrexate has been seen.

Treatment of CNS Disease

Most treatments use cranial radiation and intrathecal methotrexate. Cranial radiation at a total final dose of 24 Gy is used for children and adults. Methotrexate is given intrathecally once weekly for a minimum of 6 weeks, or until the CSF is completely clear of blasts. The treatment may result in learning problems in children and occasionally retardation of growth and delay in the onset of puberty.

Treatment of Infants

Children under 1 year of age have a very poor prognosis. ALL at this age combines features of both ALL and AML, and specific treatments combining ALL strategies with higher anthracyclin load are common practice. Because of their particular characteristics, these children should only be treated in specialized centers.

Other Measures

Hyperuricemia

Following ALL treatment, this is less common than in AML and chronic myelogenous leukemia (CML). However, ALL patients especially at risk are those with a large leukemia cell mass or with bulky disease, particularly with mature B-cell or T-cell disease. Consequently, maintenance of good hydration is essential, and all patients should be protected against hyperuricemia during intensive chemotherapy with oral **allopurinol**, 300 mg/m^2 daily. Patients with a high blast count ($>100 \times 10^9$/l), biochemical evidence of ongoing tumor lysis syndrome, or with impaired renal function can be administered urate-lytic agents such as raspburicase or enzyme-based products that destroy urate, thereby improving overall elimination of cellular debris.[31]

Pneumocystis Carinii Pneumonia

This infection should be prevented in all cases, since the lymphoid-targeted treatment that ALL patients receive makes them more prone to this infection than in myeloid conditions. Co-trimoxazole is commonly administered on two consecutive days per week during the entire duration of maintenance treatment. Patients developing co-trimoxazole-derived neutropenia can safely receive pentamidine once a month, although its efficacy is believed to be lower.

Vincristine Toxicity

Pain in the lower limbs and difficulty in walking on tiptoe are common toxicities of vincristine that only in exceptional cases can lead to the reduction of its dosage. Carbamazepine is effective in the treatment and prophylaxis of neuralgic pain.

Febrile Neutropenia

This is a common complication. It usually occurs during the intensive induction-consolidation blocks of chemotherapy and is most often related to the colonization of central venous devices by gram-positive bacteria, although potentially lethal episodes of gram-negative septicemia can occur. Mercaptopurine in maintenance treatment is administered at monitored dosages so that the neutrophil count does not fall below 0.75. Prophylactic antibiotics or antifungals have never proven to be of any use.

TABLE 8

Prognostic Factors in Acute Lymphoblastic Leukemia

Favorable	
B-cell precursor	hyperploidy
	TELL-AML1 fusion
	trisomies 4, 10, and 17
T-cell	*HOX11* overexpression
	T(11;19) with MLL-ENL fusion
Unfavorable	
B-cell precursor	under 1 year to <10 years of age
	WBC > 50 × 10^9/l
	MLL rearrangement in infants
	Philadelphia chromosome
T-cell	Adults
	WBC > 50 × 109/l
	Mediastinal bulk tumor

Prognosis

(See Table 8.) Multicenter trials have led to the definition of a number of prognostic factors in ALL. Of these, only age, blast cell count at diagnosis, and cytogenetic abnormalities are reproducible between trials, and other factors (such as sex or race) seem to play a secondary role in determining prognosis, as these can be abrogated by modern treatment regimens.

Age

Cure rates for children between 1 and 10 years with common B-cell ALL are high (up to 85% in children). However, children less than 1 year old and above 10 years have a particularly poor prognosis using standard treatments. With these treatments, cure rates for teenagers with ALL are less than 40%, with few patients above 25 years cured. Older patients have lower remission rates and remissions are shorter than in children. Generally, these features relate directly to patient age.

Chromosomal Karyotype

Abnormalities, including reciprocal translocation, duplication, and deletions, occur as clonal markers in more than 70% of ALL of B- or T-cell lineage. The major tumor-specific translocations are shown in Table 7. Generally, the detection of these translocations in ALL (but not in **Burkitt lymphoma**) is associated with a poorer prognosis. The *BCR/ABL* rearrangement associated with the Philadelphia chromosome in 20 to 30% of adult ALL patients may represent an early stem cell disease. These studies are particularly important in determining prognosis and possible treatment. In B-cell precursor ALL, a hyperdiploid (>50) chromosomal number with *TEL/AML1* fusion, there is a good prognosis, while patients with a hypodiploid number have a poor outlook. Patients with a normal chromosomal number but with structural defects have a poor prognosis. Molecular remission with <10^{-4} cells only occurs in 40% of patients with the exception of those with Ph$^+$ *BCR/ABL* translocation, where it rises to 50% cases.

Immunophenotype

(See Table 6.) Several immunophenotypic features in both childhood and adult ALL have been associated with a poor prognosis (e.g., pre-pre-B ALL, pre-B ALL, B ALL, pre-T ALL). At least in B-cell precursor ALL, the poor prognosis of pre-pre-B and pre-B phenotypes is attributable to distinct biological and clinical features of these subgroups, and the value of immunophenotyping, *per se*, is yet to be proven. Of children with common ALL (C ALL), 80% have an excellent prognosis, with about 70% surviving for 5 years or more. About 20% of this subgroup have pre-B ALL, with a poorer outlook. Children with T-cell ALL (10 to 15%), often associated with male sex, high white blood cell count (WBC), and mediastinal disease, have a poorer prognosis, with fewer than 50% surviving for 5 years. Mature B-cell disease occurs in 1 to 3% of children, usually with FAB L3 Burkitt cells. This subgroup has low remission rates and a particularly poor prognosis. In adults, there is a lower incidence of common ALL and greater incidence of T ALL (see Table 6). Because of a generally poorer outlook for adults, prognostic differences among the immunophenotypic subgroups are not as clear, although generally common ALL carries a good prognosis.

Marrow Cell Morphology

Blast cell percentages vary from 0 to 100%. The presence of more than 10% of cells with FAB L2 morphology is associated with a poor prognosis, which worsens as the proportion of L2 cells rises. Conversely, a pure L1 morphology carries a slightly better outlook. L3 morphology, usually associated with mature B-cell Burkitt lymphoma, has an especially poor prognosis. **Hand mirror cells** are sometimes seen. These are lymphoblasts where the cell axis included in the cytoplasmic protrusion doubles (at least) the wide axis. They occur in varying numbers in up to 20% of patients with ALL, more frequently observed in female patients with normal platelet counts at the time of diagnosis. The significance of this particular configuration is unknown. Their presence has been associated with a prolonged clinical course, primary refractoriness to treatment, and a higher incidence of central nervous system relapse. The significance of this particular configuration is ultimately unknown.

Relapse Sites

Relapses can occur in the bone marrow and in peripheral tissues. As a rule, prognosis depends largely on the timing of relapse, so that whenever the disease recurs during treatment or within the first year after the end of the standard treatment, a stem cell transplant should be performed after achieving a CR with systemic treatment. Late relapses can be treated without a stem cell transplant, but systemic treatment is still warranted, since there cannot be complete reassurance that other sites have been not been invaded by the disease. The need for stem cell transplantation is, however, debated.

Bone Marrow

Where disease recurs, the bone marrow is the major site of relapse. Systemic treatment is necessary, and stem cell transplantation is recommended with early relapse (i.e., 12 months after the end of treatment). Prognosis therefore largely depends on the timing of relapse, with less than 50% of long-term survivors in cases of early relapse.

Central Nervous System

In up to 10% of children and 20% of adults, the initial relapse is in the CNS, either as an isolated event (very infrequent) or in association with bone marrow relapse. Disease presents clinically with the symptoms of raised intracranial pressure, most often headache and vomiting. Physical signs are usually absent, although papilledema and palsies of the sixth or seventh cranial nerves are seen occasionally.

CNS relapse has a poor prognosis, especially in adults. Remissions of disease are often achieved in children, although less frequently in adults, following treatment with intrathecal methotrexate, but usually these remissions are short lived. Further systemic chemotherapy should always be given to patients with CNS relapse; otherwise, bone marrow relapse almost always follows. Finally, stem cell transplantation is the current consolidation approach for these events, although its need has been repeatedly debated. CNS relapse is made very much less likely by prophylactic intrathecal methotrexate (see CNS treatment — prophylaxis, above).

Testes

The testicle is the site of relapse in about 10% of boys, most commonly within a year of completion of treatment. Relapses may be isolated or associated with relapse in bone marrow or CNS. Following testicular relapse, both marrow and CNS will require further treatment, even when not involved, if relapse at these sites is to be prevented. Testicular disease usually presents as unilateral testicular swelling, although biopsy examination of the contralateral testicle frequently reveals disease. Treatment must therefore be given to both testicles, usually with 24 Gy of radiotherapy, with induction cytotoxic therapy continuing for 2 years. This form of relapse within the first 6 months of leukemia treatment carries an especially poor prognosis. Patients should be considered for intensive reinduction chemotherapy followed by stem cell transplantation.

Occular

This may occur in 1 to 2% of patients, presenting with a painful red eye. Slit-lamp examination reveals leukemic deposits, usually in the anterior chamber or in the iris. Treatment is with local radiotherapy and retreatment of the bone marrow and CNS.

Renal

Very rarely, relapse of leukemia occurs within the renal parenchyma. Early diagnosis is difficult until hematuria occurs.

ACUTE MEGAKARYOBLASTIC LEUKEMIA
See **Acute myeloid leukemia**.

ACUTE MEGALOBLASTIC LEUKEMIA
See **Megaloblastosis**.

ACUTE MONOCYTIC LEUKEMIA
See **Acute myeloid leukemia**.

ACUTE MYELOID LEUKEMIA
(AML) Malignant transformation of myeloid cell precursors. It is associated with more than 50% of blast cells in the bone marrow upon presentation and interferes with normal **hematopoiesis**.[32]

Incidence

The incidence is approximately 10 per 100,000 population per year in those over 60 years of age. It is uncommon in childhood, but a rare familial genetic disorder has been reported. Its etiology remains unknown. The etiologic associations are:

Viral infection disorders

Ionizing radiation

Benzene toxicity

Cytotoxic agents

There was a marked increase in incidence in countries associated with nuclear fallout following the disasters at Hiroshima, Nagasaki, and Chernobyl. It sometimes follows treatment by cytotoxic agents for **myelodysplasia** or **myelomatosis**.

Classification

FAB Classification

Classification is based upon morphology and immunophenotyping (summarized in Table 9).

TABLE 9

Association of FAB Subtype with Phenotypic Expression in Acute Myeloid Leukemia

FAB Subtype	% patients positive				
	CD11b	CD13	CD14	CD33	DR
M1	35	76	8	70	70
M2	62	85	12	71	91
M3	15	76	2	94	7
M4	81	71	63	85	95
M5	89	45	57	92	92

AML M0

M0 blasts display no distinctive morphological characteristics, with peroxidase and Sudan black stains being essentially negative. Immunologically, blasts express the myeloid-associated markers CD13 or CD33 in the presence of negative lymphoid markers.

AML M1

Marrow aspirates contain 90% or more of medium to large blasts with few, if any, cytoplasmic granules, Auer rods, or vacuoles. At least 3% of blasts will be peroxidase positive.

AML M2

(Acute promyelocytic leukemia; APL) More than 10% of the nucleated cells have matured to the promyelocyte stage or beyond. Many patients show frank dysplastic changes, including nuclear-cytoplasmic asynchrony, pseudo-Pelger-Hüet appearance, and abnormalities of granulation, especially hypogranulation and giant granules. Auer rods may be found at any stage, from myeloblast to mature granulocyte. The 8;21 chromosomal translocation is found in 18% of these patients. Translocation t(15;17) fused to retinoic acid receptor α(RARα) gene to PML gene on chromosome 15 gives rise to fusion protein PML-RARα, suggesting that the disruption of RARα is the cause of APL.

AML M3

A common hypergranular type and a less usual microgranular variant are recognized. The hypergranular type has a bone marrow population of abnormal promyelocytes packed with coarse red-purple granules that often obscure the nucleus. The so-called faggot cells may contain sheaves of Auer rods. The nucleus is lobulated, reniform, or bilobed and is best appreciated in the microgranular variant in which the cytoplasm contains only a fine red dusting of granules. The peroxidase stain is strongly positive, even when few granules are present. Virtually all cases are associated with the 15;17 chromosomal translocation.

AML M4

Differentiation is seen along both myeloid and monocytic lineages. Bone marrow blasts represent >30% of nucleated cells, many showing monocytic features. The monocytic component in the peripheral blood is $>5 \times 10^9$ cells/l, including monoblasts, promonocytes, and monocytes. These cells show nonspecific esterase (NSE) positivity and, with a combined esterase procedure, cells containing both NSE and the granulocyte enzyme chloroacetate esterase can be demonstrated. M4E is a distinct subtype characterized by an increased number of abnormal eosinophilic precursors containing prominent basophilic granules and associated abnormalities (especially inversion) of chromosome.[16]

AML M5

In M5a, large monoblasts with ample basophilic cytoplasm constitute 80% or more of the bone marrow monocytic component. These blasts may contain vacuoles and small azurophilic granules. The nucleus is round or convoluted with a large nucleolus, and the cell membrane is usually irregular with pseudopodia. In M5b there are mostly promonocytes and abnormal monocytes in bone marrow and peripheral blood. The nuclei of most cells are folded, with inconspicuous nucleoli. For both subtypes, peroxidase stain is negative and the NSE reaction strongly positive. Serum and urinary lysozyme levels are significantly increased.

AML M6

Erythroid precursors are >50% and blasts >30% of nucleated erythroid cells showing moderate to marked dysplastic features, e.g., nuclear lobulation, multinuclearity, karyorrhexis, and cytoplasmic vacuoles. Intense periodic acid-Schiff (PAS) block positivity may be seen in erythroid cells and increased iron incorporation with or without ringed sideroblasts. Care is needed in differentiating M6 from MDS, where erythroid blasts are nearly always <30%.

AML M7

Diagnosis depends on either (a) demonstration of platelet peroxidase by ultracytochemistry or (b) immunophenotyping to identify CD41+ glycoprotein or the CD61+ glycoprotein on the blast surface. Blast morphology is polymorphic, with some cells resembling L1 lymphoblasts (see below) with scant cytoplasm and dense chromatin, while others mimic L2 (or M0) blasts. Sudan black and peroxidase stains are negative, with variable PAS and esterase positivity.

Immunophenotyping[26,27]

Almost all AML cells are found to be CD13+ and CD33+, but expression is variable, and some cases show only one marker. There is some correlation with FAB subtypes (see Table 9). CD34+ is usually associated with a more primitive cell type, whereas CD34– is associated with more differentiated cells, e.g., M3, M4, and M5. The M4 and M5 subtypes also show CD4+, CD11c+, CD14+, CD64+, and HLA-DR expression, with the monocytic lineage usually being CD4+, CD11c+, and sometimes associated with membrane Fc receptor expression. In the more primitive FAB subtypes, TdT and CD7 positivity occurs, TdT being strongly expressed in 20 to 30% and to a lesser extent in most cases.

Cytogenetic Studies[26]

Thirty different abnormalities have been described in 60 to 80% of the patients analyzed, and some of these show a clear correlation with a particular FAB group. The commonest findings are shown in Table 10. The presence of the t(8;21) abnormality in M2 AML and the inv(16) abnormality in eosinophilic M4 AML appears to be associated with favorable prognostic findings, although some anomalies such as monosomy 7 and abnormalities of chromosome 5 are associated with a poor prognosis. However, because these changes are mainly found in elderly AML patients, it is difficult to be certain that this association is genuine.

An inhibitor of JAK2 and FLT3 tyrosine kinases has been demonstrated.[31a] FLT3/internal tandem duplications have a poor prognosis.[31b]

TABLE 10

Association of FAB Type with Chromosomal Abnormalities in Acute Myeloid Leukemia

FAB Type	Abnormality	%
M2	t(8;21)(q22;q22)	15–18
M3	t(15;17)(q22;q21)	90+
M4E	inv(16)(p13;q22)	100
M5, M4	t del(11q)	35
All AML	+8	13
	–7	9
	–5 or del(5q)	6
	t (9;22)(q34;q11)	3

WHO Classification

The WHO classification is based upon cytogenetic abnormalities:

Acute myeloid leukemias with recurrent cytogenetic abnormalities[32–36]

- AML with t(8;21)(q22;q22), (*AML1/ETO*)
- AML with inv(16)(p13;q22) of t(16;16)(p13;q22), (*CBFβ/MYH11*)
- Acute promyelocytic leukemia (t(15;17)(q22;q12), (*PML/RARα*) and variants)
- AML with 11q23 (MLL) abnormalities

Acute myeloid leukemia with multilineage dysplasia

- With prior myelodysplastic syndrome
- Without prior myelodysplastic syndrome

Acute myeloid leukemia and myelodysplastic syndrome, therapy related

- Alkylating agent-related
- Topoisomerase II inhibitor-related

Acute myeloid leukemia not otherwise categorized

- Acute myeloid leukemia, minimally differentiated
- Acute myeloid leukemia without maturation
- Acute myeloid leukemia with maturation

- Acute myelomonocytic leukemia
- Acute monoblastic and monocytic leukemia
- Acute erythroid leukemia
- Acute megakaryoblastic leukemia
- Acute basophilic leukemia
- Acute panmyelosis with myelofibrosis
- Myeloid sarcoma

Clinical and Laboratory Features

Presenting Features

These are related to bone marrow failure, with varying degrees of **anemia**, **neutropenia**, and **thrombocytopenia**. In a minority of patients, there is infiltration of tissues with leukemia cells. Pallor and tiredness occur in the majority of patients in whom hemoglobin concentrations are below 9 g/dl. Cardiorespiratory symptoms, dyspnea, tachycardia, syncope, and angina occur, especially in the elderly. The anemia of marrow failure may be worsened by accelerated destruction of red blood cells and by hemorrhage. Frank **hemolysis** is rare. One-third of patients are neutropenic at presentation. Fever occurs commonly, infection with septicemia being the usual cause, especially where the granulocyte count is below 1.0×10^9/l. The infecting organisms are those associated with neutropenia.

Peripheral Blood and Bone Marrow Morphology

The appearances of the peripheral blood film relate directly to marrow failure and infiltration with leukemic cells. Red blood cell changes are unusual unless hemolysis or **disseminated intravascular coagulation** (DIC) is a feature. The total leukocyte count may be increased, normal, or decreased, but usually reflects the blast cell count. Neutropenia (granulocyte counts of 0.5 to 1.9×10^9/l) occurs at presentation, with few or absent cytoplasmic granules and occasionally a **Pelger-Hüet**-like abnormality of the nucleus. **Neutrophils** are usually functionally and metabolically abnormal, with poor mobilization and phagocytosis, decreased bacteriocidal capacity, and reduction in enzyme activities. Blast cells are seen in varying numbers from none to $>200 \times 10^9$/l, with a median of 15 to 20×10^9/l. There is usually a relationship between blast counts in blood and bone marrow, with the marrow often being completely replaced by leukemic cells. However, peripheral blood blasts frequently differ morphologically from marrow blasts, especially in leukemias with a monocytic component. Although seen in fewer than 20% of cases, **Auer rods** in leukemic blasts, due to coalescence of primary granules, are pathognomonic of AML. Similarly, **Phi bodies** are sometimes seen. Thrombocytopenia is variable.

Hemorrhagic Disorders

Bleeding associated with clotting-factor deficiency is uncommon and presents with bruising and hematomas at venepuncture sites. Occasionally, major hemorrhage into the gastrointestinal or genitourinary tracts may occur, the commonest cause of this type of bleeding being from DIC, especially as a complication of acute promyelocytic leukemia.

Coagulation Test Abnormalities

(See Table 11.) These include a prolonged **prothrombin time**, a prolonged **partial thromboplastin time**, reductions in **platelet count** and plasma **fibrinogen** level, and raised levels of **fibrin degradation products** (FDP). Some degree of DIC occurs in about 75% of patients

TABLE 11

Laboratory Features in Patients with Disseminated Intravascular Coagulation (DIC)
in Acute Promyelocytic Leukemia

Laboratory Test	Severe DIC (Clinical Problems)	Moderate DIC (Usually without Clinical Problems)	Mild DIC
Primary changes			
Plasma fibrinogen	1 g/l	normal	normal
Fibrin degradation products	↑	↑	↑
Platelet count	$<50 \times 10^9/l$	variable	variable
Thrombin time	↑↑	variable	normal
Secondary changes			
Prothrombin time	↑↑	variable	normal
Activated partial thromboplastin time	↑↑	variable	normal
Clotting time	low	variable	normal
Factors II, V, VII, IX–XII	low	variable	no
Plasminogen	low	low/normal	variable

with acute promyelocytic leukemia (FAB M3) and occasionally in other subtypes of AML. It is accompanied by secondary **fibrinolysis** and is thought to be related to the release of procoagulants from azurophilic granules in the leukemic blasts. Surprisingly, the severity of DIC does not seem to be directly related to the number of circulating blast cells, although it is often precipitated by the onset of treatment, presumably by releasing large amounts of procoagulant from damaged blast cells.

Leukostasis and Hyperviscosity

Ten to 15% of AML patients have a leukocyte count above $100 \times 10^9/l$ at presentation and have a poor prognosis due to **leukostasis** in the small blood vessels of the central nervous system (CNS) and lungs. Patients are often protected from **hyperviscosity** by anemia, but inappropriate transfusion may lead to CNS symptoms and potentially fatal coma. **Tumor lysis syndrome** — raised potassium, calcium, and creatinine, or frank renal failure — can be a complication following treatment with cytotoxic agents.

Metabolic Disorders

The lysis of leukemic cells spontaneously, or with treatment, liberates **purines**, which are converted first to hypoxanthine and then to xanthine and uric acid by the enzyme xanthine oxidase. **Hyperuricemia** with uric acid being deposited in various body tissues may occur, but its clinical effects are seen almost exclusively in the kidneys and joints. Gouty arthropathy due to deposit of uric acid crystals in large joints is rare at presentation, but occurs occasionally in patients not protected by the xanthine oxidase inhibitor, **allopurinol**.

Renal Tract Disorders

Renal function may also be impaired, usually manifesting as hypo- or hyperkalemia and renal failure. Children are particularly vulnerable, but in adults both are more common in monocytic leukemias due to lysozyme and direct leukemia infiltration. To protect against the effects of hyperuricemia, patients should be well hydrated and receive prophylaxis with allopurinol throughout the induction period. If urate uropathy occurs, treatment with rasburicase should be given immediately.

Cardiac Disorders

Cardiac problems include heart failure associated with severe anemia and damage to cardiac muscle from **anthracycline** cytotoxic antibiotics used in treatment. Clinical changes occur at cumulative doses above 500 mg/m², but microscopic changes are seen at lower doses. Echocardiography and radionuclide scintiangiography are the most accurate methods of monitoring clinical damage.

CNS Disorders

Neurological disturbances are unusual in AML.

Treatment

General Considerations

Modern treatment regimens for AML are very intensive and almost always produce periods of severe neutropenia and thrombocytopenia, which may also be associated with various complications, especially systemic infection and bleeding. Intensive treatments should not be attempted outside centers where hematological and microbiological expertise is concentrated. It is important that the patient have as good a performance status as possible prior to commencement of chemotherapy. Particular consideration should therefore be given to fluid balance/renal function, allopurinol therapy, antibiotics, and blood-product support. Once chemotherapy is commenced, vigilance is required to detect the earliest signs of tumor lysis syndrome so that appropriate action can be taken.

Induction Chemotherapy

It is important to identify patients suitable for intensive regimens, which will usually exclude elderly patients with major organ dysfunction. Induction treatments are now almost always with combinations of **cytotoxic agents** given to produce early complete remission (CR) and repeated for three or four cycles where possible. Most widely used are combinations of an anthracycline (usually doxorubicin, daunorubicin, or idarubicin) given for 3 days, and cytosine arabinoside for 5 to 10 days, with or without 6-thioguanine. For patients under 60 years, remission rates of 60 to 80% can be expected. Cytosine arabinoside doses vary, typically 100 to 200 mg/m²/day in divided doses by bolus or continuous IV or subcutaneous (s.c.) routes. Much larger doses of cytosine, such as 1 to 3 g/m² *b.i.d.* for 3 to 5 days are often used for consolidation therapy. The addition of 5 days etoposide, 100 mg/m² by IV infusion has increased CR rates to over 80%. A clinical cardiomyopathy occurs in many patients at cumulative anthracylinic doses above 550 mg/m², and there is evidence of microscopic damage to the myocardium at lower doses. Amsacrine and anthraquinone-mitoxantrone are being used as substitutes for anthracyclines in regimens similar to those outlined above, as there is some evidence that these drugs may be less cardiotoxic without any loss of cytotoxicity. In about 20% of younger patients, remissions are long lasting, and many of these patients will achieve cure. Combinations of vincristine and corticosteroids, prednisolone, or prednisone with IV infusions of cytosine arabinoside for 5 to 10 days give remission rates of less than 60%, but these regimens are much less intensive and are appropriate for elderly patients.

About 10 to 15% of AML patients are resistant to chemotherapy, and of these, and for the 60 to 70% of patients who relapse after a first remission, a variety of intensive treatments have been used.[36a] Although not well tolerated, combinations of high-dose cytosine (1 to 3 g/m²), with or without amsacrine or etoposide, have produced significant remission rates. The use of hematopoietic growth factors (see **Cell surface receptors**; **Colony stimulating factors**) may prove to be a useful adjunct to therapy.[37] Purine nucleoside analogs (fludarabine), pentostatin, and cladribine alone, in combination, or with other agents are under trial. Calicheamicin-conjugated anti-CD33 (gemtuzumab ozogamicin) may be useful either as monotherapy or in combination with other cytotoxic agents, e.g., cytosine.[39]

Leukostasis and Hyperviscosity

(See above.) Treatment must include early and, if necessary, repeated leukopheresis. Red blood cell transfusion should only be used when significant falls in white blood cell counts have been achieved.

Transplantation

- **Autologous stem cell transplantation**. The use of stem cells from either bone marrow or peripheral blood remains controversial, as several studies have failed to show any consistent benefit over four to five courses of intensive chemotherapy.[38]
- **Allogeneic stem cell transplantation** from matched sibling donors increases the chance of cure. There is evidence that this procedure may salvage 10 to 20% of patients with resistant AML, and a search for such a donor should be undertaken early after presentation

Specific Types of Acute Myeloid Leukemia
Acute Myeloid Leukemia with t(8;21)(q22;q22), (AML1/ETO)

One of the commonest cytogenetic aberrations found in 5 to 12% of patients, usually younger. Tumor manifestations may occur at presentation. Immunophenotyping shows coexpression of the lymphoid marker CD19 along with myeloid markers. There is usually a good response to chemotherapy, with high remission rates and long-term disease-free survival when treated with high-dose cytosine in the consolidation phase.

Acute Myeloblastic Leukemia, Minimally Differentiated (FAB: AML M0)

An acute leukemia with no evidence of myeloid differentiation morphologically, but which shows immunotypic features of myeloid markers to distinguish the cells from lymphoblasts. No specific cytogenetic abnormality has been demonstrated. The patients generally have a poor prognosis.

Acute Myeloblastic Leukemia without Maturation (FAB: AML M1)

Acute leukemia with hyperleukocytosis and a high percentage of bone marrow blasts, their myeloid nature being demonstrated by myeloperoxidase and Sudan black B staining. The occurrence is about 10% of cases, the majority being adults. The blasts express at least two myelomonocytic markers, including CD13, CD33, CD117, or myeloperoxide (MPO). CD34 is often positive. There are no specific cytogenetic characteristics. The disease typically has an aggressive course.

Acute Myeloblastic Leukemia with Maturation (FAB: AML M2)

An acute leukemia characterized by >20% blasts in the blood or bone marrow, with evidence of maturation by presence of >10% promyelocytes, myelocytes, and mature neutrophils. Eosinophil precursors and basophil/mast cell precursors may be increased. It comprises 30 to 45% of cases, occurring in all age groups. Immunophenotyping may express one or more myeloid-associated antigens, CD13, CD33, and CD15, and it may express CD117, CD34, and HLA-DR. Deletions and translocations involving chromosome

12b bands 11 to 13 are associated with bone marrow basophils. Response to aggressive cytotoxic therapy is good, but survival is variable. Cases with t(8;21) translocation have a favorable prognosis, whereas those with t(6;9)(p23;q34) have a poor prognosis.

Acute Promyelocytic Leukemia
(APL; FAB M3; AML with t(15;17)(q22;q12), PML/RARα and variants)

This type of leukemia is seen in 10 to 15% of adult AML and is characterized either by abnormal promyelocytes with distinctive large granules and multiple Auer rods (faggot cells) or, less commonly, by atypical hypogranular cells with bilobed or multilobed nuclei. High levels of Annexin II, a calcin-regulated protein and a phospholipid-binding cell-surface protein, are found. A particular clinical feature is excessive bleeding due to DIC (see hemorrhagic disorders, above), and associated with DIC, there is a high frequency of renal and respiratory failure. Treatment with all-*trans*-retinoic acid (tretinoin) is effective.[39] Calicheamicin-conjugated anti-CD33 (gezutab, gemtuzumab ozogamicin) can result in complete remission for those in relapse.[40] Arsenic trioxide can also induce remission in resistant patients.

Acute Myeloid Leukemia with inv(16)(p13;q22) or t(16;16)(p13;q22), (CBFβ/MYH11)

An acute myeloid leukemia usually showing monocytic and granulocytic differentiation and the presence of a characteristically abnormal eosinophil component (AMML Eo). It occurs in 10 to 12% of patients, usually of younger age. Myeloid sarcomas may occur at presentation or relapse. Immunophenotyping shows monocytic differentiation including CD4+, CD11b+, CD11c+, CD14+, CD36+, CD64+, and lysozome. Patients with inv(16) and t(16;16) show higher rates of complete remission when treated with high-dose cytarabine in the consolidation phase and also show high survival rates.

Acute Myelomonocytic Leukemia (FAB: AML M4)

An acute leukemia characterized by proliferation of both neutrophils and monocytes. The bone marrow shows >20% blasts. In addition to the blood and bone marrow changes, leukemic skin and gum deposits occur as small rose-colored papular lesions in up to one-third of patients. Immunophenotyping shows both myeloid (CD13, CD33) and monocytic (CD4+, CD11b, CD11c, CD14+, CD36+, CD64+, and lysozyme) markers. There are no specific cytogenetic abnormalities. Response to aggressive therapy is variable, but those with an associated chromosome 16 abnormality have a favorable outcome.

Acute Monoblastic and Monocytic Leukemia
(FAB: AML M5a and AML M5b; AML with 11q23 (MLL) abnormalities)

An acute leukemia with 80% or more cells of monocytic lineage. It comprises 5 to 8% of cases, most commonly in younger patients. Hemorrhagic disorders are frequent. There is a high incidence of extramedullary disease, especially hepatosplenomegaly, lymphadenopathy, skin deposits, and gingival infiltration. Renal failure and hypokalemia occur frequently in association with increased levels of serum lysozyme, especially in patients with high blood blast counts. These features make for a worse prognosis than in most other subtypes of AML. Leukemic cells are seen in the cerebrospinal fluid of 10 to 30% of patients. Immunophenotyping shows both myeloid and monocytic markers. There are strong associations with deletions and translocations of chromosome 11, band q23 (MLL). This type of acute leukemia carries a poor prognosis, with early death for most patients. Some treatment protocols now include CNS prophylaxis for these patients. (See **Acute lymphoblastic leukemia**.)

Acute Erythroid Leukemias (FAB: AML 6; Erythroid Myelosis; Di Guglielmo's Disease)

Erythroleukemia

The presence in the bone marrow of >50% erythroid precursors in the entire nucleated cell population and >30% myeloblasts in the nonerythroid population. It comprises approximately 5 to 6% of cases of AML and is predominantly a disease of adults. Erythroblasts and immature normoblasts may predominate, sometimes with bizarre giant and multinucleated forms, so that the changes may resemble those seen in **megaloblastosis**. The erythroblasts have no immunophenotypic markers and there are no specific cytogenetic abnormalities. The disorder follows an aggressive clinical course, with poor response to treatment, although a subset of patients have a slowly progressive disease and may survive up to a year on blood-product support alone.

Pure Erythroid Leukemia

A rare disorder confined to neoplastic proliferation of precursors of the erythroid lineage. They can be identified by their expression of glycophorin A and hemoglobin A and by the absence of MPO and other myeloid markers. The disorder has a rapid clinical downward course.

Acute Megakaryoblastic Leukemia (FAB: AML M7)

An acute leukemia with >50% of megakaryoblasts in the bone marrow. It can occur *de novo* or by transformation from a **myeloproliferative disorder**. It is more frequent in patients with an underlying **Down syndrome**. The megakaryoblasts can be identified ultrastructurally by the presence of platelet peroxidase and immunophenotypically by expression of platelet glycoproteins Ib, IIb, or IIIa and CD41$^+$ and CD61$^+$. It often presents as a form of pancytopenia. Hepatosplenomegaly is rare. Those associated with t(1;22) often have abdominal masses. The prognosis is poor.

Acute Basophilic Leukemia

A rare form of acute leukemia with primary differentiation to mature and immature basophils. The blasts express myeloid markers CD13 and CD33 and early hematopoietic markers CD34 and class II HLA-DR. A few cases present *de novo* with a Philadelphia chromosome translocation t(9;22)(q34;q11). There is usually a poor prognosis.

Acute Eosinophilic Leukemia

A very rare form of acute leukemia in which blasts show a characteristic cyanide-resistant peroxidase reaction. Two cases have been reported as showing t(10;14) abnormality.

Acute Myeloid Leukemia with Multilineage Dysplasia

An acute myeloid leukemia with >20% blasts in blood or bone marrow and dysplasia in two or more myeloid cell lines. It may be seen *de novo* or following myelodysplasia. It is more usual in the elderly and rare in children. Pancytopenia is often a presenting feature. Immunophenotyping is heterogeneous, with blasts showing panmyeloid markers (CD13,CD33) with CD34$^+$. Chromosome abnormalities are those of myelodysplasia. Patients of this group have a poor prognosis.

Acute Myeloid Leukemia and Myelodysplastic Syndrome, Therapy Related

These are acute myeloid leukemias arising as a result of cytotoxic or radiation therapy, particularly common after treatment of myelomatosis.

Alkylating Agent-Related

This usually presents 5 to 6 years following exposure to an alkylating agent. There is a panmyelosis, with no specific immunotypic or cytogenetic features. It is refractory to treatment, with a poor prognosis.

Topoisomerase II Inhibitor-Related

This usually presents from 12 to 130 months posttherapy, with a median of 33 to 34 months. The major therapeutic agents implicated are epidophyllotoxins: etoposide and teniposide. Anthracyclines have also been implicated. Most cases are monoblastic or myelomonocytic. Cytogenetics shows a balanced translocation involving 11q23 (*MLL* gene), primarily t(9;11), t(11;19), and t(6;11)(11;69). They respond poorly to cytotoxic treatment, and the prognosis is generally poor.

ACUTE MYELOSIS WITH MYELOFIBROSIS

(Acute myelofibrosis; acute myelosclerosis; acute myelodysplasia with myelofibrosis) An acute proliferation in the bone marrow of erythroid, myeloid, and megakaryocytic cells associated with **myelofibrosis**. This a rapidly progressive illness characterized by **pancytopenia** but minimal (if any) **hepatomegaly** or **splenomegaly**. It needs to be distinguished from myelofibrosis secondary to other types of **acute myeloid leukemia** (AML) or **acute lymphoblastic leukemia** (ALL) and fibrotic **myelodysplasia**. Some patients have responded to AML-type therapy, but often only supportive care, e.g., **red blood cell transfusion**, **platelet transfusion**, is indicated.

ACUTE-PHASE INFLAMMATORY RESPONSE

(Systemic inflammatory response syndrome; SIRS) The systemic counterpart of inflammation to tissue injury. This may be the result of trauma, infection, ischemia, or immunological and malignant stimuli. It arises from both acute and chronic inflammation and is mediated by cytokine release from histiocytes (macrophages) of interleukin-1, IL-6, and TNF-α. This cytokine release results in:

Pyrexia through IL-1 acting as an endogenous pyrogen on the thermosensory center of the anterior thalamus; stimulation of the vasomotor center causes vasoconstriction to reduce heat loss or muscular contraction to increase metabolic activity of the liver; IL-1 stimulates the formation of PGE2 as a signal of rise in body temperature

Innate immune response with:

- Release from **histiocytes** of C3e, a low-molecular-cleavage product of C3, C4, and interferon-γ
- Increased hepatic synthesis of proteins (through IL-6, IL-1, TNF-α, and IFN-γ) pentaxin (**C-reactive protein**), serum amyloid protein A, **complement** components, **fibrinogen**, **haptoglobin**, orasomucoid, ceruloplasmin, mannose-binding lectin, and **ferritin**

Adaptive immune response through **B-lymphocyte** activation, with increased **immunoglobulin** synthesis and consequent rise in **viscosity of plasma**, as reflected by the rise in the **erythrocyte sedimentation rate**

Release of damage-limiting protein α1-antitrypsin

Increase in skeletal muscle protein catabolism

Thrombocytosis

The level of response can be determined by measuring the rise of C-reactive protein in plasma within the first 24 h of injury and by the rise of plasma fibrinogen. There is a concomitant fall of prealbumin, albumin, and **transferrin** during the following 2 to 6 days.[41] These plasma changes are complex and are usually assessed by measuring the plasma viscosity or the increase in the aggregate formation of red blood cells by the erythrocyte sedimentation rate (ESR). Other indicators are the **platelet count** and the level of serum orasomucoid. These quantitative measurements of the acute-phase response are used as an indicator of the occurrence, extent, and response to treatment of both acute and chronic inflammation, including myocardial infarction.

SIRS can lead to a compensatory **anti-inflammatory response syndrome** (CARS) with immune deactivation and increased susceptibility to infection. This can be prevented by the administration of **granulocyte colony stimulating factor**.[42]

ACUTE PROMYELOCYTIC LEUKEMIA

(APL) See **Acute myeloid leukemia**.

ADAPTIVE IMMUNITY

See **Immunity**.

ADAMTS-13

A member of the ADAMTS family of **metalloproteases** (a disintegrinlike metalloprotease with thrombospondin type-1 motifs). It is coded for on chromosome 9q34 by a 29-exon gene and spans 37 kb. ADAMTS-13 is a 1427 amino acid single peptide containing motifs that define the ADAMTS family. ADAMTS-13 cleaves high-molecular-weight **Von Willebrand Factor** (VWF). Normally VWF binds to constituents of subendothelial connective tissue and then binds to **platelet** GpIb, permitting platelet adhesion. Shear stress in flowing blood enables ADAMTS-13 to cleave a specific Tyr-Met bond A2 domain of VWF subunits. Cleavage reduces VWF multimer size and inhibits platelet adhesion. Recent advances now suggest that deficiency of this specific VWF-cleaving protease promotes tissue injury in **thrombotic thrombocytopenic purpura** (TTP). Because VWF multimers participate in the formation of platelet thrombi, proteolytic cleavage of VWF multimers normally limits platelet thrombus growth, and failure to cleave VWF appears to encourage microvascular thrombosis. **Autoantibodies** that inhibit ADAMTS-13 cause sporadic TTP, and mutations in the *ADAMTS-13* gene cause an autosomally recessive form of chronic relapsing TTP.

ADDISONIAN PERNICIOUS ANEMIA

Deficiency of **cobalamin** (vitamin B_{12}) as a result of malabsorption secondary to **gastric disorders** in which there is deficiency of intrinsic-factor secretion due to severe atrophic gastritis with achlorhydria. This form of **anemia** was first described by Thomas Addison in 1848 and is the most common cause of cobalamin deficiency. There can be a familial tendency, but the genetic mechanism is unclear. Immunological phenomena are demonstrable, the best-recognized being circulating antibodies against gastric parietal cells and intrinsic factor. Two specific intrinsic-factor antibodies occur in 60% of patients with pernicious anemia:

Type I: the "blocking" antibody, which competes with the binding site for cobalamin on intrinsic factor

Type II: the "binding" antibody, which binds at an unrelated site

Antiparietal cell antibody directed mainly against microsomes is only a marker for atrophic gastritis and is not diagnostically specific for pernicious anemia. Other immunological associations have been described but not substantiated. These include local gastric T-lymphocyte activity, lymphocytotoxic antibody, and partial reversal of the gastric defect by steroids. The association with thyroid disorders, gastric cancer, and gastric carcinoid is well recognized.

Incidence

Pernicious anemia is not a common disease, but is seen most frequently in northern Europeans, slightly more commonly in females. Peak incidence is above the age of 60 years. Occurrence before the age of 40 years is uncommon, except in women of African descent. A rare form occurs during the first two decades of life, juvenile pernicious anemia. This may be a consequence of congenital absence of functional intrinsic factor with an otherwise normal gastric mucosa or, more commonly, a precocious variant of the adult form in which about 90% of patients have circulating antibody against intrinsic factor.

Clinical Features

Patients usually present with symptoms of anemia, but sometimes this is absent and neurological features predominate, the earliest being subjective, such as symmetrical paresthesias, numbness, and loss of sensation. **Subacute combined degeneration of the spinal cord** may follow later. Mental abnormalities ranging from irritability to dementia may be present. Classically, the skin has a lemon-yellow coloration resulting from the combination of pallor and mild jaundice. The latter reflects both **ineffective erythropoiesis** and **hemolysis**. Mouth symptoms are common, as is weight loss, anorexia, and bowel disturbances, particularly looseness of stools. Gastritis usually produces no obvious symptoms.

Laboratory Features

These include anemia, raised **red blood cell** mean cell volume (MCV), **leukopenia**, and **thrombocytopenia**, the last occasionally being severe. Although **megaloblastosis** is usually conspicuous in the marrow, it is by no means universal. In some instances, the MCV may be normal due to balancing by a population of microcytes resulting from concomitant **iron deficiency** or **thalassemia**. Occasionally, a low serum cobalamin level is the only pointer to the disorder.

The diagnosis of pernicious anemia is substantiated by the demonstration of lack of intrinsic factor. This is achieved directly by assay of gastric juice for this protein, which is highly specific and highly sensitive, constituting the most reliable diagnostic test. In practice, the **Schilling test** is most commonly used. An abnormal result must be correctable when repeated with oral intrinsic factor. Unreliable results may occur because of incomplete urine collection. Additionally, transient ileal malabsorption occurs in a significant proportion of pernicious anemia patients, with resulting failure of correction following oral intrinsic factor. An indirect diagnostic approach consists of assaying serum for anti-intrinsic factor antibody. The presence of this antibody in a cobalamin-deficient individual virtually ensures the diagnosis, provided the test is not done within 48 h of the patient receiving parenteral cobalamin.

Treatment

Intramuscular hydroxocobalamin is the usual method for replacement of cobalamin deficiency. It is given in an initial dose of 1 mg, three times a week for 2 weeks, then 1 mg every 3 months. Since the gastric defect is irreversible, treatment requires lifelong replacement of cobalamin. The prognosis is excellent.

ADENINE

A purine base of **DNA**, **RNA**, some **nucleotides**, and their derivatives.

ADENOSINE

A regulator of erythrocyte **purine** nucleotides, possibly metabolized by erythrocytes *in vivo*. It is used as an experimental blood preservative.

ADENOSINE DEAMINASE

(ADA) An enzyme of the **purine** salvage pathway that converts adenosine to inosine and deoxyadenosine to deoxyinosine. A variant of severe combined **immunodeficiency** (SCID) is associated with a deficiency of ADA. An autosomally dominant increase of ADA activity affects erythroid, myeloid, and lymphoid cell populations, causing reduced production of **adenosine triphosphate** (ATP), resulting in a **hereditary nonspherocytic hemolytic anemia**.

ADENOSINE DIPHOSPHATE

(ADP) A **nucleoside** diphosphate found in all cells. ADP is phosphorylated to **adenosine triphosphate** (ATP) during catabolic reactions for energy production and is produced by hydrolysis of ATP. Adenosine diphosphate is an important agonist in **platelet** aggregation. ADP is stored in platelet dense granules or may induce platelet aggregation. ADP from damaged tissues or secretion from previously stimulated platelets may also induce platelet aggregation. ADP receptors on platelets can be divided into two groups: G-protein-coupled receptors (P2Y) and inotropic receptors (P2X). Platelets contain two P2Y receptors: P2Y1 and P2Y12. ADP binding to P2Y1 is necessary for platelet aggregation but is in itself insufficient, while P2Y12 modulates platelet shape change and **cyclic adenosine monophosphate** (cAMP) level (reduced by ADP).

P2X-class receptors from a fast ATP gated cation channel binding ADP leads to increased calcium influx into platelets, protein phosphorylation, production of **thromboxane A2**, increased fibrinogen receptor expression (**glycoprotein IIb/IIIa**, GpIIb/IIIa), aggregation, and secretion.

P2Y12 receptors form the target of the thienopyridine-derivative antiplatelet drugs ticlopidine and clopidogrel. Metabolites of these drugs covalently (irreversibly) bind the receptor, preventing receptor-coupled inhibition of adenyl cyclase activity (reducing cAMP).

ADENOSINE KINASE

An enzyme that converts adenosine to adenosine monophosphate as part of the salvage pathway of **purine** synthesis.

ADENOSINE MONOPHOSPHATE

See also **Cyclic adenosine monophosphate**.
(AMP) A nucleoside monophosphate that is a hydrolytic product of **adenosine diphosphate** (ADP) and **adenosine triphosphate** (ATP). It is a nucleoside component of **DNA** and **RNA** in deoxyribosyl and ribosyl forms. It is converted to cyclic AMP (cAMP) by adenylate cyclase. Intracellular concentrations of cAMP rise rapidly in response to extracellular signals, especially hormonal, and fall rapidly due to activity of intracellular phosphodiesterase. The level of AMP determines the rates of many intracellular biochemical pathways, especially those concerned in **platelets** with G-proteins.

ADENOSINE TRIPHOSPHATE

(ATP) A nucleoside triphosphate that is hydrolyzed from **adenosine diphosphate** (ADP) and, from it, can be regenerated by oxidation. ATP functions in **red blood cells** include:

Pump priming of glycolysis

Providing energy for the **cation pump**, which is responsible for pumping sodium out and potassium into the cell against concentration gradients

Providing energy for the calcium pump to maintain a low level of cytoplasmic calcium against a strong concentration gradient

Phosphorylation of many cell proteins by protein kinase enzymes, including those of the membrane cytoskeleton

Maintaining red blood cell shape

Regulating **oxygen affinity to hemoglobin**

Protein complexes, ATPases, are responsible for converting electrical potential energy into ATP, catalyzing hydrolysis of ATP to ADP, and releasing energy that is used in the cell. For example, sodium/potassium ATPase is a membrane enzyme that moves sodium out of the cell and potassium in at a fixed ratio of $2K^+$: $3Na^+$, Ca^{2+}, and Mg^{2+}. ATPases are the transport enzymes of cell membranes, including those of red blood cells.

ADENYLATE KINASE

(AK) A phosphotransferase enzyme acting in the reaction ATP:AMP and essential in the salvage pathway for maintaining steady-state levels of **adenosine triphosphate** (ATP). Adenine cannot be incorporated into the adenine nucleotide pool in the absence of AK, which converts **adenosine monophosphate** (AMP) generated by AK to ATP. **Adenosine kinase** deficiency leads to accumulation in the cell of AMP, thereby promoting cell death through deamination or dephosphorylation. Although not functioning in the glycolytic pathway, deficiency of AK has been associated with **hereditary nonspherocytic hemolytic anemia**.

ADHESION MOLECULES: CELL-CELL

See **Cellular adhesion molecules**.

ADRENAL GLANDS

Endocrine glands situated superior to each kidney. Each has a cortex that synthesizes corticosteroid hormones from cholesterol. Glucocorticoids (cortisone) have anti-inflammatory actions; mineralocorticoids (e.g., aldosterone) maintain the body's electrolyte balance. The synthetic **corticosteroids**, prednisone and prednisolone, are extensively used for induction of **apoptosis** in the treatment of hematological malignant disorders. They are also used for antibody suppression in the treatment of **autoimmune disorders**.

Disorders of Adrenal Glands

These are associated with a number of hematological disorders.

Adrenal hyperplasia (Cushing's disease):
- Secondary **erythrocytosis** due to increase in erythropoietin secretion
- Reduced levels of circulating **eosinophils** and **lymphocytes**

Adrenal deficiency:
- Acute adrenal failure (Waterhouse-Friedrich syndrome) caused by **disseminated intravascular coagulation** (DIC)
- Chronic adrenal insufficiency (Addison's disease) can be an autoimmune disorder. When this is associated with the gastric disorder of intrinsic-factor deficiency, **megaloblastosis** also may be present. Otherwise, there is **normocytic anemia** with **leukopenia** and **eosinophilia**.

ADRENALINE

(Epinephrine) A hormone secreted by the **adrenal glands**, its principal action being on smooth muscle via the sympathetic nervous system. It is also a weak platelet-aggregating agent and serves primarily as a potentiator of **platelet** responses to other agonists. It binds to 2-adrenergic receptors on the platelet membrane. Concentrations of adrenaline that are required to induce platelet aggregation and secretion *in vitro* are never seen *in vivo*. It does not induce platelet shape change. Adrenaline probably functions as a modulator, stimulating other, stronger agonists. Responses of normal individuals to adrenaline in platelet-aggregation studies are very variable. However, poor responses may be associated with mild bleeding in the so-called weak-agonist-response defect (see **Platelet-function disorders**).

ADULT T-CELL LEUKEMIA/LYMPHOMA

See also **Lymphoproliferative disorders**; **Non-Hodgkin lymphoma**; **Peripheral T-cell lymphoma**.

(ATLL; Rappaport: diffuse poorly differentiated lymphoma, mixed lymphocytic-histiocytic, or histiocytic lymphoma; Lukes-Collins: T-immunoblastic sarcoma) A high-grade T-cell non-Hodgkin lymphoma characterized by its association with **human T-lymphotropic virus** (HTLV)-1, clonally integrated with the virus present in all cases. The histological appearances are variable, the pattern in lymph nodes being diffuse with a mixture of medium-sized and large atypical cells with pronounced polymorphism and nuclear pleiomorphism. Multinucleated giant cells resembling **Reed-Sternberg cells** in some may cause a morphological resemblance to **Hodgkin disease**. Cells with polylobed nuclei ("clover leaf" or "flower" cells) are common in the peripheral blood. Bone marrow infiltration, when present, is diffuse and ranges from sparse to marked. The **immunophenotype** of the tumor cells is CD25+ in most cases, CD4+ in 65% of cases, and CD7+ in some; they also express CD2+, CD3+, and CD4+ antigens. **Cytogenetic analysis** shows T-cell receptor (TCR) genes clonally rearranged and clonally integrated HTLV-1 genomes in each case. The postulated origin is a peripheral CD4+ T-cell in various stages of transformation.

Incidence

The majority of patients are adults who have antibodies to HTLV-1. Most cases occur in Japan, but there is an endemic focus in the Caribbean, and sporadic cases occur in the U.K. and U.S., where the antibody incidence reflects the level of the ethnic Caribbean population.

Clinical Features

Several clinical variants have been described:

"Acute" with lymphocytosis (50×10^9/l), **lymphadenopathy, hepatosplenomegaly** (50% of cases), hypercalcemia with lytic bone lesions (75%), skin disorders (30%); less frequently, there is infiltration of the central nervous system and the lungs. Medium survival is less than 1 year.

Lymphadenopathy without leukemia or marrow involvement; a rare form.

Chronic disorder with mild lymphocytosis, without hypercalcemia or hepatosplenomegaly, but with skin disorders. The survival time is around 1 to 2 years.

Smoldering form (rare), with a very indolent course, often with skin rashes and demonstrably clonal.

Laboratory Features

Lymphocytosis characterized by marked pleomorphism, with cells similar to those seen histologically in lymphoid follicles.

Hypercalcemia, which is rare in other T-cell malignancies, the level being parallel to the blood blast count and not associated with osteolytic lesions. It is thought that the excess calcium is produced by cytokine release from malignant cells.

Detection of HTLV-1, which confirms the diagnosis. Serum antibodies (IgM, IgG) to HTLV-1 can be detected by ELISA or radioimmune assays. Viral or proviral protein can be demonstrated in infected cells.

Staging

See **Lymphoproliferative disorders**.

Treatment

See **Non-Hodgkin lymphoma**.
Without treatment, survival time ranges from a few days to more than a year, with a protracted course in chronic forms of the disease. A limited response has sometimes been achieved with **deoxycoformacin**.

ADVERSE REACTIONS TO DRUGS

(ADR) The unwanted or unexpected reaction following the administration of an appropriate dose to a patient. They are a common occurrence and, if not recognized, can cause death. Any system can be affected and any disorder mimicked. The manifestations may or may not be dose related, and the severity varies from patient to patient. These reactions can be broadly classified into:

Type A: an exaggeration of the known primary or secondary pharmacology of the drug

Type B: no apparent dose-response relationship

The adverse effects are due to either variability in drug metabolism or to their effect on the drug targets. Environmental factors, such as alcohol or tobacco excess, diet, and the presence of other disease than that undergoing treatment, may be important. These could interact with genetic predisposition of the recipient. In an increasing number of ADRs, a genetic factor has been identified.[43,43a] Some have association with **human leukocyte antigens** (HLA).

Any drug can induce a **hypersensitivity** reaction in a susceptible person:

Type I: anaphylaxis, e.g., penicillin

Type II: drug-induced hemolytic anemia

Type III: localized Arthus reaction

Type IV: contact dermatitis

Table 12 lists Type B adverse reactions of drugs on the hematopoietic system and related tissues. Any known genetically determined enzyme deficiency or HLA association is given in parentheses. Several polymorphisms affect genes encoding cytochrome P-450 enzymes. The list is not exhaustive and only gives frequent reactions by commonly used drugs.

TABLE 12

Hematological Adverse Reactions to Drugs

Drug	Adverse Reaction
Acetazolamide	bone marrow hypoplasia
Acetanilide	**hemolytic anemia** (G6PD deficiency)
Alkylating drugs	bone marrow hypoplasia, **acute myeloid leukemia** (P-450 enzymes)
Allopurinol	**agranulocytosis**, vascular purpura
Amides (lignocaine group)	severe delayed hypersensitivity reaction
Aminophylline	**platelet**-function inhibition
Aminopyrine	agranulocytosis
Amitriptyline	agranulocytosis
Anticonvulsants	**acute intermittent porphyria** (porphobilinogen deaminase)
Antihistamines	platelet-function inhibition
Antimetabolites	bone marrow hypoplasia, acute myeloid leukemia (P-450 enzymes)
Antituberculous drugs	sideroblastic anemia
Arsenicals	vascular purpura
Aspirin	bone marrow hypoplasia, **immune thrombocytopenia**, platelet-function inhibition, vascular purpura, **Henoch-Schönlein purpura**
Atropine	vascular purpura
Azathioprine	**megaloblastosis,** bone marrow hypoplasia (thiopurine methyltransferase)
Barbiturates	vascular purpura, acute intermittent porphyria (porphobilinogen deaminase)
β-Adrenoceptor blockers	platelet-function inhibition
Bismuth	vascular purpura
Busulfan	sideroblastic anemia (P-450 enzymes)

TABLE 12 (continued)

Hematological Adverse Reactions to Drugs

Drug	Adverse Reaction
Caffeine	platelet-function inhibition
Captopril	agranulocytosis
Carbamazepine	severe hypersensitivity reactions (DR3, DQ2), agranulocytosis, immune thrombocytopenia, vascular purpura
Carbimazole	agranulocytosis
Cephalosporins	hemolytic anemia, immune thrombocytopenia
Cephalothins	platelet-function inhibition
Chloral hydrate	vascular purpura
Chloramphenicol	bone marrow hypoplasia/agranulocytosis/thrombocytopenia, hemolytic anemia (G6PD deficiency), sideroblastic anemia, vascular purpura
Chlordiazepoxide	bone marrow hypoplasia
Chloromycetin	sideroblastic anemia
Chloroquine	agranulocytosis
Chlorpheniramine	bone marrow hypoplasia
Chlorpromazine	bone marrow hypoplasia/agranulocytosis
Chlorpropamide	agranulocytosis, vascular purpura
Chlorthiazide	vascular purpura
Cimetidine	immune thrombocytopenia, vascular purpura
Clofibrate	platelet-function inhibition
Clozapine	bone marrow hypoplasia (B38, DR4, DR3)
Colchicine	bone marrow hypoplasia, **cobalamin** malabsorption
Corticosteroids	**neutrophilia**, platelet-function inhibition, thrombocytopenia
Cotrimoxazole	megaloblastosis, thrombocytopenia
Cycloserine	megaloblastosis, sideroblastic anemia
Cytosine arabinoside	megaloblastosis (P-450 enzyme)
Cytotoxic antibiotics	bone marrow hypoplasia, acute myeloid leukemia (P-450 enzymes)
Dapsone	agranulocytosis, **methemoglobinemia (methemoglobin reductase)**, hemolytic anemia (G6PD deficiency)
Daunorubicin	bone marrow hypoplasia (HFE gene of hereditary hemochromatosis), hemolytic anemia (G6PD deficiency)
Diaminodiphenylsulfone	hemolytic anemia (G6PD deficiency)
Diazepam	immune thrombocytopenia
Dichlorphenamide	bone marrow hypoplasia
Diethylstilbestrol	thrombocytopenia
Digoxin/digitoxin	immune thrombocytopenia, vascular purpura
Diphenylhydantoin	megaloblastosis
Dipyrone	bone marrow hypoplasia (A24, B7, DQ1)
Doxepin	agranulocytosis
Doxorubicin	hemolytic anemia (G6PD deficiency)
Erythromycin	Henoch-Schönlein purpura
Estrogens	vascular purpura, acute intermittent porphyria (porphobilinogen deaminase)
Ethoxzolamide	bone marrow hypoplasia
Etoposide	bone marrow hypoplasia (P-450 enzyme)
Fenbrufen	vascular purpura
Fenoprofen	platelet-function inhibition
Fluorodeoxyuridine	megaloblastosis
5-Fluorouracil	megaloblastosis, bone marrow hypoplasia (dihydropyridine dehydrogenase)
Frusemide	immune thrombocytopenia, vascular purpura
Glutethimide	megaloblastosis
GM-CSF	eosinophilia
Gold salts	bone marrow hypoplasia/agranulocytosis (DR3), immune thrombocytopenia, vascular purpura
Heparin	immune thrombocytopenia
Hepatitis A and B vaccine	immune hemolytic anemia, immune thrombocytopenia
Hydralazine	agranulocytosis, **systemic lupus erythematosus** (DR4)
Hydroxychloroquine	platelet-function inhibition

TABLE 12 (continued)

Hematological Adverse Reactions to Drugs

Drug	Adverse Reaction
Hydroxyurea	bone marrow hypoplasia, megaloblastosis (P-450 enzymes)
Ibuprofen	agranulocytosis, platelet-function inhibition (P-450 enzymes)
Imipramine	agranulocytosis
Indomethacin	platelet-function inhibition, vascular purpura
α-Interferon	immune thrombocytopenia
Interleukin-2	eosinophilia
Iodides	vascular purpura
Isoniazid	sideroblastic anemia, vascular purpura
Isoprenaline	platelet-function inhibition
Levamisole	bone marrow hypoplasia (B27)
Lincomycin	sideroblastic anemia
Mefenamic acid	vascular purpura
Meprobamate	bone marrow hypoplasia, vascular purpura
6-Mercaptopurine	megaloblastosis, bone marrow hypoplasia (thiopurine methyl transferase)
Mercury	vascular purpura
Metformin	cobalamin malabsorption
Methazolamide	bone marrow hypoplasia
Methotrexate	megaloblastosis (P-450 enzymes)
Methyldopa	hemolytic anemia, immune thrombocytopenia, vascular purpura
Methylene blue	hemolytic anemia (G6PD deficiency)
Methysergide	platelet-function inhibition
Morphine	vascular purpura
Nalidixic acid	methemoglobinemia (methemoglobin reductase), hemolytic anemia in G6PD deficiency
Naproxen	platelet-function inhibition, vascular purpura
Neomycin	cobalamin malabsorption
Niridazole	methemoglobinemia (methemoglobin reductase), hemolytic anemia (G6PD deficiency)
Nitrofurantoin	hemolytic anemia (G6PD deficiency), platelet-function inhibition, vascular purpura
Nitrous oxide	megaloblastosis
Oral contraceptives	megaloblastosis, **venous thromboembolic disease** (factor V Leiden)
Pamaquine	hemolytic anemia (G6PD deficiency)
Papaverine	platelet-function inhibition
Para-amino salicylate	immune thrombocytopenia
Paracetamol	immune thrombocytopenia, Henoch-Schönlein purpura
Penicillamine	bone marrow hypoplasia, sideroblastic anemia, thrombocytopenia
Penicillins	hemolytic anemia, immune thrombocytopenia, platelet-function inhibition, vascular purpura, Henoch-Schönlein purpura
Pentamidine isethionate	bone marrow hypoplasia
Pentaquine	hemolytic anemia (G6PD deficiency)
Phenacetin	vascular purpura, Henoch-Schönlein purpura
Phenformin	cobalamin malabsorption
Phenobarbitone	cobalamin malabsorption
Phenothiazines	platelet-function inhibition
Phenylbutazone	bone marrow hypoplasia/agranulocytosis, platelet-function inhibition
Phenytoin	agranulocytosis, immune thrombocytopenia, vascular purpura
Piperazine	vascular purpura
Piroxicam	vascular purpura
Potassium chloride	cobalamin malabsorption
Primaquine	methemoglobinemia (methemoglobin reductase), hemolytic anemia (G6PD deficiency)
Primidone	bone marrow hypoplasia, megaloblastosis
Procainamide	agranulocytosis
Procaine	platelet-function inhibition, vascular purpura
Procarbazine	bone marrow hypoplasia, acute myeloid leukemia

TABLE 12 (continued)

Hematological Adverse Reactions to Drugs

Drug	Adverse Reaction
Prochlorperazine	bone marrow hypoplasia
Proguanil	megaloblastosis
Propylthiouracil	agranulocytosis
Prostaglandins	platelet-function inhibition
Pyrimethamine	bone marrow hypoplasia, megaloblastosis
Quinidine	agranulocytosis, methemoglobinemia (methemoglobin reductase), hemolytic anemia (G6PD deficiency), immune thrombocytopenia, vascular purpura
Quinine	hemolytic anemia, immune thrombocytopenia, vascular purpura
Quinolones	vascular purpura, hemolytic anemia (G6PD deficiency)
Reserpine	platelet-function inhibition, vascular purpura
Rifabutin	agranulocytosis/thrombocytopenia
Rifampicin	immune thrombocytopenia
Ropivacaine	delayed hypersensitivity allergic reaction
Sodium valproate	immune thrombocytopenia
Streptomycin	bone marrow hypoplasia
Sulfonamides	megaloblastosis, bone marrow hypoplasia/agranulocytosis, methemoglobinemia (methemoglobin reductase), hemolytic anemia (G6PD deficiency), immune thrombocytopenia, vascular purpura, Henoch-Schönlein purpura, acute intermittent porphyria (porphobilinogen deaminase)
Sulfonylurea	immune thrombocytopenia
Tetracycline	hemolytic anemia
Thiazide diuretics	bone marrow hypoplasia, immune thrombocytopenia, vascular purpura, Henoch-Schönlein purpura
Thiazolesulfone	hemolytic anemia (G6PD deficiency)
Thioguanine	megaloblastosis (P-450 enzymes)
Thiouracil	agranulocytosis
Ticlopidine	agranulocytosis
Tolbutamide	bone marrow hypoplasia/agranulocytosis, hemolytic anemia, vascular purpura
Triamterine	megaloblastosis
Trimethoprim	bone marrow hypoplasia, immune thrombocytopenia, megaloblastosis
Valproate	immune thrombocytopenia
Vinca alkaloids	bone marrow hypoplasia (P-450 enzymes)
Vitamin K	methemoglobinemia (methemoglobin reductase)
Warfarin	vascular purpura (P-450 enzymes)
Zidovudine	bone marrow hypoplasia, megaloblastosis

Note: Genetic abnormalities are given in parentheses.

Source: Extracted from *British National Formulary*, 51, British Medical Association and Royal Pharmaceutical Society, London, 2006; and Pirmohamed, M. and Park, P.K., *Trends Pharmacol. Sci.*, 22, 298–305, 2001. With permission.

AFFINITY AND AVIDITY

Measures of the strength of interaction of an **antibody** with its **antigen**. The equilibrium between antigen (Ag), antibody (Ab), and the antigen–antibody complex (Ag:Ab) is represented by

$$Ag + Ab \leftrightarrow Ag:Ab$$

If k_1 is the rate constant for the forward reaction and $k - 1$ the rate constant for the back reaction, the affinity of antibody for antigen is defined by the association (equilibrium) constant K (measured in l mol^{-1}) in the equation, where terms in square brackets represent concentrations of reactant/product:

$$K = k_1/k - 1 = [Ag:Ab]/([Ag] \times [Ab])$$

Experimentally, K is best determined in monovalent hapten-antihapten systems.

Avidity is an ill-defined term. It depends partly on affinity, but also involves factors such as antibody/antigen valency and nonspecific factors associated with binding.

AFIBRINOGENEMIA

See **Fibrinogen**.

AGGLUTINATION

See also **Agglutinins**.

Adhesion with clumping of biological material such as bacteria or cells. It is usually an effect of an antigen–antibody reaction. **Red blood cell** agglutination can occur with **immune hemolytic anemia**. It is used as an indicator in the detection of **blood group** antigens and antibodies.

AGGLUTININS

See also **Autoimmune hemolytic anemia; Direct antiglobulin (Coombs) test; Indirect antiglobulin (Coombs) test**.

Antibodies causing agglutination, usually of **red blood cells** in a saline medium.

AGGREGATE FORMATION OF RED BLOOD CELLS

The adhesion of erythrocytes (**red blood cells**) to one another. This is largely responsible for the yield stress and the viscoelasticity of whole blood and has an important effect on blood rheology by increasing **viscosity of whole blood** at low shear rates that may affect microcirculatory flow.[44] Aggregate formation occurs only on erythrocytes in static or slow-moving blood because the adhesive forces are generally small. With rapid flow, the high shear stresses overcome the adhesive cellular interactions and break up the aggregates. The extent of aggregate formation depends upon the nature and concentration of the aggregating proteins present, plasma viscosity, **deformability of red blood cells**, and their surface-charge density. Techniques for estimating aggregate formation include direct microscopy, viscometry at low shear rate, determination of the viscoelasticity of whole blood, and optical measurement of red blood cell reaggregation after disruption.

AGGREGOMETERS

See **Platelet-function testing**.

AGGRESSIVE NK-CELL LEUKEMIA

See also **Lymphoproliferative disorders; Non-Hodgkin lymphoma**.

(Real: Large granular lymphocyte leukemia, NK-cell type) A rare aggressive natural killer (NK) cell lymphocytic leukemia. It has a strong association with the **Epstein-Barr virus** (EBV), and is more prevalent among Asians than Caucasians. It is characterized by peripheral-blood **lymphocytes** with round or oval nuclei and moderately condensed chromatin and rare nucleoli, eccentrically placed in abundant pale blue cytoplasm with azurophilic granules. The cells are acid phosphatase positive, with a granular pattern, and are nonspecific esterase negative or weakly positive. Bone marrow infiltration is usually sparse, with mild-to-moderate increase in lymphocytes as well as focal aggregates. There may be a myeloid maturation arrest or erythroid hypoplasia. Infiltration of the splenic red pulp and of the hepatic sinuses is common. The **immunophenotype** of the NK-cells is CD2$^+$, CD3$^-$, CD3ϵ^+, CD56$^+$ and is positive for cytotoxic molecules. **Cytogenetic analysis** shows germline configuration of T-cell receptor genes. The postulated cell of origin is an NK-cell.

Clinical and Laboratory Features

Patients present with fever, moderate lymphocytosis (5 to 20 $\times$ 10^9/l), and often with **anemia, neutropenia** and **thrombocytopenia,** moderate **hepatosplenomegaly,** and sometimes **lymphadenopathy**. There may be associated coagulation factor disorders, hemophagocytosis, and multiorgan failure. The course is usually aggressive, with morbidity related to the cytopenias rather than tumor burden and a fatal outcome in 1 to 2 years.

Staging

See **Lymphoproliferative disorders**.

Treatment

See **Non-Hodgkin lymphoma.**

AGNOGENIC MYELOID METAPLASIA

See **Chronic idiopathic myelofibrosis.**

AGRANULOCYTOSIS

An idiosyncratic reaction to drugs, characterized by fever, sore throat, collapse, and even death, with severe **neutropenia.** The term was initially used by Schultz in 1922. Many different classes of drugs have now been implicated,[45] with a mortality risk as high as 2% for some drugs, e.g., thiouracils (see Table 13).

TABLE 13

Drugs Commonly Causing Agranulocytosis

Antibiotics	Anti-inflammatory
Chloramphenicol	Phenylbutazone
Sulfonamides	Indomethacin
Dapsone	Gold salts
Antithyroid	Analgesics
Carbimazole	Aminopyrine
Propylthiouracil	Anticonvulsants
Phenothiazines	Carbamazepine
Chlorpromazine	Phenytoin
Promazine	Valproic acid
Antihypertensives	Hypoglycemic agents
Captopril	Chlorpropamide
Hydralazine	Tolbutamide
Procainamide	Antidepressants
Quinidine	Amitriptyline
Antipsychotics	Doxepin
Clozapine	Imipramine

Pathogenesis

Two different mechanisms have been described:

Direct toxicity to granulocytes, e.g., phenothiazines

Immunological: antibodies are formed that react alone to granulocytes or react to granulocyte/drug complexes, e.g., chlorpropamide, amidopyrine

Clinical Features

The patient may present with an obvious focus of infection, e.g., furuncle, cellulitis, pharyngitis, or acutely unwell with fever, myalgia, bone pain, and collapse. If the diagnosis is not made and the offending drug stopped, death often ensues.

The diagnosis should be considered in any patient with neutropenia who takes medication. The peripheral blood often reveals an almost complete absence of neutrophils while maintaining relatively normal lymphocyte, monocyte, and platelet counts. Bone marrow cytology is variable, with complete granulocyte aplasia or hyperplasia, depending on the stage of maturation of the granulocyte affected. For example, phenylbutazone affects very early granulocyte progenitors, whereas thiouracils affect nonsegmented and segmented neutrophils. Consequently, the duration of neutropenia can roughly be predicted from the degree of bone marrow hypoplasia. Thus, if only myeloblasts are present in the marrow, 7 to 10 days of neutropenia will ensue, but if later forms are present, the period is much shorter. Clozapine, used for treatment of schizophrenia, has a genetically determined reaction causing accumulation of the cell toxin nitrenium. The rate of recovery, however, may vary depending on the offending drug, e.g., slower with phenothiazines. Occasionally, the recovery is so vigorous that rebound neutrophilia with left shift occurs.

Treatment

Recognition that the patient is taking an offending drug and its prompt withdrawal is essential. Infections should be managed similarly to any infective episode in a neutropenic patient. Colony stimulating factors, e.g., granulocyte colony stimulating factor (G-CSF), may be beneficial in reducing the period of neutropenia once the drug has been stopped.

AIDS

See **Acquired immunodeficiency syndrome**.

ALA SYNTHETASE

See **δ-Aminolevulinate synthase**.

ALBUMIN FOR TRANSFUSION

See **Human albumin solution**.

ALCOHOL TOXICITY

The hematological effects of raised levels of ethyl alcohol in the body (see Table 14). Ethyl alcohol is the constituent of many social beverages and is oxidized by alcohol dehydrogenase, present in many tissues (particularly the gastric mucosa), to acetaldehyde. The

TABLE 14

Hematological Effects of Alcohol

Red blood cells	stomatocytosis
	spherocytosis (Zieve's syndrome)
	acute intermittent porphyria with porphobilinogen deaminase deficiency
Bone marrow	iron deficiency
	dyserythropoiesis
	macronormoblastic
	megaloblastosis, direct or via folate disturbances
	sideroblastic anemia
Liver disorders	macrocytic anemia
	acanthocyte (spur cell) hemolytic anemia
	splenomegaly

liver mucosal enzyme system converts acetaldehyde to acetate, which in turn is oxidized in peripheral tissues to carbon dioxide, fatty acids, and water.

Acute alcohol toxicity can directly damage **bone marrow** or circulating **granulocytes**, causing vacuolation of erythroblasts and neutrophils, with macronormoblastic **dyserythropoiesis** and a **leukemoid reaction**. Recovery occurs rapidly sometimes with **thrombocytosis**.

Chronic alcoholism causes direct and indirect hematological effects, anemia being present in 50% of patients. As a consequence of direct action on the bone marrow, macrocytes are present in the peripheral-blood film of 90% of chronic alcoholics. A less common effect is the presence of stomatocytes due to increased red blood cell water content. Reversible sideroblastic anemia is caused by the conversion of pyridoxine to pyridoxal or as a consequence of damage to **heme** synthesis by depression of **δ-aminolevulinic acid** (ALA) dehydrogenase and **ferrochelatase**, leading to an increase of intracellular iron. This is more frequent when alcoholism is associated with **folic acid** deficiency. While these effects are primarily on **erythropoiesis**, **granulopoiesis** and **thrombopoiesis** are affected later, with **neutropenia** and **thrombocytopenia**. Following abstention from alcohol, these effects may persist for up to 4 months.

Indirect effects of alcohol are (a) **iron deficiency anemia** due to chronic hemorrhage from peptic ulceration or esophageal varices and (b) **cobalamin** deficiency from intestinal malabsorption and the consequences of disturbed folic acid metabolism.

ALDEHYDE DEHYDROGENASE

An enzyme concerned in **methemoglobin** reduction by aldehydes as substrate.

ALDER-REILLY ANOMALY

An inherited autosomally recessive condition characterized by large primary and secondary granules. All **granulocytes** may be affected. The granules contain mucopolysaccharide, and the anomaly is typically seen in **mucopolysaccharidoses**, e.g., Hurler's and Hunter's syndromes. **Eosinophils** filled with these basophilic granules may appear to be basophils. The granulocytes appear to be functionally normal.

ALDOLASE

An enzyme with variants that operates in the Embden-Meyerhof pathway in human **red blood cell** metabolism. Inherited autosomal deficiency can occur, which results in a moderate-to-severe **hemolytic anemia** in addition to a glycogen-storage abnormality and mental retardation.

ALEMTUXIMAB

See also **Rituximab**.

An anti-CD20 antibody to **B-lymphocytes**. It is used in the treatment of **lymphoproliferative disorders** where there has been failure of response to other cytotoxic agents. It causes infusion-related **adverse drug reactions**, including **cytokine release syndrome**.

ALKYLATING AGENTS

A group of **cytotoxic agents** used in the treatment of **leukemia** and **lymphoproliferative disorders**. They probably act by cross linking **DNA**, thus interfering with replication and transcription. Most are polyfunctional and contain more than one alkyl group. Nitrogen mustard derivatives and alkyl sulfonates are widely used, but ethylerimine derivatives, trazine derivatives, and nitrosoureas such as lomustine (CCNU®) are now little used.

Chlorambucil

Used in the treatment of **chronic lymphatic leukemia** and low-grade **non-Hodgkin lymphoma**. It is administered orally either in a low-dose daily schedule or in intermittent pulses. It is generally well tolerated, but occasionally gastrointestinal (GI) tract adverse effects, **leukopenia**, **thrombocytopenia**, and, rarely, hepatotoxicity are seen.

Cyclophosphamide

A drug that is transformed to its active derivative 4-hydroxycyclophosphamide in the liver. It is used to treat a variety of tumors, including leukemia, where it is mainly an agent in combination therapy for non-Hodgkin lymphoma and **acute lymphoblastic leukemia**. It is also used as a cytoreductive and immunosuppressive agent to prepare patients for **allogeneic stem cell transplantation** and **autologou**s **stem cell transplantation** (SCT). Toxicity occurs primarily in rapidly dividing tissues, including bone marrow; GI tract side effects are common. Its metabolites can produce a hemorrhagic cystitis, which is often difficult to control. Survivors may suffer bladder fibrosis and bladder carcinoma. This adverse reaction is very much less likely if a high fluid intake is maintained and high doses of intravenous cyclophosphamide are preceded by the free-radical scavenger **mesna**. Other adverse reactions include sterility, cardiac damage, and late carcinogenesis.

Melphalan

A phenylalanine derivative of nitrogen mustard used orally in a daily dosage or in intermittent pulses. Its main use is in the treatment of **myelomatosis** stem cell transplantation programs.

Lomustine (CCNU®)

A lipid-soluble nitrosourea previously used in the treatment of **Hodgkin disease** and non-Hodgkin lymphoma. A related compound is carmustine (BiCNU®), used similarly and for myelomatosis and Hodgkin lymphoma, but having toxicity to pulmonary tissue.

Mechlorethamine (Nitrogen Mustard)

The original mustine, now little used due to its highly adverse reaction of vomiting and its association with **bone marrow hypoplasia** and secondary **acute myeloid leukemia**.

Busulfan

A drug that is rapidly absorbed from the gut and has an extremely short plasma half-life of 3 to 5 min. Excretion is mainly via the kidneys. It acts predominantly on myeloid cells and was used mainly in the treatment of **myeloproliferative disorders**. **Busulfan** and **hydroxyurea** are inferior to newer treatments in the management of **chronic myelogenous leukemia**. Busulfan has been associated with irreversible bone marrow hypoplasia. Its other adverse reactions include hyperpigmentation, gynecomastia, and rarely, cataracts and pulmonary fibrosis. Together with cyclophosphamide, high-dose oral busulfan is used as conditioning treatment for SCT, but convulsions can occur, and prophylactic antiepileptic drugs are required.

ALLERGIC PURPURA

(Anaphylactoid purpura; Henoch-Schönlein purpura) A form of **vasculitis** characterized by an acute onset with arthritis, melena, and a diffuse purpuric rash of the lower limbs. Thirty percent of cases show renal and cerebral involvement. Involvement of the pleura and pericardium may also occur. Ankle edema results from proteinuria due to deposition of IgA complexes in the glomerular vessel mesangium, which may lead to glomerulonephritis. The syndrome is most common in children under the age of 10 years. A history of an upper respiratory tract infection is often elicited, and approximately one-third of cases will demonstrate elevated antistreptolysin-O (ASO) titers due to previous infection with b-hemolytic streptococci. Other cases are associated with drug **hypersensitivity** or food allergy. Treatment consists usually of penicillin, but for more severe cases **immunosuppression** by steroids, **azathioprine**, and treatment with **cytotoxic agents** may be required.

ALLOANTIBODY

An **antibody** that recognizes allotype determinants. Examples are

- "Irregular" antibodies directed against blood cell antigens, formed following the transfusion of blood components or a fetal-maternal hemorrhage. These antibodies may cause **blood transfusion complications**, **hemolytic disease of the newborn**, and **neonatal alloimmune thrombocytopenia**.
- Antibodies directed against antigens that are common to bacteria and blood cells. These antibodies are sometimes said to be "naturally occurring," such as anti-A and anti-B (see **ABO (H) blood groups**). These may also cause blood transfusion complications.
- Antibodies of human origin, or **monoclonal antibodies**, used in the laboratory to determine the **phenotype** of blood cells.

ALLOGENEIC STEM CELL TRANSPLANTATION

(Allogeneic SCT) The transplantation of **bone marrow** (BM) or peripheral blood (PB) **stem cells** (BMSC/PBSC) from a matched sibling donor or volunteer unrelated donor (VUD).

TABLE 15

Indications for Allogeneic Stem Cell Transplantation, by Disease

Disorder	Adults	Children
Acute myeloid leukemia	1st CR	1st or 2nd CR
Acute lymphoblastic leukemia	1st CR	2nd CR
		1st CR for B-cell ALL
Chronic myelogenous leukemia	1st CR	1st CR [a]
Chronic lymphoblastic leukemia	NI	b
Myelodysplasia (RAEB)	younger patients	b

Note: CR, complete remission; NI, not generally indicated.

[a] Rare in children.

[b] Does not usually occur in children.

This is now part of the treatment of a wide variety of malignant and nonmalignant hematological disorders. Age is a limiting factor, as results deteriorate with age, but SCT can be successfully performed in those up to 70 years of age.

Indications

These are summarized in Table 15. For most children with **acute lymphoblastic leukemia** (ALL), the prognosis is so good with conventional treatment that SCT is not appropriate in first complete remission (CR) except for patients with a known poor prognosis, such as those with B-cell ALL. For children with **acute myeloid leukemia** (AML), the decision is difficult, as 5-year survival rates >50% have been reported in a few studies. For adults with ALL or AML, it is appropriate in first CR, while in **chronic myelogenous leukemia** (CML), results in the first chronic phase surpass those obtained subsequently. For all leukemias, SCT in relapse should be avoided if possible, as results are poor, although 10 to 20% of patients with primary resistant AML may obtain long-term survival with allogeneic SCT. Long term survival and possible cures are now being reported in **lympho-proliferative disorders**. (For advantages and disadvantages of autologous and allogeneic SCT, see **Autologous stem cell transplantation**).

Donation of Stem Cells

Bone Marrow Stem Cell Donation

Seven to 14 days before the procedure, a unit of blood is withdrawn and stored for replacement during donation. Bone marrow is removed by multiple punctures, using heparinized syringes, from the posterior iliac crests. Umbilical cord and neonatal blood have been used.[45a]

Peripheral-Blood Stem Cell Donation

Allogeneic PBSC donors are typically primed with **granulocyte colony stimulating factor** (G-CSF) injections (10 µg/kg per day) for 4 to 6 days. This usually achieves a rapid rise in the peripheral-blood CD34+ stem cells by day four or five, at which time stem cells are harvested using a cell separator programmed to collect low-density mononuclear cells or lymphocytes. This results in a collection of hematopoietic progenitors in addition to lymphocytes, monocytes, and natural killer cells. Continuous flow machines (e.g., Cobe Spectra, Denver, CO) collect faster than intermittent flow machines and yield a purer product. Usually, one to three collections over 2 to 4 h will give a yield adequate for PBSC transplantation. In addition, both CD34+ cells and T-lymphocytes can be stored for possible use later for primary engraftment failure or as donor lymphocyte infusions.

Choice of Stem Cell Source

Bone marrow stem cell harvesting requires collection under general anesthesia and usually results in a limited (lower) number of CD34$^+$ cells than that obtained by PBSC harvesting. Although PBSC harvesting does not require a general anesthetic, the donor receives G-CSF injections and therefore may experience adverse reaction from these. Also, although at present there is no convincing evidence that G-CSF injections increase the risk of developing AML or CML, the donor should be aware of this theoretical possibility. PBSC collections result in about a 1-log increase in contaminating T-lymphocytes compared with BMT. Studies have shown that there is virtually identical engraftment between PB and BM SCT, but the former may be associated with a lower risk of chronic **graft-versus-host disease** (GVHD) and a higher risk of relapse, e.g., AML, CML. Thus, at present, neither source of stem cells is superior, but if GVHD is a concern, then BMSCT may be a better option, whereas if relapse is the major concern, then PBSCT may be preferred. Also, PBSCT is indicated in haploidentical transplants and in some HLA (human leukocyte antigen)-mismatched transplants where very high numbers of CD34$^+$ stem cells are required for a successful outcome.

Donor Matching

The current strategy is to type for **human leukocyte antigens** HLA-A and -B antigens by serology and for HLA-DR and -DQ using polymerase chain reaction-single stranded conformational polymorphisms (PCR-SSCP) or -single stranded polymorphisms (PCR-SSP). In sibling/related donors, an CTLP frequency may be of value. Mixed lymphocyte culture (MLC) does not add any further information. In unrelated donor matching, it is usual to select the best possible match, but for patients with aggressive diseases, e.g., acute myeloid leukemia, it may be permissible to mismatch one or two alleles if this is clinically acceptable. Mismatching increases the incidence of GVHD and graft rejection. The use of cytotoxic T-lymphocyte precursor (CTLP) frequency analysis may help to define donor recipient pairs in whom there is a particularly high risk of either GVHD or graft rejection. Clinical strategies can then be modified appropriately.

Transplant Procedure

Pretransplant Period

For patients fulfilling the indications shown in Table 15, a search should be made among available siblings for a donor who should be compatible at the HLA-A, -B, and -DR loci (see **Human leukocyte antigen**). The donor should be fit for BMSC or PBSC harvest. The recipient should be counseled about the risks and potential benefits of the procedure. Venous access is established via an indwelling **central venous catheter** (double-lumen) of the Hickman type. Consideration should be given to obtaining a backup remission SC harvest from the recipient, which can be stored and used as rescue should rejection of the graft occur. In addition to **hemoglobin**, **red blood cell count**, **white blood cell count**, and **platelet count**, baseline biochemistry, **blood group** serology, chest X-ray and electrocardiogram, tests for cardiovascular status and fungal precipitins, and a virology screen, including **cytomegalovirus** (CMV), should be obtained for both donor and recipient. The recipient will also require a dental assessment, lung function tests, and an echocardiogram.

Conditioning

A variety of conditioning regimens are in use, depending upon the aim of the transplant and the underlying condition. Initially, it was thought to be important to deliver myeloablative conditioning therapy, which would eradicate (hopefully) any residual malignant cells and adequately immunosuppress the recipient to allow donor stem cell engraftment. Myelo-ablative therapies are highly toxic and unsuitable for many patients with comorbidities, e.g., the elderly. We now know that many patients are not cured because of the myeloa-blative therapy but because of the graft-versus-leukemia (GVL) effect. This has led to an explosion in the development of nonablative (reduced intensity) regimens, where the primary aim is to immunosuppress the recipient adequately to allow donor cell engraftment and rely on the GVL effect to cure the patient. These regimens are therefore typically more immuno-suppressive and designed to reduce organ toxicity and hence comorbidity.[46,47]

Myeloablative regimen
- **Cyclophosphamide**, 60 mg/kg, days –6, –5
- Total body irradiation (TBI), 10 to 14.4 Gy in three to eight fractions, days –3 to –1
- **Busulfan**, 4 mg/kg, days –9 to –6
- Cyclophosphamide, 50 mg/kg, days –5 to –2

Nonmyeloablative (reduced intensity) regimen
- **Fludarabine**, 30 mg/m², days –4 to –2
- TBI, 2 Gy, day 0
- Fludarabine, 30 mg/m², days –10 to –5
- Busulfan, 4 mg/kg, days –6, –5
- Antithymocyte globulin (ATG), 10 mg/kg, days –4 to –1

Regimens using chemotherapy only (coming into increasing use)
- Fludarabine, 30 mg/m² daily, days –7 to –3
- **Alemtuximab**, 20 mg daily, days –8 to –4
- Melphalan, 140 mg/m² on day –2 only

Transplantation

After an interval of 24 h, the donor bone marrow is given via the indwelling central intravenous catheter.

Complications of Chemoradiotherapy

The metabolic products of cyclophosphamide cause cystitis, which may be severe follow-ing high-dose regimens, often with hemorrhage. Intravenous **mesna** inactivates these metabolites and should be given during and immediately after the chemotherapy sched-ule. Cystitis more than 2 weeks after chemotherapy should be investigated to exclude viral infection and GVHD. Most patients experience nausea, vomiting, and diarrhea fol-lowing TBI; parotitis and mild pancreatitis are not uncommon. Erythema and, occasionally, mild skin burning occurs. The gastrointestinal tract and skin side effects of TBI appear to be lessened when the dose is fractionated. Myeloablative conditioning regimens produce irreversible sterility, and it is important to discuss sperm or ovum storage with the recipient well before the procedure takes place. Females will suffer a radiation-induced menopause and should receive appropriate hormone replacement therapy posttransplant.

Interstitial pneumonitis is a major hazard post SCT. This is mostly caused by CMV or other viral infections and has a peak incidence at about 100 days posttransplant following myeloablative conditioning and an earlier incidence following nonmyeloablative conditioning. The nonmyeloablative regimens are particularly immunosuppressive, and reactivation of adenovirus can also become a problem.

Venous thromboembolic disease, characterized by the triad of weight gain, abnormal liver function tests, and painful hepatomegaly can occur within the first 2 weeks posttransplantation and may be fatal. Several treatments have been tried, including fluid restriction, diuretics, plasminogen activators, and defibrotide, with varying success.

Late complications of TBI include cataracts, which usually begin to form after about 12 months and may be clinically troublesome after 2 to 3 years. Cataracts are less likely and are slower to form after fractionated TBI than when single-shot TBI is used. Cataracts are also common in patients requiring steroid therapy for control of GVHD.

A higher incidence of secondary tumors is now bring reported in long-term survivors of SCT, including myeloid leukemias, sarcomas, melanoma, squamous cell carcinoma, and posttransplant lymphoproliferative disorders.

Posttransplant Lymphoproliferative Disease (PTLD)

Immunosuppression to allow donor cell engraftment results in an increased risk of reactivations of latent viruses. In PTLD, latent **Epstein-Barr virus** reactivates within **B-lymphocytes**, giving rise to rapid infiltration of lymphoid and nonlymphoid tissue by fast-proliferating B-lymphocytes — **polymorphic lymphoid proliferation**. Constitutional symptoms including fevers, sweats, and anorexia are common. Untreated, the condition is usually fatal. Therapy consists of stopping immunosuppressive drugs, chemotherapy, and **immunotherapy**. Chemotherapy similar to that used in high-grade **non-Hodgkin lymphoma** along with **rituximab** is commonly used. For resistant patients, immunotherapy can be in the form of interferon α and lymphocyte infusions (donor and autologous), with both unstimulated and virus-specific T-cells.

Transplant Period Procedures

To reduce infection, patients are usually nursed in clean environments, often with laminar air-flow facilities, which greatly reduce the risk of fungal infections. Antimicrobial prophylaxis should include a gut sterilization regimen, a "sterile" diet, oral acyclovir four times daily to protect against herpes simplex, and oral cotrimoxazole twice weekly to protect against *Pneumocystis carinii* infection.

Protection against severe GVHD is usually with oral **cyclosporin**, the dosage of which is adjusted according to blood levels, usually with twice weekly oral **methotrexate**. In addition, all blood products should be irradiated. To protect against CMV infection, all blood products should be CMV negative. Screening for CMV reactivation is essential, using a sensitive technique such as quantitative polymerase chain reaction (PCR), and preemptive therapy instituted.

Results

Matched Sibling Donor

Acute Myeloid Leukemia (AML)

The best results are achieved with SCT in first complete remission (CR). International registry data indicate long-term disease-free survival for 50 to 55% of these patients, but as this result includes data from the early experience of 15 to 20 years ago, it is probably somewhat conservative. The risk of relapse is dependent on disease status at SCT, with patients in second or subsequent CR having only a 30% long-term survival and those transplanted in relapse faring even less well. Transplant "failures" include those patients who relapse following initial success, this occurring usually within the first year post SCT, and those who fail to survive the transplant procedure or its immediate aftermath. These deaths include those due to pneumonitis or other infectious complications and those due to severe acute GVHD.

Acute Lymphoblastic Leukemia (ALL)

SCT for ALL in first CR is largely used for adults, as most children have a good or very good prognosis without SCT. Where possible, children with B-cell ALL should be transplanted in first CR, and SCT should be considered for all children in second or subsequent CR.

International pooled data show that long-term disease-free survival for patients transplanted in first CR is about 50%, about 30% for those in second or subsequent CR, and less than 15% for those transplanted in relapse. Results are poor for those patients whose disease relapse has included sanctuary sites such as the central nervous system or testis.

Chronic Myelogenous Leukemia (CML)

Although cure for CML cannot be achieved by conventional drug treatments, the majority of patients will survive in good health for up to over 5 years with **imatinib** therapy. Those patients who achieve a complete cytogenetic response may survive even longer, so that a decision to transplant is more complex than for acute leukemia. However, as best results are achieved in first chronic phase, where a matched sibling donor is available, there is presently no consensus as to how long imatinib therapy should be given in an attempt to induce a cytogenetic response before SCT is appropriate. A pragmatic approach, with a trial of **imatinib** for 12 to 24 months, is common before considering allogeneic SCT for those who have failed to have an adequate cytogenetic response.

Pooled international data show a long-term disease-free survival of less than 50%, but these results include early transplants and also those done with T-cell depletion, where relapse rates are high. Several centers have achieved long-term survivals for more than 60% of CML patients transplanted in first chronic phase, but results are not as good in second chronic phase, with few long-term survivors when SCT is done in blastic phase.

Volunteer Unrelated Donor (VUD SCT)

Leukemia patients who do not have a matched sibling donor may be considered for VUD BMT. Results are generally 10 to 30% lower than those achieved with matched sibling transplants. Long-term survivors tend to be younger than in sibling SCT, and the procedure is best reserved for patients under 40 years of age.

There may be a delay in obtaining a good matched donor, as donor banks often can only provide matching data for the HLA-A and -B loci. Even where HLA-C and -DR data are available, careful testing using DNA techniques is needed before a decision to select a potential donor is made. It is important to have as much information as possible when choosing a potential VUD, as graft failure is associated with the total number of HLA disparities, and GVHD and overall survival are associated with HLA-A, -B, -C, and -DRB-1 mismatches.

Haploidentical Transplant

The chance of finding an unrelated donor varies with the patient's ethnic origin, ranging from 60 to 70% for Caucasians to <10% for ethnic minorities. In patients without a suitable related or unrelated donor, a transplant from one's parent, half-matched sibling, or similar half-matched relative may be considered. To enable only a half-matched HLA transplant to engraft requires considerable additional immunosuppression, very large doses of stem cells (up to 10×10^6/kg CD34$^+$ cells), T-cell depletion, and vigorous antimicrobial prophylaxis and surveillance. Conditioning regimens typically include combinations of fludarabine, cyclophosphamide, thiotepa, busulfan, irradiation, and antithymocyte globulin or **alemtuximab**. The best results have been obtained in children with acute myeloid leukemia or chronic myelogenous leukemia, where 30 to 70% survive for 2 years. Many deaths are due to infective complications, as immune reconstitution is markedly slower than other types of allogeneic transplantation.

ALLOIMMUNE HEMOLYTIC ANEMIA

Immune hemolytic anemias induced by antibodies other than those of self. They occur with:

> *Hemolytic disease of the newborn.*

> *Blood transfusion complications* or infusion of **fresh-frozen plasma** in which incompatible red blood cell antibodies are introduced. These are usually of the **ABO (H) blood group**. It can also arise following injection of polyvalent pneumococcal vaccine, which provokes a rise in anti-A antibodies. This is a particular risk to the fetus of group-O mothers due to possible incompatibility.

> *Allograft-associated* **alloimmune hemolytic anemia**, a disorder that may develop in the recipients of allografts, usually renal, and that may be associated with **graft-versus-host disease** caused by donor lymphocytes in the organ graft. An alternative explanation may be the induction of autoimmunization by **ciclosporin**.

ALLOIMMUNIZATION

See also **Alloantibody**.
The development of **antibodies** to cells of the same species other than self.

ALLOLEUKOAGGLUTININS

Antibodies to **leukocytes** causing leukoagglutination, usually in pulmonary blood vessels. This gives rise to **transfusion-related acute lung injury**.

ALLOPURINOL

A xanthine oxidase inhibitor preventing the production of uric acid from xanthine and hypoxanthine. These latter two compounds are more soluble than uric acid and are readily excreted in the urine, although rarely xanthine stones may form. Allopurinol is readily absorbed from the gut with a half-life of 3 h. It is used to prevent and treat **hyperuricemia** in doses of 300 to 600 mg per day.

    Adverse drug reactions include rashes, especially if used with ampicillin, in which case the drug should be withdrawn immediately. **Hypersensitivity** reactions, **granulopenia**, **thrombocytopenia**, **bone marrow hypoplasia**, **hemolytic anemia**, and liver disorders (including **granuloma** formation) are rare. The drug may exacerbate acute gout. Allopurinol inhibits metabolism of **6-mercaptopurine** and **azathioprine**; a 75% reduction in dose of these drugs is indicated.

ALLOTYPE

Genetic variant where polymorphism of a gene (allele) occurs. Allotypic variation can result in a qualitative or quantitative difference in a blood cell surface **antigen** between different individuals. This antigenic variation may be recognized by an **alloantibody.**

ALPORT'S SYNDROME

An X-linked or autosomally recessive disorder with mutations in the *COL4A5* and *COL4A3/COL4A4* genes, respectively. These code for α5, α3, and α4 chains of type IV **collagen.**[48] It is one of the group of **giant-platelet syndromes** associated with lamellation of the glomerular basement membrane of the kidneys, nerve deafness, ocular abnormalities, and **thrombocytopenia**. It is the commonest of the hereditary nephropathies, presenting in childhood or young adulthood with benign hematuria. Diagnostic changes may be found upon renal biopsy. Platelets in Alport's syndrome may be larger but are ultrastructurally normal, as occurs in the similar Epstein's syndrome. Fetchner's syndrome is a variant of Alport's syndrome with associated leukocyte inclusions.

AMATO BODIES

See **Döhle bodies**.

AMEGAKARYOCYTIC THROMBOCYTOPENIA

Congenital or acquired **thrombocytopenia** due to **megakaryocyte hypoplasia**.

Congenital Syndromes

Thrombocytopenia with absent radii (TAR syndrome). Infants present shortly after birth with a bleeding diathesis. Early in the disease, erythropoiesis and myelopoiesis are unaffected, but frequently progression to trilineage **aplastic anemia** occurs. Examination of the bone marrow for clonal chromosome abnormality may be useful to establish the diagnosis. Therapy is mainly supportive, with **platelet transfusion**. Some patients have responded to **interleukin**-3, but this is not usually sustained upon cessation of therapy. Children with this syndrome who survive the first 2 years of life have a normal life expectancy, as the platelet count tends to rise.

Mutation of the thrombopoietin receptor (c-mpl). An autosomally recessive inheritance resulting in absence of megakaryocytes from the bone marrow. This is the more common abnormality and requires treatment by platelet transfusion. Some children respond to antithymocyte globulin (ATG) and **ciclosporin**.

Intrauterine infections associated with maternal rubella and **cytomegalovirus.**

Acquired

An adult form of megakaryocytic hypoplasia that usually progresses to **myelodysplasia** or **acute myeloid leukemia** and occasionally to **aplastic anemia**. It may be mediated by immune disorders and sometimes responds to immunosuppression by the administration of corticosteroids, antilymphocytic globulin (see **Immunoglobulins**), and ciclosporin. Regular platelet transfusion may be required.

δ-AMINOLEVULINATE SYNTHASE

See also **Porphyria**.

(ALA synthetase) The enzyme responsible for catalyzing the first enzymatic reaction in the **heme** synthesis pathway, namely, the condensation of glycine and succinyl CoA to form δ-aminolevulinic acid. ALA-S2/eALA-S is the erythroid-specific δ-aminolevulinic synthase, encoded in the *ALAS2* gene on chromosome X. Many *ALAS2* alleles have been described, most in association with X-linked **sideroblastic anemia**. Varying severity of **iron overload** occurs in many patients; responsiveness of anemia to pyridoxine therapy is common. Although most cases are detected in hemizygous males, some heterozygous women also develop *ALAS2* sideroblastic anemia due to skewed X-inactivation.

AMNIOTIC FLUID ANALYSIS

See **Hemolytic disease of the newborn**; **Hemophilia**; **Sickle cell disorder**.

AMSACRINE

A **cytotoxic agent** that is a cell-cycle-nonspecific acridine derivative, killing cells by competing for **DNA** by intercalation with base pairs. It has a short plasma half-life and is excreted mainly in bile, so that its clearance is inhibited in patients with **liver disorders**. Its main use has been in the treatment of relapsed and persistent **acute myeloid leukemia** (AML) and occasionally for **acute lymphoblastic leukemia** (ALL) at dosages of 150 to 200 mg/m^2 daily for 2 to 5 days, often combined with an **anthracycline**. Toxicities include gastrointestinal tract upsets, thrombosis of peripheral veins, hyperbilirubinemia, and cardiac arrhythmias. However, its cardiotoxicity appears to be less than for the anthracyclines.

AMYLOIDOSIS

(AL) A group of disorders characterized by the deposition in tissues of amyloid, consisting of rigid, linear, nonbranching, aggregated fibrils 7.5 to 10 nm wide and of indefinite length. The fibrils stain with Congo red, and when viewed with a polarized light source they produce an apple-green birefringence. All types of amyloid appear the same with Congo red staining and upon electron microscopy, which reveals the fibrillar pattern. The fibrils in primary amyloidosis (AL) consist of the variable portion of a monoclonal light chain (k or l). Molecular mechanisms have recently been reported.[49]

Clinical and Laboratory Features

The median age at presentation is 64 years.[50] Weakness, fatigue, and weight loss are the most frequent symptoms. Paresthesias, light-headedness, syncope, change in voice, macroglossia, dyspnea, and pedal edema often occur. The liver is palpable in about one-fourth of patients, and **splenomegaly** is found in 5%. Macroglossia is present in almost 10% of patients at diagnosis. **Purpura** is common and often involves the neck and face, particularly the upper eyelids. The skin may be fragile and easily traumatized. Signs of congestive heart failure, nephrotic syndrome, peripheral neuropathy, carpal tunnel syndrome, and orthostatic hypotension must be sought during the history and physical examination. Nephrotic syndrome or renal insufficiency is the presenting symptom in more than one-fourth of patients, and carpal tunnel syndrome is a presenting feature in one-fifth of patients. Congestive heart failure is the major feature at diagnosis in one-sixth of patients, and peripheral neuropathy is present as the major manifestation in about 15%. The presence of one of these syndromes and an M-protein in the serum or urine is a strong indication of the presence of AL and requires appropriate biopsies for diagnosis.

Anemia is not a prominent feature and, when present, is usually due to renal insufficiency or **myelomatosis**. The latter is present in 10 to 15% of patients. **Thrombocytosis** (platelets more than $500 \times 10^9/l$) occurs in almost 10% of patients. Some degree of renal insufficiency occurs in about half of patients at diagnosis. The alkaline phosphatase level is increased in 25%. The serum protein electrophoretic pattern shows a localized band or M-peak in almost half of patients. However, it is of modest size. Hypogammaglobulinemia is present in one-fifth of patients at diagnosis. Immunoelectrophoresis or immunofixation of the serum reveals an M-protein in 72% of patients. The urine contains a monoclonal light chain in 73% of patients.[19] An M-protein is found in the serum or urine in almost 90% of patients with AL at diagnosis. The **bone marrow** is characterized by a monoclonal proliferation of **plasma cells**. Most patients (60%) have fewer than 10% plasma cells.

Diagnosis[51]

This depends on the histologic demonstration of amyloid deposits in tissue. A bone marrow biopsy specimen should be stained for amyloid because it is positive in more than half of patients. An abdominal-fat aspirate is positive in almost 80%. Ninety percent of patients with AL will have amyloid in the fat aspirate or in the bone marrow at diagnosis. If these tissues are negative, a rectal biopsy or biopsy of a suspected involved organ such as the kidney, liver, heart, or sural nerve should be performed. Echocardiography is a valuable technique for the detection and evaluation of amyloid heart disease.

Prognosis

It is a progressive disease with a median survival of 13 months. Only 7% of patients survive for 5 or more years. The median survival is 4 months for those presenting with congestive heart failure.

Treatment

Melphalan with **prednisone** or **prednisolone** is superior to colchicine for therapy. The overall response rate is approximately 20%. A combination of **alkylating agents** may be tried in an effort to improve therapy. **Autologous stem cell transplantation** is useful in selected patients, but the mortality of the procedure is about 10%.[52]

ANABOLIC STEROIDS

(Androgens) A group of drugs with androgenic activity and protein-building properties, but whose mode of action is unclear. Their hematological activity affects:

Hemostasis. Stanozolol is known to increase protein C levels, although it is of questionable benefit in patients with inherited deficiencies. It also stimulates **fibrinolysis**, resulting in a decreased **euglobulin lysis time**, a reduction in plasma **fibrinogen** concentration, and a reduction in the **viscosity of whole blood**. A single intramuscular dose of stanozolol preoperatively has been shown to prevent postoperative hypofibrinolysis, although this does not reduce the incidence of postoperative venous thromboembolic disease. Stanozolol is also known to increase the synthesis of C1-esterase inhibitor and is of value in patients with hereditary **angioedema** in which there is a deficiency of the enzyme. Danazol may be of benefit in patients with **immune thrombocytopenic purpura** (ITP). It is of no value in patients with hemophilia A or B.

Bone marrow stimulation, especially erythropoiesis in **bone marrow hypoplasia**, **pure red cell aplasia**, **myelofibrosis**, immune thrombocytopenic purpura, and **myelo-dysplasia.** Nandrolone, testosterone, fluoxymesterone, and oxymethalone have been used for treatment, but the outcome is controversial. In **Fanconi anemia**, oxymethalone is the drug of choice for those without a suitable marrow donor. Side effects limit the use of all these androgens.

Adverse drug reactions:

All anabolic steroids can cause sodium retention with edema.

Virilization occurs when they are given in high dosage. The particular effects include hirsutism, hoarseness of the voice, and aggressive behavior.

Acne is common.

Liver disorders can occur, as identified by disturbed liver function tests.

ANAL CANAL DISORDERS
See **Intestinal tract disorders**.

ANAMNESIS
(Anamnestic reaction) The memory or secondary response of **lymphocytes** to **antigen** stimulation. It is the positive immunological memory in which a second or later exposure to a specific antigen causes a heightened **immune response**.

ANAPHYLACTOID PURPURA
See **Allergic purpura**.

ANAPHYLAXIS
An acute **hypersensitivity** reaction (type I) that occurs upon exposure of IgE-sensitized **mast cells** to **antigens** (allergens). These are usually pollens, house mite dust, animal dander, foods (e.g., peanuts), or more acutely by insect stings or injection of animal protein. Mild effects are urticaria, rhinitis, or asthma; severe effects are **angioedema** and circulatory collapse (anaphylactoid shock).

Mild symptoms respond to parenteral or oral **prednisolone**, parenteral antihistamine, or nebulized salbutamol for bronchial constriction. Life-threatening attacks require rapid removal of the patient from the allergen when possible, followed immediately by intra-venous **adrenaline** (1 mg to 10 ml of 1/10,000 for adults; appropriately reduced dosage for children). This may induce a cardiac arrhythmia, which itself may require treatment. If the attack does not quickly subside, an antihistamine such as **chlorpheniramine** and/ or a **corticosteroid** such as hydrocortisone should be given.

ANAPLASTIC LARGE-CELL LYMPHOMA
See also **Histiocytosis**; **Lymphoproliferative disorders**; **Non-Hodgkin lymphoma**; **Peripheral T-cell lymphoma**.

(ALCL; Luke-Collins: T-immunoblastic sarcoma; sinusoidal large-cell lymphoma; regressing atypical histiocytosis) A high-grade T-cell non-Hodgkin lymphoma characterized by expression of the Ki-l (CD30) antibody. Tumors are composed of large **lymphoblasts** with pleomorphic, often horseshoe-shaped or multiple nuclei, with multiple (usually) or single prominent nucleoli and having abundant cytoplasm. Multinucleated forms may resemble **Reed-Sternberg cells**. There is a variable admixture of **granulocytes** and **histiocytes** (macrophages). The lymphoma involves **lymph nodes**, as well as extranodal sites such as soft tissue, bone, and skin. A primary cutaneous disorder also occurs. The **immunophenotype** of tumor cells is CD30$^+$, EMA$^{+/-}$, with other T-cell-associated antigens variable. Cytogenetic studies have shown t(2;5). It is postulated that the tumor arises from extrafollicular CD30$^+$ blasts. (NPM-1 gene rearranges with anaplastic lymphoma kinase (ALK) gene.)[36]

Clinical Features

These occur in various forms:

Systemic. A relatively rare tumor, but occurring in all age groups, with 15 to 30% of cases being under 20 years of age. **Lymphadenopathy** and **skin disorders** are the presenting features. Some patients have a previous history of other lymphomas, including **mycosis fungoides** or **Hodgkin disease**.

Primary cutaneous. Predominantly in adults; indolent, with the possibility of complete or partial spontaneous regression accounting for approximately 25% of the T-cell lymphomas that arise in the skin. These lesions are multicentric in 20% of patients.

Staging

See Lymphoproliferative disorders.

Treatment

In cases of limited unicentric disease, skin lesions may respond to surgical resection or superficial b-ray radiotherapy. For therapy of solid tumors, see **non-Hodgkin lymphoma**. Multiagent chemotherapy is indicated for extracutaneous disease.

ANCROD

An enzyme purified from the venom of the Russell pit viper (*Vipera russelli*). (See **Russell viper venom time**.) It cleaves **fibrinopeptide A** (FpA) but not fibrinopeptide B (FpB), resulting in a fibrin that is very sensitive to endogenous **fibrinolysis**. Therapeutically, hypofibrinolysis is maintained by the daily administration of ancrod. Defibration is accompanied by a marked elevation in the levels of fibrinogen-fibrin degradation products, peaking 12 to 18 h after the first dose. High levels of fibrinogen-fibrin degradation products also inhibit platelet aggregation. Ancrod results in a rapid decrease in plasma fibrinogen and a reduction in blood **viscosity.** Resistance to the effects of ancrod develops due to the formation of neutralizing antibodies when repeated doses are given. Ancrod is administered as an initial loading dose of 1 to 2 U/kg subcutaneously or as an infusion over 6 h. A maintenance dose of 1 to 2 U/kg every 24 h is administered to maintain the fibrinogen at between 0.5 to 1.0 g/l. Ancrod has been of value in treating patients with

heparin-induced/associated thrombocytopenia (HIT/HAT). It has also been shown to be of value in preventing postoperative **venous thromboembolic disease** in patients undergoing hip replacement or with fractured neck of femur. There are anecdotal reports of its value in myocardial infarction, **sickle cell disorder** crises, and renal transplant rejection.

ANEMIA

The reduction of **hemoglobin** concentration below the reference range in health, usually but not invariably accompanied by a reduction in the red cell mass. The hemoglobin reference range in health, and hence the diagnosis of anemia, is affected by gender (women have lower levels during the reproductive period of life) and by age. The healthy male has a reference range of 13.0 to 17.0 g/dl and the female 12.0 to 15.0 g/dl. For levels for infants and children, see **Reference Range Tables II**, **IV**, and **V**. In common parlance, it is confused with pallor, and scientifically there is debate as to whether or not it represents a reduction in the number of circulating red blood cells or the level of hemoglobin concentration in peripheral blood.

It is the most common worldwide hematological disorder, affecting around 1.3 billion people, its true incidence being difficult to assess, as it is mostly diagnosed and treated at primary centers of health care. It is therefore an important cause of chronic debility affecting social and economic well-being as well as physical health and is of major public-health importance.

Etiology and Pathogenesis

Anemia develops when the rate of bone marrow red cell production fails to keep pace with destruction or loss of cells due to the following causes:

Acute **hemorrhage**: bleeding from a diseased organ or as a consequence of a **hemorrhagic disorder** may be external or internal into a pleural sac, peritoneum, joints, a muscle mass, or a cyst. Common causes are blood loss from:
- Traumatic accident
- Epistaxis (see **Nasal disorders**)
- Hematemesis (see **Esophageal disorders**; **Gastric disorders**; **Duodenal disorders**)
- Melena (see **Colonic disorders**; **Rectal disorders**; **Anal canal disorders**)
- Hemoptysis (see **Respiratory tract disorders**)
- Hematuria (see **Renal tract disorders**)
- Menorrhagia (see **Gynecological disorders**)
- Antepartum (accidental hemorrhage) or postpartum (see **Pregnancy**)

Chronic hemorrhage: usually occult bleeding from a site of ulceration or tumor formation. Common causes are blood loss from:
- Gastrointestinal tract: hemorrhoids, diverticulitis, ulcerative colitis, carcinoma of stomach or colon
- Renal tract: hypernephroma, Wilm's tumor of kidney, chronic cystitis, schistosomiasis, papilloma, carcinoma of bladder
- Lungs: idiopathic pulmonary hemosiderosis, **Goodpasture's syndrome**
- Nasopharyngeal: ulcerated pouches, carcinoma

Accelerated destruction: **hemolytic anemia** due to:

- **Red blood cell membrane disorders**
- **Red blood cell enzyme deficiencies**
- **Hemoglobinopathies**
- **Immune hemolytic anemias**
- **Physically induced disorders**: heat, trauma
- **Chemical toxic disorders**, including **adverse drug reactions**
- Biological lysins, e.g., **snake venom disorders**
- **Bacterial infection disorders**, e.g., **bartonellosis**, **tuberculosis**, *Mycoplasma pneumoniae*
- Viral infection disorders, e.g., **cytomegalovirus, Epstein-Barr virus, human immunodeficiency virus**
- **Protozoan infection disorders**, e.g., **malaria, leishmaniasis, babesiosis**

Splenomegaly

Failure of red blood cell production. This arises from:

Deficiency of substances necessary for **erythropoiesis**

- **Iron deficiency** as a result of dietary iron deficiency, chronic hemorrhage, malabsorption, pregnancy
- **Cobalamin** (vitamin B_{12}) deficiency, giving rise to impaired DNA synthesis (**megaloblastosis**), as occurs in **Addisonian pernicious anemia**, intestinal malabsorption due to Crohn's disease, and rarely due to dietary deficiency
- **Folic acid** deficiency, giving rise to megaloblastosis, particularly in pregnancy, malabsorption due to gluten enteropathy, and dietary deficiency
- Deficiency of general protein, **ascorbic acid** (vitamin C), **vitamin B group**, and trace metals

Primary disturbance of **erythropoiesis**

- Deficiency or failure of hemoglobin formation due to disturbances of globin chain synthesis (**thalassemias**)
- **Bone marrow hypoplasia**, either congenital or acquired, as in **idiopathic aplastic anemia**, or adverse drug reactions and **cytotoxic agents**
- **Myelodysplasia**, including the refractory anemias
- **Sideroblastic anemias**, either hereditary or acquired, as a form of myelodysplasia or due to **alcohol toxicity**, **lead toxicity**, or adverse drug reactions
- **Congenital dyserythropoietic anemias**

Secondary disturbance of marrow production

- **Anemia of chronic disorders**
- Renal tract disorders
- **Acute leukemia**
- **Myeloproliferative disorders**
- **Lymphoproliferative disorders**
- **Infiltrative myelopathies**

Dilution anemias

- **Pregnancy**
- Fluid overload as a complication of saline or blood

TABLE 16

Morphologic Classification of Anemia

Microscopy	Red Cell Indices	
	Mean Cell Volume	**Mean Cell Hemoglobin**
Normochromic/normocytic anemia	normal	normal
Hypochromic/microcytic anemia	low	low
Normochromic/macrocytic anemia	high	normal

TABLE 17

Common Causes of Anemia by Morphology

Normochromic/Normocytic	**Hypochromic/Microcytic**	**Normochromic/Macrocytic**
Acute hemorrhage	iron deficiency	cobalamin deficiency
Hemolytic anemia	thalassemia	folic acid deficiency
Marrow hypoplasia	hemoglobinopathy	posthemorrhage or posthemolysis
Anemia of chronic disorders	sideroblastic anemia	liver disorders
		renal disorders

Adaptive Responses

With the development of anemia, compensatory responses occur:

Red blood cells: increased production of 2,3-diphosphoglycerate (2,3-DPG) reduces **oxygen affinity of hemoglobin**, with increased release of oxygen to the tissues.

Heart: increase of stroke volume and heart rate, resulting in increase of output at rest.

Classification of Anemias

Apart from functional classification, anemia is, for practical clinical purposes, classified morphologically based on the red blood cell indices produced by **automated blood cell counting** or **peripheral-blood film examination** (see Table 16 and Table 17).

Clinical Features

These are very nonspecific and are determined by a number of factors:

Degree of change in **blood volume**

Reduction in oxygen transport by hemoglobin

Rate at which above change in blood volume and reduction in oxygen transport have developed (see **Hemorrhage**)

Manifestations of the underlying disorder

Ability of cardiorespiratory system to compensate; this is particularly affected by coexistent disease; as a general rule, the young compensate more readily than the elderly

Each system of the body can be affected with appropriate features:

Heart: breathlessness on effort, palpitations, systolic murmur, worsening of cardiac ischemia, congestive cardiac failure

Skin: pallor, jaundice, chronic leg ulcers

Nails: brittle, spooning (koilonychia)

Hair: premature graying

Skeletal: bony tenderness

Neuromuscular: headache, vertigo, tinnitus, faintness, loss of mental concentration, restlessness and irritability, weakness, papilledema/retinal hemorrhages, paresthesias

Gastrointestinal: anorexia, nausea, flatulence, abdominal pain, vomiting, constipation, diarrhea, glossitis, oral ulceration, dysphagia, splenomegaly, hepatomegaly, lymphadenopathy

Renal: proteinuria, hematuria, hemoglobinuria

Metabolic: increased basal metabolic rate, fever

Diagnostic Investigation

This is usually followed in the sequence:

1. Clinical history, with particular attention to familial disorders; past illnesses with hemorrhage, jaundice, or gastrointestinal tract disorders; or recent drug administration (see **Adverse drug reactions**)

2. **Automated blood cell counting** with indices; if not available, use hemoglobin concentration and **packed-cell volume**

3. **Peripheral-blood film examination** for changes in red blood cell morphology and abnormalities of white blood cell and platelets

The necessity and order of further investigations will depend upon the outcome of these basic procedures. Flowcharts can be used, depending upon whether it is normochromic/**macrocytic anemia**, hypochromic/**microcytic anemia**, or normochromic/**normocytic anemia** (see Figure 84, Figure 86, and Figure 95, respectively).

4. **Reticulocyte** count for normochromic/normocytic and normochromic/macrocytic anemias to determine the activity of bone marrow production

5. **Bone marrow** examination by aspirate and trephine biopsy; Romanowsky staining will show either normoblastic or megaloblastic erythropoiesis, infiltration by malignant cells, and any proliferation by granulocytes, lymphocytes, or megakaryocytes; iron stains will indicate excess or deficiency of iron stores together with presence of ring sideroblasts indicating a sideroblastic anemia

6. Serum iron/ferritin for hypochromic/microcytic anemias; cobalamin/folic acid for normochromic/macrocytic anemias; if these tests are not available, therapeutic response to the appropriate hematinic by reticulocyte and hemoglobin concentration should be carried out

7. Serum bilirubin, **haptoglobin**, and **methemalbumin** for normochromic/normocytic and normochromic/macrocytic anemias with a raised reticulocyte count to confirm a hemolytic anemia

8. **Direct antiglobulin (Coombs) test** if the reticulocyte count is raised; if a positive result is obtained, further tests for **warm autoimmune hemolytic anemia** and **cold autoimmune hemolytic anemia** should follow

9. **Hemoglobin electrophoresis** for any anemia with morphology suggestive of thalassemia or hemoglobinopathy

10. Clinical investigation of blood loss for confirmed iron deficiency anemias, including tests for occult blood in stools and urinalysis for hemoglobin and red cells

11. Liver enzymes, bilirubin, and alkaline phosphatases in serum for normochromic/macrocytic anemia with no apparent cause; serum urea/creatinine and renal function tests for normocytic/normochromic anemia with no apparent cause

12. Tests for specific hematological disorders as confirmation of suspected diagnosis:

    - **Acidified serum lysis test, sucrose lysis test (paroxysmal nocturnal hemoglobinuria)**
    - Osmotic fragility test
    - **Acidified glycerol lysis test (hereditary spherocytosis)**
    - **Glucose 6-phosphate dehydrogenase** and tests for other red cell enzyme deficiencies

Treatment

This depends entirely upon the cause, apart from anemia due to acute hemorrhage, where **red blood cell transfusion** may be indicated.

ANEMIA OF CHRONIC DISORDERS

Normocytic/normochromic anemia to mildly hypochromic/**microcytic anemia** of variable severity associated with acute and chronic infections of many types.[53] These include:

Chronic inflammatory disorders (especially connective tissue) and **autoimmune disorders: rheumatoid arthritis, systemic lupus erythematosus**, inflammatory intestinal tract disorder

Malignancy

Hypometabolic states (thyroid or pituitary disorders)

Renal insufficiency

Chronic rejection of solid-organ transplants

Pathogenesis

The pathophysiological hallmarks of anemia of chronic disorders include hypoferremia, hyperferritinemia, decreased delivery of **iron** to developing erythrocytes (**red blood cells**), and sequestration of iron in storage cells (**histiocytes**). Anemia of chronic disorders appears to be mediated largely by the effects of **hepcidin**, an oligopeptide produced by hepatocytes that is a potent regulator of iron absorption and internal iron homeostasis. Upregulation of hepcidin synthesis appears to be mediated by **interleukin-**6 (IL-6) in a variety of chronic disorders, and the IL-6/hepcidin sequence accounts for the **ferrokinetic** abnormalities in anemia of chronic disorders. The expression of **transferrin** receptors on the surface of erythroid precursors and the rate of iron transported into developing erythrocytes is partly regulated by **erythropoietin**. Thus, in patients with impaired renal function or decreased renal mass, decreased production of erythropoietin may contribute to anemia due to decreased erythropoietin binding to receptors on **colony forming units,**

BFU-E and CFU-E, which are precursors of erythropoiesis. Altogether, these abnormalities result in functional **iron deficiency** of developing erythroid cells; absolute iron deficiency may also be present in some patients. **Tumor necrosis factor** (TNF), **interleukin** (IL)-1, and α- and γ-**interferon** from **T-lymphocytes** also decrease erythropoietin production in many conditions associated with anemia of chronic disorders.

Clinical Features

The severity of anemia of chronic disease is usually mild to moderate, and often corresponds to that of the underlying condition and other health-related factors such as cardiovascular and pulmonary health and age. The onset of anemia may be unexpectedly brisk, especially in patients with acute or unusually intense inflammation or infection. If the underlying condition responds to treatment, the severity of anemia lessens or the anemia may resolve.

Laboratory Features

The erythrocytes are normocytic and normochromic to mildly hypochromic and microcytic, and there is reticulocytopenia. Serum iron measures typically reveal subnormal serum iron concentration, subnormal total iron-binding capacity, low-normal or decreased transferrin saturation, and elevated serum ferritin level. In the **bone marrow**, iron deposition in histiocytes (macrophages) is increased, and iron-positive granules in developing erythroid cells are markedly reduced or undetectable. In contrast, the marrow in iron deficiency reveals little or no stainable iron.

Treatment

Successful treatment of the underlying disorder is the preferred treatment for anemia of chronic disorders. This is sufficient for management of many patients with infections, inflammatory conditions, or renal insufficiency of limited duration. Successful anti-inflammatory or immunosuppressive treatment of chronic inflammatory or autoimmune disorders may result in amelioration of anemia. Erythropoietin therapy is appropriate for patients with **acquired immunodeficiency syndrome**, malignancies treated with chemotherapy or radiation therapy, and chronic renal insufficiency, with some patients receiving immunosuppressive medications. Iron therapy, either oral or intravenous, is needed in patients who have coexisting iron deficiency and in those who undergo chronic hemodialysis. **Red blood cell transfusion** is indicated in some patients, although it is associated with risks of blood transfusion complications such as **alloimmunization, iron overload**, or **transfusion-transmitted infection**. Patients who require repeated transfusions should be treated with blood that has undergone **leukocyte depletion** and be monitored regularly for the possible onset of iron overload or transfusion-related infections.

ANGIOCENTRIC LYMPHOMA

See also **Diffuse large B-cell lymphoma**; **Extranodal NK/T-cell lymphoma, nasal type**; **Lymphomatoid granulomatosis**; **Lymphoproliferative disorders**; **Non-Hodgkin lymphoma**; **Peripheral T-cell lymphoma**.
(Angiocentric immunoproliferative lesion; giant lymph node hyperplasia; lethal midline granuloma; midline malignant reticulosis; nasal T-cell lymphoma; polymorphic reticulosis) A high-grade T-cell non-Hodgkin lymphoma particularly occurring in the nose and

characterized by lymphoid invasion of vascular walls. The angiocentric and angioinvasive infiltrate is usually composed of a mixture of normal-appearing small **lymphocytes**, a variable number of **atypical lymphocytes**, **immunoblasts**, **plasma cells**, and occasionally **eosinophils** and **histiocytes**. The invasion of a vascular wall is accompanied by occlusion of the lumina by lymphoid cells showing varying degrees of cytologic abnormality. Ischemic necrosis of both tumor and normal tissue occurs. The **immunophenotype** of the atypical cells expresses pan-T antigens, particularly CD2$^+$, most cases being also CD3$^-$ and CD56$^+$. **Cytogenetic analysis** shows Ig and T-cell receptor (TCR) genes to be germline; **Epstein-Barr virus** (EBV)-encoded genes are usually present. Pulmonary lymphomatoid granulomatosis may be an EBV-associated variant of this lymphoma. Some cases of the **aggressive NK-cell leukemia** may be related to this disorder. The postulated origin is from NK T-cells.

Clinical Features

This is a rare disorder in the U.S. and Europe, but more common in Asia, Mexico, Central America, and South America. It can affect children or adults. Extranodal sites are invariably involved, including nose, palate, soft tissue, gastrointestinal tract, lymph nodes, and skin. Hemophagocytic syndromes may occur (see **Histiocytosis**). The clinical course appears to depend on the proportion of large cells; it can be indolent or aggressive.

Staging

See **Lymphoproliferative disorders**.

Treatment

See also **Non-Hodgkin lymphoma**.
The responses to chemotherapy are variable, but some patients may respond well.

ANGIODYSPLASIA

See also **Senile vascular lesions**.
A disorder affecting the vasculature of the bowel and a common cause of melena, predominantly in the elderly. Lesions are usually multiple, small, and concentrated in the cecum and ascending colon. Such dysplastic lesions are believed to be associated with aging of the bowel and years of muscular contraction. Characteristically, angiodysplasia presents with recurrent bouts of gastrointestinal bleeding, overt or occult. Isotope-labeled red blood cell scans may demonstrate the site of bleeding if losses exceed 30 1/h. Definitive diagnosis is usually made via colonoscopy. Multiple lesions are typically found. Rarely is bleeding brisk enough for clear angiographic demonstration. Hemicolectomy may be required in recurrent bleeding. Although usually seen in the elderly, angiodysplasia is often manifest in patients with **Von Willebrand Disease** (VWD) at a younger age. Angiodysplasia occurs independently of VWD, but the underlying bleeding disorder exposes the vascular anomaly for gastrointestinal and mucosal hemorrhage. Operative intervention is usually avoided in patients with VWD, and the recurrent bleeding episodes are managed conservatively. Estrogen therapy may be effective.

ANGIOEDEMA

Acute vasodilatation with increased permeability and edema. It can occur as:

Local reaction to allergy (**Hypersensitivity** Type 1).

Generalized angioedema of the skin, laryngeal, and intestinal mucosa caused by deficiency of the inhibitor of the first component of **complement** (C1 inhibitor). The attacks develop over a few hours and resolve spontaneously over the next 2 days. Laryngeal edema may result in an airway obstruction. Intestinal edema causes abdominal pain and vomiting with diarrhea when the colon is affected. It is caused by deficiency of C1 inhibitor allowing the early classical pathway to activate inappropriately after minimal stimulation. The activation of complement is aborted at the level of C3 because an appropriate surface for complement activation is missing. Both C4 and C2 are cleaved with production of excessive amounts of C2a kinin. As a result, there is generalized increase in capillary permeability. The disorder can be:

- Hereditary autosomally dominant deficiency of C1 inhibitor with attacks of angioedema in childhood and exacerbations in adolescence, which then continue throughout life[54]

- Acquired disorder in association with **lymphoproliferative disorders, autoimmune hemolytic anemias, monoclonal gammopathies, cryoglobulinemia**, or where mononuclear cells activate C1, leading to consumption of C1 inhibitor

Diagnosis of both the hereditary and acquired disorders is by detection of low serum levels of C1 inhibitor or reduced levels of C4 due to cleavage by excess C1, which acts as a screening test.

Treatment for hereditary angioedema is by giving purified C1 inhibitor prophylactically or for acute attacks. Alternative treatments are to give **anabolic steroids** such as stanozolol to stimulate the production of C1 inhibitor. **Tranexamic acid** minimizes C1 inhibitor consumption by a general lowering of complement enzyme activity, so this can be given as a preventive measure. **Danazol** may be effective in the long-term management of hereditary angioedema.[54]

ANGIOFOLLICULAR LYMPH NODE HYPERPLASIA

(Castleman's disease; lymph node hematoma; hyaline-vascular lymph node hyperplasia) Originally described by Castleman in 1956 as a localized mediastinal **lymphadenopathy** in otherwise asymptomatic patients, but now recognized as possibly presenting with other forms of lymphadenopathy.[55] The **lymph nodes** are characterized by vascular proliferation with penetrating small hyalinized vessels. Over the past 50 years, both multicentric and plasmablastic types have been described, which may be variants of several disorders.[56,57] An association with HHV-8, **Kaposi's sarcoma** virus, and augmentation by **human immunodeficiency virus** (HIV) is becoming established.

Localized

Hyaline Vascular Type

A disorder of children and young adults, with a large solitary mediastinal mass without constitutional symptoms. The follicles have widened mantle zones with concentric rings of small lymphocytes ("onion rings") that surround one or more germinal centers. These have lymphoid depletion, vascular proliferation with penetrating small hyaline vessels, and follicular **dendritic reticulum cells**. Between the nodules, the stroma is highly vascular, and there may be sheaves of small polyclonal T-cells, plasmacytoid dendritic cells of phenotype CD68+, CD45RA+, and CD123+. Upon stimulation, these dendritic cells produce type 1 **interferon** (IFN-α). The cause of the disorder is unknown, but it may progress to **dendritic reticulum cell sarcoma**. Treatment is surgical removal. **Radiotherapy** has been used for some patients.

Plasma Cell Type

A disorder of children and young adults, with lymphadenopathy of the mediastinum or abdomen, 50 to 90% associated with **anemia**, **thrombocytosis**, and hypergammaglobulinemia. The hyperplastic follicles contain sheets of **plasma cells** that secrete **interleukin-6**, the level of which is raised in plasma and produces an **acute-phase response** similar to that of **rheumatoid arthritis**. The interfollicular plasma cells produce vascular endothelial growth factor (VEGF). Progression to a **lymphoproliferative disorder** or **plasmacytoma** is rare. Surgical removal of the affected nodal masses is curative, with reduction of the plasma levels of interleukin-6.

Multicentric

A disorder of middle age with generalized lymphadenopathy, **splenomegaly**, and hepatomegaly with anemia, leukocytosis, and thrombocytosis. In addition to follicular hyperplasia, there is dilatation of the lymph node sinuses and fibrosis. It is associated with **immunodeficiency**, particularly **acquired immunodeficiency syndrome**. A plasmablastic variety occurs with **plasmablasts** in the follicular mantle zone that express HHV-8 latent nuclear antigen. Most cells express IL-6 receptor. Many progress to **plasma cell neoplasms** and are associated with **Kaposi's sarcoma**. Treatment with **rituximab** reduces the tumor burden but does not cure the disease.

Associated Disorders

Some patients with polyneuropathy, organomegaly, endocrinopathy, M-protein, and skin changes (**POEMS syndrome**) have lymphoid follicles with morphological features similar to those of the plasmablastic variety. **Cutaneous plasmacytosis** is similarly associated.

ANGIOGENESIS

The development of blood vessels, mediated by migration and division of **vascular endothelium** cells. Angiogenic stimulation occurs through secretion of the vascular endothelial growth factor (VEGF), modulated by **plasminogen** activators, **fibrin**, fibroblast growth factors, **corticosteroids**, and **heparin**. In **malignancy**, increased vascularity of tumors facilitates their growth and metastatic potential. A monoclonal antibody, bortezomib, binds with VEGF to reduce microvascular growth and could be developed for the treatment of solid tumors.

ANGIOIMMUNOBLASTIC T-CELL LYMPHOMA

(Angioimmunoblastic lymphadenopathy with proteinemia; AILD) A systemic benign-appearing **lymphoproliferative disorder** characterized by the appearance of endothelial venules surrounded by small **lymphocytes**, **plasma cells**, **immunoblasts**, and atypical "clear" cells mixed with epithelioid cells, **histiocytes**, and **eosinophils**. Around the proliferating vessels there are clusters of **dendritic reticulum cells** (follicle dendritic cells, FDC). Progression to a high-grade lymphoma of T- or occasionally B-cell type occurs.

Immunophenotyping of the tumor cells shows T-cell-associated antigens[+] and usually CD4[+]. Cytogenetic analysis shows T-cell receptor (TCR) genes rearranged in 75%, IgH in 10%, and EBV genomes in many cases; trisomy 3 or 5 may occur. The probable origin of the lymphoma is a transformed peripheral T-cell, as most tumors show clonal rearrangement of T-cell antibody receptors.

Clinical Features

A rare disease, now seen less frequently than in the past, that is more common in the elderly and particularly following drug reactions from the penicillin group. It presents with acute onset of fever, night sweats, weight loss, and abdominal pain. Painful cervical or generalized **lymphadenopathy** occurs, with erythematous maculopapular rashes, sometimes with pruritus. **Hepatosplenomegaly** is common, other features being respiratory distress from pulmonary infiltration, meningoencephalopathy, peripheral neuropathy, polyarthritis, and renal insufficiency with edema and ascites. Peripheral blood changes are **normocytic anemia**, **neutrophilia**, **eosinophilia**, **lymphopenia** but with increased numbers of **plasma cells**, and **thrombocytopenia**. **Bone marrow hypoplasia** may develop. Immunologically, there is polyclonal hypergammaglobulinemia or dysgammaglobulinemia, positive **direct antiglobulin (Coombs) test** (DAT) anti-IgG or anti-IgM, **cold-agglutinin disease** with C3 coating of **red blood cells**, **autoantibodies**, and antibodies to vimentin, the polypeptide of smooth-muscle cells and **vascular endothelium**.

Treatment

Low-grade **cytotoxic agents** are used in the early stages, with high-grade lymphoma therapy when the disorder has progressed (see **Non-Hodgkin lymphoma**). The prognosis is good in younger patients but poor in the elderly, with death due to infection or progression from lymphadenopathy to **diffuse large B-cell lymphoma** (immunoblastic lymphoma).

ANISOCHROMIA

Variation in Romanowsky staining of **red blood cells** on **peripheral-blood film examination**. Some cells are normochromic, while others show varying degrees of hypochromasia. The appearance is referred to as a "dimorphic blood picture," although "dichromic" is a more apt description. It occurs in:

  Iron deficiency anemia responding to treatment

  Hypochromic anemia following **red blood cell transfusion**

  Sideroblastic anemia

ANISOCYTOSIS

Variation in the size of **red blood cells** to a degree greater than is normally present. This may be due to the presence of cells that are larger or smaller (or both) than normal. The **red blood cell distribution width** (RDW) is increased when anisocytosis is present.

ANKYLOSING SPONDYLITIS

An inflammatory disorder of the sacroiliac and other large joints. It is related to **human leukocyte antigen** (HLA) B27, with which its incidence is roughly parallel. There is infiltration of the joints by **lymphocytes** and **plasma cells**, with local bone erosion at the attachments of intervertebral and other ligaments. Other hematological disorders include:

Anemia of chronic disorders

Elevated **erythrocyte sedimentation rate**

Leukemia, particularly if radiotherapy has been administered for treatment to the lumbosacral region in the past; local **bone marrow hypoplasia** will remain, with extensive chromosomal changes to cells as a consequence of which the risk of leukemia is ten times greater than that of the normal population

Routine treatment has been nonsteroidal anti-inflammatory drugs (NSAIDS) and physiotherapy. Anti-tumor necrosis factor -α blocking agents **infliximab**, adalimumab and **etanercept** are under trial.[58]

ANKYRIN

A protein constituent of **red blood cell** membranes (Band 2.1) that links spectrin to Band 3 protein. Rare deficiency results in heat-sensitive **fragmentation hemolytic anemia**.

ANTENATAL SCREENING

See **Hemolytic disease of the newborn; Transfusion-transmitted infections**.

ANTHRACYCLINES

See also **Cytotoxic agents**.
(Cytotoxic antibiotics) A group of drugs used for cytotoxic therapy that are natural products of the soil bacterium *Streptomyces* spp. and probably inhibit **DNA** replication by intercalation. They are cell-cycle-phase nonspecific agents that are given intravenously, usually in pulses. They are metabolized in the liver and excreted in bile, so that patients with poor liver function should receive reduced doses. The anthracyclines most often used in the treatment of **acute myeloid leukemia** (AML) are daunorubicin, doxorubicin (Adriamycin), idarubicin, and bleomycin. They are also used as part of cytotoxic regimes in the treatment of **myelomatosis** and **non-Hodgkin lymphoma**, especially **diffuse large B-cell lymphoma** treated with R-CHOP and **Hodgkin lymphoma** with ABVD. Many anthracyclines act as radiomimetics and so should be avoided when radiotherapy is also used. Cardiotoxicity is a major cumulative dose-limiting toxicity, occurring clinically above doses of approximately 450 mg/m². However, at much lower doses, microscopic cardiac muscle changes are seen. Lower doses should be used where patients have received radiotherapy to the mediastinum, which potentiates cardiotoxicity. Other **adverse drug reactions** include gastrointestinal tract upsets, especially mucositis. Alopecia is common. Bleomycin causes skin pigmentation and subcutaneous sclerotic plaques, but little **bone marrow hypoplasia**. Progressive pulmonary fibrosis becomes a problem if the cumulative dose exceeds 300,000 units. Mitoxantrone, related to daunorubicin, kills cells by intercalating with DNA and is myelosuppressive. It is given intravenously for 2 to 5 days, alone or in combination with cytarabine or vincristine. It produces remission rates in relapsed or resistant AML and in **acute lymphoblastic leukemia** (ALL) of up to 60%. In *de novo* childhood AML, remission rates of approximately 80% have been obtained when included in multidrug schedules. It has similar cardiotoxic potential to daunorubicin.

ANTIBODY

A soluble **immunoglobulin** secreted by **plasma cells**, present in serum and secretions, that specifically react with antigens and thus play a crucial effector role in the immunological elimination of these antigens. Antibody is the component of the **immune response** that mediates **humoral immunity**. Major effector mechanisms include neutralization, opsonization, activation of **complement**, and antibody-mediated cellular cytotoxicity. Depending upon their origin or function, a variety of antibody types are described:

Natural antibody: immunoglobulins secreted by B-cells, which express CD5; examples are anti-A and anti-B of the **ABO (H) blood groups**

Isoantibody: having the same genotype

Alloantibody: isoantibody raised to allotype determinants

Autoantibody: produced when failure of self-tolerance does not occur by apoptosis

Monoclonal antibody: single-type immunoglobulins that react with a single epitope of its target antigen

Heterophil antibody: IgM that reacts with a similar organ, i.e., red blood cells of a different species

Anti-idiotypic antibody concerned with the idiotypic area of an antigen

ANTICOAGULANT THERAPY

See **Aspirin**; **Coumarins**; **Dipyridamole**; **Heparin**; **Warfarin**.

ANTI-D IMMUNOGLOBULIN

See **Hemolytic disease of the newborn**.

ANTIFIBRINOLYTIC AGENTS

See **Fibrinolysis**.

ANTIGEN

The molecular structure to which a specific **immune response** develops. It should be noted that the terms "antigen" and "immunogen" are not synonymous. The term "immunogen" is used to indicate a substance capable of eliciting by itself an immune response. Not all antigens are immunogens: the classic examples are those small molecules described as haptens, which by themselves do not induce an immune response but when complexed with a larger molecule (usually protein, described as a carrier) can induce **antibodies**.

Natural antigens are most commonly proteins, although antibody (but not cell-mediated responses) may develop against other macromolecules, notably carbohydrates and nucleic acids. The nature of antigens can only be discussed in the context of the nature of the components of the immune system that specifically recognize them. These are the T-cell antigen receptors (see **Antigen presentation**) and immunoglobulin (see B-**lymphocytes**). **Immunoglobulin** may be either present in the B-cell membrane, where it is functioning as the B-cell receptor for antigen (B-CR), or as soluble antibody.

Antigen Structure

In essence, the T-cell receptor (TCR) and immunoglobulins interacting with macromolecular antigens do not recognize the entire structure but, instead, recognize smaller parts described as **epitopes**. ("Determinant" is an older term meaning essentially the same thing.) T-cell epitopes, by the nature of the process of antigen presentation, are short linear fragments of protein antigens (obviously, nonprotein antigens cannot form T-cell epitopes), and in the context of a given **human leukocyte antigen** (HLA) there may be few (one or two) or no epitopes in a given protein.

On the other hand, the epitopes with which immunoglobulins react are surface structures of the intact antigen. These may be linear peptides (say, a loop of sequence on the surface of the protein), but in practice, they are more often an array of amino acid residues on the protein surface, derived from different regions of the protein sequence. Because the antigen-combining site of the immunoglobulin reacts in a lock-and-key fashion with its epitope, the conformation of the epitope is crucial, and thus many antibodies will react only with the native protein. For this same reason, these epitopes are referred to as conformational epitopes. There are multiple, overlapping immunoglobulin-reactive epitopes on the surface of a protein, but some may be immunodominant, eliciting a greater response than others. A protein antigen therefore possesses two sets of epitopes, one for immunoglobulin and one for the TCR, which are functionally and physically quite distinct.

The nature of immunoglobulin epitopes explains why antibody responses to nonprotein antigens may develop, provided the "antigen" is present on the surface of a protein (e.g., hapten, carbohydrate moiety of a glycoprotein). The nonprotein antigen separate from its protein carrier is not immunogenic because there are no helper T-cell epitopes — which must be peptide in nature — for the B-cell to present to its helper T-cell (see B-**lymphocytes**).

ANTIGEN PRESENTATION

The means by which a protein **antigen** activates T-**lymphocyte** responses by being "processed" within a cell and thereafter "presented" to a T-cell in the context of class I or II histocompatibility antigens of the major histocompatibility complex (MHC). The details of antigen processing and presentation differ, depending on whether presentation is by class I antigens to CD8-positive T-cells or by class II antigens to CD4-positive cells (see **Human leukocyte antigens**).

Class I/CD8 Presentation

Proteins synthesized within the cell ("endogenous") are degraded, probably by **proteasomes** (large multimeric proteases), some of whose components are encoded by genes (LMP 1 and 2) in the MHC to peptides, which are nonselectively transported into the endoplasmic reticulum by a class of peptide transporters also encoded in the MHC (TAP 1 and 2). Here the peptides encounter newly synthesized class I antigens, which bind peptides of eight or nine amino acids in length and transport them via the **Golgi apparatus** to the cell surface.

Binding is to a cleft between the a1 and a2 domains, where the polymorphic residues are concentrated. Although a particular histocompatibility antigen can bind a wide range of peptides, binding is not promiscuous and there are common features displayed by binding peptides. Usually one or two "anchor" residues are conserved at particular sites in a series of peptides binding to a particular class I antigen, or at least varying in a conservative manner. These anchor residues have been shown to interact with specific amino acid residues in the peptide-binding cleft. In general, only one or two peptides from a given protein will interact with a given class I antigen, hence the restricted range of T-cell epitopes.

Class II/CD4 Presentation

Proteins taken up from outside the cell ("exogenous") are degraded to peptides in an endosomal compartment. Class II proteins, complexed with the invariant protein (CD74), travel to some ill-defined endosomal compartment where the invariant protein is degraded and the exogenously derived peptide binds. The role of the invariant protein is to prevent binding of endogenous peptides to class II antigens while within the endoplasmic reticulum; however, there is unequivocal evidence that, in some circumstances, endogenous proteins are processed via the class II pathway. There is some evidence that "professional" antigen-presenting cells have a specialized compartment for interaction between class II antigens and their peptides. The class II/peptide complex is then transported to the cell surface. The nature of peptide binding is broadly similar to that for class I antigens, with the peptide-binding cleft of class II antigens having a similar structure to the class I cleft. Slightly larger peptides bind to class II antigens.

Antigen-Presenting Cells

(APCs) Probably all cells expressing class I antigens can present endogenous antigens to CD8 T-cells, and all cells expressing class II antigens can present to CD4 T-cells. However, the efficiency with which they do so varies widely. In the case of presentation to CD4 cells, some APCs, particularly **dendritic reticulum cells** (various sorts), B-cells, and **histiocytes** (macrophages), are so much more efficient than others that they are described as professional APCs. The reasons for this include possession of specific antigen receptors (B-cells) or of extremely efficient phagocytic (e.g., macrophages) and specialized cellular compartments for processing antigen. In the case of dendritic reticulum cells, which in some assays are the most potent APCs, the reason for this efficacy is not clear, as the cells are not phagocytic and do not have antigen receptors as such, but do have receptors for antibodies and **complement** and thus can bind opsonized antigens. However, this does not explain how dendritic reticulum cells function in primary responses, which is probably their main role. Production of **cytokines** by APCs, particularly macrophages, is also important in their function.

Role of Cytokines in Controlling Presentation

Antigen presentation is, to some extent, an inducible function of cells, in that cytokines (**interferon** tumor necrosis factor, IFN-γ; **tumor necrosis factor**, TNF-α; and **interleukin**, IL-4, in particular) can upregulate expression of many of the components of the antigen-processing and -presentation machinery, including proteasome subunits, transporters in antigen presentation (TAP), and histocompatibility antigens. There is experimental evidence that this increases the efficiency of presentation to T-cells. Contrariwise, IFN-α and -β and transforming growth factor (TGF)-β downregulate class II histocompatibility antigens in at least some cell types. The importance of these cytokine effects is probably to extend the range of cells able to present antigen to T-cells and the efficacy of antigen presentation during infection.

Histocompatibility Antigen Restriction

A consequence of the dual recognition of peptide and histocompatibility antigen by the T-cell is that the histocompatibility antigen has to be correct. During ontogeny, T-cells "learn" to recognize self-histocompatibility antigens and so respond only to APCs that bear the same histocompatibility antigens, whether from the same or a different but matched individual. Restriction was first described in the context of the killing of virus-infected cells by **cytotoxic T-lymphocytes**. It was noted that killing only occurred if the T-cell and the target cell possessed identical class I histocompatibility antigens. A similar phenomenon occurs in the activation of helper T-cells. The phenomenon is important in that it helps to distinguish cellular immunological mechanisms mediated by T-cells; if the phenomenon is not restricted, then the likelihood is that T-cells are not involved.

ANTIGLOBULIN (COOMBS) TEST
See **Direct antiglobulin (Coombs) test**; **Indirect antiglobulin (Coombs) test**.

ANTIHEMOPHILIC GLOBULIN
See **Factor VIII**.

ANTI-IDIOTYPIC ANTIBODY
Naturally occurring and experimentally induced autoantibodies that react with an **antigen**-binding site or idiotype of an antibody, which is thus functioning as an **epitope**. A theory, associated with Neils Jerne, held that networks of antibody idiotypes and anti-idiotypes regulated the **immune response**; this is now disregarded.

ANTIMETABOLITES
See also **Cytotoxic agents**.
A group of drugs used for cytotoxic therapy that are analogs of normal compounds needed for cell division and function. They damage cells by interacting or competing with enzyme systems.

Methotrexate
Acts by competing for the **folic acid**-binding sites of dihydrofolate reductase, thus inhibiting the synthesis of tetrahydrofolate and resulting in decreased nucleotide synthesis with consequential reduced **deoxyribonucleic acid** (DNA) synthesis, leading to cell death. Methotrexate is used orally, parenterally, and intrathecally, chiefly in the treatment of primary and secondary **non-Hodgkin lymphoma** affecting the central nervous system (CNS) and in the prophylaxis of **acute lymphoblastic leukemia** (ALL) and **diffuse large B-cell lymphoma** associated with risk factors. Adverse reactions include skin rashes, gastrointestinal tract complications (especially mucositis), liver cell damage, **immunosuppression**, and rarely pneumonitis. **Lymphoproliferative disorders** — diffuse large B-cell lymphoma, Hodgkin lymphoma, polymorphic posttransplant lymphoproliferative disorder — have been reported, often related to Epstein-Barr virus (EBV) infection. They may regress with cessation of methotrexate therapy. The cytotoxicity of methotrexate is reversible using **folinic acid**, which converts intracellularly to reduced folate. Folinic acid should be used as "rescue" after high-dose methotrexate treatments, such as those used to treat primary non-Hodgkin lymphoma affecting the CNS, ALL, and **acute myeloid leukemia** (AML). It should be avoided if there is renal or hepatic impairment and also if pleural effusion or ascites are present. Response is particularly susceptible to genetic polymorphisms.

Cytosine Arabinoside (Cytarabine)

A pyrimidine analog with an arabinosyl sugar moiety, active during S phase of the cell cycle, killing cells by incorporation into DNA. Because of its short half-life of about 2 h, it is either used by continuous intravenous or subcutaneous infusion, or by frequent intravenous bolus injection. The drug inefficiently crosses the blood-brain barrier, so that high-dose systemic treatments lead to CNS levels capable of killing leukemia cells. More conventionally, it is used intrathecally as an alternative or complementary drug to methotrexate in the prophylaxis of the CNS in ALL. Systemically, standard dosage for AML treatments is 100 to 200 mg/m² daily for 5 to 10 days, repeated at intervals usually with an **anthracycline**. High-dose cytosine, 1 to 3 g/m² twice daily for 2 to 6 days, is useful in the treatment of relapsed and resistant AML. It may also be incorporated into regimes of treatment for relapsed non-Hodgkin lymphoma (diffuse large B-cell lymphoma). A commonly used related agent is Fludarabine, used in the treatment of **chronic lymphatic leukemia** after failure of an initial **alkylating agent**. Another, Clofarabine, is under investigation for relapsed/refractory ALL.

Fludarabine

A cytarabine derivative with similar activity to vincristine and vinblastine (see **Vinca alkaloids**). It is dephosphorylated *in vivo* and rephosphorylated intracellularly to form 2-arafluoro **adenosine triphosphate** (ATP), which inhibits DNA by blocking DNA polymerase and ribonucleotide reductase.

   It is used as a second-line agent to treat **chronic lymphatic leukemia** (CLL)/low-grade **non-Hodgkin lymphoma** at dosages of 25 to 30 μg/m² IV daily in courses lasting 5 days. Although it is associated with severe **bone marrow hypoplasia** and **immunodeficiency** with subsequent viral and other infections, but rarely with CNS toxicity or metabolic acidosis, it is increasingly used in combination therapy to treat CLL resistant to chlorambucil.

6-Mercaptopurine

A purine analog that is active in its ribonucleotide form against leukemic cells. It is used mainly in the maintenance of remissions in ALL, often with methotrexate. It is metabolized by xanthine oxidase to thiouric acid. As **allopurinol** inhibits xanthine oxidase, this drug will block this metabolic pathway, increasing the level of 6-mercaptopurine, which should be reduced to about one-third dosage in patients taking allopurinol. Azathioprine is metabolized to mercaptopurine, so if it is given simultaneously, the dose of both drugs should be reduced.

6-Tioguanine (Thioguanine)

A purine analog with a similar action, in its ribonucleotide form, to 6-mercaptopurine. It is mainly used for AML induction regimens with cytosine arabinoside, with which it is synergistic *in vitro*, although clinically results are similar with and without its addition to these regimens.

ANTINEUTROPHILIC CYTOPLASMIC ANTIBODIES

(ANCA) **Autoantibodies** that act against bactericidal permeability increasing (BPI) protein. They occur in two forms:

Cytoplasmic (c-ANCA), directed against proteinase 3 — **Wegener's granulomatosis**

Perinuclear (p-ANCA), directed against myeloperoxidase — Wegener's granuloma-tosis and, in microscopic polyangiitis, a form of **polyarteritis nodosa**

They have also been reported in patients with systemic **vasculitis** and multisystem **Behçet's disease**.

ANTINUCLEAR FACTOR

(ANF, ANA) **Antibodies** that react with nucleic acids, nuclear proteins, and cell-surface antigens (Sm antigens) to form circulating **immune complexes**. They arise in the plasma of patients with **systemic lupus erythematosus** (SLE), **Sjögren's syndrome**, **rheumatoid arthritis**, chronic hepatitis, thyroiditis, myasthenia gravis, **gastric disorders** leading to intrinsic-factor deficiency, ulcerative colitis, and **pure red cell aplasia**.

ANF can be demonstrated by:

Immunofluorescence using a section of tissue (e.g., rat liver) to which has been added fluorescein-labeled antihuman gammaglobulin

Radioimmunoassay in which isotope-labeled antigen is added to the test serum and the mixture treated with 50% saturated ammonium sulfate to precipitate the immunoglobulin; the precipitate will contain radioatoms only if an antibody-antigen reaction has occurred, the amount of which is measured as the concentration of ANF present

Latex particles coated with nuclear material, which are aggregated by ANF

L-E cell test when caused by SLE[59]

ANTIPHOSPHOLIPID-ANTIBODY SYNDROME

(APS; Hughes syndrome) A disorder in which **venous thromboembolic disease** or **arterial thrombosis** or both may occur, the serologic markers being antiphospholipid antibodies (aPL).[60,61] This is a heterogeneous group of **antibodies** that includes anticardiolipin antibodies (aCL), the **lupus anticoagulant** (LA), and antithrombotic antibodies. The aPL antibodies are directed against different phospholipid/protein complexes; LA antibodies recognize the prothrombin-phospholipid complex and in this way inhibit the phospholipid-dependent coagulation reactions. In contrast, aCL antibodies are directed against b2-glycoprotein I (b2GPI) bound to an anionic lipid surface. This increases the risk of predisposition to atherosclerosis. The aCL antibodies arising secondary to infections, e.g., tuberculosis, *Klebsiella* spp., do not have the b2GPI requirement. The measurement of anti-b2GPI antibodies identifies patients with aCL antibodies that are not associated with infection and appear to be more closely associated with a history of thromboembolic complications than aCL antibodies. It is possible that thrombosis is induced by the binding of aCL antibodies to Annexin A5 (placental anticoagulant protein).[60]

Although the antiphospholipid-antibody syndrome often occurs in patients with **systemic lupus erythematosus** (SLE), the majority of patients with the syndrome do not meet the criteria for that disease. The combination of recurrent thromboses and antiphospholipid antibodies without features of SLE is called the primary antiphospholipid syndrome. Other features of this syndrome include **thrombocytopenia**, migraine, central nervous system (CNS) demyelination, livedo reticularis, stenosis of the renal artery, recurrent spontaneous abortions, and infertility. The antiphospholipid antibodies in this syndrome can exist for many years. Thrombosis, the main complication of the syndrome,[62] can affect vessels of all sizes, including cerebral and pulmonary arteries and those of the bone marrow.[63] The risk of thrombosis in symptomatic patients with the antiphospholipid-antibody syndrome is high. Anticoagulant therapy should aim for an international nor-malized ratio (INR) of around 3.0.[64] The associated antithrombotic antibodies are of no clinical significance but can disturb monitoring of anticoagulant therapy. Anticoagulation has been claimed to reduce hypertension and to prevent irreversible renal damage.

In pregnancy, therapy with low-molecular-weight **heparin,** with or without **aspirin,** has reduced fetal loss, whereas low-dose aspirin alone seems to be relatively ineffective.[65] No improvement has been reported following high-dose **immunosuppression**.

An acute form (catastrophic APS)[66] can be precipitated by surgery, drugs, discontinua-tion of anticoagulant therapy, or infections, probably due to massive vascular endothelial cell activity. Acute renal failure or acute respiratory distress syndrome is the principal complicating disorder.

α2-ANTIPLASMIN

See **Fibrinolysis** — inactivators; **Serine protease inhibitors**.

ANTITHROMBIN III

(AT, ATIII) A single-chain glycoprotein of molecular weight 58 kDa that inhibits all of the coagulation **serine proteases** but in particular activated factor II (**thrombin**) and activated **factor X**. The rate of inhibition is accelerated 5,000- to 10,000-fold in the presence of **heparin** and other sulfated glycosaminoglycans. Heparin is not normally found in the circulation, and, physiologically, antithrombin probably binds to heparin sulfate on the vascular endothelial cells. In such a position, antithrombin is ideally positioned to inactivate free coagulation serine proteases. A deficiency or functional abnormality of antithrombin can result in an increased risk of **venous thromboembolic disease**. Such deficiencies are classified as either:

Type I: a parallel reduction in both functional and immunological antithrombin levels

Type II: the presence in the plasma of a dysfunctional protein, which may be present in normal or reduced amounts

These can be further subdivided, depending upon whether the mutation affects the heparin-binding domain (IIHBS), the reactive site (IIRS), or has multiple or pleiotropic effects (IIPL).

The risks of thrombosis are highest in individuals with type I disease, IIRS, or IIPL and lowest in cases of IIHBS. Type I deficiency is estimated to affect 1:4200 of the general population and type II approximately 1:630. In patients with thromboembolic disease, the incidence of antithrombin deficiency is estimated at between 4 and 6%.

Acquired antithrombin deficiency is seen in a variety of clinical disorders, e.g., **dissem-inated intravascular coagulation** (DIC), severe burns, **liver disorders**, **renal disorders** (nephritic syndrome); in association with various drugs (L-**asparaginase**); and in patients undergoing cardiopulmonary bypass.

The **reference ranges** for adults are 0.86 to 13.2 U/ml (function) and 0.79 to 1.11 U/ml (antigen). Those for premature infants and for full-term infants over the first 6 months of life are given in **Reference Range Tables XVII** and **XVIII**.

Antithrombin III concentrates (see **Coagulation-factor concentrates**) are of benefit in patients with congenital deficiencies of antithrombin, e.g., to cover labor where life-threatening thrombosis has occurred or where it may be inappropriate to administer heparin. The evidence that antithrombin supplementation is of value in acquired deficiencies is conflicting.

ANTITHROMBOTIC THERAPY

Therapeutic agents administered to remove or prevent thrombosis. Their mode of activity covers the whole range of **hemostasis**:

Antiplatelet-function drugs: **aspirin**, **dipyridamole**, ticlopidine, **abciximab**, clopidogrel

Anticoagulants:
- Vitamin K antagonists: coumarins (**warfarin**)
- Direct thrombin inhibitors: **argatroban**, **ximelagatran**
- Serine protease inhibitors: **heparin**, **heparin cofactor II**, **hirudin**

Fibrinolysis stimulants:
- Thrombolytic agents: streptokinase, urokinase
- Plasminogen activators: alteplase
- Fibrinopeptide A cleavage: **ancrod**

The choice of agent depends particularly on the location of the thrombus, its activity, and many other health factors including patient choice. Common treatment usage can be very broadly summarized as:

Arterial thrombosis
- Recent: fibrinolytic agents followed by aspirin or warfarin
- Preventive: aspirin or warfarin

Venous thromboembolism
- Recent: heparin followed by warfarin
- Preventive: aspirin or warfarin

APHERESIS

See **Hemapheresis**.

APLASTIC ANEMIAS

(AAs) **Pancytopenia** arising as a result of failure of hematopoietic stem cell proliferation and differentiation (see **Hematopoiesis**). The peripheral-blood examination shows **normocytic** or **macrocytic anemia**, **neutropenia** with variable **lymphopenia**, and **thrombocytopenia**. The **bone marrow** is hypocellular with normal hematopoietic marrow replaced to a greater or lesser extent by fat cells. Remaining hematopoietic cells are morphologically normal apart from mild macrocytosis and **dyserythropoiesis**, with increased granulation ("toxic granulation") in the **neutrophils**. The features may arise in a number of ways,

TABLE 18

Causes of Aplastic Anemia

Etiology	Classified Disorders	Characteristics
Idiosyncratic, acquired	idiopathic	unpredictable
	drug induced	prolonged course
	viral (hepatitis)	probably autoimmune
Inevitable	**cytotoxic agents**	dose dependent
	irradiation	predictable recovery
Genetic	**Fanconi anemia**	
	Dyskeratosis congenita	
	Schwachman-Diamond syndrome	
	Blackfan-Diamond syndrome,	
	erythroid aplasia only	
	amegakaryocytic thrombocytopenia,	
	thrombocytes only	
Immune, antibody mediated	primary **acquired AA**	multiple autoantibodies
	secondary to SLE	
Malignant	**acute leukemia**	usually ALL precedes emergence
		of leukemia
	myelodysplasia	hypoplastic myelodysplasia variant

which need to be distinguished (see Table 18). In particular, **acquired aplastic anemia** has a different pathophysiology from **bone marrow hypoplasia** and **pure red cell aplasia**. The term is also loosely used to describe the **aplastic crisis** of **parvovirus** infection where only the **erythron** is affected. A summary of causes is given in Table 18.

APLASTIC CRISIS

An abrupt fall of **hemoglobin** and **red blood cell count** due to transient erythroid bone marrow aplasia, which characteristically occurs in patients who already have **hemolytic anemia** with shortened red cell survival. Originally described in 1942, it was shown in 1981 by Pattison and his colleagues to be due to acute **parvovirus** B19 infection. Chronic pure red cell aplasia caused by persistent parvovirus infection has been associated with **immunodeficiency**, especially hypogammaglobulinemia, and **acquired immunodeficiency syndrome** (AIDS).

Pathogenesis

Parvovirus B19 is usually transmitted by droplet spread, and reticulocytes disappear about 7 to 10 days following inoculation. The B19 virus directly infects, and is cytotoxic to, proliferating erythroid progenitors within the bone marrow, leading to transient erythroid aplasia, which typically lasts 5 to 7 days. In normal, healthy individuals with a red blood cell life span of 120 days, the effect is not clinically significant, but in patients with shortened red blood cell survival — most commonly in those with congenital hemolytic anemia or sickle cell anemia — this interruption results in anemia, which can be very severe and life threatening if the underlying red cell survival is markedly reduced. Aplastic crisis has been described in patients with **sickle cell disease**, **hereditary spherocytosis**, **congenital dyserythropoietic anemia**, hereditary erythroblastic multinuclearity with positive acidified serum test (HEMPAS), **paroxysmal nocturnal hemoglobinuria**, **thalassemia**, and **pyruvate kinase** deficiency.

    The transient nature of the illness is due to the rapid development of B19-specific IgM and IgG antibodies. Neutralizing IgM antibodies arise about 10 days postinfection, and erythropoiesis recovers. IgG antibodies follow and produce lifelong immunity to further attacks. In patients who are immunocompromised, chronic infection may ensue following the acute infection and, rarely, may be associated with persistent or relapsing **pure red cell aplasia**.

Clinical and Laboratory Features

Fever may or may not precede the anemia. The typical rash of B19 infection is a later event. The severity of symptoms is proportional to the degree of anemia, with tiredness, palpitations, dyspnea, heart failure, and, rarely, death having been reported.

Peripheral blood reveals a normochromic **normocytic anemia** with a low **reticulocyte count**. Rarely, **neutropenia** or **thrombocytopenia** also occur, the cause of which is unclear, as B19 virus has no effect on colony forming units-granulocyte macrophage (CFU-GM). **Bone marrow** examination reveals erythroid precursor hypoplasia/aplasia, but giant pronormoblasts (100 μm) with vacuolation may occasionally be seen.

The patients are usually viremic at the time of presentation, and therefore the diagnosis can be made by dot-blot hybridization (if tested within 7 days of onset of symptoms) or by **polymerase chain reaction** (which remains positive for a long time). Diagnosis should be suspected on clinical grounds and is usually confirmed by finding specific IgM antibodies arising as the anemia improves. The presence of IgG antibodies excludes the diagnosis.

Treatment

In view of the self-limiting nature of aplastic crisis, **red blood cell transfusion** of red cells and/or **platelet transfusion** are usually all that is required. In immunocompromised patients with chronic infection, intravenous **immunoglobulin** is of benefit.

APOFERRITIN

See also **Transferrin**.

An iron-binding glycoprotein free of bound iron. It is synthesized by hepatocytes and cells of the monocyte-macrophage system.

APOPTOSIS

An energy-dependent mechanism of cell death induced by a variety of environmental and endogenous stimuli. The term "apoptosis" derives from the Greek "falling of leaves from the tree" and was first used to describe the mechanism of cell death of epithelial tissue. This cell death mechanism, distinct from necrosis, is often considered (inappropriately) synonymous with "programmed cell death." A key distinction between apoptosis and necrosis is that the former does not result in **inflammation**: the apoptotic cell degrades to a number of apoptotic bodies that retain the integrity of the plasma membrane, and these are phagocytosed by neighboring cells.

Apoptosis is crucial in a number of physiological processes, such as embryogenesis and the maintenance of cell numbers in tissues, and also in the killing of target cells by **cytotoxic T-lymphocytes**. It is important for removing cells where **DNA** has been irreparably damaged by either chemicals or irradiation. Subversion of apoptosis resulting in increased cell number is an important mechanism in carcinogenesis. Increasingly, apoptosis is recognized as a pathological mechanism of cell death in a variety of human diseases as diverse as motor neurone disease and **myelodysplasia**.

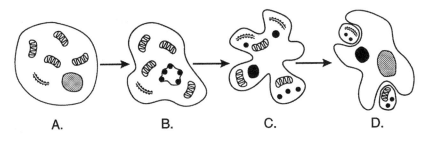

A.   B.   C.   D.

FIGURE 4
Morphological changes in apoptosis: (a) intact cell; (b) early apoptosis: nuclear chromatin condensation; (c) nucleus disintegrates, plasma membrane intact, organelles intact, cytoplasm forming into apoptotic bodies; (d) apoptotic bodies ingested by tissue macrophages.

Necrotic cell death involves cellular swelling and early disruption of the plasma membrane and cytoplasmic organelles without organized nuclear damage, and it occurs in groups of cells. In contrast, apoptosis proceeds via a series of programmed events, with little variation and within single cells. Activation of endogenous endonucleases, which cleave DNA, is an early event. Several endonucleases are recognized functionally by their dependence upon distinct cationic conditions for activity, although these have yet to be fully characterized in human cells. The earliest endonuclease-mediated DNA cleavage occurs at **topoisomerase** II binding sites and uncoils supercoiled DNA. Such 30-kb fragments are detectable by pulsed-field gel electrophoresis (see **Molecular genetic analysis**). Later events involve cleavage at internucleosomal sites, resulting in DNA fragmentation to multiples of 180 bp, detectable by agarose gel electrophoresis as "DNA ladders." Nuclear-membrane disruption then occurs, with (a) nuclear condensation into apoptotic bodies within an intact plasma membrane and (b) morphologically normal cytoplasmic organelles. Finally, the plasma membrane disrupts, and many apoptotic bodies are formed from an individual cell and rapidly ingested *in vivo* (within approximately 15 min) by **histiocytes** (see Figure 4).

Triggering of apoptosis is a combination of negative and positive signals. Negative signals include absence of growth factors and signals via **cellular adhesion molecules**; these act via the Akt kinase, which is activated through the phospholipase C cleavage and which serves to inhibit apoptosis. Positive signals include primarily DNA damage and binding of ligands to certain "death receptors."

There are two pathways leading to apoptosis: the intrinsic pathway (involving the mitochondrion and triggered by intracellular events) and the extrinsic pathway (death receptor). The intrinsic pathway involves the function of the Bcl2 family of proteins, which form heterodimers and can be pro- or antiapoptotic. Depending on the balance of the two types, apoptosis is promoted or inhibited. Key members are Bcl2 itself and Bad. The former is upregulated by DNA damage via p53, and the latter is activated by Akt. When Bcl2 predominates, it binds to the mitochondrial membrane, causing it to become permeable to cytochrome, which leaks out and activates a cascade of proteolytic enzymes called "caspases." These are, likewise, activated by the extrinsic death receptor pathway. Death receptors include CD120a (the TNF-α receptor) and CD95 (FAS, APO-1, involved in cytotoxic T-lymphocyte killing).

The activated caspases lead to digestion of structural proteins in the cytoplasm and the degradation of DNA as described above, with the formation of apoptotic bodies. A feature of apoptosis is that phosphatidylserine, normally intracellular, is exposed on the external surface of the plasma membrane, where it provides a signal for phagocytosis.

Apoptosis can be detected morphologically by electron microscopy, molecularly *in situ*, or by fluorochrome-labeling **flow cytometry**. Pulsed-field gel electrophoresis detects the earliest stage, whereas agarose gel DNA ladders represent a later stage. *In situ* end-labeling and flow-cytometric techniques employ the principles of end labeling with deoxyuridine triphosphate (dUTP) of endonuclease-induced free 3′ hydroxyl groups at DNA strand breaks by terminal deoxytidyl transferase. This technique has been shown to distinguish apoptosis from necrosis and is now widely used, and at least the flow-cytometric method has the advantage of providing simultaneous cell-cycle data (e.g., propidium iodide DNA content). This is important for studies aimed at identifying the phase of cell cycle in which apoptosis is occurring. A very convenient flow-cytometric technique takes advantage of the exposure of phosphatidylserine, which can be labeled through its affinity by Annexin V. Other flow-cytometric techniques for apoptosis detection are less sensitive and include identification of the "sub Go" peak, detecting reduced DNA content in apoptotic cells upon fixation, and Hoechst dye retention by necrotic cells and exclusion by apoptotic cells.

The role of apoptosis in hematological disease centers principally on the therapeutic induction of cell death by a variety of chemical agents, including **anthracyclines**, topoisomerase II inhibitors, and nucleoside analogs. However, apoptosis is the most likely pathological process of cell death in the **ineffective erythropoiesis** associated with **megaloblastosis**, **thalassemia**, and **myelodysplasia**, and failure to undergo apoptosis contributes to cell immortality in **follicular lymphomas**.

APROTININ

See also **Fibrinolysis** — antifibrinolytic agents.
A polypeptide extracted from bovine lung that inhibits certain serine proteases by binding to their active site. At low concentrations, it is a powerful inhibitor of plasmin, with *in vitro* molar potency 100 and 1000 times that of tranexamic acid and epsilon-aminocaproic acid (EACA), respectively. At high doses (150 to 200 kIU), it also inhibits **kallikrein**, which is formed during the activation of coagulation by cardiopulmonary bypass and has a central role in the activation of the inflammatory response. Inhibition decreases the activation of **complement**, renin, **bradykinin**, coagulation, and **fibrinolysis**. Aprotinin may have platelet effects, in that it may have a minor effect on preserving platelet function and platelet membrane receptors, possibly by inhibiting plasmin-mediated degradation. It is a bovine protein and thus can provoke a hypersensitivity reaction. There are theoretical (but unproven) prothrombotic effects, and there is a potential for the patient to develop anti-aprotinin antibodies.

ARACHIDONIC ACID

A 20-carbon unsaturated fatty acid, derived from the metabolism of membrane phospholipids by the action of phospholipase A2 and C. It has a short half-life, and the main products of its metabolism are **prostaglandins** and **leukotrienes**. Arachidonic acid metabolites are not stored and are all freshly synthesized prior to secretion. Metabolism is via two major routes: cyclooxygenase (prostaglandins) or lipoxygenase (leukotrienes).

ARGATROBAN

A synthetic N2-substituted arginine derivative of molecular weight 526 that acts as a reversible, competitive, univalent, direct thrombin inhibitor. It is used mainly in the management of **heparin**-induced thrombocytopenia (HIT). Argatroban has been used as an alternative anticoagulant in patients with HIT in various clinical conditions, including interventional cardiovascular procedures that require anticoagulation. Satisfactory clinical outcomes with acceptable complications have been reported in these patients. Whether argatroban offers additional clinical advantage over conventional heparin therapy in patients without HIT remains unclear.

ARSENIC TOXICITY

The effect on hematopoietic tissues is only replacement of phosphorus by binding to sulfhydryl groups on proteins, thus inhibiting oxidative phosphorylation and pyruvate metabolism. In the **red blood cell** membrane, this interferes with the **cation pump** control of cell volume, thus causing **hemolytic anemia**. It can also interfere in erythropoiesis, giving rise to **dyserythropoiesis** with **pancytopenia**.

Arsenic is a chronic poisoning agent readily absorbed by the gastrointestinal tract. Toxicity is rarely industrial in origin, but it still arises from domestic paint dust and chronic homicidal poisoning. The more important origin for poisoning is as arsine, a colorless toxic gas produced by the action of water on metallic arsenide. Arsine is produced during galvanizing, soldering, etching, and lead plating. Following exposure (2 to 24 h), there is abdominal pain, nausea, vomiting, dark red urine, jaundice, and anemia. Arsine is fixed by **hemoglobin** within the red blood cells, following which **hemolysis** occurs. The mortality rate following exposure approaches 25%. Treatment is by **hemapheresis**.

ARTERIAL EMBOLISM

Emboli arising from cardiac or **arterial thrombosis**. Arterial plaques with thrombi loosely attached are a causal condition, particularly in carotid and coronary arteries. **Cardiac disorders** giving rise to arterial emboli include inflamed or damaged cardiac valves, particularly with infected endocarditis; endocardium adjacent to myocardial infarction; cardiac myxoma; dilated or dyskinetic cardiac chambers, as in atrial fibrillation or on prosthetic heart valves; and low-output states with right-heart failure.

ARTERIAL THROMBOSIS

See also **Thrombosis**.

The pathogenesis, risk factors, and treatment of thrombus formation in the arterial system. The incidence of arterial thrombosis is highest in the coronary and cerebral arteries, vessels that are particularly subject to preexisting vascular disease (atherosclerosis). Thrombus forms in areas of disturbed blood flow and at sites of vessel wall rupture by an atherosclerotic plaque. The pathogenesis is variable, with probably more than one disorder operating at any one time. The current hypotheses[68–70] of atherosclerosis and subsequent thrombosis are:

> *Deposition of low-density lipoprotein* (LDL) in vessel walls, forming the "fatty streak." Phospholipids are essential components of lipoproteins. These are susceptible to free-radical or enzymatic oxidation by myeloperoxidase, lipoxygenase, and other enzymes present in the vessel wall. Oxidized phospholipids accumulate during viral infections and other inflammatory states such as **rheumatoid arthritis**. These oxidized phospholipids induce (a) chronic **inflammation** with **cell-adhesion molecules** (CAM) involving **monocytes** and **T-cells** and (b) accumulation of lipid-filled **histiocytes** (macrophages), i.e., plaque formation.[71] Release of **cytokines** and proteases and calcification of the plaque follows. Platelets are not directly implicated in this process. These lesions arise at sites of increased turbulence of blood

flow and hemodynamic stress. The calcified plaque later becomes covered by a thin fibrous cap beneath the **vascular endothelium**. Plaques enlarge by attracting more monocytes and by proliferation of underlying smooth muscle and fibrous tissue due to cell stimulation by platelet-derived growth factor (PDGF). Hemorrhage into an atherosclerotic plaque and any regrowth of endothelium over a damaged plaque adds to thickening of the vessel, with narrowing of the vascular channel.

Plaque rupture or fissuring exposes underlying **collagen** and **tissue factor**. Plasminogen activator inhibitor levels are also increased when endothelial cells are exposed to oxidized LDL. Activation of platelets with aggregation is enhanced by liberation of **adenosine diphosphate** (ADP) from damaged platelets, all leading to formation of a thrombus, which further occludes the vessel. In larger arteries with high vascular flow rate, the thrombus is composed mainly of platelet aggregates bound by strands of fibrin. Nonocclusive thrombi may become incorporated into the vessel wall and lead to further or accumulated atherosclerosis.

Vascular occlusion by the thrombus and also by emboli — **arterial embolism** — which results in distal infarction.[72]

Risk Factors for Arterial Thrombosis

These risk factors are based upon epidemiological studies rather than proven scientific research on pathogenesis.[73] A full list is given in Table 19. Common associations are:

Familial history of arterial disease

Male prevalence greater than female (currently changing)

Hyperlipidemia, particularly when associated with other metabolic disturbances such as **diabetes mellitus**, where it interacts with platelet-endothelial functions

Hypertension

Obesity with inactive lifestyle

Tobacco smoking, possibly due to increased blood levels of carboxyhemoglobin

Cardiac arrhythmias, particularly atrial fibrillation

Erythrocytosis and **essential thrombocythemia**, which increase blood viscosity (see **Viscosity of whole blood and plasma**)

Thrombophilia, particularly elevated blood levels of **fibrinogen** and plasma coagulation **factor VII**

Chronic inflammatory states such as subclinical chronic infections (chronic bronchitis, chronic dental sepsis), infection with *Chlamydia pneumoniae* or **cytomegalovirus**, and rheumatoid arthritis

Hereditary Risk Factors

These can be identified in some patients that present with arterial thrombosis, but overall, the incidence of venous thrombosis is more common than arterial thrombosis in those with hereditary thrombophilia. There are most often multifactorial risk factors, the inherited component increasing when the individual is exposed to additional acquired risk factors. For instance, there is an added increased risk of thrombosis in women taking high-dose estrogens as an oral contraceptive or as hormone-replacement therapy. A number of patients may be identified as asymptomatic as part of a family screening, once one member of the family has suffered from a thrombosis.[74]

TABLE 19

Risk Factors for Arterial Thrombosis

Hereditary	Acquired
Antithrombin/heparin disorders	Circulatory stasis
Resistance to activated protein C	**Hyperviscosity** syndrome
(factor V Leiden present)	Polycythemia rubra vera
Prothrombin 3'UTR variant gene mutation	Essential thrombocythemia
Protein C deficiency	Chronic myelogenous leukemia
Protein S deficiency	Metabolic factors
Myeloproliferative disorders	**Antiphospholipid antibody** and
Elevated factor VIII, IX, X, or XI	**lupus anticoagulants**
Fibrinolytic disorders	**Diabetes mellitus**
Hypoplasminogenemia	High-dose estrogens for oral contraception and
Dysplasminogenemia	hormone-replacement therapy
Defective or deficiency of tissue plasminogen	Acquired protein S deficiency
activator release	Lifestyle related
Dysfibrinogenemias	Active or passive **tobacco** smoke
Increased plasminogen activator inhibitor 2-I levels	Inactivity, low exercise
Other disorders	Overweight and obesity
Factor XII deficiency	Fruit and vegetable deficiency
Hypercholesterolemia	Social deprivation
Hemoglobinopathies	Low educational attainment
Thrombomodulin deficiency	Associated clinical disorders
Increased or decreased histidine-rich glycoprotein	Hypertension
Diabetes mellitus	**Paroxysmal nocturnal hemoglobinuria**
	Malignancy
	Klinefelter's syndrome
	Thrombotic thrombocytopenic purpura
	Myeloproliferative disorders
	Hemolytic uremic syndrome
	Acute infections (e.g., pneumonia)
	Chronic inflammatory disorders
	Chlamydia pneumoniae infection
	Inflammatory bowel disease
	Nephrotic syndrome
	Behçet's disease
	Cushing's disease
	Acromegaly
	Thyroid disease
	Drugs
	Drug-induced hepatic veno-occlusive disease
	(secondary to stem cell transplantation)
	Atypical antipsychotics
	Cox-2 inhibitors
	Ergotamine
	Antifibrinolytic agents
	Corticosteroids

Clinical Features

Clinical features suggestive of an inheritable factor for thrombosis include:

First thrombosis at an early age

Recurrent venous thromboembolic disease

Spontaneous thrombosis with low level of acquired risk factors identified

Family history of thrombosis

Location in unusual sites (e.g., mesenteric vessels; venous sinus thrombosis)

Recurrent thrombophlebitis

Laboratory Features

Once suspected, investigation for supporting laboratory evidence should be sought by the following tests. It is important to exclude the possibility of the patient receiving heparin or warfarin, as these will greatly affect the results.

Detection of factor V Leiden (or resistance to activated protein C)

Immunological and functional assay of protein C and protein S

Point mutation identification in the 3'UTR of the prothrombin gene

Screening for dysfibrinogenemia: defective fibrinogen polymerization

Functional assay of antithrombin III (heparin cofactor assay)

Management of Hereditary Risk Factors

Management of an immediate thrombosis is the standard care for the type and its location. However, long-term management is controversial. In all cases, the risk:benefit ratio has to be considered to help determine the duration of anticoagulation. Short-term anticoagulation (or infusion of the deficient factor such as antithrombin) should be considered to cover periods of acquired high risk, such as pregnancy or immobilization for any cause. Family screening may be of help. However, screening of girls at puberty for thrombophilia, which has been advocated, is only justified where a familial tendency exists.

Acquired Risk Factors

These are mainly disorders associated with thrombosis based on epidemiological evidence or anecdotal reports, albeit well established, as in the case of malignancy.[75] Here, thrombosis is due to either increased tissue-factor activity, as in gastric or pancreatic carcinomas, or to tumor infiltration by macrophages. Chemotherapy may further increase this activity. Epidemiological studies show that prolonged raised levels of fibrinogen and factor VII are associated with an increased incidence of coronary artery disease, stroke, and peripheral vascular disease.[76] Specific laboratory tests are those for **lupus anticoagulant**, anticardiolipin antibodies, anti-b2 glycoprotein antibodies, and **protein S**. Other tests are confirmatory for the suspected disorder, such as **hemoglobinopathies** and **paroxysmal nocturnal hemoglobinuria**. A more recently established risk factor is increased activated **protein C** resistance in women with factor V Leiden who are taking the third-generation contraceptive pill.[77] Prevention depends mostly on lifestyle, such as avoidance of cigarette smoking, moderate regular exercise, dietary prudence to avoid high-calorie or excess saturated fats, and plentiful fruit and vegetables.[78] Long-term low-dose aspirin is of benefit for those with known vascular disease. Widespread prophylaxis to cover short-term stasis, such as immobilization during aircraft journeys, is often used but remains an area of some considerable uncertainty.

Treatment

Acute Myocardial Infarction/Ischemia[79]

The currently accepted treatment for acute infarction is undergoing change. The well-established regime of an immediate dose of 150 to 300 mg of **aspirin** followed by thrombolytic therapy,[80] unless contraindicated, might be replaced by percutaneous coronary intervention (PCI), with or without stenting, followed by aspirin with a platelet membrane **glycoprotein IIb/IIIa** inhibitor (**Abciximab**) or adenosine 5'-diphosphate to block ADP receptors.[81] Aspirin (75 to 160 mg daily) has been shown to significantly reduce the risks of subsequent reinfarction and sudden postmyocardial infarction.[81] Short-term anticoagulation with **heparin** or possibly **hirudin** is often given to prevent immediate reocclusion in certain patients. Platelet glycoprotein IIb/IIIa inhibitors such as clopidogrel, in addition to low-dose aspirin for patients with the acute coronary syndrome without S-T segment elevation, have been shown as advantageous.[82] By preventing platelet-mediated coronary thrombosis, aspirin, with or without heparin, is also the treatment of choice in patients with unstable angina. The addition of **warfarin** to aspirin may reduce the risk of recurrence by half. The benefit of long-term warfarin is controversial, although there is a reduction in mortality with high-intensity anticoagulation (international normalized ratio [INR] 2.5 to 5). However, these benefits have to be weighed against the risks of hemorrhage. Oral anticoagulation, keeping INR 2 to 4, is of benefit in patients with rheumatic heart disease, artificial valves, and nonrheumatic causes of atrial fibrillation by preventing cardiac thromboembolism.[83]

Thrombotic Occlusion of Arterial Grafts

A combination of aspirin and clopidogrel or dipyridamole has been shown to be of benefit in preventing occlusion of vascular grafts and following the use of stents in angioplasty.

Retinal Thromboembolism

Antithrombotic drugs do not appear to be of any value in patients with retinal thromboembolism, although if the disorder is transient, such patients should be managed as for transient ischemic attacks (TIAs); see below.

Transient Ischemic Attacks

The majority of TIAs are probably due to platelet or platelet-rich thromboemboli originating in the vessels of the neck. TIAs are a major risk for stroke. If the source of the platelet emboli is cardiac in origin, such patients appear to benefit from long-term oral anticoagulation. If the source is noncardiac, then aspirin may be of benefit. The value of surgical intervention in noncardiac causes for selected patients with high-grade stenosis is probably beneficial when performed in centers with wide experience of the techniques involved. Carotid artery stenting with associated antiplatelet drugs is under trial.

Stroke

Minor strokes are routinely managed with aspirin or antiplatelet drugs alone, but the benefits are unproven. For more major strokes, confirmed by computed tomography (CT) and thought to be thrombotic or embolic in origin, thrombolytic therapy with recombinant t-PA (tissue plasminogen activator) followed by oral anticoagulation may be of value.

There is a risk of hemorrhage into an infarcted brain. If the cause is cardiac platelet emboli, as with atrial fibrillation, long-term oral anticoagulation may be of benefit.

Peripheral Arterial Occlusion

Acute occlusion probably occurs through thrombus formation on an atherosclerotic plaque. However, it is important to exclude secondary causes, e.g., mural thrombus following myocardial infarction and valve disease. The mainstays of treatment are anticoagulation with heparin and surgery. Thrombolytic therapy may be of benefit and should be considered. In chronic arterial disease, a variety of treatments have been used, including antiplatelet drugs, oral and intravenous anticoagulants, dextrans, and defibrinating agents, but with little benefit. Such patients are, however, at increased risk of both myocardial infarction and stroke and may benefit from appropriate prophylactic treatment (see below).

Prevention

A lifestyle that reduces the risk factors for arterial thrombosis — low-fat diet, moderate exercise, avoidance of smoking — must be recommended. For those patients who have raised levels of cholesterol, increasing the levels of inhibitors of 3-hydroxy-3-methylglutaryl coenzyme A reductase (statins),[84,85] e.g., simvastatin, appreciably reduces the cardiovascular-related morbidity and mortality. Long-term administration of antiplatelet-aggregating drugs such as aspirin, dipyridamole, ticlopidine, or clopidogrel reduces risk,[86,87] particularly for prevention of secondary myocardial infarction.

ARTHUS REACTION

A localized painful erythema and edema occurring after around 12 h at the site of injection of an immune substance such as a vaccine. It is a type III **hypersensitivity** reaction to an immune complex.

ARTIFICIAL BLOOD

See **Red blood cell transfusion**.

ASCORBATE CYANIDE TEST

A screening test for **glucose-6-phosphate dehydrogenase** (G6PD) deficiency.[88] It is based on the inability of G6PD-deficient red blood cells to detoxify H_2O_2. With cyanide as a catalase inhibitor, H_2O_2 is generated by interaction between ascorbic acid and oxyhemoglobin. In cells deficient in G6PD, H_2O_2 oxidizes hemoglobin to a brown pigment readily detected by the naked eye. Normal cells detoxify H_2O_2 via G6PD-linked reduced glutathione peroxidase. The test can be used for the detection of heterozygotes, but is not specific for G6PD deficiency, as it gives positive results in **glutathione reductase** deficiency, **unstable hemoglobin** disorders, and **pyruvate kinase** deficiency.

ASCORBIC ACID

(Vitamin C) A cofactor for propyl and lysyl hydroxylases, causing hydroxylation of proline and lysine residues in collagen, thereby stabilizing its triple helical structure. It is present at relatively high concentration in citrus fruits, berries, tomatoes, raw cabbage, green peppers, leafy vegetables, and fortified juices. In the human red blood cell (RBC), it occurs in a concentration of 0.0199 ± 0.0023 μmol/ml RBC or 0.059 ± 0.0069 μmol/g hemoglobin (Hb). Normal plasma values of ascorbic acid are 1.0 to 1.5 mg/dl, falling to <0.2 mg/dl in those with a deficiency.

It is not clear if ascorbic acid has a direct role in **hematopoiesis** or if the anemia noted in scurvy results from interactions with **folic acid** and **iron** metabolism. Attempts to induce anemia in human volunteers have failed. The **anemia** observed can be normocytic, macrocytic, or hypochromic, and the bone marrow can be hypocellular, normocellular, or hypercellular. About 10% of deficient subjects have **megaloblastosis** that responds only to the combination of ascorbic acid and folic acid. Ascorbic acid is necessary for the maintenance of folic acid reductase in its reduced and active forms. Reduced folic acid reductase prevents the conversion of folic acid to tetrahydrofolic acid, the metabolically active form. Therapy with ascorbic acid will only produce a hematological response if there is sufficient folic acid with which it can interact. Dietary iron deficiency often occurs in association with dietary ascorbic acid deficiency, and ascorbic acid deficiency itself may cause iron deficiency because of external bleeding. Iron balance may be further compromised because of the lack of ascorbic acid, which facilitates iron absorption. Many patients with scurvy do have a normocytic normochromic anemia that responds to ascorbic acid alone.

Deficiency in animals also leads to breakdown and degeneration of connective tissue within vessel walls and in the perivascular area, probably due to defects in basement collagen synthesis. Such collagen wasting in humans hinders wound healing and weakens vascular resistance to physical stress, causing petechial **hemorrhage**. Similar weakness of collagen may result in gingival disorders, with loosening of teeth due to perifollicular hemorrhage and edema. Easy bruising, particularly over the legs and forearms, with intramuscular bleeding occurs. In children, subperiosteal bleeds are common due to extravasation of blood from the epiphysial vessels. Deformation of hair follicles with keratotic plugging, perifollicular hemorrhage, and short broken corkscrew hairs are also seen.

While originally a disorder of sailors on long voyages, scurvy now occurs mainly in the elderly and in the recluse living on an inadequate diet without fresh fruit or vegetables. In addition, patients maintained for more than 6 weeks on unsupplemented intravenous fluids will become ascorbic acid depleted.

Treatment is either orange juice supplements for mild deficiency or 100 mg ascorbic acid daily for those severely affected. Those with anemia may also require iron and folic acid. Claims that ascorbic acid ameliorates upper respiratory tract infection or actively promotes wound healing have not been substantiated.

Ascorbic acid in pharmacological doses can reduce **methemoglobin**, but this action is not as effective as methylene blue for treatment of **methemoglobinemia**. Its physiological contribution to the maintenance of methemoglobin below a concentration of 1% of total hemoglobin is minor.

L-ASPARAGINASE

A high-molecular-weight enzyme used as a chemotherapeutic agent in **lymphoproliferative disorders**. Asparaginase hydrolyzes the serum amino acid asparagine to nonfunctional aspartic acid and ammonia, thus depriving tumor cells of this essential amino acid. Tumor-cell proliferation is then blocked by the interruption of asparagine-dependent protein synthesis. This drug appears to be most active in G1 phase of the cell cycle.

Asparaginase is not absorbed orally, and it is currently administered in Europe as a deep intramuscular injection, although IV asparaginase is also available and is used in the U.S. Asparaginase is used in the induction treatment of **acute lymphoblastic leukemia** (ALL) and **non-Hodgkin lymphoma**. Its efficacy is, however, debatable. ALL cells are known to develop resistance to this drug by the third month of its use, so no further doses are usually administered beyond consolidation. In addition, recent multicenter trials examining the outcome of patients where asparaginase was withdrawn due to hypersensitivity (see below) did not find any significant differences in the rate of complete remission or relapse between patients receiving the drug and those who stopped treatment. Current approaches investigating efficacy and resistance in the U.S. are exploring the possibility of rotating between the three forms of asparaginase available in three monthly periods to cover all of the high-intensity treatments prior to maintenance, after which asparaginase administration is definitely stopped. **Hypersensitivity** to asparaginase is a major problem, as cases of anaphylactic shock have been reported. It is usually recommended that the first dose be given in a hospital environment and that an intradermal test dose be administered prior to treatment. However, this test does not completely rule out hypersensitivity. Cases where hypersensitivity has been shown with *E. coli*-derived asparaginase can still be treated with the *Erwinia carotovora* derivative. An **adverse reaction to drugs** is more frequent in adults than in children. Nausea and vomiting have been reported in nearly 30% of cases. Other side effects are hypoalbuminemia, confusion, and liver toxicity, rarely leading to withdrawal. However, virtually all patients experience a fall in prothrombin, antithrombin III, and fibrinogen. These are due to a lack of protein synthesis and can lead to hemorrhage in exceptional cases. (The occurrence is so infrequent that intrathecal chemotherapy is still administered in the U.K., regardless of the asparaginase-derived clotting abnormalities.) Acute pancreatitis leading to hemorrhagic pancreatitis has also been reported in up to 15% of adults.

ASPIRIN

A widely used nonopioid analgesic, with effects on **platelet function** that may be either an **adverse drug reaction** or the primary therapeutic aim. Aspirin causes this platelet-function disorder by irreversibly inhibiting platelet cyclooxygenases (COX 1 and COX 2), resulting in a failure in thromboxane production. A single 75-mg dose of aspirin can totally inhibit thromboxane production and thus impair the function of a cohort of platelets for their life span of 7 to 10 days. Aspirin also inhibits cyclooxygenase in cells of vascular endothelium, thereby blocking the synthesis of prostacyclin (PGI2), a potent inhibitor of platelet function. However, endothelial cells are less sensitive to the effects of aspirin in low dose and can resynthesize cyclooxygenase. Aspirin induces a characteristic pattern of abnormal platelet aggregation with a primary reversible wave of aggregation with collagen, low-dose **adenosine diphosphate** (ADP), **ristocetin**, **thrombin**, and **arachidonic acid**.

Low-dose aspirin (75 mg or lower daily) is of value in reducing the risk of reinfarction in patients with a history of ischemic heart disease and in reducing the risk of stroke in patients with a history of cerebrovascular disease. It is also of value in patients with a **thrombocytosis** or **essential thrombocythemia** and can also be used as prophylaxis against arterial thrombosis, particularly in those with paroxysmal atrial fibrillation and venous thromboembolic disease in high-risk patients. It is also used in the initial treatment of acute myocardial infarction and may be of help with acute stroke.

The major adverse drug reaction of aspirin is gastric erosion causing **hemorrhage** (see **Gastric disorders**), so it is contraindicated in patients with a bleeding diathesis. Aspirin prolongs the bleeding time, and its use in patients on oral anticoagulants requires close supervision. It also crosses the placenta and can cause neonatal bleeding. Aspirin should be discontinued for at least 5 days preoperatively, as its use may be associated with an increased risk of hemorrhage. These adverse reactions question the recommendation that low-dose aspirin could be beneficial as a prophylactic for thrombosis.

ASPLENIA
See **Splenic hypofunction**.

ATPases
See **Adenosine triphosphate**.

ATRANSFERRINEMIA
Absence of plasma **transferrin** that occurs either as an inherited or an acquired disorder. In three patients with the rare hereditary form of atransferrinemia, missense mutations were detected in the transferrin gene. In the absence of transferrin, impaired **hemoglobin** synthesis and iron-deficient erythropoiesis occur, although patients develop **iron overload** due to increased iron absorption and **red blood cell transfusion**. Total serum iron-binding capacity (TIBC) and absent or very low levels of immunoreactive transferrin are characteristic. Infusion of either normal plasma or purified transferrin is followed in 10 to 20 days by reticulocytosis and increased hemoglobin concentration. The use of purified transferrin rather than plasma reduces the risk of hepatitis. A spontaneous form of atransferrinemia in the mouse has been traced to a splice-site mutation in the transferrin gene. Acquired atransferrinemia has been described in association with nephrotic syndrome, in which transferrin is presumably being lost through the kidney and in association with the occurrence of an antitransferrin autoantibody. **Anemia of chronic disorders** is usually associated with a reduction in mild or moderate serum transferrin levels.

ATYPICAL CHRONIC MYELOID LEUKEMIA
(aCML; subacute myeloid leukemia) A leukemic disorder with myeloblastic and myeloproliferative features. It is characterized by **neutrophilia** comprising immature and mature neutrophils that are dysplastic. The neoplastic cells are Philadelphia-chromosome negative, and there are no *BCR/ABL* fusion genes. There are about one to two cases for every 100 cases of Philadelphia-chromosome-positive **chronic myelogenous leukemia**. Most patients are over 60 years of age, with a slight preponderance of females. The etiology is unknown, but the postulated cell of origin is a bone marrow stem cell. Changes in the peripheral blood and bone marrow are always found, often with involvement of the spleen and liver.

   The clinical features are those related to **anemia**, **thrombocytopenia**, and **hepatosplenomegaly**. Survival is less than 2 years, with many cases evolving to **acute myeloid leukemia**. The response to **cytotoxic agents** is inconclusive due to the rarity of incidence.

ATYPICAL LYMPHOCYTES
(Atypical mononuclear cells; variant lymphocytes) Large blastoid **lymphocytes** with lack of nuclear chromatin condensation, multiple nucleoli, nuclear lobulation, mitotic figures, irregular nuclear outline, cytoplasmic basophilia, and vacuolation. They are a common feature of **infectious mononucleosis**, **viral infection disorders**, **rickettsial infection**

disorders, and drug-induced **hypersensitivity**. They are transient apart from their appearance with **adult T-cell leukemia/lymphoma**. Specific abnormalities are:

Convoluted nuclei with **mycosis/fungoides/Sézary syndrome, hairy cell leukemia**, and infections associated with **acquired immunodeficiency syndrome** (AIDS)

Villous lymphocytes (polar cytoplasmic villi) with **splenic marginal-zone lymphoma**

Binucleated cells following low-dose **ionizing radiation** or **radiotherapy**

AUER RODS

Fusion of primary granules or ellipsoidal cytoplasmic inclusions of **myeloblasts**, which stain red-purple by Romanowsky staining. They are pathognomonic of **acute myeloid leukemia** (AML), occurring in up to 20% of patients. They occur most often in the differentiating blast cells of M1, M2, M3, and M4 AML, but can be seen in myeloid-derived blasts or rarely in the myelocytes, metamyelocytes, and mature neutrophils in any of the acute myeloid leukemias, except true M5 AML. They are especially prominent in promyelocytes of M3 AML, where they may occur multiply in bundles. In hypogranular M3, they are difficult to identify, except with the naphthol AS-D chloroacetate esterase reaction. They have limited use in the morphological identification of AML.

AURORA KINASE

A novel family of serine/threonine kinases that have been identified as key regulators of **mitosis**. The three members of this kinase family, identified so far, are referred to as Aurora-A, Aurora-B, and Aurora-C kinases. Aurora kinases are localized at the centrosomes of interphase cells, at the poles of the bipolar spindle, and in the midbody of the mitotic apparatus. They are implicated in the centrosome cycle, spindle assembly, chromosome condensation, microtubule-kinetochore attachment, the spindle checkpoint, and cytokinesis. Aurora kinases are regulated through phosphorylation, the binding of specific partners, and ubiquitin-dependent proteolysis. Several aurora substrates have been identified and their roles are being elucidated. Aurora-A and -B are both overexpressed in a wide range of different human tumors. Aurora-A has also been shown to be an oncogene in *in vitro* transformation assays.

Several aurora-kinase inhibitors have recently been described. These drugs are not "antimitotic" agents in that they do not directly inhibit cell-cycle progression. Rather, following an aberrant mitosis, activation of the p53-dependent postmitotic checkpoint induces pseudo-G1 cell-cycle arrest.

AUTOAGGLUTINATION

Agglutination of an individual's **red blood cells** by his or her own plasma. It characteristically occurs at temperatures below 37°C. The phenomenon is particularly characteristic of **cold-agglutinin disease**. Autoagglutination is readily apparent on the **peripheral-blood film**, but must be distinguished from **rouleaux formation of red blood cells**.

AUTOANTIBODIES

The **antibodies** produced when failure of self-tolerance to self-antigens occurs. Most self-reactive clones are efficiently eliminated during B- and T-cell ontogeny. Some self-**antigens** or **epitopes** may never be presented, so that while potentially self-reactive **lymphocytes** occur, they never encounter self-epitopes. For other antigens, there may be tolerance of self-reactive T-cells but not B-cells. Because specific T-cells are required to help B-cells make high-affinity antibody, autoantibodies are never produced or are of low affinity and are biologically irrelevant. Tolerance may be broken in either of these situations. A state of autoimmunity is then present, with the occurrence of **autoimmune disorders**.

AUTOERYTHROCYTE SENSITIZATION

A disorder characterized by recurrent spontaneous **bruising**. It gained its name by reproduction in patients by intradermal injection of autologous whole blood, washed **red blood cells**, red cell stroma, or extracted phosphatidylserine. Although these suggest an allergic etiology, this is no longer accepted. The lesions are frequently preceded by pain at the site of the bruise, localized swelling, and warmth. Lesions most commonly occur on the thighs, then the arms, and, only rarely, the trunk. Crops of lesions occur at times of emotional stress. Affected individuals are more often female. Psychological profiles have demonstrated widespread abnormalities, raising the possibility of self-induced trauma. Historically, the diagnosis was made by the injection of 0.1 ml of homologous blood, which will generate a typical lesion. A variety of drug treatments have been used, but with limited effect. It is suggested that psychotherapy is beneficial, as this may be a psychogenic purpura.

AUTOHEMOLYSIS TEST

A method of detecting the development of spontaneous **hemolysis** in blood incubated at 37°C for 48 h by measurement in a colorimeter or in a spectrophotometer at 625 nm.[89] The reference range in health without added glucose is 0.2 to 2.0% and with added glucose 0 to 0.9%. While nonspecific, the test does produce some information about the metabolic competence of red blood cells and helps to distinguish membrane defects from enzyme defects, for example in **hereditary spherocytosis** without added glucose. This increase disappears when glucose is added. Likewise, in pyruvate kinase deficiency, autohemolysis is increased without glucose, but on this occasion the addition of glucose does not correct the anomaly. Varying incidence also occurs in **autoimmune hemolytic anemia**, **paroxysmal nocturnal hemoglobinuria**, **glucose-6-phosphate dehydrogenase** (G6PD) deficiency, and other hemolytic anemias.

AUTOIMMUNE DISORDERS

Conditions in which an individual generates an **immune response** to self-tissues by **autoantibodies**, which may result in more-or-less severe disease, with all arms of the immune system involved in tissue-damaging inflammation. It is a situation in which self-tolerance breaks down. The main mechanism of tolerance probably is in the **thymus**, where **T-lymphocytes**, which respond to self-peptides, are eliminated (central tolerance). However, self-reactive T-cells may escape to the periphery, although these are normally anergic, i.e., do not respond to the target antigen despite possessing the appropriate antigen receptor. When tolerance breaks down, an immune response to self-antigens develops.[90]

The etiology of autoimmune disease involves both inherited and environmental factors, demonstrated by the finding that disease concordance in identical twins is high, but usually much less than 100%. For example, in **rheumatoid arthritis**, concordance is about 20%, while population prevalence is 1 to 2%. A major risk factor is the inheritance of specific human leukocyte antigen (HLA) types (e.g., inheritance of DR4 confers a relative risk of four for rheumatoid arthritis), though other gene loci, e.g., cytokine genes, may also be involved. The role of HLA antigens in autoimmunity is still unclear, but the assumption is that they present specific peptides triggering the response. However, the peptides responsible for this triggering are not definitively identified. The environmental factor may be a nonself-antigen that "mimics" a self-antigen so that the self-antigen cross-reacts ("molecular mimicry"), with the response to the nonself-antigen spreading to the self-antigen. Alternatively, it may be that a strong immune response to a pathogen results in the development of an environment where peripheral tolerance breaks down, perhaps through production of cytokines, which enhance the presentation of self-antigens to previously anergic autoreactive T-cells.

Autoimmune diseases are classified into organ-specific diseases, e.g., diabetes mellitus, **multiple sclerosis**, and multisystem diseases, e.g., systemic lupus erythematosus. Several of these diseases involve the blood and circulatory systems:

Addisonian pernicious anemia (intrinsic-factor deficiency)

Ankylosing spondylitis

Autoimmune hemolytic anemia

Celiac disease (see **Intestinal disorders**)

Immune thrombocytopenic purpura

Insulin-dependent **diabetes mellitus**

Inflammatory bowel disease (see **Intestinal disorders**)

Polyarteritis nodosa

Polymyalgia rheumatica (including temporal arteritis)

Polymyositis (including dermatomyositis)

Psoriatic arthritis (see **Skin disorders**)

Reactive arthropathy (Reiter's disease)

Rheumatoid arthritis

Scleroderma

Sjögren's syndrome

Systemic lupus erythematosus

Thyroid gland disorders

Vasculitis

All components of the immune system are involved in the pathogenesis of autoimmune disease, but the contributions of the different components vary. For example, in diabetes mellitus, the main mechanism of damage is probably **cytotoxic T-lymphocyte** destruction of pancreatic islet cells. While antibodies to islet cells and their components are produced, they probably play no great role in what is essentially a type IV **hypersensitivity**. Rheumatoid arthritis is also a type IV hypersensitivity, but this time largely mediated by TNF-α production by CD4 lymphocytes; again, antibodies are present (rheumatoid factor, reacting with the Fc component of Ig). On the other hand, autoimmune hemolytic anemia is due to lysis of **red blood cells** by **complement** activated by autoantibodies.

Since autoantibodies are present in all autoimmune diseases, whether or not involved in the pathogenesis, they are valuable for diagnosis, and many tests have been developed to detect them, e.g., the **direct antiglobulin (Coombs) test** to detect antibodies to red blood cells, in which the antibody bound to red blood cells is detected by the use of fluorescent antihuman Ig and indirect immunofluorescence microscopy.

The treatment is usually some form of **immunosuppression**. A recombinant monoclonal antibody, **Natalizumab**, is undergoing clinical trial. Severely affected patients may be considered for **allogeneic stem cell transplantation**.

AUTOIMMUNE HEMOLYTIC ANEMIA

A group of anemias induced by antibodies produced by the body's **lymphocytes** against its own **red blood cells**.[91,92] They are classified in Table 20.

TABLE 20

Classification of Autoimmune Hemolytic Anemias

Warm Autoimmune Hemolytic Anemias

Primary
Secondary
  Acute infections
  Lymphoproliferative disorders
  Autoimmune disorders: **systemic lupus erythematosus**,
    rheumatoid arthritis, **scleroderma**, inflammatory
    bowel disease
  Malignancy: ovarian teratomas, **Kaposi's sarcoma**
  Immunodeficiency

Cold Autoimmune Hemolytic Anemia

Cold-agglutinin disease
  Primary
  Secondary
    Acute infections: mycoplasmas, EBV infectious
      mononucleosis, cytomegalovirus, human
      immunodeficiency virus
    Lymphoproliferative disorders
    Malignancy
Paroxysmal cold hemoglobinuria
  Primary
  Secondary
    Acute infections
    Chronic infections: syphilis

Mixed Autoimmune Hemolytic Anemia

Primary
Secondary
  Lymphoproliferative disorders
  Autoimmune disorders: **systemic lupus erythematosus**

Drug-Associated Immune Hemolytic Anemia

Drug absorption (high-affinity hapten)
Neoantigen (low-affinity happen)
Autoantibody induction

AUTOIMMUNE LYMPHOPROLIFERATIVE SYNDROME

(ALPS) See **Neutropenia**.

AUTOIMMUNITY

See **Autoantibodies; Autoimmune disorders**.

AUTOINFLAMMATORY SYNDROMES

Disorders characterized by recurrent episodes of inflammation in the absence of infection or autoantibodies. Examples are inflammatory bowel disease (Crohn's disease; see **Intestinal disorders**), **familial cold-associated acute inflammation** (familial Mediterranean fever), and **sarcoidosis**. They are all probably associated with genetic mutations.

AUTOLOGOUS BLOOD TRANSFUSION

(ABT) The transfusion of a patient's own red cells. ABT can be undertaken by:

Autologous predeposit donation

Preoperative isovolemic hemodilution

Intraoperative cell salvage

Postoperative cell salvage

The procedure is advocated to reduce hazards associated with transfusion of allogeneic blood.[94,95,(1)] These include viral transmission (see **Transfusion transmitted infection**), **alloimmunization**, **immunodeficiency**, and **graft-versus-host disease**. It does not, however, prevent the commoner complications of transfusion arising from clerical errors, bacterial infection, and circulatory overload.[(1)] ABT may be indicated where patients have formed multiple alloantibodies, particularly those to high-frequency antigens, so that selection of compatible allogeneic units is not practical. Under these circumstances, autologous units may be cryopreserved in the event of an emergency. In general, only a minority of patients who receive transfusions can receive ABT as an alternative (see Table 21). The use of allogeneic blood may be more effectively reduced by careful review of the need for transfusion.[96]

TABLE 21

Indications for Autologous Blood Transfusion Procedures

Intra-abdominal vascular procedures
Radical prostatectomy
Open-heart surgery
Total knee replacement
Total hip replacement
Placenta previa [a]
Scoliosis repair (fusion)
Multiple gestations [a]

[a] These procedures do not lend themselves to intraoperative salvage.

Autologous Predeposit

Like allogeneic blood, units of autologous blood are stored at 4°C for 35 days. Units are collected in the 4- to 5-week period preceding surgery, at intervals of 7 to 10 days. The maximum number of units to be collected depends upon the type of surgical procedure planned and the patient's level of hemoglobin. The volume of blood that can be collected preoperatively is higher for patients who receive erythropoietin, but this is expensive and its use may not be justified. Patients who donate blood preoperatively must be in good general health, without significant cardiovascular or respiratory disease. Ideally, each unit of autologous blood collected should offset the need for transfusion of an allogeneic unit. In practice, this does not happen because:

Autologous blood may be collected and not transfused.

Insufficient collection of autologous blood results in the use of additional allogeneic units.

In general, the population predepositing blood is likely to be elderly and less fit. A fall in red blood cell mass during the predeposit period may cause silent myocardial ischemia. Patients with preexisting heart disease will be less tolerant of aggressive phlebotomy schedules. Autologous predeposit donation has not been shown to be more or less safe than allogeneic blood donation. Many publications have questioned the cost effectiveness and safety of such an approach, as operations may be canceled, leading to wastage, bacterial contamination, and clerical error. The popularity of this technique has therefore waned recently.[97–99]

Preoperative Isovolemic Hemodilution

Blood is withdrawn into sterile transfer packs in the anesthetic room and replaced with colloid or crystalloid. Red blood cell loss during surgery is therefore less, since the initial hematocrit is lower. In addition, intraoperative blood viscosity is reduced, and this leads to better tissue perfusion. Autologous units are returned to the patient after completion of surgery. The advantage of hemodilution lies in the reduction of red cell loss, and this is directly proportional to the volume of surgical blood loss. Therefore, it is only useful in procedures where losses may amount to 35% of the blood volume, where on average the transfusion of two units of allogeneic blood would be avoided. Studies have shown that hemodilution compares favorably with autologous predeposit.

Intraoperative (and Postoperative) Blood Salvage

(IBS) Two different types of system are in use (see Table 22):

1. **Hemapheresis** technique of centrifugal washing. Machines such as the Haemonetics Cell Saver (Haemonetics, Leeds, U.K.) are automated devices operating by semicontinuous flow centrifugation. They are equipped with a disposable bowl and tubing that fill with anticoagulated blood drawn through a suction wand. Plasma and cellular debris are removed; the red blood cells are washed and resuspended in saline solution to a hematocrit of 0.50 to 0.60 l/l and are available for retransfusion within 10 min of aspiration. Surgical procedures such as liver transplantation or major vascular surgery require high flow of up to 250 ml per minute. These machines are therefore most effectively used where a large amount of blood is to be salvaged. In liver transplantation, large amounts of platelet transfusion concentrates and fresh-frozen plasma (FFP) are also required, so that the role of intraoperative salvage is primarily to reduce the demand on the blood bank inventory rather than to minimize donor exposure.

2. **Canister systems**. Blood is collected by vacuum into a disposable bag lining a reusable plastic canister. The blood may then be washed on a blood cell processor

TABLE 22

Comparison of Intraoperative Blood Salvage (IBS) Systems

	Centrifugal Washing	Canister
Speed	fast	slow
Hematocrit (l/l)	0.50	0.40
Risk of thromboplastin	low	higher
Cost	high	low

contained within the operating theater suite and infused via a filter. Using canister systems where the red blood cells are not washed results in a lower hematocrit and elevated plasma hemoglobin levels. In addition, thromboplastins may be released, and fatal coagulopathies have been recorded.

Intraoperative blood salvage is indicated in patients undergoing coronary artery bypass grafts, aortic aneurysm repair, orthopedic surgery, including hip and spinal operations, and liver transplantation. Between 30 and 60% of all blood transfused may be provided in this way. Intraoperative salvage is not contraindicated in patients with malignancy, since it has not been demonstrated that there is a correlation between circulating tumor cells and the subsequent development of metastasis. It has been used safely and successfully in surgery for bladder cancer, recycling on average 1.5 l of blood. Theoretically, it may be advantageous to salvage blood rather than administer allogeneic transfusion, since the immunosuppressive effect of the latter would be avoided. This subject is still controversial. Intraoperative salvage has been used in patients with infections when the red cells are washed, as this reduces the load of any bacteria present. However, it is probably not acceptable to salvage from an infected site. Despite concerns regarding cell salvage during obstetric hemorrhage leading to amniotic fluid embolism, experience is growing in this area of practice.[100]

AUTOLOGOUS STEM CELL TRANSPLANTATION

(Autologous SCT; autologous bone marrow transplantation; BMT) The transplantation of the patient's own peripheral blood or bone marrow stem cells that have been removed previously and stored during myeloablative therapy. This procedure has found increasing use in the treatment of **acute myeloid leukemia** (AML), **acute lymphoblastic leukemia** (ALL), **lymphoproliferative disorders**, and severe **autoimmune disorders**.[93] Whereas **allogeneic stem cell transplantation** (allogeneic SCT) is generally restricted to patients under 70 years, autologous stem cells can be used for patients up to 75 years if patients are selected carefully. The advantages and disadvantages of autologous SCT and allogeneic SCT are compared in Table 23.

Stem cells obtained solely from autologous bone marrow are now seldom used alone, as it has become clear that stem cells mobilized from peripheral blood (PBSC) confer the important advantage of more rapid recovery of neutrophil and platelet counts posttransplant. However, bone-marrow-derived stem cells are occasionally used to supplement PBSC when harvests of these cells are not optimal. In addition to hematopoietic progenitors, the mononuclear cell fraction in PBSC harvests is rich in immunocompetent lymphocytes and natural killer cells. Reconstitution of the immune system is also quicker with PBSC than after autologous SCT.

TABLE 23

Advantages and Disadvantages of Autologous and Allogeneic Stem Cell Transplantation

Advantages	Disadvantages
Autologous	
Applicable to older patients	No graft-vs.-leukemia effect
No limitation on donor availability	Possible residual leukemia in graft
No graft rejection	
No graft-vs.-host disease	
Allogeneic	
Infused marrow leukemia free	Age restricted
Graft-vs.-leukemia effect may increase chance of "cure"	Limited donor availability
	Possibility of serious graft-vs.-host disease

TABLE 24

Conditioning Protocols for Autologous Stem Cell Transplantation Used in Acute Myeloid Leukemia

Total Body Irradiation (TBI)-containing
   Cyclophosphamide, 60 mg/kg for 2 days plus:
      TBI (12.0–14.4 Gy over 3 days)
      TBI (9.50 Gy as a single dose)

Non-TBI-containing
   Bu Cy: Busulfan, 4 mg/kg, days −9 to −6; Cyclophosphamide, 50 mg/kg, days −5 to −2
   BACT: BCNU, 200 mg/m^2, day −6; Cytosine, 200 mg/m^2, days −5 to −2; Cyclophosphamide, 50 mg/kg, days −5 to −2; Thioguanine, 200 mg/m^2, days −5 to −2
   LACE: CCNU®, 200 mg/m^2, day −7, Etoposide, 1g/m^2, day −7; Cytosine, 2 g/m^2, days −6 and −5; Cyclophosphamide, 1.8 g/m^2, days −4 to −2

Transplant Procedure

Pretransplant Harvesting

Bone Marrow Transplantation

(BMT) The technique is identical to that described for allogeneic SCT donors. Approximately 2×10^8 nucleated cells/kg of patient weight are needed. Unlike allogeneic SCT, where donor cells are used within a short time period, it is usual to freeze and store autologous BMT-cells in liquid nitrogen for later use.

Peripheral-Blood Stem Cells

The procedure is very similar to that used to collect allogeneic peripheral-blood stem cells, except a combination of chemotherapy and granulocyte colony stimulating factor (G-CSF) is used to allow better disease control and increase stem cell yield. A typical regimen is a single intravenous dose of cyclophosphamide, 1.5 to 7 g/m^2, followed by 5 days of stimulation with G-CSF, 3 to 10 µg/kg, beginning about 1 week later at the nadir of the neutrophil count.

Conditioning

Initial protocols were often identical to those for allogeneic BMT, but increasingly a variety of chemotherapy-only regimens have been used to obtain myeloablation. A selection of protocols is shown in Table 24. Whether all these regimens are truly myeloablative is debated.

Purging

Considerable effort has been made to remove residual leukemia cells hypothetically present in the harvest of blood or marrow stem cells. Whether or not this approach is logical is controversial, and most relapses following transplant are probably due to residual leukemic cells in the patient rather than in the stored harvest. In acute myeloid leukemia (AML), the usual purging agent is a drug or chemical, and most experience has been obtained using the cyclophosphamide metabolite 4-hydroxyper-cyclophosphamide (4HC). Immunological techniques have also been used to purge harvests from AML patients, but these techniques are generally applied to the treatment of acute lymphoblastic leukemia (ALL). Monoclonal antibodies are used alone or in combination with complement. Less frequently, antibody-linked ricin conjugates or immunomagnetic methods have been used.

Results

Acute Myeloid Leukemia

Using autologous SCT (commonly a combination of BM and PBSCs) in AML in first remission, a survival plateau of 45 to 55% of patients has been achieved. These results seem to be independent of the conditioning regimen used, and probably relate more to good remission status at the time of transplant. There is debate, however, as to whether autologous SCT in first complete remission (CR) is superior to four to five courses of chemotherapy if the patient is in CR after one to two courses of chemotherapy. There is relatively little experience of autologous SCT in second remission, but results following conditioning with busulfan and cyclophosphamide show a plateau at about 25%. Results of purged autologous SCT are uncontrolled and do not suggest that these methods have significant advantage over unpurged autologous SCT. Most patients not surviving autologous SCT succumb to relapsed disease during the first 12 months posttransplant. In relatively few patients, death is related to infection or the toxicity of the conditioning regimen. Occasional deaths have been due to veno-occlusive disease in the liver.

Acute Lymphoblastic Leukemia

Selection of patients for autologous SCT is more difficult in ALL because of the greater variety of the results of conventional treatment. Consequently, in the past, patients receiving autologous BMT have often been in second or subsequent remission, with results that have appeared to be less successful than for AML. However, it seems that when used in first remission, autologous BMT produces disease-free long survivals for 40 to 50% of adults. Generally, autologous BMT is not used in the management of childhood ALL, as even in second or subsequent relapse, chemotherapy alone will give relatively good results. Experience with purged transplants is limited, and these have usually been with monoclonal-antibody-based techniques. Results seem similar to those with unpurged material.

AUTOMATED BLOOD CELL COUNTING

See also **Calibration; Flow cytometry**.
The blood parameters that can be measured by instruments using complex microprocessor devices. They generate a standard eight parameters:

Total **white blood cell count**
Red blood cell count

Hematocrit

Hemoglobin

Red blood cell indices:

- Mean cell volume (MCV)
- Mean cell hemoglobin (MCH)
- Mean cell hemoglobin concentration (MCHC)

Platelet count

Others also provide:

Three- or five-part **differential leukocyte counts** (DLC) and extended DLC including **nucleated red blood cells** (NRBCs) and immature granulocyte counts

Automated **reticulocyte counting** and parameters of reticulocyte immaturity, e.g., immature reticulocyte fraction (IRF)

Immature platelet fraction (IPF) or reticulated platelet

Systems of red cell, white cell, and platelet flags. These flags indicate the need for further study but are associated with quite high false-positive rates (15 to 18%) but commendably low false-negative rates; false-positive rates vary with the population studied, being lowest in healthy individuals and highest in hospital populations

Additional investigations most frequently required are **peripheral-blood film examination** and manual differential leukocyte count to confirm or refute a flagged parameter.

AUTOMATED BONE MARROW EXAMINATION

The examination of **bone marrow** using various impedance and light-scatter modalities. Flow-cytometric bone marrow analysis can generate cell differential counts, potentially reducing the need for microscopy examination. However, until recently, routine multiparameter blood count analyzers have not been able to accomplish this due to problems in erythroblast identification, contamination by fat particles, and heterogeneity of cell types. This is changing. While automation will not replace the microscope in the foreseeable future, advanced automated blood cell analyzers — capable of **nucleated red blood cell counting** (NRBC), **reticulocyte counting**, and measurement of the reticulocyte indices (IRF) — can have a role in bone marrow examination, provided that fat-particle interference can be avoided. First, the instrument-rated total nucleated cell count correlates well with microscopy assessment of marrow cellularity, particularly with hyper- and hypocellular marrows. Second, a useful myeloid/erythroid (M/E) ratio can be generated. Third, measurement concentration gradients between marrow and peripheral-blood counts, particularly involving the reticulocyte, have been noted. Perturbation of these gradients may occur in disease and represent an area worthy of study. Finally, three different scattergram patterns are readily visible:

Normal marrow

Quantitative abnormalities with preserved erythrogranulopoiesis

Infiltration by abnormal cells

These instrument parameters may be useful prior to microscopy.

AUTOSENSITIVITY TO DNA

A syndrome, identical in presentation and background to **autoerythrocyte sensitization**, in which painful **purpura** occurs in response to intradermal injection of buffy coat extracts. Chloroquine may be of value in the treatment of some individuals. The etiology as an allergic reaction to DNA is no longer fully accepted.

AUTOSPLENECTOMY

See **Splenic hypofunction**.

AZATHIOPRINE

A synthetic thiopurine used to induce immunosuppression. It is metabolized to an antimetabolite, mercaptopurine (see **Cytotoxic agents**), which causes **bone marrow hypoplasia** and is also hepatotoxic. The principal hematological use is for disorders that are not responding to treatment with corticosteroids. The dosage is 2 to 2.5 mg/kg by mouth daily, but it is essential to monitor patients for myelotoxic effects such as **pancytopenia**. This can be avoided by measuring before treatment the enzyme responsible for azathioprine inactivation, thioprine methyltransferase. **Immunodeficiency** can also occur, which, with bone marrow suppression, encourages opportunistic infection. Acute allergic reactions can occasionally occur. Other adverse drug reactions include dose-related alopecia, pancreatitis, and pneumonitis.

B

BABESIOSIS

Infection by small protozoa, *Babesia* spp., which gives rise to a **hemolytic anemia**. It is transmitted from feral deer mice by hard ticks of the Dermacentor and Ixodes genera. Infection is restricted to island and coastal areas of the northeastern U.S., California, France, Ireland, and Scotland in early August. It can be a **transfusion transmitted infection**. In the presence of properdin and the complement receptors C3, C5, and Mg^{2+}, the sporozoites activate the alternative pathway of **complement** and thereby become coated with C3b. The organisms then adhere to C3b receptors on the **red blood cell** membrane, which they penetrate. Once inside the red blood cell, the sporozoite becomes a trophozoite, assuming a ring shape. It does not recycle.

Clinical symptoms occur 1 to 4 weeks after the tick bite, with an acute febrile illness followed by the features of hemolytic anemia. In patients with **splenic hypofunction**, the disorder can be severe, with fulminant **intravascular hemolysis, disseminated intravascular coagulation**, hypovolemia, and renal failure. Subclinical illness can occur. Diagnosis is by detection in peripheral blood smears of babesia in red blood cells.

Treatment is by the administration of clindamycin with quinine, reducing the dosage on resolution of the fever, which usually occurs in 4 to 5 days. Severe hemolytic anemia may require **red blood cell transfusion.**

BACTERIAL INFECTION DISORDERS

See also **Babesiosis; Bartonellosis; Ehrlichiosis; Tuberculosis**.
The hematological changes associated with bacterial infections.[101] Their frequency, severity, and associated changes are much increased in those with **immunodeficiency** or **splenic hypoplasia**. These include:

Neutrophilia with metamyelocytes (**leukemoid reaction**) caused by mobilization of granulocytes from the bone marrow storage pool (Some neutrophils will show toxic granulation or **Dohle bodies**. The **neutrophil alkaline phosphatase score** is usually raised. Such changes are severe with *Diplococcus pneumoniae* or *Hemophilus influenzae* infections, mild with streptococcal or Gram-negative organisms, and minimal with staphylococcal infection.)

Neutropenia associated with infection by *Salmonella* spp., bacillary dysentery, and brucellosis, occurring around 7 to 10 days postinfection

Anemia of chronic disorders

Hemolytic anemia from:

- Direct toxic action on the **red blood cell** membrane by the toxin of *Clostridium perfringens* and of leptospires causing **intravascular hemolysis**
- Red blood cell enzyme deficiency causing increased sensitivity to hemolysis by an unknown process (The particular infectious agents causing this are *Salmonella* spp., *Shigella* spp., coliform bacilli, hemolytic streptococci, *Diplococcus pneumoniae, Hemophilus influenzae, Mycobacterium tuberculosis, Vibrio cholera,* and *Yersinia enterocolitica*.)

- **Paroxysmal cold hemoglobinuria** associated with chronic syphilis
- **Cold autoimmune hemolytic anemia** due to cold hemagglutinins of anti-**I blood group** arising from infection by *Mycoplasma pneumoniae* and other bacteria
- **Warm autoimmune hemolytic anemia** due to hemagglutinins following injection of polyvalent pneumococcal vaccine, which provokes a rise in anti-A antibodies (This is a particular risk to the fetus in blood group O mothers due to possible ABO incompatibility.)
- Intracellular invasion by *Plasmodium vivax*, *P. ovale*, *P. malariae*, *P. falciparum*, and by *Bartonella bacilliformis* (see **Malaria**; **Bartonellosis**)

Pure red cell aplasia (This can be caused by an opportunistic infection with *Mycobacterium avium intracellulare* occurring with **immunodeficiency**, especially in those with **acquired immunodeficiency syndrome** [AIDS].)

Lymphocytosis associated with *Bordetella pertussis* infection due to release of a factor from lymphocytes, which promotes their migration from lymph nodes to the circulation

Intestinal-tract heavy-chain disease; *Campylobacter jejuni* infection may lead to immunoproliferation in the small intestine of normal and abnormal plasma cells, synthesis of α-heavy chains with V-region deletion, and absence of light chains (see **Intestinal tract disorders**)

Marginal zone B-cell lymphoma (MALT lymphoma), associated with *Helicobacter pylori* and *Borrelia burgdorferi* infection (see **Gastric disorders**)

Autoimmune disorders Guillain-Barré syndrome and **reactive arthropathy**, associated with chronic infection by *Campylobacter jejuni*

Vascular purpura associated with meningococcal and streptococcal infections, which may proceed to **purpura fulminans** and **microangiopathic hemolytic anemia** with **disseminated intravascular coagulation**

Elevated **erythrocyte sedimentation rate**

Changes associated with **liver disorders** and **renal tract disorders** when these organs are damaged by bacterial infection

Coronary artery thrombosis has been associated with infection by *Chlamydia pneumoniae*

BAND-FORM POLYMORPHONUCLEAR NEUTROPHILS
See **Neutrophils**.

BANDS 1–5 OF THE RED BLOOD CELL MEMBRANE
See **Red blood cell** — red blood cell membrane.

BARR BODIES
Metachromatic masses of about 1-μm diameter located at the periphery of female cells that are at the interphase stage of **mitosis**. They are the pyknotic-inactivated remains of one of the X chromosomes of normal female cells (see **Lyonization**). These sex chromatin bodies can be detected from scrapings of buccal mucosa or in 2 to 3% of female **neutrophils**, where they are a solid round condensed chromatin mass connected to a lobe of the

nucleus by a thin chromatin strand ("drumstick"). X-chromosome inactivation, as first postulated by Mary Lyon, initiates from a single physically defined region on the X chromosome, termed the X-inactivation center (XIC). The early stages of the X-inactivation process depend critically on a gene located within the XIC that is subsequently expressed exclusively from the inactive X chromosome (Xi), referred to as the "X-inactive specific transcript" or XIST. XIST encodes a large untranslated **ribonucleic acid** (RNA) that, at interphase, associates with the Xi and initiates a series of epigenetic events that lead to the formation of the silenced state on the Xi. Judicious use of the detection of Barr bodies in the buccal mucosa, or a blood smear, can be used for sex testing or in forensic pathology.

BARTONELLOSIS

(Carrión's disease; Oroya fever) Infection by a gram-negative bacillus, *Bartonella bacilliformis*, that causes an acute **hemolytic anemia**. The bacillus is transmitted by a bite from a female sand fly of the genus *Phlebotomus verrucarum* that resides mainly on the slopes of the Andes in Peru, Colombia, and Ecuador. Following injection, the bacilli multiply over the surfaces of most **red blood cells**, which become spherocytic and are removed from the circulation by the spleen. A hemolytic anemia ensues with a leukocytosis. The infection probably gives rise to **immunosuppression**, which leads to secondary opportunistic infections, particularly by salmonella, malaria, tuberculosis, amoebae, and coliform organisms.

The clinical features are a febrile illness occurring around 3 weeks after the bite, with symptoms of an acute hemolytic anemia. **Intravascular hemolysis** can occur which, if untreated, results in a 40% mortality. In those that survive, a second phase with cutaneous **hemangiomas** follows in 3 to 6 months, persisting for months or years.

Treatment is by tetracycline and chloramphenicol. **Red blood cell transfusion** may be necessary to counteract the anemia.

BASKET CELLS

Degenerate **granulocytes** appearing as a basketlike network of pale staining material in smears of peripheral blood and bone marrow aspirate.

BASOPHIL

Circulating **leukocytes** with characteristic black/purple granules in the cytoplasm upon Romanowsky staining. First described by Paul Ehrlich in 1878, basophils and their tissue equivalent, **mast cells**, still arouse great interest, as they play a central role in several pathological states.[102]

Morphology

These cells are derived from the hematopoietic stem cell with identical maturation stages to the **neutrophil**. The characteristic black/purple granules start to appear at the promyelocyte stage.

The mature basophil (size 14 to 16 μm) has a bilobed nucleus and pink cytoplasm. The coarse, large granules (see Table 25) fill the cytoplasm, overlapping the nucleus, and contain many inflammatory mediators. There is little glycogen. The basophil has condensed chromatin, with occasional mitochondria and little if any rough endoplasmic reticulum.

TABLE 25

Granules of Basophils and Mast Cells

Basophils	Mast Cells
Histamine	Histamine
Kallikrein	Kallikrein
Myeloperoxidase	Mucopolysaccharides
Esterase	Acid phosphatase
Mucopolysaccharides	Protease
Chondroitin sulfate	
Dermatin sulfate	
Heparin sulfate	
SRS-A	
Leukotrienes	

A little over 1% of **bone marrow** cells are basophils or their precursors. The normal peripheral blood basophil count is 0.02 to $0.1 \times 10^9/l$, i.e., <1% of nucleated cells.

Kinetics and Regulation

Due to the very low numbers of basophils in bone marrow and peripheral blood, the regulation of basophil production and their subsequent release from the marrow have not been fully elucidated, but the process is probably similar to that of the neutrophil and the **eosinophil**. As with neutrophils, several **receptors to growth factors** are involved in their production and differentiation, e.g., interleukin (IL)-3, stem cell factor (SCF), granulocyte macrophage colony stimulating factor (GM-CSF), and IL-5. Basophils express the **immunophenotypes** CD40L and CCR3 and produce IL-4 and specifically IL-13α chain (CD123).

Function

The exact role of basophils is unclear. They are not very motile but can migrate into tissues and are capable of phagocytosis. Basophils and mast cells have high-affinity receptors for IgE and, when IgE/antigen complexes are bound, degranulation results. The major constituents of basophil granules, i.e., histamine and **leukotrienes**, cause the immediate **hypersensitivity** reaction as well as acting as chemotaxins for neutrophils and eosinophils. Histamine also upregulates C3b expression by neutrophils; it contracts bronchial and gastrointestinal smooth muscle.[103]

Prior to discovery of the basophil, a similar cell, the mast cell, had been discovered in tissues. This cell was heavily granulated, mononuclear, capable of mitosis, and the granules contained slightly different enzymes (see Table 25). It is probable that basophils and mast cells have a common bone marrow precursor, but basophils specifically produce IL-3α (CD123), whereas mast cells express c-Kit (CD117). Their physiological role remains undetermined.

BASOPHIL HYPERSENSITIVITY

A reaction of basophils classically associated with immediate **hypersensitivity**. Basophils also play a role in the development of delayed hypersensitivity reactions to soluble antigens that have been injected into the epidermis with Freund's incomplete adjuvant. Histological examination reveals infiltration by basophils, T-lymphocytes, and eosinophils. It is a T-cell-dependent reaction.

TABLE 26

Causes of Basophilia

Physiological	pregnancy
Malignancies	**myeloproliferative disorders, chronic myelogenous leukemia, polycythemia rubra vera**
	Hodgkin disease
	acute myeloid leukemia
	carcinoma
Endocrine	hypothyroidism
	diabetes mellitus
Infections	chickenpox
	tuberculosis
Inflammation	inflammatory bowel disease
	rheumatoid arthritis
Drugs	estrogens

BASOPHILIA

An increased level of circulating **basophils** ($>0.1 \times 10^9$/l) associated with many heterogeneous conditions (see Table 26). It is commonly associated with immediate **hypersensitivity** reactions and elevated IgE. In **chronic myelogenous leukemia**, it is sometimes the primary indication of a metamorphosis to blast crisis.

BASOPHILIC ERYTHROBLAST

(Early normoblast) See **Erythropoiesis**.

BASOPHILIC STIPPLING

(Punctate basophilia) Multiple basophilic deposits of clumped ribosomes in **red blood cells** visible on Romanowsky-stained blood films. It ranges from fine (stippling) to coarse (punctate) deposits. The causes include:

Chemicals (aniline, dinitrobenzine)

Dyserythropoiesis, including **megaloblastosis** (cobalamin, folic acid deficiencies)

Thalassemias

Unstable hemoglobins

Liver disorders

Heavy metal toxicity (**arsenic**, bismuth, **copper**, **lead**)

Prosthetic heart valves (see **Cardiac disorders**)

Pyrimidine 5'-nucleotidase deficiency

BASOPHILOPENIA

A reduced level of circulating **basophils**, $<0.02 \times 10^9$/l. Various conditions, e.g., acute infection, hyperthyroidism, irradiation, chemotherapy, or corticosteroids, can result in basophilopenia, but no adverse effects are observed. There is a very rare hereditary form.

BASSEN-KORNWEG SYNDROME

See **Abetalipoproteinemia**.

B-CELL
See **Lymphocytes**.

B-CELL CHRONIC LYMPHATIC LEUKEMIA
See **Chronic lymphatic leukemia**.

B-CELL LYMPHOMAS

Precursor B-cell neoplasm
- Precursor B-lymphoblastic leukemia/lymphoma

Mature B-cell neoplasms
- **Chronic lymphocytic leukemia**/small lymphocytic lymphoma
- **B-cell prolymphocytic leukemia**
- **Lymphoplasmacytic lymphoma**
- **Splenic marginal-zone lymphoma**
- **Hairy cell leukemia**
- **Plasma cell** myelomatosis
- Solitary **plasmacytoma** of bone
- Extraosseous plasmacytoma
- **Extranodal marginal-zone B-cell lymphoma** of mucosa-associated lymphoid tissue (MALT-lymphoma)
- Nodal marginal-zone B-cell lymphoma
- Follicular lymphoma
- Mantle-cell lymphoma
- Diffuse large B-cell lymphoma
- Mediastinal (thymic) large B-cell lymphoma
- Intravascular large B-cell lymphoma
- Primary effusion lymphoma
- Burkitt lymphoma/leukemia

B-cell proliferations of uncertain malignant potential
- Lymphomatoid granulomatosis
- **Post-transplant lymphoproliferative disorder**, polymorphic

B-CELL PROLYMPHOCYTIC LEUKEMIA
(B PLL) A **lymphoproliferative** disorder where B prolymphocytes exceed 55% of lymphoid cells counted. It is an extremely rare disorder. Most patients are over 60 years of age, with a male predominance. In addition to a **lymphocytosis** of over 100×10^9/l with **anemia** and **thrombocytopenia** in 50% of cases, prolymphocytes are present in the bone marrow and spleen. Most patients have **splenomegaly** without peripheral lymphadenopathy. The prolymphocytes on **immunophenotyping** strongly express surface IgM, IgD$^{+/-}$, and B-cell antigens CD19$^+$, CD20$^+$, CD22$^+$, CD79a$^+$, CD79b$^+$, FMC7$^+$, and CD5$^+$ in one-third of cases, along with CD23$^-$. **Cytogenetic analysis** has shown abnormalities as break points involving

14q32, particularly t(11;14)(q13;q32), in 20% of cases. The cell of origin remains unknown. Response to **cytotoxic agents** (CHOP, fludarabine) is poor and the survival time is short. For relief of symptoms, both **radiotherapy** to the spleen and **splenectomy** have been used

BEHÇET'S DISEASE

An inflammatory disorder with recurrent oral ulceration. Associated disorders include genital ulcers, uveitis, a variety of skin disorders, and intestinal, renal, central nervous system, and pulmonary lesions. There is a link with HLA-B55 allele. The only other hematological association is raised levels of the **erythrocyte sedimentation rate**.

BENCE-JONES PROTEINURIA

See **Biclonal gammopathies**.

BENZENE TOXICITY

The hematological effects of exposure to benzene. This affects 4 to 5 million workers worldwide in the coke, rubber, transportation, and petroleum/benzene production industries. Great reduction of risk has been made over the past 30 years by changes in industrial practice and industrial hygiene. Toxicity mainly affects rapidly dividing cells of the **bone marrow**, being both cycle specific and phase specific in G2 and M phases of **mitosis**. It therefore gives rise to **bone marrow hypoplasia**, which may progress to **acute myeloid leukemia**. There is a latent period of 2 to 50 years.

BERNARD-SOULIER SYNDROME

A rare congenital platelet disorder characterized by a prolonged **bleeding time, thrombocytopenia**, and **giant platelets**. It is due to a deficiency or dysfunction of the heterodimeric glycoprotein Ib/IX/V complex (GpIb/IX/V) on the surface of platelets. Due to deficiency of GpIb/IX complex, Bernard-Soulier platelets are unable to interact with subendothelial **Von Willebrand Factor** (VWF) and adhere via VWF to subendothelial matrix. In most cases, the deficiency of GpIb/IX/V is absolute or almost absolute. Bernard-Soulier syndrome is inherited as an autosomally recessive condition and is a variable, though often severe, bleeding disorder. The bleeding symptoms are disproportionate to the thrombocytopenia. Thrombocytopenia is thought to arise as a result of decreased platelet survival.

Bernard-Soulier syndrome usually presents in infancy or childhood, with bleeding characteristic of defective platelet function, such as bruising, epistaxis, and gingival bleeding. Menorrhagia, posttraumatic, and surgical bleeding and gastrointestinal hemorrhage are common in older patients. Hemarthrosis and muscle hematomas are rare.

Laboratory Features

A bleeding time in excess of 20 min with a variable thrombocytopenia, as low as $20 \times 10^9/l$. Stained peripheral blood film morphology shows the platelets to be large, frequently with mean diameter in excess of 3.5 μm. Other formed elements of the blood are morphologically normal. **Platelet-function** studies show that Bernard-Soulier platelets are unable to aggregate in the presence of **ristocetin** (due to the inability of the GpIb/IX-deficient platelets to interact with VWF). Aggregation to other agonists (adenosine diphosphate, adrenaline, collagen) is normal. **Flow cytometry** analysis shows reduced GpIb/IX complex. Differential diagnosis includes other giant platelet disorders such as **May Hegglin anomaly, Epstein's syndrome, gray-platelet syndrome**, and **myelodysplasia**.

Treatment

There is no specific treatment for Bernard-Soulier syndrome. Local measures are of major importance, as is adjunctive antifibrinolytic therapy. Treatment of bleeding in Bernard-Soulier syndrome requires the use of **platelet transfusion**. These platelets should be HLA-matched to prevent the development of ant-HLA antibodies. **DDAVP** (1-desamino-8-d-arginine vasopressin) may be transiently useful in a subgroup of patients for minor bleeding episodes. **Recombinant VIIa** may also be useful. There is no role for splenectomy or corticosteroids in attempting to increase platelet count. Hormonal control of menses is an important component to the management of menorrhagia.

BICLONAL GAMMOPATHIES

Disorders in which two paraproteins occur, present in 3 to 4% of patients with **monoclonal gammopathies**. The clinical findings are similar to those of MGUS (monoclonal gammopathy of undetermined significance). Two-thirds of patients are asymptomatic, and the remainder have **myelomatosis, amyloidosis**, macroglobulinemia, or other **lymphoproliferative disorders**. The serum protein electrophoretic pattern often shows only a single band, and the biclonal component is not recognized until immunoelectrophoresis or immunofixation is performed. Several patients with triclonal gammopathy have been recognized.

Idiopathic Bence-Jones Proteinuria

Although Bence-Jones proteinuria is most frequently associated with myelomatosis, primary amyloidosis, or **Waldenström macroglobulinemia**, it may be benign. Patients with idiopathic Bence-Jones proteinuria excrete more than 1 g of Bence-Jones protein per 24 h but have no evidence of a malignant plasma cell proliferative disorder. However, myelomatosis or amyloidosis will develop in most patients, but this may not occur for up to 20 years. Consequently, patients with apparently idiopathic Bence-Jones proteinuria should be observed indefinitely.

BILE PIGMENT METABOLISM

See **Hemoglobin** — degradation.

BILIARY TRACT DISORDERS

The interrelations of hematological disorders with those of the biliary tract. **Hemolytic anemia** causes hyperbilirubinemia and increased bile pigment in the biliary system, with consequent cholelithiasis. Biliary obstruction results in circulating **red blood cells** having a thin appearance on microscopy (see **Leptocytes**).

BILIRUBIN

See **Hemoglobin** — degradation.

B-IMMUNOBLASTIC LYMPHOMA
See **Diffuse large B-cell lymphoma**; **Non-Hodgkin lymphoma**.

BIOMICS
A generic term used to describe a variety of emerging technologies that include **genomics** (the study of genome expression, the mRNA transcripts that reflect the activity of the genome in a cell, as opposed to genetics that studies the genome or "DNA information in a cell"), **proteomics** (the study of protein), and **metabonomics** (or metabolomics, which involves the analysis of metabolites produced by a cell, organ system, or whole individual). Some of these techniques have the potential to transform biomedical research due to their ability to undertake high-throughput analyses of large numbers of samples for genomic or metabolite profiles. The technologies associated with proteomics are more laborious, although emerging approaches will also enable large numbers of proteins to be measured from a given cell or tissue sample. The technologies enable the comparison of the genetic and molecular pathways associated with disease pathology and normal cells. This will enable a more precise molecular classification of disease as well as better definition of prognostic factors and of predictive markers for response to specific drug therapies. Biomics approaches are also being used to identify novel drug targets and identify genomic "signatures" that predict specific **adverse drug reactions** (see **Warfarin**).

BITE CELLS
See **Pyknocytes**.

BLACKFAN-DIAMOND SYNDROME
(Congenital red blood cell aplasia) A rare dominant inherited autosomal disorder characterized by **pure red cell aplasia** presenting in the first year of life. Most familial cases show germline mutations in the *RPS19* gene, which encodes ribosomal protein S19. The characteristic features are:

Normochromic anemia, usually **macrocytic anemia** but occasionally **normocytic anemia**

Reduced **reticulocyte count**

Raised **hemoglobin F**

Increased i-antigen density of **I blood group**

Bone marrow normocellular with selective deficiency of erythrocyte precursors

Normal or slightly decreased **neutrophil** count

Normal or slightly increased **platelet** counts

Corticosteroids (prednisolone or prednisone) improve anemia in about 70% of patients. **Red blood** cell transfusion and **iron chelation** are reserved for those who fail to respond to corticosteroids. **Interleukin**-3 has also benefited some steroid-unresponsive patients. **Stem cell factor** is being evaluated. **Allogeneic stem cell transplantation** has been used with success. Patients with this disorder have an increased risk of **myelodysplasia** and **acute myeloid leukemia** later in life.

BLASTIC NK-CELL LYMPHOMA

See also **Non-Hodgkin lymphoma**.
(NK-cell lymphoma) A **lymphoproliferative disorder** where the cells have a lymphoblast-like morphology and evidence of commitment to the NK-cell lineage. It is a rare lymphoma without evident racial predilection. No age is exempt, but it tends to occur more frequently in the middle-aged or elderly. It presents in multiple sites, particularly the skin and lymph nodes. Involvement of the peripheral blood and bone marrow is seen. **Immunopheno-typing** shows surface CD3$^-$, CD56$^+$CD4$^+$, and CD43$^+$CD68$^-$. **Cytogenetic analysis** shows germline T-cell receptor but EBV negative. The clinical course is aggressive, with a poor response to non-Hodgkin lymphoma **cytotoxic agent** therapy.

BLEEDING TIME

A simple screening test of **platelet-function disorders**. It is prolonged in disorders that affect platelet function or the interaction of platelets with the vessel wall (e.g., **Von Willebrand Disease**, VWD). It is not prolonged in most disorders that result in a prolonged prothrombin time or activated partial thromboplastin time, such as coagulation-factor deficiencies. The bleeding time is performed on the forearm. A disposable template is used to produce small standardized incisions. A sphygmomanometer cuff is inflated to 40 mm Hg around the arm to standardize intracapillary pressure. Filter paper is used to remove blood emerging from the incisions every 30 sec. The bleeding time is the time for bleeding to cease from the incisions. The reference range is 2.5 to 9.5 min. Bleeding time is influenced by a number of factors, including operator, skin thickness, depth, and location of incisions, imprecision of the end point, and platelet count. Although bleeding time is independent of platelet count above $100 \times 10^9/l$, it is prolonged in parallel to platelet counts below this level.

BLEOMYCIN

See **Anthracyclines**.

BLIND-LOOP SYNDROMES

See **Intestinal tract disorders**.

BLOOD

The fluid circulating through the vascular system of the heart, arteries, capillaries, and veins. It is composed of **red blood cells**, white blood cells (**granulocytes**, **monocytes**, and **lymphocytes**), and **platelets**, all suspended in liquid **plasma**.

BLOOD BLISTERS

See **Hemorrhagic bullae**.

BLOOD COMPONENTS FOR TRANSFUSION

Knowledge of **blood groups** resulted in the development of blood transfusion. Initially, after ABO grouping compatibility had been assured, the procedure was performed by direct transfusion from donor to patient. The requirement of blood for mass casualties in the 1939–1945 war accelerated the need for organization of blood donation, storage, and distribution. From this arose the Red Cross Blood Transfusion Service, the American Association of Blood Banks, and the British National Blood Transfusion Service.

Blood donation has become a worldwide activity, but its organization differs from one country to another, mainly arising from the use of either volunteer donors or paid donors. Detailed donor questionnaires exclude donors that may pose a risk to recipients or to themselves by donation. Testing includes measurement of the donor's peripheral blood **hemoglobin** (Hb) concentration and screening for antibodies to the most pertinent infectious agents. Collection of specific components from donors by **hemapheresis** is being used increasingly.

Blood component materials include:

Red blood cell concentrates

Fresh-frozen plasma (FFP)

Washed red blood cells

Cryoprecipitate

Frozen red blood cells

Platelet concentrates

Leukocyte-depleted blood components

Blood Component Preparation and Storage

Techniques of blood component separation changed dramatically with the introduction of PVC containers and tubing in the 1960s. This has allowed easy separation by sedimentation of plasma and other blood components from multipack containers. Equipment is universal, based on specifications prepared by the International Standards Organization (ISO). As a consequence, blood components can be readily transmitted from one country to another when major disasters occur. Whole blood for transfusion, even for major trauma or cardiopulmonary bypass surgery, has now been largely replaced by red blood cells suspended in optimal additive solutions such as SAG-M (saline, adenine, glucose, and mannitol), allowing storage at 4°C for up to 35 days. Product specification varies with requirements for **fetal and neonatal transfusions**, particularly for exchange-transfusion procedures. Current procedures for separation of whole blood into components is shown in Figure 5.

Changes in Blood Components during Storage

Red Blood Cells

These contain **adenosine triphosphate** (ATP), much of which is derived from glycolysis. Utilization of ATP is diminished 40-fold during storage at 4°C. Additive solutions containing adenine enhance ATP levels and red cell survival. Usually, ATP levels are maintained at 50% during 35-day storage. In contrast 2,3-DPG levels decline quickly and are <10% beyond 14 days of storage. Following transfusion, 2,3-DPG levels recover to normal within 48 h; there is no evidence that this is clinically deleterious. The sodium/potassium pump is inhibited at 4°C, and potassium leaks into the surrounding medium. Red blood cell morphology changes during storage, with a proportion of cells becoming **echinocytes** and finally **spherocytes**.

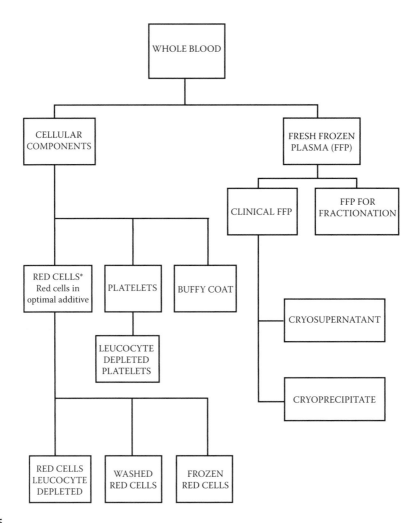

FIGURE 5
Separation of whole blood into components. (From Pamphilon, D.H., Ed., *Modern Transfusion Medicine*, CRC Press, Boca Raton, FL, 1995.)

Platelets and White Blood Cells

Platelets require optimal conditions for storage (see below). At 4°C they rapidly lose viability and undergo a shape change from discoid to spherical. These changes are irreversible after 8 h, and their *in vivo* survival is reduced to 2 days. Granulocytes and monocytes do not retain function at 4°C or higher temperatures. Microaggregates of white cells and platelets increase in numbers with storage. Leukodepletion of blood products prevents microaggregate formation, and fragmentation of leukocytes occurs progressively with time during storage.

Preparation of Red Blood Cells for Transfusion

Blood components are separated by centrifugation in one of two ways:

Platelet-rich (PRP) or platelet-poor plasma (PPP) method. Hard or soft centrifugation of whole blood yields PPP and PRP supernatant, respectively. By applying carefully controlled pressure to the main collection pack, supernatant plasma can be decanted into a satellite pack. SAG-M (120 ml) or other additive solution is then run onto the red blood cells. The hematocrit is 0.55 to 0.65 1/l, and removal of plasma and replacement with additive solution produces better flow characteristics and red blood cell viability.

Buffy coat method, also called the bottom and top (BAT) method. Units of blood are centrifuged at high velocity so that the platelets are spun down onto the buffy coat. The units are then placed in an automated extractor that applies even pressure across the unit, displacing plasma upward and red blood cells downward into additive solution while optical monitoring ensures that the buffy coat interface remains static until approximately 60 ml remains. This has a hematocrit of roughly 30 1/l. Red blood cells prepared in this way contain 1 log fewer leukocytes than conventional SAG-M red blood cell units.

Newly developed solutions that contain adenine, mannitol, phosphate, glucose, and citrate or ammonium maintain ATP levels (50%) and acceptable *in vivo* survival (75%) for up to 14 weeks, and 2,3-DPG levels are maintained for 4 to 6 weeks or longer. This is achieved by raising the pH, omitting chloride, and optimizing phosphate buffering. In the future, red blood cell storage periods may be extended even more.

Washed Red Blood Cells

These units are depleted of leukocytes and plasma proteins by washing donated blood in up to 2 l of normal saline in an automated washer. The preparation is given a 24-h shelf life. These may be indicated for patients with IgA deficiency where blood products from IgA-deficient donors are not available.

Frozen Red Blood Cells

Freezing of red cells is indicated in the following circumstances:

Rare red blood cells lacking antigens that commonly sensitize patients

Autologous red blood cells for patients with multiple red blood cell alloantibodies or rare red blood cell phenotypes

Group O red blood cells stored for military casualties

Red blood cells should be frozen within 6 days of storage at 4°C; glycerol is added in a stepwise fashion, and the cells are stored in liquid nitrogen.

Platelet Concentrates

These are made by one of three methods:

Platelet-rich plasma (PRP) method. Platelet-rich plasma obtained either by apheresis (see **Hemapheresis**) or from soft centrifugation of units of whole blood is subjected to a hard centrifugation to pellet the platelets. The majority of the supernatant plasma is expressed into a secondary container, and the platelet button/pellet is allowed to gently resuspend at room temperature in the remaining plasma.

Buffy coat method. Four or more buffy coats prepared as described above are pooled aseptically, using a sterile docking device, into a 500-ml container. They are then gently centrifuged, and the supernatant platelet concentrate is expressed into a transfer bag.

Apheresis (see **Hemapheresis**).

TABLE 27

Composition of Platelet Concentrates

	PRP-Derived	Buffy Coat-Derived	Apheresis
Platelets × 10^{11} (mean)	3.4	3.2	3.2
WBC × 10^6 (mean)	365	6 [a]	0.3

[a] WBC contamination varies in the range of 1×10^6 to 1×10^8.

Source: Values derived from data collected at the Bristol Transfusion Centre, England.

The resulting composition of the platelet concentrate by these methods is summarized in Table 27.

Storage of Platelet Concentrates

Platelets deteriorate in both quantity and quality upon storage. This platelet storage lesion is influenced by the following factors:

Storage temperature below 18°C

Method of platelet preparation

Gas-transfer properties of the platelet storage container

Concentration of platelets

White cell contamination

Volume and type of suspending medium

Agitation of the platelet concentrate

Platelets are therefore stored at 22°C using continuous agitation. The platelet density should not exceed 2000×10^6/ml, and storage containers should be made of either poly-olefin, such as PL732, or citrate-plasticized PVC. Although platelets prepared by the buffy coat and apheresis techniques have a lower number of contaminating white blood cells, there is no evidence that this influences their clinical performance. Optimal additive solutions for platelets are available that may allow a longer shelf life when used in conjunction with bacterial monitoring, but their use adds to the cost of preparation. A reduction in the volume of resuspended plasma may also reduce the risk of associated lung injury (TRALI).

Leukocyte-Depleted Blood Components[107]

These can be prepared by a variety of techniques.

Centrifugation

This is a relatively ineffective and labor-intensive method. It is most effective when the buffy coat method (see above) is used. In this method, the leukocyte content (usually 10^9) of whole blood is reduced by a factor of 1 log. Platelet concentrates contain 10^7 leukocytes per unit.

Leukocyte Filtration

This is currently the method of choice for rigorous (3 log) filtration of red blood cells and platelets. The two types of filter commonly used are column and flatbed filters. These are polycarbonate tubes and flat plastic containers filled with cellulose acetate/cotton wool or polyester fibers, respectively. Filters remove leukocytes by three principal mechanisms:

Adhesion, whereby granulocytes, monocytes, and platelets adhere directly onto the fibers via pseudopod formation, having undergone shape change indicating cellular activation.

Mechanical sieving, in which cells may become trapped within the pores of column or flatbed filters; aggregated cells are more likely to be trapped, or cells can be further captured within areas that are becoming blocked.

Cell/cell interactions, where it has been observed that granulocyte-platelet interactions occur in red blood cell filters; platelets in red blood cell transfusions adhere to the fibers, become activated, and undergo shape change with pseudopod formation, which can cause increased binding of leukocytes.

Timing of Leukodepletion

Bedside filtration may not be effective in reducing nonhemolytic transfusion febrile reactions (NHFTRs), alloimmunization, and refractoriness. The levels of a range of cytokines such as interleukin (IL)-1, IL-6, IL-8, and tumor necrosis factor-a increase during the storage of leuko-replete platelet concentrates, as they are synthesized by, and released from, disintegrating leukocytes. Prestorage leukodepletion prevents this as well as the fragmentation of leukocytes that occurs in storage, and is therefore preferred. Worldwide, many transfusion services now routinely leukodeplete blood components prestorage to $<5 \times 10^6$ leukocytes per component. Leukodepletion $<1 \times 10^6$ per component is specified by some regulatory authorities.

Plasma Components

Fresh-Frozen Plasma (FPP)

Plasma is removed from whole blood following centrifugation as described above and should be separated within 8 h of collection, rapidly frozen, and stored at −30°C. Units of FFP should contain >0.7 U/ml of factor VIII. See Figure 6.

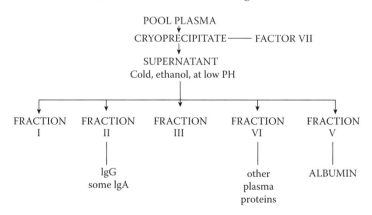

FIGURE 6
Simplified scheme of plasma fractionation. (From Pamphilon, D.H., Ed., *Modern Transfusion Medicine*, CRC Press, Boca Raton, FL, 1995.)

Cryoprecipitate

This is material precipitated from plasma when it is thawed slowly at 4 to 6°C. Cryoprecipitates are rich in coagulation **factor VIII**, **factor XIII**, **Von Willebrand Factor**, **fibronectin**, and **fibrinogen**. The majority of cryosupernatant plasma is decanted, and the remaining protein is redissolved by warming to give a small volume of cryoprecipitate solution.

Irradiation of Blood Components

All blood products to be transfused to patients undergoing **allogeneic stem cell transplantation** should be exposed to gamma irradiation. This procedure reduces the risks of **graft-versus-host disease**.[104]

Indications for Blood Component Transfusion[105,106]

See **Red blood cell transfusion**; **Platelet transfusion**; **Leukocyte-depleted blood components for transfusion**.

Complications of Blood Component Transfusion

The adverse reactions that occur following the transfusion of blood components include:

*Reactions directly related to red blood cell **alloimmunization**.* See **Hemolytic transfusion reaction**

Adverse effects of transfused leukocytes. Transfusion-associated **graft-versus-host disease** (GVHD), **immunosuppression**, alloimmunization, and platelet refractoriness

Reactions as a result of transfused platelets. **Post-transfusion purpura** (PTP) and platelet transfusion refractoriness

Reactions potentially related to antigens in plasma. Nonhemolytic febrile transfusion reactions (NHFTRs) and **anaphylaxis**

Transfusion-transmitted infection. Bacterially infected blood components; viral and prion transmission

Other complications of transfusion. **Iron overload**, circulatory overload, and transfusion-related acute lung injury

Red Blood Cell Alloimmunization: Hemolytic Transfusion Reactions

Intravascular Hemolysis (IVH)

This is invariably due to IgM anti-A and anti-B antibodies; rarely other specificities, e.g., anti-P, P1, pk, and anti-Kidd, produce IVH. It is often the result of a clerical error or mistaken sample identity, and this must be avoided, since up to 10% may be fatal. The clinical signs and symptoms are those of shock, a feeling of listlessness or oppression, chest pain, back and loin pain, rigors, and voiding of dark urine (**hemoglobinuria**). If such a reaction is suspected, the unit of blood should be taken down immediately and returned to the laboratory along with EDTA and clotted samples for urgent investigation. In addition, the first sample of urine passed should be inspected for the presence of free hemoglobin. The EDTA sample can be centrifuged to determine whether there is obvious hemolysis present.

Laboratory features include evidence of **hemoglobinemia** and hemoglobinuria, positive **direct antiglobulin (Coombs) test**, low or absent **haptoglobin**, and the presence of **methemalbuminemia** (Schumm test).

Management of IVH involves infusion of crystalloids and colloids to maintain the circulating blood volume and the correct hypotension. If a diuresis is not established with fluids, then frusemide should be given. Patients often develop **disseminated intravascular coagulation** (DIC), and any resultant **hemorrhage** should be managed with fresh-frozen plasma (FFP) and cryoprecipitate. If there is anuric/oliguric renal failure, then dialysis may be required.

Extravascular Hemolysis

Immediate. The host has a red blood cell alloantibody, and blood incompatible for the corresponding antigen is infused. There is immediate (<24 h) destruction of the incompatible red cells. If the antibody is unable to fix complement, then cells coated with immunoglobulin are removed from the circulation by FcR-gamma receptors in the **lymphoid system**. Antibodies involved are directed against **Kell blood groups**, **Duffy blood groups**, and other red blood cell antigens. The clinical findings are of variable severity and include fever, rigors, hypotension, jaundice, and hemoglobinuria.

Delayed. This occurs when there is no detectable antibody present but the recipient has been sensitized by a prior pregnancy or transfusion. The incompatible red blood cells provoke a secondary (**anamnestic**) response with a rise in antibody titer, which then causes hemolysis of the transfused-incompatible cells. This normally presents 5 to 10 days after transfusion with a fall in hemoglobin and jaundice.

Leukocyte Transfusion Adverse Effects

These may be responsible for febrile nonhemolytic transfusion reactions, refractoriness to random donor platelet transfusions, organ graft rejection, and shortened red cell survival. The following factors are important:

Leukocyte dose. Alloimmunization and subsequent refractoriness to platelet transfusions are reduced when leukocytes are depleted to $<5 \times 10^6$ per transfusion.

Prior sensitization by pregnancy. Human leukocyte antigen (HLA) antibodies occur in 10 to 30% of females. Even in patients who do not have detectable antibody, pregnancy is a high-risk factor for the development of multispecific HLA antibodies.

The antigenic stimulus. Class I HLA antigens are presented to immunocompetent cells in the recipient by class II-bearing antigen-presenting cells such as dendritic reticulum cells. Modification of antigen-presenting cells, e.g., by UV-B irradiation, prevents allosensitization.

Recipient immune responses:

- HLA type: response to foreign HLA antigens is high in individuals of HLA-DR2 phenotype. HLA-DRW6 identifies individuals who are generally high antibody responders against a variety of antigens.

- Immunosuppressant treatment: it is likely that intensive chemotherapy regimes containing immunosuppressant drugs result in a lower overall incidence of alloimmunization in patients with leukemia. In solid organ transplantation, HLA antibody formation is reduced by treatment with a**zathioprine** or **cyclosporine**.

- Sharing of **major histocompatibility complex** (MHC) determinants: it is thought that sharing of DR antigens may result in a lower incidence of HLA antibody production. This may be important, since half of the random population shares a DR antigen.

Transfusion-Associated Graft-versus-Host Disease (TA-GVHD)

In immunocompromised patients, transfusion on non-gamma-irradiated blood components may result in GVHD. This occurs in the following circumstances:

TABLE 28

Patients at Risk for TA-GVHD

Recipients of **allogeneic and autogeneic stem cell transplantation**
Patients with **Hodgkin disease** and **chronic lymphatic leukemia** treated with **fludarabine**
Recipients of intrauterine transfusions (IUT)
Recipients of exchange transfusions where there has been a prior IUT
Patients with **immunodeficiency**
Recipients of HLA-matched platelets from homozygous donors
Recipients of directed donations from first- and second-degree relatives

---

Blood components containing a significant number (greater than or equal to 10^4) of viable leukocytes transfused into immunocompromised recipients.

Patients who have received blood donated by first- or second-degree relatives who are homozygous for an HLA haplotype shared with the recipient. In this situation transfused leukocytes are not rejected but recognize one recipient HLA haplotype as foreign and may then mediate a graft-versus-host (GVH) reaction. Those at risk are listed in Table 28.

TA-GVHD can occur after the infusion of red blood cells, platelets, or even fresh plasma, but not FFP. Since the critical leukocyte load for inducing TA-GVHD is not known, its prevention by **leukodepletion of blood components** is inadvisable. All blood components should be irradiated with 2500 cGy and should bear a sticker indicating that this has been done.[40]

Immunomodulatory Effects

Transfused leukocytes may also be responsible for activating viruses in host cells; this is of particular relevance in HIV-positive patients, in whom transfused allogeneic leukocytes may accelerate the disease. Finally, immunosuppression of cellular (NK and T-cell) function may be involved in an increased incidence of cancer recurrence and postoperative wound infection seen after surgery when patients are transfused. This area is still controversial and is under study.

Platelet Transfusion Reactions

See **Platelet transfusion**.

Posttransfusion purpura

Alloimmunization and platelet refractoriness

Reactions Potentially Related to Antigens in Plasma: Nonhemolytic Febrile Transfusion Reactions (NHFTRs)

These comprise fever and/or rigors occurring 30 min to 2 h after the beginning of a transfusion in 1 or 2% of patients receiving red cell transfusions. They occur in up to 20% of recipients of multiple platelet transfusions. Pathogenesis is as follows:

Reaction of transfused leukocytes with HLA- or granulocyte-specific antibodies in the recipient

The release of **cytokines** such as interleukin (IL)-1, IL-6, IL-8, and tumor necrosis factor-alpha from donor leukocytes in stored blood components

The transfusion should be slowed or discontinued. Antipyretics will reduce fever. Where reactions have a clearly allergic component, chlorpheniramine or another antihistamine should also be given. Hydrocortisone is rarely required.

Prevention

Single NHFTRs are unlikely to recur; however, leukodepletion of red cell and platelet transfusions is recommended in patients who have recurring NHFTRs. Reducing leukocytes to $<5 \times 10^8$ prevents NHFTRs in most recipients.

Transmission of Microorganisms

See **Transfusion transmitted infection** (TTI).
The more common organisms in TTI are cytomegalovirus (CMV), human T-cell leukemia virus (HTL)-I and -II, *Toxoplasma gondii*, and *Yersinia enterocolitica*.

Apart from TTI, blood components can become infected by skin flora or bacteria that enter blood packs during component preparation. It is important that the environment in which blood products are prepared is adequately controlled to minimize microbial contamination. Blood or apheresis donors should not have marked eczema or obvious skin sepsis, as this can be a source of infection. The antecubital fossa vein must be rigorously cleaned with alcohol prior to venepuncture. A variety of organisms may contaminate blood products by this route, including *Staphylococcus aureus* and *S. epidermidis*, diphtheroids, *Serratia marcescens*, and *Pseudomonas aeruginosa*. The clinical features of transfusion reactions due to bacteria are variable and range from an elevation of temperature and rigors to severe and fatal septicemia caused directly by the bacteria or by transfusion of toxins. Severe reactions may be associated with circulatory collapse and death within 8 h.

Iron Overload

This occurs in recipients of repeated red cell transfusions, e.g., patients with **thalassemia**, **myelodysplasia**, and **sickle cell disease**.

Circulatory Overload

This occurs in

- Some elderly patients, often those with profound preexisting anemia who receive transfusions that may be inappropriate
- Patients who are overtransfused after hemorrhage

Transfusion-Related Acute Lung Injury (TRALI)[108–118]

The passive transfer of high-titer leukocyte antibodies in the plasma of some donors (usually multiparous women) may cause leukoaggregation with respiratory disorders due to disruption in the pulmonary circulation with reduced pulmonary compliance (stiff lungs). This is associated with severe reactions characterized by fever, a nonproductive cough, and dyspnea. It is diagnosed by the presence of new bilateral diffuse pulmonary infiltrates on chest radiography with gas-exchange abnormalities in the form of a normal $PaCO_2$ with PaO_2/FiO_2 ratio of less than 40 kPa/300 mm Hg.[107–109]

The main differential diagnosis is cardiogenic pulmonary edema, so to make the diagnosis of acute lung injury (ALI), there must be no evidence of a raised left atrial pressure. The "gold standard" way of excluding cardiogenic pulmonary edema is by demonstrating that the pulmonary artery wedge pressure (PAWP) is less than 18 mm Hg. PAWP is infrequently measured; therefore, other signs of cardiac failure should be absent before entertaining a diagnosis of ALI (raised jugular venous pressure [JVP]/central venous pressure [CVP], peripheral edema, ECG, or biochemical evidence of myocardial infarction and echocardiographic evidence of left ventricular dysfunction).

The international consensus view is that TRALI should be defined as an otherwise unexplained acute lung injury occurring within 6 h of a blood component transfusion.[110,(2)] The diagnosis of TRALI is rarely definitive, as ALI may occur as a complication of many pathological processes such as multiple trauma, sepsis, pancreatitis, and massive hemorrhage. Clearly, these disorders are often treated with blood transfusions, so even if a diagnosis of ALI is confidently made, it will not be clear whether it was precipitated by the disease or its treatment by transfusion.

Antibodies in donor plasma against HLA class I, class II, and granulocyte-specific antigens have been associated with recipients developing TRALI. It has been proposed that these antibodies are pathogenic and bind to leukocytes and pulmonary vascular endothelium. This leads to the sequestration of leukocytes and the release of inflammatory mediators in the lungs, the first microvascular bed encountered following the infusion of the blood component into a large vein.[111] In keeping with TRALI being an immunological reaction, the exudative pulmonary edema encountered is associated with a fever and hypotension. Agents other than antibodies that can "prime" leukocytes and cause endothelial damage, such as vasoactive lipids in stored blood, would be expected to cause a similar phenomenon.[112]

The reported incidence of TRALI is 1:5,000 to 1:10,000 transfusions.[113–115] However, the incidence has also been reported to be as high as 1:1,120 transfusions.[116] The mortality rate is approximately 25%.[117]

Treatment is supportive to maintain oxygenation with oxygen at adequate flow rate and pressure until spontaneous recovery occurs; mechanical ventilation may be required. The administration of diuretics may be associated with clinical deterioration, further accentuating the need to differentiate TRALI from cardiogenic pulmonary edema. Intravenous fluids may be required to reverse hypotension. The role of steroids is unclear. Preventive measures to reduce the risk of antibody-mediated TRALI have been introduced in some centers, including the use of male donors for single-donor FFP and the resuspension of platelets in platelet additive solution or, if pooled platelets are made using more than one donor, using only male donor plasma to resuspend the platelets. Antibody screening of donors is costly and should not be undertaken unless a diagnosis of cardiogenic pulmonary edema has been firmly excluded. Any antibodies detected may be incidental.[118] If a donor is clearly implicated in a case of TRALI by the presence of antibodies in donor plasma and the cognate antigen in the recipient or by leukocyte cross-match incompatibility, that donor should be excluded from donating therapeutic blood components.

BLOOD DONATION

The procedures and complications concerned with the collection of blood for patient transfusion from volunteer donors.[119,120]

Volunteer Blood Donor Selection

Eligibility to Donate

The recommendations and practices of blood donor eligibility vary from country to country. In general, donors are aged 18 to 65 years and in good health. Significant cardiovascular, respiratory, neurological, endocrine, multisystem autoimmune disease, or disease of uncertain cause, e.g., multiple sclerosis, contraindicate blood donation. Donors are also deferred if they are confirmed positive in microbiologic testing, if they recurrently fail the **hemoglobin** test, or if they have a severe donation-related faint or fit. Temporary deferral occurs when the donor has failed the hemoglobin screening test or is awaiting information regarding a medical condition discovered during predonation screening. They may also be awaiting reinstatement after an unconfirmed positive microbiology test from their last donation. Temporary deferral also results from recent major surgery, infection, or antibiotic treatment and following acupuncture, ear piercing, tattooing, electrolysis, and blood transfusion. Donors may be deferred for 1 year if they have traveled to a tropical area. Specific criteria vary between countries and can usually be accessed via Web sites, e.g., in the U.K.[2] and the U.S.[3] To avoid transmission of **prion infections**, U.K. donors who have previously received a blood transfusion are rejected.

In most countries, potential blood donors are screened for hepatitis B virus surface antigen, as well as for antibodies to **human immunodeficiency viruses** (HIV-1 and HIV-2) and to hepatitis C virus. More-sensitive tests to narrow the window period for detection of these infections, such as PCR for hepatitis C RNA, are added in many developed countries. In some countries, including Canada, France, Japan, the U.K., and the U.S., blood donors are also screened for antibodies to **human lymphotropic viruses** (human T-cell leukemia virus, HTLV-I and HTLV-II). There is no practical screening test for prion infection at present.

Predonation Testing and Screening

Hemoglobin. Donating 450 ml of blood results in the loss of 250 mg of iron, and this may lead to anemia if iron stores are compromised. Hemoglobin levels are estimated from a capillary sample drawn from the finger, which is then dropped into copper sulfate solution of a predetermined specific gravity. Alternatively, hemoglobinometry can be used. If the donor fails this test, a venous blood sample is taken for a full blood count and the donor referred to a doctor if appropriate.

Screening for illness. It is important at blood donor sessions that prospective donors answer a number of questions to confirm that they are eligible to donate blood. Donors should complete a questionnaire giving a yes/no answer to a list of important questions regarding medical suitability to donate. These include questions about high-risk behavior associated with transmission of HIV and other viruses. A member of the donor team should discuss the questionnaire with all first-time donors and, in some cases, with repeat donors to ensure that the material has been properly understood. Screening may be difficult where there is lack of privacy, and some donors may feel forced to donate because of peer pressure. In some centers, medical assessment of donors is carried out by interview.

Procedure of Blood Donation

If the hemoglobin and predonation screening is satisfactory, the donor lies supine on a couch and the skin of the antecubital fossa is cleaned. At this point, the blood pressure may be checked and the donor deferred if hypertension is present. An intradermal injection of 1% lidocaine is given adjacent to an antecubital vein in some centers, and a 16-G needle attached to a blood collection pack is inserted and taped into place. The collection pack contains citrate anticoagulant, which is mixed with the blood either manually or using an automated rocking blood weigher. Donation usually takes 5 to 7 min, but a maximum of 12 min is permissible.

Complications

Fainting. This is fairly common and results from hypotension occurring due to fluid loss and/or anxiety. Delayed fainting after leaving the donation session can be dangerous and has led to fatalities. It is an indication for permanent exclusion from donation.

Hypoxic fits. These are uncommon. The donor should be protected from injury and moved to the recovery position when the fit has stopped.

Hematomas. If a hematoma at the venepuncture site is noted, the donation should be stopped and a firm dressing applied.

Brachial artery puncture. This is manifest as bright red blood spurting into the collection bag. The donation should be stopped immediately and manual pressure applied to the venepuncture site for 10 min. The donor should be referred to the local emergency department.

Median nerve damage. This presents as pain or tingling in the distribution of the median nerve. The donor should be referred to a hospital.

Cardiac arrest. If the donor collapses with absent vital signs, immediate cardiopulmonary resuscitation should be commenced.

After donation is completed, donors should be advised to drink plenty of fluids and avoid strenuous activity for the next 2 h. For those with hazardous occupations, such as underwater diving, specified periods of time must elapse.

BLOOD GROUPS

The colloquial term that was used for red blood cell (RBC) surface antigens long before their biochemical nature was known. While the term can be used for leukocyte and platelet antigens or any form of polymorphism occurring in the blood, it is generally limited to antigens on RBCs and their related antibodies. Much is known about lymphocyte surface antigens, but these are regarded as part of the immune system's major histocompatibility antigens (see **Human leukocyte antigens**). There are also granulocytes and platelet surface antigens.

The recognition that circulating blood cells could have antigens on their surface capable of stimulating antibodies arose out of late-19th-century concepts on immunology. Landsteiner in 1900 showed that the red blood cells of certain persons were agglutinated by the sera of certain other persons but not by their own sera. He recognized that the population was divided into three "groups," which he termed O, A, and B types, the fourth, AB, being added later. From this beginning arose the ABO (H) blood groups. In the same year, Ehrlich and Morgenroth found that when certain goats were injected with blood from other goats, antibodies appeared in their sera that hemolysed the red blood cells of the donor goats. In 1905, Dienst found that in pregnancies where the fetus was of a blood group not that of the mother, active immunization occurred and the amount of corresponding antibody in her serum increased. The significance of any of this knowledge was not associated or appreciated until 10 years later, when Dungern and Hirzfeld made the further discovery that blood group systems were related to Mendelian genetics. Landsteiner was then able to claim that the three A, B, or O antigens are controlled by three allelomorphic genes, with two genes for each individual, the frequency in a given population remaining constant from generation to generation. He continued his research by injecting human blood into rabbits, resulting in the recognition of the MNS blood groups and P blood groups. Further progress had to wait until 1940 when, in preparing antibody to anti-M, Landsteiner and Weiner injected red blood cells from rhesus monkeys into rabbits and found that they had produced "anti-rhesus," which agglutinated 85% of human red cells. The specificity of this antibody appeared to be identical to that made in some people following a blood transfusion, but the human antibody was later found to be different and is now called anti-D, the D antigen being one of the Rh blood groups. Levine and his colleagues soon postulated that D-negative mothers could be immunized by their own D-positive fetus, the resulting antibody being the cause of "icterus gravis neonatorum" and "erythroblastosis fetalis," i.e., **hemolytic disease of the newborn**.

Some 250 blood group antigens have now been described, with a considerable understanding of their membrane biochemistry and molecular genetics.[119,120] Some of these antigens have a close genetic relationship with others (i.e., they are governed by a single gene locus or by two or more very closely linked genes) and form a "blood group system." In some systems the gene product is the antigenic determinant, whereas in others (notably the ABO (H) blood groups), the product is a glycosyl transferase. Some blood groups, e.g., the Lewis blood groups, are not intrinsic to the RBC membrane, but adsorbed from the plasma. Little is known about the functions of most blood group antigens, although some intriguing relationships have been found.[120]

The major red cell blood groups of clinical importance or interest include:

ABO (H) blood groups

Duffy (Fy) blood groups

I blood groups

Kell blood groups

Kidd blood groups

Lewis blood groups

Lutheran blood groups

MNS blood groups

P blood groups

Rh blood groups

Minor blood groups (Table 29)

Blood group antigens are either protein or carbohydrate. Carbohydrate-dependent antigens, such as ABH blood groups and Lewis blood groups, are covalently attached to proteins and sphingolipids. Determinant structures on these antigens are indirectly regulated by blood group genes, the gene products being transferase enzymes, which catalyze the transfer of monosaccharides onto carbohydrate precursors.

The RBC antigens of the blood group systems are expressed on RBC membrane proteins; the amino acid sequences of most are known. Many of these proteins have been shown to be associated with certain functions; for example, Duffy blood group antigens are carried on a glycoprotein that is a receptor for several cytokines. The number of antigenic determinants per red cell varies between about 10^3 (Knops antigens) and 10^7 (P antigen) copies.

Similar antigens can be found on the cell surfaces of some tissues outside the blood. These are most likely to be related to ABO, P, and I blood group antigens.

Nomenclature

Most of the genes governing the blood group systems have been cloned and sequenced, so that a logical nomenclature has been formulated. Genes are italicized, and RBC antigens

TABLE 29

Properties of Some Less Commonly Encountered Antigens and Antibodies of Blood Group Systems

Antigen Name	Antigen Frequency in Caucasians (%)	Ability to Cause Hemolytic Transfusion Reactions	Ability to Cause Hemolytic Disease of the Newborn	Other Comments
Ch	96	no	no	Belongs to the Chido/Rodgers (CH/RG) system, with nine antigens residing on the C4 component of complement; antibody gives variable strength reactions
Co^a	99.9	yes	mild	Belong to the Colton (CO) system (3 antigens
Co^b	10	yes	mild	described)
Cr^a	>99.9	yes	no	Belongs to the Cromer (CROM) system (13 antigens)
Di^a	0.01	yes	yes	Belong to the Diego (DI) system (21 antigens
Di^b	>99.9	yes	mild	described); Dia is more common in Native Americans and those of East Asian descent
Do^a	67	yes	no	Belong to the Dombrock (DO) system (5 antigens
Do^b	82	yes	no	described)
Gy^a	>99.9	yes	no	
Ge	>99.9	yes	no	Belongs to the Gerbich (GE) system (8 antigens)
Gil	100	yes	no	The only antigen in the GIL system
In^a	0.1	decreased red cell survival	no	Comprise the Indian (IN) system; the Ina antigen has an incidence of 4% in Asian Indians and
In^b	99	yes	no	12% in Arabs
JMH	>99.9	no	no	Anti-JMH gives variable strength serologic reactions
Kn^a	98	no	no	Belong to the Knops (KN) system (8 antigens
McC^a	98	no	no	described); the antibodies give variable
Yk^a	92	no	no	strength serologic reactions
Kx	100	yes	not applicable	Anti-Kx is a typical component of immunized males with the McLeod phenotype and chronic granulomatous disease
LW^a	100	mild	mild	Belongs to the Landsteiner-Wiener system (3 antigens); anti-LWa may be confused with anti-D
MER2	92	no	no data	The only antigen in the RAPH system
Ok^a	100	decreased red cell survival	no	The Ok(a-) phenotype has only been found in Japanese
Sc1	99	no	no	Belong to the Scianna (SC) system (5 antigens
Sc2	<1	no	no	described)
Xg^a	66 (males) 89 (females)	no	no	Belongs to the XG system (2 antigens); the Xga gene is carried on the X chromosome
Yt^a	>99	yes	no	Yt^a and Yt^b form the Cartwright (YT) system
Yt^b	8	no	no	

Source: Data from Daniels, G., *Human Blood Groups*, Blackwell, Oxford, 1995.

are usually denoted by one or two letters, the first of which is capitalized. An additional integer, or superscript letter, may be used to distinguish antigens arising from allelic variation within a blood group system; e.g., Fya and Fyb are the antigens produced by the genes *Fya* and *Fyb* of the Duffy blood groups, and Sc1 and Sc2 are the antigens produced by *Sc1* and *Sc2* of the Scianna system. Notable exceptions to this rule include the ABO blood group antigens A and B, the Kell blood group antigens K and k, and the 47 Rh blood group antigens. The International Society of Blood Transfusion (ISBT) terminology uses capitalized alphanumerics for blood group systems, e.g., KEL, RH, and FY.

Linkage with Disease

There are weak associations between some blood groups and some diseases. For example, there is a tendency for bacterial infections to attack group A people; viral infections tend to be associated with group O; cancers and clotting diseases are associated with group A, while autoimmune disorders and hemorrhagic disorders are associated with group O.[120a] The Duffy blood groups act as a receptor for *Plasmodium vivax*; RBCs of the phenotype Fy(a⁻b⁻) resist parasitic invasion. There is an association between the Kell blood groups, the Kx RBC antigen, and chronic granulomatous disease. RBC blood group antigens may also become depressed in leukemia or autoimmune hemolytic anemia.

Paternity and Forensic Testing

Blood group phenotyping was used extensively in the past for identification purposes, but has now been replaced by DNA techniques.

BLOOD PRODUCTS FOR INFUSION

Therapeutic agents derived from blood. They include:

Human albumin solution (see **Colloids for infusion**)

Coagulation-factor concentrates

Immunoglobulins

Plasma fractions are initially separated using Cohn cold ethanol extraction. By varying pH, temperature, and ethanol concentration, five different fractions are obtained (see Figure 6).

Certain protein fractions can be further purified by ion exchange, affinity or size-exclusion chromotography, or monoclonal antibody (immunoaffinity) purification. Fractionated blood products can then be treated to inactivate viruses.

BLOOD VESSEL WALL

See **Vascular endothelium**.

BLOOD VOLUME

The total circulating volume of blood made up of both cells and plasma. Over 99% of the cellular content is the red cell mass, which, like the plasma volume, can be measured by dilution techniques.[88,121] Due to high cost and the lack of personnel trained in the use of **radionuclides**, these measurements are rarely performed in current practice. Reference ranges must now be based on estimated values.[121a]

Total Blood Volume

This is determined by the addition of the red cell volume to the plasma volume when measured simultaneously. Calculation from either single determination by using the **packed-cell volume** (PCV) is unreliable. Reference ranges for adults are 65 to 85 ml/kg for males and 60 to 80 ml/kg for females. Those for children have been estimated.[121a]

Plasma Volume

This is measured by injecting a known concentration of either an azo dye, Evans blue (T-1824), which binds to albumin, or a radionuclide, e.g., ^{125}I-albumin. After 10 min, samples of plasma are removed for estimation of the dilution. When using a radionuclide, further samples have to be taken at 20 and 30 min to correct for its decay after injection. This correction is made by graphically extrapolating back the activity measured in counts per second to the time of injection. From the figures obtained, the result can be calculated from the equation:

$$\text{Plasma volume} = \frac{S \times D \times V}{A}$$

where
S = concentration of dye/radioactivity (cpm/l)
D = dilution factor of injected material
V = volume of fluid injected
A = concentration of dye/radioactivity (cpm/l) of plasma at time of injection

The results are inherently inaccurate due to the variable binding to albumin and to albumin leakage from the circulation. Reference ranges for adults are 40 to 50 ml/kg for both sexes. Those for children have been estimated.[121a]

Posture decreases the plasma volume by around 2 ml/kg, as does prolonged bed rest. There is no difference with age after infancy, where the higher total blood volume is related to the higher PCV.

During pregnancy the plasma volume increases during the second trimester, rising to a peak at the 36th week. There is a smaller rise in the red cell volume, resulting in apparent anemia (**dilutional anemia**). All these parameters return to normal by the second week of the puerperium.

Reduced plasma volume (**hemoconcentration**) occurs acutely following sudden hemorrhage, burns, intestinal stasis from obstruction, and with salt loss as a result of exercise or excessive heat. Prolonged reduction (relative **erythrocytosis**) is sometimes associated with hypertension and other forms of vascular disease, the mechanism being unknown.

Red Blood Cell Volume (Red Blood Cell Mass)

This can be measured by radionuclide labeling of red blood cells by ^{51}Cr or ^{99m}Tc in an aliquot of anticoagulated blood, which, after washing with buffered saline and resuspension, is reinjected. All procedures must be carried out with sterile precautions. Sampling is done after 5, 10, and 15 min. The dilution in the red blood cells is calculated by extrapolating back the radioactivity level to the time of injection. Calculations are made similar to those for measuring plasma volume. In patients with **splenomegaly**, sampling should be delayed until 30 min after injection of labeled red blood cells to allow for sequestration throughout that organ, with repeat sampling at 60 and 90 min. Normal levels are 25 to 35 ml/kg for males and 20 to 30 ml/kg for females. Those for children have been estimated.[121a]

Raised levels of the red cell volume and the total blood volume occur with absolute erythrocytosis from any cause. For routine diagnosis and management of patients, the measurement is now less frequently made.

Proportional increase of red cell volume either by a true increase in circulating red cells or by a reduction in plasma volume leads to raised blood viscosity with a decrease in cerebral blood flow.

Replacement

See **Blood component transfusion; Human albumin solution**.

BLOOM'S SYNDROME

A rare autosomally recessive inherited disorder in which there is a defect in replication and spontaneous chromosome breakage, resulting in a predisposition to malignancy, including **acute myeloid leukemia**, **acute lymphoblastic leukemia**, and **lymphoproliferative disorders**, these being particularly common (20% of affected individuals). Various nonhematopoietic (breast, tongue, cervix) malignancies may occur.

A clinical triad of growth retardation, facial telangiectasia, and sun-sensitive rashes is indicative of Bloom's syndrome, but these may not be apparent at birth. Other physical abnormalities, e.g., *café au lait* spots, absent or extra digits, or hypospadias, are often present. The diagnosis is confirmed upon **cytogenetic analysis** showing an increase in sister chromatid exchange.

B-LYMPHOBLASTIC LEUKEMIA/LYMPHOMA

(B-LBL; lymphoblastic lymphoma) See **Acute lymphoblastic leukemia; Precursor B-lymphoblastic lymphoma/leukemia**.

BOHR EFFECT

See also **Hemoglobin; Oxygen affinity for hemoglobin**.
The shift of the oxygen-dissociation curve to the right. The position of this curve is determined by the rate of tissue metabolism, CO_2 production, and the blood pH. Increased production of tissue CO_2 and acid metabolites results in acidosis, which shifts the curve to the right, allowing the release of more oxygen for the level of tissue Po_2 and opening hemoglobin to receive additional CO_2. It occurs instantaneously and can be highly localized to a single site. It is given a numerical value $\Delta\log p50\ O_2$, where $\Delta\log p50\ O_2$ is the change in the p50 O_2 produced by a change in pH (ΔpH).

BONE DISORDERS

See **Skeletal disorders**.

BONE MARROW

The structural organization of tissue occupying the medullary cavity of bones, its cellular content, function, methods of examination, and disorders.

Structural Organization

The bone marrow cavity is occupied largely by fat cells, with hematopoietic cells embedded in stroma of connective tissue in the long bones of adults and the flat bones of infants (see Figure 7). The marrow cavity or medulla is surrounded by a cortex of compact bone with trabeculae of cancellous bone. The trabeculae and the inner surface of cortical bone are lined by osteocytes, with osteoblasts forming new bone and osteoclasts concerned with the resorption of bone. Osteoblasts and fibroblasts are derived from mesenchymal progenitor cells, but osteoclasts are derived from hematopoietic progenitor cells. Endothelial vascular

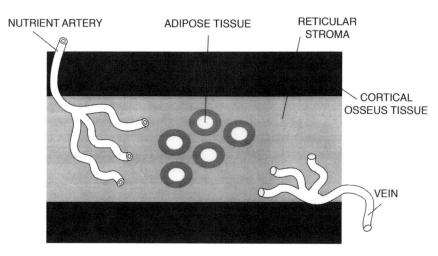

NUTRIENT ARTERY   ADIPOSE TISSUE   RETICULAR STROMA

CORTICAL OSSEUS TISSUE

VEIN

FIGURE 7
Diagrammatic representation of the structure of bone marrow at a diaphysis.

cells are derived from endothelial progenitor cells, which are most probably derived from bone marrow stem cells. The cellular content of bone marrow changes from infants through childhood to adults.[122] At birth and for the first few years of life, it is occupied by proliferating hematopoietic cells (see **Fetal hematopoiesis**), whereas the long bones of adults are largely occupied by fat cells interspersed by strands of hematopoietic tissue. The blood supply is from an osseous nutrient artery that enters through the diaphysis and bifurcates to ascending and descending branches, with lateral branching to the bone cortex. At the corticomedullary interface, some capillaries are continuous with the marrow sinuses. These converge on larger collecting sinuses arranged radially around a longitudinally oriented central sinus. Thus most of the blood reaching the sinuses has first entered the osseous tissue. The vascular sinuses are 50 to 75 μm in diameter, with a thin endothelial lining without a continuous basal lamina.

Hematopoietic Cells
The nucleated cells found in the bone marrow of adults include[122,123]:

Myeloblasts	0.3–5.0%
Promyelocytes	1.0–8.0%
Myelocytes	5.0–19.0%
Metamyelocytes	4.0–15.0%
Band forms	12.0–34.0%
Neutrophils	7.0–30.0%
Eosinophils	0.5–4.0%
Basophils	0.0–0.7%
Lymphocytes	3.0–17.0%
Monocytes	0.0–6.0%
Plasma cells	0.0–2.0%
Proerythroblasts	1.0–8.0%
Basophilic erythroblasts	0.0–4.0%
Polychromatic erythroblasts	4.0–8.0%
Orthochromatic erythroblasts	1.0–5.0%
Megakaryocytes	1–38/50 per l-p field
Myeloid/erythroid ratio	2:1–30:1

The bone marrow cell population of normal infants is given in **Reference Range Table XII**.

Cellular Stromal Composition

Central vascular sinuses are covered by endothelial cells with distinct interdigitating cell junctions. These endothelial cells secrete a variety of hematopoietic growth factors and possess receptors for others. Surrounding the vascular sinuses is a network of reticular cells that are the dominant stromal cell type *in vivo*. During the development of fetal hematopoiesis, high numbers of fibroblast colony forming cells (CFU-F) are found as migratory cells preceding the increase and migration of CFU-S or CFU-GM in the liver, spleen, and femoral marrow. The *in vitro* counterpart may be the pre-adipocyte or marrow fibroblast. These reticulum cells synthesize reticular fibers, which provide a reticular network to physically support progenitor cells, with granulocytic precursors in close proximity to these cells. They must be distinguished from phagocytic reticulum cells, which are macrophages derived from hematopoietic progenitor cells.

Marrow adipocytes are closely related to fibroblasts and are also located in the parasinal space. Their functional role is unclear. Macrophages and lymphocytes are usually considered part of the stroma, with cell-cell interaction between macrophages and erythroblasts providing evidence for the importance of cellular adhesion in hematopoietic development. Lymphoid cells and macrophages release soluble hematopoietic growth factors essential to the regulation of cellular differentiation and proliferation, and local concentrations of these, together with membrane-bound growth factor isoform expression (e.g., stem cell factor), provide the appropriate chemical growth milieu for regulation of hematopoiesis.

Finally, osteoid cells and hematopoietic cells may be derived from a common progenitor, and this, together with the localization of neutrophil precursors adjacent to osteoblasts or osteoclasts and the localization of lymphoid cells adjacent to trabeculae, implies a significant role for bone in hematopoietic development. Chondrogenesis and endochondral bone formation precede local hematopoiesis, so that osteoblastic activity may have a protective function.

Cell-Matrix Interactions

The extracellular matrix (ECM) consists of several proteins with adhesive properties enmeshed around stromal cells. These include:

Proteoglycans and glycosaminoglycans. These are produced by both stromal cells and hematopoietic cells. These proteins are adhesive to receptors on hematopoietic progenitors (Bl-CFC), other matrix proteins (laminin, fibronectin, and collagen), and hematopoietic growth factors (GM-CSF). Heparin sulfate is the principal member of this family expressed on the cell surface, while chondroitin sulfate is secreted mostly into the medium *in vitro*. *In vitro* stimulation of murine proteoglycan synthesis increases the number of hematopoietic progenitor cells.

Fibronectin. Erythroid progenitor cells bind strongly to fibronectin, and studies using the murine erythroleukemia cell line, MEL, have demonstrated fibronectin to be essential in the differentiation process and also in release of reticulocytes. This has been confirmed in human and other mammalian systems where loss of fibronectin receptors from the reticulocyte surface coincides with release. Fibronectin also binds lymphoid progenitors and megakaryocytes, with only weak attachment to granulocyte and macrophage progenitors.

Thrombospondin (TSP). This protein is secreted into the microenvironment by a variety of cells, including normal bone marrow precursors. TSP mediates adhesion of a subpopulation of hematopoietic progenitors of pluripotent, erythroid, granulocytic, and megakaryocytic lineages. TSP immobilized on plastic augments the proliferative activity of stem cell factor (SCF) upon human hematopoietic progenitors.

Hemonectin. This 60-kDa glycoprotein specifically mediates adhesion of granulocyte macrophage progenitors and immature myeloid cells.

Collagen. The precise role of types I, III, IV, and V collagen synthesized into the bone marrow extracellular matrix *in vivo* is uncertain.

CD44. This is a proteoglycan, recently identified as a receptor for hyaluronate and expressed in two forms, an epithelial isoform upregulated in carcinomas and a hematopoietic variety upregulated in certain lymphomas.

Cell–Cell Interactions

See also **Cellular adhesion molecules**.

Membrane adhesion proteins and their cellular receptors are classified into three broad groups, the immunoglobulin superfamily, the integrins, and the selectins. ICAM-1 (immunoglobulin superfamily) is present on 20 to 30% of erythroid and 50% of granulocytic progenitors and is lacking on more mature erythroid cells, but not granulocytic cells. VLA-4 (integrin family) appears central to pluripotent progenitor cell marrow homing, and antibodies to the α-chain of the VLA-4 molecule abolish lymphopoiesis in Whitlock-Witte culture.

Release of Cells to Peripheral Blood

Hematopoiesis occurs in extravascular spaces within the marrow cavity, but the mechanisms by which mature cells enter the blood vessels are unclear. Granulocytes are motile and can probably migrate through the sinusoidal wall; megakaryocyte cytoplasmic processes project between endothelial cells; but the mode of entry of the nonmotile red cells is not known. The destruction of the sinusoidal architecture in **myelofibrosis** or following invasion by metastatic carcinoma cells may explain the frequent occurrence of circulating immature blood cells in these circumstances, so-called leukoerythroblastosis.

Function

See **Hematopoiesis**; **Normoblasts** — erythropoiesis; **Granulocytes** — granulopoiesis; **Monocytes** — monopoiesis; **Lymphocytes** — lymphopoiesis; **Megakaryocytes** — thrombopoiesis.

Examination

Two techniques exist for bone marrow examination, namely aspirate and trephine biopsy. These are frequently performed together, since in many instances the findings are complementary. As a general rule, where cell detail is important, examination of an aspirate smear is most helpful, whereas if marrow architecture is important, a biopsy is indicated.
  Indications for bone marrow examination include:

  Unexplained cytopenia of any kind (see **Anemia**; **Leukopenia**; **Thrombocytopenia**)
  Suspected **leukemia** or other primary marrow malignancy

Suspected marrow replacement (see **Myelofibrosis**; **Infiltrative myelophthiasis**; **Malignancy**; **Lipid-storage disorders**)

Unexplained **lymphadenopathy, hepatosplenomegaly**

Staging of **lymphoproliferative disorders**

Diagnosis in fever due to an infectious disorder of undetermined origin

Evaluation of marrow iron stores or **iron overload**

Assessment of cellularity before and after **cytotoxic agent** therapy

As with all surgical procedures, informed consent is necessary. It may be necessary to give preoperative medication. A wide variety of needles are available. One of the most popular is the Jamshidi needle with a conical tip that can be used for both biopsy and aspiration. The iliac crest is the most frequently used site for an aspiration biopsy procedure in both adults and children. The posterior iliac spine is the most frequently used landmark with the patient's back to the operator. The sternum was formerly the preferred site, but only an aspirate could be performed. Patient anxiety and discomfort were drawbacks, as was the unavoidable although short-lived suction pain. In addition there was a remote risk of overpenetration, with injury to the heart and great vessels. In the infant under 1 year of age, the proximal anterior aspect of the tibia just below the greater tuberosity is a suitable site for marrow sampling.

The aspirate smear is stained using the Romanowsky method. Under low power (×40 or ×100), the smear is examined for clumps of tumor cells or granulomata, for adequacy of megakaryocytes, and to find the area of optimal spreading. Any areas appearing to be monomorphic are examined under high power (×400) to check for an infiltrate. No attempt should be made to assess marrow cellularity from the smear, since this can be most misleading. The marrow is then systematically examined under high power. A total of 400 to 500 nucleated cells is necessary to obtain a significant marrow differential count and estimate the myeloid:erythroid (M:E) ratio. Each cell line should then be examined for orderly maturation, and finally macrophages should be examined for inclusions (e.g., erythrophagocytosis, infectious agents such as histoplasmosis, evidence of lipid-storage disorders such as Gaucher or Niemann-Pick).

Microscopic examination of fixed, embedded sections, either marrow clot or trephine biopsy, is best for determining overall cellularity, the presence of infiltrates such as carcinoma metastases (see **Malignancy**) or neuroblastoma, granulomata, and for observing abnormal marrow architecture.

Special Techniques

Cytochemical and histochemical stains (see Table 30)

Immunophenotyping

Immunohistology (immunocytochemistry)

Cytogenetic analysis

Molecular genetic analysis

Bone marrow culture

Disorders[124]

Due to its high rate of proliferative activity, the bone marrow is susceptible to disorders that:

TABLE 30

Cytochemical and Histochemical Stains and Their Identifications

Stain	Identification
Alcian blue	cryptococci; stromal mucins
Chloro-acetate esterase	granulocyte differentiation and mast cells
Congo red	amyloid deposits
Gamori's	reticulin fibers for myelofibrosis
Gordon and Sweet's	reticulin fibers for myelofibrosis
van Giesen	collagen for myelofibrosis
Grocott's metheneamine silver	fungi
Martius scarlet blue	collagen for myelofibrosis
Periodic acid-Schiff (PAS)	plasma cells; fungi
Perl's **Prussian blue**	hemosiderin (iron)
Toluidine blue	mast cells
Ziel-Neelson	mycobacteria (tuberculosis)

Interfere with its anatomical structure

Suppress the production of hematopoietic colony stimulating factors

Stimulate the production of inflammatory cytokines

Reduce red blood cell survival

Disorders can therefore be caused by:

Genetic disorders
- **Fanconi anemia**
- **Dyskeratosis congenita**
- **Schwachman-Diamond syndrome**
- **Blackfan-Diamond syndrome**
- **Amegakaryocytic thrombocytopenia**

Infectious disorders
- **Bacterial infections**
- **Metazoan infections**
- **Protozoan infections**
- **Rickettsial infection disorders**
- **Viral infections** (including **pure red cell aplasia**)

Ionizing radiation, e.g., **aplastic anemia**

Chemical toxins, e.g., **benzene, alcohol**

Drugs, e.g., **cytotoxic agents; adverse drug reactions**

Autoantibodies
- **Acquired aplastic anemia**
- **Paroxysmal nocturnal hemoglobinuria**
- **Myelodysplasia**

Dyshematopoiesis, particularly **megaloblastosis** and **myelodysplasia**

Replacement by cellular proliferation
- **Leukemias**
- **Infiltrative myelophthiasis**

- **Myeloproliferative disorders**
- **Histiocytosis**
- **Lymphoproliferative disorders**

BONE MARROW CULTURES

The techniques used for culture of hematopoietic cells. They are mainly used for research and have a limited role in diagnostic assessment of bone marrow disorders.

Short-Term Culture

These techniques are used for assessment of harvested bone marrow or peripheral stem cells prior to their transplantation. The cells are suspended at a known starting concentration in methyl cellulose or agar supplemented with culture medium, fetal bovine serum, and growth-promoting substances such as GM-CSF, erythropoietin, or thrombopoietin. They are incubated at 37°C for 14 days in a humidified atmosphere containing 5% CO_2. The starting cells are seeded at a sufficiently low concentration so that individual colonies localized around each single multipotential cell can be seen separately from neighboring colonies and counted.

Long-Term Bone Marrow Culture (LTBMC)

The Dexter system for the long-term culture of murine bone marrow was first established in the mid-1970s. This *in vitro* model simulating the interaction between hematopoietic stem cells and stroma remains a useful biological tool. Bone marrow cells are inoculated into flasks in tissue culture medium and establish an adherent stromal monolayer (see **Bone marrow**) consisting of three cell types: phagocytic mononuclear cells (comprising 80 to 85% of adherent cells), preadipoietic fibroblasts, and some evidence for endothelial cells. An initial fall in **colony-forming units** (usually CFU-GM) is followed by a rise at about 4 weeks of culture as stem cells begin to generate their progenitor progeny. Murine progenitor production can be sustained for several months (CFU-GM, CFU-meg, HPP-CFC, CFU-S).

In 1979/1980, Gartner and Kaplan[125] reported a modification of the Dexter system for human marrow culture using tissue culture medium containing fetal bovine serum, with weekly feeding of cultures, by replacing half of the nonadherent layer (5 ml). Granulocyte macrophage (GM) colony production from normal human bone marrow could be sustained for approximately 20 weeks.

The adherent layer supports the formation of "cobblestone" areas consisting of macrophages and myeloid colony forming cells and devoid of erythroid activity. These cells are overlain by "blanket cells," probably of fibroblast origin. Murine CFU-S or CFU-Mix (stem cells/pluripotent progenitors) associate closely to or under the adherent layer. Time-lapse photography demonstrates the migration and division of these cells beneath the adherent layer, migration through that layer, and release into the nonadherent layer as progenitor cells.

Intimate contact with the stromal layer (via adhesion molecules) is essential for maximal progenitor production, and addition of exogenous growth factors does not enhance or deplete this. Separation of stromal cells from progenitors in a diffusion chamber or thin-layer agar results in reduced myelopoiesis. Stromal cytokine production is markedly stimulated by refeeding, which induces expression of GM-CSF and G-CSF mRNA, not detectable in quiescent cells, in contrast to the expression of interleukin-6, stem cell factor, and colony stimulating factor (CSF)-1.

Johnson and Dorshkind[126] have also modified the original "myeloid" Dexter conditions to preferentially support lymphopoiesis ("Whitlock Witte" conditions), and indeed cultures can be switched from myeloid to lymphoid by changing the conditions.

LTBMC has been used to preferentially grow "normal" (Ph negative) hematopoietic cells from bone marrow harvested from patients with **chronic myelogenous leukemia** (CML) following the observation that CML progenitors adhere less to stroma than normal progenitors. This marrow has been used for autotransplantation, but initial encouraging results appear to have been unsustainable.

BONE MARROW HYPOPLASIA

(Hypoplastic anemia; bone marrow suppression) A histological descriptive term applied to bone marrow with reduced hematopoietic cells of varying degree compared with stromal cells. Generally, the term is used when the cellularity is reduced to less than 25% of that seen in a normal adult trephine biopsy. The loss of hematopoietic tissue may be generalized or patchy. The term does not indicate a particular etiology; bone marrow hypoplasia may arise from a number of causes, including those that produce **aplastic anemia**, as shown in Table 31. The peripheral blood may show **pancytopenia** or loss of only one or two cell lineages, at least in the early stages. In many instances there is red blood cell **macrocytosis**, thought to be an indication of marrow stress.

TABLE 31

Causes of Bone Marrow Hypoplasia

Cause	Description	Examples	Comment
Old age	generalized reduction	over 70	not universal but progressive
Aplastic anemia	hematopoietic tissue replaced by fat cells	acquired or inherited aplastic anemias	see Table 18
Ionizing radiation	variable necrosis/fibrosis depending	therapeutic	hypoplasia distributed according to field of radiation; recovery depends on dose and duration
		accidental acute	death usually due to other effects of radiation
		chronic	may progress to myelodysplasia or acute leukemia
Cytotoxic agents	dysplasia and apoptosis	chemotherapy	drug and dose dependent
		immunosuppression	individual susceptibility to repeated doses
Infections	may contain granulomas abnormal megakaryocytes	AIDS	13–33% cases affected
		dengue fever	neutropenia and low platelets during viremic phase
Starvation	gelatinous transformation; acid mucopolysaccharides replace fat cells	anorexia nervosa cachexia in chronic disorders (TB, cancer)	
Benzene	usually proliferative but aplasia may occur	industrial exposure	may progress to more malignant phase
Myelofibrosis	severe fibrosis with loss of cellularity	malignant myelofibrosis	end stage

Bone marrow hypoplasia is a common consequence of **cytotoxic agent** therapy. A particular association is exposure to **ionizing radiation**. Marrow hypoplasia has been described following exposure to radiation, including accidental (e.g., Chernobyl, Ukraine,

Brazil), industrial (e.g., radium poisoning in watch-dial painters), and therapeutic (e.g., treatment of **ankylosing spondylitis**) exposures. Doses exceeding 1.5 to 2 cGy produce marrow hypoplasia, but recovery occurs after 2 to 4 weeks. The lethal dose depends on the medical care available to manage the cytopenia. In acute exposure leading to death, the causes are usually related to effects on other systems. When the patient survives, the acute effects of accidental radiation bone marrow recovery are usual with good transfusion support and management of infections. Doses of total body radiation above about 750 cGy in a single dose, as may be used in **allogeneic stem cell transplantation**, produce irreversible marrow depression unless stem cells are given.

Clinical and Laboratory Features

These depend on the underlying cause of the condition. The hematological features may be insignificant compared with other manifestations of the cause, or they may dominate through the cytopenias. Anemia is normochromic, and many cases have mild macrocytosis (mean cell volume 100 to 108 fl). Serum iron is usually high, with reduced ^{59}Fe clearance and utilization. Despite high erythropoietin levels, the **reticulocyte count** is low in relation to the degree of anemia. More than one biopsy may be necessary if the hypoplasia is patchy. The assessment of the biopsy includes the measurement of **reticulin** and **collagen** as well as identification of abnormal infiltrates. Differential diagnosis includes **mucopolysaccharide storage disorders**, **bone marrow necrosis**, and **lymphoproliferative disorders**, especially **hairy cell leukemia** and **acute leukemias**.

Management

This mostly involves the treatment of the underlying cause, when possible, and support for the peripheral blood cytopenias, such as anemia, **neutropenia**, and **thrombocytopenia**. Cytokines, particularly **granulocyte colony stimulating factor** (G-CSF) and **erythropoietin**, may be useful either to accelerate recovery of neutrophils and red blood cells, respectively, for example in postaccidental irradiation, or to raise counts just sufficiently to overcome the need for **red blood cell transfusion** or **platelet transfusion**. **Anabolic steroids** have been used to stimulate marrow activity, but side effects, and better results in most cases with cytokines, make their use rarely justified.

BONE MARROW NECROSIS

(BMN) A rare pathological state in which widespread areas of **bone marrow** have undergone necrosis. It is usually associated with either a hematological malignancy or **infiltrative myelophthiasis**, particularly metastatic carcinoma or tuberculosis, and as a consequence of **cytotoxic agent** therapy. It is presumed to be a result of extravascular obstruction by compression of rapidly proliferating cells, by intrinsic vascular obstruction from tumor microthrombi, or by vasoconstriction in response to release of toxins or cytokines. It can also occur with disorders giving rise to microthrombi, such as **sickle cell disorder** or **antiphospholipid syndrome**.[63] Patients suffering from involuntary starvation or anorexia nervosa have gelatinous transformation of the ground substance.[127] Clinical features and treatment are those of **aplastic anemia**.

BONE MARROW TRANSPLANTATION

(BMT) See **Allogeneic stem cell transplantation**; **Autologous stem cell transplantation**.

BORTEZOMIB

A boronic acid dipeptide that acts as a proteasome inhibitor. In **myelomatosis**, it degrades proteins that regulate cell cycle progression and causes proteolysis of the endogenous inhibitor of nuclear factor IκB (NF-κB). Degradation of IκB by proteasomes activates NF-κB which, in turn, upregulates the transcription of proteins that promote cell survival, stimulates growth, and reduces susceptibility to apoptosis. NF-κB activation also induces drug resistance in myeloma cells and upregulates the expression of adhesion molecules involved in the resistance of myeloma cell drugs. In addition, it modulates the secretion by bone marrow stromal cells of cytokines that affect growth, survival, and migration of myeloma cells. It is therefore effective in the treatment of relapsed, refractory myelomatosis. It is under investigation for the initial treatment of myelomatosis[128,129] and renal carcinoma.

B-PROLYMPHOCYTES

See also **Lymphocytes**.
Medium-sized, round lymphoid cells with prominent nucleoli. They are seen in the peripheral blood and bone marrow in cases of **B-cell prolymphocytic leukemia**.

BRADYKININ

A powerful vasoactive peptide and mediator of the response to inflammation. It is produced by release from its precursor protein, high-molecular-weight **kininogen**. Upon exposure to negatively charged surfaces, high-molecular-weight kininogen and **prekallikrein** are activated by **factor XIIa**. Prekallikrein is converted to kallikrein, and the high-molecular-weight kininogen is itself digested by kallikrein to release bradykinin, a nine-amino-acid peptide. As bradykinin is such an active peptide, its activity is closely controlled both in generation by C1 inhibitor α_2 macroglobulin and, once generated, by carboxypeptidase N.

BRUISING

See also **Ecchymoses**; **Petechiae**; **Purpura**.
Extravasation of blood into the skin or mucous membranes, which appears as a purple discoloration. In the absence of trauma, it is a clinical indication of a **hemorrhagic disorder**, particularly due to **thrombocytopenia** or **vascular purpura**.

Simple Easy Bruising

Many individuals, in particular young women, give a history of simple easy bruising for which there is often no explanation. In some cases there may be a family history. Patients should be reassured and told to avoid **aspirin** and aspirinlike drugs.

Factitious Purpura

Self-induced purpura, usually seen in easily accessible parts of the body. Nonaccidental injury ("battered child" syndrome) should be considered if there are associated abrasions. Some individuals, especially health-care workers and those working in allied professions, may knowingly self-administer anticoagulants. Patients with self-induced bleeding require prompt psychiatric counseling.

Psychogenic Purpura

Bizarre bleeding episodes and purpura in patients with psychological disorders. The diagnosis is often suspected only after the results of extensive investigation.

B-SMALL LYMPHOCYTIC LYMPHOMA

(B-SLL) See **Chronic lymphatic leukemia**.

BUFFY COAT

The layer of **leukocytes** and **platelets** lying above the red blood cells following centrifugation of whole blood. Spread on a glass slide, buffy coat films are prepared to examine microscopically a concentrate of leukocytes and view abnormal cells present in low numbers. To perform this laboratory examination, defibrinated venous blood is centrifuged in a Wintrobe hematocrit tube at 3000 rpm (1500 g) for 15 min, the supernatant serum removed, and smears made from the platelet and leukocyte layers.

BUHOT CELLS

Mononuclear cells with cytoplasmic granules, variable in size and shape and often surrounded by clear vacuoles. They are found in all **mucopolysaccharidoses**, but especially Sanfilippo syndrome, and are composed of mucopolysaccharide.

BURKITT LYMPHOMA

(Rapaport: undifferentiated large-cell lymphoma, Burkitt type; Lukes-Collins: small, non-cleaved follicle center cell lymphoma) See **Lymphoproliferative disorders**; **Non-Hodgkin lymphoma**.
A high-grade B-cell non-Hodgkin lymphoma originally described by Denis Burkitt as an endemic disease of equatorial Africa, mainly in children, who developed tumors in multiple organs and particularly in the region of the jaw. The isolation of the **Epstein-Barr virus** (EBV) from cell lines derived from Burkitt lymphomas, together with the epidemiological association of the lymphomas with *Plasmodium falciparum* malaria, has led to the speculation that high-level antimalarial immunoglobulin and EBV antibodies induce proliferation of B-lymphocytes with suppression of T-lymphocytes.

All the tumor cells are similar, being medium-sized with round central nuclei, multiple (two to five) nucleoli, and relatively abundant basophilic cytoplasm with lipid vacuoles, usually seen on imprints or smears. This tumor has an extremely high rate of proliferation and a high rate of spontaneous cell death. A "starry sky" pattern is usually present, imparted by numerous benign macrophages that have ingested tumor cells undergoing apoptosis. The **immunophenotype** of the tumor cells is SIgM$^+$, B-cell-associated antigens (CD19$^+$, CD20$^+$, CD22$^+$), CD10$^+$, CD5$^-$, and CD23$^-$. The tumor cells are Bc16$^+$ and Ki67$^+$. **Cytogenetic analysis** shows that most cases have a translocation of *c-myc* from chromosome 8 to the Ig heavy-chain region on chromosome 14 or, less commonly, to light-chain loci on chromosomes 2 or 22. EBV genomes can be demonstrated in most African cases and in 25 to 40% of those associated with **acquired immunodeficiency syndrome** (AIDS). The lymphoma probably arises from a differentiating B-cell.

Clinical Features

There are three clinical variants:

> *Endemic Burkitt lymphoma.* This occurs in equatorial Africa, affecting children aged 4 to 7 years. The male-to-female ratio is 2 or 3 to 1. In African (endemic) cases, the jaws and other facial bones are often involved.
>
> *Sporadic Burkitt lymphoma.* Seen throughout the world in children and young adults and is associated with EBV and low economic-social status. In non-African (non-endemic) cases, jaw tumors are less common. The majority present with an abdominal mass arising from a lymphoma of the distal ileum, cecum, and/or mesentery; ovaries, kidneys, or breasts may be involved. Rare cases present as acute lymphoblastic leukemia with circulating Burkitt's tumor cells (L3-ALL).
>
> *Immunodeficiency Burkitt lymphoma.* Associated with AIDS.

Burkitt lymphoma is highly aggressive, associated with bulk disease, hyperbilirubinemia, elevated **lactic dehydrogenase** (LDH), and **tumor lysis syndrome**.

Staging

See **Lymphoproliferative disorders**.

Treatment

See **Non-Hodgkin lymphoma**.

Chemotherapy cures 90% of early-stage and 60 to 80% of advanced disease patients, managed as for **acute lymphoblastic leukemia**.

BURNS

The hematological consequences of severe burns are:

> **Fragmentation hemolytic anemia**. This is a consequence of exposure of **red blood cells** to heat greater than 47°C. Morphologic changes include the appearance of many **microspherocytes**, fragmented red blood cells, and budding (pyrospherocytes), together with an increase in osmotic fragility tests. These changes are particularly striking if the blood film is examined immediately after the burn (second or third degree) is sustained. In severe burn injury, this is accompanied by **hemoglobinemia** and **hemoglobinuria**. Acute hemolysis occurring within 24 h of the burn is a direct effect of heat upon circulating red cell membranes; hemolysis occurring later may result from infusion of isoagglutinins in pooled plasma.
>
> **Neutrophil** nuclei with radial segmentation — "botyroid nucleus" or bunch of grapes. This is due to contraction of microfilaments radiating from the centriole. **Döhle bodies** are sometimes seen.
>
> **Leukemoid reaction**.

BURR CELLS

See **Echinocytes**.

BURST-FORMING UNITS OF ERYTHROID CELLS
(BFU-E) See **Colony forming units**.

BUSULFAN
See **Alkylating agents**.

C

CALCIUM

(Ca²⁺) A metallic element essential for many reactions involved in a number of different physiological processes within a coordinated **hemostasis** response. Calcium ions are required for the normal conversion of **factor IX** to IXa, **factor X** to Xa, **prothrombin** to thrombin, as well as for normal **platelet function** and platelet aggregation/secretion, **protein C** activation, **protein S** binding, and intracellular signaling. Calcium ions are bound to vitamin K-dependent coagulation proteins via carboxylated glutamic acid residues. Proteins formed in the absence of vitamin K are dysfunctional because they cannot bind calcium ions.

Although calcium has a crucial role in hemostasis, hypocalcemia does not give rise to any form of hemorrhagic disorder. Removing calcium from blood *in vitro* by chelating agents such as **ethylenediaminetetraacetic acid** (EDTA) and sodium citrate prevents coagulation. This is used to advantage to prevent coagulation occurring in blood for transfusion and laboratory analysis. Citrate-based anticoagulants are used in the preparation of **blood transfusion components**.

CALCIUM PUMP

An **adenosine triphosphate** (ATP)-driven Ca-Mg-ATPase pump by which **calcium** is pumped out of **red blood cells** against a very strong concentration gradient. A low-cytoplasmic free-calcium level is necessary to prevent cellular and membrane damage.

CALIBRATION

See also **Automated blood cell counting; Cytometry**.
Determination of a bias conversion factor for an analytical process under specified conditions, in order to obtain true results. The accuracy over the operating range must be established by appropriate use of **reference methods, reference materials**, or calibrators.[130] Most automated multiparameter blood cell counters are linear comparators and not absolute measurement devices, and therefore require careful calibration. Instrument calibration ultimately depends on the existence of reference materials to which "true" values must be assigned. This process requires reference methods.

Calibrators are of two types:

1. Preserved blood or artificial materials
2. Fresh blood to which values have been assigned[131]

CALMODULIN

See also **Cell locomotion**.
(CALcium MODULated proteIN; CaM) The major intracellular sensor of calcium, which is a crucial intracellular mediator of cell function. The normal intracellular concentration

of free Ca^{2+} is some 10^{-4} lower than in the extracellular environment; it is released from cell stores upon activation of the cell by a range of stimuli, e.g., growth factors, and when ion channels are opened to allow influx. CaM is a dumbbell-shaped molecule with two globular domains linked by a flexible region. It binds up to four Ca^{2+} ions, two to each domain, and this binding results in CaM binding and activating other proteins. These include a bewildering range of proteins important in multiple intracellular functions (depending on the particular cell type):

Protein kinases activated by Ca^{2+} CaM:
1. The CaM kinase cascade: CaM kinase kinase (CAMKK) is first activated, and this in turn phosphorylates and activates other kinases:
   - CaM kinases I and IV, which activate transcription factors involved in cell cycle and so is a part of the process initiating cell division
   - Protein kinase B, thus inhibiting apoptosis
2. CaM kinase II, which phosphorylates synapsin, facilitating neurotransmitter release
3. Phosphorylase kinase, which activates release of glucose from glycogen
4. Myosin light-chain kinase, involved in smooth-muscle contraction

Protein phosphatases activated by Ca^{2+} CaM. These include calcineurins, which activate IL-2 synthesis in T-lymphocytes.

Other signaling molecules regulated by CaM. These include nitric oxide synthase, adenylate synthase, and inositol trisphosphate 3 kinase.

CANNISTER SYSTEMS
See **Autologous blood transfusion**.

CAPILLARY FRAGILITY TESTS
A crude method of assessing **vascular endothelium** integrity, **platelets**, adhesive proteins (such as **Von Willebrand Factor**, fibronectin), and **collagen**. It can be performed by using the tourniquet test or applying a calibrated suction cup to the forearm. Capillary fragility is usually increased in vascular **purpura** and is always increased in significant **thrombocytopenia**. In contrast, the **bleeding time** is normal in vascular purpura and increased in thrombocytopenia.

CAPILLARY HEMANGIOMAS
See **Hemangiomas**.

CARBONIC ANHYDRASE
An enzyme that catalyzes the equilibrium between carbon dioxide and carbonic acid, thus aiding oxygen and carbon dioxide transport by **red blood cells**. Inhibitors to carbonic anhydrase (e.g., acetazolamide) are sometimes used as weak diuretics. They also inhibit the formation of aqueous fluid and are therefore highly effective in controlling intraocular pressure in the treatment of glaucoma. Acetazolamide has also been used as a prophylactic for mountain sickness. All drugs of this group can cause **bone marrow hypoplasia**.

CARBOXYHEMOGLOBIN

(HbCO) The product of the attachment of carbon monoxide to **hemoglobin**, thereby replacing oxygen. Since carbon monoxide has an affinity for hemoglobin about 200 times that of oxygen, even at low concentrations there is rapid formation of HbCO. In normal individuals, less than 1% is present, whereas in heavy smokers up to 10% can exist. Less hemoglobin is therefore available for oxygen transport and, in addition, the hemoglobin-oxygen dissociation curve shifts to the left, thus oxygen release to the tissues is less effective. Anemic patients suffer carbon monoxide poisoning sooner and more severely than normal individuals. A concentration of carbon monoxide of 0.1% in the inspired air results in 50% HbCO in 1 h. A concentration of 0.4% is fatal in 1 h. Rapid and accurate estimates of HbCO concentrations greater than 5 to 10% can be made using reversion spectroscopy. Direct vision spectroscopy only detects concentrations greater than 30%.

CARDIAC BYPASS HEMOSTASIS

Changes in **hemostasis** occurring as a result of cardiac bypass during cardiac surgery, giving rise to bleeding with the loss of more than 1 l of blood. Physiologically, there is a fall in **platelets** in the circulation by 25 to 60% within the first 5 min of the procedure due to increased platelet adhesion to the artificial surfaces of the bypass machine. At the same time, there is a reduction in **Von Willebrand Factor** (VWF).

Hemorrhage may be due to:

Aspirin therapy preoperative

Coumarin therapy preoperative and uncorrected

Cardiac angioplasty with the use of **abciximab** (c7E3, ReoPro®)

Congenital heart disease with acquired Von Willebrand Factor deficiency

Hemophilia

Thrombocyte levels <$100 \times 10^9/l$

Treatment is by **platelet transfusion** of 2 UK units (equivalent to platelets $100 \times 10^9/l$). The use of **fresh-frozen plasma** (FFP) or **cryoprecipitate** is debatable.[131a] Likewise is the use of an antifibrinolytic agent, aprotinin, given prophylactically. The dose is 2×10^6 KIU as bolus to the patient, 2×10^6 KIU bolus to the blood prime, followed by 0.5 KIU/h to a total of 6×10^6 KIU.

CARDIAC DISORDERS

The association of cardiac abnormalities with hematological disorders.

Cardiac disorders as a consequence of hematological disorders:

Angina as a consequence of **anemia**

Myocardial infarction from coronary **arterial thrombosis** and **polyarteritis nodosa**

Cardiomyopathy with eventual cardiac failure due to:

- Chronic anemia and pulmonary infarction in **sickle cell disorders**
- Iron deposition as hemosiderin in **iron overload** and in **hereditary hemo-chromatosis**
- Infiltration of the myocardium in **amyloidosis, sarcoidosis, scleroderma,** and **lymphoproliferative disorders** (rare)
- **Anthracycline** treatment of hematological malignancy

Mural thromboses, endocardial fibrosis, and valvular damage with cardiomegaly and arrhythmias due to **hypereosinophilic syndrome** (Loeffler's fibroblastic parietal endocarditis) and **systemic lupus erythematosus** (Liebman-Sach's endocarditis)

Acute pericarditis, hypertension, and arrhythmias associated with polyarteritis nodosa, systemic lupus erythematosus, **rheumatoid arthritis**, **ankylosing spondylitis**, and **Epstein-Barr virus** infection

Arrhythmias following anthracycline therapy, amyloidosis, rheumatoid arthritis, sarcoidosis, and **scleroderma**

Hematological disorders due to cardiac disease:

Erythrocytosis from cyanotic hypoxia of congenital heart disease

Normocytic anemia in subacute bacterial endocarditis and atrial myxoma

Splenic hypoplasia associated with congenital heart disease

Macroangiopathic hemolytic anemia due to a prosthetic heart valve, aortic stenosis, mitral regurgitation, rupture of the sinus of Valsalva, rupture of chordae tendinae, and coarctation of the aorta or aortic aneurysm

Leukocytosis, raised **erythrocyte sedimentation rate**, **rouleaux formation of red blood cells**, and elevated **immunoglobulin** levels with atrial myxoma

Thrombocytopenia, impaired clot retraction, and low **fibrinogen** and **factor V** levels associated with congenital heart disease

Arterial thromboembolism arising from thrombi on inflamed or damaged valves, particularly with infected endocarditis, endocardium adjacent to myocardial infarction, cardiac myxoma, dilated or dyskinetic cardiac chambers, as in atrial fibrillation or on prosthetic heart valves, and low-output states with right heart failure; prophylactic **warfarin** is administered for all of these disorders

CARDIAC HEMOLYTIC ANEMIA
See **Macroangiopathic hemolytic anemia**.

CARRIÓN'S DISEASE
See **Bartonellosis**.

CASTLEMAN'S DISEASE
See **Angiofollicular lymph node hyperplasia; Osteosclerotic myeloma**.

CATALASE
A **red blood cell** enzyme that catalyzes the decomposition of hydrogen peroxide to water and oxygen. Catalase destroys hydrogen peroxide (H_2O_2), which is produced by **neutrophils** or bacteria, e.g., streptococci. A rare metabolic deficiency disorder, inherited as an autosomally recessive trait, can occur, usually in patients of Asian origin. Deficiency results in recurrent ulceration of the **oral mucosa**, which manifests itself by 10 years of age and, in its severest form, causes extreme ulceration and ultimately gangrene. Neutrophils, attracted to the site of **inflammation**, merely compound the problem.

CATION PUMP

(Sodium-potassium-ATPase pump) An **adenosine triphosphate** (ATP)-driven Na-K-ATPase pump that is responsible for pumping sodium out of and potassium into the **red blood cell** against concentration gradients.

CAVERNOUS HEMANGIOMAS

See **Hemangiomas**.

CD (CLUSTER OF DIFFERENTIATION) ANTIGENS

The set of monoclonal antibodies reacting with a particular antigen and, by extension, the antigens themselves, which are defined by their reaction with monoclonal antibodies. These CD antigens are much used in definition of leukocyte types and so are important in immunophenotyping of hematopoietic malignancies. Detailed listings of CD antigens are available.[85–87]

CELL-ADHESION MOLECULES

(CAM) Three families of adhesion proteins — the integrins, selectins, and immunoglobulin superfamily — and a few additional others not belonging to these families, e.g., CD44, mediate the major cell–cell and cell–matrix adhesive interactions. They are responsible for diverse properties, from endothelial inflammatory response and platelet-endothelial contact to the homing of primitive hematopoietic cells to an appropriate microenvironmental niche. In addition, it is now clear that these molecules form complexes with intracellular signaling molecules and so mediate functions such as proliferation, motility, and resistance to **apoptosis**.

Integrins[5,6]

These are transmembrane heterodimeric proteins consisting of noncovalently linked α and β chains that uniquely pair to form at least 20 different integrin receptors. The receptors are transmembrane structures in which the cytoplasmic domain initiates intracellular signaling involving the phosphorylation of cytoplasmic proteins and modulation of cell proliferation by activation of the ras and other pathways. The cytoplasmic portion of the integrin interacts with cytoskeletal elements (talin) that in turn influence the development of focal adhesion contacts between the cell and the extracellular matrix (ECM).

Their principal function is to mediate cell–matrix interactions, but cell–cell contact is also mediated by some members of this receptor family. Integrins are expressed on the surface of a wide variety of cells, and most cell types usually express more than one integrin. An individual integrin may also bind more than one ligand and, indeed, alternative splicing of sequences within the cytoplasmic tail may activate different signal pathways upon binding of the same ligand to the same integrin. Integrin-ligand binding is low affinity (dissociation constants between 10^{-6} and 10^{-9} mol/l), but this is compensated for by many integrin–ligand interactions between the surfaces of two cells or cell–matrix. In addition, if cells are migrating, adhesion interactions must be weak.

To date, a total of 14 different α-subunits and 8 different β-subunits have been identified. A single β-subunit combines with various α-subunits, hence the families of integrins are defined by their β-subunits. There are two main β integrin families involved in hematopoiesis, the $\beta1$ and $\beta2$ integrins. Binding specificity of the integrins is largely conferred by the particular α chain.

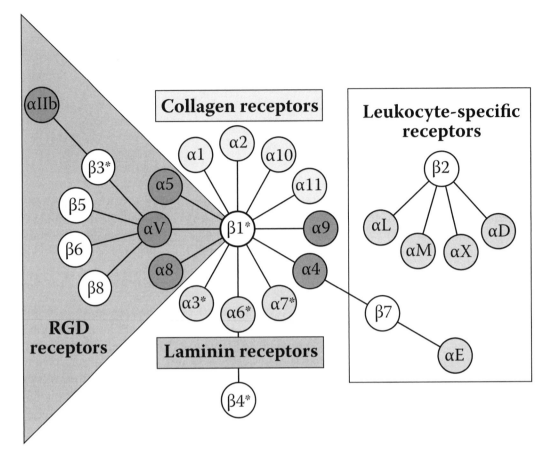

FIGURE 8
The integrin receptor family. The figure depicts the mammalian subunits and their αβ associations. (From Hynes, R.O., Integrins: bidirectionals, allosteric signalling machines, *Cell*, 110, 673–687, 2002. With permission.)

TABLE 32

The β1 (CD29, very late antigen [VLA]) Integrin Family

Associated α Chains	Alternative Nomenclature	Ligand	Leukocyte Expression
α1 (CD49a)	VLA 1	laminin, collagen	activated T-cells, monocytes
α2 (CD49b)	VLA 2 (GPIa/IIa on platelets)	laminin, collagen, fibronectin	activated B-cells, monocytes, platelets
α3 (CD49c)	VLA 3	laminin, collagen, fibronectin, thrombospondin (RGD)	B-cells
α4 (CD49d)	VLA 4	fibronectin, thrombospondin, VCAM-1 (CD106)	B-cells, thymocytes
α5 (CD49e)	VLA 5	fibronectin, fibrinogen (RGD)	memory T-cells, monocytes, platelets
α6 (CD49f)	VLA 6	laminin	memory T-cells, thymocytes, monocytes

Note: RGD = tripeptide arg gly asp; VCAM = vascular cell adhesion molecule.

β1 Integrins

See Table 32. The β1 common chain (CD29) combines with different α chains to form a variety of VLA (very late antigen) molecules that mediate the adhesion of hematopoietic cells to ECM components and ligands on stromal and endothelial cells. CD34+ cells express

TABLE 33

The β2 (CD18, leukocyte integrins) Integrin Family

Associated α Chains	Alternative Nomenclature	Ligand	Cellular Distribution
αL (CD11a)	LFA-1	ICAM-1 (CD54), ICAM-2 (CD102), ICAM-3 (CD50)	lymphocytes, granulocytes, monocytes, and macrophages
αM (CD11b)	CR3, Mac-1	ICAM-1 (CD54), complement C3b, and ECM	myeloid and natural killer cells
αX (CD11c)	CR4	fibrinogen	myeloid cells

Note: ICAM = intercellular adhesion molecule.

the integrin receptor VLA-4 (4β1) (CD49d), whose ligand is VCAM-1 (vascular adhesion molecule-1) on marrow stromal cells and fibronectin in the ECM. VCAM-1 is variably expressed on marrow stromal and endothelial cells and can be upregulated by several cytokines, including interleukin-1 (IL-1). Because VLA-4 and its ligands are widely distributed, specificity is most likely conferred by the coexpression of other adhesion molecules and can be modulated by hematopoietic cytokines. The VLA-4/VCAM-1 interaction is a critical component of the complex process of stem cell homing. Early precursors also express VLA-5 (5β1) (CD49e), which can bind to ECM fibronectin.

β2 Integrins

See Table 33. Of the three β2 integrins (CD18), the best characterized is LFA-1 (lymphocyte function antigen-1) (CD11a), which is associated with the adhesion of more mature leukocytes to the ligand ICAM1 on the endothelium, but is also found on early hematopoietic precursors. Mac-1 (CD11b), another β2 integrin, is found on mature monocytic and granulocytic cells but not on early hematopoietic precursors.

Integrins are under trial as therapeutic targets in autoimmune disease.[132]

Immunoglobulin Superfamily

These receptors also play a key role in the inflammatory response and contain from two to six immunoglobulin domains in their extracellular structure. Present on the cells of the vascular endothelium and T-lymphocytes, among others, they bind integrins (see Table 34) to promote the adhesion component of inflammation. ICAM-2 is constitutively expressed on endothelial cells, while ICAM-1 expression is induced by mediators of the **acute-phase inflammatory response**, such as tumor necrosis factor (TNF)-α, interleukin (IL)-1, and interferon (IFN)-γ. VCAM-1, expressed on endothelium, is a receptor for integrin b1a4 (VLA-4) and promotes lymphocyte/monocyte adhesion. LFA-2 (CD2) on T-cells binds a fellow family member LFA-3 (CD58) expressed on a wide variety of cells to promote T-lymphocyte adhesion.

TABLE 34

The β3 (CD61, cytoadhesin) Integrin Family

Associated α Chains	Alternative Nomenclature	Ligand	Cellular Distribution
αIIb (CD41)	platelet GPIIb	fibrinogen, fibronectin, Von Willebrand Factor, and thrombospondin	platelets, megakaryocytes, endothelium
αV (CD51)	vitronectin receptor	vitronectin, Von Willebrand Factor, fibrinogen, and thrombospondin (RGD)	platelets, megakaryocytes, endothelium

TABLE 35

The Selectin Family

Selectin	Alternative Nomenclature	Ligand	Cellular Distribution
E-selectin (CD62E)	endothelium leukocyte adhesion molecule (ELAM)	sialyl-Lewis-X antigen	activated endothelium
L-selectin (CD62L)	leukocyte adhesion molecule (LAM, LECAM)	GlyCAM (CD34), sialyl-Lewis-X antigen on endothelium	leukocytes
P-selectin (CD62P)	PADGEM	CD162 on leukocytes	platelets, activated endothelium

Selectins CD62

Currently composed of three members, each with a degree of tissue specificity, this family of lectins binds carbohydrate counterreceptors with Ca^{2+} dependence (see Table 35). L-selectin is found on lymphocytes, neutrophils, and monocytes and allows homing of lymphocytes to endothelial venules to mediate lymphocyte trafficking and neutrophil-endothelial adherence as part of inflammation. P-selectin is found on platelets and endothelial cells binding the sialated Lewis-X receptor on neutrophils, monocytes, and tumor cells. E-selectin also mediates neutrophil-endothelial binding via the sialated Lewis-X receptor. A form of leukocyte adhesion deficiency (LAD) has been described with deficient sialated Lewis-X receptor structure.

E- and P-selectins on cytokine-activated endothelium mediate binding of activated leukocytes, the first step in extravasation. The L-selectin on leukocytes plays a similar role.

CD44 (Hermes antigen, Pgp1)

This includes a number of isoforms generated by alternative splicing of a common gene, expressed widely by leukocytes. The ligands are principally hyaluronic acid and other components of the extracellular matrix. The most abundant isoform, CD44, is expressed by leukocytes, and CD44s and variants are upregulated upon activation. CD44s and the variants are particularly involved in migration of lymphocytes and **dendritic reticulum cells**.

CELL CYCLE

A descriptive concept compartmentalizing a continuum of processes leading to and including cell division and differentiation. The phases of the cell cycle are:

G0 phase, a period in the **cell cycle** where cells exist in a quiescent state

G1 phase, the first growth phase

S phase, during which the **DNA** is replicated, where S stands for the synthesis of DNA

G2 phase, the second growth phase, also the preparation phase for the cell

M phase or **mitosis** and cytokinesis, the actual division of the cell into two daughter cells

A surveillance system, so-called checkpoints, monitors the cell for DNA damage and failure to perform critical processes. Checkpoints can block progression through the phases of the cell cycle if certain conditions are not met (see **Checkpoint kinases**). Similarly, the spindle checkpoint blocks the transition from metaphase to anaphase within mitosis if not all chromosomes are attached to the mitotic spindle (see Figure 9).

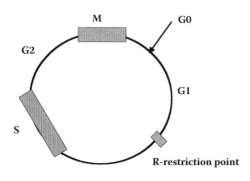

FIGURE 9

Phases of the cell cycle: restriction point. Cells will proceed to the S phase only when all steps of G1 have been completed. The restriction point R represents the time up to which growth factors are required for cell cycle progression. After this point, G1 progression is growth-factor independent, i.e., the program will continue in an autonomous manner.

If this system senses a problem, a network of signaling molecules instructs the cell to stop dividing. They can let the cell know whether to repair the damage or initiate **apoptosis** via p53. Programmed cell death ensures that the damaged cell is not further propagated.

In prokaryotic cells, continuation of cycle progression is determined by advance through a variety of checkpoints, initially the G1 "start" checkpoint. No eukaryotic "start" checkpoint has been identified, but G1 progression events are now clearer. When a cell enters gap 1 (G1), it has recently undergone mitosis. It starts to grow and perform different, specific metabolic activities. The 23 pairs of chromosomes or diploid (somatic) number of chromosomes are uncoiled, and gene transcripts (mRNA) are made from them. This is then translated into proteins to maintain cell functions. This stage can either last up to 11 h or continue throughout the cell's life span if the cell does not divide again. Cells only divide if necessary; for example, mature lymphocytes do not divide unless they are stimulated.

DNA synthesis (S phase) lasts about 8 h and is the phase in which DNA is replicated. Now each chromosome consists of two sister chromatids, each incorporating a full copy of the DNA double helix. G1 chromosomes consist of a single chromatid; after S phase, they duplicate themselves and have sister chromatids with identical DNA content. At this point, the cell has 46 pairs of chromosomes. This replication will allow daughter cells to have the same genetic material and information as the mother cell. G2 and mitosis will be delayed or aborted if DNA duplication is not complete or if important errors during the S phase have occurred (see Figure 10 and Figure 11).

During gap 2 (G2), gene transcription and translation occurs rapidly (see Figure 12). Because many proteins are needed for mitosis, the cytoplasm grows and cellular organelles duplicate. The cell is ready to divide and enters mitosis (M) (see Figure 13).

After mitosis, cells enter G1 and prepare for another cell division, but cells that do not divide enter a different stage, called G0, remaining metabolically active and waiting for signals to divide. For example, fibroblasts remain in G0 until a platelet growth factor activates them to start the healing process after a person is wounded. Nerve cells and lens cells can remain in G0 forever.

There are three key parts of the cell cycle:

1. The mechanisms that produce regular fluctuations in the levels of proteins (cyclins) and the activity of their associated kinases (cyclin-dependent kinases, Cdks); these fluctuating proteins enable the cell to go through the different cell cycle phases:

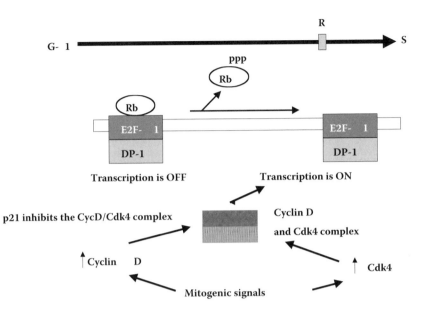

FIGURE 10
Phases of the cell cycle: early G1-S transition.

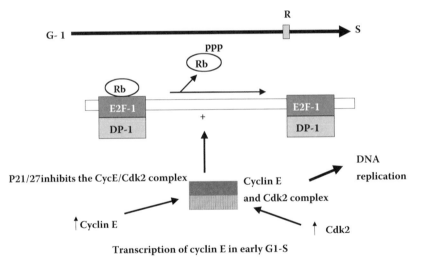

FIGURE 11
Phases of the cell cycle: late G1-S transition.

- Different cyclins are expressed during different phases of the cell cycle
  G1 cyclin: cyclin D
  S-phase cyclins: cyclins E and A
  mitotic cyclins: cyclins B and A
  other cyclins — C, F, G, H, and T — have been described
- Cyclin-dependent **kinases** (Cdks)
  G1 Cdk: Cdk4
  S-phase Cdk: Cdk2
  M-phase Cdk: Cdk1

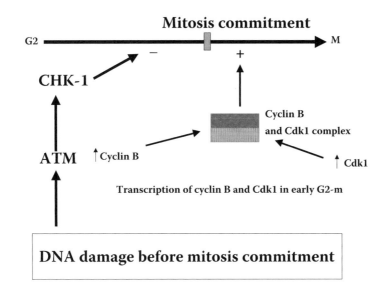

Mitosis commitment

G2 ———————————————————→ M

CHK-1

ATM ↑Cyclin B

Cyclin B
and Cdk1 complex

↑ Cdk1

Transcription of cyclin B and Cdk1 in early G2-m

DNA damage before mitosis commitment

FIGURE 12
Phases of the cell cycle: G2-M transition.

To be catalytically competent, the catalytic subunit (Cdks 1–9) and a regulatory subunit (cyclins A–H and cyclin T) must combine and be in the correct phosphorylation state. The Cdks integrate signals from the cell and its environment to either initiate or inhibit cell-cycle progression

2. The downstream events such as (a) cell-cycle arrest at particular checkpoints, which allows repair of damaged DNA, thereby preventing DNA mutation or chromosomal misalignment and consequent cell death and (b) mitosis (or **meiosis**), enabling correct numbers of chromosomes to be segregated to daughter cells

3. Signaling pathways that regulate the cell-cycle machinery in response to events outside and inside cells

A great deal of our understanding of the cell-cycle mechanisms in humans has been derived from studying hematopoietic cells, in particular **lymphocytes**. Cells in the G0 phase (quiescence) have lower rates of transcription, translation, and metabolism and reduced cell size. Quiescence of lymphocytes acts to reduce the resources (energy and space) required to maintain a vast repertoire of T and B-cells, only a small fraction of which will be clonally selected by antigen during the lifetime of the host. Quiescence might also protect cells from accumulating metabolic damage as well as genetic changes that could result in malignancy. Quiescent cells (phase G0) enter the cell cycle in response to exogenous factors such as **cytokines** and interactions with stromal cells and the extracellular matrix (ECM) in the **bone marrow** (BM) microenvironment. In addition, intrinsic transcription factors expressed in quiescent cells, including c-Myb, GATA-2, HOX family proteins, and Bmi-1, also control their growth through their effect on gene transcription. In quiescent hematopoietic stem cells, the cyclin inhibitors, p21WAF1 (p21) and p27KIP1 (p27), maintain the quiescence of HSCs and of progenitor cells, respectively, thereby governing the available pool sizes. There is also emerging evidence that the transition from the G0 state to G1 requires the cyclin inhibitors p16INK4A (p16) and p15INK4B (p15).

The core mechanism for progression from G1 to S phase, which is commonly altered in lymphoid malignancies, is referred to as the pRb pathway. Briefly, D-type cyclins synthesized in response to mitogenic signals form active complexes with Cdk4 or Cdk6, leading

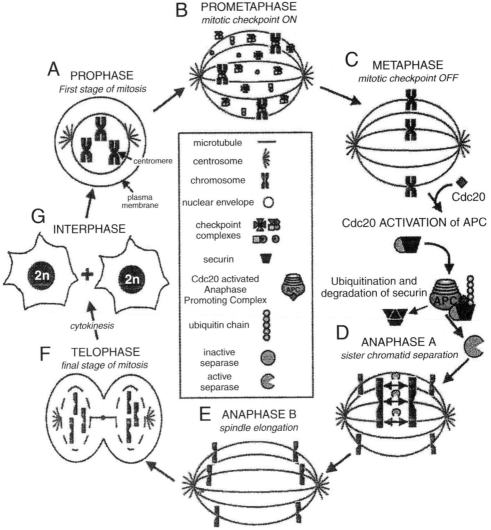

FIGURE 13

The stages of mitosis. Chromosomes enter mitosis as pairs of replicated sister chromatids that are linked by proteins known as cohesins.

A: The chromatids condense during prophase and are released into the cytoplasm by nuclear envelope breakdown, which marks the transition into prometaphase and also represents the first irreversible transition into mitosis.

B: During prometaphase, the initially unattached chromatids make connections to the microtubules of the mitotic spindle and the mitotic checkpoint is active, which means that the kinetochores assembled at the centromeres of unattached chromosomes generate a different "wait anaphase" inhibitor. Antimitotic drugs delay cells in prometaphase by producing unattached kinetochores.

C: At metaphase, every chromosome has made proper attachments to the mitotic spindle and has congressed to a central position. Production of the diffusible "wait anaphase" inhibitor has been silenced by stable kineto-chore-microtubule interactions. As the checkpoint inhibitors decay, the anaphase-promoting complex (APC), on the E3 ubiquitin ligase, becomes active and recognizes securin and cyclin B, provoking their degradation.

D: Loss of securin activates protease, separase, that cleaves the cohesins, triggering sister chromatid separation and chromosome segregation during anaphase A.

E: At anaphase B, the spindle elongates.

F: At telophase, the now-segregated chromosomes begin decondensing, and the nuclear envelopes re-form.

G: Cytokinesis separates the nuclei into two daughter cells that reenter interphase.

(From Beth, A.A., Weaver, J., and Cleveland, D.W., Decoding the links between mitosis, cancer and chemotherapy, *Cancer Cell*, 8, 7–12, 2005. With permission.)

to phosphorylation and inactivation of pRb. This in turn degrades complexes of pRb with members of the E2F family of transcription factors and associated chromatin-modifying enzymes, allowing transcription of genes required for S phase. One of these target genes encodes cyclin E1, a major G1 partner of Cdk2, which can both further phosphorylate pRb and also phosphorylate substrates required for proper DNA replication, centrosome duplication, and histone biosynthesis (see Figure 10 and Figure 11).

As a mammalian cell transits from the G1 phase of the cell cycle to the S phase, dramatic molecular changes occur as the cell begins the task of duplicating each of the 3.2 billion base pairs that make up its DNA. Replication factories scattered throughout the genome are activated and begin unwinding the double helix and building new DNA molecules at a rate of 500 nucleotides per minute, with an error rate of one nucleotide in a billion. In prokaryotes, replication begins from a single site and continues until it terminates at the end of the genome. If eukaryotes were to use an identical strategy, however, it would take on the order of several days to completely replicate its significantly larger genome. Therefore, eukaryotic cells initiate replication from multiple locations referred to as replication origins throughout each chromosome. Research in the past two decades has identified an ensemble of more than 20 proteins involved in the process of replication initiation, illustrating the complexity involved in coordinating initiation from hundreds or thousands of origins. The replication process begins with the ordered assembly of a multiprotein complex called the prereplicative complex (pre-RC). Pre-RCs are assembled on DNA at origins in a highly ordered and regulated fashion prior to S phase. During S phase, pre-RCs initiate replication by promoting origin unwinding and facilitating recruitment of the replicative DNA polymerases. The DNA unwinding is facilitated by helicase proteins called MCMs, and this process is stimulated by the cyclin E/Cdk 2 complex and a protein called Cdc6. This process is switched off by the cyclin A/Cdk1 complex and by a protein called geminin that enables the transition of the cell to the G2 and M phases.

The rapid activation of cyclin B/Cdk1 during the terminal phase of G2 commits the cell to mitosis, ultimately leading to the breakdown of the nuclear envelope. The G2 transition is an important checkpoint for DNA repair prior to mitosis. If DNA damage is induced before this commitment point, the extrinsic ATM kinase-mediated DNA damage checkpoint is triggered. Other related G2 pathways, including one mediated by the ATR kinase, can also be activated (see **Checkpoint kinases**). The ATM pathway acts throughout G2 on a range of cellular targets, even during the early stages of chromosome condensation. The principal targets of the transduction cascade that signal DNA damage in late G2, and which prevent progression into mitosis, are factors like cyclin B and Cdc25 that regulate the Cdk1 kinase. It is important to emphasize, however, that there are other, downstream kinases, including for example **aurora kinase**, polo, and cyclin A/Cdk, whose activation is also required for mitosis and that can also serve as substrates for DNA damage signaling. When activated early in G2, ATM can delay cell-cycle progression via a complex series of transcriptionally mediated responses involving p53 and the CDK inhibitor, p21. When activated during the terminal phase of G2, ATM prevents cyclin B/Cdk1 activation by inhibiting the Cdc25 phosphatase. ATM is mutated in aggressive forms of **chronic lymphatic leukemia** and may be a primary event in patients with the 11q23 deletion.

Mitosis (see Figure 13) is nuclear division plus cytokinesis, and this produces two identical daughter cells during prophase, prometaphase, metaphase, anaphase, and telophase. Interphase is often included in discussions of mitosis, but interphase is technically not part of mitosis, but rather encompasses stages G1, S, and G2 of the cell cycle.

In prophase, the chromatin in the nucleus begins to condense and becomes visible in the light microscope as **chromosomes**. The nucleolus disappears. Centrioles begin moving to opposite ends of the cell and fibers extend from the **centromeres**. Some fibers cross the cell to form the mitotic spindle.

In prometaphase, the nuclear membrane dissolves, marking the beginning of prometaphase. Proteins attach to the centromeres, creating the kinetochores. Microtubules attach at the kinetochores and the chromosomes begin moving.

In metaphase, the spindle fibers align the chromosomes along the middle of the cell nucleus. This line is referred to as the metaphase plate. This organization helps to ensure that in the next phase, when the chromosomes are separated, each new nucleus will receive one copy of each chromosome.

In anaphase, the paired chromosomes separate at the kinetochores and move to opposite sides of the cell. Motion results from a combination of kinetochore movement along the spindle microtubules and through the physical interaction of polar microtubules.

In telophase, the chromatids arrive at opposite poles of the cell, and new membranes form around the daughter nuclei. The chromosomes disperse and are no longer visible under the light microscope. The spindle fibers disperse, and cytokinesis or the partitioning of the cell may also begin during this stage. We now have a great deal of understanding of the molecular mechanisms associated with mitosis, so that proteins such as aurora kinases, polo kinases, and eg-5 are now potential targets for antimitotic therapy.

CELL-MEDIATED IMMUNITY

(CMI) The arm of the immune response mediated by cellular components of the immune system, in contrast to the humoral immunity, which is mediated by antibodies (**immuno-globulins**). The essential experimental definition of a cell-mediated response is that it should be transferable to a näive (nonimmune) individual by washed cells from an immune individual. CMI responses include delayed-type **hypersensitivity**, graft rejection, and immunity to viruses, and they are mediated largely by **cellular cytotoxicity** and the production of **cytokines**. The effector cells involved include (a) T-cells, which function in an antigen-specific manner, with cytotoxic T-cells directly lysing target cells and helper T-cells releasing cytokines and (b) other cell types, such as macrophages and natural killer cells, which, upon activation by cytokines released by antigen-activated T-cells, carry out a range of effector functions in a nonspecific manner. It should be noted that although cells are, by definition, essential for CMI, they may function via soluble mediators (cytokines), but these are not specific and so are clearly distinct from the antigen-specific mediator of the humoral response, which is antibody.

CELL-SIGNAL TRANSDUCTION

The biological and biochemical consequences after a ligand has undergone binding (usually at high affinity) with its cognate receptor. Transmission of messages from cell-membrane receptors results in an activation of a cascade of molecules, which activate nuclear genes by the binding of the final DNA-binding proteins (or transcription factors) to promoter regions of target genes within the nucleus. At the same time, pathways that inhibit and act to dampen the activating signal are also brought into play. There are a large number of cell-surface receptors but only a limited number of signaling pathways, with even more-limited functional readouts, which include: cellular proliferation, inhibition or activation of **apoptosis,** cellular differentiation, cellular adhesion and motility, and initiation of specialist functions such as **platelet** degranulation. Individual families of membrane-associated receptors are found to utilize more than one signal-transduction pathway. The combination of genes activated thus promotes mitogenesis, differentiation, or apoptosis, although little is yet known about which specific gene combinations are responsible for which function within any cellular system.

The bewildering number of receptors, receptor families, shared signal-transduction pathways, and uncertainty about which particular pathways give rise to which functional response makes this a challenging area. However, there are a number of common themes emerging.

Process of Signal Transduction

This occurs as a cascade of events via secondary messengers that include the *GTPase* (guanosine triphosphatase*) switch proteins, **tyrosine kinases***, and adaptor proteins.* A large group of GTP (guanosine 5′-triphosphate)-binding proteins act as molecular switches in signal transduction pathways. These proteins are turned "on" when bound to GTP and turned "off" when bound to GDP (Figure 14). In the absence of a signal, the protein is bound to GDP. Signals activate the release of GDP, and the subsequent binding to GTP over GDP is favored by the higher concentrations of GTP in the cell. The intrinsic GTPase activity of these GTP-binding proteins hydrolyzes the bound GTP to GDP and Pi, thus converting the active form back to the inactive form. The kinetics of hydrolysis regulates the length of time the switch is "on." There are two classes of GTPase switch proteins:

Trimeric G proteins, which as noted already are directly coupled to certain receptors

Monomeric Ras and Ras-like proteins

Both classes contain regions that promote the activity of specific effector proteins by direct protein-protein interactions. These regions are in their active conformation only when the switch protein is bound to GTP. G proteins are coupled directly to activated receptors, whereas Ras is linked only indirectly via other proteins. The two classes of GTP-binding proteins also are regulated in very different ways.

Activation of all cell-surface receptors leads to changes in protein phosphorylation through the activation of protein kinases (Figure 14). In some cases, kinases are part of the receptor itself, and in others they are found in the cytosol or associated with the plasma membrane. Animal cells contain two types of protein kinases: those directed toward tyrosine and those directed toward either serine or threonine. The structures of the catalytic

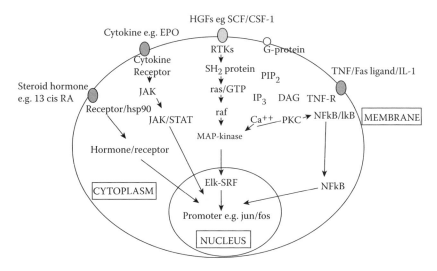

FIGURE 14
Intracellular signal transduction pathways.

core of both types are very similar. In general, protein kinases become active in response to the stimulation of signaling pathways. The catalytic activities of kinases are modulated by phosphorylation, by direct binding to other proteins, and by changes in the levels of various second messengers. The activity of protein kinases is opposed by the activity of protein phosphatases, which remove phosphate groups from specific substrate proteins. Many signal-transduction pathways contain large multiprotein signaling complexes, which often are held together by adapter proteins. Adapter proteins do not have catalytic activity, nor do they directly activate effector proteins. Rather, they contain different combinations of domains, which function as docking sites for other proteins. For instance, different domains bind to phosphotyrosine residues (SH2 and PTB domains), proline-rich sequences (SH3 and WW domains), phosphoinositides (PH domains), and unique C-terminal sequences with a C-terminal hydrophobic residue (PDZ domains). In some cases, adapter proteins contain arrays of a single binding domain or different combinations of domains. In addition, these binding domains can be found alone or in various combinations in proteins containing catalytic domains. These combinations provide enormous potential for complex interplay and cross talk between different signaling pathways.

Specificity of Signal Responses

The importance of mechanisms of signal specificity has been highlighted by the investigation of how cytokines play pivotal roles in the regulation of various biological responses, such as **immune responses**, **inflammation**, and **hematopoiesis**. The combination of cytokines and their effective concentrations at specific sites can differ, and the action of a given cytokine on a given tissue or cell type can be affected by the microenvironment of the target cell. Microenvironments can reflect different physiological conditions, e.g., a steady, stressed, inflamed, or tumor-bearing state. For example, a typical pleiotropic cytokine, **interleukin** IL-6, induces B-**lymphocyte** differentiation into antibody-producing **plasma cells**, T-lymphocyte growth and differentiation, differentiation of the **acute myeloid leukemia** cell line M1 into **histiocytes** (macrophages), and the neural differentiation of PC12 cells. The mechanisms by which a single cytokine can exert functional **pleiotropy** and how the cell specificity of cytokine action is determined include: expression patterns of receptors and subunits, including secondary signaling molecules; restriction by the spatial compartmentation of proteins that could interact with receptors; expression of different isoforms of the same receptor family; and the duration and intensity of the ligand stimulus.

Redundancy for Some Signaling Pathways

Overlapping actions for different cytokines and redundancy can be explained by similar cellular distributions of specific receptors for different cytokines as well as by the sharing of signaling pathways, which particularly occurs when different receptors share similar motifs that mediate the coupling to the same pathways. In addition, redundancy can be at least partially explained by the ability of certain cytokines to signal via more than one type of receptor complex and by the sharing of an individual receptor component by more than one cytokine.

Cross Talk Occurs between Different Signaling Pathways

One definition of cross talk is: the process by which one receptor complex and its downstream effector molecules can influence the effect of another receptor signaling response. This influence can be additive, synergistic, or inhibitory. As evidence of cross talk at the receptor level, an interaction between EPO receptor and c-kit has been described. In nonhematopoietic cells, IL-6 induces the formation of a complex between gp130 and ErbB2, the activation of ErbB2 and MAP kinase, and the growth of prostate carcinoma cells, indicating that the IL-6 receptor can recruit different signal transducers in different cells to generate varied responses.

Beyond the level of the receptors, many signaling molecules, such as kinases, phosphatases, adaptor molecules, and transcription factors, are commonly used in the signaling pathways of distinct cytokines. Thus, these signaling molecules could also be points of cross talk.

Kinase "Addiction"

This may account for dependence on specific signaling pathways in malignant transformation. Cancer cells are often "addicted to" (that is, physiologically dependent on) the continued activity of specific activated or overexpressed oncogenes and signaling pathways for maintenance of their malignant phenotype. Transgenic mice expressing the *myc* oncogene in hematopoietic cells develop **T-cell chronic lymphatic leukemia** and **acute myeloid leukemia** (AML). However, when this gene was switched off, the leukemic cells underwent proliferative arrest, differentiation, and apoptosis. Transgenic mice expressing an inducible form of the *H-ras* oncogene readily developed melanomas; when the *ras* gene is switched off, the melanomas rapidly undergo apoptosis and regress. Clinically, this is relevant to the clinical activity of small-molecule kinase inhibitors such as **Imatinib**. Upregulation of *abl* kinase activity in **chronic myelogenous leukemia** (CML) renders these cells exquisitely sensitive to kinase inhibition. Therefore, the cause of the addiction is also the Achilles heel by which the tumor might be specifically targeted by therapy. Other targets potentially are FLT3 activity in AML, Notch-1 in T-ALL, and JAK-2 in **myeloproliferative disorders**.

Specific Signaling Pathways

There are specific intracellular pathways associated with cell survival, cell proliferation, differentiation, cell-cell contact, and cellular motility. There are also important pathways that are relevant to hematological development and may be upregulated in certain leukemias. There are many signaling pathways; the main ones relevant to hematology are:

Ras/Raf/MEK/ERK pathway

PI3K/Akt pathway

JAK/STAT pathway

Wnt pathway

Notch pathway

Ras̃/Raf̃/MEK̃/ERK Pathway

This is a central signal-transduction pathway that transmits signals from multiple cell-surface receptors to transcription factors in the nucleus. This pathway is frequently referred to as the MAP kinase pathway, as MAPK stands for mitogen-activated protein kinase, indicating that this pathway can be stimulated by mitogens, cytokines, and growth factors. This pathway can be activated by Ras protein stimulating the membrane translocation of Raf. This pathway also interacts with many different signal-transduction pathways, including PI3K/Akt and JAK/STAT.

Ras is a small GTP-binding protein that is the common upstream molecule of several signaling pathways, including Raf/MEK/ERK and PI3K/Akt. The switch regions of the Ras proteins are in part responsible for the switch between the inactive and active states of the protein. Switching between these states has been associated with conformational

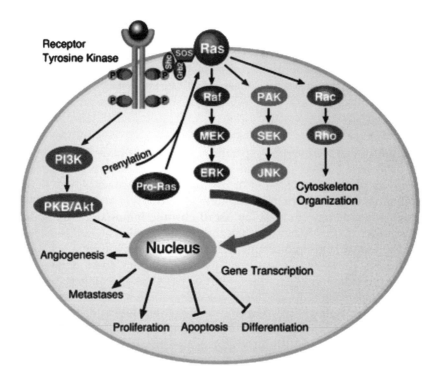

FIGURE 15

Ras-mediated signal transduction pathways. Abbreviations: PKB/Akt, protein kinase B; ERK, extracellular signal-regulated kinase; Grb2, growth factor receptor binding protein; JNK, c-JUN amino-terminal kinase; MEK, mitogen-activated protein kinase; P, phosphate; PAK, p21-activated kinase or JNK kinase; p13K, phosphotidylin-ositol 3-kinase; Shc, Src homology domain cytosol; SEK, stress-activated protein kinase; SOS, son-of-sevenless exchange factor. (From Beeram, M., Patnaik, A., and Rowinsky, E.K., Raf: a strategic target for therapeutic development against cancer, *J. Clin. Oncol.*, 23, 6771–6790, 2005. With permission.)

changes in the switch regions, which allows the binding of GTPase-activating proteins (GAPs, of which the tumor suppressor gene NF-1 is a member) and guanine nucleotide exchange factors (GEFs). When Ras is active, GTP is bound, whereas when Ras is inactive, GDP is bound. GTPases inactivate the Ras proteins, while GEFs activate the Ras proteins by either stimulating the removal of phosphate from GTP or addition of GTP (Figure 15).

RAS proteins require several posttranslational modifications (e.g., prenylation, proteol-ysis, carboxymethylation, and palmitoylation) for membrane binding and full biological activity. Protein prenylation is catalyzed by prenyl transferases that covalently attach either a 15-carbon farnesyl moiety (by farnesyl transferase, FTase) or a 20-carbon geranylgeranyl moiety (by geranylgeranyl transferase I, GGTase I) to carboxyl terminal cysteines. The preferred recognition motif for FTase and GGTase I is a carboxyl-terminal CAAX (where C = cysteine, A = aliphatic amino acid, X = any amino acid). Although the X position of the CAAX motifs determines whether a protein is a substrate for FTase (X = methionine, serine, cysteine, and glutamine) or GGTase I (X = leucine or isoleucine), these two enzymes have some degree of cross specificity.

There are three different Ras family members: Ha-Ras, Ki-Ras, and N-Ras. The Ras proteins show varying abilities to activate the Raf/MEK/ERK and PI3K/Akt cascades, as Ki-Ras has been associated with Raf/MEK/ERK, while Ha-Ras is associated with PI3K/Akt activation. Amplification of Ras proto-oncogenes and activating mutations that lead to the expression of constitutively active Ras proteins are observed in approximately 30% of all human cancers. Different mutation frequencies have been observed between Ras genes in human cancer. Oncogenic *RAS* mutations, which occur with a frequency of approximately 30 to 40% in AML and MDS, are known to provide the signals for enhanced cell proliferation and survival. N-RAS mutations are the most common, occurring in 20 to 25% of AML patients, with K-RAS mutations being found in 10 to 15% of cases. These activating mutations usually involve single amino acid substitutions at codons 12, 13, and 61, which abrogate intrinsic RAS GTPase activity and lead to constitutive RAS activation. In addition to FGFR3, c-MAF, MUM1, Cyclin D1, p53, and Rb, deregulation of RAS signal transduction either directly, via activating mutations, or indirectly, through mutations of other oncogenes or tumor suppressor genes, has been implicated as the most common single lesion reported in multiple myeloma (MM). Reported incidences of RAS mutations in MM vary from 39 to 55% at presentation to 80% at relapse, implying a role for RAS in disease progression. Due to the prevalence of Ras mutations in **acute myeloid leukemia** (AML)/**myelodysplasia** (MDS) and **myeloma**, **farnesyl transferase** inhibitors have been included in clinical trials. Interestingly, the presence of an activating Ras mutation has not predicted response to therapy with fornesyl transferase inhibitors (FTIs) in MDS and AML. There are, however, some clinical responses, and the precise role of these agents requires further study.

MEK *Genes (Members of the Extracellular Signal-Regulated Kinase Family)*

MEK proteins are the primary downstream targets of Ras and Raf. The *MEK* family of genes consists of five genes: *MEK1*, *MEK2*, *MEK3*, *MEK4*, and *MEK5*. This family of dual-specificity kinases has both S/T and Y kinase activity. The structure of *MEK* consists of an amino-terminal negative regulatory domain and a carboxy-terminal MAP kinase-binding domain, which is necessary for binding and activation of ERKs. Deletion of the regulatory *MEK1* domain results in constitutive *MEK1* and ERK activation.

Studies with this construct demonstrated that activated *MEK1* could abrogate cytokine dependency of certain hematopoietic cells. Pharmaceutical companies are developing inhibitors of *MEK*, and trials are underway in some hematological cancers.

Extracellular Signal-Regulated Kinases

The main physiological substrates of *MEK* are the members of the extracellular signal-regulated kinase (ERK) or of the mitogen-activated protein kinase (MAPK) family of genes. The ERK family consists of four distinct groups of kinases: ERK, Jun amino-terminal kinases (JNK1/2/3), p38MAPK, and ERK5. In addition, there are ERK3, ERK4, ERK6, ERK7, and ERK8 kinases, which, while related to ERK1 and ERK2, have different modes of activation, and their biochemical roles are not as well characterized. The ERK1 and ERK2 proteins are the most studied with regard to Raf signaling in hematopoietic cells.

Downstream targets of ERK include the p90Rsk kinase and the CREB, c-Myc, and other transcription factors. ERK and p90Rsk can enter the nucleus to phosphorylate transcription factors, which can lead to their activation.

PI3K/Akt Pathway

Activation of the cytokine receptors such as IL-3 stimulates the PI3K/Akt pathway. PI3K is a family of proteins that catalyze transfer of ATP to the D3 position of phosphoinositides. PI3K is a multisubunit protein consisting of an approximately 110-kDa catalytic subunit and a regulatory subunit of 50 to 100 kDa. PI3K activity is associated with cytoskeletal organization, cell division, inhibition of **apoptosis**, and glucose uptake.

Activation of class I PI3K requires translocation to the plasma membrane and association with an activated receptor tyrosine kinase or its substrates. The localization of PI3K to the membrane is activated by the IL-3 receptor, Ras, or the proline-rich regions of Shc, Lyn,

Fyn, Grb2, v-Src, Abl, Lck, Cbl, or dynamin. These interactions activate PI3K, producing phosphoinositide-3-phosphate and/or phosphoinositide-3,4,5-triphosphate. The phosphoinositide-3-phosphate product is degraded by phosphatases, including the PTEN tumor suppressor. The kinase activity of PI3K is inhibited by cAMP. The phospholipid products of PI3K activate downstream targets, including PDK, Akt, and PKC. PI3K inhibitors that show specificity have been developed. These and their derivative inhibitors show promise in suppressing this pathway in human cancer.

Akt *Oncogenes*

The downstream target of PDK1 and likely PDK2 is Akt or protein kinase B (PKB). Activated Akt was originally isolated from cells of the leukemia- and lymphoma-prone AKR strain of mice. Multiple clonings of the gene resulted in several different names for the Akt family of genes, including Akt, PKB, and RAC-PK (related to A and C protein kinase). The Akt family of genes in mammals consists of three genes: Akt1, Akt2, and Akt3. Akt1 has been mapped to chromosome 14q32, a region frequently involved in translocations in leukemia and lymphoma. Akt2 has been mapped to 19q13.1-q13.2, a chromosome region known to be amplified in ovarian and pancreatic cancers. Akt2 is amplified in 12.1% of ovarian and 2.8% of breast cancers. Akt1 and Akt2 are expressed in all tissues, but the highest expression is in brain, thymus, heart, and lung. Akt3 expression is high in brain and testes.

The most studied biological activities of Akt are the regulation of glucose transport metabolism and apoptosis. PI3K is required for glucose transport. Moreover, inhibition of PI3K decreases glucose transport. Elevated Akt activity increases glucose transport. Akt phosphorylates glycogen synthase kinase-3 (GSK-3) and 6-phosphofructose-2-kinase (PFK2) *in vitro*, thus regulating glucose flux to glycogen and glycolysis.

The antiapoptotic effects of Akt occur through its phosphorylation of a wide variety of targets. The first antiapoptotic target identified was Bad, a member of the Bcl-2 family. Phosphorylation of Bad by Akt allows phosphorylated Bad to interact with 14-3-3 proteins, promoting cell survival.

Regulation of the PI3K Pathway by Phosphatases

The PI3K pathway is negatively regulated by phosphatases. PTEN (phosphatase and tensin homolog deleted on chromosome 10, also known as MMAC1, mutated in multiple advanced cancers) is considered a tumor suppressor gene. Ras can activate PI3K, and some Ras mutations result in deregulated PI3K and downstream Akt activation. PTEN negatively regulates Akt activity; hence mutations that result in PTEN loss may lead to persistently increased levels of Akt activity. Mutations and hemizygous deletions of PTEN have been detected in some primary forms of **acute leukemia** and **non-Hodgkin lymphoma**. Increased Akt expression has also been linked with tumor progression; the Akt-related Akt-2 gene is amplified in some cervical, ovarian, and pancreatic cancers as well as in non-Hodgkin lymphoma. In cancers where there are mutations in the PI3K/Akt/PTEN pathway, there are usually mutations at PTEN or Akt but not both. A diagram of the effects of the PTEN and SHIP phosphatases on the regulation of PI3K/Akt pathway is given in Figure 15.

JAK/STAT Pathway

This pathway consists of three families of genes: the JAK, or Janus family of tyrosine kinases, the STAT (signal transducers and activators of transcription) family, and the CIS/SOCS family, which serves to downregulate the activity of the JAK/STAT pathway. The JAK/STAT pathway involves signaling from the cytokine receptor to the nucleus. JAKs

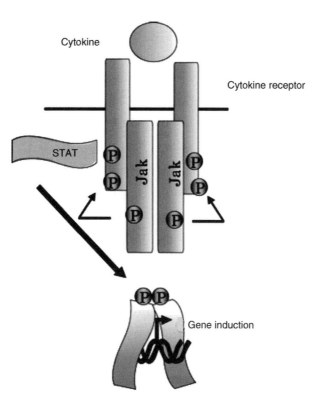

FIGURE 16
Role of Janus kinases (JAKs) and signal transducers and activators of transcription (STATs) in cytokine signaling.
Type I and Type II cytokine receptors associate with JAKs. Cytokine binding activates the JAKs, which phosphorylate the receptors, allowing the recruitment of STATs. STATs are phosphorylated and accumulate in the
nucleus, where they regulate gene expression. (From O'Shea, J.J., Targeting the JAK/STAT pathway for immunosuppression, *Ann. Rheum. Dis.*, 63, Suppl. ii, 67–71, 2004. With permission.)

are stimulated by activation of a cytokine receptor. Stimulation of JAKs results in STAT
transcription factor activity.

JAKs are a family of large tyrosine kinases, having molecular weights in the range of
120 to 140 kDa (1130 to 1142 aa). Four JAKs (JAK1, JAK2, JAK3, and Tyk2) have been
identified in mammals. JAK proteins consist of seven different conserved domains (JH1-
JH7 [JAK homology]). The JH1 constitutes a kinase domain, while JH2 is a pseudokinase
domain. Many possible roles have been proposed for the different domains of the JAK
proteins: (1) JH2 inhibits JH1; (2) JH2 promotes STAT binding; (3) JH2 is required for
kinase activity of JH1; and (4) JH6 and JH7 are necessary for association of JAKs with
cytokine receptors. A diagram illustrating the key features of the JAK, STAT, and SOCS
(suppressors of cytokine signaling) proteins is presented in Figure 16.

Loss of JAK1 produces prenatal lethality due to neurological disorders, while JAK2 results
in embryonic lethality due to defects in **erythropoiesis**. JAK3 expression is limited to hematopoietic cells, and JAK3 knockout mice have developmental defects in lymphoid cells.

Aggregation of cytokine receptors following activation allows formation of receptor
homodimers and heterodimers. Receptor aggregation allows transphosphorylation of
receptors and activation of the associated JAKs and STATs.

The STAT gene family consists of seven proteins (STAT1, STAT2, STAT3, STAT4, STAT5a,
STAT5b, and STAT6), ranging in molecular weights from 75 to 95 kDa (748 to 851 aa). The
structure of the STAT family proteins consists of an amino-terminal oligomerization
domain, a DNA-binding domain in the central part of the protein, an Src homology 2

(SH2) domain, and a transactivation domain near the carboxy l terminus. The transactivation domain is the most divergent in size and sequence and is responsible for activation of transcription. The oligomerization domain contains a tyrosine that is rapidly phosphorylated by JAKs, allowing the phosphotyrosine product to interact with the SH2 domains of other STAT proteins. Formation of STAT dimers promotes movement to the nucleus, DNA binding and activation of transcription, as well as increased protein stability.

The roles of the STAT and JAK proteins in hematopoiesis have been investigated by the creation of knockout strains of mice. STAT3 mice have severe developmental problems, resulting in fetal death. Cytokine-signaling abnormalities are associated with other STAT knockout models, but all mice are viable.

Constitutive STAT activity is associated with viral infections, but only STAT3 is known to have oncogenic properties. v-Abl and BCR-ABL induce constitutive STAT4 activity. STAT transcription factors can induce antiapoptotic gene expression, including Bcl-XL.

The JAK/STAT pathway is negatively regulated by the suppressors of cytokine signaling (SOCS) and the cytokine-induced SH2-containing (CIS) family of proteins. SOCS1 directly binds and inhibits the kinase domain (JH1) of JAK2. Gene ablation studies have indicated that the SOCS proteins have important roles in regulating the effects of **interferons**, growth hormone, and **erythropoietin**. SOCS1 and SOCS3 knockout mice are lethal, whereas SOCS2 knockout mice are 30% larger than their wild-type counterparts.

An association between an activating JAK2 mutation (JAK2(V617F)) and *BCR/ABL*-negative **myeloproliferative disorders** was recently reported.

A number of pathways, which are developmentally regulated, are now also known to be important for regulation of stem cells and during repair of end organ damage. These pathways include the Wnt, Notch, and Hedgehog systems (Figure 15, Figure 16, Figure 17, Figure 18, and Figure 19).

Wnt Pathway

Wnt signaling is initiated by the binding of a Wnt protein to the cysteine-rich domain of a receptor of the frizzled (FZ) family and a coreceptor of the low-density lipoprotein-receptor-related protein family, as shown in Figure 17. The ultimate events that are triggered by the canonical Wnt pathway are the nuclear translocation of β-catenin and its physical binding to and activation of TCF or LEF transcription factors. This pathway is important for stem cell self-renewal and the development of T and B-lymphocytes. Dysregulated Wnt signaling has become a hallmark of some types of solid tumor, most notably colon carcinomas. Given that Wnt signals are important for the survival and expansion of lymphocyte progenitors, it has been suggested that dysregulated Wnt signaling could be one mechanism underlying **leukemogenesis**.

Notch Pathway

Members of the Notch family (e.g., Notch1 and Notch3) have recently been described to play a critical role in T-**lymphocyte** development, and their constitutive activation has been related to **T-cell chronic lymphatic leukemia** in both animal models and human disease. Nevertheless, whether they act as redundant molecules, by affecting the same molecular mechanisms, or play distinct roles in T-cell differentiation or **leukemogenesis** is not clear. Altered Notch signaling impairs the developmentally regulated interplay between pre-T-cell receptor (TCR) signaling, NFkappaB, and E2A activities, thus identifying the crucial role of Notch receptors at the crossroads of disrupted lymphoid differentiation and neoplastic transformation.

Notch was originally implicated in human disease by its involvement in **T-lymphoblastic leukemia** (T-ALL). Cells from these patients (Sup-T1 cells) harbor a chromosomal translocation that fuses the TCR-β locus with a portion of the Notch gene, resulting in the constitutive

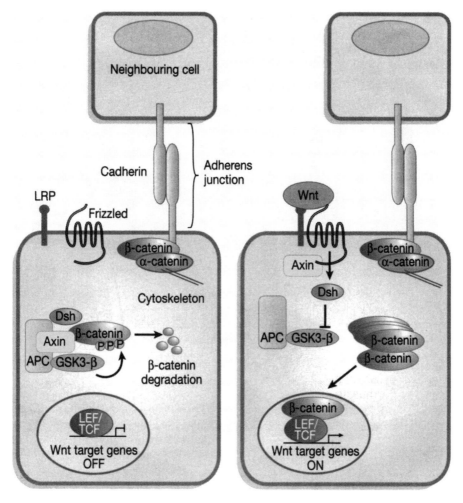

FIGURE 17
The canonical Wnt signaling pathway. In the absence of Wnt signaling (left panel), β-catenin is in complex with axin, APC, and GSK3-β and gets phosphorylated and targeted for degradation. β-Catenin also exists in a cadherin-bound form and regulates cell-cell adhesion. In the presence of Wnt signaling (right panel), β-catenin is uncoupled from the degradation complex and translocates to the nucleus, where it binds Lef/Tcf transcription factors, thus activating target genes. (Adapted from Rattis, F.M., Voermans, C., and Reya, T., Wnt signalling in the stem cell niche, *Curr. Opin. Hematol.,* 11, 88–94, 2004; from Reya, T. and Cleves, H., Wnt signalling in stem cells and cancer, *Nature,* 434, 843–850, 2005. With permission.)

expression of the intracellular domain of Notch (NIC). Notch is a single-pass transmembrane receptor that is located at the plasma membrane. Notch signaling has been shown to directly affect numerous cellular programs, including proliferation, differentiation, and apoptosis. The current model for Notch signaling suggests that, following the interaction with a ligand presented on an adjacent cell, NIC is released from the membrane and subsequently translocates to the nucleus. In the nucleus, NIC interacts with the DNA-binding transcription factor CSL (CBF-1/suppressor of hairless/Lag-1). In the absence of Notch signaling, CSL is a transcriptional repressor by virtue of association with **histone** deacetylases and corepressor proteins. It is thought that NIC displaces these corepressors and recruits coactivators to activate gene transcription. The expression of the intracellular domain of all four mammalian Notch family members has been linked to the initiation of tumorigenesis in several experimental models. Although Notch signaling is well studied, the exact mechanisms by which Notch induces cellular transformation are not well understood. The current hypothesis is

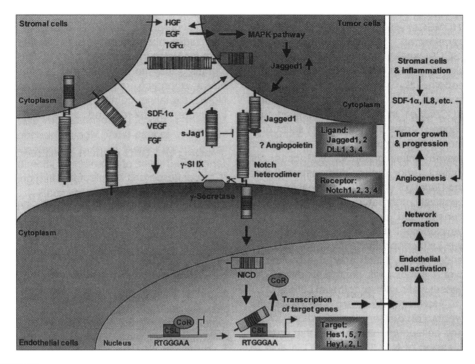

FIGURE 18

Cross talk between tumor cells, endothelial cells, and stromal cells is modulated by the Notch pathway and stimulates tumor angiogenesis. The tumor-associated growth factors HGF, EGF, and TGF-α, produced by tumor and stromal cells, induce Jagged1 expression in tumor cells (probably in endothelial cells as well as through MAPK pathway). Jagged1 on tumor cells binds to Notch receptors on endothelial cells. Upon ligand binding, the Notch intracellular domain (NICD) is released from the endothelial cell plasma membrane by the γ-secretase-dependent proteolytic cleavages of the Notch receptor. The NICD then translocates to the nucleus, where it interacts with the CSL (named for mammalian CBF1/RBP-Jkk, *Drosophila Su(H)*, and *C. elegans Lag-1*) transcription repressor to activate the transcription of target genes. The basic-helix–loop-helix proteins (bHLH), hairy/ enhancer of split (Hes1, -5, and -7) and Hes-related proteins (Hey-1, -2, and -L), are the best-characterized downstream targets. The Notch signaling from the tumor cells is able to activate endothelial cells and thus initiate tumor angiogenesis. Accordingly, the neovasculature is able to stimulate tumor growth and progression. The γ-secretase inhibitor IX (γ-SI IX) and soluble Jagged11 can protect the proteolytic cleavage by inhibiting the γ-secretase activity and blocking the interaction of Jagged11 and Notch, respectively, and therefore preventing the endothelial activation. Notch cross-signaling between tumor cells, stromal cells, and endothelial cells is likely to regulate the interaction of Notch ligands on tumor cells with receptors on endothelial cells, and vice versa. Tumor-associated stromal fibroblasts and inflammation stimulate tumor angiogenesis and progression. VEGF secreted by tumor cells and FGF generated by fibroblasts promotes tumor angiogenesis. (From Li, N.L. and Harris, A.L., Notch signaling for tumor cells: a new mechanism of angiogenesis, *Cancer Cell*, 8, 1–3, 2005. With permission.)

that Notch mediates transformation by disturbing the fine balance of activation and repression of certain genes. Very rare cases of human T-ALL harbor chromosomal translocations that involve NOTCH1, a gene encoding a transmembrane receptor that regulates normal T-cell development. It has recently been shown that more than 50% of human T-ALLs, including tumors from all major molecular oncogenic subtypes, have activating mutations that involve the extracellular heterodimerization domain or the C-terminal PEST domain of NOTCH1. These findings greatly expand the role of activated NOTCH1 in the molecular pathogenesis of human T-ALL and provide a strong rationale for targeted therapies that interfere with NOTCH signaling.

Other proteins that play a role in hematopoietic stem cell function include hedgehog, nucleostemin, and possibly nanog, and their role in leukemogenesis is being actively pursued.

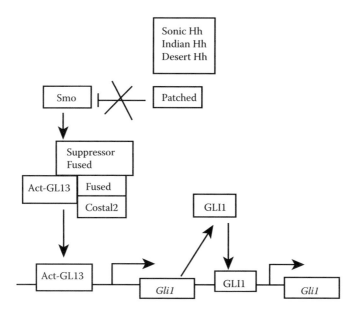

FIGURE 19

Proposed mechanisms of the Hh pathway activation. In the absence of Hh ligands, Patched suppresses Smo. Binding of Hh ligands to Patched suppresses the inhibitory function of Patched, activates the function of Smo, induces release of full-length GLI3 (Act-GLI3) from a large protein complex with Costal2/Fused/Suppressor Fused, permits translocation of Act-GLI3 to nucleus, and induces transcription of target genes such as *Gli1*. The expressed GLI1 then activates specific target genes, including *Gli1* gene itself. Thus, GLI1 is a marker of the Hh pathway activation. (From Katano, M., Hedgehog signaling pathway as a therapeutic target in breast cancer, *Cancer Letter*, 227, 99–104, 2005. With permission.)

TABLE 36

Cytokine Signaling Abnormalities

	Gene	Enzyme	Abnormality
Human	*Xq25*	CD104	hyper-M immunodeficiency
	Xq22	tyrosine kinase	SCID
		ZAP-70	agammaglobulinemia
	Xq13	γ-chain IL2 receptor	T-cell immunodeficiency
	4-bp	IFN-γR1	autoimmune disorders, susceptibility to mycobacterial disease
Mouse	*lpr*	Fas (CD95)	lymphocyte accumulation
	me/me	tyrosine phosphatase	impaired T-cells → autoimmune disorders

Cytokine Signaling Abnormalities

A number of abnormalities with their gene and encoding enzyme have been identified in humans and in mice (Table 36). Those identified result in a form of **immunodeficiency** or **autoimmune disorder.**

CELL-SURFACE RECEPTORS

The different types of cell-surface receptors that interact with water-soluble ligands. Binding of ligand to some of these receptors induces second-messenger formation, whereas ligand binding to others does not. Cell-surface receptors can be sorted into four classes:

G-protein receptors

Ion-channel receptors — Types 1 and 2

Tyrosine kinase receptors

Receptors with intrinsic enzyme activity

G-Protein-Coupled Receptors

Ligand binding activates a G protein, which in turn activates or inhibits an enzyme that generates a specific second messenger or modulates an ion channel, causing a change in membrane potential (Figure 20). The receptors for epinephrine, serotonin, and glucagon are examples.

Ion-Channel Receptors

Ligand binding changes the conformation of the receptor so that specific ions flow through it; the resultant ion movements alter the electric potential across the cell membrane. The acetylcholine receptor at the nerve-muscle junction is an example.

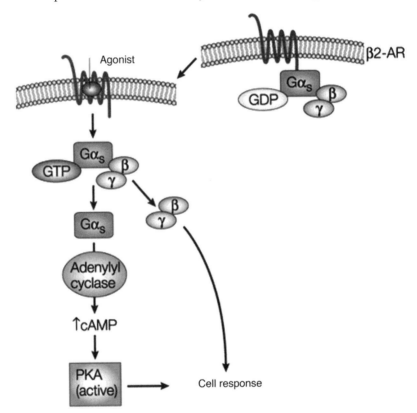

FIGURE 20

G-protein-coupled receptors. Classical examples of seven-transmembrane (7TM)-receptor signaling. In the absence of agonist, 7TM receptors such as the β2 adrenergic receptor (β2-AR) are in low-affinity state. After agonist binding, a transient high-affinity complex of agonist, activated receptor, and G protein is formed. GDP is released from the G protein and is replaced by GTP. This leads to dissociation of the G-protein complexes into its subunits and βγ dimers, which both activate several effectors. Gα2, for instance, activates adenyl cyclase, which leads to an increase in cyclic AMP (cAMP). This increase in cAMP in turn activates protein kinase A (PKA), which is a serine/threonine kinase that phosphorylates many different substrates, including 7TM receptors, other kinases, and transcription factors. (From Pierce, K.L., Premont, R.T., and Lefkowitz, R.J., Seven-transmembrane receptors, *Nat. Rev.: Molecular Cell Biol.*, 3, 639–650, 2002. With permission.)

Tyrosine Kinase-Linked Receptors

Tyrosine kinase-linked receptors (Figure 21) may have intrinsic catalytic activity and are discussed below. A subgroup of the tyrosine kinase-linked receptors consists of the cytokine-receptor superfamily. These receptors lack intrinsic catalytic activity, but ligand binding stimulates formation of a dimeric receptor, which then interacts with and activates one or more cytosolic protein-tyrosine kinases. The receptors for many **cytokines,** the interferons, and human growth factor are of this type. Based on notable amino acid homologies and conservation of characteristic sequence motifs, these receptors can be grouped into a gene family and subfamilies.

Based on details of their structural organization, the cytokine superfamily is divided into types 1 and 2.

Type-1 Cytokine Receptor Family

The family of type-1 (class-1) cytokine receptors includes those for **interleukins** (IL-2 [β-subunit], IL-3, IL-4, IL-5, IL-6, IL-7, IL-9, IL-11, IL-12), **erythropoietin** (EPO), **colony stimulating factors** (GM-CSF, G-CSF), leukocyte inhibitory factor (LIF), ciliary neurotrophic factor (CNTF), as well as the receptors for **thrombopoietin** (TPO), growth hormone, and prolactin. Further members of this protein family are some commonly shared **cell signal transduction** components such as gp130 and the γ chain of the IL-2 receptor. Upon binding of a ligand to the extracellular domain of the type-1 receptors, the receptor molecules form homodimers or heterodimers (in a few cases also trimers), and the intracellular receptor domains become associated with a variety of signaling molecules, in particular cytoplasmic tyrosine kinases and latent cytoplasmic transcriptional activators. The conserved extracellular domain of these receptors has a length of approximately 200 amino acids (type-1 family domain), which contains four positionally conserved cysteine residues in the amino-terminal region and a Trp-Ser-X-Trp-Ser (WSXWS) motif located proximal to the transmembrane domain. The four cysteines appear to be critical to the maintenance of the structural and functional integrity of the receptors. The WSXWS consensus sequence is thought to serve as a recognition site for functional protein–protein interaction of cytokine receptors. The differential expression of receptor components and signaling molecules and the selective recruitment of different intracellular signaling components explains, at least in part, some of the overlapping biological activities of the corresponding cytokine ligands. Sharing of different receptor components has also been used to further subdivide the type-1 receptor family into three different subgroups according to the critical signal-transduction component used.

Type-2 Cytokine Receptor Family

Members of the family of type-2 (class-2) cytokine receptors are only distantly related to members of the type-1 receptors. The type-2 receptor family includes receptors for **interferons** (IFN-α, IFN-β, IFN-γ), interleukin (IL-10), and **tissue factor** (coagulation factor-3). Type-2 receptors are multimeric receptors composed of heterologous subunits

The ligand-binding subunit of a receptor is referred to as the low-affinity α-chain. Other signal-transducing subunits are named β-chains or γ-chains.

Receptors with Intrinsic Enzymatic Activity

Several types of receptors have intrinsic catalytic activity (Figure 21), which is activated by binding of ligand. For instance, some activated receptors catalyze conversion of GTP to cGMP; others act as protein phosphatases, removing phosphate groups from phosphotyrosine residues in substrate proteins, thereby modifying their activity. The receptors for

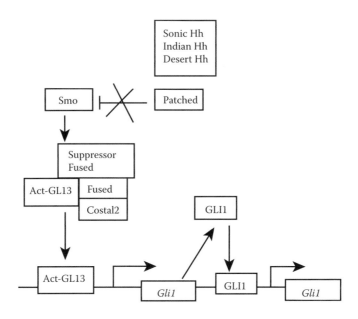

FIGURE 21
Cytokine receptor superfamily type 2. Subclassification of group 2 β-sheet cytokines. Note: see "Abbreviations" in the front of this volume. (From Nicola, N.A., An introduction to cytokines, in *Guidebook to Cytokines and Their Receptors*, Nicola, N.A., Ed., Oxford University Press, Oxford, 1994. With permission.)

insulin and many growth factors are ligand-triggered protein kinases; in most cases, the ligand binds as a dimer, leading to dimerization of the receptor and activation of its kinase activity. These receptors (often referred to as receptor serine/threonine kinases or receptor tyrosine kinases) are autophosphorylate residues in their own cytosolic domain and can also phosphorylate various substrate proteins. Receptors in this family include stem cell factor receptor or c-kit, M-CSFR, and Flt-3.

CELLULAR CYTOTOXICITY
See also **Antigen presentation**; **Lymphocytes**.
The death of cells induced by other cells. Cytotoxicity, mediated by cells of the immune system, is a major mechanism of **cell-mediated immunity** (CMI). The term is usually reserved for the killing of altered self-cells rather than bacteria and other cellular pathogens. The major role for cellular cytotoxicity is to eliminate cells infected by viruses; it may also be important in the elimination of cancer cells and in autoimmunity. **Lymphocytes** are the main mediators of cellular cytotoxicity, either CD8 cytotoxic T-cells (Tc, CTL) or natural killer (NK) cells.

CELLULAR LOCOMOTION
(Cell migration) The physiological process of cellular motility. All mammalian tissue cells have the property of movement by cellular protrusions of their cell membrane. This is a particularly well-developed function of **leukocytes**. Movement occurs by sliding filaments of **actin** and **myosin** over cell surfaces. Myosin is responsible for contraction and retraction of the cytoplasm, and the energy requirements of the myosin filaments are provided by **adenosine triphosphate**.

These two cytoplasmic proteins assemble as long filaments in the hyaline cortex at the "front end" of the cell. They are bound together by **spectrin**, which has affinity to **calmodulin**, the calcium-binding protein that is necessary for the contractile phase of locomotion.

Cellular activation results in actin filaments producing elongated stiffened bundles, termed **uropods**, located at the cellular leading edge. Here, they accumulate to form a cap and are the means of attachment to other cells and cell debris. Rhythmic propagation and dissolution of actin moves the cell in the direction of attraction by chemotactic or cellular forces. Movement or migration occurs in response to stimuli from bacteria or growth factors (see **Neutrophils**).

CENTRAL VENOUS CATHETERS

A plastic tube inserted into a vein and secured to allow regular infusion of drugs, usually **cytotoxic agents**. The ability to have long-term venous access for blood sampling as well as therapeutic intervention has made a major contribution to patient welfare over the last two decades.[133] Lines are coated with silicone to reduce infective and thrombotic complications. Insertion can be through either a central vein, e.g., internal or external jugular or subclavian vein, or a peripheral vein, e.g., femoral or antecubital fossa vein, with the tip of the catheter placed in the superior vena cava/right atrium. Tunneling of the catheter reduces the risk of dislodgement and infection. Potential complications include cardiac arrhythmias (if the line is inserted too far), infection (colonization of the line, bacteremia/septicemia, endocarditis), and **thrombosis** (within the line or the blood vessel). Strict antiseptic precautions are required whenever using the line. Colonization of the line by bacteria or fungi will necessitate line removal. Long-term anticoagulation is usually not required, with randomized studies showing marginal, if any, benefit.

CENTROBLAST

See also **Diffuse large B-cell lymphoma**.
Large B-lymphoid cells with basophilic cytoplasm and round vesicular nuclei with fine chromatin containing one to three peripheral nucleoli. They show γ-**immunoglobulin** chains on their cell surface and intracytoplasmic μ chains.

CENTROBLASTIC-CENTROCYTIC DIFFUSE LYMPHOMA

See **Follicle center lymphoma** — diffuse; **Non-Hodgkin lymphoma**.

CENTROBLASTIC-CENTROCYTIC FOLLICULAR LYMPHOMA

See **Follicle center lymphoma** — follicular; **Non-Hodgkin lymphoma**.

CENTROBLASTIC-FOLLICULAR LYMPHOMA

(Follicular, predominantly large-cell lymphoma) See **Follicle center lymphoma** — follicular; **Non-Hodgkin lymphoma**.

CENTROBLASTIC LYMPHOMA

(Monomorphic, polymorphic, and multilobed subtypes; large-cell immunoblastic lymphoma) See **Diffuse large B-cell lymphoma**; **Non-Hodgkin lymphoma**.

CENTROCYTE

See also **Lymph nodes**; **Lymphoid system**.

Transformed medium-sized B-lymphocytes with cleaved nuclei. They show μ- and δ-**immunoglobulin** chains on their surfaces.

CENTROCYTIC LYMPHOMA

See **Mantle-cell lymphoma**; **Non-Hodgkin lymphoma**.

CENTROMERE

The region in eukaryote **chromosomes** where daughter chromatids are joined together. The kinetochore, to which the spindle chromosomes are attached, lies adjacent to the centromere. The centromeric **DNA** codes for the kinetochore.

CERULOPLASMIN

See also **Hereditary hemochromatosis**.

(Ferroxidase) A blue alpha-2-glycoprotein that binds 90 to 95% of plasma **copper** and has six or seven cupric ions per molecule. It is involved in peroxidation of Fe(II) **transferrin** to form Fe(III) transferrin. Hereditary aceruloplasminemia is an autosomally recessive disorder due to mutations in the Cp gene (3q23-q24). Affected persons, usually diagnosed at middle age, have low serum ceruloplasmin and copper levels; elevated serum ferritin levels; and progressive dementia, extrapyramidal disorders, cerebellar ataxia, diabetes mellitus, and hepatic iron loading.

CHAGA DISEASE

See **Trypanosomiasis**.

CHECKPOINT KINASES 1 AND 2

These are critical kinases in **cell cycle** regulation. Damaged **DNA** in humans is detected by sensor proteins that transmit a signal via ATR to ChK1, or by another sensor complex, the signal of which is relayed by ATM to ChK2. Most of the damage signals originated by the sensor complexes for the G2 checkpoint are conducted to Cdc25C, the activity of which is modulated by 14-3-3. While the ATM-Chk2 pathway is activated primarily by **ionizing radiation** (IR)-induced DNA damage and is nonessential for cell viability, the ATR-Chk1 pathway is induced primarily by stalled DNA replication and is essential, at least for early embryogenesis. Although structurally very different, Chk1 and Chk2 phosphorylate many common substrates and have partly overlapping roles.

   The cell-cycle checkpoint pathways ultimately inhibit cyclin-dependent kinases (Cdks), thereby delaying or arresting the cell cycle at specific stages. Cdks are maintained in an inactive state by their association with cyclin subunits through the phosphorylation of Thr 14 (T14) and Tyr 15 (Y15) in their amino-terminal ATP-binding regions. The rate-limiting step in the activation of Cdks is dephosphorylation of these residues by Cdc25 dual-specificity phosphatases. Cdc25 phosphatases have a key role in the checkpoint response to unreplicated or damaged DNA. Cdc25 represents a major and crucial target for Chk1 and Chk2 in cell-cycle checkpoints. In vertebrates, there are three isoforms of Cdc25: Cdc25A, Cdc25B, and Cdc25C. Cdc25B is thought to function at the G2/M transition of the cell cycle, and Cdc25C acts during M phase to sustain the activity of Cdk1-cyclin B, a key regulator of the M phase. Checkpoint-mediated phosphorylation of Cdc25A

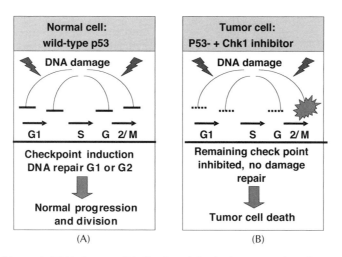

FIGURE 22

Role of checkpoint kinases in DNA damage. (A) The G_1 and G_2 checkpoints in the cell cycle will mediate DNA repair as a consequence of radiation or cytotoxic DNA damage. Once the cell DNA damage is repaired progression to cell division will occur. (B) In cells lacking p53, repair in the G_1 S phase is not possible. Therefore, repair only occurs in the G_2 M phase. If ChK1 kinase is inhibited, then DNA repair is not possible and the cell dies.

results in its inhibition and the sustained phosphorylation and inhibition of cyclin-dependent kinase 2 (Cdk2), thus leading to G1 arrest. Differences between actions in normal cells and in tumor cells are shown in Figure 22.

CHEDIAK-HIGASHI SYNDROME

A rare autosomally recessive disorder (originally described in the early 1940s) characterized by ocular and cutaneous albinism, an increased risk of **hemorrhage**, and pyrogenic infections associated with abnormal cytoplasmic granulation. This is associated with **anemia**, **neutropenia**, and **thrombocytopenia**.

Pathogenesis

The underlying defect is unknown, but abnormalities in the ability of cells to regulate cytoplasmic membranes have been observed. Granules fuse and defective degranulation occurs. The abnormal granulation — described in neutrophils, monocytes, and lymphocytes — can also occur in neuron Schwann cells, melanocytes, hepatocytes, and renal tubular cells. In neutrophils, the giant granulocytes are due to fusion of primary and secondary granules.

Clinical Manifestations

These usually appear early in childhood.

Infection: the increased risk of bacterial and fungal infections, commonly affecting the skin and the respiratory and gastrointestinal tracts. The pathogenesis is multifactorial, with defective chemotaxis, degranulation, and delayed intracellular killing all described. The neutrophil granules lack cathepsin G and elastase, and surface expression for complement C3b receptor is reduced. Neutropenia may occur late in the disease.

Hemorrhage: this may be secondary to thrombocytopenia or, more usually, defective platelet aggregation secondary to dense-body storage-pool deficiency.

Albinism: this may affect hair follicles, the skin, and the retina (manifest as photophobia) with defective melanin production and distribution.

Central nervous system manifestations, particularly peripheral neuropathy and ataxia.

Malignancies: there is an increased risk of **leukemia** and **lymphoproliferative disorders**, some of which are probably **Epstein-Barr virus** (EBV) proliferations, possibly secondary to defective natural killer (NK) cell function.

Treatment

There is no effective therapy, and prophylactic antibiotics are unhelpful. **Allogeneic stem cell transplantation** can be curative.

CHEMICAL TOXIC DISORDERS

Hematological disorders induced by adverse effects of chemicals, other than drugs, administered either accidentally, by poisoning, or from the environment.[134]

Aplastic anemia, **myelodysplasia**, **acute myeloid leukemia**: benzene compounds

Megaloblastosis: alcohol, arsenic

Sideroblastic anemia: alcohol, **lead**

Erythrocytosis: cobalt

Basophilic stippling: aniline, dinitrobenzene, lead

Methemoglobinemia: acetanilide, aniline, antimony compounds, chlorates, nitrites, nitrobenzene, permanganates, sulfonyl

Porphyria cutanea tarda: aromatic hydrocarbons in herbicide manufacture, hexachlorobenzene, wheat seed fungicide

Hemolytic anemia:
- Arsine (arsenic hydride)
- Heavy metal toxicity — arsenic, **copper**, **mercury**, lead
- Potassium chlorate
- Pyrogallic acid
- Stilbene
- Distilled water (>0.6 l causes hemolysis due to hypo-osmolarity)
- Oxygen exposure to 100% (>4 h; peroxidation of red cell membrane lipids)
- Spider venom (brown recluse spiders; acute intravascular hemolysis)
- **Snake venom disorders**: cobra (phospholipase acting directly on red cells)

Carboxyhemoglobinemia: carbon monoxide

Oxygen affinity to hemoglobin increase: cyanide

Neutrophilia: digitalis, lead toxicity, foreign proteins, insect venoms

Immunosuppression: benzene, asbestos

Hypersensitivity: nickel, di-isocyanates

Autoimmune disorders: trichloroethylene, epoxy resins

Non-Hodgkin lymphoma: 2,4-dichlorophenoxyacetic acid

Hemangiosarcomas: inorganic arsenic, thorotrast, vinyl chloride

CHEMOKINES

Cytokines that control **leukocyte** migration and may also activate cells. These are low-molecular-weight ($\approx$10 kDa) polypeptides, all related by sequence, falling into several groups. These are distinguished through possession of characteristic cysteine motifs in their sequence: the CXC family (two cysteines separated by another amino acid), CC (adjacent cysteines), C (one cysteine only), and CXXXC. These are sometimes referred to as α, β, γ, and δ, respectively. Chemokines are further distinguished numerically, this nomenclature replacing the older trivial names so that RANTES (regulated on activation, normal T-cell expressed and secreted) becomes CCL5, IL-8 CXCL8, etc. (L for ligand to distinguish from the corresponding receptor, R).

There are many chemokines (some 50 in total), and a range will be present at sites of inflammation, where they are produced by activated cells of various types.[135] The different chemokines tend to be active on different cell types, e.g., CXC on neutrophils but not macrophages, and CC the inverse. All chemokines act via receptors, which tend to be redundant (i.e., the same receptor binds multiple chemokines). Evidently, the role of these molecules is to attract and detain leukocytes to sites of inflammation. The detailed role of all individual chemokines has not been yet established.

Histiocytes (macrophages) at the site of infection release cytokines, such as IL8, that modify neutrophil integrins to make the neutrophils more adherent. This allows them to bind to endothelium and undergo diapedesis. A receptor for chemokines occurs on hematopoietic, glial, and **dendritic reticulum cells** (CXCR4), and another on **monocytes** and **lymphocytes** (CCR5). Macrophage inflammatory proteins 1α and 1β (MIP-1α, MIPβ) are the natural ligands for CCR5. Binding of these to CRC5 mediates chemotaxis and cell activation. Chemotaxis in cell locomotion is by movement along a series of chemokines

Some viruses possess genes for chemokine-binding proteins, emphasizing their importance in infection. The human immunodeficiency virus (HIV-1) utilizes chemokine receptors (CCR5 and CXCR4) as coreceptors (with CD4).

CHLORAMBUCIL

See **Alkylating agents**.

2'-CHLORODEOXYADENOSINE

(CDA) A **purine** analog that is resistant to **adenosine deaminase** (ADA) and accumulates selectively as a 5'-triphosphate metabolite in cells rich in deoxycitidine kinase. It thereby inhibits **DNA** synthesis, with a particular *in vitro* cytotoxicity to **lymphocytes** and **monocytes**, being effective for treatment of **B-cell chronic lymphatic leukemia** and **hairy cell leukemia**. It is the treatment of choice in hairy-cell leukemia at a dose of 0.1 mg/kg continuous infusion for 7 days. It is administered either as a continuous infusion over 5 to 7 days or as a 2-h bolus daily for 5 to 7 days.

CHLOROMA

(Extramedullary myeloid tumor; granulocytic sarcoma) See **Myeloid sarcoma**.

CHRISTMAS DISEASE
See **Hemophilia B**.

CHROMATIN
Stainable material of interphase nucleus consisting of nucleic acid and associated histone protein packed into nucleosomes. In eukaryotes, chromatin is the natural form of **DNA** in the nucleus. For hundreds of millions of years, NNA-binding factors have evolved with chromatin. The structural organization of DNA into chromatin is of key importance in regulating genome function and stability. Maintenance of such an organization is thus crucial to preserve cellular identity. At each cell cycle, during S phase, this is achieved by duplication of chromatin structure in tight coordination with DNA replication. Such a coordinate process requires histone synthesis and consequent deposition onto DNA by chromatin assembly factors that are efficiently coupled to DNA synthesis. Interphase chromatin is dispersed throughout the nucleus and consists of (a) euchromatin, which is relatively dispersed, occupying most of the nucleus, and (b) heterochromatin, more dense and usually adjacent to the nuclear membrane or the nucleolus. Heterochromatin consists of constitutive heterochromatin (short-repeat DNA sequences) and facultative heterochromatin (those whole chromosomes that are inactive in certain tissues, such as mammalian X-chromosome inactivation). Heterochromatin is never actively transcribed, whereas a small proportion of euchromatin is ($\approx$10%). On the basis of recent data, a number of chromatin-associated factors have been identified as putative proliferation markers for cancer diagnosis and prognosis.

CHROMOSOMES
The **DNA** of eukaryotes is subdivided into chromosomes. This is presumably for convenience of handling, as each has a long length of DNA associated with various proteins. Each chromosome has a characteristic length and banding pattern; chromosomes stained with certain dyes, commonly quinacrine (Q banding) or Giemsa (G banding), show a pattern of transverse bands of light and heavy staining that is characteristic for the individual chromosome. The chromosomes become more tightly packed at **mitosis** and become aligned on the metaphase plate.

Single unpaired chromosomes are termed haploid and are the product of meiosis. The number of haploid chromosome sets in a nucleus is termed its ploidy, i.e., diploid, triploid, tetraploid, etc.; with rapid proliferation, a variety in ploidy occurs. Chromosomes are easily identified in their most condensed state during mitosis (five to ten times more condensed than during interphase).

Mitotic Chromosomes
Following DNA synthesis and replication at S-phase, the double complement of genetic material becomes oriented into a pair of sister chromatids held together at their **centromeres** and covered by a layer of ribonucleoprotein. DNA within mitotic chromosomes is transcriptionally inactive. A 5-cm length of DNA is condensed in this state to 5 μm, and histone H1 components are phosphorylated. For full function, three chromosome components are essential: the centromere, **telomere**, and replication origin. Yeast sequences for these components are known, and insertion of these sequences into large lengths of human DNA allows replication of many copies of human DNA segments within yeasts, so-called yeast artificial chromosomes.

Centromere: essential for the attachment of the mitotic chromosomes to the mitotic spindle and for the correct chromosomal orientation at mitosis. All functional sequences are contained within 120 bp (base pairs).

Replication origin: there are many in eukaryotic DNA (see **Deoxyribonucleic acid**).

Telomere: this short terminal DNA sequence maintains the chromosome length upon cell division and seals the chromosomal ends, preventing the "stickiness" that promotes translocations of chromosomal material. During DNA replication, double-stranded DNA synthesis results in the lagging-strand synthesis of a shortened telomere. Telomeric sequences are a long series of tandemly repeated C3TA2 (human) nucleotides and are maintained in length by the action of the enzyme telomerase, which adds nucleotides to the lagging strand. Telomerase is a ribonucleoprotein whose RNA component acts as a template for the synthesis of the final nucleotides of the DNA strand. **Lymphocyte** telomerase activity is reduced with aging, and telomeric shortening has been identified in many late-stage tumors, e.g., ovarian cancer and leukemia. This may promote the stickiness of telomeric ends and the high chromosomal-translocation frequency observed in these pathological states.

Interphase Chromosomes

Chromosomes adopt a less compact structure during interphase to allow transcription of actively transcribing genes. Chromosomal components retain a nuclear location and, often, an orientation (polar centromere and centrally orientated telomeres) specific to an individual chromosome, and these are bound to the matrix-by-matrix attachment regions. Nucleoli consist of DNA loops transcribing large quantities of ribosomal RNA.

CHRONIC EOSINOPHILIC LEUKEMIA

(CEL) A **myeloproliferative disorder** in which an autonomous clonal proliferation of eosinophilic precursors results in persistently increased numbers of **eosinophils** in the blood, bone marrow, and peripheral tissues. It is not a distinct entity, and there may be overlap with other conditions in which eosinophils are increased, such as other forms of myeloproliferative disorders, **myelodysplasia**, **acute myeloblastic leukemia**, and acute eosinophilic leukemia. Organ damage occurs as a result of leukemic infiltration or the release of **cytokines**, enzymes, or other proteins by the eosinophils (see **Hypereosinophilic syndrome**). There should be evidence for clonality of the proliferating eosinophils. The commonest cytogenetic abnormalities identified involve chromosome 5, e.g., t(2;5), t(5;11), t(5;12), although other abnormalities have been reported (see below). Philadelphia chromosome is negative, and cells must be *BCR/ABL*-negative for diagnosis.

Clinical Features

The clinical features are predominantly those of the hypereosinophilic syndrome, but about 10% are asymptomatic. **Hepatosplenomegaly** is present in 30 to 50% of patients.

Laboratory Features

The peripheral blood shows eosinophils $>1.5 \times 10^9/l$, with small numbers of eosinophilic myelocytes or promyelocytes. Some eosinophils show abnormalities such as sparse granulation with clear areas of cytoplasm, cytoplasmic vacuolation, nuclear hypersegmentation or hyposegmentation, and enlarged size. **Neutrophilia, monocytosis,** and **basophilia** may accompany the eosinophilia.

Bone marrow examination shows hypercellularity, mainly due to eosinophilic proliferation. Charcot-Leyden crystals are often present. Erythropoiesis and megakaryopoiesis are usually normal. There is usually an increased number of myeloblasts, some with dysplastic features, but fewer than 20% of cells may be seen in some cases. The eosinophils show cyanide-resistant myeloperoxidase activity, but the peroxidase content of the eosinophils is normal. No specific abnormality of the eosinophil **immunophenotype** has been reported. **Mast cell** infiltration is often found in association with FIPIL1-PDGFRA fusion (see below).

Cytogenetic analysis shows a variety of abnormalities:

Trisomy 8, monosomy 7, I(17-20q) with platelet-derived growth factor (PDGFR) A and B, and *FGFRI*

t(5;12)(q33;p13), usually with chronic myelomonocytic leukemia

8p11 translocations — t(8;13)(p11;q12), t(8;9)(p11;q32-34), and t(6;8)(q27;p11) — related to the *FGFR1* gene, which is fused with different partner genes

Interstitial deletion of chromosome 4, with generation of fusion protein tyrosine kinase between PDGFRα and FIPI L1; there is an associated male preponderance

PDGFRA-FIPI L1 fusion with tryptophase-positive subset

Prognosis and Treatment

Survival is variable. Most cases respond to **imatinib** therapy, particularly those with FIPL1-PDGFRα rearrangement.[136] Although conditions with the 8p11 abnormality may initially respond to therapy, relapse is rapid and prognosis poor.

CHRONIC GRANULOMATOUS DISEASE

A rare (1:106) inherited group of disorders characterized by the failure of **granulocytes** to produce oxidative radicals and thus to kill phagocytosed bacteria. Most are inherited as an X-linked recessive (60%, type I), but autosomally recessive (30%, types II and III) and a very rare autosomally dominant form (1%) have been described. Mutations are for the genes for NAPD oxidase or its regulatory proteins.

The X-linked mutation results in defective production of cytochrome b558, which is an oxidase important in the production of superoxide. Various abnormalities in the cytochrome b558 gene, including failure of its transcription or mRNA translation, have been described. There is usually **neutrophilia** with normal recruitment to a site of infection. Here pus accumulates, the term "granulomatous" being a misnomer.

Clinical manifestations appear shortly after birth with recurrent suppurative infections, ultimately with granuloma formation. Skin infections and lymphadenitis are typical. Multiple abscesses (lung, liver, epidural) and osteomyelitis are common. The commonest pathogens are *Staphylococcus aureus*, *S. serratia*, and *Salmonellae* spp. The autosomally recessive form may run a slightly milder clinical course, but life expectancy is still only 5 to 10 years. The diagnosis is made either by chemiluminescence or the **nitroblue tetrazolium test** (NBT) failing to reduce yellow NBT to a blue/black precipitate.

Prophylactic antibiotics (trimethoprim-sulfamexazole) may be beneficial, but the mainstay of therapy is vigorous therapeutic antibiotic usage and surgical removal of problematic abscesses. **Interferon-α** increases production of cytochrome-b mRNA and is therapeutically beneficial in many patients. A few patients have been successfully treated by **allogeneic stem cell transplantation**.

CHRONIC IDIOPATHIC MYELOFIBROSIS

(CIMF; agiogenic myeloid metaplasia; myelofibrosis and myeloid metaplasia; primary myelofibrosis; primary osteomyelofibrosis) A clonal disorder involving a multipotent hematopoietic stem cell. It is characterized by bone marrow fibrosis, **splenomegaly**, and **myeloid metaplasia** (extramedullary hematopoiesis). Characteristically, there is a **leuko-erythroblastic anemia**, with teardrop **poikilocytes**. At diagnosis, anemia is usual, but the leukocyte and platelet counts can be low, normal, or high. Bone marrow aspiration is usually unsuccessful due to fibrosis. Marrow histology reveals three distinct phases as the disease progresses:

Cellular: hyperplasia of all cell lineages, with prominent but dysplastic megakaryo-cytes, often in nests, sinusoidal dilatation, neoangiogenesis, and slightly increased reticulin

Fibrotic: increased reticulin with megakaryocytic hyperplasia still prominent, but other cell lines reduced

Osteosclerotic: gross fibrosis with new bone formation and markedly disturbed archi-tecture; all cell lines reduced; however, the histologic abnormalities can be patchy

Pathogenesis

The **granulocytes, erythrocytes**, and **platelets** are monoclonal in idiopathic myelofibrosis (IMF), but fibroblasts are polyclonal. Thus, the myelofibrosis is a reaction to the neoplastic clone. **Cytogenetic analysis** in IMF shows del 13q; del 20q; trisomy 8, 9, or 21; and partial trisomy 1q. These, however, are not specific for IMF but are also seen in **polycythemia vera** and **essential thrombocytosis**. Megakaryocytes are involved in the myelofibrotic process, since, experimentally, excessive thrombopoietin production produces a histolog-ically identical syndrome, but **monocytes** also appear to be necessary for myelofibrosis by producing the transforming growth factor-β (TGF-β). Both cause **collagen** proliferation, but TGF-β also promotes **angiogenesis** and reduced collagenase activity. Osteoprotegerin is thought to be responsible for the osteosclerosis. Another a-granule enzyme, platelet factor 5, also reduces collagenase activity.

Pathophysiology

The clinical features, blood and marrow abnormalities, result from a combination of hematopoietic cell hyperplasia, increasing marrow fibrosis, myeloid metaplasia, and hypersplenism. *In vitro* colony assays show increased numbers of circulating **colony form-ing units**, CFU-MK, CFU-GEMM, CFU-GM, CFU-E, and BFU-E, in patients with IMF. Marrow fibrosis is due to increased deposition of collagen types I, III, IV, and V. **Ferroki-netics** show that myeloid metaplasia can occur in virtually any organ, but the spleen is the predominant site. Extramedullary hematopoiesis is usually ineffective and contributes little to the peripheral blood count. **Pancytopenia** results from a combination of marrow failure and hypersplenism.

Clinical Features

Nonspecific symptoms, e.g., fatigue, anorexia, weight loss, fever, and weakness, are com-mon. Abdominal pain results either from hepatomegaly, splenomegaly, or splenic infarc-tion. Symptomatic anemia and thrombocytopenia are not unusual. Splenomegaly and hepatomegaly (98 and 50% of patients, respectively) may be massive, but lymphadenopathy is uncommon. Platelet-function abnormalities appear to be more severe in IMF than in its companion myeloproliferative disorders, and bleeding is more common than thrombosis. An increase in circulating CD34$^+$ cells distinguishes IMF from its companion myeloproliferative disorders, polycythemia vera and essential thrombocytosis. A normal hemoglobin level in an IMF patient with significant splenomegaly should raise the possibility of polycythemia vera.

Prognosis

The clinical course is variable, with survival of more than 10 years possible. Prognosis can be estimated using simple scoring systems employing the hemoglobin level, leukocyte count, presence of constitutional symptoms, and circulating blast cell number. Younger patients with no constitutional symptoms and relatively normal blood counts have the best prognosis. The cause of death is variable, with infection, hemorrhage/thrombosis (cerebro/cardiovascular), and leukemic transformation most common. **Portal hypertension** appears in a minority of patients, secondary to increased portal blood flow, hepatic (Budd-Chiari syndrome) or portal vein thrombosis, or hepatic myeloid metaplasia.

Treatment

This is largely supportive (e.g., **red blood cell transfusions** or **platelet transfusions**, **folic acid**, **pyridoxine**, **allopurinol**), as no therapeutic agent has been shown to be consistently beneficial, although **hydroxyurea**, **busulfan**, and α-**interferon** have all been reported to control splenomegaly in some patients. **Corticosteroids** have also been effective in controlling constitutional symptoms and alleviating anemia in some patients. Recombinant **erythropoietin** has been effective in some anemic patients, but may promote splenomegaly. The role of **splenectomy** is controversial. Indications for splenectomy are progressive weight loss, severe thrombocytopenia, splenic pain, and portal hypertension, but postoperative complications occur in up to 50% of patients, particularly intra-abdominal thrombosis, and hepatomegaly may be progressive postsplenectomy. Splenic irradiation has a high incidence of complications due to myelosuppression and cannot be recommended except under unusual circumstances. Recently, low-dose thalidomide and prednisone have been shown to be effective in alleviating anemia and reducing splenomegaly. In younger patients with a matched donor, **allogeneic stem cell transplantation** is appropriate, and newer low-intensity approaches may extend this treatment to older patients.

CHRONIC LEUKEMIA

Excessive proliferation of a clone of a hematopoietic cell at an advanced stage of maturation that also fails to undergo apoptosis (see **Oncogenesis**). This results in an increased number of mature and immature leukocytes in the peripheral blood, usually associated with organ infiltration, particularly of the liver and spleen. The underlying clonal change has the potential to progress, resulting in bone marrow failure and acute blastic leukemia. They are largely disorders of adults, with a peak frequency in the fifth to seventh decades. The following types occur:

   Chronic myeloproliferative disorders
   Chronic myelogenous leukemia (CML)
   Atypical chronic myeloid leukemia
   Chronic neutrophilia leukemia

Chronic eosinophilic leukemia/hypereosinophilic syndrome

Chronic myelomonocytic leukemia

Juvenile myelomonocytic leukemia

Mast cell leukemia

Polycythemia vera

Chronic idiopathic myelofibrosis

Essential thrombocythemia

Myelodysplastic/myeloproliferative diseases, unclassifiable

Chronic lymphoproliferative disorders

Mature B-cell neoplasms

- Chronic lymphocytic leukemia/small lymphocytic lymphoma
- B-cell prolymphocytic leukemia
- Lymphoplasmacytic lymphoma
- Splenic marginal-zone lymphoma
- Hairy cell leukemia
- Plasma cell lymphoma
- Solitary plasmacytoma of bone
- Extraosseous plasmacytoma
- Extranodal marginal-zone B-cell lymphoma of mucosa-associated lymphoid tissue (MALT-lymphoma)
- Nodal marginal-zone B-cell lymphoma
- Follicular lymphoma
- Mantle-cell lymphoma
- Diffuse large B-cell lymphoma
- Mediastinal (thymic) large B-cell lymphoma
- Intravascular large B-cell lymphoma
- Primary effusion lymphoma
- Burkitt lymphoma/leukemia

Mature T-cell and NK-cell neoplasms

- T-cell prolymphocytic leukemia
- T-cell large granular lymphocytic leukemia
- Aggressive NK-cell leukemia
- Extranodal NK/T-cell lymphoma, nasal type
- Enteropathy-type T-cell lymphoma
- Hepatosplenic T-cell lymphoma
  Subcutaneous panniculitis-like T-cell lymphoma
  Mycosis fungoides and Sezary syndrome
  Primary cutaneous anaplastic large-cell lymphoma
  Peripheral T-cell lymphoma, unspecified
  Angioimmunoblastic T-cell lymphoma
  Anaplastic large-cell lymphoma

Hodgkin lymphoma
- Nodular lymphocyte-predominant Hodgkin lymphoma
- Classical Hodgkin lymphoma
- Nodular sclerosis classical Hodgkin lymphoma
- Lymphocyte-rich classical Hodgkin lymphoma
- Mixed-cellularity classical Hodgkin lymphoma
- Lymphocyte-depleted classical Hodgkin lymphoma

CHRONIC LYMPHOCYTIC LEUKEMIA/LYMPHOMA

See also **Lymphoproliferative disorders; Non-Hodgkin lymphoma**.
(CLL; B-CLL; small lymphocytic lymphoma; B-SLL; REAL/FAB B-cell chronic lympho-cytic leukemia; Luke-Collins: small lymphocytic B-CLL; Rapaport: well-differentiated lym-phocytic, diffuse) A low-grade non-Hodgkin lymphoma characterized by monomorphic small, slightly larger than normal, and round B-lymphocytes with clumped nuclear chro-matin and occasional small nucleolus seen in the peripheral blood, bone marrow, and lymph nodes. Larger lymphoid cells (prolymphocytes and paraimmunoblasts) are present in lymphoid follicles, usually clustered in pseudofollicles (proliferation centers) or, less often, distributed evenly throughout the lymph node. In some cases, the small lymphoid cells show moderate nuclear irregularity or cleavage. Most patients whose lymph nodes contain the characteristic infiltrate will have a **lymphocytosis** (10 to >100 × 10⁹/l) and bone marrow involvement. The lymphocytes are more fragile than normal cells, smear cells being frequent when blood or bone marrow is spread on a glass slide. Some patients are nonleukemic throughout their illness.

Immunophenotyping reveals a characteristic pattern of CD5⁺, CD19⁺, CD23⁺, with weak CD20, CD22, CD79b, and clonally restricted surface Ig expression and negative FMC 7. **Cytogenetic analysis** shows a heterogeneous disorder with five common abnormalities.

Deletions of 13q (up to 50% of cases)
Trisomy 12 (≈20% of cases)
Deletions of 11q (≈20% of cases)
Deletions of 6q (≈10% of cases)
Deletions of 17p (4 to 15% of cases)

The postulated cell of origin is a CD5⁺ CD23⁺ peripheral B-cell.[137]

Clinical and Laboratory Features[26,27,138]

A disorder of older adults with a male:female ratio of 2:1. Patients present with peripheral blood lymphocytosis, but only 50 to 70% will have **lymphadenopathy** at diagnosis. A mild anemia invariably develops at some stage. About half of patients are found by chance on routine peripheral blood examination. Other patients will present with **lymphadenop-athy, splenomegaly**, hepatomegaly, or extranodal infiltration, particularly of the skin. **Autoimmune disorders**, particularly **autoimmune hemolytic anemia** (10 to 20%) and **immune thrombocytopenia** (1 to 2%) with **purpura**, can complicate later stages of the disease. **Monoclonal gammopathy** may be found to a minor degree, associated with hypogammaglobulinemia (about 10% of patients) and mild **immunodeficiency**.

Transformation may occur due to **Richter syndrome**, where the histology is similar to **diffuse large B-cell lymphoma**, or rarely, to **acute lymphoblastic leukemia** or **myeloma-**

TABLE 37

Staging of Chronic Lymphocytic Leukemia

Rai	Binet
Stage 0 Absolute lymphocytosis >15 × 10⁹/l	Stage A 0–2 lymph node areas
Stage 1 As stage 0 + enlarged nodes	Stage B 3–5 lymph node areas
Stage II As stage 0 + enlarged liver or spleen	Stage C Hb <10 g/dl
Stage III As stage 0 + Hb <10 g/dl	or
Stage IV As stage 0 + Plt <100 × 10⁹/l	Stage C Plt <100 × 10⁹/l

tosis. Survival from Richter syndrome is only 6 to 12 months, whereas overall survival is 10 to 12 years. Many patients have a normal life expectancy and may never require treatment with **cytotoxic agents**, as the condition remains stable.

Prognosis

Five prognostic parameters have been identified:

Clinical staging, Rai or Binet (see Table 37).

Lymphocyte doubling time.

CD38 expression high, indicates an aggressive course.

Cytogenetics: deletions at 11q22-23, 17p13, and 6q21 without mutation indicate a poor prognosis. Mutated lymphocytes indicate an indolent course.

Immunoglobulin VH gene status: few or no V-gene mutations indicates an aggressive course. An easily determined indicator of this state is obtained by using flow-cell cytometry to measure the presence or absence of a tyrosine kinase zeta-chain-associated protein-70 (ZAP-70). Median survival has been shown to be significantly shorter in those who are ZAP-70 positive.[139]

Treatment

Chlorambucil was the main chemotherapy agent used up until the mid-1990s. More recently, **fludarabine** alone or in combination with an **alkylating agent** and/or **rituximab** has resulted in response rates of over 70% and longer periods of remission. CHOP chemotherapy is also as effective.[140] Patients who have p53 mutations/deletions respond best to **corticosteroids** and **alemtuximab**.

CHRONIC MYELOGENOUS LEUKEMIA

See also **Juvenile chronic myelocytic leukemia**.

(CML; chronic myeloid leukemia; chronic granulocytic leukemia) A **myeloproliferative disorder** that originates in a pluripotential bone marrow stem cell and is consistently associated with the Philadelphia (Ph) chromosome and the *BCR/ABL* fusion gene. It is characterized by anemia, granulocytosis, thrombocytosis (usually), and splenomegaly. There is an incidence of about 1 to 1.5 cases per 100,000 population per year.

Pathogenesis[141]

There is no known cause for CML, but ionizing radiation (atomic bomb and Chernobyl disaster survivors) has been implicated. In 1960, Nowell and Hungerford described an apparent deletion of the long arm of chromosome 22 — the Philadelphia chromosome (see **Cytogenetic analysis**). In 1973, Rowley discovered that the deleted part of chromosome 22 had in fact been translocated to chromosome 9 and that a part of the long arm of chromosome 9 had translocated to chromosome 22 — t(9;22)(34;11). Later studies revealed that the *ABL* oncogene, which usually resides on chromosome 9, is translocated to a region of chromosome 22 called the break-point cluster region (BCR). The *C-SIS* oncogene is similarly translocated from chromosome 22 to chromosome 9. Various break points on chromosome 9 have been described and may be up to 40 kb upstream of the *ABL* gene. Similarly, the breakpoints on chromosome 22 may vary, but most are within the 5:8 kb *M-bcr*, which is part of the much larger *BCR* gene. Within the *M-bcr*, the break points are usually between exons 2 and 3 or exons 3 and 4. The 3′ portion of the translocated *ABL* oncogene fuses with the 5′ portion of the remaining *BCR* gene to form a hybrid gene that gives rise to an 8.5-kb RNA transcript, which is in turn translated to give a 210-kDa protein. This protein has tyrosine kinase activity and in mouse models is capable of inducing leukemia. The translocated *C-SIS* gene is not thought to play a role in the pathogenesis of CML. In 90 to 95% of cases, the Philadelphia chromosome is readily identified, but in the remaining 5 to 10%, although not karyotypically obvious, molecular techniques reveal that the *BCR/ABL* hybrid gene is usually present.

The exact mechanism of leukemogenesis is unknown, but the p210 protein does inhibit normal **apoptosis** and prolong **granulocyte** survival. In the initial phase of the disease, the proliferating cells have an advantage over the normal cells but show no difference in behavior. Organization of the cell membrane, the cytoskeleton, and the pathways between integrin receptors with the nucleus is probably disturbed. *BCR/ABL* may induce degradation of inhibitory proteins.

Clinical Features

Although CML can occur at any age, it is commonest between the ages of 30 and 60 years. It occasionally occurs in children apart from J-CML. The onset is usually insidious, with nonspecific symptoms due to:

Anemia: tiredness, lethargy, dyspnea

Hyperviscosity syndrome: headaches, dizziness, coma, priapism, abdominal pain

Organ infiltration: abdominal pain, bone pain

Increased metabolic rate: fever, weight loss, sweats

Splenomegaly (80 to 90%): abdominal pain from splenic infarction

Hepatomegaly (50%)

Thrombocytopenia: bleeding and bruising

Hyperuricemia: gout

Acute febrile neutrophilic dermatosis (Sweets syndrome) is a rare complication

Laboratory Features[26,27]

Leukocytosis (20 to $500 \times 10^9/l$) with a marked left shift of **neutrophils**, with myeloid cells at all stages of differentiation, is present. Blasts can make up to 12% of peripheral blood cells. **Basophilia** occurs in all patients, and **eosinophilia** is also very common. The **neutrophil alkaline phosphatase** (NAP) score is characteristically low. Anemia is present in over 80% of cases, with occasional nucleated red cells seen, but anisopoikilocytosis is not marked. The platelet count is raised in 50 to 60% and lowered in 10 to 15% of cases. Platelet morphology may be bizarre, and macrothrombocytes are usual.

Bone marrow is markedly hypercellular, with left-shifted myelopoiesis and an increase in the myeloid:erythroid ratio. Megakaryocytes are plentiful, with immature forms often seen. Eosinophils and basophils are increased. Marrow fibrosis is increased in about 50% of cases at diagnosis. Cytogenetic analysis shows the characteristic chromosomal abnormalities of Philadelphia chromosomes, e.g., trisomy 8, loss of the Y chromosome, 17q+, in 10 to 30% of patients at diagnosis. **Immunophenotyping** shows weak expression of normal neutrophil antigens CD15 and HLA-DR. Hyperuricemia and increased serum **cobalamin** levels are usual.

Course and Prognosis

Three distinct clinical phases occur:

Chronic phase (CML-CP): blasts <5% in blood/marrow and clinical control is relatively easy. The length of the chronic phase without treatment is highly variable.

Accelerated phase (CNL-AP): blasts plus promyelocytes in marrow/blood at 5 to 30%, and new cytogenetic abnormalities, with disease control becoming more difficult. The accelerated phase is usually superseded by progression and new cytogenetic abnormalities within 6 to 12 months.

Blast crisis (CML-BP): blasts plus promyelocytes in marrow/blood >30%, with resistance to therapy and ensuing marrow failure. It represents enhanced myeloid proliferation and cell survival, with arrest of differentiation. Of untreated patients, 75% develop this change after a median of 3.5 years from diagnosis. Additional chromosomal abnormalities typically appear, particularly deletions of the p53. Other changes occur to genes *RB1*, *MYC*, *p16^{ink4a}*, *RAS*, *AML-1*, and *EVI-1*. The blast cells have myeloid and monocytic antigens (CD13, CD14, CD15, CD33) and megakaryocytic antigens (CDw41, CD61) and/or erythroid differentiation with glycophorin or hemoglobin A. Precursor B or T-lymphoblasts show their characteristic antigenic structure. Rarely, lymphoid and myeloid lineage blasts are present simultaneously. It causes **bone marrow hypoplasia** with neutropenia and thrombocytopenia, often secondary to increased marrow fibrosis.

Historically, median survival of patients with CML was 3.5 years, with 80% eventually transforming to blast crisis, which was either myeloid (80 to 90%) or lymphoid (10 to 20%). Death was usually due to overwhelming infection or hemorrhage. However, new therapeutic interventions have changed the natural history of the disease, and the prognosis has considerably improved.

Management

If hyperviscosity is present, initial treatment with **hydroxyurea** and **leukopheresis** to quickly lower the white count over 1 to 2 weeks will bring rapid symptomatic relief.

The earliest form of treatment was either radiotherapy or splenectomy, followed by the introduction of busulfan and, later, hydroxyurea. In 1987, Yoffe and his colleagues described two patients who became completely Philadelphia-negative and polymerase chain reaction (for *BCR/ABL*)-negative after treatment with α-interferon. Prospective studies showed that interferon induced a hematological response in about 70% of cases and a cytogenetic response in 20 to 30%. Patients with <30% Philadelphia cells could expect a 5-year survival of 90%. The cytogenetic response to interferon-α took 6 to 18 months. Combination therapy with **cytarabine** increased the response rate and was widely used. In 1999 a study using **imatinib mesylate**, an oral protein-**tyrosine kinase** inhibitor, resulted in an excellent hematological (95%), major cytogenetic (60%), and complete cytogenetic (41%) response in patients who had failed to respond to interferon-α/cytarabine therapy.[142,142a] Further studies (400 mg daily) showed an improved major cytogenetic (83%) and complete cytogenetic (61%) response in previously untreated chronic-phase CML patients. Molecular studies have shown, however, that only about 3% of patients become PCR negative for *BCR/ABL*, indicating that in the overwhelming majority of patients, imatinib mesylate does not eradicate the condition, self-renewing leukemic cells still being present in the bone marrow after therapy. It is not exactly clear how long imatinib therapy will control the disease, as imatinib resistance is being increasingly recognized. These patients may respond to increasing the dose of imatinib or adding further therapies, e.g., cytosine or interferon-α. Further studies are presently being conducted using larger doses of imatinib and combining imatinib with other therapies in newly diagnosed previously untreated patients. Imatinib, usually at higher doses (600 to 1000 mg daily) either alone or in combination with other agents, also has activity in blast crisis. Adverse drug effects of imatinib are generally mild, with nausea, vomiting, diarrhea, edema (including periorbital), muscle pain, and headache. Liver function tests need to be monitored, but hepatotoxicity is rarely severe enough to warrant discontinuation of therapy. Myelotoxicity is problematic in about 10% of patients, necessitating lower doses or even discontinuation.

Allogeneic stem cell transplantation with a matched sibling donor may cure up to 80% of patients, but as age increases, mortality from **graft-versus-host disease** also increases. This form of therapy is likely to be reserved for those who fail to respond initially to chemotherapy. Radiotherapy, splenectomy, and autologous bone marrow transplantation have all been used in the past, but these are not universally beneficial, and so their use will probably decline as results from chemotherapy continue to improve.[142b]

Blast crisis in those who relapse or fail to respond to imatinib should continue to be treated with the same high-dose chemotherapy as employed for **acute myeloblastic leukemia** (AML) With lymphoid blast crisis, maintenance therapy as used in **acute lymphoblastic leukemia** (ALL) should also be given.

CHRONIC MYELOID LEUKEMIA

See **Atypical chronic myeloid leukemia**; **Chronic myelogenous leukemia**.

CHRONIC MYELOMONOCYTIC LEUKEMIA

See also **Myelodysplasia**.

(CMML) A clonal disorder of a bone marrow stem cell in which **monocytosis** is a major feature. Typically, the blood contains more than $1 \times 10^9/l$ monocytes (usually 5 to $10 \times 10^9/l$) and sometimes also moderate numbers of miniature **granulocytes**. The marrow is compatible with myelodysplasia (MDS) and shows a mixture of myeloid precursors, most often not blasts. The monocytes have abnormal nuclei, which are often convoluted or hypersegmented. The results are negative for Philadelphia chromosome and *BCR/ABL* fusion genes. Clonal cytogenetic abnormalities occur, but none are specific or recurring. Raised serum and urinary **lysozyme** and increased amounts of polyclonal **immunoglobulin** are sometimes seen. The prognosis is extremely variable, with a median survival time of 20 to 40 months.

CHRONIC NEUTROPHILIC LEUKEMIA

(CNL) A rare **myeloproliferative disorder**, characterized by sustained peripheral blood **neutrophilia**, bone marrow hypercellularity due to neutrophil granulocyte proliferation, and hepatosplenomegaly. There is no Philadelphia chromosome or *BCR/ABL* fusion gene present.

TABLE 38

Diagnostic Criteria for Chronic Neutrophilic Leukemia

	Positive	Negative
Peripheral blood	leukocytosis > 25 × 10⁹/1	no acute or chronic infection
	segmt./bands >80% WBC	no malignant disease
	immature granulocytes <10% WBC	—
	myeloblasts, 1% WBC	—
	NAP score ↑	—
Bone marrow	hypercellular, increased neutrophils	no myelodysplasia
	myeloblasts <5% nucleated cells	no other myeloproliferation
	neutrophil maturation normal	no Philadelphia chromosomes
Liver/spleen	hepatosplenomegaly	no BCR/ABL fusion genes

Source: Adapted from Imbert, M., Vardiman, J.W., Bain, B. et al. in *WHO Tumours of Haematopoietic and Lymphoid Tissues*, WHO, Geneva, p. 27.

The clinical features are similar to those of **chronic myelogenous leukemia**, the diagnosis being made by exclusion based upon criteria shown in Table 38.

Hepatosplenomegaly

Survival without treatment varies from 6 months to 20 years, but the disease is progressive with increased neutrophilia, ending with myelodysplasia or **acute myeloblastic leukemia**. There is no known specific treatment, but **cytotoxic agent** therapy with **busulfan, hydroxyurea**, or **interferon**-α has been given.

CHURG-STRAUSS SYNDROME

See **Polyarteritis nodosa**.

CICLOSPORIN

(Cyclosporin) A calcineurin inhibitor that is potent for **immunosuppression** without myelotoxic effects but with marked nephrotoxic effects. It is a cyclic undecapeptide isolated from the fungus *Tolypocladium inglatnon ganus*, which suppresses **lymphocyte**-proliferative responses to mitogens and alloantigens in a dose-dependent fashion. These effects are mediated via T-cells. The mechanisms of this mediation include the prevention of precursor **cytotoxic T-lymphocyte** activation and the subsequent development of interleukin-2 responsiveness, while inhibition of B-cell stimulation appears to be due to inhibition of helper T-cells. Its particular use is in the prevention of **graft-versus-host disease** (GVHD) following transplantation, including **allogeneic stem cell transplantation** (SCT).

CIRCULATING COAGULATION-FACTOR INHIBITORS

See **Coagulation factors**.

CITROVORUM FACTOR

See **Folinic acid**.

CLONALITY

A concept derived from the establishment of a tumor from a single abnormal cell (monoclonality). It gives rise to identical progeny with a selective advantage over normal cells derived in turn from a variety of stem cells (polyclonality/oligoclonality). Whereas clonality is usually equated with **malignancy**, this has also been demonstrated in a number of nonmalignant conditions, usually autoimmune in origin (e.g., Hashimoto's thyroiditis, **immune thrombocytopenia**).

Interpretation of clonality testing requires an understanding of the principles of each test used and the sensitivity of that test. Most molecular clonality assays use the Lyon hypothesis of random somatic X-chromosome inactivation in the cells within any one tissue, and analysis is therefore restricted to female subjects (see **Lyonization**). Thus, within any given tissue, any gene with differential isoform expression, or with different polymorphisms of genomic DNA from both alleles of the maternal and paternally derived X-chromosomes, will produce a mixed pattern of both in normal material. A small proportion of tissues will by chance show a skewed lyonization pattern, and this has to be interpreted with caution using appropriate control tissue before it can be concluded that this is normal or malignant clonal expansion within a normal tissue.

Differential expression of the two **glucose 6-phosphate dehydrogenase** (G6PD) enzyme isoforms (A and B) was first used to confirm clonality of hematological malignancy (and premalignancy).

Cytogenetic analysis defines clonality as two metaphases with a structural abnormality (translocation) or three with a numerical abnormality, to distinguish these from random abnormalities, and is the most commonly used clonality test in hematological clinical practice. This technique is applicable only to dividing cells, usually of the lymphoid lineage, and it is technically difficult in hypoproliferative marrow disorders. Sensitivity is low, the limit of detection being approximately 10%. The use of fluorescence *in situ* hybridization (FISH) for detection of interphase numerical abnormalities improves the sensitivity, but it is technically difficult and largely remains a research tool.

DNA-based molecular clonality assay improves the sensitivity. X-linked probes for the *HPRT*, *PGK*, *M27b*, and, more recently, the human androgen receptor (HUMARA) genes are applied to restriction-enzyme-digested genomic DNA. Initial restriction-enzyme digestion is necessary to determine heterozygosity for each gene, and the informative rate varies from 30% for *HPRT* and *PGK* to 75% for *M27b* and >90% for HUMARA. Restriction digestion is then performed using the enzyme iso-schizomers MspI and HpaII, which both cut the same DNA recognition sequence when unmethylated, but MspI only cuts the methylated sequence. This produces characteristic band patterns for each different method. The PGK and HUMARA assays are also applicable to PCR and HUMARA for RNA analysis (RT-PCR).

CLONAL SELECTION

See also **Lymphocytes**.

The expansion of a clone of lymphocytes through selection by antigen binding to the antigen receptor specifically expressed by a cell. It depends on the phenomenon that a lymphocyte possesses a unique receptor (**immunoglobulin** or **T-cell receptor**).

CLOT RETRACTION

Shrinkage of normal blood clots following **coagulation**. It is dependent upon active metabolic processes in **platelets** that bring about contraction of actinomyosin-like protein and may be mediated by the integrin glycoprotein IIb/IIIa. It can bring together edges of small wounds and make clots less susceptible to **fibrinolysis**. Clot retraction is abnormal in **Glanzmann's thrombasthenia**, where platelets are defective in membrane glycoprotein IIb-IIIa.

COAGULATION-FACTOR CONCENTRATES

The products of plasma fractionation used in the treatment of coagulation disorders (see **Blood components for transfusion** — separation and storage). These products include concentrates of **fibrinogen, factors VIII, IX, XI,** and **XIII, Von Willebrand Factor** (VWF), **protein C, antithrombin III,** and prothrombin complex concentrates (PCC). They have had the major infectious side effects of hepatitis B and C viruses, including chronic **liver disorders, human immunodeficiency virus** (HIV), and there remains concern about variant Creutzfeldt-Jacob disease (CJD) in products from countries with known cases of bovine spongiform encephalopathy (BSE) in cattle (see **Transfusion-transmitted infection**). Historically, coagulation-factor concentrates have been associated with **immune hemolytic anemia**. Development of coagulation inhibitors (alloantibodies) in those with congenital deficiencies remains a concern. The safety of concentrates has been improved through careful donor selection; screening for hepatitis B, hepatitis C, and HIV; and viral inactivation processes.[144] Viral inactivation can be achieved by pasteurization, terminal heating of lyophilized factor concentrates (dry heating), and treatment with solvent/detergent.

Intermediate-purity concentrates have high levels of **Von Willebrand Factor, fibrinogen, fibronectin,** and **immunoglobulin** in their final product. High-purity-affinity products are prepared from a separation procedure incorporating a chromatographic step or immunoaffinity purification using monoclonal antibodies. These products have significantly lower levels of other protein components, apart from the added human albumin as a stabilizer. Recombinant factor VIII (rFVIIIc) was developed following the cloning of the factor VIII gene. Several rVIIIc products are available and are safe, effective, and well tolerated. The second-generation products added human albumin, which was replaced by sugars, but the third-generation products have no added animal protein. rFVIIIc products are free from transmissible viruses, including parvovirus B19 and hepatitis delta virus. Similarly, a single recombinant factor IX has been produced and is widely available.

Factor VIII Concentrates

Treatment with factor VIII concentrates is effective in reducing hemorrhage in patients with hemophilia A. To reduce the number of bleeds and overall morbidity, particularly in young children, thrice weekly prophylactic factor replacement therapy can be given. This will prevent chronic joint damage and other long-term effects of recurrent bleeding. Treatment with replacement therapy is also required in patients who need acute or elective surgery.

Human Factor VIII Concentrate

Such concentrates are produced by a number of different manufacturers and use either plasma derived from paid donors (largely commercial manufacturers) or volunteer donors (largely national fractionators). There is a range of technologies used to produce factor VIII concentrates, ranging from simple salt precipitation systems, through affinity chromatography, to monoclonal antibody purification systems. All of these technologies use large plasma pools (greater than 2000 l) derived from between 2,000 and 30,000 donors. The differing technologies produce products of differing purity, defined by the amount of contaminating nonfactor VIII protein in the product, expressed as specific activity in units of VIII activity per milligram total protein. Monoclonal antibody systems and affinity chromatography produce the purest factor VIII concentrates (specific activity approximately 3000 units/mg protein before formulation), and simple precipitation systems produce the least pure (1 to 10 units/mg protein). There has been a general move toward products of higher purity for clinical use, initially on theoretical grounds, as high-purity factor VIII concentrates were shown to reduce the rate of CD4$^+$ lymphocyte decline in HIV-infected hemophilic patients. The use of factor VIII concentrates has revolutionized the care of patients with severe hemophilia, enabling home care/treatment programs to be instituted.

Plasma-derived factor VIII concentrates have been associated with infectious complications, notably hepatitis B, hepatitis C, and HIV infection. Because plasma-derived concentrates are produced from large multidonor plasma pools, they were at high risk of contamination by virus-infected donors. Consequently, they became a vehicle for the transmission of blood-borne virus infections. The first to be recognized were hepatitis B, followed by non-A, non-B hepatitis (now recognized to be predominantly hepatitis C), and, of most concern, HIV. In the 1980s, donor screening and donor deferral were introduced to try to reduce contamination of plasma pools. All donors are now screened for hepatitis B, hepatitis C, and HIV infection and must be shown to be negative for each virus prior to inclusion in the donor pool. In addition, virus-inactivation procedures have been introduced to eliminate virus contamination that escapes the screening stage. Such processes include stringent terminal dry heat, pasteurization, solvent detergent treatment, and, to some extent, the technologies used to produce concentrate, which appear to exclude viruses. These processes have been extensively validated, proven to be effective in clinical study, and have essentially eliminated virus transmission by plasma-derived factor VIII concentrates that are available today. However, hepatitis A virus and **parvovirus** B19, which are nonlipid-enveloped viruses, are resistant to solvent detergent processes. Transmission of hitherto unknown infections remains a concern with the use of factor VIII concentrates.

Although infectious complications are the major concern with use of factor VIII concentrates, such concentrates are also associated with other complications, such as inhibitor development (see **Factor VIII**) and mild-to-moderate blood transfusion complications due to contaminating alloantigens present.

Porcine Factor VIII Concentrate

A polyelectrolyte fraction of porcine plasma. It was an important treatment option for patients with factor VIII inhibitors. However, it is now in very limited supply due to concerns over potential porcine *parvovirus* transmission, as it is not virus inactivated and is significantly more costly than other factor VIII concentrates. The suitability of a patient for porcine factor VIII depends partially on the degree of cross-reactivity of the patient's inhibitor with porcine factor VIII. The level of cross-reactivity is determined by modifying the Bethesda assay by substituting porcine factor VIII for human factor VIII (see **Factor VIII**). If the level of cross-reactivity is low, porcine factor VIII concentrate can be used and is likely to be effective. After treatment with porcine factor VIII, it is likely that the degree of cross-reactivity will rise. This is particularly common in high-responding patients, although in many patients porcine factor VIII can be used repeatedly with good effect. A common side effect of plasma-derived porcine factor VIII concentrate was chills/fevers after the initial infusion. **Thrombocytopenia** and more serious anaphylactic reactions have been reported occasionally. Current B-domain delta recombinant factor VIII concentrate is undergoing phase I studies.

Recombinant Factor VIII

This protein is produced in either Chinese hamster ovary cells or baby hamster kidney cells by transfection of the cell line with factor VIII-containing clones. Both full-length factor VIII protein and B-domain-deleted factor VIII concentrates are available. Following monoclonal purification, the final product was initially stabilized with human albumin (first generation). Stabilization was then achieved without the use of human albumin by using sugars (second generation). Now available are "protein-free" products that do not use adventitial animal protein material in their production and are theoretically the safest factor VIII concentrates available. Recombinant factor VIII concentrates have been shown to be efficacious and safe in clinical trial, with similar characteristics to monoclonally purified plasma-derived concentrates. As there is nowadays no protein in the material used, recombinant products offer a factor VIII concentrate free of the risk of transmission of blood-borne viral infections.

Factor IX Concentrates

These concentrates are the most effective treatment for patients with hemophilia B. Unlike factor VIII, factor IX readily diffuses into extravascular spaces, and it is necessary to give twice the amount per dose once daily to treat bleeding in these patients. Factor IX concentrates are also known as prothrombin complex concentrates (PCC), since they contain other vitamin K-dependent coagulation factors (II, X, and occasionally VII). However there is a risk of thrombosis and disseminated intravascular coagulation (DIC) due to factor Xa activation. This led to the development of high-purity factor IX concentrates using monoclonal, metal chelation, or affinity-chromatography techniques. These high-purity plasma-derived concentrates have superseded other products for the treatment of hemophilia B when recombinant factor IX is not used. They contain lower levels of adventitial activated-coagulation proteins and are safer with regard to risks of thrombosis. A recombinant factor IX concentrate is now available and is rapidly becoming the treatment of choice.

Worldwide, human plasma-derived factor IX concentrates provide the main source of concentrate for patients with hemophilia B. They are manufactured from large pools of donated plasma, are virus inactivated, and are produced by differing technologies, from modified Cohn fractionation, through affinity chromatography and metal chelates, to monoclonal antibody purification. As with factor VIII concentrates, their use in the past has been associated with transmission of blood-borne virus infections, such as HIV and hepatitis B and C.

The older-generation factor IX concentrates are produced from plasma by Cohn fractionation. These factor IX concentrates contain significant contaminating quantities of other coagulation factors, such as prothrombin (factor II), factor VII, and factor X. These factor IX concentrates are also known as prothrombin complex concentrates (PCC). In addition to significant levels of prothrombin and factor X, PCCs also contain activated factors IIa, VIIa, Xa, and IXa. These contaminant factors can induce a potentially hypercoagulable state, and factors II and X accumulate upon repeated use, as they have a longer half-life than factor IX. Use of PCC in patients with hemophilia B has been clearly associated with DIC and thromboembolic complications, such as deep-vein thrombosis, pulmonary embolism, and myocardial infarction. Such thromboembolic complications are seen particularly after surgery. Latterly, PCCs have had heparin added to the final formulation to reduce the risk of thrombosis. PCCs retain a role in the treatment of oral anticoagulant overdose and factors VIII and IX inhibitors.

Factor IX concentrates of greater purity, with significantly lower levels (100-fold or greater) of contaminating coagulation factors, have superseded PCCs. Such products (affinity chromatography, metal-chelate and monoclonal purified) are the plasma-derived products of choice for patients with hemophilia B when recombinant factor IX is not available, as the risk of thrombosis is significantly lower.

Recombinant factor IX for clinical use has been produced. This product has a reduced recovery (up to 50% lower) compared with plasma-derived factor IX concentrates due to alteration in the sulfation pattern of the recombinant protein compared with wild-type factor IX in plasma-derived products. It is hypothesized that the altered sulfation leads to lower activation of factor IX. However, the *in vivo* half-life of recombinant factor IX is the same as plasma-derived factor IX.

Factor XI Concentrate

Human plasma-derived factor XI concentrates are available from a limited number of sources. Use of such concentrates has been associated with thrombosis, including pulmonary emboli and myocardial infarction. Although previously widely used, they are now only recommended for patients with severe factor XI deficiency or in less severely affected patients when other therapies, such as fresh-frozen plasma, are not practicable or are ineffective. Thrombosis is assumed to be due to factor XI concentrates containing presumed activated coagulation factors in addition to factor XI. The maximum dose recommended is 30 units/kg. Adjuvant prophylactic heparin is often used to reduce the risk of thrombosis.

Factor XIII Concentrates

Virus-inactivated factor XIII concentrates are available for the treatment of bleeding symptoms. These are frequently given on a monthly prophylactic basis because of the long half-life of factor XIII. Levels of greater than 5 IU/dl result in secure hemostasis.

Fibrinogen Concentrates

Virus-inactivated fibrinogen concentrates are available to treat the hemorrhagic symptoms accompanying abnormalities of fibrinogen. These are generally given in a dose of 20 to 40 mg/kg every 48 h. A level of 100 mg/dl will secure hemostasis.

Antithrombin III Concentrates

Derived by fractionation of plasma, these concentrates are of benefit in patients with congenital deficiencies of antithrombin III, where life-threatening thrombosis has occurred or where it may be inappropriate to administer heparin. The evidence that antithrombin supplementation is of value in acquired deficiencies, such as disseminated intravascular coagulation (DIC) and liver cirrhosis, is conflicting.

Protein C Concentrates

An immunoaffinity-purified concentrate isolated from human plasma is available. The product includes a solvent-detergent treatment step for the inactivation of lipid-enveloped viruses. These concentrates are indicated for the treatment of homozygous protein C deficiency (when complicated by massive thrombosis), for cases of neonatal **purpura fulminans**, and for **warfarin**-induced skin necrosis.

Prothrombin Complex Concentrates

These are plasma fractions that contain factors VII, II, IX, and X. They are of two general types: inactivated concentrates and activated concentrates. Inactivated concentrates were used for the treatment of hemophilia B prior to the widespread use and availability of higher-purity factor IX concentrates and factor X deficiency. They are also used for the reversal of overdosage with oral anticoagulants, as they contain factors II, IX, and X with variable amounts of factor VII. A number of such products are now available. Their use has been shown to reverse overanticoagulated patients. Following infusion, the international normalized ratio (INR) will be corrected almost immediately.

Both inactivated and activated prothrombin complex concentrates appear to be able to bypass factor VIII and have a role in the treatment of patients with factor VIII antibodies. The substance responsible for the factor VIII bypassing activity is unknown. The activated prothrombin complex concentrates contain more of the factor VIII bypassing activity and are designed specifically for use in patients with factor VIII inhibitors. Activated prothrombin complex concentrates have proved efficacious in controlling bleeding in such patients.

Prothrombin complex concentrates have limitations: the dose is arbitrary, and they can induce a hypercoagulable state, including thromboembolism, due to the activated coagulation factors present. In addition, they contain small amounts of factor VIII and can induce an anamnestic rise in the factor VIII antibody level.

Von Willebrand Factor Concentrate

Von Willebrand Disease (VWD) is a common inherited bleeding disorder due to defects in the Von Willebrand Factor (VWF) protein. Patients with type 1 VWD and having a VWF activity of >10 IU/dl can be treated with **DDAVP** (1-desamino-8-d-arginine vasopressin), but where this is inappropriate, such as with variant type 2 VWD or where prolonged treatment is required, factor VIII concentrates containing sufficient functional VWF can be used. Although these are depleted in high-molecular-weight VWF multimers, they are effective. Most frequently, intermediate-purity human factor VIII concentrates that contain well-preserved high-molecular-weight multimers of VWF are used.

There is a lack of consensus regarding the properties that are required to make a VWF-containing concentrate suitable for the treatment of VWD. It is often stated that the presence of high-molecular-weight multimers is essential for the *in vivo* correction of the defect in primary hemostasis, but this remains controversial. **Cryoprecipitate**, for example, with a full complement of high-molecular-weight multimers, does not consistently correct the bleeding time. This is explicable, given that the bleeding time is largely dependent on platelet VWF content. In practice, many VWF/FVIII concentrates available have proven clinical efficacy despite lacking some high-molecular-weight multimers. All products used for the treatment of VWD have degraded high-molecular-weight multimers of VWF, but the products that are licensed for the treatment of VWD have relatively well-preserved multimers that make such products effective in the majority of situations. Most frequently, intermediate purity human factor VIII concentrates that contain well-preserved high-molecular-weight multimers of VWF are used. New products containing solely VWF or higher purity and better preserved VWF are currently being developed. No recombinant product is currently available. As there is an absence of a good correlation between laboratory parameters and hemostatic effect, when selecting a VWF concentrate, most reliance should be placed on products that have demonstrated reproducible clinical efficacy in clinical studies. Concern has been voiced about the very high levels of factor VIII that might be achieved when using such concentrates and the risk of **venous thromboembolism**. Therefore, it is advisable to regularly check the factor VIII in conjunction with infusion if significant or repeated quantities of such products are required. Although these products can correct the bleeding time, the results are variable. However, VWF activity and factor VIII levels are consistently and reproducibly increased after infusion.

COAGULATION FACTORS

Plasma proteins involved in the formation of fibrin clot or **hemostatic plug**. These are listed in Table 39.

TABLE 39

Coagulation Factors and Their Synonyms

Factor	Common Name	Active Form
I	fibrinogen	fibrin subunit
II	prothrombin	serine protease
III	**tissue factor** (thromboplastin)	receptor/cofactor
V	labile factor	cofactor
VII	stable factor (proconvertin)	serine protease
VIII	antihemophilic globulin	cofactor
IX	Christmas factor	serine protease
X	Stuart-Prower factor	serine protease
XI	plasma thromboplastin antecedent	serine protease
XII	Hageman (contact) factor	serine protease
XIII	fibrin stabilizing factor	transglutaminase
Prekallikrein	Fletcher factor	serine protease
High-molecular-weight kininogen	Fitzgerald factor	cofactor

Reference Ranges

See **Reference Range Tables**: Table I, reference ranges of hematological values for adults; Table XIII and Table XV for inhibitors, reference ranges for coagulation tests in healthy full-term infants during the first 6 months of life; and Table XIV, reference ranges for coagulation tests in healthy premature infants (30 to 36 weeks gestation) during the first 6 months of life and Table XVI for inhibitors.

Etiology
Hereditary Deficiencies

Factor VIII: **hemophilia A**

Factor IX: **hemophilia B**

Other factors: deficiencies occur but are rare

Acquired Deficiencies

Factor deficiencies are usually mixed for each cause.

Vitamin K deficiency: factors II, VII, IX, and X
- **Hemorrhagic disease of the newborn**
- Dietary deficiency
- Malabsorption due to **intestinal tract disorders** and chronic biliary obstruction

Liver disorders: factor V, dysfibrinogenemia

Coumarin therapy: factors II, VII, IX, and X

Antibody inhibitors: factor VIII common following treatment with factor VIII concentrate; factors V, IX, and XIII rare

Disseminated intravascular coagulation: consumption of all coagulation factors, particularly V, VIII, and IX

Massive **red blood cell transfusion**: dilution of all coagulation factors

Lupus anticoagulants bind phospholipids and so prolong the clotting times of such tests as **prothrombin time** (PT) and **activated partial thromboplastin time** (APTT).

Coagulation Factor Assays

Estimation of concentration of individual coagulation factors is required when coagulation screening tests — PT and APTT — are prolonged and suggestive of a coagulation factor deficiency. In addition, they are required to assist monitoring therapy in specific coagulation factor deficiency. Coagulation factors can be assayed using three main approaches: immunological assay, chromogenic assay, and clotting-based assay. Immunological assays provide only a measurement of antigen present and not functional activity and therefore have a more limited role in clinical use. Functional assays are the assays of predominant clinical importance. Such functional assays are the chromogenic assay and clotting-based assay. Expression of quantities is based on either a percentage of a pooled normal sample or, preferably when possible, in units based upon a **reference material**.[143]

Chromogenic Assays

These are based on the ability of the chosen coagulation factor to specifically cleave an artificial synthetic chromogenic substrate. Cleavage of the chromogenic substrate releases a chromophore that is subsequently quantitated. Specific chromogenic substrates are available for the vast majority of coagulation factors. Use of chromogenic assays is increasing, as they are relatively simple and robust.

Clotting-Based Assays

These are the most common methods of determining functional activity of a coagulation factor. These assays are based in principle upon either the PT or APTT. The principle of the clotting-based assay is measuring the ability of added plasma to correct the clotting time (PT or APTT) of a plasma known to be deficient in a specific coagulation factor. The deficient substrate plasma should be totally deficient in the coagulation factor that is to be assayed, but have all other factors in normal quantity. Ability of various dilutions of patient plasma to correct the clotting time is compared with the same dilutions of a standard plasma, enabling the patient plasma activity to be quantified. This approach to assaying coagulation factors is known as parallel-line bioassay. Factors II, V, VII, and X assays are based upon the correction of PT, while factors VIII, IX, XI, and XII assays are based upon the correction of the APTT in specific deficient plasma. Historically, deficient plasma was obtained from patients with congenital coagulation factor deficiencies. Nowadays, immunodepleted plasmas are routinely used.

Clotting-based assays are widely used but are inherently complicated assays and require care to be taken in standardization, selection of reagents, and pursuit of quality assurance. The variability seen in these assays makes the use of a reliable reference or standard plasma imperative (see **Reference Range Tables**). In the absence of a reference material, the laboratory should calibrate its own normal pool and assign it a value of 100 units/dl. Clotting-based coagulation factor assays can be modified to enable quantification of inhibitory antibodies (Bethesda assay) and rates of coagulation factor generation.

COAGULATION OF BLOOD

The processes involved in the conversion of soluble **fibrinogen** to insoluble polymers of fibrin in which blood cells become enmeshed in the fibrin strands — **hemostatic plug**. Physiological overproduction of fibrin is prevented by **antithrombin III**, **protein C**, **protein S**, and by **fibrinolyis**. There is a fine balance between intravascular coagulation and fibrinolysis.

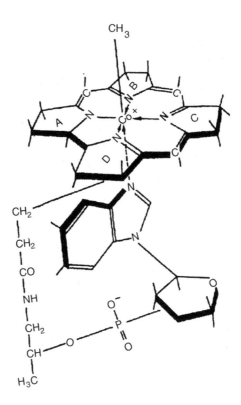

FIGURE 23
Structure of methylcobalamin. (From Chanarin, I., *Megaloblastic Anaemias*, 1st ed., Blackwell, Oxford, 1969. With permission.)

TABLE 40

Ligands of Cobalamin

Ligand	Physiological Name
–CN	Cyanocobalamin
–OH	Hydroxocobalamin
–CH$_3$	Methylcobalamin
–5'-deoxyadenosyl	5'-deoxyadenosyl coenzyme

COBALAMINS

(Vitamin B$_{12}$) A family of highly colored corrinoid compounds of high molecular weight.[145] The cobalamin molecule[146] (Figure 23) consists of:

A corrin ring (a porphyrinlike moiety) whose nucleus consists of four reduced pyrrole rings linked to a central cobalt atom. The remaining coordinate positions on the cobalt atom are occupied by an anionic group (cyanide in the therapeutic form) and a 5,6-dimethylbenzimidazole moiety.

Nucleotide bonded to the cobalt atom.

"B" group ligands attached to the cobalt atom opposite to the attachment of the nucleotide (see Table 40). These ligands provide the physiological differences.

Adenosylcobalamin and methylcobalamin are the metabolically active forms of cobalamin in mammals.

Dietary Cobalamin

Cobalamin is present in meat, eggs, and, to a lesser extent, in milk. The daily intake in a normal diet is about 5 µg, and the daily requirement is about 1 µg.

Absorption

Cobalamins are released from food in the gastric lumen, where they are adsorbed by R-binder proteins. In the duodenum, the R-binder/cobalamin complex is broken down by pancreatic proteases, thus liberating the cobalamins. The liberated cobalamin is taken up by gastric intrinsic factor (IF), and the IF/cobalamin complex passes down the small intestine to the terminal ileum, where it adheres to receptors on the brush border. The cobalamin is released from its complex and transported into the intestinal epithelial cell via specific receptors. After some hours in the cell, the cobalamins, complexed with **transcobalamin** II, are released into the portal circulation.

Transport

Cobalamins are rapidly transported by transcobalamin II via the liver to the bone marrow and other proliferating tissues. Transcobalamin II is also necessary for cellular uptake of cobalamin. Specific receptors exist for transcobalamin, which is subsequently internalized into cells by endocytosis. Excess is returned to the liver, where it is stored.

Storage

The liver is the main storage site for cobalamins. Fifty to 90% of body cobalamin is stored as methylcobalamin, the amount ranging from 1 to 10 mg. When this is within the normal range, it takes from 3 to 6 years for deficiency to be evident by hematological or neurological disturbance. There is an enterohepatic circulation for cobalamin that may equal or exceed the amount of cobalamin in the daily diet.

Metabolism

In humans, two enzymatic reactions require cobalamin. Adenosylcobalamin is required for isomerization of methylmalonyl-CoA to succinyl-CoA, particularly in the conversion of methylmalonate to succinate in myelin sheaths. Deficiency probably causes demyelination, giving rise to peripheral neuritis and **subacute combined degeneration of the spinal cord.** Methylcobalamin is an essential cofactor in the methylation of homocysteine to methionine, which is essential in the metabolism of folic acid. Deficiency is the direct cause of **megaloblastosis.** The corrinoid also maintains sulfhydryl groups in the reduced form necessary for physiological function of liver cells and red blood cells.

Excretion

Between 3 and 7 µg of cobalamin are excreted each day via the bile into the intestinal tract, most of which is reabsorbed by its binding to intrinsic factor. The daily renal excretion is up to 0.25 µg. Cobalamin not bound to transcobalamin can be cleared by the glomerular route.

Determination of Cobalamin Status[147]

Serum Cobalamin Level

Originally, these determinations were made by microbiological assay using *Euglena gracilis* or *Lactobacillus leishmanii*. These tests were cumbersome to use, took several days to perform, and were affected by antibiotics in the serum. Radioisotope dilution assays were later produced commercially in kit form and soon became the routine method of determination. Radioassay is based either on purified intrinsic factor or on intrinsic factor/ R-binder combinations in which the R-binder has been blocked by noncobalamin corrinoids.[148] The results are similar to those obtained by microbiological assay. Normal levels range from 200 to 760 ng/l, with no difference between the sexes. Assays are very sensitive for the diagnosis of clinical cobalamin deficiency and are a good measure of total body cobalamin. Normal or even elevated serum levels may be observed in cobalamin-deficient patients with concurrent liver disorders or chronic myeloproliferative disorders due to an increase in cobalamin binders in serum. Increased serum cobalamin levels occur with:

Treatment with cobalamin

Myeloproliferative disorders

Chronic myelogenous leukemia before treatment

Acute myeloid leukemia

Acute promyelocytic leukemia

Myelofibrosis

Erythroleukemia

Chronic lymphocytic leukemia (about 1/3 of cases)

Monocytic leukemia

Polycythemia rubra vera

Carcinomatosis with liver metastases

Liver disorders resulting in release of cobalamin from stores

Methylmalonic Acid (MMA)
Elevation of MMA above the normal level of 1000 ng/l is a sensitive marker for cobalamin deficiency, but is not specific. Raised levels are found with renal insufficiency and rare enzyme deficiencies.

Plasma Total Homocysteine (tHcy)
This is a very sensitive indicator of cobalamin deficiency with levels above 50 μmol/l, but of low-specificity. Raised levels occur with folate deficiency, alcoholism, renal insufficiency, hypothyroidism, and genetic enzyme deficiencies.

  Measurement of serum cobalamin levels are neither sensitive nor a specific indicator of deficiency, but the established techniques are widely used. Measurement of methyl malonic acid and homocysteine serum levels, once automated, may replace cobalamin levels for the routine diagnosis of cobalamin deficiency.

Clinical Deficiency
Causes in Adults

Nutritional disorders (rare even in strict vegans).

Lack of gastric intrinsic factor (IF) occurs with Addisonian pernicious anemia, gastrectomy, and following ingestion of corrosive materials. Assay of gastric IF is now rarely performed, but assay of IF antibodies acts as a substitute. It is usually associated with achlorhydria. In Addisonian pernicious anemia, a low cobalamin level may occur before the appearance of increased mean corpuscular volume (MCV) or anemia.

Intestinal tract disorders giving rise to ileal malabsorption of cobalamin as a consequence of bacterial overgrowth occurring with

- Diverticulosis of small bowel
- Anastomoses and fistulae
- Blind loops, pouches, or strictures

Scleroderma.

Fish tapeworm disease (see **Metazoan infection disorders**).

Alcohol toxicity.

Nitrous oxide toxicity.

Drug-induced cobalamin malabsorption (para-aminosalicylic acid, neomycin, colchicine, ethanol, potassium chloride).

Chronic pancreatic disease (rare).

Zollinger-Ellison syndrome.

Detection of intestinal malabsorption is by the **Schilling test** (now rarely available).

Causes in Children (Juvenile Pernicious Anemia)

Nutritional deficiency, usually with breast-feeding from a cobalamin-deficient mother, usually a vegan

Malabsorption of cobalamin with normal gastric acid and intrinsic factor secretion

Inborn errors of metabolism; a group of comparatively rare genetic disorders of infancy or early childhood giving rise in some to megaloblastic erythropoiesis[149,149a]

- Abnormality of receptor for IF/cobalamin complex (Immerslund-Gräsbeck syndrome)
- Enzyme deficiencies disturbing cobalamin/folate metabolism characterized by either or both methylmalonic aciduria and homocysteinuria:

  Methionine synthetase

  Methionine synthetase reductase

  Methylenetetrahydrofolate reductase

- Transcobalamin deficiencies:

  TC-II deficiency results in low serum cobalamin levels associated with raised levels of homocysteine and methylmalonic acid; the Schilling test will be abnormal

  TC-I deficiency does not give rise to cobalamin deficiency or hyperhomocysteinemia; the Schilling test will be normal

Consequences of Deficiency

Megaloblastosis with anemia.

Neurological disorders characterized by symmetrical distribution and distal onset. The initial features are sensory and include paresthesias, numbness, and loss of vibration sense. There may be visual loss from optic atrophy. Deep tendon reflexes may be lost, but hyperreflexia and spasticity only occur when the lateral columns are lost (see **Subacute combined degeneration of the spinal cord**).

Psychiatric abnormalities ranging from irritability and depression to dementia may be present.

Amblyopia associated with cigarette smoking.

Subclinical Deficiency[146]

Low to normal serum cobalamin levels (250 to 350 ng/l) occur in 10 to 20% of elderly persons and less commonly subnormal MMA levels (300 to 800 nmol/l). Causes for the low levels are usually undefined, but some progress to clinical deficiency over a period of up to 10 years.

Spuriously low values have also been recorded with folate deficiency, pregnancy, oral contraceptives, anticonvulsant therapy, human immunodeficiency virus (HIV) infection, myelomatosis, and severe transcobalamin II deficiency.

Treatment of Deficiency

Hydroxocobalamin is the most effective therapeutic agent. It is presented in ampoules containing a clear, red solution providing 1000 μg hydroxocobalamin/ml. This preparation is used intramuscularly in the treatment of pernicious anemia and other conditions associated with cobalamin deficiency, tobacco amblyopia, and Leber's optic atrophy. The preparation produces higher serum levels than the same dose of cyanocobalamin, and these levels are well maintained. The following dosage schedules are recommended:

Addisonian pernicious anemia. Initially 250 to 1000 μg intramuscularly on alternate days for 1 to 2 weeks, then 250 μg weekly until the blood count is normal. Maintenance therapy consists of 1000 μg every three months.

Neurological disorders. Initially give 1000 μg on alternate days as long as improvement is occurring. Maintenance therapy consists of 1000 μg every two months.

Prophylaxis for megaloblastic anemia resulting from gastrectomy, malabsorption syndromes, and vegans. Maintenance therapy consists of 1000 μg every three months.

Tobacco amblyopia and Leber's optic atrophy. 1000 μg daily, so long as improvement is occurring, then 1000 μg monthly.

In all patients, treatment should be continued throughout life except when the deficiency is caused by exposure to N_2O treatment or to an intestinal tract abnormality that has been surgically corrected.

The only toxic side effects are cardiac arrhythmias secondary to hypokalemia during initial treatment. Serum potassium levels should ideally be monitored initially and the patient given oral potassium supplements if low levels are found. Hypersensitivity, including anaphylaxis, very rarely occurs following an injection of cobalamin.

COBALT TOXICITY

The effects of cobalt compounds on **hematopoiesis**. This is limited to the **red blood cell**, where cobalt compounds can inactivate the enzyme 2,3-diphosphoglycerate (2,3-DPG), leading to impaired release of oxygen by red blood cells. Cobalt can also displace oxygen to form cobalt-hemoglobin, with a consequent left shift of the oxygen-dissociation curve. This results in tissue hypoxia, with increased erythropoietin production by the kidney and secondary **erythrocytosis**. It has been associated with the beer brewing industry.

CODACYTES

(Target cells) **Red blood cells** in which there is a central round stained area in addition to a rim of peripheral **hemoglobin** staining. This is a consequence of their surface area being disproportionately large compared with their volume. Codacytes are seen in:

Chronic liver disorders due to impaired cholesterol acyltransferase (LCAT) and accumulated excess cholesterol on the red cell membrane. This abnormal morphology sometimes results in a shortened cell survival. LCAT deficiency can be hereditary.

Iron deficiency.

Sideroblastic anemia.

β-**thalassemia**, **hemoglobin C**, and hemoglobin stem cell (SC) disease.

Splenectomy (post), e.g., for traumatic rupture in otherwise healthy individuals and particularly following splenectomy in patients with thalassemia.

Hereditary hypobetalipoproteinemia.

Dehydrated variant of hereditary **stomatocytosis** (hereditary xerocytosis).

COHORT LABELING OF RED BLOOD CELLS

See **Red blood cell survival**.

COLD AGGLUTININ

Agglutination of saline-suspended human **red blood cells** at low temperature, maximally at 0 to 5°C.[149] This reaction is reversible by warming. Low titers (1:32 or less) of cold-active antibodies are present in most individuals but have no deleterious effect, any autoagglutination being prevented by the thermal barrier necessary for their activity. For red blood cell injury, they require activity of the **complement** system. Most cold autoantibodies react strongly with most adult red cells and only weakly with cord blood cells, although rarely the converse occurs at temperatures below 30°C. The majority of cold agglutinins are reactive with human erythrocyte antigens of the **I/i blood group** system. I antigen complex is expressed strongly on adult red blood cells but weakly on neonatal (cord) red cells. The converse is true of i antigen complex. Thus, cold-agglutinin titers performed with adult red cells may underestimate anti-i titers. Anti-I is the predominant specificity of cold agglutinins associated with idiopathic disease and is also seen in patients with mycoplasma pneumonia and with **lymphoproliferative disorders**. Cold agglutinins with anti-i specificity are found in patients with **infectious mononucleosis** and in some patients with lymphoma. Anti-I cold agglutinins possess k light chains, whereas the anti-i variety has l light chains. Both cold-reacting IgG and IgM forms bind complement at pH 6.5 to 7.0 and lyse red cells. Neither reacts with red cells above a temperature of 32°C; however, as the temperature is lowered, the agglutinin titer increases and complement fixation occurs. The highest temperature at which agglutination occurs is referred to as the thermal amplitude. Occasionally cold autoantibodies are identified with specificity against Pr antigen. This should be suspected when the cold agglutinin reacts equally strongly with both cord and adult red blood cells and upon disappearance of agglutination with papain.

The cold-agglutinin titer should be measured using blood samples maintained and clotted at 37°C. Doubling dilutions of the serum are then made from 1 in 2 to 1 in 32 (15 tubes). Next, prepare two sets of precipitin tubes containing one volume of each dilution and one volume of a 3% suspension of group O adult (I) cells, two sets with serum dilutions and group O cord (I) cells, and another two sets with group O adult (I) cells, if available. Incubate one of each set of tubes at 4°C and the other at room temperature for 1 hour, then score the agglutination. In cold hemagglutinin disease, the titer with adult O or cord O is usually >1 in 128 at room temperature and >1000 at 4°C. Anti-I reacts more strongly with adult cells and anti-i with cord cells or with the very rare adult I cells. Most cold agglutinins are monoclonal of IgM class, as found in chronic cold-agglutinin syndrome, lymphoproliferative disorders (**angioimmunoblastic T-cell lymphoma**, **lymphoplasmacytic lymphoma/ Waldenström macroglobulinemia**, **myelomatosis**), **Kaposi's sarcoma**, and sometimes in *Mycoplasma pneumonia* infection; or they may be polyclonal when occurring in association with infections, collagen, and vascular and immune complex diseases. In infectious mononucleosis and in angioimmunoblastic T-cell lymphoma, a mixture of IgG and IgM may occur. The majority of patients with *Mycoplasma pneumoniae* infection have significant cold-agglutinin titers, but the development of hemolytic anemia is uncommon. However, subclinical red cell injury may occur. Cold agglutinins occur in over 60% of patients with infectious mononucleosis, but again, hemolytic anemia is rare.

COLD-AGGLUTININ DISEASE

Primary **autoimmune hemolytic anemias** where the autoantibodies have enhanced activity at temperatures below 37°C and usually below 31°C. They are comparatively rare disorders. No genetic or racial factors are known.[91]

Pathogenesis

Most **cold agglutinins** are monoclonal of IgM class and are unable to agglutinate red blood cells at temperatures above 30°C. The autoantibody directly agglutinates human red blood cells at temperatures below body temperature, maximally at 0 to 5°C. Fixation of **complement** to red blood cells by cold agglutinins can occur at higher temperatures but generally below 37°C. The potential of a given patient's IgM cold agglutinins to produce hemolytic anemia appears to be determined by the capacity of these antibodies to activate and bind complement to red cells at the low temperatures to which the superficial microvasculature is exposed from time to time. Although *in vitro* agglutination of red blood cells is maximal at 0 to 5°C, complement fixation by these antibodies occurs optimally at 20 to 25°C. Agglutination is not required for complement fixation to take place. The critical temperatures required for the above reactions occur at ordinary room temperatures in those patients with high-titer, high-thermal-amplitude cold antibodies. These patients tend to have a sustained hemolytic process and acrocyanosis. Patients with antibodies of lower thermal amplitude require significant chilling to initiate complement-mediated injury of red blood cells before an episode of hemolysis accompanied by **hemoglobinuria** occurs. Combinations of vaso-occlusive and hemolytic patterns also occur. Complement injures red blood cells either by direct lysis or by opsonization with subsequent sequestration, mainly by hepatic Küpffer cells and, to a lesser extent, by splenic macrophages.

Clinical Syndromes

These are all initiated or accentuated by exposure to cold temperatures.

    Chronic cold-agglutinin disease. Usually in elderly subjects with clinical features related to complement binding to red cells, with subsequent agglutination producing obstruction to blood vessels, leading to acrocyanosis, pain, numbness, and Raynaud's phenomenon. In this disorder, the hemoglobin concentration seldom falls below 7 g/dl. Laboratory diagnosis is characterized by normochromic, **normocytic anemia**, elevated **reticulocyte count**, and difficulty in spreading a peripheral-blood film. There is a decrease in the plasma **haptoglobin** level, a low-grade plasma **hemo-globinemia**, and hemosiderinuria. Cold-agglutinin titers are elevated.

Acute cold-induced intravascular hemolysis with hemoglobinuria superimposed on the chronic disease. The patient experiences fever and chills and, in extreme cases, renal failure may occur. With *Mycoplasma pneumoniae* infections, this typically occurs as the patient is recovering from pneumonia. In infectious mononucleosis, hemolysis usually coincides with the presentation of initial symptoms.

Acrocyanosis.

Laboratory Features

In chronic cold-agglutinin disease, the hemoglobin concentration seldom falls below 7 g/dl. Laboratory diagnosis is characterized by normochromic normocytic anemia, elevated reticulocyte count, and difficulty in spreading a peripheral-blood film. Clumps of red blood cells, characteristic of **autoagglutination**, may be noted on the peripheral-blood film. There is a decrease in the plasma haptoglobin level, a low-grade plasma hemoglo-binemia, and hemosiderinuria. Autoagglutination may be evident in anticoagulated blood at room temperature. This phenomenon is intensified by cooling the blood to 4°C and reversed by warming to 37°C, a property that distinguishes cold agglutination from **rouleaux formation**. In electronic cell counters, erroneous red cell indices are generated because of the autoagglutination: the mean cell volume is falsely high, and the red cell count and hematocrit are falsely low. A raised mean cell hemoglobin (MCH) and mean cell hemoglobin concentration (MCHC) are important clues to this artifact.

Cold agglutinins of high titer (>1:10,000) are found. The **direct antiglobulin (Coombs) test** with anticomplement (anti-C3 or anti-C4) is positive. The antibody reaction typically involves IgM and complement, but the IgM autoantibodies dissociate from the red blood cells subsequent to C3 binding and are therefore not generally detected *in vitro*.

Treatment

Keeping the patient warm, particularly the extremities, provides moderately effective symp-tomatic relief in patients with mild chronic hemolysis. Patients with primary cold-agglutinin disease often run a benign course, but **folic acid** supplementation is recommended. In patients with severe hemolysis, **immunosuppression** with chlorambucil or cyclophospha-mide may be beneficial, and treatment with anti-CD20 (**rituximab**) or **fludarabine** has had some success. **Corticosteroids** are only helpful for patients with either low-titer or high-amplitude IgM autoantibodies or with IgG autoantibodies. **Red blood cell transfusion** should be reserved for those patients with severe anemia of rapid onset in danger of cardiorespiratory complications. If transfusion is necessary, washed red cells should be given to avoid replenishing depleted complement components and reactivating the hemolytic sequence. The donor blood should be administered by using an in-line blood warmer at 37°C with the patient kept warm. **Plasmapheresis** may provide temporary amelioration of hemolysis, but no long-term benefit. Splenectomy is only likely to be successful in those with IgG cold autoantibodies. The postinfectious forms of cold-agglu-tinin syndrome are self-limiting.

COLD AUTOIMMUNE HEMOLYTIC ANEMIA

(Cold-induced immune hemolytic anemia disease; CHAD; cold autoimmune hemolytic disease; cryopathic hemolytic anemia; immune hemolytic anemias due to cold-reactive antibodies) The hemolytic anemia disorders arising from **autoantibodies** that have enhanced activity at temperatures below 37°C and usually below 31°C (**cold agglutinins**). These syndromes are:

Primary (idiopathic) **cold-agglutinin disease**, usually chronic.

Secondary cold-reactive antibody hemolytic anemias (*mycoplasma pneumoniae* infection, **infectious mononucleosis**, or **lymphoproliferative disorders**) either acute or chronic. Their clinical features, laboratory features, and treatment are similar to primary cold-agglutinin syndrome, but the prognosis depends upon the progress and outcome of the underlying disorder.

Paroxysmal cold hemoglobinuria, either acute (primary or secondary to infection) or chronic due to infection, particularly syphilis.

COLLAGEN

The major fibrous protein of connective tissue. The fibers are composed of masses of tropocollagen molecules, each a triple helix of collagen monomers. It is synthesized by fibroblasts, initially as a large precursor procollagen containing numerous proline and lysine residues that must be hydroxylated prior to release of procollagen from ribosomes. This molecule in the extracellular environment is cleaved from its propeptides and assembled into collagen microfibrils. These then polymerize, often with other types of collagen, to form fibrils. There are five chemical types differing in their content of hydroxyproline. Collagen types III and IV, which are richest in hydroxyproline, are the most rapidly depleted by deficiency of **ascorbic acid** (scurvy). Upon exposure, collagen activates **platelets**, but this action does not apply to monomeric collagen or to collagen type IV. As a connective tissue fiber, it is part of the **bone marrow** microenvironment. Here there is a balance between production and degradation by collagenases produced by **neutrophils**. Mutation of collagen type IV genes occurs in **Alport's syndrome**, giving rise to hematuria.[48] In **Goodpasture's syndrome**, antibodies are present against α3-chain of collagen IV, causing hemorrhage into the lungs and nephritis.

COLLOIDS FOR INFUSION

Colloids are plasma substitutes consisting of relatively high-molecular-weight molecules such as proteins and carbohydrates, which are metabolized slowly. Colloids can be used to expand and maintain blood volume in shock due to hemorrhage, trauma, septicemia, or burns.[150] Colloids commonly used include human albumin **solution,** dextrans, **gelatin,** and the etherified starches.[150]

Human Albumin Solution

A blood product manufactured by fractionation of human **plasma**. Albumin is a protein of 66-kDa molecular weight. The physiologic concentration is 35 to 50 g/l, and this accounts for 60 to 80% of the normal plasma oncotic pressure. The compositions of typical human albumin solutions (HAS) are shown in Table 41.

TABLE 41

Typical Compositions of Human Albumin Solutions

	HAS 4.5% or 5%	HAS 20%
Concentration of albumin (g/l)	45–50	200
Albumin (as % total protein)	>96	>96
Colloid osmotic pressure (mm Hg)	26–30	100–120
Sodium (mmol/l)	140–160	140–160
Potassium (mmol/l)	<2	<10
Citrate (mmol/l)	<15	<15
Bottle size (ml)	50, 100, 250, 400, or 500	50 or 100

Therapeutic Uses

Blood volume replacement: albumin is effective as a plasma substitute, but it is expensive. Many alternatives, such as polymerized gelatins, dextrans, and hydroxyethyl starch, exist; these are effective and cheaper. It may also be appropriate to resuscitate patients with crystalloids (see **Blood component transfusion**). Albumin is widely distributed, with two-thirds in the extravascular space, and infused albumin is distributed across the capillary membrane even when permeability is normal. Infused albumin may trap water in the interstitial fluid space, particularly in the lungs, and this may cause pulmonary edema. A systematic review of the use of HAS in critically ill patients suggested that these drawbacks may make crystalloid or alternative colloids a better choice.[9] A subsequent large, randomized study has suggested no difference in 28-day outcome following fluid resuscitation with 4% HAS or normal saline.[151]

Plasma exchange (see **Hemapheresis**): this often involves the removal of 3 to 4 l of plasma on several successive days. Failure to maintain plasma oncotic pressure can lead to hemodynamic instability. Ordinarily, two-thirds of removed plasma is replaced with albumin solution and the remainder with crystalloids.

Edema in patients with hypoproteinemia: this may result from loss of plasma proteins through the kidneys or gut or by underproduction in chronic liver disorders, which may lead to diuretic-resistant edema. Infusions of 20% albumin followed by the combination of frusemide and an aldosterone antagonist produces a diuresis. Twenty percent albumin is preferred, since it is hyperoncotic and has a low salt content.

Burns: there may be considerable loss of fluid from the extracellular space in badly burned patients, and microvascular permeability is increased both in injured and healthy tissue. Crystalloid solutions are equally effective in first-phase fluid replacement when compared with albumin; the latter leads to a progressive increase in lung water. Albumin solutions may be given later if plasma proteins and thus oncotic pressure remain low.

Adverse Reactions

Pyrogenic reaction causing rigors and fever, which may be due to contamination with endotoxins. Reactions of this type relate to the rate of infusion and total volume given.

Hypotensive reactions can occur as a result of contaminating kinins and kallekreins. Patients on angiotensin-converting enzyme (ACE) inhibitors are particularly prone to such reactions, and these drugs should ideally be stopped 24 h prior to plasma exchange with large volumes of albumin.

Citrate toxicity may occur if large volumes are infused during plasma exchange, associated with gastrointestinal symptoms, including abdominal pain, nausea, and vomiting, often accompanied by headache and malaise. Most of the symptoms can be attributed to hypocalcemia and can be alleviated by calcium-replacement therapy.

HAS has not been associated with the transmission of infection.

Dextrans

Dextran 70 by intravenous infusion is used predominantly for volume expansion; it has a molecular weight of 70 kDa. Dextran 40 intravenous infusion is used in an attempt to improve peripheral blood flow in ischemic disease of the limbs; it has a molecular weight of 40 kDa. Dextrans 40 and 70 have also been used in the prophylaxis of **venous thromboembolism**, but they are now rarely used for this purpose. Dextrans may interfere with blood group cross matching or biochemical measurements, and these should be carried out before infusion is begun. Dextrans have been shown to interfere with coagulation and may exacerbate hyperglycemia (see **Dextrans**).

Gelatin

This is a protein of approximately 30 kDa. It is produced from animal bone and connective tissues.

Esterified Starch

This is a starch composed of more than 90% of amylopectin that has been etherified with hydroxyethyl groups. Different products have different degrees of esterification and distribution of molecular weight — hetastarch, hexastarch, pentastarch, tetrastarch, and hydroxyethlstarch. Each solution has a mixture of molecular sizes within it, typically between 50 and 400 kDa, with a mean molecular weight of approximately 200 kDa.

Infusion Dosage

It is important to know the characteristics and any advised maximum infusional volume before treating a patient with colloid. For most preparations, an initial volume of 500 to 1000 ml is infused to adults, and the response is monitored. Some manufacturers and formularies suggest maximum infusional volumes or specific regimens in view of the individual characteristics of the colloid. Examples of these are:

RescueFlow® (Horizon) dextran 70, 6%, infuse 250 ml, followed immediately by administration of isotonic fluids. For other dextrans, the total dosage should not exceed 20 ml/kg during initial 24 h.

Voluven® (Fresenius Kabi) hydroxyethyl starch, 6%; up to 33 ml/kg may be infused daily. The total dosage should not exceed 20 ml/kg during initial 24 h.

eloHAES® (Fresenius Kabi) hexastarch, 6%; the usual daily maximum for an adult is 1500 ml; the total dosage should not exceed 20 ml/kg during initial 24 h.

For both HAES-steril® (Fresenius Kabi) and Hemohes® (Braun), it is suggested that for the pentastarch 6% preparation, up to 2500 ml daily may be given to an adult; for the pentastarch 10% preparation, the corresponding dose is up to 1500 ml daily. Depending upon the clinical condition of the patient, crystalloid solutions or blood components may also be required.

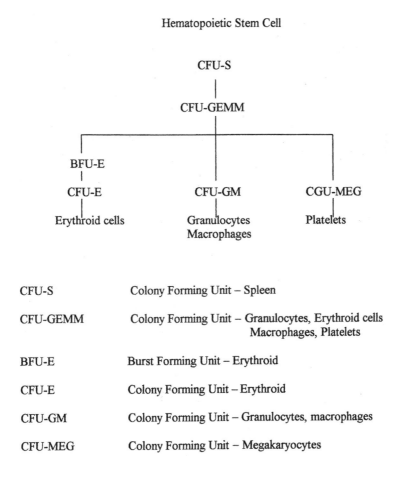

FIGURE 24
Bone marrow colony forming units (*in vitro*).

General Adverse Drug Reactions

Hypersensitivity reactions may occur, including, rarely, severe anaphylactoid reactions. Transient increase in bleeding time may occur with all colloids as a result of dilutional effects. Colloids should be used with caution in patients with raised plasma osmolality.

COLONIC DISORDERS

See **Intestinal tract disorders**.

COLONY FORMING UNITS

Hematopoietic progenitor colonies isolated from human and mouse **bone marrow**. See Figure 24.

High-Proliferative-Potential Colony Forming Cell

(HPP-CFC) An early granulocyte/macrophage progenitor cell of **hematopoiesis** more primitive than CFU-GEMM (colony forming unit-granulocytes/erythroid/macrophages/

megakaryocytes) or CFU-GM (colony forming unit-granulocytes/macrophages). Originally reported in a murine system, the HPP-CFC population is heterogeneous, with an earlier progenitor (HPP-CFU1) stimulated optimally by stem cell factor (SCF), interleukin (IL)-1, or IL-6 in untreated bone marrow, distinctly different from a later IL-3-stimulated progenitor from post-5-fluorouracil treatment and identified at 10 to 12 days of culture (HPP-CFC2). Human bone marrow assay identifies a cell capable of deriving colonies of 50,000 cells or more at 28 days of culture in a semisolid assay system supplemented with granulocyte macrophage colony stimulating factor (GM-CSF) and IL-3 for optimal growth. HPP-CFC was found to be enriched in marrow sampled from patients treated with 5-fluorouracil, although the composition of colonies was changed from predominantly monocytic pretreatment to granulocytic posttreatment.

Colony Forming Unit-Blasts

(CFU-BL) A hematopoietic progenitor capable of self-renewal, which forms small colonies composed predominantly of blast cells and thought to be more primitive than 8-day CFU-S (colony forming unit-spleen) and other committed progenitors. The assay consists of isolating an enriched progenitor population, originally using counterflow centrifugal elutriation, treating with a cycle specific chemotherapeutic agent, 5-fluorouracil, to remove committed cycling progenitors, and plating in methylcellulose with 5637 conditioned medium (or GM-CSF, IL-3, or IL-1 alone). Colonies are enumerated at day 21 to 25. These colonies can then be replated and are capable of producing single and multilineage colonies composed of every hematopoietic cell type.

Colony Forming Unit-Granulocyte/Erythroid/Macrophage/Megakaryocyte

(CFU-GEMM) The first human multipotent progenitor cell assay system isolated from bone marrow with colonies containing **granulocytes**, **erythroblasts**, **histiocytes** (macrophages), **megakaryocytes**, **eosinophils**, and **T-lymphocytes**. These primitive progenitors are found exclusively within the CD34$^+$ population in lower frequency ($\approx$0.15 to 15/10^5 mononuclear cells) and with greater variability between individuals than the more heterogeneous CFU-A progenitor compartment. CFU-GEMM represent the primitive component of the CFU-A compartment, have a low cycling status, and have limited replating (self-renewal) efficiency.

Colony Forming Unit-Granulocyte/Macrophage

(CFU-GM) A myeloid hematopoietic progenitor first identified by Bradley and Metcalf in 1996. These assays provided the model system for *in vitro* study of myeloid proliferation and differentiation. Assay systems use any of semisolid medium (agar/methylcellulose) or plasma clot for clonal colony formation. Serum-containing (batch-optimized fetal calf serum) or serum-free systems are now widely used, the latter to study individual growth factor or combination effects. Proliferation and differentiation effects vary with different growth factor additions, and indeed optimal growth factor requirements may be different for bone marrow or peripheral-blood progenitor assay. This assay is currently widely used to assess blood cell harvest yield (usually 5 to 20 $\times$ 10^4 CFU-GM/kg).

Colony Forming Unit-Type A

(CFU-A) A hematopoietic progenitor, assayable in human and murine bone marrow and analogous to the murine day-12 CFU-S in its characteristics. Representing a compartment containing more mature progenitors than CFU-GEMM and CFU-Blast, there is also greater heterogeneity of colony size and replating potential than these other primitive compartments.

Human CFU-A is assayed from low-density mononuclear cells and enriched 15-fold by CD34 selection. Cultures containing M-CSF, GM-CSF, and SCF (plus Mia PaCa-2 cell-conditioned medium) are required for optimal growth of the macroscopic CFU-A colonies (>1 mm human, >2 mm murine) scored at day 23, which contain granulocytes, macrophages, eosinophils, and normoblasts. Secondary replating efficiency (at day 13) is only 3% of primary colonies, and 7% colonies are cycling (cf. 46% for committed progenitors). The incidence of CFU-A colonies from normal human bone marrow under optimal conditions is $200/10^5$ mononuclear cells (MNC).

Colony Forming Unit-Spleen

(CFU-S) A murine assay, devised by Till and McCulloch in 1961, identified a putative stem cell capable of self-renewal. Transplantation of bone marrow into lethally irradiated mice resulted in the growth of splenic nodules, appearing initially around day 8 to 10. Initial colonies, since shown to be clonally derived (although some colonies are oligoclonal), were erythroid, with the later appearance of pure nonerythroid and then mixed-lineage foci. These 8- to 10-day colonies then disappear to be replaced by colonies of more primitive cells at day 13, with enhanced marrow repopulating activity. These colonies were therefore considered to be the progeny of a pluripotent stem cell. The potential of the day-8 CFU-S (CFU-S-8) population is estimated to be from 4,000 to 37,000 times greater than that needed in steady-state hematopoiesis. The CFU-S population can thus serve to provide a functional reserve to supply large numbers of peripheral blood cells via committed lineage-specific cells within days, whenever needed. Although most concentrated in bone marrow, CFU-S migration is well documented, and these cells can be derived from yolk sac, fetal liver, spleen, and bone marrow.

Burst-Forming Unit of Erythroid Cells

(BFU-E) The more primitive erythroid progenitor, which can be assayed in semisolid systems (agar or methylcellulose) and has been termed as a "burst-forming" erythroid unit because it produces large multifocal colonies at 10 to 14 days of culture. The primitive BFU-E matures from the CFU-S in murine models; from primitive multipotent progenitor cells (CD34+, CD38−); from human bone marrow under the influence of interleukin-1, interleukin-3, interleukin-6, and G-CSF (granulocyte colony stimulating factor); and from stem cell factor. Immature and mature BFU-Es are subsequently stimulated by interleukin-3, GM-CSF (granulocyte macrophage colony stimulating factor), and EPO (erythropoietin), resulting in several divisions before the committed unipotential colony forming unit-erythroid (CFU-E) develops. Late BFU-E becomes responsive to high concentrations of erythropoietin for optimal growth (usually 2 units/ml). BFU-E has high proliferative potential but a low proportion (≈20%) in S-phase. This progenitor circulates and can be grown from peripheral blood or bone marrow. Within the colony forming populations, there is heterogeneity of response to erythropoietin. While hypertransfusion suppresses CFU-E production, that of BFU-E expands, suggesting that CFU-E is the progeny of BFU-E. CFU-E has a greater proliferative capacity. See also **Hematopoiesis**.

Colony Forming Unit-Erythroid

(CFU-E) The committed unipotential erythroid progenitor, more mature than BFU-E and dependent *in vitro* upon erythropoietin for survival. CFU-E has a lower proliferative potential than BFU-E, but the majority are in S-phase. This progenitor, found only in human bone marrow, is also assayed with plasma clot or semisolid culture medium using relatively low concentrations of erythropoietin (cf. BFU-E requiring high concentrations). Colonies are relatively small and detected after several days of culture. CFU-E is the progeny of BFU-E and is selectively repressed by hypertransfusion (removal of endogenous erythropoietin drive) while BFU-E expands. CFU-E gives rise to the earliest morphologically recognizable erythroid cell, the proerythroblast. The more mature CFU-E forms clusters of 20 to 25 proerythroblasts within 7 days. Erythroid proliferation and maturation are stimulated by erythropoietin, the effect of which is increased as the progenitor cell matures from immature BFU-E through to CFU-E. Erythropoietin receptors are present on the surface of CFU-E. Erythropoietin is internalized into the CFU-E, where it is degraded and, in the process, activates the transcription genes, which produce α- and β-globins. This binding of erythropoietin initiates the production of cellular transcription factors, synthesis of membrane and cytoskeletal proteins, synthesis of **heme** and **hemoglobin**, and the terminal differentiation of the cells.

COLONY STIMULATING FACTORS

(CSFs) Late-acting **cytokines** concerned in hematopoietic regulation. Those identified are granulocyte/macrophage (GM-CSF), macrophage (M-CSF), and granulocyte (G-CSF) (see Figure 25).

Recombinant CSFs represent a means by which the numbers of hematopoietic cells can be modulated, thus making these agents potentially useful in treating hematological and immunological deficiencies. Recombinant CSFs are recommended to reduce the likelihood of febrile **neutropenia** when the expected incidence is >40%; after documented febrile neutropenia in a prior chemotherapy cycle to avoid infectious complications and to maintain dose-intensity in subsequent treatment cycles when chemotherapy dose-reduction is not appropriate; and after high-dose chemotherapy with autologous progenitor-cell transplantation. CSFs are also effective in the mobilization of peripheral-blood progenitor cells. Administration of CSFs after initial chemotherapy for **acute myeloid leukemia** does not appear to be detrimental, but clinical benefit has been variable and caution is advised.[152] Further research is warranted as a means to improve the cost-effective administration of the CSFs and to identify clinical predictors of infectious complications that may direct their use.

Those available are:

rhG-CSF unglycosylated (filgrastim) by subcutaneous or intravenous infusion over 30 min and glycosylated (lenograstim) by intravenous infusion

GM-CSF (Molgramostim)

Their use can be considered for:

Reduction in neutropenia following cytotoxic therapy for hematological malignancies, with the exception of **chronic myelogenous leukemia** and **myelodysplasia**

Reduction in duration of neutropenia in myeloablative therapy followed by bone marrow transplantation

Mobilization of peripheral-blood progenitor cells for harvesting and subsequent autologous or allogeneic infusion

Severe congenital neutropenia or cyclic neutropenia: molgramostim is not effective; a contraindication is **congenital neutropenia** with abnormal cytogenetics (Kostman's syndrome)

Persistent neutropenia in advanced AIDS

Throughout a course of treatment, monitoring of hemoglobin, white cell count, platelet counts, and liver enzymes is essential, and with **leukocytosis** or **thrombocytosis** it should be discontinued. For congenital neutropenias, regular bone marrow examination is recommended due to risk of myelodysplasia.

Adverse drug reactions include musculoskeletal pain, transient hypotension, abnormal liver enzymes and uric acid, thrombocytopenia, proteinuria, hematuria, allergic reactions, and cutaneous vasculitis. More generalized reactions affecting the gastrointestinal and respiratory tracts and cardiovascular system occur following treatment with molgramostim.

COLUMN-AGGLUTINATION TECHNIQUES

The methods for assessing agglutination of cells, usually **red blood cells,** by diffusion through a column of gel or beads. These methods are now widely used for **blood group** determination and red cell antibody screening in **pretransfusion testing** and **antenatal screening. Direct** and **indirect antiglobulin (Coombs) tests** can be performed using a gel in which an antiglobulin reagent has been incorporated, without the need for washing red blood cells free of antiglobulin reagent, as is the case with traditional tube techniques. The end point is also more robust and more straightforward to interpret than in tubes. In one widely available system, the antibody antigen reaction takes place at the top of a chamber that contains a Sephadex™ gel. Agglutinates are unable to pass through the gel, but in a negative reaction, unagglutinated red blood cells are centrifuged to the bottom of the reaction chamber.

COMMON LYMPHATIC ENDOTHELIAL AND VASCULAR ENDOTHELIAL FACTOR-1

(CLEVER-1) An endothelial cell protein that facilitates **lymphocyte** transmigration. It is increased in response to **inflammation** and mediates adhesion of metastatic tumor cells to the vessel walls.

COMPATIBILITY TESTING

See **Pretransfusion testing of blood.**

COMPENSATORY INFLAMMATORY RESPONSE SYNDROME

See **Acute phase response.**

COMPLEMENT

(C) An interacting multicomponent system of plasma proteins, mostly enzymes, activated in a coordinating manner. It has three functions:

Development of pores in cell membranes, thereby lysing cells

Generation of inflammatory mediators

Opsonization of antigens, thereby facilitating phagocytosis

Its major role is as an effector mechanism in **humoral immunity** to bacteria. Individuals with inherited C deficiencies are extremely susceptible to bacterial infections.

Complement is present in plasma as an inactive form with multiple, separate components that interact to generate active complement. Activation is by a cascade mechanism, in that components in a series of enzymatic reactions catalyze the activation of successively larger numbers of the next component in the cascade. There are two activation mechanisms: the classical pathway, for which the initial stimulus is antibody bound to antigen, and the alternative pathway, triggered by components of pathogenic organisms, especially bacterial cell walls. Activation takes place either on the plasma membrane of a cell or in immune complexes in solution.

Complement components are denoted by a serial notation C1, C2, etc. Activated complement components are generally notated with a bar (here underlined, e.g., C1). Activation steps are often proteolytic, and the cleavage products are notated as a and b.

Complement Activation

In the classical pathway, binding of antibody to antigen results in conformational changes in the Fc region, allowing the C1 component (itself a multimer of 18 C1q molecules and two each of C1s and C1r) of complement to bind. Only IgG1, 2, 3 and IgM can bind complement. The C1 complex must interact with at least two Fc regions; hence, IgM is particularly efficient. The binding activates proteolytic activity of C1r, which cleave C1s, thus activating its proteolytic activity. C1s cleaves both C4 and C2: the large fragments of these (C4b and C2a) combine on the antigenic surface to give an enzymic complex described as C3 convertase. Its function is to cleave C3 to generate two fragments (C3a and b). C3b binds to the C4bC2a complex to give a further enzyme activity, C5 convertase.

In the alternative pathway, C3b, generated by spontaneous cleavage, binds to a range of surfaces, particularly of bacterial cells, and they can bind factor B, which is then susceptible to cleavage by factor D so that a complex C3bBb is produced. This is able to cleave further C3 to generate C3bBbC3b, which is a C5 convertase.

Cell Lysis

In both pathways, the final step is the generation of the membrane attack complex (MAC), which punches holes in cell membranes and thereby lyses them. C5 convertase cleaves C5 to C5a and b. C5b binds C6 and C7 to give a complex C5aC6C7. The formation of this complex exposes hydrophobic regions of C7, which interacts with membrane phospholipids and thus inserts the complex into a membrane. Once inserted, C8 and multiple copies of C9 bind. C9 is a perforin-related protein (important in cellular cytotoxicity) that polymerizes in the membrane to form a pore, thus resulting in cell lysis.

Inflammatory Role of Complement

Several of the C-component fragments (C3a, C4a, and C5a) not involved in generation of the MAC degranulate basophils and mast cells, thus releasing the inflammatory mediators (serotonin and histamine) and aggregate platelets, leading to the release of their proinflammatory cytokines. C5a stimulates the extravasation of granulocytes and monocytes. Hence, C activation contributes to the development of the inflammatory response in infection.

Control of Activation

The major function of C is cell lysis, and probably the most important control is proximity to a plasma membrane. Several components of C, particularly C3b, are labile if not rapidly incorporated into complexes, and so these are inactivated if they diffuse away. This helps to ensure that only cells in the immediate vicinity are damaged. However, the forerunner of the MAC, C5aC6C7, is relatively stable: it can diffuse away from a soluble immune complex, and thus "bystander" effects do occur.

In both pathways, generation of C3b is also under control by a number of proteins collectively described as regulators of complement activation (RCA). These include factor H and decay-accelerating factor (DAF: CD55).

Complement Factor Inhibitors (CFIs)

These have effects outside the complement system. CFI inhibits C5a as well as other chemotactic factors. Deficiency, either a hereditary autosomally dominant defect or an acquired defect of the inhibitor to C1 esterase, which also inhibits plasmin and kallikrinin, allows generation of the peptides bradykinin and C2 kinin, which cause edema — **hereditary angioedema** (HAE). Uninhibited C1 cleaves C2 and C4, resulting in reduction of their serum levels, which provides a screening method for diagnosis. The normal level of the inhibitor itself is around 1.0 U/ml, and the levels for premature and full-term infants are given in **Reference Range Tables XVII** and **XVIII**. The disorder is treated by giving **danazol** and prophylactically by giving **tranexamic acid**.

Complement Deficiency

See **Immunodeficiency**.

CONGENITAL AMEGAKARYOCYTOSIS

See **Amegakaryocytic thrombocytopenia**.

CONGENITAL ANOMALIES

A number of hematological fetal disorders that may be evident either as a neonatal disorder or as a disease of childhood. They are quite distinct from hereditary hematological disorders. They arise from both maternal and intrinsic disorders of the fetus:

Maternal disorders:
- Maternal antibodies: **hemolytic disease of the newborn; immune thrombocytopenic purpura** (including **neonatal alloimmune thrombocytopenia**)
- **Parvovirus** infection: **hydrops fetalis**
- **Vitamin K** deficiency: **hemorrhagic disease of the newborn**
- Transplacental hemorrhage: anemia, hydrops fetalis

Fetal disorders
- Prematurity: anemia
- Juvenile myelomonocytic leukemia
- Vascular anomalies: **hemangiomas, hemangioblastomas**

CONGENITAL DYSERYTHROPOIETIC ANEMIAS

(CDAs) A complex, uncommon group of hereditary refractory anemias characterized by **ineffective erythropoiesis**, multinuclear erythroblasts, and secondary **hemosiderosis**.[153]

Several genetic defects may be present, since the same disorder is transmitted as an autosomally recessive trait in some families and as a dominant trait in others.

The clinical features reflect the degree of **anemia.** The diagnosis can be made at any time of life, but anemia is often detected in infancy or childhood. There is no sex predominance. Family history is frequently unrewarding. Intermittent jaundice may be the only sign. Enlargement of the spleen or liver may be present, and pigment gallstones are common. Somatic abnormalities can occur, including hyperpigmentation, mental retardation, shortened phalanges, and syndactyly. Sequelae of **iron overload**, even in the absence of transfusion or therapeutic iron, have been reported. The level of **hemoglobin** is variable. The peripheral-blood film shows marked anisopoikilocytosis with prominent macrocytes, tear-drop poikilocytes, punctate basophilia, and crenated cells. The **reticulocyte count** is low. Leukopenia and thrombocytopenia are usually attributed to splenomegaly. Increased stainable iron in sideroblasts is present in the marrow, reflected by the raised serum ferritin level. **Ferrokinetics** reveal ineffective erythropoiesis. The red cell life span is modestly shortened. The long-term prognosis in this group of disorders is good. There is no specific treatment, each case requiring management as events demand. The complications of iron overload may be significant.

Three major types of CDA have been described, although there is much overlap in diagnostic features, prognosis, and treatment, and patients have been reported presenting with a combination of syndromes.[153a]

Type I CDA

The first clinical features have been reported to appear both in early childhood and in adolescents. It is inherited in an autosomally recessive trait associated with gene *CDA1*. Patients usually present with mild anemia, moderate jaundice, and splenomegaly. No serological abnormalities have been reported. Bone marrow aspirate shows abnormalities in almost all red cell precursors, with megaloblastoid changes, macrocytosis, internal nuclear bridges, and pseudo-Gaucher cells. This condition only requires regular **iron chelation**. In most cases, **red blood cell transfusion** is not be required. **Interferon-α** may be efficient therapy.

Type II CDA or HEMPAS

(Hereditary erythroblast multinuclearity with positive acidified serum) Originally described in 1962, it is the commonest form of congenital dyserythropoietic anemia. Inherited as an autosomally recessive trait, it is most prevalent in northwestern Europe, Italy, and North Africa. The underlying defect is unknown, but recently an abnormality of red cell membrane a-mannosidase II has been described.

Clinical Features

These are variable, but jaundice (90%), splenomegaly (70%), hepatomegaly (40%), and gallstones (20%) are typical. Only about 25% of patients are sufficiently anemic to warrant red blood cell transfusion. There is marked ineffective erythropoiesis, and many patients ultimately develop cirrhosis and diabetes secondary to iron overload.

Laboratory Features

The hemoglobin and red cell morphology are variable. Anemia (which can be either absent or very severe) is typically normocytic and normochromic, with anisopoikylocytosis and basophilic stippling. Reticulocytes may be normal or slightly reduced. Ferrokinetic studies reveal a raised plasma-iron turnover and ineffective erythropoiesis due to a failure to incorporate iron into the developing red cells. The bone marrow is hyperplastic due to erythroid hyperplasia. Late erythroblasts contain 2 to 7 nuclei with otherwise normal appearance and exhibiting internuclear bridging. Megaloblastic changes are not seen, but karyorrhexis is typical.

The unique feature of HEMPAS red blood cells is their ability to be lysed by certain acidified sera of normal people (30%). These sera contain a high titer of a specific IgM anti-HEMPAS antibody. Unlike **paroxysmal nocturnal hemoglobinuria** (PNH), the **sucrose lysis test** is negative, and the patient's own acidified serum is incapable of causing lysis. As with Type III CDA, there may be increased anti-I or anti-i activity.

In patients with anemia, regular red blood cell transfusions and iron-chelation therapy may be necessary. **Splenectomy** may reduce transfusion requirements in patients where anemia is a relevant feature, but is of no benefit for any other problem. In nonanemic patients, either regular **phlebotomy** or iron chelation can be used to prevent hemosiderosis.

Type III

An autosomally dominant disorder. As in the other types of CDA, anemia is the most relevant clinical feature. A differential characteristic is the presence of gigantoblasts with up to 12 nuclei in the bone marrow and remarkable basophilic stippling. Serological data in occasional reports have shown an increased tendency for agglutination by anti-i and an increased lysis by anti-I, with no acid lysis.

CONGENITAL ERYTHROPOIETIC PORPHYRIA

See also **Porphyrias**.

A rare disorder resulting from an autosomally recessive severe deficiency of uroporphyrinogen III cosynthase. This condition does not appear to affect any particular geographic location or ethnic group. A variety of mutations of the uroporphyrinogen III cosynthase gene have been described in affected persons. The clinical features usually appear in infancy or early childhood, including severe photosensitivity, red staining of the teeth (red porphyrin fluorescence under ultraviolet illumination), hypertrichosis, and alopecia. Skin lesions appear as subepidermal vesicular or bullous eruptions on areas exposed to the sun and often heal with much scarring. **Hemolytic anemia** is associated with fluorescence of normoblasts, **splenomegaly**, **red urine**, porphyrin-rich gallstones, pathologic fractures, short stature, and other skeletal abnormalities due to erythroid hyperplasia. Massive quantities of uroporphyrin I are excreted in the urine, and coproporphyrin I is excreted in urine and feces. Patients with this disorder have been treated with topical sunscreen agents or beta-carotene to block visible light, **splenectomy**, **red blood cell transfusion** to reduce erythropoiesis, oral charcoal or cholestyramine to bind excess porphyrins, ascorbic acid or alpha-tocopherol as free-radical scavengers, and **allogeneic stem cell transplantation**.

CONGENITAL FOLATE MALABSORPTION

See **Folic acid**.

CONGENITAL INTRINSIC FACTOR DEFICIENCY

See **Gastric disorders** — intrinsic factor.

CONGENITAL NONSPHEROCYTIC HEMOLYTIC ANEMIA

(CNSHA) A term originally used to describe a group of hereditary hemolytic anemias in which the red blood cell **osmotic fragility test** was normal and that did not respond to **splenectomy**, thus distinguishing them from **hereditary spherocytosis**.[154] The group of disorders has proved to be extremely heterogeneous, both in cause and in clinical presentation. The term is reserved for those patients with no major red cell morphological abnormality. These disorders can arise from:

Red blood cell membrane disorder, either hereditary or acquired

Red blood cell enzyme deficiency, usually hereditary

Unstable hemoglobins

In many instances, the underlying lesion cannot be identified.

CONGENITAL R-BINDER DEFICIENCY

See **R-binder proteins**.

CONGENITAL RED BLOOD CELL APLASIA

See **Blackfan-Diamond syndrome**.

CONGENITAL THYMIC APLASIA

See **Immunodeficiency** — primary congenital.

CONNECTIVE TISSUE DISORDERS

(Collagen diseases) A group of clinical syndromes with widespread **inflammation** in connective tissue, with fibrinoid material seen histologically.[155] Constitutional manifestations occur in joints, blood vessels, heart, skin, and muscles. The clinical features are determined by the tissue distribution of the lesions, each syndrome having a distinct clinical pattern. Any one disorder can be associated with **vasculitis** or **microangiopathic hemolytic anemia**. The syndromes are hereditary or acquired:

Hereditary
- Ehlers-Danlos syndrome
- Pseudoxanthoma elasticum
- Marfan's syndrome
- Osteogenesis imperfecta

Acquired
- **Polymyositis**
- **Polyarteritis nodosa**
- **Polymyalgia rheumatica**
- **Scleroderma**
- **Systemic lupus erythematosus**
- **Wegener's granulomatosis**

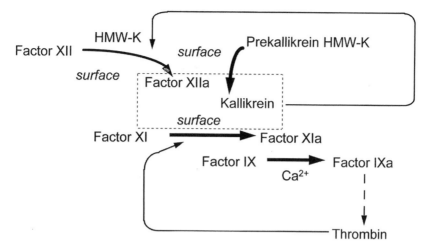

FIGURE 25
A schema of contact activation. HMW-K, high-molecular-weight kininogen.

CONSUMPTION COAGULOPATHY
See **Disseminated intravascular coagulation**.

CONTACT ACTIVATION

The reaction of a series of enzyme cascades that are initiated when plasma comes into contact with a negatively charged foreign surface. The initial activation is based around **factor XII, prekallikrein**, and **high-molecular-weight kininogen**. This method of coagulation activation is of doubtful physiological significance and is probably predominantly an *in vitro* phenomenon. Severe congenital deficiency of any of the above components produces significant laboratory abnormalities but few, if any, clinically significant effects. Most patients are discovered incidentally. There may be a weak association between severe factor XII deficiency and **venous thromboembolism**.

A schema of contact activation is shown in Figure 25. Following contact with a foreign surface, factor XII undergoes limited proteolysis to generate factor XIIa. This slowly converts prekallikrein to kallikrein and activates factors XI and VII, thereby feeding down the coagulation cascade. In turn, kallikrein releases kinins from high-molecular-weight kininogen, stimulates fibrinolysis by acting as a plasminogen activator, and fully activates factor XII to factor XIIa. This establishes a positive feedback loop. High-molecular-weight kininogen functions as a nonenzymatic cofactor and accelerator in this process. This process is not calcium dependent. Activated platelets may be involved in this process, not by generating factor XIIa, *per se*, but by providing a receptor-rich membrane for contact-activation components.

CONVOLUTED T-LYMPHOCYTIC LYMPHOMA
See **Non-Hodgkin lymphoma; Precursor T-lymphoblastic lymphoma**.

COPPER

The hematological disorders arising from deficiency or excess of copper in the body. In an adult, total body copper is approximately 80 mg and is found mainly in the liver and brain. The normal level of serum copper is 12 to 20 μmol/l. The daily requirement for copper is about 1 mg, and the average diet contains 1.5 mg.

Copper Deficiency

Etiologic factors include poor diet, inadequate total parenteral nutrition, excessive therapeutic use of zinc, and use of iron chelation. Because copper is absorbed in the duodenum, resection of the duodenum together with poor nutrition can result in copper deficiency. Copper is required for production of **ceruloplasmin** and the enzymes cytochrome oxidase and superoxide dismutase. These enzymes are used in **iron** transport, **heme** synthesis, and **red blood cell** membrane function. With copper deficiency, iron accumulates in the intestinal mucosa and liver due to a failure of iron mobilization. In the adult, there is maturation arrest of the myeloid series and failure to synthesize heme, thus giving rise to **sideroblastic anemia** and maturation arrest of the myeloid series with **neutropenia**. Both immature erythroid and myeloid cells contain cytoplasmic vacuoles. In infancy, there is additionally osteoporosis and pathological fractures.

Copper Toxicity

Inorganic copper in toxic amounts causes inhibition of -SH group enzymes of **glucose 6-phosphate dehydrogenase** and **glutathione reductase**, leading to failure in the protection of red blood cells from oxygen free radicals, with oxidation of **hemoglobin**. It also damages the red blood cell membrane. Hence, a **hemolytic anemia** can result from excess of red blood cell copper. This is rare, occurring as an accidental ingestion of a copper salt, high copper content of hemodialysis water, and as a complication of liver disorders due to sudden release of unbound copper in hepatic infarction and in hepatitis. In the early stages of hepatolenticular degeneration (Wilson's disease) with hepatic decompensation due to copper toxicity, acquired deficiency of hexokinase results in the development of hemolytic anemia.

COPROPORPHYRIN

See also **Porphyrias**.

Coproporphyrins I and III are the only two of the four possible isomeric forms of this tetrapyrrole that occur in nature. In normal humans, the type III isomer predominates over the type I isomer in urine, whereas the reverse is true in bile. Increased quantities of coproporphyrins are excreted by persons with **hereditary coproporphyria**.

CORDOCENTESIS

See **Hemolytic disease of the newborn**.

CORTICOSTEROIDS

Hormones naturally secreted by the adrenal cortex. Physiologically, both hydrocortisone, which has glucocorticoid activity with weak mineralocorticoid activity, and aldosterone, with strong mineralocorticoid activity, are secreted.

Action

Glucocorticoids are mainly anti-inflammatory:

Increased **granulocyte**-pool release, resulting in **neutrophilia**

Increased maturation of **monocytes** to **histiocytes** (macrophages), resulting in **mono-cytopenia**

Suppression of T-lymphocyte and NK (natural killer)-cell function; **lymphocyt**e death occurs by the induction of **apoptosis**; **lymphopenia** is evident

Reduced histiocyte (macrophage) activity, particularly **antigen** handling

Glucocorticoid action also increases gluconeogenesis with reduction of protein in bone, muscle, blood vessels, skin, and fat deposits. Its action is antagonized by neuropeptides.

Mineralocorticoid action is mainly sodium and water retention with potassium loss. These actions particularly apply to cortisone and hydrocortisone.

Therapeutic Indications

Pharmacological use of corticosteroids varies with the required effect, but this must be balanced with the inevitable adverse reactions. Corticosteroids are synthetic preparations, and resistance to their effects may develop over time.

Hydrocortisone

Replacement therapy for adrenal gland disorders

Acute **immunosuppression** for **anaphylaxis** shock and acute **angioedema** affecting the respiratory tract

Adult dosage by intramuscular injection or slow intravenous injection or infusion is 100 to 500 mg repeated as required, or three to four times in 24 h. The reduced dosage for children is 25 to 100 mg, depending upon age. Hydrocortisone should not be continued any longer than absolutely necessary.

Prednisolone

Long-term immunosuppression for autoimmunity disorders, such as **autoimmune hemolytic anemia**, **immune thrombocytopenia**, systemic lupus erythematosus, **polyarteritis nodosa**, dermatomyositis, **polymyalgia rheumatica**, and **Sjögren's syndrome**

Lymphocyte suppression in:

- Hematological malignancy: **acute lymphoblastic leukemia, myelomatosis, non-Hodgkin lymphoma**; it is sometimes part of a **cytotoxic agent** therapy regime
- Sarcoidosis

Stress due to severe illness or trauma, in a dose dependent upon severity

The usual adult dose is 10 to 20 mg daily by mouth, with a maximum of 60 mg for severe disorders. Due to the risk of side effects, the dose should be lowered as soon as possible commensurate with disease control. A maintenance dose of 2.5 to 15 mg daily should be achieved. Sudden reduction or removal of dosage should be avoided. A reduced-dosage schedule for children appropriate to their age and weight must be used.

Prednisone

Prednisone is similar to prednisolone, being an alternative form of corticosteroid that is less used in the U.K. It has a similar level of glucocorticoid activity, but it is only active after conversion to prednisolone in the liver.

Methylprednisolone

Aplastic anemia in which a similar response rate to that of antilymphocyte globulin can be achieved, but with increased side effects, which prolong hospitalization

Immune thrombocytopenia, where the remission rate is similar to standard prednisolone therapy, but the response is quicker in many patients

Graft-versus-host disease in doses up to 1 g/m^2/day, required to establish control; patients are weaned off over a period of days or weeks, depending on response

Non-Hodgkin lymphoma as part of a multidrug regime, e.g., etoposide, methylprednisolone, cytarabine, and cisplatin

Methylprednisolone has increased immunosuppressive activity above that of the parent compound. The adverse reactions are similar, but their severity is increased.

Dexamethazone

Used for high-dosage therapy when water retention is to be avoided, such as in cerebral edema, due to its having high glucocorticoid activity but low mineralocorticoid activity. It is also used in cytotoxic agent therapy regimes.

Adverse Drug Reactions

These vary with the predominant action of the preparation:

Mineralocorticoid effects: hypertension, edema (particularly of the face), muscle weakness, Cushing's syndrome

Glucocorticoid effects: diabetes mellitus, osteoporosis with fractures, skin striae, capillary fragility (vascular purpura), cataracts, glaucoma

Peptic ulceration (with perforation), esophageal ulceration, acute pancreatitis

Increased susceptibility to infection with risk of septicemia without clinical signs of fever; candidiasis, esophageal ulceration, and pancreatitis may be adverse reactions with long-term therapy

Suppression of growth in children

Neuropsychiatric disturbances of euphoria, paranoia, and depression; aggravated epilepsy

Musculoskeletal effects: proximal myopathy, tendon rupture, osteoporosis (fractures)

Ophthalmic effects: glaucoma, cataracts, papilledema

Skin disorders: atrophy, striae, acne, telangiectasia

COUMADIN

See **Warfarin**.

COUMARINS

A group of drugs taken for oral **anticoagulant therapy**.[156] Their action is by antagonizing **vitamin K** reductases, indirectly blocking g-glutamyl carboxylation. Dicoumarol is a naturally occurring vitamin K antagonist. Synthetic pharmacological preparations include **warfarin** sodium (Coumadin), acenocoumarol (Sinthromin), and phenprocoumon (Marcoumar).

Warfarin sodium (4-hydroxy-coumarin; Coumadin; Marevan) inhibits two vitamin K-dependent enzymes within the liver: vitamin K-dependent reductase and vitamin K-dependent quinone reductase. As a result, this prevents efficient recycling of vitamin K to its enzymatically active form and therefore limits a vitamin K-dependent carboxylase. Warfarin is the most widely used oral anticoagulant in the U.S. and in the U.K. Acenocoumarol (Sinthrome) and phenprocoumon (Marcoumar) are popular in countries other than the U.K. and the U.S.

Phenindione is an indiandione derivative rather than a 4-hydroxy derivative and appears to inhibit vitamin K reductases in a similar manner to the coumarins. It is not widely used because of potential hepatotoxicity. It can also give rise to **agranulocytosis** and **eosinophilia**.

C-REACTIVE PROTEIN

(CRP; Pentaxin) A cyclic pentameric protein derived from hepatocytes. This protein is synthesized in liver cells in response to IL-1 and IL-6 through TNF in the **acute-phase inflammatory response**. The protein forms a complex with phospholipids and C-polysaccharide on bacterial surfaces, particularly of the pneumococcus. It acts as an opsonin-stimulating phagocytosis and also activates the **complement** system through the lectin pathway. In the acute-phase response, its level in plasma increases within 6 to 10 h by up to 1000 times. Methods of measurement are by laser immunonephelometry or turbidimetry, immunoassay by homogeneous enzymes, or by fluorescein or radionuclide labeling. The results should be expressed against a WHO reference standard[157] and reported as milligrams per liter. The reference range is 0.1 to 8.0 mg/l. Sequential measurement over a 24-h period is important for effective monitoring.

The levels are raised in patients with unstable angina and, for an unexplained reason, indicate high risk of myocardial infarction, where they may be a predictor of poor outcome, but this remains unproven.[158] They are also raised to a lesser degree in cigarette smokers and those with high body mass. The levels decline rapidly once the inflammatory process has regressed.[70] Persistently raised levels are reduced by cholesterol-lowering agents such as statins or clofibrates and can be used as an indicator of reduced likelihood of recurrence.[159]

CRENATION OF RED BLOOD CELLS

Numerous tiny projections from the surfaces of **red blood cells**. It is usually an artifactual appearance due to delay in the preparation of peripheral-blood films. Scanning electron microscopy reveals that crenated cells differ from **acanthocytes**, although the distinction may be difficult to see under ordinary light microscopy.

CRISTANTABUSE

See L-**Asparaginase**.

CROSS-REACTIVE MATERIAL

(CRM) The property of a plasma that differentiates disorders of **coagulation factors** arising from absence of protein (no antigen, no activity) and reduced protein (low antigen, low activity) from a structurally abnormal protein that has reduced activity, although it may be present in relatively normal amounts (normal antigen, low activity). It is based on the detection of antigen by polyclonal antibodies. CRM⁻ samples have no antigen and no activity. CRM⁺ samples have detectable antigen with reduced activity. CRMr samples have reduced antigen and reduced activity. Within CRM⁺ and CRMr groups, there is considerable clinical phenotypic heterogeneity. This distinction is now largely of historical value due to the advance in knowledge of molecular pathology of coagulation factor disorders.

CRYOFIBRINOGENEMIA

The presence in blood of abnormal cold **precipitins** (4°C) of plasma protein complex, indistinguishable from **fibrinogen** or **fibrin**. Their presence is generally asymptomatic, but may cause multiple thromboemboli in the skin, lung, and myocardium. Cryofibrinogenemia may be idiopathic or secondary. Essential (idiopathic) cryofibrinogenemia occurs in association with **disseminated intravascular coagulation** (DIC). Secondary causes include systemic infections and malignancy. In the skin, cryofibrinogenemia usually results in **purpura** of the legs with ulceration. These lesions may be associated with Raynaud's phenomenon or livedo reticularis. It should be suspected in those with cold hypersensitivity and peripheral ischemia, especially when skin ulceration is present and no cause is apparent. Treatment includes plasmapheresis, fibrinolytic agents, warfarin, and stanozolol.

CRYOGLOBULINEMIA

See **Heavy-chain diseases**.

CRYOPATHIC HEMOLYTIC ANEMIA

See **Cold-agglutinin disease; Cold autoimmune hemolytic anemia**.

CRYOPRECIPITATE

The material precipitated from plasma when it is thawed slowly at 4 to 6°C. Cryoprecipitates are rich in coagulation **factors VIII** and **XIII, Von Willebrand factor, fibronectin,** and **fibrinogen**. The majority of cryosupernatant plasma is decanted, and the remaining protein is redissolved by warming to give a small volume of cryoprecipitate solution. Cryoprecipitate is used principally as a source of fibrinogen in the treatment of bleeding disorders, particularly **disseminated intravascular coagulation**.[160] Cryoprecipitate is no longer used in the treatment of patients with **hemophilia** or **Von Willebrand Disease**, since there is a small risk of transmitting viral infection. These patients should receive alternative factor VIII or VWF concentrates.

Cryosupernatant contains a variable amount of fibrinogen and most coagulation factors except factors VIII and XIII.[161] Cryosupernatant is occasionally effective as replacement fluid in patients with **thrombotic thrombocytopenic purpura** (TTP) who undergo plasma exchange. It is thought that depletion of factor VIII and VWF multimers prevents their deposition on endothelial walls. However, prospective studies have failed to demonstrate any significant difference in use as initial therapy. Cryosupernatants can also be used as starting material for fibrin glue.

CRYPTIC LYMPHOMAS

Lymphoproliferative disorders in which **monoclonal antibodies** of the cold-agglutinin type develop, causing **cold-agglutinin disease**. Associated hemolytic anemia is rare.

CUTANEOUS PLASMACYTOSIS

Multiple skin plaques of polyclonal **plasma cells** associated with **lymphadenopathy**. The lymphoid follicles resemble those seen in **angiofollicular lymph node hyperplasia**. It is a disorder mainly of Japanese patients.

CUTANEOUS T-CELL LYMPHOMA

See **Mycosis fungoides/Sezary syndrome**.

CYANOCOBALAMIN

See **Cobalamins**.

CYANOSIS

Bluish discoloration of the skin due to deficiency of **oxyhemoglobin** in the blood. Apart from cardiorespiratory disorders, this occurs with **methemoglobinemia**, **sulfhemoglobinemia**, **hemoglobinopathies** with low oxygen affinity, and in **erythrocytosis** and **polycythemia rubra vera**. Cyanosis of the ears and fingertips occurs after exposure to cold in individuals with **cryoglobulinemia** and **cold-agglutinin disease**. Cyanosis is a function of the total amount of reduced hemoglobin, methemoglobin, or sulfhemoglobin present. The minimum amount of these pigments causing detectable cyanosis is 5 g reduced hemoglobin, 2 g methemoglobin, and about 0.5 g sulfhemoglobin per deciliter of blood.

CYCLIC ADENOSINE MONOPHOSPHATE

See also **Adenosine monophosphate**.
(cAMP) A nucleoside produced from **adenosine triphosphate** (ATP) by the enzyme adenyl cyclase. It is an essential substance in cellular enzyme cascade. In **platelets**, it is a suppressor of platelet aggregation. Normal levels of platelet cAMP are 1 pmol/100 platelets. Elevated cAMP levels occur via stimulation of adenyl cyclase or inhibitor of cAMP phosphodiesterase (see **Phosphodiesterase inhibitors**). **Prostacyclin** stimulates cAMP production. Mechanisms involved in cAMP suppression of platelet function include inhibition of **arachidonic acid release** from phospholipids, induction of uptake of cytoplasmic calcium, and immobilization of platelet contractile apparatus.

CYCLIC NEUTROPENIA

A rare, infrequently inherited autosomally dominant disorder, characterized by **neutropenia** occurring over 14 to 35 days (usually about 21 days) and lasting 3 to 10 days. The specific abnormality is unknown, but probably resides at the stem cell level, as the defect is transplantable and other hematopoietic lineages can be affected in a cyclical manner. Heterogeneous germline mutations of the *ELA2* gene, encoding neutrophil elastase, may explain some cases. The mechanism is probably cyclical inhibition of granulocyte stem cell proliferation. The severity of neutropenia varies between patients, with some asymptomatic while others suffer severe neutropenia and infections (especially of skin and chest), becoming clinically evident within the first year of life. **Monocytes** also cycle, but in a phase opposite to that of neutrophils. Several studies have shown cyclical hematopoiesis with raised monocytes, reticulocytes, lymphocytes, and eosinophils occurring simultaneously with neutropenia. **Erythropoietin** and **colony stimulating factors** increase during the neutropenic period, probably accounting for other cell cycling. Life-threatening infections can accompany the 3- to 4-day neutropenic nadir of the cycle.

Most patients reach adulthood, and a reduction in the incidence of septic episodes at puberty has been reported. The diagnosis is made by regular **white blood cell counting**, with differential leukocyte counting at least three times per week for at least 2 months (longer periods of observation may be required if cycling of other cells is to be appreciated). A number of therapeutic approaches have been tried (corticosteroids, lithium), being of limited benefit for some patients. Early experiences showing response to G-CSF — regardless of the severity of the neutropenia, together with the absence of acquired refractoriness to this **colony stimulating factor** — have led to a unified approach based on long-term administration of standard doses of G-CSF on a daily or on an every-other-day basis. This strategy does not avoid cycling, but it reduces its duration from about 21 to 12–14 days, and the peak neutrophil count is significantly elevated above baseline levels. G-CSF reduces infections in affected patients, predominantly by decreasing the transit time of the bone marrow storage/maturation pool. However, prophylactic antibiotics such as azithromycin (or erythromycin in cases of intolerance) are commonly administered to these patients together with G-CSF.

CYCLIC THROMBOCYTOPENIA

A rare disorder in which the level of **platelets** in the peripheral blood rises and falls with a periodicity in the range 15 to 35 days. It can be regarded as a facet of cyclic **hematopoiesis**. It often occurs in women but is not related to menstrual cycle. Rarely, it is a manifestation of **myelodysplasia**. While usually self-limiting, severe fluctuations in platelet count, if associated with symptoms, can be treated with **corticosteroids**. Although not related to the menstrual cycle, response has been reported to administration of the combined oral contraceptive pill.

CYCLIN-DEPENDENT KINASES

(CDKs) Heteromultimeric kinases that phosphorylate specific Ser/Thr residues. In their simplest form, they are dimeric complexes composed of a catalytic subunit and a regulatory subunit. To be catalytically competent, the catalytic subunit (CDK1–9) and a regulatory subunit (cyclins A–H and cyclin T) must combine and be in the correct phosphorylation state. The activity of most of the CDK complexes can be associated with specific points during **cell cycle** progression. CDKs integrate signals from the cell and its environment to either initiate or inhibit cell cycle progression. CDK4/CDK6 activity is required for entry into G1 and subsequent cell-cycle progression, and CDK2/cyclin E is required for G1 progression. CDK4-cyclin D phosphorylates the Rb tumor suppressor gene and leads to its dissociation from the E2F transcription proteins. This dissociation allows the E2F-DP complexes to initiate the transcription of a large number of genes required for **DNA** synthesis. CDK1 activity is an absolute requirement for **mitosis**, and degradation of this complex is required for exit from mitosis.

As with other protein kinases, the activity of CDKs is regulated by phosphorylation and dephosphorylation events. Additionally, a large number of proteins have been identified that bind to and inhibit the CDK-cyclin complexes. These proteins are generally referred to as cyclin kinase inhibitors (CKIs) and can be divided into two groups: the INK4 (inhibitors and kinases) and cyclin inhibitor protein–kinase inhibitor protein (CIP–KIP) inhibitors. The INK4 inhibitors bind and inhibit CDK4- and CDK6-associated kinase activities specifically, and the CIP-KIP family inhibits all of the CDK complexes.

Alterations in the expression of a number of CDKs have been noted in a wide range of human tumor types; for this reason, CDK inhibitors are currently the focus of research into novel therapeutic targets for hematological malignancies.

CYCLOOXYGENASE

Enzymes found in **platelets** and cells of the **vascular endothelium** that convert **arachidonic acid** to cyclic endoperoxides. There are two isoforms, cyclooxygenase-1 and -2 (COX-1 and COX-2). COX-2 has the same activity as COX-1, but it is induced in cells by **inflammatio**n. Platelet cyclooxygenase converts arachidonate, liberated from platelet membranes by phospholipases during platelet activation, to cyclic endoperoxides prostaglandins PGG2 and PGH2. These intermediates are then converted, by thromboxane synthetase, to thromboxane A2. Thromboxane A2 is a powerful platelet activator. Endothelial cyclooxygenase converts arachidonate to identical endoperoxides, but these are then converted, by prostacyclin synthetase, to **prostacyclin**. Prostacyclin is a potent vasodilator and has significant antiplatelet effects. **Aspirin** irreversibly inactivates both platelet and endothelial cell COX-1 and COX-2.

CYCLOPHOSPHAMIDE

See **Alkylating agents**.

CYCLOSPORIN

See **Ciclosporin**.

CYTARABINE

See **Cytosine arabinoside**.

CYTOGENETIC ANALYSIS

Cytogenetics is the study of **chromosomes**. The study of human chromosomes has been going on for well over 100 years, but initial progress was impeded by difficulties associated with the preparation of high-quality chromosomes from mammalian cells. In the 1960s, the addition of phytohemagglutinin to blood lymphocyte cultures was found to stimulate mitosis, significantly increasing the number of metaphase spreads. The discovery of the Philadelphia chromosome and its association with **chronic myelogenous leukemia** was a milestone in cytogenetics. The 1970s saw the introduction and successful application of a variety of staining techniques that gave chromosomes a banding pattern. Banding greatly improved the accuracy of chromosome analysis, thereby permitting the analysis of many different tissues and diseases. High-resolution banding methodology made it possible to detect chromosomal deletions and identify small translocations.

More recently, molecular cytogenetic techniques that are based on fluorescence *in situ* hybridization (FISH) have become invaluable tools for the diagnosis and identification of the numerous chromosomal aberrations that are associated with hematological malignancies.[26,162] FISH can be used to identify chromosomal rearrangements by detecting specific DNA sequences with fluorescently labeled DNA probes. The technique of comparative genomic hybridization (CGH) involves two-color FISH. It can be used to establish ratios of fluorescence intensity values between tumor DNA and control DNA along normal reference metaphase chromosomes, and thereby to detect DNA copy-number changes, such as gains and losses of specific chromosomal regions and gene amplifications. Spectral karyotyping (SKY) is a novel molecular cytogenetic method for characterizing numerical and structural chromosomal aberrations. SKY involves the simultaneous hybridization of 24 differentially labeled chromosome-painting probes, followed by spectral imaging and chromosome classification, to produce a color karyotype of the entire genome. The use of SKY has contributed significantly to the identification of chromosomal anomalies that are associated with constitutional and cancer cytogenetics, and has revealed many aberrations that go undetected by traditional banding techniques.

Cytogenetic analysis of bone marrow and peripheral-blood metaphase mononuclear cells has provided a starting point for the molecular genetic identification of leukemia and lymphoma pathogenesis at the gene level, e.g., t(15;17) of acute promyelocytic leukemia and t(8;21) of AML M2 (see **Leukogenesis**). It is also assisting with diagnosis, e.g., **myelodysplasia**, and is useful in detecting minimal residual disease after therapy (Philadelphia chromosome detection and α-interferon therapy). Stratification of risk, and thus risk-directed therapy, is now available for **acute lymphoblastic leukemia** (ALL). Abnormalities include reciprocal translocation, duplication, and deletions, which occur as clonal markers in more than 70% of ALL of B or T lineage (Table 7). In **acute myeloid leukemia** (AML), over 30 different abnormalities have been described, and some show a clear correlation with a particular FAB group (Table 10). Many examples of cytogenetic abnormalities have been found in **lymphoproliferative disorders** (Table 99).

CYTOKINE-RELEASE SYNDROME

A severe **adverse drug reaction** characterized by fever, rash, angioedema, bronchospasm and severe dyspnea. This syndrome is a not infrequent 1 to 2 h following infusion of **Rituximab**. Patients should be given an analgesic and antihistamine before each infusion to reduce the incidence of these effects. Premedication with **corticosteroids** should also be considered.

CYTOKINES

See also **Acute-phase inflammatory response**; **Angiogenesis**; **Apoptosis**; **Cell-signal transduction**; **Cell-surface receptors**; **Chemokines**; **Colony stimulating factors**; **Hematopoiesis**; **Interferons**; **Interleukins**; **Lymphocytes**; **Transforming growth factor**; **Tumor necrosis factor**.

Polypeptide factors produced transiently or constitutively by a range of cell types, acting usually locally, altering the physiology of target cells by binding to cell-surface receptors and activating the expression of specific genes. T-cells in particular are important producers of cytokines. They are therefore mediators of intercellular communication. Cytokines are related to hormones, but hormones are produced by specific organs and travel to their target cells via the vascular circulation. In general, cytokines are produced locally by a variety of cell types and act locally (i.e., are paracrine factors); indeed, cytokines often are thought of as short-range hormones, produced locally and acting locally. Certainly those cytokines produced exclusively by activated T-cells tend to be produced at the site of the T-cell activation and, with few exceptions, are found in very small amounts in the circulation. Cytokines may act in a juxtacrine manner, i.e., surface-bound cytokines acting on cells in physical contact. But there are exceptions, e.g., erythropoietin, which is produced by the kidney and liver, acting on the bone marrow, and IL-6, which is produced locally in response to inflammation but travels via the circulation to activate the acute-phase response in the liver.

An important feature of cytokines is that their production is modulated by pathological rather than physiological stimuli. Because pathological stimuli are, at least at the start, local — even if subsequently disseminated — then local responses are essential. The initial response will activate mechanisms that will repair the damage or eliminate the pathogen and these will be both local and systemic. A reason for transient production (other than economy) is that many cytokines, especially those involved in inflammatory responses,[163] activate damage to tissue and thus contribute to autoimmune disease and to the classic symptoms of infection — fever, malaise, etc. So their production is strictly controlled to those situations where they are required.

Nomenclature and Classification[10,11,12]

The nomenclature of cytokines is generally confused. The term "lymphokines" was introduced in the 1970s, and these were given names based on the biological activity that led to their recognition, e.g., interferons for those that interfere with virus replication, and **interleukin** (IL), which was coined for those cytokines that are produced by and act on leukocytes. But the terms are not used systematically, as several interleukins are produced by nonleukocytes (e.g., IL-1), and some interleukins can act on nonleukocytes (e.g., IL-6). Many cytokines have functional names, which may be misleading (e.g., transforming growth factor b is growth inhibitory for most cell types). Those that induce chemotaxis or direct migration are sometimes referred to as **chemokines**.[135] Updated lists of cytokines are published by R & D Systems.[11,12]

Families of cytokines can be defined on the basis of their receptor usage: class (or type) I and II cytokine receptors are distinct families of molecules. Subsequently, it was shown that the cytokines utilizing these distinct receptor types themselves are structurally related, and so by extension, these became class I and class II cytokines. Class I cytokines are further subdivided into *short chain*, of which about 12 are defined, including IL-2, IL-4, IL-5, IL-7, IL-13, IL-15, and granulocyte/monocyte colony stimulating factor (GM-CSF), and *long chain*, again about 12, including IL-6, IL-11, oncostatin M (OSM), ciliary neurotrophic factor (CNTF), cardiotrophin (CT-1), leptin (LPT), and G-CSF. Class II cytokines include the growing family of interferons (α, β, γ, κ, λ, ω) as well as IL-10 and its relatives.

Cytokines produced by helper T-cells are also grouped into Th1 and Th2 cytokines (see **Lymphocytes**).

Cytokines are often redundant in their effects, i.e., different cytokines can have the same effect; they often have multiple effects on the same cell and are often pleiotropic in their actions, i.e., they have different effects on different cells. This complexity of actions is due to the activation of different subsets of genes in different cells, presumably as a consequence of different signaling pathways downstream of the receptor, as there is no evidence that different cell types have different receptors for individual cytokines. Different cytokines may interact, either synergistically or antagonistically or merely in an additive manner.

Tables 42 through 47 list the major cytokines with their usual abbreviations and common synonyms. They are not totally inclusive, but represent those most likely to be encountered in matters concerning hematology.

Cytokine Receptors

Cytokines act on target cell-surface receptors. These are usually specific for a particular cytokine, but there are examples of related cytokines binding the same receptor. Cells lacking a receptor for a particular cytokine will not respond to that cytokine. The binding

TABLE 42

Cytokine Nomenclature: Early Interleukins

Usual Name and Abbreviation	Common Synonyms
Interleukin 1, IL-1-α and IL-1-β [a]	lymphocyte activating factor (LAF), endogenous pyrogen (EP)
Interleukin 1 receptor antagonist, IL-1ra	none
Interleukin 2, IL-2	T-cell growth factor (TCGF)
Interleukin 3, IL-3	multicolony stimulating factor (multi-CSF)
Interleukin 4, IL-4	B-cell growth (or stimulatory) factor I (BCGF I or BSF I)
Interleukin 5, IL-5	eosinophil differentiation factor (EDF) or B-cell growth factor (BCGF) II
Interleukin 6, IL-6	B-cell stimulatory factor (BSF) II, hepatocyte stimulating factor (HSF)
Interleukin 7, IL –7	lymphopoietin
Interleukin 8,[b] IL-8, and many related factors (the chemokines)	monocyte-derived neutrophil chemotactic factor (MDNCF), neutrophil activating factor (or peptide) (NAF or NAP), etc.
Interleukin 9, IL-9	none
Interleukin 10, IL-10	cytokine synthesis inhibitory factor (CSIF)
Interleukin 11, IL-11	none
Interleukin 12, IL-12	natural killer cell stimulatory factor (NKCSF)
Interleukin 13, IL-13	none
Interleukin 14, IL-14	none
Interleukin 15, IL-15	none

[a] Two closely related proteins with separate genes.
[b] Supergene family, i.e., multiple, closely related proteins with related biological properties.

TABLE 43

Cytokine Nomenclature: Hematopoietic Factors

Usual Name and Abbreviation	Common Synonyms
Monocyte colony stimulating factor, M-CSF	CSF-1
Granulocyte monocyte colony stimulating factor, GM-CSF	none
Granulocyte colony stimulating factor, G-CSF	none
Stem cell factor, SCF	c-kit ligand, mast cell growth factor, steel factor
Erythropoietin, EPO	none

TABLE 44

Cytokine Nomenclature: Nonhematopoietic Growth Factors and Other Growth Regulatory Cytokines

Usual Name and Abbreviation	Common Synonyms
Fibroblast growth factors 1–7,[a] FGF-1–7	FGF-1, acidic aFGF; FGF-2, basic bFGF; FGF-7, keratinocyte FGF
Nerve growth factor, NGF	none
Platelet-derived growth factor, PDGF	none
Insulinlike growth factors,[b] IGF-I and IGF-II	none
Epidermal growth factor, EGF	none
Transforming growth factor-α, TGF-α [c]	none
Transforming growth factor-β, TGF-β [c]	none
Hepatocyte growth factor, HGF	scatter factor
Oncostatin-M (OSM)	none

[a] Supergene family, i.e., multiple, closely related proteins with related biological properties.
[b] Two closely related proteins with separate genes.
[c] TGF-α and TGF-β are distinct proteins with differing biological activities, despite similarity in names.

TABLE 45

Cytokine Nomenclature: Antiviral Cytokines — Interferons (IFNs)

Usual Name and Abbreviation	Common Synonyms
Interferon-α and -β, IFN-α and -β [a]	type 1 IFNs (α and β), leukocyte IFN, L-IFN-α, fibroblast, F-IFN-β
Interferon-γ, IFN-γ [b]	type 2, immune IFN, macrophage activating factor (MAF)

[a] Supergene family, i.e., multiple closely related proteins with related biological properties.
[b] A protein distinct from IFN-α and IFN-β, with overlapping properties.

TABLE 46

Cytokine Nomenclature: Tumor Necrosis Factors

Usual Name and Abbreviation	Common Synonyms
Tumor necrosis factor-α, TNF-α	cachectin
Tumor necrosis factor-β, TNF-β	lymphotoxin (LT)

TABLE 47

Cytokine Nomenclature: Cytokines Involved in Pregnancy and Proliferation of Embryonic Cells

Usual Name and Abbreviation	Common Synonyms
Leukemia inhibitory factor, LIF [a]	hepatocyte stimulating factor (HSF), differentiation inducing factor (D-factor, DIF, DIA, etc.)
Ovine and bovine trophoblast protein-1a	antiluteolytic protein

[a] Related to interferon-α (IFN-α).

of a cytokine to its receptor will activate physiological changes in the target cell through activation of cellular genes. This is through intracellular signaling systems targeted on transcription factors. The receptors for cytokines are now in many cases cloned and defined in molecular terms; many receptors for cytokines belong to a superfamily distinct from both the G-protein receptors typical of polypeptide hormones and most growth factor receptors.

Cytokine receptors can be homodimeric (e.g., IL-6R, G-CSFR) or heterodimeric (most others). The heterodimeric receptors include one of three common subunits: gp 130 (IL-6, LIF, and IL-11 receptors); common b (IL-3, IL-5, and GM-CSF receptors), and common g (IL-4, IL-7, IL-9, and IL-13 receptors). IL-2 differs in that it has a third component, IL-2Ra, in addition to IL-2Rb (not the common b) and common g. It is the unique component of the receptor that confers specificity.

The transmission of the signal to the gene involves the activation of **tyrosine kinases** associated with the receptor; these phosphorylate a class of proteins described as signal transducers and activators of transcription (STATS), which thereupon translocate to the nucleus and there activate their target gene.

Physiological Roles of Cytokines

Table 48 indicates (a) the major physiological and pathological processes in which cytokines are implicated and (b) the individual cytokines involved in each.

TABLE 48

Summary of Major Cytokine Functions

Process	Cytokines Involved
Regulation of lymphocyte activation	IL-1, IL-2, IL-4, IFN-γ
Regulation of inflammation and the acute-phase response	IL-1, TNF-α, IL-6, IL-8, TGF-β
Activation of effector mechanisms in the immune response to infection	IFNs (antiviral and activation of NK-cells), IFN-γ (macrophage activation)
Control of hematopoiesis	SCF, IL-3, CSFs
Stimulation of wound healing and angiogenesis	PDGF, EGF, TGF-β, angiogenic factor
Control of cell growth and survival: proliferation and apoptosis	growth factors and others, e.g., IFNs, TNFs

It is extremely difficult to ascribe biological roles to cytokines. The situation is confused by the pleiotropy (multiple seemingly unrelated effects of the same factor); pleiotypy (multiple seemingly unrelated responses of the same cell to a single factor); redundancy (the same response can be elicited by different factors); and interactions of cytokines. Many of the biological activities of cytokines are defined *in vitro* and may not relate to function *in vivo*. The only ways definitively to illustrate experimentally the importance of a cytokine *in vivo* is to remove it and observe the consequences. This can be achieved by gene knockout and antibody depletion. Unfortunately, this approach is unsubtle (knockout or depletion may not define which of many activities of a particular cytokine are crucial to a particular phenotype); can be done only in animal models, which may not reflect the human situation; and cannot discount the fact that another cytokine may compensate for the loss of a cytokine in what is a many-layered system (redundancy). The animal experiments are, however, now supplemented by the discovery of a number of inherited human conditions in which cytokines may be deficient and by the effective use in therapy of anticytokine antibodies.

Such experiments have demonstrated, for example, that IL-1 and IL-2 are not essential to the health of mice, as are IFNs. The discovery that humans inheriting a defect in the IFN-γ system are especially susceptible to infection by *Mycobacteria* spp. confirms mouse data that this cytokine's major physiological role is activation of the oxidative burst in macrophages, but surely this is not the only role of IFN-γ (Table 45).

Another experimental approach to the determination of cytokine roles is the administration of cytokines to observe their *in vivo* effects on, for example, the development of disease. This approach has two main disadvantages: a cytokine administered via the circulation may not reach the appropriate site of action, but more subtly, the administration of a cytokine may make no difference where that particular cytokine is already perhaps present in excess through local production.

A superior experimental design is to eliminate a cytokine from an animal and so find how this affects an animal's physiology or response to disease. This can be done by the use of antibodies to deplete cytokines from tissues. This has the disadvantage that the antibody administered into the circulation again may not reach the site of cytokine production and action. More recently, the gene "knockout" approach, where a cytokine gene is eliminated by homologous recombination to generate ("nullizygous") animals, has been used; subsequent derangements in their physiology or response to disease can be very informative.

Cytokines in Pathogenesis of Disease

Autoimmune disorders: the breakdown of tolerance to self-antigens results in activation of self-reactive T-cells that produce cytokines, triggering effector mechanisms that damage cells expressing these antigens and also amplifying the autoimmune process.[164]

Infectious disease: production of cytokines contributes to symptoms (headache, fever, malaise, etc.) and tissue damage by activating inflammation and the acute-phase response. A number of viruses, in particular large DNA viruses, such as the pox and herpes viruses, produce proteins that modulate immune responses in favor of the viruses by interfering with cytokine function. Many of these have been hijacked from the cellular genome. Some are homologs of the cellular cytokine, and their nomenclature follows that of the cellular cytokine, e.g., vIL-10, and they are often described as virokines. Others are soluble homologs of cytokine receptors (viroceptors) or cytokine-binding proteins. These factors include vIL-6, vIL-8, vIL-10, IFN-γ receptor homologs and IFN-αβ binding proteins, TNF receptor-homologs, chemokine-binding proteins, etc.

Atherosclerosis: the developing fibrous plaque contains proliferating histiocytes (macrophages) and activated T-cells that produce a range of cytokines that are mitogenic for smooth-muscle cells and chemotactic for other inflammatory cells.

Malignancy: more-subtle effects of cytokines may include TGF-β functioning as a tumor promoter and inhibiting potential immune responses to the tumor. Cytokines are also important in angiogenesis, invasion, and metastasis.

Cytokines in Therapeutic Strategies

The therapeutic potential of cytokines has long been recognized, but this potential has not yet been achieved. The first cytokine to be characterized was IFN-α, whose broad spectrum of antiviral activity suggested it would be an effective antiviral drug, and whose inhibitory effects on tumor cell growth suggested that it might be the answer to all malignant disease. However, the clinical trials of the IFNs through the 1970s and 1980s showed that, both as an antiviral and as an antitumor drug, IFN-α would not have the major therapeutic role that had been predicted for it. Similarly, as other cytokines have been tested, the same has generally been found, and cytokines as yet have only limited uses in the treatment of disease.

IFNs have found a role in some chronic virus infections, particularly hepatitis B and C, especially in conjunction with other antivirals, and in some cancers, such as **chronic myelogenous leukemia**. IL-2 or TNF-α act as single agents in cancer therapy, but they have marginal effects on tumor growth, and the **adverse drug reactions** are severe. Hematopoietic growth factors are finding a role in the treatment of **neutropenia**, an adverse reaction of **cytotoxic agent** therapy. In time it may become possible to mobilize hematopoietic stem cells for transplantation. **Erythropoietin** is in use for certain types of anemia. Because some cytokines are involved in the pathological effect in some diseases, their elimination, e.g., by monoclonal antibody therapy, may become important for treatment of the toxic shock syndrome and in diseases such as **rheumatoid arthritis**.

CYTOMEGALOLOVIRUS

(CMV) A virus of the herpes group commonly present in humans but causing disease in those with immunodeficiency.[165] This virus infects epithelial cells and monocytes, giving rise to enlargement of the cell with perinuclear and nuclear inclusions (owl-eyed cell). After primary infection, the virus remains latent in the infected lymphocytes. Antibodies are produced that are detectable by passive hemagglutination, latex agglutination, enzyme immunoassay, and by indirect immunofluorescence. Here, a monoclonal antibody specific for 72-kDa protein of CMV, synthesized during an early stage of viral replication, identifies CMV-infected fibroblasts by their dense homogeneous staining of the nuclei.

The consequences of infection differ with age and immune status:

Fetal. The infection is acquired *in utero* from a symptomatic mother and remains symptomless after birth. Those whose mothers have had a primary CMV infection during the first trimester are prone to congenital disorders, particularly hearing defects and mental retardation.

Normal infants and children. 60% normally acquire the virus from either the mother or institutional contacts.

Normal adults. Usually asymptomatic but may cause an infectious mononucleosislike disorder.

Immunodeficiency *subjects*:

Neonates: They may develop **thrombocytopenia,** which rapidly resolves, bleeding being rare. Choroiditis is the main clinical effect. **Splenomegaly** is common.

Adults and older children: They usually receive the virus from contaminated blood, with symptoms 1 to 3 months posttransfusion (see **Transfusion-transmitted infection**). It is a common complication of **acquired immunodeficiency syndrome** (AIDS) and stem cell transplantation.. The clinical and hematological effects are:

- **Infectious mononucleosis**like syndrome with splenomegaly and atypical lymphocytes, but without **heterophil antibodies**

- Transient **cold-agglutinin disease**, antibodies to rheumatoid agglutinins and antinuclear factor, and cryoglobulins; anti-i cold hemolytic anemia may occur

- **Bone marrow hypoplasia** with reduced granulopoiesis, large basophilic lymphoid cells, and reduced thrombopoiesis with cytoplasmic vacuolation; this causes **neutropenia** and thrombocytopenia

- Pneumonia, gastrointestinal disease, hepatitis, retinitis, and encephalitis, particularly in those with clinical AIDS

- Graft-versus-host disease following transplantation

Treatment with ganciclovir may be effective.

CYTOMETRY

See also **Flow cytometry**.
Counting of cells, which in hematological practice includes, *inter alia*, measurement of volume, nuclear size, and cytoplasmic features. There are a number of methods that can be used.

Counting Chambers (Hemocytometers)

The improved Neubauer counting chamber is most commonly used. Meticulous technique must be observed; discard if the chamber is incompletely filled, overflows into the moat, or if debris or air bubbles are present. Edge rules must be strictly followed. There are two main types of error:

Technical errors resulting from poor equipment, indifferent technique, and unrepresentative sampling

Inherent or field errors due to varying distribution of cells in the counting area, the effect of which is diminished only by counting a very large number of cells

The technique is time consuming and has largely been replaced by electronic counters except in some laboratories for white blood cell counting and platelet counting. Many Third World laboratories still have to rely on these methods[165a] of red blood cell counting, which are intrinsically inaccurate.

Visual Microscopy
This is used for **differential leukocyte counting, reticulocyte counting,** and **bone marrow** examination.

Pattern Recognition
Pattern recognition (image processing) is used for differential leukocyte counting.

Flow Cytometry
The physical and chemical characteristics of cells (particles) are analyzed as cells (particles) pass in a fluid stream, ideally in single file, through a measuring apparatus. The principle is used in automated blood cell counting for **red blood cell counting**, reticulocyte counting, **white blood cell counting**, differential leukocyte counting, **platelet counting, immunophenotyping**, and determination of fetal-maternal hemorrhage in **hemolytic disease of the newborn.**

CYTOSINE ARABINOSIDE
See **Antimetabolites**.

CYTOSKELETON OF BLOOD CELLS
See **Lymphocytes; Neutrophils; Red blood cells** — membrane.

CYTOSOL CALCIUM
Calcium ions (Ca^{2+}) that serve as intracellular second messengers, affecting calcium-dependent enzyme activities and protein-protein interactions. They are central to the control and regulation of platelet activation. **Platelets** maintain a low intracellular cytosolic Ca^{2+} concentration, but there is a steep concentration gradient across the plasma membrane. Low cytosolic Ca^{2+} concentrations are maintained by expelling Ca^{2+} across the plasma membrane and into the membrane-bound dense tubular system, which acts as a repository for Ca^{2+} ions. Strong agonists such as **thrombin** and **collagen** cause a tenfold increase in cytosol-free Ca^{2+} concentration, whereas weaker agonists such as **adenosine diphosphate** (ADP) have a smaller effect. The increase in cytosolic Ca^{2+} concentration is due to discharge of Ca^{2+} ions from the platelet dense tubular system and increased Ca^{2+} influx.

CYTOTOXIC AGENTS
Drugs used to inhibit or prevent nuclear division of rapidly dividing cells. Their action is by interference with **DNA** replication and by repair or disruption of mitochondrial function leading to **apoptosis** in slowly dividing cells. Many agents have both properties.

TABLE 49

Cytotoxic Combination Therapy

Name	Drug	Dose	Route [a]	Schedule (days)	Interval Cycle (days)
ABVD	adriamycin	25 mg/m^2	IV	1 and 15	28
	bleomycin	10 U/m^2	IV	—	—
	vinblastine	6 mg/m^2	IV	—	—
	dacarbazine	375 mg/m^2	IV	—	—
BEACOPP	bleomycin	10 U/m^2	IV	8	28
	etoposide	650 mg/m^2	IV	1–3	—
	adriamycin	25 mg/m^2	IV	1	—
	cyclophosphamide	650 mg/m^2	IV	1	—
	vincristine	1.4 mg/m^2	PO	8	—
	procarbazine	100 mg/m^2	PO	1–7	—
	prednisone	40 mg/m^2	PO	1–14	—
CHOP	cyclophosphamide	750 mg/m^2	IV	1	21
	doxorubicin	50 mg/m^2	IV	1	—
	vincristine	1.4 mg/m^2	IV	1	—
	prednisone	100 mg	PO	1–5	—
COPP	cyclophosphamide	400–650 mg/m^2	IV	1 and 8	28
	vincristine	1.4 mg/m^2	IV	1 and 8	—
	procarbazine	100 mg/m^2	PO	1–14	—
	prednisone	40 mg/m^2	PO	1–14	—
CVP	cyclophosphamide	400 mg/m^2	IV	1 and 8	21
	vincristine	1.4 mg/m^2	IV	1	—
	prednisone	100 mg/m^2	OP	1–5	—
MOPP	nitrogen mustard	6 mg/m^2	IV	1 and 15	28
	vincristine	1.4 mg/m^2	IV	—	—
	procarbazine	100 mg/m^2	PO	—	—
	prednisone	40 mg/m^2	PO	—	—

[a] IV = intravenous, PO = *per os*, by mouth.

Source: Compiled from Horning, S.J., Hodgkins Disease[267] and Foon, K.A. and Fisher, R.I., Lymphomas[397] in *Williams Hematology*, 6th ed., Beutler, E., Lichman, M.A., Coppler, B.S., Kipps, T.J., and Selligsohn, U., Eds., McGraw-Hill, NY, 2001, pp. 1224, 1250. With permission.

It follows that they are liable to damage normal tissue to some degree, particularly the rapidly dividing cells of the hematopoietic, gastrointestinal, and reproductive systems. The pharmacological groups are:

L-Asparaginase   Epipodophylloxin
Alkylating agents   Hydroxyurea
Amsacrine   Procarbazine
Anthracyclines   Retinoids
Antimetabolites   Vinca alkaloids
Deoxycoformacin

The drugs can be used singly, but more often they are used in combinations that act at different phases of the cell cycle (see **Mitosis**). Common combinations are shown in Table 49.

In all combinations, the dose of vincristine can be capped at a dose of 2 mg. To these combinations, another single drug may be added, such as:

Rituximab, 375 mg/m^2/day at weekly intervals, i.e., r-CHOP

Methotrexate, 200 mg/m^2 on days 8 and 15

Acronyms of other less common combination therapies are:

ACE   cytosine arabinoside, cisplatin, etoposide
BACOP   bleomycin, leucovorin, doxorubicin, cyclophosphamide, vincristine,
                dexamethasone
BACT   (BiCHu®) Carmastine, cytosine, cyclophosphamide, thioguanine
BEAM   carmustine, etoposide, cytosine arabinoside, melphalan
CAP   cyclophosphamide, doxorubicin, prednisone
CDE   cyclophosphamide, dexamethasone, etoposide
CAVP   cyclophosphamide, doxorubicin, vincristine, prednisone
ESHAP   etoposide, methyl prednisolone, cytarabine, prednisone
DHAP   dexamethasone, cisplatin, cytarabine, prednisone
LACE   Lomustine CCNU®, etoposide, cytosine, cyclophosphamide
MACOP   methotrexate, doxorubicin, cyclophosphamide, oncovin, bleomycin,
                leucovorin
VAD   vincristine, doxorubicin, dexamethasone
VACOP-B   vinblastine, doxorubicin, cyclophosphamide, oncovin, bleomycin, leucovorin
VAPEC-B   vinblastine, doxorubicin, prednisone, etoposide, cyclophosphamide,
                bleomycin
VBM   vinblastine, bleomycin, methotrexate

Adverse drug reactions are common to all groups of cytotoxic drugs. Common reactions
include:

Nausea and vomiting, particularly with alkylating drugs. This requires antiemetic
    therapy using dopamine and serotonin antagonists. The latter are the treatment
    of choice for the more emetic chemotherapy regimens, as they are more potent
    and do not usually cause central effects such as sedation and dystonic reactions,
    which are common with dopamine antagonists. Steroids, e.g., dexamethazone,
    may also be useful.

Bone marrow hypoplasia, which is common to all cytotoxic drugs with the exception
    of vincristine and bleomycin. It occurs, usually at 7 to 10 days postadministration,
    with **thrombocytopenia** and **neutropenia**. The main consequence is opportunistic
    infections (see **Immunodeficiency**).

Alopecia that, while being reversible, usually requires the wearing of a wig tempo-
    rarily.

Hyperuricemia resulting in renal dysfunction if not treated with allopurinol.

Neuropathy: central with purine analogs, peripheral with vinca alkaloids.

Impaired reproductive function, particularly after treatment with alkylating drugs.

Teratogenicity affecting the fetus when the drugs are given during pregnancy.

Secondary malignancies, particularly with alkylating agents and epipodophylloxin.

Immunosuppression, especially with purine analogs.

These drugs are used in the treatment of the following hematological malignancies:

Acute myeloid leukemia (AML)
Acute lymphoblastic leukemia (ALL)
Chronic myelogenous leukemia (CML)

Chronic eosinophilic leukemia
Chronic neutrophilic leukemia
Chronic lymphatic leukemia (CLL)
Hodgkin disease
Myelomatosis (plasma cell leukemia)
Non-Hodgkin lymphoma
Plasmacytoma

CYTOTOXIC ANTIBIOTICS
See Anthracyclines.

CYTOTOXIC T-LYMPHOCYTE-PRECURSOR FREQUENCY
See Human leukocyte antigens.

CYTOTOXIC T LYMPHOCYTES
See Lymphocytes.

D

DACROCYTE

(Tear drop poikilocyte) **Red blood cells** with a single elongated or pointed extremity due to the action of splenic macrophages on abnormal cell membrane. Dacrocytes are found with:

Myelofibrosis of bone marrow

Bone marrow infiltration by metastatic carcinoma

Heinz body **hemolytic anemia**

Dyserythropoiesis, particularly **megaloblastosis**

Thalassemia

DANAZOL

A synthetic androgen/progesterone that inhibits pituitary gonadotrophins. It is commonly administered for menstrual disorders but also in the long-term management of hereditary **angioedema**. Danazol has a relatively low incidence of **adverse drug reactions**, but these include masculinization and mild weight gain.

DAUNORUBICIN

See **Anthracyclines**.

DDAVP

(1-desamino-8-D-arginine vasopressin) A synthetic vasopressin analog that increases endogenous **factor VIII** and **Von Willebrand Factor** (VWF) by their release from endothelial stores without the pressor effects of vasopressin. It does, however, retain the antidiuretic effects of the natural hormone and also stimulates release of tissue plasminogen activator. The mechanism for the rise in factor VIII was thought to be due to its consequent stabilization in plasma. DDAVP is used to correct the hemostatic defect in mild **hemophilia A** and **Von Willebrand Disease**. Typically, it is administered as an infusion, but may be given subcutaneously or intranasally (in a special concentrated preparation). Responses to DDAVP are variable among individuals but relatively consistent within individuals. With repeated dosing there is a progressively diminishing response. It will raise the levels of factor VIII and Von Willebrand Factor by two- to fivefold, with a peak 60 min after the completion of intravenous infusion of DDAVP (90 to 120 min after subcutaneous and intranasal application). Vasodilatation, hypotension, and facial flushing are not uncommon and may necessitate an intravenous infusion being slowed. Fluid retention is a significant potential side effect. DDAVP should be avoided in young children and those with underlying cardiovascular disease. DDAVP is not contraindicated in uncomplicated pregnancy, although, like all drugs, it should be used with caution. No teratogenic effect has been observed in animals. DDAVP has a thousandfold greater affinity for vasopressin type 2 (V2) receptors over the type 1 (V1) receptors and so has very little oxytocic effect.

D-DIMERS

Plasma D-dimers are generated when the endogenous **fibrinolysis** degrades fibrin. Because 2 to 3% of plasma **fibrinogen** is degraded to fibrin, small amounts are found in healthy individuals. They are rarely elevated in a healthy population. D-dimer levels are elevated approximately eightfold after **venous thromboembolism** (VTE). D-dimer levels are a highly sensitive marker for thrombosis but have low specificity. D-dimer is elevated in a range of conditions other than VTE, notably **disseminated intravascular coagulation**, inflammatory disease, malignancy, and after trauma or surgery. A positive test, therefore, does not confirm the presence of VTE. However, a negative test can be used as strong evidence that clotting has not occurred. When the D-dimer result is used in conjunction with pretest probability scoring systems for deep-vein thrombosis (DVT) and pulmonary embolism (PE), a negative (i.e., below preordained cutoff value) D-dimer test can reduce the number of imaging investigations required for the diagnosis of VTE. However, if there is a high clinical suspicion of VTE, diagnostic tests should proceed in spite of a normal D-dimer. It should be noted that a D-dimer assay is of no use in those with high clinical probability of VTE. In elderly or inpatients, D-dimer retains a high negative predictive value, but it is normal in less than 20% of patients, and, hence, not very useful.[166]

Multiple D-dimer assays have been developed, with sensitivities ranging from 80 to 100%, but with no standardization among these assays. Not all D-dimer testing methods have sufficient sensitivity, but some of the rapid assays with a negative predictive value approaching 100% are comparable with the reliable but labor- and time-consuming conventional ELISA (enzyme-linked immunosorbent assay) tests. Clinicians must be aware of the characteristics of the test used in their hospital.

DEEP-VEIN THROMBOSIS

See **Venous thromboembolism**.

DEFEROXAMINE MESYLATE

See **Desferrioxamine; Iron chelation**.

DEFORMABILITY OF RED BLOOD CELLS

The changes in the shape of **red blood cells** to enable passage through capillaries. It is a physiological phenomenon that is disturbed in many red blood cell disorders, increasing with raised levels of mean cell hemoglobin concentration (MCHC), as occurs with aging red blood cells due to the decline in their viscoelasticity.

Measurement[17] is by using a rotational viscometer at a high shear rate (≈ 200 s^{-1}) to break up aggregate formation of red cells and so provide an indication of their deformation. Other more sophisticated methods use an ektacytometer (a combination of a viscometer and a laser diffractometer). Here, during rotational shearing, the test red cells diffract a helium-neon laser beam; the resulting ellipsoid diffraction pattern provides an index of the degree of red cell elongation or deformability. The osmolarity of the suspending buffer can be altered progressively during shearing so that elongation is measured when the cells are either swollen and spherocytic (hypotonic buffer), normal and discocytic (isotonic buffer), or dehydrated and echinocytic (hypertonic buffer). This versatility provides a rheological assessment of cell geometry (ratio of surface area to cell volume), membrane properties, and cytoplasmic viscosity, but cell elongation is also influenced by temperature and cell volume.

Red cell filtration by passing whole blood through narrow pores of 3- to 5-μm diameter gives a poor assessment of deformability because the flow rate is also influenced by contaminating leukocytes, heparin-induced aggregating platelets, and fibrinogen-induced aggregation of red cells. Micropipette aspiration gives an assessment of red cell membrane mechanical properties, including viscoelasticity, but is limited by the small number of cells that can be studied and by the preparation of standardizing pipettes with a bore of 1 to 2 μm.

DELTA ASSAYS

A variety of secondary replating assay systems used to identify primitive stem cells. Delta assays include long-term **bone marrow culture** systems — colony forming unit-blasts (CFU-Bl), high-proliferative-potential colony forming cell (HPP-CFC), and day-12 colony forming unit-spleen (CFU-S) — and the plastic/stroma (P/S) adherence assay. This assay is often thought to be synonymous with the delta assay, although this is not mutually exclusive. The P/S assay consists of a plastic-adherence step and a stromal-adherence step with progenitor and stem cell assay of the adherent and nonadherent cells. Plastic- and stromal-adherent cells represent a population of stem cells (approximately 10% of CD34+ cells from human bone marrow), while cells exhibiting only stromal adherence are mainly CFU-Bl and nonadherent cells are committed progenitors.

DENDRITIC RETICULUM CELLS

Bulky mononuclear cells with elongated stellate processes found particularly in association with **lymphoid follicles**. They have a different morphological identification from tissue-supporting cells such as fibroblasts, which have a different genealogy and function (Figure 26). Their precise origin is disputed, some claiming that the progenitor cell is the hematopoietic stem cell, others arguing that it is the mesenchymal stem cell or both. They have genealogical relationships with myeloid, lymphoid, and histiocytic cells. For normal maturation and function they rely on the Rho guanosine triphosphatase (GTPase), Rac 1.

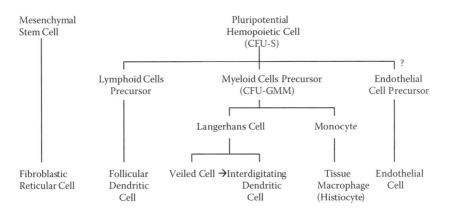

FIGURE 26
Hypothetical genealogy of hematopoietic tissue-supporting cells.

They are present in tissues such as skin and mucosae, adjacent to the environment, where they capture and process **antigens**. Dendritic cells are motile and migrate to the blood and lymph from peripheral organs such as the skin and epithelial mucosa en route to lymphoid follicles.[167]

Upon stimulation by **inflammation**, they mature and migrate to secondary lymphoid tissue. Their function is to link innate (preformed) and adaptive or cognate (specificity and memory) immune systems by **antigen presentation** to CD8+ T-lymphocytes. This action is probably in conjunction with the major histocompatibility complex (MHC) class II antigen (**human leukocyte antigen**) costimulatory modules such as CD80+. They retain MHC-peptide antigen on their surfaces for long periods, thereby enhancing the likelihood of T-cell binding and subsequent activation. They initiate both T-cell CD 4+ and T-cell CD8+ responses. Active movement of membrane extensions has been shown, with a simultaneous disappearance of actin-rich podosomes. This suggests a coordinated redeployment of **actin** to increase endocytosis induced by pathogens.[168]

Dendritic reticulum cells are the most potent of professional antigen-presenting cells, and it is thought that they may be the only cells able to activate naïve T-cells, and thus are crucial to the primary immune response They also produce cytokines and influence the type of T-cell response (Th1 or Th2) that develops. They have been proposed as the principal protagonist in allograft rejection.

There are five types of dendritic reticulum cells recognized:

Langerhans cells. Mononuclear phagocytes of the skin, identifiable using cytochemical, e.g., gold, or immunohistochemical stains. The nucleus is lobulated, and the plentiful cytoplasm contains characteristic rod-shaped Birbeck granules. The cytoplasmic membrane forms long dendritic processes that "infiltrate" the skin. Although, like macrophages, they derive from the hematopoietic stem cell, there are many differences from tissue macrophages, which suggests a different evolution. These cells are essentially tissue-resident myeloid dendritic cells. Their primary function is **antigen presentation** following absorption through the skin. They also probably play a role in delayed **hypersensitivity** reactions, e.g., contact dermatitis and skin graft rejection. Upon activation, they translocate to the secondary lymphoid tissues, where they present antigen in conjunction with MHC class II antigen to T-cells.

Veiled cells. Langerhans cells that have migrated to peripheral blood and have a convoluted single nucleus and long "veils" of motile cytoplasm. They can be isolated from peripheral blood by leukapheresis followed by differential centrifugation through Percoll gradients.

Interdigitating dendritic cells. Located within T-cell-dependent areas of lymphoid tissue, e.g., paracortex of lymph nodes. They are derived from Langerhans cells and have the same phenotype. Veiled cells, having reached the lymph node, become interdigitating cells. Their dendritic processes envelop the surrounding T-lymphocytes and are involved in antigen presentation and T-cell regulation.

Follicular dendritic cells. These are now considered distinct from myeloid dendritic cells. These cells are located in the primary and secondary lymphoid follicles. They bear **complement** and Fc receptors and retain antigen antibody complexes on their surfaces, but apparently they do not internalize these. They appear to contribute to B-cell activation and survival.

Plasmacytoid dendritic cells. These are designated by their morphology being similar to **plasma cells** with intensely basophilic cytoplasm and eccentric nucleus. They express Toll-like receptors and MHC-II. Stimulation results in secretion of interferon-α and interferon-β. Their function is the proliferation and stimulation of allogeneic helper T-cells in response to inflammation, autoimmune disease, and neoplasia. There is some evidence that they prime CD8[+] T-cells for viral antigens. They probably are derived from CD34[+] hematopoietic cells with the help of thrombopoietin and FLT-3 and precursor plasmacytoid monocytes and plasmacytoid T-cells.

DENDRITIC RETICULUM CELL SARCOMAS

Rare neoplasms of **dendritic reticulum cells** located within **lymphoid tissue**.[169] Under the WHO Classification of Tumors of Hematopoietic and Lymphoid Tissues (see Appendix), the group includes:

Follicular dendritic sarcoma

Langerhans cell sarcoma

Interdigitating dendritic cell sarcoma

There are also some rare indeterminate tumors of Langerhans cell phenotype but without the characteristic granules. They are regarded as transitional cells between a Langerhans cell and an interdigitating dendritic cell.

DENGUE

An infectious disease due to flaviviruses transmitted through the bite of an infected arthropod, particularly the mosquito *Aedes aegypti* (Asian tiger moth). The infection is common in the tropics, with an estimated incidence of 50 million to 100 million cases annually and over 10,000 deaths. With global warming, this mosquito could become resident in Europe and the U.S. The virus enters the host plasma and spreads to the **lymph nodes**, where it replicates in **histiocytes** (macrophages) and **dendritic reticulum cells**, inducing the formation of α-**interferon**. There is increased production of **immunoglobulin** M and also a **cell-mediated immune response**. The presenting clinical features are fever with facial flushing and hemorrhagic manifestations due to **thrombocytopenia**. A severe manifestation is dengue hemorrhagic fever due to increased vascular permeability — dengue hemorrhagic fever (DHF). Most patients have serological evidence of a preexisting infection. No specific antibiotic therapy is at present available, treatment being entirely symptomatic. Eradication of the disease depends upon mosquito control.[170]

DENTAL HEMORRHAGE

Persistent **hemorrhage** following dental extraction. This may be due to local trauma but is often associated with **hemorrhagic disorders**. Avoidance can be achieved by correcting the disorder when known either by **platelet transfusion**, by the appropriate **coagulation factor**, or by its release stimulus, such as **DDAVP**. If hemorrhage has occurred, restoration of the deficiency to the normal level is essential. In addition, a fibrinolytic inhibitor such as **tranexamic acid** should be given while observing the usual precautions when administering these drugs. A light pack should also be inserted.

2'-DEOXYCOFORMACIN

A **cytotoxic agent** that occurs naturally as a by-product from the culture of *Streptomyces antibioticus*. It acts by enzyme inhibition, preventing the conversion of **adenosine** to inosine. Intracellular accumulation of adenosine occurs, although cell death may be due to concomitant inhibition of RNA synthesis.

2'-Deoxycoformacin is efficacious in the treatment of **hairy cell leukemia**, but it has recently been replaced by 2-CDA (2'-chlorodeoxyadenosine). It is still used in combination therapy for treatment of **chronic lymphatic leukemia** (CLL) and **non-Hodgkin lymphoma**, particularly for **mycosis fungoides/Sézary syndrome**. The usual dose is 2 to 5 mg/m^2 by intravenous injection or infusion for 2 to 5 days. At these doses, toxicity is mainly **bone marrow hypoplasia**, with especially **thrombocytopenia**. At higher doses, it is toxic to the liver, kidney, and central nervous system.

DEOXYHEMOGLOBIN

See **Oxygen affinity to hemoglobin**.

DEOXYRIBONUCLEIC ACID

(DNA) A sequence of nucleotides, usually in two helical chains, that in turn encodes for amino acids that are assembled into functional proteins. Genetic information is passed from parent to daughter cells via the **genes** encoded within DNA. Critical to the structure of DNA is the principle that this structure is independent of the sequence of its component nucleotides. The sequence of nucleotides is crucial to provide the specificity for an individual polypeptide synthesis. The relationship between the sequence of DNA and the corresponding protein is called the genetic code.

Structure

Each nucleic acid comprises a nitrogenous base, a pentose sugar, and a phosphate group. A base linked to a sugar is called a **nucleoside** and, upon addition of the phosphate group, a **nucleotide**. Nitrogenous bases are either purines (arginine and guanine) or pyrimidines (cytosine, uracil, and thymine). Purines have fused five- and six-member rings, while pyrimidines have only a five-member ring (Figure 27). DNA contains arginine (A), guanine (G), cytosine (C), and thymine (T), while within **ribonucleic acid** (RNA), thymine is substituted by uracil. The pentose sugar in DNA is 2-deoxyribose, and this is linked at position 1 to the nitrogenous base. Nucleotides can carry phosphate in the 5' or 3' positions.

The nucleotides are linked by a series of 5' to 3' sugar-phosphate (phosphodiester) bonds (Figure 27). The nitrogenous bases protrude from the backbone. Nucleosides exist as energy-rich triphosphates (e.g., ATP), providing energy for many cellular functions. The nucleotide backbone of DNA is synthesized from triphosphate components with the b and g phosphates cleaved off as pyrophosphate.

Paired, complementary antiparallel nucleic acid strands are organized into a double helix, allowing a double-check system for maintenance of faithful sequence and held together by noncovalent bonds facilitating strand separation for DNA replication under physiological conditions. Complementary nucleic acids are A to T and C to G. The C-G bond is stronger, linking three hydrogen bond pairs, while A-T links only two. Hydrophobic base stacking interactions also occur along the middle of the double helix. Under physiological conditions, the helix is right-handed, with a turn every 3.4 nm and 10.5 base pairs (bp) per turn. The turns run clockwise looking along the helical axis. Each base is rotated 34° around the axis of the helix relative to the next, such that 10.5 bases = 360°. The helix has a major groove (2.2 nm) and a minor groove (1.2 nm) (Figure 28c) whose geometry is critical in determining the interaction between DNA and proteins. This form of DNA structure is B-DNA. Local variations in sequence can cause DNA to have intrinsic

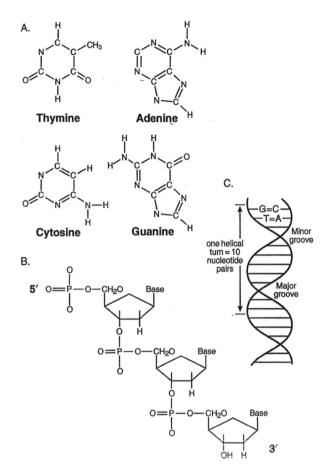

FIGURE 27
Structure of DNA. (a) Nucleotide chemical structures; (b) sugar/phosphate backbone; (c) DNA double helix.

rigidity, flexibility, or curvature. For example, oligo (dA) or oligo (dT) tracts within DNA are found to be straight and rigid, whereas sequence changes producing periodicity changes greater or smaller than 10 to 11 bp will cause normally straight DNA to take up a corkscrewlike path.

Higher-order DNA structure consists of supercoiling. Although usually thought of as open-ended, most DNA strand ends are closed. This is a prerequisite to maintain super-coiling, which is further rotation of the double helix, analogous to the progressive twisting of a rubber band to ultimately form a ball. Supercoiling is usually negative, relaxing the double helix and providing the energy required for processes such as strand separation upon initiation of DNA synthesis.

Supercoiled DNA is then wrapped around a histone octamer to form the nucleosome, each with 146 bp of DNA helix and attached to adjacent nucleosomes by linker DNA flanked by histone H1. More than 80% of DNA is contained within nucleosomes, and it is these structures that give chromatin the bead-on-a-string electron microscopic appearance.

Replication of DNA

The process of faithful replication of the genetic code in a parent cell for transmission to its progeny is remarkable for its complexity and reliability. S-phase of the cell cycle is part

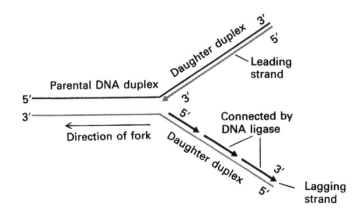

FIGURE 28

Replication of DNA. During replication of DNA, the duplex is sequentially unwound, exposing a single-stranded region (the growing fork), where complementary daughter strands are synthesized. Since nucleic acid chains can grow only in the 5' → 3' direction (indicated by arrowheads on daughter strands), only one new strand, the leading strand, is synthesized continuously. Synthesis of the other new strand, the lagging strand, proceeds discontinuously, initially forming fragments, because replication must begin at a new (downstream) start site on the parental strand as each new region of DNA is exposed at the growing fork. DNA ligase connects the lagging-strand fragments to form a continuous strand. (From Lodish, H., Baltimore, D., Berk, A., Zipura, S. L., Matsudaira, P., and Darnell, J., Nucleic acid synthesis, in *Molecular Cell Biology*, 3rd ed., Scientific America Books, W.H. Freeman & Co., New York, 1995, p. 117. With permission.)

of interphase, during which the cell doubles its DNA content. The stimulus to activate cell division in hematopoietic tissue derives from an appropriate combination and concentration of mitogenic growth factors.

In any cell, initiation of DNA synthesis commits the cell to division, but daughter cells are unable to separate until DNA replication is complete (Figure 29). In eukaryotic cells, DNA replication is initiated from many origins, and the units of replication are called replicons. Each replicon fires only once per cell division, but not all replicons fire simultaneously. DNA synthesis is bidirectional from any one origin and ceases upon contact with an oncoming replicon. From the estimated speed of replication, the entire mammalian genome could be replicated in 1 h if all replicons are used, but a typical S-phase duration is >6 h. This implies that only 15% replicons are active at any given moment.

The DNA template is prepared in the proper structural condition (at least in phage DNA) prior to the initiation of synthesis by a helicase enzyme that separates strands of DNA and by a single-strand binding protein that prevents DNA from re-forming into the duplex state. DNA daughter-strand synthesis always occurs in the 5' to 3' direction by addition of bases complementary to the parent strand to the free 3'-OH ends. Thus, synthesis of the leading strand is continuous, while that of the lagging strand is discontinuous, consisting of a series of Okazaki fragments ≈1000 to 2000 bases in length. DNA synthesis is initiated at a replication origin by synthesis of a short primer of RNA. A replisome is then assembled, consisting in eukaryotes of DNA polymerase a (lagging-strand synthesis), DNA polymerase d (leading-strand synthesis), PCNA (elongation), replication factor C (elongation; ATPase), replication factor A (single-strand binding), and topoisomerases (maintenance of DNA topology). Bases are converted from the nucleotide triphosphate to monophosphate form, and the energy for addition is provided by ATP.

It is clear that actively transcribing genes are replicated earlier in S-phase than transcriptionally repressed genes. Following synthesis of the daughter strand, the nucleosome complexes are immediately reconstituted and DNA tertiary structure rapidly reestablished, although a short delay in histone assembly allows nontranscribed genes to become transcriptionally active if exposed to the appropriate transcription-factor binding.

Damage and Repair

The cytogenetic abnormalities of many hematological disorders have already been characterized (e.g., t15:17 in acute promyelocytic leukemia and t9:22 in acute myeloid leukemia), and for some disorders, this knowledge has allowed the characterization of genes affected and the effects of their dysregulation. The etiologic processes leading to these molecular abnormalities are unknown, although environmental toxic damage to DNA may clearly contribute to this. In addition, a number of inherited disorders of DNA repair associated with hematological disease, such as Fanconi anemia and ataxia-telangiectasia, result from an inability to respond appropriately to DNA-damaging agents.

DNA Damage

This is mediated in a number of ways:

Spontaneous. The most common form of DNA damage, which most frequently results from misincorporation of bases during semiconservative DNA replication. The mutation frequency is very low (one in $\approx 10^6$ to 10^9 nucleotides) due to the highly efficient repair mechanisms indicated below. Tautomeric shifts in bases, such as a change of the amino group (NH_2) to the imino (NH) tautomeric form, encourages mispairing during DNA replication. Deamination of bases may also occur spontaneously, resulting in cytosine deaminating to uracil, adenine to hypoxanthine, and guanine to xanthine. Uracil incorporation into DNA is mutagenic, resulting in G.C to A.T base-pair transitions, but this is normally rapidly corrected by uracil DNA-glycosylase. Deamination of 5-methylcytosine to thymine (T.G mispairing) is potentially more harmful to the cell, as no specific pathway exists to remove these particular thymine residues in preference to native thymine. Depurination and depyrimidation also occur spontaneously.

Environmental. Chemical damage to DNA can produce DNA-DNA inter- or intrastrand cross-links that impair DNA replication. This is commonly seen with alkylating agents that modify bases such as 6O of guanine. Ultraviolet irradiation commonly produces pyrimidine dimers formed from the covalent linkage of adjacent pyrimidine bases in the same polynucleotide chain. DNA polymerization ceases at these points. Ionizing radiation, in contrast, produces direct DNA damage, such as strand breaks and indirect damage by free radicals.

The influence of chromatin structure is important, as certain forms of chemical damage affect only linker DNA, whereas pyrimidine dimers are formed in nucleosomal and linker DNA.

Repair of DNA Damage

Reversal of damage. Enzymatic photoreactivation, repair of 6O-alkylguanine, purine insertion, ligation of DNA strand breaks.

Excision of damage. Mediated by DNA glycosylases with or without AP endonucleases (base-excision repair); excision mediated by direct-acting damage-specific endonucleases (nucleotide excision repair); mismatch excision repair.

Tolerance of damage. Replicative bypass of template damage with gap formation, translesion DNA synthesis.

DEOXYURIDINE (dU) SUPPRESSION TEST

Preincubation of normal **bone marrow** with an appropriate concentration of dU to suppress the incorporation of tritiated thymidine (3H-TdR) into DNA. Suppression is reduced when there is deficiency of either **cobalamin** or **folic acid** due to failure of the thymidylate synthesis reaction in which deoxyuridine monophosphate (dUMP) is methylated to thymidine monophosphate (dTMP), the methyl donor being the folate coenzyme 5,10-methylene tetrahydrofolate in its polyglutamate form.

The test can be helpful in the differential diagnosis of **megaloblastosis** by measuring the efficiency of a cobalamin- and folate-dependent step within bone marrow cells. In cobalamin- or folic acid-deficiency states, the ability of deoxyuridine to suppress the incorporation of [3H] thymidine into the DNA of bone marrow cells is impaired.

DERMATOMYOSITIS

See **Polymyositis**.

DESAMINE-8-d ARGININE VASOPRESSIN

See **DDAVP**.

DESFERRIOXAMINE

See also **Iron chelation**.

(Deferoxamine Mesilate) Desferrioxamine is a naturally occurring hexadentate iron chelator derived from *Streptomyces pilosus*; once combined with **iron**, the compound is known as ferrioxamine. Desferrioxamine is used for chelation therapy of persons with **iron overload** who are unable to tolerate therapeutic phlebotomy due to chronic anemia. Desferrioxamine also forms stable complexes with Ni^{2+}, Cu^{2+}, Zn^{2+}, Al^{3+}, La^{3+}, and Yb^{3+}, and thus may be used for chelation therapy of persons with aluminum intoxication.

DEXTRANS

A group of bacterial branched-chain polysaccharides of varying molecular weight (MW). Dextrans with MWs of 40 kDa and 70 kDa have been used as antithrombotic agents. Dextrans increase the bleeding time and inhibit platelet function, with the maximum effect observed 4 to 8 h after starting the infusion. The precise mechanism by which the dextrans operate is unclear, although they are known to adsorb onto the surface of platelets and, in addition, to decrease the plasma levels of Von Willebrand Factor (VWF). Patients may exhibit allergic reactions to dextran, and in some individuals they may precipitate renal failure by physically blocking the renal glomeruli. Bleeding is seen more commonly with dextran 70 than dextran 40. Dextrans are of benefit in patients undergoing various vascular surgical operations, e.g., femoropopliteal bypass. Their effect on increasing blood volume may be unacceptable, especially if there is a compromised cardiovascular system. Although the dextrans increase the bleeding time, the therapeutic effect is achieved in the majority of patients without an increased risk of bleeding.

Dextran 40 and dextran 70 are used for plasma expansion in shock syndromes (see **Colloids for infusion**). High-molecular-weight dextrans (>100 kDa) increase aggregation of red blood cells and raise blood viscosity, which may interfere with **pretransfusion testing**.

DIABETES MELLITUS

A common syndrome characterized by chronic hyperglycemia and relative insulin deficiency, resistance, or both. Primary diabetes is predominantly an **autoimmune disorder** with pancreatic β-islet cell antibodies, resulting in insulin deficiency (type 1A). Secondary diabetes has a genetic basis with lifestyle components but no immune involvement in its pathogenesis.

The hematological disorders associated with diabetes mellitus include:

Arterial thrombosis, particularly coronary artery disease and peripheral vascular disease, probably due to glycation end products damaging **vascular endothelium** with platelet aggregation or increased blood viscosity. This is exacerbated by an increase in capillary pressure associated with hypertension. Prophylactic treatment with **aspirin** may reduce the incidence of stroke, transient ischemic attacks, myocardial infarction, and angina.

Addisonian pernicious anemia due to intrinsic factor antibodies as an associated autoimmune disorder.

Hemolytic anemia in G6PD-deficient patients with ketoacidosis.

Hereditary hemochromatosis (bronzed diabetes) due to iron deposition in the pancreatic islets of Langerhans.

The level of glycosylated hemoglobin (HbA1c) is used both for monitoring of treatment and for diagnosis.

DIAMOND-BLACKFAN SYNDROME

See **Blackfan-Diamond syndrome**.

DIAPEDESIS OF NEUTROPHILS

See **Inflammation and Neutrophils**.

DIFFERENTIAL LEUKOCYTE COUNTING

Counting of the various types of leukocytes in the peripheral blood or bone marrow. It is performed either by visual microscopy (see **Peripheral-blood film examination**), by pattern recognition (image processing), or by automated **flow cytometry**. Visual differential leukocyte counting is extremely labor intensive, and good performance requires skill and motivation. The procedure is error prone, the three principal sources of difficulty being cell-distribution errors, cell recognition, and statistical sampling errors.

The current status of pattern-recognition instruments is problematic. These include statistical sampling errors (still too few cells counted), continuing need for blood film preparation (time and cost), and the need for all abnormalities identified by the instruments to be reviewed by a human operator.

The introduction of flow-cytometry-based systems provided the ability to screen for significant abnormalities with a false-negative rate lower than that of an experienced technologist and the requirement to make films only on those specimens flagged by the instrument. The accuracy of the flow cytometry systems for differential leukocyte counting can only be judged against the visual method.[171] The reference ranges in health of differential leukocyte counts ($\times 10^9$/l) are given in Table 50.

TABLE 50

Reference Ranges in Health of Differential Leukocyte Counts ($\times 10^9/l$)

Cell Type	Adult	Neonate	Infant (day 7)	Children (6 years)
Neutrophil	2.0–7.0	6.0–26.0	1.5–10.0	1.5–8.0
Eosinophil	0.02–0.5	0.4 (mean)	0.5 (mean)	0.2 (mean)
Basophil	0.02–0.1	—	—	—
Monocyte	0.2–1.0	1.1 (mean)	1.1 (mean)	0.4
Lymphocyte	1.0–3.0	2.0–11.0	2.0–17.0	1.5–7.0

Source: Derived from **Reference Range Tables I** and **IX**.

For details of ranges in neonates, infants, and children, see peripheral-blood leukocytes reference ranges in **Reference Range Table IX**. These ranges are based upon a manual method. Modern automated methods suggest lower ranges.

With improved technology, a number of routine analyzers are offering an increased capability in enumerating cell types not normally found in the blood, the extended differential count (EDC). These include immature granulocytes, nucleated red blood cells (NRBCs), atypical lymphocytes, and hematopoietic progenitor cells.

DIFFUSE HISTIOCYTIC LYMPHOMA

See **Diffuse large B-cell lymphoma**.

DIFFUSE INFILTRATIVE LYMPHOCYTOSIS SYNDROME

(DILS) A painful neuropathy due to **lymphocyte** infiltration of peripheral nerves. It is a specific complication of **acquired immunodeficiency syndrome** (AIDS), probably due to a secondary viral infection.

DIFFUSE LARGE B-CELL LYMPHOMA

See also **Angioimmunoblastic T-cell lymphoma**; **Burkitt lymphoma/leukemia**; **Lymphoproliferative disorder**; **Non-Hodgkin-lymphoma**; **Primary mediastinal (thymic) large B-cell lymphoma**.

(Rappaport: diffuse histiocytic lymphoma, occasionally diffuse mixed lymphocytic-histiocytic lymphoma; Lukes-Collins: large cleaved or large noncleaved cell follicle center B-immunoblastic lymphoma) An aggressive B-cell non-Hodgkin lymphoma constituting 30 to 40% of adult non-Hodgkin lymphomas worldwide. It is characteristically composed of large cells with large vesicular nuclei, prominent nucleoli, and basophilic cytoplasm. In most cases, the predominant cell resembles either a **centroblast** (large noncleaved cell) or an **immunoblast**; the most common appearance is a mixture of centroblastlike and immunoblastlike cells. Other cell types include large cleaved or multilobated cells and anaplastic large cells morphologically identical to those of **anaplastic large-cell lymphoma**. The **immunophenotype** of the tumor cells is $SIg^{+/-}$, $CIg^{-/+}$, B-cell-associated antigens$^+$, $CD20^+$, $CD45^{+/-}$, $CD5^{-/+}$, $CD10^{-/+}$. **Cytogenetic analysis** shows a rearrangement in about 30% of cases:

Germinal center B-cell signature: LMO2 and BCL6

Activated B-cell signature: BCL2, CCND2, SYA3

Lymph node signature: FN1 associated with mesenchymal cells and therefore probably reactive

c-myc rearranged in some cases

Clinical Features

A lymphoma of adults, the median age being in the sixth decade, but the range is broad, including children. Patients typically present with **lymphadenopathy**, sometimes accompanied by **hepatosplenomegaly**. Involvement of the **bone marrow** is less common at diagnosis.

Prognosis

This is an aggressive lymphoma, but patients have a good prognosis with therapy. A relationship of outcome exists with the cytogenetic analysis. Those with *LMO2*, *BCL6* rearrangements show prolonged survival, whereas those with *BCL2*, *CCND2*, and *SCYA3* rearrangements show a short survival time.

Staging

See **Lymphoproliferative disorders**.

Treatment

The initial treatment of choice is six to eight cycles of R-CHOP (see **Cytotoxic agents**). The addition of Rituximab has improved the outcome.[172] In selected patients with limited-stage disease, fewer cycles of chemotherapy accompanied by **radiotherapy** may be effective. In relapsed patients, **allogeneic stem cell transplantation** is the treatment of choice.

DIFFUSE LARGE-CELL IMMUNOBLASTIC LYMPHOMA

See **Adult T-cell leukemia/lymphoma; Anaplastic large-cell lymphoma; Non-Hodgkin lymphoma; Peripheral T-cell lymphoma**.

DIFFUSE LARGE CLEAVED CELL LYMPHOMA

See **Diffuse large B-cell lymphoma; Non-Hodgkin lymphoma**.

DIFFUSE, MIXED SMALL- AND LARGE-CELL LYMPHOMA

See **Angiocentric lymphoma; Angioimmunoblastic T-cell lymphoma; Diffuse large B-cell lymphoma; Follicular lymphoma; Intestinal T-cell lymphoma; Lymphoplasmacytic lymphoma/Waldenström macroglobulinemia; Mantle-cell lymphoma; Marginal-zone B-cell lymphoma; Non-Hodgkin lymphoma; Peripheral T-cell lymphoma**.

DIFFUSE POORLY DIFFERENTIATED LYMPHOCYTIC LYMPHOMA

See **Follicular lymphoma; Non-Hodgkin lymphoma**.

DIFFUSE, SMALL CLEAVED CELL LYMPHOMA

See **Extranodal marginal B-cell lymphoma**; **Follicular lymphoma**; **Non-Hodgkin lymphoma**; **Peripheral T-cell lymphoma**; **Splenic marginal-zone lymphoma**.

DI GEORGE SYNDROME

See **Immunodeficiency** — primary congenital.

DI GUGLIELMO'S DISEASE

See **Acute myeloblastic leukemia**.

DIHYDROFOLATE REDUCTASE

An essential enzyme in **folic acid** metabolism. Deficiency of this enzyme is a rare cause of **megaloblastosis** in infancy.

DILUTIONAL ANEMIA

A **normocytic anemia** that results from raised plasma volume disproportionate to the red cell mass (see **Blood volume**). It is a physiological anemia of **pregnancy.** It is also associated with **splenomegaly**, where the degree of anemia is proportional to the size of the spleen, which can be corrected by **splenectomy**.

2,3-DIPHOSPHOGLYCERATE

(2,3-DPG) A diester essential for modulating **oxygen affinity to hemoglobin**. It binds to a specific site on the β-chain and decreases oxygen affinity. 2,3-DPG is also part of the 2,3-diphosphoglycerate pathway (Rapoport-Luebering shunt) of **red blood cell** metabolism.

DIPHOSPHOGLYCEROMUTASE

(DPGM) An enzyme in the 2,3-diphosphoglycerate pathway of **red blood cell** metabolism. The rare autosomally recessive deficiency of this enzyme leads to marked decrease in red blood cell 2,3-DPG levels, with the consequent left shift of the oxygen-dissociation curve resulting in **erythrocytosis**. Severe deficiency of 2,3-DPGM results in virtual absence of intracellular 2,3-DPG, but although there is interference in the delivery of oxygen to the tissues, this is offset by the increase in red cell mass, and red blood cell life span is unimpaired.

DIPYRIDAMOLE

See also **Platelet-function disorders**.
A phosphodiesterase inhibitor that inhibits platelet function by augmenting the effects of coadministered antiaggregates such as **aspirin**. It is used therapeutically as an antithrombotic agent and specifically for the prevention of thrombi formation on prosthetic heart valves. **Adverse drug reactions** include gastrointestinal disturbances, myalgia, headache, hypotension, and tachycardia.

DIRECT ANTIGLOBULIN (COOMBS) TEST

(DAT) The laboratory technique used to determine the *in vivo* coating of **red blood cells** (RBCs) with globulins.[173] A polyspecific or monospecific antihuman globulin antiserum interacts with IgG or C3 on red blood cell surfaces, promoting **agglutination**.

Uses of DAT

Diagnosis of **warm autoimmune hemolytic anemia** (AIHA). The RBCs of patients with warm AIHA almost invariably have a positive DAT. Usually IgG, often complement, and sometimes IgA or IgM components are detectable on the RBCs. If desired, antibody can be eluted from the RBCs to determine its serologic specificity.

Investigation of **drug-induced autoimmune hemolytic anemia**. Drugs may induce the formation of antibodies against the drug itself or against RBC antigens. A number of different mechanisms have been proposed. The DAT is usually positive, but depending upon the mechanism, the causative globulin may be IgG or complement or both.

Diagnosis of **cold autoimmune hemolytic anemia** or **paroxysmal cold hemoglobinuria** (PCH). The RBCs of these patients are usually strongly DAT-positive using anticomplement (C3d) reagents.

Investigation of **hemolytic disease of the newborn** (HDN). RBCs of infants affected by HDN will almost always have a positive DAT, showing IgG coating. An eluate prepared from the RBCs can be used to confirm the specificity of the causative antibody. In ABO HDN, the DAT may be negative or weakly positive. If the DAT is positive, maternal antibody is not detectable, and ABO HDN has been excluded, the cause may be an antibody to a low-frequency antigen.

Investigation of a suspected hemolytic **blood transfusion complication**. The value obtained from performing a DAT in this circumstance is limited because antibody-coated transfused RBCs are likely to have been removed from the circulation.

Technical Aspects/Quality Assurance

The DAT is technically similar to the **indirect antiglobulin (Coombs) test**, but no incubation phase with plasma is required. A fresh anticoagulated sample should be used whenever possible, kept at 37°C in the case of cold AIHA or PCH. Tube- or **column-agglutination techniques** are usually used, with polyspecific or monospecific (e.g., anti-IgG, -IgM, -IgA, -C3d) reagents. A reagent control should be used with cold AIHA to control for autoagglutination.

DISCOCYTE

See also **Normocytes**.
The normal biconcave disc form of the mature **red blood cell**.

DISSEMINATED INTRAVASCULAR COAGULATION

(DIC) A pathological activation of **hemostasis** within the circulation in response to injury. Disseminated intravascular coagulation is characterized by the widespread activation of coagulation, which results in the intravascular formation of **fibrin** and, ultimately, thrombotic occlusion of small and midsize vessels. Intravascular coagulation will compromise the blood flow and organ perfusion and, in conjunction with hemodynamic and metabolic derangements, may contribute to multiple organ failures. At the same time, the use and subsequent depletion of **platelets** and coagulation proteins resulting from the ongoing coagulation may induce severe hemorrhage. Fibrin deposition within vessels results in physical damage to circulating erythrocytes and a **microangiopathic hemolytic anemia** (MAHA). Clot formation stimulates **fibrinolysis**, and the high levels of fibrin and fibrinogen degradation products (FDPs) further exacerbate the bleeding tendency by interfering with platelet function and fibrin polymerization.

Etiology
Multifactorial, but depending on whether the disorder is acute or chronic.

Acute DIC
Gram-negative and gram-positive bacterial infections, particularly by meningococci

Viral infections

Malaria

Leukemias, especially acute promyelocytic leukemia

Physically induced disorders: brain injuries, burns, drowning, ingestion of corrosive poisons, heat stroke

Liver disorders resulting in liver failure

Snake venom disorders

Pregnancy: placental abruption, amniotic fluid embolus, pre-eclampsia and eclampsia, sepsis, or a dead fetus

Malignancy

Hemophilia A treated with **coagulation factor concentrate** — prothrombin complex concentrate (PCC)

Intravascular hemolysis from **blood component transfusion** complications and **paroxysmal nocturnal hemoglobinuria** (PNH)

Chronic DIC

Vascular anomalies

- **Hemangiomas** — cavernous
- Dissecting aneurysms

"HELLP" syndrome, comprising Hemolysis, Elevated Liver enzymes, Low platelets, and Pregnancy

Pathogenesis
All of these causes act by a multiplicity of reactions as a result of an initiating factor, which becomes complicated by consequential factors (see Table 51).

Initiating Factors

Vascular endothelium injury by proteolytic enzymes arising from microorganisms or by excessive heat

TABLE 51

Hemostatic Abnormalities in DIC

	Abnormality	Mechanism
1. Platelets	Thrombocytopenia	Aggregation in the microcirculation
	↓ Adhension/aggregation	Premature platelet activation
		↑ FDPs
2. Clotting factors	↓ Fibrinogen	Consumed during coagulation or lysed by plasmin during fibrinolysis
	↓ Factors V and VIII	Consumed during coagulation or lysed by plasmin during fibrinolysis
	↓ Factor XIII	Activation and consumption
	↓ Antithrombin	Consumed during inactivation of activated clotting factors
3. Coagulation system activation products	↑ FDPs	Increased degradation of fibrin and fibrinogen by plasmin
4. Fibrinolysis	↓ Plasminogen	Consumed during fibrinolysis
	↑ Plasmin-antiplasmin complexes	Reflects increased fibrinolysis

Platelet activation by circulating particles, e.g., intravenous fluids and immune complexes, including endotoxins

Release of tissue factor (thromboplastin) from tissue trauma, malignancy, damaged erythrocytes, snake venoms

Consequential Factors

Endothelial injury causing contact activation with release of kallikrinins, plasminogen activators, and complement (an intact complement system is essential for DIC to occur)

Phospholipid from tissue factor stimulates intravascular coagulation activation, with consumption coagulopathy of both coagulation factors and platelets

Excess of fibrin with coincident release of plasminogen activators, which stimulates excessive fibrinolysis

All of these processes result in spontaneous **hemorrhage.** The occurrence of circulating fibrin damages red blood cell membranes, causing **fragmentation hemolytic anemia** and **intravascular hemolysis**. Fibrin deposition in the microvasculature leads to local infarction with the characteristic histological appearances of DIC. It is important to appreciate that the initial process is thrombosis due to widespread pathological activation of coagulation. Multiple organ damage is an early feature of DIC, with hemorrhage occurring later.

Clinical Features

A multisystem disorder associated with a wide spectrum of clinical abnormalities (see Table 52). Two specific syndromes that share many common features are:

Hemolytic-uremic syndrome

Thrombotic thrombocytopenic purpura

Most cases are secondary to other disorders, the diagnosis of DIC being made by clinical observations and the results of laboratory investigation.

TABLE 52

Clinical Manifestations of DIC

Organ	Manifestations
Cardiovascular system	Hypotension, acidosis, shock, myocardial ischemia and infarction, thrombosis, adrenal failure
Renal	Acute renal failure, acute tubular necrosis, hematuria
Skin	Purpura, bleeding, skin necrosis
Liver	Jaundice, hepatic failure
Lung	Adult respiratory distress syndrome (ARDS)
Gastrointestinal tract	Bleeding, ulceration
Central nervous system	Stupor, coma, convulsions, intracranial bleeding

Laboratory Features

Prothrombin time, **activated partial thromboplastin time**, and **thrombin time**, all usually prolonged

Fibrinogen titer reduced

Platelet count reduced, the most common abnormality

Fibrin/fibrinogen degradation products and **D-dimer** levels increased

Peripheral blood film examination may show fragmented red blood cells

Hemoglobinemia and **hemoglobinuria** are present occasionally

Treatment

The basis is to treat the cause and to support the patient. Supportive measures include the replacement of any deficient clotting factors or platelets by means of **platelet transfusion**, **fresh-frozen plasma**, and **cryoprecipitate**. **Heparin** may be useful in cases of chronic DIC or where the clinical features are predominantly thromboembolic. Fibrinolytic inhibitors may be useful in patients with DIC due to primary fibrinolysis, but are contraindicated in most other cases because of the risks of thrombosis. Hemofiltration may be necessary to treat renal failure. Significant data exist in animal models supporting the potentially beneficial role of activated **protein C** in DIC/sepsis models. Recent reports of the use of recombinant human activated protein C in similar clinical settings have shown significantly improved survival. Recombinant activated protein C is now indicated for use in severe sepsis and may be an important adjuvant therapy.

DIVALENT METAL TRANSPORTER-1

(DMT1/Nramp2; Natural resistance-associated transporter-2) An intestinal and erythroid protein responsible for active transport of **iron**, zinc, manganese, cobalt, cadmium, copper, nickel, and lead. The transporter is upregulated by dietary **iron deficiency**. To date, an Nramp2 mutation (12q13) has been described in one person, a woman with hypochromic, **microcytic anemia** and **iron overload**.

DNA

See **Deoxyribonucleic acid**.

DÖHLE BODIES

(Amato bodies) Small elliptical bodies found in the cytoplasm of **neutrophils**, about 5 μm long and 1 to 2 μm thick. Upon Romanowsky staining, they appear bright blue. They occur with:

May-Hegglin anomaly (see **Leukocyte anomalies**)

Pyogenic infections

Burns

Pregnancy

Myelodysplasia and **acute myeloid leukemia**

G-CSF therapy

Kwashiorkor

DONATH-LANDSTEINER TEST

A method for detecting **paroxysmal cold hemoglobinuria** by an IgG autoantibody (anti-P specificity) that binds to **red blood cells** in the cold, but only activates **complement** and causes lysis at or near 37°C. This test[2] involves incubating red blood cells with the patient's serum, first in the cold and then at 37°C with added fresh complement. When the autoantibody is present, visible **hemolysis** appears when the cells are brought back to 37°C.

DOPPLER ULTRASONOGRAPHY

See **Venous thromboembolic disease**.

DOWN SYNDROME

(DS) A complex multisystemic congenital syndrome involving bone and joint abnormalities, metabolic complications, and intellectual deficit. DS is the result of trisomy 21 (the most frequent chromosomopathy, with 0.06 to 0.1% newborn babies affected worldwide), an abnormality believed to be originated by a defect in the meiotic chromosome disjunction. Its phenotypical repercussion in the newborn is variable, and it appears to depend on the proportion of cells presenting the trisomy in cases of mosaicism.

Children with DS may present several hematological abnormalities that, in many cases, have a confusing appearance and can mislead clinicians. Newborn babies usually present exaggerated responses to stress, with transient **leukemoid reactions** that resemble congenital leukemia, with very high white cell counts (up to $400 \times 10^9/l$), myeloblasts in peripheral blood, anemia, thrombocytopenia, hepatosplenomegaly, and skin infiltrates. These cases achieve spontaneous remission within a few months. However, they can be morphologically indistinguishable from true acute leukemia, and cases of transient leukemoid reaction in children with DS have been found to have cytogenetic abnormalities, suggesting **clonality**. To avoid misdiagnosis, it is generally recommended to closely monitor the child, avoiding any active antileukemia treatment, and watch evolution.

Children with DS are known to have a 20-fold higher incidence of acute leukemia, with **acute myeloid leukemia** (specially FAB M7 or megakaryoblastic leukemia) being the most common type. However, **acute lymphoblastic leukemia** is also more frequent than in the average population. Its pathogenesis remains unknown, but genotypically normal relatives of children with DS have been found to have a higher than normal tendency to develop leukemia. This finding has led to formulating a hypothesis of the existence of a familial tendency for meiotic nondisjunction leading to an increased risk of leukemia.

Acute leukemia arising in children with DS does not have a worse prognosis than in normal children. In fact, congenital leukemia in DS appears to have better prognosis in various clinical trials than a similar leukemia in typically normal newborn babies. However, children with DS are especially susceptible to drug toxicity (especially antifolate agents), which in many ways conditions their final outcome, as it leads to a systematic reduction of dosage of agents such as methotrexate.

DOXORUBICIN

See **Anthraclines**.

DREPANOCYTE

(Sickle cells) An anomaly of **red blood cells** due to the cells containing polymerized **hemoglobin S**. They show varying shapes, from bipolar spiculated forms to holly-leaf and irregularly spiculated forms, and occur in **sickle cell disorders**, Hb C-Harlem, and Hb Memphis/S.

DRUG-INDUCED AUTOIMMUNE HEMOLYTIC ANEMIA

The mechanisms, clinical and laboratory features, and treatment of **immune hemolytic anemias** caused by **adverse drug reactions**. Drug-related immune hemolysis must be distinguished from spontaneous **autoimmune hemolytic anemia** (AIHA) and from drug-induced nonimmune hemolytic reactions.[245] It is usually a relatively benign process, but fatal cases do occur.

Pathogenesis

Low-molecular-weight substances such as drugs are not immunogenic in their own right. Firm chemical coupling of drug to a protein carrier is necessary. Proteins serving this carrier function may or may not be a component of the red blood cell (RBC) that the antidrug antibody may ultimately damage.

General mechanisms of drug-mediated immunological injury to red blood cells are:

Drug adsorption type, where drug-RBC membrane complex acts as **antigen** to induce **autoantibody** (high-affinity hapten)

Neoantigen (immune complex) type, where drug binds weakly to the RBC surface and the immune system reacts as if a foreign antigen had been formed by production of autoantibodies (low-affinity hapten)

Autoantibody induction

Distinguishing among these mechanisms is not always easy, and some cases may involve a combination of mechanisms. Drug-related nonimmunological protein adsorption may result in a positive direct antiglobulin reaction without actual RBC injury, and this further complicates the issue.

Drug-Adsorption Type

This applies to drugs that bind firmly to proteins, including those on the RBC surface. Penicillin is the prime example. Most individuals develop IgM antibodies directed against the benzylpenicilloyl determinant, but no immune injury to RBCs occurs. The antibody responsible for hemolytic anemia is of the IgG class and is directed more commonly against nonbenzylpenicilloyl determinants. Other manifestations of penicillin sensitivity are usually not present.

With high-dose penicillin, if the patient has an IgG antipenicillin antibody, the antibody binds to the RBC-bound penicillin molecules, and the **direct antiglobulin test** with anti-IgG becomes positive. Antibody eluted from RBCs or present in the serum of such patients reacts (in direct antiglobulin tests) only against penicillin-coated RBCs, a critical step in distinguishing these drug-dependent antibodies from autoantibodies.

Destruction of RBCs coated with penicillin and IgG antipenicillin antibody occurs mainly through sequestration by splenic histiocytes (macrophages). Hemolytic anemia due to penicillin is subacute in onset, occurs typically only after the patient has received the drug for 7 to 10 days, and ceases a few days to weeks after discontinuation of the drug.

Cephalosporins and semisynthetic penicillins have antigenic cross-reactivity with penicillin. Hemolytic anemia is rare. Rarely, tetracycline and tolbutamide may cause hemolysis by this mechanism.

Neoantigen Type

Many drugs induce immune injury to RBCs, **platelets**, or **granulocytes**. This immune reaction has severe characteristics; small doses of drug trigger the process, and cellular injury is mediated chiefly by complement activation at the cell surface. The process has previously been termed the immune complex mechanism, but such terminology is now less appropriate. Recent studies suggest that blood cell injury is actually mediated by a cooperative interaction of three reactants to generate a ternary complex involving:

Drug (or drug metabolite in some cases)

Drug-binding membrane site on the target cell

Antibody

Binding of the drug to the target cell membrane is weak until stabilized by attachment of the antibody to both drug and cell membrane.

RBC destruction by this mechanism may occur intravascularly after completion of the whole complement sequence. Some destruction of C3bp-coated but unlysed RBCs may be mediated by splenic and liver sequestration. The direct antiglobulin test is usually positive only with anticomplement reagents. Quinine and quinidine are the classical offenders in this group, but many other drugs are also implicated, such as chlorpropamide and rifampicin.

Autoantibody Induction

Several drugs induce formation of antibodies against the patient's own RBCs. The principal offender has been α-methyldopa. Positive direct antiglobulin reactions (with anti-IgG reagents) vary in frequency from 8 to 36%. Patients taking higher doses of the drug develop antiglobulin positivity with greater frequency. There is a lag period of 3 to 6 months between the start of therapy and development of a positive antiglobulin test.

The direct antiglobulin reaction is usually positive only for IgG. Occasionally, weak anticomplement reactions are encountered as well. Patients have strongly positive direct antiglobulin reactions as well as serum antibody, evidenced by the positive indirect antiglobulin reaction. The antibodies frequently show evidence of specificity for determinants of the Rh complex.

Less than 1% of patients taking α-methyldopa exhibit hemolytic anemia. Hemolysis is usually mild to moderate and occurs chiefly by splenic sequestration of IgG-coated red cells. The antibodies appear to be true autoantibodies reactive with native red cell antigens. It is not possible to distinguish these antibodies from similar warm-reacting autoantibodies in idiopathic AIHA.

A small proportion of patients receiving cephalosporin antibiotics develop positive antiglobulin reactions due to nonspecific adsorption of plasma proteins to red cell membranes without the development of hemolytic anemia. The clinical importance is the potential to produce difficulties in cross-match procedures unless the drug history is taken into account.

Clinical Features

These are quite variable and depend on the rate of hemolysis. Patients with high-affinity hapten and autoimmune types exhibit mild-to-moderate RBC destruction, with insidious onset over a period of days to weeks. Many patients with hemolysis mediated by the ternary complex mechanisms have sudden onset of severe hemolysis with **hemoglobinuria** after only one dose of the drug, particularly if the patient has been previously exposed. Acute renal failure may occur.

Serological Features

Drug-induced immune hemolysis must be distinguished from AIHA, congenital hemolytic anemia, and drug-mediated hemolysis due to disorders of RBC metabolism, e.g., G6PD deficiency. Patients with drug-related immune hemolytic anemia exhibit a positive direct antiglobulin test.

With the high-affinity hapten mechanisms, there is a superficial resemblance to warm antibody AIHA. In penicillin-related immune hemolytic anemia, antibodies in the serum or eluted from the patient's RBCs react only with penicillin-coated red cells. The IgG antibodies in warm-type AIHA react with unmodified human RBCs. Such a serological distinction, plus the history of exposure to high blood levels of penicillin, should be decisive.

Hemolysis mediated by the low-affinity hapten (ternary complex) mechanism is associated with a positive direct antiglobulin test with anticomplement serum. This pattern is similar to that encountered in AIHA mediated by **cold agglutinins** and the brisk hemolysis also seen in certain cases of cold-antibody AIHA. In the drug-induced cases, cold-agglutinin titer and the **Donath-Landsteiner test** are normal.

In patients with autoimmune hemolytic anemia due to α-methyldopa, the direct antiglobulin reaction is strongly positive for IgG only. Anti-red-cell autoantibody is regularly present in the serum of these patients and mediates a positive indirect antiglobulin reaction with unmodified human red cells. The evidence must, however, be circumstantial, with the knowledge that discontinuation of drug has consistently permitted a slow but definite recovery. Finally, helpful information may be gained by stopping the suspect drug.

Treatment

Discontinuation of the offending drug is often the only treatment needed. This may be life saving in patients with the ternary complex mechanism. **Red blood cell transfusion** may be necessary in severe, life-threatening anemia. The finding of an isolated direct positive antiglobulin test in the absence of hemolytic anemia is not an absolute indication to stop therapy. Future episodes should be prevented by avoidance of the offending drug, particularly when alternative therapy is available. With severe hemolysis, **corticosteroid** therapy may help, but RBC transfusion will only lead to hemolysis of any incompatible donor blood.

DRUG-INDUCED PURPURA

The occurrence of petechial and purpuric adverse drug reactions.[174] Drug-related **purpura** may occur as a result of specific antibody to **vascular endothelium**, **immune complex** development, or changes in vessel permeability. Drugs known to cause vascular purpura are listed in Table 53. Treatment is withdrawal of the candidate drug. Patients should be advised to avoid similar drugs.

TABLE 53

Drugs Associated with Vascular Purpura

Aspirin	Mefenamic acid
Allopurinol	Mercury
Arsenicals	Meprobamate
Atropine	Methyldopa
Barbiturates	Morphine
Bismuth	Naproxen
Carbamazepine	Nitrofurantoin
Chloral hydrate	Penicillins
Chloramphenicol	Phenacetin
Chlorthiazide	Phenytoin
Chlorpropamide	Piperazine
Cimetidine	Piroxicam
Digoxin	Procaine
Estrogens	Quinine
Fenbrufen	Quinidine
Frusemide	Reserpine
Gold	Sulfonamides
Indomethacin	Tolbutamide
Iodides	Warfarin
Isoniazid	

DRUGS USED FOR HEMATOLOGICAL DISORDERS

The groups of drugs acting directly on the hematopoietic system and its related tissues. Response is subject to variation.[143a] Inappropriate use of these drugs, either by overdosage or administration by the incorrect route, can cause deleterious effects. Commonly used drugs are given in Table 54.

TABLE 54

Commonly Used Drugs in Hematological Disorders

Drug Group	Disorder Treated
Abacavir	retroviral infection (AIDS)
Abciximab	antiplatelet aggregation
Acetazolamide	high-altitude (mountain) sickness
Adrenaline (epinephrine)	severe anaphylaxis
Alemtuximab	follicular lymphoma
Allopurinol	hyperuricemia
All-transretinoic acid (tretinoin)	acute promyelocytic leukemia
Amsacrine	acute myeloid leukemia; acute lymphoblastic leukemia (ALL)
Anabolic steroids	bone marrow hypoplasia; hemostasis

TABLE 54 (continued)

Commonly Used Drugs in Hematological Disorders

Drug Group	Disorder Treated
Aprotinin	antifibrinolysis
Ascorbic acid	ascorbic acid deficiency
L-Asparaginase	acute lymphoblastic leukemia
Aspirin	antiplatelet aggregation
Azathioprine	immune thrombocytopenia; autoimmune hemolytic anemia; allogeneic stem cell transplantation (SCT)
Bevacizumab	reduction of microvascular growth
Bleomycin	lymphoproliferative disorders
Busulfan	allogeneic BMT; chronic myelogenous leukemia
Chlorambucil	chronic lymphatic leukemia
Chlorpheniramine (chlorphenamine)	mild anaphylaxis
Ciclosporin (cyclosporin)	allogeneic SCT; severe autoimmune disorders
Cladribine	acute myeloid leukemia
Cobalamins	cobalamin deficiency
Colony stimulating factors (G-CSF; GM-CSF)	bone marrow hypoplasia and neutropenia following cytotoxic agents; dyskeratosis congenita; agranulocytosis
Corticosteroids	anaphylaxis; acquired aplastic anemia; autoimmune hemolytic anemia; immune thrombocytopenia; immune neutropenia; leukemia/lymphoma; major blood transfusion complications
Cotrimoxazole	prophylactic antibiotic in immunosuppression/immunodeficiency
Coumarins	venous thromboembolic disease; prevention of arterial thromboembolism in atrial fibrillation and heart valve prostheses
Cyclophosphamide	acute lymphoblastic leukemia; allogeneic SCT; lymphoproliferative disorders; acquired aplastic anemia
Cytarabine	acute myeloblastic leukemia
Danazol	hereditary angioedema
Daunorubicin	leukemia/lymphoma
L-Desamino-8-d-arginine vasopressin (DDAVP)	mild hemophilia; Von Willebrands Disease
Desferrioxamine (deferoxamine)	iron overload
Dexamethazone	immunosuppression
Didanosine	retroviral infection (AIDS)
Dipyrimadole	antiplatelet aggregation
Doxorubicin	leukemia/lymphoma
Efavirenz	retroviral infection (AIDS)
Epoprostenol (prostacyclin)	inhibition of platelet aggregation; pulmonary hypertension
Etanercept	ankylosing spondylitis
Etoposide	leukemia
Erythropoietin	anemia of chronic renal failure
Factor VIII	hemophilia A
Factor IX	hemophilia B
Fludarabine	chronic lymphatic leukemia; acquired aplastic anemia
Folic acid	folate deficiency
Folinic acid	cytotoxic therapy to counteract antimetabolite methotrexate
Granulocyte colony stimulating factor (G-CSF)	severe neutropenia
Granulocyte macrophage colony stimulating factor (GM-CSF)	severe neutropenia post SCT
Hematin	acute intermittent porphyria
Heparin	venous thromboembolic disease
Heparinoids	prophylaxis of venous thromboembolic disease
Hydroxyurea (hydroxycarbamide)	chronic myelogenous leukemia
Idarubicin	acute myeloid leukemia
Imatinib	chronic myelogenous leukemia

TABLE 54 (continued)

Commonly Used Drugs in Hematological Disorders

Drug Group	Disorder Treated
Immunoglobulins	immunodeficiency; immune thrombocytopenia; hemolytic disease of the newborn (prophylaxis)
Indinavir	retroviral infection (AIDS)
Interferons	lymphoproliferative disorders
Infliximab	ankylosing spondylitis
Iron	iron deficiency
Lamivudine	retroviral infection (AIDS)
Lenalidomide	myelodysplasia
Lopinavir	retroviral infection (AIDS)
Melphalan	myelomatosis
Mercaptopurine	acute leukemias
Mesna	prevention of cyclophosphamide toxicity
Methotrexate	leukemia/lymphoma
Mitoxantrone (mitozantrone)	lymphoproliferative disorders
Nelfinavir	retroviral infection (AIDS)
Nevirapine	retroviral infection (AIDS)
Pentamidine	pneumocystis infection (prophylactic in ALL)
Pentostatin	acute myeloid leukemia
Plasminogen activators	arterial thrombosis; venous thromboembolic disease
Procarbazine	Hodgkin lymphoma
Protamine sulfate	excess heparin
Pyridoxine hydrochloride	sideroblastic anemia
Rasburicase	acute hyperuricemia
Retinoids	acute promyeloblastic leukemia
Ritonavir	retroviral infection (AIDS)
Rituximab	follicular lymphoma
Saquinavir	retroviral infection (AIDS)
Stavudine	retroviral infection (AIDS)
Stanozolol	hereditary angioedema
Streptokinase	arterial thrombosis; pulmonary embolism
Thioguanine (thioguanine)	acute myeloblastic leukemia
Tranexamic acid	antifibrinolysis
Vinblastine	leukemia/lymphomas
Vincristine	leukemia/lymphomas
Zaletabine	retroviral infection (AIDS)
Zidovudine	retroviral infection (AIDS)

DUFFY (Fy) BLOOD GROUPS

See also **Blood groups**.

A specific antigen–antibody system located on red blood cells (RBCs).

Genetics and Phenotypes

Six antigens have been described within the Duffy (Fy) blood group system. The most important of these are the Fya and Fyb antigens, regulated by the *Fya* and *Fyb* genes. In blacks there is a high frequency of the Fy gene, which produces neither Fya nor Fyb on RBCs. Individuals with the Fy(a−b−) phenotype have increased resistance to infection by the malarial parasite *Plasmodium vivax*. The common phenotypes and genotypes are shown in Table 55.

TABLE 55

Phenotypes and Genotypes of the Duffy (Fy) Blood Group System

Phenotype	Caucasians		Blacks	
	Genotype	Frequency (%)	Genotype	Frequency (%)
Fy(a⁺b⁻)	Fya Fya	20	Fya Fya Fya Fy	10
Fy(a⁺b⁺)	Fya Fyb	48	Fya Fyb	3
Fy(a⁻b⁺)	Fyb Fyb	33	Fyb Fyb Fyb Fy	20
Fy(a⁻b⁻)	Fy Fy	0	Fy Fy	68

Antibodies and Their Clinical Significance

Duffy antibodies are almost always immune in origin and of the IgG class.

Duffy antibodies are found in about 11% of transfusion recipients who have antibodies other than anti-D. Anti-Fya is the most common Duffy antibody specificity.

Fy(a⁻b⁻) persons can make anti-Fy3, which reacts with Fy(a⁺b⁻), Fy(a⁻b⁺), and Fy(a⁺b⁺) RBCs.

Blacks with Fy(a⁻b⁻) appear to be less susceptible to producing Duffy antibodies than Caucasians with Fy(a⁺b⁻) or Fy(a⁻b⁺).

Duffy antibodies can give rise to severe hemolytic **blood transfusion complications** if incompatible RBCs are transfused, although these are unusual. These complications include delayed as well as immediate hemolytic transfusion reactions. The selection of blood for patients with anti-Fya or anti-Fyb should not be difficult, and compatible blood should be provided in all but life-threatening situations. The provision of blood for patients with anti-Fy3 may be more difficult, depending on the ethnic mix in the local blood donor population.

In patients who are likely to receive multiple transfusions over a long period of time, consideration should be given to the selection of blood that is matched for Duffy antigens. Patients with **sickle cell disorder** are likely to be Fy(a⁻b⁻), and although these patients appear to be particularly susceptible to producing RBC antibodies, this risk must be weighed against the difficulty of providing blood that is Fy(a⁻b⁻).

Duffy antibodies are occasionally implicated in hemolytic disease of the newborn, but this is usually mild.

Anti-Fya and anti-Fyb are not detected in pretransfusion testing using **enzyme techniques for antibody detection**.

DUODENAL CYTOCHROME b

(Dcytb) An intestinal brush border enzyme that reduces dietary ferric **iron** and thus enhances its absorption. In **iron deficiency**, Dcytb is stimulated via enhanced protein expression. In **hereditary hemochromatosis** due to HFE C282Y homozygosity, Dcytb is upregulated posttranslationally.

DUODENAL DISORDERS

See **Intestinal tract disorders**.

DUTCHER BODIES

Intranuclear invaginations of **immunoglobulin-**containing cytoplasm. The inclusions are periodic acid-Schiff-positive and consist of IgM. They are found in neoplastic plasmacytoid **lymphocytes** and **plasma cells** in patients with **Waldenström macroglobulinemia** and other **lymphoproliferative** disorders. Dutcher bodies can also be found in benign conditions such as chronic synovitis.

DYSCONVERTINEMIA

See **Factor VII** — deficiency.

DYSERYTHROPOIESIS

The production by the bone marrow of abnormal cells of the **erythron**, which may be congenital, inherited, or acquired. As well as morphological abnormalities, there may be functional abnormalities, usually manifesting in shortened red cell survival resulting in anemia (i.e., **ineffective erythropoiesis**). The underlying defects may be abnormal DNA synthesis, failure of hemoglobin production, enzyme deficiency, membrane defects, or a combination of these, as shown in Table 56.

TABLE 56

Causes of Dyserythropoiesis

Congenital/Inherited

Abnormal **DNA** synthesis
    Congenital dyserythropoietic anemia
    Megaloblastosis — transcobalamin II deficiency; methylmalonicaciduria
Failure of hemoglobin production
    Thalassemia
    Sideroblastic anemia
Red blood cell enzyme deficiencies
    Pyruvate kinase deficiency
Red blood cell membrane disorders
    Hereditary infantile pyropoikilocytosis
    Abetalipoproteinemia

Acquired

Abnormal **DNA** synthesis
    Megaloblastosis
    Myelodysplasia
    Acute myeloid leukemia — M6
    Paroxysmal nocturnal hemoglobinuria
    Myeloproliferative disorders
Severe protein-calorie malnutrition
Failure of hemoglobin production
    Sideroblastic anemia
    Iron deficiency
Combination of acquired causes
    HIV infection
    Drugs — cytosine arabinoside; zidovudine
    Alcohol toxicity

Diagnosis

Peripheral Blood

The blood film reveals anisopoikilocytosis, which can be microcytic, normocytic, or macrocytic. The underlying conditions may reveal specific morphological abnormalities, e.g., a dimorphic picture in sideroblastic anemia.

Bone Marrow

Nuclear abnormalities include multinuclearity, nuclear/cytoplasmic asynchrony, internuclear bridging, nuclear lobulation or budding, and **karyorrhexis**. The cytoplasm may be vacuolated or contain inclusions such as siderotic granules, e.g., sideroblastic anemia, or precipitated hemoglobin molecules, e.g., hemoglobin C. The associated ineffective erythropoiesis is detectable by **ferrokinetic** studies. Serum **iron** and **ferritin** are often increased.

Treatment

The management is that of the underlying disorder with **red blood cell transfusion** and/ or **iron chelation**.

DYSFIBRINOGENEMIA

See **Fibrinogen**.

DYSGRANULOPOIESIS

The production of **granulocytes** with abnormal morphology. Abnormalities may affect the nucleus, cytoplasm, or both and are found in many different inherited and acquired hematopoietic disorders, with **myelodysplasia** (MDS) the most common cause. The abnormalities are rarely singular.

Nuclear abnormalities include:

Hyposegmentation (**Pelger-Huët anomaly** with pince-nez appearance) — hereditary; myelodysplasia; preceding **chronic myelogenous leukemia**; **cytotoxic agent** therapy

Hypersegmentation (with twinned nuclei and bizarre shapes) — hereditary; **megaloblastosis**; **renal tract disorders** with uremia

Cytoplasmic abnormalities include:

Hypogranulation (surplus ribosomes form a dense basophilic rim) — myelodysplasia

Myeloperoxidase deficiency — hereditary; acquired idiopathic

Hypergranulation — **Alder-Reilly anomaly**; infections

Giant granules — **Chediak-Higashi syndrome**

Vacuolation — **Jordan's anomaly**

Inclusions — **Döhle bodies**; May-Hegglin anomaly

Monocytes are often abnormal in MDS, with sausage-shaped nuclei or nuclear lobulation. Both **basophils** and **eosinophils** may look abnormal in **myeloproliferative disorders**, especially chronic myelogenous leukemia.

DYSHEMATOPOIESIS

See **Dyserythropoiesis**; **Dysgranulopoiesis**; **Dysmegakaryopoiesis**.

DYSKERATOSIS CONGENITA

A rare hereditary condition characterized by the presence of abnormalities of integument (nail dystrophy, leukoplakia, skin pigmentation) and **neutropenia**. In most cases, presenting features are mild, but patients develop severe neutropenia or severe pancytopenia in the third or fourth decades of life, with **bone marrow hypoplasia** or aplasia. Inheritance is usually X-linked recessive, but both autosomally dominant and autosomally recessive pedigrees have been described.

The pathogenesis is not fully elucidated. Although early reports proposed the existence of immune mechanisms leading to bone marrow failure, more recent evidence suggests that punctual mutations in Xq28, present in a large proportion of patients, can lead to chromosomal fragility in various marrow cell types. Of these, fibroblasts appear to be particularly affected, as suggested by the incapacity of the marrow environment to maintain long term **bone marrow cultures**.

Therapy is unsatisfactory, but oxymethalone, granulocyte colony stimulating factor (G-CSF), and granulocyte/macrophage colony stimulating factor (GM-CSF) are all capable of producing transient improvements in blood counts. **Allogeneic stem cell transplantation** has been successful but, as in other chromosomal fragility syndromes, some apparently cured patients have suffered late epithelial cancers. Present experimental tendencies are focused on reduced-conditioning allografts, avoiding radiotherapy.

DYSMEGAKARYOPOIESIS

The production of **megakaryocytes** or **platelets** with abnormal morphology. Megakaryocyte abnormalities are typically seen in **myelodysplasia**, with micromegakaryocytes characteristic in mononuclear and polynuclear forms. Platelets may be abnormal in size as well as appearance. A list of these abnormalities is given in Table 57.

TABLE 57

Platelet Abnormalities in Dysmegakaryopoiesis

Giant platelets	**Alport's syndrome**
	May-Hegglin anomaly
	Myeloproliferative disorders
	Bernard-Soulier syndrome
	Montreal platelet syndrome
Small platelets	**Wiskott-Aldrich syndrome**
Cytoplasmic abnormalities	**Myelodysplasia**

DYSPLASTIC ANEMIA

See **Myelodysplasia**.

DYSPROTEINEMIC PURPURA

Hemorrhage and **purpura** associated with dysproteinemia, including macroglobulinemia, hyperglobulinemia, cryoglobulinemia, and **myelomatosis**. Lesions may be macular or papular, discrete or confluent. Cryoglobulinemia-associated purpura classically occurs after cold exposure. The mechanisms by which dysproteinemias cause purpura are not clear. Possible mechanisms include coating of platelets, direct inhibition of platelet function, reduction of different coagulation factors, failure of fibrin polymerization, and excess calcium binding. Damage to **vascular endothelium** may also occur secondary to coating by abnormal protein and hypoxia due to sludging in small vessels. Temporary correction may be seen after **plasmapheresis**.

E

ECARIN CLOTTING TIME

(ECT) A test of blood coagulation based on the cleavage of **prothrombin** to meizothrombin by the snake venom from *Echis carinatus*. Meizothrombin binds direct thrombin inhibitors but exhibits only 5% of the clotting activity of thrombin. Residual meizothrombin converts **fibrinogen** to fibrin, and the resulting fibrin can be measured as a clotting time.

For many direct thrombin inhibitors, the **activated partial thromboplastin time** (APTT) is nonlinear at high concentrations of the inhibitors. The ECT is proposed as an alternative to the APTT for the measurement of direct thrombin inhibitors, as it overcomes some of the APTTs limitations.

ECCHYMOSES

Discoloration of the skin caused by extravasation of blood, which usually results from trauma. Ecchymoses are of greater significance if the trauma that led to the lesion was trivial. They persist for a greater length of time than **petechiae**, and undergo a series of color changes from blue to green, yellow, and brown as the extravasated blood is slowly broken down. This may take between 10 days and 3 weeks, dependent upon the size of the ecchymosis. Spontaneous ecchymosis or ecchymoses occurring after trivial injury may be indicative of a defect of **hemostasis** and require full investigation.

ECHINOCYTES

(Burr cells) An anomaly of **red blood cells** in which they lose their disc shape and become spiculated with 10 to 30 short, equally spaced projections over the entire surface of the cell. The principal associations are:

Doxorubicin chemotherapy

Neonatal hypoxia

Phosphoglycerate kinase deficiency

Aldolase deficiency

Pyruvate kinase deficiency

Splenectomy (post)

Renal disorders — hemolytic uremic syndrome

Burns hemolytic anemia

Upper gastrointestinal hemorrhage

Liver disorders

Nutritional phosphate deficiency

Decompression phase of deep-sea diving

Cardiopulmonary bypass surgery
Post-blood transfusion
Storage defect in banked blood
Prolonged storage of blood in EDTA

ECULIZUMAB

A monoclonal antibody acting against terminal complement protein C5. In patients with **paroxysmal nocturnal hemoglobinuria** (PNH), it reduces intravascular hemolysis, hemoglobinuria, and the need for red blood cell transfusion.

EHLERS-DANLOS SYNDROME

A rare inherited disorder arising from an abnormality of **collagen**, which may lead to **hemorrhage** because of an increased fragility of the subcutaneous tissues. There are more than ten separate forms of the syndrome recognized. Four clearly have a dominant inheritance, and one is X-linked. Hyperextensibility of joints is a common finding. Patients may exhibit a prolonged **bleeding time** and **platelet-function testing** abnormalities, but only with collagen from the patient. Occasional patients may also have associated **factor XII** deficiency.

EHRLICHIOSIS

Tick-borne infections by bacteria of the genus *Ehrlichia*, mainly found in North America. These bacteria parasitize **granulocytes** and **monocytes** as inclusions termed morulae, which replicate and are released into the blood circulation.[175] Fever, myalgia, malaise, and headaches are the presenting symptoms. Fatal outcomes have followed involvement of the central nervous system. Diagnosis is made by examination of Giemsa-stained peripheral blood at the peak of the fever for purple inclusions in granulocytes or monocytes. The presence of infection can be confirmed serologically. Treatment is by the administration of doxicyclin.

ELLIPTOCYTES

See also **Red blood cells** — structure.
(Ovalocytes) Oval-to-elongated cells found in **hereditary elliptocytosis** (hereditary ovalocytosis), **thalassemia**, **iron deficiency**, and **megaloblastosis**. The greatest numbers are found in hereditary elliptocytosis in which more than 90% of adult red blood cells assume this form.

EMBDEN-MEYERHOF PATHWAY

See **Red blood cells** — metabolism.

EMBRYONIC HEMATOPOIESIS

See **Hematopoiesis**.

EMPERIPOIESIS

The engulfment by **megakaryocytes** of **red blood cells**. This is a common feature of **essential thrombocythemia**.

ENDOCYTOSIS

The processes of **phagocytosis** and **pinocytosis** by which particulate and soluble material, respectively, are taken up into membrane-bound vesicles (collectively, endosomes).

ENDOPLASMIC RETICULUM

A eukaryotypic cytoplasm organelle comprising a complex system of membranous stacks (cisternae); often continuous, with the outer of the two nuclear membranes bearing attached **ribosomes.** Endoplasmic reticulum is not continuous with the **Golgi apparatus,** but it is functionally integrated.

ENDOTHELIUM

See **Lymphatic system**; **Vascular endothelium**.

ENDOTOXIC SHOCK

(Septic shock) An acute state of hypotension caused by the effects of bacterial toxins. Endotoxin triggers the innate **immune response**, as **histiocytes** (macrophages) are activated through Toll-like receptors. Macrophage activation includes the secretion of tumor necrosis factor-α (TNF-α), prostaglandins, and nitric oxide. TNF triggers more nitric oxide secretion from smooth muscle and endothelial cells, with a resulting decrease of vascular tone and impaired cardiac output. Endothelial-cell activation may also trigger the coagulation pathway, leading to **disseminated intravascular coagulation** with multiple organ failure.

ENOLASE

An enzyme of the Embden-Meyerhof pathway of **red blood cell** metabolism. A rare deficiency has given rise to a mild **hemolytic anemia** following ingestion of furantoin.

ENTEROPATHY-TYPE T-CELL LYMPHOMA

(IE) A form of **non-Hodgkin lymphoma** specifically located in the small intestine. It occurs as swellings with ulceration of the jejunum or cecum, often leading to perforation. Histologically the tumors are of small, medium, and large anaplastic cells with intraepithelial **T-lymphocytes** and **histiocytes** in the adjacent mucosa. The **immunophenotype** of the tumor cells is CD3+, CD7+, CD8+/−, CD4−, CD103+ (MLA: HML-1, LFG1; Bly7, Ber-ACT8). Cytogenetic analysis shows that T-cell receptor genes are clonally rearranged.

Acute intussusception, obstruction, or perforation may precipitate the diagnosis. Predisposing causes are gluten enteropathy, Crohn's disease, common variable **immunodeficiency**, selective Ig deficiency, and **acquired immunodeficiency syndrome** (AIDS).[176] There is a strong association with gluten enteropathy, and a gluten-free diet is therefore recommended. Surgical removal is indicated for lesions limited to the jejunum presenting as intestinal enteropathy (IE) disease and followed by **cytotoxic agent** therapy using an **anthracycline**-based combination (CHOP), but the results are poor. Prognosis depends upon the extent of the lesion and the degree of malabsorption

ENZYME TECHNIQUES FOR ANTIBODY DETECTION

Methods for the detection of irregular antibodies to **red blood cells** (RBCs) using enzymes.[177] RBCs treated with proteases such as papain, bromelain, and ficin are sometimes used in **pretransfusion testing** for the detection of irregular antibodies. These enzymes cleave negatively charged, hydrophilic residues from the RBC membrane. Enzyme methods are often used in automated blood grouping systems. **Rhesus (Rh) blood group** antibodies may be detected at very low concentrations using enzyme methods. However, the routine use of enzyme methods for antibody screening is not recommended because antibodies detectable only by enzyme methods are only rarely of clinical significance, although "enzyme-only" antibodies such as anti-c may become reactive by antiglobulin and clinically important following a **red blood cell transfusion** of antigen-positive blood. Anti-Fya, -Fyb, -M, and -N are not detected using enzyme techniques. ELISA (enzyme-linked immunosorbent assay) techniques have found only a very limited application in blood-group serology, primarily because of the cost and complexity of ELISA compared with traditional hemagglutination methods.

EOSINOPHIL

Circulating **granulocytes** with characteristic reddish-orange granules in the cytoplasm upon Romanowsky staining, first described by Paul Ehrlich in 1879. The circulating level in the peripheral blood is a balance of production, tissue egress after margination and emigration through postcapillary venules, apoptosis, and cytolytic degradation. They are derived from myeloid progenitor cells stimulated by IL-3 and IL-5.

Morphology

The mature eosinophil (size 12 to 16 µm) has a bilobed nucleus with large specific granules filling the cytoplasm. A smaller nonspecific granule is also produced (see Table 58). The structure of the mature eosinophil is similar to that of the neutrophil with condensed nuclear chromatin, glycogen, and vesicles, but has a better developed endoplasmic reticulum, mitochondria, and Golgi apparatus. Only 3 to 6% of **bone marrow** cells are eosinophils or their precursors.

TABLE 58

Content of Eosinophil Granules

Specific	Nonspecific
Major basic protein	Aryl sulfatase
Eosinophil myeloperoxidase	Acid phosphatase
Eosinophil cationic protein	
Lysophospholipase	
Eosinophil-derived neurotoxin	
Collagenases	

The normal peripheral blood eosinophil level is 0.02 to 0.5 × 10^9/l, i.e., up to 8% of nucleated cells. For levels in children, see peripheral blood leukocyte counts in **Reference Range Table IX**.

Metabolism

Glucose provides the main energy source for eosinophils and is metabolized largely by glycolysis and occasionally also by oxidation.

Kinetics and Regulation

The transit time from myeloblast to mature eosinophil is estimated to be 9 days, i.e., similar to that of the **neutrophil**. The blood transit time may be as short as 5 h.

As with neutrophils, several receptors to growth factors are involved in the production and differentiation of eosinophils, but *in vitro* studies suggest that interactions between **interleukin** (IL)-3, granulocyte macrophage **colony stimulating factor** (GM-CSF), and IL-5 are pivotal. The release of eosinophils from the marrow storage pools is also regulated by these **cytokines**, but especially by IL-5.

Functions

Antimicrobial

The eosinophil is a motile phagocytic cell that, upon activation by cytokines (e.g., IL-5) or chemotaxins (e.g., lymphocyte chemotactic factor, FMLP, C5a), is able to leave the circulation. Adherence and migration through endothelium are very similar to those of neutrophils. In addition to the **cell adhesion molecules** used by neutrophils, e.g., CD11a/CD18, CD11b/CD18, CD54, CD62L, CD62P, CD62E, the eosinophil expresses CD49d, which binds to endothelial VCAM-1. Eosinophil expression of CD11b is enhanced by IL-5, whereas endothelial expression of VCAM-1 is enhanced by IL-4. Interestingly, both IL-4 and IL-5 are produced by the T-helper subset 2 of CD4 lymphocytes.

Eosinophils express receptors for IgG, IgE, IgM, C3b, and C5a, and although capable of phagocytosing opsonized microbes, they play a central role in the defense against large, nondigestible organisms such as helminthic parasites, e.g., *Schistosoma mansoni* and *Trichinella* spp., apparently by depositing their granules on the surface of the organism. Major basic protein and, to a lesser extent, eosinophilic cationic protein account for the majority of the antihelminthic activity; halide production (generated by eosinophil peroxidase) and superoxides, hydroxyl radicals, and singlet oxygen (oxidative metabolism) may play a very minor role.

Production of Inflammatory Mediators

The production of mediators of **inflammation** by eosinophils (Table 59) causes direct and indirect effects (through the recruitment of other cells) that are important in the pathogenesis of acute (especially allergic) and chronic inflammatory reactions. Also, eosinophils may be able to act as antigen-presenting cells (via CD4 expression) to T-cells.

Usually, the role of the eosinophils in inflammation is beneficial, but they play a central role in asthma. Activated eosinophils accumulate in the lungs after antigen challenge and degranulate. Major basic protein, eosinophil peroxidase, eosinophil cationic protein, and eosinophil-derived neurotoxin are all toxic to respiratory epithelium. Thus, it now seems that late-phase asthmatic reactions are a delayed-type **hypersensitivity** response and hence are T-cell mediated with reactive **eosinophilia**.

TABLE 59

Inflammatory Mediators

Cytokines	IL-3, IL-5, GM-CSF (e.g., autocrine)
	IL-1, IL-6, IL-8, tumor necrosis factor, TNF-α
	Transforming growth factor, TGF-α, TGF-β
Chemotaxins	Eotaxin (from epithelial cells)
	Leukotriene C4 (from mast cells) platelet-activating factor
Granules	Major basic protein and eosinophil peroxidase
	Stimulate release of histamine from basophils and mast cells

EOSINOPHILIA

Defined as elevation of **eosinophil** level in the peripheral blood $>0.5 \times 10^9/l$. If persistent, the level is usually $>1.5 \times 10^9/l$, but it can be as high as $30 \times 10^9/l$. It can also be associated with tissue infiltration and damage.[178,178a]

Etiology

Reactive Eosinophilia

Eosinophil proliferation is stimulated by **interleukins** (IL)-2 and IL-5 secreted by Th2 cells. Granulocyte macrophage **colony stimulating factor** (GM-CSF) and IL-3 are eosinophil growth factors, but IL-5 is the most likely agent responsible for their release and subsequent activation. The commonest causes of reactive eosinophilia are:

Allergy: asthma, hay fever, urticaria, eczema

Adverse drug reaction causing allergy: GM-CSF, IL-2

Metazoan infection disorders: hydatid, ascaris, filariasis, schistosomiasis

Bacterial infection disorders: scarlet fever, tuberculosis

Protozoan infection disorders: malaria

Skin disorders: dermatitis herpetiformis, pemphigus, psoriasis

Intestinal tract disorders: inflammatory bowel disease, celiac disease, Crohn's disease, tropical sprue, milk precipitin disease

Vasculitis: polyarteritis nodosa, Wegener's granuloma

Clonal

Chronic eosinophilic leukemia

Hypereosinophilic syndrome

Chronic myelogenous leukemia

Myeloproliferative disorders

Acute myeloid leukemia

Myelodysplasia

Hodgkin disease

Non-Hodgkin lymphoma (especially T-cell)

Local Tissue Infiltration

Tissues may be locally infiltrated by eosinophils. These cells can contain intracellular and extracellular deposits of lysophospholipase, so-called Charcot-Leyden crystals. Specific tissues are:

Lungs

Pulmonary eosinophilia (Loeffler's syndrome), where mild pulmonary symptoms and transient pulmonary infiltrates are characteristic. This acute self-limiting illness is usually associated with parasitic infections and occasionally with drugs.[179]

Pulmonary eosinophilic granuloma (see **Histiocytosis**).

Pulmonary infiltrates with eosinophilia (PIE syndrome).

Fascia (Eosinophilic Fasciitis)

Symmetrical inflammation and sclerodermalike sclerosis of the dermis, subcutis, and deep fascia associated with eosinophilia.[180] It often occurs following strenuous or unusual exercise and also following the ingestion of L-tryptophan — used in treatment of insomnia, premenstrual syndrome, and depression — and the ingestion of rapeseed oil. Hypergammaglobulinemia with raised **erythrocyte sedimentation rate** (ESR) and circulating **immune complexes** is frequently present; antibody-mediated **bone marrow hypoplasia** is also seen. Spontaneous resolution may occur after 3 to 5 years.

Treatment

Symptomatic relief and reduction in the level of eosinophilia may be obtained by the administration of **corticosteroids**. Intense and persistent eosinophilia can give rise to general tissue damage — mediated by major basic protein due to degranulation and cytokine products (see **Hypereosinophilic syndrome**) — that requires more-intensive therapy.

EOSINOPHILIC GRANULOMA

See **Histiocytosis**.

EOSINOPHILOPENIA

A reduced level of circulating **eosinophils** $<0.02 \times 10^9$/l. Reduced eosinophil counts commonly occur during infections and **inflammation** due to reduced marrow release or increased margination and migration. Patients have been described as having thymoma or adrenal insufficiency and eosinophilopenia. It has also been associated with prolonged **corticosteroid** therapy.

EPIDOPHYLLOTOXINS

See **Etoposide**.

EPINEPHRINE

See **Adrenaline**.

EPITOPES

The part of an **antigen** that interacts with **immunoglobulins** and **T-lymphocyte** receptors. In the case of immunoglobulins, epitopes of native antigens comprise surface features of the molecule and may be amino acids (not necessarily contiguous) or other structures, e.g., carbohydrates. In the case of the T-cell receptor, the antigen epitope comprises a short peptide derived from a part of the protein component of an antigen. At least in the case of immunoglobulins, the terms "epitope" and "determinant" are essentially synonymous.

EPOPROSTINOL
See **Prostacyclin**.

EPSTEIN-BARR VIRUS

(EBV) A virus of the herpes group present in the saliva of 10 to 20% of normal adults and transmitted by oral contact. It was first identified in Africa from cultured lymphoblasts from **Burkitt lymphoma** tissue. Transmission is mainly salivary, particularly by kissing, but it can be a **transfusion transmitted infection**. Virus replication occurs within B-cells, the outcome of which depends upon the immune status of the person infected. The virus is capable of persisting for long periods in the host without clinical evidence of infection. **Antibody** to EBV is present in 90% of all adults, 25% of healthy students, 35% of Caucasian children, and 54% of African children. It is probable that the infection causes an oligoclonal expansion of a few dominant clones of CD8T-cells, which were originally generated in the **thymus** and clonally expanded upon contact with the virus in the periphery. EBV infects its target B-cell for life and proliferates without production of free virus.

Consequences of Infection
These depend upon age and immune status of the person infected.

Normal Children

Small dose: virus genome dormant in B-cells, with subclinical seroconversion and immunity

Large dose: virus replication, with oropharyngitis, release of virus particles in buccal fluid, and seroconversion; **hemophagocytosis** may be associated

Normal Adolescents
The responses vary with the dose and the immune status. In the nonimmune receiving a large dose, the responses are:

Intense antibody production
- EBV seroconversion
- Heterophil antibodies of IgM cross-reacting autoimmune type appearing at 2 to 3 weeks and declining after 6 weeks of infection
- Autoantibody production:

    Anti-i IgM, usually inactive and transient, but **autoimmune hemolytic anemia** and autoimmune hepatitis can occur

    Rheumatoid arthritis factor, ANF, and Wasserman factor

    Tumor necrosis factor (TNF)

Cellular reaction

- Intense CD4$^+$ T-cell response in paracortical areas of lymph follicles (see **Infectious mononucleosis**), with production of IL-6, IFN-γ, and TNF contributing to the fever and patient fatigue
- Some CD4$^+$ and CD8$^+$ T-cells become memory cells and help in any future response to infection by the virus

Immunodeficient Persons

Lymphoblastic transformation to Burkitt lymphoma or, more rarely, **non-Hodgkin lymphoma**[181,182]

Epithelial transformation to nasopharyngeal carcinoma

Acute fatal infectious mononucleosis, sometimes with splenic rupture

Lymphoid interstitial pneumonitis in children with **acquired immunodeficiency syndrome** (AIDS)

Those with X-linked recessive lymphoproliferative immunodeficiency show an increased prevalence to the first two effects.

EPSTEIN'S SYNDROME

An autosomally dominant syndrome of mild **thrombocytopenia** with giant platelets, associated with nephritis and sensorineural deafness. **Platelet function** may be normal. In others, platelet aggregation may be reduced in response to weak agonists (**adenosine diphosphate** [ADP], iron deficiency, collagen). It is one of a group of giant-platelet syndromes, similar to **Alport's syndrome**.

ERYTHREMIC MYELOSIS

See **Acute myeloblastic leukemia** — erythroleukemia.

ERYTHROBLASTEMIA

See **Normoblastemia**.

ERYTHROBLASTOPENIA OF CHILDHOOD

See **Transient erythroblastopenia of childhood**.

ERYTHROCYTE

See **Red blood cell**.

ERYTHROCYTE SEDIMENTATION RATE

(ESR) The measurement of the rate of fall of **red blood cells** through a column of **plasma**. Increased rates are due to increased levels of large plasma proteins such as **fibrinogen**

TABLE 60

Erythrocyte Sedimentation Rate Ranges in Health

Age (years)	95% Upper Limit
Men	
17–50	10
51–60	12
61–70	14
>70	~30
Women	
17–50	12
51–60	19
61–70	20
>70	~35

Source: Data from Lewis, S.M., Miscellaneous tests, in *Dacie and Lewis Practical Haematology*, 10th ed., Lewis, S.M., Bain, B.J., and Bates, I., Eds., Churchill Livingstone, Edinburgh, 2001, p. 529. With permission.

and **immunoglobulins**, which cause rouleaux formation and clumping of red blood cells. The original method using Westergren open-ended glass pipettes has now limited availability due to biological hazard, but under strictly standardized conditions and with care to avoid spillage of blood, it is reserved as the primary reference method. A secondary reference method,[183] traceable to the primary reference method, uses undiluted venous blood of **packed-cell volume** 0.35 l/l or lower, anticoagulated with EDTA to give a dilution of <1%. The blood is then well mixed under standardized conditions and drawn into a standardized sedimentation tube, which is held upright by a rigid holding device. The blood sample is suspended under standardized conditions at 20 ± 3°C for 60 min, when the height of the red cell column is read. For routine practice, the venous blood is collected into EDTA and then diluted with trisodium citrate before pipetting. This allows the blood to be drawn up by vacuum in a closed system with reduced biohazard. With all methods, once collected, the blood must be tested within 4 h. Decreased levels occur with **erythrocytosis** and increased levels with reduced red blood cell concentration, but mathematical corrections for anemia have no value. The ranges in normal health are given in Table 60.

The ESR is a nonspecific test for the assessment and monitoring of the **acute-phase response**, particularly used for monitoring progress and response to therapy. Despite the wide range of alternative methods of assessing the acute-phase response, particularly plasma **viscosity**, the ESR remains a widely used test, mainly due to its low cost and convenience, it being easily performed at many sites of clinical practice, both laboratory-based and point-of-care testing (near patient testing) sites. A normal result helps to exclude organic disease, whereas a raised ESR indicates the need for further investigation.

ERYTHROCYTIC PROTOPORPHYRIN

See **Heme** — synthesis; **Hemoglobin**.

ERYTHROCYTOSIS

An increase in the concentration of **red blood cells** within the circulation associated with a rise in **hemoglobin** and **packed-cell volume** (PCV). The normal range of red blood cell counts is 3.8 to 4.8×10^{12}/l for females and 4.5 to 5.5×10^{12}/l for males. Erythrocytosis

TABLE 61

Causes of Erythrocytosis

Absolute (True)

Primary
Idiopathic **erythrocytosis**
Polycythemia rubra vera

Secondary
Appropriately increased erythropoietin levels
Hypoxic lung disease
Congenital cyanotic heart disease
High altitude
High-O$_2$-affinity Hb
Carbon monoxide intoxication
Sleep apnea
Right-to-left cardiac or vascular anomalies
Tobacco abuse
Hepatopulmonary syndrome
Congenital **methemoglobinemia**
Inappropriate increased erythropoietin levels
Renal disease: cysts, renal transplantation, renal artery stenosis, focal sclerosing glomerulonephritis, and Bartter's
    syndrome
Tumors: hepatocellular carcinoma, hypernephroma, cerebellar hemangioma, meningioma, uterine leiomyoma,
    pheochromocytoma, and other adrenal tumors
Familial: erythropoietin receptor mutations, VHL mutations (Chuvash polycythemia), and 2,3-BPG mutation
Drugs: androgens, recombinant erythropoietin

Relative (Pseudo)

Androgens
Tobacco abuse
Hypertension
Burns
Diuretics
Dehydration from any cause
Alcohol abuse
Diuretic therapy

may be absolute (true) if there is an increase in red cell mass, or relative (pseudo) if there is a fall in plasma cell volume resulting in an apparent erythrocytosis (see **Blood volume**).

Absolute Erythrocytosis

This can be primary (**polycythemia rubra vera**) or secondary, involving appropriate or inappropriate increase in **erythropoietin** levels (see Table 61).

Secondary Erythrocytosis
Appropriate Increase in Erythropoietin Levels

Hypoxic lung disease. Arterial hypoxia due to many different lung pathologies, e.g., chronic obstructive airways disease, pulmonary fibrosis, hypoventilation syndromes, etc., results in erythrocytosis. If the PCV rises above 0.55 l/l, the increase in blood viscosity may reduce cerebral blood flow. **Phlebotomy** (venesection) down to a PCV of 0.50 to 0.52 l/l is desirable if the patient is symptomatic.

Congenital cyanotic heart disease. A marked left-to-right shunt causes arterial hypoxemia and high PCV values (0.7 to 0.8 l/l). Patients usually have clubbing and **cyanosis**, with occasional **thrombocytopenia**. Phlebotomy to a level of a PCV less than 0.65 l/l may alleviate symptoms due to **hyperviscosity** and improve blood flow.

Altitude erythrocytosis. This occurs upon ascending to heights at which inspired oxygen tension falls, causing excessive antidiuretic hormone and adrenal steroid secretion. This results in a reduced plasma volume and relative erythrocytosis. The low-inspired-oxygen tension stimulates respiration, but the tachypnea causes a left shift in the oxygen-dissociation curve due to hypocapnia and alkalosis. However, hypoxia stimulates **2,3-diphosphoglycerate** (2,3-DPG) production, giving a compensatory right shift in the oxygen-dissociation curve, allowing increased release of oxygen to tissues. Hypoxia also stimulates erythropoietin secretion and secondary erythrocytosis. Overall, there is a slight right shift in the oxygen-dissociation curve. Red cell counts up to $8 \times 10^{12}/l$ with hematocrits over 0.60 l/l are not unusual. If the ascent is rapid, arterial hypoxia leads to acute mountain sickness. Anorexia, vomiting, headache, and irritability occur within 6 to 72 h. Rarely, convulsions, coma, and even death can occur. At heights greater than 15,000 ft above sea level, a defective physiological response may occur (chronic mountain sickness or Monge's disease), whereby excessive erythrocytosis occurs secondary to alveolar hypoventilation. Chronic hypoxia and carbon dioxide retention causes lethargy, headache, somnolence, and coma. Patients are cyanosed, with finger clubbing and peripheral edema. This condition is more likely to occur in older (>40 years of age) patients. Both forms of mountain sickness rapidly improve with descent to sea level.

*High **oxygen affinity to hemoglobin** disorders.* Abnormalities in both α- and β-globin chains cause impaired oxygen release. The tissue hypoxia stimulates compensatory erythrocytosis. Over 80 different abnormalities have been described with an autosomally dominant inheritance, e.g., Hb Malmo, Hb Chesapeake, Hb Heathrow. The oxygen-dissociation curve is left-shifted. Most patients are asymptomatic, although a few have suffered from thromboses. Phlebotomy therapy is usually not necessary.

Methemoglobinemia. This condition can be inherited or acquired. Methemoglobin has high affinity for oxygen, and the dissociation curve is left-shifted. Acquired causes are due to drugs, e.g., sulfonamides, phenacetin, primaquine.

Vascular anomalies. Large atrioventricular malformations may result in arterial hypoxia and erythrocytosis.

***Tobacco** excess.* Heavy smokers have increased levels of carbon monoxide, with a resultant left shift in the oxygen-dissociation curve. Smoking can also lead to a reduction in plasma volume.

Pickwickian syndrome of gross obesity and somnolence causes central and peripheral hypoventilation but more common is sleep apnea syndrome due to upper airway obstruction.

Inappropriate Increase in Erythropoietin Levels

Renal tract disorders, where erythrocytosis secondary to increased erythropoietin production has been described in a wide range of renal diseases, including tumors, parenchymal disease, and renal-artery stenosis. The mechanism is usually renal ischemia, and treatment of the underlying disease usually reverses the erythrocytosis.

Tumors, particularly those associated with the von Hippel Lindau syndrome, may secrete erythropoietin, and upon removal, resolution of erythrocytosis occurs. Often, however, the serum erythropoietin level is not increased outside the normal range.

Familial erythrocytosis. Mutations of the von Hippel Landau protein (VHL) have now been recognized as a common inherited cause of erythrocytosis. Initially identified in the Chuvash people of Russia, this form of erythrocytosis has been found worldwide.

Relative Erythrocytosis

Many different terms have been used to describe relative erythrocytosis, including Gaisbock's syndrome, stress, and apparent and pseudo-erythrocytosis, in which there is a raised packed-cell volume (PCV) but a normal red cell mass (RCM). Relative erythrocytosis is due to a reduced plasma volume, which can be caused by many differing conditions (see Table 61). The risk of **venous thromboembolic disease** is less than that seen in absolute erythrocytosis, but the risk is not trivial. Where possible, the cause of the relative erythrocytosis should be corrected. If unsuccessful, phlebotomy should be instituted to maintain a PCV below 0.45 l/l, since phlebotomy expands the plasma volume.

Differential Diagnosis[184]

To differentiate absolute from relative erythrocytosis, simultaneous measurement of the red cell volume (using ^{99m}Tc-labeled red cells) and plasma volume (using ^{125}I-labeled albumin) is necessary. The normal range of red cell volume is 25 to 35 ml/kg for males and 20 to 30 ml/kg for females, whereas the normal range for plasma volume is 40 to 50 ml/kg for both sexes. The differentiation of cause can be determined by a diagnostic algorithm using:

  Measurement of the red cell volume

  Measurement of oxygen saturation

  Measurement of serum erythropoietin

ERYTHROKINETICS

The movement of **red blood cells** through the body from the bone marrow to the peripheral organs. The size of the red blood cell mass defines **anemias** and erythrocytosis, while the kinetics of red blood cell production and their destruction determines pathogenesis. The three main components of red cell kinetics — the size of the red cell mass, the rate of red cell production (**erythropoiesis**, as quantified by **ferrokinetics** and the **reticulocyte count**), and the rate of red blood cell destruction (**red blood cell survival**) — can be measured.

ERYTHROLEUKEMIA

(Di Guglielmo's disease; Erythemic myelosis) See **Acute myeloid leukemia** — M6.

ERYTHROMELALGIA

A syndrome of erythremia and burning pain in the hands and feet occurring in a variety of disorders:[185]

Thrombocytosis with platelet-mediated arteriolar **inflammation** and **thrombosis**, which responds to treatment with **aspirin**

Primary disorder from childhood of bilateral distribution upon exposure to warmth or exercise, probably of genetic origin but refractory to drug therapy

Secondary to peripheral vascular disease of all forms

ERYTHRON

The collective term for progenitor and adult **red blood cells** as a functional organ. The erythron has three cell components:

The pool of early erythroid progenitors characterized by their capability to give rise to erythroid colonies *in vitro*

An intermediate compartment comprising proerythroblast-to-marrow reticulocyte

Mature red blood cells

ERYTHROPHAGOCYTOSIS

See also **Histiocytosis**.

Ingestion of **red blood cells** by histiocytes (macrophages), **monocytes**, or **neutrophils**. Physiologically, the red blood cells are removed in the liver and spleen with the mediation of **complement**. Pathologically, it occurs with:

Complement-fixing antibodies in **immune hemolytic anemias** (particularly **paroxysmal cold hemolytic anemia**)

Protozoal infection disorders

Bacterial infection disorders

Viral infection disorders, usually herpetic

Chemical toxic disorders

Some forms of **histiocytosis**, e.g., Rosai-Dorfman histiocytosis; familial erythrophagocytic lymphohistiocytosis is a rare, usually fatal, disorder that is inherited as an autosomally recessive trait

To a mild degree, erythrophagocytosis commonly occurs at the margins of tumors, particularly lymphomas, but here it is not of sufficient degree to account for any anemia. It is an uncommon appearance in peripheral-blood films and usually presents as a cytopenia. It can be diagnosed by bone marrow aspiration or, occasionally, by lymph node biopsy, where histiocytes that have ingested red blood cells (and sometimes associated leukocytes or platelets) can be readily identified.

ERYTHROPOIESIS

See **Hematopoiesis; Hematopoietic regulation; Ineffective erythropoiesis; Normoblasts; Reticulocytes**.

ERYTHROPOIETIC PROTOPORPHYRIA

See also **Porphyrias**.

The clinical disorder resulting from an autosomally dominant partial deficiency of ferrochelatase due to mutations on Ch18q; penetrance is variable. Splicing mutations are most frequent, although a variety of missense and other mutations have been described. Increased levels of free protoporphyrin occur in red blood cells, plasma, and feces. Abnormalities usually occur first in childhood and are associated with cutaneous photosensitivity, including sensations of burning, itching, edema, erythema, onycholysis, thickening of the skin, and scarring. Some patients develop anemia or progressive liver injury. Children with mild disease can be managed by avoiding exposure to direct sunlight and by using topical sunscreen products. Oral beta-carotene (120 to 180 mg/day) may reduce photosensitivity in 1 to 3 months. Cholestyramine reduces photosensitivity by decreasing hepatic protoporphyrin content. If severe hemolysis is present, splenectomy may be helpful. The benefit of liver transplantation was temporary in children with hepatic failure.

ERYTHROPOIETIN

(EPO) A glycosylated α-globulin with a molecular mass of 38 kDa. The gene responsible for EPO is located on chromosome 7. The site of production is the kidney (several renal cell types may be involved) in response to hypoxia.[186] Extrarenal tissues, particularly the liver, have some capacity for EPO synthesis in response to severe hypoxia. During fetal life, the liver is the main site of production. It is now possible to synthesize recombinant EPO. The half-life of EPO, both natural and recombinant, is 5 to 6 h, and it is cleared predominantly by the liver (particularly when desialated) and excreted by the kidney when not utilized within the **erythron**. EPO receptors are found on CD34+ bone marrow cells and on all morphologically identifiable erythroid precursors to orthochromatic erythroblasts. EPO acts principally as a survival factor, preventing **apoptosis** of erythroid cells, from late **colony forming unit** BFU-E (burst-forming unit-erythroid) onwards, with the highest density of receptors per cell at the CFU-E (colony forming unit-erythrocytes)/proerythroblast stage in those cells most responsive to EPO.

It acts on progenitor rather than precursor cells, and its actions can be summarized as:

Induction of transformation of erythroid-committed CFU-E to proerythroblasts

Action in consort with various growth factors such as burst-promoting activity (BPA) to enhance proliferation of BFU-E

Increase in transition from proerythroblasts to basophilic erythroblasts, thus shortening marrow transit time

Possible control over the release of **reticulocytes** from the bone marrow (release of stress reticulocytes)

Many positive regulatory cytokines synergize with EPO to promote erythroid differentiation. The most potent of these is stem cell factor (SCF), although IL-3, IL-11, granulocyte/macrophage colony stimulating factor (GM-CSF), and G-CSF produce similar effects. Indeed, BFU-E units are IL-3 dependent.

Recombinant EPO, produced commercially, is used for the treatment of end-stage **renal tract disorders**, antiretroviral-associated anemia in **acquired immunodeficiency syndrome** (AIDS), **aplastic anemia** both primary and secondary, and **bone marrow hypoplasia**, especially when due to **cytotoxic agent** therapy. EPO also has pleiotropic properties that can provide protection against acute ischemic injuries in several organs and tissues. The main **adverse drug reaction** is a dose-dependent increase in blood pressure. A rise in platelet count may occur, but thrombocytosis is rare. Another rare reaction is the development of **pure red blood cell aplasia**.

ESOPHAGEAL DISORDERS

The effects of esophageal disease on the hematopoietic system. These are all due to **hemorrhage**, either acute or chronic B from hiatus hernia (associated with esophagitis or ulceration), varices, telangiectases, or carcinoma.

ESSENTIAL THROMBOCYTHEMIA

(ET; Hemorrhagic thrombocythemia; Primary thrombocythemia) A rare chronic clonal disorder of the stem cell, characterized by **megakaryocyte** hyperplasia and **thrombocytosis**. It is one of the **myeloproliferative disorders** and shares many features, especially with **polycythemia rubra vera** (PRV).

Clinical Features

The disorder is uncommon under the age of 50 years, with men and women equally affected. Patients may be asymptomatic at diagnosis or present with **hemorrhage** or **venous thromboembolic disease**. Epistaxis and gastrointestinal hemorrhage are the usual bleeding sites, but any part of the body may be affected. Postoperative bleeding is common. Both arterial (skin, central nervous system) and venous (legs, hepatic) thrombosis can occur. Thrombosis is usually microvascular, secondary to platelet plugging. Patients classically present with ischemic lesions of digits, which may progress to gangrene. Cerebral symptoms such as transient ischemic attacks or amaurosis fugax are common. **Splenomegaly**, usually mild to moderate, is present, with hepatomegaly less common. **Splenic atrophy** due to splenic vein thrombosis is well described. As with PRV, there is an increased incidence of peptic ulceration.

Laboratory Features

Thrombocytosis is universal (600 to $3000 \times 10^9/l$), with platelet production as much as 15 times above normal. There is platelet anisopoikilocytosis with abnormal granulation, and megakaryocyte cytoplasm may appear in the peripheral blood. **Mean platelet volume** (MPV) is typically increased, and macrothrombocytes are common. In virtually all patients, there is abnormal platelet aggregation to epinephrine, with loss of both the primary and secondary wave. Fewer patients have abnormal **adenosine diphosphate** (ADP), arachidonic acid, and collagen aggregation. In some cases, spontaneous aggregation *in vitro* is demonstrable. **Anemia** is usually due to blood loss, but **erythrocytosis** occurs in 30% of cases.

    Granulocytosis (12 to $30 \times 10^9/l$) with left shift is present in 30 to 70% of patients, with **basophilia** and **eosinophilia** also not uncommon. The **neutrophil alkaline phosphatase** (NAP) score is usually normal or high.

    Bone marrow examination reveals marked atypical megakaryocyte hyperplasia with clumping. Immature forms are conspicuous and bizarre megakaryocyte morphology is usual. There may also be mild erythroid and myeloid hyperplasia. Marrow **reticulin** is usually normal, but may be slightly increased. *In vitro* **colony forming unit** assays reveal increased BFU-Mk formation, some of which may be spontaneous colonies. A few patients also have increased BFU-E and CFU-GM. Results of **cytogenetic analysis** of bone marrow cells are usually normal, but various abnormalities have been described. Patients with Philadelphia-chromosome-positive thrombocythemia represent cases of **chronic myelogenous leukemia** (CML). **Hyperuricemia**, pseudohyperkalemia, and increased serum **cobalamin** levels are often found.

Differential Diagnosis

Reactive thrombocytosis occurs in many conditions, but platelet counts rarely exceed 1200 $\times 10^9$/l, and resolution occurs with successful treatment of the underlying disorder. Difficulties can arise in distinguishing ET from other myeloproliferative disorders, but marrow cytogenetics and the application of the polycythemia rubra vera study group diagnostic criteria[187] aid in distinction from CML and PRV, respectively.

Course and Prognosis

Essential thrombocythemia is a chronic disorder and, provided that life-threatening thrombosis or hemorrhage does not occur at diagnosis, the survival curve with treatment is the same as normal age-matched controls. However, most patients do ultimately succumb to thromboembolic complications, although transformation to **acute myeloid leukemia** (5 to 10%) and **myelofibrosis** (10 to 25%) also occurs.

Treatment

The risk of thrombosis/hemorrhage increases with increasing platelet count, especially in the elderly. In urgent situations, e.g., digital or cerebrovascular ischemia, plateletpheresis (see **Hemapheresis**) and **aspirin** are both useful when used alone or, preferably, in combination. The long-term aim is to achieve a platelet count as near normal as possible without inducing serious adverse side effects.[187a] Myelosuppression can be achieved with various **cytotoxic agents**, e.g., busulfan, melphalan, chlorambucil, α-interferon, and anagrolide, but the safety, cost, efficacy, and tolerability of **hydroxyurea** (hydroxycarbamide) make this the agent of choice. **Radioactive phosphorus** (^{32}P) is useful in elderly patients, but hydroxyurea is also often required for a few weeks until its maximum effect has occurred.

ETAMSYLATE

A hemostatic agent used orally to reduce capillary bleeding in the absence of **thrombocytopenia**. Its action is by correction of abnormal platelet adhesion. It is also used for prophylaxis and treatment of periventricular hemorrhage in low-birth-weight infants given by intramuscular or intravenous injection.

ETANERCEPT

A drug that inhibits the activity of **tumor necrosis factor-α**. It is used for the treatment of highly active **rheumatoid arthritis** and **ankylosing spondylitis**. **Adverse drug reactions** include **bone marrow hypoplasia** and demyelination in the central nervous system.

ETHYLENEDIAMINETETRAACETIC ACID

(EDTA) A chemical that effectively chelates **calcium** in blood and is used as such as an anticoagulant for blood cell counting and other procedures involving cells. The lack of solubility of the free acid in aqueous solution makes the sodium and potassium salts preferable for use, the latter being most popular. The dipotassium salt is used in dry form, whereas the tripotassium salt is generally used in liquid form. The recommended range for adequate anticoagulation for both K2 and K3 salts is 3.7 to 5.4 µmol (1.5 to 2.2 mg) per ml of blood.[188] EDTA is particularly useful in specimen collection, since it best preserves the cellular components of blood. Three problem parameters are the **white blood cell count** (method differences), the **red blood cell mean cell volume** (MCV), and the mean platelet volume (MPV). The International Committee for Standardization in Hematology (ICSH) has recommended the dipotassium salt of EDTA as the anticoagulant of choice for blood cell counting and sizing.[188]

ETOPOSIDE

(VP16) A synthetic epipodophyllotoxin that is a phase-specific **topoisomerase**-inhibiting agent active in the G2 phase of the **cell cycle**. Etoposide is active, and increasingly used, in the treatment of **acute myeloid leukemia** (AML), especially for the monocytic subtypes. It is also effective for **acute lymphoblastic leukemia** (ALL), especially in combination with cytosine arabinoside. **Adverse drug reactions** include gastrointestinal tract upsets, alopecia, fever, and, less commonly, peripheral neuropathy (see **Cytotoxic agents**).

EUGLOBULIN LYSIS TIME

(ECLT; ELT) See **Fibrinolysis**.

EVAN'S SYNDROME

Auto-immune hemolytic anemia associated with **neutropenia** and **thrombocytopenia**.

EXCHANGE TRANSFUSION

See **Fetal/neonatal transfusion**; **Hemapheresis**.

EXERCISE

The effects of exercise or exertion on hematological values and of hematological disorders on exercise. These are:

Leukocytosis, mainly of **granulocytes**, with strenuous exercise over a short period, probably due to mobilization of the granulocyte pool.

Eosinophilic fasciitis, attributed to unusual or excessive physical activity.

Increased **fibrinolysis**, probably due to release of plasminogen activators from the vascular endothelium.

In the presence of **anemia**, the oxygen debt per unit of activity increases, leading to slower recovery of heart rate and respiratory minute volume.

Increase in circulating **T-lymphocytes** as a consequence of catecholamine release. There is a concomitant decrease in their integrin molecules, so that adherence to endothelial cells is reduced. Type I **hypersensitivity** reactions are also reduced.

Prolonged exercise such as in marathon runners results in excess **corticosteroid** production, which affects the immune system by inhibition of macrophage function and T-cell function, thereby inducing a mild **immunodeficiency**.

March hemoglobinuria occurs following walking or running on hard surfaces for a prolonged period of time.

Intense exercise in a warm climate can cause death from **disseminated intravascular coagulation**.

EXOCYTOSIS

The cellular process, particularly concerning **granulocytes**, whereby a vesicle (e.g., secretory vesicle), often budded from the endoplasmic reticulum or Golgi apparatus, fuses with the cell membrane for the release of vesicle contents into plasma. When restricted to the anterior region of the cell, it becomes an important stage in **cellular locomotion**.

EXTRAMEDULLARY HEMATOPOIESIS

See **Myeloid metaplasia**.

EXTRAMEDULLARY HEMOLYSIS

See **Extravascular hemolysis**

EXTRAMEDULLARY MYELOID TUMOR

See **Myeloid sarcoma**.

EXTRAMEDULLARY PLASMACYTOMA

See **Plasmacytoma**.

EXTRANODAL MARGINAL-ZONE B-CELL LYMPHOMA OF MUCOSA-ASSOCIATED LYMPHOID TISSUE

(MALT-lymphoma) See **Marginal-zone B-cell lymphoma; Non-Hodgkin lymphoma**.

EXTRANODAL T-CELL LYMPHOMA, NASAL TYPE

(REAL: angiocentric T-cell lymphoma; Others: malignant midline reticulosis, polymorphic reticulosis; Angiocentric immunolymphoproliferative lesion; Lethal midline granuloma). An aggressive **lymphoproliferative disorder** occurring in adults and more commonly in males, with an extranodal presentation. This includes the nose with the surrounding area and other extranodal sites. The histology is characterized by a diffuse angiocentric and angiodestructive lesion with a broad range of cell size. The characteristic **immunophenotype** is $CD2^+$, $CD56^+$, and $CD3^-$, $CD4^-$, $CD8^-$, and $CD57^-$. The cell of origin is usually an activated natural killer (NK) cell. Patients may respond well to systemic **cytotoxic agents** with or without radiation therapy (see **Non-Hodgkin lymphoma**).

EXTRAVASCULAR HEMOLYSIS

(Extramedullary hemolysis)
The destruction of **red blood cells** within tissues, usually by **histiocytes** in the spleen or other areas of the lymphoid system. It is associated with many forms of **hemolytic anemia** and as a complication of **red blood cell transfusion**. It may be compensated by increased bone marrow erythropoiesis.

EYE DISORDERS

See **Ophthalmic disorders**.

F

FAB CLASSIFICATION OF ACUTE LEUKEMIA

See **Appendix**.

FAB CLASSIFICATION OF MYELODYSPLASIA

See **Appendix**.

FABRY'S DISEASE

A **lipid-storage disorder** inherited as a sex-linked recessive trait, originally described in 1898. Deficiency of α-galactosidase results in the accumulation of α-galactosyl-lactosylcer-amide in the skin and other epithelia. Male hemizygotes have the full syndrome, but female heterozygotes may exhibit some manifestations. Presentation is in childhood or adolescence, with skin lesions distributed over the scrotum, umbilicus, thighs, and buttocks. Histologically, the lesions are angiokeratoma, which increase in number with time. Deposition within the kidneys or heart ultimately results in organ failure, which along with strokes usually results in death in the fourth or fifth decade. There is no effective treatment. Diagnosis is made by the identification of the characteristic skin lesions, the demonstration of vacuolated **histiocytes** (macrophages) within the **bone marrow**, or by enzyme estimation.

FACTITIOUS PURPURA

See **Bruising**.

FACTOR V

See also **Coagulation factors**; **Hemostasis**.

A 2224-amino acid plasma glycoprotein of molecular weight 330,000. It is a critical cofactor in coagulation, which in its activated form facilitates the conversion of factor II (**prothrombin**) to factor IIa. It acts as a catalyst to this reaction in the prothrombinase complex and increases the rate of conversion 200,000- to 300,000-fold. It circulates as a single-chain protein in a precursor inactive form. It has a domain structure that is very similar to that of **factor VIII** (see Figure 29).

It is coded for by a complex 25-exon gene on chromosome 1 (1q21-25) and encodes a 6.6-kb mRNA. Upon activation by thrombin or **factor X**a, it is converted into its active two-chain form. Thrombin cleaves factor V at Arg709-Ser710, Arg1018-Thr1019, and Arg1545-Ser1546.

Following cleavage, the two chains are linked via a divalent metal ion bridge. Factor V binds to phospholipid surfaces via binding sites in the light chain. Inactivation of factor V occurs via activated **protein C** and its cofactor **protein S** by cleavage at Arg506.

Although it is predominantly synthesized in the liver (plasma factor V), **megakaryocytes** also synthesize factor V, which is stored in platelet a-granules (platelet factor V). Platelet

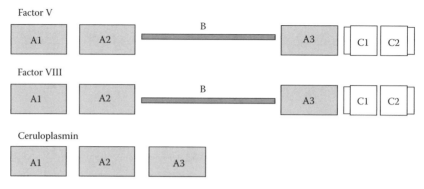

FIGURE 29

Domain structure of **factor V** and factor VIII. Both are composed of triplicated A domains, a duplicated C domain, and a large B domain. The A domains of ceroplasmin are homologous to those of factor V and factor VIII. The A and C domains of factor V and factor VIII share a 40% sequence homology. The B domains do not share a sequence identity.

factor V is secreted upon platelet activation and accounts for approximately 20% of the body mass of factor V. Factor V has a binding-protein multimerin that acts in a similar manner to **Von Willebrand Factor** and factor VIII.

Plasma concentration of factor V is 0.5 to 1.0 mg/dl (0.5 to 2.0 U/ml). It has a half-life of approximately 12 h. Values for premature and full-term infants during the first 6 months of life are given in **Reference Range Tables XIII** and **XIV**.

Factor V Leiden

Factor V Leiden is a variant form where an Arg506-Glu mutation occurs. This leads to a factor V molecule that is resistant to cleavage by activated protein C (APC resistance). As a consequence, activated factor V persists, rather than being inactivated by protein C. About 5% of white populations are heterogeneous for this variant, with about 1 in 1000 homozygous. The factor V Leiden mutation is thought to have arisen ≈30,000 years ago as a founder effect. The evolutionary advantage of factor V Leiden is unknown, but it has been proposed that it reduces blood loss after trauma, increases fertility, and reduces the likelihood of postpartum hemorrhage. More than 95% of those with activated protein C resistance on plasma testing will exhibit this mutation. It is the most common genetic risk factor for **venous thromboembolic disease**. The relative risk for thrombosis with factor V Leiden is two- to eightfold. As the defect is so common, it accounts for 20 to 50% of venous thrombosis, dependent upon the population studied. The importance of factor V Leiden is its frequency and the fact that it acts synergistically with other acquired thrombosis risk factors, particularly when associated with taking oral contraceptive pills.[189] Here, the thrombosis risk rises 35-fold, although the absolute risk of thrombosis remains at significantly less than 0.5% per year for any single individual.

Mutations at Arg306, the second APC cleavage site, have also been described. These may account for some of the non-factor V Leiden cases of phenotypic APC resistance.

Factor V Deficiency

A rare autosomal disorder, consanguinity being frequently seen in affected kindred. Factor V deficiency presents with a bleeding disorder due to the direct lack of plasma and platelet factor V.

Combined factor V and factor VIII deficiency is also seen. Acquired factor V deficiency is frequently seen in **liver disorders.**

Treatment is via local measures, and replacement therapy with **fresh-frozen plasma** is the mainstay of therapy, as no commercial factor V concentrate is available. Levels of approximately 25 units/dl are thought to be hemostatic in mild and moderate bleeding episodes. In severe bleeding, **platelet transfusion** may be used, as platelets contain factor V. Platelet concentrate is not the treatment of choice due to the risk of the development of platelet antibodies.

FACTOR VII

See also **Coagulation factors; Hemostasis**.

A 406-amino acid plasma glycoprotein and **serine protease** of molecular weight 50,000. It is a component in the initiation of blood coagulation that forms a complex with **tissue factor** to generate an enzyme complex that activates **factor X** and **factor IX**. It is coded for by a 13-kb, nine-exon gene on chromosome 13. It is a **vitamin K**-dependent protein and has 10 N-terminal glutamic acid residues that are terminal gamma carboxylated to form the Gla domain. Calcium-binding properties of factor VII are crucial to its normal function and biological activity. Factor VII is activated by cleavage of the Arg153-Ile153 peptide bond. Activators include **thrombin**, activated **factor X**, and activated **factor IX**. Autoactivation of factor VII may occur when bound to tissue factor or a positively charged surface. Of the activators, factor IX is the most potent.

In contrast to other coagulation factors, it is suggested that factor VII may have low but significant levels of activity in proenzyme form. This characteristic would be important in the initial amplification in the coagulation cascade, but it remains controversial.

Laboratory evaluation of factor VII can be variable, depending upon the source of thromboplastin (tissue factor) used in the assay, as different species have differing affinities for human factor VII. It circulates at a concentration of around 1.0 mg/dl (0.5 to 2.0 U/ml). It has a half-life of 4 to 6 h. Levels are notably low in the newborn due to liver immaturity (see **Reference Range Tables XIII** and **XIV**). There is a rise in level with age, and long-term epidemiological studies in healthy persons have shown increased levels in those who develop coronary artery disease. High levels may therefore be a risk factor for **arterial thrombosis**.

Factor VII Deficiency

This is a rare autosomal disorder. Consanguinity is frequently seen in affected kindred. Factor VII deficiency presents with a bleeding disorder due to the direct lack of plasma factor VII. Factor VII levels of less than 2 U/dl are associated with severe bleeding (including **hemarthroses** and intracranial hemorrhage) comparable with that seen in classic **hemophilia A**. It is the only hereditary coagulation factor deficiency that causes isolated prolongation of the **prothrombin time**. Factor VII deficiency can be seen as part of an inherited multiple coagulation factor deficiency. As factor VII has such a short half-life, acquired factor VII deficiency is frequently seen in **liver disorders** and **vitamin K** deficiency.

Treatment is via local measures and replacement therapy using **recombinant factor VIIa** at 10 to 15 µg/kg, 6 to 12 h as required until the hemorrhage is arrested. Alternatively, therapy with **fresh-frozen plasma** or prothrombin complex concentrate containing factor VII can be used (see **Coagulation factor concentrates**). Levels of approximately 25 U/dl are thought to be hemostatic in mild and moderate bleeding episodes. Factor VII deficiency may be seen as part of an inherited multiple coagulation factor deficiency. As factor VII

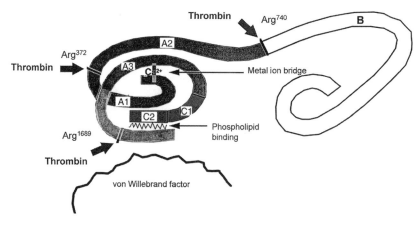

FIGURE 30
Model of factor VIII and thrombin cleavage sites.

has such a short half-life, acquired factor VII deficiency is frequently seen in liver disorders and vitamin K deficiency.

FACTOR VIII

See also **Coagulation factors**; **Hemostasis**.

(Antihemophilic globulin) A 2351-amino acid plasma glycoprotein of approximately 360,000 molecular weight. A 19-amino acid signal peptide is removed during secretion. It has a domain structure that is very similar to that of **factor V** (see Figure 29) and is related to the copper protein ceruloplasmin. The large B domain is of unknown function, has no known homology to other proteins, and is not required for coagulant activity. B-domainless mutants show normal factor VIII clotting activity and may be expressed in recombinant systems at a higher level than full-length factor VIII. Factor VIII is coded for by a complex 26-exon, 186-kb gene on the X chromosome and codes a 9-kb mRNA. It is one of the largest and least stable coagulation factors, with a complex polypeptide composition, circulating in plasma in a noncovalent complex with **Von Willebrand Factor** (VWF).

Plasma concentration of factor VIII is <0.01 mg/dl (0.5 to 2.0 IU/ml) for normal healthy adults. Values for premature and full-term infants are given in **Reference Range Tables XIII** and **XIV**. Levels of factor VIII may rise as an **acute phase response** protein and in response to stress or exercise. It is frequently elevated in **pregnancy**, **liver disorders**, and with **vasculitis**.

It has a half-life of approximately 12 h. VWF functions to protect factor VIII from premature proteolytic degradation and concentrate factor VIII at sites of vascular injury. Factor VIII is a critical protein procofactor in coagulation for factor IX. In activated form, factor VIII facilitates the conversion of factor X to factor Xa. It acts as a catalyst to this reaction in the Xase complex and increases the rate of conversion 200,000-fold.

Although synthesized in the liver as a single-chain molecule, factor VIII is cleaved shortly after synthesis so that it circulates as a heterodimer. The heterodimer comprises an 80-kDa light chain linked through a divalent metal cation bridge to a heavy chain (90 to 200 kDa) that contains variable amounts of the B domain. Upon activation by thrombin (or factor Xa), factor VIII is cleaved at Arg372, Arg740, and Arg1689 (see Figure 30).

Factor VIII circulates as a two-chain heterodimer, heavy (A1-A2-B) and light chains (A3-C1-C2) linked by a divalent metal ion bridge, as shown. Thrombin cleaves factor VIII at Arg372, Arg740, and Arg1689 to yield a series of smaller chains. The Arg372 cleavage is the rate-limiting step, which yields 50- and 40-kDa fragments from the heavy chain, both of which are essential for catalytic activity. The Arg740 cleavage removes any remaining B-domain remnant, to yield a 90-kDa heavy chain. At the same time, a small fragment is cleaved that removes VWF from factor VIII. Activated factor VIII, factor VIIIa, is very unstable and rapidly loses cofactor function, probably due to subunit disassociation. Inactivation of factor V also occurs via activated protein C and its cofactor protein S by cleavage at Arg336 and Arg562.

Factor VIII Deficiency

This is the cause of **hemophilia A**. Combined deficiencies of factor VIII and other factors (such as factor V and factor VIII) are also seen, but these are rare.

Antibodies to Factor VIII

(Factor VIII inhibitors) Antibodies to factor VIII may develop in patients with hemophilia A (alloantibodies) or previously normal patients who develop acquired hemophilia (autoantibodies). The origin of these antibodies is speculative,[189a] but it includes:

Patients who have no cross-reactive material (CRM) and who have had no factor VIII from birth

Mutations in the Az domain or fraction C1 and C2 domains

Reaction to factor VIII concentrate due to variable manufacturing processes

Such antibodies interfere with the coagulant function of factor VIII, inhibiting its biological efficacy. They are generally IgG antibodies with a predominance of IgG4 subclass. They do not fix complement and do not lead to immune complex disease. Inhibitors are most frequently directed against the A2 and C2 domains of the factor VIII molecule. The reaction between the antibody and factor VIII is both time and temperature dependent.

Laboratory tests for antibody rely on the neutralization of factor VIII activity of normal plasma. Levels of inhibitor are quantified using the Bethesda assay, where one Bethesda inhibitor unit (Bu) is the amount of antibody that reduces the activity of a given sample of plasma by 50% after 2 h of incubation.[189b]

Inhibitors have a prevalence of approximately 5 to 10% of all hemophilic patients and approximately 15% of patients with severe hemophilia A. However, the incidence of severe hemophilia A inhibitors is up to 25%, with approximately 10% being transient. The vast majority occur in patients with a factor VIII level of less than 3 IU/dl. The time of development of an inhibitor to factor VIII is not predictable but most often occurs in childhood after a limited number of exposures (median ≈10/day) to factor VIII. Failure of standard replacement therapy to treat a bleeding episode is often the first indication of the presence of an inhibitor.

Hemophilic patients with inhibitors fall into two groups: those in whom the antibody does not rise upon further exposure to factor VIII (low responders) and those in whom the antibody rises dramatically after further exposure (high responders). Inhibitor titers rise within a few days of exposure but only decline very slowly. A level of above 10 Bu is generally regarded as a high titer inhibitor.[19] High-level inhibitors significantly complicate treatment of bleeding episodes. These features are important in determining therapeutic options for these patients.

Management of acute bleeding will depend upon the level of the inhibitor and how well the patient responds to treatment. Low-responder patients may be treated using high-dose factor VIII concentrate sufficient to overcome the inhibitor. High-responder patients are best treated with (activated) prothrombin complex concentrate or **recombinant factor VIIa**. Porcine factor VIII concentrate, if available, may be used, provided that the inhibitor shows low cross reactivity (see **Coagulation factor concentrates**). Long-term treatment to reduce the inhibitor may be performed using immune-tolerance-induction regimens. Such regimens may be low-dose or high-dose factor VIII concentrate regimes. High-dose regimes involve infusions twice daily for upwards of 12 months at a dose of 100 IU/kg or greater. Such regimes usually give factor VIII concentrate alone, but may occasionally be given along with **corticosteroids, alkylating agents, plasmapheresis**, and **immunoglobulin** in the more complex regimens (e.g., Malmo regime). All high-dose regimens are extremely expensive to institute. With these high-dose regimens, immune tolerance is instituted in approximately 80% of patients, who may show significant improvement in their inhibitor titer.

Inhibitors to factor VIII may also arise outside of hemophilia and lead to the development of acquired hemophilia. Acquired factor VIII inhibitors may be seen in association with **malignancy, pregnancy, systemic lupus erythematosus** (SLE), **rheumatoid arthritis, adverse drug reactions**, inflammatory bowel disease, and old age. About 50% of all patients have no underlying disorder. Acquired factor VIII inhibitors (**autoantibodies**) often have more-complex reaction kinetics than inhibitors seen in patients with hemophilia (**alloantibodies**). Such inhibitors usually present with widespread bleeding (often subcutaneous) in patients without any previous history of bleeding. Bleeding episodes can be treated using different blood products, much as outlined above for alloantibodies. Long-term treatment for acquired hemophilia involves immunosuppression to eliminate the inhibitor (using corticosteroids, **cyclophosphamide, azathioprine**, and intravenous immunoglobulin or **rituximab**) and treatment of any underlying disease.

FACTOR IX

See also **Coagulation factors**; **Hemostasis**.
(Plasma thromboplastin component) A **vitamin K**-dependent **serine protease** that is essential for blood clotting. It circulates as a single-chain polypeptide of 415 amino acids with a molecular weight of 57,000. It is coded by a 34-kb gene on the long arm of the X chromosome and is the largest of the family of vitamin K-dependent proteins. Factor IX is synthesized in the liver. It comprises several functional domains, including Gla domain (calcium binding), epidermal growth factor-like domain, and trypsin-like domain (catalytic site). Twelve N-terminal glutamic acid residues are terminal gamma carboxylated to form the Gla domain. Calcium-binding properties of factor IX are crucial to its normal function and biological activity.

Activation of factor IX occurs via cleavage of two peptide bonds, Arg145-Ala146 and Arg180-Val181. This activation can be achieved by either active factor XI, factor XIa, or by activated **factor VII**, factor VIIa, complexed to **tissue factor**. Cleavage of the Arg145-Ala146 occurs rapidly, whereas the Arg180-Val181 cleavage is rate limiting. Cleavage into factor XIa generates a protein with a heavy and light chain bound together via a single disulfide bond. A 24-amino acid activation peptide is removed during cleavage. Together with **factor VIII**, factor IXa can then proceed to activate **factor X**. In addition, factor IXa may also activate **factor VII**.

The plasma factor IX concentration in a healthy population is around 0.01 mg/dl (0.4 to 1.6 IU/ml). It has a half-life of approximately 24 h. It partitions between both the intravascular and extravascular spaces. Levels of factor IX are low at birth due to hepatic immaturity, being only 20 to 50% of normal. Levels increase to normal by the age of 6 months (see **Reference Range Tables XIII** and **XIV**). There is a small increase in level seen in **pregnancy** and in women taking estrogen-containing contraceptives. Congenital deficiency of factor IX results in **hemophilia B**.

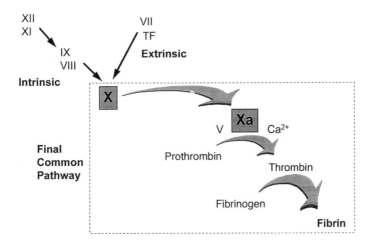

FIGURE 31
Central role of factor X in the final common pathway of coagulation.

FACTOR X

See also **Coagulation factors**; **Hemostasis**.
A plasma glycoprotein and **serine protease** of molecular weight 59,000. It is the pivotal component in the common pathway of blood coagulation (see Figure 31). Factor X is coded for by a 22-kb gene on chromosome 13. It is a **vitamin K**-dependent protein and has 11 N-terminal glutamic acid residues that are terminal gamma carboxylated to form the Gla domain. Calcium-binding properties of factor X are crucial to its normal function and biological activity. It is synthesized as a single chain but exists in plasma as heavy and light chains linked by a single disulfide bond. Factor X is activated by cleavage of the Arg51-Ile52 peptide bond. Activators include activated **factor VII/tissue factor** complex and activated **factor IX/factor VIII** complex in the presence of calcium ions.

Factor Xa in conjunction with **factor V** forms a complex on the membrane surface, **prothrombinase complex**, which converts **prothrombin** to **thrombin**. Factor X is inhibited by **antithrombin** and α_2-macroglobulin.

Factor X circulates at a concentration of around 0.75 mg/dl (0.5 to 2.0 U/ml). For values of premature and full-term infants, see **Reference Range Tables XIII** and **XIV**. It has a half-life of ≈36 hours.

Factor X Deficiency

This is an autosomal disorder. Factor X deficiency may be seen as part of an inherited multiple coagulation-factor deficiency. Consanguinity is seen in affected kindred. Acquired factor X deficiency is seen in **liver disorders** and with vitamin K deficiency. It presents with a bleeding disorder due to the direct lack of plasma factor X. Factor X levels of less than 2 U/dl are associated with severe bleeding, similar to those seen in classic hemophilia A, but often not as severe. Individuals with levels above 15 U/dl have few bleeding symptoms, although bleeding may occur with major surgery and trauma. Diagnosis is made by specific assay following identification of a prolonged **prothrombin time** and **activated partial thromboplastin time**. Treatment is via local measures and replacement therapy with **fresh-frozen plasma** or intermediate-purity human factor IX concentrate (see **coagulation factor concentrates**). Such intermediate-purity factor IX concentrates contain approximately 1 unit of factor X activity per unit of factor IX present. Levels of approximately 15 U/dl are thought to be hemostatic in mild and moderate bleeding episodes.

FACTOR XI

See also **Coagulation factors; Hemostasis**.

(Plasma tissue thromboplastin antecedent) A zymogen of a **serine protease** of molecular weight 160,000 that is involved in the **contact activation** phase of blood coagulation. Factor XI is a homodimer, comprising two identical subunits bound together by a disulfide bond, that circulates bound to high-molecular-weight kininogen. It is coded by a 15-exon, 23-kb gene on chromosome 4 (q32-35). It has a plasma half-life of approximately 72 h.

Factor XI is cleaved to active factor XIa by active **factor XII**, factor XIIa, in the presence of high-molecular-weight kininogen. Activation cleavage occurs within each subunit at Arg369-Ile370 in a region bound by a disulfide linkage, thus yielding two heavy chains and two light chains in the active molecule (see Figure 32).

Factor XIa activates **factor IX** in the presence of calcium. No specific additional cofactors are required for this reaction. Both factor XI and factor XIa bind to platelets. The role of the factor XI-platelet interaction in physiological terms is unknown.

It circulates at a concentration of around 1.2 mg/dl (0.4 to 1.6 IU/ml). For levels in premature and full-term infants, see **Reference Range Tables XIII** and **XIV**.

Factor XI Deficiency

This is the only contact factor deficiency that is known to be associated with a clinical bleeding tendency. The condition is particularly noted in Ashkenazi Jews, and because of the gene frequency of $\approx$4%, appreciable numbers of homozygous patients (i.e., severe cases) are to be expected. Three point mutations of the factor XI gene are described in these populations (splice junction, stop, and missense) that appear to account for most cases of factor XI deficiency.

Factor XI deficiency is inherited as an autosomal disorder. Severe factor XI deficiency is defined as a factor XI level of below 15 IU/dl. Bleeding is frequently relatively mild and predominantly seen only after surgery or significant trauma. Spontaneous bleeding is rare. Bleeding severity does not correlate particularly well with the plasma level of factor XI. This makes treatment and defining adequate levels for hemostasis difficult. The best predictor of bleeding is past history of hemostatic challenge. The diagnosis is suggested by an isolated prolongation of the **activated partial thromboplastin time** (APTT), other screening tests of coagulation being normal. The diagnosis is confirmed by a specific coagulation-factor assay. Other screening tests of coagulation are normal. Bleeding time is normal, although there are a few reported cases of prolongation. Treatment is via local measures or replacement of factor XI, dependent upon the extent of the bleeding problem. Traditionally, **fresh-frozen plasma** has been used for replacement of factor XI, but now factor XI concentrates are available in limited supply (see **Coagulation factor concentrates**). These concentrates facilitate the management of patients with factor XI deficiency, but they may be associated with venous thromboembolic disease, much as prothrombin complex concentrates were, and should therefore be used with caution. **Recombinant factor VIIa** has been reported to be of value in the management of factor XI deficiency.

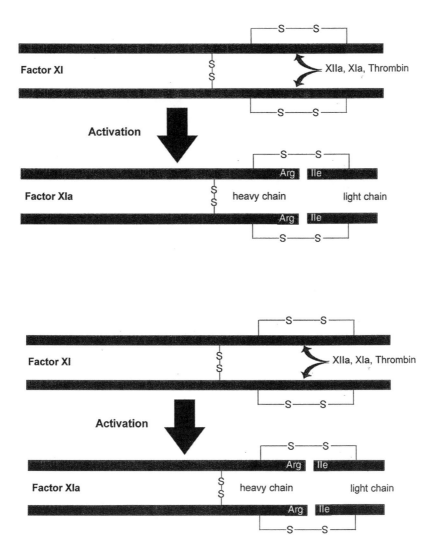

FIGURE 32
Activation of factor XI. Factor XI can be activated by factor XIIa, factor XIa, or thrombin. Only the light chain possesses catalytic activity.

FACTOR XII

See also **Coagulation factors**; **Hemostasis**.

(Hageman factor) A single-chain **serine protease** of molecular weight 80,000 that is the first component of the intrinsic pathway of blood coagulation. It is involved in **contact activation**. The factor XII gene is 12 kb in size and located on chromosome 5. Factor XII has a half-life of approximately 2 days.

In the process of contact activation factor XII is absorbed onto negatively charged surfaces and undergoes limited proteolysis at specific sites to yield active factor XII. This slowly converts **prekallikrein** to **kallikrein**, which specifically cleaves factor XII to yield fully active factor XIIa. In addition, factor XIIa can autoactivate factor XII. Factor XIIa can activate factor XI to promote downstream activation of the coagulation cascade.

Factor XII circulates at a concentration of around 0.4 mg/dl (0.3 to 1.5 IU/ml), with slightly lower levels for both premature and full-term infants (see **Reference Range Tables XIII** and **XIV**). Deficiency is only very rarely associated with excessive bleeding. It often presents incidentally as an isolated prolongation of the **activated partial thromboplastin time**. Paradoxically, there may be a weak association between factor XII deficiency and thrombosis, although this is controversial. The mechanism of increased thrombotic risk is thought to be impaired contact activation of **fibrinolysis**.

FACTOR XIII

See also **Coagulation factors**; **Hemostasis**.

(Fibrin-stabilizing factor) A cysteine transglutaminase enzyme operating in the final step in coagulation: the formation of a stable fibrin clot. Factor XIII circulates in plasma as an inactive tetramer consisting of two "a" subunits (molecular weight 75 kDa) coded 6p24-25 and two "b" subunits (molecular weight 80 kDa), coded 1q31-32, i.e., a2b2. Factor XIII is also found in platelets and megakaryocytes, the placenta, uterus, and macrophages, but in these tissues only the "a" subunit is present. The "a" subunit is synthesized in several tissues, including macrophages, whereas the "b" subunit is synthesized only in the liver. The plasma concentration of the "a" subunit is 15 µg/ml and that of the "b" subunit is 14 µg/ml, suggesting the formation of a stoichiometric complex.

Activation of factor XIII to XIIIa involves cleavage by thrombin of the "a" subunit (at Arg37-Gly38) followed by its separation from the "b" subunit to give the active transglutaminase. The active site is located on the "a" subunit (at cysteine 314) and the "b" subunit appears to act solely as a noncatalytic carrier for the "a" subunit. After activation by thrombin, Factor XIIIa catalyzes the formation of cross-links between the χ-chain of fibrin in an antiparallel manner, resulting in covalent dimerization of the χ-chains. At a much slower rate, polymerization of the α-chains occurs. Factor XIIIa also cross links fibronectin to fibrin and collagen, which has been shown *in vitro* to enhance fibroblast proliferation. This may explain the delayed wound healing observed in factor XIII-deficient individuals. In addition, factor XIIIa also cross links a$_2$-antiplasmin into the α-chain of the fibrin clot, thereby increasing the resistance of the clot to lysis by plasminogen. The activation of factor XIII by thrombin is facilitated by the presence of both fibrinogen and fibrin polymers.

Factor XIII circulates at a concentration of around 2.5 mg/dl (0.4 to 1.7 U/ml), with slightly lower values for premature and full-term infants (see **Reference Range Tables XIII** and **XIV**) and has a half-life of ≈9 days.

Factor XIII Deficiency

Affected individuals have significantly lower levels of factor XIII, shown clinically by prolonged bleeding following trauma and after surgery (including dental extractions) and delayed wound healing.

Hereditary Deficiency

An autosomally recessive deficiency state often presenting as prolonged bleeding from the umbilical stump. Other significant symptoms include intracranial hemorrhage and recurrent soft-tissue hemorrhage with a tendency to form cysts. Hemarthroses are rare occurrences. In most cases, relatives of the propositus are heterozygotes with reduced levels of factor XIII, and frequently asymptomatic. Consanguinity is not uncommon. In the homozygous state, affected individuals have less than 1 U/dl factor XIII antigen or activity. Women with factor XIII deficiency have an increased risk of recurrent spontaneous abortions, and replacement treatment may be required during pregnancy.

Hereditary factor XIII deficiency is classified as:

Type I: reduced/absent subunits "a" and "b"

Type II: absent subunit "a" but normal/reduced levels of subunit "b"

Type III: reduced level of subunit "a" but absent subunit "b"

Type II deficiencies appear to be more common than either type I or type III. Because the "b" subunit of factor XIII appears to act as a carrier for the "a" subunit, deficiencies of the "b" subunit are likely to associated with a secondary deficiency of the "a" subunit. Families with a dysfunctional factor XIII have been reported.

Acquired Deficiency

This may occur in some forms of **leukemia**, of **liver disorders**, and with **disseminated intravascular coagulation**. A screening test for factor XIII can be performed by determining the solubility of the patient's recalcified plasma in either $5M$ urea or 1% monochloroacetic acid. Individuals with less than 1 U/dl factor XIII activity show an increased solubility. Specific factor XIII assays to determine the precise level of factor XIII activity are available. Treatment is ideally with factor XIII concentrate given ≈4 times weekly, as factor XIII has a long half-life and minimal factor XIII activity (≈5 U/dl) may be sufficient to prevent bleeding complications. Alternatively, **fresh-frozen plasma** or **cryoprecipitate** may be used when factor XIII concentrate is not available.

FAMILIAL COLD-ASSOCIATED AUTOINFLAMMATORY SYNDROME

See also **Autoinflammatory syndromes**.

(FCAS; familial Mediterranean fever) An autosomally dominant disorder characterized by recurrent episodes of rash, arthralgia, and fever after exposure to cold temperature. **Neutrophilia** occurs 4 to 8 h later. The disorder generally resolves spontaneously within 24 h. It is caused as a result of mutation of the CIAS1 gene that encodes the protein cryopyrin that activates caspase 1, resulting in the release of **interleukin**-1. Treatment with interleukin-1 receptor (IL-1 Ra) has been claimed to be effective.[190]

FAMILIAL ERYTHROCYTOSIS

See **Oxygen affinity to hemoglobin** — disorders.

FAMILIAL HEMATOLOGICAL DISORDERS

See **Hereditary anomalies**.

FAMILIAL LECITHIN; CHOLESTEROL ACYLTRANSFERASE DEFICIENCY

See **Abetalipoproteinemia; Acanthocytosis**.

FAMILIAL SELECTIVE MALABSORPTION TO VITAMIN B₁₂

See **Immerslund-Gräsbeck syndrome**.

FANCONI ANEMIA

(FA) An inherited **aplastic anemia** associated with skeletal and skin abnormalities. Originally described in 1927, inheritance is autosomally recessive with variable penetrance.

Pathogenesis

The precise genetic abnormality remains unknown. Chromosomal fragility with a defect in **DNA** repair has been demonstrated. Cultures of lymphocytes, marrow cells, or skin fibroblasts from patients with FA reveal increased nonspecific chromosomal damage when stressed by agents (mitomycin C or diepoxybutane), causing cross bindings between DNA chains. Similar events can occur with nonstressed cultures. However, a number of cases with negative fragility tests have been reported. Antenatal diagnosis is possible using chorionic villus cells or fetal blood sampling. Bone marrow hypoplasia appears to result directly from stem cell failure with reduced CFU-GM and BFU-E that precede increasing **pancytopenia**.

Clinical Features

In addition to marrow hypoplasia, there are typical physical abnormalities (see Table 62), but these may be absent in over 25% of patients. Bone marrow failure occurs at any time between birth and <30 years (mean 8 years), with platelet counts being usually first affected. Bone marrow aspiration may reveal **megaloblastosis** and increased macrophage activity (including **hemophagocytosis**) in the early stages, but eventually the picture is indistinguishable from other forms of marrow hypoplasia.

Growth retardation occurs in about 75% of patients, and most of them remain below the tenth percentile. The skeletal abnormalities of the face give rise to an elflike appearance. Variable upper-limb abnormalities, including triphalangeal or absent-hypoplastic thumb and absent radii, have been described.

Median survival is about 25 years in untreated patients. The bone marrow hypoplasia is progressive, requiring **red blood cell transfusion** and **platelet transfusion** in support. In 10% of patients, **acute myeloid leukemia** supervenes (average age 15 years), and this is the presenting feature in up to 25% of affected individuals. Carcinomas (especially squamous cell) occur in about 5% of patients (mean age 23 years), with the oropharynx and the gastrointestinal and urogenital tracts the most frequently affected sites. Hepatocellular carcinoma is also more common than in the general population, but this probably relates to androgen therapy.

TABLE 62

Clinical Features of Fanconi Anemia

Low birth rate
Growth retardation
Short stature
Microcephaly
Micro-ophthalmia
Microstomia
Skeletal abnormalities, e.g., thumb, wrist, forearm
Skin pigmentation
   Generalized hyperpigmentation
   Café au lait
   Depigmentation
Genitourinary
   Horseshoe or pelvic kidney
   Cryptorchism
Strabismus
Mental retardation

Management

Apart from **red blood cell transfusion**, most patients benefit from **anabolic steroids**. Oxymethalone (2.5 mg/kg/day) improves all of the hematological indices in most patients. However, side effects are problematic, with hyperactivity and aggressive behavior, along with virilization of females and the development of secondary sexual characteristics in boys. Hepatic complications, with peliosis hepatis, cholestatic jaundice, hepatic adenoma, and ultimately carcinoma, are not infrequent. Hepatocellular carcinoma may temporarily regress after withdrawal of androgen therapy. **Allogeneic stem cell transplantation** is the only curative approach for bone marrow failure. In about 80% of cases hematopoiesis is restored, but the other skeletal abnormalities and the risk of malignancy remain. Potential donors include siblings and unrelated donors, using bone marrow or umbilical cord cells. An increasing number of successful haploidentical grafts has also been reported. Siblings who are apparently heterozygous for Fanconi's are suitable. Fanconi cells are supersensitive to conditioning chemoradiotherapy; therefore, traditional conditioning schedules should be avoided; cases of lethal results after radiotherapy or standard-dose cyclophosphamide have been reported. Because a small percentage of patients with FA do not present the typical chromosomal abnormalities after stress cultures, a practical approach to bone marrow failure of suspected constitutional (familial) origin is to administer a nonmyeloablative reduced-dose conditioning schedule based on the immunosuppressive effect of CAMPATH (humanized monoclonal specific human CD52), ALG, or ATG, together with low-dose (50 mg/kg) cyclophosphamide, prior to transplant. It is also advisable to harvest and cryopreserve bone marrow from these patients as soon as the diagnosis is made and before the bone marrow fails, whenever this is possible. This would then be a potential source for autologous hematopoietic regeneration, should the allograft fail.

FARNESYL TRANSFERASE

(FT) An enzyme involved in the posttranslational modification of ras protein. The ras family of proto-oncogenes is an upstream mediator of several essential **cell-signal transduction** pathways involved in cell proliferation and survival. Point mutations of ras oncogenes result in constitutively active RAS and have been shown to be oncogenic. However, ras activation can occur in the absence of ras mutations secondary to upstream receptor activation.

Farnesyl transferase is involved in posttranslational modification of the ras proteins by covalently linking a farnesyl group to the ras protein. This permits the ras protein to be translocated to the surface membrane, allowing the protein to be involved in signaling for increased proliferation and inhibition of **apoptosis**.

The first important step in RAS activation is farnesylation by farnesyl transferase, and inhibitors of this enzyme have been demonstrated to inhibit RAS signaling and to have antitumor effects. However, it is now clear that farnesyl transferase inhibitors (FTIs) have activity independent of RAS, most likely due to effects on prenylated proteins downstream of RAS, which explains their activity in several malignancies where ras mutations are rare. Several FTIs are in clinical development for the treatment of hematological malignancies, but these have not yet completed the trial stage.

FAS

See also **Cellular cytotoxicity**.

Fas/CD95/Apo-1 is a cell-surface molecule of the **tumor necrosis factor** (TNF) group. Cell activation triggers its expression on **lymphocytes**. Prolonged activation makes Fas-expressing cells sensitive to death by **apoptosis** once the Fas ligand (Fas-L), expressed by activated T-lymphocytes, has interacted with Fas. These interactions control the life span of peripheral lymphocytes.

Deficiency of Fas gives rise to a massive nonmalignant lymphoproliferation of CD4⁻ and CD8⁻ T-cells. The deficiency can be associated with hypergammaglobulinemia and autoimmunity to peripheral-blood cells.[191]

FAVISM

See **Glucose 6-phosphate dehydrogenase** — deficiency.

Fc RECEPTORS

(FcRs) Cell membrane receptors specific for the Fc portion of **immunoglobulin** (Ig). Binding of the immunoglobulin confers upon the cell the specificity of the immunoglobulin. When antigen binds to Ig complexed with the FcR, cross-linking of the receptors occurs and activates the cell. Cells bearing FcRs include **histiocytes** (macrophages), **lymphocytes**, **mast cells**, and **basophils**. There are several FcRs, with differing specificities for immunoglobulins IgG, IgA, and IgE (indicated Fcg, etc.) present on different cell types, with differing functions and with differing affinities for the appropriate Ig (low, medium, high) (see Table 63).

Disorders mediated by Fcγ receptors include **immune thrombocytopenic purpura** and **warm autoimmune hemolytic anemias**.

TABLE 63

Properties of Fc Receptors

Receptor	CD No.	Cell Types	Function
g RI, high affinity	CD64	monocyte/macrophage	ADCC;[a] triggers phagocytosis, oxidative burst, cytokine release
g RII, low affinity	CD32	monocyte/macrophage, granulocyte	ADCC; endocytosis
g RIII, low affinity	CD16	natural killer cells, macrophages, granulocytes, some T-cells	ADCC
aR, medium affinity	CD89	monocytes/macrophages, neutrophils	phagocytosis, oxidative burst
eRI, high affinity		mast cells, basophils	degranulation
eRII, low affinity	CD23	activated B-cells	antigen presentation?

[a] ADCC, antibody-dependent cellular cytotoxicity.

FELTY'S SYNDROME

The association of **rheumatoid arthritis** with **neutropenia** and **splenomegaly**. It sometimes occurs with a form of T-cell **lymphocytosis** of the large granular type. The neutropenia is probably a consequence of accelerated **apoptosis** of granulocyte precursors.

FEMALE REPRODUCTIVE ORGAN DISORDERS

See **Gynecological disorders**.

FERRITIN

A water-soluble complex of ferric hydroxide and the protein apoferritin. This is a 440-kDa protein consisting of 243 subunits arranged to form a hollow sphere, the central cavity of which can store about 4500 **iron** atoms in the form of electron-dense particles. Because the plasma ferritin concentration is correlated with the total body iron stores, serum ferritin measurements are important in the diagnosis of disorders of iron metabolism. Hyperferritinemia is also common in a variety of liver disorders unassociated with **iron overload**. The reference range in adult males is 15 to 200 ng/ml, with a median of 100 ng/ml, and in adult females is 12 to 150 ng/ml, with a median of 30 ng/ml. The levels rise from 25 ng/ml at birth to 200–600 ng/ml at 1 month, falling to around adult levels by 6 months of age.

FERROCHELATASE

(Heme synthase) A mitochondrial enzyme in the final step in **heme** synthesis. Mutations in the corresponding *FECH* gene (18q21.3) cause **erythropoietic protoporphyria**.

FERROKINETICS

The measurement of **iron** movement throughout the body.[25] Three isotopes of iron (^{59}Fe [$T^{1/2}$ 45 days], ^{55}Fe [$T^{1/2}$ 2.16 years], and ^{52}Fe [$T^{1/2}$ 8.2 h]) have been used in clinical practice to measure:

Absorption of iron from an oral dose

Distribution of iron after intravenous injection

Imaging of iron uptake in organs

Absorption of Iron from an Oral Dose

Iron absorption is measured using single doses of inorganic iron (usually radiolabeled) or food substances labeled with intrinsic or extrinsic radioiron tags. Iron absorption is quantified subsequently by measuring retained radioiron in a whole-body counter, radioiron that becomes incorporated into hemoglobin or bound to transferrin, or radioiron that is excreted (primarily in stool).

Distribution of Iron after Intravenous Injection

After intravenous injection of ^{59}Fe complexed to transferrin *in vitro*, the rate of radioiron clearance from the plasma (^{59}Fe plasma $T^{1/2}$) and subsequent uptake in erythrocytes are measured. From these data, the plasma iron concentration and plasma volume, the rate of formation of erythrocytes, and the red blood cell iron turnover can be calculated.

The initial clearance of iron is exponential, and sampling during this period can be used to calculate the $T^{1/2}$. In normal individuals, the $T^{1/2}$ is about 90 min (range 60 to 140 min). In patients with erythroid hyperplasia, the $T^{1/2}$ is shorter; in patients with marrow hypoplasia, the $T^{1/2}$ is longer. When plasma iron clearance is related to the plasma iron concentration, a value can be obtained for the plasma iron turnover (PIT). The reference range in healthy subjects is 70 to 140 µmol/l/day (4 to 8 mg/l/day). Increased PIT occurs in **iron deficiency, hemolytic anemias, myelofibrosis**, and **ineffective erythropoiesis** (especially **thalassemia**). In **bone marrow hypoplasia**, PIT is normal or reduced. However, PIT values in health and disease often overlap.

Incorporation of radioactive iron into developing erythroid cells occurs within a few days and reaches a maximum at 10 to 14 days after injection. Normal utilization is 70 to 90% by days 10 to 14 after injection. Decreased erythroid incorporation of radioiron suggests that:

Mature erythrocytes are destroyed soon after their release from the marrow

Immature red cells are destroyed in the marrow before release (ineffective erythropoiesis)

Serum iron is diverted to nonerythropoietic tissue as with bone marrow hypoplasia due to slow uptake by the **erythron**

An early, steep rise in the red cell radioiron utilization curve (rapid marrow transit time) suggests the presence of erythroid hyperplasia or a high erythropoietin level. Early maximum utilization with a subsequent falloff suggests the occurrence of hemolysis. Using the plasma iron clearance and utilization of iron, the red cell turnover (in units of mg/dl blood for 24 h) can be calculated; the normal value is 0.30 to 0.70 mg/dl of blood for 24 h.

Imaging of Iron Uptake in Organs

^{59}Fe is a gamma emitter, and thus its radioactivity can be measured *in vivo* by scintigraphy, and sites of distribution of the administered ^{59}Fe and the sites of erythropoiesis can be determined. ^{59}Fe activity is measured by placing a collimeter over the heart, liver, spleen, and upper part of the sacrum of a prone patient. Counts at these sites should be performed as soon as possible after intravenous ^{59}Fe administration, and again after 5, 20, 40, and 60 min, and then hourly for 6 to 10 h. Subsequent measurements are then made daily or on alternate days for the next 10 days. Initial counts are expressed as 100%, and subsequent counts are expressed proportionately after correction for decay. Although laborious, the technique is informative in patients thought to have bone marrow hypoplasia, myelofibrosis, or refractory anemia, conditions in which specific ^{59}Fe counting patterns are observed. It may also be helpful to determine sites of extramedullary erythropoiesis when splenectomy is contemplated. Whole-body scanning can also be performed using ^{52}Fe.

FERROPORTIN-1

See also **Hereditary hemochromatosis**.

(Iron-regulated transporter-1) An **iron**-responsive exporter of iron located in relatively large quantities in the basal aspect of syncytiotrophoblasts, where it transports iron from mother to embryo; in the basolateral surface of duodenal enterocytes, where it transports iron from enterocyte to blood; and in the histiocytes (macrophages), where it exports stored iron outside the cell. Mutations in the corresponding *FPN1* gene (2q32) are associated with **iron overload**. In many cases, transferrin saturation is normal, and iron accumulation predominates in macrophages in various organs.

FETAL HEMATOLOGICAL DISORDERS

The disorders of the fetus of hematological origin.

Hemolytic disease of the newborn (HDN) due to rhesus and other blood group incompatibilities with the mother.[192] The fetus may develop hydrops fetalis, often leading to abortion, or have icterus gravis neonatorum with kernicterus after birth.

Hemoglobinopathies, particularly hemoglobin Barts and hemoglobin H disease, both usually resulting in abortion.

Thalassemias and **sickle cell disorders**. These can be detected *in utero* by fetal blood sampling, either by amniocentesis or by chorionic villous biopsy. DNA techniques using the polymerase chain reaction on fetal cells in the maternal blood will, when developed, aid prenatal diagnosis.

Anemia due to massive fetomaternal **hemorrhage** or to **parvovirus** infection.

Hemophilia and related disorders. These can be diagnosed by fetal sampling.

Thrombocytopenia due to:

- **Immune thrombocytopenic purpura** (ITP) as a consequence of maternal platelet antibody transmission across the placenta, their origin being either maternal ITP antibodies or alloimmune antibodies — **neonatal alloimmune thrombocytopenia** (NAIT)
- Congenital infection (e.g., cytomegalovirus, rubella)
- Chromosomal disorder such as trisomy 21 mutation or thrombocytopenia with absent radii (TAR)

Hypercoagulable states (**thrombophilia**) of familial origin with increased risk of fetal death or stillbirth:

- **Antithrombin III** deficiency
- **Protein C** deficiency
- **Protein S** deficiency
- Resistance to activated protein C in those with **factor V** Leiden

Acute lymphoblastic leukemia. In some children there is evidence of fetal origin.

Diagnosis of fetal disorders using fetal cells in maternal blood is being used increasingly.[192] Techniques are available using samples obtained by venepuncture, so-called non-invasive techniques.[193,194]

FETAL HEMATOPOIESIS

See **Hematopoiesis**.

FETAL/NEONATAL TRANSFUSION

The procedures for blood transfusion of donor red blood cells, platelet transfusion, and infusion of fresh-frozen plasma to the fetus or neonate.[195]

Intrauterine Transfusion of Red Blood Cells (IUT)

These are used to treat immune-mediated **hemolytic disease of the newborn** (HDN) most commonly, but also fetal disorders with anemia, particularly when due to massive fetomaternal hemorrhage or parvovirus infection.

Product Specification

If IUT is undertaken for HDN due to anti-D, group O RhD-negative blood is used. Where HDN is due to other antibodies in the rhesus system, select group O blood negative for the appropriate antigen. Where antibodies are directed against antigens other than RhD, select O RhD-negative blood negative for the appropriate antigen. Blood should be **cytomegalovirus** (CMV) seronegative or leukodepleted ($<5 \times 10^6$ WBC), even if the mother is herself seropositive, since transplacental transfer of some specificities of IgG antibodies may be low in the second trimester. Blood should be <5 days old and, immediately prior to transfusion, the unit is concentrated by centrifugation to a final hematocrit of over 70%. All units for IUT should be irradiated with 2500 cGy and transfused within 24 h.

Clinical Considerations

Before considering IUT, the following procedures should be observed:

Ascertain the father's phenotype for the corresponding antigen; if he is negative, then the fetus is not at risk of HDN.

Ensure that the antibody involved is associated with moderate or severe HDN.

Refer the patient to a center specializing in fetal medicine.

Consider noninvasive assessment (e.g., cranial Doppler ultrasound) for cases with fetal anemia.

Exchange Transfusions

Product Specification

Heparinized whole blood may be used. Its advantage is a high level of 2',3-DPG, provision of coagulation factors and platelets, and no risk of hypocalcemia resulting from citrate infusion. However, as it can only be stored for a maximum of 24 h, it may be difficult to complete the necessary viral testing within this time, and there is also a risk of hemorrhage from infusion of heparin. Heparinized blood is still used in Europe but not in North America. Citrated group O RhD-negative blood is usually the most convenient to use for all exchange transfusion procedures. The donor red blood cells should be compatible upon **pretransfusion testing** with maternal serum, and if the mother is known to have red cell antibodies, appropriate antigen-negative blood should be selected. Blood should be <5 days old. Babies undergoing exchange transfusion for treatment of ABO hemolytic disease of the newborn (HDN) should receive blood that has been tested to exclude units containing hemolytic and/or high-titer anti-A/-B. The **hematocrit** should be 50 to 60%, and blood should be screened for the presence of Hb S. Prior to transfusion, blood should be warmed to 37°C.

Clinical Considerations

Exchange transfusion is indicated in HDN to correct anemia and hyperbilirubinemia or to treat hyperbilirubinemia due to nonimmune causes (often prematurity).

Neonatal Direct Infusion of Red Blood Cells

Product Specification

Group O blood is used, since this reduces wastage and, in addition, each unit can be divided into several aliquots — a multipack. If the serum does not contain red cell alloantibodies and units of donated blood are screened, then pretransfusion testing is unnecessary in the first 4 months of life.

Blood for neonatal transfusion can be up to 35 days old and suspended in optimal additive solutions such as SAG-M (saline, adenine, glucose, and mannitol) and Adsol. The volume to be transfused is usually 5 to 15 ml/kg given over a period of 2 to 3 h. The hematocrit should be 0.55 to 0.75 l/l. Small-volume red cell transfusions do not cause hyperkalemia, even when blood is 35 days old and the amount of adenine transfused is less than in blood anticoagulated with CDP-A (citrate-dextrose-phosphate-adenosine); in addition, only a small amount of mannitol, well below the theoretical limit of toxicity, is given.

Clinical Indications

Shock associated with surgical or pathologic blood loss

Replacement **venesection** losses ("bleeding into the laboratory"); usually, losses of 10% of the total blood volume in acutely ill infants

Maintenance of **hemoglobin** >12 g/dl in ill neonates with cardiac and/or respiratory disease that requires assisted ventilation or added oxygen and in neonates who have had recurrent apneic attacks

Maintenance of hemoglobin >8 g/dl in other neonatal disorders

The level of hemoglobin at full term is 19.0 ± 2.2 g/dl (lower in preterm babies) and falls to reach a physiological nadir 8 to 12 weeks after birth. The refractory anemia of the premature infant is due to an inappropriately low erythropoietin level for the degree of anemia. The marrow is cellular, and normal *in vitro* growth of erythroid colonies is seen. The hemoglobin or hematocrit measurements may not be reliable indicators of neonatal anemia in preterm babies, since these do not accurately reflect reduction in the red cell mass (RCM, see **Blood volume** — red cell volume). This is because the **plasma volume** may also be reduced.

Platelet Transfusions

Product Specification

Where possible, transfusions to neonates should be ABO- and Rh-group identical. If ABO-identical platelets are not available, the donor plasma of the platelet concentrate should be ABO-compatible with the recipient's red cells. A dose of 10 ml/kg of either a random donor platelet unit or an apheresis unit usually increases the platelet count by $50 \times 10^9/l$.

For IUT, group O RhD-negative platelets are used, since the ABO group of the fetus is usually unknown, and donations with hemolysins and/or high titers of anti-A/-B should be avoided. The volume of platelets to be transfused is given by the following formula:

$$\text{Volume} = 2 \times \text{desired increment } (\times 10^9/l) \times \text{fetoplacental blood volume (ml)}$$
$$\div \text{ platelet count of the concentrate } (\times 10^9/l)$$

The aim is to raise the posttransfusion platelet count to 300 to $500 \times 10^9/l$. Platelet concentrates are volume-reduced by additional centrifugation prior to transfusion to obtain a platelet count of $3000 \times 10^9/l$. In addition, the products for IUT should be CMV seronegative or leukodepleted. They should also be c-irradiated with 2500 cGy.

Clinical Indications

Intrauterine platelet transfusions are indicated for fetomaternal alloimmune thrombocytopenia (FMAIT).

Platelet transfusions to neonates are given at platelet counts of $>100 \times 10^9/l$ if there is intracranial hemorrhage or neurosurgery is required. They are given at platelet counts $<50 \times 10^9/l$ where there is major bleeding and prophylactically to sick preterm babies where the platelet count is $<30 \times 10^9/l$. Thrombocytopenia is common in neonatal intensive-care units and may be caused by septicemia, disseminated intravascular coagulopathy, perinatal asphyxia, hyperbilirubinemia, and intrauterine viral infections.

FIBRIN

See **Fibrinogen**.

FIBRIN DEGRADATION PRODUCTS

(FDP) See **Fibrinogen**.

FIBRINOGEN

(Factor I) The plasma protein that constitutes the final part of the coagulation cascade (see **Hemostasis**) and is responsible for the formation of the fibrin clot, which reinforces and stabilizes the **platelet plug**. Activated platelets bind fibrinogen via the platelet membrane **glycoprotein IIb/IIIa** complex.

   Fibrinogen is the most abundant plasma protein at 2 to 4 g/l and circulates in the plasma as a hexamer consisting of three pairs of chains — Aα, Bβ, and χ2 — held together by disulfide bonds located toward the N-terminus of the protein. The Aα chains consist of 610 amino acid residues, the Bβ chain 461 residues, and the χ2 chain 411 residues. Approximately 10% of the χ2 chains have an additional 20 residues at their C-terminus due to alternative **RNA** splicing. Although the three chains are similar in sequence, suggesting a common ancestral origin, the differing properties of each chain are conferred by their individual sequences. The genes for the three chains that constitute fibrinogen have been cloned and sequenced, and are clustered together within a 50-kb span of **DNA** on chromosome 4 (4q23–32). The Aα gene is located in the middle of the fibrinogen gene cluster downstream of the χ-chain and upstream of the α-chain and consists of five exons spanning 5.4 kb of DNA. Alternative splicing of a sixth exon leads to the formation of an extended α-chain (aE). The *Aα* gene encodes a 625-amino acid polypeptide and a signal peptide of either 16 or 19 residues. The *Bβ* gene consists of eight exons spread over ≈8 kb of DNA and is located downstream of both the *Aα* and *χ2* genes within the fibrinogen gene cluster. Transcription of the *Bβ* gene occurs in the opposite direction to both the *Aα* and *χ2* genes and encodes a 411-amino acid mature protein and a signal peptide of 26 residues. The *χ2* gene is located ≈10 kb upstream of the *Aα* gene and 35 kb upstream of the *Bβ* gene. It consists of ten exons spanning at least 10.5 kb of DNA. Two different forms of χ-chains exist (a major form χA and a minor form χB) as a result of alternative splicing at the 3′ end of the gene. The major form is found in both hepatocytes and platelets, whereas the minor form is found only in hepatocytes.

   Fibrinogen is synthesized primarily by the liver, although it has also been shown to be synthesized by **megakaryocytes**. Platelets are also capable of endocytosing fibrinogen that has been adsorbed onto its surface. Approximately 25% of fibrinogen is found extravascularly. The rate of fibrinogen synthesis can increase 25-fold in response to increased demands, e.g., increased **fibrinolysis**.

Measurement

A variety of methods are available:[196]

Dry Clot Weight

Fibrinogen in plasma is converted into fibrin by the action of thrombin and calcium. The clot is collected on wooden sticks or glass beads, from which it can be easily removed and weighed, the level being expressed as grams per liter of plasma. The normal range by this method is 2.0 to 4.0 g/l.

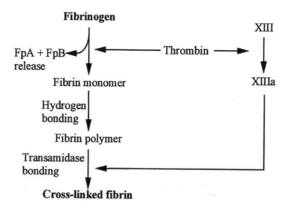

FIGURE 33
Conversion of fibrinogen to cross-linked fibrin.

Claus Technique[196]

Diluted plasma is clotted by a strong thrombin solution, the plasma being diluted to reduce the level of inhibitors, such as fibrin degradation products and heparins. A calibration curve is prepared for each batch of thrombin reagent, with the clotting time in seconds plotted against the fibrinogen concentration in grams per liter on log/log graph paper. From this curve the results of a patient's sample can be read, the value being in grams per liter, with a normal range of 2 to 4 g/l. For levels in premature and full-term infants, see **Reference Range Tables I, XIII**, and **XIV**.

Micronephelometric Method

Here, a nephelometer is used to measure the light-scattering intensity of test plasma diluted in buffered saline, the turbidity of which is measured both before and after heating at 56°C for 15 min. The difference between the nephelometer readings is multiplied by a constant factor corresponding to the dilution used.

The dry clot weight method is the most reliable, but laborious. The Claus method is subject to interference by heparin or warfarin, so the nephelometric method is used for comparative studies, being both economical and rapid. Quality control procedures using the dry-weight method are essential.

Fibrinogen is an **acute-phase response** protein and is largely responsible for raised levels of the **erythrocyte sedimentation rate** (ESR). Prolonged raised fibrinogen levels have been shown in epidemiological studies to be a risk factor for coronary **arterial thrombosis**. Furthermore, a rapidly rising level after myocardial infarction is a prognostic sign for a poor outcome in those not receiving thrombolytic therapy.[76]

Conversion of Fibrinogen to Fibrin

This occurs in three steps (see Figure 33):

1. Thrombin binds to fibrinogen in the region of the N-terminus of the Aα and Bβ chains, leading to cleavage of the Aα chain between Arg16 and Gly17 and the release of a small peptide termed fibrinopeptide A (FpA: Ala1-Arg16). Somewhat slower cleavage of the Bβ chain at Arg14-Gly15 releases a second peptide, fibrinopeptide B (FpB: Gly1-Arg14). Cleavage of the Aα and Bβ chains by thrombin exposes binding domains in the central E domain that interact with sites on the χ chain.

2. Fibrin monomers form **protofibrils**.

3. **Factor XIIIa** catalyzes cross linking of the polymerized fibrin, creating links between adjacent lysine and glutamine residues, making the chain stronger and relatively resistant to lysis by **plasmin**.

Fibrinopeptides A and B

(FpA/FpB) These constitute less than 2% of the mass of fibrinogen. FpA can also be generated by batroxobin, a protease isolated from the venom of *Bothrops atrox*. Similarly, FpB can be liberated by the venom of *Agkistrodon contortrix* (see **Snake venom disorders**). Increased levels of FpA and FpB may be found in a number of clinical situations, e.g., venous thromboembolic disease and coronary artery disease.

Degradation of Fibrinogen and Fibrin

This is induced by plasmin, which hydrolyzes arginine and lysine bonds in a variety of substrates, although its major physiological effect is upon fibrin and fibrinogen. Degradation of noncross-linked fibrin is identical to that of fibrinogen, whereas that of cross-linked fibrin is significantly different and gives rise to a number of characteristic fragments.

Degradation of Fibrinogen and Non-Cross-Linked Fibrin by Plasmin

The globular domains of fibrinogen comprise the two D domains, the single E domain, and the long Aα chain extensions from the D domains (see Figure 34). Digestion of fibrinogen or non-cross-linked fibrin involves an initial cleavage of several small peptides

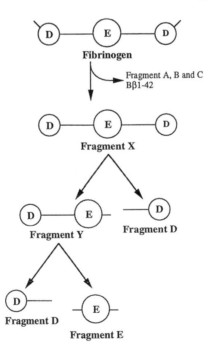

FIGURE 34
Plasmin digestion of fibrinogen and non-cross-linked fibrin by plasmin.

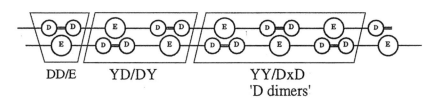

FIGURE 35
Plasmin digestion of cross-linked fibrin.

(termed fragments A, B, and C) from the C-terminal portion of the Aα chain, followed rapidly by removal of the N-terminal 42 amino acids from the Bβ chain. The residual fragment, known as fragment X, consists of all three of the domains but lacks the long Aα chain extension. Assay of the Bβ1-42 fragment generated at this stage provides a sensitive index of fibrinolytic activity. Asymmetrical digestion of fragment X then occurs, with the release of fragment D (in which the chains remain linked by disulfide bonds) and the residue of fragment X, termed fragment Y. Fragment Y, therefore, consists of the central E domain and either of the terminal D domains. Further digestion of fragment Y by plasmin results in the cleavage of the second D domain to give a second fragment D. The residue of fragment Y, consisting of the disulfide-linked N-terminal ends of all six chains, is termed fragment E. Fragments X, Y, and D are able to bind to the fibrin monomer, inhibiting polymerization and thereby interfering with clot formation. Fragments Y, D, and E also increase the rate of conversion of plasminogen to plasmin, thereby increasing the rate of fibrinolysis once initiated.

Degradation of Cross-Linked Fibrin by Plasmin

The unique cross-linked structure of fibrin results in the generation of a series of specific fragments during lysis by plasmin, namely D dimers, fragment E (both free and complexed to a D dimer complex), and YD/DY fragments (see Figure 35).

Afibrinogenemia

The total absence of fibrinogen in plasma is a rare disorder, and affected individuals are presumed to be homozygotes or compound heterozygotes for mutations that result in a failure of fibrinogen synthesis. Consanguinity of the parents is common. Afibrinogenemia is associated with a prolonged **bleeding time**, and *in vitro* platelet-aggregation tests show no response to agonists that operate through the release mechanism. Affected individuals also show a prolonged **activated partial thromboplastin time** (APTT) and **thrombin time** (TT), and the ESR is characteristically very low. Many patients demonstrate mild **thrombocytopenia**. Clinically, the disorder is associated with a severe bleeding diathesis with spontaneous bleeding into muscles, joints, and mucous membranes. Recurrent miscarriages are common in women. Prolonged bleeding from the umbilical stump or after circumcision or after eruption of the teeth may also occur.

Hypofibrinogenemia

This is thought to be the heterozygous form of afibrinogenemia.

Dysfibrinogenemias

Characterized by a dysfunctional fibrinogen that may be present in normal or increased amounts. Some patients previously diagnosed as hypofibrinogenemic have subsequently been found to have trace amounts of a dysfunctional fibrinogen. Dysfibrinogenemias arise from structural abnormalities in the fibrinogen molecule that lead to dysfunction in one or more of the stages that are involved in the conversion of soluble fibrinogen to insoluble cross-linked fibrin, i.e., reduced release of FpA and FpB (37% of cases), defective fibrin monomer polymerization (71% of cases), or defective fibrin cross linking (6% of cases). Most cases are heterozygotes, although ≈5% of cases are homozygous. Approximately one-quarter of patients exhibit a bleeding tendency. The risk of bleeding is increased in homozygous patients. Approximately 250 families have been described (with 25 mutations) with dysfibrinogenemia, 60% remaining asymptomatic, 28% with associated hemorrhage, 20% with thromboses (see Hyperfibrinogenemia, below), and 2% with hemorrhage and thrombosis.

Some 50 families have been described with dysfibrinogenemia in which the variant fibrinogen is associated with an increased risk of thrombosis, both arterial and venous thromboses. In this latter group, the variant fibrinogen frequently demonstrates abnormal polymerization, and it is suggested that this renders it more resistant to lysis by plasmin.

An acquired dysfibrinogenemia is seen in patients with liver disorders and in those with a low serum albumin, e.g., nephrotic syndrome. In this latter group, correction of the abnormal coagulation tests can be demonstrated if the patient's plasma is supplemented with albumin *in vitro*.

Treatment of Fibrinogen Deficiencies

Those with hypofibrinogenemia and dysfibrinogenemia may be asymptomatic and require no treatment. For minor bleeding problems, fibrin glue or antifibrinolytic therapy, e.g., tranexamic acid (see **Fibrinolysis** — antifibrinolytic agents) and **DDAVP**, may be useful. Fibrinogen concentrates are the mainstay of treatment for more severe bleeds (see **Coagulation-factor concentrates**).

Hyperfibrinogenemia

This occurs transiently, with inflammation as part of the acute-phase response. Persistent high levels are associated with age, familial tendency, smoking, oral contraceptives, menopause, obesity, diabetes mellitus, and "stress." There are seasonally higher levels in winter months. The consequences of hyperfibrinogenemia are increased blood **viscosity** and platelet aggregation, with acceleration of atherosclerosis. It is therefore a risk factor for **thrombosis**[80] and, as such, is strongly associated with an increased mortality rate from cardiovascular causes in patients with intermittent claudication or venous embolic disease; it may also contribute to age-related macular degeneration and reocclusion after coronary artery bypass surgery or angioplasty. High levels of plasma fibrinogen are a poor prognostic feature for myocardial infarction. The only known fibrinogen-lowering agent is intravenous **ancrod**.

FIBRINOLYSIS

The principal effector of clot removal by which degradation of fibrin into smaller fragments occurs through the action of **plasmin** (see Figure 36).

The components of the fibrinolytic pathway are plasminogen with endogenous and exogenous activators to form plasmin.

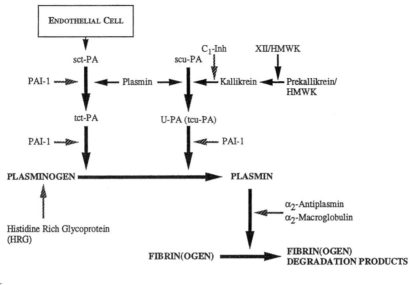

FIGURE 36
Normal fibrinolysis.

Plasminogen

The inactive zymogen form of the active enzyme plasmin. Plasminogen is synthesized in the liver, circulates in plasma at a concentration of $2.4\mu M$ (200 mg/l), with a $T^{1/2}$ of 2 h. Plasminogen contains five homologous looped structures called "**kringles**," four of which contain lysine-binding sites through which the molecule interacts with its substrates and its inhibitors. Plasminogen is synthesized as a single-chain molecule consisting of 790 amino acids and with a molecular weight of 92 kDa. The N-terminus contains a glutamic acid residue, and this molecule is known as Glu-plasminogen. Internal autocatalytic cleavage occurs during activation of plasminogen, with the release of an activation peptide. The N-terminus of the plasminogen now contains a lysine residue and is therefore known as Lys-plasminogen.

Conversion of plasminogen to plasmin can occur via two routes (see Figure 37).

Most activators cleave plasminogen at arginine 560 to generate a two-chain protein termed Glu-plasmin, which comprises a light chain and a heavy chain linked by a single disulfide bridge. The light chain is derived from the C-terminus of the protein and contains the active serine catalytic site, whereas the heavy chain is derived from the N-terminus and contains the kringle domains and the four lysine-binding sites. Glu-plasmin, despite being a serine protease, is functionally inactive, since its lysine-binding sites are masked. It is only when it is converted to Lys-plasmin by autocatalytic cleavage between Lys76-Lys77, with the release of an activation peptide (residues 1–76), that the lysine-binding sites on the four kringle domains are exposed and the affinity of the protease for fibrin is dramatically increased. Both Glu-plasmin and Lys-plasmin attack the Lys76-Lys77 bond to form Lys-plasminogen. This is capable of binding to the fibrin clot before it develops protease activity, and it is, therefore, brought into close proximity with the physiological activators. Plasminogen is known to bind to a number of proteins, including histidine-rich glycoprotein (HRG), tetranectin, and **thrombospondin.** Tetranectin is known to increase plasminogen activation by tissue plasminogen activator (t-PA), whereas platelet-derived thrombospondin is a noncompetitive inhibitor of plasminogen activation by t-PA.

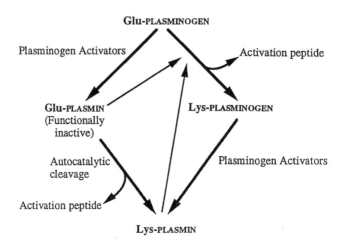

FIGURE 37
Conversion of plasminogen to plasmin.

HRG modulates the activity of plasminogen by binding to the lysine-binding sites, blocking subsequent binding to the fibrin clot.

Normally, free t-PA is rapidly inactivated because of an excess of plasminogen activator inhibitor type 1 (PAI-1), and any free plasmin generated is rapidly inactivated by α_2AP (antiplasmin). In the presence of a fibrin clot, t-PA released locally binds to the clot, the affinity of t-PA for plasminogen being increased around 400-fold. Thus, both plasminogen and t-PA are bound to the clot, activation occurs, and plasmin is generated.

Plasminogen is measured by a functional chromogenic assay based on the full transformation into plasmin by activators. The normal range is 0.75 to 1.35 U/ml. Levels are low in the full-term infant and may be very low in the preterm infant with **liver disorders** or **disseminated intravascular coagulation** and during or after thrombolytic therapy. Deficiencies and variants of plasminogen have been described, but in the heterozygous form, these appear to be of little clinical significance. As plasminogen is an **acute-phase response** protein, levels are increased with infection, trauma, myocardial infarction, and malignant disease. Its levels are also raised in **pregnancy**, with age, and with the use of the oral contraceptive pill.

Endogenous Activators of Fibrinolysis

Tissue Plasminogen Activator

(t-PA) These glycoproteins are synthesized primarily by the cells of the **vascular endothelium**, although many other cells are also capable of its synthesis. The concentration of t-PA in plasma varies in response to stress, injury, exercise, and a number of physiological and pharmacological stimulants. t-PA is synthesized as a single-chain glycoprotein (sct-PA) and contains two kringle domains through which it is thought to bind to fibrin and lysine analogs. Although sct-PA has significant proteolytic activity, its biological activity is small until bound to a fibrin clot, whereupon its affinity for plasminogen is increased ≈400-fold. The plasmin generated by the activation of plasminogen is capable of cleaving sct-PA into a two-chain molecule (tct-PA) with a significant increase in activity. This cleavage occurs rapidly when sct-PA is bound to a fibrin clot. t-PA has a short half-life (5 min) and is rapidly cleared from the circulation by the liver.

sct-PA and tct-PA are inhibited by the serine protease inhibitor PAI type 1 (PAI-1). A second inhibitor of t-PA, PAI-2, is found in plasma in significant amounts during pregnancy. t-PA (both sct-PA and tct-PA) has been expressed in cell culture and has been widely used in thrombolytic studies.

Urokinase

(UK) This was originally isolated from urine, but it is synthesized by a range of normal and pathological cell types. Urokinase is synthesized as an inactive (or with very little activity) single-chain zymogen (scu-PA, also known as pro-urokinase) and must be converted to the two-chain form (tcu-PA or U-PA) before it is functionally active. scu-PA is converted to tcu-PA (U-PA) by plasmin and kallikrein. tcu-PA activates plasminogen to plasmin by proteolytic cleavage at Arg560-Val561. Inhibition of the active enzyme occurs via PAI-1, PAI-2, and also by protease nexin 1. Although urokinase can activate plasminogen in plasma, it is thought that its major role is as an extravascular activator of plasminogen, especially where tissue destruction or cell migration occurs. UK is also known to bind to a specific cellular receptor present on the surfaces of monocytes, fibroblasts, and endothelial cells, among others.

Exogenous Activators of Fibrinolysis

A number of exogenous activators of fibrinolysis exist and have been widely used in thrombolytic studies.

Recombinant t-PA

(rt-PA) In both its single-chain and two-chain form, rt-PAs have been used for clinical thrombolysis. They are relatively fibrin-specific, have relatively little systemic activity, and have short half-lives ($\approx$5 min). They do not provoke an immune response and are useful in individuals with significant antibody titers to streptokinase. The bleeding complications observed with rt-PA are similar in severity and frequency to those observed with streptokinase and urokinase, suggesting that fibrin specificity does not confer protection against hemorrhage.

Streptokinase and Urokinase

(SK, UK) Streptokinase is derived from b-hemolytic streptococci. It has no intrinsic activator activity, but forms a 1:1 complex with plasminogen that leads to the conversion of plasminogen to plasmin, and the complex is then capable of activating other plasminogen molecules. Urokinase is commonly isolated from tissue culture and is capable of directly activating plasminogen to plasmin. Both SK and UK have little affinity for fibrin, and their use is associated with significant hyperplasminemia resulting in proteolytic degradation of fibrinogen and other plasma proteins. UK and SK in particular have been widely used for thrombolysis in both venous and arterial thromboembolic disease. SK, however, induces an immune response, limiting its use. The $T^{1/2}$ for SK is $\approx$30 min and for UK is $\approx$10 min. UK is considerably more expensive than SK.

Acylated Plasminogen SK Activator Complex

(APSAC) A chemically modified SK derivative in which SK is complexed to plasminogen to provide fibrin specificity. The plasminogen moiety binds through its kringle domains to the fibrin clot, deacylation of the SK occurs, rendering it active, and it then operates in a manner identical to SK alone. However, it has a long $T^{1/2}$ ($\approx$70 min), resulting in sustained fibrinolysis.

Inactivators of Fibrinolysis

Plasminogen Activator Inhibitor Types 1 and 2

(PAI-1, PAI-2) PAI-1 is the major inhibitor of t-PA and U-PA. PAI-1 is synthesized by the endothelial cells and hepatocytes. Platelets are known to contain significant amounts of PAI-1. The synthesis of PAI-1 is stimulated by various cytokines. PAI-1 is an acute-phase protein and its levels vary widely. A familial increase of PAI-1 has been reported, and molecular analysis of such families has shown the presence of a common polymorphism within the promoter region of the PAI-1 gene, 675 bp upstream of the transcription initiation site. A congenital absence of PAI-1 has also been reported, and although such cases are extremely rare, they are associated with a severe bleeding diathesis. PAI-2 is produced by the placenta, and plasma levels therefore increase during pregnancy. PAI-2 is not detectable in normal plasma, although PAI-2 is found in monocytes. The precise role of PAI-2 is unclear, but because levels increase in pregnancy, it is reasonable to assume that it has a role in hemostasis, possibly in the maintenance of placental function. PAI-2 has a much lower affinity for either t-PA or U-PA than PAI-1. It has been suggested that excess PAI-1 activity is related to hyaline membrane disease of infant lungs.

Plasminogen Activator Inhibitor Type 3

(PAI-3) This has a low affinity for both t-PA and U-PA and is probably unimportant in the regulation of fibrinolysis. PAI-3 is primarily an inhibitor of activated protein C (APC) and appears identical to protein C inhibitor (PCI).

α_2-Antiplasmin

(α_2AP) The major inhibitor of plasmin and a member of the serpin family of inhibitors. α_2AP binds irreversibly to plasmin, forming a stable 1:1 bimolecular complex that is then removed from the circulation by the liver. α_2AP has a plasma concentration of around 1.0 U/ml; the levels for premature and full-term infants are given in the **Reference Range Tables**. It exists in two forms: a plasminogen-binding form ($\approx$70%) and a nonplasminogen-binding form ($\approx$30%). The latter has less inhibitory activity than the plasminogen-binding form of α_2AP, from which it may be derived by proteolysis.

Miscellaneous Inhibitors of Fibrinolysis (e.g., α_2-macroglobulin)

These have a relatively low affinity for plasmin, although they have a high plasma concentration and may become functionally important if plasmin levels are high.

Evaluation of the Fibrinolytic Pathway

Fibrinolysis can be studied by using global screening tests, which provide an index of the overall efficiency of the fibrinolytic mechanism, or by assaying the specific proteins involved in fibrinolysis. Of crucial importance when evaluating fibrinolysis is adequate preparation of the patient and correct handling of blood samples. Patients should rest for at least 20 min before samples are collected, and blood samples must be collected without the use of a tourniquet. Both the global screening tests and individual assays for t-PA and PAI-1 can be assessed in relation to venous occlusion or following the administration of pharmacological agents such as **DDAVP**. In general, an increase in fibrinolytic activity is observed following venous occlusion.

Global Screening Tests of Fibrinolysis[19]

Euglobulin Clot Lysis Time

(ECLT/ELT) This provides a global measure of fibrinolysis. The euglobulin fraction is isolated from plasma by acidification and cooling and is rich in fibrinogen, plasminogen, and the plasminogen inactivators but deficient in the inhibitors of fibrinolysis, e.g., α_2AP. The euglobulin fraction is dissolved in buffer, clotted with thrombin, and the time to clot lysis measured. The normal range lies between 60 and 270 min but varies with age, sex, smoking, and alcohol consumption. The test is very sensitive to low levels of fibrinogen and plasminogen, which may give rise to false results.

Fibrin Plate Lysis

A plasminogen–fibrinogen-rich solution is clotted onto petri dishes using thrombin. The euglobulin fraction of plasma is added, the plate is incubated at 37°C for 24 h, and the zones of lysis measured. The area of lysis reflects the concentration of plasminogen activators.

Specific Tests of Fibrinolysis[196a]

Plasminogen, t-PA, PAI-1, and α_2AP may provide a more accurate assessment of fibrinolysis than global screening tests.

Disorders of Fibrinolysis

Disordered fibrinolysis is recognized to be associated with both bleeding (see **hemorrhagic disorders**) and thrombotic complications. Clinical disorders arise from hyperfibrinolysis, which occurs when fibrinolytic activity is greater than fibrin formation (see Table 65). In addition to breakdown of fibrin, increased fibrinolysis reduces platelet aggregation by degradation of receptor glycoproteins 1b and platelet fibrinogen receptor **glycoprotein IIb/IIIa**. Furthermore, hyperfibrinolysis inhibits fibrin polymerization, resulting in a weak hemostatic plug.

TABLE 64

Clinical Disorders Related to Abnormal Fibrinolysis

Bleeding

Increased activation
   Iatrogenic
   Exogenous activators
   Defective clearance–liver disorders
Local excess activation
   Menorrhagia
   Malignancy, e.g., prostate
Decreased inhibition
   Defect in plasma protein
   $\downarrow$ PAI-1, α_2-AP

Thrombosis

Decreased inhibition
   Abnormal plasma protein
   $\downarrow$ Plasminogen; abnormal fibrinogen
Increased inhibition
   Iatrogenic $\uparrow$ inhibition
   EACA/tranexamic acid
   Hereditary $\uparrow$ PAI-1

Primary Fibrinolyis

Liver disorders due to reduced clearance of t-PA and reduced α_2-antiplasmin from
   diminished protein synthesis

Cardiopulmonary bypass due to increased t-PA concentration

Thrombolytic drug therapy

Neoplasia, particularly of prostate and pancreas

Prostatic surgery

Secondary Hyperfibrinolysis

This is a consequence of **disseminated intravascular coagulation** (DIC), where activation
of the coagulation cascade occurs with thrombin generation, which stimulates the vascular
endothelium to secrete t-PA. Treatment is by antifibrinolytic drugs (see below). These are
contraindicated where hyperfibrinolysis is due to DIC because here it is the mechanism
for removal of fibrin from pathological microthrombi.

Antifibrinolytic Agents[196b]

Aprotinin

Aprotinin is a substance isolated from bovine lung and other bovine organs. It inhibits
plasmin and other serine proteases and is of value in reducing blood loss in patients
undergoing cardiopulmonary bypass surgery.

Synthetic Lysine Analogs

Tranexamic acid (*trans*-p-aminomethyl-cyclohexane carboxylic acid). *In vitro*, tranexamic
acid accelerates plasminogen activation to plasmin by binding to plasminogen and altering
its conformation such that it is more susceptible to activation. However, subsequent lysis
by plasmin is prevented by the conformational changes that take place, which results in
steric inhibition of the binding of lysine residues on fibrin (and fibrinogen) to the lysine-
binding sites on plasmin(ogen). Tranexamic acid is excreted by the kidney and inhibits
renal urokinase. Urinary levels are approximately 75 to 100 times the plasma levels. It is
therefore contraindicated in patients with upper genitourinary bleeding, where its use can
lead to the development of thromboses within the renal tract. Antifibrinolytic agents have
been associated with thrombosis, but the risks appear to be small unless there is some
ongoing thrombogenic stimulus.

   Tranexamic acid is widely used to control blood loss in a number of clinical situations, e.g.,
after tonsillectomy; in women with menorrhagia; for the treatment of recurrent epistaxes,
particularly if there is an associated bleeding diathesis; after dental extractions in patients
with a bleeding diathesis; and in the treatment of acute promyelocytic leukemia (M3).

FIBRINOPEPTIDES A and B

See also **Fibrinogen**.

Peptides liberated from the A and B chains of fibrinogen, respectively, following its
activation by **thrombin**. FpA can also be generated by batroxobin, a protease isolated from
the venom of *Bothrops atrox*. Similarly, FpB can be liberated by the venom of *Agkistrodon
contortrix* (see **Snake venom disorders**). FpA and FpB constitute less than 2% of the mass
of fibrinogen. Assays for both FpA and FpB are available and allow an accurate assessment
of the action of thrombin, especially in thrombotic states. The presence of these small

peptides implies cleavage of fibrinogen by thrombin and not fibrinolysis. Assays for FpA and FpB are technically demanding and require close attention to venepuncture technique and sample processing. Increased levels of FpA and FpB may be found in a number of clinical situations and may be of value in the diagnosis or monitoring of a number of disorders, e.g., deep venous thrombosis and coronary artery disease. *In vivo* plasmin activity can be measured by quantitating B1-42 and B15-42 peptides, which are released by the proteolytic activity of plasmin on fibrinogen and fibrin.

FIBRIN PLATE LYSIS
See **Fibrinolysis**.

FIBRIN STABILIZING FACTOR
See **Factor XIII**.

FIBRONECTIN
See **Bone marrow** — microenvironment.

FISH TAPEWORM DISEASE
See also **Metazoan infections**.
Carriers of fish tapeworm, *Diphyllobothrium latum*, develop **megaloblastosis**. It only occurs in Finland. Deficiency of **cobalamin** (vitamin B_{12}) results from impaired absorption as a result of competition between the worm and the host for dietary cobalamin. The resulting macrocytic anemia responds to expulsion of the worm, but often the response is suboptimal in the absence of treatment with cobalamin.

FITZGERALD FACTOR
See **Kininogens**.

FLAME CELLS
(Thesaurocytes) Giant multinuclear **plasma cells** that appear carmine red or vermilion upon Romanowsky staining. They often possess fiery fringes of devitalized margins containing immunoglobulin precipitates. They occur in **myelomatosis,** particularly IgA myeloma.

FLETCHER FACTOR
See **Prekallikrein**.

FLOW CYTOMETRY
The physical and chemical characteristics of cells (particles) analyzed as cells (particles) as they pass in a fluid stream, ideally in single file, through a measuring apparatus.[197] The principles are used in hematological diagnosis for:

Automated blood cell counting
- Red blood cell parameters
- Reticulocyte parameters
- White blood cells
- Differential leukocytes
- Platelet parameters

Fluorochrome-labeling cytometry
- **Immunophenotyping**
- **Reticulocyte counting**
- Fetomaternal hemorrhage

More than one flow technique can be used in the same instrument. All techniques produce a rapid result, often of several variables simultaneously, e.g., cell size and cytochemical staining reaction. These techniques can therefore measure a large number of cells in a single sample. Results are presented digitally and as frequency distribution curves, histograms, or scattergrams.

Aperture Impedance

A measured volume of diluted blood cell suspension in buffered electrolyte solution flows through an aperture of specific dimensions. Electrodes are located at both ends of the aperture, with a direct current from a constant source of electricity flowing between them. A blood cell flowing through the aperture displaces conductive fluid and increases electrical resistance, producing a corresponding change in potential between the electrodes, with the amplitude of the pulse generated being approximately proportional to the volume of the cell (see Figure 39). The pulses generated are distorted unless the cell passes through the center of the aperture. In Figure 38, (A) represents the ideal, whereas means must be adopted to prevent or eliminate (B) and (C). Unless such aberrant pulses are detected, inaccuracies in size assessment will arise. A process of electronic editing avoids this problem. Alternatively, the flow system can be modified by sheathed flow to ensure that all cells pass through the center of the aperture, a process called hydrodynamic focusing (see Figure 39).

Light Scattering

The diluted blood suspension flows through a focused light source with cells in single file scattering light. The incident wave is converted to a forward light scatter, side scatter, or fluorescence scatter (Figure 40). The newer light-scatter instruments employ laser light that possesses depth of focus and less diffusion of light.

To separate background interference from cells and distinguish one cell type from another, discrimination thresholds must be set. These are specific for each instrument type. Red blood cells are counted in diluted blood that also contains leukocytes and platelets. The error in the red cell count due to the presence of leukocytes is negligible, but will increase when the white count increases ($>50 \times 10^9$/l). Platelets are so small that they do not usually introduce errors into the red cell count, but the reverse is not true. Extremely microcytic red cells and red cell fragments introduce an error into the platelet count.

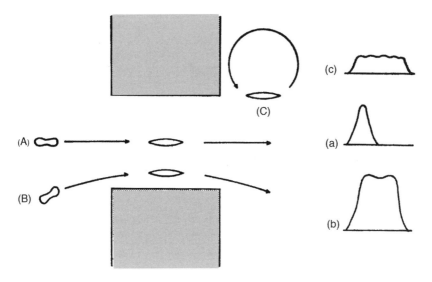

FIGURE 38
Diagrammatic representation of cell-counting and cell-volume measurement by aperture impedance and the effect of nonaxial flow. (A) = particle in axial flow producing an electronic pulse with the characteristics shown in (a); (B) = particle in nonaxial flow producing an electronic pulse with the characteristics of (b); (C) = a recirculating particle producing an electronic pulse with the characteristics of (c).

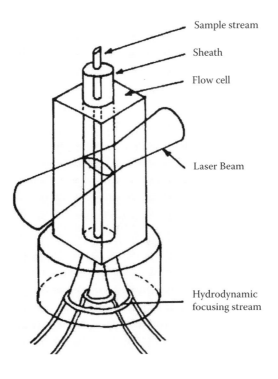

FIGURE 39
Schematic illustration of a hydrodynamic-focused (sheathed flow) counter using the light-scattering principle.

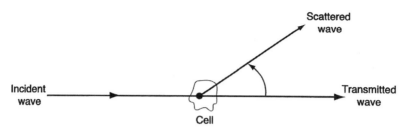

FIGURE 40

Diagrammatic representation of cytometry by light scattering.

TABLE 65

Common Fluorochromes

Fluorochrome	Excitation Laser / Wavelength (nm)	Peak Emission (nm)	Fluorescence Channel/Filter (nm)
Fluorescein (FITC)	argon/488	510	FL1/530
Green fluorescent protein	argon/488	—	FL1/530
Phycoerythrin (PE)	argon/488	580	FL2/585
Propidium iodide (PI) (for DNA)	argon/488	620	FL3/630
Peridinin–chlorophyll–protein (PerCP)	argon/488	680	FL3/630
Activated protein C (APC)	helium/633	660	FL4/685

Fluorochrome-Labeling Cytometry

A technique used to measure light emitted by a fluorochrome that has reacted stoichiometrically with some component of a cell, so that the intensity of light emission is a measure of the amount of that component.[198] Fluorochromes are substances that will emit light when excited by light of a shorter wavelength. These may themselves react directly and specifically with cellular components (e.g., propidium iodide, which complexes with double-stranded DNA) or may be rendered specific by conjugation with an antibody, e.g., the commonly used green fluorescein isothiocyanate (FITC) or red phycoerythrin (PE). The use of labeled antibodies may be direct, in which case the primary antibody (i.e., that which reacts with the protein of interest) is labeled, or indirect, in which case the primary antibody is unlabeled but the secondary antibody that reacts with the primary is labeled. Obviously, quantification of the target molecule is possible only when the fluorochrome (or fluorochrome-labeled antibody) is present in saturating concentration: this must be determined by titration. More recently, the use of fluorescent green (FGP) protein has been exploited to identify cells that have been successfully transfected: cells transfected with a vector containing the *GFP* gene will fluoresce green. There are many other such proteins becoming available. Table 65 shows some common fluorochromes.

In practice, a suspension of cells is passed in a stream of fluid that intersects a monochromatic light beam (usually generated by a laser). Light that is scattered by the cell gives two sorts of information: low-angle forward scatter (i.e., in the same direction as the light beam and at a few degrees to it) is dependent on the size of the particle, and side scatter (90° to the beam) is related to the opacity of the particle. The scattered light is collected by lenses, amplified, and the intensity measured. Light emitted by fluorochromes is likewise collected: a series of filters is used to separate emitted light of different wavelengths so that light that is scattered or emitted by fluorochromes can be distinguished. With sophisticated instruments with more than one laser (and thus more than one activation wavelength) and advanced fluorochrome technology, four or more wavelengths of emitted light may be analyzed.

The information generated is conventionally stored in a computer memory in list mode, that is, with all the measured parameters correlated for individual cells so that the information can be recalled and the whole cell population analyzed for all parameters. The data can be analyzed in a number of ways using appropriate software either built into the instrument or run on a separate computer.

Data Analysis

The simplest analysis is the single histogram. The magnitude of each parameter is stored in digital form, and the histogram displays the number of cells in each light-intensity channel, of which there may be up to 1024 on modern instruments. Obviously, only one parameter can be displayed at a time in a particular histogram.

More-complicated displays allow correlated displays, of which the simplest is a correlated dot plot: the two axes are different parameters, and each cell is represented as a single dot on the display, its position defining the value of the two parameters for that cell. Contour plots and mock three-dimensional displays can be used to give a visual impression of the number of cells in each region of the plot.

The actual numbers of cells in each region of a plot can be determined by electronically placing markers about a population of cells and using the software to count the number within that region. An important technique is gating, which is useful when analyzing mixtures of cells: by electronically "drawing" a box about a set of cells in a display, those cells can be selected and analyzed for other parameters separately from the other cells.

A simple illustrative example is the analysis of CD4 and CD8 cells in blood. Red blood cells (RBC) are first removed, e.g., by selective lysis, and the leukocytes then stained with, for example, FITC anti-CD4 and PE anti-CD8. The sample is run through the cytometer, and correlated data for forward and side scatter and green and red fluorescent light are stored.

The first step in analysis is to display the data as a correlated plot of side against forward scatter. In a good preparation, four populations will be seen: debris and residual RBCs, with very low forward and side scatter; lymphocytes, somewhat larger than RBCs and thus with greater forward scatter but still low side scatter; monocytes, larger and somewhat granular, and thus with greater forward and side scatter; and finally granulocytes, with the greatest scatter parameters. A gate is drawn around the lymphocytes, and using that gate, a correlated plot of green (FITC) vs. red (PE) fluorescence is displayed. Three populations will be visible. At the origin will be non-T-cells (chiefly B-cells). The CD4 cells will be a discrete population of high green, low red fluorescence and the CD8 as low green, high red. The number of cells in each region is then determined, thus giving the CD4:CD8 ratio.

Cell Sorting

Here, the first step is analysis — but under "live" conditions, i.e., while the machine is running — and up to two live gates, left or right, are defined. The stream containing the cells is broken into charged droplets, which ideally contain single cells, by a high-frequency audio field. After passing through the light beam, the droplet is deflected to the left or right by an electric field according to whether the cell it contains was gated left or right, and collected in a test tube, thereby separating two populations — as it might be, CD4 and CD8 cells. The technique is laborious and not suitable for large numbers of cells. Alternatives such as panning have always been available, which, though less discriminating, are more convenient (and much cheaper); this technique is now being replaced by magnetic-bead technology.

Applications

Fluorochrome-labeling flow cytometry is used to study cells in suspension, so it is not surprising that its use is most common in the study of blood disorders. However, tissues such as solid tumors can also be processed into single-cell suspensions and then studied by flow cytometry. By far the most important application is the use of fluorescent antibodies for the identification and quantitation of cellular proteins, and in this application the only limitation is the availability of suitable antibodies. The most common application is in **immunophenotyping** using a panel of monoclonal antibodies to surface antigens to identify cell types, for example in the diagnosis of leukemias and lymphoproliferative disorders.[199] Permeabilization of cell membranes using, for example, the mild detergent saponin allows the quantitation of intracellular antigens, such as products of oncogenes. Use of DNA-specific fluorochromes allows the estimation of DNA in cells for studies of ploidy and **apoptosis**. Fluorochromes specific for ions such as Ca^{2+} can be used for their estimation. Fluorochromes can also be used to measure fetomaternal hemorrhage as the predisposing cause of **hemolytic disease of the newborn**. This technique is also an established automated method for **reticulocyte counting**.

FOAM CELLS

Histiocytes (macrophages) laden with small uniform droplets of lipid throughout the cytoplasm. They are large cells (20 to 100 µm) and usually mononuclear. The cytoplasmic droplets stain faintly with periodic acid-Schiff, are birefringent in polarized light, and consist mostly of sphingomyelin, although some ceroid is also present. Under the electron microscope, the droplets have a lamellar structure. They most commonly occur in the spleen, liver, lymph nodes, and CNS, characteristically in **Niemann-Pick disease**, where there is accumulation in histiocytes of sphingomyelin. They may also be seen in generalized gangliosidosis.

FOLIC ACID

(Pteroylglutamic acid) A compound of a pteridine derivative and an l-glutamic acid (see Figure 41, Figure 42, Figure 43, Figure 44).[145,147] Folic acid exists in dihydro- and tetrahydro-forms, reduction occurring by the action of ascorbic acid or by enzymes — folate reductases — that are widely distributed in animal tissue, bacteria, and viruses. Biological activity of folate in tissue metabolism is by one-carbon transfer of the tetrahydro-form, for which dihydrofolate reductase is essential. Folate antagonists (2,4-diaminopyrimidines) act by competitive blocking of dihydrofolate reductase. All folates are highly labile and sensitive to oxidation. Folates are polyglutamates, glutamic acid conjugates of up to seven molecules attached through the l-carboxyl group. They are catabolized by enzymes that are present in plasma and intestinal secretions — folic acid conjugates.

Folate compounds arise from substitution at the N5, N10 positions of the pteroyl rings. Metabolic activity occurs with transfer of one carbon atom at various levels of oxidation (see Figure 44 and Table 66).

Specific folate coenzymes are concerned with oxidation-reduction interconversion and others with single-carbon transfer.

Dietary Folate

There are many sources of folic acid, the richest of which are fresh vegetables, particularly asparagus, broccoli, endive, spinach, lettuce, and lima beans. The best fruit sources are

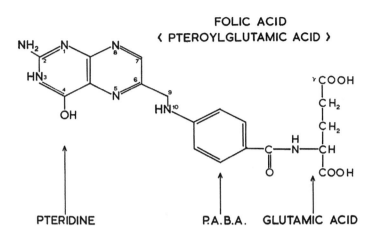

FIGURE 41
Structure of folic acid. (From Shinton, N.K., *Br. Med. J.*, 1, 556–559, 1972. With permission.)

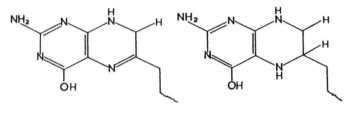

FIGURE 42
Reduced forms of folic acid. (From Shinton, N.K., *Br. Med. J.*, 1, 556–559, 1972. With permission.)

TABLE 66

Folates and Their Activities

Group	Formula	Activity
Formimino	HCNH	histidine catabolism
Formyl	HCO	purine synthesis
Hydroxymethyl	$HOOCO_2–$	methionine synthesis
Methenyl	–CH=	purine synthesis
Methylene	$–CH_2–$	thymidilate synthesis
Methyl	$–CH_3$	methionine synthesis

bananas, lemons, and melons. Folic acid is also found in liver, kidney, mushrooms, and yeast. Prolonged cooking destroys food folate. The average daily Western diet contains 400 to 600 µg folate. The normal adult has a requirement of about 50 µg daily. Dietary folate is mainly in the polyglutamate form.

Absorption and Transport

The principal site of folate absorption is the proximal jejunum. Folates are readily absorbed as a monohydrofolate, but dietary folate is in the polyglutamate form. Reduction by hydrolytic enzymes (deconjugases) occurs in the brush border of the intestinal epithelial

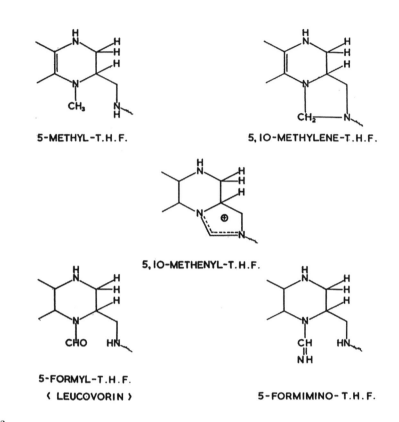

FIGURE 43
Compounds of folic acid. (From Shinton, N.K., *Br. Med. J.*, 1, 556–559, 1972. With permission.)

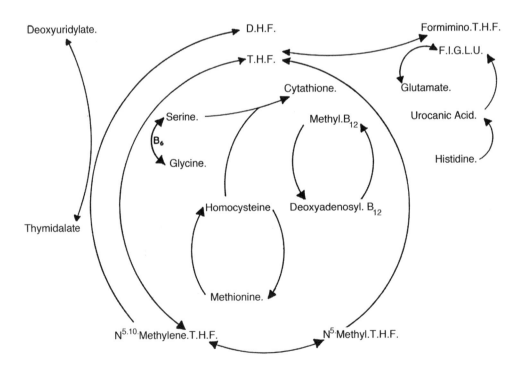

FIGURE 44
Folates in metabolic pathways. (From Shinton, N.K., *Br. Med. J.*, 1, 556–559, 1972. With permission.)

cells and is maintained in this form by similar intracellular lysomes. Intracellular conversion is made to ^{5}N-methyl tetrahydrofolate. In this form, it enters the plasma before transportation by specific carriers to metabolically active cells. Here it releases its methyl group to homocysteine.

Folate-binding proteins can be detected in the plasma of about 15% of healthy individuals. In these persons, the protein is found to be about 70% saturated. Increased levels of folate-binding protein have been reported in pregnancy, in women on oral contraceptives, in alcoholics who are folate deficient, and in patients with uremia, cirrhosis, and chronic myelogenous leukemia. Folate-binding protein with similar properties has been found in normal granulocytes. The function of these proteins in granulocytes remains unknown, although it has been suggested that they may deny access to folate by invading bacteria.

Storage

The adult body contains approximately 5 mg folate, about one-third of which is stored in the liver. When folate intake is prevented, megaloblastosis as an indication of deficiency occurs within 3 to 4 months. The levels of red blood cell folate reflect the body stores and can readily be assayed.

Metabolism

Before folate can be metabolically active, it must be reduced to the tetrahydrofolate form, with dihydrofolate being an intermediate and dihydrofolate reductase catalyzing the reaction. Folates are coenzymes in many mammalian metabolic pathways in which there is a one-carbon-unit transfer. These include:

Formylation of glycinamide ribonucleotide and 5-amino-4-imidazole carboxamide ribonucleotide in early purine synthesis

Methylation of deoxymandelic acid to thymidylic acid in pyrimidine nucleotide biosynthesis

Amino acid conversions:

- Serine and glycine, which also require pyridoxine as coenzyme
- Histidine to glutamic acid through formiminoglutamic acid (FIGLU)
- **Homocysteine** to methionine, which also requires cobalamin as coenzyme; here, there is a transfer of a methyl group from N^5-methyl tetrahydrofolate to cobalamin to form methyl cobalamin and a subsequent transfer of this methyl group to homocysteine to form methionine (see Figure 44)

Excretion

Twenty percent of food folate is not absorbed and passes into feces. There is a daily urinary loss of 2 to 5 mg, which is much increased if the tissues are saturated. With deficiency, there is an increased excretion of formiminoglutamic acid, which is accentuated by an oral dose of histidine and forms the basis for the FIGLU-excretion test as an indicator of folate deficiency.

Determination of Folate Status[147]

The amount of folate present in body fluids, food, or tissue is determined by biological assay; prior to assay, folate must be released from any binding protein. The methods available are:

Microbiological assay using *Lactobacillus casei.*

Radioisotope-dilution assay using a commercially available kit. Most clinical laboratories use this method, where the amount of folate in serum (methyltetrahydrofolate) is determined by its ability to compete with a known amount of radiolabeled vitamin for its binding protein.

Serum Folate

The normal adult range is 3 to 20 μg/l; the values for premature infants and normal children are given in **Reference Range Table XI**. Reduced levels may only indicate reduction in folate intake over the preceding few days. A low serum folate level will rise quickly upon refeeding. Serum folate levels are low in patients with megaloblastic anemia secondary to folic acid deficiency. They may also be low in the absence of megaloblastic anemia in patients with alcoholism, low dietary intake, in normal pregnancy, and in those on anticonvulsant therapy. Therefore, a low serum folate level must be interpreted with caution. The measurement is most useful when interpreted with the **cobalamin** level of patients with proven **megaloblastosis**. The reference range in health is 5 to 30 ng/ml. Levels for normal premature infants and children are both lower (see **Reference Range Table XII**). Normal DNA synthesis is not sustained with levels below 4.0 ng/ml.

Red Blood Cell Folate

These are largely polyglutamate concentrations determined on red cell hemolysates, but these are more reliable indicators than serum levels of folate-induced megaloblastic change. Additionally, they are not affected by dietary changes, but they are low in about 50% of patients with cobalamin deficiency. The red cell folate level may be normal, despite folate deficiency, when there is an increased reticulocyte count and following recent blood transfusion. The reference range in health is 160 to 640 μg/l; the values for infants are 74 to 995 μg/l and for children 96 to 364 μg/l (see **Reference Range Table XI**).

Clinical Deficiency
Causes

Decreased intake due to:
- Poor nutrition, particularly in the elderly living alone
- Poverty, sometimes associated with chronic alcoholism
- Destruction of folate due to excessive cooking of vegetables
- Hyperalimentation
- Premature birth (particularly with infection, diarrhea)
- Synthetic diets (e.g., for phenylketonuria)
- Goat's milk as mode of feeding
- Intensive-care parenteral feeding

Malabsorption
- Congenital with inability to absorb folate or transport it into the cerebrospinal fluid
- Acquired malabsorption due to **intestinal tract disorders** such as:
  Gluten enteropathy
  Tropical sprue

Crohn's disease (regional enteritis)

Extensive resection

Enteropathy-type T-cell lymphoma

Whipple's disease

Scleroderma

Amyloidosis

Dermatographic enteropathy

Alcohol abuse

In many patients, there is a mixture of causes

Increased Requirements

Pregnancy: folate requirements increase five- to tenfold because the growing fetus draws on the mother's folate stores even when those are severely deficient. Multiple fetuses, a poor diet, infection, coexisting hemolytic anemia, or anticonvulsant medication all increase the problem, as does lactation. Folate deficiency is very common in pregnancy and is the major cause of the megaloblastic anemia of pregnancy.

Increased cell turnover from:

- Chronic **hemolytic anemia**
- Exfoliative dermatitis
- Psoriasis treated with methotrexate

Hemodialysis (folate lost in dialysis fluid)

Chronic **liver disorders** with diminished storage, which accentuates dietary deficiency, in particular with **alcohol toxicity**. Adverse effects of alcohol consumption, such as increased risk of breast cancer, may be mitigated by an adequate dietary intake of folate.

Disturbances of Folate Metabolism
Hereditary[149,149a]

Deficiency of methylenetetrahydrofolate reductase (MTHFR), the most common hereditary folate deficiency, giving rise to neuropathy but no megaloblastic anemia. MTHFR gene polymorphisms cause mild-to-moderate reductions in MTHFR activity, but no direct clinical manifestations.[200]

In infancy, folate malabsorption disorders present with severe anemia, delayed development, and low serum folate levels.

Deficiency of glutamate formiminotransferase causes variable disorders, from megaloblastic anemia of childhood with developmental issues to merely increased excretion of formiminoglutamate (FIGLU)

Acquired

Dihydrofolate reductase inhibition can occur due to intake of alcohol and drugs such as methotrexate, aminopterin, pyrimethamine, trimethoprim, sulfasalazine, proguanil, and triamterene.

Autoantibodies[201] causing:

- Blockage of 5-methyl THF transport through the choroid plexus into the central nervous system (CNS), giving rise to infantile irritability, psychomotor retardation, ataxia, dyskinesis, and pyramidal tract disturbances
- Blockage of placental folate receptors

Clinical Features of Deficiency/Disturbance

Megaloblastosis by disturbance of DNA synthesis via nucleoside biosynthesis. This occurs within 3 to 4 months of any cause for deficiency.

Increased incidence of neural-tube defects in fetus (see **Pregnancy**), a consequence of the need, in some, for increased folate by the fetus to facilitate remethylation of homocysteine to methionine and 5-adenosyl methamine.

Hypercoagulable state (**thrombophilia**) in those with deficiency of 5,10-methylene tetrahydrofolate reductase, leading to increased blood levels of **homocysteine**. This may be present in 5 to 10% of the population, but is of doubtful significance as a thrombotic risk factor.

Association with Alzheimer's dementia is under investigation.

Diagnosis of Folate Deficiency/Disturbance

Appropriate clinical features

Laboratory evidence of low folate status

Full clinical response to physiological test doses of folic acid

Diagnosis can be difficult in pregnancy. The "physiologic" anemia of pregnancy develops as a result of expansion in plasma volume, only partly offset by an increase in red cell mass (see **Dilution anemia**). The hemoglobin level falls and is associated with macrocytosis. Serum and red cell folate levels fall steadily during pregnancy, even with adequate nutrition. The only finding that will reliably distinguish physiologic anemia of pregnancy from folate deficiency is a megaloblastic bone marrow.

Treatment

Folate deficiency is treated by oral folic acid in doses of 1 to 5 mg daily. A parenteral preparation containing 5 mg/ml is also available, which is indicated for those in whom deficiency is due to malabsorption, e.g., congenital folate malabsorption.

Folinic acid is required in the treatment of severe toxicity caused by folate antagonists that block dihydrofolate reductase. A dose of 3 to 6 mg daily intramuscularly (already in the tetrahydro-form) bypasses the reductase blockade.

Therapeutic doses of folate will partly correct the hematopoietic abnormalities in cobalamin deficiency, but **nervous system disorders** can be induced, with disastrous results. If the need for treatment is urgent and the nature of the deficiency is unclear, both folate and cobalamin must be given after obtaining samples for assay.

It is recommended that folic acid supplements in doses of 200 to 500 µg/day be given during pregnancy, not only because folate requirements are increased, but also because of an association between severe folate deficiency and premature delivery, spontaneous abortion, abruptio placentae, toxemia, and neural-tube defects.

In the U.S. and Canada, grain and pasta are fortified with folic acid. This provides folic acid before the pregnancy has been diagnosed and thus reduces the incidence of neural-tube defects and may also help those with hyperhomocysteinemia. As the individual dose of folic acid is low, the potential danger of causing neurological manifestations in those with cobalamin deficiency is remote, but the extent of this problem is not yet known.[202,202a]

Adverse drug reactions of folate medication, apart from inducing cobalamin deficiency, are rare. Pruritic rash and bronchospasm have been recorded.

FOLINIC ACID

(Leucovorin) N^5-formyl tetrahydrofolate form of **folic acid** used to counteract the folate antagonistic action of methotrexate, thus avoiding methotrexate-induced mucositis and bone marrow suppression. It does not counteract the antibacterial activity of folate antagonists such as trimethoprim (cotrimoxazole). Folinic acid is given as calcium folinate 15 mg orally every 6 h for two to eight doses, depending upon the dose of methotrexate to be counteracted against. An alternative antagonist is calcium levofolinate, a single isomer of folinic acid. This is given by intramuscular or intravenous injection 24 h following methotrexate.

FOLLICLE CENTER LYMPHOMA

(Rappaport: diffuse poorly differentiated lymphocytic lymphoma; Lukes-Collins: diffuse small-cleaved follicle center cell lymphoma) See **Follicular lymphoma**.

FOLLICULAR DENDRITIC CELL

See **Dendritic reticulum cell**.

FOLLICULAR DENDRITIC CELL SARCOMA

A rare neoplastic proliferation of spindled-to-ovoid cells showing morphologic and phenotypic features of **dendritic reticulum cells** located in **lymph nodes** near the germinal centers, associated with B-cell proliferation. They have the **immunophenotyp**e CD23, CD21, CD35, complement receptors, and 5′-nucleotidase activity. They give rise to cervical and axillary **lymphadenopathy**. Treatment is by surgical excision followed by **radiotherapy**.

FOLLICULAR LARGE-CELL LYMPHOMA

See **Follicular lymphoma**; **Diffuse large B-cell lymphoma**; **Non-Hodgkin lymphoma**.

FOLLICULAR LYMPHOMA

(Rappaport: nodular poorly differentiated lymphocytic, mixed lymphocytic-histiocytic, histiocytic, or undifferentiated lymphoma; Lukes-Collins: small-cleaved, large-cleaved, or large-noncleaved follicular center cell (follicular) lymphoma; REAL: follicle center lymphoma) See also **Histiocytic lymphoma**; **Lymphoproliferative disorders**; **Non-Hodgkin lymphoma**.

A B-cell non-Hodgkin lymphoma (NHL) characterized by proliferation of follicle center cells, which are usually a mixture of **centrocytes** (cleaved follicle center cells) and **centroblasts** (large noncleaved follicle center cells). There are two morphological types:

Follicular, where the pattern is at least partially follicular, but diffuse areas may be present, often with sclerosis. Centrocytes typically predominate over centroblasts.

Diffuse, where the cells resemble centrocytes, with a minor component of centroblasts, all entirely diffuse. This is a rare lymphoma.

The **immunophenotype** of the tumor cells is B-cell-associated antigen+, CD20+CD10+/−, CD5−, CD23−/+, CD43−, CD11c−. Cytogenetic analysis shows t(14;18), with rearrangement of bcl-2 in 70 to 95% cases. It probably originates from germinal center B-cells.

Clinical Features

This is a lymphoma occurring predominantly in adults, with an equal male:female incidence. They constitute 40% of adult NHLs in the U.S., the incidence being lower in other parts of the world. Most patients present with multicentric **lymphadenopathy**, **bone marrow** infiltration, and occasionally with **splenomegaly**.

The median survival is nine to ten years. With the exception of selected patients with limited stage I disease, it has rarely been curable, but now patients respond to a large range of interventions both at the time of initial treatment and at relapse.

Staging

See **Lymphoproliferative disorders**.

Treatment

The initial treatment is dependent on a number of factors, some patients initially requiring observation only. The initial standard treatment has been the CVP combination of **cytotoxic agents**. To this combination, **rituximab** (R-CVP) is now added either initially or as maintenance. In patients who have relapsed, monoclonal antibodies conjugated to radioimmunoconjugates ([131]iodine tositumomab and [90]yittrium ibritumomab tiuxetin) are efficacious. For Stage III or Stage IV, treatment is via a one-week course of ^{131}I-tositumomab by 1-h infusion of 450 mg tositumomab followed by 20 min infusion of ^{131}I-tositumomab. This is repeated in 7 to 14 days using 35 mg tositumomab and a dosimetric dose of ^{131}I-tositumomab to total body irradiation of 75 cGy. This results in 90% response and 75% complete in 4.3 to 7.7 years.[202a,202b] A clinical trial is necessary for confirmation.[90] Yttrium is under investigation for therapy as is autologous stem cell transplantation.[90a]

FOLLICULAR, SMALL-CLEAVED CELL LYMPHOMA

See **Mantle-cell lymphoma; Non-Hodgkin lymphoma**.

FOLLICULAR SMALL-, MIXED-, OR LARGE-CLEAVED CELL LYMPHOMA

See **Follicle center lymphoma; Non-Hodgkin lymphoma**.

FONDAPARINUX

(Fxa)

Synthetic analog of **antithrombin**-binding pentasaccharide. It consists of chemically modified (hypersulfated) pentasaccharide sequences that are native to **heparin**(s), with the chemical modification increasing affinity for antithrombin. They are uniform in size, and the chain length is too short to bridge antithrombin to thrombin, which is necessary for inhibition of thrombin by antithrombin (AT). Consequently, the pentasaccharide derivatives are specific inhibitors of FXa only. They avidly bind antithrombin, inducing a conformational change that increases AT activity to selectively neutralize FXa ≈300-fold. Upon the activated antithrombin binding, fondaparinux is released and available to bind to more antithrombin.

Fondaparinux is 100% bioavailable upon subcutaneous injection, with a rapid onset of action, and does not bind to other plasma proteins, platelets, or endothelium. It is eliminated unchanged by the kidneys and has minimal or no effect on **activated partial thromboplastin time** (APTT), **prothrombin time** (PT), or thrombin time (TT).

In clinical trials, fondaparinux has been shown to be more effective at preventing **venous thromboembolic disease** than low-molecular-weight heparin (LMWH), but in other settings it is otherwise equivalent to LMWH.

Idrapinux is a hypermethylated analog of fondaparinux that binds to antithrombin with very high affinity such that antithrombin is fully activated. It is administered by once-weekly subcutaneous injection. It may have a role in therapeutic anticoagulation where **warfarin** use is difficult, but it has a long half-life and there is no antidote.

FRAGMENTATION HEMOLYTIC ANEMIA

Red blood cell damage with loss of membrane and cytoplasm, leading to marked changes in red blood cell morphology and **anemia**. It occurs as a consequence of physical trauma to the red blood cells within the circulation. Following injury, the cell reseals, but adopts a bizarre shape (**schistocytes**). The morphologic changes include crescent forms, helmet cells (keratocytes), triangular cells, and **microspherocytes**. Ultimately, these cells are removed from the circulation by **histiocytes**, mainly in the spleen. The causes of this abnormality are either mechanical trauma in large vessels (**macroangiopathic hemolytic anemias**; traumatic or mechanical hemolytic anemia) or disorders of the microvasculature (**microangiopathic hemolytic anemias**).

FRATAXIN

A protein involved in mitochondrial **iron** metabolism; encoded by *FRDA* (9q13). Most cases of Friedreich ataxia (hypertrophic cardiomyopathy and ataxia) are due to trinucleotide (GAA) repeat expansions in the *FRDA* gene; more-severe disease is associated with more repeats. Other cases are due to missense mutations. Frataxin deficiency can cause decreased mitochondrial iron storage, which can impair iron metabolism, promote oxidative damage, and lead to progressive **iron storage**.

FREE ERYTHROCYTE PROTOPORPHYRIN

(FEP) See also **Porphyria**.
Unbound protoporphyrin in **red blood cells**. Any limitation to the supply of iron to erythrocytes that synthesize **hemoglobin** results in an increase in FEP. As an indicator of **iron deficiency**, FEP measurement is less sensitive than that of **transferrin** saturation. However, FEP is stable and thus is a reliable indicator of iron deficiency that responds to treatment. FEP can distinguish between iron deficiency (increased FEP) and α-**thalassemia** (normal FEP). FEP is increased in anemia associated with **lead toxicity**, and FEP measurement can be used to screen for this disorder. The reference range of FEP levels is 22 to 48 μg/l red cells; in severe iron deficiency, levels of ≈200 μg/l red cells may occur.

FRESH-FROZEN PLASMA

(FFP) Plasma separated from whole blood by differential centrifugation, which is then frozen to below −30°C within 8 h of collection.[160] It can then be stored for up to 24 months. FFP can be treated by the addition and removal of methylene blue to inactivate viruses without evidence of significant reduction in efficacy. Virally inactivated plasma is also available in the solvent-detergent (SD)-treated FFP. For cost-effective SD treatment, most manufacturers pool 100 to 1000 donations, which are then split into packs for transfusion of approximately 200-ml volume following treatment. **Factor VIII** is the most labile plasma constituent of clinical importance and must be present at a minimum level of 0.7 IU/ml in thawed FFP. The pooling for SD treatment may reduce the incidence of transfusion-related acute lung injury by dilution of pathogenic antibodies and inactivation of human leukocyte antigen (HLA) antibodies by soluble HLAs. The disadvantages of pooling are increased donor exposure and possible reductions in anticoagulant proteins. Before transfusion, units are thawed in a 37°C water bath or microwave incubator, and these should be infused within 2 h when **factor V** and factor VIII are required, as the concentration of these factors declines rapidly upon thawing. Storage up to 24 h in a blood bank refrigerator after defrosting is not associated with significant reductions in vitamin K-dependent coagulation **factors II, VII, IX,** and **X**. An initial dose of 10 to 15 ml/kg body weight is usually required, but this will vary according to the indication. Its effect should be monitored using the **prothrombin time** (PT) and **activated partial thromboplastin time** (APTT) tests. FFP should be ABO compatible, and since hemagglutinating anti-A and anti-B are present, group O FFP should only be given to group O recipients; group A or B FFP can be given to group O recipients as well as to patients of the same group. FFP contains a small amount of red blood cell stroma, which may boost the immune response to Rh blood group D (RhD) in patients who have already been sensitized by prior pregnancy or transfusion. Rh-compatible FFP should therefore be given to women of childbearing potential; if this is not possible, then appropriate prophylaxis with anti-D should be administered. FFP should not be transfused at a rate >30 ml/min.

Clinical Indications
Definite Indications

Reversal of severe overdose with **coumarin** *anticoagulants*: FFP is indicated if the patient is at risk from life-threatening hemorrhage. Alternatively, prothrombin complex concentrate (PCC) (containing factors II, VII, IX, and X) can be given, and this might be regarded as the treatment of choice. However, some PCCs do not contain factor VII, and if the correct concentrate is not available, FFP should be infused at an initial dose of 10 to 15 ml/kg.

Vitamin K *deficiency.*

Acute **disseminated intravascular coagulation** (DIC): this can be triggered by a number of disease processes, and treatment should aim primarily to remove the underlying cause. However, deficiencies of factor V, **fibrinogen, fibronectin**, and **platelets** due to activation of both coagulation and fibrinolytic pathways give rise to both abnormal bleeding and thrombosis. In severe DIC, FFP is given at the doses stated above and the results of the coagulation tests are monitored. Cryoprecipitate is also indicated if fibrinogen levels are <0.8 g/l.

*Thrombotic **thrombocytopenic purpura*** (TTP): FFP has been shown in clinical trials to be the treatment of choice in TTP. A large daily dose approximating one plasma volume (around 40 ml/kg) is required and is usually given during a therapeutic plasma-exchange procedure.

Inherited single-coagulation-factor deficiency in cases where a specific or combined coagulation-factor concentrate is not available. Most coagulation factors and inhibitors of coagulation are now available for clinical use, including factor XIII, antithrombin III, and protein C. If these concentrates are unavailable, use FFP.

Complement-C1 *esterase-inhibitor deficiency*: a partially purified concentrate is available and should be used wherever possible to treat severe attacks of **angioedema** in patients with either hereditary or acquired deficiency. FFP is not ordinarily recommended for treating angioedema, since it provides complement, which exacerbates further inflammation and tissue edema. FFP is effective in prophylaxis against angioedema in susceptible individuals at doses of 5 to 10 ml/kg.

Conditional Uses

Massive acute **hemorrhage**: FFP is an inappropriate routine adjunct to replacement of large volumes of blood. Disordered coagulation is associated more with the clinical circumstances surrounding the transfusion than the volume of blood lost. Abnormal bleeding following massive transfusion may occur with hypofibrinogenemia, **thrombocytopenia**, and DIC. If the PT or APTT are significantly prolonged, or if it is felt that treatment should not wait for the results of these tests, then FFP at a dose of 10 to 15 ml/kg will replace both fibrinogen and other critical clotting factors. If severe hypofibrinogenemia is present (plasma fibrinogen <0.8 g/l), then **cryoprecipitate** should be given.

Liver disorders: FFP is indicated if there is active bleeding or if bleeding is suspected. Large volumes of FFP (≥30 ml/kg) may be required. This volume may not be readily tolerated, and PCC is then indicated. However, the latter carries a risk of inducing thrombosis or DIC in patients with liver disease and should only be used if this risk is justified by the severity of the patient's clinical situation.

Cardiopulmonary bypass surgery: nonsurgical bleeding during cardiopulmonary bypass is usually caused by platelet-function disturbances. FFP may be needed for patients with clinical bleeding and proven abnormalities of coagulation that are not attributable to **heparin**.

Special **neonatal disorders**: FFP is given to neonates to reduce the likelihood of intraventricular hemorrhage and to treat DIC-complicating sepsis. In neonates with necrotizing enterocolitis and associated erythrocyte T-antigen activation, infusion of FFP may cause **intravascular hemolysis**.

Contraindications

FFP is not indicated in the management of hypovolemia or as a part of nutritional support. It has no place in the treatment of immunodeficiency states for which immunoglobulin preparations should be used.

Adverse Effects

See also **Blood components for transfusion** — complications.

Hypersensitivity reactions: urticaria occurs in 1 to 3% of recipients of FFP, and occasionally **anaphylaxis** occurs. FFP should not be transfused at >30 ml/min.

Transfusion-transmitted infection: FFP may transmit HIV, hepatitis B, hepatitis C, and parvovirus.

Hemolysis: may be due to ABO incompatibility or found in neonates with necrotizing enterocolitis and associated T-lymphocyte activation.

Circulatory overload: whenever possible, specific factor concentrates are preferred to avoid fluid overload.

Transfusion-related acute lung injury: leukocyte antibodies can cause damage to pulmonary tissue.

FRUCTOSE

An alternative sugar to glucose as a source of energy for **red blood cell** metabolism.

FRUCTOSE-1,6-DIPHOSPHATE

A sugar of the glycolytic Embden-Meyerhof pathway of **red blood cell** metabolism. (See also **Glyceraldehyde-3-phosphate dehydrogenase**.)

FUNGAL INFECTION DISORDERS

The association of fungal infection and hematology. Hematological changes occur with all disseminated fungal infection, commonly following suppression of the T-cell-mediated **immune response** such as occurs with post-**allogeneic stem cell transplantation** or in **acquired immunodeficiency syndrome**. The common fungal organisms are *Candida albicans*, *Aspergillus* spp., and *Cryptococcus* spp. There is an increased **transferrin** saturation and raised **ferritin** concentrations with reduction in total iron-binding capacity (TIBC). **Anemia, leukopenia**, and **thrombocytopenia** are associated with **myeloid hyperplasia** and erythroid/megakaryocytic **bone marrow hypoplasia**. **Lymphadenopathy** and **splenomegaly** are common. *Aspergillus fumigatus* and other fungi produce gliotoxin, which has immunosuppressive properties that facilitate immune evasion by the host. This is achieved by inhibition of **histiocytes**, inhibition of **cytotoxic T-lymphocytes**, and induction of **apoptosis** by B-cells.[203] There is a high mortality rate in those with **neutropenia**. Every effort should be made to identify the infecting fungus. Serodiagnostic assays including the detection of genomic DNA sequences of fungal antigens may help. Macrophages in the marrow can be a source for culture of the infecting fungus. Treatment is by administering triazole antifungals: fluconazole, itraconazole, vericonazole, or amphotericin B.[204] There are few evidence-based recommendations for therapy, but isolated reports suggest that combination therapy may be beneficial. Fluconazole is used prophylactically in high-risk patients. **Adverse drug reactions** include nephrotoxicity and cardiotoxicity.

G

GAISBÖCK'S SYNDROME

See **Erythrocytosis** — relative erythrocytosis.

GALACTOSE

An alternative sugar to glucose as a source of energy for **red blood cell** metabolism.

GASSER CELLS

Lymphocytes (and **monocytes**) that contain large basophilic granules seen in **mucopolysaccharidoses** (except Morquio's syndrome). The granules are themselves made up of mucopolysaccharide.

GASTRECTOMY SYNDROMES

See **Cobalamins**; **Gastric disorders**; **Intestinal tract disorders**.

GASTRIC DISORDERS

The changes in the stomach leading to hematological disorders.

Chronic Gastritis — Type A Inflammation

Type A (**autoimmune**) inflammation is caused by antibodies to the parietal cells and to the intrinsic-factor binding sites. Progression leads to atrophic gastritis, common in the elderly,[205] which may be associated with infiltration by **lymphocytes** and **eosinophils**. There is an increased risk of atrophic gastritis when the gastritis has been treated with omeprazole. The consequences are decreased gastric secretion:

Hydrochloric acid (achlorhydria; reduced hydrogen ion secretion) <15 ml/h (10% normal) and no increase upon stimulation by 40 µg/kg body weight histamine. Gastric acid facilitates the absorption of ferric **iron** and food iron but has little effect on absorption of ferrous iron or heme iron. Of those with **iron deficiency**, 30% have parietal-cell autoantibodies. Low hydrogen ion concentration in the small intestine leads to bacterial overgrowth, which accentuates malabsorption of iron, **folic acid**, and **cobalamin**.

Pepsinogen with lack of pepsin for food digestion; there are associated low levels of serum and urine pepsinogen.

Intrinsic factor (IF) deficiency results in cobalamin malabsorption, with all the features of **Addisonian pernicious anemia**.

R-binder proteins: cobalamin-binding proteins, which differ from IF by their rapid electrophoretic mobility. They are a family of immunologically similar substances synthesized by gastric mucosal cells.

Peptic Ulceration and Gastric Carcinoma

Both are associated with infection by *Helicobacter pylori*.[206] They lead to acute **hemorrhage** (hematemesis) or chronic hemorrhage with consequent **iron deficiency**. The red blood cells sometimes take the shape of **echinocytes**. Hemorrhage from these or other causes can be detected insensitively by fecal occult blood tests or, more accurately, by labeling the patient's red blood cells with ^{51}Cr (see **Erythrokinetics**) followed by reinjection intravenously. The subsequent fecal radioactivity is counted.

Hereditary Hemorrhagic Telangiectasia

This causes acute or chronic hemorrhage.

Gastric Erosion

Acute or chronic hemorrhage due to gastric erosion occurs as an **adverse drug reaction** by:

Nonsteroidal anti-inflammatory drugs (NSAID)

Corticosteroids

Serotonin reuptake inhibitor antidepressants

Gastric Surgery

Total gastrectomy: IF-secreting cells will have been removed so that cobalamin deficiency occurs in all patients within 5 years unless parenteral cobalamin therapy is given. Iron supplementation is also necessary.

Partial gastrectomy: malabsorption of cobalamin occurs in 30 to 40% of patients, with **megaloblastosis** in 18%. Folate and iron deficiency may follow. Iron deficiency is common when the duodenum has been bypassed in Billroth II and Polya procedures. All deficiencies can be avoided by prophylactic therapy.

Gastric-bypass operations: cobalamin deficiency is common due to exposure of IF to pepsin and trypsin.

Vagotomy and pyloroplasty with gastroenterostomy: iron deficiency is common, particularly when the duodenum has been bypassed in Billroth II and Polya procedures.

Zollinger-Ellison Syndrome

A triad of fulminating peptic ulceration, gastric acid hypersecretion, and multiple gastrinomas of the pancreas. If there is inability of cobalamin release from R-binder proteins to intrinsic factor, cobalamin malabsorption will ensue.

Extranodal Marginal-Zone B-Cell Lymphoma

These tumors are probably a consequence of *Helicobacter pylori* infection. The lymphoid cells of the lamina propria — mucosa-associated lymphoid tissue (MALT) — proliferate laterally along the submucosal planes, giving a characteristic rigid appearance radiologically by sonography and computerized tomography (CT) scanning. The lymphoma later penetrates the mucosa, becoming polypoid and fungating with hemorrhage. Due to the lack of distinction from some carcinomas, surgical biopsy is required for a definitive diagnosis. There are three genetic abnormalities in MALT lymphoma: t(1;14), t(11,18), and t(14;18). These likely affect a common molecular pathway, *TRAF6*. They may transform to a **diffuse large B-cell lymphoma**.

Treatment of low-grade gastric MALT lymphomas by antibiotics and proton-pump inhibitors with subsequent irradiation of *H. pylori* may result in remission. Eighteen months should be allowed for the full effect. If this does not occur, **radiotherapy** results in about a 90% overall survival. **Cytotoxic agent** therapy with chlorambucil and CVP has been effective. The monoclonal antibody **rituximab** is under evaluation. Local surgery is not commonly indicated. Ten-year survival is likely for 90% of those treated from an early stage of the disease. Cytotoxic agent therapy is now the initial treatment of choice in primary gastric diffuse large B-cell lymphoma. Surgery is indicated for gastric perforation.

GASTROINTESTINAL DISORDERS

See **Gastric disorders; Intestinal tract disorders**.

GAUCHER'S DISEASE

A rare autosomally recessive **lipid-storage disorder** of sphingolipid, originally described by Gaucher in 1882. Deficiency of the enzyme b-glucocerebrosidase results in the accumulation of b-glucocerebroside within tissues.[207] The clinical severity of Gaucher's disease depends on the amount of residual b-glucocerebrosidase activity:

Type 1, adult (chronic nonneuropathic)

Type 2, juvenile (subacute neuropathic)

Type 3, infantile (acute neuropathic)

All nationalities may be affected, but there is a particularly high incidence of type 1 disease among Ashkenazi Jews (1:2500 births). Various mutations of the glucocerebrosidase gene have been described, and homozygote and double heterozygote patients have been identified. Prenatal diagnosis is often possible.

In the adult form (glucocerebrosidase level 12 to 44%), glucocerebroside largely accumulates in internal organs, whereas in the infantile form (glucocerebrosidase level 0 to 9%), neural tissue is the target. Clinical manifestations start to appear in childhood, with **hepatosplenomegaly**, skin pigmentation, brown pingueculae, and orthopedic problems. Abdominal pain is usually due to splenic infarction, but bony pain may result from pathological fractures or osteonecrosis (especially femoral head). X ray of the lower femur may reveal a classic flasklike expansion — Ehrlenmeyer deformity. **Anemia, thrombocytopenia**, or **leukopenia** can occur, either as a result of **bone marrow** infiltration by lipid-laden macrophages or **splenomegaly**.

In the infantile form, along with hepatosplenomegaly, neurological defects start to appear in the neonatal period. Fits, developmental retardation, cranial nerve neuropathies, and spasticity are common. Death under 2 years of age is usual.

The juvenile form is intermediate between the other two forms, but by the second decade of life, neurological problems and even death are not unusual.

The diagnosis is usually made by **bone marrow** aspiration, where classical large (20 to 80 μm) glucocerebroside-laden macrophages (Gaucher cells) are seen. Identical cells are seen in liver, spleen, or lymph node biopsies. The cytoplasm is faintly blue, and numerous fibrillae can be seen, giving a "whirl-like pattern." The single eccentric nucleus is small and often becomes denser as the cytoplasm fills up. Gaucher cells are strongly PAS-positive and TRAP-positive. Confirmation is by blood b-glucocerebrosidase levels or molecular analysis.

A variety of recent treatments are under investigation. These include:

Intravenous β-glucoside, particularly for adults[208]

Enzyme replacement by recombinant Cerezyme® (Genzyme Corp., Cambridge, MA)

Substrate reduction by N-butyl deoxynojirimycin (Zavesca®, Actelion Pharmaceuticals, Allschwill, Switzerland)

Gene therapy (in the future)

Allogeneic stem cell transplantation

Symptomatic treatment such as **splenectomy** for splenic infarction or hypersplenism is often necessary.

GEL TECHNIQUES

See **Column-agglutination techniques**.

GENES

Mendel's units of inheritance, occurring in allelic pairs, one being contributed from each parent. Alleles can be inherited in a dominant, recessive, or incomplete (partial dominant) manner. The molecular definition of the gene is straightforward: "Each fragment of DNA contain[s] the entire nucleic acid sequences necessary to encode a functional polypeptide or RNA molecule."[13]

Eukaryotic genes are monocistronic, encoding only one protein, although isoforms can subsequently be generated by alternative RNA splicing or by differential use of promoters, terminators, or splice sites upon transcription. On average, a eukaryotic gene is five times longer than its mRNA due to the presence of noncoding DNA sequences, called introns. These are composed of repetitious DNA, which in turn consists of simple-sequence DNA: Alu repeats (10 to 15% of total DNA), mobile intermediate-repeat DNA (short [SINES] and long [LINES] retrotransposons, 25 to 40% of total DNA), and unclassified spacer DNA (single-copy sequences not encoding protein, 45 to 50% of total DNA).

Protein-encoding DNA accounts for only 5% of total DNA in humans and consists of single-copy genes, functionally duplicated (pseudogenes and gene families) and tandemly repeated genes encoding pre-rRNA (≈250 copies), 5S-rRNA (≈2000 copies), and tRNA (1300 copies). Gene characterization usually begins with identification and sequencing of cDNA and subsequent screening of a genomic library (see **Molecular genetic analysis**). Identification of introns follows a comparison of genomic and cDNA sequences. Introns vary in number and size, but within one gene are invariant upon cell replication. The average exon number in mammals is seven, average gene length is 16.6 kb, and average mRNA length is 2.2 kb. Exons tend to be short (average 150 to 200 bases), whereas average intron length is 200 to 2000 bases. The total number of human genes is estimated to be 125,000.

Gene duplication to produce families of related genes results in retention of the basic gene structure, with similar exon size, sequence, and intron-exon boundaries but variation in intron size. Gene families may be clustered on the same chromosome (homologous recombination) or scattered among chromosomes following evolutionary translocation.

GENE THERAPY

The manipulation of human genomes as a form of treatment for hematological disorders. Gene therapy is based on the early principle that if aberrant genes causing disease can be replaced with "correct" versions, the disease might be controlled, prevented, or cured. However, at present, gene therapy is most likely to be clinically effective as a preventative or curative treatment only for single-gene defects, such as adenosine deaminase (ADA) deficiency. Several clinical trials of gene therapy protocols have already been completed, but the effectiveness of the protocols was generally disappointing, predominantly due to the inefficiency of the gene transfer vectors used. In addition, some safety issues have come to light after development of leukemia in some children after gene therapy for ADA.

The term "gene therapy" applies to protocols that involve an element of gene transfer, either *in vivo* or *ex vivo*. *In vivo* gene transfer is the introduction of genes to cells at the site they are found in the body, for example, lung epithelial cells following inhalation of the gene transfer vector. *Ex vivo* gene transfer involves the transfer of genes into viable cells that have been temporarily removed from the patient and are then returned following treatment (e.g., bone marrow cells). Gene therapy can be subdivided into somatic-cell and germline gene transfer (transfer to haploid sperm or egg cells of the reproductive system). While the potential benefits of germline gene therapy in humans are significant, the potential for abuse and eugenics merits circumspection about the use of technique. Germline gene therapy is being widely used for the production of transgenic animals for research and, increasingly, for agriculture and biotechnology, but the long-term effects of each transferred gene in animals will need to be carefully monitored and analyzed.

Routes of Administration and Cell Targets for Vectors

Intrathymic and hepatic approaches have been used for administration of either live cells transfected with DNA or naked DNA itself. Bone marrow cells, however, remain the most widely utilized mode of gene therapy administration. Their potential of hematopoietic stem cells (HSCs) for self renewal and differentiation into all hematopoietic lineages makes them very attractive as targets for gene transfer, especially in those situations (such as genetic disorders) where long-term transgene expression is required. Unfortunately, because HSCs occur at extremely low frequencies in the bone marrow and peripheral blood, it is difficult to obtain sufficient numbers of cells to transduce *ex vivo* for a subsequent *in vivo* biological effect. Stem cell mobilization protocols and positive- and negative-enrichment methods using cell-surface markers that are potentially specific for stem cells may be used to increase the yield of HSCs.

Forms of Gene Transfer Vectors

Vectors are the vehicles that are used in gene therapy to transfer the gene of interest (transgene) to the target cells, which will then go on to express the therapeutic protein encoded by the transgene. All of the vectors that are currently available have both advantages and disadvantages. Many factors must be taken into consideration when choosing a vector. The most important are:

- Length of time that the transgene needs to be expressed
- Dividing state of the target cells
- Type of target cell
- Size of the transgene

- Potential for an immune response against the vector to be induced, and whether this is deleterious
- Ability to administer the vector more than once
- Ease of production of the vector
- Facilities available
- Safety issues
- Regulatory issues

Viral Gene Delivery

The general principle involved in the development of most viral vector systems is that an intact wild-type virus is modified for safe use and effective gene transfer; for example, the specific genes that are involved in viral replication can be modified or deleted, thus rendering the new recombinant virus "replication defective" and safer for use in gene therapy protocols. Usually, the transgene that is to be delivered by the virus must be inserted into the viral genome, using molecular biological techniques. In general, the more severely attenuated the viral vector is from its wild-type state (i.e., the greater the number of virulence-associated genes that have been removed), the safer the virus is for use in gene therapy protocols. The size of the transgene has to be matched to the potential space in the viral genome. Because many of the viruses that are used as vectors lack replication genes and therefore cannot replicate in normal cells, the recombinant virus with its transgene must be grown up to higher titers in a packaging cell line, which includes the complementary genes that the virus requires to replicate. The recombinant viral particles can then be purified as live infectious virus from the packaging cell line and used to infect (transduce) cells or tissues *in vivo* or *ex vivo*.

Retroviral vectors, such as Moloney-murine-lentivirus-related viruses (MMLV) and lentivirus, are the most widely used viral vectors in clinical trials at present. Retroviruses will only transduce cells that are actively undergoing mitosis, and are therefore well suited for gene transfer protocols to pluripotent HSCs. Retroviral vectors give good gene expression over the long term and are technically easy to produce. However, low viral titers are yielded (generally up to 1×10^7 colony forming units per ml) and, although very rare, contamination with helper virus is a possibility, which needs to be monitored.

In contrast to MMLV vectors, lentivirus vectors are able to transduce nondividing cells as well as those that are actively dividing, thereby considerably broadening their usefulness as gene transfer vehicles. Because they integrate their genetic material into the genome of the host cell, lentivirus vectors have the potential to result in the long-term, stable gene expression of transgenes. The prospects for using lentivirus vectors as gene therapy vectors for immunological purposes are very exciting because of their inherent tropism for CD4+ T-cells, macrophages, and HSCs. Genetic modifications have been used to widen the tropism of the vector, and they have since been used to target airway epithelial cells for the gene therapy of cystic fibrosis.

Adenoviruses are nonenveloped, icosahedral, double-stranded DNA viruses with a capsid diameter of 70 to 100 nm comprising 252 capsomeres (240 hexons and 12 pentons). They are not incorporated into the genome of the target cell (nonintegrating) but remain as an extrachromosomal entity in the nucleus of the host cell. Replication-defective recombinant adenoviruses are the second most commonly used viral vectors in clinical trials today. Their transduction efficiency is high compared with that of other viral vectors. A major disadvantage of using adenovirus as a vector *in vivo* is the **cytotoxic T-lymphocyte** (CTL) response that is induced against capsid-derived peptides; this response can cause the destruction of vector-transduced cells and also leads to local tissue damage and inflammation. The period of expression of an adenovirus-encoded transgene is also relatively short. Expression is reported to last at a "reasonable" level for ≈14 days *in vivo*, but can be prolonged by manipulation of the **immune response**. Furthermore, because the adenoviral genome does not integrate into the genome of the target cell, only one of the daughter cells (if the target cells are dividing) will contain the transgene, thereby halving the total number of cells that contain the transgene.

Adenoviral gene delivery is ideally suited to those situations that require only a one-off delivery of a transgene, for example, in growth-factor therapy, in which transient, as opposed to long-term, expression of growth factor is required. In protocols that are aimed at inducing transplantation tolerance, the delivery of an adenoviral vector to a recipient before transplantation should be sufficient to induce a regulatory population of T-lymphocytes that will impart long-term immunological tolerance to that recipient.

The adeno-associated virus (AAV) vectors offer many of the same advantages as adenovirus vectors, including a wide host-cell range and, in some situations, a relatively high transduction efficiency. However, unlike adenovirus, AAV is minimally cytopathogenic. AAV also stably integrates into the host-cell genome at specific sites (in chromosome 19 of humans), thereby increasing duration of transgene expression. However, in primary-cell transductions, most of the AAV vector DNA does not integrate into the host genome but remains extrachromosomal, and this inefficiency might limit its use for *in vivo* application.

Herpes simplex virus (HSV) vectors are being developed for several gene therapy applications. At present, the main problem associated with the use of HSV as a gene therapy vector is concern about its safety for clinical use, owing to reports of the wild-type virus replicating lytically in the human brain and resulting in potentially serious encephalitis. Although vaccinia virus vectors are not currently in use for transplantation studies, they are in development for cancer gene therapy. Vaccinia virus vectors do not integrate into the genome of the host cell; however, they can accommodate large transgenes and are extremely immunogenic. Most transgenes are expressed at a high level *in vivo*, eliciting a specific immune response against the tumor antigen, which otherwise would not have occurred at levels that are sufficient to kill cancer cells. If required, more than one gene can be cloned into the vector, owing to its large capacity.

Nonviral Gene Delivery

Nonessential genes can be removed from viral vectors to make more room for transgenes, to reduce inflammatory responses, or to increase their safety. This involves the virus being simplified, sometimes to an extreme. What remains can be an artificial "vector shell" that has been designed to allow the gene of interest to be expressed at high levels, in a highly regulated specific manner, and for a controlled period of time (either short term or long term). Another approach to achieve the same result is to produce a system that can simply introduce genetic material to the nucleus of cells. This has been the focus of intensive research over the past few years and has resulted in the development of several nonviral vector systems.

Liposomes

In their most basic form, liposomes consist of two lipid species: a cationic amphiphile and a neutral phospholipid. Liposomes bind to and condense DNA spontaneously to form complexes that have a high affinity for the plasma membranes of cells. This results in the uptake of liposomes to the cytoplasm by the process of endocytosis.

Fusigenic Virosomes

Recently, some of the advantages of viral delivery vectors have been combined with the safety and "simplicity" of the liposome to produce fusigenic virosomes. Virosomes have been engineered by complexing the membrane fusion proteins of hemagglutinating virus of Japan (HVJ, which is also known as Sendai virus) with either liposomes that already encapsulate plasmid DNA or oligodeoxynucleotides for antisense applications. The inherent ability of the viral proteins in virosomes to cause fusion with cell membranes means that these hybrid vectors can be very efficient at introducing their nucleic acid to the target cell, resulting in good gene expression. Each viral vector has a limit on the size of the transgene that can be incorporated into its genome.

DNA-Ligand Conjugates

DNA-ligand conjugates have two main components: a DNA-binding domain and a ligand for cell-surface receptors. The transgene can therefore be guided specifically to the target cell, where it is internalized via receptor-mediated endocytosis. Once the DNA-ligand complex is in the endocytic pathway, the conjugate is likely to be destroyed when the endosome fuses with a lysosome.

Naked DNA

One of the simplest ideas for nonviral gene delivery techniques is the use of purified DNA in the form of plasmids. This approach is being used for DNA vaccines, among other protocols, and has been tried in many situations for gene therapy. However, despite the simplicity of this approach, studies have revealed that the transfection efficiency is very low and will therefore limit its application.

Applications of Gene Therapy in Hematology

Inherited single gene disorders, e.g., severe combined **immunodeficiency** (SCID) associated with a defect in the $\gamma\chi$ chain of the IL-2 receptor; hemophilia

Strategy for adoptive **immunotherapy** to enhance immune responses against CMV or leukemic blasts[209]

Protocols to restore the function of tumor suppressor genes, e.g., p53

Protocols to secrete antiangiogenic agents

Adverse Reactions

There remain safety concerns regarding the retroviral gene therapy protocols. Two children in the French SCID-XI study who appeared to be cured of their immunodeficiency by gene therapy developed T-cell leukemia nearly 3 years after transplantation with gene-modified cells .The malignant T-cells stemmed from a single transduced cell, in which the oncoretroviral genome was inserted in the vicinity of the LIM domain 2 (*LMO-2*) locus in a manner that enabled the retroviral promoter to mediate aberrant expression of *LMO-2* in these cells. The *LMO-2* gene encodes a transcription factor that is required for normal hematopoiesis; however, when aberrantly expressed, it is associated with childhood leukemia. This serious adverse effect of gene therapy may be confined to selection pressures associated with the clinical condition, but it highlights the need for caution before this therapy can be widely applied.[209]

GENETIC COUNSELING
See also **Hemophilia**; **Sickle-cell disorders**; **Thalassemia**.

The involvement of an individual or family in the evaluation to confirm, diagnose, or exclude a genetic condition, malformation syndrome, or isolated birth defect. The process includes discussion of the natural history of the patient's condition and the role of heredity; identification of medical management issues; calculation and communication of genetic risks; and provision of or referral for psychosocial support.

During genetic counseling, the reason for referral and the diagnoses under consideration is generally considered first, together with the patient's or family's perception of disease status or risk, beliefs about the cause of disease, and their perception of disease burden. The patient's birth history, past medical history, and current status should be documented, and a directed family history constructed in pedigree form to include: the patient's first-degree, second-degree, and further-removed relatives as appropriate; the status of current pregnancies; ethnic background; and presence of consanguinity. Additional medical records, including diagnostic testing, of the patient and affected family members should be obtained and reviewed as needed, together with the family's social history, education, employment, and social functioning. Sources of psychosocial support (community, religious, family) should also be considered, and potential ethical issues, such as confidentiality, insurability, discrimination, and nonpaternity, should be identified. Genetic counseling consultations also incorporate a physical examination of the patient and other family members as needed.

Such a consultation aims to increase the patient's understanding of genetics and describe risks to family members compared with general population risks. In this setting, the health-care professional shares information with the patient about the expected course of the disease, management issues, and possible treatments or interventions, as well as reproductive options, if and when appropriate, which may include: pregnancy with prenatal testing, pregnancy without prenatal testing, remaining childless, parenting by adoption, pregnancy by egg or sperm donation, and pregnancy following preimplantation genetic diagnosis.

The genetic counseling session should also, if appropriate, explore strategies for communicating information to others, especially family members who may be at risk, and provide written materials and referrals to support groups, other families with the same or similar condition, and local and national service agencies.

GENOMICS

A DNA chip technology that enables the analysis of thousands of genes or entire genomes. It is being used cautiously in clinical oncology.[210] There are many forms:

Oligonucleotide Arrays, e.g., Affymetrix (Figure 45)

These consist of short single-stranded **DNA** segments, oligonucleotides (or oligos), which are built to order by chemical synthesis. They are integrated or spotted using inkjet technology into a chip measuring approximately 1.25×1.25 cm. Recent arrays have used 11 oligonucleotides, each of approximately 25 bp per gene, and with the completion of the Human Genome Project, analysis expression of >47,000 transcripts derived from 38,500 human genes can be analyzed. Messenger RNA (mRNA) is extracted from target tissue and hydrolyzed with the chip. The multiple independent oligonucleotides hybridize to different regions of the same RNA and give rise to a fluorescent signal that can be measured. The use of multiple interdependent oligonucleotides improves the signal-to-noise ratios and the dynamic range of detection. Furthermore, the use of single-base-mismatch (MM) control proteins identical to the perfect match (PM) partners serves as a

348

GENOMICS

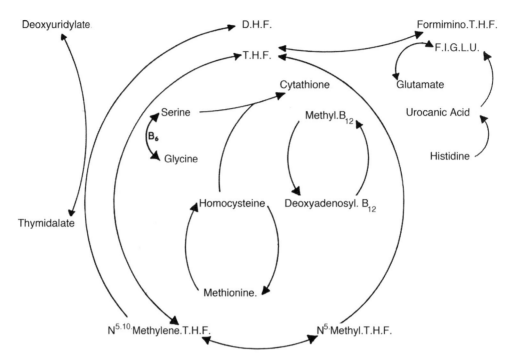

FIGURE 45
Gene-expression analysis using oligonucleotide microarray. Up to half a million distinct oligonucleotides are synthesized on the microarray by photolithography and act as probes in individual "features" on the microarray surface. About 30 distinct oligonucleotides, printed in individual features, represent the partial sequence of one gene. Fluorescent-labeled cDNA derived from a single test sample is hybridized to the microarray, allowing the expression level of up to 15,000 genes to be measured in the test sample. (From Aitman, T.J., DNA microarrays in medical practice, *Br. Med. J.*, 323, 611–615, 2001. With permission and courtesy of Affymetrix, Santa Clara, CA.)

control for hybridization and facilitates the direct saturation of background and cross-hybridization signals.

cDNA Arrays, e.g., Agilent (Figure 46)

This system uses 60-bp cDNA molecules per gene (derived from PCR amplicons). The format of 60-bp cDNA provides greater sensitivity over the 25-bp format and only requires one 60-mer per gene or transcript. The Agilent system employs a two-color system for mRNA hybridization, in which the same array is hybridized with two different mRNA samples. Reference mRNA labeled with one fluorescent probe, e.g., Cy3, is compared with a second mRNA fluorescent-labeled probe, e.g., Cy5 isolated from treated cells or disease tissue under investigation.

Both techniques give a large amount of expression information, require complex normalization and evaluation algorithms, and therefore require sophisticated bioinformatics approaches to evaluate the data. They have been used to develop molecular markers for acute leukemia and have also enabled classification of diffuse large B-cell lymphoma (DLBCL) into germ cell and activated B-cell categories.

While there is no doubt that genomics offers a huge potential, there are still a number of technical issues, such as tissue sampling and standardization of analyses and technology platforms, that require refinement over the next 4 to 5 years. To facilitate such an organized compilation of microarray data, the Microarray Gene Expression Database (MGED) Group has created guidelines for the submission of microarray data, referred to as minimal information about a microarray experiment (MIAME).

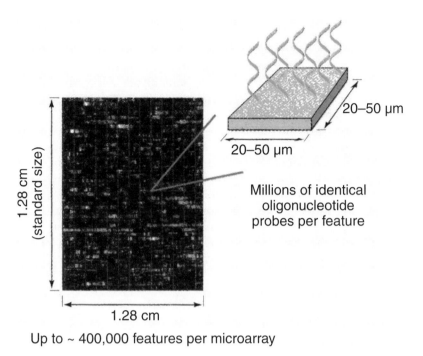

20–50 μm

20–50 μm

Millions of identical
oligonucleotide
probes per feature

1.28 cm
(standard size)

1.28 cm

Up to ~ 400,000 features per microarray

FIGURE 46

Gene-expression analysis in two tissue samples using spotted DNA microarray. RNA extracted from samples 1 and 2 is labeled with red or green fluorescent dyes. The dye-labeled RNA populations are mixed and hybridized to the microarray, on which has been spotted cDNA from thousands of genes, each spot representing one gene. The RNA from each sample hybridizes to each spot in proportion to the level of expression of that gene in the sample. After hybridization, the red and green fluorescent signal from each spot is determined, and the ratio of red to green reflects the relative expression of each gene in the two samples. For example, the gene TEPP1 is shown to be expressed at a higher level in sample 2 than in sample 1. (From Aitman T.J., DNA microarrays in medical practice, *Br. Med. J.*, 323, 611–615, 2001; and as adapted from Brown, P.O. and Botstein, D., Exploring the new world of the genome with DNA microarrays, *Nature Genet.*, 21, Suppl. i, 33–37, 1999. With permission.)

Genomic Imprinting

This concerns the unequal functional activity of some genes or chromosomal regions, dependent on whether they have been inherited maternally or paternally. Each somatic cell of the human body contains 46 chromosomes consisting of two sets of 23; one inherited from each parent. These chromosomes can be categorized as 22 pairs of autosomes and two sex chromosomes; females are XX and males are XY. Two copies of each autosomal gene exist at the molecular level; one copy derived from each parent. Until the mid-1980s, it was assumed that each copy of an autosome or gene was functionally equivalent, irrespective of which parent it was derived from. However, it is now clear from classical experiments in mice and from examples of human genetic disease that this is not the case. The functional activity of some genes or chromosomal regions is unequal and is dependent on whether they have been inherited maternally or paternally. The activity or silence of an imprinted gene or chromosomal region is set during gametogenesis.

Imprinted genes show three primary characteristics: monoallelic expression, clustering in evolutionarily conserved imprinted domains, and an association with parental-allele-specific methylation. Monoallelic expression refers to the transcription of a gene from a single parental allele and is the fundamental characteristic of imprinted genes. Imprinted genes tend to occur in discrete clusters in the genome. This is most probably because the allelic expression, or silencing, of imprinted genes within a cluster is coordinately regulated by a cis-acting element (or elements) referred to as an "imprinting center" or "imprinting control element." DNA can be modified by methyl groups that attach to cytosine residues within specific CpG (cytosine phosphate diester guanine) dinucleotide pairs. CpG islands at the 5′ end of genes are generally unmethylated to allow constitutive expression of the genes. Methylation of a single parental allele is a hallmark of imprinted genes. Methylation of an allele usually correlates with its transcriptional inactivation. Parental-allele-specific methylation is the primary candidate for the gametic imprint.

Imprinted expression can be detected by several methods, including exploitation of an expressed polymorphism occurring within an exon of a gene. Because imprinting is important for normal fetal development, and might occur only during gestation, fetal tissue samples are needed to determine which parental allele a gene is transcribed from. If the fetal genomic DNA is heterozygous for an exonic polymorphism within the gene under study, then imprinted expression from one parental allele can be detected by visualization of a single allele in the RNA. If a corresponding maternal DNA sample is also available for study, then the parental origin of the active allele can be determined. If the gene is not imprinted, i.e., it is biallelic in some tissues, then both alleles will be represented in the RNA.

GENOTYPE
The genetic constitution of an individual cell or organism. In practice, the term is often used to describe the patient-specific expression of **antigens** such as red blood cell antigens (e.g., **blood groups**) or **human leukocyte antigens** (HLA), but correctly should be used in relation to the characteristics of **DNA** encoding these proteins, e.g., restriction enzyme digestion patterns.

GIANT-CELL ARTERITIS
See **Polymyalgia rheumatica**.

GIANT-PLATELET SYNDROMES
A group of autosomally dominant syndromes with modest increases in **platelet** size. These include **May-Hegglin anomaly**, **Bernard Soulier syndrome**, **Alport's syndrome**, **Mediterranean thrombocytopenia**, **Epstein's syndrome**, and a range of other rare inherited disorders. Leukocyte anomalies such as inclusions may also be seen.

GINGIVAL DISORDERS
See **Oral mucosa disorders**.

GLANDULAR FEVER
See **Infectious mononucleosis**.

GLANZMANN'S THROMBASTHENIA
A rare inherited platelet disorder characterized by a prolonged **bleeding time**, normal platelet count, and absence of macroscopic platelet aggregation. It is due to a defect in the platelet surface membrane **glycoprotein IIb/IIIa** (GpIIb/IIIa) complex. Because of the

deficiency of functional GpIIb/IIIa complex on the platelet surface, Glanzmann thrombasthenic platelets contain an insufficient number of GpIIb/IIIa-related binding sites for **fibrinogen** and **Von Willebrand Factor** to provide adequate support for platelet aggregation. In addition, Glanzmann platelets do not spread normally after contact with the subendothelial matrix. Deficiency of GpIIb/IIIa complex is variable and may be either quantitative or qualitative. Patients with Glanzmann's thrombasthenia divide into two main groups: those with less than 5% normal levels of GpIIb/IIIa and those with a residual 10 to 20% GpIIb/IIIa activity, which may allow some residual normal function. There is no clear relation between bleeding severity and residual GpIIb/IIIa activity.

Glanzmann's thrombasthenia is an autosomally recessive disorder, with clustering seen in populations with high degrees of consanguinity. It presents in infancy or childhood with bleeding characteristic of defective platelet function, such as bruising, epistaxis, and gingival bleeding. Later in life menorrhagia, posttraumatic and surgical hemorrhage, gastrointestinal hemorrhage, and hematuria may be recurrent problems. Hemarthrosis and muscle hematomas are rare. Severity of bleeding is very variable, even within affected kindreds.

Laboratory Evaluation

Bleeding time in excess of 20 min with normal platelet count and platelet morphology is highly suggestive. Platelet-function studies show failure to aggregate with agonists **adenosine diphosphate** (ADP), **collagen**, **thrombin**, and **arachidonic acid**. Aggregation may be normal with ristocetin. Glanzmann platelets bind the agonists normally and undergo normal shape change from discs to spheres in the presence of agonists, but due to the deficiency of GpIIb/IIIa fail to bind fibrinogen and hence fail to aggregate.

Heterozygotes have approximately 50% normal levels of GpIIb/IIIa complex, and this provides a method of carrier detection.

Treatment

No specific treatment is available for Glanzmann's thrombasthenia. Local measures to arrest bleeding and antifibrinolytic drugs are of major importance. Significant bleeding requires **platelet transfusion**. As bleeding is lifelong, use of human leukocyte antigen (HLA)-matched platelet transfusion concentrates should be considered to reduce alloimmunization and refractoriness to platelet transfusion. Occasionally, alloantibodies to GpIIa/IIIa complex develop, severely limiting the effectiveness of transfused platelets. Hormonal control of menses is an important component to the management of menorrhagia. **DDAVP** has not proved efficacious. There is no place for splenectomy or corticosteroids. A number of reports suggest that recombinant factor VIIa may be useful in arresting hemorrhage, particularly in patients that have developed GpIIb/IIIa antibodies and are therefore refractory to platelet transfusion.

GLOBIN-CHAIN GENES

See **Hematopoiesis** — globin switching.

GLUCOSE-6-PHOSPHATE DEHYDROGENASE

(G6PD) A key enzyme in the hexose monophosphate shunt (pentose phosphate pathway; phosphogluconate pathway) of **red blood cell** metabolism. G6PD generates nicotinamide-adenine dinucleotide (NADPH) to maintain **glutathione** in reduced form (GSH). GSH preserves the sulfhydryl groups in the cell membrane, thus preventing oxidative injury to the cell. This can be demonstrated cytochemically.[211] Screening tests are based on the inability of cells from deficient subjects to convert an oxidized substrate to a reduced state.[89]

G6PD Deficiency

A common inherited disorder in which reduced activity of the enzyme results in **hemolytic anemia**, particularly after the administration of certain drugs. It is estimated that at least 300 million people are affected by G6PD deficiency. The gene required for expression of G6PD is located on the X chromosome; thus the deficiency is fully expressed in males. The heterozygote female is clinically normal, although G6PD activity in the latter is intermediate between the normal subject and the affected male. This deficiency is found in all parts of the world, but its greatest frequency is in tropical and subtropical areas. The geographic distribution coincides with that of tropical malaria, and the hypothesis that G6PD-deficient red cells might be resistant to falciparum malaria is supported by a number of studies. G6PD deficiency may coexist with **thalassemia** and **sickle cell disorders**, as these conditions appear in similar parts of the world. There is wide variation in clinical expression.

G6PD-Deficiency Variants

As G6PD is a polymorphic enzyme, an enormous number of variants exist (more than 300). Mutant isoenzymes are classified on the basis of enzyme activity and clinical expression:

Class I, severe deficiency with chronic hemolysis

Class II, severe deficiency with acute episodic hemolysis

Class III, moderate deficiency with acute episodic hemolysis

Class IV, normal activity

Class V, increased activity

The normal enzyme is designated B or B+, the + indicating normal enzyme activity (65% to 150%). This is the most common enzyme found in all population groups. Another common variant is A+, which is especially prevalent in persons of African descent. B and A are the only letter designations, the other variants having mainly geographical names. A and B differ in amino acid composition at position 376, where asparagine is substituted for aspartic acid. Enzyme A moves faster than enzyme B upon electrophoresis. Some 16% of African-Americans have enzyme A, and the most common clinical variant enzyme is A− (possibly resulting from a second amino acid substitution), found in 11% of African-Americans. The A− enzyme activity is only 5 to 15% of normal. The electrophoretic mobilities of A and A− are identical; however, separation can be achieved chromatographically. The most common variant in Caucasians is B−, seen most often in the Mediterranean area. The mutation involves a single amino acid substitution at position 563. Enzyme activity of B− is usually less than 1% of normal.

The A− variant is the most common and is highly prevalent in the African population. Enzyme activity becomes more severely deficient as the red cell ages. Reticulocytes and younger red cells can, however, cope with normal oxidative stresses. When oxidative stress increases, such as with fever, infection, diabetic ketoacidosis, or exposure to certain drugs and toxins, inadequacy of G6PD activity results, with accumulation of oxidized glutathione, reaction with hemoglobin, loss of heme, and formation of **Heinz bodies**. The end stage resembles the Heinz body hemolytic anemia that occurs with unstable hemoglobin disorders. If the oxidative stress is severe, **intravascular hemolytic anemia** results. Usually, hemolysis is self-limiting, since only the old red cells are destroyed, the younger cells resisting the challenge.

The B-variant occurs in individuals with the Mediterranean isoenzyme variant. These individuals inherit a very unstable isoenzyme that survives in the red cell for only a few hours. This clinical variant is characterized by moderate to severe chronic hemolysis that is exacerbated, perhaps even to a potentially fatal level, upon exposure to oxidants. The classical example is the syndrome of favism, which results in profound hemolysis following the ingestion of fava beans.

Clinical Features

The severity and course of a hemolytic episode depend on the type of variant and the type of oxidative stress. Hemolysis is usually milder in the A⁻ than in the B⁻ type. Many patients with G6PD deficiency have no symptoms or signs. Others show the features of chronic hemolytic anemia, while the remainder suffer from episodes of acute hemolysis precipitated by certain drugs, infection, metabolic upset, and sometimes even spontaneously in infancy. The main clinical types are:

Drug-induced hemolytic anemia: many drugs precipitate attacks of hemolysis in both A⁻ and B⁻ types. Typically, hemolysis commences 1 to 3 days after exposure to the drug. The hemoglobin falls rapidly, the urine turns dark brown to black, and "bite" cells are noted in the peripheral-blood smear. The laboratory features of intravascular hemolysis appear. Within 4 to 6 days, the hemolysis gradually subsides and the reticulocyte count starts to rise. The following drugs and chemicals may precipitate an episode of hemolysis in patients with G6PD deficiency:

Acetanilide	Pamaquine
Chloramphenicol	Pentaquine
Daunorubicin	Phenylhydrazine
Diaminodiphenylsulfone	Sulfacetamide
Doxorubicin	Sulfamethoxazole
Methylene blue	Sulfanilamide
Nalidixic acid	Sulfapyridine
Naphthalene	Thiazolesulfone
Niridazole	Toluidine blue
Nitrofurantoin	Trinitrotoluene

Favism: eating fava beans (*Vicia faba*) or even inhalation of its pollen by susceptible individuals may precipitate an attack of hemolysis. This occurs most commonly in the Mediterranean countries and the Middle East. Symptoms commence within a few hours, with nausea, vomiting, malaise, and dizziness, followed by the classical laboratory features. Favism does not occur in the A⁻ variant.

Icterus neonatorum: neonatal jaundice is reported mainly from the Mediterranean countries. Again, neonates with the A⁻ variant seem to be less at risk.

Hereditary nonspherocytic anemia, which can occur in patients with the common Mediterranean type. The entity is usually first noticed in infancy or early childhood. Hemolysis is exacerbated by infection or by the administration of drugs. The spleen is usually enlarged; however, splenectomy has not proved beneficial.

Diabetic ketoacidosis may precipitate severe hemolysis in patients with G6PD deficiency.

G6PD Testing
Indications

Prior to treatment with certain antimalarial drugs

Nonimmune hemolytic disease of the newborn

Hemolysis associated with oxidant drugs

Red cell morphology suggestive of oxidant change

Unexplained hemolytic anemia

Favism

Hemoglobinuria

Sickle cell disease, since both may occur in the same individual

Family history of G6PD deficiency or favism

Measurement of G6PD Activity[211]

All tests depend on detecting the rate of reduction of NADP to NADPH and are based on one of the following properties of NADPH:

Absorption of light at 340 nm

Fluorescence produced by long-wave UV light

Ability to decolorize or lead to the precipitation of certain dyes

Tests carried out during or soon after an episode of hemolysis may be difficult to interpret because of the much higher activity level in young as opposed to more mature red cells. If the diagnosis needs to be made rapidly, the parallel assay of 6-phosphogluconate dehydrogenase activity, which mirrors G6PD, will take into account the effect of older cells lost by hemolysis. Screening tests (fluorescence or dye decolorization) are useful to differentiate between normal and grossly deficient samples,[83] but neither can reliably detect G6PD deficiency in heterozygous women. Samples from patients with anemia and those with a high leukocyte count can also give misleading results with the screening tests. Fresh samples are required for the decolorization test, and heparin should not be used, as it may affect the decolorization time.

If G6PD deficiency is suspected clinically, the patient should be tested initially by the fluorescence or dye-decolorization test. If the result is abnormal or equivocal, the quantitative assay should be performed to confirm or refute. If a female has an equivocal result in the quantitative assay, then the cytochemical assay should be performed. Following a hemolytic episode, all tests may give a normal result, even in the presence of marked deficiency. Retesting 4 months after the hemolytic episode may be the only way to avoid missing the diagnosis. The various procedures are as follows:

Fluorescent screening test

Dye-decolorization screening test

Cytochemical assay

Quantitative assay: the serum values vary with the measurement method but usually range from 4.3 to 11.8 U/g (SI, 0.28 to 0.76 mU/mol) of hemoglobin

These tests are not difficult to perform and can be undertaken in most laboratories. The reagents are stable and no specialized equipment is required. The limitations of the tests must be appreciated.

Treatment of Deficiency Disorders

In most patients, simple avoidance of oxidant drugs is all that is required. Although infection must be treated promptly, antibiotic drug dosage must be kept at the lowest effective concentration for the shortest possible time. During a hemolytic attack, the precipitating drug must be stopped immediately. **Red blood cell transfusion** from donors not deficient in G6PD may be necessary. It is important to use intravenous fluids and diuretics to maintain a good urine flow, thus preventing renal damage by hemoglobinuria. In infants with severe jaundice, exchange transfusion (see **Fetal/neonatal transfusion**) may be considered. The results of **splenectomy** have been disappointing, although it can be beneficial in carefully selected cases.

c-GLUTAMYL CYSTEINE SYNTHETASE

An enzyme of the phosphogluconate pathway of **red blood cell** metabolism. Homozygote deficiency causes a well-compensated lifelong **hemolytic anemia**, but heterozygote deficiency has no clinical effects.

GLUTATHIONE

(GSSG) The principal redox buffer of **red blood cell** metabolism guarding **hemoglobin** from oxidation to **methemoglobin**. It is part of the hexose monophosphate shunt (phosphogluconate pathway) that couples oxidative metabolism with nicotinamide-adenine diphosphate (NADP) and glutathione reduction, counteracting environmental oxidants and preventing denaturation of globin. The key enzymes are **glucose-6-phosphate dehydrogenase** (G6PD) and **glutathione reductase** (GSH), deficiency of either leading to the precipitation of denatured hemoglobin on the inner surface of the red cell membrane, resulting in membrane damage and hemolysis.

Glutathione Stability Test

Normally, incubation of red cells with acetylphenylhydrazine (an oxidizing drug) has little effect on GSH content, since its oxidation is reversed by glutathione reductase, which in turn relies on G6PD for a supply of NADPH. In normal individuals, GSH is only lowered by about 20%, whereas in G6PD-deficient subjects it is lowered by substantially more than this; in hemizygotes, almost all may be lost, and in heterozygotes the fall may be up to 50%.

Reduced Glutathione (GSH)

This constitutes almost all of the nonprotein sulfhydryl compounds present in the mature erythrocyte. Its function is as a source of reducing potential for the cell. It can be assayed by a spectrophotometric method in which the slit width must be less than 2 nm; otherwise, the accuracy of the method cannot be guaranteed. Glutathione is reduced or absent in the rare inherited defects of glutathione synthesis. It is also used as part of the glutathione stability test in the diagnosis of G6PD deficiency. Glutathione levels may be increased with **dyserythropoiesis**, **myelofibrosis**, and **pyrimidine-5'-nucleotidase** deficiency. The reference range in health is 75 to 100 mg/dl RBC. Deficiency is associated with a poor prognosis in **acquired immunodeficiency syndrome** (AIDS).

GLUTATHIONE REDUCTASE

An enzyme of the phosphogluconic metabolic pathway of **red blood cell** metabolism. The gene encoding this enzyme is located on chromosome 8. Deficiency is rare and its association with **hemolytic anemia** uncertain, although in one family there was clear evidence of severe deficiency, glutathione instability, and susceptibility to oxidant-induced hemolysis.

GLUTEN ENTEROPATHY

See **Intestinal tract disorders**.

GLYCERALDEHYDE 3-PHOSPHATE DEHYDROGENASE

(Phosphotriose dehydrogenase) An enzyme involved in **red blood cell** metabolism, which catalyzes the conversion of glyceraldehyde-3-phosphate to 1,3-diphosphoglycerate in the Embden-Meyerhof pathway. Deficiency of this enzyme is rare, any **hemolytic anemia** being well compensated.

GLYCEROL LYSIS TEST

See **Acidified glycerol lysis test**.

GLYCOPHORINS

Four intrinsic **red blood cell** membrane proteins, rich in carbohydrate, that have a large number of sialic acid residues on the outer surface and contribute in a major way to the negative charge on the outer surface of the red cell. The glycophorins — designated A, B, C, and D — are identified using sodium-dodecyl-sulfate polyacrylamide gel electrophoresis (SDS-PAGE) stained with periodic acid-Schiff. A mutation of the glycophorin gene (GYPC) gives rise to the Leach phenotype, where there is lack of expression of the Gerbich blood group antigen.

GLYCOPROTEIN IIb/IIIa

(GpIIb/IIIa) Cellular membrane receptors that are a key component to **platelet** function, to binding **fibrinogen**, and to **Von Willebrand Factor**.[196b] As the **integrin** GpIIb/IIIa receptor is the final common pathway to platelet aggregation, GpIIb/IIIa receptor antagonist reduces acute ischemic complications following plaque fissuring or rupture. Pharmaceutical preparations of these inhibitors (**abciximab**, tirofiban, and eptifibatide) are widely used in percutaneous coronary intervention (PCI). Given in combination with traditional regimens, they are superior to placebo in management of non-ST-elevation acute myocardial infarction.

Abciximab is a novel antibody that blocks GpIIb/IIIa on platelets and consequently inhibits platelet function. It is a Fab fragment of a human–mouse chimeric monoclonal antibody, and it has a very high affinity but a slow dissociation rate from the GpIIb/IIIa. Although it has a short half-life of 10 to 30 min, it has a long biological half-life due to its strong affinity to the glycoprotein IIb/IIIa receptor, remaining bound in circulation for up to 15 days. However, this prolonged presence has minimal residual activity, and platelet function returns to baseline 12 to 36 h after therapy. More than 80% of GpIIb/IIIa must be blocked to inhibit platelet function. Abciximab has provided robust, consistent, and highly significant reductions in death or myocardial infarction (MI) in several large PCI trials, and it is a preferred drug used during coronary interventions to inhibit platelet activity.

Tirofiban — a nonantibody GpIIb/IIIa inhibitor — is a synthetic nonpeptide derivative of tyrosine. It has an intermediate affinity for the platelet receptor, a long plasma half-life, and a short biological half-life, resulting in a rapid recovery of platelet activity approximately 4 h following therapy cessation. Tirofiban is effective in treating unstable angina in combination with aspirin and heparin, at reducing myocardial infarction, and in PCI.

Eptifibatide is also a nonantibody glycoprotein IIb/IIIa inhibitor. It is a cyclic heptapeptide with lower affinity to platelet receptors than other agents. Plasma half-life is 2 to 3 h. Platelet activity normalizes about 4 h after therapy. This agent also provides improved outcome in PCI and unstable angina compared with control.

GLYCOSYLATED HEMOGLOBIN

(Hemoglobin A_{ic}; Total fasting hemoglobin) Measurement used for monitoring treatment of diabetic mellitus disorders. The level of HbA_{ic} indicates the average glucose level during the preceding 2 to 3 months. The value is reported as a percentage of total hemoglobin, normally 4 to 7% but rising to 10% or more with poor glucose control.

GOAT'S MILK ANEMIA

An **anemia** of infants raised on goat's milk, who may become **folic acid** depleted and develop **megaloblastosis**.

GOLGI APPARATUS

A cell constituent located within the cytoplasm that functions to modify proteins transported from the **endoplasmic reticulum** (ER) and to direct the transport of these proteins to the appropriate location. Located centrally within the cell, the Golgi apparatus divides functionally into three regions: the cis-Golgi, with which protein-containing vesicles from the ER fuse, and the medial region, through which proteins are transported to the trans-Golgi, from which appropriate transport vesicles take proteins to either membrane, lysosomes, or secretory vesicles. Proteins are mainly modified by either N- or O-glycosylation within the Golgi apparatus.

GOODPASTURE'S SYNDROME

See also **Respiratory tract disorders; Renal tract disorders**.
An **autoimmune disorder** (hypersensitivity reaction-type II) in which IgG autoantibodies bind a glycoprotein of **collagen** IV in basement membranes.[48] It affects the lung alveoli and glomeruli, causing both hemoptysis and hematuria followed by renal failure.

GRAFT-VERSUS-HOST DISEASE

(GVHD) A reaction caused by donor-derived cells against recipient tissue. It may occur within the first 100 days of transplantation (acute GVHD) or after the first 3 months (chronic GVHD). It is a troublesome complication of **allogeneic stem cell transplantation** (allogeneic SCT) and can follow **blood component transfusion** if the component contains **T-lymphocytes** (see Table 28).

Acute GVHD

The great majority of sibling recipients will experience some degree of acute GVHD at 7 to 14 days, and when this is mild, it is often a useful clinical indicator of the success of grafting. This reaction is due to alloreactive T-cells in the transplanted material reacting against the recipient's foreign antigens. Removal of T-cells prevents this, but is associated with graft rejection and subsequent relapse. Almost certainly, donor T-cells involved in GVHD also possess an antileukemic effect, which explains the poorer results obtained in identical twin and T-cell-depleted SCT.

Acute GVHD usually involves the skin, with erythema or a mild maculopapular rash. Often, the first sign of this is tenderness in the tips of the fingers or in the soles of the feet, both of which later become erythematous. More-severe reactions include generalized erythroderma, bullae, and desquamation.

In addition, there is often mild diarrhea and, less commonly, disordered liver function. Early clinical-stage GVHD is usually controlled by treatment with high-dose intravenous methylprednisolone, which can be rapidly reduced and withdrawn as GVHD fades. Later stages are often difficult to treat and can be fatal. A commonly used staging system is shown in Table 67.

TABLE 67

A Grading System for Chronic Graft-versus-Host Disease

Type of Disease	Extent of Disease
Limited	Localized skin involvement, liver dysfunction, or both
Extensive	Generalized skin involvement
	Localized skin involvement or liver dysfunction plus any of the following:
	Chronic aggressive hepatitis, bridging necrosis, or cirrhosis
	Eye involvement (Schirmer tear test <5 mm)
	Mucosalivary gland involvement
	Mucosal involvement, e.g., mouth, vagina
	Other target organ involvement, e.g., lung, gut

Chronic GVHD

Among recipients of allogeneic SCT, 25 to 50% develop some degree of chronic GVHD, which occurs 100 days or more post-SCT and is characterized by scleroderma-like changes, which may progress. Its incidence is higher in patients over 30 years old and in those who have experienced significant acute GVHD. Often, there is a sicca syndrome, with dry eyes and mouth and vulval and vaginal dryness in females. The skin is dry, with hair loss, often with tethering over bony prominences, sometimes with restricted movement at joints, leading to contractures. Hyperpigmentation and hypopigmentation, which may have been seen in acute GVHD, becomes patchy and accentuated.

Chronic GVHD of the gut is associated with weight loss due to malabsorption. GVHD is seen less frequently in other organ systems, but the lung is occasionally involved, manifesting as bronchiolitis obliterans.

Pathogenesis of GVHD

Three phases of GVHD are now recognized:

Conditioning therapy phase

Afferent therapy stage

Efferent therapy stage

Host tissues will have been damaged by a combination of the underlying disease, infections, and the conditioning therapy. For example, the lining of the gut is damaged, allowing bacterial products including endotoxin to cross into the circulation, resulting in activation of host cells with secretion of proinflammatory cytokines such as interleukin-1 and tumor necrosis factor . As a consequence, expression of MHC antigen (human leukocyte antigen, HLA) and **cell-adhesion molecules** is increased.

During the afferent phase, T-lymphocyte activation occurs following contact with antigen presented in conjunction with HLA class I and with HLA class II molecules by antigen-presenting cells (APCs). The APCs also secrete interleukin-1, which acts as an activator of naïve T-lymphocytes. The T-lymphocytes then secrete interleukin-2, which binds to CD25 on the very same T-lymphocytes, giving rise to an autocrine (positive feedback) growth loop. This leads to clonal proliferation and differentiation.

During the efferent phase, multiple cytokines are released, giving rise to both stimulatory and inhibitory signal to accessory hematopoietic cells, including B and T-lymphocytes (cytotoxic, helper, suppressor, and regulatory), monocytes, macrophages, and natural killer cells. These so-called secondary effector cells release cytokines, including tumor necrosis factor α, tumor necrosis factor β, interleukin-1, interleukin-2, interleukin-6, interleukin-10, interferon-γ, and tumor growth factor β, leading to tissue damage. There are many polymorphisms of the various cytokines and their receptors, leading to enhanced or diminished GVHD reactions

Prevention[104]

The most effective preventative measure for GVHD is good donor-recipient matching. Methods employing DNA analysis should be used in conjunction with the basic serological tests. These include DNA restriction-fragment-length polymorphism typing (DNA-RFLP) and polymerase chain reaction (PCR) sequence-specific oligonucleotide-probe typing. T-cell depletion of donor bone marrow is a very effective way of preventing GVHD, the effect being directly proportional to the number of T-cells removed. Unfortunately, T-cell depletion is associated with significant rates of graft rejection (10 to 20%) and leukemia relapse (30 to 60%).

Ciclosporin, **methotrexate**, or a combination of the two is effective in GVHD prevention and is used for the great majority of patients undergoing allogeneic SCT. Oral cyclosporin (or IV in cases where retention of the drug or malabsorption are problems) alone is effective, but the addition of oral methotrexate is an advantage. Methotrexate alone is now seldom used, as the higher dosages used are cytotoxic, and better results are achieved with cyclosporin-methotrexate combinations. Tacrolimus as a single agent or in combination with other agents is also effective as prophylaxis for acute GVHD.

Treatment

Mild acute GVHD may not require treatment, but some patients with grade 1 disease and all patients with GVHD of stage 2 and above will require IV high-dose **corticosteroid**, methyl prednisolone, at a dose of at least 1 g/m² daily. Initially, it is possible to halve this dose each day, but after a few days, reductions should be made less frequently to avoid recrudescence of GVHD. These large steroid doses render patients more susceptible to bacterial and fungal infection, and particular care with patient observations and cleanliness is needed. Other complications include increased risk of gastrointestinal hemorrhage and diabetes mellitus. Other measures used to treat acute GVHD include *in vivo* T-cell depletion, azathioprine, mycophenolate, and tacrolimus. Anti-thymocyte globulin may be beneficial in resistant acute GVHD, but the increased immunosuppression may be a problem.

Chronic GVHD can often be difficult to eradicate, and the best that may be achieved is suppression. In time, chronic GVHD may resolve. Patients with good engraftment will often respond to continuous low-dose oral prednisolone or cyclosporin/tacrolimus with or without oral azathioprine/mycophenolate. Some patients will respond to thalidomide or ecodronate, which seems particularly useful for chronic GVHD of the skin.

GRANULOCYTES

White blood cells (leukocytes) that contain large granules in their cytoplasm: **neutrophils**, **eosinophils**, and **basophils**. To maintain a peripheral circulating level of 2.0 to $7.5 \times 10^9/1$ — mainly neutrophils — requires the production in excess of 10×10^{10} granulocytes per day, making the **bone marrow** one of the largest and most active organs of the body.

Granulocyte Compartments

The granulocyte pool can be divided into several compartments (Figure 47).

Bone marrow mitotic compartment. This contains 2×10^9 cells/kg. Only myeloid cells up to the myelocyte stage of maturation are able to undergo mitosis, with an estimated four to five mitoses necessary to reach the mature myelocyte stage. The transit time from myeloblast to mature myelocyte is estimated to be 4 to 6 days.

Bone marrow maturation-storage compartment. This contains 6 to 9×10^9 cells/kg. The transit time from nonsegmented to mature neutrophil is 6 to 8 days, but may be considerably shorter, i.e., 2 days, if infection is present. This compartment contains over ten times the number of late-maturing granulocytes compared with the total blood granulocyte pool.

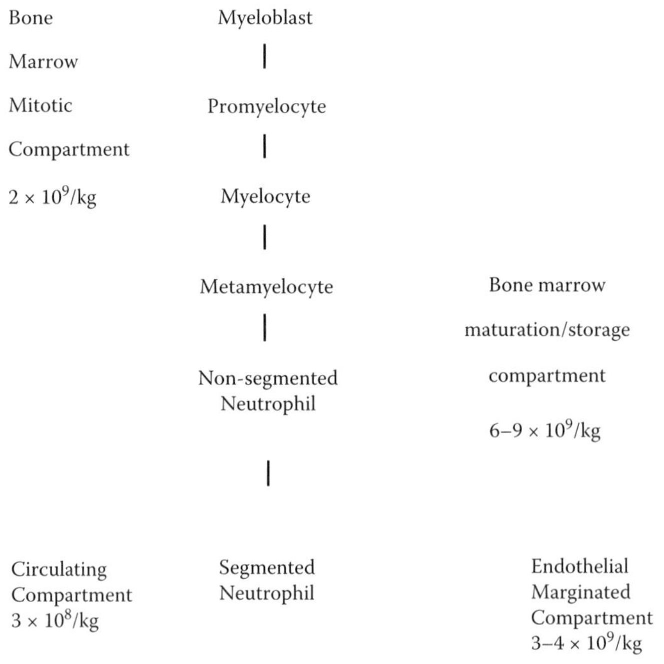

FIGURE 47
Granulocyte compartments.

Endothelial marginated compartment. This contains 3 to 4×10^9 neutrophils/kg at any given time.

Circulating neutrophil compartment. This contains 3×10^8 neutrophils/kg. The mature neutrophil spends only 6 to 10 h in the peripheral circulation.

Granulopoiesis

This results from interactions in **bone marrow** between hematopoietic stem cells, T-lymphocytes, the extracellular matrix, and stromal cells, which produce many growth factors and differentiation cytokines that positively and negatively regulate hematopoiesis. Pluripotential stem cells are usually in the G0 stage and must be induced into division and, ultimately, granulocyte maturation. The **cytokines** required for granulopoiesis are now being identified, but due to their multiplicity of action, their overlapping spectrum of activity, and their interactions with each other, the information is far from complete. Many cytokines not only have a direct effect, but also cause the release of secondary cytokines. *In vitro* studies have shown that stem cell factor (SCF) and interleukin (IL)-3 can lead to self-renewal of the pluripotential stem cells and, along with granulocyte/macrophage colony stimulating factor (GM-CSF) or G-CSF, allow commitment to granulopoiesis. In addition, IL-3, GM-CSF, and G-CSF are necessary for survival and differentiation of neutrophil progenitors. Other cytokines, e.g., IL-1, IL-4, IL-6, IL-11, IL-13, and M-CSF, have also been shown to induce granulopoiesis, usually in conjunction with IL-3, SCF, or GM-CSF. These cytokines promote the phagocytic and cytotoxic capacity of terminally differentiated neutrophils and monocytes. G-CSF promotes expansion of the postpromyelocytic pool of myeloid cells to produce rapid increase in neutrophils and has found clinical application in, for example, the accelerated restoration of hematopoiesis postchemotherapy and in congenital neutropenia.

As well as these stimulatory cytokines, several others have been shown to be inhibitory, e.g., α-interferon and transforming growth factor-β. The exact role of each cytokine should become clearer in the future due to the development of knockout mice, which have mutations or deletions of the gene encoding a particular cytokine. Recent studies have shown that mice with no GM-CSF production have normal granulopoiesis, whereas those with no G-CSF production have decreased granulopoiesis. Thus, GM-CSF does not appear to be essential *in vivo*, whereas G-CSF does (for the mouse).

Morphology of Maturing Cells

Neutrophils

Of the **neutrophils**, the earliest morphologically recognizable myeloid cell is the myeloblast (size 10 to 20 μm), which comprises less than 3.5% of bone marrow cells (Figure 48). The nucleus is round or oval and occupies up to 80% of the cell. Nuclear chromatin is fine, and up to six nucleoli may be present (typically two or three). The cytoplasm is basophilic and can be either agranular or contain a few primary (azurophilic- or myeloperoxidase-positive) granules, depending on maturation. The next morphologically recognizable cell in myeloid maturation is the promyelocyte (size 22 to 25 μm), which makes

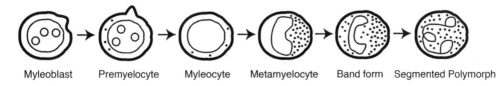

FIGURE 48
Changes in nuclear shape during neutrophil maturation.

up 25% of marrow cells. Nuclear chromatin is slightly coarser than in the myeloblast, but nucleoli are still prominent. The cytoplasm is more abundant, and secondary (specific, myeloperoxidase negative) as well as primary granules start to appear. The loss of nucleoli and further chromatin condensation indicates further maturation to the myelocyte (size 12 to 20 µm), which makes up about 20% of marrow cells. The cytoplasm contains increased numbers of secondary granules and becomes less basophilic, ultimately ending pink in color.

Mitosis is possible up to the myelocyte stage, but further nuclear and cytoplasmic maturation occurs in nondividing cells. It is estimated that four to five mitoses occur between the myeloblast and the myelocyte stage.

Indentation of the nucleus (now kidney shaped) heralds further maturation to the metamyelocyte (size 14 to 20 µm), which makes up 25 to 30% of marrow cells. Secondary granules increase in prominence, but tertiary granules also start to appear within the pink cytoplasm. The nucleus, with coarse/condensed chromatin, now elongates, becoming horseshoe shaped, which is initially nonsegmented (often referred to as band neutrophils or stab neutrophils) before developing the typical lobulated appearance of the mature neutrophil (size 12 to 14 µm), which makes up about 30% of marrow cells.

Transmission-electron microscopic studies of myeloblasts show that their cytoplasm contains small mitochondria and initially a poorly developed Golgi apparatus. As the myeloblast develops, the Golgi apparatus becomes more active, producing round dense (primary) granules. Promyelocytes have a moderate number of mitochondria, a rough endoplasmic reticulum, and an active Golgi apparatus that now produces less-dense (secondary) granules. A third granule (tertiary) starts to appear after the myelocyte stage that is heterogeneous in size with intermediate density. As maturation to neutrophils occurs, the mitochondria and rough endoplasmic reticulum diminish in number, with the Golgi apparatus becoming less active. From the metamyelocyte stage onward, glycogen appears, which ultimately acts as an energy source for the mature neutrophil.

The release of neutrophils from the bone marrow seems to be under the influence of humoral factors, several of which have been identified in animal studies. Iron deficiency, endotoxin, etiocholanolene corticosteroids, and pharmacological doses of GM-CSF and G-CSF can all result in neutrophil release, but their physiological role (if any) is unclear. Glucocorticosteroids increase circulating neutrophils by allowing release of neutrophils from the marrow, demargination of neutrophils, and preventing from neutrophils entering tissues.

Eosinophils

Eosinophil cells are derived from the hematopoietic stem cell, with identical maturation stages to the neutrophil. The characteristic specific reddish-orange granules are identifiable from the promyelocytic stage.

Basophils

Basophil cells are derived from the hematopoietic stem cell with identical maturation stages to the neutrophil. The characteristic black/purple granules start to appear at the promyelocyte stage.

GRANULOCYTE COLONY STIMULATING FACTOR

See also **Colony stimulating factors**; **Hematopoietic regulation**.

(G-CSF) A glycoprotein acting as a **cytokine** in the late stages of myeloid cell differentiation. It is produced by **neutrophils**, **vascular endothelium** cells, fibroblasts, and **bone** marrow stromal cells after stimulation by endotoxin, tumor necrosis factor, interleukin-1, or granulocyte macrophage (GM)-CSF. A recombinant form can be given as a bolus to stimulate granulopoiesis after intense cytotoxic agent therapy, to aid in treatment of **myelodysplasia** and other causes of severe **neutropenia**, and to improve neutrophil function. A wide variety of **adverse drug reactions** have been reported, including musculo-skeletal pain, transient hypotension, liver enzyme disturbances, **thrombocytopenia**, and allergic reactions.

GRANULOCYTE/MACROPHAGE COLONY STIMULATING FACTOR

See also **Colony stimulating factors; Hematopoietic regulation**.
(GM-CSF) A **cytokine** acting at the intermediate stage of myeloid differentiation. **Adverse drug reactions** are common, and it is ineffective for congenital **neutropenia**. Any therapeutic advantage over G-CSF has yet to be determined.

GRANULOCYTE SARCOMA

See **Myeloid sarcoma**.

GRANULOCYTOSIS

See also **Neutrophilia; Eosinophilia; Basophilia**.
An increased level of **granulocytes** in the circulating peripheral blood above $8.00 \times 10^9/l$.

GRANULOMA

A pathological formation that arises when acute **inflammation** fails to destroy the noxious agent, e.g., bacteria, suture. These lesions must be distinguished from eosinophilic granulomas (see **Histiocytosis**). The causes are listed in Table 68.

TABLE 68

Causes of Granulomata Formation

Foreign body	Epithelioid
Beryllium	Bacteria
Zirconium	tuberculosis
Silica	leprosy
Talc	syphilis, i.e., gumma
Sutures	brucellosis
Freund's adjuvant	listeria
Asbestos	
Fungal	histoplasmosis
	blastomycosis
	coccidioidomycosis
	cryptococcus
Helminth	schistosomiasis
	filariasis
Unknown	sarcoidosis
	Crohn's disease
	Wegener's granulomatosis
	Hodgkin disease
	allopurinol
	primary biliary cirrhosis

Granulomata are 0.5- to 2-mm nodules of collections of transformed **histiocytes** (macrophages) termed epithelioid cells, since they resemble epithelial cells, often surrounded by **lymphocytes**, **plasma cells**, and giant (syncytial) cells. It is a form of Type IV **hypersensitivity** reaction in which the invading microorganisms/substance activate T-lymphocytes, which in turn release substances that transform macrophages into epithelioid cells. These cells have abundant rough **endoplasmic reticulum**, a prominent **Golgi apparatus**, and numerous cytoplasmic vesicles and vacuoles. The vacuoles contain mucopolysaccharide. Macrophages may fuse to form giant cells with up to 200 nuclei. When the nuclei are arranged peripherally, the term "Langerhans giant cell" is used. Alternatively, the term "foreign body giant cell" is appropriate when the particular foreign body, e.g., silica, is identified within the cytoplasm. Lymphocytes at the center of the granuloma are usually CD4 type with CD8 type at the periphery. The secondary products of the histiocytic response include:

Hydrolytic enzymes: lysozyme, collagenase, elastase, plasminogen activator, lysosomal acid hydrolases

Arachidonic acid metabolites: prostaglandins, leukotrienes

Complement components

Oxygen metabolites: hydrogen peroxide, superoxide, hydroxyl radicals

Cytokines: IL-1, interferons α and β

Immunogenic granuloma often result in the development of tissue necrosis, but peculiar to granulomata associated with **tuberculosis** is caseous ("cheesy") necrosis.

GRANULOPENIA
See also **Neutropenia**.
A decreased level of granulocytes in the circulating peripheral blood, below $2.5 \times 10^9/l$.

GRANULOPOIESIS
See **Granulocytes**.

GRAY-PLATELET SYNDROME
See **Storage pool disorders of platelets**.

GROWTH FACTORS
See **Cell-surface receptors; Cytokines**.

GYNECOLOGICAL DISORDERS
The interrelationship of hematological disorders with those affecting the female reproductive organs.

Iron deficiency resulting from menorrhagia of all causes

Polycythemia rubra vera in those with uterine fibroids

Hemostatic disorders causing menorrhagia

- Von Willebrand Disease
- Factor XI deficiency
- Thrombocytopenia

Vulval ulceration due to **agranulocytosis**, either primary or secondary to **cytotoxic agents**

Postpartum endometritis in **sickle cell disorders**

Impaired ovarian function from **radiotherapy** or cytotoxic agents for hematological malignancy

H

HAGEMAN FACTOR
See **Factor XII**.

HAIR DISORDERS
The changes in hair associated with hematological disorders:

Premature graying associated with **Addisonian pernicious anemia**, probably due to an associated autoimmune disorder

Diffuse alopecia with normal scalp skin associated with severe **iron deficiency**

Alopecia as a consequence of **cytotoxic agent** therapy for hematological malignancies

HAIRY CELL LEUKEMIA
(HCL) A **lymphoproliferative disorder** characterized by small B-cells (rarely T-cells) with oval, convoluted, or indented nuclei, chromatin slightly less clumped than that of a normal B-lymphocyte, and abundant pale cytoplasm with "hairy" projections on smear preparations. The cells may be phagocytic and usually contain acid phosphatase isoenzyme 5, which is resistant to tartaric acid (tartrate-resistant acid phosphatase). **Bone marrow** is always involved, the infiltrate being interstitial or diffuse and characterized by widely spaced small nuclei separated by clear areas of expanded cytoplasm, giving a halo or "fried egg" appearance. **Reticulin** is increased, often resulting in a "dry tap" on marrow aspiration. The diagnosis is best made on a marrow trephine biopsy. In cases of minimal involvement, immunostaining with anti-B-cell antibodies such as DBA44 or L26 may be useful. In the **spleen**, the tumor involves the red pulp, the white pulp being usually atrophic. Lymph node involvement is uncommon. The **immunophenotype** of the tumor cells is SIg$^+$ (M$^{+/-}$D, G, or A), B-cell-associated antigens$^+$ (CD19$^+$, CD20$^+$, CD22$^+$, and CD7A$^+$), CD5$^-$, CD10$^-$, CD23$^-$, CD11c$^+$ (strong), CD25$^+$ (strong), FMC7$^+$, and CD103$^+$ (MLA: mucosal lymphocyte antigen, recognized by HML-1, B-ly7, Ber-ACT8, LF61). Molecular genetic studies show Ig heavy and light chains rearranged with Ig-variable-region genes mutated. The most probable origin is a peripheral B-cell at an unknown stage of differentiation.

Clinical Features
Patients are adults with **splenomegaly** and those with pancytopenia with **monocytopenia**. A few circulating "hairy" lymphocytes may be seen in peripheral-blood films. Rare spontaneous remissions are reported. There is increased susceptibility to infections, especially in those with granulocytopenia or impaired-cell-mediated immunity.

Treatment

Interferon, deoxycoformycin, or **2′-chlorodeoxyadenosine** can induce long-term remissions. 2′-Chlorodeoxyadenosine is the treatment of choice, with very high remission rates and long survival times. The disorder does not respond to conventional lymphoma chemotherapy.

HAM'S TEST
See **Acid hemolysis test**.

HAND MIRROR CELL
The name given by Norberg in 1974 to **lymphoblasts** and **myeloblasts** presenting with a cytoplasmic protrusion giving the resemblance to the cells of a hand mirror. More commonly, hand mirror cells is the name given to lymphoblasts presenting in varying numbers in up to 20% of patients with **acute lymphoblastic leukemia**, where the cell axis included in the cytoplasmic protrusion doubles (at least) the wide axis. The significance of this particular configuration is ultimately unknown, although immune mechanisms promoting the survival of these cells have been argued, since their presence has been associated with a prolonged clinical course, primary refractoriness to treatment, and a higher incidence of central nervous system relapse. They are more frequently observed in female patients with normal platelet counts at the time of diagnosis.

HAND-SCHÜLLER-CHRISTIAN DISEASE
See **Histiocytosis**.

HAPTEN
See also **Antigen**.
Typically a small artificial molecule (e.g., dinitrophenol) bound to the surface of a protein and functioning as an antibody epitope. **Antibodies** can be generated to the hapten only in association with the carrier protein. The study of hapten and carrier functions was germinal in showing that both T- and B-lymphocytes were essential in antibody production, with experiments by Mitchison and others showing that the T-cell was carrier specific but that the hapten was a B-cell. The rationalization now is that the T-cell generates help in response to linear epitopes from the protein, while the B-cell generates antibody in response to the surface epitope, hapten.

HAPTOGLOBIN
See **Hemoglobin** — degradation.

HASSELL CORPUSCLES
See **Thymus**.

HEART DISORDERS
See **Cardiac disorders**.

HEAT INSTABILITY TEST

See **Hemoglobinopathies**.

HEAT STROKE

A disorder induced by exposure to high temperatures. As a consequence, **neutrophils** show "botyroid" nuclei due to contraction of **chromatin** microfilaments.

HEAVY METAL DISORDERS

See **Arsenic toxicity**; **Cobalt toxicity**; **Copper**; **Lead toxicity**; **Manganese**; **Mercury toxicity**.

HEAVY-CHAIN DISEASE

(HCD) **Lymphoproliferative disorders** of lymphoplasmacytic cells characterized by the presence of an M protein consisting of an incomplete heavy chain.[212,213] The three major types are γ, α, and μ.

γ-Heavy-Chain Disease (γ-HCD)

This lymphoplasmacytic cell proliferative disorder produces a monoclonal γ-heavy-chain in the serum or urine. The γ-chains have significant amino acid deletions. The median age is approximately 60 years, although it has been noted in young persons. γ-HCD usually presents with a lymphoma-like illness, but the clinical manifestations are diverse and range from an aggressive lymphoproliferative process to an asymptomatic state. Weakness, fatigue, and fever are common. **Hepatosplenomegaly** and **lymphadenopathy** occur in 60% of patients. A normochromic, **normocytic anemia** is found in 80%. The serum protein electrophoretic pattern is unimpressive, with a normal appearance, a broad-based increase, or a small localized band. The urinary heavy-chain protein concentration ranges from traces to 20 g/24 h. Bence-Jones proteinuria is not found. Examination of the **bone marrow** and **lymph nodes** reveals an increased number of lymphocytes, plasma cells, and lymphoplasmacytic cells. Osteolytic lesions are rare. Treatment with **cytotoxic agents** should be limited to symptomatic patients. Therapy with **cyclophosphamide**, **vincristine**, and **prednisone** or **prednisolone** is a reasonable approach. If there is no response, **doxorubicin** (adriamycin) should be added.

α-Heavy-Chain Disease (α-HCD)

This is the most common type of heavy-chain disease. It usually develops in the second or third decade of life and is characterized by severe malabsorption with loss of weight, diarrhea, and steatorrhea. The term "immunoproliferative small intestinal disease" (IPSID) is restricted to patients with small-intestinal lesions that have the same pathologic pattern as that of α-HCD, but with no identifiable a heavy chains. IgA linear deposition at the interface of skin and mouth epithelium with connective tissue may be present.

The serum electrophoretic pattern is normal or shows a broad band in the α2 or β regions. The amount of monoclonal α-chain in the urine is small, and Bence-Jones proteinuria is present. The bone marrow is also normal. The small bowel is infiltrated with mature plasma cells or lymphoplasmacytic cells. The mesenteric lymph nodes are usually affected, but the liver, spleen, and peripheral nodes are not involved. Treatment with antibiotics may produce benefit. If patients do not respond to antibiotics or if more advanced disease is present initially, they should be treated with cyclophosphamide, doxorubicin (Adriamycin), vincristine, and prednisone or prednisolone.

μ-Heavy-Chain Disease (μ-HCD)

This is characterized by the presence of a monoclonal μ-chain fragment in the serum. Most patients have a chronic lymphoproliferative disorder similar to **B-cell chronic lymphocytic leukemia.** The serum protein electrophoretic pattern shows a monoclonal spike in about 40% of patients. Two-thirds have Bence-Jones proteinuria, usually of the κ-type. The bone marrow reveals an increase in plasma cells, lymphocytes, and lymphoplasmacytic cells. Vacuolization of the plasma cells may be prominent. Treatment with alkylating agents and corticosteroids is often beneficial.

Cryoglobulinemia

This disorder is characterized by precipitation of protein when cooled and dissolving when heated.[214] It can be classified as:

Type I (monoclonal — IgG, IgM, IgA, or, rarely, monoclonal light chains)

Type II (mixed — two or more immunoglobulins, one of which is monoclonal)

Type III (polyclonal — no M protein is found)

Many patients with large amounts of type I cryoglobulin are completely asymptomatic, but if the cryoglobulin precipitates at higher temperatures, pain, purpura, Raynaud's phenomenon, and ulceration or sloughing of the skin may occur. Mixed cryoglobulinemia (type II) most commonly consists of monoclonal IgM and polyclonal IgG, but other combinations are seen. Patients with this type frequently have vasculitis, glomerulonephritis, lymphoproliferative disease, or chronic infectious processes. Hepatitis C has frequently been associated with mixed cryoglobulinemia. Type III cryoglobulinemia is often asymptomatic. Successful treatment has been claimed using **rituximab.**

Pyroglobulins

Pyroglobulins precipitate when heated to 56°C and do not dissolve when cooled. They are usually associated with **monoclonal gammopathy** of undetermined significance (MGUS), multiple myeloma, or macroglobulinemia. They are asymptomatic and are of no clinical consequence.

HEINZ BODIES

Small **red blood cell** inclusions seen as refractile objects in dry unstained peripheral blood smears, by dark-ground illumination or phase-contrast microscopy. They are more usually demonstrated by staining with methyl violet, cresyl violet, new methylene blue, brilliant cresyl blue, brilliant green, or rhodanile blue, where they appear as small basophilic inclusions beneath or adjacent to the red blood cell membrane. They are not visualized on Romanowsky-stained blood films but may be suspected by the presence of "bite" cells produced by splenic pitting. They are a late sign of oxidative damage and represent an end product of **hemoglobin** degradation.

TABLE 69

Disorders Causing Heinz Body Formation

Unstable hemoglobins
Glucose-6-phosphate dehydrogenase deficiency
Chemical toxic disorders
   Arsine (arsenic hydride)
   Benzene derivatives
   Heavy metals — **arsenic, copper, mercury, lead**
   Potassium chlorate
   Naphthalene
   Toluidine blue
   Trinitrotoluene
   Oxygen exposure to 100% > 4 h by peroxidation of membrane lipids
Adverse Reaction to Drugs (usually in association with G6PD deficiency)

Acetanilide	Pamaquine
Dapsone	Pentaquine
Daunorubicin	Quinidine
Diaminodiphenylsulfone	Sulfacetamide
Methylene blue	Sulfamethoxazole
Nalidixic acid	Sulfapyridine
Niridazole	Thiazolesulfone
Nitrofurantoin	

"Heinz body hemolytic anemia" is a generic term used to encompass all hemolytic disorders caused by oxidative damage to red blood cells. Because of their role in oxygen transport of hemoglobin, red blood cells are selectively menaced by redox intermediates. Protection is physiologically provided by **glutathione** enzymes and by membrane thiols. Glutathione stability depends upon the enzyme **glucose-6-phosphate dehydrogenase** (G6PD), and deficiency of this enzyme increases the risk of damage by oxidants. Oxidation of membrane thiols loosens the membrane skeletal structure, thereby disturbing the **cation pump**, which allows swelling of the cell and preferential clearance by the spleen. The second effect when the protective mechanism is overcome is a change in hemoglobin tetramers to mixed disulfides and sulfhemoglobins. Sustained oxidant effects result in the precipitation of aggregates of insoluble hemochromes — Heinz bodies. Within the cytosol, these are harmless, but many adhere to the inner aspect of the cell membrane, which is then deformed, with increased cell permeability giving rise to hemolysis. Deformability of red blood cells is reduced, making them more easily trapped in the spleen and hence removed, giving rise to a hemolytic anemia. Heinz bodies are found particularly with **hyposplenism** or postsplenectomy in the disorders listed in Table 69.

HELMET CELLS

See **Fragmented red blood cells**.

HELMINTH INFECTION DISORDERS

See **Metazoan infection disorders**.

HELPER CELL

See **Lymphocytes**.

HEMANGIOBLASTOMA

(Hemangioendotheliomas) Vascular tumors seen primarily in older children and young adults, frequently in the cerebellum, accounting for ≈2% of all intracranial tumors. Most are single, although they can be associated with retinal, renal, or pancreatic cysts. There may be a familial element, suggesting a genetic component to the disorder. Occasionally they may produce ectopic erythropoietin, leading to **erythrocytosis**. They can arise in the liver and migrate to bones, spleen, and subcutaneous and retroperitoneal tissues. The tumor cells secrete plasminogen activators. It is a highly invasive tumor, having a mortality of around 60%.

HEMANGIOENDOTHELIOMA

See **Hemangioblastoma**.

HEMANGIOMA

A group of congenital vascular anomalies that can occur in any organ or tissue. Most are harmless and involute within 2 to 4 years.

Capillary Hemangiomas

(Port wine stain; Nevus flammeus) Pink or deep burgundy vascular nevi present at birth and usually segmental. The majority persist for life and may increase in size with age. They may be cutaneous manifestations of other vascular anomalies in which the viscera as well as the skin are involved. Treatment is usually unsatisfactory. Plastic surgery may be successful but can be associated with extensive scarring. Laser treatment has produced encouraging results.

Cavernous Hemangiomas

(Strawberry nevi) Superficial soft vascular malformations, usually appearing within the first few months of life. They are almost never present at birth, although the affected area can be preceded by an area of pigmentation. The lesion can grow rapidly but will stabilize at 9 to 12 months of age. Gradual resolution by fibrosis usually occurs.

Kasabach-Merrit Syndrome

A rare disorder consisting of one or more giant hemangiomas in the skin or elsewhere, causing bleeding defects due to local consumption of clotting factors and platelets. Most of the hemangiomas are large and occur in the limbs alone or on part of the trunk, with involvement of the adjacent limbs. Large, giant hemangiomas are frequently associated with consumption coagulopathy (**disseminated intravascular coagulation**, DIC) due to the local fixation of platelets in the vessels of the mass, leading to the activation of the clotting cascade. The DIC can be very severe and is the most frequent cause of morbidity and mortality (intracranial bleeding) in these children. Strict monitoring of the clotting activity is therefore required.

Treatment varies with the clinical state. In the absence of relevant clotting abnormalities, observation with careful monitoring of **fibrinogen** and **D-dimer** is a cautious approach. Should there be any need for coagulation factor or platelet replacement, a curative approach should be tried, as factor replacement will only be effective for 12 or 24 h and will effectively increase the size of the hemangioma, as most of the factors will be deposited and consumed in the tumor mass.

A variety of eradicative approaches have been tried. Steroids (mainly dexamethasone) for 2 to 4 weeks at cytostatic doses have been reported to reduce the size of the hemangioma, although cessation of steroids may be associated with rapid regrowth, and further courses of treatment may be necessary. More recently, **vincristine** (standard doses, every 2 to 4 weeks for a total of four to six courses) has been used. Vincristine has been shown to reduce the size of the mass without later regrowth, and its platelet-genetic effect also prevents transfusions. The use of **aspirin** or **dipyridamole** to reduce platelet consumption has also been reported, and this can be tried when a conservative approach has been decided. Other reports abrogate the use of tranexamic acid (see **Fibrinolysis**), as cases of localized thrombosis with later mass reduction have been described.

HEMANGIOSARCOMA

Rare chemically induced angioformative tumors of the liver associated with previous exposure to vinyl chloride, thorotrast, and inorganic arsenic.

HEMAPHERESIS

The separation and planned removal of a specific blood component (or components) from a patient or volunteer donor with the immediate return of all other blood components. The term is derived from *heme* and *apheresis* "to take away" (Greek and Latin). Hemapheresis is performed routinely using automated blood cell separators.[215] Disposable plastic harnesses installed on the machine connect the donor **venepuncture** site, bags of anticoagulant and saline solutions used to prime the machine, and a cell-separation chamber. Blood is normally anticoagulated with citrate. The whole blood/anticoagulant ratio is about 10:1. Centrifugal systems used are either a spinning bowl (Latham bowl, Haemonetics Corp., Braintree, MA; see Figure 49), a circular plastic separation channel (Gambro BCT, Lakewood, NJ; Dideco, Mirandola, Italy; Fresenius Kabi AG, Bad Homburg, Germany), or two rotating separation chambers (Baxter Healthcare Corp., Round Lake, IL). The same principle applies to all types — that of centrifugal separation of blood components. The manner in which each is harvested varies from machine to machine. Some machines operate by continuous flow so that two venepunctures are required. Other intermittent-flow centrifugal machines process blood in "batch fashion," drawing blood into the machine from one venepuncture site, harvesting the specified component(s), and returning the residue (usually red blood cells) through the same needle.

There are a number of adverse effects that are common to all forms of hemapheresis. These are given in Table 70.

Plasmapheresis

Donor Plasmapheresis

This is used mainly for the harvesting of **fresh-frozen plasma** (FFP). Plasma may be recovered from the differential centrifugation of units of whole blood, but additional plasma is also collected by automated apheresis. Volunteer apheresis donors should be fit and have ordinarily given one or two uneventful blood donations. Selection criteria vary from country to country, but usually the donors should be between 18 and 60 years old. They should not be <50 kg in weight. A donor should not undergo a total of more than 24 plasma/plateletpheresis procedures per annum, including not more than 12 leukopheresis

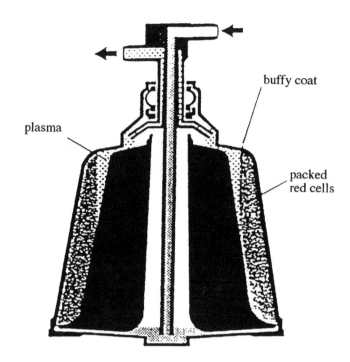

FIGURE 49

The Latham bowl. Whole blood is pumped down the feed tube and enters the bowl at the bottom. Centrifugal force spins the dense cellular components to the outside, leaving platelet-rich plasma (PRP) or plasma in the inner band. As the bowl fills, plasma or PRP flows out of the effluent tube into the collection bag. (Reproduced by permission of Haemonetics (UK) Ltd.)

TABLE 70

Adverse Effects of Apheresis Donation

Dizziness and fainting
Hematoma formation
Citrate toxicity
Red blood cell loss if procedure aborted
Chilling during reinfusion on the return cycle

procedures per annum. There should normally be a minimum of 48 h between procedures, and a donor should not normally undergo more than two procedures within a 7-day period. Not more than 15 l of plasma should be donated by one donor in a year. Not more than 2.4 l of plasma should be donated by one donor in any 1-month period. After a whole-blood donation, or the loss of an equivalent number of red cells during an apheresis procedure, a donor should not normally donate plasma, platelets, or leukocytes for a period of 8 weeks. The possible adverse effects of donation are given in Table 70.

Therapeutic Plasmapheresis

These procedures are performed mainly for the removal of unwanted plasma constituents. In principle, they are similar to donor apheresis except that several liters of plasma are removed and replaced during the procedure with albumin and saline. The procedure is termed plasma exchange (PEX). Patients with problems of venous access may require preliminary insertion of a central line. PEX provides an immediate way of removing unwanted substances present in the plasma or substances bound to plasma proteins. These include:

Paraproteinemia immunoglobulins present in **myelomatosis** and **lymphoplasma-cytic lymphoma/Waldenström macroglobulinemia** that may cause **hyperviscosity** syndrome and impairment of renal function. These patients often have a significant anemia and can be transfused once their viscosity has been lowered to <4.0 mPa.s. Some patients with acute oliguric renal failure associated with myeloma kidney improve if plasma exchange is instituted early. PEX also plays a significant role in the treatment of acute nephritis, e.g., **Goodpasture's disease**, vasculitis in **Wegener's granulomatosis**, and microscopic polyarteritis.

Autoantibodies such as Guillain-Barré syndrome, where autoantibodies damage the outer myelin sheath of nerves, thereby causing severe paralysis and sensory loss or myasthenia gravis in patients whose clinical condition is difficult to control by conventional methods. It is also effective for chronic inflammatory demyelinating polyneuropathy.

Cryoglobulinemia (see Heavy-chain disease)

Lipids, e.g., removal of cholesterol in familial hypercholesterolemia.

Toxins or drugs.

Thrombotic thrombocytopenic purpura (TTP). PEX provides a convenient way to infuse large volumes of FFP.

Many patients who undergo therapeutic plasma exchange are unwell, such as ventilated patients with Guillain-Barré syndrome who in addition have disturbed autonomic nervous function and therefore fluctuations in blood pressure, which may in turn prevent adequate blood flow. Febrile, urticarial, or anaphylactic reactions may occur when FFP is used as replacement fluid. Associated problems with a central line include infection, bleeding, and pneumothorax.

Plateletpheresis

Donor Plateletpheresis

This is performed by a variety of procedures:

Platelet-rich plasma (PRP) collection. This is performed on intermittent-flow centrifugal cell separators. Centrifugal forces allow separation of whole blood into packed red cells, buffy coat, and supernatant PRP (see Figure 50). This is harvested in three to five cycles, to a maximum volume of 600 ml. PRP requires secondary centrifugation in the laboratory to leave a platelet pellet, which is then resuspended in approximately 150 ml of residual plasma. This process yields 1.5 to 2 $\times 10^{11}$ platelets; two units constitute an adult therapeutic dose.

Surge plateletpheresis. Blood components are separated as described in Figure 50. The centrifugal speed is higher so that platelets and leukocytes lie close to the bulk of red cells. When the platelet layer nears the exit point of the bowl, blood flow from the donor temporarily ceases, and platelet-poor plasma collected during the filling phase of the bowl is recirculated through it at 200 ml/min. The surge of plasma elutriates the platelet layer, which passes into the collection bag. Collection ceases before excessive numbers of white cells are collected. Optimizing the amount of plasma collected and performing about eight machine cycles yields about 4×10^{11} platelets.

Continuous flow apheresis. Cell separators continually extract platelets from a dynamic interface between red cells and plasma into a collection bag along with a predetermined volume of plasma. Much higher yields (6.0×10^{11}) may be obtained if the duration time is 90 to 120 min. Many centers routinely divide these products equally into two adult therapeutic doses, e.g., a double donor product (DDP). Leukocyte contamination is lower using continuous-flow cell separators than during surge plateletpheresis or PRP collection, e.g., 10^6 vs. 10^7 or 10^8 white blood cells per product.

Therapeutic Thrombocytopheresis

Therapeutic apheresis for the removal of platelets is very rarely performed. Gross overproduction of platelets occurs in **essential thrombocythemia** (ET) and **chronic myelogenous leukemia** (CML), where the platelet count may exceed $1000 \times 10^9/l$. Although such patients are usually treated with chemotherapy, platelet removal as an emergency procedure at the time of diagnosis may help reverse vascular insufficiency caused by thrombotic vents.

Leukapheresis

Donor Leukapheresis

This procedure is occasionally performed to collect large numbers of granulocytes for transfusion to septic neutropenic patients with proven infection unresponsive to appropriate antibiotics.

Therapeutic Leukapheresis

This is performed in the following circumstances:

Leukocytosis in **myeloproliferative** or **lymphoproliferative disorders**. Microaggregates of leukocytes form in the circulation at counts $>50 \times 10^9/l$ and can cause neurologic, pulmonary, and other complications. Rapid reduction of the leukocyte count by automated apheresis is effective in this situation and is most commonly performed in CML. It may also be indicated occasionally in patients with **acute leukemias**. The leukocytes removed from patients with CML can be cryopreserved as a source of hematopoietic stem cells for subsequent autotransplant.

Peripheral-blood stem cell (PBSC) collection. This procedure offers an alternative to bone marrow stem cells in **autologous stem cell transplantation** for patients with **Hodgkin disease**, **non-Hodgkin lymphoma**, **myelomatosis**, **acute lymphoblastic leukemia** (ALL), **acute myeloid leukemia** (AML), CML, and certain adult and pediatric solid tumors. They are capable of long-term hematopoietic reconstitution. Recently, PBSCs have been used as an alternative to **allogeneic stem cell transplantation**. They are capable of inducing long-term pediatric solid tumors. Ordinarily, hematopoietic stem cells capable of long-term bone marrow repopulation are found only in small numbers in the peripheral blood. Their numbers may be increased by maneuvers such as exercise, steroid administration, adrenalin, and growth factors. The number of circulating stem cells is increased following bone marrow recovery after chemotherapy and following the administration of cytokines, such as **granulocyte colony stimulating factor**. Chemotherapy and cytokine treatment may be combined to cause a considerable expansion of PBSCs. Usually one to four apheresis procedures are performed, each lasting 3 to 4 h, depending on the amount of prior chemotherapy and the efficacy of the mobilizing regimen. Advantages of PBSC transplantation include reduced likelihood of tumor cell contamination and more rapid hematopoietic and lymphoid reconstitution. Acceptable yields of PBSC are defined as follows:

- Mononuclear cells 3×10^8/kg recipient body weight
- CFU-GM 10×10^4/kg recipient body weight
- CD34$^+$ 2×10^6/kg recipient body weight

HEMARTHROSES

Bleeding into joints, which is common in severe **hemophilia A**. The most frequently affected joints are knees, elbows, ankles, shoulders, hips, and wrists. The first episode usually occurs in childhood. Hemarthrosis is usually spontaneous or associated with minimal trauma. Initial symptoms are vague warmth/tingling progressing through discomfort and limitation of movement to pain, overt swelling, cutaneous warmth, and severe limitation of movement. Once bleeding has stopped the blood is slowly resorbed, and the joint returns toward normal over a period of days to weeks. Hemarthrosis stimulates synovial proliferation and chronic inflammation. Chronic synovitis predisposes to further episodes of hemarthrosis, ensuing progressive joint damage, chronic arthritis, and deformity. Hemarthrosis should be treated at the earliest opportunity to prevent joint damage.

HEMATIN

Oxidation product of **heme**. It is seen in **red blood cells** invaded by **malaria** plasmodia.

HEMATOCRIT

See also **Packed-cell volume**.

The relative volume occupied by red blood cells in capillary or venous samples of whole blood. There is considerable variation in values between venous and capillary blood and particularly in the spleen, where it can be as high as 0.8 l/l.

HEMATOLOGY

The study of hematopoietic cells and their related components and products both in health and disease. The subject has largely developed over the latter half of the 20th century but, apart from intense scientific investigation, blood has always had emotional connotations, so that ethical considerations remain in the background of any advance in diagnosis or treatment.

The ancient Greeks considered that excess of blood, plethora, was the cause of all ills, and in support of this belief they introduced the practice of bloodletting supported by starvation to induce anemia. The introduction of the microscope in the 17th century allowed the study of circulating blood cells, and interest arose in their possible function. In the early 19th century, studies were made on the causes of anemia and its deleterious effects on the patient. This was followed later in that century by descriptions of eponymous blood diseases such as **Addisonian pernicious anemia**, **Hodgkin disease**, and **Henoch-Schönlein purpura**. Therapy remained palliative until the early 20th century, when Castle first treated patients with Addisonian pernicious anemia by giving them raw liver followed later by liver extract. Measurement of hemoglobin and microscopic examination of peripheral-blood cells hence became essential to medical practice as a means of diagnosis.

Hematology subsequently developed as a branch of clinical pathology performed by physicians in a side room of their hospital ward or clinic. The widespread requirement for patients' blood counting was followed by coagulation-factor screening and pretransfusion blood testing, leading to the opening of central hematology laboratories. Out of this came considerable laboratory research into the causes of anemia, the identification of blood coagulation disorders, and blood transfusion serology. The study of the lymphoid system, antigens, and antibodies slowly led to the development of the separate but related discipline of immunology. The advent of **cytotoxic agent** therapy for leukemia and lymphomas moved the emphasis of hematology back to the clinician, so that, in many centers, hematology became a combined clinical and laboratory discipline. Research into the genetics of hemoglobin and immunoglobulins paved the way for deeper knowledge of genetics, culminating in the definition of the human genome.

Hematological studies are made at all periods of life — the embryo, fetus, neonate, infant, child, and adult, including the changes induced by **pregnancy**. General physiological matters include **blood volume, viscosity**, and **rheology**. These normal physical, chemical, and biological processes are affected by variations arising from the effects of posture, exercise, and altitude. Some disorders are more specific to a particular age or state, e.g., fetal disorders, pregnancy. Overall, most hematological disorders can affect any age and can therefore be considered as individual entities.

Disturbances of the hematopoietic system affect most organs of the body, and in so doing their own characteristics may change, e.g., **monocytes** in the peripheral blood becoming **histiocytes** or tissue macrophages. Disorders of each specific organ may therefore produce hematological effects, and equally, hematological disorders may have interactions with specific organs or systems.

The genetic basis of hematology is fundamental, and most disorders can have a genetic or hereditary basis. Well-recognized examples are **hemophilia, sickle cell disease**, and **thalassemia**. Environmental and lifestyle factors equally can affect the hematological system. **Physically induced disorders** include the effects of **ionizing radiation, traumatic disorders**, particularly **burns**, and **hemorrhage** from accidents. **Chemical toxic disorders**, particularly **heavy metal disorders** and **lead toxicity**, have effects on all blood cells and on the **bone marrow**. Excessive exposure by abuse of **tobacco** and **alcohol** has effects on both **hematopoiesis** and **hemostasis**. Common causes of **bone marrow hypoplasia** are **adverse drug reactions**, and these may induce **hemolysis**, particularly in those with deficiency of the red cell enzyme **glucose 6-phosphate dehydrogenase**. Drug **hypersensitivity** may cause **vasculitis** and **purpura**. Biological toxins, such as those of **snake venom disorders**, affect the **coagulation/fibrinolysis** mechanisms.

Infection frequently induces hematological change, e.g., **bacterial disorders, viral disorders, fungal disorders, metazoan disorders**, and **protozoan disorders**. Certain infections produce specific effects, e.g., **Epstein-Barr virus** causing **infectious mononucleosis; parvovirus** causing **pure red cell aplasia; human immunodeficiency virus** leading to **acquired immunodeficiency syndrome**; plasmodial invasion of red blood cells occurring in **malaria**.

Deficiency disorders are common, particularly **iron deficiency** anemia, with **megaloblastosis** from **cobalamin** or **folic acid** deficiency being less frequent.

Autoantibodies can damage red blood cells (**autoimmune hemolytic anemias**) and cause gastric **intrinsic-factor** deficiency with failure of cobalamin absorption, leading to megaloblastosis. Many immunological disorders such as **rheumatoid arthritis** or systemic **lupus erythematosus** have concomitant hematological effects.

Oncogenesis results in **myelodysplasia, acute leukemias, myeloproliferative disorders**, and **lymphoproliferative disorders**.

The worldwide prevalence of blood disorders is difficult to determine, the available statistics being for specific disorders, whose incidence varies considerably with region and ethnicity. Malaria is the world's commonest infectious disease, with around 300 million cases. Human immunodeficiency virus disease is estimated at over 20 million persons infected, but only 1 million reside in the developed countries of Australasia, Europe, or North America; most of those infected will in time progress to acquired immunodeficiency syndrome. Cardiovascular disease, particularly **thrombosis** in those over 65 years of age, accounts for about half of deaths in Europe and North America. It has been estimated that around 7 million persons or 3% of the population in the U.S. have some form of athero-sclerotic disorder.[216] The Framingham study has shown that 2 per 1000 between 35 and 65 years of age will develop stroke or some form of transient ischemic attack, the incidence rising to 11/1000 for those 65 to 94 years of age.[217] In the U.S., **venous thromboembolism** is given as the reason for admission to hospital of 260,000 patients each year, the true incidence being much higher, as many patients are treated outside hospital.[218] Anemia due to malnutrition is common in Africa and to a lesser extent in South Asia; iron deficiency anemia is the commonest hematological disorder in Europe and North America, particu-larly in women during the childbearing years. Sickle cell disease in some form affects 45% of the population in West Africa, with similar large numbers in Saudi Arabia and east-central India; 10% of the American black population are affected,[219] compared with 14,000 persons affected in the U.K. Thalassemia is the most common worldwide genetic disorder, with a high frequency in Mediterranean countries, India, Burma, and Southeast Asia. In some Asian and black populations the α-thalassemia gene frequency exceeds that of the normal genotype. This is probably the result of selection due to adaptation to plasmodial infection. In black Americans the α-thalassemia trait is common but rarely of clinical significance. β-Thalassemia has a prevalence of 10% in black Americans, of whom 1 to 2% will have the homozygous form with anemia.[220]

In comparison, hematological malignancy is rare.[221,222,223] The American Cancer Society figures for 1994 give an age-adjusted incidence for leukemia of 5/100,000, similar figures being given by the number of patients admitted to National Health Services (NHS) hos-pitals in the U.K.[224] The U.K. Leukemia Research Fund data collection survey from January 1984 to December 1988 gave estimates of 1/100,000 for **chronic myelogenous leukemia**,[225] 15/100,000 for **chronic lymphatic leukemia**,[226] and 6/100,000 for **myelomatosis**.[227] The International Agency for Research on Cancer for the years 1978 to 1982 gave figures of 2.38/100,000 for Hodgkin disease[228] and 10/100,000 for **non-Hodgkin lymphoma**.[229] Wide regional variations have been reported, there being low incidences in Japan and India apart from the **adult T-cell leukemia** in Kyushu Island of Japan. Even lower-incidence figures are quoted for hemorrhagic disease. In the U.S., 1/5000 males have hemophilia, 85% having hemophilia A.[230] All other hematological disorders are rare.

The study and practice of hematology is at present broadly divided into hematopoiesis and oncology, hemostasiology, transfusion medicine, and immunohematology.

> *Hematopoietic system and oncology.* This includes the structure of bone marrow, hemato-poiesis for each individual cell line (**red blood cells**, **neutrophils**, **eosinophils**, **basophils**, **mast cells**, **monocytes**, and **histiocytes** [tissue macrophages] with their products and interacting substances [**hemoglobin**, **cytokines**]), together with pro-cedures involved in the diagnosis and treatment of their disorders.

> *Hemostasiology.* This discipline includes the processes of blood coagulation and their disorders, including the function of **platelets**; plasma **coagulation factors**, with their inhibitors and interactions; and **fibrinolysis**, together with the structure and function of **vascular endothelium**. The diagnosis and treatment of **hemorrhagic diseases** and thrombosis is the clinical component of the specialty.

> *Transfusion medicine.* This is the study of **blood group** antigens and antibodies (blood group serology) and of **blood components for transfusion**.

Immunohematology. This is the study of **lymphopoiesis** and the **lymphoid system**, including its products (**immunoglobulins**, cytokines) and disorders. Those disorders due to autoimmunity may be specific to certain organs such as the thyroid, pancreas, or joints, but some are general involving a variety of tissues, e.g., systemic lupus erythematosus. Their clinical study can therefore be dispersed to organ-specific specialties such as endocrinology and rheumatology. Lymphoid oncology for therapeutic reasons is merged with other hematological malignancies.

The organization of practice varies widely, dependent upon the economic circumstances of the geographical area. In the poorer regions, it is limited to hemoglobin measurement and examination of a **peripheral-blood film**, with treatment by hematinics. In the economically advantaged countries, the patient attends a hospital clinic or a specialist's office for investigation and diagnosis, with **point-of-care testing** by the physician. Tests in these clinics include measurement of hemoglobin, **red blood cell counts** with indices, **white blood cell** and **platelet counts**, peripheral-blood-film examination, and **erythrocyte sedimentation rate** (ESR). Patients with hematological malignancy, immunodeficiency, or a hemorrhagic disease are in some regions referred to specialist centers, but in others the hematologist remains both the laboratory pathologist and the physician. There is an increasing practice for clinical laboratories to join with clinical chemistry and to be overseen by a clinical scientist. The range of investigation between clinics, general hematology laboratories, and specialist centers varies greatly. Some specialist centers act as reference laboratories for specialized techniques such as **cytogenetic analysis, immunophenotyping, ferrokinetics**, immunoserology, **red blood cell enzyme deficiencies**, or coagulation-factor assays. Bone marrow trephine biopsies and lymph node biopsies are usually referred for examination by a histopathologist. Hematological oncologists carry out specific treatments, including cytotoxic agent therapy for leukemias and lymphomas, and **stem cell transplantation** from bone marrow or peripheral blood. Specialists in transfusion medicine perform **hemapheresis** and match donor/recipient for **human leukocyte antigens** prior to all forms of transplantation. Patients with hemophilia, **thrombophilia**, and other uncommon coagulation deficiencies also attend specialist centers for investigation and treatment. Patients requiring **splenectomy** are transferred to a surgeon and those requiring **radiotherapy** to a radiation oncologist.

Hematology continues to evolve. Increasing automation and point-of-care testing will provide quicker and more reliable results, thereby changing the management of laboratory practice. Molecular biology is providing greater diagnostic and prognostic accuracy in hematological malignancy, and new therapeutic maneuvers such as **gene therapy** are developing. This is having the effect of bringing immunology and hematology ever closer together as specialties. The introduction of blood substitutes, when fully developed, will have a marked effect on transfusion medicine. Unfortunately, alongside better knowledge of the subject, new disorders are emerging. These are particularly the consequence of mutations in viruses that can infect *homo sapiens*, particular examples being the HIV viruses and **prion infections**, both of which can be transmissible by transfusion from donors. New infective agents continue to develop, particularly in Africa and Asia, posing the threat of worldwide pandemics. Such a development would have great consequences for the practice of hematology.

Hematology, though better understood, is becoming such a complex discipline that no single hematologist can be competent in all aspects. However, whether practicing clinical or laboratory hematology or confined to a specialist area, all hematologists will continue to require a general and basic knowledge of the subject.

HEMATOPOIESIS

(Hemopoiesis) The formation of blood cells at the various stages of maturation and their ontology.

Embryonic Hematopoiesis

Hematopoiesis and vasculogenesis in the mammalian embryo begin in the blood islands of the yolk sac and continue, somewhat later, within the embryo proper.[231] Blood islands are formed from mesodermal aggregates that have migrated from the primitive streak. The outer cells differentiate into endothelial cells and the inner to primitive blood cells. The close developmental association between hematopoietic and endothelial cell lineages has led to a hypothesis that they share a common progenitor, the hemangioblast. The mechanisms that control formation of hemangioblast and embryonic hematopoietic and endothelial (angioblastic) stem/progenitor cells are still not well understood. Formation of these cell types from nascent mesoderm requires signals from an adjacent outer layer of primitive (visceral) endoderm. Indian hedgehog (Ihh), a member of the hedgehog family of extracellular morphogens, is secreted by visceral endoderm and alone is sufficient to induce hematopoiesis and vasculogenesis in explanted embryos. Murine gene-targeting studies indicate involvement of hedgehog **cell signal transduction** in these processes *in vivo*, but suggest that additional molecules (perhaps, for example, Wnt proteins) are required for induction and patterning of hematopoietic and vascular mesoderm.

The earliest detectable human hematopoietic tissue, termed hemocytoblasts (stem cells), is found about the third week in the embryonic yolk sac. Erythropoiesis takes place here within the primitive circulation (intravascular) and is megaloblastic in type. Multipotent progenitors derived from the mesoderm are present at early stages of gestation. In mice, embryonic stem cells obtained from the blastocyst can be maintained in an undifferentiated state by leukemia inhibitory factor (LIF) and differentiate upon withdrawal of LIF. This has provided an excellent model for the study of early hematopoiesis, at least in the murine system.

During vertebrate embryogenesis, hematopoiesis occurs in different sites (see Figure 51). This reflects the different demands of the developing organism. The earliest site is the yolk sac, which contains mainly erythrocytes, macrophages, and megakaryocytes. These cells are different from their adult counterparts. Erythrocytes are still nucleated; macrophages bypass the monocyte stage and synthesize a different set of enzymes; and megakaryocytes show an accelerated production of platelets and reduced ploidy. These differences enable the developing embryo to ensure a necessary quick supply of oxygen, the clearance of dead cells, and protection from possible infection and bleeding. No hematopoietic stem cell (HSC) repopulating activity is seen at this stage following transplantation of yolk sac cells in lethally irradiated adult recipient mice. However, HSCs, capable of multilineage long-term reconstitution, are present when yolk sac cells are transplanted in conditioned newborn recipients, suggesting that yolk sac HSCs have special microenvironmental requirements for graft survival.

A second wave of hematopoietic development occurs at the dorsal level, which contains the aorta and the developing urogenital system, in the aorta–gonado–mesonephros (AGM) region. Clusters of hematopoietic cells are associated with the endothelium of the ventral wall of the dorsal aorta. These cells present multipotent potential for erythroid, myeloid, and lymphoid lineages and are capable of multilineage engraftment in adult recipients. The presence of hematopoiesis is only transient in the AGM and is mainly represented by HSCs. Soon after their appearance in the AGM region, cells with repopulating activity in

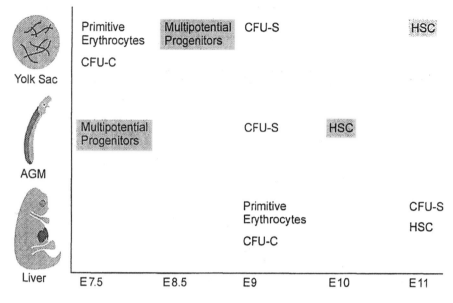

FIGURE 50

Temporal appearance of different types of assayable hematopoietic cells: colony forming units-culture (CFU-C), multipotent progenitors, colony forming units-spleen (CFU-S), and hematopoietic stem cells (HSC) in the yolk sac (top), para-aortic splanchnopleure/aorta-gonad-mesonephros (PAS/AGM) region (middle), and fetal liver (bottom). Note that the more-differentiated progenitors can be detected at earlier stages than the hematopoietic stem cells. The lineage relationships between these cells remains to be established. Also, note that multipotential progenitors and HSCs are detected 1 day earlier in the PAS/AGM than in the yolk sac. (From Dzerziak, E. and Oostendorp, R., Hematopoietic stem cell development in mammals, in *Hematopoiesis: A Developmental Approach*, Zon, L.I., Ed., Oxford University Press, Oxford, 2001, p. 211. With permission.)

adult recipients are seen both in the yolk sac and in the liver. For many years, investigators have believed that yolk sac blood islands contained HSCs capable of primitive hemato-poiesis and of migration to the developing liver to initiate definitive hematopoiesis. Challenging the idea of a singular origin of hematopoiesis in the yolk sac, it has been proposed that there is a more potent intraembryonic HSC site in the AGM region. HSCs arise for the first time in the AGM region and migrate to the yolk sac and fetal liver, the main source of hematopoietic cells in fetal life. Around the time of birth, HSCs migrate from the liver to the bone marrow, to be responsible for adult hematopoiesis (see Figure 52). However, this sequential migration has not been directly demonstrated *in vivo* in mammalian species. Instead, evidence for this view is based on the quantitative temporal measurement of hematopoietic activity in the yolk sac and liver (using adult recipients) following establishment in the AGM region. Migration is inferred because at this stage precursors become detectable in significant numbers in the blood stream. In this model, the presence of an HSC pool in the yolk sac, with ability to engraft in neonatal hosts before the onset of the AGM region, is not taken into consideration. Figure 50 and Figure 51 summarize the current view of embryonic hematopoiesis.

It is possible that HSCs arise from at least two independent sites and that endothelial cells of both the dorsal aorta and of the yolk sac can be the direct precursors of hemato-poietic cells. Indeed, close juxtaposition and temporally parallel onset of the endothelial and hematopoietic lineages are seen in the yolk sac and AGM regions. The idea of a common precursor, the "hemangioblast," for the endothelial and the hematopoietic lineage first emerged some years ago, and it has been revived in the last decade. Evidence that emphasizes the close relationship between those two lineages includes the observation that genes affecting primitive hematopoiesis encode receptor tyrosine kinases, such as

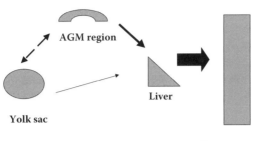

FIGURE 51
Migration of progenitors that leads to definitive hematopoiesis.

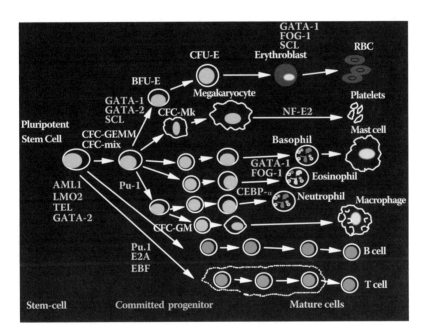

FIGURE 52
Schematic view of transcription factors and hematopoiesis.

Flk-1, that are involved directly in endothelial-cell proliferation and angiogenesis. Furthermore, these cells express markers that are common to both endothelial and hematopoietic stem cells, e.g., CD31, CD34, c-kit, and VE-cadherin endothelial/hematopoietic cluster marker. Recent studies have used sorted cell populations with endothelial markers from murine yolk sac and AGM. Those cells formed blood cells in culture in the presence of stromal cells. This has been confirmed in humans, and vascular endothelial cells isolated from fetal liver and fetal bone marrow have been shown to also be capable of multilineage hematopoiesis.

Considerable attention has been focused on the mechanisms that regulate the induction, differentiation, and maintenance of the hematopoietic system during development (see Figure 52), but there are similarities and important differences with adult hematopoiesis. Hematopoietic **cytokines** and their associated **cell signal transduction** pathways have been well studied, but less is known about hematopoietic functions of other classes of signaling molecules. The Notch, Wnt, and Hedgehog pathways play important roles in a variety of developmental processes and have also been shown to control self-renewal or differentiation of HSCs and more-committed progenitors.

Another group of molecules important for the development of hematopoiesis are the on-bone morphogenetic protein (BMP) members. These are part of the transforming growth factor-β (TGF-β) superfamily of extracellular signaling molecules. Although BMPs were originally discovered on the basis of their ability to induce ectopic bone formation, it rapidly became evident that these peptides have much broader functions during development. Genetic studies have demonstrated that BMP-4, in particular, plays critical roles in formation and patterning of mesoderm. Hematopoietic specification of ventral mesoderm is sensitive to the concentration of BMP-4, with embryonic globin expression occurring within a narrow range. BMP-4 synergizes with vascular endothelial growth factor (VEGF) in the generation of HSCs from embryoid bodies, pointing to a possible role for it at the level of hemangioblast specification. b-Fibroblast growth factor (b-FGF) has been implicated in commitment to hemangioblast development.

Thrombopoietin, recently shown to play an important role in maintenance and proliferation of the HSC and in yolk sac hematopoiesis, synergizes with, and can actually replace VEGF, indicative of an important role in hemangioblast development. Other molecules involved in fetal hematopoietic commitment decisions include the Wnt pathway.

By analyzing embryonic stem (ES) cell lines carrying various mutations and using colony assays to determine the growth factor requirements of ES cells as they differentiate from a pluripotent to differentiated state, it has been possible to dissect some aspects of the genetic regulation of hematopoietic commitment.

The basic helix-loop-helix transcription factor stem cell leukemia gene SCL is expressed in all embryonic hematopoietic sites. It is absolutely required for embryonic hematopoiesis; in addition, it is expressed in the embryonic vasculature and is required for proper vascular development, being critical for the development of the hemangioblast.

Flk-1 is a receptor tyrosine kinase that is activated by VEGF. Loss of Flk-1 blocks endothelial development and day-8.5 yolk sac hematopoietic development. While SCL is important for the specification of hemangioblasts, Flk-1 enables migration of the hemangioblasts to sites that would allow their survival and proliferation. The precise relationship and cross talks between SCL and Flk-1 remain to be elucidated.

Another genetic regulator, *Runx1*, also known as *Cbfa2* or *AML1*, is strongly expressed at all hematopoietic sites of the day-8.5 embryo, but its expression is maintained strongly only in intraembryonic sites. Mutation of *Runx1* blocks definitive but not primitive hematopoiesis, leading to embryonic death by day 12.5.

In serum-free, chemically defined medium, activin A and BMP-4 are able to induce dorsal or ventral mesoderm formation in ES cells, respectively. Recently, BMP-4 has been shown to enhance the self-renewal of the earliest hematopoietic progenitors. A hemangioblast colony assay has not yet been documented for differentiating human ES cells.

Colony forming unit-granulocytes/macrophages (CFU-GM) is detected as early as 5 weeks of gestation. Erythroid progenitors are also present in the yolk sac at this stage. Initially, only nucleated erythroid cells are morphologically identifiable within the yolk sac, but lymphoid cells and megakaryocytes appear later at this stage. Yolk sac erythropoiesis has ceased by week 10 of gestation.

Fetal Hematopoiesis

Hematopoiesis changes throughout an individual's life — from embryonic through fetal life and childhood before finally reaching adult maturation — with regard to both its site and cellular composition (Figure 53). In erythropoiesis, there is also a specific change in globin-chain synthesis from embryonic life onward (Figure 54).

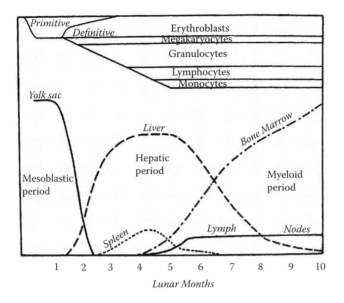

FIGURE 53
Stages of hematopoiesis in the embryo and fetus, indicating the comparative participation of the chief centers of hematopoiesis and the approximate times at which the different types of cells make their appearance. (From Rothstein, G., in *Wintrobe's Clinical Hematology*, Lee, G.R. et al., Eds., Williams & Wilkins, Baltimore, 1993, p. 80. With permission.)

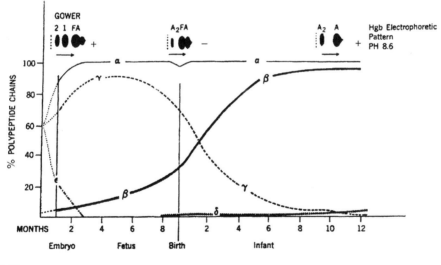

FIGURE 54
Proportions of the various human hemoglobin polypeptide chains through early life. The hemoglobin electrophoretic pattern (Gower 2, Gower 1, HbF, HbA, HbA2) typical for each period is also shown. (From Pearson, H.A., *J. Pediatr.*, 69, 473, 1966. With permission.)

Fetal Liver and Spleen Hematopoiesis

At about week 6 of gestation, erythropoiesis begins in the fetal liver extravascularly, with mature cells entering the fetal circulation. Erythropoiesis is also detectable in the spleen by week 12, this remaining the primary site of erythropoiesis until week 24. Circulating platelets are detectable at 8 to 9 weeks.

Fetal Bone Marrow Hematopoiesis

This commences at around week 16 to 18, as fetal liver hematopoiesis is challenged by hepatocyte proliferation, and assumes the primary role from week 24 onward. Mature **neutrophils** first appear in the peripheral blood at this time.

Control of Fetal Erythropoiesis

Fetal primitive multipotent progenitors (murine CFU-S and CFU-GEMM) proliferate more rapidly than newborn or adult. Fetal liver CFU-Ss have greater marrow repopulating ability than adult CFU-Ss. Committed fetal progenitors show a similar proliferative pattern, both in the myeloid and erythroid lineages, although CFU-GM show reduced sensitivity to GM-CSF and are reduced in number compared with adults. Erythroid progenitors predominate in fetal marrow in contrast to adults. Fetal BFU-E produces larger colonies and shows increased erythropoietin (EPO) sensitivity and earlier maximal colony growth in culture than newborn or adult human BFU-E. The proportional synthesis of γ-globin within the BFU-E colonies is also twofold greater in fetus than newborn. Maximal BFU-E numbers are seen early in the second trimester, and at mid-gestation the BFU-E number is three times that of newborns and ten times that of adults.

EPO production is low up to 20 weeks, then remains constant throughout gestation, with values well below adult serum concentrations (mean 1.6 ± 2.5 mIU/ml).

Hematopoietic ontogenic control mechanisms have been little studied in humans,[232] but early murine embryonic data have demonstrated expression of erythroid control genes (EPO-receptor, GATA-1, and α-globin) in day-6.5-postconception embryos (mesodermal tissue) prior to the morphological identification of erythroid cells. EPO and EPO-r are also expressed in embryonic stem cells *in vitro*, derived from the blastocyst. EPO expression is not detected *in vivo* until the yolk sac stage. The homeobox genes *HOX-2.2* and *HOX-2.3* are also expressed at day 6.5 postconception and are part of a family of genes encoding DNA-binding proteins with a major role in embryogenesis. *HOX-2.2* and *HOX-2.3* are particularly associated with erythroid development and differentiation.

Globin Switching

(See Figure 55 and Table 71.) An orderly sequence of production of different globin proteins occurs during fetal development in response to changes in requirements for red blood cell oxygen-carrying capacity. The earliest globin chains detectable in the embryo yolk sac are zeta (ζ), an α-type chain with locus on chromosome 16, and epsilon (ε), a β-type chain with locus on chromosome 11. The earliest fetal hemoglobin is thus HbGower1 (ζ2ε2), and it is the major form at 5 to 6 weeks. HbGower2 (α2ε2) is present from 4 to 13 weeks of gestation. HbPortland (ζ2γ2) also persists from 4 to 13 weeks but is found in infants with homozygous α-thalassemia. Synthesis of ζ- and ε-chains ceases at the time the liver takes over from the yolk sac as the site of erythropoiesis. At that time, α- and γ-chains become dominant. HbF (α2γ2) is the major fetal form and accounts for 90 to 95% of the total hemoglobin until 34 to 36 weeks gestation. Adult hemoglobin (HbA; α2/β2) is

TABLE 71

Temporal Expression of Globin Chains

ζ2ε2	Hb Gower 1	embryonic
α2ε2	Hb Gower 2	embryonic
ζ2γ2	Hb Portland	embryonic
α2γ2	HbF	fetal
α2δ2	HbA2	adult
α2β2	HbA	adult

detectable from week 11 of gestation, after which time the proportion of HbA increases as HbF declines. The amount of HbF in neonates varies from 50 to 90%, but thereafter declines at a rate of 3% per week and is generally less than 2 to 3% by the age of 6 months.

Gene switching at the β-globin locus is accomplished at the transcriptional level and is regulated at the level of chromatin structural changes, exposing DNAse I-hypersensitive sites within the long-terminal-repeat regions (LTRs) located 5′ to the ε-globin gene-promoter regions. Silencing of γ-globin transcription is accomplished by transacting factors and not, as originally thought, by direct competition for the β-globin gene. These silencing factors progressively silence the influence of the locus control region (LCR) upon e, Gg, and Ag gene transcription as human embryonic erythropoiesis develops from the yolk sac through fetal liver to bone marrow.

Increased proportions of HbF occur in infants small for gestational age who have experienced severe fetal anoxia, who have trisomy 13, and in infants dying from sudden infant death syndrome. Persistence of embryonic hemoglobins has been reported in some infants with developmental abnormalities. Decreased levels are found in trisomy 21.

Fetal Blood Groups

These are the same as those of adults, apart from the I blood group, where "i" antigen predominates on fetal red cells to be replaced by "I" antigen on adult red cells. There are no blood group antibodies in the absence of immunoglobulin formation.

Fetal Blood Cell Values

Hemoglobin concentration rises from a mean of 11.7 g/dl at 18 weeks to 13.6 g/dl at >30 weeks, with a steady rise in hematocrit (0.37 l/l to 0.43 l/l) and concomitant fall in mean cell volume (131 fl to 114 fl). Circulating normoblasts constitute 45% of nucleated cells at 18 weeks, falling to 17% at >30 weeks. Lymphocyte percentage falls from 88% to 68%, with neutrophils only rising significantly after 30 weeks (8% at 26 to 29 weeks to 23% at >30 weeks). Eosinophil, monocyte, and basophil percentages remain reasonably constant throughout. Platelet concentration also remains constant.

Adult Hematopoiesis

Adult hematopoiesis is the process by which HSCs divide and differentiate to maintain a supply of mature blood cells so as not to exhaust the HSC compartment within the lifetime of the individual. HSCs are pluripotent cells able to give rise to at least ten different functional cell types (neutrophil, **monocytes/histiocytes** (macrophages), **basophils, eosinophils**, erythrocytes (**red blood cells**), **platelets, mast cells, dendritic reticulum cells**, and B and T-lymphocyte**s**). Two types of stem cells have been defined. The long-term repopulating cells (LTRC) are capable of producing all blood cell types for the entire life span of the individual and of generating progeny that display similar potentiality on secondary transplant. The short-term repopulating cells (STRCs) reconstitute myeloid and lymphoid compartments for a short period of time. The process by which stem cells give rise to terminally differentiated cells occurs through a variety of committed progenitor cells, often overlapping in their hematopoietic capacity. During commitment, cells can undergo extensive proliferation and sequential differentiation, accompanied by a decrease in self-renewal capability to produce mature cells. The primary function of this transit population is to increase the number of mature cells produced by each stem cell division.

The clonal succession model proposed by Kay in 1965 suggests that one or a small number of HSC clones give rise to mature blood cells as needed, and the remaining HSCs remain quiescent and do not contribute to hematopoiesis until the proliferative capacity of the already engaged stem cells has been exhausted. This hypothesis has been supported by data from retrovirally marked donor-transplant studies in lethally irradiated mice, which have indicated that only one or very few clones contribute to hematopoiesis at any given time. Furthermore, the data from clonal studies of human hematopoiesis in the elderly would support this evaluation. On the other hand, studies that have analyzed steady-state hematopoiesis in an alternative model using bromo-2'-deoxyuridine (BrdU) incorporation kinetics suggest that up to 8% of HSCs enter the **cell cycle** per day, and although at any given time over 75% of HSCs are in G0, all HSCs are recruited into the cell cycle on average every 57 days.

Hematopoiesis is regulated by a complex interaction of secreted **cytokines**, stromal cell interactions, which in turn regulate transcription factors and the cell cycle machinery. The HSCs divide to contribute to hematopoiesis either in a stochastic process or are directed by microenvironmental cues to differentiate down particular lineages.

Hematopoietic stem cells are defined by four distinguishing features:

Self-renewal, defined as the production of exact duplicates with the maintenance of all attributes of the original

Pluripotentiality, enabling differentiation into all mature hematopoietic lineages

Quiescence, such that at a given time point the majority of stem cells will be in G0

Expression of p glycoprotein-like pumps that extrude dyes such as rhodamine-123

In normal steady-state hematopoiesis, the size of the stem cell population is maintained at a constant level by the balance of stem cell production by cell division and stem cell loss via differentiation. This is a tightly regulated process of controlled self-renewal together with the provision of differentiated cells to meet demand, but with considerable capacity to expand the stem cell population when necessary, for example following myelo-suppressive chemotherapy or infections.

The concept of an undifferentiated stem cell giving rise to the spectrum of blood cells via an intermediate state of progenitor cells was postulated by Pappenheim as early as 1917. In the 1950s, Miklem and coworkers demonstrated the existence of the hematopoietic stem cell in the bone marrow by the rescue of irradiated recipients by bone marrow injection in mice. Till and McCulloch later characterized such stem cells following the discovery that murine bone marrow contained single cells that could give rise to myelo-erythroid colonies in the spleen of a transplant recipient. In these experiments, random chromosome markers were produced by irradiating donor bone marrow. Following trans-plantation into conditioned recipients, colonies of daughter cells, each derived from a single clonogenic precursor, were found in the spleen. These colonies were shown to contain differentiated myeloerythroid cells together with more primitive cells that could themselves both self-renew and differentiate. They had the ability to produce more spleen colonies and to reconstitute hematopoiesis in lethally irradiated secondary recipients. This observation formed the basis of the widely used "colony forming unit-spleen" (CFU-S) quantitative assay of stem cell activity.

The teams of Bradley and Metcalf and of Pluznik and Sachs independently performed experiments seeding adult murine spleen cells onto a soft agar medium in the presence of a feeder layer. These produced clones of cells constituting two types of hematopoietic colonies that could be analyzed morphologically as the **colony forming units**.

Further research into hematopoietic stem cell biology required the development of techniques for cell purification and the refinement of **hematopoietic stem cell assays** capable of investigating properties of multilineage differentiation as well as self-renewal. It became apparent that day-8 CFU-Ss were not in fact formed by HSC but by more-mature

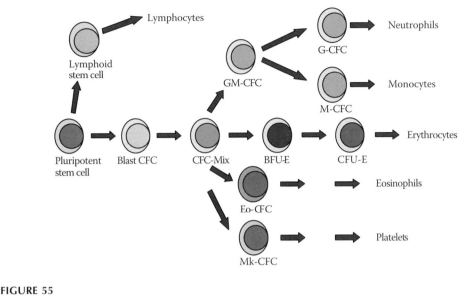

FIGURE 55
Proposed hierarchy of hematopoietic colony forming potential.

committed progenitors, and more-primitive day-12 CFU-Ss were described, with the ability to rescue irradiated recipients. *In vitro* clonogenic assays were also refined. It was demonstrated that the colony forming cells (CFCs) could be subdivided into different classes according to the differentiated progeny they produced in response to various known growth factors. These included the multipotent CFC-Mix, which could produce all of the different types of myeloid cells but not T- and B-lymphocytes. In turn, these underwent differentiation to produce bipotent and unipotent progenitors, such as the granulocyte and macrophage CFCs, eosinophil CFCs, erythroid progenitors called burst-forming units-erythroid (BFUs-E), and the more mature colony forming units-erythroid (CFUs-E) that were able to respond to erythropoietin. This hierarchy of hematopoietic progenitors is shown in Figure 55.

Though stem cell assays provided a means of identifying the hematopoietic progenitor subpopulations, their purification and detailed characterization were greatly facilitated by the development of the fluorescence-activated cell sorter (FACS). This provided a rapid means of subdividing cellular populations according to their innate size and granularity profile, together with their expression of specific cell-surface markers. (See **Ineffective erythropoiesis**; **Normoblast**; **Reticulocyte**.)

Measurement of Erythropoiesis

Normal red blood cell production is extremely effective, and most red blood cells live, or have the potential to live, a normal life span. Under certain conditions, however, a fraction of red blood cell production is ineffective, with destruction of nonviable red blood cells either within the marrow or shortly after the cells reach the blood. Effective erythropoiesis is most simply estimated by determining the reticulocyte count. This count is usually expressed as the percentage of red blood cells that are reticulocytes, but it can also be expressed as the total number of circulating reticulocytes per liter of blood (reticulocyte % × RBC per l). An elevated reticulocyte count may give an erroneous impression of the actual rate of red cell production because of premature release of reticulocytes into the circulation. To correct for this premature release, some workers calculate a reticulocyte index to compensate (see **Reticulocyte count**).

Ineffective Red Blood Cell Production

Ineffective erythropoiesis is suspected when the reticulocyte count is low or is normal or only slightly increased in the presence of erythroid hyperplasia in the bone marrow. In certain disorders, such as **Addisonian pernicious anemia**, **thalassemia**, and **sideroblastic anemia**, ineffective erythropoiesis is a major component of total erythropoiesis. This can be quantified by **ferrokinetics**. Using ferrokinetic methods, ineffective erythropoiesis is calculated as the difference between total plasma iron turnover and erythrocyte iron turnover plus storage iron turnover.

Total Erythropoiesis

This is the sum of effective and ineffective red cell production and can be estimated from marrow examination by first determining the relative content of fat and hematopoietic tissue. The myeloid/erythroid ratio is then determined. These, taken in conjunction with the red cell count and the reticulocyte count, will usually provide quantitative information on the rate and effectiveness of red blood cell production.

Erythropoiesis can be demonstrated by imaging marrow, liver, and spleen with ^{99m}Tc sulfur colloid or ^{111}indium, even though these isotopes primarily label the monocyte-macrophage system. Their uptake is similar to ^{59}Fe, and they can be used to demonstrate erythroid tissue, but accurate quantitation of total erythropoiesis is made by measuring the rate of red blood cell production (see **Ferrokinetics**).

The identification and cloning of an array of growth factors whose activities *in vivo* and *in vitro* have marked effects on the growth and function of hematopoietic progenitors was enabled by the combination of hematopoietic assays of protein purification from conditioned media and application of cDNA cloning.

HEMATOPOIETIC REGULATION

The maintenance of hematopoiesis in steady state by a balance of negative and positive **cytokine** regulators (Table 72). A variety of the cytokines that include the hematopoietic

TABLE 72

Hematopoietic Growth Factors

Factor	Major Target Cell or Precursor
EPO	erythrocytes
G-CSF	neutrophil, also precursors of myeloid lineage
GM-CSF	erythrocyte, neutrophil, eosinophil, basophil, monocyte, megakaryocyte
IL-1a and b	primitive precursor cells — lymphocyte activating factor
IL-2	T-cells
IL-3	neutrophil, eosinophil, basophil, monocyte
IL-4	T-cells, B-cells
	cofactor in granulopoiesis
IL-5	B-cells, eosinophils
IL-6	B-cells and precursors, neutrophils, monocytes, megakaryocytes, early precursor cells
IL-7	pre-B-cells, pre-T-cells
IL-9	erythroid precursors
IL-11	B-cells, megakaryocytes, mast cells
IL-12	early precursors, NK lymphocytes
M-CSF	monocytes
SCF	stem cells
TPO	megakaryocytes

Note: See "Abbreviations" in the front of this volume; for synonyms, see **Cytokines**.

TABLE 73

Hematopoietic Regulators

Cell Lineage	Transcriptional Regulators	Cytokine Regulators
s	Pu.1; CEBP-u.1; CEBP-α; CEBP-ε	G-CSF; GM-CSF; IL-3; M-CSF; SCF; IL-6
Macrophage	Pu.1	GM-CSF; IL-3; M-CSF
Eosinophil	Pu.1; GATA-1; Fog-1	IL-5; GM-CSF; IL-3
Mast cell	Pu.1; GATA-1; Fog-1	SCF; IL-3; IL-9; TPO
Megakaryocyte	GATA-1; Fog-1; GATA-2; SCL; NF-E2	TPO; IL-3; LIF; SCF; IL-6; IL-11; EPO
Erythroid	GATA-1; Fog-1; GATA-2; SCL	EPO; IL-3; SCF
T-cell	Ikaros; Ets1; GATA-3; NFATc; TCF1; LEF1; sox4; NF-kB; LKLF	IL-2 SCF; IL-7; IL-12; FL
B-cell	Ikaros; EBF; E2A; RAG1; RAG2; Pax-5; Vav	IL-7 SCF; IL-5; IL-12; FL

growth factors (HGFs) (see Table 73), stem cell factor (SCF), flt3 ligand (FL), thrombopoietin (TPO), **interleukin**-3 (IL-3), granulocyte/macrophage colony stimulating factor (GM-CSF), and IL-6 have been shown, in various combinations, to promote the growth and differentiation of hematopoietic stem cells (HSCs) (see Table 73 and Table 74). The role of these cytokines is largely as survival factors for HSCs, and their role in the *in vitro* self-renewal of HSCs remains controversial. This has important clinical implications, since it is unlikely that the current repertoire of cytokines will result in stem cell expansion for therapeutic purposes. In contrast, TGF-β1 inhibits growth of HSCs. This 25-kDa protein is produced by the BM stroma as well as progenitors and therefore regulates HSCs in a paracrine/autocrine manner. It has been shown that antisense or antibody inhibition of TGF-β1 releases stem cells from quiescence. TGF-β1 induces quiescence through the p21/27 pathways.

HSC proliferation is intimately linked to the stromal cells and extracellular matrix (ECM) in distinct microenvironmental niches (see **Bone marrow**). The ECM is composed of a variety of molecules, including fibronectin, laminin, **collagens**, and proteoglycans. Some components of the ECM bind to cytokines produced by the stroma, immobilizing them within the microenvironmental niches and thus creating zone in which HSCs and cytokines can coalesce. More recently, it has been shown that mouse HSC cells (specifically long-term HSC cells) are tethered to N-cadherin-expressing, spindle-shaped osteoblastic cells lining trabecular bone. Consequently, increasing trabecular bone surfaces increases the number of niches and HSC cells to fill them. Osteoblast activity and trabecular bone is regulated by parathyroid hormone (PTH), which enlarges the HSC pool by:

Increasing the amount of trabecular bone and with it the available niche space

Stimulating bone-lining cells to make large amounts of a ligand called Jagged1, which activates the Notch receptors on the attached HSC cells

Directly stimulating the HSC/CFU-S cells to start replicating DNA

Without endogenous PTH, the HSC/CFU-S pool shrinks as the cells terminally differentiate and hematopoiesis declines. There is also evidence that key regulators of **angiogenesis** also regulate the **bone marrow** (BM) microenvironmental niche. HSCs expressing the receptor **tyrosine kinase** Tie2 adhere to osteoblasts in the BM niche. The interaction of Tie2 and its ligand angiopoietin-1 (Ang-1) leads to tight adhesion of HSCs to stromal cells, resulting in maintenance of long-term repopulating activity of HSCs. Thus, Tie2/Ang-1 signaling pathway plays a critical role in the maintenance of HSCs in a quiescent state in the BM niche (see Figure 56).

Other signaling pathways are also intimately linked with HSC self renewal (see **Cell signal transduction**).

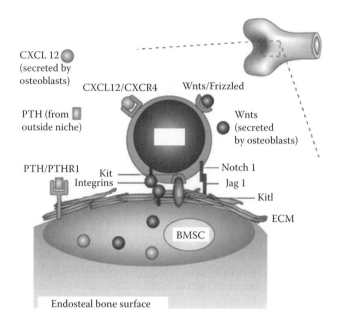

FIGURE 56

Molecular regulation of hematopoietic stem-cell niche. The functional interactions between stem cells and their niches have been conserved from flies to mammals. Within the extracellular matrix (ECM) at the endosteal surface of the bone marrow cavity, osteoblasts and other bone marrow stromal cells (BMSCs) regulate hemato-poietic stem cells (HSCs) through segregated signals as well as by cell-cell and cell-matrix interactions. Chemokines such as CXCL12 provide a signal to recruit CXCR4-expressing HSCs to the appropriate niche, whereas ECM components interact with HSC-expressed integrins to retain the stem cells. Niche cells also provide hematopoietic cytokines such as Kitl. Further regulation of HSCs by the HSC niche depends upon activation of Notch signaling by Jagged ligands (Jag1) and Wnt signaling by secreted Wnt ligands. The HSC niche itself is regulated by paracrine factors, such as parathyroid hormone (PTH) and its receptor (PTHR1), that can alter the cellular composition and size of the niche, and thereby regulate HSC numbers. (From Eckfeldt, C.E., Mondenhall, E.M., and Verfaille, C.M., The molecular repertoire of the "almighty" stem cell, *Nat. Rev.: Molecular Cell Biol.*, 6, 726–737, 2005. With permission.)

In addition to external factors, HSCs are also regulated by a complex network of **transcription regulation** factors that include *c-Myb*, GATA family of transcription factors, *AML1/Runx*, SCL, *Hox* family of transcription factors, and *Bmi-1*. The relationship between the external factors such as growth factors, distinct signaling pathways, and downstream transcription machinery remains to be elucidated. A significant amount of information regarding hemato-poietic lineage commitment is derived from knockout-mouse models. While these models have given us insights into the critical role of these transcription factors in the early stages of embryonic development, they offer limited insights into their role in established adult hematopoiesis. This is highlighted by observations of GATA2 knockout mice, which are embryonically lethal at day 11 with a defect in HSC maintenance. However, the specific role in established hematopoiesis is controversial, with some experiments suggesting a role in HSC growth suppression and others indicating enhancement. Improved models of adult hematopoiesis are required to define the role of transcription factors in hematopoiesis. Nonetheless, Table 73 and Figure 52 summarize our current state of understanding.

HEMATOPOIETIC STEM CELL ASSAYS

Assays developed to evaluate *in vitro* the self-renewal and differentiation potential of human hematopoietic stem cells (HSCs). The basic colony forming cell (CFC) assay remains an important measurement of progenitor cells, and a more primitive progenitor population known as high proliferative potential CFC (HPP-CFC) has been described using the same semisolid culture method. This HPP-CFC population has been described in human bone marrow with a frequency of 2 per 10^5 mononuclear cells. In the presence of stem cell factor and IL-3, HPP-CFC demonstrated fivefold expansion and replating capability suggestive of self-renewal. HPCs have also been demonstrated *in vitro* in "cobblestone assays," where bone marrow cells are added to preformed stromal layers, producing a cobblestone appearance of adherent hematopoietic colonies consisting of undifferentiated blast cells. Replating of these colonies produced CFU-Mix, CFU-GM, and BFU-E, but only limited self-renewal.

Another widely used assay, one that reproduces the interrelationships between HPC and the stromal cells of the **bone marrow** microenvironment, is the long-term culture initiating cell (LTC-IC) assay. This assay detects primitive cells that can give rise to CFC, which is released into the culture supernatant for many weeks when cultured on supportive marrow stromal layers. Such an assay can therefore separately and quantitatively assess LTC-IC self-renewal versus altered output of clonogenic progeny. LTC-IC represents a biologically heterogeneous population with differing proliferative and multipotency characteristics. In mice, they are able to reconstitute hematopoiesis, but in the human setting, adaptations of this assay have demonstrated that single lineage-negative CD34+ (Lin⁻CD34+) cells can differentiate *in vitro* into cells with myeloid, natural killer (NK), B-cell, or T-cell phenotypes, but these are unable to self-renew. Supplementing stromal-based culture with early-acting cytokines has defined a more-primitive human progenitor, the myeloid-lymphoid initiating cells that can generate secondary progenitors that do have the ability to reinitiate long-term hematopoiesis.

However, there is no definite proof that *in vitro*-defined human progenitors correlate with LTC-ICs, and *in vitro* assays cannot evaluate homing ability. The most compelling evidence for stem cell activity remains the ability of cells to reconstitute the hematopoietic system of patients following myeloablative chemotherapy or radiation. Surrogate, *in vivo* animal models have therefore been developed using xenogenic transplant recipients. One such model measures the ability of human HSC to initiate multilineage hematopoiesis in the bone marrow of severely combined immunodeficient (SCID) mice. Intravenous injection of human bone marrow or cord blood resulted in engraftment of primitive cells that proliferated and differentiated in the murine bone marrow, producing large numbers of LTC-ICs, CFCs, and immature CD34+CD38⁻ cells, as well as mature myeloid, erythroid, and lymphoid cells. The numbers of CFCs and LTC-ICs fell soon after transplant but expanded significantly over the next 4 weeks, implying that they were being produced from more-primitive cells, termed SCID repopulating cells (SRCs). This assay was subsequently refined with the creation of a new mouse strain, crossing the SCID gene onto the nonobese diabetic (NOD) background. These NOD/SCID mice had reduced immunological function, enabling engraftment of lower cell doses, thereby making it possible to test HSC purification strategies.

Another *in vivo* assay involves transplantation of selected human HSC into fetal sheep, which lack the immunological capacity to reject them. The pre-immune status of the early-gestational-age sheep fetus permits the long-term engraftment of human HSCs in the absence of irradiation or other myeloablative therapies. The human HSCs engraft host marrow and persist for long periods, showing multilineage expression and responsiveness to human cytokines. This assay is relatively specific for stem cells: only the CD38⁻ fraction of CD34+ cells exhibited long-term persistence of human cells and were able to secondarily engraft a further recipient, indicating the presence of long-term repopulating cells. This assay has the advantage of prolonged follow-up study; however, high cost precludes its widespread use.

Murine Hematopoietic Stem Cell Surface Markers

Methods of excluding cells bearing lineage markers of mature lymphocytes, granulocytes, monocytes/macrophages, erythroid cells, natural killer cells, etc., play an important role in the enrichment for progenitor cells. Using a cocktail of antibodies detecting these lineage markers, followed by lineage-positive cell removal by bead adsorption or flow cytometric sorting, murine bone marrow can be effectively divided into lineage-positive (Lin$^+$) and lineage-negative (Lin$^-$) fractions. Adding analysis for the expression of Thy-1 or Sca-1, which are cell-surface glycophosphatidyl inositol-linked immunoglobulin superfamily molecules, refines further for HSC. In mice, the Sca-1$^+$Lin$^-$Thy-1^{lo} cell subset is enriched for all clonogenic assays and for radioprotective cells, while the Sca-1$^-$ subset was enriched for day-8 CFU-S, which represented more-committed myeloid progenitors. The total population of Sca-1$^+$Lin$^-$Thy-1^{lo} cells represents approximately 1 per 2000 cells in the bone marrow. In competitive repopulation assays, irradiated mice injected with just a single Sca-1$^+$Lin$^-$Thy-1^{lo} cell, together with 2×10^5 host marrow cells, show multilineage reconstitution. Sca-1 positivity is now included in the majority of purification protocols for murine HSC. The function of the Sca-1 antigen is unclear, but it appears to play a role in lymphocyte activation and possibly in the activation of hematopoietic stem cells. However, Sca-1 knockout mice are healthy, with normal numbers and percentages of all hematopoietic lineages.

More recently, it was recognized that the tyrosine kinase receptor c-kit is expressed on primitive HSC and progenitor cells, providing an additional marker. In combination with lineage depletion and Sca-1 expression, such cells possessed day-12 CFU-S and pre-CFU-S activity, could form colonies on stromal layer culture, and had the capability to rescue irradiated transplant recipients, giving rise to long-term multilineage repopulation. The c-kit$^+$Thy-1^{lo}Lin$^{-/lo}$Sca-1$^+$ (KTLS) phenotype, representing approximately 0.05% of murine bone marrow cells, is now widely used to define bone marrow HSC in mice. However, they may be less useful in the setting of G-CSF-induced mobilization of progenitors, as c-kit expression has been found to be selectively reduced on Lin$^-$Sca-1$^+$c-kit$^+$ cells, but not on Lin$^-$Sca-1$^-$c-kit$^+$ cells in the bone marrow following G-CSF treatment.

Rhodamine efflux capability may also be added to divide c-kit$^+$Lin$^-$Sca-1$^+$ cells into more-primitive Rho^{lo} and more-mature Rho^{hi} subsets. Although both populations confer similar levels of radioprotection and have similar content of day-12 CFU-S, most if not all long-term repopulating cells are found in the Rho^{lo}c-kit$^+$Lin$^-$Sca-1$^+$ fraction. Rho^{hi}c-kit$^+$Lin$^-$Sca-1$^+$ cells displayed *in vivo* repopulation kinetics resemble those of the short-term repopulating cells. The c-kit$^+$Lin$^-$Sca-1$^-$ cells were largely Rho^{hi} and could be stimulated to form colonies and clusters *in vitro* in the presence of single cytokines, which is a characteristic of committed progenitor cells. Wheat-germ-agglutinin-positive Rho^{lo} cells have been further divided by a second rhodamine incubation in the presence of verapamil, an inhibitor that blocks the multiple drug resistance (MDR) efflux pump. Rho^{lo}/Rho(VP)$^-$ cells exhibited better long-term repopulating activity (LTRA), whereas Rho^{lo}/Rho(VP)$^+$ cells demonstrated better short-term repopulating activity (STRA). However, transplanting just 30 Rho^{lo}/Rho(VP)$^-$ cells could produce sufficient cells to rescue 60% of lethally irradiated recipients; therefore they did possess some STRA. It was unclear whether these purified cell fractions contained homogeneous populations of cells with both LTRA and STRA or mixed populations of cells with only LTRA or STRA.

Human HSCs are frequently characterized as CD34-positive, CD38-negative. Whereas fetal murine progenitors are also CD34-positive (Ito T Exp Haem 2000), in normal adult murine bone marrow the reverse appears to be true, with the majority of stem cells expressing CD38, but not CD34. Interestingly, further experiments have demonstrated that the expression of these two antigens appears to be reciprocal and alters according to the activation status of the cells. For example, cells from chemotherapy-treated mice or progenitors mobilized into the peripheral blood by G-CSF proved to be primarily CD34$^+$CD38$^-$.

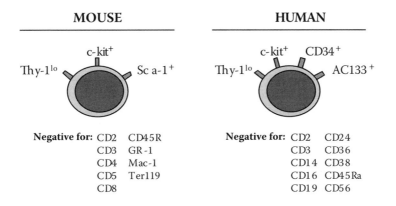

FIGURE 57
Surface phenotype of mouse and human hematopoietic stem cells.

Human Hematopoietic Stem Cell Markers

Attempts to purify human HSC have involved a variety of techniques, including density centrifugation, cell-cycle status, and dye efflux. However, the most frequently utilized method involves the characterization of cell-surface antigen expression. Studies on these markers have shown that human HSC shares some of the cell markers found on murine HSC (Figure 57). Additionally, as in mice, human HSCs do not express many of the lineage markers characteristic of terminally differentiating hematopoietic cells. Removal of such lineage-positive cells therefore enriches for more-primitive cells. Recognition that the sialomucin CD34 could be used as a positive selection marker for HSC was a significant advance. In clinical practice, it has become the basis of the enumeration, isolation, and manipulation of human HSC. Transplant experiments in species including baboons and mice have demonstrated long-term repopulation by CD34+-selected cells. However, the physiological function of the CD34 molecule remains unclear, with potential roles in cell adhesion and homing.

CD34 is not specific for HSC, as it is also expressed on clonogenic progenitors and some lineage-committed cells. Both human and murine CD34 are also expressed outside the hematopoietic system on vascular endothelial cells and some fibroblasts. The frequency of CD34+ cells has been estimated at 1 to 4% of nucleated cells in adult bone marrow and <1% in umbilical cord blood. However, isolated CD34+ cells are heterogeneous, and the stem cell content constitutes only a small fraction. CD38 is a transmembrane glycoprotein expressed on B and T-lymphocytes, NK-cells, and myeloid cells at various stages of development. In contrast to CD34 expression, increases with differentiation and the most primitive human progenitors, as demonstrated by CFC, LTC-IC, and xenogenic transplantation assays, were found in the subset (1%) of CD34+ cells that were CD38−. Hence the surface phenotype of human HSC was refined as lineage-negative CD34+CD38−.

Patterns of expression of other markers, including Thy-1, c-kit, HLA-DR, and CD71, have all distinguished populations enriched in progenitors. More recently, a novel five-transmembrane antigen recognized by the monoclonal antibody AC133 has been reported. CD34 and AC133 antigens are coexpressed on primitive cells from human bone marrow, mobilized peripheral blood, and umbilical cord blood.

Despite this experimental evidence, recent studies have appeared to confound the assumption that HSC must be CD34-positive. There is minimal clonogenic CFC or LTC-IC activity in the human Lin−CD34− population, suggesting limited stem cell properties; however, xenogenic transplant assays have told a different story. Using the fetal sheep model, Lin−CD34− cells proved capable of long-term repopulation and multilineage engraftment *in vivo*. These cells could reconstitute secondary recipients, demonstrating their extensive self-renewal potential. Large numbers of CD34$^+$ cells were found in CD34$^-$ recipients, suggesting that the Lin$^-$CD34$^-$ fraction was more primitive than CD34$^+$ cells. Both CD34$^+$ and CD34$^-$ cells from primary/secondary hosts could engraft secondary/ tertiary recipients, indicating that CD34 expression on human HSC is reversible. Further reports of CD34$^-$ human repopulating cells utilized the NOD/SCID murine assay. Transplantation of human Lin$^-$CD34$^-$ cell subpopulations into NOD/SCID mice demonstrated repopulating activity comparable with that seen for CD34$^+$ SRCs. The repopulating activity of the Lin$^-$CD34$^-$ cells appears to reside in the CD38$^-$AC133$^+$ fraction.

Further data regarding CD34-negative HSC populations has been provided by the identification of "side population" (SP) cells. Experiments using the fluorescent vital dye Hoechst 33342 as an indicator of DNA content, and hence of cell cycle in murine bone marrow, have isolated a distinct subpopulation of cells with a high level of dye-efflux activity. These SP cells were highly enriched for, but not entirely composed of, KTLS cells, lacked murine CD34 expression, and were capable of long-term marrow repopulating activity. This group subsequently demonstrated the existence of SP cells in the human, rhesus monkey, and pig hematopoietic systems; in each species, the cells expressed low or undetectable amounts of CD34.

Further dye-efflux studies have suggested that human SP cells may possess a more heterogeneous surface phenotype than murine SP cells. Indeed, CD34 expression has now been demonstrated within this subpopulation in umbilical cord blood. Functional assays on adult human CD34$^-$CD38$^-$ SP cells have so far failed to detect CFC or LTC-IC, and engraftment has not been obtained in NOD/SCID mice; in contrast, CD34$^+$ SP cells have shown stem cell activity in these assays. SP cells have been isolated from normal peripheral blood, but the majority was shown to be lineage-committed. However, even the Lin$^-$SP fraction did not demonstrate growth in LTC-IC or cobblestone assays and failed to engraft NOD/SCID mice, but it did appear to contain lymphocyte/dendritic cell precursors. However, G-CSF mobilized peripheral blood has been demonstrated to contain a significantly increased proportion of SP cells, including lineage negative SP cells that are more likely to represent a clinically relevant progenitor cell population.

The molecular mechanism defining the SP phenotype has recently been attributed to the **adenosine triphosphate** (ATP)-binding cassette transporter *ABCG2*. High levels of *ABCG2* mRNA were found in murine bone marrow SP cells, and enforced expression of *ABCG2* in an epithelial cell line conferred an SP phenotype. However, the continued presence of SP cells in an *ABCG2* knockout mouse implies that more than one dye-efflux pump is responsible for the SP phenotype.

The variety of emerging cell-surface markers emphasize the considerable heterogeneity within the "stem cell" compartment. A hierarchy of stem cells appears to exist, based on varying capacities for self-generation and differentiation.

HEMATOPOIETIC SYSTEM

The cells and their associated substances that arise from the **bone marrow** (see **Hematopoiesis**). The myeloid-lymphoid progenitor pluripotential stem cells, probably arising from the yolk sac and migrating to the liver and bone marrow, differentiate to myeloid progenitor cells, which mature in the bone marrow, and lymphoid progenitor cells, with B-**lymphocytes** maturing in the bone marrow and T-lymphocytes in the **thymus.** These lymphoid cells migrate to the **spleen, lymph nodes**, skin and epithelial mucosa. **Dendritic reticulum cells** are derived directly from the hematopoietic stem cell, mature in the skin, and migrate via the blood and lymph to the lymphoid follicles of the spleen and lymph nodes.

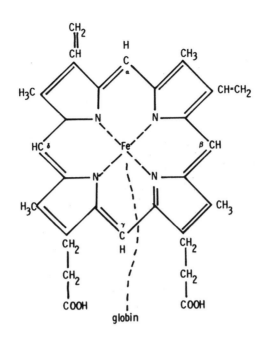

FIGURE 58
Structure of heme. (From Hoffband, A.V. and Pettit, J.E., *Essential Haematology*, 2nd ed., Blackwell, Oxford, 1980.
With permission.)

HEME

The prosthetic group for **hemoglobin**, but which is essential for the function of all aerobic
cells. Approximately 85% of heme is synthesized in the erythropoietic marrow, the remainder being produced mainly by the liver. Heme is composed of an **iron** atom coordinated
to four pyrrole rings of porphyrin through the nitrogen atom on each pyrrole ring (see
Figure 58).

The steps in heme synthesis are:

Glycine + Succinyl-CoA

⇓   1

δ-Aminolevulinic acid

⇓   2

Porphobilinogen

⇓   3

Hydroxymethylbilane

⇓   4

Uroporphyrinogen III

⇓   5

Coproporphyrinogen III

⇓   6

Protoporphyrinogen IX

⇓   7

Protoporphyrin

⇓   8

Heme

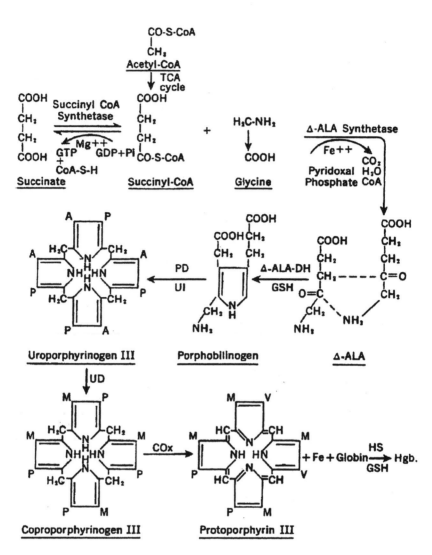

FIGURE 59
Chemical steps in the biosynthesis of hemoglobin. (Prepared with the help of Dr. G.W.E. Plaut and Dr. G.E. Cartwright. Reproduced from Wintrobe, M.M., *Clinical Hematology*, 6th ed., Henry Kimpton, London, 1967. With permission.)

Enzymes are:

1. δ-Aminolevulinic acid synthase
2. δ-Aminolevulinic acid dehydratase
3. Porphobilinogen deaminase
4. Uroporphyrinogen III cosynthase
5. Uroporphyrinogen decarboxylase
6. Coproporphyrinogen oxidase
7. Protoporphyrinogen oxidase
8. Ferrochelatase

These are shown diagrammatically in Figure 59.

The rate of heme synthesis in the liver is regulated largely by the enzyme ALA synthase, which in turn is under the control of heme. Any substance that disturbs the liver heme concentration potentiates the action of ALA synthase. The mode of regulation of heme production in the marrow is less clear. The formation of porphobilinogen, uroporphyrin, and coproporphyrin takes place in the cytoplasm, and the final assembly of the protoporphyrin ring occurs in the mitochondria. The final step is made with a mitochondrial enzyme **ferrochelatase** (heme synthetase). Iron is then incorporated to form heme.

HEME OXYGENASE-1

An essential enzyme in **heme** catabolism, cleaves heme to form biliverdin, which is subsequently converted to bilirubin by biliverdin reductase and carbon monoxide. Activity is induced by its substrate heme and by various nonheme substances. Mutations in the *HMOX1* gene (22q12) have been described in a 6-year-old boy who had severe growth retardation, persistent **fragmentation hemolytic anemia**, and **intravascular hemolysis**.

HEMIN

The Fe^{3+} oxidation product of **heme**. Hemin is a feedback inhibitor of **δ-aminolevulinic acid** (ALA) **synthase**, inhibits δ-ALA synthase transport from cytosol into mitochondria, and decreases δ-ALA synthase synthesis.

HEMOCONCENTRATION

The increase, usually rapid, in the relative **red blood cell volume** of blood. It occurs with dehydration due to plasma loss, as in burns, and with fluid transfer from the circulation to tissues, as in shock syndromes.

HEMOGLOBIN

The protein that transports oxygen in **red blood cells** from the lungs to the tissues. Hemoglobin molecules are composed of four globin chains to which an **iron**-containing porphyrin, **heme**, is attached.

Structure

The hemoglobin molecule consists of four polypeptide subunits, two of α-type and two of β-type (68 kDa), each containing an active heme group (see Figure 60, Figure 61, and Table 74) located within a hydrophobic crevice.

The α-chain contains 141 amino acids and the β-chain 146 amino acids. The γ- and δ-chains also have 146 amino acids. The γ-chain differs from the β-chain by 39 amino acids and contains isoleucine, which is absent in the other chains; the δ-chain differs from the β-chain by ten amino acid substitutions. In the normal adult, there are two α-chains and two β-chains, designated a2b2 or hemoglobin A. During **fetal hematopoiesis**, hemoglobin F (HbF; a2g2) is the predominant form. There are two types of γ-chain in HbF that differ from the amino acid at position 136: glycine (Gg chain) or alanine (Ag chain). Hemoglobin A2 lacks β-chains, which are replaced by δ-chains; a variety of fetal/embryonic hemoglobins are formed: Hb Gower 1 ζ2ε2, Portland ζ2λ2, and Gower 2 α2ε2. At birth, β-globin synthesis has started to replace γ-chain synthesis, which decreases to adult level by the age of 6 months. This high synthetic rate is accomplished because of the continuous

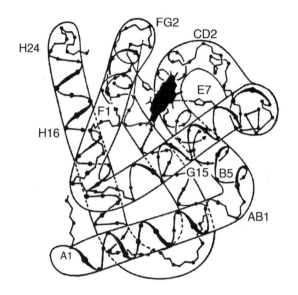

FIGURE 60
Diagrammatic representation of the tertiary configuration of the myoglobin of sperm whale, showing helical segments A–H, interhelical segments AB, etc., and the position of heme group (black) in heme pocket. (Adapted from Dickerson, R.E., in *The Proteins*, Vol. II, Neurath, M., Ed., Academic Press, New York, 1964. With permission.)

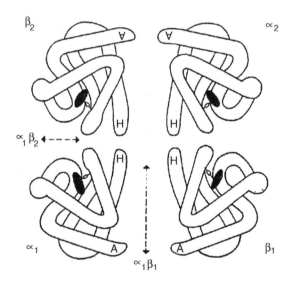

FIGURE 61
Diagrammatic representation of the relationship between α- and β-globin chains in the hemoglobin tetramer. The a1b1(a2b2) contact is the stabilizing contact; the a1b2(a2b1) is the contact across which the β chains slide during oxygenation and deoxygenation. (From Hoffbrand, A.V. and Lewis, S.M., *Tutorials in Postgraduate Medicine: Haematology*, William Heinemann, London, 1966. With permission.)

transcription of the relevant genes, the stability of the subsequent mRNA, and the efficient translation of these messages.

The four polypeptide subunits fit together primarily by hydrophobic bonding. Those amino acid residues on the surface of the tetrameric molecule are largely hydrophilic. By this means, exclusion of water from the interstices of the molecule causes the chains to

TABLE 74

Structure of Physiological Hemoglobin Chains

Hemoglobin A	α2β2
Hemoglobin A2	α2δ2
Hemoglobin F	α2Gγ2 or Aγ2
Hemoglobin Gower 1	ζ2ε2
Hemoglobin Gower 2	α2ε2
Hemoglobin Portland	ζ2γ2
Hemoglobin Barts	γ4
Hemoglobin H	δ4

bond, and the attraction of water to the surface prevents further polymerization. In **hemoglobin S**, the substitution of a hydrophobic for a hydrophilic residue on the outer surface of the β-chain allows polymerization of more than one tetramer. Identification of the various forms is performed by **hemoglobin electrophoresis**.

The heme group, which binds hemoglobin, requires protoporphyrin. Protoporphyrin synthesis commences with the formation of **δ-aminolevulinic acid** (ALA) from glycine and succinyl-coenzyme A. The formation of porphobilinogen, uroporphyrin, and coproporphyrin then takes place in the cytoplasm, and the final assembly of the protoporphyrin ring occurs in the mitochondria. Heme transport to the cytosol is aided by an **adenosine triphosphate** (ATP)-binding cassette, **ABC7**. Iron is then incorporated.

As soon as the red blood cell begins to differentiate, it rapidly synthesizes hemoglobin, some 90% of the dry weight of the mature cell consisting of hemoglobin.

Synthesis

Globin chain synthesis is regulated by globin chain genes (see **Hematopoiesis** — globin switching). The controlling genes are in:

Chromosome 11-linked cluster 5′ to 3′: ε-, Gλ-, Aλ-, δ, β

Chromosome 16-linked order 5′ to 3′: ζ, α2, α1

The genes are arranged in the order on each chromosome in which they are switched on during intrauterine life. Promoters and enhancers are present on the upstream flanking regions that are recognized by nonspecific transcription factors.

Heme biosynthesis is dependent on a series of enzymes that convert glycine and succinyl-CoA to protoporphyrin. Steps in the biosynthesis of hemoglobin are shown in Figure 63.

Oxygen Transport

The uptake of oxygen in the lungs and its release to the tissues involves a specific change in the molecular structure of hemoglobin. Each heme group is capable of binding an oxygen molecule. At the time of deoxygenation, certain residues (Bohr groups) lose a proton (**Bohr effect**), and this permits the formation of salt bridges between the charged groups of different chains, thus increasing the rigidity of the tetrameric structure. **2,3-Diphosphoglycerate** (2,3-DPG) binds in a cavity in the central part of the molecule in the deoxy configuration (see Figure 63).

As oxygenation progresses, this central cavity closes, preventing DPG binding. The higher the oxygen tension, the more likely oxygen will bind, driving off bound DPG. The higher the DPG level, the more likely the tetramer will have DPG bound to it, forcing it into the deoxy configuration and expelling oxygen. Thus, it is possible for DPG to regulate

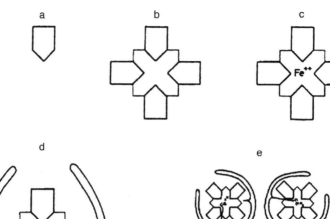

FIGURE 62
Steps in the development of hemoglobin shown diagrammatically. (a) Pyrrole unit; (b) porphyrin ring; (c) heme; (d) heme-globin complex; (e) four heme-globin units forming one complex hemoglobin molecule. (From Lehmann, H. and Huntsman, R.G., *Man's Haemoglobins*, Elsevier, Amsterdam, 1966. With permission.)

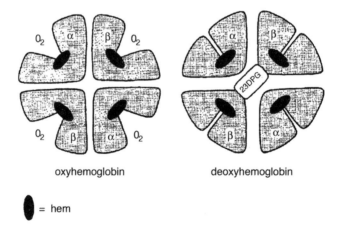

FIGURE 63
Oxygenated and deoxygenated hemoglobin molecule: 2,3-DPG; α- and β-globin chains of normal adult hemoglobin (HbA). (From Hoffband, A.V. and Pettit, J.E., *Essential Haematology*, 3rd ed., Blackwell, Oxford, 1980. With permission.)

oxygen affinity within the red blood cell. As hemoglobin moves from its deoxy to its oxy configuration, carbon dioxide and DPG are expelled from their position between the β-chains, opening up the molecule to receive oxygen and increasing its oxygen affinity. This is responsible for the sigmoid shape of the oxygen-dissociation curve, which is the arithmetic plot of hemoglobin oxygen saturation (ordinate) against the partial pressure of oxygen (abscissa) (see Figure 64). The **oxygen affinity to hemoglobin** is expressed as P_{50},

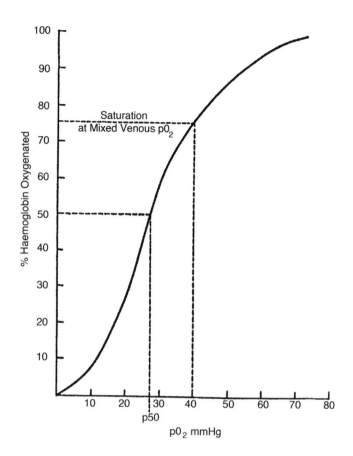

FIGURE 64

Oxygen-dissociation curve. (From Hoffbrand, A.V. and Pettit, J.E., *Essential Haematology*, 2nd ed., Blackwell, Oxford, 1984. With permission.)

which is the oxygen tension at which hemoglobin is half-saturated and is taken from the midpoint of the oxygen dissociation curve. This value is 26 mm Hg in normal cells, which compares with:

Partial pressure of oxygen
In room air   =   100 mm Hg
Pulmonary alveoli   =   95 mm Hg
Systemic arterial blood   =   90 mm Hg

Delivery of oxygen is determined by pO_2 of the tissues. The steep portion of the oxygen-dissociation curve allows a relatively large amount of oxygen to be unloaded for a small decrement in pO_2. With increasing oxygen affinity, the oxygen-dissociation curve is "shifted to the left." High values for the pO_2 indicate a lower oxygen affinity for hemoglobin. A right shift in the dissociation facilitates oxygen delivery. The three primary determinants of the pO_2 are temperature, pH, and 2,3-diphosphoglycerate (2,3-DPG) concentration. The Bohr effect is an important buffer system of the body. When blood reaches the tissues where the oxygen tension is lower and the hydrogen ion concentration is increased by lactic acid and carbon dioxide, the Bohr shift of the oxygen-dissociation curve makes more oxygen available.

TABLE 75

Reference Ranges in Health of Hemoglobin
in Peripheral Blood

Men	13.0–17.0 g/dl
Women	12.0–15.0 g/dl
Infants (full term)	17.9–21.5 g/dl
Children (1 year)	10.5–13.5 g/dl
Children (2–6 years)	12.5–13.5 g/dl
Children (6–12 years)	11.5–13.5 g/dl

Source: Derived from **Reference Range Tables IV** and **V**.

Measurement of Hemoglobin (Hemoglobinometry)

In circulating blood, hemoglobin is a mixture of hemoglobin, oxyhemoglobin, carboxyhemoglobin, methemoglobin, and small amounts of other forms of the pigment. To measure the compound accurately, it must be converted into a single stable derivative. The cyanmethemoglobin derivative can be prepared easily and reproducibly and is widely used for hemoglobin estimation. All forms are readily converted, with the exception of sulfhemoglobin, which is seldom present in significant amounts. Cyanmethemoglobin can be measured spectrophotometrically by its absorbance at 540 nm or by Whatman paper color scale.[235] Errors in the measurement are those of dilution and estimation of color intensity. Turbidity in the sample leads to falsely high levels. Turbidity arises from improperly lysed red cells, high concentration of white blood cells, and the presence of nucleated red blood cells, paraprotein, or lipids during the reproductive period of life. A summary of reference ranges is given in Table 75. The details of premature and full-term infants and for children are given in **Reference Range Tables II** through **V**.

Hemoglobin Degradation

Aged red blood cells are removed from the circulation by phagocytes in the spleen. In these cells, hemoglobin is split into globin and heme. Globin is hydrolyzed into its component amino acids, which return to the body amino acid pool. Iron is split from the heme moiety and reused for hemoglobin synthesis or stored in ferritin and hemosiderin. The heme moiety is cleaved by heme oxygenase-1 to biliverdin and carbon monoxide, and the biliverdin is rapidly reduced to bilirubin, which is excreted into the bile. The process is represented in Figure 65.

Increased rates of hemoglobin degradation occur in **hemolysis**, **ineffective erythropoiesis**, and resorption of hematoma. There are two sites of red blood cell destruction: intravascular and extravascular. In **intravascular hemolysis**, hemoglobin is released into the

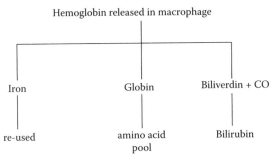

FIGURE 65
Degradation of hemoglobin.

plasma and bound to **haptoglobin**, transported to the liver as a haptoglobin-hemoglobin complex, and converted to bilirubin. In the extravascular variety, hemoglobin breakdown occurs within phagocytes that degrade heme directly to bilirubin, although there is also some liberation of free hemoglobin into plasma.

Haptoglobin is a glycoprotein consisting of α- and β-polypeptide chains that migrate as an α-globulin upon serum protein electrophoresis. The plasma concentration of haptoglobin is expressed as its hemoglobin-carrying capacity (reference range 0.8 to 2.7 g/l by the radial immunodiffusion method; 0.3 to 2.0 g/l by the hemoglobin-binding-capacity method). With active intravascular hemolysis, the level of haptoglobin falls rapidly.

The haptoglobin-hemoglobin complex is too large to pass through the renal glomerulus; however, when the haptoglobin mechanism is saturated, hemoglobin readily passes into the urine in the form of αβ-dimers. Whether or not there is **hemoglobinuria**, the presence of hemoglobin in the glomerular filtrate results in the excretion of **hemosiderin** in urine. In addition to the haptoglobin mechanism, some hemoglobin is probably cleared directly into the parenchymal cells of the liver. Another mechanism exists that becomes operative, particularly when the haptoglobin-hemoglobin process is saturated. Hemoglobin is oxidized to **methemoglobin**, from which the oxidized heme moiety (hemin or ferriheme) is readily dissociated and becomes tightly bound to **hemopexin**, a glycoprotein of the β-globulin class that avidly binds **hemin**. The resulting hemopexin-hemin complex is slowly cleared by the hepatic parenchymal cells. This mechanism, too, can become saturated, and when this occurs free hemin becomes associated with albumin to produce **methemalbumin**. Both methemalbumin and the hemopexin-hemin complex impart a brownish color to plasma, which can be detected spectrophotometrically. The enzymatic conversion of heme to bilirubin takes place, *inter alia*, in phagocytic cells of the monocyte-macrophage system in spleen, marrow, and parenchymal cells of liver and kidney. The enzymes involved include microsomal heme oxygenase and biliverdin reductase. Carbon monoxide is carried in the blood as carboxyhemoglobin and excreted via the lungs.

Plasma contains both unconjugated and conjugated bilirubin, the latter being regurgitated from the liver after its conjugation. Conjugated bilirubin is measured in plasma by the "direct" van den Bergh reaction. "Indirect" bilirubin corresponds roughly to the unconjugated form. Both forms are carried in plasma bound to albumin. Free bilirubin is easily displaced from albumin by fatty acids, salicylates, sulfonamides, and acidosis, a point that has relevance in the neonate. These substances must be avoided in severe unconjugated hyperbilirubinemia. Otherwise, there is a risk that free bilirubin will cross the blood–brain barrier and cause **kernicterus** (bilirubin toxicity to the brain). The excretion of bilirubin into bile is mediated in the liver by the three steps of uptake, conjugation, and secretion. Uptake is accomplished by the enzyme glutathione S-transferase (ligandin), which binds the unconjugated bilirubin in the hepatic cell cytosol. Another protein, called Z protein, acts as a secondary cytoplasmic binder. Conjugation with glucuronic acid is catalyzed by glucuronyltransferase. This enzyme is completely lacking in Crigler-Najjar syndrome, which is characterized by severe unconjugated hyperbilirubinemia and kernicterus. Gilbert syndrome may be the partial expression of this disorder. Secretion of conjugated bilirubin is an active energy-dependent process. Dubin-Johnson and Rotor syndromes are inherited benign disorders of the secretory process.

Once excreted into bile, conjugated bilirubin traverses the intestinal tract. After reaching the colon, bacteria convert the conjugated bilirubin to a series of compounds, collectively known as **urobilinogen**. The glucuronic acid residues are hydrolyzed by β-glucuronidase. Urobilinogen compounds are largely excreted in the feces as stercobilinogen. However, 20% is reabsorbed and enters the portal circulation, reaching the liver, where it is remetabolized and reexcreted in the bile (enterohepatic recirculation).

HEMOGLOBIN A

(HbA) The principal **hemoglobin** of adults (96 to 97% of total circulating hemoglobin). Hemoglobin A molecules consist of two α-globin chains and two β-globin chains.

HEMOGLOBIN A2

(HbA2) A normal minor component of adult **hemoglobin** that consists of two α-globin chains and two δ-globin chains. Hemoglobin A2 is quantified by either electrophoretic or column chromatography techniques (see **Hemoglobinopathies**). It comprises 2.2 to 3.5% of total adult hemoglobin; the values for infants are given in the **Reference Range Table III**.

HEMOGLOBIN BART'S

(g4) Abnormal **hemoglobin** that occurs in a form of α-**thalassemia**. Hemoglobin Bart's molecules consist of a tetramer of γ-globin chains because no α-chains are formed. Hence, patients with Hemoglobin Bart's have no hemoglobin A, F, or A_2. This tetramer transports oxygen ineffectively. Thus, in hydrops fetalis (thalassemia 1 homozygosity), Bart's hemoglobin comprises 80 to 90% of the total hemoglobin; the remainder is tetramers of β-globin chains (β4), termed hemoglobin H. This abnormality is lethal because hemoglobin H is both relatively unstable and an ineffective oxygen transporter.

HEMOGLOBIN C

(Hb C) A common hereditary variant of **hemoglobin** arising as a result of substitution of lysine for glutamic acid at the 6 position of the β-globin chain. Hemoglobinopathy C occurs predominantly in persons of African descent (4% of African Americans are heterozygotes). Heterozygotes are asymptomatic, although **codacytes** (target cells) are present in the blood. Most hemoglobin-C homozygotes have mild or moderate **hemolysis**, reticulocyte count of 2 to 6%, **anemia**, and **splenomegaly**. Spontaneous splenic rupture sometimes occurs. Some erythrocytes contain tetrahedral crystals (Hb-C crystals), and others are possibly induced by dehydration of the red cells *in vitro* between slide and coverslip. Hypertonic dehydration of cells in 3% NaCl buffer for 4 to 12 h can also induce their formation. Prevalence of the intraerythrocytic crystals increases after splenectomy. No specific therapy is required or available for hemoglobinopathy C. Hemoglobin C/β-thalassemia occurs in North and West Africa (see **Thalassemia**).

HEMOGLOBIN D

(Hb D) A hereditary structural variant of **hemoglobin** due to substitution of glutamine for glutamic acid at the 121 position on the β-globin chain. Hemoglobinopathy-D homozygosity is characterized by mild **hemolytic anemia** and mild or moderate **splenomegaly**. The heterozygotes are asymptomatic. The electrophoretic and solubility phenotypes of Hb G and Hb D are so similar that the two states are usually not differentiated. Each can be associated with Hb S and is usually silent. Hb S/D-Punjab presents as mild **sickle cell disorder**.

HEMOGLOBIN E

(Hb E) A hereditary structural variant of **hemoglobin** arising due to the substitution of lysine for glutamic acid at the 26 position on the β-chain. Hemoglobinopathy E homozygosity is characterized by mild **hemolytic anemia** with many **codacytes** (target cells) in the peripheral blood film. Some patients have **splenomegaly**. Hemoglobinopathy E heterozygosity is clinically silent. Hemoglobin E is most prevalent in Southeast Asia, although hemoglobin E/β-**thalassemia** occurs worldwide.

HEMOGLOBIN ELECTROPHORESIS

The separation of the structural variants of hemoglobin based upon their different mobilities in an electric field generated by direct current through a buffer. The technique is used to identify the different types of hemoglobins by a variety of methods.[236,237]

Cellulose acetate electrophoresis at alkaline pH. This is used for distinguishing common **hemoglobinopathies** as occur with **hemoglobins A, C, F,** and **S,** provided that there is no disproportionate amount of any one hemoglobin.

Citrate agar electrophoresis at pH 6.0. This technique permits differentiation between Hbs S, D, and G or between Hbs C, E, and O. The method also permits demonstration of small quantities of HbF that are clearly separated from HbA.

Cellulose acetate electrophoresis at pH 6.5. This method permits distinction of Hb H and Hb Bart's from other fast-migrating Hb variants.

Starch-gel electrophoresis. The migration of common hemoglobin variants on this medium is similar to that obtained on cellulose acetate at alkaline pH.

Starch block electrophoresis. Although this is a sensitive medium for separating hemoglobins, the process is very time consuming and unsuitable for routine use.

Globin-chain electrophoresis. This method is used either at acid or alkaline pH for refined systematic identification of certain hemoglobin variants following screening by cellulose acetate electrophoresis at alkaline pH and citrate agar electrophoresis at acid pH. It is particularly valuable for the identification of Hb D (Punjab), Hb O Arab, and Hb G Philadelphia.

HEMOGLOBINEMIA

The presence of free **hemoglobin** in plasma. The free hemoglobin level is increased in **hemolytic anemias** when hemolysis is sufficiently severe to saturate the hemoglobin-haptoglobin mechanism. It occurs predominantly during **intravascular hemolysis**, and therefore marked increases (with or without **hemoglobinuria**) occur in:

Blackwater fever of **malaria**

Cold-agglutinin disease

March hemoglobinuria

Macrovascular hemolytic anemias

Paroxysmal cold hemoglobinuria

Paroxysmal nocturnal hemoglobinuria

Minor increases in free hemoglobin level occur in **autoimmune hemolytic anemias, sickle cell disease,** and **thalassemia.** Care must be taken to ensure that hemolysis has not occurred during **venepuncture** to collect specimens for free hemoglobin determination. A plasma hemoglobin level greater than 50 mg/dl appears pinkish to the naked eye and suggests intravascular hemolysis. Plasma that contains a free hemoglobin level of more than 150 mg/dl is bright red, and such patients will also have hemoglobinuria.

HEMOGLOBIN F

(HbF) The principal fetal **hemoglobin**, the level of which usually declines during early childhood to less than 1% of the total hemoglobin. (For methods of detection and measurement, see **Hemoglobin electrophoresis** and **Hemoglobinopathies**.) The reference levels for the first year of life are given in **Reference Range Table III**.

Increased levels of HbF occur in adult life in a variety of acquired and inherited disorders.

Inherited disorders
- Primary: hereditary persistence of fetal hemoglobin; δβ-**thalassemia**
- Secondary: **sickle cell disorder**; β-thalassemia

Acquired disorders
- Acute **hemorrhage**
- **Acquired aplastic anemia**
- **Blackfan-Diamond anemia**
- **Fanconi anemia**
- **Myelodysplasia** — refractory anemia
- **Paroxysmal nocturnal hemoglobinuria**
- Post-**allogeneic stem cell transplantation**
- Pregnancy

The increased levels appear to be a secondary effect, either due to a reduction in the number of progenitor cells or an acute demand for increased red cell production.

HEMOGLOBIN H

A tetramer of β-globin chains formed when there is no hemoglobin A, F, or A2, due to this type of α-**thalassemia**.

HEMOGLOBIN LEPORE

The **hemoglobin** present in of one of a group of δβ-**thalassemias**. Hb Lepore results from deletion of a part of the linked b and d genes, producing a fusion gene. Homozygotes have very severe transfusion-dependent anemia, whereas heterozygotes have a clinical phenotype of thalassemia minor. The diagnosis is indicated by the peripheral-blood-film appearances of thalassemia, normal iron stores, high HbF level, and low levels of an abnormal hemoglobin — hemoglobin Lepore.

HEMOGLOBIN M

A group of abnormal **hemoglobins** arising from autosomally dominant amino acid substitutions that permits autooxidation of heme iron, resulting in **methemoglobin** formation and **cyanosis**. In most of the hemoglobins M, tyrosine is substituted for either the proximal or distal histidine, permitting the formation of an iron-phenolate that resists reduction by the normal red blood cell metabolic systems. Four such M hemoglobins are recognized (Boston, Saskatoon, Iwate, Hyde Park). A fifth Hb M (Milwaukee) results from substitution of glutamic acid for valine residue 67 of the β-chain. Although spectroscopy is the simplest method for identification in theory, the mixed spectra of methemoglobin A and hemoglobin

M may be difficult to interpret. For this reason, all hemoglobin M samples should be converted to methemoglobin so that any differences produced on electrophoresis will be due to the amino acid substitution and not to the different charge of the iron atom. Electrophoresis at pH 7.1 is most useful for the separation of Hbs M.

All of these hemoglobinopathies are characterized by cyanosis, the appearance of other clinical features being determined by the globin chain affected. In α-chain variants (Hb M-Boston, Hb M-Iwate), the dusky coloration is present at birth. In β-chain variants (Hb M-Hyde Park, Hb M-Milwaukee, Hb M-Saskatoon), features appear at the age of 6 months. The chief effect of both varieties is cosmetic.

HEMOGLOBIN O-ARAB

(Hb O-Arab) A hereditary structural variant of **hemoglobin** due to substitution of lysine for glutamic acid at the 121 position of the β-chain. The disease is characterized by **splenomegaly**, mild **hemolytic anemia**, and **codacytes** (target cells) seen in the peripheral blood film. The double heterozygote with Hb S is clinically indistinguishable from patients with **sickle cell disease**.

HEMOGLOBINOMETRY

The measurement of **hemoglobin**, usually after its conversion to cyanmethemoglobin.

HEMOGLOBINOPATHIES

Clinical syndromes arising from disorders of **hemoglobin**. They are a vast group of genetically determined disorders.

Categories

Structural variants of the hemoglobin molecule:
- **Hemoglobin S** disease (**Sickle cell disease**)
- **Hemoglobin C** disease
- **Hemoglobin D** disease
- **Hemoglobin E** disease
- **Hemoglobin M** disease
- **Hemoglobin O-Arab** disease
- **Unstable hemoglobins**: Heinz body hemolytic anemia
- **Oxygen affinity to hemoglobin** disorders
  Low affinity: **methemoglobinemia**
  High affinity: **erythrocytosis**

Failure of globin-chain synthesis — **thalassemias**
- α-Chain synthesis: α-thalassemias
  Homozygotes: **hemoglobin Bart's** or **hemoglobin H** disease
  Heterozygotes: silent or trait
- β-Chain synthesis: β-thalassemias
  Homozygotes: thalassemia major

Heterozygotes: thalassemia intermedia or thalassemia minor

Mixed hemoglobin variant/thalassemia, e.g., hemoglobin C/thalassemia

Clinical Features

Many are associated with anemia, some with a **hemolytic anemia** and **splenomegaly**. Variants resulting in diminished oxygen affinity with methemoglobinemia have **cyanosis**; others are associated with increased oxygen affinity have the features of erythrocytosis.

Laboratory Investigation

Investigation must follow a definite but flexible plan that involves performing screening tests to recognize any hemoglobinopathies (structural variants of hemoglobin) or thalassemias (including failure of neonatal switch from HbF to HbA). Evaluation of possible thalassemia requires the quantitative measurement of hemoglobins, especially in geographic areas in which there is a high incidence of double heterozygotes for hemoglobinopathies and thalassemias.

The necessary tests[238,239] are

Initial tests

- **Red blood cell counting** with red blood cell indices
- Peripheral blood film examination
- Cellulose acetate electrophoresis at alkaline pH
- Hemoglobin S sickling and solubility tests
- Hemoglobin A2 estimation
- Hemoglobin F estimation
- Demonstration of inclusion bodies

Further hemoglobin electrophoresis

- Citrate agar electrophoresis at pH 6.0
- Cellulose acetate electrophoresis at pH 6.5
- Electrophoresis at neutral pH for Hb H and Hb Bart's
- Globin-chain electrophoresis
- Starch block electrophoresis
- Starch gel electrophoresis

Special tests

- Detection of unstable hemoglobins
- Detection of hemoglobin M
- Detection of altered oxygen affinity hemoglobins

Reference laboratory procedures

- *In vitro* globin-chain synthesis ratio (see **Hematopoiesis** — globin switching)
- Isoelectric focusing
- High-pressure liquid chromatography (HPLC)
- Analysis of novel variants
- Molecular genetic evaluation (selected cases)

Red Blood Cell Counting with Red Blood Cell Indices and Peripheral Blood Film Examination

The red blood cell count, red blood cell indices, and red blood cell morphology give valuable information. Examination of the blood film may show changes characteristic of certain structural hemoglobin variants, e.g., target cell production or hemoglobin crystal formation in association with Hb C, or sickle cells with Hb S. In thalassemia, the mean corpuscular volume (MCV) and mean corpuscular hemoglobin (MCH) are reduced, often out of proportion to the degree of anemia. Thalassemia and **iron deficiency** both cause microcytosis and hypochromia. A number of discriminant functions based on red blood cell count, hemoglobin concentration, MCV, MCH, and, more recently, red cell distribution width (RDW) have been elaborated to aid this distinction. In thalassemia, RDW is usually normal, whereas in iron deficiency both anisocytosis and anisochromia cause increased RDW. Target cells are more frequently seen in thalassemia; pencil-shaped elliptocytes are often present in iron deficiency anemia. Serum **iron** measurement readily separates the two conditions. In thalassemia, serum indirect bilirubin and lactate dehydrogenase levels are elevated due to increased cell turnover.

Cellulose Acetate Electrophoresis at Alkaline pH

This method distinguishes the common hemoglobin structural variants. Adequate separation of hemoglobins A, C, F, and S is achieved, but it is not possible to distinguish hemoglobins D, G, and Lepore from S, or to distinguish hemoglobins E, O, and A2 from C. With this technique, it can be difficult to detect small amounts of HbF when large amounts of HbA exist, and vice versa.

Tests for Hemoglobin S

Homozygous sickle cell disease can usually be diagnosed from the clinical features and the presence of sickle cells in the blood film. In Hb S heterozygotes, sickle cells are usually not observed upon examination of peripheral blood film, but are usually visible after 1-h incubation with metabisulfite. The Hb S solubility test, which depends upon the decreased solubility of Hb S at low oxygen tensions, is used to distinguish Hb S from other hemoglobins having the same cellulose acetate electrophoretic mobility. False negative results may be obtained when outdated reagents are used, when blood from a child under 6 months of age is tested, after recent blood transfusion, and in heterozygotes when the proportion of HbS is <20%. False-positive results can occur in association with unstable hemoglobins, particularly after splenectomy and sometimes by other "sickling" hemoglobins (HbC-Harlem, HbS-Travis). Whenever a solubility test is positive, hemoglobin electrophoresis must be performed before a definitive diagnosis is made.

Quantitation of Hemoglobin A2 Levels[240,241]

Two methods are available:

- Elution after cellulose acetate electrophoresis (although accurate cutting of the A2 band may be difficult, and the technique cannot be used in the presence of hemoglobins of similar electrophoretic mobility, e.g., Hbs C, E, O). Values less than 3% are considered to be normal and those above 3.5% abnormal. Elevated levels occur in thalassemia and in the β-unstable hemoglobinopathies. Some overlap occurs between normal persons and subjects with β-thalassemia trait.
- Microcolumn chromatography (Tris HCl and glycine KCN). With the first technique, values below 3% are considered normal and those above 3.5% abnormal.

Elevated levels occur in thalassemia and in the β-unstable hemoglobinopathies. Again, some overlap occurs between normal subjects and those with β-thalassemia trait. The reference range in health is 1.5% to 3.5%. β-Thalassemia subjects have HbA2 levels of 4.0 to 7.0%, except for δβ-thalassemia subjects, in which levels are low. Patients with b-unstable hemoglobinopathies have the same reference range as the thalassemia patients. Values in excess of 10% suggest the presence of another structural variant.

HbA2 levels are increased in **Addisonian pernicious anemia** and decreased in **iron deficiency** anemia, **sideroblastic anemia**, and **aplastic anemia**.

Quantitation of Hemoglobin F Levels

Methods available are:

Resistance to denaturation at alkaline pH (Betke method[242] is reliable for amounts of HbF below 15%; however, for levels over 50%, and in cord blood, the method of Jonxis and Visser[243] is preferred. No single method is suitable for measuring HbF levels across the entire range.)

HPLC (see below)

Immunodiffusion methods (commercial kits)

Enzyme-linked immunoassay (ELISA)

The reference range in health for adults is 0.2 to 1.0%. In cord blood at term, the level is 60 to 80%. Increased levels are found in β- and dβ-thalassemias, hereditary persistence of fetal hemoglobin, and in sickle cell disease. Raised levels often appear in a number of acquired disorders, including aplastic anemia, leukemias, other neoplastic diseases, and any form of red cell expansion.

Red Blood Cell Inclusions

These are frequently found in such hemoglobinopathies as:

Hemoglobin H inclusions (stained by 1% new methylene blue; they appear as multiple greenish-blue bodies)

α-Chain inclusions in β-thalassemia (stained supravitally by methyl violet; they appear as irregularly shaped bodies close to the normoblast nuclei or in the peripheral blood after splenectomy)

Heinz bodies in unstable hemoglobin diseases (stained by methyl violet)

Detection of Unstable Hemoglobins

The diagnosis may be suspected clinically due to the presence of cyanosis (due to methemoglobinemia), anemia, jaundice, and the passage of brown to almost black urine (due to dipyrromethenes derived from partially degraded heme molecules). The presence of Heinz bodies in the blood of a patient with hemolytic anemia after splenectomy makes the diagnosis almost certain.

The autohemolysis test can be useful. When unstable hemoglobin is present, the serum will appear brown after incubation for 48 h due to the presence of methemoglobin and Heinz bodies.

Hemoglobin instability can be demonstrated on the basis of sensitivity to heat or precipitation by isopropanol. In the heat-instability test, while a normal control may show minimal cloudiness at 1 h, major unstable hemoglobin will have undergone marked precipitation and will be grossly flocculent at 2 h. In the isopropanol precipitation test, clinically significant unstable hemoglobin will undergo precipitation at 5 min and be grossly flocculent at 20 min (normal control remains clear at 20 min).

Detection of Hemoglobin M
Methemoglobin variants can be detected spectrophotometrically or by starch-gel electrophoresis at pH (see **Hemoglobin M**).

Detection of Altered-Affinity Hemoglobins
See **Oxygen affinity to hemoglobin** — disorders.

Measurement of Globin-Chain Synthesis Ratios
See **Thalassemias** — diagnosis.

Isoelectric Focusing
The commercial availability of precast, thin agarose gels and high-quality buffers containing ampholytes represents a significant advance. These thin gels permit high resolution of hemoglobins within reasonable electrophoresis times. Because many samples can be run on a single gel, some laboratories use this method for the initial analysis of unknown hemoglobin variants. Under well-controlled conditions, hemoglobins migrate to a characteristic position in the gel and can be identified accurately, provided high-quality controls are run in parallel. Such controls are now available commercially.

High-Pressure Liquid Chromatography (HPLC)
The method[244] is based upon the interchange of charged groups on the hemoglobin molecule. Automated cation-exchange HPLC for the identification of variant hemoglobins is currently being used as the initial diagnostic method in hemoglobinopathy laboratories with a high workload. In addition to identification, individual components can also be quantified.

Analysis of Novel Variants
This must be undertaken in a reference laboratory experienced in identification and characterization of variant hemoglobins. Some laboratories will purify the variant globin and undertake amino acid analysis on isolated peptides. Conclusive identification evidence is provided by peptide sequencing. Another approach combines peptide analysis with DNA technology (see molecular genetic approach below) to rapidly identify a variant. Peptide analysis is used to identify the region containing the suspected substitution and, subsequently, polymerase chain reaction (PCR) DNA amplification is used to amplify, clone, and sequence the DNA of the mutation.

Molecular Genetic Evaluation
The techniques being used are Southern blotting with DNA filter hybridization methods and with PCR DNA-amplification techniques. The clinical applications include:

Definitive diagnosis of thalassemias and hemoglobinopathies

Prenatal diagnosis using analysis of DNA of fetal cells obtained by amniocentesis or chorionic villous sampling — procedures that entail some risk to the fetus — or concentration of erythroblasts by magnetic activated cell sorting that can avoid this risk when it is available

Neonatal screening

Screening for Hemoglobinopathy[238]

Screening for sickle hemoglobin and for thalassemia is necessary to detect the presence and to determine the significance of the inherited condition. **Genetic counseling** should be available. Criteria for screening will depend on local disease prevalence and circumstances. Three important groups of circumstances require consideration in specific populations:

Screening for sickle hemoglobin of specific patient subgroups:
- Preoperative
- Specific disease subgroups — aseptic necrosis of hips or shoulders, certain types of retinopathy, priapism

Broad-based population antenatal screening for sickle cell disease and hemoglobinopathy variants for couples at risk of having a child with Hb Bart's hydrops fetalis, β-thalassemia major

Neonatal screening to detect babies with β-thalassemia or sickle cell disease

HEMOGLOBIN S

(Hb S) A hereditary structural variant of **hemoglobin** due to substitution of valine for glutamic acid at position 6 on the β-chain. It gives rise clinically to **sickle cell disease**. The combination of Hb S with deficiency of β-globin chains gives rise to the disorder hemoglobin S/β-**thalassemia**. Double heterozygosity of Hb S/C, Hb S/D-Punjab, or Hb S/O-Arab, all less common than Hb SS, gives rise to less severe forms of sickle cell disease. (For methods of detection, see **Hemoglobinopathies**.)

HEMOGLOBINURIA

The presence of **hemoglobin** in urine. This occurs when hemoglobin degrades rapidly, as in acute **intravascular hemolysis** once **haptoglobin** has been saturated. It is diagnosed by urine being red or brown after centrifugation to remove intact red blood cells. A qualitative measure can be obtained using Hemastix, but this will not distinguish between hemoglobinuria and myoglobinuria (this requires electrophoresis or differential solubility in ammonium sulfate).

HEMOJUVELIN

See also **Hereditary hemochromatosis**.

A potent regulator of **iron** absorption, the precise mechanism of action of which is presently unknown. Hemojuvelin contains multiple protein motifs consistent with a function as a membrane-bound receptor or secreted polypeptide hormone. Mutations in the corresponding gene *HJV* (1q21) cause an autosomally recessive form of hemochromatosis characterized by early age of onset and severe iron overload, often complicated by hypogonadism, cardiomyopathy, and hepatic cirrhosis.

HEMOLYSINS

External agents, either chemical or biological, causing damage to **red blood cell** (RBC) membranes. This damage leads to premature removal of RBC by **histiocytes**, particularly in the **spleen**. Surface immune reactions act similarly, e.g., lytic antibodies. Membrane phospholipases may be destroyed by **snake venom**. Berries, fruits, and seeds sometimes contain hemolysins, the most potent being the saponins of corn cockle (*Agrostemma githago*) and bouncing bet (*Saponaria officinalis*). A rapidly absorbed ricinlike protein, abrin, from the jequirity bean, sometimes used in necklaces, is a powerful hemolysin.

HEMOLYSIS

The lysis of **red blood cells** with release of extracellular **hemoglobin**. This may be physiological, pathological, or technical.

Physiological Hemolysis

This occurs slowly as a physiological process of red blood cell removal, the red cell stroma being phagocytosed by the macrophage-monocyte system with hemoglobin degradation.

Pathological Hemolysis

Pathogenesis

Accelerated red cell destruction within the circulation — **intravascular hemolysis** — with **hemolytic anemia**, **hemoglobinemia**, and **hemoglobinuria.** There is also release of potassium into the plasma.

Extravascular hemolysis by **histiocytes** (macrophages), particularly in the **spleen**. There is also increased urinary excretion of **hemosiderin.** It may be compensated by increased bone marrow erythropoiesis. This is the process occurring in chronic hemolytic anemia.

Etiology

Many factors and mechanisms contribute to accelerated red cell removal in disease. These can be broadly divided into four groups, depending on the abnormality present.

Abnormalities of rheology and viscosity
- Abnormal red blood cell shape: spherocytosis, elliptocytosis, pyropoikilocytosis
- Increased **viscosity:** abnormal hemoglobins such as **sickle cell disease** and precipitated hemoglobin (Heinz body disorders)
- Dehydration

Red blood cell membrane defects
- Inherited and acquired membrane instability, such as **hereditary spherocytosis**, **hereditary elliptocytosis**, or **hereditary infantile pyropoikilocytosis**
- Abnormalities of transmembrane proteins, e.g., sialoproteins, ankyrin (deficiency of β and γ sialoglycoprotein results in elliptocytosis)

External factors
- Attachment of **immunoglobulins** to cell surface

- Oxidation of red blood cell membrane
- **Complement** lysis
- Enzyme-induced hemolysis, e.g., phospholipase activity produced by *Clostridium welchii* or cobra venom
- Heat-induced spherocytosis
- Trauma, particularly to the skin and subcutaneous tissues, as in **march hemoglobinuria**

Dysfunction of vasculature or organs

- Microangiopathy
- Abnormal shear forces, e.g., prosthetic heart valves
- Splenomegaly
- **Ineffective erythropoiesis**: intramedullary destruction, particularly in **Addisonian pernicious anemia** and **thalassemia**

Red blood cell survival measurement can be used to estimate the degree and site of red cell removal.

Technical Hemolysis

This occurs in blood sampling, particularly due to the use of too small a needle or its bore or to poor **venepuncture** technique. It can also occur in the laboratory due to the use of an impure **anticoagulant** or to infection within the sampling container. It is essential that this error be appreciated, as it will lead to incorrect values being reported, particularly those concerning plasma potassium concentration.

HEMOLYTIC ANEMIAS

The pathogenesis, diagnosis, and clinical and laboratory features of anemias resulting from an increase in the rate of **red blood cell** (RBC) destruction, irrespective of the cause.[91,154,245] Because of the various compensatory mechanisms (erythroid hyperplasia and anatomical extension of the bone marrow), red cell destruction may be increased manyfold before anemia develops — compensated hemolytic disease. Depending upon the rapidity or chronicity of the condition, hemolysis occurs as either **extravascular hemolysis** in the spleen and other sites of red cell destruction or within the circulation — **intravascular hemolysis**.

Pathogenesis

Hemolytic disease has many causes, some intrinsic to the red blood cell membrane or its cytosol, which are usually inherited defects, and others being extrinsic disorders, which are usually acquired, that directly or indirectly affect the plasma environment.

Inherited intrinsic **red blood cell membrane disorders** affect the membrane's permeability in a variety of ways, all leading to swelling and increased liability of early removal from the circulation by phagocytes, particularly in the spleen. A multiplicity of **red blood cell enzyme deficiencies** can be inherited, many of which cause disturbance to cellular metabolism and consequent changes in the morphology of the cell as well as the development of insoluble hemochromes (**Heinz bodies**), all of which increase the rapidity of removal by phagocytes. The presence of **hemoglobinopathies** can induce crystallization

— sickling, precipitation of hemoglobin as Heinz bodies, or secondary disturbance of the red cell membrane (**codacytes**).

Externally induced red cell membrane damage may be caused by **physically induced disorders** such as trauma, leading to fragmented red blood cells. **Chemical toxic disorders** and drug-induced hemolysis are mainly a result of failure of protective enzyme mechanisms of the phosphogluconate pathway of red blood cell metabolism, which remove superoxides and peroxides. Oxidation of the membrane thiols exposes them to redox agents that generate oxygen free radicals, which loosen the membrane skeletal structure and so disturb the **cation pump**. Thiol oxidation also induces hemoglobin denaturation, leading to intracellular precipitation (Heinz bodies), which deforms the membrane further, thereby increasing its permeability. This can also be induced by hypo-osmolarity following water intoxication from drowning or by surgical irrigation with intake of more than 0.6 l of distilled water. The red cell membrane can also be damaged by immune reactions on the surface (**immune hemolytic anemias**), by biological **hemolysins** affecting phospholipases, and by **bacterial**, **viral**, and **protozoan infection disorders** or their toxins. Red blood cells deficient in **glucose-6-phosphate dehydrogenase** (G6PD) are particularly vulnerable to these external agents. All of these actions give rise to preferential clearance of these red blood cells by the spleen and so produce a hemolytic anemia. Overactivity of the spleen due to **splenomegaly** with red blood cell sequestration has similar consequences. Bone marrow **aplastic crises** (usually caused by associated parvovirus infection) or **megaloblastosis** due to folic acid deficiency may be a feature. A classification is given in Table 76.

Diagnosis of Hemolytic Anemia

A detailed history — including family, occupational, hobby, and drug history — is required, followed by clinical examination and carefully selected laboratory tests. The absolute test to confirm the presence of hemolytic disease is measurement of the ^{51}Cr-labeled red cell survival, although this may not be required in every patient.

Clinical Features

General features include some or all of:

  Pallor of mucous membranes

  Fluctuating jaundice

  Splenomegaly

  Urine turns black upon standing (excess urobilinogen)

  Pigment gallstones

  Ankle ulcers

Laboratory Features

The first step is to diagnose that hemolytic disease exists. This procedure is conveniently divided into:

  Tests indicating increased red blood cell breakdown
  - Serum bilirubin increased (unconjugated and bound to albumin)
  - Urinary **urobilinogen** increased

TABLE 76

Classification of Hemolytic Anemias

Red blood cell membrane disorders	**hereditary spherocytosis**
	hereditary elliptocytosis
	abetalipoproteinemia
	paroxysmal nocturnal hemoglobinuria
	rhesus null
Red blood cell enzyme deficiencies (congenital	Embden-Meyerhof pathway
nonspherocytic hemolytic anemias)	pentose phosphate pathway
	methemoglobin reductase pathway
	glutathione pathway
	2,3-diphosphoglycerate
	nucleotide enzyme
	adenylate kinase
Hemoglobinopathies	**thalassemias**
	sickle cell disorders
	Hb C, D, E, M, etc.
	unstable hemoglobins
Immune hemolytic anemias	**warm autoimmune hemolytic anemia**
	cold autoimmune hemolytic anemia
	alloimmune hemolytic anemia
	drug associated immune hemolytic anemia
	hemolytic blood transfusion complications
	hemolytic disease of the newborn
Physically induced disorders (traumatic,	**macroangiopathic hemolytic anemia**
mechanical, heat)	**microangiopathic hemolytic anemia**
	march hemoglobinuria
	burns
	hereditary infantile pyropoikilocytosis
Chemical toxic disorders (including drugs)	**Heinz body hemolytic anemia**
	hypo-osmolarity
Biological hemolysins	spider venom (brown recluse)
	snake venom disorders (cobra)
	saponins from *Agrostemma githago* and *Saponaria officinalis*
	jequirity bean
Bacterial infections	*Bartonella bacilliformis*
	Clostridium perfringens (welchii)
	Diplococcus pneumoniae
	Escherichia coli
	Hemophilus influenza
	Mycobacterium tuberculosis
	Mycoplasma pneumoniae
	Salmonella spp.
	Shigella spp.
	Streptococcus spp.
	Vibrio cholerae
	Yersinia enterocolitica
Viral infections	coxsackie
	cytomegalovirus
	Epstein-Barr virus
	herpes simplex
	human immunodeficiency virus
	influenza A
	rubella
	varicella
Protozoan infection	babesiosis
	malaria, *Plasmodium* spp.
	Toxoplasma gondii
	leishmaniasis (kala-azar)
Splenomegaly	

- Fecal stercobilinogen increased (test seldom performed)
- Serum **haptoglobin** absent

Tests indicating increased red blood cell production

- **Reticulocyte count** increased
- Erythroid hyperplasia in the **bone marrow**

Evidence of damage to red cells

- Abnormal red cell morphology seen upon examination of peripheral blood film, e.g., microspherocytes, RBC fragmentation
- Shortened red blood cell survival measurement

Recognition of the nature of the hemolytic mechanism

- **Direct antiglobulin test** (DAT)
- Tests for intravascular hemolysis: **hemoglobinemia**, **hemoglobinuria**, **hemo-siderinuria**, **methemalbuminemia** (Schumm's test)

The final step is to determine the precise diagnosis. This will depend on the results of tests carried out according to the following schedule:

Suspicion of hereditary hemolytic disease

- **Osmotic fragility test**
- **Autohemolysis test**
- **Glucose 6-phosphate dehydrogenase** (G6PD) screening test
- **Pyruvate kinase** assay
- Assay of other enzymes involved in glycolysis
- Red cell **glutathione**

Suspicion of acquired hemolytic disease

- Direct antiglobulin (Coombs) test (DAT) using anti-Ig and anticomplement sera
- Autoantibody tests to red blood cell **blood group** antigens
- **Cold-agglutinin** titer
- Donath-Landsteiner test
- Serum protein electrophoresis

Suspicion of drug-induced hemolytic disease

- G6PD screening test
- **Glutathione** stability test
- Heinz bodies
- **Methemoglobinemia/sulfhemoglobinemia**
- Drug-dependent antibodies

Diagnosis remains obscure

- **Acid hemolysis test** for paroxysmal nocturnal hemoglobinuria
- Sucrose lysis test
- **Flow cytometry** for evidence of PNH cells

HEMOLYTIC DISEASE OF THE NEWBORN

(HDN) An **alloimmune hemolytic anemia** of the fetus and the newborn characterized by anemia with **extramedullary hematopoiesis** and hyperbilirubinemia. **Hydrops fetalis** with fetal or neonatal death occurs in 20 to 25% of cases. Severe neonatal jaundice (icterus gravis) with risk of brain damage (**kernicterus**) occurs in another 25 to 30%.[246]

Etiology of HDN

The disorder is a consequence of alloimmunization in the mother to paternal red blood cell antigens expressed in the fetus. Sensitization of fetal cells with production of antibodies is commonly caused by **blood group** antigens of the **Rh ("Rhesus") blood group** and **ABO blood group**, less frequently by those of other blood groups.

Rh HDN

Anti-RhD HDN is the most prevalent form, but others of clinical significance include antibodies to antigens C, c, E, and e (after anti-D, anti-c is the most frequent and severe). In common parlance, the presence or absence of D-antigen determines the RhD-positive or RhD-negative status of an individual. Each parent transmits a set of the three antigens (Cde, c[d]e, cDE are the most common) to the fetus, who can therefore be RhD-negative (dd), RhD-positive heterozygous for D(Dd), or RhD-positive homozygous for D(DD). If the father is homozygous for D and the mother is Rh negative, all of their children will be D positive; if the father is heterozygous, in each pregnancy, the chances are equal that the fetus will be D positive or D negative. Only the D-positive fetus can provoke RhD immunization, and only the D-positive fetus will be affected by the anti-D produced. The frequency of D-negative status varies from 15% in Caucasians to 8% in North American blacks, 2% in Asians, and 0.3% in Chinese. The RhD antigen is developed by the sixth week of gestation, and by 8 to 10 weeks is found in the liver and spleen (see **Fetal hematopoiesis**). Transplacental passage of RhD antigen anti-D causes the production of anti-D in the RhD-negative woman, and passage of anti-D into the circulation of the RhD-positive fetus causes hemolytic disease of the newborn. HDN due to anti-c is less common than that due to anti-D, but is of similar severity. HDN due to anti-C or anti-E is usually milder.

ABO HDN

ABO sensitization is serologically much more common than Rh sensitization and often occurs in first pregnancies, but rarely produces severe anemia. ABO hemolytic disease (**see ABO (H) blood groups**) usually occurs in group A or group B babies born to group O mothers.

HDN Due to Other Blood Group Antigens

Anti-Kell (see **Kell blood groups**) HDN is rarer than that due to anti-D, but does not differ in degree of severity or outlook. Kpa, k, Fya, and S and other antibodies directed at antigens within the **Kidd blood groups**, **MNS blood groups**, **P blood groups**, and minor blood groups have been associated with moderate or severe HDN. It has also occurred with some independent public and private antigens.

Pathogenesis

Asymptomatic transplacental passage of fetal red blood cells occurs in 75% of pregnancies, parturition, or delivery. RhD immunization, or other blood group sensitization, may occur as a result of fetal transplacental hemorrhage (TPH) of less than 0.1 ml of maternal red blood cells, with the dose of Rh antigen influencing the risk. Spontaneous or therapeutic abortion carries a risk of 2 to 4% of Rh immunization. Preeclampsia and obstetrical procedures such as amniocentesis, external version, cesarean section, and manual removal of the placenta increase the risk of transplacental passage of fetal red cells and the risk of Rh immunization. IgG anti-D traverses the placenta and coats fetal Rh-positive red cells.

Noncomplement-fixing antibodies such as anti-D mediate red blood cell destruction through attachment to macrophages bearing FcR c-receptors in the spleen. Erythrocytes are either phagocytosed initially or lose a portion of their membrane, causing spherocytosis and an increased likelihood of subsequent red cell lysis and phagocytosis of cell fragments. Antibody-sensitized red cells may also be lysed by natural killer (NK) lymphocytes, which adhere to receptors for the Fc portion of IgG and release lysosomal enzymes. Anti-D antibodies involved in HDN are of IgG1 or IgG3 subclass. IgG1 and IgG3 anti-D in combination cause more severe HDN than either antibody alone. IgG3 anti-D is more likely to cause fetal **hemolysis** than IgG1 anti-D, but it usually occurs in significant concentration only when both are present.

The primary response is often weak, and IgM is produced. Although it may develop as early as 4 weeks after sensitization, it may not be detectable for 6 months. A second exposure to Rh-positive red blood cells produces a rapid increase in IgG anti-D. RhD immunization occurs within 6 months after delivery of a first D-positive child. In 1967, Clark recognized the relationship between ABO incompatibility of mother and fetus with the occurrence of Rh-induced HDN. As a consequence of ABO incompatibility, transplacental hemorrhage induces Rh sensitization in only 8 to 9% of mothers. Some women who have no detectable antibodies are nonetheless immunized and may mount a secondary immune response in the next D-positive pregnancy, giving a true incidence of immunization of 16%. The risk in a second D-positive ABO-compatible pregnancy is similar, but this decreases thereafter, as there is a greater residual number of nonresponders to the D-antigen. ABO incompatibility partially protects against Rhesus immunization. This is due to rapid intravascular hemolysis of ABO-incompatible RhD-positive red cells. RhD immunization may also occur as a result of incompatible blood transfusion.

Hemolysis occurs in the spleen, leading to **anemia, erythropoietin** production, compensatory medullary and extramedullary erythropoiesis, and **hepatosplenomegaly**, with an outpouring of immature nucleated red cells. Anemia and intrahepatic circulatory obstruction cause hepatocellular damage, hypoalbuminemia, and generalized edema.

In utero, unconjugated bilirubin is mainly cleared across the placenta, conjugated by the maternal liver, and excreted. Despite placental clearance, the total bilirubin level in the severely affected fetus at birth may be very high. After birth, the infant's immature hepatic Y-transport and glucuronyl transferase mechanism is unable to conjugate the large amounts of bilirubin produced by red cell hemolysis. Untreated, the bilirubin-binding capacity of albumin is rapidly exceeded. Free, unconjugated bilirubin interferes with vital intracellular metabolic processes in the human and causes cell death (kernicterus).

Anti-Kell antibodies bind to erythroid progenitors, suppressing erythropoiesis. Fetal anemia due to anti-Kell may be severe in the absence of erythroblastosis, reticulocytosis, and hyperbilirubinemia. Levels of bilirubin in amniotic fluid may be misleadingly low. The Kell antibody titer is not predictive of severity.

Clinical Features

The most severely affected fetuses are grossly hydropic, the majority *in utero*. The occasional hydropic infant born alive presents with extreme pallor, marked hepatosplenomegaly, **petechiae**, and modest edema. Moderate disease and hydrops fetalis grade into each other. Mothers carrying severely affected erythroblastic fetuses not infrequently present with polyhydramnios and a preeclampsia-like syndrome. If the disease is ameliorated by intrauterine fetal transfusion, the maternal syndrome disappears.

Mild-to-moderate anemia affects 25 to 30% of fetuses that are born near term. They may or may not appear slightly jaundiced at birth. Hepatosplenomegaly is present, but edema is absent. Untreated, jaundice progresses rapidly (icterus gravis neonatorum), and within 2 to 4 days of birth there is evidence of brain damage. The neonates become hypotonic and refuse to suck, lose their neurovegetative reflexes, become spastic, lie in a position of opisthotonus, and may suffer from convulsions. Respiratory failure supervenes, and most infants die. The 10% who survive have severe brain damage.

About 50% of fetuses are only mildly affected. Mild-to-moderate hyperbilirubinemia and anemia occur, but recovery is complete.

HDN due to anti-c, anti-Kell, and anti-k (Cellano) does not differ in clinical expression from anti-D erythroblastosis. Anti-C and anti-E only rarely produce hydrops. Jaundice and the risk of kernicterus are the most significant clinical problems in anti-A, anti-B, anti-C, and anti-E hemolytic disease.

Assessment of Risk of HDN

Past History

HDN worsens with successive pregnancies.

Maternal Antenatal Screening

In all pregnancies, a sample of maternal blood is taken by venepuncture at first attendance, usually at 12 to 18 weeks of pregnancy, and examined for the presence of irregular antibodies by the **indirect antiglobulin (Coombs) test**. A further screen is undertaken at 28 and 34 weeks.

Maternal Alloantibody Titer

HDN requiring intervention is unlikely to occur at levels of anti-D <4 IU/ml in maternal serum. At levels 4 to 10 IU/ml, further intervention may be indicated if other adverse features are present. A level >10 IU/ml usually indicates that intervention is required. For other antibodies, a titer of 1:32 is usually taken as significant, but this must be considered in conjunction with other parameters. Generally, samples should be tested every 4 weeks before 24 to 28 weeks and every 2 weeks thereafter.

Determination of Fetal Blood Group

This can be ascertained by:

Chorionic villous sampling. This invasive procedure carries the risk of abortion.

Polymerase chain reaction of fetal cells in maternal blood (see **Molecular genetic analysis**). This recent development is under investigation for wide use and shows great future potential.[193]

Amniotic fluid analysis. The deviation in optical density from linearity at 450 nm, i.e., the delta-OD 450 reading, or "bilirubin bulge" first described by Liley, allows disease to be classified as mild, moderate, or severe according to the gestational age. The accuracy of prediction is greatest between 27 and 34 weeks, but is often inaccurate before 27 weeks of gestation. Molecular analysis of amniocytes using polymerase chain reaction (PCR) is used in some centers to determine the fetal RhD, Kell, and Fya type.

Ultrasound. High-quality ultrasound (US) allows an estimate of the placental and hepatic size and the presence or absence of edema, ascites, and other effusions. The placenta can be accurately located, which is of benefit when fetal transfusion is to be given. Doppler US scanning can be used to assess the velocity of fetal blood flow in a vessel, typically the middle cerebral artery. The velocity is inversely proportional to fetal Hb.

Cordocentesis (percutaneous umbilical blood sampling). In experienced hands, fetal blood sampling cordocentesis carries a risk on the order of 0.2 to 0.5%, but a very accurate profile of the severity of hemolytic disease can be obtained. It allows measurements of hemoglobin, hematocrit, blood group, serum bilirubin, platelet, and leukocyte counts. An antiglobulin test can also be performed. This is the most accurate way of determining the severity of HDN in the absence of hydrops. The traumatic fetal mortality rate is <1%. Cordocentesis is likely to cause fetomaternal hemorrhage and should, therefore, only be taken when it is felt that there is a high chance of moderate to severe HDN. Sampling may be started as early as 18 weeks gestation, and the most suitable site is the insertion of the umbilical vessel into the placenta. Transfusions may be required from the outset, and compatible phenotypically matched blood must be available for transfusion as well as for intravascular fetal transfusion (see below).

Management of HDN

Fetal Treatment

Early delivery: it is usual to deliver at 36 to 38 weeks gestation if:

- The risks of prematurity are outweighed by the greater risk of HDN
- Fetal transfusions have been given and further invasive procedures can be avoided

Intraperitoneal fetal transfusions: red blood cells placed in the peritoneal cavity are absorbed and function normally. Intraperitoneal fetal transfusion is of no value in moribund hydropic fetuses. The procedural mortality is as high as 7%. This has largely been superseded by intravascular fetal transfusion.

Intravascular fetal transfusion (IVT): this is done using a 20- or 22-gauge spinal needle introduced into an umbilical blood vessel (preferably the vein but occasionally the artery) under ultrasound guidance. It allows direct measurement of fetal blood parameters (see above) and increases the circulating maternofetal red cell mass immediately. Red cells are transfused in 10-ml aliquots while turbulence in the fetal blood vessel is observed. A mean of 50 ml/kg estimated nonhydropic fetal weight is given. Transfusions are discontinued if there is significant bradycardia or marked ventricular dilatation.

Neonatal Treatment

Exchange transfusion is the cornerstone of management (see **Fetal/neonatal transfusion**). At delivery, cord-blood findings in conjunction with the clinical appearance of the infant will determine the need for and the type of treatment. The direct antiglobulin test is invariably positive in all forms of erythroblastosis. Cord-blood hemoglobin and unconjugated bilirubin estimations reflect the severity of disease. Prompt treatment is indicated when the cord-blood hemoglobin is <11 to 11.5 g/dl or when the indirect bilirubin level is >75 to 85 µmol/l. If the infant is premature, prompt treatment should be carried out at hemoglobin levels <12.5 g/dl and bilirubin levels greater than 60 μmol/l. Reticulocyte counts are not of much value in management; however, nucleated red cell counts can be helpful in assessing the degree of disease and the need for treatment. Cord-blood levels of conjugated bilirubin greater than 45 μmol/l or the presence of heme (red-brown discoloration) is always indicative of severe hemolytic disease. A two-blood volume-exchange transfusion removes about 85 to 90% of coated hemolyzing red cells, preventing anemia and further bilirubin overproduction. Bilirubin removal can be increased by addition of 5 to 6 g of human albumin to each unit of blood used. The blood should be group O or group-specific if mother and baby are ABO-compatible but negative for the red blood cell antigen to which the mother is immunized. It must be tested for compatibility against maternal serum for the initial exchange and against the previous postexchange transfusion blood sample for subsequent exchange transfusions.

Whether treatment has been required initially or not, serial hemoglobin and unconjugated bilirubin measurements at 8- to 12-h intervals help determine management. Persistent elevation of conjugated bilirubin to 340 to 540 μmol/l is due to hepatocellular damage. However, if the infant survives, the conjugated bilirubin levels subside over the first few weeks or months, leaving no residual hepatic damage. Transient marrow inactivity with anemia is universal following the acute phase of erythroblastosis. The anemia is well tolerated, and no treatment should be undertaken unless the hemoglobin level drops below 7 g/dl.

Visual light in the blue spectrum bypasses the glucuronyl transferase mechanism, converting toxic unconjugated bilirubin into water-soluble nontoxic isomers. Phototherapy is a valuable adjunct in the management of all forms of neonatal hyperbilirubinemia, reducing the need for repeated exchange transfusions.

Administration of albumin reduces the risk of kernicterus and the need for exchange transfusion, but this should not be undertaken in the severely anemic infant in borderline cardiac failure.

Phenobarbital induces enzyme maturation when given to the mother (30 mg, three times daily) or to the infant (5 mg/kg per day) after birth. However, since there is a 48-h latent interval, it is not very effective once hyperbilirubinemia has developed.

Throughout the management of these very sick infants, every effort must be made to prevent hypoxia and acidosis, which, if present, materially increase the risk of kernicterus, cerebral hemorrhage, and cardiac and renal failure.

Prevention

Immunoprophylaxis of Rh HDN[247,248]

Injections of anti-D **immunoglobulin** (100 to 300 μg intramuscularly or 120 μg intravenously) should be given within 72 h of delivery of an RhD-positive infant to an unsensitized RhD-negative mother. If transplacental fetomaternal hemorrhage, estimated by **Kleihauer test** or **flow cytometry**,[249] exceeds the calculated protection of the dose administered, extra anti-D should be given as required, i.e., 25 μg (125 IU) for every 1 ml of fetal blood.

Anti-D immunoglobulin should also be given:

When RhD-incompatible platelets have been given

Following spontaneous or induced abortion

Following chorionic villus sampling

Following amniocentesis or cordocentesis (if procedures are repeated, the dose should be repeated every 6 weeks)

Following external cephalic version or vaginal bleeding during pregnancy (may be associated with fetomaternal hemorrhage)

In addition, prophylaxis has been recommended for all Rh-negative women antenatally as well as postnatally, either 100 μg (500 IU) intramuscularly at 28 and 34 weeks or 300 μg (1500 IU) at 28 weeks.[250,251] The mechanism whereby RhD immunoglobulin prevents immunization is unknown, although it mediates accelerated clearance of Rh-positive red cells.

Immunoprophylaxis has reduced the incidence of Rh-immunized women in the U.K. to around 1% and deaths from HDN of 46/100,000 births in 1969 to 1.6/100,000 births in 1990. Further reduction depends upon greater compliance with recommended procedures, particularly for those exposed to sensitization during pregnancy. Investigations are proceeding into the safety of using a single dose of rhesus (D) immunoglobulin at 28 to 30 weeks only. Determination of fetal blood groups could reduce the need for Rh immunoprophylaxis where there is an Rh-negative or ABO-incompatible fetus. This procedure, if widely used, could reduce the requirement for RhD immunoglobulin by around 40%.

Intensive Maternal Plasma Exchange

This is a costly procedure, and its benefits are equivocal. Although it reduces anti-D levels by about 75%, it should be reserved for the pregnant woman with a history of hydropic fetal death before 26 weeks gestation.

HEMOLYTIC TRANSFUSION REACTION

See **Blood components for transfusion** — complications.

HEMOLYTIC UREMIC SYNDROME

(HUS) Acute renal failure combined with **thrombocytopenia** and **microangiopathic hemolytic anemia** (MAHA). It is predominantly a disease of children, less frequently adults or postpartum, characterized by extensive damage to the vascular endothelium. The presenting features are bloody diarrhea with hematuria, anuria, or oliguria.

Etiology

The infecting organisms are multiple, but most are the consequence of:

Infections with Escherichia coli, *producing verocytotoxin*: the most frequently identified is *E. coli 0157:H7*.[252] Verocytotoxins (VT1 and VT2) are potent exotoxins that produce an irreversible effect upon certain cell lines (Vero and HeLa). *E. coli 0157:H7* is widely distributed in the guts of animals, and outbreaks are frequently associated with hamburger meats and unpasteurized milk.

Infection with Shigella dysenteriae *type I, producing verrotoxin VT1*: this appears to be the commonest cause of acute renal failure in children under 5 years of age in the Indian subcontinent. The shiga toxins damage the renal and endothelial cells by inhibiting protein synthesis or by damaging the nuclear genetic material.

Infections with neuraminidase-producing organisms: e.g., pneumococcus; a rare cause of HUS.

Miscellaneous viral infections: e.g., echovirus, adenovirus, occasionally HIV.

Pathogenesis

Damage to the **vascular endothelium**, probably by toxin, augments **thrombin** generation and impairs **fibrinolysis**, giving rise to **disseminated intravascular coagulation**. This leads to the formation of microthrombi, particularly affecting the renal glomeruli. The production of circulating **fibrin** causes mechanical damage to circulating red blood cells, **fragmentation hemolytic anemia**, and **intravascular hemolysis**. There may be, in addition, consumption coagulopathy.

Clinical Features

Renal damage with micro or gross hematuria is common together with proteinuria. Urine volumes may be decreased or increased. In the D^+ form of HUS, the blood pressure is usually normal, but in the D^- form the blood pressure is often markedly elevated. Involvement of the central nervous system is uncommon, although in the very ill individual there may be CNS (central nervous system) signs and symptoms. Patients usually give a history of diarrhea.

Laboratory Features[252a]

Anemia with raised **reticulocyte count** and fragmented red blood cells, i.e., schistocytes (helmet cells, burr cells), upon peripheral-blood film examination

Neutrophilia

Thrombocytopenia

Prolonged **prothrombin time** (PT), prolonged **activated partial thromboplastin time** (APTT), prolonged **thrombin time** (TT), and low **fibrinogen**; the **Euglobulin clot lysis time** may be prolonged

Hemoglobinemia, hemoglobinuria, methemalbuminemia, reduced **haptoglobin**, and raised **bilirubin; hemosiderinuria** occurs later

Treatment

Treatment is essentially supportive. Individuals who are anuric or oliguric may require dialysis until renal function returns. **Plasmapheresis** or the infusion of plasma appears to be of little benefit. The value of antibiotic therapy when infection is thought to be the cause remains uncertain. A recovery rate of 95% is to be expected.

Prevention

This depends upon safe food handling and cooking at temperatures sufficient to kill any contaminating organisms. The U.S. Food and Drug Administration (FDA) has approved the irradiation of red meat.

HEMOPEXIN

See **Hemoglobin** — degradation.

HEMOPHAGOCYTIC HISTIOCYTOSIS

See **Histiocytosis**.

HEMOPHAGOCYTOSIS
See **Histiocytosis**.

HEMOPHILIA A

(Factor VIII deficiency) An X-linked disorder causing a deficiency of coagulation-**factor VIII** in males, the most common of the inherited bleeding disorders. About one-third of all patients have no family history and result from a spontaneous mutation somewhere within the kindred. About 1 in 5000 males are affected. There is no evidence to suggest a variation in incidence among different ethnic groups. Numerous mutations of the factor VIII gene have been identified that lead to hemophilia A of varying degrees of severity. Approximately 50% of all cases of severe hemophilia A are due to a large inversion of the factor VIII gene (focusing on intron 22) on the X chromosome.

The level of factor VIII determines the clinical severity of the disorder, the normal range being 50 to 150 U/dl. Severe hemophilia A (VIII:C <2 U/dl) accounts for about 50% of all cases of hemophilia A and is characterized by spontaneous bleeds, notably into joints and muscles. Recurrent painful hemarthroses and muscle hematomas dominate the picture and, untreated, lead to progressive arthritis, deformity, and crippling. Internal hemorrhage may occur. Although uncommon, spontaneous cerebral hemorrhage is an important cause of death in patients with severe disease. In moderate hemophilia A (VIII:C 2 to 5 U/dl), bleeding usually occurs after minor trauma, although these patients may have occasional spontaneous bleeds. Mild hemophilia (VIII:C >5 u/dl) is a very broad group with a range of clinical features determined by the residual factor VIII level. In its mildest form, bleeding occurs only after surgery (including dental surgery) or trauma. It is important to note that operative and posttraumatic bleeding may be life threatening in both mild and severe hemophilia if uncontrolled. Severe hemophilia A usually presents in early childhood. Mild hemophilia may not present until adult life after a notable hemostatic challenge.

Laboratory Features

The diagnosis is suggested by an isolated prolongation of the **activated partial thromboplastin time** and confirmed by a low factor VIII level on a specific assay. The **prothrombin time**, **thrombin time**, and **bleeding time** are all normal, as is **Von Willebrand Factor**, the factor VIII carrier protein.

Management

Patients with hemophilia should be registered and attend specialist hemophilia centers. The World Federation of Hemophilia states that the aim of therapy is to "minimize disability and prolong life, to facilitate general and social and physical well-being and to help each patient achieve full potential." Bleeding problems are treated by elevating the factor VIII level toward normal. This can be done by the administration of **coagulation-factor concentrates**, with factor VIII as replacement therapy or, for mild patients, by administration of **DDAVP**. To control spontaneous hemorrhage, the factor VIII level must be raised to above 20% of normal, although generally more than this is given. For severe hemorrhage or to cover surgical procedures, the level must be raised to 100% of normal and maintained above 60% until the hemorrhage has stopped or healing has occurred. As the half-life of factor VIII is approximately 12 h, replacement therapy may need to be given twice daily. Antifibrinolytic therapy, such as **tranexamic acid**, is a useful adjunctive treatment.

The availability of coagulation-factor concentrates for factor VIII has revolutionized the treatment of hemophilia A. Patients are taught to self-administer factor VIII at home at the earliest sign of bleeding. Such on-demand therapy has significantly reduced morbidity associated with chronic and recurrent joint and muscle hemorrhage. Many patients are now in adulthood with little joint damage. With the increased safety and availability of factor VIII concentrates today, prophylaxis regimens are regularly used, particularly in young children, and are the standard of care. Factor VIII concentrate is administered three times per week at a dose of 25 to 40 U/kg. Using such a regimen, bleeding episodes are significantly reduced in number, as is concomitant joint damage. Prophylaxis is particularly encouraged in children to preserve normal joint function. However, problems with venous access in young children may necessitate insertion of a subcutaneous implantable venous-access device to allow regular administration of factor VIII concentrate.

Plasma-derived factor VIII concentrates have been associated with complications such as **transfusion-transmitted infections** and inhibitor formation. Many patients with hemophilia A have been infected with either hepatitis C253 or HIV. Progressive **HIV** infection and progressive liver cirrhosis leading to hepatocellular carcinoma are problems in patients previously infected. As of 2005, more than 60% of those infected with HIV during the 1980s have died of **acquired immunodeficiency syndrome** (AIDS). The long-term survivors are today effectively managed by highly active antiretroviral therapy (HARRT), but coinfection with hepatitis C continues to cause deaths. For the future, these infections can be avoided by the administration of recombinant factor VIII (see **Coagulation-factor concentrates**), but at a considerably higher economic cost.

The presence of inhibitors — antibodies to factor VIII — may be a most serious complication requiring a high level of treatment (see **Factor VIII** — antibodies to Factor VIII).

Advances in treatment now allow patients to lead a relatively normal lifestyle, although hemophilia can impose significant social and physiological burdens on the patient and his family. Families often require extensive help and support in these areas as well as advice in terms of seeking appropriate future employment. Female members of such families will also have concerns related to bearing sons with the same affliction. For parents dealing with the condition, the discomfort and disruption it causes their son during childhood, and the ensuing treatment, can be notably difficult. A normal schooling is encouraged, although contact sports are to be avoided.

Hemophilia A is a disruptive and potentially disabling condition. However, advances in treatment, improved safety of therapeutic products, and availability over recent years allow for an optimistic prognosis. It is an ideal target for somatic gene therapy.

Carrier Status

Although hemophilia A is an X-linked disorder and manifests only in males, it is important to identify the female members of the kindred that carry the hemophilia-A gene. Carrier females may exhibit low levels of factor VIII activity due to **Lyonisation** of the X chromosome. Levels in some carrier females may be <10 U/dl. Carrier females with low levels of factor VIII activity may bleed and require factor VIII therapy after trauma and surgery, as would a comparable male with mild hemophilia. Consequently, all females in a kindred with hemophilia should have their factor VIII level determined.

Identification of carrier status is important for female members of the affected family. Women should be offered genetic counseling. Prenatal diagnosis is possible, and women should be nondirectionally counseled about these issues. Evidence for carrier status can be obtained by measuring factor VIII level, as outlined above. Von Willebrand factor can be used as an additional discriminant, but such phenotypic data cannot rule out carrier status. The principle of linkage using **DNA** markers has been widely applied to the definitive determination of carrier status. **Restriction-fragment-length polymorphisms** within or close to the factor VIII gene allow the mutant allele to be tracked with confidence through the kindred, enabling carrier status to be confidently determined in the majority of women. Nowadays, with the improvement of knowledge and **molecular genetic analysis**, direct sequence detection of carrier defects is the normal method for diagnosis, and this technology can also be used for antenatal testing. Samples of chorionic villus, taken at around 10 weeks gestation, can be similarly analyzed for the presence of the mutant allele, enabling selective termination of affected males.

HEMOPHILIA B

(Factor IX deficiency, Christmas disease) An X-linked disorder causing a deficiency of coagulation-**factor IX** in males. It is approximately one-sixth as common as hemophilia A, with an incidence of around 1 in 30,000 live male births. The factor IX gene also lies in relatively close proximity to the factor VIII gene at the tip of the X chromosome. Hemophilia B resembles hemophilia A to the extent that they cannot be distinguished clinically (for clinical features, see **Hemophilia A**), and indeed they were not differentiated historically. The two conditions can only be distinguished by specific factor assays. The marked phenotypic resemblance is due to a common final pathology, that is, failure of the factor IX/VIII complex responsible for generation of factor Xa.

The clinical features of hemophilia B show the same relationship to level of residual factor IX activity as does hemophilia A to residual factor VIII activity (see hemophilia A). The diagnosis is suggested by an isolated prolongation of the **activated partial thromboplastin time** (APTT) with a low factor IX activity upon specific assay.

Management

The mainstay of treatment is **coagulation-factor concentrates** — factor IX used in on-demand or prophylaxis programs as with hemophilia A. Factor IX has a longer half-life than factor VIII (24 vs. 12 h). Replacement therapy is usually administered on a daily basis for acute bleeding episodes and twice weekly for prophylaxis. Recombinant factor IX is now widely available. High-purity plasma-derived factor IX concentrates are also available. Both preparations reduce the risk of thromboembolism seen with previous prothrombin complex concentrates. Complications due to therapy and general principles of management are the same as for hemophilia A (see **Hemophilia A**).

Inhibitors to factor IX occur with a significantly lower frequency (approximately 2% to 3%) than factor VIII inhibitors in hemophilia A. They are often, but not invariably, associated with a significant gene deletion or mutation with the first 10 to 15% of coding sequence.

Hemophilia B is a disruptive and disabling condition. However, advances in management, improved safety of therapeutic products, and availability allow for an optimistic prognosis. As the factor IX gene is smaller and simpler than the factor VIII gene, hemophilia B is the ideal model for somatic gene therapy.

Carrier Status

The management of carriers is also similar in principle to hemophilia A, as is carrier detection and antenatal diagnosis. Discriminant analysis of phenotypic carrier status employs factor IX activity and factor IX antigen levels. As the factor IX gene is significantly smaller and less complex than the factor VIII gene, there is a much greater understanding of the molecular mechanisms and identification of mutations responsible for causing hemophilia B. This has enabled antenatal diagnosis and carrier detection to employ direct mutation analysis, which in many cases provides an earlier diagnosis than **restriction fragment length polymorphism** (RFLP) analysis. Direct mutation detection using **molecular genetic analysis** is now the normal method of diagnosis.

HEMORRHAGE

(Bleeding) The loss of blood from the circulation, which may be sudden (acute hemorrhage) or occur over a prolonged period (chronic hemorrhage), always resulting in anemia.

Acute Hemorrhage

This occurs as a result of traumatic accidents or from bleeding from a diseased organ:

Epistaxis: see **Nasal disorders**

Hematemesis: see **Esophageal disorders**, **Gastric disorders**, **Intestinal tract disorders**

Melena: see **Intestinal tract disorders**

Hemoptysis: see **Respiratory tract disorders**

Hematuria: see **Renal tract disorders**

Menorrhagia: see **Gynecological disorders**

Acute hemorrhage may complicate pregnancy either antepartum (accidental hemorrhage) or postpartum. Bleeding may be internal into a pleural sac, peritoneum, joints, a muscle mass, or into a cyst. It may be the consequence of a **hemorrhagic disorder**.

The immediate effect of acute hemorrhage is a fall in total **blood volume**, the consequences of which vary according to the age of the patient. Young, otherwise healthy individuals tolerate a loss of 30% (1500 ml) of blood volume with only mild postural hypotension. Older subjects or those with concomitant cardiac or respiratory disease quickly develop peripheral circulatory failure (shock) with smaller losses. The rapid loss of 750 ml causes a substantial fall in cardiac output and blood pressure, with peripheral **vasoconstriction**. Hypovolemic shock occurs when the loss exceeds 40% of blood volume (2000 ml). The signs of hypovolemic shock include anxiety and confusion, rapid weak pulse, cold clammy skin, and air hunger. With losses up to 50% blood volume, cardiac failure and death are imminent unless plasma volume expanders are administered immediately. It is the hypovolemia causing inadequate tissue perfusion with acute renal failure and cerebral ischemia, rather than the depletion of red blood cells, that poses the immediate threat, death occurring with a sudden loss of more than 33% or a prolonged loss of more than 50% over 24 h.

Physiological restoration of the plasma volume is slow, so that **hemoconcentration** occurs and the degree of **anemia** is masked. With survival, and provided that no further bleeding occurs, the plasma volume rises in 4 to 6 h, with consequent reversal of the hemoconcentration, which is usually complete in 24 h but with a temporary dilution of blood causing a fall in the **red blood cell count**. The acute anemia induces marrow stimulation, leading initially to a raised **reticulocyte count**, the degree of which is an indicator of erythropoietic activity. **Thrombocytosis**, followed by **neutrophilia** and sometimes a **leukemoid reaction**, occurs. The reticulocytosis with macrocytosis occurs within 1 to 2 days, reaching a peak of 5 to 15% between the fourth and seventh day. Leukocytes return to normal in 3 to 4 days and reticulocytes in 5 to 6 weeks.

Treatment is to stop the bleeding immediately by whatever means appropriate to the site and cause, and to rapidly replace the blood volume by infusion of fluids, followed by transfusion of red blood cells (see **Red blood cell transfusion**). Prolonged bleeding can only be controlled by massive blood transfusion, following which **platelet transfusion** will become necessary to overcome further bleeding from **thrombocytopenia**. Later, treatment of any consequent **iron deficiency** anemia may hasten recovery.

Chronic Hemorrhage

This occurs from any site of ulceration, particularly in the gastrointestinal or renal tracts. Chronic pulmonary hemorrhage leading to pulmonary hemosiderosis is a rare cause of chronic blood loss. The outcome of all these events is iron deficiency anemia.

HEMORRHAGIC BULLAE

(Blood blisters) Elevated circumscribed lesions filled with blood. Technically, a bulla is a lesion of greater than 0.5 cm diameter, with vesicles being lesions smaller than 0.5 cm. The colloquial term blood blister is loosely used for all such lesions. Hemorrhagic bullae occur in extreme **thrombocytopenia** (platelet often less than $10 \times 10^9/l$) and signify the need for intervention.

HEMORRHAGIC DISEASE OF THE NEWBORN

A neonatal disorder with bleeding due to a deficiency of **vitamin K**. Three types are recognized:

Classical form. This occurs in breast-fed, full-term infants, usually on the second or third day of life, and presents with widespread bruising, gastrointestinal bleeding, bleeding from venepuncture site, and intracranial hemorrhage. Investigations show a grossly prolonged **prothrombin time** and **activated partial thromboplastin time**. The deficiency arises because of the low transplacental passage of vitamin K during pregnancy, the low levels of vitamin K in breast milk, the sterile gut of the neonate, and because breast-fed neonates often drink very little in the first few days of life. Bottle-fed infants rarely suffer from vitamin K-deficient bleeding as they appear to absorb sufficient vitamin K from the milk. The treatment is essentially preventative and prophylactic administration of vitamin K (1 mg *i.m.*) to the neonate, usually immediately after delivery. Vitamin K may also be given orally, but multiple doses are required, so there are risks that the full dose will not be given or that it is vomited. Some concerns have been expressed over a possible link between intramuscular administration of vitamin K and an increased risk of childhood cancer that is now discounted. There is general agreement that those at particular risk due to prematurity, complicated delivery, sepsis, maternal anticonvulsant therapy, and infants requiring surgery should receive i.m. vitamin K.[254]

Bleeding within the first 24 h of life. This form arises because the mother has taken drugs during the last trimester of the pregnancy that interfere with vitamin K metabolism, e.g., anticonvulsants, warfarin. Women on anticonvulsants in pregnancy should receive additional vitamin K and, similarly, infants at high risk of developing vitamin K-deficient bleeding may require additional prophylaxis.

Bleeding after the first week of life. Usually is secondary to some malabsorptive disease, e.g., celiac disease, cystic fibrosis, a1-antitrypsin deficiency. Treatment depends upon the underlying condition. Parenteral vitamin K may be of help.

HEMORRHAGIC DISORDERS

These diseases are characterized by recurrent hemorrhage occurring spontaneously or as a result of minor trauma. They arise from:

Coagulation-factor-deficiency disorders

Thrombocytopenia

Platelet-function disorders

Vascular purpura

Excessive **fibrinolysis**

A clinical diagnosis can be made based upon the form of the hemorrhage but requires confirmation by laboratory investigation. **Hemarthroses** and large muscle bleeds are characteristic of coagulation-deficiency disorders, such as **hemophilia A**. Similar episodes occur in **hemophilia B**. Screening tests of **prothrombin time** (PT) and **activated partial thromboplastin time** (APTT) will differentiate between deficiencies of **factors II, V, VII, and X**, such as occur with excessive **anticoagulant therapy** or **liver disorders** and deficiencies of **factor VIII** (hemophilia A) and **factor IX** (hemophilia B).

Multiple purple spots seen following extravasation of blood from capillaries as **purpura** is seen in the skin, mucous membranes, and retinas can be caused by thrombocytopenia, vascular disorders, or disturbance of platelet function. Platelet counting will differentiate between thrombocytopenia and vascular or platelet-function disorders. Vascular anomalies are often diagnosed by the clinical history and dermatological examination, but for platelet function, special tests after the patient has been fasting are necessary.

Excessive fibrinolysis is comparatively rare, being characterized by an acute onset of generalized bleeding from the orifices and into the skin. It is always an emergency state, confirmation by laboratory tests being difficult. A prolonged thrombin time can be a screening test indicator, but the euglobulin clot lysis is the best definitive test if readily available.

Coagulation Factor Deficiency Disorders
Hereditary

Factor VIII: hemophilia A

Factor IX: hemophilia B

Other coagulation-factor deficiencies occur but are rare.

Acquired

Factor deficiencies are usually mixed for each cause.

Vitamin K deficiency: factors II, VII, IX, and X

- **Hemorrhagic disease of the newborn**

- Dietary deficiency
- Malabsorption due to intestinal tract disorders and chronic biliary obstruction

Liver disorders: factor V, dysfibrinogenemia

Coumarin therapy: factors II, VII, IX, and X

Antibody inhibitors: factor VIII common following treatment with factor VIII concentrate; factors V, IX, and XIII rare

Disseminated intravascular coagulation: consumption of all coagulation factors, particularly V, VIII, and IX

Massive **red blood cell transfusion**: dilution of all coagulation factors

Lupus anticoagulants bind phospholipids and so prolong the clotting times of such tests as PT and APTT.

Thrombocytopenia

This can be diagnosed when a platelet count below the lower limits of the **reference range**, i.e., $140 \times 10^9/l$, is found. This must allow for the normal lower levels found in neonates and infancy. Thrombocytopenia occurs when platelets are lost from the circulation faster than they can be replaced. **Pseudothrombocytopenia** must be excluded by examination of the blood film as a cause of thrombocytopenia before considering other causes.

True thrombocytopenia may result from failure of platelet production, an increased rate of destruction of platelets, or loss/removal from the circulation. A simplified classification of thrombocytopenia is given in Table 146. A combination of mechanisms may account for thrombocytopenia in some situations. For example, in **chronic lymphocytic leukemia**, thrombocytopenia may arise from a combination of defective marrow production due to marrow infiltration, immune-mediated destruction due to circulating autoantibodies, and splenic sequestration secondary to splenomegaly. However, as a general rule, when thrombocytopenia is due to defective marrow production, there is a reduction in the number of megakaryocytes in the bone marrow, whereas in thrombocytopenia due to decreased platelet survival, there is an increase in the number of marrow megakaryocytes. Bone marrow examination is an integral part of the investigation of thrombocytopenia unless the cause is obvious.

Thrombocytopenia rarely leads to bleeding unless the platelet count is below $75 \times 10^9/l$. There is no clear relationship between severity of thrombocytopenia and severity of bleeding symptoms. However, spontaneous hemorrhage may be expected when platelet counts are below $10 \times 10^9/l$. The **bleeding time** is prolonged in parallel to the platelet count when this is below $100 \times 10^9/l$.

After defective marrow production of platelets, decreased platelet survival due to immune mechanisms form a large group. There is a range of mechanisms involved in immune destruction of platelets, from classical **immune thrombocytopenic purpura** to nonspecific immune mechanisms. Raised platelet-associated immunoglobulin may be seen in many of these other immune-destructive mechanisms. These may be due to platelet alloantibody or autoantibody bound to the platelet surface or to passive adsorption of antigen–antibody immune complex or antibody to a nonplatelet antigen that has been adsorbed onto the platelet surface. Any of these mechanisms will lead to removal of platelets by the reticuloendothelial system. Conditions that are associated with raised platelet-associated immunoglobulin and nonspecific immune destruction include **malaria**, **lymphoproliferative disorders**, **viral infections**, **acquired immunodeficiency syndrome** associated thrombocytopenia, **systemic lupus erythematosus**, preeclampsia, and some

drug-induced purpuras. Treatment of thrombocytopenia in such cases is variable, dependent upon the cause, but broadly is directed at the underlying cause and immunomodulation with steroids and/or immunoglobulin, if appropriate.

Platelet-Function Disorders

These are both inherited and acquired (see Table 128).

Vascular Purpura

These lesions represent breakdown in primary hemostasis unassociated with primary defects in blood coagulation or platelets. A classification of vascular purpuras is given in Table 152.

Excessive Fibrinolysis

Disordered fibrinolysis is recognized to be associated with both bleeding and thrombotic complications (see Table 65).

HEMOSIDERIN

The intracellular, yellowish, iron-containing pigment visible under light microscopy in iron-loaded cells or tissues. Electron microscopy reveals it as a composition of dense clusters of **ferritin** that has been converted into hemosiderin by partial degradation of its protein shell by lysosomal enzymes. Hemosiderin is found mainly in cells of the monocyte-macrophage system located in bone marrow, spleen, and liver (Küpffer cells).

HEMOSIDERINURIA

This is a sequel to the presence of **hemoglobin** in the glomerular filtrate. This hemoglobin is reabsorbed by the renal tubular cells and broken down into hemosiderin. The hemosiderin-containing tubular cells are then shed in the urine and can be demonstrated by performing a Prussian blue stain on a centrifuged urine sediment. A true positive shows blue-staining granules within intact tubular cells. Free iron outside tubular cells merely indicates contamination. A positive test is a valuable sign of **intravascular hemolysis**; however, hemosiderinuria is not found at the onset of a hemolytic attack, even if **hemoglobinemia** and **hemoglobinuria** are present. Hemosiderinuria persists for several weeks following a hemolytic episode.

HEMOSIDEROSIS

See also **Iron overload**.

Hemosiderosis indicates the presence of systemic iron overload due to excessive **red blood cell transfusion** or, less frequently, to the chronic ingestion of supplemental **iron** or receipt of excessive parenteral iron supplements. Systemic hemosiderosis can cause functional and structural damage to the heart, liver, and endocrine tissues. This term also connotes localized accumulation of excess iron in tissues identified using light microscopy and the Perl's stain (see **Prussian blue reaction**).

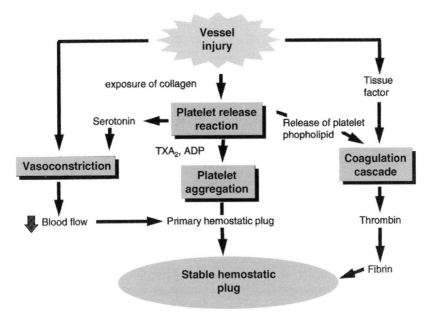

FIGURE 66
Diagrammatic representation of the overall hemostatic response following vessel wall injury.

HEMOSTASIS

After tissue injury, protection of the integrity of the vascular system is maintained by this host defense mechanism. It works in conjunction with the mechanisms of **inflammation** and repair to produce a coordinated response. Hemostatic systems are normally quiescent, but following tissue injury or damage they become rapidly activated, involving **coagulation-factor** proteins, **platelets**, **vascular endothelium**, and **leukocytes**. The coordinated hemostatic response ultimately produces a **platelet plug**, fibrin-based clot, deposition of white cells at the point of injury, and activation of inflammatory and repair processes.

All components of the hemostatic mechanism exist under resting conditions in an inactive form. A diagrammatic representation of the overall hemostatic response is shown in Figure 66.

Following injury there is immediate **vasoconstriction** and reflex constriction of adjacent small arteries. This slows blood flow into the damaged area. The reduced blood flow allows contact activation of **platelets**. Upon activation by tissue injury (or other agonists), platelets undergo a series of physical, biochemical, and morphological changes. Platelets adhere to exposed connective tissue, mediated in part by **Von Willebrand Factor**. Exposure of **collagen** and local **thrombin** generation (see below) lead to secretion of platelet granule contents. Secretion of platelet granule contents, which include **adenosine diphosphate** (ADP), serotonin, and **fibrinogen**, further enhances platelets to be activated, form aggregates, and interact with other platelets and leukocytes. This forms the initial platelet plug. In addition, the vascular endothelium undergoes a series of changes to move from its resting phase (with predominantly anticoagulant properties) to a more active procoagulant and repair phase (see **Vascular endothelium**). In concert with these cellular changes, inactive plasma coagulation factors are converted to their respective active species by cleavage on one or two internal peptide bonds. In sequence, these active factors generate

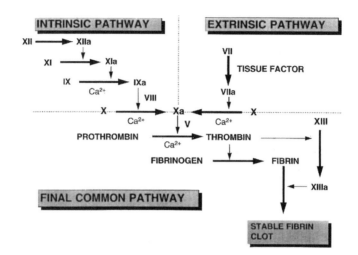

FIGURE 67
Classical "waterfall" hypothesis of coagulation.

thrombin, which leads to formation of fibrin from fibrinogen (to stabilize the platelet plug), cross linking of the fibrin formed (via activation of **factor XIII**), further activation of platelets, and also activation of fibrinolytic pathways (to enable **plasmin** to dissolve fibrin strands in the course of wound healing). Additionally, thrombin interacts with other nonhemostatic systems to promote cellular chemotaxis, fibroblast growth, and wound repair.

Coagulation Cascade

The classic "waterfall" hypothesis for coagulation proposes the so-called intrinsic and extrinsic pathways (see Figure 67). The intrinsic system assumes that exposure of contact factors (**factor XII**, high-molecular-weight **kininogens, prekallikrein**) to an abnormal/injured vascular surface leads to activation of **factor XI**, which in turn activates **factor IX**. Activated factor IX, in the presence of its cofactor **factor VIII**, then activates **factor X** to factor Xa in the presence of phospholipid. In turn, factor Xa with its cofactor **factor V** together form the prothrombinase complex, which converts **prothrombin** to thrombin. Thrombin then converts fibrinogen to fibrin. The extrinsic system assumes that **factor VII** and **tissue factor**, released from damaged vessels, directly activate factor X and coagulation factor lying below factor X in the final common pathway.

The division into extrinsic and intrinsic systems and the ability to test these two systems in the laboratory (the **prothrombin time** and **activated partial thromboplastin time**, respectively) has been valuable in understanding clinical bleeding problems, but it fails to represent accurately what happens *in vivo* in hemostasis. Findings concerning direct activation of factor IX by tissue factor–factor VII complex have led to revision of the coagulation cascade, with tissue factor–factor VII complex, **tissue factor pathway inhibitor** (TFPI), and factor X central to the model (see Figure 68).

Coagulation is initiated when tissue damage exposes blood to tissue factor beneath the endothelium. Factor VII binds tissue factor and the tissue factor–factor VII complex directly activates factor X to factor Xa and some factor IX to factor IXa. In the presence of factor Xa, TFPI then inhibits further generation of factor Xa and factor IXa by inhibiting tissue factor–factor VIIa complex. Under these conditions, further factor Xa can only be generated by the factor IXa/factor VIIIa pathway. Enough thrombin exists at this point to

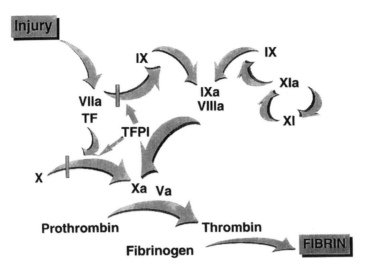

FIGURE 68
A revised coagulation cascade.

activate factor VIII to factor VIIIa, and the initial generation of factor IXa (by tissue factor–factor VIIa) allows this route to factor Xa generation to proceed. Further factor IXa is produced via thrombin activation of the factor XI pathway.

The revised cascade assumes that tissue factor–factor Xa complex is responsible for the initial generation of factor Xa and thrombin, sufficient to activate factor V, factor VIII, and platelet aggregation locally. Following inhibition by TFPI, the amount of factor Xa produced is insufficient to maintain coagulation, and therefore factor Xa generation must be amplified using factor IXa and factor VIIIa to allow hemostasis to progress to completion.

Unlike the waterfall hypothesis, the revised hypothesis does not assume that initial generation of factor Xa and thrombin is the end of the hemostatic process. Rather, it assumes that following initial generation, the hemostatic response must be reinforced or consolidated by a further progressive generation of factor Xa and thrombin. This allows the hypothesis to encompass the competing influences of inhibitors of coagulation, blood flow washing away activated coagulation factors, and thrombin-activated fibrinolysis. Additionally, it does not include all factors known to be involved in blood coagulation.

The revised hypothesis also allows a better explanation of bleeding seen in **hemophilia A** and **hemophilia B**. In these two conditions, bleeding occurs both spontaneously (intrinsic system) and after trauma (extrinsic system), which cannot easily be reconciled on the classical waterfall hypothesis. Using the revised schema, it is clear that without factor VIII or factor IX, bleeding will ensue because the amplification and consolidating generation of factor Xa is insufficient to sustain hemostasis.

HEMOSTATIC PLUG

See **Hemostasis**.

HENOCH-SCHÖNLEIN PURPURA

(Anaphylactoid purpura) A form of **leukocytoclastic vasculitis** with lesions affecting the skin, joints, intestines, and kidneys. It most frequently occurs in children following streptococcal infection, drug hypersensitivity (e.g., aspirin, carbidopa/levodopa,[255] erythromycin, penicillin, paracetamol, phenacetin, sulfonamides, thiazide diuretics), or food allergy. Purpura usually evolves from erythematous macules of the lower extremities. It is associated with abdominal pain, melena, arthritis, and ankle edema due to proteinuria from deposition of IgA immune complexes in the glomerular vessel mesangium.[255a] It is usually self-limiting, but with persistence, especially of glomerulonephritis, **immunosuppression** with **corticosteroids** or **azathioprine** may be effective.

HEPARIN

A sulfated glycosaminoglycan isolated from the intestinal mucosa of various animals (cattle and pigs) or from bovine lung tissue. The major source is hog intestinal mucosa. Heparin consists of alternating chains of uronic acid and glycosamine. The anticoagulation properties of heparin reside in a five-sugar (pentasaccharide) sequence, which binds tightly to argine and lysine sites on antithrombin. In its unfractionated form, heparin consists of chains of varying molecular weights. Heparin binds to **antithrombin III**, inducing a conformational change in the protein that results in increased inhibitory activity of antithrombin toward many of the coagulation serine proteases (at least 5000-fold). Physiologically, free heparin is not found in the circulation, but heparan sulfate, located on the surface of vascular endothelial cells, appears to serve the same role. Heparin is widely used as an anticoagulant both therapeutically and prophylactically. It has a rapid onset of action and a short $T^{1/2}$, but it can only be administered parenterally, as it is rapidly destroyed by enzymes within the gut. Heparin can be enzymatically fractionated or chemically cleaved to enrich for smaller, low-molecular-weight species with a mean molecular weight of 5 kDa. Unfractionated heparin (UFH) has both anti-factor Xa and anti-factor IIa activity, but low-molecular-weight heparin (LMWH) has primarily anti-factor Xa activity, as there are fewer longer species that can effectively cross link the heparin antithrombin complex to thrombin.[255b]

Unfractionated Heparin

(UFH) This form of heparin is extremely heterogeneous and comprises 10 to 100 saccharide units with molecular weight range 3 to 30 kDa (mean 15 kDa). Only about 25% of UFH has anticoagulant activity. The remainder may be responsible for the deleterious side effects, particularly heparin-induced **thrombocytopenia** (HIT). Conventionally, UFH is given therapeutically in a loading dose of 5000 U followed by a continuous infusion of 1000 units/h (24,000 U/day), and its anticoagulant action is monitored by measuring the **activated partial thromboplastin time** (APTT) 6 h after starting treatment. The APTT should be maintained at 1.5 to 2.5 times the midpoint of the normal range. Because of the capacity of heparins to bind nonspecifically to a number of plasma proteins, the pharmacokinetics of unfractionated heparin are dose dependent. These components vary significantly between individuals and with time. As a consequence, intravenous infusions of heparin require close monitoring and should be checked at least twice daily. The half-life of the anticoagulant activity of heparin in humans is generally regarded as being approximately 1 h. Heparin can also be given prophylactically in a dose of 5000 U subcutaneously twice or thrice daily. In this case, no prolongation of the APTT should be obtained. Heparin is safe to administer in pregnancy, as it does not cross the placenta.

In cases of overdose, the heparin infusion is discontinued, and in cases of life-threatening hemorrhage, it can be reversed by the administration of protamine sulfate. Protamine is a basic protein derived from fish sperm that binds to heparin to form a stable salt. The precise dose of protamine to administer can be calculated from a protamine sulfate correction test. Protamine in excess dosage acts as an anticoagulant. When there are fewer than 18 saccharide units, heparin is progressively more resistant to neutralization by protamine. Low-molecular-weight heparin is, therefore, less well antagonized by protamine and may not be completely neutralized.

The major side effect of heparin is **hemorrhage**. Osteoporosis may occur with prolonged heparin administration (e.g., during pregnancy, renal dialysis). Recent animal work suggests the osteoporosis side effects may be less with low-molecular-weight heparins. The other major complication of heparin therapy is heparin-induced (associated) thrombocytopenia (HIT/HAT).[255c] Thrombocytopenia is a common occurrence in patients treated with heparin. A transient drop in platelet count is seen in approximately 5 to 30% of patients beginning on the second or third day of treatment, but rarely does the count fall below $100 \times 10^9/l$. This is type 1 heparin-induced thrombocytopenia. It distinguishes it from the more important type 2 HIT where, in a small number of patients, the platelet count may fall to much lower levels ($<20 \times 10^9/l$) and paradoxically may be associated with the formation of platelet thrombi and clinical thrombosis. This rare complication is due to the presence of a platelet-activating antibody directed against platelet factor 4 (PF4) complexed to heparin. PF4/heparin complexes express multiple neoepitope sites and bind to platelet surfaces. HIT-IgG antibodies recognize the neoepitope sites on PF4, forming multimolecular PF4/heparin/IgG complexes on platelet surface. The IgG Fc regions bind and cross-link the platelet FcgIIa receptors, leading to platelet activation. Activated platelets release additional PF4 from granules, leading to a vicious cycle. Clinically, it presents 5 or more days after exposure to heparin. Platelet counts frequently are in the range 20 to $100 \times 10^9/l$, but a drop of 50% or more in the resting count may constitute HIT, even if the count remains above $150 \times 10^9/l$. Thrombosis is common, often with major vessel occlusion, and can be both arterial and venous.

Various assays are available to detect HIT antibody, but they are time consuming, and it should be noted that the diagnosis of HIT is predominantly a clinical one. *In vitro* platelet-aggregation tests may detect the presence of such antibodies and can also be used to screen other heparins and heparinoids for cross-reactivity. An ELISA (enzyme-linked immunosorbent assay) for PF4 complexed to heparin is also available and is more sensitive and specific than platelet-aggregation-based tests. Once the diagnosis is made, heparin must be stopped. Despite stopping heparin, there remains a high incidence of thrombosis that warrants treatment at therapeutic levels of anticoagulation. Paradoxically, warfarin is contraindicated, as it leads to an increased risk of microvascular thrombosis. Low-molecular-weight heparin cannot be used due to very high (80%+) cross-reactivity. Alternatives to heparin include heparinoids (danaparoid), recombinant hirudin, or argatroban. HIT antibodies may persist for up to 6 weeks after the cessation of heparin.

Low-Molecular-Weight Heparins

(LMWH) Low-molecular-weight heparins contain 10 to 20 saccharide units and have a mean molecular weight in the range of 4 to 6 kDa. The smaller, more uniform size of LMWHs, their reduced nonspecific protein binding, and their principally renal excretion give them a more predictable dose–response relationship compared with UFH. LMWHs have a longer $T^{1/2}$ than UFH and in general only need to be administered once daily. LMWHs do not cross the placenta and can be safely administered during pregnancy. Although osteoporosis and HIT have been reported with LMWHs, the incidence appears to be less than with the use of UFH. When given therapeutically (e.g., in pregnancy), LMWHs are monitored by an anti-factor Xa assay. LMWHs may be of benefit in reducing the thrombotic complication of orthopedic surgery, particularly hip-replacement surgery and other orthopedic indications.

Protamine sulfate can be used to correct overdoses of LMWHs, but it is less effective than in its reversal of unfractionated heparin and is, therefore, usually given in twice the amount. LMWHs are now being used as first-line therapy for uncomplicated deep-venous thromboses and pulmonary embolus. In such cases, LMWH is administered subcutaneously once or twice daily (depending upon the precise heparin preparation) in a dose based upon the patient's body weight. No monitoring of the anticoagulant effect is necessary, and in some centers such treatment is carried out on an outpatient basis. The efficacy of such treatment is as good and probably better than conventional treatment with unfractionated heparin, and the complications of treatment are no greater.

HEPARIN COFACTOR II

(HCII) A member of the serpin family of **serine protease** inhibitors, but it only inhibits **thrombin**. It is a single-chain glycoprotein of molecular weight 65,600 Da. It forms a 1:1 stoichiometric complex with thrombin and has no anti-factor Xa activity. Antithrombin exists in twofold molar excess over HCII. The inhibitory activity is increased by **heparin**, by heparan sulfate, and by dermatan sulfate, which is found in high concentration on the surface of subendothelial cells. HCII may be important in inhibiting free thrombin once the **vascular endothelium** is breached, but in the presence of an intact endothelium, antithrombin is physiologically more important. HCII can be assayed by comparing activation of test and control plasma by dermatan sulfate incubated with human thrombin. The residual, uninhibited thrombin is then measured by chromogenic assay. Calibration curves are drawn, the normal concentration being around 1.0 U/ml, with a range of 55 to 145%. HCII deficiency in patients with **venous thromboembolic disease** is found in approximately 0.5% of cases and is of doubtful clinical significance.

HEPARINOIDS

Synthetic substances with **heparin**like activity. Lomoparin (Organon) contains primarily heparan sulfate (80%), dermatan sulfate, and chondroitin sulfate. It enhances **antithrombin** and **heparin cofactor II** and has both anti-factor IIa (antithrombin) and anti-factor Xa activity. Lomoparin may be of value in patients who develop heparin-induced thrombocytopenia.

HEPATIC DISORDERS

See **Liver disorders**.

HEPATOSPLENIC T-CELL LYMPHOMA

(REAL: Hepatosplenic γδ T-cell lymphoma) A rare extranodal and systemic neoplasm derived from cytotoxic T-**lymphocytes** usually of γδ T-cell receptor type. The cells are of medium size, showing sinusoidal infiltration of the spleen, liver, and bone marrow. It has a peak incidence in male adolescents and young adults. Patients usually present with marked **thrombocytopenia**, mild **anemia**, and **leukocytosis**. The abnormal cells in the **bone marrow** can only be demonstrated by **immunohistology** staining for CD3. **Immunophenotyping** of the neoplastic cells are CD3+, usually TCRδ1+, TCRαδ−, CD56−/+, CD4−, CD8−, and CD5−. The cells also express the cytotoxic-granule-associated protein TIA-1. **Cytogenetic analysis** shows the cells to have rearranged TCRγ genes. The occurrence of this lymphoma is more common following immunosuppression for solid organ transplantation.

The course is aggressive, but patients may respond to **cytotoxic agent** therapy (see **non-Hodgkin lymphoma**).

HEPATOSPLENOMEGALY

See also **Liver disorders**; **Splenomegaly**.
Combined enlargement of the liver and spleen.

HEPCIDIN

An antimicrobial 25 amino acid peptide formed in hepatocytes of the liver that is a key regulator of **iron** metabolism; substantial amounts are present in urine. Hepcidin production appears to be upregulated predominantly by interleukin-6 and is thus stimulated by **inflammation**.[256] High hepcidin levels cause decreased iron absorption and increased retention of iron by **histiocytes** (macrophages), largely through the interaction of hepcidin and **ferroportin**. Production is increased in all forms of **iron overload**; it is hence a negative regulator of iron metabolism. The gene that encodes hepcidin is *HAMP* (19q13), mutations of which cause autosomally recessive **hereditary hemochromatosis**, often severe and of early onset.

HEPHAESTIN

A **copper**-containing homolog of **ceruloplasmin** found predominantly in the intestinal mucosa at the villous tips. Hephaestin is a ferroxidase necessary for iron release from intestinal epithelial cells that may interact with an iron-transport protein to facilitate transmembrane movement of **iron**. Mutations of the mouse homolog of the human *HEPH* gene (Xq11-q12) cause sex-linked hypochromic, **microcytic anemia**.

HEREDITARY COPROPORPHYRIA

See also **Porphyrias**.
An uncommon disorder resulting from a partial deficiency of coproporphyrinogen oxidase that is inherited as an autosomally dominant trait. The few persons who were homozygous or compound heterozygous for pertinent mutations of the coproporphyrinogen oxidase gene had severe enzyme deficiency and were severely affected. Clinical manifestations include neurologic, psychiatric, and abdominal symptoms similar to those that occur in persons with **acute intermittent porphyria** (AIP) and **variegate porphyria** (VP). Attacks are often precipitated by drug therapy (especially phenobarbital), but they also occur after **pregnancy**, menses, and contraceptive steroids. Approximately one-third of patients have photosensitivity. During attacks, urinary and fecal porphyrins are increased, with a predominance of **coproporphyrin** III; unlike AIP, porphobilinogen deaminase activity is normal. Prevention of acute attacks depends on avoidance of precipitating factors; management of acute attacks is similar to that for AIP and VP.

HEREDITARY ELLIPTOCYTOSIS

(HE) An inherited autosomally dominant **red blood cell membrane disorder** characterized by the presence of significant numbers of oval or elliptical red cells in the peripheral blood. Most patients are heterozygous; however, a small number of homozygous patients have been described. In some families, the gene for the disorder is linked to **Rh blood group**.

Although this disorder is reported worldwide, it seems to be more frequent in Southeast Asia, Melanesia, France, Africa, and some Mediterranean countries. A Southeast Asian variant is associated with a marked increase in red cell membrane rigidity, a phenomenon that may be responsible for the comparative resistance to penetration of these cells by **malaria** plasmodia and consequently, although indirectly, to the high endemic gene frequency of elliptocytosis in this region.

Red blood cell membrane protein abnormalities are known to occur in 30 to 40% of patients:

Abnormal spectrin self-association

Abnormal spectrin interaction with band 2.1

Abnormal binding of band 2.1 with band 3

Partial or complete deficiency of band 4.1

Deficiency or absence of glycophorin C

The remaining patients await identification of abnormality. The glycolytic pathways are intact, and no abnormal hemoglobins have been identified. HE is clearly a red cell membrane skeleton disorder, since isolated HE ghosts and membrane skeletons, in which lipid bilayers and integral proteins are totally removed, retain their elliptical shape.

Red blood cell morphology shows various types, broadly grouped as:

Ovalocytosis: an intermediate form of the process from discocyte to rod-shaped cell. Marked hemolysis can occur, associated with increased sodium permeability.

Rod-shaped red cells: considered to be an extreme form of elliptocytosis.

The majority of patients have no clinical symptoms or signs. Only a small number have persistent chronic **hemolytic anemia**. In the latter, anemia is generally mild, jaundice is intermittent, and **splenomegaly** may be present. Other features of chronic hemolysis such as gallstones, leg ulcers, and skeletal abnormalities are seldom present.

The characteristic finding is of 25 to 90% elliptocytes in the peripheral-blood film. The **hemoglobin** concentration seldom falls below 12.0 g/dl, and the red blood cell indices are usually normal. The **reticulocyte count** is increased. In anemic patients with overt hemolysis, both **osmotic fragility** and **autohemolysis** are increased, the latter being partially corrected by glucose addition. Red blood cell survival is only slightly decreased in asymptomatic patients; however, when overt hemolysis exists, there is significant shortening of red cell life span.

The majority of patients do not require any treatment; however, in those who do, **splenectomy** may be beneficial; elliptocytes will persist following splenectomy.

HEREDITARY ERYTHROBLASTIC MULTINUCLEARITY
WITH POSITIVE ACIDIFIED SERUM TEST

(HEMPAS) See **Congenital dyserythropoietic anemias**.

HEREDITARY HEMATOLOGICAL ANOMALIES

(Familial hematological disorders, genetic hematological disorders) Hematological disorders of genetic origin. From family studies of most disorders, the type of chromosome has been determined, i.e., autosomal or sex-linked, dominant or recessive. In a few disorders,

TABLE 77

Hereditary Hematological Disorders

General cellular disorders	DNA replication (**Bloom's syndrome**)
	Pyrimidine metabolism (**orotic aciduria**)
	Nucleotide synthetase (**Lesch-Nyhan syndrome**)
	Cell locomotion (**Wiskott-Aldrich syndrome**)
	Lipid-storage disorders
	Mucopolysaccharidoses
Bone marrow	**Aplastic anemia**
	Congenital dyserythropoietic anemia
Red blood cells	**Red blood cell membrane disorders**
	Red blood cell enzyme disorders
	Oxygen affinity and **hemoglobin disorders**
Hemoglobin	Hemoglobinopathies, including **sickle cell disorders** and **thalassemias**
	Porphyrias
	Hereditary sideroblastic anemia
Leukocytes	**Leukocyte anomalies**
Platelets	**Platelet-function disorders** (including storage pool disease)
Coagulation factor	Deficiencies of **fibrinogen, factors II, V, VII, VIII, IX, X, XI, XII** causing hemophilia and hemorrhagic disorder
	Factor V Leiden and deficiencies of **antithrombin III, protein C**, and **protein S** causing thrombophilia
Blood vessel wall	**Hereditary hemorrhagic telangiectasia**
Intestinal tract	**Folic acid** malabsorption
	Cobalamin malabsorption
	Hereditary hemochromatosis
Specific protein deficiencies	Apoprotein B (**abetalipoproteinemia**)
	Complement C1 inhibitor (**angioedema**)
	Transferrin (atransferrinemia)

the precise chromosomal abnormality, i.e., deletion, translocation, has also been determined. This particularly applies to coagulation-factor deficiencies and to **hemoglobinopathies.** The disorder may affect cells and tissues generally or may be specific to cells of the **hematopoietic system**. The disorders can be grouped according to the anomaly (Table 77).

Prenatal diagnosis, particularly of hemoglobin disorders, can be performed in the first trimester by chorionic villous sampling. Routine neonatal screening for sickle cell disease and thalassemia is advocated in some regions, particularly where the ethnic minority population at risk is >15%.

HEREDITARY HEMOCHROMATOSIS

(Genetic hemochromatosis) A group of heritable disorders in which **iron overload** can occur due to a primary defect in **iron** absorption or its control.

Types of Hemochromatosis
HFE Hemochromatosis

Hemochromatosis affects approximately 0.5% of Caucasians of Western European decent. Iron absorption is inappropriately high for body iron content in some homozygotes for this autosomally recessive condition, leading to the development of iron overload. "Classical" hemochromatosis is linked to the **human leukocyte antigen** (HLA) region on *Ch6p*, particularly with HLA-A3, and a missense mutation in a major histocompatibility (MHC)-like gene on *Ch6p* (*HFE* exon 4: nt 845GA; C282Y) that occurs in 60 to 100% of unrelated hemochromatosis patients.[257] Approximately 14% of Caucasians are heterozygous for the *C282Y* mutation. A second common *HFE*-coding-region mutation (*HFE* exon 2: nt 187 CG; *H63D*) occurs in approximately 20% of western Caucasians. *H63D* is uncommonly associated with a hemochromatosis clinical phenotype, although some compound heterozygotes (*C282Y/H63D*) have mild iron overload with little or no iron-associated organ injury. *HFE*-coding-region mutations other than *C282Y* and *H63D* have been identified in persons with hemochromatosis and in members of general Caucasian and Vietnamese populations. A *Cys282/Tyr* mutation is associated with sporadic porphyria cutanea tarda.[258]

Uncommon Types of Hemochromatosis

A rare but especially severe type of heritable iron overload that occurs in children, adolescents, or young adults (juvenile hemochromatosis), usually due to inheritance of two mutations in the hemojuvelin (*HJV*) gene. Other uncommon forms of autosomally recessive hemochromatosis are associated with mutations in the alternate **transferrin** receptor gene (*TFR2*) or the hepcidin gene (*HAMP*). Mutations in the ferroportin gene (*FPN1*) are associated with an autosomally dominant form of hemochromatosis. Mutations in the erythroid-specific ALA synthase gene (*ALAS2*) are associated with X-linked **sideroblastic anemia** and a hemochromatosis-like picture in some families. African iron overload (Bantu siderosis) may have an as-yet-unidentified genetic component, although the consumption of large quantities of iron in traditional beer is an important contributing factor in the pathogenesis of this form of iron overload. Iron overload in African-Americans is common but often unrecognized clinically; this appears to be a phenotypically and genotypically heterogeneous group of disorders.

Clinical Features

Approximately 1% of *HFE C282Y* homozygotes develop overt evidence of progressive iron deposition that injures the liver, joints, pancreas, heart, and other organs and reduces longevity. Males predominate in most clinical series; onset of symptoms due to iron overload is rare before age 30 years. Subjective complaints that are common in persons with severe iron overload due to hemochromatosis include: weakness and fatigue, arthralgia, abdominal pain, impotence, loss of libido, amenorrhea, and symptoms of cardiac dysfunction or diabetes mellitus. Physical examination may reveal brownish (melanin) or grayish (iron) skin hyperpigmentation (especially in areas unexposed to the sun), hepatomegaly, loss of secondary sexual characteristics, muscle wasting, or weight loss. Some patients have osteoarthropathy of the second and third metacarpophalangeal and interphalangeal joints and of the knee or hip, although these findings are not diagnostic. Complications of hepatic cirrhosis (including primary liver cancer) and diabetes mellitus are the leading causes of death. Panhypopituitarism or congestive heart failure are typical of early age-of-onset forms of hemochromatosis due to mutations in genes other than *HFE*. Approximately 99% of persons who are homozygous for *HFE* C282Y do not have signs or symptoms due to iron overload, although many of them (especially men) have biochemical abnormalities of serum transferrin saturation, **ferritin**, testosterone, luteinizing hormone, follicle-stimulating hormone, or hepatic transaminases.

Laboratory Diagnosis

An elevated transferrin saturation value (60% for males, 50% for females) on at least two occasions in the absence of other known causes strongly suggests the diagnosis of hemochromatosis. Because affected individuals are assumed to be homozygotes or compound heterozygotes for *HFE* missense mutations, the demonstration of *C282Y* homozygosity in an individual with elevated serum transferrin saturation values also establishes the diagnosis. Persons who are heterozygous for the *C282Y* mutation, those who are compound heterozygotes for the *C282Y* and *H63D* mutations, and those who are *H63D* homozygotes sometimes have elevated values of transferrin saturation. Iron overload in these persons, if any, is usually mild. In some persons who are suspected to have hemochromatosis (particularly those who have clinical evidence of hepatic abnormalities), biopsy of the liver may be indicated. This procedure is most helpful to evaluate persons for the occurrence of hepatic cirrhosis due to iron overload, thereby helping to define both diagnosis and prognosis. HLA-identical siblings of a proband who has "classical" hemochromatosis are considered to have hemochromatosis, regardless of their transferrin saturation values. HLA typing is not otherwise useful for diagnosis.

The severity of iron overload is assessed most frequently using the serum ferritin concentration. Persons suspected of having hemochromatosis who have elevated serum concentrations of hepatic enzymes, hepatomegaly, or other evidence of a liver disorder should undergo liver biopsy. These specimens should be evaluated for iron by histochemical (Perl's Prussian blue staining) and quantitative (atomic absorption spectrometric or biochemical) methods, including computation of the hepatic iron index. However, hepatic biopsy is of limited utility in diagnosing hemochromatosis before iron overload has occurred. The quantity of iron removed by **phlebotomy** is a retrospective indicator of the severity of iron overload and cannot be used alone for diagnosis. Liver-imaging techniques can visualize hepatic iron deposits, but these are too insensitive to evaluate most young, asymptomatic persons. The sequelae of iron overload — hepatopathy, arthropathy, diabetes mellitus and other endocrine disorders, and cardiac abnormalities — may require additional evaluations. Some patients have coincidental abnormalities that augment iron absorption and increase iron overload, including excessive dietary iron supplementation or ethanol ingestion, porphyria cutanea tarda, or hemolytic anemia. Viral hepatitis with (and without) hemochromatosis is often associated with elevated serum iron parameters, elevated concentrations of hepatic transaminases, and increased stainable iron in the liver.

Population screening using serum transferrin saturation or unsaturated iron-binding capacity, followed in positive cases by tests for *C282-Y* homozygosity, has been advocated, but this carries the risk of inducing illness or stigmatization.[259]

Treatment

Most persons with hemochromatosis who have iron overload benefit from phlebotomy. Typically, females of childbearing years whose serum ferritin is 200 ng/ml and other adults with serum ferritin of 300 ng/ml should be treated. Therapeutic phlebotomy removes 200 to 250 mg of elemental iron per unit of blood (1 unit = 450 to 500 ml). Children and adolescents with early-age-of-onset hemochromatosis sometimes have severe iron overload (often with cardiac and anterior pituitary failure) and need aggressive phlebotomy. Delaying therapy until symptoms of iron overload develop is not recommended. Patients with advanced hepatic cirrhosis, severe atherosclerosis, or malignancy may not be candidates for phlebotomy treatment.

The depletion of body-iron stores typically involves the removal of one unit of blood weekly. Many females, smaller persons, the elderly, and those with coexisting anemia or cardiac or pulmonary problems can sustain the removal of only 0.5 units of blood weekly or 1 unit every two weeks. The hemoglobin concentration or hematocrit should be quantified before each treatment; subjects whose values are <11.0 g/dl or <33.0%, respectively, often experience undue fatigue and other consequences of hypovolemia and anemia. The serum ferritin concentration (in the absence of inflammation) and the hepatic iron con-centration permit an estimation of the amount of phlebotomy needed. Men require twice as many units of phlebotomy as women, on the average. Iron depletion is complete when the serum ferritin concentration is 10 to 20 ng/ml. Phlebotomy prevents progression of liver injury due to iron overload in many patients. Weakness, fatigue, skin hyperpigmen-tation, cardiac symptoms, and arthralgias often improve after iron depletion; in contrast, joint deformity, hypogonadotrophic hypogonadism, and control of diabetes mellitus improve infrequently.

Lifelong phlebotomy should be performed to keep the serum ferritin at 50 ng/ml. This requires, on average, 3 to 4 units of phlebotomy per year in men and 1 to 2 units per year in women. Some persons, particularly the elderly, appear to require no maintenance phlebot-omy. Sustaining overt iron deficiency by phlebotomy is not desirable. Hyperferrinemia and elevated transferrin saturation — essential attributes of hemochromatosis — recur quickly after iron depletion. Thus, the serum iron concentration and transferrin saturation are unsuit-able as measures of body iron stores and the progress of phlebotomy therapy.

Persons with hemochromatosis should avoid supplemental or medicinal iron, consume red meats and alcohol in moderation, and limit supplemental vitamin C to 500 mg daily. Because the absorption and retention of many nonferrous metals is also increased in hemochromatosis, mineral supplements should be used only for demonstrated deficien-cies. *Vibrio vulnificus*, a spiral bacterium, can cause fatal infection when ingested in raw or improperly cooked shellfish, or by entering open wounds of those who handle seafood or bathe in warm seas. Persons with hemochromatosis, chronic liver disease, or immune-deficient states should consume only thoroughly cooked seafood items from these waters and take other measures to prevent *V. vulnificus* infections. Dietary regimens do not enhance iron excretion or prevent iron reaccumulation.

Morbidity and Mortality

The longevity in persons with hemochromatosis who have hepatic cirrhosis or diabetes mellitus is reduced. The occurrence of these two complications of iron overload in the same patient may interact in an additive manner to reduce life expectancy further. The cardiac effects of hemochromatosis can also be life threatening, although these serious complications are uncommon and do not significantly affect the survival statistics in large-case series. Significant morbidity in hemochromatosis also occurs due to arthropathy (especially in women), complications of hepatic cirrhosis and diabetes mellitus, and endo-crine dysfunction. In persons whose hemochromatosis is diagnosed before the develop-ment of severe iron overload, longevity is normal, and hemochromatosis-associated morbidity is minimal or nondetectable when therapeutic phlebotomy is performed rou-tinely to maintain body iron stores at a low normal level.

HEREDITARY HEMORRHAGIC TELANGIECTASIA

(HHT, Osler-Rendu-Weber syndrome) An autosomally dominant disorder of the blood vessel wall (see **Vascular endothelium**) that affects both sexes equally and is the most common of the hereditary vascular disorders. The telangiectatic lesions are dilated arte-rioles and capillaries lined by a thin endothelial monolayer. The lesion is characteristically dark red, slightly raised, and has one or more radiating "vascular" legs. Common sites for lesions are nasal mucosa, lips, oral mucosa, tongue, esophagus, and gastrointestinal tract. Rarely, lesions may occur in the respiratory tract, urinary tract, eye, gynecological structures, spleen, or liver. Lesions typically are 1 to 3 mm and are red, flat, and blanch upon pressure. They are usually first seen in the second to third decades of life, the number of lesions often increasing with age. Hemorrhage occurs from lesions in the skin or mucosal surfaces, either as overt recurrent bleeding from superficial lesions or persistent recurrent iron-deficiency anemia from occult blood loss. Prolonged **hemorrhage** after dental surgery, epistaxis, and menorrhagia are common. Associations include pulmonary arteriovenous fistulas (from which emboli can arise), cirrhosis, **splenomegaly**, and "mini" Kasabach-Merrit syndrome (see **Hemangiomas** — Kasabach-Merrit syndrome). Treatment of bleeding episodes is nonspecific. Treatment involves local cautery of the lesion, which may require endoscopy for telangiectases affecting the gut. Oral estrogens may be of benefit in some patients, with variable success, but are associated with significant **adverse drug reactions** with long-term use. Chronic **iron deficiency** anemia is common and requires oral iron therapy.

HEREDITARY INFANTILE PYROPOIKILOCYTOSIS

(HPP) A rare but severe congenital **hemolytic anemia** inherited as an autosomally recessive trait. The instability of HPP red cells upon exposure to heat results from a defective spectrin dimer-dimer association. The effect is increased susceptibility to shear stress and thermal denaturation. Similar abnormalities have been described in **hereditary elliptocytosis** (HE). HPP and HE may, in fact, be part of a spectrum rather than distinct entities. The disorder occurs mainly in infants of African descent who present with hemolytic anemia and concomitant erythroid marrow expansion, giving rise to skull and facial deformities. Moderate to severe anemia is present, with an elevated **reticulocyte** count. The blood film shows microcytosis, hypochromia, poikilocytosis, spherocytosis, and elliptocytosis, sometimes with fragmentation. The **osmotic fragility test** and **autohemolysis test** are both increased and are not corrected by addition of glucose. Complete response to **splenectomy** is rare.

HEREDITARY MACROTHROMBOCYTOPENIA

Large **platelets** occurring in a range of unrelated inherited disorders. Platelets appear large on stained blood films and are often associated with functional abnormalities. Hereditary macrothrombocytopenias also fall under the broad grouping of **giant platelet syndromes**.

HEREDITARY NONSPHEROCYTIC HEMOLYTIC ANEMIA

See **Congenital nonspherocytic hemolytic anemia**.

HEREDITARY OROTIC ACIDURIA

An autosomally recessive disorder of **pyrimidine** metabolism. It is characterized by **megaloblastosis** unresponsive to **cobalamin, folic acid**, or other vitamin therapy. Growth impairment and the urinary excretion of large quantities of orotic acid are diagnostic.

HEREDITARY PERSISTENCE OF HEMOGLOBIN F

(HPFH) A heterogeneous condition characterized by the persistence of **hemoglobin F** into adult life. It was first described in black (15 to 30% HbF in heterozygotes and 100% in homozygotes) and Greek populations. Other variants have since been described. Deletion and nondeletion forms exist. It does not give rise to any red blood cell abnormalities.

HEREDITARY SPHEROCYTOSIS

(HS) An autosomally dominant hereditary **red blood cell membrane disorder** characterized by **hemolytic anemia**, intermittent jaundice, **splenomegaly**, and the presence of
microspherocytes in the peripheral blood. Multiple generations and several members in
a family are usually affected. Chromosomal studies suggest that a gene deletion on the
short arm of chromosome 8 is responsible. The disease occurs primarily in persons of
Northern European descent. The sexes are equally affected. The disorder is due to marked
reduction in red cell membrane spectrin. Possible mechanisms in its development are:

Reduced spectrin synthesis

Synthesis of unstable spectrin

Impaired binding of spectrin to the membrane

Band 4.1 deficiency

Band 4.2 deficiency

A correlation exists between the level of spectrin and the severity of the disease. Patients
with less than 50% of normal spectrin have severe disease.

The microspherocyte is the characteristic red blood cell with a decreased surface:volume
ratio. While **reticulocytes** appear normal, there is progressive red blood cell membrane
loss with each passage of the cells through the spleen. Eventually, these microspherocytic
cells are unable to pass through the splenic microvasculature, where they die prematurely.
In vitro, glucose reduces hemolysis by preventing sodium loss through the leaky membrane. It has been postulated that glucose deprivation, perhaps because of stasis and
plasma skimming, results in an adverse environment for the red blood cell in the splenic
circulation.

Clinical Features

The severity of the disease varies greatly from patient to patient and even in the same
patient during the course of the disease. The principal features are:

Anemia: generally mild, but sudden drops in hemoglobin level occur (aplastic crisis),
usually precipitated by bacterial or virus infections (often human **parvovirus** B19).
Aplastic crises must be distinguished from megaloblastic crises due to folate
depletion resulting from the hemolytic process. True hemolytic crises occur mainly
in children, but anemia occurs at all ages.

Jaundice: fluctuating, usually mild in adults but can be severe in children during
hemolytic crisis.

Gallstones: pigment stones occur in 50% of patients.

Ulceration of the leg: intractable and not uncommon.

Splenomegaly: present in most patients.

Laboratory Features[260]

Peripheral-blood count and blood film: mild anemia, normal or slightly increased mean
cell volume (MCV), often very high mean cell hemoglobin count (MCHC), reticulocyte count increased, microspherocytes and polychromasia on blood film.

Biochemistry: evidence of extravascular hemolysis — increased unconjugated bilirubin, reduced or absent haptoglobin, increased urinary urobilinogen.

Bone marrow: erythroid hyperplasia — normoblastic or macronormoblastic.

Direct antiglobulin test: negative.

Osmotic fragility: increased, but may require incubation for 24 h at 37°C to become obvious. (The degree of hemolysis in the test roughly parallels the number of microspherocytes in the peripheral blood.)

Autohemolysis test: increased, but corrected by glucose addition.

^{51}Cr *red blood cell survival measurement*: used to confirm reduction in red blood cell life span and to demonstrate splenic red blood cell destruction.

Treatment

No treatment is required in the mild case. **Splenectomy** is the treatment of choice in the moderate to severe case. At the time of splenectomy, it is important to remove any accessory splenic tissue. Pneumococcal and hemophilus vaccination should be given before splenectomy is performed, and oral penicillin is administered postsplenectomy. If possible, splenectomy should be avoided during childhood. The prognosis following splenectomy is good, with reduction in hemolysis and healing of any leg ulceration.

HEREDITARY STOMATOCYTOSIS

See **Stomatocytes**.

HEREDITARY X-LINKED SIDEROBLASTIC ANEMIA

See **Sideroblastic anemia**.

HERMANSKY-PUDLAK SYNDROME

An autosomally recessive syndrome comprising oculocutaneous tyrosinase-positive albinism, platelet dense granule deficiency, and presence of ceroidlike pigment inclusions in cells of the hematopoietic system. Heterozygotes are unaffected. The storage disorder may lead to pulmonary fibrosis and inflammatory bowel disease. The predominant platelet defect is that of platelet storage-pool disease (p-SPD, see **Storage-pool disorders of platelets**). Symptoms include bruising, epistaxis, and menorrhagia. The platelet count is usually normal. Bleeding episodes can be controlled by transfusion of **platelets** or **DDAVP**.

HESS'S TEST

See **Tourniquet test**.

HETEROPHIL ANTIBODIES

IgM **agglutinins** that cause agglutination or hemolysis of heterologous **red blood cells**, particularly those of sheep, cattle, and horses.[261] They arise in response to **T-lymphocyte** transformation during the second week of **Epstein-Barr virus** (EBV) infection, where infected lymphocytes contain viral capsid antigens that give rise to the heterophil antibodies. Titers remain high for 1 to 2 months, after which they decline within 4 to 6 weeks. Tests are available using formalized horse red blood cells as indicator (Monospot) or, for batch testing, sheep red blood cells using the Paul-Bunnell microtest.

When the heterophil antibody test is negative in patients with clinical **infectious mononucleosis**, an EBV-specific antibody test should be performed with the considered possibility of infection by **cytomegalovirus**, rubella, adenovirus, *Toxoplasma gondii,* or a hepatitis virus.

HEXOSEKINASE

(HK) The enzyme catalyzing the first step in both the Embden-Meyerhof pathway and the hexose monophosphate pathway in **red blood cell** metabolism.

Hexokinase Deficiency

In view of its central position in both the Embden-Meyerhof pathway and hexose monophosphate shunt, it has been generally assumed that hereditary defects of this enzyme would have lethal consequences. Four distinct tissue-specific enzymes of hexokinase have been identified (HK1–HK4). HK1 has been identified in red blood cells and its deficiency has been reported. Inheritance usually conforms to the autosomally recessive mode. The principal manifestation of severe deficiency is mild-to-severe hemolytic anemia. Heterozygous deficiency states appear to have no effect on cell function or on longevity. In severe cases with marked **red blood cell transfusion** requirement, **splenectomy** may be of some benefit. Hexokinase deficiency has been reported as part of **Fanconi anemia**. An acquired form of the deficiency is found in Wilson's disease (hepatolenticular degeneration) due to inhibition of the enzyme by the high levels of **copper**.

HEXOSE MONOPHOSPHATE SHUNT

See **Red blood cells** — metabolism.

HIGH-GRADE B-CELL LYMPHOMA, BURKITT-LIKE

(Rappaport: undifferentiated non-Burkitt lymphoma/leukemia; Lukes-Collins: small non-cleaved follicle center lymphoma) See also **Burkitt lymphoma/leukemia**; **Lymphoproliferative disorders**; **Non-Hodgkin lymphoma**.

A high-grade B-cell non-Hodgkin lymphoma characterized by cell size and nuclear morphology intermediate between that of typical Burkitt lymphoma and typical **diffuse large B-cell lymphoma**, with a high proliferation rate. Some show a "starry sky" pattern. The **immunophenotype** of the tumor cells is B-cell-associated antigens+, CD5−, and usually CD10−. **Cytogenetic analysis** shows uncommonly a *c-myc* rearrangement, but the *b-cl-2* is rearranged in 30% of cells.

Clinical Features

These are relatively uncommon lymphomas occurring mostly in adults, with or without a history of **immunodeficiency**. They present with **lymphadenopathy** but rarely extranodal involvement. The tumor is aggressive, being usually fatal in adults, but the prognosis in children is good with treatment.

Staging

See **Lymphoproliferative disorders**.

Treatment

See **Non-Hodgkin lymphoma**.

HIGH-MOLECULAR-WEIGHT KININOGEN

(HMW-K, Fitzgerald factor) See also **Contact activation**; **Kininogens**.
A cofactor facilitating the attachment of **prekallikrein** and **factor XI**, with which it circulates as a complex. *In vitro* platelets can provide the necessary negatively charged surface for this mechanism. The concentration in plasma is 2.5 mg/dl, but it is of doubtful physiological significance and is not essential for hemostasis.

HIGH-PROLIFERATIVE-POTENTIAL COLONY FORMING CELL

See **Colony Forming Units**.

HILL'S CONSTANT

The number of molecules of oxygen that combines with one molecule of **hemoglobin**. This figure is 2.6 and not 4, the value that would be expected due to the effect of binding one molecule of oxygen by hemoglobin based on the affinity for binding further oxygen molecules (see **Oxygen affinity to hemoglobin**).

HIRUDINS

65-Amino acid proteins originally isolated from the medicinal leech *Hirudo medicinalis* although now produced by recombinant engineering. Hirudins are direct inhibitors of **thrombin** and are capable of inhibiting both bound and soluble thrombin. They are capable of interrupting **platelet** recruitment in **aspirin**- and **heparin**-resistant thrombotic processes developing at sites of arterial damage. In experimental thrombotic models, hirudins are more effective than heparin in preventing rethrombosis following tissue plasminogen activator (t-PA)-induced thrombolysis. Hirudins also prolong the bleeding time through their effects on platelet function. Hirugen is a 12-residue synthetic peptide that mimics the C-terminus of hirudin, thereby blocking the anion-binding exosite of thrombin. Hirulog is a bivalent peptide connected by a linker sequence such that inhibition of thrombin is achieved by occupation of both the anion binding site and the catalytic site.

Recombinant hirudins available are lepurudin and desirudin. Lepurudin can be used for anticoagulation in patients with Type II (immune) heparin-induced thrombocytopenia who require parenteral antithrombotic treatment. The dose is adjusted according to the **activated partial thromboplastin time** (APTT). Desirudin can be used for prophylaxis of **venous thromboembolic disease** in patients undergoing hip and knee replacement. **Hypersensitivity** reactions can occur.

HISTIOCYTES

(Mononuclear tissue macrophages) Large (15 to 18 μm) mononuclear phagocytic cells found in connective tissues and body fluids, the term being originally used by Metchnikoff in 1891. The shape is variable, with abundant cytoplasm and a cytoskeleton consisting of microfilaments largely composed of **actin**.[262] They have one or two oval/round nuclei, eccentric with uncondensed **chromatin** and a single nucleolus. In response to chronic infection, some become multinucleated giant or epithelioid cells. Histiocytes are mobile, with pseudopodium formation contributing to both movement and function. The sky-blue cytoplasm may contain a few azurophilic granules and vacuoles. Electron microscopic studies reveal the macrophage to be an active cell with a well-developed rough **endoplasmic reticulum**, **Golgi apparatus**, and plentiful **mitochondria**. Granules are typically myeloperoxidase-negative. Macrophages are capable of **DNA** synthesis, but rarely undergo **mitosis**, and readily form syncytia. They have a diffuse activity for lysosomal enzymes, including phosphatases and nonspecific esterases. Phenotypically, they are CD68+. Phagocytosed red blood cells and leukocytes may be seen.

Origin and Distribution

Macrophages are derived from blood **monocytes** that have migrated from the circulation to enter the body tissues or cavities. In liquid culture, blood monocytes slowly transform into macrophagelike cells, a process that can be hastened by the addition of various agents, e.g., c-**interferon** (IFN). However, the exact physiological mechanism resulting in monocyte transformation to macrophages is still largely unknown.

The turnover rate of macrophages is estimated to be 20 to 40 days, depending on the animal model used and organ studied. The renewal of tissue macrophages is largely achieved by monocyte migration.

General Properties

Macrophages are highly mobile, responding to various chemotaxins, e.g., C5a, formyl-methionyl-leucyl-phenylalanine (FMLP), **collagen**, and elastin. They have several key roles in the **immune response**, particularly triggering **inflammation** through production of **cytokines** and **arachidonic acid** metabolites, phagocytosing and destroying particulate material, and **antigen presentation** to T-lymphocytes. They have an important link between the adaptive and innate immune systems.

Macrophages are stimulated by cytokines, e.g., IFN-γ and bacterial products, particularly lipopolysaccharide (endotoxin), which transform a relatively inert cell into a highly active one. Other mediators inhibit macrophage activation, e.g., interleukin (IL)-4, IL-10, and transforming growth factor (TGF)-β. The activated macrophage is much more potent in all its functions.

Specific Properties
Motility

Macrophages are highly mobile, responding to various chemotaxins, e.g., C5a, FMLP, collagen, and elastin.

Phagocytosis

This is by direct binding of particles through specific **Fc receptors**, **fibronectin**, and **complement** (C3b, C3bi, C4b), with or without **opsonin** intervention. Phagocytosis is much more efficient when the particle is opsonized by being coated either with antibody or complement; the macrophage possesses receptors for both of these. The engulfed particle enters a phagolysosome within the cell and is there destroyed; the "oxidative burst" generates highly oxidizing compounds, e.g., peroxide, that will kill bacteria, etc., and proteolytic enzymes (cathepsins) digest the material within the phagolysosome. The oxidative burst depends on activation of the macrophage, and where the IFN-γ system is defective it is very inefficient. Histiocytes also destroy old/damaged **red blood cells** and release **iron** from **hemoglobin**. The iron is then transported by **transferrin** to the iron stores or **erythron**. They also remove necrotic tissue, allowing healing and remodeling to occur.

Cytotoxicity

Macrophages can produce, via oxidative metabolism, reactive oxygen species that are bactericidal. Nonoxidative enzymes are also produced, e.g., **lysozymes**, acid hydrolases, and proteases. Both microbes and tumor cells are killed.

Antigen Presentation

Phagocytosed (or endocytosed) antigen can be processed (if necessary) in the cytoplasm and then presented to T-cells on the membrane surface in conjunction with **human leukocyte antigen** (HLA) class II molecules. Upregulation of HLA class II molecules and increased antigen presentation are mediated by IFN-γ.

Cytokine and Growth-Factor Production

Interleukin-1, a cytokine that plays an important role in neutrophil migration and fibroblast, epithelial, and endothelial proliferation. It is a major regulator of hematopoiesis, an important activator of T-cells, and a stimulator of B-cell proliferation. It is the principal endogenous pyrogen.

Tumor necrosis factor (TNF)-α upregulates endothelial expression of CD62E and CD54, allowing leukocyte migration. It is the major mediator of endotoxic shock and is released from macrophages secondary to the binding of endotoxin to CD14.

Interleukin-6 induces an **acute-phase response**, e.g., C-reactive protein release, and may be a B-cell growth factor.

Transforming growth factor (TGF)-α stimulates fibroblast proliferation.

Angiogenesis factor stimulates endothelial proliferation with production of nitric oxide.

Granulocyte **colony stimulating factor** (G-CSF), macrophage (M)-CSF, and GM-CSF stimulate granulocyte production, differentiation, and ultimately activation.

Cyclooxygenase (COX)-2 induction enhances production of **prostaglandins**.

Location

Histiocytes (macrophages) are distributed throughout all organs and cavities of the body. They have differing morphology and roles, depending on the ultimate location, and have specialized roles within their residential organs (see Table 78).

Lung Histiocytes (Macrophages)

These play a central role in protecting the respiratory system from inhaled antigens or irritants. Most macrophages are alveolar, but a few remain interstitial. They express Fc and complement receptors, but have an inefficient respiratory burst. Smoking increases the number of alveolar macrophages that contain higher levels of elastase. The release of

TABLE 78

Specialized Organ Role of Histiocytes

Site	Role
Liver (Küpffer) cells	phagocytosis of blood-borne particles, e.g., viruses, bacteria ± opsonins, red cells
	neutralization of endotoxin
	production of IL-1
	acts as antigen-presenting cells
	recycles red cell iron
Spleen	phagocytosis of blood cells
	acts as antigen-presenting cells
Brain (microglial cells)	phagocytosis of dying cells and debris
Bone marrow	regulation of hematopoiesis
Bone (osteoclast)	bone resorption
Peritoneum	cytotoxic to tumor cells and bacteria

metalloprotease, elastase, and collagenase causes tissue damage, ultimately resulting in emphysema. Lung histiocytes can also act in antigen presentation, interacting with T-cells.

Splenic Histiocytes (Macrophages)

These present in both red and white pulp of the **spleen**. They play a major role in the destruction of aged or damaged red and white blood cells. They are also involved in antigen presentation and interactions with T and B-cells. In **autoimmune hemolytic anemia** and **immune thrombocytopenia**, the spleen is the major site of destruction via their Fc and complement receptors.

Peritoneal Histiocytes (Macrophages)

These express Fc and complement receptors and are particularly apt at killing bacterial and tumor cells.

Intestinal Histiocytes (Macrophages)

Underlying the basement membrane of the intestinal epithelial cells, histiocytes play a role in the presentation of absorbed antigen to T-cells and the development of immune tolerance. The expression of HLA class II molecules is increased in inflammatory bowel disease (perhaps due to c-IFN release) and is intimately involved in the development of granuloma in Crohn's disease and tuberculosis.

Liver Histiocytes (Macrophages) — Küpffer cells

Tissue macrophages residing in the sinusoids of the liver act as a "filter" for blood-borne microbes/particles, etc. They express both Fc and complement receptors (in particular), and phagocytosis is markedly increased by opsonins (Ig, C3, fibronectin). Gut-derived bacteria or endotoxins are efficiently removed by Küpffer cells, as are circulating aged/damaged red blood cells. Küpffer cells also release cytokines such as IL-1, which acts as an endogenous pyrogen and a mediator in the acute-phase response. Similar to other macrophages, liver histiocytes can interact with T-cells and act as antigen-presenting cells.

Bone Marrow Histiocytes (Macrophages)

These are essential components of the bone marrow stroma, secreting hematopoietic growth factors, e.g., IL-1, GM-CSF, G-CSF, and M-CSF.

Bone Histiocytes (Macrophages) — Osteoclasts

Highly motile, multinucleated giant cells that, although lacking macrophage/monocyte-specific antigens, appear to be a highly specialized macrophage. Their maturation requires the macrophage colony-stimulating factor (M-CSF) and is accelerated by cytokines, such as IL-6 and the systemic calciotropic hormones, parathyroid hormone (PTH) and vitamin D. Osteoclasts reside in bone and play a major role in bone resorption. For this process, the motile osteoclast alights upon a bone surface and seals off an area to form an adhesive ring. The osteoclast then develops, above the surface of this ring, an invaginated plasma membrane structure termed the "ruffled border." This is an organelle, but it acts as a lysosome that dissolves bone mineral by secreting acid onto the isolated bone surface while breaking down bone matrix by secretion of catheptic proteases. Osteoclast activation and bone resorption occur with **myelomatosis**.

Nerve Histiocytes (Macrophages) — Microglial Cells

Brain macrophages mostly reside in the gray matter, with a perivascular distribution. They congregate around dying neurons as a result of monocyte recruitment from the blood. Although their function is not fully understood, they seem to be involved in the phagocytosis of dead cells and necrotic tissue.

HISTIOCYTIC MEDULLARY RETICULOSIS

See **Histiocytosis**.

HISTIOCYTIC SARCOMA

A rare malignant proliferation of cells, showing morphologic and immunophenotypic features similar to those of mature **histiocytes**. Phenotypically, these are CD68, lysosome, CD11c, and CD14. They present either in the skin, intestinal tract, or in multiple sites (malignant **histiocytosis**). Hepatosplenomegaly is common. The bone may show lytic lesions, and **bone marrow hypoplasia** is not uncommon. It is an aggressive neoplasm with a poor response to **cytotoxic agent** therapy.

HISTIOCYTOSIS

See also **Langerhans histiocytosis**.
Reactive proliferation of **histiocytes** (macrophages).

Hemophagocytic Syndrome

(HPS) This is not a single disorder but a group, probably a consequence of defects in the immune-effector cells.

Infection-Associated Hemophagocytic Histiocytosis

This aggressive, often fatal, disorder is most commonly seen in childhood. It is precipitated by **Epstein-Barr virus** (EBV) or other **viral infections**. Histiocyte proliferation leads to a "**cytokine** storm," with release of **interferon** (INF)-δ, **tumor-necrosis factor** (TNF)-α, **interleukin** (IL)-6 and IL-10, and intense **phagocytosis**. Presentation is usually acute, with fever, **lymphadenopathy**, **hepatosplenomegaly**, and nonspecific symptoms, e.g., malaise and anorexia. **Pancytopenia** is usual, but one lineage may be more affected than others. The **bone marrow** aspirate reveals hemophagocytosis by histiocytes, with 40% of patients having **myelofibrosis**. It is fatal in up to 40% of cases, but therapy directed at the underlying cause plus aggressive supportive therapy may lead to recovery. **Corticosteroids, ciclosporin** A, and **etoposide** have been successfully used.

Familial Hemophagocytic Histiocytosis

(FHH) This rare autosomally recessive disorder of **perforin** gene defects results in deregulated proliferation and accumulation of cytotoxic **T-lymphocytes** and histiocytes, with overproduction of inflammatory cytokines.[262a] It presents in early childhood with fever, cachexia, lymphadenopathy, and hepatosplenomegaly. Pancytopenia due to hemophagocytosis by histiocytes is seen. The prognosis is very poor, with only 10% of patients alive at 1 year. Combination cytotoxic agent therapy and **allogeneic stem cell transplantation** may benefit a few patients.

Sinus Histiocytosis with Massive Lymphadenopathy

See also **SHML; Rosai-Dorfman disease**

This is a proliferation of histiocytes and of lymphocytes within the lymph node sinuses and in the lymphatics at extranodal sites. Lymph node histology reveals lymphophagocytosis by sinus histiocytes, although plasma cells and neutrophils may also be phagocytosed (emperipolesis). Fibrosis may be prominent. It is a rare worldwide condition that presents usually in children and young adults with fever, neutrophilia, and massive lymphadenopathy. Extranodal disease affects the genitourinary tract, respiratory tract, and CNS. Laboratory features include a normocytic anemia that may be **autoimmune hemolytic anemia** and polyclonal hypergammaglobulinemia. The cause is unknown, but is probably an immune disturbance following a viral infection. The condition is usually benign and the treatment symptomatic.

HISTOCOMPATIBILITY

See **Human leukocyte antigens** — typing.

HISTONES

Proteins concerned in the packaging of eukaryotypic **DNA**. The nucleosome core contains DNA tightly wrapped around a central histone octamer comprising two molecules of each of the four core histones (H2A, H2B, H3, and H4). These core histones are subject to a variety of enzyme-catalyzed posttranslational modifications that modulate **gene** expression. Of these modifications, acetylation of lysine on the histones has been extensively studied. Lysine and arginine account for 30% of amino acids on histones, and the N-terminal (tails) of lysine is responsible for 50% of the positive charge on the histone. The overall charge on the histone is responsible, in part, for the relaxation or tight binding of histones to DNA nucleosome. The neutralization of the positive charge by acetylation leads to loosening of histone-DNA contacts, which facilitates the accessibility of a variety of factors to DNA.

Histone Deacetylation

See also **Transcriptional regulation**.

Several lines of evidence suggest that inappropriate transcriptional repression mediated by histone deacetylases (HDACs) is a common molecular mechanism that is used by oncogenes. Acetylation status of histones is controlled by activities of two families of enzymes, the histone acetyltransferases (HATs) and histone deacetylases. Histone acetylation neutralizes the histone charge and therefore is thought to facilitate transcription by loosening the interactions between histones and DNA. Conversely, removal of acetyl groups from histones by HDACs is thought to cause tighter nucleosomal packaging of DNA, preventing transcription factors access to DNA. Accordingly, gene-silencing events and oncogenes such as *Myc/Mad* and *RB/E2F* that repress or silence transcription recruit HDACs.

HDACs are a diverse family of enzymes acting on diverse substrates. They are characterized by their ability to remove acetyl groups from conserved lysine residues in histones and nonhistone proteins such as transcription factors, *GATA-1*, *NF-YA*, *YY1*, *PLZF*, *E2Fs*, *myoD*, and p53.

HDACs comprise a large group of proteins divided into three major classes based on their homology to yeast proteins, and each HDAC class consists of several isoforms. HDACs vary in their molecular size, pattern of expression tissue distribution, and ability to complex with corepressors.

Class-1 HDACs (HDAC1, HDAC2, HDAC3, and HDAC8) are small molecules (most of them with a molecular weight less than 400), are homologous of the yeast transcriptional regulator RPD3, and are found almost exclusively in the nucleus and ubiquitously distributed.

Class-2 HDACs have molecular weight in the range between 1084 and 1215 (HDAC4, HDAC5, HDAC6, HDAC7, HDAC9, HDAC10, and HDAC11), are homologs of yeast HDA1 deacetylase, shuttle between nucleus and cytoplasm, and are tissue specific (almost exclusively found in heart and skeletal muscles and brain). HDAC6 is unique in having two active deacetylase catalytic domains and functions as a tubulin deacetylase as well, and its inhibition results in tubulin acetylation.

Class-3 HDAC members are a recent discovery and hence not well known.

The function of HDACs is complex. In addition to deacetylation histones, HDACs also deacetylate non-histone protein such as p53, E2F, ∝-tubulin, and myoD. Most HDACs have nuclear localizing signals, allowing them to locate in the nucleus where their predominant substrate is found. Some have nuclear export signals that allow them nuclear and cytoplasmic locality.

The general effects of HDAC inhibitors (HDAIs) include (a) increased acetylation of histones as well as other proteins such as nuclear factors and (b) modulation of the expression of genes that play a role in the control of cell growth, differentiation, and **apoptosis** in cancer cells *in vitro* and *in vivo*. HDAIs function by displacing the **zinc** ion, a necessary component of the charge-delay system, which is responsible for removal of the acetyl group from N-terminal lysines.

HIV

See **Human immunodeficiency virus**

HODGKIN DISEASE

(HD) A **lymphoproliferative disorder** of uncertain cellular origin, first described by Thomas Hodgkin in 1832.

Incidence and Etiology

The occurrence is worldwide,[228] representing around 1% of all tumors. It has a bimodal age distribution of 9 to 17% of cases at 15 to 34 years, with a second peak of incidence in the sixth and seventh decades. The highest incidence is in the U.S., Europe, and Israel, especially in economically advantaged communities, with lowest incidence in China and Japan. There is little difference in incidence between the sexes. Familial aggregation occurs. The risk of developing the disease is increased in siblings of young adult patients, but the risk of disease among siblings is small. Space-time clustering has suggested a transmissible agent, the most likely possibility, based upon epidemiological studies, being viral. **Epstein-Barr virus** (EBV) is the strongest risk factor, the genome being expressed in **Reed-Sternberg cells** in about one-half of cases. The capacity of EBV to transform **lymphocytes** and the increased incidence of HD in those with a past history of **infectious mononucleosis** supports this hypothesis.[181,263] Malignant cells in EBV-positive Hodgkin disease express several viral genes, one of which is *LMP-1*, a known oncogene. The association of patients with **human leukocyte antigen** (HLA) class I polymorphisms suggests a possible susceptibility of their immune system to the disorder.

Pathogenesis

The cellular origin of the malignant cells is a B-lymphocyte. The characteristic histological feature of HD is the presence in lymphoid tissue of giant cells, some with a bilobed mirror-image nucleus: Reed-Sternberg cells (RS cells). A derangement of immunological homeostasis occurs, with lymphocytopenia particularly affecting T-cells. Humoral immunity remains normal but with delayed cutaneous hypersensitivity. Lymphocyte production following mitogen or antigen stimulation is impaired, with increased susceptibility to infection by opportunistic microorganisms (see **Immunodeficiency**).

Clinical Features

Patients present at all ages, adults being more commonly affected than children. It is more common in males than females. Hodgkin lymphoma usually presents as locally limited **lymphadenopathy**, but virtually any tissue outside of the central nervous system may be involved. One-half of the patients have mediastinal involvement, spreading in a centripetal fashion and becoming disseminated; 75% of patients are curable. Late relapses may occur. Two to 5% of patients are at risk to develop **diffuse large B-cell lymphoma**. Some patients have "B" symptoms (night sweats, fevers, and greater than 5% weight loss), pruritus, extranodal masses, glomerulonephritis, and other symptoms.

Staging

The Costwold staging classification of Hodgkin lymphoma is the most recent staging system, based on anatomic distribution of disease. Standard staging studies (see **Lymphoproliferative disorders**) include history, physical examination, complete blood count with differential, sedimentation rate, chemistry analysis (liver and renal studies, electrolytes, and **lactic dehydrogenase** [LDH]), chest X ray, and CT scans of the chest, abdomen, and pelvis. PET scans are being evaluated.

Prognostic Factors

Seven factors are independently associated with a worse prognosis:

Serum albumin level of less than 4 g/dl

Hemoglobin level of less than 10.5 g/dl

Male gender

Age greater than 45 years

Stage IV

Total WBC count ≥15,000/µl

Absolute lymphocyte count of <600/µl

Histological Types of Hodgkin Disease

The World Health Organization (WHO) Classification of Tumors: Tumors of the Hemato-poietic and Lymphoid Tissues (see **Lymphoproliferative disorders**) divides the disease into six types:

Nodular-lymphocyte-predominant Hodgkin lymphoma

Classical Hodgkin lymphoma

Nodular-sclerosis classical Hodgkin lymphoma

Lymphocyte-rich classical Hodgkin lymphoma

Mixed-cellularity classical Hodgkin lymphoma

Lymphocyte-depleted classical Hodgkin lymphoma

Nodular Lymphocyte-Predominant Hodgkin Lymphoma

Characterized by large numbers of small mature lymphocytes with a variable component of histiocytes. There is a nodular pattern, with or without diffuse areas, the nodularity being more easily recognized using immunohistology stains with anti-B-cell antibodies. Progressively transformed germinal centers are often seen, with atypical cells having vesicular polylobulated nuclei with small nucleoli, probably a lacunar variant of RS cells (lymphocytic/histiocytic cells; L&H cells; "popcorn cells"). These cells have **immunophe-notype** CD45$^+$, CD15$^-$, and B-associated antigens$^+$. No diagnostic RS cells are seen. A prominent meshwork of **dendritic reticulum cells** suggests a B-cell origin, but plasma cells are infrequent. Eosinophils and neutrophils are rare; fibrosis is absent.

Clinical Features

Early-stage disease presents in 75 to 90% of patients; "B" symptoms or bulky disease are not common. The mediastinum is involved less than 5%. Compared with lymphocyte-rich classical Hodgkin lymphoma, lymphocyte-predominant Hodgkin lymphoma exhibits a pattern of multiple relapses after remission induction.

Nodular Sclerosis Hodgkin Lymphoma

Of patients with Hodgkin lymphoma, 40 to 70% have nodular sclerosing histology. It is characterized by a partially nodular pattern, with fibrous bands separating the nodules; diffuse areas are common, as is necrosis. The characteristic cell is the lacunar RS cell, diagnostic RS cells being also present. These tumor cells show immunophenotype CD15$^{+/-}$, CD30$^+$, and CD45$^-$. Other cells are lymphocytes, histiocytes, plasma cells, eosinophils, and neutrophils. Tumor cells are EBV$^+$ in about 50% of cases.

Clinical Features

Most patients are adolescents and young adults, but it can occur at any age; females equal or exceed males. The mediastinal lymph nodes are commonly involved. There is a good prognosis with treatment.

Mixed-Cellularity Hodgkin Lymphoma

This type is characterized by diffuse or slightly nodular infiltration of typical RS cells (some of lacunar type), with immunophenotype CD15$^{+/-}$, CD30$^+$, and CD45$^-$. There is no band-forming sclerosis, although fine interstitial fibrosis may be seen.

Clinical Features

A lymphoma occurring predominantly in males, with staging at diagnosis usually more advanced than in nodular sclerosis or lymphocyte-predominance types. The prognosis is less good in patients greater than 60 years of age. The frequency of mixed-cellularity Hodgkin lymphoma increases to 50% in males with Hodgkin lymphoma who are older than 70 years.

Lymphocyte-Depleted Hodgkin Lymphoma

This type is characterized by hypocellularity of the lymphoid tissue involved, infiltration being mainly by RS cells with bizarre variants. These tumor cells show immunophenotype CD15$^{+/-}$, CD30$^+$, and CD45$^-$. There is diffuse fibrosis and necrosis.

Clinical Features

This is the least common type of HD, representing less than 1% of cases, and usually occurring in older persons. It is particularly common in those with Hodgkin lymphoma associated with **acquired immunodeficiency syndrome** (AIDS). It presents with abdominal lymphadenopathy, hepatosplenomegaly, and bone marrow involvement, but without peripheral lymphadenopathy. At diagnosis, the disease is usually at an advanced stage, but the prognosis with treatment is no worse than for other types.

Lymphocyte-Rich Classical Hodgkin Lymphoma

A diffuse tumor with classical RS cells, some of lacunar type, relatively infrequent, in a background of lymphocytes with occasional eosinophils or plasma cells.

Treatment

This depends on the staging (see above).

Radiotherapy

High-dose supervoltage techniques — megavoltage linear accelerators, telecobalt, or betron procedures — have been effective when applied to involved areas as appropriate for treatment of stage IA or IIA disease, and these were the treatments of choice in limited-stage disease.[107] With this treatment, 90% of stage IA patients and 80% of stage IIA remain in remission after 5 years, but adverse reactions were significant. Radiation therapy is now the treatment of choice only in localized and locally recurrent nodular-lymphocyte-predominant Hodgkin lymphoma. Apart from reducing tumor burden, radiotherapy is not the appropriate form of treatment for stages IB, IIB, III, or IV.

Cytotoxic Agent Therapy

Cytotoxic agent therapy[264–269] is now the treatment of choice for all cases. For nonbulky IA and IIA disease, the treatment is now three cycles of ABVD chemotherapy followed by reduced doses of involved-field radiation therapy at a dose of 20 to 40 Gy. The preferred regimens are ABVD and BEACOPP.[108,109] ABVD is administered for six to eight cycles, followed by involved-field radiation therapy to previous sites of bulky disease. With ABVD in stage III and IV disease, the complete response rate is 80%, and the partial remission rate is 60%. A major problem is nausea and vomiting, for which a number of regimens are available. Neutropenia is a frequent complication, often with severe infections.[269] These complications may require reduction in dosage.

Second malignancies, such as breast carcinoma in women, sometimes occur.[265] Such malignancies are related to alkylating-agent chemotherapy, radiation therapy, splenectomy, age, and advanced stage, particularly occurring in females. Solid tumors represent 55 to 75% of the therapy-related tumors.

Autologous or Allogeneic Stem Cell Transplantation

Autologous stem cell transplantation (peripheral-blood [or bone marrow] stem cell transplantation [PBSCT]) plus G-CSF is the treatment of choice in relapsed disease after chemotherapy.[111] Response can be expected in only 40 to 50% of patients. **Allogeneic stem cell transplantation** combined with chemotherapy may be applicable for treatment of multiple relapse.[270]

Surgery

Apart from an incisional or excisional lymph node or extranodal tissue biopsy for diagnosis, surgery now has no role in treatment. Laparotomy followed by splenectomy as a staging procedure is no longer advocated due to the inevitable complications, particularly the long-term increased susceptibility to infection, myelodysplasia, and secondary acute leukemia.

AIDS-Associated Hodgkin Disease

The mortality risk is increased fivefold, particularly when associated with intravenous drug abuse. The histological types are usually mixed cellularity or lymphocyte depleted. They present clinically at an advanced stage. With CD4 counts greater than $200 \times 10^9/l$, aggressive chemotherapy can be administered. Approaches have included dose-adjusted CDE and R-CDE.

HOMOCYSTEINE

An amino acid in the pathway linking the metabolism of methionine, cysteine, **cobalamin**, **folic acid**, and **vitamin B$_6$**. It is normally present in plasma at concentrations of 5 to 15 μmol/l. Hereditary deficiency of the enzyme cystathionine-β-synthase, which converts methionine to cystathionine, is an uncommon autosomally recessive disorder of worldwide distribution associated with hyperhomocysteinemia (plasma levels >50 μmol/l) and homocysteinuria. Polymorphisms of methylenetetrahydrofolate reductase (MTHFR) may be associated. A single patient has been reported with deficiency of N^5-methyl FH4: hemocysteine methyl transferase. He presented with anemia, megaloblastosis, and mental retardation. Both serum and red cell folate levels were revised.[149]

Acquired deficiency results in plasma levels of 25 to 50 μmol/l. This occurs with deficiencies of cobalamin, folate, and vitamin B$_6$, with chronic renal disorders, and with hypothyroidism.

Hyperhomocysteinemia is an established risk factor for thrombosis, both **arterial thrombosis** — coronary artery disease and cerebrovascular disease — and **venous thromboembolism**. The mechanism by which elevated levels of homocysteine results in vascular damage is unclear, although studies both *in vivo* and *in vitro* have shown damage to both **vascular endothelium** and **platelets**, with a reduction in the levels of various **coagulation factors**.[271] Raised levels of homocysteine are also associated with osteoporosis.[272] Some patients with hyperhomocysteinemia respond to treatment with vitamin B_6, others to dietary restriction of methionine. Supplementing the diet with exogenous folate reduces plasma homocysteine levels; fortification of bread and pasta with folate may reduce the incidence of hyperhomocysteinemia. Elevated levels of homocysteine are associated with the genetic variant MTHFR 677C→T that also has an association with coronary artery disease, but lowering of the homocysteine level has a doubtful effect on the incidence of myocardial infarction.[273],[9]

HORN CELL
See **Keratocytes**.

HOWELL-JOLLY BODIES
An anomaly of **red blood cells** in which nuclear remnants, usually removed by the splenic macrophages, are seen in these cells of Romanowsky-stained films as deeply basophilic, pyknotic, spherical inclusions. They are usually single, but in **dyserythropoiesis** they can be multiple. They are found in association with:

> **Cytotoxic agent** therapy (cyclophosphamide, chlorambucil, nitrosoureas)
>
> Dyserythropoietic anemias, including **megaloblastosis**
>
> **Splenic hypofunction** (inherited or acquired)
>
> **Glucose-6-phosphate dehydrogenase** deficiency and other **red blood cell enzyme deficiencies**
>
> **Splenectomy** (post)

HUGHES SYNDROME
See **Antiphospholipid-antibody syndrome**.

HUMAN ALBUMIN SOLUTION
See **Colloids for infusion**.

HUMAN HEMATOPOIETIC IMMORTALIZED CELL LINES
In vitro immortalized human hematopoietic cells propagated in culture. These have provided models for the study of **oncogenesis** and differentiation in **hematopoiesis**. Establishment of human tumor cell lines is difficult but has been achieved by prolonged liquid culture of mononuclear cells (usually bone marrow but occasionally blood) from patients with hematopoietic malignancies. A population of cells ultimately obtains a growth advantage with reduced or absent growth-factor requirements.

In Vitro Growth Characteristics of Transformed Cells

Focus formation, a feature of transformed murine fibroblasts, which no longer exhibit contact inhibition of the adherent layer. This change can be induced by transfection of transforming human oncogenes into mouse NIH3T3 cells.

Tumorigenicity after inoculation into Nude (nu/nu) mice (impaired cell-mediated immunity).

Growth-factor independence: immortalized cell lines grow with reduced or absent growth-factor requirements due to upregulated signal-transduction pathways, e.g., RAS oncogene ("gain of function") mutations, increased surface growth-factor-receptor expression, or loss of tumor-suppressor genes ("loss of function").

Genetic Mechanisms Associated with Immortalization

Cooperating oncogenes: *in vitro* tumor cell transformation often requires cooperation between "transforming" oncogenes (e.g., *v-src*, mutant *RAS*) and "immortalizing" oncogenes (*myc*, *bcl-2*, and mutant p53). Oncogenic DNA viruses may possess more than one cooperating oncogene, e.g., *v-erbA* and *v-erbB* in avian erythroblastosis.

Genomic instability: immortalization is usually associated also with multiple karyotypic abnormalities, both numerical (loss more than gain) and structural (translocations and partial deletions). This may result in homozygous deletion of tumor-suppressor genes such as p53 or amplification of others, e.g., *myc*, promoting immortalization.

Epstein-Barr Virus-Transformed B-Cells

Infection of resting B-cells by **Epstein-Barr virus** (EBV) promotes cellular lymphoblastic transformation with an increased size and active cycling. This system is used to create an immortalized cell line capable of propagating a genetic abnormality identified in B-cells from patients with hematopoietic malignancy for further study, e.g., uncharacterized gene rearrangement.

HUMAN IMMUNODEFICIENCY VIRUSES

(HIV) Retroviruses of the lentivirus subgroup that generate information from **RNA** to **DNA**. The original infection in chimpanzees of sub-Saharan Africa has been estimated by mathematical modeling to have occurred about 1930. Human infection was first detected in 1981. They all have the same genomic structure, share at least 40 to 50% homology, and infect T-lymphocytes through the CD4 receptor. They are related to the simian immunodeficiency virus (SIV) and to the simian T-cell leukemia virus (STLV).

Types of Virus[274]

HIV-1 has nine genetic subtypes of HIV-1 (A–H and O), subtype B being most common in Europe and North America, subtype E in Southeast Asia, and subtypes A, C, and D in Africa. This virus is highly infectious for humans and is the cause of **acquired immunodeficiency syndrome** (AIDS).

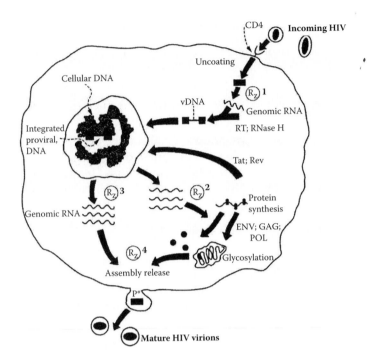

FIGURE 69
Human immune deficiency virus (HIV) life cycle. (From Bridges, S.H. and Sarver, N., *Lancet*, 345, 427–432, 1995.
With permission.)

HIV-2 occurs in wild African monkeys and has been detected in West African pros-
titutes. It has a slow rate of infection and few humans develop clinical AIDS.

Structure

Each virus is made up of gp120 molecules and contains two groups of RNA-encoding
structure genes (see Figure 69):

Pol coding for reverse transcriptase (RT) and RNaseH, which convert RNA to DNA
    in the host cell
Env, which codes for envelope proteins

These are bound by a gag-derived protein, p24.
    The concentration of HIV RNA strands, as occurs in plasma, directly reflects the titer
of HIV particles. This can be determined by a modified PCR technique. The level of protein
p24 can be measured in serum.
    The viruses can be inactivated by heat at 60°C for 10 h and are destroyed by chemical
viricidal agents, e.g., sodium hypochlorite bleach.

Transmission

Semen via the rectal or vaginal mucosa, including artificial insemination. Subtype E
    is particularly taken up by Langerhans epithelial cells of the oral and genital
    mucosa. This may explain the predominance of heterosexual transmission in

Southeast Asia. Homosexual transmission of subtype B is prevalent with institutional living such as occurs in prisons. The simultaneous infection with other sexually transmitted organisms appears to facilitate HIV-1 transmission, while other organisms such as hepatitis C virus have an inhibitory effect.

Blood components or blood products, as by therapeutic infusion, or contamination by needles or syringes, as in drug abusers. Transmission was by contaminated plasma to patients with **hemophilia** prior to heat treatment of plasma. Needle-stick injury is an uncommon route of transmission for health-care workers.

Transplacental from mother to fetus, occurring in 15% of infected mothers. Other routes of transmission are intrapartum or via breast milk. Vertical transmission is less frequent with HIV-2 than HIV-1. Measurement of HIV-1 RNA levels may predict progression of the infection, but in early pregnancy the risk of transmission can be reduced by maternal antiretroviral therapy.

Organ donation by transplantation.

Pathogenesis

Following infection, rapid attachment of the virus occurs to CD4 antigen-receptor sites on T-cells. This protein is highly concentrated on helper T-cells. The virus also uses the CCR5 and CXCR4 **chemokine** receptors on **monocytes/histiocytes** and **dendritic reticulum cells**. Deficiency of the allele for this cytokine, as is present in around 10% of North Americans and northern Europeans, may account for lower incidence of AIDS in these populations relative to exposure in Africans. Other cells vulnerable to infection of a lesser degree include fibroblasts and neurotropic glial cells. Genetic alterations in the genes controlling cell receptors may explain differences in progression of HIV disease.[275] Attachment of the virus occurs through a constellation of surface gp120 envelope glycoprotein spikes, covalently linked to transmembrane gp41 glycoprotein and disulfate isomerase. Uninfected CD4$^+$ lymphocytes are sensitive to an actor termed TNF-related apoptosis-inducing ligand (TRAIL), the presence of which contributes to CD4$^+$ cell depletion.

Following entry, two viral enzymes — reverse transcriptase (RT) and RNaseH — convert viral RNA genome to a double-stranded DNA copy (see Figure 70). The DNA migrates to the nucleus, where it is integrated into cellular DNA through another HIV enzyme (integrase). Viral RNA synthesis and transport of partly spliced RNAs to the cytoplasm depends upon two viral regulatory proteins, Tat and Rev. Following synthesis and processing of the viral structural proteins (ENV, POL, GAG), the mature peptides and two copies of the genomic RNA assemble and form a virus particle that is then released from the surface of the cell. Final maturation of the virus is presumed to occur extracellularly. Steps where ribozymes (Rz) can inhibit HIV replication are:

Rz-1: incoming viral RNA (pre-integration)

Rz-2: *de novo*-synthesized mRNAs

Rz-3: *de novo*-synthesized genomic RNA

Rz-4: genomic RNA during assembly

Inside infected CD4 lymphocytes, rapid multiplication of the virus particles occurs within 36 h and leads to death of the host cell. The resultant viremia allows them to be identified in serum and cerebrospinal fluid at an early stage of infection by its core antigen (p24). In association with this virus replication, there is a massive proliferation of the invaded lymphocytes. The level of viremia can only be detected for as long as it is in

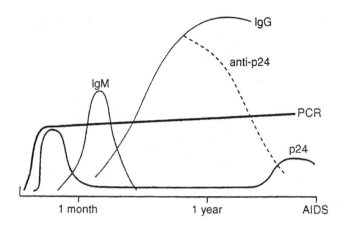

FIGURE 70

The evolution of plasma laboratory markers of the naturalization of HIV infection (x-axis is not to scale). The course of HIV can now be modified by combined antiviral treatment, which will suppress HIV PCR reactivity but will not usually modify the anti-HIV response. (From Mortimer, P.P. and Loveday, C., The virus and the tests, in *ABC of AIDS*, 5th ed., Adler, M.W., Ed., British Medical Journal Publishing Group, London, 2001, Fig. 2.5. With permission.)

excess of its antibody (anti-core). The virus load is increased by the presence of another infection, particularly by malaria plasmodia.

Stimulation of B-lymphocyte proliferation by proliferating T-cells leads to hypergammaglobulinemia, with the production of IgM antibody to HIV, which is detectable in serum from 4 to 6 weeks postinfection. This is followed by persistence of IgG antibody production (see Figure 71).

The effects of the virus on CD4 lymphocytes are either immediate cytolysis, occurring in around 90%, or symbiosis within the cell nucleus (see Figure 71). Viral replication is reduced by specific cytotoxic T-lymphocytes, but cell lysis is increased if the cytotoxic T-lymphocytes are activated by other viruses or by cytokines from B-cells or by antigen-presenting cells. Neither antibody offers protection to the host, but HIV suppressor factors may naturally halt progression. The HIV-infected macrophage/monocyte remains in a chronic nonlytic state predominantly in the brain. Within days or weeks, a transient fall in CD4 lymphocytes may occur with a rise in CD8 cytotoxic/suppressor lymphocytes until a steady state is reached with a reservoir of infected cells. This proliferation in lymph nodes accounts for the clinically detected generalized lymphadenopathy. Failure of CD8[+] T-cells to prevent progression of immunodeficiency is probably due to defective antigenic presentation to them by interdigitating dendritic cells.

Over the next 5 to 10 years, persistent T-cell proliferation continues. Gradually, the T-cells fail to overcome the HIV multiplication, ending in lymphopenia with reduction in cellular immunity and the changes of immunodeficiency. With the progressive fall in CD4 cells, there is a parallel fall in p24 antibody with reappearance of p24 antigen.

B-cell proliferation in the bone marrow with increase of **interleukin**-1 (IL-1) and other inflammatory cytokines stimulates the production of growth factors — granulocyte colony stimulating factor (G-CSF), granulocyte/macrophage CSF (GM-CSF), and macrophage CSF (M-CSF) — during the early phases of HIV-1 infection. With advancing disease, there occurs a reduction of **erythropoiesis**, **granulopoiesis**, and **thrombopoiesis**, resulting in **anemia**, **neutropenia**, and **thrombocytopenia**. Opportunistic infection gives rise to an increase in marrow macrophages and atypical lymphocytes. Tumor necrosis factor (TNF-α) production is increased and has a variable effect upon hematopoietic growth factors. B-cell proliferation also leads to an increase of **plasma cells** with the production of **autoantibodies**

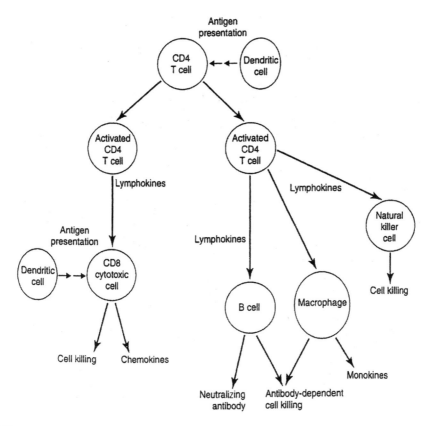

FIGURE 71
Induction of an immune response. (From Beverley, P. and Helbert, M., Immunology of AIDS, in *ABC of AIDS*, 5th ed., Adler, M.W., Ed., British Medical Journal Publishing Group, London, 2001, Fig. 3.1. With permission.)

and **immune complexes**, which gives rise to **immune thrombocytopenia**. The **direct antiglobulin (Coombs) test** becomes positive in 60 to 80% of patients, but immune hemolytic anemia is rare. The appearance of the **antiphospholipid antibody** (lupus anticoagulant) in 20 to 70% of patients and reduced levels of active **protein C** leads to thromboembolic disorders. These complications may not occur for 5 to 10 years postinfection, but the rate of progression can be increased by concomitant infections such as other sexually transmitted diseases, pregnancy, and malnutrition. Opportunistic infections are the predominant complication, sometimes associated with malignancy such as **Kaposi's sarcoma** or **lymphoproliferative disorders**. Dementia can follow transmission by infected monocytes to glial cells of the central nervous system.

Maternal vertical transmission of virus and HIV antibody leads to similar immunological changes in the fetus. HIV antibody testing is not reliable for diagnosis during the first 15 months of life due to transplacental passage of maternal antibody.

Laboratory Detection
Antibody

Enzyme-linked immunosorbent assay (ELISA). This provides the best indicator of infection by HIV-1, with seroconversion occurring early. There is a high degree of false positivity when using commercial kits. With a negative result in a clinically suspect case, the test should be repeated in 3 to 6 months.

TABLE 79

Estimated Regional Incidence of HIV Infection by 2005

Region	Million/cases
Sub-Saharan Africa	25.80
South and S.E. Asia	7.50
Latin America	1.90
E. Europe and Central Asia	1.70
North America	1.20
East Asia	1.00
Western Europe	0.80
Caribbean	0.50
Oceania	0.40

Source: Data from UNAids, Report on Aids Epidemic Update 2005. Joint United Nations Programme on HIV/Aids.[275a,b]

HIV-1 Western blot. This is the best confirmatory test and is necessary when screening low-risk populations. If the result is indeterminate, a repeat test should be performed after 4 to 6 weeks.

Antigen

Viral culture of limited value due to lack of sensitivity and expense

PCR techniques to amplify chosen genome sequences

p24-antigen-capture assay

Incidence

Infection of persons by HIV-1 was estimated in 2005 to have reached 40 million, with the incidence rate continuing to rise.[14a] The estimated worldwide incidence distribution is shown in Table 79. The incidence in women has now caught up with that in men, thus accounting for half of the persons infected. It is hoped that the proportion of people living with HIV will continue to rise due to population growth and life-prolonging effects of antiretroviral therapy.

HUMAN LEUKOCYTE ANTIGENS

(HLA) A group of polymorphic glycoproteins present on the surface of nucleated cells occurring in multiple different allelic types. They are often referred to as histocompatibility antigens because they determine acceptance or rejection of tissue grafts between individuals; however, their biological role is to enable antigen recognition by T-lymphocytes. The genes for HLA molecules are located in a region of the genome known as the major histocompatibility complex (MHC), and these are often loosely referred to as MHC antigens.

Biochemistry

Two classes of HLA antigens occur: I and II. Of these, three class I antigens (A, B, and C) and three class II (DR, DP, and DQ) are immunologically important. Other HLA antigens exist but have distinct roles. All are dimeric: class I antigens comprise a heavy chain that varies with the allele and a constant light chain (β-2 microglobulin). Class II antigens comprise two equal chains, both of which vary. All of these HLA antigens are membrane proteins. An important feature of their three-dimensional structure is that the membrane distal portion of the molecule exhibits a groove; X-ray crystallography has shown that short peptides are located here. This feature — the peptide-binding groove with its bound peptide — is key to the function of these proteins.

Major Histocompatibility Complex

(MHC) The human MHC is located on the long arm of chromosome 6. Gene cloning and physical mapping have provided a very clear idea of the genes present in the MHC. There are multiple loci — over 400, of which 250 are expressed (the remainder being pseudo-genes) — in about 4 million base pairs of **DNA**. The complex is divided into three regions — class I, class II, and class III — wherein are located the genes for the corresponding antigens, plus many others.[276] Some of the other genes present encode key immunological roles, but many do not.

Because the genes in the MHC are physically close together, they tend to be inherited as a unit (linkage disequilibrium) so that particular combinations of HLA-A, -B, and -DR occur much more frequently than would be expected if segregation were random. Such a unit is termed a haplotype, i.e., a collection of alleles along a particular **chromosome**.

Class I Region: Genes

Here are encoded the heavy chains of HLA-A, -B, and -C antigens, as are the so-called nonclassical (or class Ib) antigens -E, -F, and -G, which in contrast to the classical genes are not polymorphic. β-2 Macroglobulin is encoded on chromosome 15. The structure of all class I heavy-chain genes is essentially the same. The genes are split into six exons corresponding to the five mature protein domains α1, α2, α3, the transmembrane region, and cytoplasmic domain. The sixth exon encodes a signal peptide whose function is to direct the newly synthesized protein to the endoplasmic reticulum: it is cleaved off as the protein travels to the cell surface and so is not present in the mature surface-expressed histocompatibility antigen. The α helices of the α1 and α2 domains lie parallel to one another, forming the sides of the peptide-binding groove for which a section of β-pleated sheet forms the floor. The polymorphisms of the A, B, and C genes are mostly located in the gene region corresponding to the peptide binding groove.

Class II Region: Genes

There are paired genes for the α- and β-chains (indicated by suffixed A and B) of the classical class II antigens DP, DQ, and DR. The pairs are generally closely associated together. The DRA gene is essentially nonpolymorphic, and the two dozen or so DR specificities correspond to polymorphisms in the DRB genes. Both A and B genes of DP and DQ are polymorphic. Again, as with class I genes, the exons of the genes correspond to protein domains, but with one less exon and domain, the mature protein being a heterodimer. The structure of the peptide-binding groove is very similar to that of class I antigens, but both α- and β-chains are involved.

The products of two further, nonpolymorphic class II antigens, DM and DO, are accessory molecules for class II function (see below). Also present in the class II region are genes for accessory molecules for class I function: TAP1 and TAP2 (transporters in antigen presentation) and LMP1- and LMP2 (large multimeric proteases)-encoding proteasome (see below).

Class III Region Genes

None of these is directly involved in HLA function, but the product of at least one, **tumor necrosis factor** (TNF), influences expression of HLA antigens by cells.

Biological Role of HLA/MHC

T-Cell Antigen Presentation

(TAP) The T-cell receptor for antigen (see **Antigen presentation**), unlike the B-cell receptor (i.e., **immunoglobulin**), does not bind intact antigen in the soluble phase. Instead, it recognizes peptide fragments located in the groove of HLA antigens on the surface of cells. The consequence of recognition of antigen "presented" to the T-cell in this fashion is activation of the T-cell.

HLA class I antigens present peptides derived from proteins in the cytoplasm of the cell. These are "processed" by digestion through **proteasomes** and transport into the rough **endoplasmic reticulum** (RER) through TAP. Once within the RER, they associate with class I antigens synthesized there, binding to the peptide-binding groove, and are then transported to the cell surface via the **Golgi apparatus**. The class I/peptide complex is recognized by CD8$^+$ cytotoxic T-lymphocytes (CTL). This is because the CD8 molecule binds to class I antigens.

Class II antigens, on the other hand, present peptides derived from the exterior of the cell taken up into specific vesicles. Within these vesicles, proteins are degraded, and the peptide fragments bind to the groove of class II antigens and thence are transported to the cell surface, where the complex is recognized by CD4$^+$ helper T-cells. This is because the CD4 molecule binds to class II antigens.

All proteins, self and nonself, are dealt with in this way. The binding of peptides to the groove is more-or-less specific in that a particular HLA protein binds a range of different peptides, which tend to have common features. Generally, only one or a few peptides derived from a particular protein will bind a particular HLA protein.

Obviously, only antigens for which T-cells bear specific T-cell receptors (TCRs) will activate an immune response. Evidently, this is a mechanism for sampling proteins both within cells (class I/CD8) and in the extracellular environment (class II/CD4), thus generating CTL or Th cells as necessary. All nucleated cells can express class I antigens, and thus there is surveillance, e.g., for virus infection. On the other hand, relatively few cell types express class II antigens, of which the most important probably are **dendritic reticulum cells**, and these provide surveillance, e.g., for the presence of exotoxins, etc., for which an antibody response is appropriate.

Immune Recognition of Allografts

A feature of recognition by T-cells of foreign antigen is that this occurs in the context of self-HLA, i.e., the T-cell and the target cell must be HLA-identical (MHC antigen restriction). However, in the case of allografts, nonself-HLA is recognized by T-cells, and a very strong rejection reaction occurs. Hence, so far as possible in allografting, donors are chosen whose HLA antigens match those of the patient. It appears that the main HLA antigens mediating rejection are A, B, and DR; bearing in mind that the alleles on both chromosomes are expressed, this means that up to six antigens need to be matched. The better the match, the more likely the graft is to survive. However, given the high level of polymorphism of HLA genes, six antigen matches are rare. Nevertheless, lesser matches are common, e.g., between first-degree relatives and because of linkage disequilibrium (see above). Hence, typing of HLA antigens to match donor and recipient is extremely important in transplantation, most especially for stem cell/bone marrow, where there is a risk of **graft-versus-host disease**.

Methods of HLA Typing (Tissue Typing)

Improved understanding of the gene sequences of HLA antigens has complicated the nomenclature by demonstrating that there are many more alleles than were envisaged in premolecular times. On the other hand, this understanding has largely simplified HLA typing through polymerase chain reaction (PCR) techniques, which has made all other techniques essentially obsolete except as research tools.

Serological Typing

Peripheral-blood mononuclear leukocytes or purified T-cells are used for serological definition of HLA-A, -B, and -C specificities, whereas cell suspensions enriched for B-cells are used for typing HLA-DR and -DQ. The conventional method of typing is microlymphocytotoxicity: cells are first mixed with specific antisera, complement is then added, and cell viability is determined; cell lysis indicates a positive reaction. HLA-specific sera (alloantisera) are obtained from immunized individuals, from previously pregnant women (who generate antibodies to paternal HLA on the fetus), or from transfused subjects, and then screened for monospecific sera (i.e., those that respond to one HLA type only). **Monoclonal antibodies** are now available.

Molecular Genetic Analysis

Molecular genetic analysis methods have the great advantages that cell type and viability are not important, screening for alloantisera is unnecessary, and less-subjective interpretation is involved.

> **Restriction-fragment-length polymorphism** (RFLP) technique was used for several years, but it has now been largely superseded by PCR-based tests. RFLP suffers from the disadvantage that the analysis is lengthy and that certain specificities give identical DRB RFLP patterns and require DQ RFLPs for interpretation. In addition, some heterozygous combinations are indistinguishable.

PCR-based techniques:

- Sequence-specific oligonucleotide techniques: PCR primers are used to amplify target HLA sequences. Subsequently, the product of the PCR is hybridized to a range of single-strand oligonucleotide (SSO) probes specific for particular HLA types to determine which particular sequence is present.

- Sequence-specific primers: DNA is amplified using a series of primer pairs that are sequence-specific for different alleles. Amplification only occurs when the allele specific to a particular primer pair is present.

Cellular Assays

Mixed-Lymphocyte Culture

(MLC) This is a bidirectional assay performed using a mixture of allogeneic mononuclear cells from a potential donor/recipient pair. It tests the ability of cells from the donor to respond to irradiated stimulator cells from the recipient (graft-versus-host [GVH] response) and the ability of recipient cells to respond to irradiated donor stimulator cells (host-versus-graft [HVG] rejection, response). The MLC is an assay in which cells from both the donor and recipient are alternatively irradiated or treated with mitomycin-C so that they cannot themselves proliferate. Response to HLA class I difference is minimal, and therefore the test mainly detects different class II molecules. Stimulation is primarily induced by potent antigen-presenting cells such as dendritic reticulum cells, which cluster with responding allogeneic lymphocytes. The latter incorporate [3H]-thymidine, and it is usual in human MLCs to measure proliferation at days 5 to 7. This is then compared with the appropriate autologous (negative) and third-party (positive) controls. The results are normally expressed as a relative response (RR). This is given by the formula:

$$RR = \frac{\text{cpm test} - \text{cpm autologous}}{\text{cpm max} - \text{cpm autologous}} \times 100$$

This procedure is now little used for donor matching.

Cytotoxic T-Lymphocyte Precursor Frequency

(CTLP) This is essentially a quantitative cell-mediated lympholysis assay incorporating limiting-dilution analysis. Responder mononuclear cells are cocultured with limiting numbers of irradiated stimulator cells for 10 days and are fed with interleukin (IL)-2 and fresh medium on days 3 and 6 of culture. Phytohemagglutinin (PHA)-stimulated targets (i.e., stimulator cells from the same patient) are cultured simultaneously and after 10 days are labeled with ^{51}Cr and added. Chromium release is measured and compared with spontaneous release and a 100% control (all cells lysed). Because the assay is performed with limiting dilutions of responder cells, their frequency can be calculated. It is possible to use either donor or patient cells as the stimulator cells in this assay, and this will give an indication of the possibility of HVG and GVH responses between the donor and recipient. The assays are used mainly in **allogeneic stem cell transplantation** (SCT). The frequency of CTLPs may correlate with GVH/HVG response in unrelated-donor allogeneic SCT. Similarly, the frequency of helper T-cell precursor cells may be assayed by limiting dilution. The culture supernatants are assayed for IL-2 production by measuring the proliferation of an IL-2-dependent cell line. This assay is oversensitive in unrelated-donor SCT, but it may be useful in defining unsuspected HLA differences in sibling/related SCT and thus individuals at particular risk of GVHD.

Nomenclature of HLA Alleles

Initially, HLA types were defined using serologic techniques. By such means, roughly 20 HLA-A, 42 HLA-B, 8 HLA-C, 18 HLA-DR, and 6 HLA-DQ antigens were identified. Using molecular techniques, it can be shown that a very much higher degree of polymorphism exists and that some serologically defined antigens may exist in many molecular forms. (However, many of the alleles differ in nucleotide sequence not resulting in amino acid substitution — so-called synonymous substitutions.) This has led to the adoption of an initially four-digit code by the World Health Organization (WHO) Nomenclature Committee for Factors of the HLA System, a code that is now up to eight digits. The first two digits are equivalent to the old serological type; thus HLA A2 is now a family of some 80 molecular types such as HLA A*022001. The complete nomenclature in the year 2005 can be downloaded from the Anthony Nolan Trust Web site.[14]

The clinical importance of matching or mismatching for the finer specificities is not yet clear.

Clinical Applications

Disease Associations

A number of disorders are linked to HLA types. These include:

Ankylosing spondylitis	B27
Diabetes mellitus — insulin dependent	DQ2
Goodpasture's syndrome	DR2
Hereditary hemochromatosis	A3
Multiple sclerosis	DR2
Rheumatoid arthritis	DR4
Systemic lupus erythematosus	DR3

Matching for Transplantation

Matching for **transplantation** involves:

Solid organs: kidney, pancreas, liver, heart

Allogeneic stem cell transplantation

Platelet transfusion

HUMAN T-LYMPHOTROPIC VIRUS

(HTLV) A retrovirus endemic in Japan, the Caribbean, parts of Central and South America, and Africa (up to 5% are carriers in some regions). It is transmitted sexually, via blood, and vertically from mother to child. The target cell for HTLV replication is the CD4 **T-lymphocyte**. Most carriers are asymptomatic, but in about 5%, disease may develop after many years of incubation. In equatorial regions, disease manifests as a progressive neurological disorder — tropical spastic paraparesis (TSP, also known as HTLV-associated myopathy [HAM]) — due to inflammation of the spinal cord, which appears to be an autoimmune condition triggered by the virus. More widespread geographically is **adult T-cell leukemia/lymphoma**. This is an aggressive disease with lesions in skin, bone, and viscera and a characteristic hypercalcemia. The pathogenesis is probably related to the product of the viral gene *tax*, which upregulates the cellular IL-2 receptor, resulting in proliferation of the infected cell.

HUMORAL IMMUNITY

Specific immunity mediated by soluble components of the immune system — specific antibodies (**immunoglobulins**) — in distinction to **cell-mediated immunity** (CMI). The experimental definition of humoral immunity is that it should be transferred by serum. Clear examples include resistance to virus infection and antitoxin responses. **Cytokines** — products of CMI — may also be transferred by serum, but their effects, e.g., in inducing inflammation, are nonspecific and are not a part of humoral immunity. In some senses, the term "humoral immunity" has become obsolete as we gain a fuller understanding of immunology and realize that antibodies are but one effector component of a system that is essentially cellular.

HUNTER'S SYNDROME

A **mucopolysaccharidosis** due to deficiency of iduronate-sulfate sulfatase with accumulation of dermatan and heparan sulfate in tissues. It is inherited as a sex-linked recessive trait. Clinical manifestations are almost identical to **Hurler's syndrome**, with gargoyle facies, hepatosplenomegaly, growth retardation, and mental retardation. Corneal clouding never occurs, and the clinical course is slower, with survival up to 30 years possible. **Alder-Reilly** bodies, **Gasser cells**, and **Buhot cells** are seen in the peripheral blood and bone marrow. There is no specific treatment, but **allogeneic stem cell transplantation** has benefited a few patients.

HURLER'S SYNDROME

A **mucopolysaccharidosis** due to deficiency of α-L-iduronidase, with the accumulation of dermatan and heparan sulfate in tissues. It is inherited as an autosomally recessive trait, with clinical manifestations appearing 6 to 12 months after birth. Typical facies consist of coarse features, enlarged tongue, depressed nasal bridge, and prominent eyebrows (gargoylism). Corneal cloudiness, hepatosplenomegaly, and skeletal deformities (claw hand, joint deformities) are common. All patients have mental and physical retardation, with death usually occurring in adolescence secondary to congestive cardiac failure. Accumulation of mucopolysaccharide in neutrophils (**Alder-Reilly** bodies), lymphocytes (**Gasser cells**), or bone marrow mononuclear cells (**Buhot cells**) is seen. There is no specific treatment, but **allogeneic stem cell transplantation** has benefited some patients.[277]

HYDROCORTISONE
See **Corticosteroids**.

HYDROPS FETALIS
A disorder of the fetus characterized by ascites and generalized edema together with hepatosplenomegaly and, in some cases, cardiomegaly. Edema is attributed to a low serum albumin, and it is thought that hypoxic damage to vascular endothelium may be important in lowering the serum albumin. Complications include **disseminated intravascular coagulation** (DIC), pulmonary hemorrhage, and subarachnoid hemorrhage. Hydrops fetalis is usually caused by severe **hemolytic disease of the newborn** (HDN), but may also be due to massive transplacental hemorrhage (TPH) and **parvovirus** infection.

HYDROXOCOBALAMIN
See **Cobalamins**.

HYDROXYCARBAMIDE
See **Hydroxyurea**.

4-HYDROXYCOUMARIN
See **Warfarin**.

HYDROXYUREA
(Hydroxycarbamide) A ribonucleotide reductase inhibitor that prevents the conversion of ribonucleotide diphosphates to the deoxyribonucleotide form, this being essential for **DNA** synthesis. There is, therefore, killing of cells in the S-phase of the cell cycle (see **Mitosis**). Gut absorption is very good, with excretion mostly via the kidneys. The plasma half-life is relatively short, at about 3 h.

Hydroxyurea is used in the initial treatment of leukocytosis in acute and **chronic myelogenous leukemia** and to control thrombocytosis in the chronic **myeloproliferative disorders**, particularly **essential thrombocythemia**, because of its relatively rapid effect and because its short half-life is associated with rapid recovery of bone marrow suppression when it is discontinued. It is also used chronically in **sickle cell anemia** to reduce the frequency of painful crises. Adverse effects include nausea, vomiting, diarrhea, skin rash, fever, mouth ulcers, painful leg ulcers, hyperpigmentation of nails or skin, granulomatous lung disease, and hyperuricemia. Hydroxyurea is mutagenic in rodents and in combination with alkylating agents and irradiation in humans; its mutagenicity as a single agent in humans is still a matter of debate. Reversible **bone marrow hypoplasia** with anemia, leukopenia, and thrombocytopenia is the usual dose-limiting toxicity. **Macrocytes** and **neutrophil** hypersegmentation are characteristic peripheral-blood smear appearances with chronic administration.

HYPERCHROMASIA

(Hyperchromia) Unusually deep uniform staining of **red blood cells** with a lack of central pallor. It is seen in **macrocyte**s where there is increased thickness of the cells and when the cells are abnormally rounded, as with **microspherocytes**.

HYPERCOAGUABLE STATES

See **Thrombophilia**.

HYPEREOSINOPHILIC SYNDROME

(HES) The clinical features arising from persistent **eosinophilia** ($>1.5 \times 10^9/l$), with infiltration of multiple organs, including heart, lungs, central nervous system (CNS), skin, joints, pancreas, esophagus, and peritoneum.[178,278,278a,278b] The causes are either clonal (**chronic eosinophilic leukemia**), secondary to **parasitic infections** or vasculitides (**eosinophilic granuloma**), or idiopathic. Activated-T-cell release of **interleukin** (IL)-5 results in reactive eosinophilia with tissue damage due to degranulation.

Clinical Manifestations

These may be nonspecific, e.g., fever, weight loss, or specific symptoms related to tissue infiltration.

Cardiac Disorders

The heart is affected in over 90% of cases and is usually the cause of demise. Three stages of cardiotoxicity have been described; these initially affect the endocardium, but later the myocardium becomes involved.

Necrotic: acute inflammatory reaction

Thrombotic: mural thrombi secondary to endothelial damage or arteritis

Fibrotic: scarring of the myocardial muscle, valves, chordae tendinae (endomyocardial fibrosis)

Heart failure, secondary to valvular regurgitation or myocardial failure, and arrhythmias due to neural bundle damage are typically seen.

Pulmonary Disorders

The patient may present with cough, but ultimately dyspnea (secondary to eosinophil infiltration impairing gas exchange) supervenes. The chest X-ray often shows pulmonary infiltration.

Skin Disorders

Both nonspecific papulo/nodular rashes and urticaria are seen, with biopsy revealing eosinophilic infiltration.

Other Organ Disorders

Neurological involvement, e.g., encephalopathy and peripheral nerve palsy (usually sensory), can occur in about 50% of patients, with renal, hepatic, splenic, and ocular manifestations less common. **Anemia** and **thrombocytopenia** can also occur.

Prognosis and Treatment

Patients have a variable prognosis, with survival of about 1 year if untreated. The extent of organ involvement (especially cardiac) at diagnosis is prognostically important. Organ failure should be treated by conventional therapy. Prednisolone/prednisone (see **Corticosteroids**), with or without **hydroxyurea**, results in excellent response in the majority of patients, but **vincristine** or **6-mercaptopurine** may be of benefit in resistant cases. Alternatively, immunosuppressive therapy directed at T-lymphocytes, e.g., **ciclosporin**, may prove to be effective. Psoralen + ultraviolet A radiation therapy (PUVA) is useful for localized skin manifestations. In patients with severe refractory disease, **allogeneic stem cell transplantation** may be curative. If this is not available or is contraindicated, **immunosuppression** directed at T-lymphocytes, e.g., ciclosporin, may be effective.

HYPERHISTAMINEMIA

Raised plasma histamine levels occurring in up to 70% of patients with **polycythemia rubra vera** (PRV). It has been suggested that histamine, which is released from **basophils** or tissue **mast cells**, is responsible for the pruritus seen in PRV, but the relatively poor response to histamine antagonists (both H1 and H2) has led some authors to question this theory. Hyperhistaminemia has also been described in patients with gastric carcinoid.

HYPERREACTIVE MALARIAL SPLENOMEGALY

(Tropical splenomegaly) A disorder occurring in chronic **malaria** associated with massive **splenomegaly**. This enlargement is caused by an increased phagocytic activity and accompanying lymphocyte response. There is elevation of serum IgM and plasmodium-specific IgM, both arising from the spleen. There is a marked decline in CD8+ **lymphocytes**. It is often accompanied by hepatic sinusoidal lymphocytosis. Treatment is vigorous antimalarial therapy, but if this fails to reduce splenic size, **splenectomy** may be indicated.

HYPERSEGMENTED NEUTROPHILS

See **Neutrophils** — maturation.

HYPERSENSITIVITY

An exaggerated or inappropriate **immune response** that causes tissue damage. The reaction is a characteristic of the individual concerned, occurring upon the second or subsequent contact with a particular **antigen**.

Type I: Immediate Hypersensitivity

See also **Anaphylaxis**.

An IgE (immunoglobulin E) response to innocuous environmental antigens (i.e., allergens such as pollens, dust from house mites, and animal dander). Antigen stimulates Toll-like receptors (TLRs) on antigen-presenting cells that cross link with IgE at a receptor on the surface of **mast cells**, causing them to degranulate and produce histamine with an acute inflammatory reaction, commonly rhinitis or asthma and sometimes **angioedema**. Tumor-necrosis factor (TNF) activates local endothelium to promote diapedesis of **neutrophils** and **monocytes**. Interleukin (IL)-4 activates Th2 cells and B-cells to produce Th2 cytokines and IgE, IL-3, and IL-5, the latter stimulating **eosinophil** production and activation. Positive feedback by IL-4 sustains the T-cell response as long as antigen (allergen) is available.

Type II: Antibody-Dependent Cytotoxic Hypersensitivity

Antibody is directed against **antigen** on a cell surface or a tissue, antigen with natural killer (NK) **lymphocytes**, or **complement** initiating cell lysis. It occurs as a consequence of IgG or IgM binding to cells. Antibodies have the following reactions:

Cytolysis due to activation of complement or opsonization, e.g., **hemolysis** of red blood cells

Binding to cell receptors for hormones, possibly linked with a **human leukocyte antigen** (HLA) allele to stimulate cell activity, e.g., thyrotoxicosis or **Wegener's granulomatosis**

This is the basis for autoimmunity and **autoimmune disorders** when this reaction is exaggerated. Examples are **autoimmune hemolytic anemia**, **immune thrombocytopenic purpura**, **systemic lupus erythematosus**, celiac disease, and insulin-dependent **diabetes mellitus**.

Type III: Immune Complex Disease

Immune complexes become deposited in tissues, with complement activation, phagocyte migration, and cell damage. Antigen can be from an infectious origin (e.g., hepatitis B), an innocuous environmental antigen (e.g., fungal spores causing Farmer's lung), or an autoantigen (e.g., DNA in systemic lupus erythematosus). The reaction is dependent upon the antigen being in excess of antibody and being present long enough to elicit an antibody response that fails to remove the immune complex, e.g., serum sickness.

Type IV: Delayed-Type Hypersensitivity

Antigen-sensitized Th1 cells mediated by **histiocytes** (macrophages) to release **cytokines**. It usually occurs following secondary contact with a specific antigen. The cytokines induce an inflammatory reaction and activate macrophages, which in turn release mediators (as in graft rejection) or cause an allergic contact dermatitis with **vascular purpura**. This cell-mediated immune response can be monitored by exposure to new antigens in a sensitizing dose followed later by a challenge dose. Dinitrochlorobenzene can be used in this way to determine the prognosis following treatment of malignancy. This reaction is responsible for granuloma formation following infection by *Mycobacterium tuberculosis* and hepatitis B virus, pathogens that are hard to eradicate. The lesions of **multiple sclerosis** (MS) are of this type.

HYPERSENSITIVITY VASCULITIS
See **Vasculitis**.

HYPERSPLENISM
See also **Splenic hyperfunction**.
The term introduced by Damashek in 1955 for excessive destruction of circulating blood cells by the spleen. He defined the syndrome as:

Cytopenia of one or more peripheral-blood cell lines

Bone marrow hyperplasia commensurate with the cytopenia

Splenomegaly

Correction of the cytopenias following **splenectomy**

The term is now rarely used, as few disorders completely fulfill all criteria at the same time.

HYPERURICEMIA
A raised level of plasma uric acid above 90 mg/l. Uric acid is the major end product of **purine** metabolism (see Figure 72). It is excreted by the kidney at a rate of about 500 mg/day, but the pKa is only 5.7, and therefore crystallization occurs at urinary pH < 5.5. This typically happens in the distal renal tubule and collecting duct, resulting in renal impairment. Precipitation can also occur within other tissues, e.g., skin (tophi), joints (gout), and in hyperuricemic conditions.

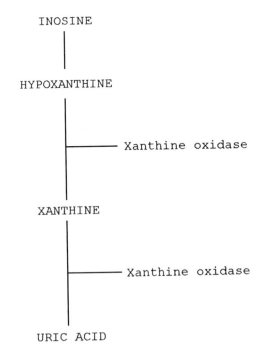

FIGURE 72
Purine metabolism to uric acid.

Hyperuricemia may be found at presentation of any hematopoietic malignancy, especially those that are widespread and have a rapid cell turnover, e.g., **acute leukemia** and **myeloproliferative disorders**. Hyperuricemia is very likely to occur once chemotherapy is initiated in these tumors, either in isolation or as part of a generalized **tumor-lysis syndrome**. Treatment with **allopurinol** (typically 300 to 600 mg/day) and vigorous intravenous hydration should be started prior to the administration of chemotherapy, as the level of uric acid excretion may increase five- to tenfold. If renal impairment has already occurred, urinary alkalinization with sodium bicarbonate to maintain a urinary pH > 7.0 should be given in addition to allopurinol. Patients with a high blast count (>100×10^9/l), with biochemical evidence of ongoing tumor-lysis syndrome or with impaired renal function, can be administered urate-lytic agents such as uricase or raspburicase, enzyme-based products that destroy urate and improve overall elimination of cellular debris. Hemodialysis may be required for patients with progressive renal failure. In patients with a myeloproliferative disorder and acute gout, the administration of allopurinol may exacerbate symptoms, in which case nonsteroidal anti-inflammatory drugs (platelet count and platelet function allowing) or colchicine are indicated.

HYPERVISCOSITY SYNDROME

A syndrome of fatigue, headache, sluggish mentality, and visual disturbances caused by slowing of the microcirculation due to increased blood viscosity. It occurs in association with:

Lymphoplasmacytic lymphoma/Waldenström macroglobulinemia

Myelomatosis

Polycythemia rubra vera

Chronic myelogenous leukemia

A characteristic feature seen on ophthalmoscopy is "sausage-shaped" dilatation of the retinal vessels.

Hyperviscosity is treated by **hemapheresis** as appropriate to the underlying disorder.

HYPERVOLEMIA

The increase in whole **blood volume** usually associated with raised plasma volume.

HYPOCHROMIA

Pale staining of **red blood cells** (RBC), the area of central pallor exceeding one-third of the total surface of the cell. Hypochromia results from:

Failure of heme synthesis (**iron deficiency** anemia, **sideroblastic anemia**)

Failure of globin synthesis (**thalassemia**)

Counts are reduced in both the **RBC mean cell volume** and the **RBC mean cell hemoglobin**.

HYPOFIBRINOGENEMIA
See **Fibrinogen**.

HYPOPHOSPHATEMIA

Levels of plasma (serum) phosphate below 0.2 mg/dl, leading to impaired glycolysis, including that of **red blood cell** metabolism. It occurs with:

Excessive antacid therapy

Total parenteral nutrition that does not include phosphate supplements

Starvation

Clinically, the patient presents with anorexia, confusion, malaise, and weakness. Red blood cell **adenosine triphosphate** (ATP) and 2,3-DPG are decreased, resulting in decreased membrane deformability and increased **oxygen affinity to hemoglobin**. The administration of phosphate results in cure.

HYPOSPLENISM
See **Splenic hypoplasia**.

I

I BLOOD GROUPS

A specific antigen–antibody system located on **red blood cells** (see **Blood groups**). I and i are carbohydrate antigens. The RBCs of almost all adults express I antigen but only weakly express i antigen. Conversely, red cells taken from umbilical cord samples, and from adults with the rare i phenotype, express i but only weakly express I. Genetic conversion of i to I phenotype requires 1-6-N-acetyl glucosamyl-transferase and offers a choice of exons. Mutation of the third exon is associated with congenital cataracts.[279] This occurs in Asians but not in Europeans.

A weak, cold, active auto-anti-I (or cross-reacting anti-HI) is often present in normal plasma. This autoantibody is of no clinical significance but occasionally gives unwanted positive results in **pretransfusion testing.** However, auto-anti-I active at temperatures above 25°C is the most common cause of **cold-agglutinin disease.**

Idarubicin

See **Anthracyclines**.

IDIOPATHIC APLASTIC ANEMIA

See **Acquired aplastic anemia**.

IDIOPATHIC MYELOFIBROSIS

See **Chronic idiopathic myelofibrosis.**

IDIOPATHIC THROMBOCYTOPENIA

See **Immune thrombocytopenia; Thrombocytopenia.**

IMATINIB

(Glivec; imatinib mesylate) An orally administered competitive inhibitor of **adenosine triphosphate** (ATP) binding to **tyrosine-kinase** receptors, including KIT protein, *BCR/ABL* fusion protein, and PDGF receptor. It is used for the treatment of gastrointestinal stromal tumors and for **chronic myelogenous leukemia** (CML).[142,142a] Initially, it was given to CML patients after failure of interferon-α or for blast crisis, but following successful trials, it is becoming used as an initial therapy agent. Adverse drug reactions include gastrointestinal disturbances, cardiac disorders, myalgia, and headache.

IMMATURE GRANULOCYTES

Metamyelocytes, myelocytes, and promyelocytes present in peripheral blood (see **Granulocytes**). These cells are identified in peripheral-blood film examination, but criteria for instrument identification varies between instrument manufacturers. Standardization is necessary, but initiatives will be difficult.

IMMATURE PLATELET FRACTION

(IPF) See **Platelet indices**.

IMMATURE RETICULOCYTE FRACTION

(IRF) See **Reticulocyte indices**.

IMMERSLUND-GRÄSBECK SYNDROME

(Familial selective malabsorption of vitamin B_{12}) A rare inherited disorder (autosomally recessive inheritance) characterized by the appearance of **megaloblastosis** with **anemia** and proteinuria during the first five years of life.[148] The original patients were reported from Finland and Norway, but more recently from the Middle East. The Finnish patients have shown mutations in the cubulin (CUBN) gene on chromosome 10p12.1, which encodes for a receptor for the intrinsic factor-**cobalamin** complex. This results in failure of enterocytes to absorb the complex. In the Norwegian patients, a candidate gene was located on the long arm at 14q32, mutations being found in the AMN gene. It is postulated that the 5′ end of the gene product is needed for cobalamin absorption. The consequent anemia responds completely to cobalamin therapy, although the proteinuria persists.

IMMUNE COMPLEXES

High-molecular-weight complexes of **antigen** and **antibody**, often many molecules of each, sometimes including **complement** components. These are present in the circulation during infections (e.g., hepatitis) and with autoimmune disorders (e.g., **systemic lupus erythematosus**) and may be deposited at various sites, including kidney, joints, and capillaries. A consequence of the deposition is local activation of complement and thus tissue damage, resulting in conditions such as glomerulonephritis.

IMMUNE DISORDERS

Pathological states of the immune system giving rise to tissue injury and illness.

Immunodeficiency: primary and secondary

Hypersensitivity reactions
- Type I: allergic
- Type II: autoimmunity
- Type III: immune complex disorders
- Type IV: delayed hypersensitivity, e.g., contact dermatitis, graft rejection of transplanted tissue

Lymphoproliferative disorders

IMMUNE DOMINANCE

See also **Antigen**.

The preferential immune response to one epitope rather than another, i.e., a hierarchical response to different epitopes. Immunodominance of T-cell epitopes is easy to understand: it depends on their ability to interact with appropriate HLA antigens. Immunodominance of B-cell epitopes is less easy to understand, and it is less clear that it exists. One problem with the latter is the confusion between immunodominance and neutralization epitopes, i.e., that region of an antigen to which antibody binding results in inactivation

IMMUNE EVASION BY PATHOGENS

The failure of the immune system to clear pathogens (see **Immunodeficiency**). Detailed explanation will depend on the individual situation, but a number of general mechanisms can be described.

General Mechanisms

Antigenic Variation

The classic example is the causative organism of sleeping sickness, *Trypanosomas brucei*. The major host response is through **antibodies** to trypanosomal surface glycoprotein **antigens**, which destroy the protozoa either by **complement**-mediated mechanisms or through **opsonins**. However, trypanosomes possess multiple genes (hundreds) for the surface antigen, which are expressed singly. The progeny of the initially infecting protozoa all express the same surface antigen; the antibody response develops and kills the majority of the protozoa. However, a few will switch expression to a second surface antigen that is not recognized by antibodies to the first antigen, and so repeated waves of replication occur.

Another example is infection by **human immunodeficiency virus** (HIV). Upon initial infection, this virus replicates, and an antibody response to the envelope protein develops that, together with cell-mediated responses, initially controls the infection. However HIV, in common with some other RNA viruses, has a high rate of mutation, and so variant viruses with altered envelope protein epitopes arise that are not recognized by preexisting antibodies.

Production of Immunosuppressive Factors

Many parasitic infections depress **cell-mediated immunity** (CMI) in an ill-defined manner. Certain viruses produce factors that suppress immunity; for example, an interleukin-10-like factor produced by **Epstein-Barr virus** and certain cancers, notably gliomas, produce the immunosuppressive cytokine-transforming growth factor-β.

Downregulation of Major Histocompatibility Antigens

Several viruses, e.g., adenoviruses, prevent the transport of class I major histocompatibility antigens to the cell surface, thus reducing susceptibility to T-cells mediating cellular cytotoxicity. Downregulation of histocompatibility antigens is a common feature of tumor cells, possibly with the same result.

Downregulation of Target Antigens

In the case of **Burkitt lymphoma/leukemia**, the Epstein-Barr virus antigens in infected B-cells, which would normally result in a vigorous **cytotoxic T-lymphocyte** response (as in **infectious mononucleosis**), are not expressed. Another example is infection by herpes simplex virus. After the primary infection, this virus may persist (as DNA) in cells with no expression of viral proteins until it is reactivated, when replication and the characteristic blistering are repeated.

Destruction of Components of the Immune System

HIV targets CD4 cells, thereby destroying the key cell of the immune system.

Inhibition of Phagocyte Function

Some bacteria produce factors that inhibit phagocytosis, e.g., the polysaccharide capsule of *Streptococcus pneumoniae*. Other bacteria, particularly the mycobacteria, are able to escape from the phagolysosome of macrophages, thus avoiding the bacteriocidal mechanisms and dividing within the cytoplasm of the macrophage.

IMMUNE HEMOLYTIC ANEMIAS

A group of anemias induced by **antibodies** that react with **red blood cell** antigens to cause a reduction in red cell survival.[91,280] These disorders are classified as follows:

Autoimmune hemolytic anemia (AIHA). The antibody is produced by the body's own lymphocytes against its own red blood cells. It is characterized by a positive **direct antiglobulin (Coombs) test**. The disorders are further classified according to the optimum temperature at which the reaction occurs:

- **Warm autoimmune hemolytic anemia** reacting at 37°C
- **Cold autoimmune hemolytic anemia** reacting below 37°C
- **Mixed-type autoimmune hemolytic anemia**
- **Drug-induced immune hemolytic anemia**

Alloimmune hemolytic anemia. The antibody is produced by a different person of the same species against the recipient's red blood cells. It occurs with:

- **Blood transfusion complications** — hemolytic reactions
- **Hemolytic disease of the newborn**
- Allograft-associated alloimmune hemolytic anemia

IMMUNE RESPONSE

Activation of the immune system either through body fluids (**humoral immunity**) or by cellular activation (**cell-mediated immunity**). It can be activated by infection of foreign antigens, tissue damage, especially necrosis, or by neoplastic proliferation. It is also activated by neuropeptides through **lymphocytes** and **histiocytes**. Much of the response acts through the **lymphoid system**. There are three types of response:

Innate or natural immunity. Here an immediate response occurs to antigenic stimulation involving **neutrophils**, natural killer (NK) lymphocytes, **complement**, **interferon**, and histiocytes (macrophages). Examples are:

- Activation of interferon-α production by virus-infected cells (the proximal inducer for interferon-α is probably double-stranded RNA, with no question of any sort of antigenic specificity), which will then activate NK-cells to lyse cells in a non-antigen-specific manner.

- Induction of **inflammation** triggered by the release of mediators from necrotic cells and by components of pathogens, classically bacterial lipopolysaccharides. This results in migration of neutrophils, phagocytosis, and lysis.

Adaptive immunity. Cytokines derived from histiocyte stimulation in the innate immunity involve T-cells and B-cells by T-cell receptors (T-CRs) and B-cell receptors (B-CRs), maybe in specific antigen–antibody reactions. The term perhaps is best reserved for the situation where the response is specifically directed against a foreign antigen and immunological memory develops. The term "memory" in this context indicates a modified response (often augmented, but sometimes depressed, as in the phenomenon of tolerance) upon a second exposure to the same stimulus. The response occurs in phases:

- Cognitive. Recognition of antigen by T- and B-CRs upon presentation by plasma complement, MCH complex molecules, adhesion molecules, and dendritic cells.

- Activation. Proliferation of cells with matching receptor following signal transduction

- Effector. Elimination of antigen by antibody dissolution or phagocytosis.

Hypersensitivity. Exaggerated immune responses giving rise to tissue damage. Here there are four types of reaction:

- Type I: allergic response
- Type II: autoimmunity
- Type III: immune complex disease
- Type IV: delayed cell-mediated immunity, examples being chronic inflammation with granulomas, graft rejection after transplantation, and vasculitis

IMMUNE SURVEILLANCE

See also **Tumor antigens; Immune evasion by pathogens**.

Observation of the tissues of an organism for the presence of deviations and elimination of such deviations when detected. Where the deviations are the presence of non-self components (microorganisms, etc.) there is no difficulty with the concept, and the immune system is well adapted for this through its circulatory nature.

In practice, the term "immune surveillance" is more usually restricted to surveillance for cancer. The idea is that cells transform to a potentially malignant phenotype with a relatively high frequency, but that the majority of such cells are destroyed by the immune system, particularly the cell-mediated arm, recognizing tumor antigens. This simple hypothesis is probably vitiated by the observation that immune suppression (e.g., in organ transplantation, or in animal experimental models) does not lead to a massive increase in the prevalence of cancers. However, some cancers are much increased, particularly B-cell **lymphoproliferative disorders** and skin cancers. These are exceptional in that the lymphomas are viral in origin (nearly all are **Epstein-Barr virus** positive), and so what we are seeing is essentially the failure of antiviral immunity, and the skin cancers (commonly squamous cell and basal cell carcinomas) are recognized from experimental systems as being unusually immunogenic. Most common cancers show very modest augmentations in frequency.

A modified concept of immune surveillance is tenable if we accept (1) that most tumors are multifactorial in cause and that the failure of surveillance is but one factor and (2) that many cancers develop mechanisms for evading surveillance.

IMMUNE THROMBOCYTOPENIC PURPURA

(ITP) A disorder caused by interaction of **immunoglobulins** with the **platelet** surface, leading to their accelerated destruction. The platelets become coated with **autoantibodies**, usually IgG, most frequently directed against **glycoprotein IIb/IIIa** complex. The normal lifespan of a platelet is 7 to 10 days; in ITP, this is dramatically reduced to a few hours. By a negative feedback loop, platelet destruction leads to an increased total **megakaryocyte** mass, with increased numbers of megakaryocytes of early, low ploidy in the bone marrow. Total platelet production and turnover may be increased up to fivefold before thrombocytopenia ensues. Sensitized platelets are prematurely removed from the circulation by cells of the reticuloendothelial system. Although the spleen is the primary site of platelet removal, heavily coated platelets are predominantly removed later by the liver. Little intravascular lysis of platelets occurs.

Etiology

ITP may occur at any age, but it is most common in children and young adults. In children, the sex incidence is equal, but after adolescence females are affected three to four times more frequently than males. Two clinical types are recognized:

Acute, self-limiting
Chronic, recurrent, and/or persistent

There is considerable overlap between these two groups. The majority of cases of thrombocytopenia seen in clinical practice are secondary. The diagnosis of acute or chronic ITP can only be made after careful evaluation and consideration/exclusion of other causes, particularly drug-induced thrombocytopenia.

Acute Immune Thrombocytopenic Purpura

This occurs most commonly in children between the ages of 2 and 6 years. It has a relatively acute onset with mucocutaneous bleeding. Epistaxis, bruising, and scattered purpura or petechiae are common presenting features. There may be a history of a recent viral infection, including varicella zoster or measles, in the 2 to 3 weeks preceding presentation, or recent vaccination, although in many cases no specific infectious association can be identified. Bleeding may be severe at presentation. Serious life-threatening hemorrhage is rare (approximately 1%). Intracranial hemorrhage only occurs in less than 1%, usually associated with platelet counts of $<20 \times 10^9/l$.

Chronic Immune Thrombocytopenic Purpura

This is more commonly a disease of young to middle-aged females. The onset is more insidious than acute ITP. Symptoms at presentation vary from mild, easy bruising and petechiae to relatively severe mucocutaneous bleeding, including menorrhagia. A proportion of cases may be picked up incidentally on routine blood counting. In chronic ITP, symptoms run a variable and intermittent course. Remissions may occur and last for months to years. In other cases, the symptoms remain but vary in severity. For a given degree of thrombocytopenia, severity of bleeding in ITP may appear less than that seen in thrombocytopenia from marrow failure. This is attributed to the circulation of predominantly young, functionally superior platelets. Splenomegaly is not a usual feature of ITP.

Drug-Induced Immune Thrombocytopenia

This is often related to an allergic mechanism. Thrombocytopenia may be seen in association with a range of drugs, including quinine, quinidine, heparin, *para*-aminosalicylates, sulfonamides, rifampicin, and digoxin. The antibody is usually directed against the drug bound to a plasma protein. Antigen–antibody complexes are passively adsorbed by platelets. The platelet is then damaged as an innocent bystander. Coated platelets are removed by the reticuloendothelial system due to the bound immunoglobulin or associated **complement**. Immediate cessation of the suspected drug(s) is essential. Recovery is then usually rapid, dependent upon half-life and elimination of the offending drug from the circulation.

Fetal/Neonatal Immune Thrombocytopenic Purpura

As the predominant antibody in ITP is IgG, it may cross the placenta. The fetus may therefore become severely thrombocytopenic. The origin of the antibody may be maternal or alloimmune (see **Neonatal alloimmune thrombocytopenia**). Antenatal fetal hemorrhage is rare. In milder cases, the infant may become thrombocytopenic at or shortly after birth. Children born of mothers with ITP should be closely monitored for thrombocytopenia and cerebral hemorrhage

Laboratory Features

The outstanding abnormality is thrombocytopenia with a **platelet count** of 10 to $50 \times 10^9/l$ or less in drug-associated immune thrombocytopenia.[281] Hemoglobin and white cell counts are normal. **Peripheral-blood film examination** shows reduced platelet numbers, with those present being larger than normal. **Bone marrow** examination shows a normal or increased number of megakaryocytes, which are often relatively immature. The bone marrow is otherwise normal. Antiplatelet IgG may be detected either alone or with complement, on the platelet surface. IgM may also be detected. Platelet-associated IgG may be 5 to 10 times above the normal. The quantity of platelet-associated IgG per platelet correlates well with severity of disease and is inversely proportional to platelet count and mean platelet survival. In acute ITP, serum IgG and IgM are both raised. Platelet-associated Ig in acute ITP may be 4 to 5 times that seen in chronic ITP. In chronic ITP, platelet autoantibodies are often platelet specific; in acute ITP, much of the platelet-associated Ig may be of **immune complex** origin, bound through platelet **Fc receptors**.

Occasionally a positive **direct antiglobulin (Coombs) test** and **autoimmune hemolytic anemia** coexist (Evans's syndrome).

Treatment

This is, in part, dependent upon clinical type.[282–289] Over 80% of acute ITP of childhood will remit permanently without intervention. Treatment is not indicated other than avoidance of injury. The remaining 20% will take up to 6 months to remit or may develop into recurrent, chronic ITP. Duration of thrombocytopenia of greater than 6 months classifies ITP as chronic. Although other therapies may need to be employed, even chronic ITP of childhood is a relatively benign condition, with eventual remission, although this may not be for a few years. Should the platelet count fall below $20 \times 10^9/l$, a short, high dose of prednisone/prednisolone should be considered rather than platelet transfusion, which is unlikely to be effective. Splenectomy should only be considered when there is no remission after 12 to 24 months.

In contrast to acute ITP of childhood, less than 10% of adults with chronic ITP recover spontaneously. Treatment is aimed at reducing the level of antibody and the rate of platelet destruction. First-line treatments include steroids. Eighty percent of patients remit upon high-dose **corticosteroids**. The dose is gradually reduced after remission. A partial or complete relapse commonly occurs upon reduction of steroid dosage. If an unacceptably high dose is required to maintain the platelet count, alternative therapies should be considered.

Infusion of high-dose **immunoglobulin** is able to rapidly raise platelet counts in the majority of patients. A dose of either 400 mg/kg/day for 5 days or 1 g/kg/day for 2 days is used. This is useful in selected patients, such as those with hemorrhage, steroid-refractory ITP, pregnancy, and prior to surgery. The mechanism of action may involve blocking reticuloendothelium Fc receptors, although other mechanisms have been proposed.

Splenectomy is recommended in patients who require an unacceptable dose of steroids or who fail to remit by other means after 3 months. Over 75% of patients show a good response to splenectomy, with sustained clinical remissions. The remaining 25% of patients do not respond or have a poorly maintained response, with counts returning to presplenectomy levels in a few weeks. Only rarely does relapse occur late, after a number of years. Accessory spleens should then be excluded as a possible cause by **radionuclide** scanning. Addition of steroids or other immunosuppressive therapies may be required.

Immunosuppression by drugs can be used in refractory ITP. These include **vincristine**, **cyclophosphamide**, **ciclosporin**, and **azathioprine** (reduction of antibody generation). Many refractory cases show a useful response to immunosuppression. Alternative therapies with anti-D immunoglobulin or α-**interferon** may also have some place in refractory cases. Use of **platelet transfusions** is very limited. Transfused platelets will be destroyed as rapidly as native platelets. However, platelet transfusion may temporarily control a life-threatening hemorrhage.

The management of maternal immune thrombocytopenia is controversial.[117] Treatment with corticosteroids or immunoglobulin may be required to maintain and support the level of maternal platelets prior to delivery.

Neonatal immune thrombocytopenia can be treated with corticosteroids, immunoglobulin, or exchange transfusion (see **Fetal/neonatal transfusion**).

IMMUNITY

Ability of an organism to resist infection using host defense mechanisms (see **immune response**). This involves **neutrophils**, **monocytes**, **histiocytes** (macrophages), **lymphocytes** — T-cells and B-cells producing antibodies (**immunoglobulin**), natural killer (NK) cells, and **dendritic reticulum cells** with the essential support of plasma **complement** — **major histocompatibility complex** (MHC) molecules, **cell-adhesion molecules**, and messenger **cytokines** and **chemokines**. Immunity is either innate (nonspecific) or adaptive (specifically acquired immunity). Innate immunity is immediately active, using physical and chemical barriers followed by the immune response. Adaptive immunity follows innate immunity when it has failed to remove the "foreign" material by the action of specific T-cell receptors. Passive immunity can be achieved by the maternal acquisition across the placenta or by intravenous injection. Active immunity can be induced by vaccination. When a population has achieved a high incidence of immunity to a specific infection (e.g., smallpox), it is referred to as having herd immunity.

IMMUNOASSAY

The determination of amounts of particular proteins (and other **antigen** materials) by the use of specific **antibodies**. Essential for the technique is a monospecific antibody with a good affinity for the target antigen. Generally, **monoclonal antibodies** are used, but polyclonal antibodies may be suitable or, indeed, may be preferable, as they react with a wide range of **epitopes** on the target antigen.

In essence, the technique depends on the specific and quantitative interaction between an antibody and its antigen. Where the antibody is present in excess, all the combining sites of the antigen will be saturated. Removal of excess antibody and estimation of the amount of combined antibody allows determination of the amount of the target antigen. Immunoassays can be carried out in the liquid phase (in which case there must be a step to separate combined from uncombined antibody) or with one or other components of the reaction attached to a solid phase, in which case separation is achieved by washing.

Estimation of amounts of antibody is done by labeling techniques. The antibody can be conjugated with a **radionuclide** (e.g., ^{125}I) in the case of radioimmunoassay (RIA), where bound radioactivity is proportional to the amount of antibody, or with an enzyme (e.g., phosphatase) in enzyme-linked immunosorbent assay (ELISA) catalyzing a reaction generating a colored compound, in which case the amount of color generated is proportional to the amount of antibody. The latter, for reasons of safety and convenience, is now far more common than the former.

Solid-phase ELISAs are often carried out in clear plastic plates with 96 wells of 200-μl capacity arranged in an 8×12 array for which there are many automated readers available for measurement of the absorption of the contents of the wells.

The most common design is "sandwich" or "capture" ELISA, requiring two antibodies. The first is coated to the wells of the plate, and dilutions of the unknown sample added. After incubation, the plate is washed, and the conjugated antibody (which must react with a different epitope of the antigen) is added in excess. After a further period for reaction, the plate is again washed and the color substrate added and the amount of color subsequently determined. "Competition" ELISAs rely on a preliminary step in which the unknown antigen sample is incubated in solution with a known amount of antibody. The mixture is then added to a plate to which the antigen has been coated, and unreacted antibody reacts with the antigen. The excess is washed away, and the amount of bound antibody is determined by a conjugated antibody reacting with the primary antibody. In this technique, only one antibody to the antigen is required. A third type — "indirect" — may be used to measure specific antibody. Plates are coated with the antigen, dilutions of unknown antibody added, and the amount reacted is determined as with competition ELISA. In every case, standard curves have to be constructed with reagents of known concentration to measure concentrations of unknowns by interpolation.

ELISAs are now commonly used for determination of a wide range of substances, including such nonproteins as drugs and insecticides as well as proteins. Many ELISAs are commercially available in kit form, providing all necessary reagents and protocols.

IMMUNOBLAST

Large **lymphocytes** (18- to 25-μm diameter) arising from "virgin" T and B-cells upon exposure to **antigen** by antigen-presenting cells (APC), commonly for the first time in the **lymph nodes** or **spleen**. With Romanowsky dyes, the nucleus is light staining, with a fine **chromatin** pattern and with large nucleoli, which are often solitary in the middle of the nucleus or as an indentation of the nuclear membrane. The cells have a broad rim of strongly basophilic-staining cytoplasm.

IMMUNOBLASTIC LEUKEMIA/LYMPHOMA

See **Diffuse large B-cell lymphoma**; **Non-Hodgkin lymphoma**.

IMMUNOCYTOMA, LYMPHOPLASMACYTIC TYPE

See **Lymphoplasmacytic lymphoma/Waldenström macroglobulinemia**; **Waldenström macroglobulinemia**.

IMMUNODEFICIENCY

Reduction in the function of the normal **immune response** for defense. This usually implies protection from infection, but can include malignancy associated with depletion of either B- or T-lymphocytes.[290] Immunodeficiency is the major aspect of the **immune evasion by pathogens**. It occurs infrequently as a primary disorder where lymphocyte collections, such as in the white pulp of the spleen, are depleted. With increasing frequency, it is a secondary complication of infection, the most prominent example being **acquired immunodeficiency syndrome**.

Primary (Congenital) Immunodeficiency

There are around 130 distinct entities, many of which overlap T-cell- and B-cell-deficiency disorders. T-cell-deficiency disorders result in loss of **cell-mediated immunity**. The effects are increased susceptibility to infection by viruses, fungi, bacteria, and protozoa. B-cell-deficiency (**humoral immunity** deficiency) effects include impaired protection against bacterial infection and, to a lesser extent, fungi and viruses. Because B-lymphocyte function in humans is T-lymphocyte dependent, T-cell deficiency results in some degree of humoral immunodeficiency, a combined immunodeficiency. The disorders are usually detected during infancy by the recurrence of pyogenic and opportunistic infections.

Etiology

Genetic Mutations

Severe Combined Immunodeficiency — (SCID; "Swiss" type) An inherited T-cell deficiency that results in failure of B-cell production of **immunoglobulins**. This leads to failure of both cell-mediated immunity and humoral immunity, allowing opportunistic infections to occur from early infancy. Nine different molecular defects are known. The commonest include:

X-linked (XL-SCID) absence or defective function of *JAK3*, resulting in the failure is of several T-cell receptors (IL-2, IL-7, and IL-4), the subunit being in the λ chain. It results in impaired intrathymic T-cell maturation and function. Abnormal immunoglobulin synthesis follows with only 96 or 128 amino acids. Children with this disorder develop fulminant **infectious mononucleosis**, which may proceed to a **lymphoproliferative disorder**.

Autosomally recessive deficiency of genes for **adenosine deaminase** (ADA) or for **purine** nucleoside phosphorylase (PNP). This deficiency allows the formation of metabolites toxic to lymphocyte stem cells. Deficiency of ADA in lymphocytes results in an accumulation of **adenosine triphosphate** (ATP) and deoxyATP, thus inhibiting **DNA** synthesis. Such children are normal at birth but develop progressive immunological impairment as deoxyATP accumulates. **Red blood cell transfusion** cells rich in ADA or the use of polyethylene glycol-modified ADA has temporarily restored immunity in some instances.

Protein kinase *ZAP-70* abnormality leads to a deficiency of circulating CD8 T-cells.

Recombinase-activating gene deficiency (*RAG-1, RAG-2*) *Artemis* mutation is expressed early and effectively terminates lymphocyte maturation with defective innate resistance to both vial and bacterial infection from birth.

Translocation of part of chromosome 22 (Di George syndrome), resulting in a decreased number of CD3⁺ T-cells. This not usually inherited and leads to only limited opportunistic infections.

Reticular dysgenesis with associated phagocytic deficiency.

Hyper-IgM Syndrome — An X-linked mutation at X-q26 that can be inherited as an autosomally recessive disorder that causes T-cells to fail to express CD154, the ligand for CD40. IgA and IgG cannot be expressed because T-cells are unable to offer appropriate help to induce B-cells to switch immunoglobulin class. Furthermore, T-cells are unable to communicate normally with antigen-presenting cells (APCs). **Neutropenia** is an associated disorder. It gives rise to recurrent pyogenic infections, **autoimmune disorders**, and **lymphoproliferative disorders**. **Allogeneic stem cell transplantation** may offer the only form of treatment.

X-Linked Hypogammaglobulinemia — (Bruton's disease) Here, the arrest of differentiation of pre-B-cells is caused by mutations of the genes encoding *ZAP-70* and Bruton's **tyrosine kinase** (Btk), which is involved in phosphoinisotide hydrolysis during *BCR* signaling. This causes impaired B-cell development. B-cells are not present in the circulation, lymph nodes, or gastrointestinal tract, with only pre-B-lymphocytes in the bone marrow. Affected males have no circulating IgA, IgM, IgD, or IgE and only small amounts of IgG. This leads to recurrent pyogenic infections once maternal IgG has been exhausted after 5 to 6 months of life. It is treated by regular injections of immunoglobulin.

Genetic Polymorphisms

Human leukocyte antigen (HLA) alleles of the same gene at a single locus are common and thus give rise to a variety of disorders:

Inability to bind some viral peptides.

Abnormalities of promoter for **tumor necrosis factor** (TNF) increase the risk of cerebral malaria and the risk of developing septic shock.

Deficiency of **cytokines** or their receptors, such as interferon-γ receptor (IFN-γR1). These deficiencies may increase the consequences of infections such as by HIV-1.

Mannin-binding lectin disturbances lead to failure to activate the classical **complement** pathway.

Polygenic Disorders

These are caused by the interaction of several genes.

Common Variable Immunodeficiency — (CVID) Here there are low serum levels of IgA2 and IgA4, IgM, reduced B-cells, and mild cell-mediated deficiency.[291] This may be caused by failure of B-cells to terminally glycosylate and secrete immunoglobulins, a failure of helper T-cell factor production, or an increase in specific suppressor T-cell effects. The immunological defect in this disorder is not limited only to B-cells but also includes macrophages and immunoregulatory T-cells. The defects in T-cell immunity include abnormalities of activation and lymphokine production, which usually progress with age. Older patients commonly have chronic diarrhea due to associated malabsorption.

Autoimmune Disorders — Autoimmune disorders can occur, particularly gastric atrophy, hepatitis, and gastric carcinoma. A relatively common occurrence is persistent generalized **lymphadenopathy** with absence of lymphoid follicle germinal centers. Nodular intestinal lymphoid hyperplasia and hepatosplenomegaly are additional features. Pyogenic and opportunistic infections occur, particularly giardiasis. Management includes immunoglobulin replacement therapy, antimicrobial therapy, pulmonary drainage, and immunomodulatory therapies, including recombinant human IL-2 conjugated with polyethylene glycol and cimetidine.

Selective IgA Deficiency and IgG Subclass Deficiency — These are due to failures of terminal B-cell ontogeny. IgA deficiency is the most common primary immunodeficiency and causes a predisposition to allergic and autoimmune disorders, e.g., autoantibodies to rheumatoid arthritis factor, **antinuclear factor** (ANF), and reticulin. Those who also have a deficiency of IgG2 and IgG4 are susceptible to pyogenic infection. Some patients have a first-degree relative with CVID deficiency. Susceptible genes may be located within the major histocompatibility complex class III region of chromosome 6. Treatment is difficult due to the short half-life of IgA and the presence of autoantibodies.

Specific T-Cell-Deficiency Disorders

Hypoplasia of the Thymus — (Congenital thymic aplasia; Di George syndrome) This is caused by failure of a population of neural crest cells to migrate and interact with endodermally derived cells of the third, fourth, and fifth pharyngeal pouches, resulting in absence of **thymus** and parathyroid glands. The consequences are **lymphopenia** as a result of monosomy of chromosome 22q-11 in addition to hypoparathyroidism with convulsions and tetany; cardiovascular anomalies such as Fallot's tetralogy and right-sided aortic arch; characteristic abnormal facies with low-set notched ears, micronathia, and nasal clefts; recurrent opportunistic infections. Severely affected patients usually succumb within the first few weeks or months of life unless a fetal thymus transplant or allogeneic stem cell transplant can restore immunological function.

Ataxia-Telangiectasia — An autosomally recessive disorder affecting mitogenic signal transduction, meiotic recombination, and cell-cycle control. The defect gives rise to cerebellar ataxia, oculocutaneous telangiectasia, gonadal dysgenesis, and immunodeficiency causing malignancy. A large proportion of patients have a selective IgA deficiency and some an associated IgG2 subclass deficiency. Most have depressed or absent IgE levels. T-cell abnormalities include lymphopenia, decrease in helper T-cell/suppressor T-cell ratio, and an overall decrease in cytotoxic T-cells.

Wiscott-Aldrich Syndrome — (WAS) **Wiscott-Aldrich syndrome** is an X-linked recessive defect causing **thrombocytopenia**, eczema, and immunodeficiency of IgM but increased IgA with normal IgG. Malignancy often follows.

Major Histocompatibility Complex (MHC) Class II Deficiency

An autosomally recessive defect causing failure of expression of MCH class II molecules on antigen-presenting macrophages and B-lymphocytes. This is necessary for positive selection of CD4 cells in the thymus; otherwise, a CD4 deficiency occurs. Recurrent infections follow, particularly affecting the gastrointestinal tract.

Complement Deficiency

These defects are mostly autosomally recessive inherited states.

C1q, C1r, C1s, C4, or C2 deficiency. These result in immune complex disorders, particularly systemic lupus erythematosus but without ANF antibodies. Activation of the alternative pathway may be sufficient to provide normal immune evasion of pathogens.

C3 deficiency. Either a primary defect or secondary to deficiency of C3b inhibitor, which inactivates factor I or factor H (b1H globulin). It results in recurrent pyogenic infections due to failure of opsonization.

C5, C6, C7, C8, factor D, and properdin deficiency (membrane attack complex) all result in recurrent neisseria infections but are otherwise asymptomatic.

Neutrophil Functional Deficiencies

Chronic granulomatous disease

Leukocyte-adhesion deficiency

Complications of Primary Immunodeficiency
Infections

In addition to the common infections, fatal infectious mononucleosis with hemophagocytosis sometimes occurs (see Table 82).

Lymphoproliferative Disorders

The lymphoproliferative disorders include:

Diffuse large B-cell lymphoma

Hodgkin disease

Precursor T-lymphoblastic leukemia/lymphoma

Management of Primary Immunodeficiency

This involves the following procedures.

Genetic counseling with prenatal diagnosis if possible.

Avoidance of immunization with live vaccine due to risk of inducing overwhelming infection by the specific virus.

Transfusion of blood avoided unless essential, in which case it must be irradiated prior to use.

Appropriate antimicrobial therapy only as necessary.

Permanent grafting of immunocompetent tissue — allogeneic bone marrow transplantation, fetal thymic graft, fetal liver graft. Transplantation by infusion of T-cell-depleted bone marrow is the only treatment for children with SCID and has resulted in lymphoid chimeras with normal immunological growth and development.[292]

Replacement of missing factors:
- Cell extracts: thymic hormones, transfer factor
- Enzymes: SCID, PNP, ADA by bone marrow transplantation
- Immunoglobulin replacement therapy either by intravenous, intramuscular, or subcutaneous injections[292]

- **Allogeneic stem cell transplantation** by infusion of T-cell-depleted bone marrow: the only treatment for children with SCID. It has resulted in lymphoid chimeras with normal immunological growth and development[293]

Gene therapy to correct lymphocyte disorders by retroviral-mediated gene transfer into hematopoietic precursor cells. Gene therapy has been initially successful for XL-SCID but may result in the development of a leukemia or lymphoma.[294]

Secondary Immunodeficiency

Etiology

Transient low levels of immunoglobulin occur in infancy as an extension of "physiological hypogammaglobulinemia" beyond 6 months of birth due to either reduced maternal IgG transfer across the placenta or an abnormally slow rate of IgG synthesis by the infant. Complement deficiency occurs due to consumption during infections and active autoimmune disorders. Other secondary immune deficiencies are caused by a multiplicity of disorders.

Malnutrition of protein leads to defects in cell-mediated immunity, antibody production, phagocytic function, and complement activity.

Proteinuria (nephrotic syndrome), protein-losing enteropathy, and burns cause reduced levels of immunoglobulin, with recurrent infections.

Immunosuppression, usually by drug therapy (see Table 80).

Lymphoproliferative disorders:

- **Myelomatosis** and **plasmacytoma**s: B-lymphocyte deficiency
- **Hodgkin disease** and **non-Hodgkin lymphoma**: T-lymphocyte deficiency
- **Chronic lymphatic leukemia**: combined T and B-cell deficiency

All forms may produce an anti-idiotype to their overproduced immunoglobulin with an idiotype–anti-idiotype reaction, which consumes C1, C4, C2, and C1 inhibitor. This may cause an acquired **angioedema**.

Acquired immunodeficiency syndrome (AIDS). A T-lymphocyte deficiency disorder associated with **HIV infection**.

TABLE 80

Action of Drugs on Immune Functions

Immune Function Affected	Drug
T- and B-lymphocyte function	azathioprine
	corticosteroids
	cyclophosphamide
	ciclosporin
Neutrophil chemotaxis	gentamicin
	amikacin
	tobramycin
Cell-mediated immunity	cotrimoxazole
	rifampicin
	tetracycline
Primary and secondary antibody responses	chlorambucil
	cotrimoxazole
	rifampicin

TABLE 81

Relationship of Immunodeficiency with Common Infections

Antibody (B-Cell)	Cellular (T-Cell)	Complement	Neutrophil Phagocytosis
Pyogenic bacteria	Viruses	Pyogenic bacteria	Bacteria
staphylococcal	cytomegalovirus		staphylococcus
streptococcal	vaccinia	*Neisseria*	Gram negative
hemophilus	herpes		
Viruses	Measles	Some viruses	Fungi
enterovirus			*Aspergillus*
polio	Fungi		*Candida*
echovirus	*Candida*		
	Bacteria		
	mycobacteria		
	Listeria		
	Protozoa		
	Pneumocystis carinii		

Allogeneic stem cell transplantation and allogeneic blood transfusion, causing dys-regulation of lymphopoiesis, which can increase the risk of postoperative infec-tion. As a blood-transfusion complication, this can be avoided by using **leukocyte-depleted blood components for transfusion**. The disturbed lymphopoiesis also contributes to the risk of colonic carcinoma recurring following colectomy.

Ionizing radiation can cause both T- and B-lymphocyte deficiency.

Adenovirus infection is an emerging pathogen for immunodeficiency.

Complications

The main consequences of secondary immunodeficiency are opportunistic infections. The relationship of the types of immunodeficiency with common infections is given in Table 81.

Management of Secondary Immunodeficiency

This is treatment of the complicating opportunistic infections[9] by the appropriate antibi-otic, as shown in Table 82.

IMMUNOELECTROPHORESIS

A separation technique used for the identification of **antigens** based upon their differences in the mobility in an electric field that is generated by direct current through a buffered solution. Antigens are placed in wells of an agar gel. After separation, a trough is cut between the wells, which are then filled with **antibody**, and diffusion is allowed to take place. Where an antigen–antibody reaction occurs, arcs of precipitation form (precipitin reaction). The technique is used for the diagnosis of **hemoglobinopathies, monoclonal gammopathies**, and **biclonal gammopathies**.

IMMUNOFLUORESCENCE

See **Flow cytometry; Immunohistology.**

IMMUNOGEN

See also **Antigen.**

TABLE 82

Antibiotics Appropriate to Opportunistic Infections

Opportunistic Infection	Antibiotic
Pneumocystis carinii	cotrimoxazole
	pentamidine
Toxoplasmosis	sulfadiazine
	pyrimethamine
Cryptosporidosis	none
Tuberculosis	isoniazid
	rifampin
	pyrazinamide
	streptomycin
	ethambutol
Cytomegalovirus	ganciclovir
	foscarnet
	cidofovir
Mycobacterium avium complex	rifampin/rifabutin
	clofazamine
	ethambutol
	amikacin
	macrolides: clarithromycin, azithromycin
	ciprofloxacin
Candidiasis	fluconazole
Penicillium marneffei	amphotericin B
Cryptococcus neoformans	fluconazole/itraconazole

Data derived from *British National Formulary*, 51, British Medical Association and Royal Pharmaceutical Society, London, Sept. 2006.

A substance that is by itself able to elicit an **immune response**. Some substances, e.g., haptens, are unable to generate an immune response by themselves, but **antibodies** can be made if the hapten is attached to a complex protein carrier. Hence the hapten is antigenic but not immunogenic. Immunogens are generally large, complex molecules and are almost always proteins.

IMMUNOGLOBULINS

(Ig) Glycoproteins of a large superfamily with a similar structure and **antibody** activity. Immunoglobulins have the ability to combine with determinants on antigens, i.e., **epitopes**. Immunoglobulins play two crucial roles in the **immune response**: as receptors (membrane immunoglobulin [mIg]) on the surface of B-**lymphocytes** for specific antigens (B-cell receptor [BCR]) and as effector molecules, i.e., antibodies, secreted by B-cells. They perform the receptor function by virtue of interacting with signaling molecules within the B-cell such that, upon engagement with antigen, the B-cell is activated (other signals are required). They perform effector functions by combining with the antigen; this may inactivate it directly, e.g., as in toxin neutralization, or else indirectly through interaction with **complement** or with cells, e.g., **neutrophils**.

General Structure of Immunoglobulins

There are multiple secreted Ig classes (isotypes, originally defined antigenically) that perform different functions (see below). Which antibodies are produced in an immune response depends on the maturation of the response (IgM early, others late) and whether the response is of the Th1 or Th2 type, the latter predisposing to greater amounts of antibody and production of the IgE type.

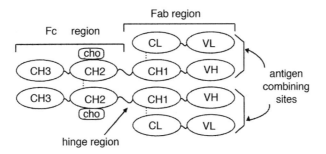

FIGURE 73

Schematic diagram of an immunoglobulin. Oval boxes represent protein domains, swung lines represent random coil linking regions, and dotted lines represent disulfide bonds. CH1, -2, and -3 are constant regions of the heavy chains; CL is a constant region of light chains; and VH, VL are corresponding variable regions. The box labeled "cho" represents carbohydrate. Note that the molecule is flexible around the "hinge" region and that, for steric reasons, it adopts a Y conformation unless otherwise constrained, e.g., by combining with polyvalent antigen.

The structure of all Igs is broadly the same: a combination of paired heavy (H) and light (L) polypeptide chains with molecular weights of about 50 kDa and 25 kDa. The basic unit consists of two pairs and so has the formula H2L2, forming a Y-shaped molecule (see Figure 73).

Sequencing studies of human myeloma proteins revealed that the N-terminal half of light chains (about 110 amino acids) is unique to each myeloma, whereas the C-terminal half falls into one of two classes (l or k). These are designated variable (VL) and constant (CL) regions. Similarly, heavy chains show a unique variable region (VH) of about the same size as the light chain V region and one of five possible isotype-specific constant (CH) regions (g, a, μ, d, or e). Both chains show repeating units of 110 amino acids — two in the light chains (VLCL) and three or four for heavy chains (VHCH1CH2CH3CH4) — that show a degree of sequence similarity. Structural studies have shown that these represent folded globular units, domains, joined by short linkers. It is thought that these Ig domains represent archetypal gene products that may have played a role in molecular recognition in primitive eukaryotic organisms. Related domains are found in other members of the immunoglobulin supergene family, which includes the T-cell receptor and histocompatibility antigens (see **Human leukocyte antigens**), among many others.

The VHCH1 domains are linked to the rest of the heavy chain by a flexible length of polypeptide, described as the hinge. This is susceptible to proteolytic digestion, which cleaves the molecule into two fragments, one (Fab) retaining antigen-binding activity and the other (Fc) crystallizable.

More-detailed structural analysis has shown that both the VL and the VH regions contribute to the antigen-combining site of the Fab component of the molecule. In particular, three hypervariable regions (also known as complementarity-determining regions CDR-1–3) — which are located approximately 25 to 35, 50 to 60, and 85 to 95 residues from the N-terminus, respectively — actually constitute the combining site. The folding of the domain brings these regions together at the tip of the molecule, where they are available to combine lock-and-key fashion with the antigen. The remaining parts of the V regions are relatively invariant and are described as framework regions. Their function is to maintain the structure of the combining site. The variable region bears antigenic determinants, termed idiotypes, that are unique to individual antibody molecules. Effector functions (other than direct neutralization) of the Ig molecule are performed by the CH regions — the Fc fragment.

TABLE 83

IgG Subtype Properties

IgG Subtype	Serum Half-Life (day)	Complement-Fixing Macrophage	FcR Binding	IgG (% total)	Crosses Placenta
1	20–25	+	++	66	+
2	20–25	±	±	22	±
3	7–10	++	++	8	+
4	20–25	-	+	4	+

Functions and Structures of Different Ig Isotypes

IgG

This, the dominant serum Ig (about 14 mg/ml, 80% of total Ig), plays the major role in neutralization of pathogens and their products, fixation of complement, uptake of antigen by macrophages via their receptor for the Fc component of the IgG (FcRg), and crossing the placenta to protect the newborn. It has a long half-life in serum and persists for many months after immunization, and it is probably the primary factor in resistance to secondary infections by many pathogens. Efficacy of vaccination often correlates with IgG titer. IgG synthesis commences later than IgM in a primary infection (see below and **Lymphocytes — B-cells**) and persists longer. As the immune response matures, the affinity of IgG produced increases. IgG of high affinity is produced rapidly after a second challenge with an immunogen.

The IgG molecule is a monomer of the basic H2L2 unit, the H-chain consisting of one V and three C domains. There are four IgG subtypes, 1, 2, 3, and 4, with differing properties (see Table 83).

IgA

This is the next most abundant isotype in the serum (about 4 mg/ml), with a half-life of a few days. It is the major secreted Ig, present on mucous membranes and secretions, including milk. Its role is to combat infections at the site of entry: the reason vaccination is so inefficient against many gut infections is thought to be that present vaccination strategies do not give good IgA responses, and IgG responses are irrelevant to mucosal immunity. It does not fix complement, pass the placenta, or bind to FcRs. IgA structure is more complex than IgG. In the serum, it consists of one basic unit (the H-chain with three C domains), but in secretions, it is a dimer or tetramer linked by a J (for joining) peptide. IgA-secreting plasma cells are located in the submucosa; the IgA binds to poly-Ig receptors on the inner surface of the epithelial cells of the mucosa and is transported across. During transport, a further peptide (the secretory component) is added to the IgA to mask protease-sensitive sites in the hinge region. There are two subtypes, with similar properties.

IgM

The third major Ig serum type, at about 1.5 mg/ml, with a half-life of a few days. It is the first Ig to be produced in a primary infection and is not produced in a second challenge. It has relatively low affinity for antigen (i.e., low-equilibrium binding constant), but by virtue of its structure it has high avidity (see **Affinity and avidity**). It strongly activates complement and binds to macrophage FcR. Secreted IgM is a pentamer of the basic Ig unit (membrane IgM is a monomer), with the H-chain and four C domains linked via pairs of disulfide linkages between adjacent CH3 and CH4 domains and, in one case, a single J-chain (same as IgA). The molecule is therefore polyvalent, and when reacting with antigens with repeating antigenic sites, it is able to form multiple bonds and thus achieve strong binding — hence the high avidity.

IgE

This is present in the serum in trace amounts. Its role is to mediate immediate **hypersensitivity** reactions. It does this by binding mast cell and basophil FcRs, thereby sensitizing the cell; binding of the appropriate antigen triggers release of the factors such as histamine, giving rise to the well-known response. The structure of IgE is similar to IgG, except there are four CH domains.

IgD

Although present in the serum in trace amounts, it is probably more important in its membrane-bound form as an antigen receptor for B-cells. IgD does not fix complement or bind to FcRs.

Membrane Immunoglobulin

See **Lymphocytes** — B-cells.

Ig Genes and Generation of Antibody Diversity

The problem of how unique antibodies could be generated to any antigen was solved by cloning the structures of Ig genes. Genes for the variable and constant regions of both heavy and light chains are separate in the germline and in all somatic cells, except B-cells, where they are brought close together to form a single transcriptional unit.

There are multiple genes for V regions (human Vl, about 100 on chromosome 22; Vk, about 100 on chromosome 2; VH, about 200 on chromosome 14), but unique (or a small number) genes for each subtype heavy chain. In addition, there are smaller components — D (for diversity, H-chains only) and J (for junction) — that are located between the V and C regions. There are up to a few tens of J-region genes and a few D-region genes. During B-cell ontogeny, the chromosomal DNA is rearranged by recombination to bring the three L or four H components together at random, thereby generating a range of V(D)JC combinations. The (D)J region of the mature gene corresponds to CD3. Additional diversity arises from imprecise joining, from random addition of nucleotides at the junction, and by somatic mutation (clustering at the CD regions) during maturation of the antibody response. Thus, in the mature B-cell, the genes present on one chromosome are rearranged to generate a unique V region adjacent to the cluster of C-region genes. The genes of the other chromosome are not rearranged, thus ensuring that a single immunoglobulin is produced by the B-cell (allelic exclusion). Finally, diversity is generated from the combination of distinct L and H chains in the mature Ig.

The structure of the genes for heavy and light chains of mature B-cells is shown in a simplified form in Figure 74.

Control of IG Gene Expression: Class Switching

The structure of the V-Cµ-C region of the heavy-chain genes of the mature B-cell is shown in Figure 75.

In the näive cell, the entire region is transcribed to give a long primary transcript that is alternately spliced to generate mRNAs encoding either IgM or IgD, both incorporating the exons M1 and M2 without the S region of the fourth exon. M1 and M2 encode the

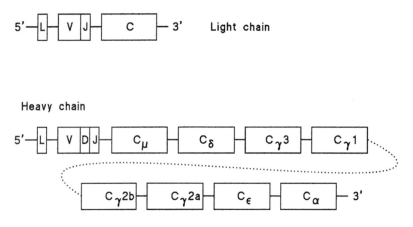

FIGURE 74
Schematic diagram of light- and heavy-chain genes in mature B-cells (see also Figure 76). L represents the common leader sequence of the mRNA; V(D)J is the common variable region generated by recombination of V, D, and J segments; C represents constant-domain-coding regions, one for each of the light chains and eight for the heavy chain.

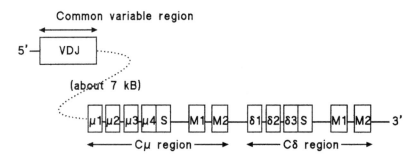

FIGURE 75
The structure of the V-Cμ-Cδ region of the heavy-chain genes of the mature B-cell. This shows part of Figure 74 in more detail. Each box represents an exon (coding unit), and the connecting lines represent introns: μ1–4 correspond to the domains of the constant region of the heavy chain of IgM; d1–3 corresponds to IgD; S is the sequence contiguous with the last of the heavy-chain exons that encodes the signal sequence; M1 and M2 are the exons that encode the signal sequence corresponding to membrane-spanning domains.

membrane-spanning region of the mature mIg molecule; hence this mRNA generates the B-cell receptor. In the plasma cell, however, the splicing is different, so that S rather than M1 and M2 is incorporated in the mRNA; S encodes the signal for secretion of the Ig. What controls these alternate splicings is unknown.

The first Ig to be secreted by plasma cells in a primary immune response is IgM. As the immune response matures, the Ig H-chain genes in the proliferating B-cells undergo further rearrangements so that, progressively, μ, d, and e genes are deleted, and the resulting plasma cells secrete IgG (different subclasses), IgE, or IgA. This class switching is under the control of cytokines; in particular, IL-4 favors production of IgG1 and IgE, whereas IFN-γ favors IgG-2a and IL-5 and TGF-β favor IgA.

Disorders of Immunoglobulins

Deficiency

See **Immunodeficiency**.

Excess Production

Monoclonal gammopathies
- **Heavy-chain disease**: λ, α, μ, cryoglobulinemia
- **Monoclonal gammopathies** of undetermined significance (MGUS)
- **Waldenström macroglobulinemia**
- Light-chain disease (primary **amyloidosis**)

Myeloma and **plasmacytoma**

Therapeutic Preparations

These are prepared by fractionation of pooled plasma. They contain mainly immunoglobulins of the IgG isotype and must be given parenterally.

Intramuscular Immunoglobulin (IMIg)

This can passively transfer specific antibody to protect against hepatitis B, rabies, and tetanus when it is accompanied with the appropriate vaccine. *Varicella zoster* IMIg is given to prevent primary infection in nonimmune immunosuppressed patients; anti-RhD immunoglobulin is used for the prevention of Rh **hemolytic disease of the newborn**. In addition, normal immunoglobulins are effective against hepatitis A and measles. IMIg is no longer used in the treatment of immunodeficiencies, since it has been superseded by intravenous immunoglobulin (IVIg) or, more recently, by subcutaneous injections.[292]

Intravenous Immunoglobulin (IVIg)

This is indicated in the following conditions:

Primary immunodeficiency:
- Common variable **immunodeficiency**
- X-linked agammaglobulinemia
- X-linked immunodeficiency with hyper-IgM
- **Wiskott-Aldrich** syndrome
- Severe combined immunodeficiency (SCID) syndromes
- Di George syndrome
- Certain cases of IgG-subclass deficiency
- Antibody deficiency with normal/near-normal immunoglobulins

Secondary immunodeficiency, usually for patients with **lymphoproliferative disorders**. Serious bacterial infection, such as septicemia and pneumonia, can be prevented by IVIg infusions. Not all patients require treatment. Patients who fail to make specific IgG responses to pneumococcal vaccine are most likely to benefit.

Allogeneic transplantation: IVIg has been shown to reduce the risk of bacterial sepsis and fungal infection. The incidence of cytomegalovirus interstitial pneumonia is also reduced.

Autoimmune antibody disorders:
- Acute **immune thrombocytopenic purpura** (ITP): IVIg has been shown to be both effective and safe, and a proportion of patients with chronic ITP also respond. It is also effective in treating ITP associated with HIV infection. Other

autoimmune disorders in which IVIg is or may be effective include dermato-myositis, Guillain-Barré syndrome, **Kawasaki disease**, **multiple sclerosis**, my-asthenia gravis, **polymyositis**, **antineutrophilic cytoplasmic antibody** (ANCA) associated systemic **vasculitis**.

- Rhesus (D) immunoglobulin: can be used prophylactically for hemolytic disease of the newborn.
- **Acquired aplastic anemia**: antithymocyte globulin (ATG), prepared by immunizing horse or rabbit with thymocytes, is used for conditioning prior to allogeneic stem cell transplantation and treatment when a suitable donor is not available. It has also been used in the treatment of amegakaryocytic hypoplasia. Its action is by reduction of **cytotoxic T-lymphocytes**.

Adverse reactions, particularly **anaphylaxis** and **transfusion-transmitted infections**, may occur.

IMMUNOHISTOLOGY

The use of **antibodies** as specific staining reagents to visualize particular proteins in cellular preparations, either sections of solid tissues or cell suspensions deposited upon slides by smearing or by centrifugal techniques ("cytospins").[295] There are many variations upon the basic technique. For "direct" techniques, the antibody must be conjugated with either a fluorochrome (fluorescein, rhodamine red, or phycoerythrin is commonly used) to allow visualization by emitted light upon illumination with appropriate ultraviolet (UV) light (immunofluorescence) or with a suitable enzyme that will generate a colored deposit after incubation with a suitable substrate (peroxidase or phosphatase is commonly used). Alternatively, the antibody can be conjugated with biotin and reacted with a streptavidin-enzyme complex and subsequently with the substrate; this technique gives a degree of amplification and is more sensitive. The use of alternative conjugation techniques (e.g., different fluorochromes whose emissions can be distinguished by the use of different filter systems) allows dual labeling. Combinations with conventional histology and techniques such as *in situ* hybridization are also possible.

For "indirect" techniques, the primary antibody (i.e., that which reacts directly with the target antigen) is unconjugated and subsequently must react with a secondary antibody conjugated as above. This technique, while more flexible (not requiring conjugated primary antibodies) and more sensitive, is more subject to artifacts, and dual-labeling techniques are more difficult.

Immunohistological techniques depend on the preservation of the antigenic structure of the tissues examined, and this is best done by the use of acetone-fixed frozen sections. However, conventional formalin-fixed paraffin sections can be used with certain antibodies, and antigens can be rescued by procedures such as microwave treatment. Many antibodies are now available that react with antigens in formalin sections that hitherto had to be visualized in frozen sections, which give much poorer morphology.

In addition to its use for the detection of tissue antigens, immunohistology can be used for the detection of **autoantibodies.** The patient's serum is incubated with tissue sections as the primary antibody, and the reaction is visualized by indirect (usually fluorescent) techniques. The staining pattern indicates the specificity of the autoantibodies.

IMMUNOLOGY

The scientific study of the immune system. It is closely related to **hematology** by its study of the biology of hematopoietic cells and with microbiology and infectious diseases by causal relationship. It is taught and practiced as an independent scientific discipline.

IMMUNOPHENOTYPING

The use of immunological markers to determine the **phenotype** of cells. The techniques are particularly useful for:

Determining type of **acute myeloid leukemia** (AML) according to FAB classification

Distinguishing **acute lymphoblastic leukemia** (ALL) from other **lymphoproliferative disorders**

Determining the clone from which B-**lymphocyte** proliferations descend and distinguishing them from nonneoplastic reactive **lymphocytosis**; aberrant T-cell phenotypes can be suggestive of T-cell clonal proliferation

Confirming or refuting a histological diagnosis that has typical or atypical clinical features.

Methodology[296–298]

Peripheral-blood specimens are more suitable than bone marrow, which should only be used when it is infiltrated and when there are few or no abnormal cells in the peripheral blood.

Immunofluorescence of unfixed cells in suspension marked by **monoclonal antibodies**. Cells are identified by phase microscopy or by **flow cytometry**.

Staining of surface **immunoglobulins** (SmIg) by anti-immunoglobulins.

Immunoalkaline phosphatase antialkaline phosphatase on fixed cells spread on a glass slide.

Monoclonal Antibodies

Recommended panels of markers are available for distinguishing B-cells from T-cells, ALL, and AML subtypes and for identifying possible clonality of B and T lymphomas.

Reporting Results

An estimate is made of the proportion of abnormal cells in the specimen under test. The cutoff point to consider a marker as positive varies with the number of expected cells and thus differs with each marker. This can be expressed as weak/strong or as a percentage, interpreted as +, +/−, −/+, −.

Specific Applications

Acute Myeloid Leukemia[296]

Almost all AML is CD13$^+$ or CD33$^+$, but expression is variable, and some cases show only one marker. There is some correlation with FAB subtypes (see Table 9). CD34 positivity is usually associated with a more-primitive cell type, whereas CD34 negativity is associated with more-differentiated cells, e.g., M3, M4, and M5. The M4 and M5 subtypes also show CD14 positivity and human leukocyte antigen (HLA)-DR expression, with the monocytic lineage being usually CD4 positive, CD11c positive, and sometimes associated with membrane Fc receptor expression.

In the more-primitive FAB subtypes, TdT and CD7 positivity occurs, TdT being strongly expressed in 20 to 30% and to a lesser extent in most cases. In addition, there may be a smaller subpopulation of blasts, which is TdT-positive or CD13- or CD33-positive.

Acute Lymphoblastic Leukemias[296]

(ALL) The availability of a range of monoclonal antibodies has allowed precise classification of ALLs along their B- or T-cell lineages. Myeloid leukemia can be separated from ALL by pan-myeloid surface antigens (including CD13, CD33, and CDw65).

The lymphoblast stage can be determined along B- or T-cell lineages (see Table 6). These lineages are based on patterns of expression rather than on the presence or absence of any single antigen. Several subgroups of B- or T-cell ALL can be separated by further immunophenotype studies, supplemented by studies of Ig and T-cell-receptor gene patterns.

Lymphoproliferative Disorders[297]

The differential diagnosis using the WHO classification of tumors (see Appendix) depends upon morphological, cytogenetic, and immunophenotyping features. The advances in using immunohistology methods furthers this approach, which may in time have therapeutic implications. The immunophenotypes of these disorders are summarized in Table 103.

IMMUNOPROPHYLAXIS

See **Hemolytic disease of the newborn**.

IMMUNOSUPPRESSION

The deliberate inhibition of the **immune response**. This is done by a variety of drug groups for specific effects:

Corticosteroids (e.g., prednisone).

Anti-inflammatory analgesics (nonsteroidal anti-inflammatory drugs [NSAID]): both steroids (e.g., prednisone) and NSAIDs (e.g., aspirin) function largely by inhibiting **cyclooxygenase** (COX), of which there are two isoforms: COX-1 and COX-2. COX-2 is particularly associated with inflammation through generation of **prostaglandins**. Specific COX-2 inhibitors such as Celebrex have been developed that do not have the undesirable side effects associated with inhibition of COX-1.

Calcineurin inhibitors such as **ciclosporin** and tacrolimus inhibit IL-2 synthesis by activated T-cells and, thus, T-cell proliferation.

Antimetabolites such as azathioprine and methotrexate block DNA synthesis of cells so that activated T-cells cannot proliferate and generate effector cells.

Monoclonal antibodies directed toward immune cells or inflammatory cytokines. Examples are basiliximab (specific for CD25 and thus for activated T-cells) and **infliximab** (antibody to the inflammatory **cytokine, tumor necrosis factor**-α [TNF-α]). These are effective drugs, but they need to be "humanized" by the replacement of murine epitopes with human equivalents from the original murine monoclonal antibody, using genetic engineering to prevent the development of neutralizing antibodies.

Drug usage depends on the clinical situation: mild chronic asthma is treated with antihistamines and with low-dose inhaled steroids; autoimmune disorders (e.g., rheumatoid arthritis) with the corticosteroids and antimetabolite methotrexate, perhaps with infliximab; and transplantation with a combination of prednisone, azathioprine, and ciclosporin or tacrolimus, adding in perhaps basiliximab to suppress episodes of acute rejection.

All immunosuppressive regimes have the defect in that they predispose the patient to infections and potentially virus-induced cancers, particularly in those transplant patients on long-term high-dose therapy.

IMMUNOTHERAPY

Treatment of chronic inflammatory and neoplastic disorders by immune reactions. An increasing number of recently developed therapies are using these processes.

Passive immunotherapy using **monoclonal antibodies** (see Table 110). Monoclonal antibody therapy may be complicated by allergic and anaphylactoid pulmonary reactions such as rhinitis and asthma. Secondary stimulation of antibodies to the infused monoclonal antibody may occur later.

Active immunotherapy using cytokines:

- **Interferon**-α and interferon-γ for **hairy cell leukemia**, chronic hepatitis B; interferon-β for MS
- Recombinant **interleukin**-2 (aldesleukin), which activates T-**lymphocytes** and natural killer (NK) cells for metastatic renal carcinoma

Anti-lymphocyte **immunoglobulin**

Anti-thymocyte globulin used for **acquired aplastic anemia**

Vaccines

- **Epstein-Barr virus** (EBV)-specific cytotoxic T-cells for posttransplant lympho-proliferative disease
- Dendritic cells: DNA, CMV, HIV-1 vaccines are being developed

Gene therapy is under investigation.

INBORN ERRORS OF COBALAMIN METABOLISM

See **Cobalamins**.

Indirect antiglobulin (Coombs) test

(IAT) A technique to detect **antibodies** to **antigens** of **red blood cells** (RBCs).[173] In **pretransfusion testing** and prenatal screening, this is the most suitable test for the detection of clinically significant antibodies to red cell antigens because of its high sensitivity and specificity. The IAT may also detect antibodies that are of no clinical significance. Antibody screening, antibody identification, and phenotyping can be performed.

The usefulness of the IAT lies in its ability to detect IgG RBC antibodies such as anti-D, which can cause **hemolytic disease of the newborn** and blood transfusion complications (see **Blood components for transfusion** — complications). IgG antibodies are, in general, not direct **agglutinins**, because the maximum distance between **immunoglobulin** Fab binding sites on a single molecule is less than the minimum distance that RBCs (with surface negative charge) can usually approach each other in a saline medium (see Figure 76). In the IAT, antiglobulin reagent (usually anti-IgG) forms IgG–anti-IgG complexes that can bridge the distance between RBCs. Because the final phase of the IAT (except in **solid-phase**

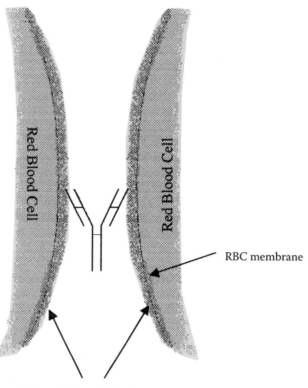

Red Blood Cell

Red Blood Cell

RBC membrane

Extracellular carbohydrate

FIGURE 76
An IgG antibody positioned between two RBCs at the minimum distance of approach. The Fab arms of IgG can span the distance between extracellular RBC carbohydrate antigens, but not the distance between RBC membrane protein antigens such as RhD.

techniques) is agglutination of RBCs, the IAT will also detect most agglutinating antibodies (which can be IgM, IgG, or IgA). Further sensitivity in the IAT is possible by using a polyspecific antiglobulin reagent containing anti**complement** as well as anti-IgG, because some RBC antibodies initiate the classical pathway of complement after uptake onto RBCs.

Methodology

Liquid-phase tube. The IAT can be performed by mixing plasma and RBCs suspended in physiologic saline, but the use of a **low-ionic-strength saline** (LISS), polyethylene glycol, or **low ionic polycation** increases sensitivity. Whichever method is used, following incubation with plasma, the RBCs must be washed free of plasma globulins, and anti-human globulin reagent is then added to cross link antibody-coated RBCs. Results are interpreted by gentle agitation or tipping or rolling of the tube, or by transfer onto a microscope slide.

Solid-phase techniques.

Column-agglutination techniques.

Whichever technique is used, a gradation of results from negative to very weak positive to strong positive is possible. It is essential that antiglobulin techniques be quality assured as part of a comprehensive quality assurance program (see **Quality management in the hematology laboratory**).

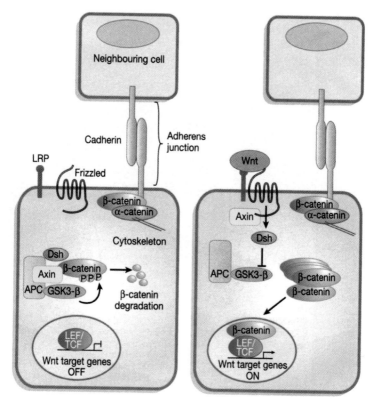

COLOR PLATE 1
Wnt signalling in stem calls and cancer.

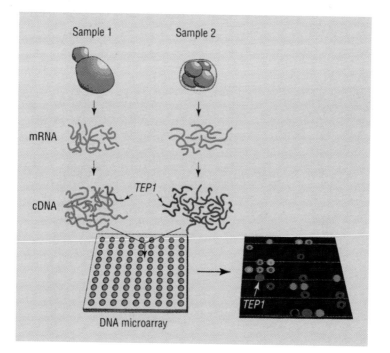

COLOR PLATE 2
DNA microassay.

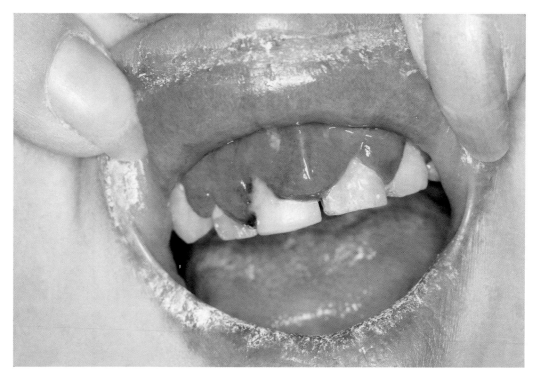

COLOR PLATE 3
Lead deposition in gums in a patient with lead poisoning.

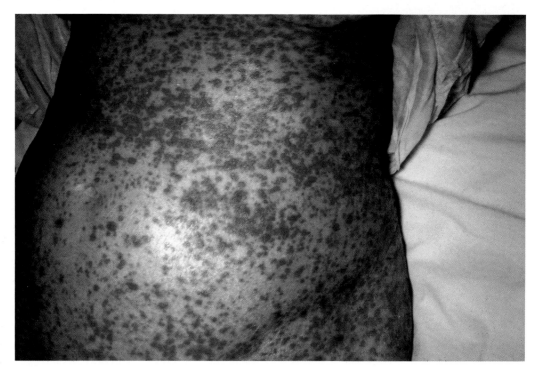

COLOR PLATE 4
Bruising and purpura in a patient with immune thrombocytopenic purpura.

COLOR PLATE 5
Abdominal purpura in a patient with varicela who was immunosuppressed by steroid therapy.

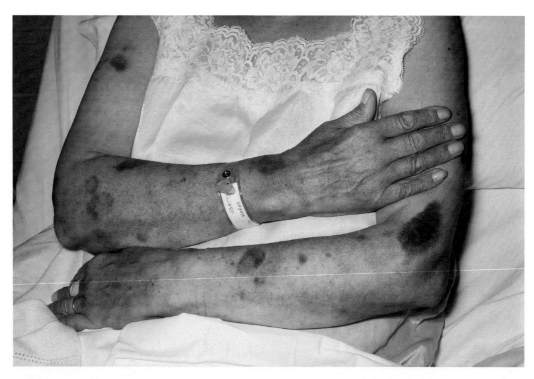

COLOR PLATE 6
Bruising and skin necrosis of the hands in a patient with disseminated intravascular coagulation due to *B.klepsiella* septicemia complicating cytoxic therapy for non-Hodgkin lymphoma.

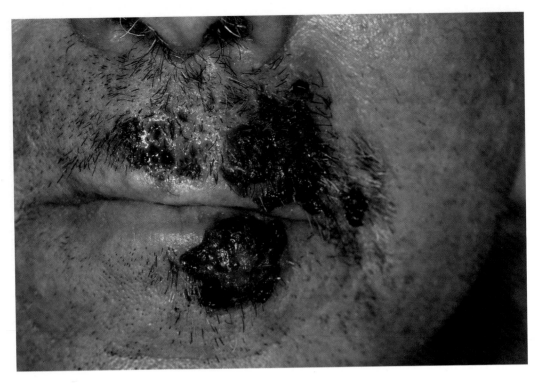

COLOR PLATE 7
Herpes zoster of the lips in a patient following allogeneic stem cell transplantation for acute myelogenous leukemia.

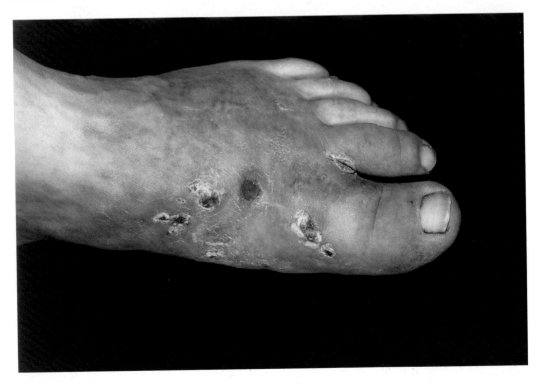

COLOR PLATE 8
Skin necrosis of the foot in a patient with "organ failure" complicating acute monocytic leukaemia.

TABLE 84

Causes of Ineffective Erythropoiesis

Megaloblastosis
Congenital dyserythropoietic anemia
Defective **heme** synthesis
Iron deficiency
Sideroblastic anemia
Thalassemia
Myelodysplasia — refractory anemia
Myelofibrosis
Acute myeloid leukemia — especially FAB-type M6

INDUCER T-CELLS
See **Lymphocytes**.

INEFFECTIVE ERYTHROPOIESIS
Failure to produce adequate numbers of mature **red blood cells**, despite plentiful erythroid progenitors. Normal **erythropoiesis** is 80 to 90% efficient, with the remaining 10 to 20% of erythroid cells undergoing cell death, probably by **apoptosis** within the marrow cavity. Effete erythroid cells are engulfed by marrow macrophages, which may ultimately contribute to **iron overload** such as occurs in some patients with **thalassemia** or **sideroblastic anemia**.

There are many conditions associated with ineffective erythropoiesis, either due to defective nuclear maturation, hemoglobin formation, or a combination of both (see Table 84). Ineffective erythropoiesis is usually suspected by the combination of bone marrow erythroid hyperplasia with a reduced or inappropriately low **reticulocyte count**. Peripheral-blood-film examination often reveals anisocytosis, poikilocytosis, in addition to the specific changes associated with the underlying diagnosis, e.g., pencil cells in iron deficiency.

Ferrokinetics provides more-detailed information. Radioactive iron (^{59}Fe) bound to transferrin is injected into the patient and taken up by proliferating erythroid progenitors at a rate determined by the degree of erythroid activity (normally 60 to 140 min). The plasma iron turnover (PIT) can be calculated using the formula:

$$\text{PIT} = \frac{\text{plasma iron concentration } (\mu g/l) \times 10^3}{T^{1/2}} \times \frac{100 - 0.9 \times \text{hematocrit}}{100}$$

where $T^{1/2}$ = plasma half-clearance time.

In normal subjects, 70 to 90% of the injected ^{59}Fe is demonstrable in the peripheral blood within 10 to 14 days. The PIT normally is 4 to 8 mg/l/day. In patients with ineffective erythropoiesis, there is accelerated plasma clearance (25 to 50 min), elevated iron turnover (9 to 55 mg/l/day), and reduced ^{59}Fe reappearance in blood (30 to 50%). Body-surface counting of ^{59}Fe also allows estimation of the contribution of the various organs to erythropoiesis. Ferrokinetic studies are less accurate if erythropoiesis is not in a steady state, the serum iron is elevated, or red blood cell survival is sufficiently shortened.

Indirect methods of diagnosing ineffective erythropoiesis include measurement of serum lactate dehydrogenase or *in vitro* counting of BFU-E (burst-forming unit-erythroid).

INFANTILE POLYARTERITIS NODOSA
See **Kawasaki disease**.

INFECTIOUS DISORDERS
See **Bacterial infection disorders**; **Metazoal infection disorders**; **Protozoal infection disorders**; **Rickettsial infection disorders**; **Viral infection disorders**.

INFECTIOUS MONONUCLEOSIS
(Glandular fever) A syndrome characterized by fever, sore throat, **lymphadenopathy**, and the presence of **atypical lymphocytes** in the peripheral blood. It is a response to infection, commonly by the **Epstein-Barr virus** (EBV), but occasionally by toxoplasma, **cytomegalovirus** (CMV), and other viruses that enter B-**lymphocytes**, provoking the formation of atypical mononuclear cells similar to those seen during reactive lymphocyte transformation.[165] **Antibodies** are produced against the specific infecting agent after 2 to 3 weeks as well as **heterophil antibodies** that react with sheep, horse, and bovine red blood cells.

Subclinical infection is common in early life, but the infectivity rate is low. It is a disease of young people clustered together in schools and colleges.

Clinical Features
The clinical syndrome is preceded by a prodromal period of 7 to 10 days with lethargy, malaise, headaches, stiff neck, and dry cough. The following features are characteristic but do not necessarily all occur in each patient.

Bilateral discrete tender cervical lymphadenopathy in 75%, with generalized lymphadenopathy in 50%

Sore throat with inflamed oral and pharyngeal surfaces in 50%; follicular tonsillitis with edema and exudate in 25%; a few cases developing pharyngeal obstruction — edema glottidis

"Hairy" leukoplakia of the tongue and mouth

Pyrexial illness with headache and photophobia

Splenomegaly in about 50%

Hepatomegaly in 15%, with jaundice in 10%; all cases have raised alkaline phosphatases with raised LDH, GPT, and GOT; hyperbilirubinemia is present in 50%

Morbilliform rash on trunk in 5%

Purpura and epistaxis from thrombocytopenia during first week of illness

Conjunctivitis and periorbital edema

Rare features are:

- Tachycardia with ECG changes
- Convulsions, stupor, coma
- Palsies, including cranial nerves
- Peripheral neuropathy
- Abdominal pain due to mesenteric adenitis

In those with **immunodeficiency**, a lymphomalike proliferation of B-cells carrying the EBV can occur, sometimes resulting in **non-Hodgkin lymphoma**.

Laboratory Features
Peripheral Blood

Pleomorphic atypical lymphocytosis

Early: round or oblong large granular lymphocytes, the nuclei being large, eccentric, often indented and lobulated with condensed chromatin. There is abundant basophilic cytoplasm with azurophilic granules and a few vacuoles. These cells are probably damaged natural killer (NK) lymphocytes.

Peak of illness: large lymphocytes (18- to 25-μm diameter) with copious agranular cytoplasm and hyperdiploid nuclei similar to mitogen-transformed T-cells.

During second week: blastoid lymphocytes, probably activated suppressor T-cells and antigen-specific cytotoxic T-cells that display S-phase features.

Late stage: large pale amoeboid cells with a tendency to adhere to neighboring red blood cells, causing a curved indentation ("kissing cells").

Any form of atypical lymphocytes may persist for 1 to 2 months.

Thrombocytopenia in 50%

Neutropenia

Serology

Presence of heterophil antibodies against sheep red blood cells (Paul-Bunnell test) or formalized horse red blood cells (Monospot test). Highest titers are reached at 2 to 3 weeks, persisting for up to 6 weeks. These antibodies only occur with EBV infections.

Autoimmune hemolytic anemia with IgM antibodies showing I blood group specificity (see **I blood groups**).

Treatment
This is symptomatic apart from **corticosteroids** for edema glottidis. Ampicillin should be avoided, as it causes an erythematous rash. Acyclovir decreases EBV replication but has no advantageous clinical effects.

Prognosis
Resolution occurs after 4 to 6 weeks, but may be slow, being associated with lethargy and malaise. Isolated fatalities have occurred from encephalitis, edema glottidis, hepatic necrosis, splenic rupture, and renal failure.

INFILTRATIVE MYELOPHTHIASIS
(Infiltrative myelopathy) Infiltration of the **bone marrow** by ectopic cells displacing and disorganizing hematopoietic cells. The origins of these ectopic cells are most commonly from

a neuroblastoma, Ewing's sarcoma, oat cell carcinoma of lung, prostatic carcinoma, breast carcinoma, squamous cell carcinoma of lung, rhabdomyosarcoma, and melanoma.[298a]

The frequency of bone marrow infiltration depends on the clinical stage of the disease, i.e., presentation, relapse, and autopsy. The rate of detection can be increased with multiple sampling and the use of bone marrow biopsies.

Bone marrow infiltration can be indicated by **anemia**, **thrombocytopenia**, or **neutropenia**, but classically there is a **leukoerythroblastic anemia**. However, even in patients with proven bone marrow infiltration, there is not always an abnormality in the peripheral blood. Rarely, some metastatic tumors, e.g., breast and prostate, may result in marked secondary **myelofibrosis**, and distinction from primary myelofibrosis may be difficult.

INFLAMMATION

A local response to injury resulting in vascular dilatation and enhanced permeability of capillary walls, facilitating efflux of fluid and cells from the blood into the tissues. The **acute-phase response** is the systemic counterpart of local inflammation. The cardinal outward signs of acute inflammation include swelling, redness, and warmth, consequences of increased local blood flow and vascular permeability. Histologically, it is typified by the accumulation of **leukocytes**, including activated T-lymphocytes and phagocytic cells (**histiocytes** and **neutrophils**). Inappropriate manifestations of inflammation include pathologic conditions such as asthma, contact dermatitis, and the toxic-shock syndrome, which is essentially a systemic inflammatory response resulting in circulatory collapse and death. Inflammation may be acute, resolving when the stimulus is effectively removed, or chronic, when continued irritation by a pathogen is not cleared, as with mycobacterial infection or certain cancers.

Acute Inflammation[299]

This is almost certainly triggered by components of pathogens activating production of a range of pro-inflammatory mediators, **cytokines, prostaglandins**, and activated **complement** by myeloid **dendritic reticulum cells** and macrophages at the site of infection. These pathogenic components include bacterial lipopolysaccharide (LPS, endotoxin), bacterial DNA containing unmethylated CpG not found in mammalian DNA, both of which induce interleukin (IL)-1, tumor necrosis factor (TNF)-α, and IL-12; and dsRNA, produced by many viruses, which induces interferon (IFN)-$\alpha\beta$. These cytokines will trigger further cytokine production, of which IL-6 and chemokines are particularly important, by other cells, including cells such as endothelial cells, fibroblasts, and keratinocytes. LPS will also activate expression of COX-2 in macrophages and epithelial cells, thereby augmenting the production of prostaglandins and leukotrienes, which also have inflammatory roles. The presence of bacteria will activate complement, generating anaphylatoxins (chemotaxins), which also are pro-inflammatory. At this stage, the process is purely innate. However, as the infection progresses and T-cells are activated, they will also contribute to the inflammatory process by producing cytokines, including those listed but also others such as IL-3 and IFN-γ, which play a wider role in the activation of the immune response.[300]

These pro-inflammatory factors will act on endothelial cells to increase the expression of **cell-adhesion molecules**: CD62P, CD62E, and CD54. This will augment the adherence of neutrophils via CD11a/CD18, which then migrate across the endothelium (diapedesis) under the influence of anaphylatoxins and chemokines and become activated, avidly phagocytosing bacteria, foreign particles, etc. At the same time, capillary permeability increases, aiding the efflux of cells and molecules such as antibodies and, of course, plasma,

TABLE 85

Roles of the Tissue Macrophages (Histiocytes) in Inflammation

Cytokine release, e.g., IL-1, PDGF, FGF, TGF-β, inducing fibroblast proliferation and collagen synthesis
Production of inflammatory mediators, e.g., PAF-1, prostaglandin metabolites
Phagocytosis and killing of bacteria using oxidative and nonoxidative mechanisms
Release of proteases and hydrolytic enzymes responsible for liquefaction of tissues
Antigen presentation, thus forming a link with both cell-mediated and humoral immune mechanisms
Growth-factor release, e.g., FGF, TNF-α, which encourages angiogenesis
Stimulation of epithelial cells via EGF, TGF-α, FGF

Note: EGF, epidermal growth factor; FGF, fibroblast growth factor; IL-1, interleukin-1; PAF-1, platelet-activating factor; PDGF, platelet-derived growth factor; TGF-α and -β, transforming growth factor-alpha and -beta; TNF-α, tumor-necrosis factor-alpha.

resulting in the swelling, redness, and warmth. IL-6 travels to the liver, where it activates the acute-phase response.

Experimental models of inflammation show that the migration of neutrophils peaks at 3 to 4 h and has almost completely ceased by 6 h. In contrast, **monocytes**, having started to migrate at the same time as neutrophils, peak at 12 to 24 h, so that by 48 h the inflammatory infiltrate consists almost entirely of mononuclear cells. Monocytes traverse the endothelium in a way similar to that of neutrophils. Monocytes, in addition to responding to general chemotaxins, also seem to be able to respond alone to certain ones, e.g., MCP1 (monocyte chemoattractant protein 1). The activity of tissue macrophages (histiocytes) is summarized in Table 85.

At later stages, neutrophils (and slightly less mature forms) are released from their storage sites under the influence of hematopoietic growth factors, resulting in **neutrophilia** (and left shift). These growth factors are released from macrophages, fibroblasts, and endothelial cells under the influence of IL-1, TNF-α, bacterial products, or activated T-cells. Additionally, platelets adhere and degranulate, releasing their load of growth factors, which will contribute to tissue repair processes.

When there is immunological memory, there is activation of **mast cells** by antigens binding to IgE complexed with high-affinity receptors, FceRI. The mast cells degranulate, releasing cytokines and prostaglandins, but also the vasoactive substances histamine, serotonin, and bradykinin.

The inflammatory process is probably largely terminated by elimination of the stimulus, but the cytokine transforming growth factor (TGF)-β may also be involved, as mice lacking the gene for TGF-β are subject to uncontrolled inflammation.

Pus Formation

The destruction of bacteria, irritants, or necrotic tissue is the result of granulocyte and complement activation. Neutrophils and monocytes/macrophages are able to phagocytose opsonized (with Ig, complement, or defensins) and nonopsonized bacteria through specific receptors on their membrane surface (Fc or C3b). The neutrophils envelop the bacteria in a membrane, forming a phagosome into which their granules are rapidly discharged. Unfortunately, many neutrophils die, releasing their cytoplasmic contents into the surrounding tissues. The neutrophil granules, with highly reactive radicals, cationic proteins, proteases, lipases, and collagenases, cause further tissue damage.

The complement system can be activated via the classical or alternative pathway, depending on the stimulus. Complement components act as vasoactive amines (C3a, C5a), chemotaxins (C5a), opsonins (C3b), and cytolytic agents (C3b-9 complex).

The dead neutrophils, bacteria (dead or alive), and liquefied necrotic tissue make up pus. Monocytes/macrophages, as mentioned previously, are the predominant granulocytes during the later stages of acute inflammation and, as well as phagocytosing and destroying bacteria, are involved in wound healing.

Chronic Inflammation

This usually occurs as a continuum of acute inflammation, but *de novo* chronic inflammation can occur in response to **viral infection disorders**, **parasitic infection disorders**, autoimmune antibodies, and malignancy. Acute inflammation is followed by chronic inflammation when the damaging agent is not destroyed and a balance therefore develops between ongoing tissue damage and wound healing.

The principal cells of chronic inflammation are monocytes/macrophages, plasma cells, lymphocytes, eosinophils (especially in allergic responses), neutrophils, and fibroblasts. Monocytes and lymphocytes leave the systemic circulation under the influence of cytokines, e.g., MCP-1, IFN-γ, and IL-4. Migration of monocytes and lymphocytes involves utilizing the CD49d/CD29, CD54, and CD14 adhesion molecules. Having entered the tissues, the monocyte transforms into a macrophage, where it has a central role in the development, maintenance, and possible healing of chronic inflammation (see Table 87).

Within 3 to 5 days, fibroblast and endothelial proliferation starts to occur, resulting in granulation tissue. Occasionally, macrophages fuse, forming giant cells, which may or may not be associated with **granuloma** formation. If destruction of the noxious agent occurs, wound healing may follow, again with macrophages and fibroblasts central to the process.

Chronic inflammatory disorders include **tuberculosis**, Crohn's disease (see **Intestinal tract disorders**), **sarcoidosis**, and syphilis.

INFLIXIMAB

See also **Tumor necrosis factor**.

One of a number of drugs developed to inhibit the activity of TNF-α. They have been shown to be effective in the treatment of **autoimmune disorders**, **rheumatoid arthritis**, and Crohn's disease, but they carry the risk of adverse reactions affecting the central nervous system. The immunosuppression they induce makes opportunistic infection possible. A particularly adverse reaction is fibrosing alveolitis.

INNATE IMMUNITY

See **Immunity**.

INOSINE

The deamination product of **adenosine** when catalyzed by adenosine deaminase. Inosine is a component of a particular preservative solution used in **blood transfusion components**.

INOSINIC ACID

A **purine** nucleoside, which is a precursor of **adenosine monophosphate** (AMP) and guanine monophosphate. It is a rare monomer in nucleic acids, where it pairs with cytosine.

INTEGRINS

See **Cellular adhesion molecules**.

INTERDIGITATING DENDRITIC CELL SARCOMA

A neoplastic proliferation of spindle to ovoid cells with folded nuclei and having pheno-typic features similar to those of interdigitating **dendritic reticulum cells**.[169] These are located in T-**lymphocyte** rich areas of the mononuclear-phagocytic system. Their immu-nophenotypic expression is: CD45, HLA-DR, and S-100. They give rise to generalized **lymphadenopathy** and show rapid progression. Treatment is by **cytotoxic agent** therapy.

INTERDIGITATING RETICULUM CELLS

See **Dendritic reticulum cells**.

INTERFERONS

(IFNs) A group of **cytokines** with antiviral and other functions.[301] Originally defined as a single virus-induced factor that inhibited virus replication, there are now multiple IFNs that all share the antiviral activity of the prototype but otherwise are disparate in their properties. IFN-α (about a dozen subtypes) -β, -γ, -δ, -κ, -λ (three subtypes, also known as IL-28A, IL-28B, and IL-29), -τ, and -ω have all been recognized. All but IFN-γ are related by sequence and are referred to as type I IFNs, while IFN-γ is type II. IFN-δ and -τ appear to be expressed in trophectoderm (ovine/bovine and pig, respectively) and presumably are involved in the establishment of pregnancy. IFN-α, -β, -κ, and -λ are all induced in a variety of cell types by viral infection (proximal inducer: dsRNA); IFN-α and -β share a common receptor, while IFN-κ and -λ share a separate receptor. On the other hand, IFN-γ is expressed by activated lymphocytes and natural killer (NK) cells and utilizes a third receptor type. Of the different IFNs, α, β, and γ appear to be the most important.

Properties of Type IFN-$\alpha\beta$

The defining cell property induced by IFNs is the antiviral state. This takes some 6 to 8 h to develop and results in the failure of the production of viral proteins upon subsequent infection. The mechanism of the antiviral state involves the production of a novel oligo-meric nucleotide, 2',5'-oligoadenylate, the activation of a ribonuclease that degrades viral mRNA, and the inactivation of a protein-synthesis-initiation factor, thus inhibiting protein synthesis. Other major properties include enhanced expression of **major histocompatibility complex** class I (but not class II), which will modify susceptibility to cellular cytotoxicity, and the activation of NK-cells and inhibition of cellular proliferation. Experiments with knock-out mice show type I IFNs to be of key importance in various viral infections.

Properties of IFN-γ

IFN-γ induces the expression of class II histocompatibility antigens and other cellular components of antigen presentation, activates macrophages, and regulates immunoglob-ulin class switching during B-cell maturation. Experiments with knockout mice indicate IFN-γ to have a key role in resistance to intracellular bacteria, a finding confirmed by the identification of inherited defects in the IFN-γ system leading to defects in resistance to *Mycobacteria* spp.[302]

TABLE 86

Commonly Used Therapeutic Interferons and Their Brand Names

Nonproprietary Name	Source	International Brand Name (Manufacturer)
rIFN-α2α	Recombinant bacterial cell	Roferon (Roche)
rIFN-α2β	Recombinant bacterial cell	Intron A (Schering-Plough)
rIFN-α2β	Recombinant bacterial cell	Viraferon (Schering-Plough)

Therapeutic Uses of IFNs

Type I IFNs are the only interferons to have found a significant therapeutic role. Interferon-α is much used in the treatment of hepatitis B and C, in some **lymphoproliferative disorders**, such as **hairy cell leukemia**, and in a few solid neoplasms, such as renal cell carcinoma. Interferon-β use is limited to relapsing, remitting **multiple sclerosis**. The therapeutic interferons are often referred to by their manufacturer's brand name (see Table 86).[303] Clinical results and adverse drug reactions vary between preparations, so it is essential to consult the information supplied with the precise preparation that is to be administered.

INTERLEUKINS

See also **Cytokines**.

Glycoproteins produced by **leukocytes**, which act as cytokines in the communication between **lymphocytes** and other cells concerned in **hematopoiesis.** The range of knowledge of interleukins is now vast. Those important in hematology are described briefly here, with a summary of sources, stimuli, and biological activity concerning the majority provided in Table 87.

Interleukin-1

(IL-1; hematopoietin-1) IL-1a and IL-1b are the products of two separate genes on chromosome 2q12-21, with only 25% structural protein homology. A natural receptor antagonist exists (IL-1ra), with promising *in vivo* activity against **graft-versus-host disease** associated with bone marrow transplantation.

Functionally, IL-1 is a cytokine acting directly on primitive progenitors (early acting) and indirectly on progenitors by stimulating the release of other cytokines, such as granulocyte/macrophage colony stimulating factor (GM-CSF), G-CSF, M-CSF, and IL-6 from macrophages or T-cells. IL-1 is also part of the **acute-phase response** to toxins, such as endotoxin, and triggers synthesis of acute-phase reactant proteins from hepatocytes and endothelial cells. It therefore induces **C-reactive protein** and is found in the synovia of patients with **rheumatoid arthritis**.

Interleukin-2

(IL-2) Produced only from activated T-cells, this 133-amino acid peptide, whose gene is located on chromosome 4q26-28, binds to its receptor on T-cells (and B-cells and natural killer [NK] cells) to promote rapid clonal expansion of antigen-specific clones. These activated T-cells secrete further IL-2 and multiple other cytokines. With IL-5, it is concerned in the activation and proliferation of **eosinophils**. Promising therapeutic uses are in solid-tumor therapy using IL-2-stimulated lymphokine activated killer (LAK) cells or tumor-infiltrating lymphocytes (TIL).

TABLE 87

Summary of Sources, Stimuli, and Biological Activity of Interleukins

Interleukin	Other Names	Main Cell Sources	Stimulus for Production	Main Biological Activities	Comments
IL-1α, IL-1β	lymphocyte activating factor, etc.	keratinocytes, monocytes/macrophages, etc.	LPS and other cytokines stored as inactive precursor in macrophages	coactivator for T-cells; activator of endothelial cells; activator of Langerhans cells: pro-inflammatory	—
IL-2	T-cell growth factor, etc.	CD4 T-cells	cognate antigen and T-cell mitogens	supports T-cell proliferation; activates NK-like activity	IL-2 has been used to expand *in vitro* populations of so-called lymphokine-activated killer (LAK) cells, which have been used in cancer therapy (largely unsuccessfully)
IL-3	multi-CSF, hematopoietin, etc.	T-cells, keratinocytes, endothelial cells, etc.	cognate antigen and T-cell mitogens (T-cells)	supports growth of hematopoietic progenitor cells	see **Hematopoiesis**
IL-4	B-cell stimulating factor or growth factor, etc.	CD4 (Th2) T-cells	cognate antigen and T-cell mitogens	drives proliferation and differentiation of B-cells and antibody class shift; downregulates cytokine production by macrophages	key cytokine in the humoral immune response
IL-5	eosinophil colony stimulating/differentiation factor, etc.	CD4 (Th2) T-cells	cognate antigen and T-cell mitogens	—	(in humans) stimulates growth and differentiation of eosinophils (in mice) also B-cell growth and differentiation
IL-6	B-cell differentiation factor, etc.	monocytes/macrophages, fibroblasts, endothelial cells	IL-1, TNF, LPS, IL-17, etc.	maturation of B-cells to plasma cells; activation of acute-phase response; pro-inflammatory	prototype of IL-6 family: LIF, CNTF, OSM, IL-11, and CT-1
IL-7	pre-B-cell growth factor, lymphopoietin, etc.	bone marrow stromal cells	none known, i.e., probably constitutive	stimulator of lymphopoiesis	see **Lymphocytes**
IL-8 (several subtypes)	chemotaxin, leukocyte-inhibitory factor, etc.	monocytes/macrophages, endothelial cells, keratinocytes, etc.	IL-1, TNF, dsRNA, etc.	chemotactic for all leukocytes; activates neutrophils	one of the **chemokine** family

TABLE 87 (continued)

Summary of Sources, Stimuli, and Biological Activity of Interleukins

Interleukin	Other Names	Main Cell Sources	Stimulus for Production	Main Biological Activities	Comments
IL-9	mast cell growth factor	CD4 (Th2) T-cells	cognate antigen and T-cell mitogens; IL-1	potentiates Ig production with IL-4; hematopoietic growth factor	see **Hematopoiesis**
IL-10	cytokine-synthesis inhibitory factor	CD4 (Th2) T-cells	cognate antigen and T-cell mitogens	inhibits cytokine production of Th1 but not Th2 cells, thus downregulating cell-mediated immunity	Epstein-Barr virus encodes an IL-10 homologue, vIL-10
IL-11	megakaryocyte-differentiation factor	bone marrow stromal and other mesenchymal cells; endometrium	none known, i.e., probably constitutive	inducer of acute-phase response: hematopoietic growth; differentiation factor	member of IL-6 family
IL-12	T-cell-stimulating factor	macrophages, dendritic cells	bacterial, parasitic, and viral products, e.g., LPS, dsRNA	stimulates cytokine production by Th1-type CD4 cells and inhibits IgE production	IL-12 is widely regarded as a key activator of cell-mediated immunity: its production by cells of the innate immune system aids Th1 function
IL-13	—	CD4 (Th2) T-cells	cognate antigen and T-cell mitogens	similar to IL-4: drives proliferation and differentiation of B-cells and antibody class shift; downregulates cytokine production by macrophages	related to IL-4 in amino acid sequence
IL-14	—	—	—	—	very little literature available on this cytokine, which may relate to IL-4
IL-15	—	macrophages, muscle, and many other cell types NOT T-cells	unknown	similar to IL-2; may be important in generation of memory	may function as membrane-bound cytokine ("juxtacrine" factor)
IL-16	lymphocyte-attractant factor	CD8 (preformed) and CD4 cells; many other cell types	histamines (CD8 cells), IL-1 (fibroblasts)	binds CD4 and is chemoattractant for CD4 cells; inhibitor of Th2 cytokine production: anti-inflammatory; inhibits HIV growth	potential in AIDS therapy
IL-17	cytotoxic T-cell-associated antigen 8	CD4 T-cells	IL-23; cognate antigen and T-cell mitogens	pro-inflammatory through stimulating release of chemokines, etc., from endothelial cells, etc.	6 subtypes (A–F); a homolog is encoded by *Herpesvirus saimiri*

Interleukin	Other Names	Main Cell Sources	Stimulus for Production	Main Biological Activities	Comments
IL-18	IFN-γ-inducing factor	macrophages, dendritic cells, etc.	microbial products, e.g., LPS	growth and differentiation factor for Th1 CD4 cells, thereby enhancing cell-mediated responses	related to IL-1 in sequence
IL-19	—	T-cells	cognate antigen and T-cell mitogens	unclear	IL-10 family member
IL-20	—	macrophages	bacterial products	keratinocyte activation to produce pro-inflammatory cytokines	IL-10 family member
IL-21	—	CD4 (Th2) T-cells	cognate antigen and T-cell mitogens	IL-4 antagonist action on B-cells? Activator of CTL?	related to IL-2 and IL-15
IL-22	—	T-cells	cognate antigen and T-cell mitogens	pro-inflammatory?	IL-10 family member
IL-23	—	macrophages, dendritic cells	bacterial, parasitic, and viral products, e.g., LPS, dsRNA	activation of memory T-cells?	IL-12 family member
IL-24	melanoma-differentiation-associated gene 7	melanoma; monocytes and T-cells	cytokines	induces apoptosis selectively in cancer cells	IL-10 family member
IL-25	—			—	same as IL-17E
IL-26	AK155	T-cells	Herpesvirus saimiri	—	IL-10 family member
IL-27	—	macrophages, dendritic cells	microbial products, e.g., LPS	promoter of Th1 responses?	IL-12 family member
IL-28	IFN-λ1, IFN-λ3	many cell types	virus infection (dsRNA)	similar to INF-α	two subtypes, A and B: member of the type I interferon family
IL-29	interferon-λ2	many cell types	virus infection (dsRNA)	similar to INF-α	member of the type I interferon family
IL-30	—				uncertain status
IL-31	—	CD4 (Th2) T-cells	T-cell mitogens	pro-inflammatory	—
IL-32	—	T-cells, NK-cells?	IL-2, IL-15?	activator of cytokine production in macrophages	—
IL-33	—			—	not yet discovered

Interleukin-3

(IL-3; burst-promoting activity; multi-CSF; hematopoietin-2) The IL-3 gene is located within a cluster of growth factor and growth factor receptor genes on 5q23-31, commonly deleted in **myelodysplasia** and **acute myeloid leukemia**. A highly glycosylated monomeric protein, IL-3 displays species specificity in stimulating proliferation of committed progenitors and acts at an intermediate stage of commitment. IL-3 is mainly synthesized in activated T-cells. IL-3 alone is unable to support single-lineage proliferation, but in combination with GM-CSF, G-CSF, erythropoietin (EPO), IL-11, and IL-6 synergizes to enhance monocyte/macrophage, neutrophil, erythroid (red blood cell), and megakaryocyte colonies. Its potent activity upon megakaryocyte colonies leads to therapeutic use in the treatment of **amegakaryocytic thrombocytopenia**. A recombinant IL-3/GM-CSF fusion protein (PIXY 321) has activity that is synergistic compared with either factor alone.

Interleukin-4

(IL-4; B-cell differentiation factor; B-cell growth factor) The gene for IL-4, located within the cytokine locus on chromosome 5q, encodes a glycosylated protein of 19 kDa. IL-4 is produced by a subset of T-helper cells designated Th2, although secretion from mast cells and bone marrow stromal cells has also been described. The activities of IL-4 are diverse:

Synergy with intermediate-acting cytokines to promote erythroid and myeloid progenitor growth

T- and B-cell ontogeny, stimulation of B-cell proliferation, selective secretion of IgG1 and IgE, and enhancement of B-cell antigen presentation by upregulation

Mast cell stimulation, enhancement of NK and LAK cell proliferation, and involvement in Type I **hypersensitivity** reactions

Enhancement of monocyte/macrophage antigen processing and presentation

Interleukin-5

(IL-5; B-cell growth factor 2; eosinophilic CSF) The gene for interleukin-5 is also located within the cytokine cluster on chromosome 5q and encodes a 40- to 45-kDa protein requiring homodimerization for activity and produced by T-cells. IL-5 stimulates proliferation and differentiation of eosinophils *in vitro* and *in vivo*, with antibodies to IL-5 blocking the murine eosinophilia induced by parasitic infections. Mice transgenic for the IL-5 gene produce 70-fold more eosinophils than their controls. IL-5 also promotes selective IgA secretion from lipopolysaccharide-stimulated B-cells. An anti-interleukin-5 antibody has been used in the hypereosinophilic syndrome.[304]

Interleukin-6

(IL-6; B-cell stimulatory factor 2; interferon-β2; hepatocyte-stimulating factor, 26-kDa protein) The gene for IL-6 is located on chromosome 7p21-22 and encodes a protein of molecular mass 21 to 28 kDa. IL-6 is produced by many tissues, including T-cells, macrophages, myeloma cells, endothelial cells, fibroblasts, thrombocytes, and hepatocytes. It follows that there is release in the acute-phase response of **inflammation**, where it stimulates release of the hepatic protein **hepcidin**. Release is also associated with **myelomatosis** and with **angiofollicular lymph node hyperplasia**.[305]

Interleukin-7

(IL-7; lymphopoietin-1; Pre-B-cell growth factor) The IL-7 gene is located on chromosome 8q12-13 over 6 exons and encodes a 25-kDa protein secreted mainly from bone marrow stromal cells. IL-7 stimulates T-cells (mature and immature cells derived from nonmarrow epithelial cells) and B-cell progenitor proliferation, induces the generation of cytotoxic T-lymphocytes (CTL) and lymphokine-activated killer (LAK) cells, and may synergize with other cytokines such as IL-2 and IL-12 in this action. (It is essential for thymic production of T-lymphocytes by promoting survival of T-cell receptor rearrangement of "double negative" pro- and pre-T-cells.)

Interleukin-8

(IL-8; neutrophil-activating protein-1) The IL-8 gene is located on chromosome 4q12-21, among the chemokine-a-gene superfamily gene cluster. IL-8 is produced by peripheral-blood mononuclear cells, endothelial cells, and fibroblasts and in response to tumor necrosis factor (TNF)-α stimulation by human renal epithelial cells and human pulmonary fibroblasts. IL-8 acts predominantly as a neutrophil chemoattractant, with additional neutrophil-degranulating activity and chemoattractant activity for basophils and lymphocytes. IL-8 production can be partially blocked *in vivo* by anti-TNF-α antibodies, but is not always dependent upon stimulation by this pro-inflammatory cytokine.

Interleukin-9

(IL-9; p40; T-cell growth factor-II; mast cell enhancing activity) IL-9 is produced from activated T-cells and monocytes following stimulation by mitogens or cytokines and acts to promote erythroid and mast cell colony growth. The erythroid-growth-promoting effect is directly mediated, with stimulation of burst forming unit-erythroid (BFU-E) and colony forming unit-erythrocyte (CFU-E) growth in the presence of EPO.

Interleukin-10

(IL-10; cytokine synthesis inhibitory factor) The IL-10 gene is located on human chromosome 1 and encodes an 18-kDa protein monomer showing homology with an **Epstein-Barr virus** (EBV) gene BZRF1. IL-10 is produced by a subset of T-helper cells (Th2), macrophages, and Ly-1-positive B-cells.

IL-10 activity was first described as a product of Th2 cells that was capable of suppressing the production of cytokines such as IL-2, IL-3, GM-CSF, TNF-α, and transforming growth factor (TGF)-β by Th1 cells as mediators of the inflammatory response. IL-10 also inhibits antigen presentation by macrophages to Th1 cells by downregulation of MHC class II expression and inhibits macrophage production of TNF-α, IL-1a, IL-1b, IL-6, and IL-8. IL-10 also increases production of IL-1 receptor antagonist.

IL-10 also stimulates B-cell proliferation, differentiation, and immunoglobulin synthesis, particularly in combination with IL-4. Thus, while inhibiting T-cell and macrophage components of the acute inflammatory response, IL-10 upregulates B-cell function. The viral homolog of IL-10 (vIL-10, BZRF1) appears to play a role in the pathogenesis of EBV and other viral infections.

Interleukin-11

(IL-11; adipogenesis inhibitory factor) The IL-11 gene is located on chromosome 1q13.3-13 and encodes a nonglycosylated, 23-kDa protein. Produced by a bone-marrow-derived stromal cell line and human fetal lung fibroblasts, IL-11 is an early-acting cytokine promoting the exit from G0 of multipotent human hematopoietic progenitors in synergy with IL-3, stem cell factor (SCF), or combinations of IL-3 and G-CSF to expand primitive LTC-IC populations (see **Hematopoietic regulation**) capable of sustained production of large numbers of hematopoietic cells and generation of high-proliferative-potential colony-forming cell (HPP-CFC). IL-11 combined with SCF and EPO promotes human CFU-E growth, and with IL-3 it promotes BFU-E growth, even in the absence of EPO. *In vivo*, IL-11 promotes cycling of erythroid progenitors. IL-11 in combination with IL-3 enhances megakaryocyte progenitor growth and increases size and ploidy of **megakaryocytes**. In addition, IL-11 stimulates T-cell-dependent B-cell immunoglobulin secretion, and in fibroblasts and adipocytes it is a potent inhibitor of adipogenesis. IL-11 also promotes secretion of acute-phase proteins in a similar manner to IL-6.

Interleukin-12

(IL-12; natural killer stimulatory factor; cytotoxic lymphocyte maturation factor) The gene is located on chromosome 5q and the product is a heterodimer of p40 and p35 chains linked by disulfide bonds. Produced by macrophages and B-cells, IL-12 promotes the proliferation of T-cells and NK-cells and enhances NK/LAK activity, particularly in synergy with low IL-2 doses. IL-12 also stimulates IFN-γ production from NK and T-cells. IL-12 produced early in the defense of infection promotes the generation of Th1 CD4-positive cells and cytotoxic T-lymphocytes (CTL). An anti-interleukin 12 antibody has been shown to be active in the treatment of Crohn's disease.[306]

Interleukin-13

(IL-13) The gene for IL-13 is located on chromosome 5 within the cytokine gene cluster and encodes a 10-kDa protein produced principally by activated Th1 cells. IL-13 shares many properties with IL-4. IL-13 is an early-acting cytokine, synergizing particularly with SCF to promote colony formation from murine Lin⁻, Sca-1⁺ marrow cells, which are highly enriched for marrow repopulating cells. On more-mature murine progenitors, addition of IL-13 redirects SCF- and G-CSF-stimulated granulocytic colony growth to macrophage colony production. IL-13 also stimulates B-cell proliferation and immunoglobulin secretion and upregulates monocyte class II MHC antigen expression.

Note that for some of the higher interleukins (from about 20 onward), there may be little literature, and a definitive understanding of their production, roles, etc., is lacking.

Pleiotropic Activities

Promotion of **immunoglobulin** secretion from stimulated B-lymphocytes without isotype specificity.

Activity upon early progenitor cells, stimulating entry into cell cycle and synergizing with intermediate-acting factors such as IL-3 to promote granulocyte and granulocyte/macrophage progenitor growth. IL-6 promotes growth arrest and differentiation in a variety of human hematopoietic immortalized cell lines.

A potent *in vitro* and *in vivo* stimulatory effect on megakaryocyte colonies and platelet production, particularly in combination with IL-3. Thus, the combination of IL-3 and IL-6 *in vivo* promotes rapid hematopoietic reconstitution after chemotherapy by stimulating early progenitors, granulocyte/macrophage colony growth, and the megakaryocyte lineage (see **Thrombopoiesis**).

A major mediator of acute inflammation response, stimulating hepatocyte production of C-reactive protein, serum amyloid A, and fibrinogen.

Synonymous with interferon-β2, IL-6 is thought to possess antiviral activity.

Inhibitory effects upon erythroid colony growth (*in vitro* and *in vivo*); promotion of a normocytic anemia in monkeys despite an increase in reticulocyte counts.

T-cell activation and differentiation of **cytotoxic T-lymphocytes**

Suppresses establishment of pulmonary and hepatic metastases of syngeneic methylcholanthrene-induced sarcomas in mice. rhIL-6 also inhibits the growth of human breast carcinomas transplanted into athymic mice.

Stimulation of the hypothalamic-pituitary axis.[135]

Pathological Manifestations

One strain of IL-6 transgenic mice has developed a lymphoproliferative disorder similar to Castleman's disease (angiofollicular lymph node hyperplasia).

Stimulation of replication of the **human immunodeficiency virus** (HIV) and the proliferation of **Kaposi's sarcoma** cells *in vitro*.

Possible association with the pathogenesis of autoimmunity and proliferative glomerulonephritis.

Autocrine stimulation of some multiple myeloma cells that secrete IL-6 and possess receptors for IL-6. Antibodies to IL-6 block myeloma colony growth from some patients.

Role in pathogenesis of **Hodgkin disease**.

INTERMEDIATE OR POORLY DIFFERENTIATED LYMPHATIC, DIFFUSE, OR NODULAR LYMPHOMA

(ILL, IDL, PLL) See **Mantle-cell lymphoma**; **Non-Hodgkin lymphoma**.

INTERNATIONAL NORMALIZED RATIO

(INR) The **prothrombin** ratio which, by calculation, would have been obtained had the original primary human reference thromboplastin been used to perform the **prothrombin time**. It is derived from either of the following calculations:

INR = (prothrombin time ratio)^{ISI}

INR = antilog (log [prothrombin time ratio] × ISI)

Both calculations are based upon the **international sensitivity index** (ISI). The INR is the index used for international control of anticoagulant therapy (see **Warfarin**).

INTERNATIONAL SENSITIVITY INDEX

(ISI) The slope of the calibration line obtained when the logarithms of the **prothrombin time** obtained using the World Health Organization **reference material** for thromboplastin, which has been assigned a value of 1.0, are plotted on the vertical axis of log-log paper and the logarithms of the prothrombin time obtained by the test thromboplastin are plotted on the horizontal axis. Aliquots of the same normal and anticoagulated patients' plasma samples must be used. As thromboplastins are used from a variety of sources and usually give different prothrombin times for the same plasma sample, due to differing potencies and responsiveness to levels of **vitamin K** coagulation factors, all thromboplastins for such purposes should be assigned their ISI.

INTESTINAL CELL LYMPHOMA

(Diffuse mixed lymphocytic-histiocytic lymphoma; diffuse large-cell lymphoma; diffuse small cleaved cell lymphoma; histiocytic lymphoma; malignant histiocytosis of the intestine; T immunoblastic sarcoma; T, pleomorphic small-, medium-, and large-cell lymphoma; ulcerative jejunitis) See **Intestinal tract disorders**; **Lymphoproliferative disorders**; **Non-Hodgkin lymphoma**.

INTESTINAL-TRACT DISORDERS

The pathological changes of the duodenum, jejunum, ileum, colon, rectum, and anal canal, with associated hematological disorders. These are discussed in the following subsections.

Disorders of Malabsorption

Gluten enteropathy (celiac disease, nontropical sprue, idiopathic steatorrhea):[307,308,309a] permanent intolerance of the intestinal mucosa to gluten — a water-insoluble, glutamine-rich protein fraction of wheat, barley, and rye grains — and to the related gliadin. Ninety percent of patients have associated HLA-DQ2, a heterodimer encoded by HLA-DQA1*05 and HLA-DQB1*02 genes with an inherited genetic abnormality on chromosome 14. Most patients carry IgA autoantibodies to reticulin, gliadin, transglutaminase, and endomycin. Activation of T-cells is by a peptide in gliadin, which leads to an immune response that damages enterocytes and causes villous atrophy. Celiac disease is sometimes associated with dermatitis herpetiformis, and there is a predisposition to enteropathy-type T-cell intestinal lymphoma. Jejunal biopsy, either by Crosby capsule or fiber-optic endoscopy, shows subtotal villous atrophy, cryptic hypoplasia, and mononuclear-cell infiltration of the submucosa. These changes cause malabsorption, leading to weight loss, glossitis, abdominal discomfort, and steatorrhea; a negative calcium balance, **iron deficiency** anemia, and **folic acid**-deficiency anemia follow. **Vitamin K** malabsorption may cause deficiency of vitamin K-dependent coagulation factors. Occasionally, ileal involvement causes **cobalamin** malabsorption, leading to megaloblastosis. A gluten-free diet results in reversal of the intestinal changes and of any malabsorption in 80% of patients.

Tropical sprue. A malabsorption disorder following enteric infection acquired in tropical countries, particularly the West Indies, southern India, parts of southern Africa, and Southeast Asia. The clinical and pathological manifestations are similar to those of gluten enteropathy, except that it affects the distal rather than the proximal small intestine and almost always leads to cobalamin deficiency, which may not become manifest until 20 years or more after relocation. Malabsorption of folate also occurs, possibly because the diseased intestine fails to deconjugate folate polyglutamates. Megaloblastic anemia is therefore common in these patients. Therapy consists of daily oral administration of 1 to 3 mg folic acid plus cobalamin if deficiency is present. Because patients with this disease tend to relapse without supplementation, maintenance therapy is recommended for at least 2 years. Broad-spectrum antibiotics are a useful therapeutic adjunct, but alone fail to correct the disorder. There is no improvement with a gluten-free diet.

Extensive surgical resection of the bowel, causing cobalamin and folate malabsorption.

Bacterial overgrowth syndrome. This occurs when gastric acid secretion and peristalsis fail to maintain intestinal bacteria at a physiological level; stomach disorders with achlorhydria, diverticulae, fistulae, strictures, surgical blind loops, and pouches from anastomoses, resulting in stasis or recirculation, cause this overgrowth. Competition for available cobalamin and folate then makes these no longer available to the host, leading to deficiency and consequent megaloblastosis. It is associated with weight loss, steatorrhea, and diarrhea.

Fish tapeworm (*Diphyllobothrium latum*). Infestation causing cobalamin deficiency from impaired absorption due to competition for dietary cobalamin (see **Metazoan infection disorders**).

Inflammatory Disorders[310]

Crohn's disease (regional ileitis): is a chronic inflammatory disorder of the small and large intestines, similar histologically to **sarcoidosis**, associated with recessive mutations in a gene on chromosome 16 resulting in the formation of a nucleotide-binding oligomerization domain protein (NOD2).[311] This is a Toll-like receptor (TLR) that upon activation induces through **interleukin**-12 a Th1 inflammatory response. Crohn's disease causes iron deficiency anemia due to chronic **hemorrhage**, cobalamin deficiency when affecting the ileum, and folate deficiency if it affects the jejunum.[312] Treatment with an anti-tumor necrosis factor (TNF) monoclonal antibody (**infliximab**), an antibody to interleukin-12, and another against α4-integrins are all under evaluation.[132,306]

Ulcerative colitis: causes iron-deficiency anemia due to chronic hemorrhage, folate malabsorption from the jejunum, and rarely, cobalamin malabsorption from the ileum, leading to megaloblastic anemia.

Apart from these inflammatory disorders, in those patients with type 1 **human leukocyte antigens** (HLA), there is associated arthritis of the spine and peripheral joints. This association also occurs in postdysenteric-reactive arthritis. It is proposed that these extra-intestinal manifestations are a consequence of homing by memory lymphocytes.

Vascular Disorders

Ischemic bowel disease.

Atherosclerosis of intestinal arteries results in stenosis, thrombosis, and emboli arising from cardiac disorders. Treatment may require anticoagulant therapy with heparin or warfarin or possibly acute surgical intervention.

Sickle cell disorders complicated by infarction of mesenteric vessels, often during an acute crisis and causing abdominal pain.

Angiodysplasia of the colon and hemorrhoids of the anal canal causing iron-deficiency anemia from chronic hemorrhage.

Scleroderma affecting the microvasculature of the intestinal wall.

Polyarteritis nodosa causing hemorrhage, ischemia, and perforation of the bowel.

Leukoclastic vasculitis (Henoch-Schönlein purpura) causing melena and abdominal pain.

Angioedema of intestinal wall due to deficiency of C1 inhibitor of complement.

Hemorrhoids of the anal canal causing iron-deficiency anemia from chronic hemorrhage.

Neoplastic Disorders

Intestinal polyps or telangiectases causing chronic hemorrhage and iron-deficiency anemia; detection of blood loss can be made insensitively by tests of fecal occult blood or more accurately by labeling the patient's red blood cells with ^{51}Cr (see **Red blood cell survival measurement**), reinjecting them intravenously, and counting the fecal radioactivity.

Carcinoma of the colon or rectum, giving rise to melena or iron-deficiency anemia.

Lymphoproliferative disorders:

- Intestinal T-cell lymphoma, now recognized as **enteropathy-type T-cell lymphoma** in the WHO Classification of Tumors of the Hematopoietic and Lymphoid Tissues (see Appendix).

- Primary small-intestinal lymphoma (PSIL), a focal B- or T-cell chronic lymphatic leukemia/lymphoma without involvement of other areas of the intestinal tract. Local surgical removal is indicated.

- **Heavy-chain disease** (Mediterranean lymphoma), a B-cell proliferation (normal and abnormal plasma cells) involving a wide segment of the intestinal wall associated with hypergammaglobulinemia. There is increased synthesis of α-heavy chains with the V-region deleted and absence of light chains. Cobalamin malabsorption occurs in 10 to 15% of patients. It is associated with infection by *Campylobacter jejuni*.

- **Marginal-zone B-cell lymphoma** (MALT type) associated with *Helicobacter pylori* infection of the small intestine, rarely the colorectum.[176]

- Complications of **cytotoxic agent** therapy for hematological malignancies, including:

  Intestinal cell damage

  Graft-versus-host disease

  Both conditions, uncommonly causing diarrhea

INTRACELLULAR SERPINS

See **Serpins**.

INTRAUTERINE TRANSFUSION

(IUT) See **Fetal/neonatal transfusion**.

INTRAVASCULAR FETAL TRANSFUSION

(IVT) See **Fetal/neonatal transfusion**.

INTRAVASCULAR HEMOLYSIS

(Intravascular hemolytic anemia) The destruction of **red blood cells** (RBCs) in the intravascular space associated with **hemoglobinemia** and **hemoglobinuria**. It may result from microangiopathy (e.g., **disseminated intravascular coagulation** and **thrombotic thrombocytopenic purpura**), immune-mediated destruction (e.g., **blood transfusion complication** — hemolytic transfusion reaction and **paroxysmal cold hemoglobinuria**), or the effects of drugs or bacterial toxins on susceptible RBCs. Rapid RBC destruction may cause hypovolemia and renal failure.

INTRAVASCULAR LARGE B-CELL LYMPHOMA

(Lukes-Collins: angiotropic large-cell lymphoma) See also **Diffuse large B-cell lymphoma**.

Clinical Features

A rare intravascular, diffuse, large B-cell lymphoma that is usually disseminated with involvement of cutaneous, pulmonary, renal, bone marrow, and central nervous system.

Staging

See **Lymphoproliferative disorders**.

Treatment

See **Diffuse large B-cell lymphoma; Non-Hodgkin lymphoma**.
This is an aggressive lymphoma that is initially managed with **cytotoxic agent** therapy (R-CHOP) or other aggressive chemotherapy.

INTRINSIC FACTOR

(IF) A glycoprotein of molecular weight 45 kDa, synthesized and secreted by the parietal cells of the gastric fundus. Secretion of IF usually parallels the secretion of HCl and is enhanced by the presence of food in the stomach, by vagal stimulation, and by histamine and gastrin. This is one of a number of protein cobalamin binders (see **R-binder proteins**). Antibody production to IF occurs in some stomach disorders and is the cause of its deficiency in **Addisonian pernicious anemia**.

Congenital Deficiency

This is a rare inborn error of cobalamin metabolism that results from production of functionally abnormal IF. Three variants are recognized:

Formation of unstable IF

Formation of IF that binds cobalamin, but the complex cannot be transported across the ileal mucosa

Failure completely of IF production

These disorders are transmitted as autosomally recessive traits, with patients presenting between the ages of 6 months and 2 years with irritability, glossitis, and **megaloblastosis**.

Acquired Deficiency

This arises from:

IF blocking autoantibody at the B12 binding site

IF precipitating antibody that inhibits ileal binding

Functionally abnormal IF; this is immunologically identifiable as IF, but without the presence of IF antibodies; it is unable to bind cobalamins, including the ileal receptor

In both congenital and acquired disorders, the abnormal **Schilling test** is corrected by oral intrinsic factor. The effect of the disorder is failure of cobalamin absorption, and treatment is by conventional doses of cobalamin.

IONIZING RADIATION

The effects on blood cells and blood-forming tissues of radiation that remove electrons from atoms or molecules.[313–317] These depend upon the type of radiation, the amount and rate of energy absorbed and distributed, and the extent to which the dose is split or fractionated. Cell damage arises from exposure to radiant energy from electromagnetic rays — X-rays and cosmic rays — and from energetic particles (protons, fission nuclei, α-particles [helium nuclei], and β-particles [electrons], or from the production of free radicals). Neutrons do not ionize directly but impart energy to protons.

All body cells are susceptible to damage by ionizing radiation. The effects depend upon the physical nature and quantity of radiant energy, the mitotic activity of the cell population exposed, and the duration of radiation exposure. The radiant energy of neutrons, X-rays, and most cosmic rays are highly penetrating, enabling the radiation to ionize tissue components at distances ranging up to many meters. The potential tissue penetration by β-particles (electrons emitted during cosmic-ray decay) is much shorter, and penetration by most α-particles is limited to a fraction of 1 mm.

The unit of measurement of ionization is the rad (radiation absorbed dose) or, with X-rays, the roentgen (R). A dose of 1 rad is the radiation dose that will cause the formation of 1.61×10^{12} ion pairs per gram of air or tissue. Rems (radiant equivalent man) are used for protection purposes, being biologically equivalent to the rad. For radiotherapy-dosage purposes, the unit is the Gray (Gy), where 1 Gy = 100 rads. The effects of different doses are shown in Table 88.

A single dose of 750 rads to the whole body is lethal due to widespread **DNA** breaks in all body cells. The effects of lower doses vary with the dose and the length of exposure. The susceptibility of cells in G0 (resting) phase **mitosis** (i.e., mature erythroid or myeloid

TABLE 88

Dose-Related Effects of Penetrating Ionizing Radiation

Effects	Rads
No symptoms	0 –125
Reversible symptoms and signs	125 –250
50% mortality	500
100% mortality	700
Doubling of mutations	5 –150
Maximum cumulative occupational exposure	5 per year

Source: From Jandl, J.H., *Blood: Textbook of Hematology,* 2nd ed., Little, Brown, Boston, 1996, chap. 4. With permission.

cells) to radiation damage is several orders of magnitude less than active cycling cells. Vulnerability varies widely, in proportion to the fraction of cells in the S phase and the rate of DNA synthesis. The cytokinetic effect of a single dose of radiation to hematopoietic cells is the prompt killing of cells, with recovery depending upon the fractions of multi-potential and pluripotential cells that survive. A single brief exposure may cause extensive destruction of circulating blood cells, differentiated bone marrow cells, and gut mucosal cells, resulting in **agranulocytosis**, **thrombocytopenia**, and bleeding from intestinal ulcer-ation, with death resulting in a few days. If the individual survives the acute cellular destruction, complete recovery usually follows within 2 to 3 weeks, as the stem cells in their resting phase will have been resistant to radiation damage. Prolonged exposure to moderate doses leads to **myelodysplasia** with **ineffective erythropoiesis, dyserythropoie-sis, dysgranulopoiesis**, and **dysmegakaryopoiesis**. An increasing rate of stem cell death leads to **aplastic anemia**. Doses of less than 150 rads have little or no immediate effect upon hematopoietic tissue, but chromosomal aberrations can occur upon recovery. These will increase with repeated low-dose exposure. Double-strand breaks in DNA are respon-sible for chromosome breaks. Those that do not rejoin, that rejoin incorrectly, or that rejoin with other breaks to form translocations (but without causing the cell to die or to cease proliferating) are most likely to lead to malignant transformation, i.e., **leukemia**. The multistep mechanism of **oncogenesis** can be rapidly triggered by multiple concurrent hits by neutrons, causing the sequence of mutation, promotion, and progression to malignant cloning.

Aplastic anemia or **bone marrow hypoplasia** has been long associated with nuclear fallout, as seen from victims of Hiroshima and Nagasaki and the more recent explosion at the Chernobyl nuclear power station. Chronic exposure effects have been seen following treatment of **ankylosing spondylitis**, where delayed aplastic anemia, acute leukemia (both myeloid and lymphoblastic), myelodysplasia, and **lymphoproliferative disorders** have occurred. Patients given thorium dioxide (thorotrast) as an intravenous contrast medium also suffered similar late complications. Chronic radium poisoning also occurred in work-ers who painted watch dials with luminous paint when they moistened their brushes orally.

Uncertainty persists concerning the dose-effect curve at low levels of ionizing radiation. The most pervasive population exposure comes from natural background sources, including cosmic rays (0.045 rad/yr at sea level), isotopes such as uranium in natural deposits, and assimilated **radionuclides**. Epidemiological studies have failed to reveal an excess cancer rate in stable populations exposed to three times the average levels of natural radiation, but concern continues for those living near nuclear energy centers and radioactive waste dumps.[318] A small excess risk has been claimed for those working in the nuclear industry. The main source of harnessed radiation is medical diagnostic X-rays. For persons exposed to this hazard by occupation, health screening by film badges is more effective than by serial **automated blood cell counting**.

Ionizing radiation is used for the treatment of some hematological malignancies (see **Radiotherapy**).

IRON

The metal essential for synthesis of energy-storage and -transport molecules. The total body iron content of the average adult male is 50 mg per kilogram of body weight, and for the adult female it is 35 mg/kg. Most iron is present in the **heme** compounds — **hemoglobin** (70%) and myoglobin (5%) — with a small amount bound to plasma **trans-ferrin**. Small but important amounts of iron are incorporated in cellular cytochromes, in other heme-containing enzymes (e.g., catalase and peroxidase), and in nonheme iron enzymes (e.g., xanthine oxidase and ribonucleotide reductase). Iron is stored in **ferritin** (and partially degraded ferritin, termed **hemosiderin**). The average daily intake of iron in Western diets is 10 to 30 mg. Worldwide, however, there is great variation. Approximately 5 to 10% of dietary iron is absorbed in healthy adults. During periods of rapid growth, pregnancy, or when iron stores become depleted, absorption is increased three- to fivefold. Conversely, in states of iron overload, absorption may be decreased. Men unavoidably lose 1.0 mg iron daily in sweat, urine, or feces. Women of reproductive age lose additionally through menstruation a total daily loss of about 2 mg. In pregnancy, the average daily loss is about 3 mg.

Iron Absorption

Entry and Processing by Enterocytes

Iron is absorbed in the small intestine in a gradient duodenum-jejunum-ileum. Nonheme iron is converted by **duodenal cytochrome** to inorganic ferrous form before absorption. Certain iron complexes are poorly available for absorption (e.g., cereal protein iron complexed with phytate; egg yolk iron complexed with phosphates). Inorganic iron absorption is facilitated in the presence of animal proteins, breast milk, and reducing agents such as ascorbic acid and cysteine. At the microvillous membrane of absorptive enterocytes, iron in physiologic concentrations is reduced to the ferrous form and transported across the apical (microvillous) membrane by **divalent metal transporter-1** (DMT1). At high concentration, iron may cross the mucosa passively. Within enterocyte cytoplasm, some iron appears to occur as unbound iron salts or as low-molecular-weight complexes ("labile iron pool"). Another absorptive pathway appears to be specific for heme iron derived from heme proteins in foods of animal origin. Heme may be absorbed via specific absorptive cell-surface receptors, or may be oxidized to the ferric state and converted to hemin in the gut lumen before cellular uptake, where heme oxygenase cleaves the porphyrin ring to release iron from the molecules. Excess iron may be incorporated into ferritin within the enterocyte and subsequently lost into the gut lumen by desquamation. A protein, **hemojuvelin**, is a potent regulator of iron absorption. Iron release from the intestinal epithelial cells requires the enzyme **hephaestin**.

Transfer from Enterocytes to the Circulation

Iron is exported from the basolateral aspect of absorptive cells via **ferroportin** with the participation of hephaestin (a multicopper oxidase) and bound to transferrin via transferrin receptors on the basolateral surfaces of absorptive cells. The affinity of transferrin and transferrin receptors in this location may be modulated by HFE protein (see **Hereditary hemochromatosis**). A variable proportion of the absorbed iron appears in the plasma, bound to transferrin, within a few hours after ingestion. The amount of storage iron and the overall rate of erythropoiesis are important regulators of iron absorption. This effect may be mediated at the molecular level by **hepcidin**, an oligopeptide produced primarily in the liver.

Body Iron Compartments

There are six iron compartments in the human body based on anatomic distribution, chemical characteristics, and function. These are:

Hemoglobin (about 70%)

Myoglobin (about 5%)

Labile iron pool

Tissue iron compartment

Transport compartment

Storage compartment (ferritin/hemosiderin in macrophages about 30%)

The labile iron pool contains 80 storage compartments (ferritin/hemosiderin) for up to 90 mg of iron in the healthy individual. Iron temporarily leaves the plasma and enters interstitial and intracellular fluid compartments for a short time before being incorporated into heme or a storage compound. Tissue iron normally amounts to 6 to 8 mg, including iron in cytochromes and a variety of enzymes (cytochrome C, catalase, cytochrome oxidase, and succinic dehydrogenase). Although small, this compartment is important. Some of its components reflect changes in the total body iron content. Although the transport compartment is the smallest, kinetically it is the most active and serves as the common intermediate pathway by which iron in the other compartments can be interchanged. **Frataxin** is an essential protein involved in mitochondrial iron metabolism.

Assessment of Iron Status[319]

There is no single test of iron status suitable for all clinical circumstances. The available tests include the serum iron, the total iron-binding capacity and its percentage saturation, the serum ferritin, the urine ferrioxamine iron, and the liver iron concentration. In addition, there are ferrokinetic measurements using **radionuclides** to estimate iron absorption and iron distribution (clearance, turnover, utilization), for surface counting, and for whole-body scanning (see **Ferrokinetics**).

Measurement of Serum Iron

Serum iron concentration, total iron-binding capacity (TIBC), and calculation of percentage saturation of the TIBC permit assessment of iron repletion and tissue iron delivery in health, in iron deficiency, and in iron overload. The serum iron concentration and transferring saturation fluctuate from day to day in the same subject and exhibit diurnal variation (highest values occur in the morning). These measurements reflect iron supply to the tissues at the time of sampling and do not provide any information on the level of body iron stores. In iron deficiency, the serum iron concentration is low and the TIBC becomes elevated and thus the percentage transferrin saturation is decreased (see **Iron deficiency**). Serum iron concentration and transferring saturation levels are decreased in many acute or chronic inflammatory or neoplastic disorders, although the percentage saturation of TIBC is usually reduced less than in iron deficiency (see **Anemia of chronic disorders**). Increased serum iron concentration and transferring saturation level are usually due to hepatocyte injury, ineffective erythropoiesis, or hemochromatosis or other primary overload disorders (see **Hereditary hemochromatosis**). The adult reference range in health for serum iron is 11 to 28 μmol/l; for TIBC it is 47 to 70 μmol/l; and for transferring saturation of TIBC it ranges from 16 to 30% (see **Reference Range Tables VI and VII**.

The serum ferritin concentration parallels the magnitude of body iron stores and is therefore useful in the diagnosis both of iron deficiency and iron overload. Reference ranges for serum ferritin vary with age and sex. In childhood and adolescence, reference ranges of 20 to 30 μg/l are found. Adult male levels are 30 to 300 μg/l, with a mean value of about 100 μg/l. In women, however, values remain around 30 μg/l (range 15 to 200 μg/l). Significantly higher values of serum ferritin occur in healthy persons of Asian or sub-Saharan African descent than in Western whites. Serum ferritin levels are increased due to increased body iron stores (hemochromatosis, secondary iron overload), increased ferritin protein synthesis (inflammatory disease, certain tumors, hyperthyroidism, chronic ethanol ingestion), and release of tissue ferritins (hepatic necrosis, chronic liver disease, spleen or bone marrow infarction, certain tumors). Serum ferritin is an acute-phase reactant. Thus, elevated values of serum ferritin may occur in persons with conditions such as **rheumatoid arthritis**, renal failure, and other inflammatory or neoplastic processes, even in the presence of iron deficiency. Hereditary hyperferritinemia and congenital cataract is a syndrome due to mutations in the iron-responsive element of the L-ferritin gene; affected patients do not have iron overload. Serum ferritin less than 15 µg/l indicates depletion of iron stores, although most adults with serum ferritin <50 µg/l have little or no stainable marrow iron. Subnormal ferritin levels occur in patients with iron deficiency, pregnancy, and sometimes occur in patients with vitamin C deficiency. Although elevated serum ferritin levels are usually due to race/ethnicity or disorders other than **iron overload**, an elevated level of serum ferritin is also found in association with increased body iron stores (hereditary hemochromatosis, secondary iron overload).

Estimates of the liver iron concentration using a specimen obtained by biopsy provide important data on the severity of iron overload in persons with hemochromatosis and other iron overload disorders; concurrent histological assessment of liver biopsy provides information about the extent of hepatic injury. The liver biopsy is dried to constant weight, dissolved in acid, and analyzed using atomic absorption spectrometry. The reference range in health is 6 to 24 µmol Fe/g dry liver (0.03% to 0.13% of the dry weight). In patients with untreated hereditary hemochromatosis, the hepatic iron level may be greater than 1% of the dry weight. An age-adjusted hepatic iron concentration (hepatic iron index) may be especially useful in distinguishing HFE hemochromatosis from other disorders.

Oral Iron Absorption Test

This is of very limited value as an indicator of the ability of the gastrointestinal tract to absorb iron. A small amount of isotope-labeled inorganic iron is given by mouth and fecal radioactivity measured to permit indirect quantitation of absorbed iron.

Iron Excretion Measurement

The level of urine iron excretion following a test dose of **desferrioxamine** can be used as a screening test for parenchymal iron overload. When normal iron stores are present, the 24-h urinary iron excretion following a single test dose of desferrioxamine (500 mg) will not exceed 1.1 mg (20 µmol), whereas in subjects with hereditary hemochromatosis, excretion is usually greater than 4.5 mg (80 µmol). Lower excretion values are found immediately following blood transfusion.

IRON-BINDING CAPACITY — TOTAL

(TIBC) See **Iron**.

IRON-BINDING PROTEIN

See **Transferrin**.

IRON CHELATION

The removal of excess body **iron** by chemical means. This is a standard therapeutic maneuver in patients with certain forms of severe **thalassemia** and in **hereditary hemochromatosis**. **Desferrioxamine** is the only currently approved effective iron chelator for clinical use.[320] Its two main disadvantages are expense and the need for parenteral administration. Desferriox-amine must be given subcutaneously or by continuous intravenous infusion because it is poorly absorbed when given by mouth and has a very short half-life. Desferrioxamine therapy to remove excess iron is much less efficient than **phlebotomy**. The adverse drug reactions of desferrioxamine include erythema, swelling and discomfort at the site of subcutaneous injection, depletion of certain nonferrous metals, dose-dependent optic neuropathy (increased retinal pigmentation, central scotoma, night blindness), high-tone deafness, and an increased susceptibility to infection by *Yersinia* and *Staphylococcus aureus*.

IRON DEFICIENCY

See also **Iron.**

The consequences of the body iron being less than normal. This state exists as a continuum with three recognizable subdivisions:

Iron depletion: reduced body iron content but normal levels of **hemoglobin**, serum iron, and **transferrin** concentrations

Iron deficiency without anemia: absent storage iron and reduced serum iron and transferrin concentrations (sideropenia)

Iron deficiency with anemia

As iron deficiency progresses, changes occur in a number of body tissues and systems, including the iron-containing proteins, red blood cells, muscle, nervous system, immune system, gastrointestinal tract, and bone.

Causes of Iron Deficiency

The numerous causes of iron deficiency can be subdivided according to mechanism.

Dietary deficiency, particularly of heme iron, rarely occurs alone except in infancy and childhood. Dietary fads or weight-loss diets, if followed for long periods, sometimes cause iron deficiency.

Malabsorption, an uncommon cause of iron deficiency except following gastrointestinal surgery (especially bariatric surgery) and in malabsorption syndromes (see **Intestinal tract disorders; Gastric disorders**).

Chronic hemorrhage:

- Stomach disorders, e.g., peptic ulceration, carcinoma
- Intestinal tract disorders, e.g., inflammatory bowel disease, carcinoma, **hereditary hemorrhagic telangiectasia**
- Respiratory tract disorders, e.g., idiopathic pulmonary hemosiderosis, **Goodpasture's syndrome**
- Renal tract disorders, including carcinoma, bladder fluke infection, nephrolithiasis
- Nosocomial anemia from overly frequent or excessive blood donation, excessive laboratory-test blood loss ("blood test anemia"), postsurgical blood loss (especially after major thoracic surgery, total hip or knee replacement)
- Factitious anemia
- **Pregnancy**, with diversion to the fetus and placenta and loss amounts of approximately 900 mg

Intravascular hemolysis: **paroxysmal nocturnal hemoglobinuria, paroxysmal cold hemoglobinuria**, and **fragmentation hemolytic anemia**.

Neonatal disorders: deficiency of iron stores at birth. This deficiency can be prevented by delay in clamping of the umbilical cord.

Clinical Features

The clinical features are those of the causative condition, in addition to those common to patients with all forms of anemia. Clinical features specific to iron deficiency include atrophy or spooning of nails (koilonychia), painful atrophic tongue, angular stomatitis, dysphagia and esophageal webbing as part of the Plummer-Vinson syndrome, increased susceptibility to oral or esophageal carcinoma, and pica (especially a craving for ice [pagophagia] in women). Iron deficiency without anemia may decrease physical endurance, work capacity, learning, memory, and immune function. In infancy, it may delay growth and development.

Diagnosis

In patients with anemia, hypochromia, and microcytosis, pencil-shaped poikilocytes are typically present in peripheral-blood films. The **red blood cell count** is normal or reduced, and **thrombocytosis** is common, although a minority of subjects have **thrombocytopenia**. The differential diagnosis includes **thalassemia**, the **anemia of chronic disorders**, and **sideroblastic anemia**. The serum iron level is decreased, the total iron-binding capacity increased, and the percentage saturation reduced. The serum ferritin level is reduced and serum transferrin receptor (sTfR) increased. Stainable iron is absent from the bone marrow.

Treatment

Management depends on the severity of the anemia, its cause, and the ability of the patient to tolerate and absorb medicinal iron.

Oral Iron Therapy

Where possible, oral iron therapy is preferred. Several oral preparations exist, of which ferrous sulfate tablets or solution are the least expensive. Each dose should contain between 50 and 100 mg elemental iron, and most adult patients should be treated with 150 to 200 mg Fe daily. Doses for children should be reduced according to body size; oral preparations that stain teeth should be avoided. Iron medication is best taken before meals and at bedtime. This sometimes causes unpleasant adverse effects, especially nausea, gastric irritation, diarrhea, constipation, or metallic taste. Enteric-coated and delayed-release ferrous sulfate preparations are available. The disadvantage of using such preparations is that there is delay in releasing iron, which is then presented to a segment of intestinal mucosa where iron uptake is less efficient than in the proximal small intestine. When there is intolerance to ferrous sulfate, as occurs in 10 to 20% of patients, ferrous gluconate or ferrous fumarate can be used, but these should not be first-choice drugs, since they are more expensive than ferrous sulfate. Ascorbic acid, succinate, and fructose all enhance gastrointestinal absorption of iron, but iron preparations containing them are more expensive than ferrous sulfate, and they produce a greater frequency of adverse effects. Carbonyl iron preparations are generally well tolerated but are often expensive. Preparations containing multiple hematinics should not be used. To replenish iron stores, treatment should be continued for 2 to 3 months after correction of the hemoglobin level and mean cell volume. Antacids, H_2-receptor antagonists, proton pump blockers, and calcium supplements are potent inhibitors of iron absorption and are therefore common causes of apparent failure of oral iron therapy.

Parenteral Iron Therapy

Indications for parenteral iron therapy include intolerance to oral iron, noncompliance by the patient with severe iron deficiency, and malabsorption. Iron dextran (a ferric hydroxide-dextran compound prepared as a viscous solution containing 50 mg of elemental iron per milliliter) is the commonly used preparation. The preferred method of administration is by bolus intravenous infusion in doses from 500 to 2000 mg diluted in 250 ml of normal saline. True anaphylaxis is uncommon, but it is prudent to premedicate patients with intravenous H_1- and H_2-receptor antagonists and a small dose of **corticosteroids**. After premedication, an intravenous test dose of 0.5 ml iron dextran solution is administered over 5 to 10 min. If the patient experiences no immediate adverse effects, the infusion should commence and continue over at least 2 h. The infusion must be stopped immediately if itching, breathlessness, chest pain, back pain, or hypotension develop. Although many of these are "rate-related" reactions that resolve spontaneously, epinephrine should be immediately available if true anaphylaxis occurs. Larger "single-dose replacement" quantities of iron are more likely to be associated with arthralgias and bone pain of delayed onset than 500-mg doses. Iron sucrose and other iron saccharides are used commonly in patients undergoing chronic hemodialysis, but iron quantities per dose are low (typically 50 mg). The intramuscular injection of iron dextran is not recommended for many reasons.

Red Blood Cell Transfusion

In patients with a very severe anemia, emergency transfusion of packed red blood cells may be required (see **Red blood cell transfusion**), particularly if the patient is in cardiac failure or requires urgent surgery. Care must be taken to avoid fluid volume overload during transfusion in such patients. It is always important to ascertain the cause of the iron deficiency as soon as possible, because this will significantly influence the treatment of such patients.

IRON OVERLOAD

See also **Hereditary hemochromatosis; Iron; Iron chelation**.
The causes and consequences of excess body iron.[321] Iron overload occurs in a number of conditions:

Primary absorptive defect: hereditary hemochromatosis is the most important cause.

Secondary generalized or organ-specific disorders:

Anemia with **ineffective erythropoiesis**
- Transfusion-dependent **thalassemi**a
- **Sideroblastic anemia** requiring regular transfusion
- **Bone marrow hypoplasia** requiring chronic transfusion

Liver disorders
- Alcoholism
- Postportacaval shunt
- Chronic hepatitis C

Porphyria cutanea tarda

Excessive oral iron intake

- Prolonged excessive iron medication (rare)
- African iron overload (associated with one or more genetic traits that promote iron absorption)

Hereditary atransferrinemia

Hereditary aceruloplasminemia

   The causes of secondary iron overload are usually obvious, with the exception of liver disease with or without a history of alcoholism, which may be difficult to distinguish from early hereditary hemochromatosis. Liver biopsy provides a definitive diagnosis. Iron deposition in the Küpffer cells is often predominant in iron overload resulting from certain **ferroportin** mutations, African or African-American iron overload, repeated blood transfusions, or alcohol-induced increased absorption. In HFE, HAMP, or HJV-associated hemochromatosis, iron deposition in hepatocytes is more prominent than that in Küpffer cells. Typically, hepatic cirrhosis does not develop until iron stores exceed 20 g, although chronic ethanol ingestion or chronic hepatitis C may act in synergy to cause cirrhosis at lower levels of hepatic iron overload.

IRON UTILIZATION
See **Erythrokinetics; Ferrokinetics.**

IRREGULARLY CONTRACTED RED BLOOD CELLS
Small, densely staining **red blood cells** with irregular margins and lacking central pallor. They are formed when there is oxidant damage or damage to the red cell membrane by precipitation of an **unstable hemoglobin** or of free α- or β-chains. In drug- or chemical-induced **hemolytic anemia** they are associated with **Heinz bodies**. In films from patients with unstable hemoglobins, a form of **keratocyte** with membrane deficiency is seen ("bite" cells).

ISOANTIBODY
An antibody raised in an individual of the same species from which the antigen is derived. If there are allotype determinants, **alloantibodies** are generated. Otherwise, such antibodies will in effect be anti-self antibodies and normally will only be generated if tolerance is broken, e.g., by the use of powerful adjuvants.

ISOIMMUNIZATION
See also **Hemolytic disease of the newborn; Neonatal alloimmune thrombocytopenia.** A state induced by antigens that are "foreign" to the subject. It is caused by transfused cells, either by donation or by fetal cells crossing the placenta and giving rise to maternal antibodies.

ISOPROPANOL PRECIPITATION TEST
See **Hemoglobinopathies; Unstable hemoglobins.**

J

JEJUNAL DISORDERS
See **Intestinal tract disorders**.

JOB'S SYNDROME
A rare disorder characterized by recurrent infection, eczema, raised serum IgE, and **eosinophilia**. Bacterial, fungal, and viral infections can occur, with staphylococcal skin abscesses and respiratory tract (both upper and lower) infections typical. The **leukocytes** have defective chemotaxis, but in other respects are functionally normal. Prophylactic antibiotics have been shown to be beneficial.

JOINT DISORDERS
The disorders of joints in hematological disease.

Transient polyarthritis occurring in:

- **Sickle cell disorders** (SCD) with ischemia of periarticular structures
- **Thalassemia**, which in the long term induces osteoporosis, resulting in crush fractures and secondary spondylosis
- Hematological malignancy of all types

Acute dactylitis in SCD due to infarction of the carpal and tarsal bone marrow of phalanges

Hemarthroses of **hemophilia A**

Joint pain with **acute lymphoblastic leukemia**, particularly in children

"Frozen shoulder" due to **amyloidosis** affecting the glenohumeral articulation

Joint effusion, sometimes hemarthrosis, in chronic SCD

Autoimmune disorders: ankylosing spondylitis, relapsing polychondritis, rheumatoid arthritis, systemic lupus erythematosus

JORDAN'S ANOMALY
An autosomally recessive hereditary disorder in which cytoplasmic vacuolation of **granulocytes** (and, to a lesser extent, **lymphocytes**) due to dissolution of lipid is seen as described by Jordan in 1953. In the two families described to date, one had hereditary muscular dystrophy and the other had hereditary ichthyosis.

JUVENILE MYELOMONOCYTIC LEUKEMIA
(J-MML) A rare clonal disorder of **hematopoiesis** affecting stem cells and characterized by proliferation, principally of cells of granulocytic and monocytic lineage. The cells are

Philadelphia (Ph) chromosome-negative and there are no *BCR/ABL* fusion genes. It is etiologically, morphologically, and clinically distinct from classical adult **chronic myelogenous leukemia** (CML) and must be differentiated from a small number of children (usually age 5 or greater) who have adult-type CML, including Philadelphia (Ph) chromosome positivity.

Clinical Features

Peak incidence is at age 1 to 2 years, about 10% being associated with neurofibromatosis type-1 (NF-1), suggesting a congenital origin. Presentation is with failure to thrive, fever, or recurrent infections. **Hepatomegaly** and **lymphadenopathy** are common. Dermatitis (typically facial) or other skin manifestations — e.g., xanthomata, café au lait spots — occur in over half the cases. Sequelae of **thrombocytopenia** or **neutropenia** may also be evident.

Laboratory Features

Peripheral blood counts show **neutrophilia** (20 to 200 $\times$ 10^9/l) with left shift, **basophilia** and **eosinophilia**, **neutrophil alkaline phosphatase score** (NAP score) low, and marked **monocytosis** (1 to 100 $\times$ 10^9/l). Occasional blasts and nucleated red blood cells are often seen. **Anemia** is present in virtually all cases, with a raised **hemoglobin F** (perhaps up to 85% of all Hb) occurring in two-thirds of patients. Other features of fetal erythropoiesis, e.g., absent I antigen and low HbA2, may also be found. The platelet count is variable, but usually low. *In vitro* colony assays usually show monocytic growth. **Bone marrow** cytology reveals myeloid hyperplasia with reduced erythroid and megakaryocytic activity.

Treatment and Prognosis

Prognosis is extremely poor, with most patients dying within one year from bleeding or overwhelming sepsis. Typically, there is resistance to **cytotoxic agent** therapy, but some patients have responded to 6-mercaptopurine and cytosine arabinoside. **Allogeneic stem cell transplantation** is the treatment of choice.

JUVENILE PERNICIOUS ANEMIA

See **Cobalamin**.

K

KALLIKREIN

An active two-chain protein that is involved in **contact activation** of blood coagulation. It is produced by the catalytic cleavage of its single-chain precursor, **prekallikrein**. The active site in kallikrein is located in the light chain (molecular weight 33,000 or 36,000), while its **kininogen**-binding site is located in the heavy chain (molecular weight 52,000). Kallikrein activates factor XII to factor XIIa, a reciprocal of the prekallikrein-to-kallikrein activation mediated by factor XIIa. Kallikrein also releases a small vasoactive peptide, **bradykinin**, from high-molecular-weight kininogen. In addition, kallikrein can participate in mediating several aspects of acute **inflammation**.

KAPOSI'S SARCOMA

(KS) A multicentric hemorrhagic tumor of the skin, with or without visceral involvement. Atypical endothelial cells line the vascular channels, which are distended by blood or organized thrombi. It is a disease of homosexual and bisexual men and of African men and women, rare among hemophiliacs, drug abusers, and recipients of infected blood transfusions, but often associated with **immunodeficiency**. It is due to transmission of human herpes virus-8 (HHV-8), which induces the *GPCR* gene to express Ephrin B2. This protein induces growth of **vascular endothelium** and **angiogenesis** by increasing the activity of **heme oxygenase** (HO-1), probably through the vascular endothelial growth factor (VEGF). It commonly presents as a violaceous skin lesion, although lesions affecting the mucous membranes, e.g., hard palate, are not uncommon. Visceral disease may occur, affecting the lungs and gut, and patients may therefore present with hemoptysis or gastrointestinal bleeding. Excision, local radiotherapy, particularly with retinoids, and local intralesional chemotherapy may be beneficial. In cases with symptomatic visceral disease, systemic **cytotoxic agent** chemotherapy with single agent doxorubicin, bleomycin, vincristine, or liposomal doxorubicin may be of benefit, but patients frequently relapse unless treatment is given continuously. Inhibition of angiogenesis by captopril has been claimed,[322] as has human chorionic gonadotropin.[323] No known treatment is ever curative.

KARYORRHEXIS

The breakdown of nuclear **chromatin**. This is particularly seen during **erythropoiesis** at the time of development from an orthochromatic erythroblast to a mature red blood cell. The nuclear chromatin becomes condensed, pyknotic, and moves to the side of the cell and is later extruded through the cell membrane.

KASABACH-MERRIT SYNDROME

See **Hemangiomas**.

KAWASAKI DISEASE

(Infantile polyarteritis nodosa; mucocutaneous lymph node syndrome) An acute **vasculitis** of infants and young children presenting with fever, erythema of skin and mucous membranes, desquamation of the hands and feet, and cervical **lymphadenopathy**. Other hematological changes are **thrombocytosis** and a raised **erythrocyte sedimentation rate**. If untreated, the disorder can lead to coronary artery aneurysms or ectasia-causing coronary artery thrombosis and myocardial infarction. It is treated with **aspirin** and intravenous **immunoglobulin**, 2 g/kg in a single infusion.[324,325]

KELL BLOOD GROUPS

A specific antigen–antibody system located on red blood cells (see **Blood groups**).

Genetics and Phenotypes

Twenty-five antigens have been described within the Kell (KEL) blood group system. The most important of these are the K (Kell) and k (Cellano) **antigens**, regulated by the *K* and *k* genes. There are four other sets of antithetical antigens: Kpa/Kpb/Kpc, Jsa/Jsb, K11/K17, and K14/K24. The Jsa antigen is found on the red cells of about 20% of black donors, but is very rare in white donors. Some phenotypes and genotypes are listed in Table 89.
   Other phenotypes include:

Ko: absence of all Kell system antigens in a rare recessive trait. Immunized persons may produce anti-Ku, reacting with normal red blood cells (RBCs) but not with their own RBC.

McLeod: an X-linked disorder of males, all Kell antigens being expressed weakly. Mild congenital **hemolytic anemia** ensues, with **acanthocytes** and neuromuscular abnormalities with elevated serum creatinine phosphokinase. It is sometimes associated with **chronic granulomatous disease** because of the deletion of a common part of the X-chromosome. Female carriers have a dual population of normal and acanthocytic RBCs.

TABLE 89

Common Phenotypes and Genotypes of the Kell Blood Group System

Phenotype	Genotype	Frequency in Caucasians (%)
K⁺k⁻	KK	0.2
K⁺k⁺	Kk	9
K⁻k⁺	kk	91

Antibodies and Their Clinical Significance

Usually immune in origin, and of the IgG class, although saline-reactive IgM naturally occurring antibodies are sometimes encountered.

Found in about 29% of transfusion recipients who have antibodies other than anti-D. Anti-K is the most common Kell antibody specificity, and is the most common immune red cell antibody specificity outside the ABO and Rh systems. Anti-k and anti-Kpa are the next most common Kell system antibodies found in Caucasians, but are much less common than anti-K.

Persons of the very rare phenotype Ko lack expression of all Kell antigens on red cells and can make anti-Ku, which reacts with all red cells except those of the Ko phenotype.

Kell antibodies can give rise to severe hemolytic blood transfusion complications (see **Blood components for transfusion** — complications) if incompatible red blood cells are transfused. These complications include delayed as well as immediate hemolytic transfusion reactions.

The selection of blood for patients with anti-K, anti-Kpa, or anti-Jsa is not difficult because of the low incidence of the corresponding antigens in the general donor population. However, selection of blood for patients with the much less common anti-k, anti-Kpb, or anti-Jsb will be problematic because of the high incidence of the antigens. **Autologous blood transfusion** should be considered.

RBCs that are K negative are often chosen for patients who are likely to receive multiple transfusions, because of the relatively high immunogenicity of the K antigen.

Kell antibodies can cause severe **hemolytic disease of the newborn**. The cause of the disease appears to be different from that caused by other RBC antibodies in that fetal anemia is caused by suppressed erythropoiesis rather than by hemolysis of fetal RBCs. Antibody titer does not correlate well with the severity of the disease.

Pretransfusion testing may fail to detect the presence of anti-Kpa, anti-Jsa, and other antibodies to low-frequency Kell system antigens, as the RBCs used in testing protocols are often negative for these antigens.

KERATOCYTE

(Horn cells) Fragmented **red blood cells** with pairs of spicules formed by fusion of opposing membranes to form a pseudovacuole giving the appearance of a half moon. They occur as a result of mechanical damage by prosthetic heart valves or by fibrin causing **disseminated intravascular coagulation.**

KERNICTERUS

The deposition of bilirubin in the basal ganglia of the brain. At autopsy, this deposition appears as yellow staining. It occurs when severe jaundice develops in infants, usually due to **hemolytic disease of the newborn**. Seventy percent of infants with kernicterus die between days 2 to 5 of life, and survivors have permanent cerebral damage, which is manifests as choreo-athetosis and spasticity. High-frequency deafness may be the only sign.

KIDD (JK) BLOOD GROUPS

A specific antigen–antibody system located on red blood cells (see **Blood groups**).

Genetics and Phenotypes

Three **antigens** have been described within the Kidd (JK) blood group system (see Table 90). The Jka and Jkb antigens are regulated by the *Jka* and *Jkb* genes.

TABLE 90

Phenotypes and Genotypes of the Kidd (Jk) Blood Group System

Phenotype	Genotype	Frequency in Caucasians (%)
Jk(a⁺b⁻)	Jka Jka	28
Jk(a⁺b⁺)	Jka Jkb	49
Jk(a⁻b⁺)	Jkb Jkb	23

Antibodies and Their Clinical Significance

Always immune in origin, and usually of the IgG class. They are often **complement binding**.

These are found in about 4% of transfusion recipients who have antibodies other than anti-D, but more often in patients who have had an immediate or delayed hemolytic blood transfusion complication (see **Blood components for transfusion** — complications). Anti-Jka is the most common Kidd antibody specificity.

Because of the relatively high frequency with which Kidd antibodies are associated with hemolytic transfusion reactions, it is important that antigen-negative, cross-match-compatible blood be provided if Kidd antibodies are present or have previously been detected in the patient's plasma. Kidd antibodies that are only weakly detectable in **pretransfusion testing** are often capable of causing delayed hemolysis if antigen-positive blood is transfused. Once anti-Jka or anti-Jkb has been identified, the selection of antigen-negative blood should not be difficult unless additional antibodies are present.

In patients who are likely to receive multiple transfusions, consideration should be given to the selection of blood that is matched for Jka and Jkb antigens.

Kidd antibodies are rarely implicated in **hemolytic disease of the newborn**.

It is important to use **pretransfusion testing** methods that are sufficiently sensitive to facilitate the detection of weak Kidd antibodies. Although polyspecific anti-globulin reagents containing anti-complement may give improved detection of some Kidd antibodies, low-ionic-strength **indirect antiglobulin (Coombs) test**, **column agglutination**, and **solid-phase techniques** using anti-IgG only are sufficiently sensitive for routine use. There is general agreement that the use of screening red blood cells having homozygous expressions of the Kidd antigens is highly desirable.

KININOGENS

Precursors of the active kinins **bradykinin** and lysyl-bradykinin. They exist in plasma in two distinct molecular-weight forms: high-molecular-weight kininogens and low-molecular-weight kininogens. High-molecular-weight kininogen is the nonenzymatic cofactor for prekallikrein in the contact-factor phase of blood coagulation. Its function is to act as a surface-binding protein to link **prekallikrein** to negatively charged surfaces. High-molecular-weight kininogen (MW 120,000) also binds **factor XI** and presumably also functions to localize factor XI. Low-molecular-weight kininogens have no known role in blood coagulation. Kininogens carry the bradykinin moieties as the central part of their polypeptide chains. Deficiency of high-molecular-weight kininogen is associated with significant prolongation of the **activated partial thromboplastin time** but with few clinical consequences.

KLEIHAUER TEST
See **Hemolytic disease of the newborn**.

KOSTMANN'S SYNDROME
A rare autosomally recessive condition, presenting within the first month after birth with severe **neutropenia** and recurrent infections (otitis, gingivitis, pneumonitis, enteritis, and peritonitis have all been reported). All other cell lineages appear unaffected, but **monocytosis** and **eosinophilia** may be present. Mild **splenomegaly** has been reported. Immune assessments are usually normal, with occasional cases showing hypergammaglobulinemia. **Bone marrow** cellularity is variable, with no maturation beyond the myelocyte stage. The underlying defect is unknown, and no chromosomal abnormality has been described. Although spontaneous remissions have been reported, the condition, if untreated, is usually fatal early in life, either due to overwhelming sepsis or the development of **acute myeloid leukemia** (typically with a monosomy 7 karyotype). Infections should be vigorously treated with broad-spectrum antibiotics. Recently, administration of granulocyte **colony stimulating factor** (G-CSF) has been shown to benefit many patients by increasing neutrophil counts with a reduction of infections. Macrolide-based antibiotic prophylaxis is common policy in these cases. **Allogeneic stem cell transplantation** has been successfully tried in early cases with severe repeated infections.

KRINGLES
Domains found in many proteins concerned with **fibrinolysis** and named after a Danish breakfast roll. They have a specific structure consisting of ≈80 amino acids held together by three disulfide bonds. Proteins may contain one or more kringle domains, e.g., plasminogen has four kringles, whereas tissue plasminogen activator (t-PA) has only two. The kringle-1 domain in plasminogen is responsible for binding to fibrin.

KÜPFFER CELL
See **Histiocytes**.

L

LABILE FACTOR
See **Factor V**.

LACTATE DEHYDROGENASE
(LDH) An enzyme that catalyzes the reversible reduction of pyruvate to lactate by (nico-tinamide-adenine dinucleotide) NADH in the Embden–Meyerhof pathway of **red blood cell** (RBC) metabolism. Five isoenzymes exist, of which 1, 2, and 3 are the predominant forms. Hereditary deficiency of this enzyme is a benign condition not producing any clinical features. RBCs are rich in lactate dehydrogenase, so levels are increased when there is increased **hemolysis**. The **reference range** in health is 300 to 600 IU/l, whereas in hemolytic anemia, levels in excess of 1000 IU/l are found, but without abnormality in other liver enzyme levels.

LACTOFERRIN
(Lf) A 78-kDa iron-binding glycoprotein homologous to **transferrin**. Lactoferrin belongs to a family of iron-binding proteins that modulate iron metabolism, **hematopoiesis**, and immunological reactions; all of these proteins are encoded by genes on 3q. Lactoferrin is found in large quantities in **neutrophil** granules, milk, and tears. Like transferrin, lacto-ferrin binds iron in a 2:1 molar ratio, but unlike transferrin, lactoferrin maintains iron binding at low values of pH. Iron-free lactoferrin (apolactoferrin) is secreted into plasma with **inflammation**, where it becomes bound to specific receptors on **histiocytes** (mac-rophages) and thus functions as a bactericidal agent. Increased lactoferrin-receptor activity results in cellular-iron overload, which can arise in nervous tissue.

Deficiency of Lactoferrin
A rare congenital neutrophil disorder of unknown inheritance pattern, in which specific cytoplasmic granules are few, and those present are devoid of lactoferrin. It results in lack of neutrophil response to chemotactic signals and inability of the cell to reduce surface-charge density, with resulting diminution of adhesiveness to the surfaces of particles. Hence, there is disturbance of phagocytosis and release of microbicidal O_2 metabolites. Clinical features are recurrent pyogenic infections, particularly deep-seated skin abscesses. Management entails the use of palliative antimicrobial drugs and prevention of infection.

LANGERHANS CELL
See **Dendritic reticulum cells**.

LANGERHANS CELL SARCOMA

A rare neoplastic proliferation of **Langerhans cells** with overtly malignant cytologic features. It can progress from **Langerhans cell histiocytosis**, with the cells having the same phenotype. There is multiorgan involvement of lymph nodes, spleen, liver, lung, and bone. It is an aggressive tumor with short survival time.

LANGERHANS HISTIOCYTOSIS

(LCH) See also **Histiocytosis**.

Neoplastic proliferation of Langerhans **dendritic reticulum cells**, with expression of CD1a, S-100 protein, and presence of Birbeck granules seen upon ultrastructural examination. The cells are large and oval, with voluminous cytoplasm and grooved or lobulated nuclei. The cytoplasmic Birbeck granules have a "tennis racket" shape, about 200 to 400 nm long and 33 nm wide.

Systemic LCH

Although historically the systemic Langerhans cells proliferation (Hand-Schuller-Christian disease) has been regarded as a separate entity from eosinophilic granuloma, there is an overlap, and the two conditions probably represent a continuum. The term "histiocytosis X" was used by Lichenstein in 1953 to group eosinophilic granuloma, Hand-Schuller-Christian disease, and Letterer-Siwe syndrome.

Adults

A slow progressive disease where systemic proliferation of **histiocytes** is usually accompanied by T-lymphocytes and **eosinophils** and is associated with a "cytokine storm" of GM-CSF, interferon-γ, IL-2, IL-4, IL-5, IL-7 IL-10, and TNF-α. The disorder is more common in Hispanics than Caucasians. It presents with multiple bony lesions of the skull, spine, and long bones; gingival hypertrophy; scaly erythema and popular rashes; ulcerative intertrigo; and classically with diabetes insipidus and exophthalmos. Invasion of the pituitary may also result in growth-hormone deficiency. A "high risk" form involves the liver, spleen, lungs, and bone marrow. Treatment is with **cytotoxic agents**, such as corticosteroids, 6-mercaptopurine, methotrexate, or vinblastine, but no evidence-based recommendations are available.

Children

Aggressive acute systemic histiocytosis (previously called Letterer-Siwe histiocytosis or reticulocytosis) presents in infancy with seborrheic dermatitis, **lymphadenopathy**, **hepatosplenomegaly**, and bony lesions. **Thrombocytopenia**, due to bone marrow infiltration or splenomegaly, can render the rash hemorrhagic.

Local LCH

Bone

A benign proliferation of Langerhans cells, classically affecting bones, with skull, femur, and ileum the commonest sites (eosinophilic granuloma). Histology is variable, with a mixture of Langerhans cells, macrophages, fibroblasts, lymphocytes, and eosinophils. The disease usually begins in childhood, presenting as mild discomfort in the affected bone. Usually, only one bone is affected, but multifocal disease can occur. Radiographs of the affected bone show a "punched out" lesion. Organ involvement, e.g., of the central nervous system (CNS), occurs by contiguous spread from bony lesions.

Lungs

Pulmonary eosinophilic granuloma lesions are confined to the lungs, but the histologic changes are similar to systemic histiocytosis. It is usually a mild sudden respiratory disorder of young males associated with smoking. Radiographs reveal a diffuse small nodular infiltrate with small cystic areas creating a honeycomb appearance. Longstanding disease leads to pulmonary tissue destruction, but the lesions usually resolve spontaneously, leaving only fibrosis. Diagnosis is only by lung biopsy, with electron microscopy demonstrating typical Langerhans cells. Corticosteroid therapy may hasten recovery.

Skin

Congenital self-healing histiocytosis (CSHH) is a rare form of histiocytosis, present at birth, with solitary or multiple skin nodules. There is an equal sex occurrence. The lesions undergo spontaneous regression at 3 to 4 months and do not usually recur. The prognosis depends on the extent of the systemic infiltration, with eosinophilic granuloma being highly curative. Remission in most patients with systemic disease is achieved with curettage or radiotherapy and with single or multidrug chemotherapy (corticosteroids, etoposide). Relapse may occur, and multiple courses of chemotherapy may be necessary.

LARGE ANAPLASTIC B-CELL LYMPHOMA

See **Anaplastic large-cell lymphoma**; **Non-Hodgkin lymphoma**.

LARGE-CELL IMMUNOBLASTIC LYMPHOMA

See **Diffuse large B-cell lymphoma**; **Primary mediastinal (thymic) large B-cell lymphoma**; **Non-Hodgkin lymphoma**.

LARGE GRANUALR LYMPHOCYTIC LEUKEMIA, NK-CELL TYPE

See **Aggressive NK-cell leukemia**.

LAZY LEUKOCYTE SYNDROME

A rare, poorly characterized condition presenting with **neutropenia**, a failure to develop neutrophilia when infected, and impaired **neutrophil** mobility upon testing. Presentation is usually in childhood, with recurrent bacterial infections. The **bone marrow** is cellular, with normal granulocyte maturation. Various abnormalities of **actin** filaments or response to chemotaxins have been described.

LEAD TOXICITY

(Plumbism; saturnism) The hematological effects of raised levels of lead in the body.[326] Being a natural component of earth, a small amount of lead is present in most humans. The body content is reflected and expressed by the quantity present in blood, which is normally 10 to 15 µg/dl. This rises with atmospheric pollution up to 30 µg/ml and in exposed workers up to 60 μg/ml. Lead toxicity usually arises in occupations such as lead smelting, battery manufacture, brass founding, radiator repair, paint stripping (especially when power drivers are used), construction work, and pottery and ceramics manufacture, especially when associated with poor industrial hygiene. It is now more common in industry of developing countries. Historically, lead poisoning occurred with wine drinking, but it is now associated with the sport of gun use at shooting galleries. Children are at risk from flaking paint, lead-soldered cans, lead-glazed toys and ceramics, and environmental exposure to lead-contaminated soil as well as airborne lead from industrial emissions and leaded gasoline fumes.

Clinically, the acute toxic effect of lead is encephalopathy, and the chronic disorder is either intermittent abdominal colic or peripheral neuropathy. These effects occur with blood levels above 80 μg/dl, but hematopoietic effects occur with levels above 50 μg/ml, and sometimes even as low as 20 μg/dl. Lead toxicity is due to its high affinity for sulfhydryl (–SH) groups, which inhibit the activity of 5-aminolevulinic acid (ALA) dehydrogenase and ferrochelatase (see **Heme**). This results in increased amounts of ALA in urine and protoporphyrin in circulating **red blood cells** (RBCs). The suppression of ferrochelatase impairs the transport of iron through the mitochondrial membrane; consequently, the protoporphyrin chelates with zinc instead of with iron, leading to increased zinc protoporphyrin. A lead **porphyria** ensues. Protoporphyrin levels rise with coproporphyrinogen III accumulating in the cytosol, but most is excreted as the oxidized derivative coproporphyrin as well as some uroporphyrin. Impaired globin synthesis occurs as a consequence of both impaired heme synthesis and deranged amino acid transport. As a result of this combined effect, hemoglobin synthesis is inhibited, giving rise to a microcytic anemia. **Iron overload** occurs in the bone marrow, presenting as erythroid hyperplasia, **sideroblastic anemia**, and hyperferrinemia.

Lead also disturbs the RBC membrane, with inhibition of Na-K-ATPase and consequent leakage of potassium. Cell shrinkage occurs with increased mechanical and **osmotic fragility** and reduced RBC life span, leading to a **reticulocyte count** of 10 to 15% of RBC. Acute severe lead toxicity can cause intravascular hemolysis with grossly distorted RBCs.

Finally, lead also interferes with the breakdown of **RNA** by inhibiting the enzyme **pyrimidine 5'-nucleotidase**, causing accumulation of denatured RNA in red cells and the appearance of **basophilic stippling** upon Romanowsky staining. The changes in blood are summarized in Table 91.

Upon suspicion of lead toxicity, the person must be immediately removed from exposure. For those with mild toxicity, this may be sufficient, but for those with more severe effects or a blood lead level of over 70 μg/ml, treatment with chelating agents such as penicillamine, dimercaprol (BAL), or ethylenediamine tetraacetic acid (EDTA) — given as the calcium disodium salt (sodium calcium edetate), either singly or in combination — is

TABLE 91

Summary of Hematological and Biochemical
Changes Induced by Lead Toxicity

Cellular	microcytic anemia
	basophilic stippling
	reticulocytosis, 10 to 15%
	red cell protoporphyrin raised
	ring sideroblasts
Biochemical	raised serum iron
	raised urine 5-aminolevulinic acid
	raised urine porphobilinogen
	raised urine coproporphyrin III

required. Repeated courses may be necessary due to mobilization of stored lead. With chelation therapy, urinary levels of ALA and coproporphyrin return to normal within days, but the hemoglobin level takes about 1 month to recover.

LE CELL

See **Lupus erythematosus cells**.

LECTINS

Proteins or glycoproteins of nonimmune origin that have specificities for carbohydrate structures. Some lectins have specificity for carbohydrate **blood group** determinants on the red blood cell membrane. These lectins can be used in direct-agglutination blood-grouping methods. The most useful lectins in blood group serology are:

Anti-A1, derived from the seeds of *Dolichos biflorus* (specificity for N-acetylgalac-tosamine; see **ABO(H) blood groups**)

Anti-H, derived from the seeds of *Ulex europaeus* (specificity for L-fucose; see **ABO(H) blood groups**)

Anti-T, derived from the peanut *Arachis hypogaea* (specificity for D-galactose; see **Polyagglutinability**)

Anti-Tn, derived from the seeds of *Salvia sclarea* (specificity for N-acetylgalactosamine; see **Polyagglutinability**)

LEISHMANIASIS

(Kala-azar) The pathogenesis, clinical and laboratory features, and treatment of infection by the protozoa genus *Leishmania donovani*. Visceral leishmaniasis is endemic in northern China, northeast India, east Africa, the Middle East, and eastern South America. It is transmitted by bites from the female sandfly of *Phlebotomus* spp., which resides on dogs, jackals, wild Canidae, and humans. Upon entry through the skin, *L. donovani* is in the flagellate (promastigote) form, which transforms to a round (amastigote) Leishman-Donovan body of 2-μm diameter. The protozoa pass from the circulation into **histiocytes** (macrophages) of the liver, spleen, and bone marrow, within which they multiply as mastigotes. Release of mastigotes from these cells leads to invasion of other macrophages, the disease becoming systemic in 4 to 10 months. **Immune complexes** develop, with **complement** deposition on red blood cells.

Clinical Features

Initial features are skin lesions at the site of the bite, with local scarring. There is then an incubation period of 3 to 12 months without symptoms. Clinical kala-azar presents with **hepatosplenomegaly** and fever, which persists and progresses to **amyloidosis**, wasting, congestive cardiac failure, edema, and renal failure. Mortality is 90% if untreated.

Laboratory Features

Hypergammaglobulinemia with IgG/IgM 3 to 6 g/dl; circulating immune complexes; **cold agglutinins**; antibodies to rheumatoid factor; antibodies specific to *L. donovani* can be demonstrated by immunofluorescence or enzyme-linked immunosorbent assay (ELISA). Serological tests show high sensitivity.[326a]

Pancytopenia

Autoimmune hemolytic anemia: positive **direct antiglobulin (Coombs) test**

Bone marrow biopsy: macrophages with Leishman-Donovan bodies, which can be cultured to develop the flagellate form

Severe **anemia** in late stages, with raised levels of erythropoietin[327]

Treatment

An organic pentavalent antimony compound, sodium stibogluconate, 20 mg/kg daily by intravenous or intramuscular injection for 7 to 30 days. This is successful in 90% of cases. Alternative therapies use liposomal amphotericin B or pentamidine isethionate. Oral miltefosine is under investigation. There are no prophylactic measures.

LENALIDOMIDE

A thalidomide analog used in the treatment of **myelodysplasia**. It has the advantage of avoiding neurologic toxicity. It has been found to be particularly effective in patients with a pure erythroid form of the disorder. It is under trial for treatment of **myeloma** and chronic idiopathic myelofibrosis.

LEPTOCYTE

(Thin cells) Flat **red blood cells** of normal volume and hemoglobin concentration, **hemoglobin** being located at the periphery. They are seen in blood smears from patients with **thalassemia**, **biliary disorders**, and **iron deficiency**.

LESCH-NYHAN SYNDROME

A sex-linked hereditary disorder that causes deficiency of the enzyme hypoxanthine-guanine phosphoribosyltransferase (HGPRT), leading to failure of **nucleotide** synthesis. It is characterized by hyperuricemia and neurological disorder with choreoathetosis, spasticity, mental retardation, and often self-mutilation. There is growth failure, sometimes **megaloblastosis**, and impaired **B-lymphocyte** function. The only available treatment is **allopurinol** to remove excess purines. In the future, **gene therapy** may become available.[328]

LETTERER-SIWE HISTIOCYTOSIS

See **Histiocytosis**.

LEUCOVORIN

See **Folinic acid**.

LEUKEMIA

See also **Acute leukemias**; **Lymphoproliferative disorders**, **Myeloproliferative disorders**. Aberrant **hematopoiesis** initiated by rare leukemic stem cells that have maintained or reacquired the capacity for indefinite proliferation through accumulated mutations or epigenetic changes.

Leukemic cells have certain defined characteristics:

Monoclonal origin

Acquired gene mutations

Genetic instability, clonal diversification, and subclonal selection

Dysregulation or uncoupling of certain cellular factors, proliferation, differentiation, and cell death

Net growth advantage, clonal dominance, and vascular and extravascular spread with compromising of normal tissue fractions

LEUKEMOGENESIS

Leukemogenesis encompasses the causes and mechanisms that lead to the development of the different types of **leukemia**.[329] In contrast to solid tumors that typically contain point mutations, gene amplifications, or deletions, leukemias are commonly associated with chromosomal translocations that juxtapose two unrelated genes and their products, leading to aberrant expression or function of the fusion protein. These chromosome abnormalities give an insight into key molecular events that are associated with the development of leukemia. Combining these with a greater understanding of how congenital/hereditary leukemias develop has further highlighted potential molecular pathways. The common karyotypic abnormalities and molecular/functional correlates are given in Table 92.

Studies of human **acute myeloid leukemia** (AML) samples indicate that most translocations involve transcription factors or components of the transcriptional activation complex (see **Cytogenetic analysis**). Recurring examples include translocations involving both subunits (*AML1 and CBFβ*) of core-binding factor (CBF), *MLL* gene rearrangements, *HOX* family members, and the coactivators CBP and p300. Specifically, the following chromosomal translocations involving transcriptional elements have been detected in and cloned from human AML samples: *CBFβMYH11, MLLCBPMLLENL, Nup98HOXa9* (*Nup98HOXd13*), and, frequently, *AML1ETO*. Significantly, when these oncogenic translocations are expressed in murine bone marrow experimental models, a myeloid leukemia is usually not evident for 6 to 12 months. In these models of leukemia, this is taken as evidence to support the requirement of additional mutations to cause full acute leukemic disease.

The study of hereditary retinoblastoma led Knudson to propose the "two-hit hypothesis" for cancer. This suggests that most somatic tumors require at least two mutations (e.g., one germline and one somatic, or two somatic mutations) for the development of **malignancy**. Some genetic events are associated with dominant mutations in two genes, e.g., a t(8;21) translocation with mutations in the kinase domain of c-kit, or recessive mutations in two alleles of a single gene. This is the case for the tumor-suppressor genes, e.g., *RB*,

TABLE 92

Common Karyotypic Abnormalities and Molecular/Functional Correlates

Disease	Karyotypic Abnormality	Gene Implicated	Gene Product Function
Chronic myeloid leukemia	t(9;22)(q34;q11)	*BCR/ABL*	tyrosine kinase
Acute myeloblastic leukemia	t(8;21)(q22;q22)	*ETO*	transcription factor
Acute myeloblastic leukemia	t(15;17)(q22;q12)	*PML/RAR-a*	fusion protein
Acute myeloblastic leukemia	inv(16)(p13;q24)	*CBFβ*	transcription factor
T-cell acute lymphoblastic leukemia	t(11;14)(p32;q11)	*RHOM2*	"LIM" protein
T-cell acute lymphoblastic leukemia	t(10;14)(q24;q11)	*HOX11*	homeobox factor
B-cell lymphoma	t(14;18)	*BCL-2*	anti-apoptosis
B-cell lymphoma	t(11;14)(q13;q32)	*PRAD1/cyclin D1*	cell cycle; G1

P53, WT1, NF1. Mutations of both alleles of the tumor-suppressor genes lead to genetic instability.

For example, *TP53* mutations result in a failure of **apoptosis**. Normally, **DNA** lesions are repaired by processes such as recombinational repair, which leave the chromosome intact. When the induction of this repair is compromised, apoptosis ensues. This process fails in the presence of *TP53* mutations, and florid karyotypic changes emerge abruptly. This is the state of chromosomal instability (CIN).

A second form of genomic instability — mutational microsatellite instability (MIN) — is not associated with CIN but with mutations in mismatch repair (MMR) genes such as *MSH2* or *MLH1*. These have been shown to occur in AML and may be an early event in the development of leukemias. Thus the accumulation of CIN and MIN confer unique properties to a cell that lead to a progression from a normal state to malignancy. These accumulated genetic alterations include, but are not limited to, independence in growth-factor signaling (e.g., mutations in the kinase domain of *FLT3* or *c-kit*), escape from apoptosis, and an endless ability to replicate.

The current "two-hit" model of AML cites mutation in a cellular kinase that results in a constant growth-promoting signal and mutation in a hematopoietic transcription factor that leads to disrupted developmental potential as the important elements driving leukemogenesis. Clinically, support for the acquisition of cooperating mutations in the development of AML is well documented. In nearly all cases of **chronic myelogenous leukemia** (CML), a translocation exists that constitutively activates a **tyrosine kinase**. Common translocations associated with CML include *BCR/ABL, TEL/ABL, TEL/PDGFβR,* and *TEL/ JAK2.* Expression of these constitutively active kinases results in increased proliferation or survival of affected cells without a block in cellular differentiation. It is the subsequent gain of mutations involving transcription factors that provides the differentiation block and causes the onset of acute disease. Examples of CML progression to AML (or CML blast crisis) provide supporting evidence for this hypothesis. Cases of CML patients with *BCR/ABL*-positive disease (activated kinase) progressing to acute leukemia with ensuing acquisition of either the *Nup98HOXA9* or *AML1ETO* translocations (transcription factors) are documented. The acquisition of *TP53* mutations is also an important event.

Further evidence for the "multihit hypothesis" is derived from the 8;21 chromosomal translocation, which encodes the *AML1ETO* fusion protein. This fusion gene is detectable in the **stem cells** of AML patients, but interestingly, it also remains detectable in stem cells from patients in long-term remission. This indicates that other genetic events, besides the *AML1ETO* translocation, are necessary for progression and lethality of leukemia.

Finally, and probably most compelling, are data obtained in a study of twins who developed AML due to the *TELAML1* translocation. Upon analysis of this unique biological phenomenon, it was determined that one *TELAML1* clone was transmitted between the twins *in utero.* Despite this shared, identical mutation, neither twin was born with leukemic disease. In fact, both twins developed AML later in life, but at greatly differing times. All of these data point to the need for multiple cooperating mutations during leukemogenesis. Similar data are also available for **acute lymphoblastic leukemia** (ALL).

It should be noted, however, that translocations involving tyrosine kinases are rare in acute leukemias. While the *BCR/ABL* fusion is responsible for the majority of cases of CML, similar fusions of activated kinases are not usually found in AMLs. The signal for increased growth and//or survival is derived from point mutations in certain tyrosine kinases that can cause constitutive activation of these growth regulators. In particular, Flt3 tyrosine kinase internal tandem duplications (ITDs) and activating mutations are widely seen; in fact, Flt3 activation is described as the most common alteration in AML. Interestingly, the ITD mutations are related, with poor prognosis in AML, while the point mutations have not been shown to be of prognostic significance.

In addition, up to 5% of all cases of AML contain mutations that cause constitutive activation of the kinase c-kit. So, the presence of one of these active kinases in addition to the commonly seen transcription-factor translocations corresponds well with the two-hit theory of leukemogenesis.

The precise temporal relationship between genetic instability, acquisition of the balanced chromosome abnormalities, and mutations in tumor-suppressor genes remains to be elucidated. Furthermore, it is unclear which event represents the "first hit" for cancer and what are the precise predisposing causes. There is increasing evidence that, in most adult leukemias, the stem cell is the target for some of the early genetic events, but more studies are required to define whether cellular compartments are predisposed to specific genetic lesions and how these impact on differentiation of the different progenitor compartments.

Causes of Leukemia

The causes of a vast majority of leukemias remain unknown. However genetic and epidemiological studies give an insight into some of the known causes and associations and also offer an insight into potential cellular pathways that contribute to leukemogenesis:

Congenital

Bloom's Syndrome

Bloom's syndrome is a rare autosomally recessive inherited disorder in which there is a defect in replication and spontaneous chromosome breakage, resulting in a predisposition to malignancy, including acute myeloid leukemia, acute lymphoblastic leukemia, and **lymphoproliferative disorders**, these being particularly common (20% of affected individuals).

Down Syndrome

Down syndrome is associated with transient **myeloproliferative disorder** (TMD) and acute **megakaryoblastic leukemia** (AMKL). Children with Down syndrome have a 10- to 20-fold elevated risk of developing leukemia. Individuals with Down syndrome-related AMKL have mutations in the *GATA1* gene. One class of mutations imparts a myeloproliferative or survival advantage, as illustrated by activating mutations in *FLT3*, encoding a receptor **tyrosine kinase**, or the increased dosage of genes in chromosome 21 in persons with Down syndrome. To generate overt leukemia, a second class of genetic alterations must produce lineage-specific blocks in differentiation. The mutations responsible for this step have been demonstrated mainly in genes encoding chimeric transcription factors produced by chromosomal translocation. *GATA1* is a transcription factor that plays a pivotal role in myeloid lineage commitment (see above).

As many as 10% of infants with Down syndrome present with TMD at or shortly after birth. TMD is characterized by an abundance of blasts within peripheral blood and liver, and undergoes spontaneous remission in a majority of cases. TMD may be a precursor to AMKL, with an estimated 30% of TMD patients developing AMKL within 3 years. Mutations in *GATA1* play a critical role in the etiology of TMD, and mutagenesis of *GATA1* represents a very early event in myeloid leukemogenesis in Down syndrome.

Familial Platelet Disorder

When associated with myeloid malignancy, this disorder is related to mutations in the *Runx* (AML1) gene. Family studies indicate that those with the familial platelet disorder have a propensity to acute myeloid leukemia.

Schwachman-Diamond Syndrome

The **Schwachman-Diamond syndrome** is characterized primarily by exocrine pancreatic insufficiency, hematological abnormalities (including increased risk of malignant transformation), and skeletal abnormalities. Recently, the α-mutated gene on chromosome 7 has been found in patients with Schwachman-Diamond syndrome. This gene has been named *SBDS* (Schwachman-Bodian-Diamond syndrome). *SBDS* is a member of a highly conserved protein family. Orthologs exist in species ranging from Archaea to vertebrates and plants. Indirect lines of evidence indicated that the orthologs could function in **RNA** metabolism. The predicted protein is 28.8 kDa. The amino acid sequence had no homology to any functional domain known at that time, and no signal peptides were detected. It is suggested that Schwachman-Diamond syndrome may arise from a defect in RNA metabolism. Manifestations of the disease must reflect the loss or perturbation of a cellular function that is critical for the development of pancreatic acini, myeloid lineages, and chondrocytes at growth plates of bones.

Neurofibromatosis Type I

(NF1) This disorder is caused by mutation in the neurofibromin gene. The risk of malignant myeloid disorders in young children with NF1 is 200 to 500 times the normal risk. Several young children with some features of NF1 and hematological malignancies have been identified with homozygous mutations in the mismatch-repair genes *MLH1*. Data from the study of children with myeloid disorders have provided evidence that NF1 may function as a tumor-suppressor allele in malignant myeloid diseases and that neurofibromin is a regulator of *RAS* in early myelopoiesis.

Fanconi Anemia

Fanconi anemia is an autosomally recessive disorder affecting all bone marrow elements and is associated with cardiac, renal, and limb malformations as well as dermal pigmentary changes. Leukemia is a fatal complication and may occur in family members lacking full-blown features.

Ataxia-Telangiectasia

(AT) An autosomally recessive disorder characterized by cerebellar ataxia, telangiectases, immune defects, and a predisposition to malignancy. Chromosomal breakage is a feature. AT-cells are abnormally sensitive to killing by **ionizing radiation** (IR) and abnormally resistant to inhibition of DNA synthesis by ionizing radiation. The latter trait has been used to identify complementation groups for the classic form of the disease. At least four of these (A, C, D, and E) map to chromosome 11q23 and are associated with mutations in the ATM gene. Patients with AT have a strong predisposition to malignancy, in particular lymphomas and leukemias. In general, lymphomas in AT patients tend to be of B-cell origin (B-**chronic lymphocytic leukemia**, B-CLL), whereas the leukemias tend to be of the T-CLL type.

Other Hematological Congenital Conditions

Other hematological congenital conditions with a predisposition to leukemia include **Blackfan-Diamond syndrome**, **Kostmann's syndrome**, and **Chediak-Higashi syndrome**.

Environmental Causes

Many investigators have attempted to correlate environmental risk factors with the development of acute leukemia, but the evidence has not been strong for any one factor. Exposure to ionizing radiation has been shown to be a risk factor for leukemia. Other factors for which there is evidence (some of it weak or lacking) include: residential or occupational nuclear exposure by child or parent to nonionizing radiation, pesticides, **benzene**, **tobacco** (cigarette) smoke, **vitamin K**, cod-liver oil, natural and human-made **topoisomerase II inhibitors**, maternal **alcohol** abuse, dietary nitrates (hot dogs), and drinking water contaminated with trichloroethylene.

Radiation

The association of ionizing radiation and leukemogenesis comes from data derived from the study of individuals exposed to the Hiroshima and Nagasaki atomic bombs, from patients receiving radiation therapy, and from individuals inadvertently exposed to radiation. Less than 5 years after the atomic bombings of Hiroshima and Nagasaki, it was established that there was an excess of leukemia among the atomic bomb survivors. When examined by leukemia subtype, there was a substantial excess of acute forms of leukemia and chronic myeloid leukemia, but no excess of **chronic lymphocytic leukemia** (CLL). A few years later, results of studies of mortality among adult British males who had received X-ray therapy for **ankylosing spondylitis** by Court-Brown and Doll in 1965 and later by Darby and his colleagues in 1987 also found an excess of deaths from acute leukemia and chronic myeloid leukemia but no excess of CLL. These findings indicate that there are differences in the radiogenicity of leukemia by subtype, with CLL being much less readily inducible by exposure to ionizing radiation than other types of leukemia.

Clusters of childhood leukemia have been noted, some in close geographical proximity to nuclear power stations. While this is a controversial area, the generally accepted view is that these clusters are a rare response to a common infection and that population mixing between infected and susceptible individuals in these secluded areas produces localized epidemics of the relevant infection and a consequent increased risk of childhood leukemia. Such an explanation would also account for the increased levels of childhood leukemia that have been found after other instances of pronounced urban/rural population mixing in areas with no enhanced exposure to radiation.

Virus Infection

There is a clear causal relationship between certain infectious agents and leukemia. For example, **human lymphotropic virus-1** (HTLV-1) in **adult T-cell leukemia/lymphoma** (ATLL) and **Epstein-Barr virus** (EBV) in **Burkitt lymphoma** and **posttransplant lympho-proliferative disorders** (PTLD). In addition to these associations with chronic infection, data from childhood leukemia studies have suggested that infectious agents may be involved in the etiology of some leukemias, but no specific agents have yet been identified.

Epidemiological studies have demonstrated that the relative percentage of malignant lymphoid proliferations varies widely according to geographical location and ethnic populations. HTLV-1 is the etiological agent of ATLL and is also associated with **cutaneous T-cell lymphoma** (CTCL). However, a definite role of HTLV-1 in mycosis fungoides (MF)/Sézary syndrome (SS) remains controversial. While most HTLV-1-infected individuals remain asymptomatic carriers, 1 to 5% will develop ATLL, an invariably fatal expansion of virus-infected $CD4^+$ T-cells. This low incidence and the long latency period preceding occurrence of the disease suggest that additional factors are involved in development of ATLL.

The incidence of PTLD after **allogeneic stem cell transplantation** (SCT) is ≈1%, with the majority of cases developing during the first year after transplantation. The incidence of PTLD is significantly increased by risk factors such as

(a) underlying **immunodeficiency**

(b) T-cell-depleted donor cells

(c) the use of closely matched unrelated donors or HLA-mismatched family members,

(d) intensive **immunosuppression** with T-cell antibodies for the prophylaxis and treatment of **graft-versus-host disease** (GVHD).

In contrast to the incidence after T-cell depletion, PTLD has a much lower frequency when both T and B-cells are depleted. This supports the concept that the malignant outgrowth of EBV-positive B-cells is favored by an imbalance between EBV-infected B-cells and EBV-specific T-cells. For example, SCT recipients treated with a monoclonal anti-CD52 antibody (Campath-1), which removes T and B-cells, have a PTLD incidence of <2%. The incidence of PTLD after allogeneic umbilical cord blood transplant is also <2%, and T-cell suppressive therapy for GVHD is an important risk factor. Few cases of PTLD have been described following autologous SCT; risk factors include the use of CD34-selected stem cell products and the use of antithymocyte globulin in patients with **autoimmune disorders**.

Chemical Toxicity

Benzene and its metabolites (derived from cigarette smoke and pollution), bioflavonoids, herbal medicines, anthraquinone laxatives, podophyllin resin, quinolone antibiotics, pesticides, and most phenolic chemicals and their metabolites have some association with leukemias. In particular, the use of DNA topoisomerase inhibitors is associated with the breakage and recombination of the *MLL* gene with one of several potential partner genes. This is a common feature of infant acute leukemia as well as a number of patients with adult leukemia. Because *MLL* rearrangements are frequently seen in adult cases of leukemia following **cytotoxic agent** treatment with topoisomerase II inhibitors, it is possible that *in utero* exposure to topoisomerase inhibitors could also be an important risk factor in infant leukemia. The list of topoisomerase inhibitors is extensive and includes not only chemotherapeutic agents but also benzene metabolites and bioflavonoids. Studies of occupational groups with varied chemical exposures (e.g., farmers, petroleum workers, and rubber workers) have reported excess risk for **non-Hodgkin lymphoma** (NHL), **myelomatosis**, and other **B-cell lymphomas**. While not conclusive, these studies raise questions about the effects of chemical exposures on the lymphocytic versus myeloid cell lines.

LEUKEMOID REACTION

The appearance of a **peripheral-blood film** resembling leukemia from a patient who does not have this disease. This reaction may be granulocytic (more common) or lymphocytic. Many causes have been identified (see Table 93). The **white cell count** may be very high, 50 to $100 \times 10^9/l$, with the patient presenting with features of the underlying cause. The peripheral-blood film may show reactive **neutrophils** with left shift, toxic granulation, and **Döhle body** formation. Anemia may or may not be present, depending on the underlying cause, and a normal platelet count is usual. Bone marrow aspiration often reveals hyperplasia, but seldom to the extent of leukemia. The **neutrophil alkaline phosphatase score** is normal or raised, which is useful in distinguishing myeloid leukemoid reactions from **chronic myelogenous leukemia**.

Pathogenesis

This has only been studied in animal models, and various possible mechanisms have been described. Tumors have been shown to stimulate granulocyte production and release via the production of growth-factor receptors. Lymphocytic leukemoid reactions can be produced by pertussis infections where lymphocyte migration from the circulation is prevented.

TABLE 93

Causes of Leukemoid Reactions

Infections	Myeloid	Lymphocytic
Bacterial	pneumococcal	whooping cough
	meningococcal	tuberculosis
	diphtheria	—
	tuberculosis	—
Viral	—	infectious mononucleosis
	—	cytomegalovirus infections
	—	measles
	—	chickenpox
Malignancy	Hodgkin disease	stomach
	lung	melanoma
	breast	ovary
	adrenal	—
Reactive	eclampsia	eclampsia
	agranulocytosis (rebound)	exfoliative dermatitis
	hemorrhage	drugs
	burns	—

LEUKOAGGLUTININS

Antibodies to **leukocyte**s causing **agglutination** *in vitro*. Typically seen in granulocyte agglutination testing. They may be **human leukocyte antigen** (HLA) or **granulocyte** specific. They may also be found in the serum of some donors implicated in transfusion-related acute lung injury (see **Blood transfusion components** — complications).

LEUKOCYTE

(White blood cells) Colorless amoeboid cells in circulating blood, urine, and other body fluids or tissues. They include **granulocytes**, **monocytes** with **histiocytes** (tissue macrophages), and **lymphocytes**. Any of these cells may be degenerate, particularly when in pus or in an abscess cavity. Like all nucleated cells, their surface carries antigens that are most easily detected on lymphocytes (see **Human leukocyte antigen**).

LEUKOCYTE-ADHESION DEFICIENCY

(LAD) Reduced or absent expression in families of **cell adhesion molecules** CD11a/CD11b/CD11c and CD18, which play an important part in granulocyte migration from the circulation. Inheritance is autosomally recessive, and consanguinity is not unusual. LAD was first described nearly 25 years ago in patients with delayed separation of the umbilical cord, **neutrophilia**, neutrophil defects, and systemic bacterial infections. Leukocyte functional studies confirmed a neutrophil adhesion defect; flow cytometric and protein analysis indicated the absence of the leukocyte integrins on the cell surface; and molecular analysis demonstrated heterogeneous molecular defects in the leukocyte integrin CD18 subunit.

Recurrent bacterial infections are the hallmark of LAD. A severe infection involving the umbilical stump usually represents the initial presentation. Episodes of severe bacterial infection then ensue throughout the perinatal, childhood, and young-adult years. These infections take the form of severe gingivitis and periodontitis as well as recurrent, cutaneous, nonhealing wounds. These infectious episodes in LAD are accompanied by a leukocytosis ranging from 15,000 to 100,000 cells/µl. Despite the marked leukocytosis, the inability of **neutrophils** to migrate to the site of infection in LAD results in the absence of pus at the inflammatory or infectious sites. LAD is inherited in an autosomally recessive manner, and children with LAD are usually compound heterozygotes with a different mutation on each allele of the CD18 gene (also known as the β2 subunit of the leukocyte integrin family). Because the CD11 subunits, or alpha subunits, of this adhesion receptor family require the CD18 subunit for heterodimer formation and surface expression, deficiency of CD18 results in failed, decreased, or aberrant surface expression of the CD11/CD18 complex in LAD. The majority of LAD cases involve single point or missense mutations in the CD18 subunit, leading to an altered CD18 precursor that is unable to dimerize with the CD11 subunits. Prophylactic and therapeutic antibiotics are useful forms of treatment. **Allogeneic stem cell transplantation** (if available) may be curative.

LEUKOCYTE ALKALINE PHOSPHATASE SCORE
See **Neutrophil alkaline phosphatase score**.

LEUKOCYTE ANOMALIES
Abnormalities of leukocytes, mainly of **neutrophils**, that are either stem cell deficiencies of granulocytes, acquired dysplastic changes, or defects of microbiocidal function.

Hereditary Stem Cell Deficiencies

Cyclic hematopoiesis (**cyclic neutropenia**)

Congenital agranulocytosis (**Kostmann's syndrome**)

Neutropenia and pancreatic insufficiency (**Schwachman-Diamond syndrome**)

Pelger-Huët anomaly

May-Hegglin anomaly

Alder-Reilly anomaly

Jordan's anomaly

Hereditary hypersegmentation of neutrophils; normal neutrophils have five or fewer nuclear lobes, but here there may be as many as eight to ten lobes

Hereditary giant neutrophils (autosomally dominant)

Chediak-Higashi syndrome

Acquired Dysplasias

Giant neutrophils: cells that are occasionally seen in infections and a characteristic neutrophil (macropolycyte) of megaloblastosis.

Hypersegmented neutrophils: in megaloblastic anemia, up to ten lobes may be present, this being one of the most sensitive and reliable indicators of megaloblastic anemia. It persists for up to 14 days after the commencement of therapy. Hypersegmentation has also been described in patients with renal tract disorders and iron deficiency. The mechanism by which nuclear segmentation occurs is unclear, as is its purpose, and therefore the pathogenesis of hypersegmentation in megalo-blastic anemia is also unknown.

Toxic vacuolation: vacuolation of neutrophils is usually the result of bacterial infections, but can also occur with ethanol abuse and diabetic ketoacidosis. The vacuoles contain fat or remnants of phagocytosed bacteria. Similar changes are seen in neutrophils after prolonged storage in EDTA (ethylenediaminetetraacetic acid).

Toxic granulation: an increase in the number and size of azurophilic (primary) granules in neutrophils usually indicates bacterial infection, and other neutrophil changes, such as **Döhle bodies** and vacuolation, may be seen. Staining at pH 5.4 with Giemsa makes the granules more prominent. Several granules may fuse to form secondary lysosomes.

Döhle bodies.

Pelger-Huët anomaly.

May-Hegglin anomaly.

Defects of Microbicidal Function

There are large numbers of extremely rare disorders that predispose to pyogenic infection, the more common anomalies being

Chronic granulomatous disease

Myeloperoxidase deficiency

Lazy leukocyte syndrome

Job's syndrome

Lactoferrin deficiency

Leukocyte-adhesion deficiency

LEUKOCYTE-DEPLETED BLOOD COMPONENTS FOR TRANSFUSION

The removal or reduction of leukocytes from volunteer donor whole blood or separated red blood cell and platelet preparations[107] (see **Blood transfusion components**). The clinical indications are:

Prevention of alloimmunization occurring with:

- Prospective organ graft recipients
- Patients with severe **bone marrow hypoplasia**, particularly if **allogeneic or autologous stem cell transplantation** (SCT) is planned

Prevention of **cytomegalovirus (CMV)** transmission by:

- Intrauterine transfusion to fetus
- Premature neonates
- CMV-negative pregnant women
- Allogeneic SCT where both donor and recipient are CMV seronegative
- Autologous SCT in CMV-negative patients
- Organ transplant patients where both recipient and organ donor are CMV-negative

- CMV-seronegative **HIV**-positive patients
- Reduction of nonhemolytic febrile blood transfusion reactions (NHFTR)
- Inhibition of possible spread of **prion infection**

It remains controversial as to whether leukodepletion of blood components prevents nonhemolytic febrile transfusion reactions.

LEUKOCYTE ESTERASES

Specific esterases of **neutrophils** and their precursors (e.g., chloroacetate esterase), or nonspecific esterases, reacting with both neutrophils and **monocytes** (α-naphthyl acetate or α-naphthyl butyrate esterases). The esterase reactions are used to distinguish various types of **acute leukemia** — purely myeloid (FAB M1–M3) from myelomonocytic or monocytic leukemias (FAB M4 and M5). A dual esterase (chloroacetate esterase plus α-naphthyl esterase) reaction has been developed for laboratory ease, with each esterase resulting in a different colored reaction, depending on the substrate used.

LEUKOCYTE FILTRATION

See **Leukocyte-depleted blood components for transfusion**.

LEUKOCYTE INHIBITORY FACTOR

(LIF) A **cytokine** involved in the regulation of the development of immature hematopoietic cells.

LEUKOCYTOCLASTIC VASCULITIS

(Allergic vasculitis) A term used to describe **vasculitis**, in which numerous fragmented **neutrophils** are seen at the site of the damaged vessel. Vasculitis mediated by **immune complexes** (type III hypersensitivity) is most likely to be leukocytoclastic. **Adverse drugs reactions** associated with the disorder follow treatment with allopurinol, penicillin, sulfonamides, thiazide diuretics, quinolones, and bosentan. It may be localized to the skin, with eosinophilic infiltration, but without systemic involvement, either as erythema elevatum diutinum or granuloma faciale. **Henoch-Schönlein purpura** is a form of the disease.

LEUKOCYTOSIS

Increase in the total **white cell count** above the physiological range in health, i.e., $10 \times 10^9/l$.

LEUKOERYTHROBLASTIC ANEMIA

The term used to describe the changes seen when immature **granulocytes** (usually myeloblasts) and **normoblasts** (usually restricted to bone marrow) circulate in the peripheral blood in association with **anemia**. The causes are listed in Table 94. It is thought to be due to either damage to the marrow/blood barrier or premature release as part of a "fight or flight" **acute phase response**.

TABLE 94

Causes of Leukoerythroblastic Anemia

Malignancy	Nonmalignancy
Metastatic cancer	**Myelofibrosis**
Breast	Osteopetrosis
Prostate	**Gaucher's disease**
Lung	Tuberculosis
Myelomatosis	Acute **hemolysis**
Hodgkin disease	Acute **hemorrhage**
Non-Hodgkin lymphoma	Acute **hypoxia**
	Megaloblastosis
	Thalassemia major

LEUKOPENIA

Reduction in the total **white cell count** below the physiological range in health, i.e., $4 \times 10^9/l$.

LEUKOPHERESIS

See **Hemapheresis**.

LEUKOSTASIS

The term used to describe "sludging" of blood within arterioles and capillaries as a result of very high granulocyte counts ($>100 \times 10^9/l$), with consequent whole-blood **hyperviscosity**. It characteristically occurs in blast crisis of **acute myeloid leukemia** (AML) or **chronic myelogenous leukemia** (CML), but has on rare occasions been associated with **acute lymphoblastic leukemia** (ALL) and CML in the chronic phase. Organ dysfunction results from tissue hypoxia, with the lungs and central nervous system most commonly affected, with respiratory distress, headaches, confusion, ataxic gait, visual disturbances, coma, and death. Priapism has been described in CML.

Immediate intensive **leukopheresis** and **cytotoxic agent** chemotherapy is indicated. The correction of anemia may exacerbate symptoms and should be withheld until the white count is below $100 \times 10^9/l$.

LEUKOTRIENES

The slow-reacting substances of anaphylaxis (see **Basophil hypersensitivity**). They are synthesized from **arachidonic acid**, which is liberated by stimulation of eosinophils, basophils, and macrophages in all forms of **inflammation**, including acute asthma. Their principal action is to cause the accumulation of neutrophils and eosinophils.

LEWIS BLOOD GROUPS

A specific antigen–antibody system located on red blood cells (see **Blood groups**).

Biochemistry

The Lea and Leb blood group antigens are carbohydrates, but unlike the **ABO(H) blood group** antigens, they are not intrinsic to the red blood cell (RBC) membrane: they are

TABLE 95

Phenotypes and Genotypes of the Lewis Blood Group System

Red Blood Cell Phenotype	Genotype		Frequency in Caucasians	ABH Secretor Status
	Secretor	Lewis		
Le(a⁻b⁺)	*SeSe* or *Sese*	*LeLe* or *Lele*	75%	+
Le(a⁺b⁻)	sese	*LeLe* or *Lele*	20%	–
Le(a⁻b⁻)	*SeSe* or *Sese*	lele	4%	+
Le(a⁻b⁻)	sese	lele	1%	–

fucosylated glycosphingolipids incorporated into the RBC membrane from plasma via lipoproteins as a carrier. The antigens are widely distributed in body tissues and in secretions, including saliva. There are four additional antigens described within the Lewis blood group system.

Genetics and Phenotypes

The Lea and Leb antigens are determined by the Secretor (*Se* or *FUT2*) and Lewis (*Le* or *Fut3*) genes. As with ABH, these genes specify glycosyl transferases that catalyze the transfer of monosaccharides onto carbohydrate precursors. When the *Se* gene is present, a precursor is synthesized that allows (1) the production of the Leb antigen, following the action of the *Le* gene, and (2) the production of ABH antigens in secretions, following the actions of *ABH* genes.

In the absence of the *Se* gene, action of the *Le* gene results in the production of the Lea antigen. As a consequence of this gene interaction, the phenotype Le(a⁺b⁺) is rarely found on RBCs. The phenotypes and genotypes are shown in Table 95.

The Le(a⁻b⁻) phenotype is much more common in black populations than in white populations. Reduced expression of Lewis antigens on RBCs is also found when there is a low or relatively low plasma lipoprotein concentration, as in infants and pregnant women.

Antibodies and Their Clinical Significance

Usually naturally occurring, IgM, and **complement** binding.

Their apparent incidence is dependent on the **pretransfusion testing** methods in use, as many Lewis antibodies show greater activity at temperatures below 37°C.

Anti-Lea is found only in Le(a⁻b⁻) individuals.

Anti-Leb is found in Le(a⁻b⁻) and occasionally in Le(a⁺b⁻) individuals; anti-LebH fails to react *in vitro* with Le(a⁻b⁺) RBCs from A1 individuals.

Anti-Lea and anti-Leb often occur together as separable antibodies or as anti-Leab.

Lewis antibodies only very rarely cause hemolytic blood transfusion complications (see **Blood components for transfusion** — complications) if incompatible RBCs are transfused. They can be ignored for the purposes of transfusion if they are inactive at 37°C in direct agglutination or antiglobulin methods, and therefore provision of blood for patients with Lewis antibodies is unlikely to be problematic.

Because Lewis antibodies are complement binding, they are often lytic *in vitro* and are reactive in the **indirect antiglobulin (Coombs) test** when a polyspecific reagent containing anticomplement is used.

Lewis antibodies do not cause **hemolytic disease of the newborn** (HDN) for the same reasons that ABO HDN is rare (see **ABO(H) blood groups**).

Ganglioside

	A	B	C	D
Ceramide	→ glucose	→ galactose	→ N acetyl galactosamine	→ galactose

↓

N acetylneuramic acid

Globoside

	A	B	E	F
Ceramide	→ glucose	→ galactose	→ galactose	→ N acetyl galactosamine

Sphingomyelin

G

Ceramide → phosphate → choline

	Enzyme	*Disorder*
A	β Glucocerebrosidase	**Gaucher's Disease**
B	GL-2-β-Galactosidase	Lactosyl Ceramidosis
C	Hexosaminidase A	**Tay-Sach's Disease**
D	GM_1-B-Galactosidase	GM_1 Gangliosidosis
E	α Galactosidase	**Fabry's Disease**
F	Hexosaminidase A and B	Sandhoff's Disease
G	Sphingomyelinase	**Niemann-Pick Disease**

FIGURE 77
Degradation of sphingolipids.

LIP DISORDERS

See **Oral mucosa disorders**.

LIPID-STORAGE DISORDERS

Lysozymal enzyme disorders causing the accumulation of lipids in tissue cells, particularly **histiocytes** (macrophages). Sphingolipids have a major structural role in forming the cell wall of mammalian cells. The sphingolipid molecule is composed of ceramide and a hydrophilic side chain. The distinguishing feature of each sphingolipid is that determined by the compounds esterified to the ceramide. Gangliosides contain hexoses and N-acetyl-neuramic acid and are largely found in brain tissue. Globosides have N-acetylgalac-tosamine and are important in the antigen expression on the cell surface of hematopoietic cells. Sphingomyelin has phosphorylcholine as a side chain.

After normal cell death, the sphingolipids are degraded by a series of enzymes (see Figure 77). These enzymes are glycoproteins that have been synthesized in the rough **endoplasmic reticulum**. Posttranslational modification of the enzymes occurs in the **Golgi**

apparatus, often resulting in the production of various isoenzymes. The enzymes are released into the cytoplasm as primary **lysosomes**. The degradation of sphingolipids is largely performed by macrophages. Phagocytosed material fuses with primary lysosomes to form secondary phagosomes, resulting in lipid degradation. In the degradation of gangliosides and globosides, the hydrophilic (carbohydrate) side chain is removed before hydrolysis of ceramide can occur. The degradation of the hydrophilic side chain starts distally and proceeds proximally to ceramide.

In lipid-storage disease, there is an autosomally recessive deficiency of a particular degrading enzyme, usually due to single-gene mutation. Mutations affecting both enzyme synthesis and posttranslational modification have been described. Thus, there may be either reduced production of a normal enzyme or normal production of an abnormal enzyme. The failure of degradation results in the accumulation of the particular sphingolipid substrate within the tissues, with subsequent tissue damage.

Clinical Features

The clinical manifestations are largely determined by the distribution of the particular sphingolipid within normal tissue, with abnormalities of sphingomyelin affecting predominantly neural tissue, and with globoside and ganglioside resulting in more generalized damage (e.g., liver, spleen, and bone). The genes encoding the degrading enzymes are autosomal (except α-galactosidase A), with a normal gene usually compensating for the mutated gene. Thus, autosomally recessive inheritance is usual. These conditions are therefore particularly prevalent in consanguineous relationships or inbred populations. Indeed, for reasons largely unknown, Ashkenazi Jews have a particularly high incidence of **Gaucher's**, **Niemann-Pick**, and **Tay-Sachs diseases**. Clinical manifestations within affected individuals may be variable due to many patients being compound heterozygotes at the molecular level.

Diagnosis

The diagnosis of a lipid-storage disease should be suspected in any infant/child with progressive neurological deficit, organomegaly, or skeletal abnormality, particularly if in the setting of ethnicity or consanguinity. Tissue biopsy, e.g., **bone marrow**, may reveal pathognomonic macrophages, but definitive diagnosis by measuring specific enzyme activity in serum, leukocytes, or cultured skin fibroblasts is usually necessary.

Treatment

Adult Gaucher's disease may benefit from specific enzyme replacement, but in all other disorders no specific therapy has been shown to be beneficial. **Allogeneic stem cell transplantation** has benefited a few patients, with damaged organs slowly recovering. Genetic counseling is essential, and prenatal diagnosis via chorionic villus sampling is available for many of the disorders.

LIPOPROTEIN (a)

(Lp[a]) A low-density lipoprotein with a specific protein adjunct. It has many properties in common with low-density lipoprotein (LDL) but contains a unique protein, apo(a), that is structurally different from other apolipoproteins. The size of the apo(a) gene is highly variable, resulting in the protein molecular weight ranging from 300 to 800 kDa. Apo(a) influences to a major extent metabolic and physicochemical properties of Lp(a), and the size polymorphism of the apo(a) gene contributes to the pronounced heterogeneity of Lp(a). There is an inverse relationship between apo(a) size and Lp(a) levels. Lp(a) levels differ between populations, with blacks having higher levels than Asians and whites. Many meta-analyses support an association between Lp(a) and coronary artery disease, and the levels of Lp(a) carried in particles with smaller size apo(a) isoforms are associated with cardiovascular disease or with preclinical vascular changes. There is also an interaction between Lp(a) and other risk factors for cardiovascular disease. The physiological role of Lp(a) is unknown, but it has been suggested that Lp(a) may inhibit **fibrinolysis** and potentiate **thrombosis**.

LIPOXYGENASES

A family of enzymes that catalyze oxygenation of fatty acids to lipid hydroperoxides. They provide an alternative pathway for **arachidonic acid** metabolism. **Platelets** contain 12-lipoxygenase, whereas leukocytes contain both 12- and 5-lipoxygenases. Lipoxygenase products include a range of leukotrienes. Activation of lipoxygenase may occur in response to contact activation of blood coagulation. Lipoxygenase products have diverse effects on platelet aggregation and in general appear to enhance platelet aggregation induced by adrenaline/epinephrine and thrombin. This may be mediated via complex interactions with **prostacyclin** and **thromboxan**e. The exact role of lipoxygenase products in **hemostasis** and **thrombosis** remains unclear at present.

LIVER

The structure and hematological function of the lobular organ situated in the right hypochondrium of the abdominal cavity.

Structure

The liver weighs 1500 g in an adult and is composed of three lobes covered over much of its surface by the peritoneal mesothelium. It has anatomical relationships with the diaphragm, spleen, and stomach. At the hilum on its lower aspect, the porta hepatis, hepatic blood vessels and biliary ducts enter or leave the organ. It has a dual blood supply via the hepatic artery (25%) and from the **portal circulation** (75%), which returns blood from the spleen, pancreas, stomach, small intestine, and the greater part of the colon to the liver. This system allows the products of digestion to pass through the liver before entering the general circulation. Both hepatic and portal vessels divide along trabeculae to become hepatic sinusoids, which have direct contact with the hepatic cells. The sinusoids are lined by endothelial cells on whose surface are fixed **histiocyte**s (macrophages): Küpffer cells. These remove damaged red blood cells and bacilli entering via the portal circulation. Blood leaves the sinusoids by the hepatic veins.

The hepatic cells are arranged as acini around the branching system of ducts and vessels within fibrous connective tissue: the portal area (see Figure 79). Bile collects from the hepatic cells via biliary canaliculi to bile ducts, which lie alongside the hepatic arteries and veins and eventually form the common bile duct, which drains into the duodenum.

Function

In embryonic life, the liver from 2 to 7 months is the main site of hematopoiesis. Later, hepatic functions are multiple, including the metabolism of absorbed nutrients; storage of nutrients such as cobalamins; the production of bile for emulsification of fats in the

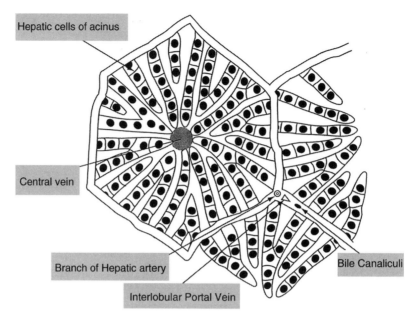

FIGURE 78
Diagrammatic representation of the structure of a liver lobule. (From Fawcett, D.N. (Figure 27-1) in A. Brown and Fawcett, A. *A Textbook of Histology,* Chapman & Hall, London, 1994.)

duodenum; degradation of toxic substances, including drugs; and the synthesis of plasma proteins. This latter function includes synthesis of **fibrinogen** and **vitamin K**-dependent **coagulation factors**.

LIVER DISORDERS

The associations of liver and hematological disorders

Hematological Effects of Liver Disorders

Acute hepatic failure

- Failure of synthesis of plasma **coagulation factors** by the hepatic cells
- Consumption coagulopathy
- Both can result in acute **hemorrhage**; Treatment by plasma exchange (see **Hemapheresis**) is preferable to infusion of **fresh-frozen plasma**

Chronic hepatitis and cirrhosis

- Hepatic cell failure with dysfibrinogenemia from surplus sialic acid in **fibrinogen** molecules, causing abnormal fibrinogen polymerization but minimal effects on **hemostasis**; reduced production of fibrinogen, **factor V**, and the **vitamin K**-dependent **coagulation factors** (factor II, **factor VII**, **factor IX**, and **factor X**) also occurs, leading to **hemorrhagic disorders**

- **Portal hypertension** with esophageal varices, causing hematemesis and **iron deficiency** from chronic **hemorrhage**; **splenomegaly** with pancytopenia from blood cell sequestration

- **Macrocytic anemia** due to increased cholesterol in the red cell membrane with characteristic **leptocytes** ("target cells") and **acanthocytes** in the peripheral blood

- **Iron overload**, particularly with alcoholic cirrhosis
- Elevated levels of serum **cobalamin**
- **Folic acid** deficiency due to lack of hepatic cell storage

Hepatoma associated with erythrocytosis, neutrophilia, and leukemoid reactions

Effects of Hematological Disorders on the Liver

Hematopoietic cell proliferation within the liver

- **Myeloid metaplasia** with **myelosclerosis**
- **Non-Hodgkin lymphomas**, particularly **adult T-cell leukemia/lymphoma, chronic lymphatic leukemia, hairy cell leukemia, mycosis fungoides/Sézary syndrome, peripheral T-cell lymphoma**
- Hemangioendotheliomas

Hepatic vein thrombosis (Budd-Chiari syndrome), a pathological change that leads clinically to abdominal pain, liver tenderness, and abrupt ascites; diagnosis is based upon portal angiography and occurs with:

- Disorders that predispose to thrombosis: **polycythemia rubra vera**, oral contraceptives, hepatocellular carcinoma
- **Paroxysmal nocturnal hemoglobinuria** (PNH), where 30% have an intravascular thrombosis of the portal tract due to the hypersusceptibility of their platelets to complement activation by circulating thromboplastic material from hemolyzed red blood cells
- Stem cell transplantation (25% allogeneic) as a consequence of damaged endothelium, with activation of the coagulation cascade

Hepatic **porphyrias**: metabolic defects affecting the cells of the liver as well as those of the erythron; **porphyria cutanea tarda** is associated with **iron overload** of the liver

Hereditary hemochromatosis: a hereditary defect of **iron** absorption that leads to iron overload of the liver, pancreas, and lymph nodes

Sickle cell disorders, giving rise to hepatomegaly and disturbed function

Graft-versus-host disease, causing hyperbilirubinemia

Effects of Combined Liver and Hematopoietic Disorders

Alcohol toxicity

- Acute: **fragmentation hemolytic anemia** with **acanthocytes** (spur-cell hemolytic anemia; Zieve's syndrome)
- Chronic: **folic acid** deficiency

Parvovirus hepatitis with **pure red cell aplasia**

Tuberculosis with **granuloma** of liver and bone marrow

Wilson's hepatolenticular degeneration with **hemolytic anemia** due to **copper** oxidation of the red cell membrane

LOMUSTINE
See **Alkylating agents**.

LOW IONIC POLYCATION TECHNIQUES

A method used in **pretransfusion testing** to increase the sensitivity of the **indirect anti-globulin (Coombs) technique** for red blood cell (RBC) antibody detection.

Polybrene® and protamine sulfate are positively charged molecules that can be used to promote the agglutination of antibody-coated RBCs. The "manual Polybrene test" (MPT) has a high sensitivity for Rh antibodies and is sometimes used in routine and some other pretransfusion testing. Like enzyme techniques for antibody detection, not all clinically important antibodies (anti-K and anti-Fya in the MPT) are detected well, so the technique should not be used to replace the indirect antiglobulin technique. The manual Polybrene test can be used as the first stage of an indirect antiglobulin technique, although conventional low-ionic-strength saline (LISS) techniques may be found to be more convenient.

LOW-IONIC-STRENGTH SALINE

See also **Indirect antiglobulin (Coombs) test**.
(LISS) A solution used for blood group serology tests.

LOW-MOLECULAR-WEIGHT HEPARIN

(LMWH) See **Heparin**.

LUPUS ANTICOAGULANTS

(LAs) **Immunoglobulins**, usually IgG, that bind phospholipids active in coagulation. Lupus anticoagulants are nonspecific **antibodies** that prolong the clotting time of phospholipid-dependent clotting assays such as the **activated partial thromboplastin time** (APTT) and **prothrombin time** (PT). They are one form of **antiphospholipid antibody** and may require binding with β2-glycoprotein 1 before they exhibit anticoagulant properties. Unlike specific factor antibodies, LAs are usually associated with **venous thromboembolic disorders**, **arterial thrombosis**, or recurrent fetal loss. LAs do not specifically inhibit individual coagulation factors; rather, they neutralize anionic phospholipid-protein complexes that are involved in the coagulation process. Prolongation of clot-based assays is highly dependent on the sensitivity of the reagent employed. Reagents with reduced amounts of phospholipids and dilute **Russell viper venom time** (DRVTT) have enhanced sensitivity for LAs. Due to the heterogeneity of LA antibodies, no single assay will identify all cases. Plasma samples for the detection of an LA must be free of platelets or platelet fragments. To make a diagnosis of an LA, a sample should show:

Prolongation of at least one phospholipid-dependent clotting test

Evidence of inhibitory activity shown by the effect of the patient's plasma on pooled normal plasma, i.e., the LA defect persists in mixtures of test and normal plasma with only a partial correction on mixing

Evidence that the inhibitory activity is dependent upon phospholipid, i.e., the LA can be neutralized by the addition of exogenous phospholipid

LAs must be distinguished from other coagulopathies that may give similar laboratory results

Tests for Lupus Anticoagulants

Laboratories should employ a screening test to detect the presence of LA and then conduct separate tests to confirm its presence. Screening tests and confirmatory tests are summarized by the International Committee for Thrombosis and Hemostasis (ICTH) Subcommittee on Lupus Anticoagulants/Antiphospholipid Antibodies.[61]

Screening Tests

Activated partial thromboplastin time/dilute activated partial thromboplastin time

Taipan snake venom clotting time/dilute Russell viper venom time (DRVVT)

Dilute prothrombin time (thromboplastin inhibition test)

Kaolin clotting time/silica clotting time

Confirmatory Tests

APTT-based assays with phospholipid correction

DRVVT with phospholipid correction

Taipan snake venom time with phospholipid correction

Dilute prothrombin time with phospholipid correction

Kaolin clotting time (KCT) with phospholipid correction

Silica clotting time with phospholipid correction

It is recommended that screening should include at least two sensitive tests. For example, the APTT can be used as a screening test for LA. If the APTT is prolonged, the performance of mixing tests with normal plasma will confirm the presence on an inhibitor. A second test should also then be used, preferably the KCT or more frequently the DRVVT, with a correction procedure to confirm the prolongation of the clotting time (basic DRVVT or KCT) and to confirm that the inhibitory activity is directed to phospholipids (correction tests). The presence of LA must be confirmed in separate blood samples collected at least 6 weeks apart to demonstrate persistent positivity.

LUPUS ERYTHEMATOSUS CELLS

(LE cells) **Neutrophils**, or more rarely **monocytes** or **eosinophil**s, that have ingested nuclear material altered by antinuclear antibody (ANF) to form a spheroidal mass that displaces the cell nucleus to the periphery. Upon Romanowsky staining, this mass is pale purple. LE cells can be artificially induced by incubating slightly traumatized normal leukocytes of peripheral blood or bone marrow with the test serum and then Romanowsky staining a film of the deposit.[59] The test is positive in 75% of patients with **systemic lupus erythematosus**, and LE cells are commonly found in those with lupoid hepatitis and occasionally in those with **rheumatoid arthritis**.

Lutheran (Lu) blood groups

A specific antigen–antibody system located on red blood cells (see **Blood groups**). Nineteen antigens have been described within the Lutheran (Lu) blood group system. The most important of these are the antithetical Lu^a and Lu^b, produced by the Lu^a and Lu^b genes. In Caucasians, 92% are Lu(a-b+) and 8% are Lu(a+b+). The phenotype Lu(a+b-) is rare, and Lu(a-b-) is very rare. Anti-Lu^a and anti-Lu^b can be IgG or IgM or both. Anti-Lu^a is not uncommon. It is more usually found in plasma containing other red cell antibodies. Anti-Lu^a does not cause blood transfusion complications. Anti-Lu^b has very rarely been the

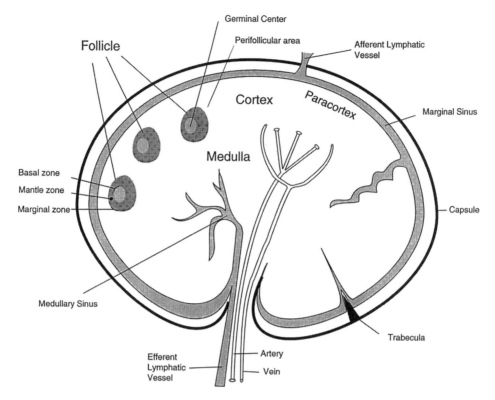

FIGURE 79
Diagrammatic representation of the lymph-node structure. (From Fawcett, D.N. (Figure 15-1) in A. Brown and Fawcett, A. *A Textbook of Histology,* Chapman & Hall, London, 1994.)

cause of a delayed hemolytic blood transfusion complication (see **Blood components for transfusion** — complications). Anti-Lua and anti-Lub have been associated only with very mild **hemolytic disease of the newborn**.

LYMPH

The fluid circulating through the lymphoreticular system of lymphatic vessels, **lymph nodes**, **spleen**, and **liver**. It usually contains **lymphocyte**s with other blood cells and occasional fat particles.

LYMPH NODE

Accumulations of lymphoid tissue along the course of the lymphatic vessels. The nodes often occur in chains, the most prominent being the cervical, axilla, inguinal, mesenteric, and retroperitoneal chains. Each node is surrounded by a fibrous capsule with trabeculae passing into the cortex (see Figure 79). Between the trabeculae, the lymphoid tissue is supported by a network of reticular fibers and associated reticular cells. The meshes of this network are filled by **lymphocytes**, **plasma cells**, **histiocytes** (macrophages), and interdigitating **dendritic reticulum cells**. The capsule is penetrated at the hilum of the node by afferent and efferent lymphatic vessels. The afferent lymphatics are continuous, with a sinus running beneath the capsule. From here, the lymph flows inward through narrow cortical sinuses to a medullary sinus and eventually outward through the efferent lymphatics.

Within the cortex, lymphocytes are grouped in lymphoid follicles.[330] Primary follicles of unstimulated lymphocytes are located at the periphery of the cortex. Secondary follicles, composed of poststimulated lymphocytes and with a germinal center of transformed lymphocytes and immature plasma cells, are located more deeply within the cortex. B-cells occupy the periphery or cortical area, while T-cells form the paracortical (predominantly CD4$^+$ cells) and perifollicular areas. The medulla of each node consists of lymphatic vessels and a medullary sinus.

For lymph node disorders, see **lymphadenopathy**.

LYMPHADENITIS

See **Lymphadenopathy**.

LYMPHADENOPATHY

Enlargement of the regional **lymph nodes**.

Etiology

Local

Infection: pyogenic bacteria, actinomyces, **tuberculosis**, viral (e.g., cat scratch fever, lymphogranuloma venereum)

Lymphoproliferative disorders: **Hodgkin disease, non-Hodgkin lymphoma**

Angiofollicular lymph node hyperplasia (Castleman's disease)

Histiocytic and dendritic neoplasms

Carcinoma metastases

General

Infections: bacterial (e.g., brucellosis, syphilis, tuberculosis, salmonelloses, bacterial endocarditis), fungal (e.g., histoplasmosis), protozoal (e.g., toxoplasmosis), viral (e.g., **infectious mononucleosis**, measles, rubella, viral hepatitis, **HIV**)

Noninfectious inflammatory diseases (e.g., **sarcoidosis, rheumatoid arthritis, systemic lupus erythematosus**, serum sickness, **Wegener's granulomatosis**)

Leukemias, especially **chronic lymphatic leukemia** (CLL), **acute lymphoblastic leukemia** (ALL), and **juvenile myelomonocytic leukemia** (J-MML)

Lymphoproliferative disorders: Hodgkin disease, non-Hodgkin lymphoma

Waldenström macroglobulinemia

Carcinoma metastases

Hemophagocytic **histiocytosis**, sinus histiocytosis with massive lymphadenopathy

Reaction to drugs and chemicals (e.g., hydantoins and related chemicals, beryllium)

Hyperthyroidism

Lymphoid hyperplasia: **autoimmune lymphoproliferative syndrome** (ALPS)

Diagnosis

Lymph nodes of the cervical, axillary, and inguinal chains are palpable, but hilar (mediastinal) lymphadenopathy requires detection by radiography of the chest, while mesenteric or retroperitoneal lymphadenopathy requires scanning by **lymphangiography**, computerized tomography (CT), or magnetic resonance imaging (MRI). Whenever possible, the cause of lymphadenopathy should be determined by histological examination of a biopsy specimen or, as a second choice, cytological examination of a needle-aspirate.

LYMPHANGIOGRAPHY

Injection of a radio-opaque dye into lymphatic vessels, thereby allowing radiological identification of size and structure of the appropriate regional **lymph nodes**. The procedure is most commonly used as bipedal injections for viewing the retroperitoneal nodes of the para-aortic, paracaval, and paraileac lymph node groups. In experienced hands, these can be identified with 80 to 90% accuracy. Although it is an invasive procedure, it has the advantage over computerized tomography (CT) and radiogallium scanning in that it shows the architectural pattern of the nodes in addition to nodal bulk. It is particularly useful for staging of **non-Hodgkin lymphoma**.

LYMPHATIC SYSTEM

See **Lymphoid system**.

LYMPHEDEMA

Edema of subcutaneous tissue due to blockage of lymphatic vessels.[331] It is essential to exclude the possibility of edema being due to disturbances of filtration rate, as occurs with cardiac failure. Lymphedema is usually unilateral, affecting one limb only and is associated with **lymphadenopathy**. Diuretics have little effect, treatment depending upon the underlying cause.

LYMPHOBLAST

Large **lymphocytes** (10- to 15-µm diameter) characterized by a low cytoplasmic:nuclear ratio, presence of nucleoli, and nuclear membrane irregularity. They can be formed *in vitro* by activation or transformation of lymphocytes by an inducer. They are characteristic of **acute lymphoblastic leukemia** (ALL) and of **precursor B-lymphoblastic leukemia/lymphoma**.

LYMPHOBLASTIC LYMPHOMA

See **Precursor B-lymphoblastic lymphoma/leukemia**; **Precursor T-lymphoblastic lymphoma/leukemia**; **Non-Hodgkin lymphoma**.

LYMPHOCYTES

Mononuclear cells found in peripheral blood, lymph, lymph nodes, and the lymphoid follicles of liver, spleen, bone marrow, and submucosa of the gastrointestinal, respiratory, and renal tracts (see **Lymphoid system**). These cells are broadly of three types, two small: B (bone marrow derived) and T (thymus derived); and one large: natural killer (NK) cells (null, non-B-cells, non-T-cells, and large granular lymphocytes [LGLs]). These different cell types have characteristic immunological markers and are functionally distinct. All are derived from a common lymphoid progenitor (CLP) cell in the bone marrow. Mature lymphocytes are found in peripheral blood, lymph, lymph nodes, and the lymphoid follicles of liver, spleen, bone marrow, and submucosa of the gastrointestinal tract (Peyer's patches). Small lymphocytes have been well recognized morphologically from the mid-19th century, but are now recognized to be functionally heterogeneous. They have long been associated with the **immune response**, but it is only in the last 30 years that their more precise functions have become appreciated. In the peripheral blood there are, in adults, 1.5 to 4.0×10^9 lymphocytes/l (for children see **Reference Range Table X**), of which the T-cell is the commonest. Concentration, apart from that in lymphoid follicles, occurs at sites of inflammation, such as infections, autoimmune responses, or allografts.

Small Lymphocytes

Morphology

In health, the great majority of small lymphocytes (6- to 10-μm diameter) are quiescent, resting in G0 of the **cell cycle**. They are about 10 μm in diameter, with a uniform round, intensely staining condensed nucleus, largely heterochromatin. There is only a rim of agranular cytoplasm containing a few ribosomes and organelles. The resting lymphocyte is spheroidal, with a surface studded by villous processes, the surface proteins and receptors being uniformly distributed along the membrane. The cytoskeleton is composed of protein fibrils similar to other cells. Little macromolecular synthesis occurs during this resting phase, and the cell is metabolically quiescent.

Lymphocytes are particularly motile cells capable of migration throughout the lymphoreticular system and other tissues. **Cell locomotion** is largely by moving over other cell surfaces, including passage through vascular endothelium cells of the capillary walls. During migration through the lymphoreticular system and tissues they can adopt a characteristic "hand mirror" configuration. (See **hand mirror cell**.)

Activation of lymphocytes during an immune response (see below) results in rearrangement of the cytoskeleton proteins. The **actin** microfilaments produce **uropods**, which come together to form a cap. Activation also results in the entrainment of a program of differentiation: over a period of some hours, the cell transforms into a lymphoblast that is much larger (10- to 15-μm diameter), with a greater cytoplasm:nucleus ratio, much more loosely packed euchromatin within the nucleus, many more **mitochondria**, and a more developed **Golgi apparatus** and **endoplasmic reticulum**. The cell becomes metabolically very active, with an organized program of expression of genes, resulting in much-enhanced **RNA** and protein synthesis and the acquisition of a range of functions, which are the essence of the effector-cell response. Above all, the cell enters the cell cycle and divides. Division is terminated, probably by the removal of the proliferative stimulus (**antigen**), and the cells either undergo **apoptosis** or differentiate to become memory cells.

Types of Small Lymphocytes

Mature naïve (i.e., have not been activated) B and T-cells are morphologically identical, but are distinguished by virtue of cell-surface markers (CD antigens) and other markers as indicated in Table 96. The two cell types are fundamentally different through the expression of distinct antigen receptors: the B-cell receptor (BCR), whose antigen-specific component is immunoglobulin, and the T-cell receptor (TCR). The genes for these are present in segmented forms in precursor stem cells, and the mature genes are generated during lymphopoiesis by rearrangement of the gene segments. There are subsets of T-cells distinguished by possession of CD4 or CD8 surface antigens. These subsets are further subdivided according to the different cytokines they produce upon activation (see below).

TABLE 96

Characterizing Markers of Mature Resting Lymphocytes

Lymphocyte	Marker
B	CD19, CD20, CD21, CD22, sIg
T	CD2, CD3, CD4 (Th only), CD6, CD8 (Tc only), TCR
NK	CD56, CD161, KIR, NCR

Note: This is not an exhaustive list.

T-cells are also divided into two sets, depending on the nature of their TCR, which can be αβ or γδ in type. However, the significance of the TCRγδ subset has not been well established, although it may have a role in mucosal (particularly gut) immunity.

Natural Killer (NK) Cells

These are large granular cells, around 20 μm in diameter, comprising 5 to 10% of total peripheral leukocytes. Although cytotoxic and thus, at a gross level, functionally similar to CD8 T-cells, they are found in detail to be very different from the other types of lymphocyte. In particular, they do not possess the *sine qua non* of an antigen-specific receptor: their cytotoxicity, unlike that of CD8 cells, is not antigen specific, and they do not possess all of the surface markers that characterize T and B-cells. For this reason, they are sometimes referred to as non-T and non-B-cells, or null cells.

Lymphopoiesis

Some of the detail of the differentiation of lymphocytes has been worked out in mice (see Figure 81, Figure 82, and Figure 83). The CLP, present in adult bone marrow, lacks all lymphoid lineage-specific markers, but is characterized by possession of the receptor for the **cytokine** IL-7, which is essential for lymphopoiesis, but without the receptor for erythropoietin (EPO-R). IL-7 and CLP are able to generate T-, B-, and NK-cells but not myeloid cells. The CLP cell is not a stem cell, lacking CD34 and incapable of unlimited renewal. B-cells and NK-cells differentiate within the bone marrow, while T-cells differentiate in the thymus after CLPs have migrated there. The decision on whether B- or T-cells develop from CLPs depends on the expression of the gene-transcription factor Notch 1: with it, T-cells develop; without it, B-cells develop. NK-cells appear to require IL-15 for their development. A key step in the differentiation of B- and T-cells is the rearrangement of segmented genes to form mature antigen receptors (BCR or TCR), and this is dependent on IL-7. This of course defines their function, as described below. Although the stages in differentiation through which the three lymphoid cell types pass are well described, the actual factors — cytokines, hormones, adhesion molecule interactions — driving the differentiation process remain ill-defined. Presumably, the different microenvironments determine which lineage is followed.

Figure 80, Figure 81, and Figure 82 indicate the main recognized stages of the different lymphocytes and the major markers that distinguish them.

Immunological Roles of Lymphocytes

The role of small lymphocytes in the immune system was demonstrated by draining lymph (and thus lymphocytes) from the thoracic duct of rats, which thereby became relatively immunosuppressed. It is now clear that lymphocytes are the key to the adaptive immune response. The function of B and T-cells is to respond to the challenge of infection by a

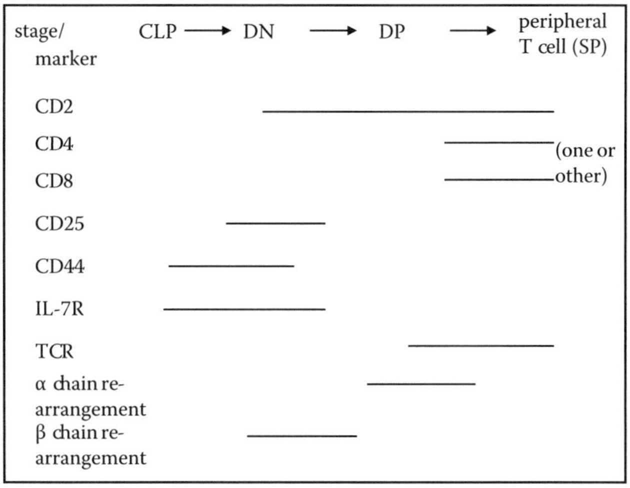

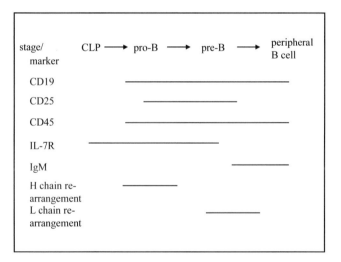

FIGURE 80
Development of T-cells.

FIGURE 81
Development of B-cells.

pathogen, identified by antigenic differences from the host, and eliminate it. In essence, these cells:

Recognize foreign antigen and proliferate to generate large numbers of effector cells to eliminate the antigen

Activate components of the innate immune system with the same end

When (if) the pathogen is eliminated or contained, the immune response comes to a halt, but memory is generated such that a second exposure is more effectively dealt with (usually).

The recognition of foreign antigen is of course through their specific antigen receptors (BCR, TCR) generated during lymphopoiesis (see above): engagement of the receptor leads

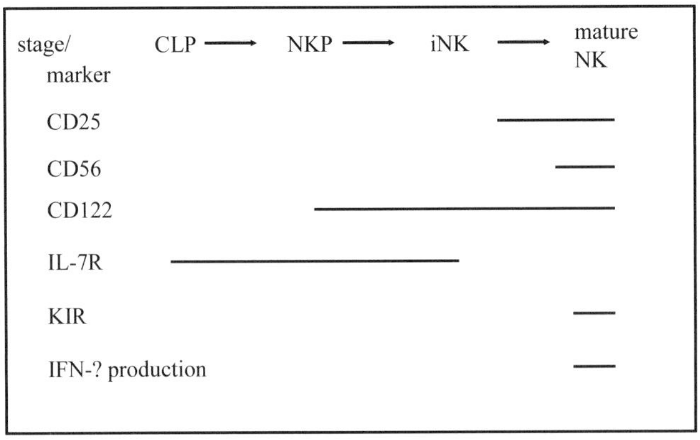

FIGURE 82
Development of NK-cells.
Abbreviations: DN, double negative (e.g., T-cell precursor lacking both CD4 and CD8 markers); DP, double positive (possessing both CD4 and CD8); SP, single positive (CD4 or CD8); NKP, NK precursor; iNK, immature NK; KIR, killer-cell immunoglobulin-like receptor. For other terms, see text. (See also **Cytokines**; **CD antigens**.)

to activation. All the progeny of an activated lymphocyte bear the same receptor (with some modification in the case of B-cells, see below), hence the term "clonal distribution" of receptors. This process of course amplifies the number of antigen-specific cells and thus increases magnitude of the response. After the proliferative phase, the cells differentiate to become effector or memory cells. Effector cells die, but memory cells do not. There are more memory cells in a primed individual than there are naïve cells, which accounts for the greater magnitude of the secondary response.

Broadly, the role of B-cells is to produce antibody and that of T-cells is to provide "help" (i.e., to take part in the activation process), particularly for B-cells to make antibody (the CD4 subset) or to generate cytotoxic cells (CD8 subset). Additionally, T-cells (both subsets) produce cytokines.

Role of B-Cells

The role of B-cells is to differentiate to **plasma cells** and produce **antibodies**. B-cells are also potent antigen-presenting cells (see **Antigen presentation**) for their cognate **antigen**, which is taken up by the BCR.

The membrane immunoglobulin (mIg) component of the BCR (in naïve B-cells, this will be IgD or IgM) is associated with other membrane proteins whose role it is to transmit the signal to the nucleus, whereupon a program of proliferation and differentiation ensues. However, without the secondary signals from T-cells, the B-cell is not activated and instead undergoes **apoptosis**. The essential costimulatory signals are engagement of CD40 and CD80 on the surface of the B-cell with CD154 and CD28, respectively, on the T-cell. The CD40/154 interaction prevents B-cell apoptosis, and the CD80/28 interaction coactivates cytokine production by the T-cell. CD54 interacting with CD11/18 stabilizes the B/T-cell conjugate. T-cell cytokines also supply activation and proliferation signals to the B-cell. The T-cell involved with the B-cell is the CD4 helper T-cell, which also needs to be activated by engagement of its TCR by processed antigen in association with class II MHC (major histocompatibility complex) antigens. Depending on the subset of CD4 cells providing help (see below), the B-cell will mature to produce either IgG or IgE (see below).

As B-cells proliferate and mature, through a process of somatic mutation, immunoglob-ulins with different antigen-binding sites (see **Immunoglobulins**) of varying affinity will be generated: B-cells bearing BCRs with higher-affinity Ig components will be selected by virtue of their superior binding of the cognate antigen. As the immune response develops ("matures"), the affinity of antibodies produced increases. Maturation also involves "class switching": the Ig produced shifts from IgM through to IgA and IgG and then IgE (depending on the cytokine environment). Finally, the plasma cell is generated.

Role of T-Cells

T-cells are activated by virtue of engagement of the TCR with antigen presented in the context of MHC (see **Antigen presentation**). CD4 cells "see" antigen in the context of MHC class II and CD8 in the context of class I. The antigen-presenting cell (APC) may be a B-cell (as described above), but in the context of a primary immune response (when antigen-specific B-cells will be very rare), the major APC initially is almost certainly the myeloid dendritic cell. Other cell types, such as histiocytes (macrophages), can act as APCs, but generally only after their activation in the context of a developing immune response. As with B/T-cell interaction, coactivation signals are essential. These are again CD40/154, CD80/28, and CD54 interacting with CD11/18 and cytokines, particularly IL-2, which is a key growth factor for T-cells.

The effector functions of T-cells include provision of help as described above (CD4 only), production of cytokines (both subsets), and cytotoxicity (CD8 only).

CD4 cells are divided into Th1 and Th2 cells according to their production of cytokines. Th1 cells produce, *inter alia*, IL-2, IFN-γ, and TNF-β and tend to promote inflammation, CD8 cell activation, and macrophage activation, i.e., cellular immune responses. Th2 cells produce *inter alia* IL-4 and IL-5 and promote Ig production, class switching to IgE, and eosinophil activation, i.e., humoral and antiparasite (and allergic) responses. Cytokine production drives processes such as **inflammation** (TNF, IL-1, etc.). Activation of other cell types, such as eosinophils (IL-5), macrophages (IFN-γ), NK-cells (IL-2), and neutrophils (chemokines, etc.), are among the most important, along with control of Th1/Th2 balance (IL-10 versus IFN-γ) and termination of the immune response (IL-10 and TGF-β).

CD8 cells upon activation become killer cells (Tc cells or CTL, **cytotoxic T-lymphocytes**), which will destroy cells presenting the specific activating antigen (peptide) in the context of MHC class I. Killing is achieved either by inducing apoptosis via the CD95/CD178 pathway or via lysis by directional secretion of **perforins**, complementlike molecules that insert into the target cell membrane and render it permeable.

Memory B- and T-Cells

Memory B-cells have undergone class switching, and the mIg component of the BCR is IgG, IgA, or IgE, and it will be of high affinity. Memory T-cells are less well defined, but they appear to have altered levels of adhesion molecules (e.g., CD62L), which control their migration patterns so that they tend to recirculate between blood and tissues more efficiently. Both B- and T-memory cells tend to express the O isoform of CD45. Memory cells respond to antigen similarly, as do naïve cells.

Role of Natural Killer (NK) Cells

These cells are not antigen-specific, as are B- and T-cells. They are cytotoxic (by the same mechanisms as CD8 cells) and produce cytokines such as IFN-γ and TNF-α. Their cytotoxicity is activated by cytokines, principally interferons and IL-2. NK-cells recognize their targets via both inhibitory and activating signals. Inhibitory signals are transmitted via killer-cell immunoglobulinlike receptors (KIRs) and the CD94:NKG2A heterodimer. The ligand for these is MHC class I antigen, and binding prevents killing. Hence NK-cells tend to kill targets that lack class I antigens. Receptors receiving activating signals are termed natural cytotoxicity receptors (NCRs). The ligands for these are generally not defined, with the exception of certain viral hemagglutinins. These different receptors are variously expressed by NK-cells; hence there is a variety of such cells. Some T-cells also express these receptors and also exhibit natural cytotoxicity. NK-cells possess Fc receptors and can also mediate antibody-dependent cellular cytotoxicity (ADCC). NK lymphocytes proliferate in the pregnant uterus alongside extravillous trophoblasts, presumably in the adaptation of the cardiovascular system to pregnancy.

The function of NK-cells in the immune system is not so clear as that of other lymphocytes. However, the observation that a number of viruses downregulate class I antigens and that activating ligands include viral hemagglutins, together with the observation that virus-induced IFNs activate NK-cells, strongly suggests that NK-cells have an antiviral role. They may also have an anticancer role; class I antigens are often downregulated in cancers. In support of these hypotheses, NK-defective individuals are known to be more susceptible to the development of lymphomas and to infection by some viruses.

LYMPHOCYTIC SMALL-CELL LEUKEMIA/LYMPHOMA

See **Chronic lymphatic leukemia**; **Non-Hodgkin lymphoma**.

LYMPHOCYTOPENIA

(Lymphopenia) A reduced level of circulating **lymphocytes** below $5.5 \times 10^9/l$ for children and $1.5 \times 10^9/l$ for adults. The common causes are:

Severe **bone marrow hypoplasia**

Hodgkin disease

Immunosuppression therapy

Ionizing radiation

Immunodeficiency, particularly **acquired immunodeficiency syndrome** (AIDS)

Thymoma/reticular dysgenesis (see **Thymus**)

Non-Hodgkin lymphoma

LYMPHOCYTOSIS

An absolute increase in the total number of circulating **lymphocytes** above $9 \times 10^9/l$ for infants, $7 \times 10^9/l$ for children, and $4 \times 10^9/l$ for adults. A relative lymphocytosis occurs with **pancytopenia**. Lymphocytosis arises either from redistribution of lymphocytes from lymphoid tissue to the circulation or as a result of lymphoproliferation. In infants and young children, it occurs in response to infections that cause **neutrophilia** in adults.

Infectious lymphocytosis is a benign disorder of children and young adults, occurring when groups are living together in schools and other institutions. It is usually subclinical, but fever, abdominal discomfort, upper-respiratory symptoms, and fleeting skin rashes may occur. Lymphocyte levels rise sharply to $20 \times 10^9/l$ during the first week of illness, declining within 3 to 7 weeks. **Eosinophilia** may follow.

Infection by *Bordella pertussis* causes peripheral-blood lymphocyte levels to rise to over $10 \times 10^9/l$ during the initial 2 to 3 weeks, due to release of a factor that prevents passage of **lymphocytes** into the **lymph nodes**.

Large granular lymphocytosis is a disorder with an increase of large granular lymphocytes (LGL), usually of **immunophenotype** CD3+, CD8+, CD57+ with TCRab. There is an associated **neutropenia**, but infection is rare. Some patients have **rheumatoid arthritis** with or without **splenomegaly** (Felty's syndrome).

Chronic polyclonal B-cell lymphocytosis of cigarette **tobacco** smokers is associated with **Epstein-Barr virus** infection.

Other causes are:

Viral infections: Epstein-Barr virus (**infectious mononucleosis**), **cytomegalovirus** (CMV), rubella, infectious hepatitis A, herpes simplex and zoster, **human immunodeficiency virus** (HIV), measles, rubella varicella

Bacterial infections: **tuberculosis,** brucellosis, syphilis

Parasitic infections: toxoplasmosis, malaria causing **hyperreactive malarial splenomegaly**

Transient inflammatory reaction with myocardial infarction, trauma, and **sickle cell disorder**

Hypersensitivity: Type IV allergic reactions, e.g., alveolitis

Lymphoproliferative disorders: **chronic lymphatic leukemia, adult T-cell leukemia/ lymphoma, Burkitt lymphoma, follicular lymphoma, hairy cell leukemia, lymphoplasmacytic lymphoma/Waldenström macroglobulinemia, mantle-cell lymphoma, mycosis fungoides/Sézary syndrome, precursor B- and T-lymphoblastic lymphomas, acute lymphoblastic leukemia**

LYMPHOCYTOTOXIN

(LT) See **Tumor Necrosis Factor**.

LYMPHOEPITHELIOID LYMPHOMA

See **Peripheral T-cell lymphoma**; **Non-Hodgkin lymphoma**.

LYMPHOID INTERSTITIAL PNEUMONITIS

Diffuse lymphoid infiltration of the lungs occurring in association with **acquired immunodeficiency syndrome** (AIDS) in children.[134] **Epstein-Barr virus** (EBV) genome is present in most cases, so it is postulated that EBV stimulates intrapulmonary B-lymphocytes, causing the lymphoid hyperplasia. Clinically, it gives rise to hypoxia with finger clubbing and eventual respiratory failure. It is associated with **lymphadenopathy** and **hepatosplenomegaly**.

LYMPHOID SYSTEM

(Lymphatic system) A whole-body system composed of lymphatics, with localized collections of lymphocytes in **lymph nodes**, **spleen**, **bone marrow**, skin, and mucosa of respiratory and intestinal tracts. The lymphatic vessels are lined by permeable endothelium with thin connective tissue walls, but they do not have nonreturn valves. They mostly follow the line of the blood capillaries and veins. The lymphatic capillaries join to form lymphatic trunks, which drain into either the left or right thoracic duct. These drain into the left or right subclavian veins. Through the neuropeptide axis, there is an interrelationship with the nervous system and the pituitary gland. Lymphocytes are located in the body, either encapsulated in follicles, as in lymph nodes and spleen, or nonencapsulated scattered lymphocytes known as mucosa-associated lymphoid tissue. The bone marrow and thymus are regarded as primary lymphoid organs, being the organs where lymphocytes are produced, with the remainder (i.e., spleen lymph nodes, skin) classified as secondary lymphoid organs, where the lymphocytes come into contact with foreign antigen, are clonally expanded, and mature into effector cells.

Lymphoid Follicles

These structures in lymphoid tissue, particularly of the lymph nodes and spleen, form following germinal-center response to antigenic stimulation.[133b] Those that do not have a germinal center are termed "primary follicles." The main function of the germinal-center reaction is proliferation and maturation of B-cells to plasma cells or memory cells. In response to the antigen-BCR complex and the TCR-MCH class II complex, immature B-lymphocytes commence replication within 18 to 24 h, forming a primary focus of B-blast cells at the boundary of the T-cell zone of the paracortex and the follicle. Within 3 to 4 days, **centroblasts** and **immunoblasts** develop. Those of the basal zone (see **Lymph nodes** — Figure 80) are intensely proliferating cells, which are joined by bulky **dendritic reticulum cells** (follicular dendritic cells), giving a "starry sky" appearance upon histological microscopy. Centroblasts gradually migrate through the mantle zone, transforming to medium-sized lymphoid cells with cleaved nuclei and pale cytoplasm: **centrocytes.** By day 7, these centrocytes accumulate above the mass of centroblasts and beneath the dense follicular mantle zone or corona, composed of mature T and B-lymphocytes. Some B-cells differentiate into plasma cells and secrete IgM; others establish the germinal center, where they further proliferate.

Mucosa-Associated Lymphoid Tissue (MALT)

Intestinal- or Gut-Associated Lymphoid Tissue

(GALT) Specialized lymphocytes having numerous microfolds on their luminal surface (M cells) overlie the lymphoid collections (Peyer's patches). They absorb, transport, process, and present antigen to the lymphoid cells lying between them. B-cells secreting IgA are abundant. IgA and smaller amounts of IgM are transported across the intestinal epithelium to prevent microorganisms from entering. Dimers of IgA bind to a receptor on the membrane of the epithelial cells. The IgA-receptor complex is endocytosed and transported across the cell into the lumen. The secretory component of the receptor remains attached to the IgA and protects it from proteolytic enzymes in the gut lumen.

Bronchial-Associated Lymphoid Tissue

(BALT) The cells of this tissue secrete IgA, which (as in the intestine) protects the bronchial mucosa from infection.

Skin-Associated Lymphoid Tissue

(SALT) This is distributed throughout the dermis, with occasional collections of lymphoid cells.

Genito-Urinary Tract-Associated Lymphoid Tissue

This is similar to intestinal and bronchial lymphoid tissue, with a secretory function to protect the mucosa from infection.

LYMPHOMA

Lymphoproliferative disorders in which cellular proliferation occurs in **lymph nodes**, **bone marrow**, extranodal tissue, and peripheral blood. There is usually a mass of regional lymph nodes as well as other sites of proliferation in the lymphoreticular system.

LYMPHOMATOID GRANULOMATOSIS

See also **Angiocentric lymphoma**; **Lymphoproliferative disorders**; **Non-Hodgkin lymphoma**.

A rare extranodal angiocentric lymphoma associated with **Epstein-Barr virus** infection that can involve the lung, kidney, liver, or cutaneous areas. The **lymphocytes** are CD20$^+$ and variably positive for CD79a and CD30. The disease is classified into stages of Grades I, II, and III. Grade III may respond to aggressive **cytotoxic agent** chemotherapy. Grades I and II disorders may respond to **interferon** alpha 2b.

LYMPHOMATOID PAPULOSIS

Transient, spontaneously regressing papulonodular inflammatory infiltration of the superficial and deep dermis by **T-lymphocytes** with cerebroform nuclei (Sézary-like) and by smaller numbers of multinucleated hyperdiploid large cells similar to **Langerhans cells** carrying Ki-1 antigen, as found in **Reed-Sternberg cells**. These cells display surface cutaneous lymphocyte antigen (CLA), an adhesion molecule that mediates in the initial tethering of T-lymphocytes to the endothelium of cutaneous postcapillary venules. **Diffuse non-Hodgkin lymphoma** develops in 10 to 20% of patients. **PUVA light therapy** and oral **methotrexat**e are effective treatments in patients with multicentric disease.

LYMPHOMATOUS MENINGITIS

Infiltration of the meninges by **lymphocytes** or **lymphoblasts** in association with a **lymphoproliferative disorder**. It particularly occurs with **acute lymphoblastic leukemia** (ALL), **diffuse large B-cell lymphoma**, and **Burkitt lymphoma**. Apart from ALL, the condition is relatively rare, so that there is little evidence-based therapy. In addition to CNS penetrating systemic therapy, intrathecal injection of **cytotoxic agents** such as cytarabine or methotrexate can be tried, particularly where there is a prospect for remission following **allogeneic stem cell transplantation**. This regime can also be used as a palliative treatment in leptomeningeal relapse and where intensive therapy with curative intent is not an option.

LYMPHOPLASMACYTIC LYMPHOMA/WALDENSTRÖM MACROGLOBULINEMIA

See also **Lymphoproliferative disorders**; **Non-Hodgkin lymphoma**.

(Immunocytoma, lymphoplasmacytic type; Rappaport: well-differentiated lymphocytic, plasmacytoid, diffuse lymphocytic, and histiocytic lymphoma; Lukes-Collins: plasmacytic-lymphocytic lymphoma; plasmacytoid lymphoma) A high-grade B-cell non-Hodgkin lymphoma characterized by diffuse proliferation of small lymphoplasmacytic cells. These cells have abundant basophilic cytoplasm with lymphocytelike nuclei. Many cells have large intranuclear inclusions (Dutcher bodies) that stain positive with para-aminosalicylic acid (PAS). The growth pattern in the lymphoid follicles is often interfollicular, with sparing of the sinuses and little stromal destruction. The cells have surface and cytoplasmic Ig, usually M type, and usually lack IgD, with **immunophenotype** B-cell-associated antigen[+], CD5[-], CD10[-], CD43[+/-], CD25, or CD11c faintly positive in some cases. **Cytogenetic analysis** shows rearrangement of heavy- and light-chain genes but no specific abnormality. The origin of the tumor is probably CD5[-] peripheral B-cell undergoing differentiation to a **plasma cell**. Transformation to a **diffuse large B-cell lymphoma** may occur.

Clinical Features

The lymphoma occurs in middle and older age groups. The sites involved are bone marrow, lymph nodes, and spleen, and less frequently peripheral blood and extranodal sites. The majority of patients have a monoclonal serum paraprotein of IgM type, and **hyperviscosity** may follow. The course is indolent.

Treatment

The disorder does not respond well to treatment with either **cytotoxic agent** therapy or **radiotherapy**. Patients may respond to newer interventions that include **rituximab** with CHOP chemotherapy; rituximab, dexamethasone, and cyclophosphamide; rituximab fludarabine; and 2-CDA (2′-chlorodeoxyadenosine), rituximab, and cyclophosphamide.

LYMPHOPOIESIS
See **Lymphocytes**.

LYMPHOPROLIFERATIVE DISORDERS
An increase in the division of the cells of the **lymphoreticular system**, leading to the formation of a tumor mass termed a lymphoma. An increase of abnormal lymphocytes in the peripheral blood is termed a lymphatic leukemia.

Causal Factors
There continues to be a worldwide increase of the incidence of these disorders, the precise cause remaining unknown, but a number of possible risk factors, based upon epidemiological associations, have been identified (see **Oncogenesis**).

Virus infection: **human T-lymphotropic virus** (HTLV)-1 is clearly associated with **adult T- cell lymphoma**, both conditions having an overlapping incidence. **Epstein-Barr virus** (EBV) has been isolated from cell lines derived from **Burkitt lymphoma**, and patients with this lymphoma have EBV antibodies. EBV is also associated with **posttransplant lymphoproliferative disorders**. Hepatitis C virus is associated with **cryoglobulinemia**, and there is also an above-average incidence of lymphoproliferative disorders that also occur with human herpesvirus (HHV)-6 or HHV-8 infection. A viral association with **myelomatosis** has recently been suggested.

Bacterial infection: *Helicobacter pylori* and *Borrelia burgdorferi* are both associated with the **gastric disorder** of mucosa-associated lymphoid tissue (MALT) lymphoma.

Ionizing radiation: following high-dosage exposure to **ionizing radiation**, as occurred in Hiroshima and Nagasaki, there has been a fivefold increased incidence of lymphomas after a latent period of 15 years. A similar increased incidence has been expected from the fallout radiation of the Chernobyl disaster of 1986, but as yet no increased incidence of leukemia has occurred.

Genetic inheritance (**Lymphoproliferative diseases associated with primary immune disorders**): An increased incidence occurs in hereditary ataxia-telangiectasia, **Fanconi's anemia**, **Wiscott-Aldrich syndrome**, common-variable immune deficiency, severe combined immunodeficiency, X-linked lymphoproliferative disorder, Nijmegen breakage syndrome, hyper-IgM syndrome, autoimmune lymphoproliferative syndrome, and **Bloom's syndrome**. Nonrandom cytogenetic change is found in some patients (see Table 97).

Immunodeficiency: both hereditary and acquired **immunodeficiencies** lead to an increased incidence of all types of lymphoproliferative disorder. This is particularly prevalent in patients with **acquired immunodeficiency syndrome**. Secondary malignancies following **cytotoxic agent** therapy have been reported.

Chronic antibody stimulation: could be the origin of **heavy-chain diseases**. This might be the factor in lymphomas following **sarcoidosis** and gastric lymphomas associated with *Helicobacter pylori* infection (see **Gastric disorders**).

Occupational (agriculture): exposure to pesticides or increased exposure to solar ultraviolet rays may be a causative factor.

Classification

The disorders are classified as either **Hodgkin disease** or **non-Hodgkin lymphoma/leukemia** (NHL). Both groups are further classified into subtypes. Over 20 classifications have been proposed, and at least eight different classification schemes have been utilized worldwide that were not derived from information in cytochemistry, cytogenetics, or molecular genetics. The most commonly applied classification systems discriminate between low-, intermediate-, and high-grade subtypes. It is not possible to compare clinical trials that utilize different classifications, because not all subcategories can be translated from one classification scheme to another. The Working Formulation for Clinical Usage (WFCU), which grouped non-Hodgkin lymphomas by natural history and response to therapy, has been the most utilized scheme in the U.S.[332] (See **Non-Hodgkin lymphoma** — Table 124.)

This and other existing schemas (e.g., the Kiel classification[333]) have been based upon morphology only. These morphologic patterns have characteristic-associated cytochemical, cytogenetic, and oncogene associations. This classification scheme broadly categorizes patients into two groups: the favorable low-grade lymphomas and the unfavorable intermediate- and high-grade lymphomas. Some patients have different histologic types of the lymphoma in the same biopsy specimen or in different biopsy specimens sampled at the same time. If both follicular and diffuse areas are involved, then the lymphomas are called follicular in the WFCU, but the disease in these patients has the significant characteristics of each histologic subtype, and patients are treated as having the more aggressive subtype. An intermediate-grade histology may be present at one site and low-grade histology at another site. The most frequent discordance is diffuse large-cell lymphoma in lymph nodes and follicular small cleaved cell in the bone marrow.

These classifications have not always overlapped with those based on immunological data, such as the Lukes-Collins classification.[334] Furthermore, several subtypes of non-Hodgkin lymphomas have recently been characterized that do not fit into the WFCU. Diffuse intermediate and mantle-zone lymphomas originate from mantle-zone B-lymphocytes, express CD5 antigen, frequently contain the t(11;14) chromosome translocation, and have a survival intermediate between that of small lymphocytic and diffuse small-cleaved-cell NHL. Anaplastic large-cell lymphomas frequently have Ki-1 antigen (CD30), are often T-cell in

TABLE 97

Genotypic and Cytogenetic Changes of Lymphoproliferative Disorders

Histological Type	Genetic Features
Hodgkin Disease	
Lymphocyte predominance	Ig and TCR genes germline, large cells EBV–
Nodular sclerosis	Ig and TCR genes usually germline, rearrangements occur, tumor cells EBV+ (40%)
Mixed cellularity	Ig and TCR genes usually germline, tumor cells EBV+ in majority
Lymphocyte depletion	Ig and TCR genes germline
Non-Hodgkin Lymphoma	
B-Cell Neoplasms	
I. Precursor B-lymphoblastic	Cytogenetics variable; Ig heavy chains usually rearranged, light chain genes may be rearranged
	T-cell receptor rearrangements present in a minority
II. Peripheral B-cell neoplasms	
1. B-cell chronic lymphatic lymphoma	Trisomy 12 in 33%; 13q up to 25%; T(11,14) and *bcl-1* reported; IgG heavy and light chains rearranged
2. Lymphoplasmacytoid lymphoma	No specific abnormalities; IgG heavy and light chains rearranged
3. Mantle-cell lymphoma	t(11,14) and *bcl-1* positive in majority
4. Follicle center cell lymphoma	*bcl-2* expression present in most; t (14,18) (q32;q21) present in 70–95%; IgG gene rearranged
5. Marginal-zone B-cell lymphoma	No *bcl-2* or *bcl-1*; trisomy 3 or t (11,18) in extranodal
6. Splenic marginal-zone lymphoma	Not well studied
7. Hairy cell leukemia	No specific abnormality; Ig heavy and light chains rearranged
8. Plasmacytoma	IgH and L chain genes rearranged or deleted
9. Diffuse large B-cell lymphoma	*bcl-2* gene rearranged in about 30%; c-myc rearranged 6q21-23 abnormalities reported
Subtype: primary mediastinal (thymic) lymphoma	IgH and L chains rearranged; no specific abnormality described
10. Burkitt's lymphoma	*c-myc* translocation t(8,14) or less common t(2,8) (p11-12;q24) or 5(8,22) (q24;q11)
11. High-grade B-cell lymphoma	c-myc rearrangement uncommon; *bcl-2* rearranged 30%
T-Cell Neoplasms	
I. Precursor T-lymphoblastic	Rearrangements of TCR genes variable
II. Peripheral T-cell and NK neoplasms	
1. T-cell chronic lymphatic lymphoma	Clonal rearrangements of TCR genes: inv 14 (q11;q32) in 75%; trisomy 8q
2. Large granular lymphoma	T-cell cases: clonal rearrangements of TCR genes
	NK-cell cases: germline and clonality not proven
3. Mycosis fungoides	TCR genes clonally rearranged
4. Peripheral T-cell lymphoma	TCR genes usually rearranged; Ig genes germline
5. Angioimmunoblastic lymphoma	TCR genes rearranged 75%; IgH rearranged 10%; trisomy 3 and/or 5 may occur; EBV genome detected in many
6. Angiocentric lymphoma	Clonal Ig gene rearranged; EBV genome present in B-cells
7. Intestinal T-cell lymphoma	TCR b genes clonally rearranged
8. Adult T-cell lymphoma	TCR genes clonally rearranged; clonally integrated HTLV-1 genomes found in all cases
9. Anaplastic large-cell lymphoma	TCR genes rearranged 50–60%

Source: Data derived from Harris, N.L. et al., *Blood*, 84, 1361–1392, 1994. With permission.

origin, typically present with extranodal disease that most commonly involves the skin, and treatment is similar to diffuse large-cell lymphoma. Monocytoid B-cell lymphoma was reported to be a novel B-cell neoplasm. Considerable antigenic heterogeneity is present in the different subgroups of non-Hodgkin lymphomas. Monoclonality in B-cell lymphomas can be defined operationally by light-chain restriction on the cell surfaces or cytoplasm, but there is not a similar immunohistochemical counterpart in T-cell lymphoproliferative diseases. Evidence supports the clonal origin of lymphoid neoplasms, and rearrangements of antigen-receptor genes are found in most malignant lymphomas of B-cells or T-cells. Previous classification schemes did not incorporate these new and evolving concepts.

To recognize new entities and refine previously recognized disease categories, the International Lymphoma Study Group reported on a revised European–American classification of lymphoid neoplasms (REAL classification) (Tables 98 and 99).[335] This classification was subsequently modified to be the World Health Organization (WHO) Classification of Tumors of Hematopoietic and Lymphoid Tissues (see Appendix). The lymphoproliferative disorders are listed in Table 100.

Anatomical Distribution

The predominant site for each type of lymphoma (see Table 101) provides an early clinical feature upon which the differential diagnosis can be based.

Staging

The Ann Arbor staging system (see Table 102) was used initially for Hodgkin disease,[336] but it is now also applied to non-Hodgkin lymphomas. The initial evaluation includes the history, physical examination, complete blood count, **erythrocyte sedimentation rate**, **lactate dehydrogenase** (LDH), liver function tests, serum creatinine, serum calcium, chest radiograph, computed tomographic scan of the abdomen (with or without the chest), and bilateral **bone marrow** studies.[337]

Immunophenotyping

The differential diagnosis using the WHO classification (see Appendix) depends upon morphological, cytogenetic, and immunophenotyping features. The immunophenotypes of these disorders[296,338] are summarized in Table 103.

Diagnosis

The precise diagnosis of lymphoma type depends upon morphological appearances (see Classification subsection, above) together with immunophenotyping (see Immunophenotyping subsection, above). The clinical features vary with each tumor type, and none of the biological markers are totally lineage specific. Molecular genetics and cytogenetics incorporate the pathologic events, but they are not routinely used by the hematopathologist. Fine-needle aspiration cytology with clonality analysis and immunophenotyping by flow cytometry can be used for screening.

Treatment

This is dependent upon the specific disorder and its staging (see Classification subsection, above, and Ann Arbor staging [Table 102]). It is, broadly, some regime of cytotoxic agent therapy supported in a few patients by radiotherapy.

LYMPHOPROLIFERATIVE DISORDERS ASSOCIATED WITH PRIMARY IMMUNE DISORDERS

See **Lymphoproliferative disorders**; **Immunodeficiency**.

TABLE 98

REAL Classification of Lymphoid Tumors

CLASSIFICATION OF LYMPHOID NEOPLASMS RECOGNIZED BY THE INTERNATIONAL LYMPHOMA
STUDY GROUP (REVISED EUROPEAN–AMERICAN CLASSIFICATION OF LYMPHOID NEOPLASMS
[REAL CLASSIFICATION])

B-Cell Neoplasms

   I. Precursor B-cell neoplasms: precursor B-lymphoblastic leukemia/lymphoma

  II. Peripheral B-cell neoplasms

     1. B-cell chronic lymphocytic leukemia/prolymphocytic leukemia/small lymphocytic lymphoma

     2. Lymphoplasmacytoid lymphoma/immunocytoma

     3. Mantle cell lymphoma

     4. Follicle center lymphoma, follicular

       Provisional cytologic grades: I (small cell), II (mixed, small, and large cell), III (large cell)

       Provisional subtype: diffuse, predominantly small-cell type

     5. Marginal zone B-cell lymphoma

       Extranodal (MALT-type +/− monocytoid cells)

       Provisional subtype: nodal (+/− monocytoid cells)

     6. Provisional entity: splenic marginal-zone lymphoma (+/− villous lymphocytes)

     7. Hairy cell leukemia

     8. Plasmacytoma/plasma cell myeloma

     9. Diffuse large B-cell lymphoma

       Subtype: primary mediastinal (thymic) B-cell lymphoma

    10. Burkitt's lymphoma

    11. Provisional entity: high-grade B-cell lymphoma, Burkitt-like

T-Cell and Putative NK-Cell Neoplasms

   I. Precursor T-cell neoplasm: precursor T-lymphoblastic lymphoma/leukemia

  II. Peripheral T-cell and NK-cell neoplasms

     1. T-cell chronic lymphocytic leukemia/prolymphocytic leukemia

     2. Large granular lymphocytic leukemia (LGL)

       T-cell type

       NK-cell type

     3. Mycosis fungoides/Sézary syndrome

     4. Peripheral T-cell lymphomas, unspecified

       Provisional cytologic categories: medium-sized cell, mixed medium and large cell, large cell, lymphoepithelioid cell

       Provisional subtype: hepatosplenic CD T-cell lymphoma

       Provisional subtype: subcutaneous panniculitic T-cell lymphoma

     5. Angioimmunoblastic T-cell lymphoma (AILD)

     6. Angiocentric lymphoma

     7. Intestinal T-cell lymphoma (+/− enteropathy associated) (See **Intestinal cell lymphoma**)

     8. Adult T-cell lymphoma/leukemia (ATL/L)

     9. Anaplastic large-cell lymphoma (ALCL), CD30+, T- and null-cell types

    10. Provisional entity: anaplastic large-cell lymphoma, Hodgkin's-like

Hodgkin's Disease (HD)

    I. Lymphocyte predominance

   II. Nodular sclerosis

 III. Mixed cellularity

 IV. Lymphocyte depletion

  V. Provisional entity: lymphocyte-rich classical HD

TABLE 99

Comparison of Commonly Used Lymphoma Classifications

Kiel Classification	REAL Classification	Working Formulation
B-lymphoblastic B lymphocytic, chronic lymphocytic leukemia (CLL) B lymphocytic, prolymphocytic chronic lymphatic leukemia	Precursor B-lymphoblastic lymphoma/leukemia	Lymphoblastic
Lymphoplasmacytoid immunocytoma Small lymphocytic, plasmacytoid	B-cell chronic lymphocytic leukemia/prolymphocytic leukemia/small lymphocytic lymphoma	Small lymphocytic, consistent with CLL
Lymphoplasmacytic immunocytoma	Lymphoplasmacytoid lymphoma	Small lymphocytic, plasmacytoid Diffuse, mixed, small, and large cell Small lymphocytic
Centrocytic Centroblastic centrocytoid subtype	Mantle cell lymphoma	Small lymphocytic Diffuse, small cleaved cell Follicular, small cleaved cell Diffuse, mixed, small, and large cell Diffuse, large cleaved cell

Note: The REAL classification scheme incorporated many lymphoma entities that were recognized by the Kiel Classification, the Lukes-Collins classification, and the REAL Working Formulation. The REAL classification included established entities, differentiated B- and T-cell disorders, incorporated subsets of B-cell disorders, and included distinct T-cell disorders. It was defined by morphologic, immunologic, and genetic techniques available in 1994.

Source: Modified from Harris, N.L. et al., *Blood*, 84, 1361–1392, 1994. With permission.

TABLE 100

WHO Classification of Tumors of Lymphoid Tissue

B-Cell Neoplasms

Precursor B-cell neoplasm
Precursor lymphoblastic leukemia/lymphoma

Mature B-cell neoplasms
Chronic lymphocytic leukemia/small lymphocytic lymphoma
B-cell prolymphocytic leukemia
Lymphoplasmacytic lymphoma
Splenic marginal-zone lymphoma
Hairy cell lymphoma
Plasma-cell lymphoma
Solitary plasmacytoma of bone
Extraosseous plasmacytoma
Extranodal marginal-zone B-cell lymphoma of mucosa-associated lymphoid tissue (MALT-lymphoma)
Nodal marginal-zone B-cell lymphoma
Follicular lymphoma
Mantle-cell lymphoma
Diffuse large B-cell lymphoma
Mediastinal (thymic) large B-cell lymphoma
Intravascular large B-cell lymphoma
Primary effusion lymphoma
Burkitt lymphoma/leukemia

B-cell proliferations of uncertain malignant potential
Lymphomatoid granulomatosis
Posttransplant lymphoproliferative disorder, polymorphic

TABLE 100 (continued)

WHO Classification of Tumors of Lymphoid Tissue

T-Cell and NK-Cell Neoplasms

Precursor-T-cell neoplasms
Precursor T-lymphoblastic leukemia/lymphoma
Blastic NK-cell lymphoma

Mature T-cell and NK-cell neoplasms
T-cell prolymphocytic leukemia
T-cell large granular lymphocytic leukemia
Aggressive NK-cell leukemia
Adult T-cell leukemia/lymphoma
Extranodal NK/T-cell lymphoma, nasal type
Enteropathy-type T-cell lymphoma
Hepatosplenic T-cell lymphoma
Subcutaneous panniculitislike T-cell lymphoma
Mycosis fungoides
Sézary syndrome
Primary cutaneous anaplastic large-cell lymphoma
Peripheral T-cell lymphoma, unspecified
Angioimmunoblastic T-cell lymphoma
Anaplastic large-cell lymphoma

T-cell proliferations of uncertain malignant potential
Lymphomatoid papulosis

Hodgkin Lymphoma

Nodular lymphocyte-predominant Hodgkin lymphoma
Classical Hodgkin lymphoma
Nodular sclerosis classical Hodgkin lymphoma
Lymphocyte-rich classical Hodgkin lymphoma
Mixed-cellularity classical Hodgkin lymphoma
Lymphocyte-depleted classical Hodgkin lymphoma

Immunodeficiency-Associated Lymphoproliferative Disorders

Primary immunodeficiency or primary immunoregulatory disorder
Human immunodeficiency virus (HIV) infection (see **Acquired immunodeficiency syndrome**)
Post-transplant lymphoproliferative disorders (see **Allogeneic stem cell transplantation**)
Methotrexate-associated disorders (see **Antimetabolites**)

LYMPHORETICULAR SYSTEM

A term introduced to unite the lymphoid system with histiocytes (macrophages) and dendritic reticulum cells. In this respect, it is the morphological counterpart of the immune system.

LYONIZATION

The inactivation pattern of the X **chromosom**e and its relationship to inheritance. It is named after Mary Lyons, a British cytogeneticist who first made the observation that female embryonic tissue contains two copies of X chromosomes per cell, one maternally (Xm) and one paternally (Xp) derived. During embryonic development, one X chromosome is randomly inactivated such that any one tissue becomes a mosaic for Xm and Xp.

TABLE 101

Predominant Sites of Selected Lymphoproliferative Disorders at Presentation

	PB	LN	S	L	BM	N/P	R	C	G/I	CNS
Hodgkin Disease										
I. Lymphocyte predominant		+	±	±	±					
II. Nodular sclerosis		+	±	±	±					
III. Mixed cellularity		+	+	+	±					
IV. Lymphocyte depletion		+	+	+	+					
Non-Hodgkin Lymphoma										
B-Cell Neoplasms										
I. Precursor B neoplasms	+	±			±			±		+
II. Mature B-cell neoplasms										
1. Small lymphocytic lymphoma	+	+	+	+	+			±		
2. Lymphoplasmacytoid	±	+	+		+					
3. Mantle-cell lymphoma		+	+		+	+			+	
4. Follicle lymphoma	±	+	+		+					
5. Extranodal marginal-zone lymphoma		+						±	+	
6. Splenic marginal-zone lymphoma	±		+		±					
7. Hairy cell leukemia	+	+	+		+					
8. Plasmacytoma	±				+					
9. Diffuse large B-cell lymphoma	±	+	+	+	±		±			
Subtype: primary mediastinal (thymic)		+					±			
10. Burkitt's lymphoma	±	+				+			+	
11. High-grade B-cell lymphoma		+								
T-Cell and NK-Cell Neoplasms										
I. Precursor T neoplasms	+	+								+
II. Mature T-cell and NK neoplasms										
1. T-cell prolymphocytic	+	+	+	+	+		±			
2. T-cell large granular lymphocytic leukemia	+	±								
3. Mycosis fungoides	+	+						+		
4. Peripheral T-cell lymphoma		+	+	+			±	+		
5. Angioimmunoblastic T-cell lymphoma		+	+	+			±			±
6. Extranodal NK/T-cell lymphoma, nasal type		+				+	+	+		+
7. Enteropathy-type T-cell lymphoma									+	
8. Adult T-cell lymphoma	+	+	+	+	+		±	+		±
9. Anaplastic large-cell lymphoma		+						+		

Note: PB, peripheral blood; LN, lymph nodes; S, spleen; L, liver; BM, bone marrow; N/P, nose/pharynx; R, respiratory tract; C, cutaneous (skin); G/I, gastrointestinal tract; CNS, nervous system; +, commonly involved; ±, rarely at presentation.

TABLE 102

Ann Arbor Staging System of Lymphomas

Stage	Criteria
IA or IB	A single nodal region or a single extranodal site
IIA or IIB	Two or more nodal regions or an extranodal site and regional nodal involvement on the same side of the diaphragm
IIIA or IIIB	Lymphatic involvement on both sides of the diaphragm
IVA or IVB	Liver and bone marrow involvement or extensive involvement of another extralymphatic organ

Note: A denotes absence of fever, night sweats, or 10% weight loss; B denotes presence of fever, night sweats, and 10% weight loss.

The decision to inactivate is not based upon any specific **DNA** sequence. The mechanism of X inactivation is unclear but may involve repression by heterochromation (position-effect variegation) or DNA **methylation**. Clonally derived cell populations show a single X-inactivation pattern, and this is used for many available clonality assays.

LYOPHILIZED COAGULATION FACTOR CONCENTRATES
See **Coagulation factor concentrates**.

LYSOSOME
A membrane-bounded cytoplasmic organelle containing a variety of hydrolytic enzymes that can be released into a phagosome or to the exterior of the cell. Release of lysosomal enzymes in a dead cell leads to autolysis. Early endosomes are located at the periphery of a cell, and late endosomes are located in the perinuclear region, with the lysosomes between. Lysosomes are composed of membrane and vesicles containing hydrolytic enzymes. Primary lysosomes are small and contain no inclusions; secondary lysosomes are larger and contain partially degraded organelles. Secondary lysosomes are phagocytic vesicles with which primary lysosomes have fused. They often contain undigested material.

Functioning at a low pH (4.8), the enzymes are a series of acid hydrolases, including proteases to degrade proteins and polypeptides, nucleases to degrade RNA and DNA, and phosphatases, among others. Their function is to mediate in the degradation of senescent membrane components and organelles and to digest endocytosed foreign material within the cell. Following contact with bacteria they form phagosomes.

Lysosomal Storage Diseases[339]
Although the first description of a lysosomal storage disorder was that of Tay-Sachs disease in 1881, the lysosome was not discovered until 1955, by Christian De Duve. The first demonstration by Hers in 1963 of a link between an enzyme deficiency and a storage disorder (Pompe's disease) paved the way for a series of seminal discoveries about the intracellular biology of these enzymes and their substrates, culminating in the successful treatment of **Gaucher's disease** with beta-glucosidase in the early 1990s. It is now recognized that these disorders are not simply a consequence of pure storage, but result from perturbation of complex cell-signaling mechanisms. These, in turn, give rise to secondary structural and biochemical changes that have important implications for therapy. Defective lysosomal acid hydrolysis of endogenous macromolecules leads to accumulation of lipids and mucopolysaccharides (see **Lipid-storage disorders**; **Mucopolysaccharidoses**). Over 40 disorders have been described. They are all single-gene, autosomally recessive disorders. Excessive accumulation occurs in macrophages, including microglia and in mesenchymal cells. The disorders are multisystem but particularly affect the liver, spleen, and central nervous system (CNS). This produces disturbances due to space occupation, but there is also macrophage activation and possibly disturbance of cellular mitochondrial function. Where the enzyme defect can be identified, enzyme-replacement therapy is the most likely effective treatment. **Allogeneic stem cell transplantation** can help, but the outcome is variable. Significant challenges remain, particularly the treatment of CNS disease. It is hoped that recent advances in understanding of lysosomal biology will enable successful therapies to be developed.

TABLE 103

Immunophenotypes of Lymphoproliferative Disorders

Hodgkin Disease

Lymphocyte predominance   CD45$^+$, B-cell-associated antigens$^+$, Cdw75a$^+$, EMA$^{+/-}$, CD115$^-$, CD30$^{-/+}$, T-cells surrounding L and H cells: CD57$^+$

Nodular sclerosis   CD15$^{+/-}$, CD30$^+$, CD45$^-$

Mixed cellularity   B- and T-cell-associated antigens$^-$

Lymphocyte depletion   EMA$^-$

Non-Hodgkin Lymphoma

B-Cell Neoplasms

I. Precursor B-lymphoblastic   leukemia/lymphoma Tdt$^+$, CD19$^+$, CD79a$^+$, CD22$^+$, CD20$^{+/-}$, CD10$^{+/-}$, HLA-DR$^+$, SIg$^-$, cMu$^{-/+}$, CD34$^{+/-}$, may coexpress CD13 or CD33$^{-/+}$

II. Peripheral B-cell neoplasms

1. Chronic lymphocytic leukemia/small lymphocytic lymphoma   faint SIgM$^+$, SIgD$^{+/-}$, CIg$^{-/+}$, B-cell-associated antigen$^+$, CD5$^+$, CD23$^+$, CD43$^+$, CD11c$^{-/+}$ (faint), CD10$^-$

2. Lymphoplasmacytic lymphoma/ Waldenström macroglobulinemia   SIgM$^+$, CIgM$^+$ (some cells), B-cell-associated antigens$^+$, CD5$^-$, CD10$^-$, CD43$^{+/-}$, CD25$^+$, or CD11c$^+$ (faint)

3. Mantle-cell lymphoma   SIgM$^+$ usually IgD$^+$ l > k, B-cell-associated antigens$^+$, CD5$^+$, CD110$^{-/+}$, CD23$^-$, CD43$^+$, CD11c$^-$

4. Follicular lymphoma   SIg$^+$ (IgM$^{+/-}$ IgD > IgG > IgA), B-cell-associated antigen$^+$, CD10$^{+/-}$, CD5$^-$, CD23$^{-/+}$, CD43$^-$, CD11c$^-$

5. Extranodal marginal-zone B-cell lymphoma (MALT lymphoma)   SIg$^+$ (IgM > IgG or IgA), CIg$^+$ (40%), B-cell-associated antigen$^+$, CD5$^-$, CD10$^-$, CD23$^-$, CD43$^{-/+}$, CD11c$^{+/-}$

6. Splenic marginal-zone lymphoma   Similar to marginal-zone B-cell lymphoma, above

7. Hairy cell leukemia   SIg$^+$ (M$^{+/-}$ IgD, IgG, or IgA), B-cell-associated antigens$^+$, CD5$^-$, CD10$^-$, CD23$^-$, CD11c$^+$ (strong), CD25$^+$ (strong), FMC7$^+$, CD103$^+$ (MLA: HML-1, B-ly7, Ber-ACT8, LF61)

8. Plasmacytoma   SIg$^-$, CIg$^+$ (IgG, IgA, rare IgD or IgE; or light chain only); most B-cell-associated antigens$^-$, CD79a$^{+/-}$, CD45$^{-/+}$, HLA-DR$^{-/+}$, CD38$^+$, EMA$^{-/+}$, CD43$^{+/-}$, CD56$^{+/-}$

9. Diffuse large B-cell lymphoma   SIg$^{+/-}$, CIg$^{-/+}$, B-cell-associated antigens$^+$ (CD19$^+$, CD20$^+$, etc.), CD45$^{+/-}$, CD5$^{-/+}$, CD10$^{-/+}$

10. Mediastinal (thymic) large B-cell lymphoma   Ig$^-$, B-cell-associated antigens$^+$, CD45$^+$, CD15$^-$

11. Burkitt lymphoma/leukemia   SIgM$^+$, B-cell-associated antigens$^+$, CD10$^+$, CD5$^+$, CD23$^-$

12. Lymphomatoid granulomatosis   CD20$^+$, CD79a variable$^+$, variable CD30$^+$, LMP1$^+$, CD15

T-Cell and NK-Cell Neoplasms

I. Precursor T-lymphoblastic   CD7$^+$, CD3$^+$, Tdt$^+$, CD1a$^{+/-}$, CD4, 8^{++} or CD4, 8^{--}, Ig$^-$, occasional cases express natural killer (NK) antigens (CD16, 57)

II. Peripheral T-cell and NK-cell neoplasms

1. T-cell chronic lymphatic   CD7$^+$, T-cell-associated antigens$^+$, CD4$^+$ (65%), CD4$^+$ CD8$^+$ (21%), CD4$^-$ CD8$^+$ (rare), CD25$^-$

2. T-cell large granular lymphocytic leukemia   *T-cell:* cD2$^+$, CD3$^+$, CD5$^-$, CD7$^-$, TCRab$^+$, CD4$^-$, CD8$^+$, CD16$^+$, CD56$^-$, CD57$^{+/-}$, CD25$^-$
NK-cell: CD2$^+$, CD3$^-$, TCRab$^-$, CD4$^-$, CD8$^{+/-}$, CD16$^+$, CD56$^{+/-}$, CD57$^{+/-}$

3. Mycosis fungoides/Sézary syndrome   T-cell-associated antigens$^+$, CD7$^+$ (33%), CD4$^+$, CD25$^-$, S-100$^-$ CD1a$^+$ interdigitating Langerhans cells present

4. Peripheral T-cell lymphoma, unspecified   T-cell-associated antigens variable, CD4 > CD8, may be CD4$^-$, CD8$^-$

5. Angioimmunoblastic T-cell   T-cell-associated antigens$^+$, usually CD4$^+$

6. Angiocentric lymphoma   pan T antigens expressed (CD2$^+$, CD5$^{+/-}$, CD7$^{+/-}$), often CD3$^-$, may be CD4$^+$ or CD8$^+$, and often CD56$^+$

TABLE 103 (continued)

Immunophenotypes of Lymphoproliferative Disorders

7. Enteropathy-type T-cell lymphoma	$CD3^+$, $CD7^+$, $CD8^{+/-}$, $CD4^-$, $CD103^+$ (MLA: HML-1, LFG 1, Bly7, Bev-ACTB)
8. Primary cutaneous CD30-positive T-cell lymphoproliferative disorders	
a: Primary cutaneous anaplastic large-cell lymphoma (C-ALCL)	$CD4^+$, $CD30^+$ (>75%), variable $CD2^-$, $CD5^-$, and $CD3^-$
b: Lymphomatoid papulosis	$CD4^+$, $CD8^-$, $CD30^+$ in type A lesions and $CD30^-$ in type B lesions, variable loss of $CD2^-$, $CD5^-$
c: Borderline lesions	
9. Anaplastic large cell	$CD30^+$, $CD45^{+/-}$, $CD25^{+/-}$, $EMA^{+/-}$, $CD15^{-/+}$, $CD3^{-/+}$, T-cell-associated antigens variable, $CD45RO^{-/+}$, $CD68^-$

Note: +, >90% cases; +/−, >50% cases; −, <10% cases; −/+, <50% cases. B-cell-associated antigens = CD19, CD20, CD22, CD79a; T-cell-associated antigens = CD2, CD3, CD5.

Source: Data derived from Harris, N.L. et al., *Blood*, 84, 1361–1392, 1994. With permission.

M

MACROANGIOPATHIC HEMOLYTIC ANEMIA

(Traumatic mechanical hemolytic anemia; cardiac hemolytic anemia; march hemoglobinuria) Anemias arising as a result of mechanical trauma to **red blood cells** circulating through the heart or blood vessels, with and without surgical intervention.

Following heart valve replacement, severe **hemolysis** does not arise from hemodynamic turbulence alone, but in a space bound by a foreign surface. Nonendothelialized surfaces are thrombogenic and may cause platelet aggregation, thrombus formation, and distant embolization. Largely to overcome these thrombogenic problems, nonthrombogenic tissue valves have been developed, which has minimized hemolysis. Mildly compensated hemolysis is usually present. Even so, the **reticulocyte count** is slightly elevated, as is the level of serum **lactate dehydrogenase**. The peripheral blood shows changes of **fragmentation hemolytic anemia** — helmet cells, triangular cells, and other fragmented forms. **Hemoglobinemia** may be present and the **haptoglobin** level reduced. There is often hemosiderinuria. **Iron deficiency** may occur. In severe cases, the only treatment may be to replace the valve. In mild cases, treatment with **iron** and **folic acid** will usually prevent decompensation.

Without surgery, macroangiopathic hemolytic anemia is usually associated with:

Aortic stenosis

Mitral regurgitation

Ruptured sinus of Valsalva

Ruptured chordae tendinae

Coarctation of aorta

Aortic aneurysm

March hemoglobinuria

MACROCYTES

Large **red blood cells** with a diameter >8.5 μm and mean cell volume usually >96 fl. There are two forms:

Oval macrocytes characteristic of **megaloblastosis, bone marrow hypoplasia, myelodysplasia**, and other dyserythropoietic states

Round macrocytes found in chronic **liver disorders**, myelodysplasia, **myelomatosis**, bone marrow hypoplasia, and **tobacco** smoking

Macrocytes are also found when there is increased **erythropoiesis**, when they are due to the presence of reticulocytes. On Romanowsky stained blood films, reticulocytes are more basophilic than adult red cells (**polychromasia**). Physiological macrocytosis occurs in the neonatal period and with pregnancy. A rare familial type has been described.

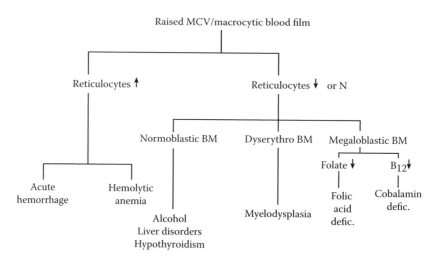

FIGURE 83

Flow diagram of investigation of macrocytic anemias. (Adapted from Bates, I. and Bain, B.J. Approach to the diagnosis and classification of blood disorders, in *Dacie and Lewis Practical Haematology,* 10th ed., Churchill-Livingstone Elsevier, 2006, Philadelphia, Figure 23.2. With permission.)

MACROCYTIC ANEMIA

The association of a reduced level of **hemoglobin** with macrocytosis, demonstrated either by automated blood cell counting or upon examination of peripheral-blood film. Differentiation is a stepwise process (see Figure 84). The initial investigation is a **reticulocyte count**; a raised count suggests either acute **hemorrhage** or some form of **hemolytic anemia**. A normal or low reticulocyte count requires a **bone marrow** aspirate/trephine biopsy to establish the presence or absence of **megaloblastosis**. Assays for **cobalamin** and **folic acid** are then indicated. A **dyserythropoietic** marrow count less than the megaloblastic count is usually due to one of the forms of **myelodysplasi**a or, less frequently, **aplastic anemia**. A normoblastic marrow suggests that the macrocytosis may be due to **alcohol toxicity**, the commonest cause of macrocytosis, or the result of a **liver disorder** or **thyroid disorder**, usually hypothyroidism.

MACROGLOBULINEMIA

See **Lymphoplasmacytic lymphoma/Waldenström macroglobulinemia**.

MACROPHAGE

See **Histiocyte**.

MACROPOLYCYTE

Giant **neutrophils** seen in **megaloblastosis**.

MAJOR HISTOCOMPATIBILITY COMPLEX

(MHC) See **Human leukocyte antigens**.

MALARIA

The pathogenesis, clinical and laboratory features, and management of infection by species of the genus *Plasmodium*, which is composed of *P. falciparum*, *P. malariae*, *P. ovale*, and *P. vivax*. On a global basis, this is the commonest infection of humans,[340] accounting for 300 to 500 million cases per annum, with a mortality of up to 2 million per annum. Transmission is by a bite from the female *Anopheles* mosquito, which resides in tropical climates of Central and South America, Asia, Africa, and Oceania. Many patients with malaria enter Europe and the U.S. as travelers from the tropics, but transmission onward in these countries is rare. Transmission of the erythrocyte cycle can be a **transfusion-transmitted infection** as a result of shared needles or syringes in drug abusers, by needle-stick injury in health-care workers, by organ transplantation, or by accidental laboratory inoculation. The species of infecting plasmodium varies from one region of the world to another, but in endemic areas, mixed infections can occur.

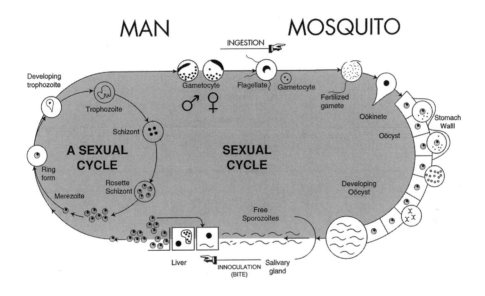

FIGURE 84
Life cycle of malaria plasmodium.

Pathogenesis

The life cycle of the plasmodia is a sexual cycle between human and mosquito, with an asexual cycle within humans (see Figure 84). Sporozoites enter the human circulation from mosquito saliva. They are rapidly cleared by liver **histiocytes** (macrophages), from where they enter the hepatic parenchymal cells — the hepatic phase of infection. Here they transform to schizonts, which subdivide rapidly during schizogony. A single sporozoite can divide to produce 2,000 to 40,000 merozoites in 1 to 2 weeks. These are released into the circulation when the schizont bursts and invades **red blood cells**. With infection by *P. falciparum* and *P. malariae*, the hepatic phase now ends, but with *P. ovale* and *P. vivax*, some schizonts can remain in the liver cells for years, from which later red blood cell invasion can occur. Entry into the red blood cells occurs by **receptor-mediated endocytosis**. For this, *P. vivax* is associated with **Duffy blood group** determinants (Fya, Fyb) so that Duffy-negative persons are naturally resistant to invasion by this parasite. *P. falciparum* receptor is associated with red blood cell membrane proteins Band 3 and glycophorin A. Malarial parasite invasion has the following effects on the red cell membrane:

Reorganization of the phospholipid distribution

New permeability pathways introduced

New surface antigens

Increased areas of attachment to endothelium associated with protrusions seen on electron microscopy; this specifically occurs with *P. falciparum* invasion and leads to cerebral microvascular infarctions — cerebral malaria

Intracellular merozoites transform to trophozoites — ring forms. These devour **hemoglobin** with the production of oxidized heme — **hematin** or "malarial pigment." For this reaction, *P. falciparum* can remove iron from **ferritin**, but the presence in the cell of **hemoglobin F, hemoglobin S,** or deficiency of the enzyme **glucose-6-phosphate dehydrogenase** confers resistance. This leads to a natural selection for survival in tropical areas of these genetic disorders. The trophozoites enlarge, taking an amoeboid shape that differs from one plasmodium to another, and by which a microscopic diagnosis can be made.

The trophozoites undergo schizogony to form a new generation of merozoites — the asexual reproduction cycle. The red blood cell is eventually lysed, with the merozoites entering the circulation to further invade red blood cells.

Some intracellular merozoites become sexual gametocytes, which are dormant and do not lyse the host cell. If these cells are ingested by a female *Anopheles* mosquito, sexual fertilization of male and female gametocytes occurs in the mosquito stomach wall. An oocyst develops within the egg cavity, with eggs eventually passing to the mosquito salivary gland. From here, the human cycle continues.

Malarial infection stimulates T-lymphocytes and histiocytes (macrophages). Release of opsonizing **antibodies** allows killing of parasites by cytotoxic T-lymphocytes and natural killer (NK) cells. This may be the mechanism for natural protection against malaria, with the parasitized cells being removed by the cells of the **reticuloendothelial system**.

Adhesion of infected red blood cells to **vascular endothelium** may be a factor in the cause of **normocytic anemia**. With massive cell lysis, **intravascular hemolysis** due to **disseminated intravascular coagulation** can occur, particularly with *P. falciparum*, resulting in **hemoglobinemia** and **hemoglobinuria** (blackwater fever). Hypovolemia and renal failure result, with a mortality of 20 to 30%. This is a particular complication in those taking quinine. Adhesion of *P. falciparum*-infected erythrocytes to cerebral vessels gives rise to the syndrome of cerebral malaria.

Extravascular hemolysis from erythrophagocytosis associated with splenic pooling of red blood cells and phagocytosis follows prolonged or recurrent infection, which can end with the syndrome of **hyperreactive malarial splenomegaly**.

Clinical Features

Incubation period from the mosquito bite to symptoms: 10 to 14 days for *P. falciparum* and *P. vivax*; 18 days to 6 weeks for *P. malariae* (see Table 104). Not all infected individuals develop symptoms, probably due to their level of reactive nitrogen intermediates (RNI) and their major histocompatibility complex (MHC) genes. Non-falciparum malaria is usually benign. The presenting illness is fever with headache. This has a periodicity depending upon the infecting plasmodium — tertian (every 3 days), quartan (every 4 days). Concomitant infection, particularly by HIV, has a deleterious and progressive effect on the malarial disease.

TABLE 104

Febrile Illness Depending upon Type of Plasmodium

Species	Clinical Disease
P. falciparum	malignant tertian malaria
P. vivax	benign tertian malaria
P. malariae	quartan malaria
P. ovale	tertian malaria

Benign Malaria

Patients develop cyclic fever with weakness, malaise, headache, and myalgias, followed by pallor with hepatosplenomegaly. The illness usually resolves without treatment but with recurrent febrile illness.

Malignant Malaria

Fever tends to be continual rather than cyclical, followed by convulsions and an impaired level of consciousness. Respiratory distress and circulatory collapse may occur. Severe pallor from hemolytic anemia with jaundice and hemoglobinuria is a common complication (blackwater fever). Renal failure may follow. Splenic rupture can complicate severe splenomegaly. There is a high mortality rate.

Laboratory Features[341,342]

Normochromic normocytic anemia **or anemia of chronic disorders**

Identification of invading plasmodium by:

- **Peripheral-blood film** stained by either Giemsa or Field's technique. Thin blood films or thick blood smears can be used, the thicker smears increasing the efficiency but in practice reducing the reliability. **Buffy coat** films may be preferred. Fluorescent microscopy using Kawamoto acridine orange or benzthiocarboxypurine is an alternative method of staining, but this requires special training and expensive equipment and reagents. Irrespective of technique, the smear should, whenever possible, be obtained at the height of the fever. The principal differentiating features are:

  P. falciparum: numerous small ring forms with double chromatin dots

  P. malariae: single ring forms with single chromatin dot

  P. ovale: ovoid ring forms; Schuffner's dots; large gametocytes

  P. vivax: single ring forms with single or double chromatin dot; few Schuffner's dots; multiple merozoites occur; large gametocytes (see Table 105); in addition, spherocytes are present

- Serological methods used for screening those suspected at **blood donation**, but these are too slow for routine diagnostic use

- Immunochromatographic technique is available to identify antibodies against a histidine-rich protein-2 synthesized by *P. falciparum* when invading red blood cells

- Molecular probes using **polymerase chain reaction** techniques can be used[343]

Increase in the **reticulocyte count**

Hemoglobinemia, hyperbilirubinemia, and hemoglobinuria

TABLE 105

Microscopic Differentiation of Malaria Plasmodia

	P. falciparum	*P. vivax*	*P. ovale*	*P. malariae*
Infected red blood cells	normocytic; Maurer's clefts	macrocytic; Schuffner's dots	macrocytic; oval and fimbriated; Schuffner's dots	normocytic or microcytic
Ring forms	2 or more, delicate peripheral; small chromatin dot	1–2 in cell; large, thick; large chromatin dot	single, thick, compact	single, small, compact
Later trophozoites	compact; vacuolated; small chromatin dot	amoeboid central vacuole; light blue cytoplasm	amoeboid	band across cell; deep blue cytoplasm
Schizonts	18–24 merozoites filling 2/3 cell	12–24 merozoites, irregular spacing	8–12 merozoites filling cell	6–12 merozoites around central pigment mass
Pigment	clumped dark mass	fine, granular, yellow-brown	coarse, light brown	distinct dark mass
Gametocytes	crescentic; diffuse chromatin; single nucleus	spherical; compact; single nucleus	oval filling 3/4 cell	round filling 2/3 cell

Source: Adapted from Bain, B.J., and Lewis, S.M., Preparation and staining methods for blood and bone marrow films, in *Dacie and Lewis Practical Haematology*, 10th ed., Churchill-Livingstone Elsevier, Philadelphia, 2006, Table 4.3. With permission.

Reduced red cell life span (see **Erythrokinetics**)

Autoantibodies to red blood cells, both IgM and IgG, giving a positive **direct anti-globulin (Coombs) test**

Bone marrow hypoplasia and **dyserythropoiesis**; multinucleate erythroblasts, karyorrhexis, and erythrophagocytosis are seen

Leukopenia due to **neutropenia**, particularly with *P. falciparum* invasion associated with hypersegmentation; eosinophils are reduced, except following antimalarial therapy; initial lymphopenia may be followed by **lymphocytosis**, particularly of B-cells

Monocytosis with vacuolation, erythrophagocytosis, malarial pigment, and hemosiderin inclusions

Thrombocytopenia in 85% of cases

Hematological Disorders Associated with Malaria

Burkitt lymphoma: high incidence in malarious areas, which may be due to stimulation of centroblast proliferation, defective cell-mediated toxicity, or enhanced **Epstein-Barr virus**-induced lymphocyte transformation.

Sickle cell disorders: reduction of asexual parasitemia by *P. falciparum* gives advantageous protection. This may arise by blocking the entry of parasites into the red cells, by restriction of parasite development due to the presence of hemoglobin S, or by an increase in antibody-mediated opsonization with premature red cell removal.

Thalassemia: red blood cells containing high levels of hemoglobin F suppress parasite development, thus offering some protection. This may be the explanation for the greater incidence of survival in those with thalassemia in tropical areas.

Glucose-6-phosphate dehydrogenase deficiency: the malarial parasite adapts well to G6PD-deficient cells, so that these disorders coexist.

Melanesian ovalocytosis: persons with this disorder are less commonly infected.

Hereditary infantile pyropoikilocytosis: affected red blood cells resist invasion by malarial parasites.

Management

Treatment of Infection[344–349]

Benign Malaria

Oral chloroquine — initial dose 600 mg (of base), then single dose 300 mg * after 6 to 8 h, followed by single daily doses of 300 mg* for 2 days. In areas with chloroquine-resistant strains of plasmodia, amodiaquine can be used, but adverse reactions may result. For *P. vivax* and *P. ovale* infection, to remove liver parasites, primaquine 15 mg daily* for 14 to 21 days should follow, but this is contraindicated for pregnant women. To avoid hemolytic anemia, the dosage should be reduced for those with G6PD deficiency to 30 mg weekly for 8 weeks.

Malignant Malaria[348]

(*P. falciparum* chloroquine-resistant strains) Oral quinine 600 mg* (of quinine salt) every 8 h for 7 days, followed by Fansider®, three tablets as a single dose. If Fansider-resistant, doxicyclin (20 mg daily) should be given every 6 h for 7 days. Mefloquine or malarone can replace quinine/Fansider, and should be used for children, where oral quinone and chloroquine or pyrimethamine with sulfadoxine are contraindicated, but resistance has been reported. Other new drug combinations include atovaquine and proguanil, artemisinin derivatives, and mefloquine.[349] An artemisinin compound (artemether) has been used as a rectal suppository. With high parasitemia or therapeutic failure, **hemapheresis** should be considered.

Prophylaxis[350,351,(15)]

Pharmacological protection, although not absolute, remains the mainstay of prophylaxis, with choice of drug and dosage* depending upon the area to be visited. For most regions, chloroquine 300 mg weekly* 1 week before entry to all endemic areas, throughout the stay, then for 4 weeks after leaving is the usually recommended regime. An alternative regime is malarone for 1 to 2 days before leaving home and continued until 1 week after leaving the malarial zone. Mefloquine 250 mg weekly can be used to cover those who are entering areas where chloroquine-resistant plasmodia are known to be present, where it is considered that the risk of infection outweighs the possible adverse drug reactions. Proguanil hydrochloride 200 mg daily* should be added to chloroquine for those entering sub-Saharan Africa, South Asia, Southeast Asia, and Oceania. Drug resistance by the plasmodia has led to the development of numerous drug combinations with the addition of doxicyclin and atovaquone. Blockade of folate synthesis by the plasmodia is the basis for two recently introduced combinations, sulfadoxine-pyrimethamine and proguanil-dapsone. Many more drug combinations   and vaccines are undergoing clinical trial.

MALIGNANCY

The hematological changes that occur with malignancy, other than those of primary bone marrow disease.[121,352] A summary of these is shown in Table 106.

---

* Dosage for the average weight adult. Reduced dosage for children should be calculated according to weight and age.

TABLE 106

Hematological Changes Associated with Malignancy

Hematological Change	Tumor or Associated Therapy
Pancytopenia	
Bone marrow hypoplasia	**cytotoxic agent therapy**; **radiotherapy**
Myelodysplasia	cytotoxic agent therapy; radiotherapy
Leukoerythroblastic anemia	metastases in **bone marrow**
Megaloblastosis	**folic acid** deficiency; **cobalamin** deficiency with gastric carcinoma
Red blood cell disorders	
Anemia of chronic disorders	most malignancies
Iron deficiency anemia	gastrointestinal, uterine carcinoma
Pure red cell aplasia	thymoma
Immune hemolytic anemia	**lymphoproliferative disorders**; ovarian carcinoma
Microangiopathic hemolytic anemia	mucin-secreting carcinomas
Polycythemia rubra vera	kidney, cerebellum, uterus, liver tumor
White blood cell disorders	
Neutrophilia	most malignancies
Leukemoid reaction	disseminated tumors and those with necrosis
Eosinophilia	**Hodgkin lymphoma**, others
Monocytosis	various malignancies
Atypical lymphocytes	mycosis fungoides/Sézary syndrome
Hemorrhagic disorders	
Thrombocytosis	gastrointestinal with hemorrhage, others
Disseminated intravascular coagulation	mucin-secreting carcinomas of prostate
Acquired inhibitors of **coagulation factors**	many malignancies
Thrombosis	
Migrating venous **thrombophlebitis**	many malignancies

Source: Adapted from Hoffbrand, A.V. and Pettit, J.E., *Essential Haematology,* 4th ed., Blackwell Scientific, Oxford, 2001. With permission.

Metastases

Although tumors are often divided into those that metastasize via blood or lymphatics, this classification is somewhat artificial, as there are many interconnections between the two systems. Also, some tumor cells have been shown to have free passage through normal lymph nodes, and free passage often occurs once the lymph node architecture has been disrupted by tumor infiltration.

Tumors may spread locally through tissue planes, but metastatic spread is a multistep process (the metastatic cascade):

Local tumor spread into surrounding lymphatic or blood vessels

Detachment of tumor cells and distal embolism

Arrest within the vessel

Egression from the vessel

Proliferation and local tissue invasion

Detachment of tumor cells is increased if there is rapid tumor growth or necrosis. Tumors that express low levels of **cell adhesion molecules**, e.g., laminin and fibronectin, have a higher rate of detachment. The arrest of tumor cells in blood vessels involves interaction with **platelets** and thrombus. **Heparin** was found to alter the distribution of experimental metastases, and the tumors that activate platelets by releasing prostaglandins are highly metastatic. Once arrested, the tumor cells adhere to the **vascular endothelium** using, initially, gangliosides and, later, adhesion molecules, e.g., CD62E. Tumor egression through the endothelium is probably similar in mechanism to neutrophil migration. Having entered the surrounding tissue, tumor growth is enhanced by the release of factors such as vascular endothelial growth factor (VEGF), causing autocrine growth and **angiogenesis.**

Metastatic spread is a very inefficient process, with <0.01% of circulating tumor cells ultimately leading to a metastatic deposit. Adhesion of the malignant cells to the endothelium of both lymphatic and vascular channels is related to a specific receptor, CLEVER-1. Animal studies have shown that both **cytotoxic T-lymphocytes** and natural killer (NK) **lymphocytes** play a major role in preventing the establishment of metastases, but platelets and fibrinogen inhibit this activity.

Hematogeneous spread often involves intermediate sites, usually the lung or liver, depending on the venous drainage of the primary tumor. Although in theory the final sites of metastases are random, typical patterns are seen, suggesting that some tumors preferentially metastasize to certain organs (tropism). Evidence to support this theory has been provided by animal models, where various organs have been transplanted prior to the injection of tumor cells. Melanoma B16 cells preferentially metastasize to both the original and the transplanted lung.[353] It is thought that some organs produce locally active growth factors and have endothelia particularly suited to a certain tumor cell, which enables adherence and tropism to occur.

MALONDIALDEHYDE

A highly reactive oxidant produced from the peroxidation of polyunsaturated fatty acids. It can also be produced in platelets as a by-product of **cyclooxygenas**e pathways. Malondialdehyde can be used as a marker of lipid peroxidation and oxidative damage in ischemia and thrombosis. Production of malondialdehyde in **platelets** is blocked by the administration of **aspirin**. This forms the basis of a nonisotopic method for measuring platelet survival, but it is not used widely.

MALT LYMPHOMA

See **Marginal-zone B-cell lymphoma**.

MANGANESE

A normal constituent in the synthesis of mucopolysaccharides. It also has a physiological action in control of **vitamin K**. Its uptake is enhanced by those with **iron deficiency** anemia, as has occurred in miners. Its effects are on the nervous system, giving rise to a disorder resembling Parkinson's disease.

MANNOSE

An alternative to glucose as a source of energy in **red blood cell** metabolism.

MANTLE-CELL LYMPHOMA

See also **Lymphoproliferative disorders**; **Non-Hodgkin lymphoma**.
(Rappaport: intermediate or poorly differentiated lymphocytic diffuse or nodular lymphoma, malignant lymphoma, diffuse small cleaved cell type) An aggressive B-cell non-Hodgkin

lymphoma characteristically involving the mantle zone of reactive lymphoid follicles.[354] The tumor pattern is diffuse or slightly nodular follicles composed of small- to medium-sized uniform lymphoid cells with scant pale cytoplasm, dispersed chromatin, and inconspicuous nucleoli. The nuclei are mostly irregular or "cleaved." In some tumors, the cells are nearly round, and in others they may be very small and resemble small lymphocytes. A small proportion of these cells have larger nuclei with more-dispersed chromatin — blastoid variant. Cells circulating in the peripheral blood are monomorphic, showing slight nuclear indentations, condensed chromatin, and occasional nucleoli. There are several variants reported, but only two blastoid variants are considered to be of potential clinical significance.

The **immunophenotype** of tumor cells is SIgM$^+$, usually IgD$^+$, l > k, B-cell-associated antigens$^+$, CD5$^+$, CD10$^{-/+}$, CD23$^-$, CD43$^+$, CD11c$^-$, bcl-6 negative. All are bcl-2 positive, and most express cyclin D1. **Cytogenetic analysis** shows a chromosomal translocation t(11;14) involving the Ig heavy-chain locus and the bcl-1 locus on the long arm of chromosome 11 in almost all cases. The most probable origin is a CD5$^+$ peripheral B-cell of the inner mantle zone of a reactive lymphoid follicle.

Clinical Features

This histological form of non-Hodgkin lymphoma represents 3 to 10% of new cases. They occur in older adults, with a high male-to-female ratio. It is usually at an advanced stage at diagnosis. The sites involved include lymph nodes, spleen, bone marrow, peripheral blood, and extranodal sites, especially the gastrointestinal tract (lymphomatous polyposis). The course is aggressive, with a poor prognosis even after treatment. The median survival ranges from 3 to 5 years; the blastoid variant is more aggressive (median survival 3 years), and about 20% of patients are alive in long-term follow-up studies.

Staging

See **Lymphoproliferative disorders**.

Treatment

See also **Cytotoxic agents**; **Non-Hodgkin lymphoma**.

The optimal treatment strategies are under investigation. R-CHOP and hyperCVAD are current initial treatments. Patients achieving a complete remission are candidates for **allogeneic stem cell transplantation**.

MARCH HEMOGLOBINURIA

See also **Fragmentation hemolytic anemia**.

The common factor in the production of march hemoglobinuria is repetitive-contact trauma between a body part and a hard surface. There is no evidence of red cell or serum abnormality in this condition. Anemia rarely develops, and no red cell fragments are seen in the peripheral blood. Laboratory features consist of reduced **haptoglobin**, mild **hemoglobinemia**, some elevation of the **lactate dehydrogenase** and, rarely, **hemosiderinuria**.

MAREVAN

See **Warfarin**.

MARFAN'S SYNDROME

An autosomally dominant disorder with abnormalities of elastin, **collagen**, and glycosaminoglycans. The collagen in affected individuals is abnormally soluble and shows defective cross-linking. Patients may show easy **bruising** and unexplained **hemorrhage** after surgery. **Platelet** function may be abnormal. Aortic dilatation and its complications are the major causes of death.

MARGINAL-ZONE B-CELL LYMPHOMA

See also **Gastric disorders**; **Lymphoproliferative disorders**; **Non-Hodgkin lymphoma**. (Lukes-Collins: small lymphocytic lymphoplasmacytoid, diffuse small cleaved cell lymphoma, small lymphocyte B, lymphocytic-plasmacytic, parafollicular B-cell; MALT lymphoma; Rappaport: well-differentiated lymphocytic, poorly differentiated lymphocytic, mixed lymphocytic-histiocytic) A group (about 7%) of non-Hodgkin lymphomas characterized by their widespread distribution of heterogeneous cells proliferating in the marginal zones of lymphoid follicles. Some cells resemble **centrocytes,** others resemble small cleaved follicular center cells with angulated nuclei but with more abundant cytoplasm, similar to **lymphocytes** of Peyer's patch, mesenteric nodes, or those of the splenic marginal zone. There are also monocytic B-cells, **plasma cells**, and a few large cells (centroblastic or immunoblastic) present in most cases. Reactive follicles are usually present, with the neoplastic marginal-zone cells or monocytic B-cells occupying the marginal zone or the interfollicular region; occasional follicles may contain an excess of marginal-zone or monocytic cells, giving them a neoplastic appearance (follicular colonization). In lymph nodes, these marginal-zone cells may have a perisinusoidal, parafollicular, or marginal-zone pattern of distribution. Plasma cells are often distributed in distinct subepithelial or interfollicular zones, and these are neoplastic (monoclonal) in up to 40% of cases. Proliferation in epithelial tissues of the marginal-zone cells typically involves infiltration of the epithelium, so-called lymphoepithelial lesions. The **immunophenotype** of the tumor cells is SIg (M > G or A)$^+$, CIg$^+$ (40%), B-cell-associated antigens$^+$, CD5$^-$, CD10$^-$, CD23$^-$, CD43$^{-/+}$, CD11c$^{+/-}$. Cytogenetic studies show no rearrangement of bcl-1 or bcl-2; trisomy 3 and t(11;18) occur in extranodal cases. The postulated origin is a marginal-zone B-cell of a lymphoid follicle with capacity to home to tissue compartments.

Extranodal Marginal-Zone Lymphoma of Mucosa-Associated Lymphoid Tissue

(MALT lymphoma; low-grade B-cell lymphoma of MALT type) These are tumors of adults, with a slight female predominance. Many patients have a history of autoimmune disease, such as **Sjögren's syndrome**, Hashimoto's thyroiditis, or *Helicobacter pylori* gastritis (see **Gastric disorders**). It has been suggested that "acquired MALT" secondary to autoimmune disease or infection in these sites may form the substrate for lymphoma development. The majority present with localized stage I or II extranodal diseases (see **Lymphoproliferative disorders** — staging) that involve glandular epithelial tissues of various sites, most frequently the stomach and also ocular adnexa, skin, lung, thyroid, salivary glands, and spleen. Dissemination occurs in up to 30% of the cases, often in other extranodal sites, with long disease-free intervals. Reported pathogens putatively associated with MALT lymphomas include *H. pylori* in gastric MALT, *Campylobacter jejuni* in immunoproliferative small intestinal disease (IPSID), *Chlamydia psittaci* in ocular adnexal MALT, and *Borrelia burgdorferi* in cutaneous MALT. Early gastric MALT lymphomas often respond to antibiotic therapy against *H. pylori*. For patients who are unresponsive after 18 months, **radiotherapy** is the treatment of choice. In patients with extragastric MALT, lymphoma radiotherapy provides excellent local control. Chemotherapy is efficacious, including doxycycline for ocular adnexal MALT, cefotaxime for cutaneous MALT, and broad-spectrum antibiotics for IPSID.

Nodal Marginal-Zone Lymphoma

The majority occur in patients with Sjögren's syndrome or with other extranodal MALT-type lymphomas. Tumors with morphologic features identical to those described for extranodal MALT-type or monocytic B-cell lymphomas have occasionally been reported with isolated or disseminated **lymphadenopathy** in the absence of extranodal disease. Other sites involved include bone marrow and, rarely, peripheral blood. The clinical course is indolent. Transformation to **diffuse large B-cell lymphoma** may occur. Disseminated tumors should be treated as high-grade non-Hodgkin lymphoma, but are generally unresponsive.

MAST CELL

Large tissue-fixed cells (15 to 18 μm) containing coarse basophilic granules occurring normally in connective tissue,[102] having a morphological similarity between them and circulating **basophils**. Mast cells originate from uncommitted and mast-cell-committed progenitors under the influence of Th2 lymphocytes and cytokines (interleukin [IL]-3 and IL-4). T mast cells located near mucosal surfaces contain the enzyme trypsin, whereas TC mast cells located in the connective tissue contain both trypsin and chymotrypsin. All mast cells express CD13 and KIT, with CD34 also expressed in cells present in both bone marrow and peripheral blood. Mast cells activate **eosinophils**, and eosinophil products activate mast cells. They have a genetic relationship with eosinophils, found with the idiopathic **hypereosinophilic syndrome**. Morphologically, they are round/oval cells with prominent basophilic cytoplasmic granules and a single nonsegmented nucleus. The granules contain many proinflammatory agents, e.g., histamine, heparin, proteoglycan, proteases, leukotrienes, platelet-activating factor, and prostaglandin D2 (PGD2). These enzymes increase mucous secretion and smooth-muscle contraction. Mast cells also produce many cytokines, e.g., IL-1, IL-2, IL-3, IL-4, IL-5, IL-6, IL-10, IL-13, tumor necrosis factor (TNF)-α, and vascular endothelial growth factor (VEGF). IL-4 activates T-helper cells, and IL-4 and IL-5 stimulate eosinophil production and activation.

Mast cells express surface receptors for the Fc portion of IgE, which upon binding triggers degranulation, resulting in the immediate (type 1) hypersensitivity reaction. Mast cells have a long life span of several months to years.

The physiological role of mast cells is thought to be innate **immunity**, particularly to parasitic infections, by chemoattraction of neutrophils to a site of infection. They also play a role in **hypersensitivity** reactions, e.g., asthma, hay fever, drug allergy, contact dermatitis, insect sting, and probably chronic **inflammation**. Here the mast cells degranulate when IgE/antigen complexes are fixed on their cell membrane. Antigens, such as pollen or drugs, can bind directly to mast cells, expressing IgE, or, alternatively, bind to IgE in the circulation and then fix to mast cells. The immediate release of proinflammatory mediators causes vasodilatation, smooth-muscle contraction, and increased capillary permeability (**angioedema**), which clinically is manifest as urticaria, laryngeal stridor, bronchoconstriction, and shock. In extreme reactions (**anaphylaxis**), death may ensue, which is particularly likely if specific IgE is already present following previous exposure to the antigen, e.g., drug. **Arachidonic acid** metabolism is activated by mast cells when exposed to antigen. This results in:

Production of prostaglandins to increase vascular permeability, vasodilatation, and constriction of smooth muscle. They also inhibit Th1 response.

Production of leukotrienes, which cause a slower smooth-muscle contraction. They act as chemotactic stimuli for neutrophils and eosinophils.

MAST CELL LEUKEMIA

See **Mastocytosis** — systemic mastocytosis.

MASTOCYTOSIS

Proliferation of **mast cells** and their subsequent accumulation in one or more organ systems — skin, bone marrow, spleen, liver, and lymph nodes.[355] The cells are derived from myeloid hematopoietic progenitors. Apart from specific syndromes (see below), mastocytosis can be reactive to all forms of hematological malignancy, particularly **acute myeloid leukemia** and **Hodgkin disease**.[356]

Cutaneous Mastocytosis

Urticaria pigmentosa: here there are local accumulations of mast cells within the dermis. It is a benign disorder, usually of childhood, due to release of histamine and heparin from mast cells. Patients complain of flushing, urticaria, diarrhea, and have a reddish/brown rash, usually trunkal in distribution, with a positive Darier's sign, i.e., lesions transform to urticaria when friction is applied. The condition usually resolves by puberty. No therapy is indicated unless pruritus is present, in which case antihistamines and **PUVA light therapy** are of benefit.

Diffuse cutaneous mastocytosis: infiltration of mast cells here occurs into the papillary and upper reticular dermis. It is a disorder of children.

Mastocytoma of the skin: a single mass of cells, usually of the trunk or wrist, is manifest.

Systemic Mastocytosis

Indolent systemic mastocytosis: a benign disorder where typically no treatment is required.

Systemic mastocytosis: this is associated with hematopoietic clonal nonmast cell lineage disease, e.g., **myelodysplasia** or **acute myeloid leukemia.** Here it is necessary to treat the underlying hematological disorder.

Aggressive systemic mastocytosis: if the lesions are slowly progressive, interferon-α +/− steroids or cladribine may be adequate. If the lesions are rapidly progressive with organ damage, treatment is required by polychemotherapy +/− interferon-α. Imatinib therapy may be beneficial, especially in cases that do not possess the Asp-816-Val mutation. Consideration of **allogeneic stem cell transplantation** must be given for resistant cases.

Mast cell leukemia: this is an extremely rare form of acute leukemia with a very poor prognosis. The bone marrow is infiltrated by well-differentiated "tissue" mast cells in which dense basophilic granules may almost obscure the nucleus. Where the nucleus is visible, it is rounded, with dense chromatin. Because of the dense cytoplasmic granules, these cells may resemble promyelocytes, but in mast cell leukemia, the granules tend to aggregate around the nucleus and the myeloperoxidase reaction is negative. Treatment similar to that used for acute myeloid leukemia is often required with possibly interferon-α maintenance. Allogeneic stem cell transplantation should be considered in all cases.

Mast cell sarcoma: an extremely rare disorder characterized by local but destructive proliferation of atypical immature mast cells. A leukemic variant may occur. The prognosis is poor.

Extracutaneous mastocytoma: localized tumors of mast cells, usually in the lung. Surgical excision is the only form of treatment.

MAY-HEGGLIN ANOMALY

A rare autosomally dominant inherited condition initially described in 1905 by May and characterized by large round or rod-shaped inclusions in granulocytes (**neutrophils, eosinophils, monocytes**). The inclusions consist of **RNA** and are morphologically identical to **Döhle bodies** (i.e., blue coloration with Romanowsky stain). About one-third of patients have associated **thrombocytopenia** with giant platelets and a hemorrhagic tendency. Platelet survival is variable, but may be shortened.

MEAN CELL VOLUME

(MCV) See **Red blood cell indices**.

MEAN CORPUSCULAR HEMOGLOBIN

(MCH) See **Red blood cell indices**.

MEAN CORPUSCULAR HEMOGLOBIN CONCENTRATION

(MCHC) See **Red blood cell indices**.

MECHANICAL PURPURA

Extravasation of blood into tissues due to a sudden increase in venous intravascular pressure. It may arise from coughing or vomiting and asphyxia and is a well-recognized complication. Characteristically, the **purpura** involves the face, neck, and periorbital area. Mechanical purpura may also arise due to sucking, especially in adolescents ("love bite"). A not uncommon cause in the parents of small children and, indeed, in small children is the sticking of sucker toys onto the forehead ("purpura cyclops").

MECHLORETHAMINE

(Nitrogen mustard) See **Alkylating agents**.

MEDIASTINAL (THYMIC) LARGE B-CELL LYMPHOMA

(Large-cell lymphoma of the mediastinum) See also **Diffuse large B-cell lymphoma**.

A subtype of diffuse large B-cell lymphoma that involves localized disease in the mediastinum. It characteristically involves females in the third to fifth decades of life. The **immunophenotyp**e is that of diffuse large B-cell lymphoma with CD19$^+$, CD20$^+$, weak CD30$^+$, CD5$^-$, and CD10$^-$. The disorder is thymic B-cell in origin.

Staging
See also **Lymphoproliferative disorders**.
PET scans are useful because many patients have a residual mediastinal mass after treatment.

Treatment
See also **Non-Hodgkin lymphoma**.
R-CHOP (**rituximab**, **cyclophosphamide**, Adriamycin, **vincristine**, and **prednisone** or **prednisolone**) followed by consolidation **radiotherapy** is the treatment of choice. Patients with disease beyond the mediastinum have a less favorable response.

MEDITERRANEAN THROMBOCYTOPENIA
Slightly reduced levels of **platelet counts** noted in patients from the Mediterranean and surrounding areas, when compared those from Northern Europeans. There is no associated bleeding disorder. Although the platelet count is low, platelets are larger than normal and the total platelet mass is unchanged. Approximately 2% of affected individuals have platelet counts of less than $90 \times 10^9/l$.

MEGAKARYOBLASTIC LEUKEMIA
See **Acute myeloid leukemia** — acute megakaryoblastic leukemia.

MEGAKARYOCYTE
The precursor cells of **platelets**, which are derived from pluripotential stem cells in the **bone marrow** (see **Hematopoiesis**). Megakaryocytes develop from the pluripotent stem cell, and the earliest identifiable megakaryocyte precursor is the promegakaryoblast, defined by its expression of receptors for thrombospondin as well as platelet glycoproteins (and sometimes erythroid markers), and presumed to be intermediate between morphologically unidentifiable megakaryocyte colony-forming cells (CFU-Meg) and recognizable forms. The first morphologically identifiable precursor is the megakaryoblast, 15 to 50 μm in size with scant cytoplasm and an oval nucleus. These cells mature with cytoplasmic expansion through the promegakaryocyte stage to mature megakaryocytes. Stages I, II, and III are terms often used to describe these maturation stages and correspond approximately to the megakaryoblast, promegakaryocyte, and mature megakaryocyte, respectively.

Morphologically recognizable megakaryocytes lack proliferative capacity but increase in size by mitotic endoreduplication. Cell volume (nuclear and cytoplasmic) approximately doubles with each endomitosis. The earliest morphologically identifiable cell is already 8N ploidy (see **Mitosis**), and endoreduplication produces cells of 16 and 32N, each of which produce progressively greater numbers of megakaryocytes per cell. The total estimated megakaryocyte maturation time in humans is 5 days. Accelerated platelet production is achieved by an initial increase in megakaryocyte nuclear and cytoplasmic size, followed later by the appearance of more-immature forms (from the "stem cell" pool) and an increase in megakaryocyte number. Total platelet production may increase up to eightfold.

Several hematopoietic **cytokines** are known to stimulate megakaryopoiesis. Interleukin (IL)-3 stimulates CFU-Meg proliferation, but has little effect upon megakaryocyte maturation. IL-6, IL-11, stem cell factor (SCF), and leukocyte migration inhibitory factor (LIF) exhibit little colony-promoting activity alone, but can augment IL-3 induced growth. IL-6 and IL-11 act predominantly upon megakaryocyte maturation. Erythropoietin (Epo) shows moderate effects upon colony growth and maturation. The recent purification of thrombopoietin (Tpo),[188] the ligand for the c-mpl receptor, and the cloning of the Tpo gene have led to significant advances in the area of megakaryopoietic control. Antisense oligonucleotides to c-mpl specifically inhibit proliferation of human CFU-Meg (HuMGDF) but not platelet count or ablation of production, suggesting that Tpo is central but not essential to platelet production.

Murine Tpo stimulates both the proliferation of committed megakaryocyte progenitor cells and maturation of megakaryocytes, and it synergizes with Epo, SCF, and IL-11 to stimulate CFU-Meg proliferation.

Cytoplasmic maturation includes development of platelet-specific granules, membrane glycoproteins, and lysosomes. As a result of the endomitotic process, there is an increase of membrane, which is accommodated by invagination. This process continues until individual platelets are clipped off (cytoplasmic fragmentation) from the main body of the megakaryocyte. It is possible that circulating megakaryocytes undergo cytoplasmic fragmentation in the pulmonary capillary bed. Megakaryocyte maturation is under humoral control, regulated by the cellular homolog of viral oncogene v-mpl (C-MPL) via thrombopoietin, which is synthesized in the liver and kidney. There is a simple negative-feedback loop. In situations where platelet production is increased, platelets are produced from megakaryocytes with rapid cytoplasmic maturation but less nuclear maturation. Octaploid, or even tetraploid, cells may produce platelets under such circumstances. Such platelets are often larger than normal and more metabolically active. Thrombopoietin has been prepared in a recombinant form (rHuMGDF) and, conjugated with polyethylene glycol (PEGylated rHuMGDF), has been used successfully in the treatment of patients with advanced cancer.[357]

MEGAKARYOCYTIC HYPOPLASIA

Absence or arrested maturation of the committed **megakaryocyte** precursor, causing **thrombocytopenia**. It may occur as an isolated defect, in association with other congenital anomalies, or as a result of congenital viral infection and drugs. Combined defects include **amegakaryocytic thrombocytopenia** with absent radii syndrome.

MEGAKARYOBLASTIC LEUKEMIA

See **Acute myeloid leukemia** — Acute megakaryoblastic leukemia.

MEGAKARYOCYTIC HYPOPLASIA

Absence or arrested maturation of the committed **megakaryocyte** precursor, causing **thrombocytopenia**. It may occur as an isolated defect, in association with other congenital anomalies, or as a result of congenital viral infection and drugs. Combined defects include **amegakaryocytic thrombocytopenia** with absent radii syndrome.

MEGALOBLASTOSIS

Disorders caused by impaired **DNA** synthesis. In most instances, megaloblastic change results from deficiency of **cobalamins** (vitamin B_{12}), **folic acid**, or both.

Megaloblasts are larger than their normal counterparts and have more cytoplasm relative to the size of their nuclei. As the cell differentiates, the **chromatin** condenses more slowly than normal. As **hemoglobin** synthesis proceeds, the increasing maturity of the cytoplasm contrasts with the immature appearance of the nucleus — a feature termed "nuclear-cytoplasmic asynchrony." Granulocyte precursors also display nuclear cytoplasmic asynchrony and enlargement: the giant metamyelocyte. **Hypersegmented neutrophils** are prominent in the peripheral blood. Ineffective thrombopoiesis also occurs. Not only are **platelets** frequently reduced in number, they also display a functional abnormality. Rapidly proliferating cells in other tissues also show megaloblastic features. Epithelial cells in the mouth, stomach, small intestine, and the uterine cervix are larger than their normal counterparts and contain atypical immature-looking nuclei. The slowing of DNA replication in the megaloblastic anemias of folate and cobalamin deficiencies has long been attributed to a decrease in deoxythymidine synthesis from deoxyuridine, resulting from operation of the "methylfolate trap" (see **Folic acid** — metabolism).

Causes of Megaloblastosis

The multiple causes are listed in Table 107.

TABLE 107

Causes of Megaloblastosis

Cobalamin deficiency
Folic acid deficiency
Combined cobalamin and folic acid deficiency
Drugs
    Anticonvulsants: diphenylhydantoin, phenobarbitone, primidone
    Antimetabolites: purine analogs, **6-mercaptopurine**, 6-thioguanine, **azathioprine**, acyclovir, pyrimidine
       analogs, 5-fluorouracil, 5-fluorodeoxyuridine, 6-azauridine, zidovudine
    Cycloserine
    Dihydrofolate reductase inhibitors: **methotrexate**, aminopterin, pyrimethamine, trimethoprim, sulfasalazine,
       proguanil, triamterene
    Cobalamin-absorption interferents: p-aminosalicylic acid, metformin, phenformin, neomycin, colchicine
    Glutethimide
    Inhibitors of ribonucleotide reduction: **cytosine arabinoside, hydroxyurea**
    Nitrous oxide toxicity
    Oral contraceptives
Chemical toxicity: **arsenic toxicity, mercury toxicity**
Inborn errors of metabolism
    Congenital deficiency of intrinsic factor
    Deficiency of IF-cobalamin receptor (**Immerslund-Gräsbeck syndrome**)
    Transcobalamin II deficiency
    Hereditary orotic aciduria
    Lesch-Nyhan syndrome
    Thiamin-responsive megaloblastic anemia
Bone marrow hyperplasia
    Myelodysplasia (refractory megaloblastic anemia)
Myeloproliferative disorders
Chronic **hemolytic anemia**

Clinical Features

Anemia: patients show severe pallor and slight jaundice, producing the characteristic lemon-yellow tint to the skin. The mild icterus results from both **ineffective erythropoiesis** and hemolysis.

Neurological features are not uncommon in cobalamin deficiency, but do not occur with folic acid deficiency.

Gastrointestinal disorders: oral and tongue soreness, anorexia, weight loss, and bowel disturbances are all common. There may be a previous history of surgery or radiotherapy to the stomach or small bowel, malabsorption, or unexplained diarrhea. There may be evidence of malnutrition, including growth impairment in children. Alcohol abuse is frequently associated with dietary neglect and malnutrition. A family history is important and should include nonimmediate relatives.

Laboratory Features

Red blood cell counting; Red blood cell indices; Peripheral-blood-film examination. The patient may or may not be anemic. If anemia is present, it may be severe. In uncomplicated cases, the mean cell volume (MCV) and the red cell distribution width (RDW) are increased. The disorder must be considered when the MCV is greater than 100 fl; however, when the value lies between 100 and 110 fl, so long as iron deficiency is not present, the most likely causes are alcoholism and liver disease. Above 110 fl, megaloblastic anemia becomes the most likely cause. The MCV may, however, be normal if the megaloblastic state coexists with either iron deficiency or the **anemia of chronic disorders**. There may also be leukopenia and thrombocytopenia. The peripheral-blood smear is usually distinctive. Macro-ovalocytes are the prime feature, but poikilocytes, both teardrop-shaped forms and fragmented cells, are also present. **Basophilic stippling** and **Howell-Jolly bodies** may be seen. Nucleated red cells, if present, show megaloblastic features. Neutrophil hypersegmentation and giant platelets may be present.

Bone marrow. The marrow aspirate is hyperplastic. Megaloblastic change may be seen in any of the hematopoietic cell lines, although the major changes involve the erythroid series. Sideroblasts are seen in increased numbers. In severe megaloblastic anemia, many of the erythroid cells are promegaloblasts containing an unusually large number of mitotic figures. Unless iron deficiency is present, the iron content of macrophages is usually increased. When megaloblastic anemia occurs in combination with iron deficiency, thalassemia minor, or the anemia of chronic disorders, many megaloblastic features may be masked. Marrow examination may reveal partially developed "intermediate" megaloblasts, which are smaller and less striking than fully developed megaloblasts. Usually, however, the megaloblastic nature of the marrow is indicated by the presence of giant metamyelocytes. With abnormalities of DNA and histone synthesis due to deficiency of thymidine production from deoxyuridine, the erythroid precursors are large, as are their nuclei; however, the nuclear chromatin is clumped and does not possess the delicate lacelike appearance of the typical megaloblast (see above). Patients misdiagnosed as iron deficient and treated accordingly will only respond incompletely, and frank megaloblastic features will emerge as iron stores become replenished. Masking of megaloblastic anemia also occurs when patients receive small amounts of folate, often in the form of a meal, but the degree of anemia does not alter. Similar masking can occur following blood transfusion. Usually clear morphologic clues to the megaloblastic nature of the marrow are found upon careful microscopic examination.

Biochemical features. Because there is both ineffective erythropoiesis and hemolysis of circulating red cells, the plasma unconjugated bilirubin level is often elevated (but does not usually exceed 2.0 mg/dl), and the serum **lactate dehydrogenase** level is raised. Methyl malonic aciduria may be present.

Differential Diagnosis

Confirmation of megaloblastic state — red blood cell (RBC) indices and bone marrow

Cobalamin or folic acid deficiency by assay of serum B_{12} and folate levels and of RBC folate level

Establishment of the cause of the deficiency, depending upon results of bone marrow and biochemical features described above

Acute Megaloblastic Anemia

Potentially fatal megaloblastosis due to severe sudden depletion of tissue cobalamin can occur in a few days. The clinical features suggest an immune cytopenia. There can be profound leukopenia and thrombocytopenia, but often there is no anemia. Diagnosis is made on bone marrow examination and confirmed by the rapid response to appropriate therapy. Exposure to nitrous oxide is the most frequent cause, but it can also occur in any severe illness associated with extensive transfusion, dialysis, total parenteral nutrition, or exposure to weak folate antagonists such as trimethoprim.

Refractory Megaloblastic Anemia

This is a form of **myelodysplasia** often culminating in **acute myeloid leukemia**. It is characterized by ring sideroblasts, excess iron, hyperplasia of mast cells, and a mixture of erythroblastic and megaloblastic erythropoiesis in the bone marrow. Occasional patients respond to pharmacological doses of pyridoxine (200 mg/day).

MEIOSIS

The division of gametocytes to produce four daughter cells, each of n **chromosome** complement. This is achieved by two cell divisions as follows. DNA replication and anaphase proceed as for **mitosis**. Chromosomes then align at the equator and sister chromatids pair as bivalents. Each bivalent contains all four of the cell's copies of one chromosome. At the first cell division, bivalents separate to form daughter cells, each containing homolog pairs ($2n$). Each daughter cell then divides again, with each of a pair of homologs separated into two more daughter cells (n).

MELANESIAN OVALOCYTOSIS

An autosomally dominant gene defect causing deletion of Band 3 proteins of the **red blood cell** membrane. It occurs in the lowland areas of Papua New Guinea. The red blood cells are oval, thought to be due to an elongated Band 3 protein with reduced membrane deformability. It is not associated with clinical symptoms or anemia, but there is a geographical association with reduced incidence of **malaria** in those affected.

MELPHALAN

See **Alkylating agents**.

MEPOLIZUMAB

A **monoclonal antibody** that neutralizes anti-leukin-5 antibody. It is under trial for use in treatment of the **hypereosinophilic syndrome** with eosinophilic dermatitis.

6-MERCAPTOPURINE

See **Antimetabolites**.

MERCURY TOXICITY

A cause of **hemolytic anemia**, due to interference with –SH group enzymes. In the **red blood cell** membrane, this disturbs the **cation pump** control of cell volume, thereby causing hemolytic anemia. It can also disturb **erythropoiesis**, giving rise to **megaloblastosis**. Chronic mercury poisoning is very rare, occurring from the inhalation of mercury vapor or ingestion of small amounts of mercuric nitrate used in felt manufacture.

MESNA

An antagonist to the metabolite acrolein, an excretory product of oxazaphosphorines cyclophosphamide and ifofamide, and used as a **cytotoxic agent** in the treatment of **lymphoproliferative disorders**. Treatment with mesna prevents urothelial toxicity manifested by hemorrhagic cystitis. Mesna itself produces such adverse drug reactions as gastrointestinal disturbances, hypotension, and tachycardia.

METABONOMICS

An emerging technology that may be regarded as the end result of **gene** and protein regulation in that it studies endogenous metabolism. It can be undertaken as a high-throughput technique using proton-NMR (nuclear magnetic resonance) spectroscopy or liquid chromatography/mass spectrometry (LC/MS). The samples are subject to external factors that can give rise to variation, such as diet, drugs, and exercise.

METALLOPROTEASE

An enzyme secreted by **histiocytes** (macrophages). This enzyme is particularly significant in chronic obstructive lung disease.

METAMYELOCYTE

See **Neutrophil** — maturation.

METAZOAN INFECTION DISORDERS

The hematological changes associated with metazoan, including helminth, infections. These organisms include:

Nematodes: *Ascaris lumbricoides, Strongyloides stercoralis, Filaria* spp., *Trichinella spiralis, Dracunculus medinensis*

Cestodes: *Diphyllobothrium latum, Taenia solium, Cysticercus cellulosea, Hymenolepis diminuta, Echinococcus* spp.

Trematodes: *Schistosoma mansoni, S. japonicum, Clonorchis sinensis*, opisthorchiasis, fascioliasis

The hematological disorders include:

Eosinophilia, particularly marked with *Trichinella spiralis* and *Ascaris lumbricoides*.

Leukocytosis, either due to **neutrophilia** or **lymphocytosis**.

Iron deficiency, when the intestinal tract is infected by hookworm, *Trichuris trichiura* or *Fasciolopsis buski*, and when the urinary bladder is infected by *Schistosoma haematobium*.

Cobalamin deficiency from intestinal bacterial overgrowth in association with infection by *Diphyllobothrium latum* (fish tapeworm disease). Deficiency of cobalamin results from impaired absorption due to there being competition between the worm and the host for the dietary content. **Megaloblastosis** has been found in a proportion of carriers of the fish tapeworm, but only in Finland. The anemia responds to expulsion of the worm but often is suboptimal in the absence of treatment with cobalamin.

Liver disorders as a consequence of infection by *Clonorchis sinensis, Schistosoma mansoni*, or *S. japonicum*.

Parasitinemia of peripheral blood by *Wucheria bancrofti* and loa-loa (filariasis). This can be prevented in a population at risk of infection by the administration of a combination of two drugs from diethylcarbazine — albendazole and invermectin — for a period of 4 to 6 years.

METHEMALBUMIN

A compound of free **hemin** (ferriheme) attached to albumin. It arises when haptoglobin and hemopexin are both saturated during **hemoglobin** degradation. It is detected in plasma by direct spectroscopy. Small concentrations are more easily detected by extracting the pigment with ether and converting it to ammonium hemochromagen (Schumm's test).

METHEMOGLOBIN

Oxidized **hemoglobin** in which **iron** is in the ferric form. Methemoglobin is brown in color and is characterized by absorption of light at 632 nm. Methemoglobin formation occurs at a rate of about 3% per day, but this process is counterbalanced by a more rapid reduction process via methemoglobin reductase. Thus less than 1% of the circulating hemoglobin is normally oxidized as methemoglobin.

METHEMOGLOBINEMIA

The abnormal state that occurs when more than 1% of hemoglobin is oxidized. Three distinct entities are recognized:

Congenital methemoglobinemia. This autosomally recessive disorder with reduced activity of NADH (nicotinamide-adenine dinucleotide) diaphorase (sometimes designated as methemoglobin reductase or cytochrome b_5 reductase) is associated with a variety of mutations of the NADH diaphorase gene identified at the nucleotide level.[358] **Cyanosis** is the usual presenting feature, although some patients have enzyme deficiency in nonerythroid cells associated with progressive encephalopathy and mental retardation. The level of methemoglobin is 8 to 40%, and NADH diaphorase activity is typically less than 20% of normal. Patients should avoid exposure to nitrites or aniline derivatives, which sometimes can precipitate methemoglobinemia in normal persons. The most satisfactory chronic treatment is ascorbic acid, 300 to 600 mg three or four times daily.

Acquired (toxic) methemoglobinemia. This can occur in normal subjects exposed to strong oxidants in dyes, drugs (dapsone, nalidixic acid, niridazole), solvents, or fertilizers; in infants fed soups or well water rich in nitrates, which are converted to nitrites by the action of intestinal bacteria; and in infants or adults after inhalation of nitric oxide. The level of methemoglobinemia is increased, but the activity of NADH diaphorase is reduced. Preferred treatment is the intravenous administration of methylene blue, 1 to 2 mg/kg body weight, over 5 min.

Hemoglobin M *hemoglobinopathies.* A group of disorders caused by mutations that affect the "heme pocket" of hemoglobin. These changes cause the production of an iron phenolate complex that prevents the reduction of ferric iron to a ferrous form. There is no known effective treatment.

METHEMOGLOBIN REDUCTASE

See **Methemoglobin**.

METHOTREXATE

See **Antimetabolites**.

METHOTREXATE-ASSOCIATED LYMPHOPROLIFERATIVE DISORDERS

See also **Immunodeficiency** — secondary immunodeficiency; **Lymphoproliferative disorders**; **Non-Hodgkin lymphoma**.

Following treatment with **methotrexate**, patients with **rheumatoid arthritis** have a two- to fourfold increased risk of developing non-Hodgkin lymphoma, particularly **diffuse large B-cell lymphoma**.

Clinical Features

The incidence of lymphoma correlates with disease activity of rheumatoid arthritis and with an **erythrocyte sedimentation rate** greater than 40 mm/h.

Treatment

Some patients may respond to withdrawal of immunosuppression if the **Epstein-Barr virus** (EBV) *in situ* hybridization studies are positive. **Rituximab** is under investigation. Treatment appropriate to the lymphoma histology is otherwise the treatment of choice.

METHYLATION

The introduction of methyl groups into cytosine residues in eukaryotic **DNA**. It is mainly thought to be associated with transcriptional gene repression in euchromatin, although genes are also methylated in heterochromatin. Two percent to 7% of cytosines in mammalian DNA are methylated, and these are concentrated at CG doublets in the genome, such that the majority of these are methylated. Demethylation of 5-methyl cytosine to thymidine occurs at these so-called mutational hotspots and is responsible for much of the DNA mutation. CpG islands are areas of the genome rich in CG doublets ($\approx$30,000 in the mammalian genome) and are often associated with promoter regions of genes. The diagnostic importance of methylation lies in its value for X-linked clonality studies.[359]

METHYLMALONIC ACIDURIA

A rare inborn error of metabolism or, more commonly, **cobalamin** deficiency, where an increased urinary excretion of methylmalonic acid is a reliable indicator. It is not increased in folic acid deficiency. The level falls toward normal after a few days of cobalamin therapy.

MICROANGIOPATHIC HEMOLYTIC ANEMIA

Anemias arising as a result of mechanical trauma to **red blood cells** from circulation through small blood vessels.[360,361] The mechanism is probably associated with **fibrin** deposition on the **vascular endothelium**. This anemia occurs with:

Disseminated intravascular coagulation of all causes

Disseminated carcinoma

Eclampsia and preeclampsia

Giant **hemangiomas**

Hemolytic-uremic syndrome

Immune complex disorders: **systemic lupus erythematosus** (SLE)

March hemoglobinuria

Malignant hypertension with glomerulonephritis

Thrombotic thrombocytopenic purpura

Drug-induced by mitomycin C, inhibitors of the Ca^{2+}-activated phosphatase, calcineurin (ciclosporin), quinine

Postallogeneic transplantation for bone marrow, kidney, liver, heart, or lung

Total-body irradiation

It is a nondiarrhea-associated illness with an insidious onset and often no prodromal illness. Severe hypertension is common, with irreversible renal damage. There may or may not be anemia; the **reticulocyte count** is elevated; the peripheral blood will show changes of **fragmentation hemolytic anemia**; and there is frequently **intravascular hemolysis**.

MICROCYTES

Small **red blood cells** with a diameter <6.0 μm. The **mean cell volume** (MCV) is usually <80 fl. Their staining is either:

Hypochromic, as in **iron deficiency** or **thalassemia**

Normochromic, as in the **anemia of chronic disorders**

Hyperchromic, as in disorders associated with **microspherocytes**

The causes of microcytosis are given in Table 108.

TABLE 108

Causes of Microcytosis

Hereditary	Acquired
β **thalassemia** trait — heterozygotes	**iron deficiency**
β thalassemia major — homozygotes	**anemia of chronic disorders**
δβ and γδβ, all forms of thalassemia	**myelodysplasia**
hemoglobin Lepore (all forms)	secondary **sideroblastic anemia**
persistence of **hemoglobin F**	hyperparathyroidism
hemoglobin H disease	**ascorbic acid** deficiency
sickle cell/hemoglobin C disease	**lead toxicity**
sideroblastic anemia	cadmium toxicity
atransferrinemia	antibody to erythroblast **transferrin** receptor

MICROCYTIC ANEMIA

The association of reduced level of **hemoglobin** with microcytosis, demonstrated either by **automated blood cell counting** or on **peripheral-blood-film examination**. Differentiation is a stepwise process, as shown in Figure 86. The initial investigation is to determine the level of serum **iron.** If raised, the most likely cause is a form of **sideroblastic anemia**. A raised or normal serum iron occurs with **hemoglobinopathies**. Low serum iron levels associated with reduced serum **ferritin** levels indicate **iron deficiency**, but a raised level suggests an **anemia of chronic disorders**, which requires clinical evaluation for diagnosis.

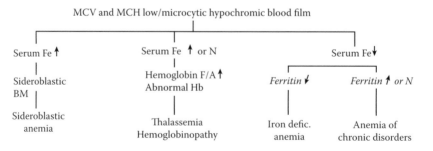

FIGURE 85

Flow diagram of the investigation of microcytic anemia. (Adapted from Bates, I. and Bain, B.J. Approach to the diagnosis and classification of blood disorders, in *Dacie and Lewis Practical Haemotology,* 10th ed. Churchill Livingstone, Elsevier, 2006. Table 23.1. With permission.)

MICROGLIAL CELLS
See **Histiocytes**.

MICROSPHEROCYTES
Red blood cell spherocytes of reduced volume with dense **hemoglobin** content. They are found in posttransfusion patients and in patients with **hereditary spherocytosis**, **immune hemolytic anemias**, Heinz body hemolytic anemias, water-dilution hemolysis, and **fragmentation hemolytic anemia**.

MITOCHONDRIA
Organelles located in the cytoplasm of eukaryotic cells. They contain enzymes necessary for cell energy production via **adenosine triphosphate** (ATP) generation. Mitochondria also contain other specialized enzymes active in protection against oxidative damage and in **heme** biosynthesis. Many mitochondrial proteins are actively imported, such as δ-aminolevulinic acid. Structurally, mitochondria consist of an outer and inner membrane, an intermembrane space, and a cytosolic matrix containing most proteins.

The sequence of the entire mitochondrial genome is known, and it possesses little noncoding **DNA**. Mitochondrial DNA is inherited uniparentally (maternal). Mutation occurs more frequently than for genomic DNA, and inherited mitochondrial DNA deletional syndromes have been associated with **sideroblastic anemia**.

MITOGEN
Substances that induce **mitosis**. In the context of the immune system, the term is usually reserved to describe substances that drive **lymphocyte** proliferation in a polyclonal manner, i.e., independent of the antigenic specificity of the lymphocytes involved. Specific antigens are mitogenic for lymphocytes bearing their specific receptors and many cytokines, and they also have nonspecific mitogenic effects, but these are not regarded as mitogens. Often mitogens are more efficient when combined with other agents, e.g., cytokine growth factors, and occasionally are mitogenic only when combined with other factors and so can be referred to as comitogens. Those factors recognized as mitogens include:

Certain plant lectins such as concanavalin A (conA), phytohemagglutinin (PHA), and pokeweed mitogen (PWM). These function by binding to the carbohydrate moieties of glycoproteins, thereby cross linking them; where the cross-linked proteins are receptors functioning in growth control (e.g., the CD3 component of the T-cell antigen receptor), this will trigger their signal-transduction functions and so drive the cell into proliferation. Lectin mitogens show some degree of cell specificity, e.g., con A and PHA are T-cell mitogens, whereas PWM is a B-cell mitogen.

Antibodies that bind receptors controlling growth again cross link and trigger signal transduction, e.g., anti-CD3 for T-cells and anti-IgM for B-cells.

Certain proteins function as so-called superantigens. These are powerful mitogens for helper T-cells; they function by cross-linking the Vb domain of the T-cell antigen receptor with any class II histocompatibility antigens on another cell. Superantigens include staphylococcal enterotoxins and some retroviral proteins.

Perhaps the most potent mitogen is the combination of a Ca^{2+} ionophore such as ionomycin and the phorbol ester PMA (phorbol myristic acetate). These increase intracellular Ca^{2+} concentrations and activate protein kinase C, both essential steps in the activation of cell proliferation. PMA is indeed a powerful comitogen in combination with, for example, con A.

MITOSIS

Cell division resulting in two daughter cells, each having the same number and type of **chromosomes** as their parent nucleus (compare **Meiosis**). Somatic cell division consists of two phases: (a) interphase to allow DNA replication within the parent cell ($2n$ to $4n$ DNA complement) and (b) mitosis for the separation of nuclear and cytoplasmic material to form two daughter cells. Each somatic cell contains two copies of each chromosome (diploid or $2n$), called homologs. **Cell cycle** progression, through S-phase to mitosis (G2/M), is controlled by cell-cycle-associated proteins, particularly cyclin B. Once DNA replication is complete, mitosis can proceed.

Nuclear **chromatin** becomes organized into chromosomes, each consisting of a pair of sister chromatids. After DNA replication, each sister chromatid contains $2n$ chromosome complement. The nucleus is then replaced by the mitotic spindle itself, derived from the microtubules comprising the cellular cytoskeleton. The spindle is attached at each pole of the cell to centrosomes containing the microtubule-organizing center. Some spindle fibers traverse the cell, while others attach chromosomes via the kinetochore, which lies within the chromosome centromere. Mitotic cell division is then divided into several phases (see Figure 86).

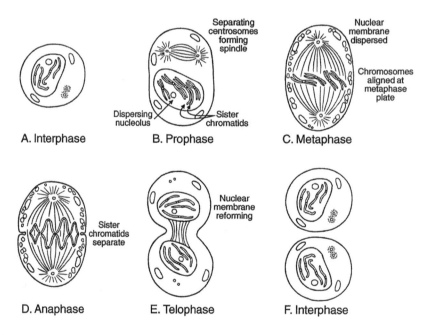

FIGURE 86

Cycle of mitotic nuclear division. (1) Prophase, when the individual pairs of sister chromatids become apparent. (2) Prometaphase, during which the chromosomes move toward the equator. (3) Metaphase, during which chromosomal alignment at the equator occurs. (4) Anaphase, where each sister chromatid is pulled apart to the poles following functional centromeric duplication. This is accomplished by shortening of the spindle microtubules. (5) Telophase, the process of chromosomal decondensation to reform the nucleus at each pole of the cell. (6) Cytokinesis, the process of daughter-cell separation following the formation of a contractile ring of microtubules to pinch the cells apart. Each daughter cell therefore receives one member of each sister chromatid pair.

MITOXANTHRONE
See **Anthracyclines**.

MIXED-LYMPHOCYTE CULTURE
See also **Human leukocyte antigens**.

(MLC) A technique in which **lymphocytes** from different (allogeneic) donors are cultured together. Where there are differences in HLA class II antigens, T-lymphocytes will be mutually stimulated to proliferate by virtue of their T-cell receptors engaging allogeneic class II antigens of the other donor. This technique was used initially to define HLA class II specificities but has been superseded.

MIXED-, SMALL-, AND LARGE-CELL LYMPHOMAS
See **Angiocentric lymphoma**; **Non-Hodgkin lymphoma**.

MIXED-TYPE AUTOIMMUNE HEMOLYTIC ANEMIA
Warm immune hemolytic anemia in association with **cold agglutinins**. While in the majority of patients the cold agglutinins are not clinically significant, occasionally they have sufficient thermal amplitude of high titer to also have the clinical features of cold-agglutinin syndrome. The condition may be primary or secondary to **lymphoproliferative disorders** or **systemic lupus erythematosus**. The disorder usually has a chronic course with acute exacerbations. The warm antibodies are usually IgG and the cold autoantibodies specific against I antigen. Treatment is similar to warm autoimmune hemolytic anemia (AIHA).

MNS BLOOD GROUPS
See also **Blood groups**.

A specific antigen–antibody system termed MNS located on **red blood cells** (RBCs).

Genetics and Phenotypes
The MNS blood groups are sited on glycophorins A and B (GPA and GPB), which are sialoglycoproteins that traverse the RBC membrane lipid bilayer. The most important are the MN antigens (on GPA) and the Ss and U antigens (on GPB). The large number (43) of antigens defined within the MNS blood group system is due to the close proximity of the genes that encode GPA and GPB, with the consequent opportunity for hybrid genes to appear. Because of linkage disequilibrium between these gene loci, the S antigen is found about twice as frequently in MN as in NN individuals. The common phenotypes and genotypes are shown in Table 109.

TABLE 109

Phenotypes and Genotypes of the MNS Blood Group System

Phenotype	Genotype	Frequency in Caucasians (%)
M+	MM or MN	78
N+	NN or MN	72
S+	SS or Ss	55
s+	ss or Ss	89

Antibodies and Their Clinical Significance

Some anti-M and most anti-S and anti-s antibodies are immune in origin and are of the IgG class. Some anti-M and most anti-N antibodies are naturally occurring and are of the IgM class.

MNS antibodies are not commonly found. Anti-M is the most frequent. Anti-S and anti-s are likely to be encountered in the presence of other red cell antibodies. Some examples of anti-M and anti-N can be detected serologically only below 37°C and are not clinically significant.

The phenotype S–s–U is sometimes found in subjects of African descent. These individuals can make anti-U, which reacts with the RBCs of almost all Caucasians.

MNS antibodies can give rise to severe hemolytic blood transfusion complications (see **blood components for transfusion** — complications) if incompatible RBCs are transfused, although these are unusual. Anti-M, anti-S, and anti-s are sometimes implicated as the cause of a delayed hemolytic transfusion reaction. The selection of blood for patients with anti-M, -N, -S, or -s should not be difficult, and compatible blood should be provided in all but life-threatening situations. The provision of blood for patients with anti-U will be problematic because of the high incidence of the U antigen in the donor population. Autologous blood transfusion should be considered.

MNS antibodies are occasionally implicated in hemolytic disease of the newborn, but this is usually mild.

Very rarely, antibodies to low-frequency MNS antigens have been shown to cause hemolytic transfusion reactions or **hemolytic disease of the newborn**.

Anti-M and anti-N are not detected using enzyme techniques for antibody detection.

MOLECULAR GENETIC ANALYSIS

The principles of the methodology used for the study of **DNA** and **RNA**. This is providing a rapid insight into the pathogenesis of hematopoietic disease. The advent of the polymerase chain reaction (PCR) has been the major advance and is used extensively to study DNA and gene expression (see **Reverse transcriptase**).

DNA-Based Techniques

Southern Blotting

Developed by Ed Southern, the principle of Southern blotting involves restriction endonuclease digestion of DNA, and hybridization with target (usually gene specific) complementary DNA probes. Restriction endonucleases are bacterially derived enzymes, which cut DNA at sequences specific for each enzyme. Some recognize the same sequence but cut only methylated or unmethylated DNA (isoschizomers; see **Clonality**). Depending upon sequence specificity, restriction endonucleases are subdivided into "frequent" (e.g., EcoR1) or "infrequent" (e.g., Not1) cutters.

DNA is digested by appropriate single-restriction enzymes or combinations, and fragments are separated by gel electrophoresis. Fragments are then blotted (transferred) to nylon membranes and hybridized with a DNA (full-length complementary DNA [cDNA] or oligonucleotide) probe of the gene of interest. The probe can be radiolabeled (usually ^{32}P) and the signal detected by autoradiography or by using a nonisotopic detection system (e.g., digoxygenin).

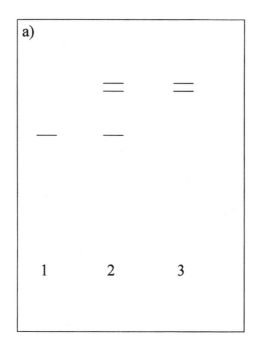

FIGURE 87
Southern blotting. Representation of autoradiograph following Southern blotting: (1) no digestion with restriction enzyme; (2) one allele digested (heterozygous for RFLP); (3) both alleles digested (homozygous for RFLP).

Abnormal restriction patterns — **restriction-fragment-length polymorphisms** (RFLPs) — indicate either a normal polymorphic base substitution (not usually changing the amino acid code) or a pathological mutation (Figure 87).

Pulse-Field Gel Electrophoresis

(PFGE) This allows separation and resolution of larger DNA fragments (50 to 5000 kb) than Southern blotting. Infrequent cutting enzymes digest DNA, which is then electrophoresed in an orthogonal electric field. Its application is for linkage analysis.

Cloning DNA

Used to obtain large quantities of genomic DNA from small numbers of cells. Several different strategies are available, depending largely upon the fragment size of DNA required. The principle of each involves digestion of genomic DNA and insertion into vectors that replicate their DNA plus the inserted fragments of interest.

Plasmid cloning: restriction digestion of genomic DNA with enzymes creating "sticky" fragment ends (<20 kb) that anneal to linearized "sticky" plasmid ends. Plasmids are then infected into transformed *Escherichia coli* and selected for by antibiotic-resistant genes. This approach is most appropriate for cloning and sequencing of known genes.

δ-cloning: higher transformation efficiency uses the bacteriophage-δ as vector and can create a genomic library consisting of most of the genome of interest. Identification of specific genes of interest is by hybridization of radiolabeled cDNA probes to colonies (clones) transferred onto membranes.

Cosmid cloning: useful for larger DNA fragments of ≈45 kb and utilizing cosmid vectors.

b)

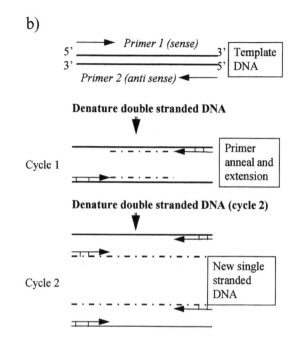

FIGURE 88
Polymerase chain reaction (see text for details).

Yeast artificial chromosome cloning (see **Chromosomes**).

Expression (cDNA) libraries: isolate poly-A mRNA and reverse transcribe to ssDNA. This is then converted to dsDNA and ligated to linker sequences before packaging in δ-phage and cloning as above.

Polymerase Chain Reaction

An alternative to cloning for producing multiple copies of short DNA fragments whose sequence ends are known. Taq polymerase enzyme elongates oligonucleotide primers (adding complementary nucleotides) annealed to target denatured ssDNA. Sequential cycles of denaturing, annealing, and primer extension allow progressive amplification of the sequence of interest (Figure 88). The high sensitivity makes amplification of contaminating DNA a major problem, although strategies to eliminate this are available. PCR is used in DNA diagnosis, e.g., for **thalassemias** and **factor V** Leiden. It is also used in clinical practice for mRNA detection (reverse transcribed prior to amplification), such as in the setting of minimal residual disease detection. The technique is also used for the detection of bacterial and viral infections.

Sequencing

Two major methods involving chain termination at specific nucleotides (Sanger dideoxy method) or alkali denaturation (Maxam-Gilbert method). Now fully automated using fluorescent labeling cytometry or manual detection systems (^{35}S radiolabeling). Allows sequence determination of fragments of 300 to 500 bp.

PCR-Single Stranded Conformational Polymorphisms Analysis

(PCR-SSCP analysis) Useful for mutational screening where genomic wild-type sequence is known; short target sequences are amplified, denatured, and electrophoresed into a nondenaturing gel. Single-stranded DNA with sequence differences exhibit mobility shifts. Shifted bands can be excised, cloned, and sequenced.

Transgenic/Knockout Mice

Random integration of exogenous genetic material into murine DNA occurs at high frequency, and addition of fully functional genes into a murine genome enables gene functional analyses *in vivo*, i.e., the transgenic model.

Targeted gene deletion allows study of the normal functional role, and this is accomplished by introducing knockout mutations in one allele of murine embryonic stem cells, which are transplanted back into early murine embryos to create chimeras for that gene. Cross breeding of chimeras produces some animals with homozygous gene deletions, the so-called knockout mice.

RNA-Based Techniques; Detection of Gene Expression

See also **Ribonucleic acid**.

Three techniques are widely used to detect expression of specific mRNA species. The most commonly employed technique is reverse-transcriptase PCR in the detection of minimal residual disease in hematological malignancies characterized by fusion mRNAs (e.g., *bcr-abl*, PML/RARa).

Reverse Transcriptase

(PCR, RT-PCR) Applicable to small quantities of RNA (<1 μg) and usually considered to be at best semiquantitative, the principle involves reverse transcription of mRNA to single-stranded cDNA by the enzyme reverse transcriptase. This cDNA provides the template for specific gene amplification by PCR. This is the most sensitive technique for detection of gene expression.

Northern Blotting/Dot (Slot) Blotting

Total RNA (usually >10 μg) is electrophoresed in a denaturing agarose gel and then transferred to nylon or nitrocellulose membranes. Specific mRNA species can then be identified and quantitated after hybridization with radioactive (or digoxygenin labeled) cDNA probes. A variant technique involves directly immobilizing total RNA into slots or dots on nylon/nitrocellulose with DNA probe hybridization. This technique is quantitative (along the linear portion of the densitometry curve in dilutional experiments) but requires large quantities of RNA.

RNAse A Protection

The most specific RNA detection technique with smaller quantities of RNA required than for Northern blotting. A radiolabeled specific DNA probe (usually 200 to 400 bp) is designed for the gene of interest and hybridized in solution with the specific target mRNA. After treatment with RNAse A, all mRNA is degraded except for the protected hybridized fragment, which can be electrophoresed and quantitated by autoradiography and densitometry.

MONOBLAST

See **Monopoiesis**.

MONOCLONAL ANTIBODIES

Molecules of a similar single immunoglobulin molecular type. They possess a unique **antibody**-combining site that reacts with a single **epitope** of its target **antigen**, and are therefore exquisitely monospecific.

Immunization of an animal with an antigen leads to a polyclonal B-**lymphocyte** response such that the resulting polyclonal antibody is a mixture of many different **immunoglobulins** with different antigen-combining sites and mixtures of different isotypes. These antibodies are not monospecific, in that they react with a range of antigens rather than a single antigen. For many purposes — e.g., diagnosis, serotyping of viruses, **immunophenotyping** of leukemias — monospecific antibodies are important. With care, and some luck, a polyclonal antibody that is relatively monospecific can be prepared. Examples include antibodies to bacterial serotypes, which can be prepared by immunization with purified bacteria and rendered effectively monospecific by cross adsorption with other serotypes. However, monospecific polyclonal antibodies to components of complex antigens, e.g., lymphocytes, cannot easily be prepared. The development of monoclonal antibody technology has solved this monospecificity problem, and at the same time provided a renewable tissue culture source of antibody, avoiding the need for repeated reimmunization of animals. A monoclonal antibody, by virtue of being a single immunoglobulin molecular type and thus possessing a unique antibody-combining site, will react with a single epitope of its target antigen, and so is exquisitely monospecific.

Preparation of Monoclonal Antibodies

In essence, B-cells are cloned and immortalized *in vitro* so that they grow and produce their monoclonal antibody indefinitely in culture.

A mouse (or some other suitable host) is immunized with the antigen; the spleen — a rich source of B-cells — is removed; and a single-cell suspension is prepared. This will be enriched for B-cells specific for the immunizing antigen. The spleen cells are then fused with a mouse myeloma cell line that is capable of indefinite growth *in vitro* but that does not itself produce immunoglobulin. A proportion of the fusion products will be capable of growth as hybridomas, which are essentially immortalized B-cells. The parental B-cells will not grow, and growth of the parental myeloma can be selected against by drugs to which the hybridomas are resistant. The hybridomas can be cloned *in vitro*, and the resulting cell clones are the products of fusions between the myeloma and single B-cells; hence each clone will produce a single immunoglobulin — a monoclonal antibody.

Obviously, not all the hybridoma clones will be producing the desired antibody, and so they need to be screened. If the original immunization is effective, and if the investigator is lucky, a relatively high proportion (on the order of a few percent) of the clones will produce the desired antibody. However, it is often the case that very large numbers of clones have to be screened to find one producer. The screening procedure is often laborious and the whole process very tedious.

The hybridoma secretes antibody into the supernatant culture medium and is easily harvested. Large-scale culture techniques allow the production of milligram or even gram quantities of antibody, which can then be used as a reagent on an industrial scale.

Human Monoclonal Antibodies

For some purposes, monoclonal antibodies derived from rodents are unsuitable, particularly for administration to humans, as immune responses develop to the murine determinants on the antibody. However, repeated attempts to develop a reliable system for generation of human monoclonal antibodies have met with failure. The "humanization" of mouse monoclonal antibodies (replacement of mouse-specific sequences in the antibody genes so that the secreted antibody is effectively human, while retaining the original antibody specificity) has met with some success, but is unlikely to become a routine procedure.

Phage Display

More recently, the phage-display technique has been adapted to generate monoclonal antibodies. In this technique, the entire antibody repertoire of an individual is, in effect, cloned into a bacteriophage culture (as cDNA); individual phages then encode the $V_H V_L$ regions of an antibody, which is expressed as a part of the phage coat so that this can be selected by binding to the appropriate antigen. The isolated phage is then grown and the antibody sequences isolated and engineered to immunoglobulin. The great advantage of this procedure is that the antibodies generated are fully human.

Diagnostic Uses of Monoclonal Antibodies[362]

Diagnostic procedures for infectious diseases, involving techniques such as enzyme-linked immunosorbent assay (ELISA) and immunofluorescence microscopy

Immunophenotyping of hematological disorders by **flow cytometry**

Pretransfusion testing to determine the ABO and Rhesus (D) blood groups of a recipient

Therapeutic Applications of Monoclonal Antibodies[362a]

See Table 110.

All of these agents can be complicated by allergic and anaphylactoid pulmonary reactions such as rhinitis and asthma. Secondary stimulation of antibodies to the infused monoclonal antibody may occur later. Lymphocyte inhibition may give rise to a variety of **viral infection disorders**.

TABLE 110

Therapeutic Use of Monoclonal Antibodies

Monoclonal Antibody	Antagonist	Disorder
Abciximab	platelet glycoprotein IIb/IIIa	percutaneous stenting
Alemtuximab	B-lymphocyte	**chronic lymphocytic leukemia**/lymphoma
Bortezomib	proteasomes	refractory relapsing **myelomatosis**
Eculzimab	terminal complement C5	**paroxysmal nocturnal hemoglobinuria**
Imatinib	tyrosine kinase from *BCR/ABL* fusion genes	**chronic myelogenous leukemia**
Infliximab	TNF-α-autoantibodies	Crohn's disease
		multiple sclerosis
		rheumatoid arthritis
Mepolizumab	antileukin-5	**hypereosinophilic syndrome**
Natalizumab	α4 integrins	Crohn's disease
		multiple sclerosis
Rituximab	CD20 B-lymphocytes	**non-Hodgkin lymphoma**
Trastuzumab (herceptin)	growth factor receptor 2	breast carcinoma
Adalimumab	TNFα	psoriatic arthritis, rheumatoid arthritis
Efalizumab	T-cell activation	chronic plaque psoriasis
Ibritumemab	CD20 B-lymphocytes	**follicular lymphoma**
Tositumemab	growth factor receptor 2	follicular lymphoma

MONOCLONAL B-CELL LYMPHOCYTOSIS

Presence of monoclonal B-lymphocytes in the circulation of otherwise normal individuals.[363] It is analogous to **monoclonal gammopathy** of undetermined significance (MGUS). Likewise, it can be a precursor state for a lymphoproliferative disorder such as **chronic lymphatic leukemia**.

MONOCLONAL GAMMOPATHIES

(Paraproteinemias) Disorders with the occurrence of monoclonal proteins without manifestations of malignancy. Each monoclonal protein (M-protein or paraprotein) consists of two heavy polypeptide chains of the same class and subclass and two light polypeptide chains of the same type. The heavy chains consist of gamma (γ) in IgG, alpha (α) in IgA, mu (μ) in IgM, delta (δ) in IgD, and epsilon (ε) in IgE. The light-chain types are kappa (κ) or lambda (λ).[364]

A paraprotein is seen as a narrow peak (like a church spire) in the γ, β, or $\alpha2$ regions of the densitometer tracing, whereas a dense, discrete band is seen on the cellulose membrane or on agarose with electrophoresis. Serum protein electrophoresis is indicated for any adult with unexplained weakness, fatigue, **anemia**, increased **erythrocyte sedimentation rate**, back pain, osteoporosis, osteolytic lesions, fractures, **immunoglobulin** deficiency, hypercalcemia, **Bence-Jones proteinuria**, renal insufficiency, or recurrent infections. It should also be performed in patients with nephrotic syndrome, refractory congestive heart failure, orthostatic hypotension, peripheral neuropathy, or carpal tunnel syndrome of unrecognized cause, because a localized spike or band strongly suggests primary **amyloidosis** (AL).

Immunoelectrophoresis or immunofixation is necessary to verify the presence of an M-protein and its type. An M-protein may be present even when the total protein concentration, β- and γ-globulin values, and quantitative immunoglobulin results are all within normal limits. A small M-protein may be concealed among the normal β or γ components. In addition, a monoclonal light chain (Bence-Jones proteinuria) is rarely seen with electrophoresis. Furthermore, the M-protein appears small or is not evident in patients with IgD myeloma or γ, μ, or a **heavy-chain disease**.

Quantitation of immunoglobulins should be performed with a nephelometer. However, the concentrations of IgM may be 1 to 2 g/dl more than expected on the basis of the densitometric tracing. IgG and IgA concentrations may also be spuriously elevated on nephelometry.

Analysis of Urine

Sulfosalicylic acid is more reliable for the detection of monoclonal light chains (Bence-Jones protein) than the usual dipstick tests. The heat test for Bence-Jones protein may produce both false-positive and false-negative reactions and is not recommended. Electrophoresis of an aliquot from a 24-h urine collection should be done in all patients with a serum M-protein and in patients with **myelomatosis**, **Waldenström macroglobulinemia**, primary amyloidosis, monoclonal gammopathy of undetermined significance, heavy-chain diseases, or suspicion of these conditions. An M-protein often appears as a dense, localized band upon electrophoresis or a tall, narrow, homogeneous peak on the densitometer tracing. The amount of M-protein secreted can be calculated from the size of the densitometer spike and the amount of protein in the 24-h specimen. Immunoelectrophoresis shows a prominent arc with the appropriate light-chain antisera. Immunofixation is more sensitive and is most helpful when monoclonal light chains occur in the presence

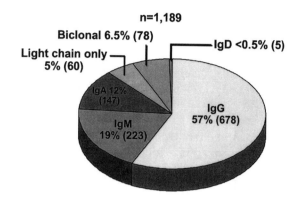

FIGURE 89
Types of serum monoclonal protein. (Data collected from Mayo Clinic, 2003.)

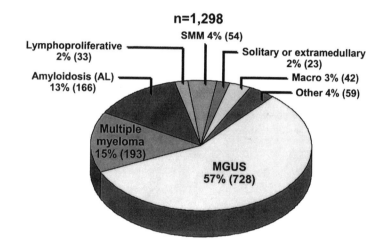

FIGURE 90
Diagnoses in patients with monoclonal protein. (Data collected from Mayo Clinic, 2003.)

of a polyclonal increase in light chains. It is also useful for detecting monoclonal heavy-chain fragments in the urine.

The most common gammopathy is IgG, followed by IgM and IgA (see Figure 89). The most common clinical diagnosis is monoclonal gammopathy of undetermined significance. The types of diagnosis in patients with monoclonal proteins are seen in Figure 90.

Monoclonal Gammopathy of Undetermined Significance[364]

(MGUS; benign monoclonal gammopathy) MGUS denotes the presence of an M-protein in persons without evidence of myelomatosis, macroglobulinemia, primary amyloidosis, or other related disorders. For many years, this disorder was considered to be benign and often was called benign monoclonal gammopathy. However, it is now known that a proportion of cases will evolve to symptomatic multiple myeloma, macroglobulinemia, or amyloidosis. For this reason, the term "MGUS" is more appropriate. The frequency of MGUS is age related. It occurs in 1% of persons older than 50 years and in 3% of those older than 70 years. Because of the high prevalence of MGUS, it is of importance to know whether the M-protein will remain stable and benign or progress to a symptomatic disorder.

At the Mayo Clinic, 241 patients with an apparently benign gammopathy have been followed for 24 to 38 years.[365] The vast majority were older than 40 years, the median age being 64 years. The heavy-chain type was mostly IgG (73%), with IgM (14%) and IgA (11%) being less common and biclonal (2%) rare. The initial M-protein level ranged from 0.3 to 3.2 g/dl (median, 1.7 g/dl). The level of uninvolved or background immunoglobulins was often reduced. A small number of patients had a monoclonal light chain (Bence-Jones protein), but the amount was small in most patients. The bone marrow contained around 3% plasma cells (range, 1 to 10%), which were nondiagnostic in appearance. During follow-up, multiple myeloma or related disorders developed in about a quarter of patients, the commonest being multiple myeloma, followed by macroglobulinemia, amyloidosis, and a malignant lymphoproliferative disorder. The actuarial rate of development of serious disease was 17% at 10 years, 34% at 20 years, and 39% at 25 years. Development of serious disease was not significantly different in IgG, IgA, or IgM gammopathies.

The interval from the recognition of MGUS to the diagnosis of multiple myeloma in the same series of patients ranged from 1 to 32 years. The diagnosis of myeloma was made more than 20 years after the detection of a serum M-protein in ten patients. The median survival after diagnosis of myeloma was 33 months, which is similar to that in the usual series of patients with myeloma. Macroglobulinemia of Waldenström occurred at a median interval of 10.3 years after the recognition of the M-protein. Primary amyloidosis occurred 6 to 19 years after the diagnosis of MGUS (median, 9 years). A **lymphoproliferative disorder** occurred 4 to 19 years after recognition of the M-protein.[19]

To confirm the findings of the 241 Mayo Clinic patients from the U.S. and other countries, a population-based study of 1384 patients with MGUS from the 11 counties of southeastern Minnesota was evaluated at Mayo Clinic from 1960 to 1994.[366] Median age at diagnosis was 8 years older than the 241 cohort (72 years vs. 64 years). Fifty-nine percent were 70 years or older, while only 2% were younger than 40 years at diagnosis. The M-protein was IgG in 70%, IgM in 15%, IgA in 12%, and biclonal in 3%. The M-protein ranged in size from 3.0 g/dl to an unmeasurable level. Thirty-eight percent of 840 patients who were evaluated had a reduction of uninvolved (normal or background) immunoglobulins. Thirty-one percent had a monoclonal light chain in the urine. The bone marrow contained a median of 3% plasma cells (range 0 to 10%) in the 160 patients who had a bone marrow examination. When anemia and renal insufficiency were present, there was no relationship to the plasma cell proliferative process.

During a follow-up of 11,900 person-years (median 15.4 years),[366] 115 patients (8%) developed multiple myeloma, Waldenström macroglobulinemia, AL amyloidosis, lymphoma, or chronic lymphocytic leukemia. The cumulative probability of progression to one of these disorders was 10% at 10 years, 21% at 20 years, and 26% at 30 years (1% per year). Even after 25 years of a stable MGUS, patients were at risk for progression. An additional 32 patients had an increase in M-protein to more than 3 g/dl or a bone marrow containing more than 10% plasma cells, but symptomatic multiple myeloma or Waldenström macroglobulinemia did not develop. The number of patients with progression to a plasma cell disorder (115 patients) was 7.3 times the number expected. The risk of developing Waldenström macroglobulinemia was increased 46-fold, multiple myeloma was increased 25-fold, and AL was increased 8.4-fold. Lymphoma was increased at 2.4-fold. This is underestimated because only lymphoma associated with an IgM protein was counted. Multiple myeloma occurred in 75 (65%) of the 115 patients with progression. In almost one-third of patients, multiple myeloma was diagnosed more than 10 years after the M-protein was recognized, while it was diagnosed after 20 years of follow-up in five patients. Characteristics of the 75 patients with multiple myeloma were comparable with those of the patients with newly diagnosed multiple myeloma referred to Mayo Clinic except that the southeastern Minnesota patients were older (72 years vs. 66 years).

The differentiation of a patient with MGUS from one with myeloma is difficult. The size of the serum and urine M-proteins, hemoglobin value, percentage of plasma cells, presence of hypercalcemia or renal insufficiency, and the presence of lytic bone lesions are helpful. An elevated plasma cell-labeling index, which is a measurement of the proliferative rate of the plasma cells, suggests multiple myeloma. Circulating plasma cells in the peripheral blood are also a good indication of active myeloma. However, active multiple myeloma may be present with a normal labeling index and no circulating plasma cells. No single factor can differentiate a patient with benign monoclonal gammopathy from one in whom a malignant plasma cell disorder will develop. The serum M-protein must be measured periodically and a clinical evaluation conducted to determine whether serious disease has developed. In patients with a recently recognized apparent MGUS, protein electrophoresis should be repeated in 3 to 6 months (depending on the size of the M-protein), and if the M-protein is stable, the test should be repeated in 6 to 12 months. If the M-protein remains stable and the patient appears to be well, electrophoresis and clinical evaluation should be performed annually thereafter.

Monoclonal gammopathies can also be associated with sensorimotor peripheral neuropathy. The patients are characterized by a slowly progressive, mainly sensory neuropathy beginning in the lower extremities. In approximately half of patients with an IgM M-protein and peripheral neuropathy, the M-protein binds to myelin-associated glycoprotein (MAG). The type and severity of neuropathy with anti-MAG are not different from those without anti-MAG activity. Amyloidosis must be considered and excluded in the differential diagnosis. Plasma exchange (see **Hemapheresis**) may produce impressive results. If the symptoms progress, treatment with **alkylating agents** may be beneficial.

MONOCYTE

Mononuclear leukocytes (size 10 to 18 µm) with irregular, often lobulated nucleus and opaque grayish-blue cytoplasm with fine azurophilic granules and vacuoles in the cytoplasm. There are no nucleoli, and the nuclear chromatin is fine. Prominent **mitochondria**, rough **endoplasmic reticulum**, and well-developed **Golgi apparatus** are present.

The monocyte is highly motile with plentiful cytoplasmic **actin** and glycogen. The granules contain peroxidase, fluoride-resistant nonspecific esterase (α-naphthyl butyrate or acetate esterase), or specific esterase (naphthol-AS-D chloroacetate) along with acid phosphatase. Hypogranular monocytes are young cells, implying rapid turnover, whereas hypergranular implies a response to **inflammation**. The normal peripheral-blood monocyte count is 0.2 to 1.0×10^9/l; the values for children are given in **Reference Range Table IX**.

Monopoiesis

Monocytes are derived from the pluripotential hematopoietic stem cell, but identification of mixed granulocyte-macrophage colonies in tissue culture suggests that there is a later common precursor (see **Granulopoiesis; Hematopoiesis; Hematopoietic regulation**).

The earliest identifiable monocyte precursor is the monoblast (size 14 to 22 µm), which is very similar in appearance to the myeloblast. The large single nucleus may be indented with fine lacy chromatin and one to two nucleoli. The agranular cytoplasm is deeply basophilic and contains few mitochondria and little rough endoplasmic reticulum. The monoblast gives rise to the promonocyte (size 11 to 20 µm), which is characterized by fine nuclear chromatin, a high nuclear cytoplasmic ratio, and gray/blue cytoplasm that may contain a few azurophilic granules consisting of myeloperoxidase, nonspecific esterase, and lysozyme. The rough endoplasmic reticulum and Golgi apparatus are well developed.

Several mitochondria and characteristic bundles of filaments are seen. The promonocyte normally only resides in the bone marrow, but the further differentiation to the monocyte allows release into the circulation and ultimately the tissues.

Monocytic growth and differentiation is regulated by various cytokines, although the exact role of each is still under study. Interleukin (IL)-3, granulocyte/macrophage colony stimulating factor (GM-CSF), and M-CSF, either alone or in combination, result in mono-cytic differentiation in tissue culture. It has been estimated that 2 to 3 days are necessary for a monoblast to mature into a bone marrow monocyte. The monocyte enters the circulation for only 8 to 17 h before migrating into tissues and transforming into macro-phages. Approximately 50% of monocytes produced ultimately reside in the liver (Küpffer cells), with the lung (15%) and peritoneum (8%) also important migratory sites. Although IL-3 and GM-CSF in pharmacological doses increase the peripheral-blood monocyte count, their role in physiological regulation is unknown.

Function of Monocytes

The migration of monocytes from the circulation is similar to that of neutrophils, with monocyte and endothelial adhesion molecules playing a central role. Upon stimulation by C5a, c-IFN (interferon), IL-1, IL-8, tumor necrosis factor (TNF)-α, etc., monocytes upregulate expression of CD11a, CD11b, CD11c, CD18, and CD49d/CD29. **Vascular endo-thelium** also upregulates expression of CD54, CD62E, and VCAM1 in response to cyto-kines. This enables monocytes to bind and migrate through the endothelium. Having bound, the monocyte spreads over the surface of the endothelium and migrates between two endothelial cells. The main role of monocytes is to enter the tissues and transform into **histiocytes** (macrophages), although they themselves may play a role in the regulation of **inflammation**.

MONOCYTE CHEMOATTRACTANT PROTEIN

(MCP; chemokines numbered MCP 1–3 of the CCL family) See **Chemokines**.

MONOCYTIC LEUKEMIA

See **Acute myeloid leukemia**; **Myelodysplasia**.

MONOCYTOID B-CELL IMMUNOCYTOMA

See **Marginal-zone B-cell lymphoma** — extranodal marginal-zone lymphoma; **Non-Hodgkin lymphoma**.

MONOCYTOID, INCLUDING MARGINAL-ZONE, LYMPHOMA

See **Marginal-zone B-cell lymphoma** — nodal marginal-zone lymphoma; **Non-Hodgkin lymphoma**.

MONOCYTOPENIA

A reduced level of circulating **monocytes**, $<0.2 \times 10^9/l$. It is always secondary to other causes such as stress, acute infections, **corticosteroid** administration, **bone marrow hypo-plasia**, **acute myeloid leukemia**, and **hairy cell leukemia**. It may predispose to infection, which will usually resolve with successful treatment of the predisposing cause.

TABLE 111

Causes of Monocytosis

Infections
    Bacterial: **tuberculosis**, brucellosis, subacute bacterial endocarditis, syphilis
    Protozoal: **malaria**, kala-azar, trypanosomiasis
    Rickettsial: Rocky Mountain spotted fever, typhus
    Viral: **cytomegalovirus**
Inflammatory reactive disorders
    Intestinal disorders: Crohn's disease, ulcerative colitis
    Systemic disorders: **rheumatoid arthritis, systemic lupus erythematosus**
Malignancy
    Carcinoma of any organ
    Hodgkin disease
    Lymphomatoid granulomatosis
    Acute myeloid leukemia: acute monocytic leukemia, acute myelomonocytic leukemias
    Chronic myelomonocytic leukemia
Cytokine agent therapy
    Granulocyte macrophage **colony stimulating factor**
    Interleukin-3
    Macrophage colony stimulating factor
Neutrophil disorders
    Recovery from **agranulocytosis** or acute infection
    Neutropenia, particularly cyclic neutropenia
Post**splenectomy**

MONOCYTOSIS

A raised level of circulating **monocytes**, $>1.0 \times 10^9/l$, usually due to increased monocyte production. There are many hematological and nonhematological causes (see Table 111). The monocyte count is usually only moderately elevated (1 to $5 \times 10^9/l$), but in **acute monocytic leukemia** and **chronic myelomonocytic leukemia**, very high counts ($>30 \times 10^9/l$) may be seen.

Causes

Hematological

Apart from the monocytic leukemias, monocytosis is observed in **chronic myelogenous leukemia**, malignant **histiocytosis**, and **polycythemia rubra vera** (see Table 111). One-third of patients with **Hodgkin disease** have monocytosis. Monocytosis may also be observed in several forms of **neutropenia**, particularly cyclical neutropenia and familial benign neutropenia. It is a feature of recovery from **agranulocytosis** and often precedes the neutrophil rise.

Nonhematological

Monocytosis is not uncommon in chronic infections, especially **tuberculosis**. Almost one-quarter of patients with subacute bacterial endocarditis have monocytosis, and a few may even have circulating **histiocytes** (macrophages).

MONOPOIESIS

See **Monocytes**.

MONTREAL PLATELET SYNDROME

A rare autosomally dominant hereditary **hemorrhagic disorder** associated with **thrombocytopenia** and the **giant-platelet syndrome**. In contrast to **Bernard-Soulier syndrome**, with which it has similarities, platelet aggregation to ristocetin is normal. Thrombin-based aggregation is low or absent. Spontaneous aggregation occurs when platelet-rich plasma is stirred. The platelet gigantism has been ascribed to an abnormal spreading reaction to **adenosine diphosphate** (ADP), thrombin, or cold. Platelet membrane glycoproteins are normal. It may be due to a calpain defect. Treatment is similar to that for Bernard-Soulier syndrome, from which it is difficult to distinguish.

MOSAICISM

Abnormality of **chromosomes** in which clones of cells with different genotypes are derived from the same zygote. Therefore, only a portion of the dividing cells are abnormal. It occurs in **glucose-6-phosphate dehydrogenase** deficiency and **chronic myelogenous leukemia**.

MOTT CELL

(Grape or morula cells) **Plasma cells** containing agglomerates of Russell bodies giving an appearance of bags of bluish marbles or grapes. They occur with hypergammaglobulinemia of all forms, including **myelomatosis** and **plasmacytoma**.

MOUTH DISORDERS

See **Dental hemorrhage**; **Oral mucosa disorders**; **Tongue disorders**.

M-PROTEIN

See **Monoclonal gammopathies**.

MUCOCUTANEOUS LYMPH NODE SYNDROME

See **Kawasaki disease**.

MUCOPOLYSACCHARIDOSES

A very rare group of inborn errors of metabolism in which deficiency of a particular lysosomal enzyme results in the accumulation of mucopolysaccharide within tissues (see Table 112). All are inherited as autosomally recessive traits, except **Hunter's syndrome**, which is sex-linked recessive. The combined incidence is about 1:10,000.

Mucopolysaccharide accumulation typically occurs in the central nervous system, skeleton, soft tissues, and visceral organs, e.g., liver, spleen, lung, and heart. Central nervous system deposition may result in hydrocephalus, cord compression, radiculopathies, and mental retardation. Skeletal abnormalities are due to hypoplastic bony formation and abnormal remodeling. Dwarfism is characteristic. Typical facies (previously called "gargoylism") are seen in Hunter's, Hurler's, and Maroteaux-Lamy disease and consist of thickened lips and face, nasal bridge depression, prominent forehead, and bushy eyebrows.

TABLE 112

Forms of Mucopolysaccharidosis

Name	Enzyme Deficiency	Mucopolysaccharide Accumulating
Hurler	α-L-iduronidase	dermatan sulfate heparan sulfate
Hunter	iduronate sulfate sulfatase	dermatan sulfate heparan sulfate
Sanfilippo A	heparan-N-sulfate sulfamidase	heparan sulfate
Sanfilippo B	N-acetyl-α-D-glucosaminidase	heparan sulfate
Sanfilippo C	α-glucosamine-N-acetyl transferase	heparan sulfate
Sanfilippo D	N-acetyl-glucosamine-6-sulfatase	heparan sulfate
Morquio	N-acetyl-galactosamine-6-sulfatase	keratin sulfate
Maroteaux-Lamy	N-acetyl-galactosamine-4-sulfatase	dermatan sulfate

In all types of mucopolysaccharidoses, granulocytes may contain large granules — **Alder-Reilly bodies**. Similar inclusions may be seen in lymphocytes (**Gasser cells**) in all types except Morquio's syndrome.

Usually, only supportive and symptomatic therapies are available, although a few patients have benefited from **allogeneic stem cell transplantation**.

MUCOSA-ASSOCIATED LYMPHOID TISSUE

(MALT) **Lymphoid tissue** located in the mucosa of the intestinal or gut-associated lymphoid tissue (GALT, Peyer's patches) and nasopharyngeal tonsils and adenoid tracts (NALT). Typically, the B-lymphocytes are in follicles surrounded by T-cells. The follicles have a specialized epithelial covering (M cells) that takes up **antigen** by pinocytosis as part of the innate immune system. These lymphoid tissues are rich in IgA-producing **plasma cells**. The lymphocytes have T-cell receptors but are without mean corpuscular hemoglobin (MCH) antigens. The M cells transport antigen into the subepithelial area containing the lymphocytes. After exposure to antigen, these lymphocytes can migrate to other mucosal tissues, including the reproductive tracts and to the mammary and salivary glands.

MULTIMERIN

A massive, soluble protein found in **platelets**, **megakaryocytes**, and **vascular endothelium**. Like **Von Willebrand Factor** (VWF), multimerin is composed of variably sized, disulfide-linked homomultimers, but it is not found in normal plasma. Multimerin is the major platelet-binding protein for **factor V**, and it could be important for regulating the factor V stored in platelet α-granules.

Multimerin's large homomultimers are assembled from a precursor protein, prepromultimerin, that is expressed in megakaryocytes and vascular endothelium. Prepromultimerin contains a signal sequence, multiple N-glycosylation sites, an RGDS (tripeptide of adhesive proteins such as fibrinogen and fibronectin) site, a central coiled-coil region, epidermal growth factor (EGF)-like and partial EGF-like domains, and a C-terminal region that resembles the globular domain in the extracellular matrix protein **collagens** type VIII and X. In both platelets and endothelial cells, prepromultimerin undergoes extensive N-glycosylation, proteolytic processing, and multimerization to generate mature multimerin.

MULTIPLE MYELOMA

See **Myeloma**.

MULTIPLE SCLEROSIS

(MS) Multiple plaques of demyelination are within the brain and spinal cord. The disease occurs worldwide and is regarded as an **autoimmune disorder**, probably induced by a nonspecific viral infection in those who are genetically susceptible. There is a weak familial incidence, but there is a firm **human leukocyte antigen** (HLA) linkage with HLA-A3, B7, D2, and DR2. **Lymphocyte** and **histiocyt**e infiltration of the central nervous system (CNS) occurs with IFN-γ activity and possibly CD4+ and CD8+ cells acting against the myelin sheaths.

Treatment is by immunosuppression with **corticosteroids**, **β-interferon**, **ciclosporin**, **azathioprine**, **cyclophosphamide**, and **plasmapheresis**. Dosage of chemotherapy is limited by **bone marrow hypoplasia**, but this may be increased following **allogeneic stem-cell transplantation** in the early stages of the disease. Clinical trials are underway for use of an antiadhesion molecule (**natalizumab**) against α-4 integrins.

MUSCLE DISORDERS

The disorders of muscles occurring with hematological disorders.

   Hemorrhage into muscle masses with **hemophilia** and related disorders. Muscles commonly affected are the anterior flexor group of the arm, the posterior soleus, and the gastrocnemius of the legs. It may lead to cyst formation and **pseudotumors**, particularly in the thigh, buttock, and psoas muscles.

   Amyloidosis with deposition in ligaments of the carpal tunnel, leading to median nerve compression. Deposition in muscle causes pseudohypertrophy.

MYCOPLASMA INFECTION DISORDERS

The hematological changes associated with *Mycoplasma pneumoniae* infection. These occur in the postinfective period and are the commonest cause of high-titer cold agglutinins (see **Cold-agglutinin disease**) to **I blood group** antigens. The agglutinins activate **complement**, giving rise to red blood cell sludging with acrocyanosis of peripheral tissue and to chronic hemolytic anemia.

MYCOSIS FUNGOIDES/SÉZARY SYNDROME

(Cutaneous T-cell lymphoma; CTCL; small-cell cerebroform lymphoma) See also **Lymphoproliferative disorder**; **Non-Hodgkin lymphoma**.

A T-**lymphocyte** non-Hodgkin lymphoma characterized by the predominance of small cells with cerebroform nuclei and a minority of larger cells with similar nuclei that infiltrate the epidermis.[367,368] In some patients, these epidermotropic infiltrates have Pautrier microabscesses in the epidermis with aggregates of cerebroform cells. When both skin and systemic involvement occur, e.g., peripheral blood and paracortex of **lymph nodes**, the term "Sézary syndrome" is used.[367] The infiltrate is invariably accompanied by **Langerhans cells**. Bone marrow is usually normal. The **immunophenotype** of the tumor cells is T-cell-associated antigens+ (CD2+, CD3+, CD5+), CD4+, CD7+, CD8+ (rare), and CD25-. **Cytogenetic analysis** shows rearrangement of T-cell receptor (TCR) genes. The cellular origin is probably a peripheral epidermotropic CD4+ T-cell.

Clinical Features

Patients are usually middle-aged adults, with a particular prevalence for black ethnicity. Prior to the diagnosis being made, nonscaly eruptions of long duration occur. Presentation is by skin lesions in the following stages:

T1: patch, papules, or plaques involving less than 10% of the skin

T2: patch, papules, or plaques involving 10% or more of the skin

T3: tumors

T4: generalized erythroderma

Peripheral-blood involvement may be slight or prominent, occurring in about 5% of cases at any clinical phase. T-cell **lymphocytosis** with hyperconvoluted cerebroform nuclei of at least $1 \times 10^9/l$ occurs. In advanced stages, the lymph nodes and other organs may become involved. The 10-year relative survival is 100% in T1, 67% in T2, and about 40% in T3 and T4 disease. The most common cause of death is infection, secondary to the skin lesions. As a terminal event, a transformed, large, involuted **T-cell lymphoma** may develop.

Staging

See **Lymphoproliferative disorders**.

Treatment

In the patch phase, this is limited to bland emollients and ultraviolet-beam therapy. Plaques require topical nitrogen mustard or psoralen plus ultraviolet A radiation **(PUVA) light therapy**. Bexartene gel is effective in the early stage of the disease. For the tumor phase, **radiotherapy** by total body electron beam therapy (4 to 7 MeV), limiting ionizing radiation effects to the outer 1 cm of the skin in doses of 3000 to 3600 cGy, is given over 8 to 10 weeks. Extracorporeal photopheresis alone or combined with **interferon-α** has been used.[369] Disseminated lymphoma requires courses of **cyclophosphamide**, hydroxydaunomycin/doxorubicin, Oncovin®, prednisone (CHOP), or **2′-deoxycoformycin** (DCF) (see **Cytotoxic agents**). Denileukin diftiox, whose binding and toxicity are dependent on target-cell expression of IL-2R, has a reported response rate of 30% in patients who are refractory.

MYELOBLASTS

See **Neutrophils** — maturation.

MYELOCYTES

See **Neutrophils** — maturation.

MYELODYSPLASIA

(Myelodysplastic syndromes; MDS; preleukemia) A form of **dyserythropoiesis**, including **ineffective erythropoiesis, megaloblastosis, dysgranulopoiesis** (including **leukocyte anomalies**), and **dysmegakaryopoiesis**, or any combination of these abnormalities. This is a common practical definition, but taken in the broadest interpretation, the term "myelodysplasia" can include all variations from normal morphology of cells derived from or maturing in the bone marrow.

Etiology

The majority of cases are idiopathic, but some are secondary to agents causing damage to the bone marrow (see **Bone marrow hypoplasia**). A few are associated with **paroxysmal nocturnal hemoglobinuria**. An association with JAK inhibitors (see **Cell-signal transduction**) has been claimed.[31a]

Classification

French-American-British (FAB) Group

The FAB classification of myelodysplasia is described in Table 113.

Although the FAB divisions are somewhat arbitrary, they were an attempt to identify those patients at high risk of transformation to **acute myeloid leukemia** (AML). A level of 30% blasts in bone marrow was seen as critical in defining this division, but patients with 20 to 30% of marrow blast cells with more than 5% blasts in blood are categorized as having refractory anemia with excess of blasts in transformation (RAEB-t). Where blood **monocyte** counts exceed $1 \times 10^9/l$, the suggested diagnosis is one of **chronic myelomonocytic leukemia**. Although clinically useful, this classification is biologically incomplete, as it does not take account of other important parameters such as karyotype.

TABLE 113

FAB Classification of Myelodysplasia

Type	Blood Findings	Bone Marrow
Refractory anemia (RA)	<1% blasts	dysplasia in one or more cell lines <5% blasts
RA with ring sideroblasts (RAS)	<1% blasts	as RA plus ring sideroblasts in >15% erythroblasts
RA with excess blasts (RAEB)	<5% blasts	as RA plus 5 to 20% blasts
RAEB in transformation (RAEB-t)	>5% blasts (or <5% with Auer rods)	as RA plus 20 to 30% blasts or as RAEB with Auer rods
Chronic myelomonocytic leukemia (CMML)	any above plus $>1 \times 10^9/l$ monocytes	any above plus promonocytes

WHO Classification

This introduces other categories and places chronic myelomonocytic leukemia in a separate group of myelodysplastic/**myeloproliferative disorders**.

Myelodysplastic syndromes

- Refractory anemia
- Refractory anemia with ring sideroblasts
- Refractory cytopenia with multilineage dysplasia
- Refractory anemia with excess blasts
- Myelodysplastic syndrome associated with isolated del(5q) chromosome abnormality
- Myelodysplastic syndrome, unclassifiable

Myelodysplastic/myeloproliferative diseases
- Chronic myelomonocytic leukemia
- Atypical chronic myeloid leukemia
- Juvenile myelomonocytic leukemia
- Myelodysplastic/myeloproliferative diseases, unclassifiable

Clinical Features

These relatively common syndromes are primarily disorders of the elderly, particularly males aged 50 to 70 years. Many patients are asymptomatic at presentation, the abnormality being detected upon routine peripheral-blood examination. Others complain of fatigue due to **anemia** and may be prone to infection as a result of **neutropenia** or to bleeding due to **thrombocytopenia**.

Laboratory Features[370]

Morphological Features of Blood and Bone Marrow

Typically, there are multiple cytopenias and a hypercellular marrow with dysplastic changes, often in all precursor cells (trilineage dysplasia), but isolated cytopenias may occur.

Erythropoiesis

Characteristically, there is ineffective erythropoiesis with erythroid hyperplasia and often macrocytosis in blood and bone marrow.[371] Patients with RAS may show red cell dimorphism and in all types of MDS, anisocytosis and poikilocytosis are common. Nonspecific features of dyserythropoiesis, such as pyknosis multinuclearity, abnormal sideroblasts, and, rarely, nuclear budding and bridging are seen; nucleated red cells in blood, often with dysplastic changes, are not uncommon.

Granulopoiesis

Neutropenia is common with or without an absolute monocytosis. Neutrophils may show nuclear and cytoplasmic abnormalities with loss, and occasionally complete absence, of primary granules. The **Pelger-Huët anomaly** is occasionally present. In CMML, there is monocytosis exceeding $1 \times 10^9/l$, occasionally with cells intermediate between monocytes and myelocytes. In bone marrow, granulocyte precursors may be atypical, and in RAEB and RAEB-t, granular and agranular blasts with or without Auer rods may be present.

Thrombopoiesis

Although thrombocytopenia is seen in more than half of all patients, occasionally thrombocytosis occurs, especially in refractory sideroblastic anemia (RSA). Giant platelets deficient in granules are common, and ultrastructure studies on these platelets show defective or absent microtubules with dilated canalicular systems. Platelet function shows abnormal adhesion and aggregation, with a prolonged bleeding time. Megakaryocytes are often abnormal, usually distorted, and smaller than normal "micromegakaryocytes," which produce small numbers of dysplastic platelets and may be mistaken for myeloid blasts. Occasionally, giant megakaryocytes are seen.

Cytogenetic Analysis

Abnormal clones are demonstrable in half of all patients, the commonest abnormalities being monosomy 7 or 7q⁻, trisomy 8, monosomy 5 or 5q⁻, and loss of the Y chromosome. Trisomy 1q, 9 and 21, 21q⁻, and iso 17q are also described.

Chromosome anomalies appear to be related to leukemic change and to disease progression, and several give independent prognostic information.

Single or multiple abnormalities of chromosomes 7 and 8 and the development of any change in patients with a previously normal karyotype indicate a poor outlook, but an isolated 5q⁻ does not affect prognosis. α-**Thalassemia** has been reported as an acquired complication. Hypermethylation of DNA with silencing of multiple genes (epigenesis), associated with many forms of malignancy, is characteristic of myelodysplasia. A recently identified change is histone acetylation. Aberrant methylation may be associated with resistance to chemotherapy.

Hematopoietic Progenitors

In vitro culture studies may give prognostic information in MDS. In RA and RAEB, **colony forming unit**-granulocytes/macrophages (CFU-GM) show a reduction in colony and cluster growth. At diagnosis, undetectable myeloid colony growth and low colony/cluster ratio in blood indicate a high risk of subsequent transformation. In RA and RAS, high cluster growth (>15 clusters/ml) is associated with shorter-than-normal survival, while survival in RAEB is independent of cluster number but is related to colony/cluster ratio.

Specific Types of Myelodysplasia

Refractory Anemia

(RA) Patients are typically over 50 years of age with anemia, reticulocytopenia, macrocytosis, and marrow changes that include erythroid hyperplasia or dyserythropoiesis, sometimes dysgranulopoiesis, but with fewer than 5% blasts. The incidence of leukemic change is less than 25%, and these patients often have minor abnormalities of granulopoiesis at diagnosis.

Refractory Anemia with Ringed Sideroblasts

(RAS) As with RA, patients who demonstrate abnormalities of granulocyte morphology at presentation have a higher chance of transforming to AML, although generally this transformation occurs in less than 15% of RSA patients and is very low in patients with abnormalities confined to the erythroid series. However, separation into these two groups may be artificial, as all gradations of these abnormalities are seen, especially if patients are carefully followed. Other factors said to have prognostic importance are thrombocytosis, seen occasionally in RSA, which is favorable, while male sex and older age are unfavorable.

About 20% of MDS patients have RSA, although the degree of sideroblastosis varies considerably. The exact genetic defect is unclear, but defective DNA synthesis and abnormal mitochondrial enzymes result in abnormal heme synthesis (see **Sideroblastic anemia** — myelodysplasia with ringed sideroblasts). Abnormalities in mitochondrial DNA have been described in a minority of cases.

Myelodysplastic Syndrome Associated with Isolated del(5q) Chromosome Abnormality

(5q-syndrome) This is a disorder mainly of middle-aged women with severe refractory anemia. There are <5% blasts in the blood and bone marrow. It is associated with long survival, and karyotypic evolution is uncommon.

Refractory Cytopenia with Multilineage Dysplasia

(RCMD) A myelodysplastic syndrome with bicytopenia or pancytopenia and dysplastic changes in 10% or more in two or more of the myeloid cell lines. There are less than 1% blasts in the peripheral blood and less than 5% in the marrow. Auer rods are not present, and the monocytes in the blood are <1 × 10^9/l. This group accounts for approximately 24% of cases. Chromosomal abnormalities include trisomy 8, monosomy 7, del(7q), monosomy 5, del(5q), and del(20q). Clinical course is variable, the prognosis depending upon the degree of cytopenia and dysplasia.

Refractory Anemia with Excess Blast Cells

(RAEB) In RAEB, the marrow contains more than 5% blast cells (see Table 113), often with erythroid hyperplasia and occasional sideroblasts. Sideroblasts are usually non-ring forms, but occasional ring sideroblasts may be seen. The few patients who have more than 5% blasts with many ring sideroblasts should be diagnosed as RAEB rather than RSA, as blasts are a better discriminator of prognosis. The blood in RAEB may show occasional blast cells, often in the presence of granulocytopenia and thrombocytopenia. Phi bodies are sometimes seen in the cytoplasm. Approximately 30 to 50% have clonal cytogenetic abnormalities, including +8–5, del(5q), –7, del(7q), and del(20q). Patients with up to 30% marrow blasts are termed RAEB-t and often progress quickly to acute myeloid leukemia. The only form of treatment is stem cell transplantation, so HLA matching at diagnosis may be lifesaving.

Chronic Myelomonocytic Leukemia

(CMML) A disorder showing features of both myelodysplasia and myeloproliferation, either of which may predominate. Typically, the blood contains more than 1 × 10^9/l monocytes, usually 5 to 10 × 10^9/l, and sometimes also moderate numbers of miniature granulocytes. The marrow is compatible with MDS and shows a mixture of myeloid precursors, most often not blasts. The monocytes have abnormal nuclei that are often convoluted or hypersegmented. The differential diagnosis is CML, and occasionally distinction is difficult, especially from Ph-negative CML. Clonal cytogenetic abnormalities occur, but none are specific or recurring. Other features of monocytic leukemia are often seen, such as raised serum and urinary lysozyme and the presence of increased amounts of polyclonal immunoglobulin. The prognosis is extremely variable, with a median survival time of 20 to 40 months.

Therapy-Related Myelodysplasia

See also **Acute myeloid leukemia** — therapy related.
(t-MDS) More-intensive **cytotoxic agent** therapy coupled with longer survival has led to the emergence of therapy-related leukemia, often following treatment for Hodgkin disease, non-Hodgkin lymphoma, and plasma cell myeloma, these patients accounting for up to 15% of all acute leukemias. In up to 50% of these patients, there is a demonstrable myelodysplastic phase. Clinical findings in t-MDS are similar to those in *de novo* MDS, although the marrow is commonly more hypocellular, and well-marked trilineage dysplasia is usual.[329] In addition, marrow fibrosis can be demonstrated upon reticulin staining in up to 50% of patients. With or without progression to acute leukemia, prognosis is poor and management difficult.

Prognosis

To compare the outcome of clinical trials, an International Prognostic Scoring System (IPSS) was devised.[372] It was derived from the analysis of data from more than 800 patients managed solely with supportive care. The model applies a score that is weighted according to the independent statistical power of each of three prognostic features affecting the bone marrow:

Blast percentage
Cytogenetic pattern
Number of cytopenias

The cumulative score enables segregation of patients into four subgroups with varying expectations for survival and the risk of, and interval to, AML progression. Based upon this model, numerous trials of drugs are in progress. It is anticipated to confirm that 35 to 40% of patients transform eventually to acute myeloid leukemia.

Treatment

The primary curative treatment, where possible, is **allogeneic stem cell transplantation** (ASCT) from an HLA-identical sibling donor.[373] Disease-free survival ranges from 29 to 40%, with a corresponding nonrelapse mortality of 37 to 50% and a rate of relapse ranging from 23 to 48%. The best outcome is for younger patients.

Patients without evidence of blast cells in the bone marrow or who are unsuitable for ASCT can be treated by supportive care for appropriate cytopenias with red blood cell and platelet transfusions accompanied by antibiotics (see **Bone marrow hypoplasia**).

Therapeutic developments are in progress using drugs aimed at specific cellular targets (see Table 114).

Lenalidomide, a thalidomide analog, avoids the neurological adverse drug reactions of thalidomide and is reported to be effective in achieving transfusion independence in those with a pure erythroid disorder. Other new therapies include methylation inhibitors: 5-azacytidune (AZA) and 5-aza-2′-deoxycytidine (DAC).

TABLE 114

Targets of Interaction and Drugs under Trial in Treatment of Myelodysplasia

Target	Drug
Antiangiogenic	thalidomide
	lenalidomide
	bevacizumab
	arsenic trioxide
Receptor tyrosine kinase inhibitor	imatinib mesylate
Farnesyl transferase	tipifarnib
	ionafarnib

MYELOFIBROSIS

An increase in the fibrotic matrix of **bone marrow**. This may be primary **idiopathic myelofibrosis** (IMF) or secondary myelofibrosis (see Table 115). **Reticulin fibrosis** is the term used when the fibers are of un-cross-linked soluble collagen (usually type III) and contain deposits of matrix protein that are argyrophilic, staining black when exposed to silver. Acute panmyelosis with **myelofibrosis** is a rapidly progressive fibrosis associated with all forms of acute leukemia. In IMF, osteosclerosis is seen in some patients, but this also occurs with secondary myelofibrosis. In IMF associated with **myeloma metaplasia** and **BCR/ABL** negative, JAK2 tyrosine kinase mutation has been shown.[373a]

TABLE 115

Causes of Secondary Myelofibrosis

Hematological malignancies
Acute myeloid leukemia (AML)
Chronic myelogenous leukemia (CML)
Polycythemia rubra vera (PRV)
Myelomatosis
Hairy cell leukemia (HCL)
Hodgkin disease (HD)
Non-Hodgkin lymphoma (NHL)
Carcinoma metastatic to the bone marrow
Nonmalignant conditions
Systemic **mastocytosis**
Myelodysplasia
Gray-platelet syndrome
Infection
HIV, Tuberculosis
Renal osteodystrophy
Hyperparathyroidism
Systemic lupus erythematosus
Vitamin D deficiency

MYELOID METAPLASIA

(Extramedullary hematopoiesis) The production of red blood cells, granulocytes, and platelets outside the marrow cavity, in the liver and spleen, and rarely in lymph nodes or other organs. It characteristically occurs in **idiopathic myelofibrosis**, but is also seen in **polycythemia rubra vera**, **chronic myelogenous leukemia**, **myelodysplasia**, syphilis, osteopetrosis, **lymphoproliferative disorders**, and carcinoma metastases in the bone marrow. For unknown reasons, extramedullary hematopoiesis also occurs with ischemic hepatic injury and with **hemolytic anemia**. If exuberant, extramedullary hematopoiesis can result in organ dysfunction, and no organ is immune from this. It can also form mass lesions in the thoracic and abdominal cavities and the brain. The diagnosis can be made by evaluation of a blood smear, which will contain circulating erythroblasts and myelocytes, or by a tissue biopsy. Extramedullary hematopoiesis is usually ineffective and contributes little to cell numbers in the peripheral blood. Teardrop **poikilocytes** are a characteristic change in red blood cell morphology that is associated with **splenomegaly** but not necessarily extramedullary hematopoiesis.

MYELOID SARCOMA

(Extramedullary myeloid tumor; granulocytic sarcoma; chloroma) A tumor mass of **myeloblasts** or immature myeloid cells occurring in an extramedullary site or in bone. They may appear green in color due to the presence of myeloperoxidase[374] (the term being first used by King in 1953), the color rapidly fading on exposure to light. It may precede or occur concurrently with **acute myeloid leukemia** (AML), **chronic myelogenous leukemia**, **myelodysplasia**, or with **myeloproliferative disorders**. In the latter group, the appearance of such an extramedullary mass may herald the onset of blast crisis or transformation to overt acute leukemia. It may also be the initial manifestation of relapse in a previously treated AML in remission.

The most common locations are subperiosteum of the skull, paranasal sinuses, sternum, ribs, vertebrae and pelvis, lymph nodes, and skin. Diagnosis is based upon immunocytochemistry for **myeloperoxidase**, **lysozyme**, and chloroacetate esterase. **Immunophenotyping** shows expression of myeloid markers CD13, CD33, CD117, and MPO. **Cytogenetic analysis** shows that associations of the AML myeloblasts are with t(8;21)(q22;q22) and abnormal eosinophil components (AMML Eo) with inv(16)(p13;q22) or t(16;16)(p13;q22).

Tumors arising with myelodysplasia or myeloproliferative disorders are a blastic transformation. As with AML, the prognosis is that of the underlying leukemia. Solitary masses without evidence of leukemia respond well to **radiotherapy**.

MYELOMA

(Multiple myeloma; plasma cell myeloma) Neoplastic proliferation of a single clone of **plasma cells** engaged in the production of a monoclonal immunoglobulin.[375] **Immunophenotype** of the majority of cells is SIg⁻, CIg⁺, CD38⁺, and PCA-1⁺. (See **Osteosclerotic myeloma; plasmacytoma**.)

Clinical Features

Myeloma accounts for 1% of all malignant disease and slightly more than 10% of hematological malignancies. The annual incidence of multiple myeloma in the U.S. is 4 per 100,000 population.[376] An apparent increased incidence in recent years is probably related to increased availability and use of medical facilities, particularly in the elderly. Its incidence in African-Americans is almost twice that in Caucasians. The median age at diagnosis is 65 years. Only 2% of patients are younger than 40 years at diagnosis.

Bone pain, particularly in the back or chest, is present at diagnosis in more than two-thirds of patients. Weakness and fatigue are common, but fever and weight loss are rare. The major symptoms may result from an acute infection, renal failure, hypercalcemia, or **amyloidosis**. The liver is palpable in about 15% of patients, but the spleen is rarely enlarged. Extramedullary plasmacytomas are uncommon and are usually observed late in the course of the disease.

Laboratory Features[377,377a]

Normocytic, **normochromic anemia** is present in almost 60% of patients at diagnosis.[378] A spurious decrease of hemoglobin and hematocrit values may occur because of an increased plasma volume from the osmotic effect of large amounts of M-protein. The serum creatinine level is more than 2 mg/dl in one-fifth of patients, and hypercalcemia (calcium value more than 11 mg/dl) is present at diagnosis in about 15% of patients. Serum protein electrophoresis reveals a peak or localized band in 80% of patients, hypogammaglobulinemia in 7%, and no apparent abnormality in the remainder. An M-protein is detected in the serum in more than 90% of patients. Approximately 52% are IgG, 22% IgA, 16% light chain only (Bence-Jones proteinuria), IgD 2%, and biclonal gammopathy 1%. An M-protein is found in the urine in 78% of patients at diagnosis. Ninety-eight percent of patients with myelomatosis have an M-protein in the serum or urine at the time of diagnosis. Conventional radiographs reveal lytic lesions, fractures, or osteoporosis in more than 70% of patients at diagnosis. ⁹⁹Tc bone scans are inferior to conventional radiographs for detecting lesions in myeloma. Computed tomography and magnetic resonance imaging are helpful in patients with myeloma who have pain but no abnormality on radiographs.

The bone marrow contains more than 10% plasma cells in 95% of patients, but the level of plasma cells can range from less than 5% to almost 100%. Bone marrow involvement may be more focal than diffuse and require repeat bone marrow examinations for diagnosis. The identification of a monoclonal population of plasma cells (k or l) by immunoperoxidase or immunofluorescence staining is helpful for differentiating monoclonal plasma cells from reactive plasmacytosis due to autoimmune disorders, metastatic carcinoma, liver disease, or infections. Plasmablastic morphology, which occurs in about 10% of cases, is associated with a short survival. **Cytogenetic analysis** has revealed abnormalities in about one-third of patients, but no diagnostic abnormalities have been identified. Aneuploidy is found in about 80% of patients.

Minimal criteria for the diagnosis of multiple myeloma are a bone marrow containing more than 10% plasma cells or a plasmacytoma plus at least one of the following:

M-protein in the serum (usually more than 3 g/dl)

M-protein in the urine

Lytic bone lesions

These findings must not be from carcinoma metastases, connective tissue diseases, chronic infection, or other **lymphoproliferative disorders**. Patients with multiple myeloma must be differentiated from those with **monoclonal gammopathy** of undetermined significance (MGUS) and smoldering multiple myeloma (see below).

Prognosis

Multiple myeloma has a progressive course; the median survival is approximately 3 years. The plasma cell-labeling index, β2-microglobulin level, **C-reactive protein** value, and soluble **interleukin**-6 (IL-6) receptor level are all statistically significant and independent prognostic factors. Increased age, plasmablastic morphology, circulating plasma cells in the peripheral blood, increased myeloma colony growth, and increased levels of IL-6 are all associated with more-aggressive disease.

Treatment

Not all patients who fulfill the minimal criteria for the diagnosis of multiple myeloma should be treated. The patient's symptoms, physical findings, and all laboratory data must be considered. The present treatment of choice is chemotherapy followed by peripheral-blood stem cell transplantation.

Chemotherapy[379]

The oral administration of **melphalan** and **prednisone** or **prednisolone** produces objective response in 50 to 60% of patients. Many combinations of chemotherapeutic agents have been used to improve survival. In a metanalysis of 18 published trials, there was no difference in overall survival in a comparison of single versus multiple **alkylating agents**. Chemotherapy should be continued for at least 1 year, and if the patient is in a plateau state (stable serum and urine M-protein and no evidence of progression), treatment should be stopped. Interferon-α2 (IFN-α2) appears to prolong the plateau state but probably does not influence survival. Patients should be followed closely and the same chemotherapy reinstituted if relapse occurs more than 6 months later.

Newer therapies include thalidomide with dexamethazone, with aspirin being given simultaneously to prevent venous thrombosis associated with thalidomide.[380] **Bortezomib** is being used,[129] and **lenalidomide** (Revlimid) is being assessed in clinical trials.

New anti-angiogenesis agents, which have less thrombogenic potential, less fatigue, and less peripheral neuropathy, are also undergoing evaluation.

Stem Cell Transplantation

Allogeneic stem cell transplantation has the advantage over **autologous stem cell transplantation** in that the graft contains no tumor cells, and there may be a graft-versus-tumor effect. However, there is an early mortality of 25% during the first 6 months, the risk of graft-versus-host disease is troublesome, and relapse of multiple myeloma is common. Furthermore, only 5 to 10% of patients with multiple myeloma are eligible because of their age, renal function, and the lack of an HLA-matched sibling donor.

Autologous peripheral-blood stem cell or bone marrow transplantation is applicable for up to 50% of patients because there is no specific age limit and a matched donor is unnecessary.[381] However, two major hurdles must be overcome:

Eradication of multiple myeloma from the patient may not occur even with large doses of chemotherapy and radiation.

Infused peripheral-blood stem cells or autologous bone marrow is contaminated by myeloma cells or their precursors and may be responsible for relapse.

Radiotherapy

Palliative **radiotherapy** should be limited to patients with disabling pain from a well-defined focal process that has not responded to chemotherapy.

Refractory Multiple Myeloma

Almost all patients with multiple myeloma who respond to chemotherapy eventually relapse. Previous studies revealed that the highest response rates have been with VAD (vincristine, Adriamycin [doxorubicin], and dexamethasone). Most of the activity of VAD is from dexamethasone, so it is often used as a single agent. Intravenously administered methylprednisolone (2 g × 3 times weekly for a minimum of 4 weeks) is helpful in patients with **pancytopenia** and has fewer side effects than dexamethasone. VBAP (vincristine, carmustine [BiCNU®], Adriamycin [doxorubicin] on day 1 and prednisone or prednisolone daily for 5 days) every 3 to 4 weeks benefits about one-third of patients. IFN-α2 produces an objective response in 10 to 20% of patients with refractory multiple myeloma. A proteasome inhibitor, bortezomib, is under clinical trial for this form of myeloma.

Clinical Variants of Classic Multiple Myeloma[375]

Smoldering Multiple Myeloma

This subgroup of multiple myeloma is characterized by the presence of a serum M-protein (paraprotein) of more than 3 g/dl and more than 10% plasma cells in the bone marrow, but with no anemia, renal insufficiency, hypercalcemia, skeletal lesions, or other clinical manifestations of multiple myeloma. The concentration of uninvolved immunoglobulins is usually reduced, and a small amount of M-protein is often found in the urine. The plasma cell-labeling index (proliferative rate) is low. Biologically, these patients have a benign monoclonal gammopathy, but they must be followed because symptomatic multiple myeloma will develop in many. Up to three lytic bone lesions without pain is termed "indolent myeloma." Recognition of smoldering or indolent myeloma subset is crucial because such patients should not be treated unless progression occurs.

Nonsecretory Myeloma

These patients have no M-component in either the serum or the urine and account for only 3% of patients with multiple myeloma. The diagnosis is established by identification of an M-protein in the cytoplasm of the plasma cells by immunoperoxidase or immunofluorescence. However, cases in which no M-protein can be found in plasma cells have been reported. The response to therapy and survival are similar to those in patients with myeloma who have an M-component, except that they have less renal involvement.

Plasma Cell Leukemia

This is defined as the presence of more than 20% plasma cells in the peripheral blood and an absolute plasma cell value more than $2 \times 10^9/l$. It is considered primary when it presents *de novo* (60% of cases) and as secondary when it is a leukemic transformation of a previously recognized multiple myeloma (40%). Patients with primary plasma cell leukemia are younger and have a higher incidence of **hepatosplenomegaly** and **lymphadenopathy**, fewer lytic bone lesions, smaller M-protein components, and a longer survival than patients with secondary plasma cell leukemia. Treatment is not satisfactory, but single or combination alkylating-agent therapy often produces short responses.

IgD Myeloma

The M-protein is small, and **Bence-Jones proteinuria** of the l type is more common than in multiple myeloma.

Osteosclerotic Myeloma

Plasmacytoma

Management of Complications
Hypercalcemia

This disorder should be suspected in the presence of anorexia, nausea, vomiting, polyuria, polydipsia, constipation, confusion, or stupor. Hydration and prednisone/prednisolone are sufficient in most cases. If these measures fail, pamidronate disodium (Aredia) or gallium nitrate is effective.

Renal Insufficiency

Patients with acute renal failure should be treated promptly with fluid and electrolyte correction and hemodialysis if necessary. Plasmapheresis (see **Hemapheresis**) may be helpful for acute renal failure. Maintenance of high urine output (3 l/24 h) is important for preventing renal failure in patients with Bence-Jones proteinuria. Allopurinol is effective for the prevention and treatment of hyperuricemia.

Infections

Prompt and appropriate therapy of bacterial infections is essential. Pneumococcal and influenza immunizations should be given to all patients. Prophylactic daily penicillin given orally often benefits patients with recurrent pneumococcal infections. Intravenously administered gamma globulin may be helpful in some patients, but it is very expensive.

Skeletal Lesions

Patients should be encouraged to be as active as possible, but they must avoid undue trauma. Fixation of long bone fractures or impending fractures with an intramedullary rod and methyl methacrylate has given excellent results. Pamidronate disodium (Aredia) or zoledronic acid (Zometa) reduce skeletal pain and fractures.

Hyperviscosity

Symptoms of **hyperviscosity** may include nasal bleeding, blurred vision, neurologic symptoms, and congestive heart failure. Most patients have symptoms when the relative serum viscosity reaches 6 or 7 cP (normal, less than 1.8 cP), but the relationship between serum viscosity and clinical manifestations is not precise. Plasmapheresis promptly relieves the symptoms and should be done regardless of the viscosity level when the patient is symptomatic.

Spinal Cord Compression

This should be suspected in patients with severe back pain, with weakness or paresthesias of the lower extremities, or with bowel or bladder dysfunction. Magnetic resonance imaging (MRI) or computed tomography (CT) must be done immediately and radiation given.

MYELOPEROXIDASE

An enzyme found in the primary granules of **neutrophils** and the specific granules of **eosinophils** and **basophils** as well as **monocytes**. It starts to appear in the rough **endoplasmic reticulum** at the promyelocyte stage. Myelocytes demonstrate myeloperoxidase in the Golgi apparatus, but mature neutrophils only exhibit myeloperoxidase in the primary granules. Other peroxidases present in neutrophils, eosinophils, and basophils, although similar in activity, are not identical.

Myeloperoxidase mediates intracellular bacterial killing by producing highly toxic halides:

$$H_2O_2 + Cl^- \quad \xrightarrow{\text{myeloperoxidase}} \quad HOCL + OH^-$$

Increased neutrophil myeloperoxidase is seen in **Addisonian pernicious anemia** and other forms of **megaloblastosis**.

Myeloperoxidase can be detected using various enzymes, e.g., Sudan black or chromogens such as 3,3'-diaminobenzidine, which are useful in distinguishing myeloid from lymphoid leukemias and various subtypes of **acute myeloid leukemia** in the **FAB classification**. It has been claimed that the serum level is raised prior to myocardial infarction.

Deficiency of Myeloperoxidase

This can be either inherited or acquired. The condition can be identified by automated differential leukocyte counting where myeloperoxidase-deficient neutrophils usually count as "large unstained cells" or "others." The hereditary form (frequency 1:4000) is an autosomally recessive trait and, despite *in vitro* testing indicating failure of intracellular bacterial and fungal killing, is a benign condition.

MYELOPROLIFERATIVE DISORDERS

The designation given by Dameshek in 1951 to a group of disorders, which he classified together, since they appeared to share many pathological, clinical, and laboratory features.[382] They include **polycythemia rubra vera** (PRV), **chronic myelogenous leukemia** (CML), **essential thrombocythemia** (ET), and **idiopathic myelofibrosis** (IMF). Subsequent studies have shown that they are all clonal hematopoietic stem cell disorders that have a growth advantage over normal polyclonal hematopoietic cells. Only CML has a distinct diagnostic marker, the Ph chromosome, and its pathogenesis differs markedly from IMF, PRV, and ET, which can be considered together as a distinct group. Developments in

genomics may confirm or refute this classification. Support arises from the identification in patients with PRV, ET, and IMF of a single-point mutation (Val-617-Phe) of the **tyrosine kinase** JAK2.[383]

Pathophysiology

Normal **hematopoiesis** is polyclonal. The monoclonal origin of the myeloproliferative disorders was first demonstrated in female patients who were heterozygous for the X-chromosome-linked isoenzymes **glucose-6 phosphate dehydrogenase** (G-6PD), hypoxanthine ribosyl transferase (HPRT), or phosphoglycerate kinase (PGK). Clonal karyotypic abnormalities were later demonstrated in many of these disorders, confirming their monoclonality; more recently, molecular techniques (e.g., *BCR-ABL* in CML) have also confirmed the clonal origin of these diseases (see **Cytogenetic analysis**). Using the same techniques, it has been shown that the various differentiated hematopoietic cells are also derived from the same hematopoietic stem cell, demonstrating that the pathogenesis of the myeloproliferative disorders resides at the stem cell level (see Figure 91). In all cases, cells of the erythroid, myeloid, and megakaryocytic lineages are clonal, but this has not always been demonstrated for **lymphocytes**. B-cells are usually clonal, but most patients have circulating nonclonal T-lymphocytes. This may be due to the persistence of long-lived polyclonal T-lymphocytes that were formed before the myeloproliferative disorder developed. Fibrosis is common in the myeloproliferative disorders, but it is reactive, since the fibroblasts are always polyclonal.

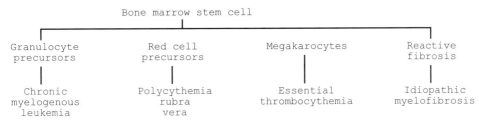

FIGURE 91
Relationship of the origin of myeloproliferative disorders.

Growth Advantage

Primitive bone marrow and peripheral-blood hematopoietic cells can be grown *in vitro* using specific exogenous growth factors and will produce colonies of varying phenotypic expression (see **Hematopoiesis** and Figure 91). Interestingly, a common feature of all the chronic myeloproliferative disorders is growth-factor-independent *in vitro* colony formation and growth-factor hypersensitivity. *In vitro* colony assays have also provided further evidence that the primary defect(s) in myeloproliferative disorders resides at the stem cell level. Patients with IMF show increased numbers of circulating CFU-Meg, CFU-GEMM, CFU-GM, CFU-E, and BFU-E, which is in keeping with the increase in circulating CD34[+] cells in this disease (see Figure 25). In ET patients, not only is there increased circulating CFU-MEG, but also increased BFU-E and CFU-GM. Similar behavior is seen in PRV. In the late 1970s, several studies revealed that granulocytes, erythrocytes, and platelets were monoclonal, but fibroblasts were polyclonal in the chronic myeloproliferative disorders. Thus, the myelofibrosis in these disorders is secondary to hematopoietic clonal proliferation. Further studies have identified both megakaryocytes and monocytes as the likely cells responsible for the myelofibrosis by releasing transforming growth factor-β (TGF-β).

This growth factor, synthesized by megakaryocytes, is stored in **platelet** α-granules. TGF-α also stimulates **angiogenesis** and inhibits collagenase activity. Another α-granule enzyme, platelet factor 5, also reduces collagenase activity. Marrow fibrosis is due to increased synthesis of types I, III, IV, and V **collagen**. It is important to note that myelofibrosis *per se* does not disrupt marrow function; marrow failure is due to abnormalities intrinsic to the clonal hematopoietic progenitor cells.

Clinical and Laboratory Features

IMF, PRV, and ET share the following common features:

Chronic course

Overproduction of one or more of the formed elements of the blood

Domination of the malignant clone over polyclonal hematopoiesis

Marrow hypercellularity and megakaryocytic dysplasia

Abnormalities of chromosomes 1, 8, 9, 13, and 20

Marrow fibrosis

Myeloid metaplasia

Thrombosis

Hemorrhage

Transformation to acute leukemia

Pancytopenia from a combination of marrow failure and hypersplenism

All three myeloproliferative disorders also have a tendency to transform into one another. Therefore, it is not surprising that they share many clinical and laboratory features, as shown in Table 116.

Constitutional symptoms and **hyperuricemia** are due to hypermetabolism secondary to increased cell turnover and largely parallel the tempo of the various diseases. **Extramedullary hematopoiesis** results in organomegaly, which can be extreme. The specific clinical syndromes manifested by myeloproliferative disease patients depend on the behavior of the malignant clone and are a consequence of overproduction of blood cells, exuberant extramedullary hematopoiesis, bone marrow failure, or transformation. The biological basis of these disorders, with the exception of CML, is unknown, and even in CML, the Ph chromosome does not appear to be the primary genetic lesion.

Treatment

This depends entirely upon the behavior of the myeloproliferative disorder. For example, in PRV, reduction of the red blood cell mass by **phlebotomy** to a normal level based on gender will prevent major vessel thrombosis and hemorrhage, while reduction of the platelet count in both PRV and ET may be necessary to control microvascular complications such as **erythromyelalgia**, ocular migraine, or hemorrhage. In IMF, apart from supportive measures such as blood transfusion and chemotherapy, the use of biological-response modifiers such as interferon-α or thalidomide or even **splenectomy** may be necessary to control extramedullary hematopoiesis. **Interferon-α**, **imatinib** mesylate, and **allogeneic stem cell transplantation** are the major treatment options for CML.

TABLE 116

Clinical and Laboratory Features of Myeloproliferative Disorders

		Chronic Myelogenous Leukemia (CML)	Polycythemia Rubra Vera (PRV)	Essential Thrombocythemia (ET)	Idiopathic Myelofibrosis (IMF)
Constitutional symptoms:	weight loss	60%	30%	rare	60–70%
	sweats	60–70%	30%	rare	10–20%
	pruritus	<10%	20–40%	rare	10–20%
Clinical signs:	hepatomegaly	50% may be massive	30–50% moderate	10–50% moderate	50–75% may be massive
	splenomegaly	85–90% may be massive	50–90% moderate	40–80% moderate	90–100% may be massive
Laboratory findings:	lymphadenopathy	60%	rare	rare	rare
	leukocytosis (>12 × 10^9/l)	Up to 200 × 10^9/l neutrophilia with prominent basophilia and eosinophilia	60–70%	30%	50%
	thrombocytosis	50%	50%	100%	50–70%
	erythrocytosis	10%	100%	30%	rare
	anemia	80%	very rare; bleeding may lead to severe iron deficiency	10–20%	90%
	NAP, LAP	low (90%)	high (75%)	variable	variable
	hyperuricemia	90%	50–70%	usual	usual
	serum B12	100%	often high	usually low/normal	often high
	marrow fibrosis	10–50% at diagnosis	10–15%	rare	90% at diagnosis

MYOGLOBIN

A monomeric globin structurally similar to **hemoglobin**. Each myoglobin molecule consists of a **heme** group almost surrounded by loops of a long polypeptide chain. Myoglobin is present in all skeletal and cardiac muscle cells, where it serves as a reservoir of oxygen to protect against cellular injury during periods of anoxia.

MYOSIN

See **Cell locomotion**.

N

NAIL DISORDERS

The changes in nails in patients with hematological disorders.

Koilonychia, concavity, or spoon-shaped nails, caused by thinning and softening of the nail plate. This classical nail disorder of **iron deficiency** is associated with postcricoid webbing. It is also seen in **hereditary hemochromatosis** and **porphyrias**.

Clubbing, possibly due to reduced **ferritin** opening up the nail-bed arteriovenous anastomoses. It is traditionally associated with suppurative lung disease and cyanotic heart disease, but also occurs with celiac disease, Crohn's disease, and **Hodgkin disease** involving the lungs.

NASAL DISORDERS

The hematological disorders that result in patients from disorders of the nose

Hemorrhage (epistaxis) due to:
- Local inflammation
- Hemorrhagic disorders
- Hereditary hemorrhagic telangiectasia
- **Hemangiomas** (angiolipomas and lymphangiomas)
- Tumors, particularly **extranodal T-cell lymphoma**, **nasal type**, and **angiocentric lymphoma**

Eosinophilic vasomotor rhinitis associated with nasal plops, aspirin sensitivity, and asthma

Nasopharyngeal carcinoma associated with **Epstein-Barr virus** infection[383a]; adaptive immunotherapy to EBV-specific T-cells may have a potential therapeutic effect

NASAL T-CELL LYMPHOMA

See **Angiocentric lymphoma**; **Extranodal T-cell lymphoma, nasal type**; **Non-Hodgkin lymphoma**.

NATALIZUMAB

A recombinant **monoclonal antibody** acting against α4 integrins.[132] It blocks the ability of α4β and α4γ integrin to bind with their respective endothelial counterreceptor vascular-cell adhesion molecule-1 (VCAM-1) and mucosal addressin-cell adhesion molecule-1 (MAdCAM-1) required by lymphocytes for entry into the CNS. It is under clinical trial for treatment of **multiple sclerosis**[384] and Crohn's disease.[385]

NATURAL KILLER (NK) CELL

See **Lymphocytes** — large granular cells.

NECROSIS

The inadvertent death of cells, usually by exposure to metabolic disturbance such as hypoxia or toxic action. It usually gives rise to **inflammation**.

NEONATAL ALLOIMMUNE THROMBOCYTOPENIA

(NAIT) **Platelet** depletion caused by alloimmunization of the mother-to-fetal platelet **antigens**. It is the platelet equivalent of **hemolytic disease of the newborn**. The mother produces an antibody against an antigen that is absent in her own platelets but is present on the fetal platelets; the maternal antibody then crosses the placenta and causes premature destruction of fetal platelets. This occurs in approximately 1 in 1000 to 3000 pregnancies and may affect the firstborn in about 50% of these, suggesting that placental passage of platelets occurs during gestation. NAIT platelet depletion frequently reappears in subsequent pregnancies. The most frequently implicated alloantigen is the human platelet antigen HPA-1A (PLA1), carried by **glycoprotein IIIa** in 98% of the population in which the antibody binds. Mothers who generate anti-HPA-1A belong to the other 2% of the population who are HPA-1A negative. This may also impair platelet function.

Clinical Features

Clinical features are predominantly those of **thrombocytopenia** presenting at birth, although a proportion of affected infants may develop significant intrauterine hemorrhage in the third trimester. The thrombocytopenia typically worsens within the first 48 h after birth but, untreated, the count becomes normal within weeks due to the natural decline in the level of circulating antibody (typically IgG, whose half-life is 2 weeks). Bleeding, which usually occurs only in cases with $<30 \times 10^9/l$ platelets, is often limited to mild purpura, but it can also be severe and include visceral or intracranial (20%) bleeding. The latter can cause neurologic damage or death. Sometimes these bleeds have already occurred *in utero* and cause porencephalic cysts. Overall, NAIT has been reported to have a mortality of 15%, with intracranial bleeding being the most frequent cause.

Diagnosis

Most cases are diagnosed on clinical basis, as specific testing requires the submission of samples to reference laboratories requiring up to 2 weeks to establish the definitive diagnosis. NAIT is suggested by the lack of an obvious alternative cause for thrombocytopenia in a child born to a mother without a history of previous immune thrombocytopenia or other autoimmune disease and who does not have thrombocytopenia herself and is not taking any drug potentially causing thrombocytopenia.

Simple testing[286] includes a serologic matrix whereby maternal serum is tested against group O HPA-1A-positive platelets, paternal platelets, and maternal platelets. This excludes the presence of **immune thrombocytopenic purpura** (ITP) and should confirm that a platelet antibody present in the maternal serum reacts with paternal platelets. More-complex testing involves screening maternal serum against phenotyped or genotyped platelets to determine the specificity of the antibody and then confirming (often by molecular techniques) the genotype or phenotype of both the mother and father. It is important to ascertain whether the father is homozygous for the antigen against which the maternal antibody could be directed in subsequent pregnancies. The platelet phenotype of the fetus can then be determined on a blood sample obtained by cordocentesis. In severely thrombocytopenic fetuses, such procedures carry a significant risk of potentially fatal bleeding, and so preparations must be made for immediate **platelet transfusion** (group O, **cytomegalovirus** [CMV]-negative, irradiated, and negative for the appropriate platelet antigen). The diagnosis is often made postnatally in the first affected fetus, although thrombocytopenia may be attributed to nonimmune causes.

Clinical Management

Mother

In a pregnancy known to be at risk from severe thrombocytopenia due to platelet alloimmunization, the following steps can be taken:

Intrauterine transfusion of platelets is preferred for severe cases. Most reported *in utero* intracranial hemorrhage occurs after 30 weeks gestation. Cord blood sampling with immediate determination of fetal platelet count and platelet transfusion can be done from 28 weeks gestation. Transfusions may be required at weekly intervals so that the nadir platelet count is ordinarily $>50 \times 10^9/l$.

In less severely affected cases, corticosteroids and intravenous immunoglobulin may be effective.

With profound fetal thrombocytopenia, delivery should be by cesarean section to avoid birth trauma.

Neonate

The need for treatment is debatable and, in the absence of complicating reasons, it is generally agreed to watch and wait for the spontaneous correction of the platelet numbers in cases where the absolute platelet count is $>50 \times 10^9/l$. Neonates with lower platelet counts should be treated. Thromboapheresis using maternal platelets is an option in case of emergency. Otherwise, platelets from an HPA-1A-negative donor is the product of choice. Prednisone (2 mg/kg/day) can be used in the immediate postdelivery period in cases with severe thrombocytopenia, although its benefit is debatable. Intravenous immunoglobulin is reported to shorten the duration of the thrombocytopenia.

NEONATAL HEMATOLOGY

See also **Premature Infant Hematology**

The hematological status of infants during the first 4 weeks of life. Neonatal whole-blood specimens can be obtained from the umbilical cord (cord blood) or from a heel prick (capillary blood), the values from each site being different. Physiological adjustments occur rapidly so that the ranges for capillary blood are wide.[386] The comparatively high levels for **hemoglobin**, **red blood cells**, and **packed-cell volume** (PCV) (see Table 117) fall within hours after clamping of the cord. These falls continue for the next 2 months or more due to slowing of **erythropoiesis**. At birth, **normoblasts** are common in the peripheral blood (about 25/100 leukocytes) with **reticulocyte counting** at 5% of red blood cells, both declining within 4 to 5 days. At this time, there is a switch of **hemoglobin F** to **hemoglobin A**. At 7 to 9 months, the hemoglobin level averages 9 g/dl (physiological anemia of infancy). The red blood cell life span is 75% that of the adult.

TABLE 117

Reference Ranges of Common Neonatal Blood Parameters

	Capillary	Cord
Hemoglobin, g/dl	14.5–22.5	13.5–19.5
Red blood cells (RBCs), $\times 10^{12}/l$	4.0–6.6	3.9–5.5
Packed-cell volume (PCV), l/l	45–67	42–62
Mean cell volume (MCV), $\times 10^{-15}l$	95–121	98–118
Mean cell hemoglobin (MCH)	31–37	31–37
White blood cells (WBCs), $\times 10^9/l$	9.0–30.0	—
Neutrophils, $\times 10^9/l$	6.0–26.0	—
Eosinophils, $\times 10^9/l$	0.4 (mean)	—
Monocytes, $\times 10^9/l$	1.1 (mean)	—
Lymphocytes, $\times 10^9/l$	2.0–11.0	—
Reticulocytes, % RBC	1.8–4.6	—
Normoblasts	500–200	—
Platelets, $\times 10^9/l$	232–378	—
Fibrinogen, g/l	1.67–3.99	—
Prothrombin time, sec	10.1–15.9	—
Activated partial thromboplastin time (APTT), sec	31.3–54.5	—
Factor V, U/ml	0.34–1.08	—
Factor VII, U/ml	0.28–1.04	—
Factor VIIIC, U/ml	0.50–1.78	—
Factor X, U/ml	0.12–0.68	—
Factor XI, U/ml	0.10–0.66	—
Factor XII, U/ml	0.13–0.93	—
Prekallikrein, U/ml	0.18–0.69	—
High-molecular-weight kininogen, U/ml	0.06–1.02	—
Factor XIIIa, U/ml	0.27–1.31	—
Factor XIIIb, U/ml	0.30–1.22	—
Plasminogen (CTA), U/ml	1.60–2.30	—

Note: Table values derived from **Reference Range Tables**.

Leukocyte levels in the peripheral blood are comparatively high, with many **metamyelocytes**, some **myelocytes**, and a few **promyelocytes**, all falling within 2 weeks. **Eosinophil** and **monocyte** levels may be higher than those of adults. **Neutrophil** function is not fully developed, particularly chemotaxis, and monocyte phagocytosis is similarly underdeveloped.

Platelets, **fibrinogen**, and coagulation **factor V** and **factor VIII** are at adult levels, but **thrombin time** is prolonged, probably due to fibrinogen being of fetal type. Vitamin K-dependent coagulation **factors II**, **VII**, **IX**, and **X** are reduced, with prolongation of both **prothrombin time** and **activated partial thromboplastin time**. **Factor XI** and **Factor XII** are 50% of adult levels. There is an increase in plasminogen activators, with increased **fibrinolysis** at birth declining rapidly. Transient deficiencies of **plasminogen**, **antithrombin III**, **heparin cofactor II**, **protein C**, and **protein S** render infant blood hypercoagulable.

Blood volume is 80 to 85 ml/kg, lower if there is delay in clamping of the cord. Falls to adult levels occur early.

Bone marrow is hyperplastic with few fat spaces. Erythroid precursors are 30 to 65% and myeloid precursors are 45 to 75% of nucleated cells. Cellularity soon decreases after birth, particularly of the red blood cell precursors, which reach adult levels at 3 months.

IgM antibodies, including **blood group** antibodies, are absent at birth, appearing after 3 to 6 months of life.

Hematological Disorders

1. Anemia

Post-**hemorrhage** anemia detectable at birth:

- Overt, from rupture of the umbilical cord, placenta previa, abruption of placenta, incision of placenta during cesarean section
- Occult, before birth, by fetomaternal transfusion or twin-to-twin transfusion
- Occult, internal, after birth — retroperitoneal, rupture of liver or spleen, intracranial

Hemolytic anemia

- Isoimmune **hemolytic disease of the newborn** (HDN)
- Congenital disorders of **red blood cell** membrane
- Congenital disorders of red blood cell metabolism
- **Hemoglobin H** disease

Congenital pure red cell aplasia (**Blackfan-Diamond syndrome**)

Repeated blood sampling

2. Polycythemia and Hyperviscosity

Polycythemia and **hyperviscosity** occur together or independently.[386] Hyperviscosity is due to raised plasma **fibrinogen**. Only symptomatic treatment is indicated, unless seizures occur, when partial-exchange transfusion should be induced (see **Fetal/neonatal transfusion**).

3. Hemorrhage[387]

Coagulation factor deficiency, usually presenting as bleeding from the umbilical stump, at circumcision, or as hematomas

- Hereditary **hemophilia** disorders
- Acquired coagulation deficiency (**hemorrhagic disease of the newborn**) due to vitamin K deficiency and immaturity of hepatic cells; now uncommon due to **vitamin K** being given prophylactically to the mother or to the infant at birth

Neonatal alloimmune thrombocytopenia (NAIT), presenting with generalized ecchymoses and petechiae as a consequence of maternal platelet antibodies

Neonatal **thrombocytopenia**

4. Disseminated Intravascular Coagulation

For **disseminated intravascular coagulation**:

Transplacental passage of tissue factor (thromboplastin) from the mother occurs with abruptio placentae or eclampsia

Bacterial or viral infections (herpes simplex, cytomegalovirus, rubella) — **purpura fulminans**

Giant **hematomas**

5. Isoimmune Neutropenia

Maternal antibodies to neutrophil-specific antigens can cross the placenta, resulting in neonatal **neutropenia** similar to red blood cell HDN.

NEONATAL THROMBOCYTOPENIA

Platelet count in the newborn below $100 \times 10^9/l$. Causes of neonatal thrombocytopenia differ from those of the adult, and many reflect intrauterine influences or complications of delivery or of the neonatal period. Thrombocytopenia may be only part of the broader syndrome[117] (see Table 118). By far, the most frequent cause is bacterial septicemia, acquired at delivery or via interventions (**central venous catheters**, cannulae, urethral catheters).

Regardless of its origin, the infants have generalized ecchymoses and petechiae, although they are frequently unnoticed, as they often present only in pressure areas (abdomen or back, in direct contact with sheets, hidden from examination). It is seldom life threatening, but the neonatal tendency to intracranial hemorrhage in circumstances of severe thrombocytopenia ($<50 \times 10^9/l$ in neonates) must be considered. If severe, **platelet transfusion** is indicated, and investigation of coagulation by screening tests to rule out disseminated intravascular coagulation (DIC) must be performed in all cases.

TABLE 118

Causes of Neonatal Thrombocytopenia

Intrauterine infections
Rubella, cytomegalovirus, herpes simplex, toxoplasmosis, syphilis
Platelet antibodies
Maternal immune thrombocytopenic purpura or systemic lupus erythematosus; neonatal alloimmune thrombocytopenia
Drugs
Disseminated intravascular coagulation
Megakaryocytic hypoplasia
Congenital
Acquired maternal drug ingestion
Hereditary thrombocytopenias
Giant hemangioma
Metabolic disorders
Postexchange transfusion

NEONATAL TRANSFUSION

See **Fetal/neonatal transfusion**.

NERVOUS SYSTEM DISORDERS

The relationships of the nervous system with hematological function and disorders.[388] A direct relationship occurs through the neuropeptide axis. Substance P, produced by sensory nerve cells, binds to receptors on endothelial cells and on **lymphocytes** to induce **inflammation**.

Demyelination of the central nervous system and peripheral nerves leading to peripheral neuropathy:

- **Subacute combined degeneration of the spinal cord** due to **cobalamin** deficiency

- Progressive multifocal encephalopathy, a subacute demyelination disorder associated with **myeloproliferative disorders** and **lymphoproliferative disorders**
- **Multiple sclerosis**: inflammatory lesions involving T-cells and histiocytes (microglial macrophages) followed by plaque formation with demyelination

Subarachnoid, subdural, epidural, and intracerebral hemorrhage:

- **Thrombocytopenia** of all causes, particularly with cytotoxic therapy and **disseminated intravascular coagulation** due to promyelocytic leukemia
- **Hemophilia** or related disorders, the commonest cause of death in these disorders
- Hyperfibrinolysis or hypofibrinogenemia, particularly with **thrombolytic therapy**
- Anticoagulant therapy
- **Sickle cell disorders** (SCD) when associated with vascular malformations

Cerebrovascular or spinal arterial **thrombosis**:

- Sickle cell disorders
- Polycythemia rubra vera
- Leukostasis of **chronic myelogenous leukemia**, particularly at the time of blast crisis, or **acute myeloid leukemia** (AML) when the white blood cell count exceeds $200 \times 10^9/l$
- **Hyperviscosity** disorders, particularly multiple myeloma and myeloma-related disorders
- Disseminated intravascular coagulation (DIC)
- Polyarteritis nodosa

Cerebral vein and sinus thrombosis:

- Hypercoagulable states (**thrombophilia**)
- Antiphospholipid-antibody syndrome
- Sickle cell disorder
- **Polycythemia** and **thrombocythemia**
- Paroxysmal nocturnal hemoglobinuria

Cranial and peripheral neuropathy:

- Herpes zoster complicating **cytotoxic agent** therapy
- **Vinca alkaloid** or intrathecal **methotrexat**e toxicity
- Compression or infiltration of nerves or nerve roots by **Hodgkin lymphoma** (HD), **myelomatosis**, **amyloidosis** (AL), **hypereosinophilic syndrome**, polyarteritis nodosa, **sarcoidosis**
- Monoclonal gammopathy
- **Radiotherapy** plexopathy
- Guillain-Barré syndrome with HD and **acquired immunodeficiency syndrome** (AIDS)
- Chronic idiopathic demyelinating polyradiculoneuropathy with HD or **non-Hodgkin lymphoma** (NHD)
- Paraneoplastic subacute sensory or subacute motor neuropathy with HD

- Restricted lumbosacral polyradiculopathy, mononeuropathies, chronic inflammatory demyelinating polyradiculoneuropathy (CIPD), vasculitic mononeuritis multiplex, **cytomegalovirus** (CMV), cauda equina neuritis, and **diffuse infiltrative lymphocytosis** (DILS), all associated with AIDS

Compression of spinal cord:

- Extramedullary mass of a lymphoproliferative disorder
- Marrow hyperplasia of SCD
- Hemorrhage due to hemophilia
- Arteriovenous angiomas

Meningeal infiltration by **acute lymphoblastic leukemia** (ALL), HD, NHL, particularly **angiocentric lymphomas**, rarely acute myeloid leukemia (AML). It is usually diagnosed by detection of "blasts" in the cerebrospinal fluid. Cell counting of cerebrospinal fluid can be performed by automated methodology (see **Automated blood cell counting**). Symptoms of meningeal irritation and cranial nerve palsy are the clinical signs of infiltration.

Space-occupying lesions by:

- NHL or HD, which are all common secondary neoplasms of treated ALL and especially as a complication of AIDS; usually they are localized tumors, but multifocal leukoencephalopathy does occur
- Hemangiomas

Encephalopathy associated with AIDS and **hypereosinophilic syndrome**

Double athetosis, gaze palsies, and deafness caused by **kernicterus** due to deposition of bile pigment in the basal ganglia, basal nuclei, and thalamus of the neonate affected with **hemolytic disease of the newborn**

Syncope following severe reduction of **blood volume** due to acute hemorrhage

Myoclonic seizures, amblyopia, spasticity, cerebellar ataxia, and dementia due to **lipid storage disorders**

Dementia due to deposition of **granulomas** including sarcoidosis or **vasculitis** in cerebral tissue

NEUROPEPTIDES

Hormones produced by the pituitary and sensory nerve cells, with widespread activity on other body systems. They include prolactin, macrophage-inhibiting factor, and substance P. Their actions include antagonism to **corticosteroids** and activation of T and B-lymphocytes and **histiocytes** (macrophages).

NEUTRALIZATION

The process of biological inactivation of pathogens such as viruses and of proteins such as toxins or cytokines. The simplest mechanism of neutralization is likely to be the physical coating of the **antigen** by the **antibody**, thus sterically hindering access to active sites. However, there is evidence that, at least for some of the more-complex targets, such as certain viruses, inactivation is by binding of antibody, resulting in conformational changes in the target. Neutralization is an important effector function of antibody, but **complement** components, particularly C3b, may enhance antibody neutralization of viruses.

Neutralization Epitopes

Studies with panels of **monoclonal antibodies** have shown that, in some situations, not all of these are capable of neutralizing; there are neutralizing and nonneutralizing antibodies. Monoclonal antibodies react with small portions of the antigen, i.e., **epitopes**, thus only when antibodies bind to certain regions of the antigen can they neutralize. This leads to the concept of neutralizing epitopes. These can be mapped, and often these epitopes fall into discrete, small regions on the surface of the antigen. For example, the major envelope protein of influenza virus, hemagglutinin, possesses three major neutralization epitopes, which together comprise only a small percentage of the surface area of the protein.

NEUTROPENIA

A reduced level of circulating **neutrophils**: mild (1 to $2 \times 10^9/l$), moderate (0.5 to $1 \times 10^9/l$), severe (0.1 to $0.5 \times 10^9/l$), or very severe ($<0.1 \times 10^9/l$). There is an increasing risk of infection as the severity of the neutropenia increases.[391]

Etiology

Several pathophysiological mechanisms (see Figure 92) have been identified, including:

Impaired cell proliferation

Ineffective granulopoiesis in the bone marrow

Marginalization in the blood vessels (pseudoneutropenia)

Reduced neutrophil survival

Increased tissue migration, either inherited or acquired

	Stem Cell Compartment	Mitotic	Maturation Storage	Marginal Pool	CBP
↓ Proliferation	↓	↓	↓	↓	↓
Ineffective Granulopoiesis	↑	↑	↓	↓	↓
↓ Neutrophil Survival	↑	↑	↓	↓	↓
↑ Margination	N	N	N	↑	↓
↑ Tissue Migration	N	N	↑	↓	↓

CBP - Circulating Blood Pool

FIGURE 92
Simplified schematic diagram of the pathophysiology of neutropenia

TABLE 119

Causes of Neutropenia

Inherited	Acquired
Diminished Proliferation	
Kostman's syndrome	Paroxysmal nocturnal hemoglobinuria (PNH)
Reticular dysgenesis	Drug induced
Schwachman-Diamond syndrome	Cyclical
Idiopathic	Idiopathic
Hereditary/aplastic anemia, e.g., Fanconi anemia	Infections
	Marrow infiltration
	Marrow failure, e.g., aplastic anemia
Ineffective Proliferation	
Chediak-Higashi syndrome	Myelodysplastic syndrome
Transcobalamin II deficiency	Megaloblastic anemia
	Alcohol
↓ *Neutrophil Survival*	
	Large granular lymphocyte (LGL) disorders
	Hypersplenism
	Felty's syndrome
	Drug induced
	Isoimmune neutropenia
	Postdialysis
	Autoimmune diseases
	Infections
	Agranulocytosis
↑ *Margination*	
	Pseudoneutropenia
	? Autoimmune disorders
↑ *Release into Tissues*	
	Sepsis (overwhelming)

It is most commonly seen as part of a pancytopenia due to **bone marrow hypoplasia** following **cytotoxic agent** therapy or **radiotherapy** and with all forms of **hypersplenism**. The causes of neutropenia are listed in Table 119.

Clinical Features

Fever, as a consequence of **inflammation**, occurs in 90% of patients, but this may not be evident in those who have concurrent **immunodeficiency**, such as occurs posttransplantation. The location of any infection in the mouth, chest, abdomen, perianal, and perivaginal tissue must be found. Identification of the infecting organism should determined by microbiological examination of any infected site, by blood culture, and by fecal culture. Respiratory symptoms should be investigated by bronchoalveolar lavage or open lung biopsy. Common infecting organisms are:

Staphylococcus aureus
Coagulase-negative *Staphylococcus* spp.

Methicillin-resistant *Staphylococcus aureus* (MRSA)

Enterococci

Viridans streptococci

Escherichia coli

Klebsiella pneumoniae

Pasteurella aeruginosa

Candida albicans, C. kruzei, C. tropicalis

Aspergillus spp.

Cryptococcus neoformans

Herpes simplex virus

Varicella-zoster virus

Pneumocystis carinii

Management

This is largely determined by the cause and severity of the neutropenia. If possible, the underlying cause should be vigorously treated. Prophylactic oral antibiotic therapy was widely practiced, but it is no longer recommended because particular antibiotics gain a growth advantage and several well-documented nosocomial outbreaks have occurred. If severely neutropenic patients become pyrexial, routine cultures should be taken and intravenous antibiotics (to cover both gram-positive and gram-negative organisms) promptly commenced (see **Immunodeficiency**). Fungal infection usually follows prolonged periods of neutropenia, and prophylactic antifungal therapy, e.g., fluconazole, nystatin, and amphotericin, has been shown to be useful.

Recombinant human **granulocyte colony stimulating factor** (rhG-CSF), either glycosylated or nonglycosylated, can be used for:

Postintensive chemotherapy or bone marrow transplantation, with the exception of myeloid leukemias

Congenital neutropenia of childhood

Recombinant human **granulocyte/macrophage-colony stimulating factor** (GM-CSF) is also available, but this has more **adverse drug reactions** and is ineffective in congenital neutropenia. For those with recurrent neutropenia from any cause, prophylactic antibiotics such as azithromycin (or erythromycin in cases of intolerance) are advocated together with G-CSF.

Selective Forms of Neutropenia

Inherited Neutropenias

Cyclic Neutropenia

A rare, infrequently inherited autosomally dominant disorder characterized by neutropenia occurring over 14 to 35 days (usually about 21 days) and lasting 3 to 10 days. The abnormality probably resides at a mutation in the stem cell level 19p13.3, causing a cyclical inhibition of granulocyte stem-cell proliferation. Monocytes also show a phasic presence but opposite to that of neutrophils. The diagnosis is made by regular **white blood cell counting** with differential leukocyte counting at least three times per week for at least 2 months. Treatment is by long-term administration of G-CSF.

Severe Congenital Neutropenia

(SCN) A genetically heterogeneous group of disorders involving genes *ELA2*, *Gfi1*, *wASP*, and *G-CSFR*. Those with *ELA2* mutations may evolve to **myelodysplasia** or **acute myeloid leukemia**. While giving rise to severe infections, the condition responds to G-CSF given in doses of 10 µg/kg/day.

Kostmann's Syndrome

A rare autosomally recessive inherited condition presenting within the first month after birth with severe neutropenia and recurrent infections. The underlying defect is unknown, and no chromosomal abnormality has been described. If untreated, the condition is usually fatal early in life.

Schwachman-Diamond Syndrome

(SDS) A rare autosomally recessive disorder with bone marrow aplasia, dwarfism, and pancreatic exocrine insufficiency. The bone marrow involvement is variable among subjects, with severe congenital neutropenia being the originally reported feature. An increasing number of patients with severe neutropenia, severe thrombocytopenia, and moderate anemia have bone marrow aspirates showing an aplastic appearance. Neutropenia is the most relevant feature; it is the most important morbidity-conditioning factor in these patients, and infections are the most frequent cause of death. **Allogeneic stem cell transplantation** is the only curative approach.

Hermansky-Pudlak Syndrome Type 2

Hermansky-Pudlak syndrome is an autosomally recessive disorder of gene *AP3B1*. There are associated platelet storage-pool defects and oculocutaneous albinism.

Chediak-Higashi Syndrome

Chediak-Higashi syndrome is an autosomally recessive disorder of gene *LYST*. There is associated impairment of platelet function and oculocutaneous albinism.

Barth Syndrome

A sex-linked genetic disorder of the *TAZ* gene. The neutropenia is often cyclic with dilated cardiomyopathy.

Cohen Syndrome

An autosomally recessive disorder of the *COH1* gene associated with mental retardation and dysmorphism.

Acquired Neutropenias

Drug-induced neutropenia

(Agranulocytosis) Toxicity to drugs, apart from cytotoxic agents given as treatment for hematological malignancies, may be dose-related or a drug-induced immune disorder, where the drug acts as antigen, giving rise to neutrophil antibodies. There is a high rate of infectious complications.

Infection-Induced Neutropenia

The exact mechanisms involved in infection-related or -induced neutropenia have not been fully elucidated, but in some infections, e.g., **human immunodeficiency virus** (HIV),

TABLE 120

Antineutrophil Antibodies

Antigens	Glycoprotein	Allele frequency (%)
HNA-1a	FcγIIIb (CD16)	58
HNA-1b	FcγIIIb (CD16)	88
HNA-1c	FcλIIIb (CD16)	5–38
HNA-2a	CD177 (gp50–64)	94
HNA-3a	Gp70–95	97
HNA-4a	CD11a	99
HNA-5a	CD11b	96

Source: International Society of Blood Transfusion. With permission.

it is multifactorial. Virtually all types of infective agents can cause neutropenia. Those commonly involved are:

Bacteria: typhoid, **tuberculosis**, brucella

Virus: HIV, hepatitis A and C, **cytomegalovirus**, **Epstein-Barr virus**, rubella, measles, chickenpox, yellow fever

Protozoa: **malaria, babesiosis**

Rickettsia: typhus, Rocky Mountain fever

In hematological practice, multiple infections from indwelling central line or urethral catheters are common.

Immune Neutropenia

This is caused by antibodies against neutrophil-specific antigens that are glycoprotein receptors. They can be alloimmune or autoimmune and can be detected by a variety of methods (see Table 120):

Granulocyte immunofluorescence test

Granulocyte indirect immunofluorescence test

Granulocyte agglutination test

Enzyme-linked immunoassays

Monoclonal antibody-specific immobilization of granulocyte antigens

Isoimmune Neutropenia — Maternal IgG antibodies to the neutrophil-specific antigens (e.g., HNa-1a, HNA-2b, NB-2a) cross the placenta, resulting in neutropenia of the newborn (analogous to **hemolytic disease of the newborn**). The severity of neutropenia is variable and can last up to 2 months. If serious infection occurs, antibiotics should be given along with G-CSF.

Primary Autoimmune Neutropenia — *Infancy and childhood*: a rare disorder with onset from 6 to 12 months of age with a moderate to severe neutropenia. Antibodies are uniformly IgG and directed primarily against HNA-1 and HNA-2. Spontaneous remission occurs in 95% of cases over a period of 2 years and rarely requires treatment with G-CSF.

Adults: a rare chronic autoimmune disorder that has minimal symptoms. It is associated with increased apoptosis of granulocyte precursors due to excessive production of interferon-γ and transforming growth factor (TGF)-α. There is no evidence of progression to **systemic lupus erythematosus, aplastic anemia**, or **acute leukemia**.

Secondary Autoimmune Neutropenia — Here the neutropenia is associated with a systemic autoimmune disorder, but it is probably of the same pathogenesis as the primary form.

Rheumatoid Arthritis (Felty's Syndrome)

Rheumatoid arthritis (Felty's syndrome) is a form of neutropenia that has a considerable morbidity to bacterial infection, is associated with **splenomegaly**, and shows high levels of rheumatoid factor, antinuclear factor, and hypergammaglobulinemia. There is no simple correlation between splenic size and the degree of neutropenia. Some patients have an immune component, with IgG and immune complexes demonstrable on the neutrophil surface. Bone marrow usually shows increased granulopoiesis, suggesting shortened white cell survival; thus, the pathogenesis appears to be a combination of **hypersplenism**, increased **apoptosis**, and autoimmunity. This probably explains why **splenectomy** fails to correct the neutrophil count in about 25% of patients. Neutropenia is usually mild/moderate, but can be severe. Splenectomy should be reserved for patients with life-threatening infections.

Systemic Lupus Erythematosus

Systemic lupus erythematosus: neutropenia occurs in about 50% of patients but is rarely severe. In addition to antineutrophil antibodies, increased apoptosis and decreased marrow granulocyte production are attributed as causal factors.

Autoimmune Lymphoproliferative Syndrome

(ALPS) A secondary immune neutropenia of childhood that is caused by heterogeneous mutations of the fas gene, leading to abnormalities of lymphoid apoptosis where the **T-lymphocytes** are CD4⁻ and CD8⁻. The syndrome is associated with **pancytopenia, lymphadenopathy**, and splenomegaly. There is a markedly increased incidence in these patients of **non-Hodgkin lymphoma**.

T-Cell Large Granular Lymphocytic Leukemia

T-cell large granular lymphocytic leukemia: neutropenia in patients with this lymphoproliferative disorder is severe. The tumor cells express FasL on their surface, which induces Fas/FasL apoptosis. Mechanisms of autoantibody destruction similar to those of rheumatoid arthritis (RA) are also active.

Splenomegaly

Splenomegaly (apart from Felty's syndrome) can cause a moderate/mild neutropenia from hypersplenism, whatever the underlying cause. Neutrophils are sequestered, but normal survival ensures that they are available if required, albeit at a slower rate. Splenectomy is rarely required for neutropenia alone, but if indicated for other reasons, it will usually return the neutrophil count to normal.

Ionizing Radiation

Ionizing radiation, usually as a form of **radiotherapy**, such as radiophosphorus for **polycythemia rubra vera**, or rarely due to accidental exposure, are rare causes neutropenia.

Nonimmune Chronic Idiopathic Neutropenia

This heterogeneous group of benign conditions is usually associated with only mild/moderate neutropenia and hence only a small increased risk of infections. Some patients have demonstrable inhibitors of granulopoiesis, but in others the defect seems to be reduced production of receptors of growth factor.

NEUTROPHIL

Polymorphonuclear neutrophils are **leukocytes** (13-μm diameter) that, upon Romanowsky staining, show characteristic dense nuclei of up to five lobes and pale cytoplasm with many pink-blue or gray-blue granules. The peripheral-blood steady-state **reference range** is 2.5 to 7.0×10^9/l. For values in infancy and childhood, see **Reference Range Table IX**.

Morphology

The nucleus of the mature neutrophil consists of up to five lobes of condensed chromatin of varying size and shape, connected by thin chromatin strands. "Polymorph left shift" refers to the presence within the blood of neutrophil precursors, i.e., blasts, promyelocytes, myelocytes, metamyelocytes, or band cells (see **Granulocytes**). It classically occurs with neutrophilia, and some responses may be so severe as to be termed a leukemoid reaction, the causes being the same. "Polymorph right shift" refers to the presence of hypersegmented neutrophils in the circulation (see **Leukocyte anomalies**).

Up to 3% of neutrophil nuclei from females have a drumsticklike appendage joined to one of the nuclear lobes. This is thought to be the inactivated X chromosome (analogous to the **Barr body**).

The cytoplasm of neutrophils contains granules classified as:

Primary granules containing lysosomal enzyme, **myeloperoxidase**, and a serine protease, elastase

Secondary granules containing **lactoferrin**

Tertiary granules containing gelatinase (see Table 121)

In addition, an agranular secretory vesicle, which appears after the metamyelocyte stage and results from **endocytosis** of the cell membrane, has become recognized. The secretory vesicle seems to act as a store of membrane-bound surface receptors, which are important in neutrophil function. Strictly speaking, neutrophil granules are **lysosomes**, with the enzymes produced by the rough **endoplasmic reticulum** prior to the addition of a lipoprotein

TABLE 121

Content of Neutrophil Granules

Primary	Secondary	Tertiary
Myeloperoxidase	Lactoferrin	Gelatinase
α-1-Antitrypsin	Collagenase	Acetyltransferase
Elastase	Lysozyme	
Azurocidin	Transcobalamin	
Bactericidal permeability		
Increasing protein		
Cathepsins		
Defensins		
Lysozyme		
β-Glucuronidase		

membrane by the **Golgi apparatus**. It is now apparent that not only are the enzymes contained by the three types of granules different, but so is the composition of the lipo-protein membrane. This enables the exocytosis of the three types of granules (and, for that matter, the secretory vesicles) to be differentially regulated.

Metabolism

Neutrophils are able to metabolize carbohydrates, lipids, proteins, nucleic acids, and nucleotides to a greater or lesser extent. They contain large amounts of stored glucose in the form of glycogen, which is mainly metabolized by glycolysis, ultimately resulting in the production of lactate. The Krebs cycle utilizes 5% of glucose, but despite the relative paucity of mitochondria within the mature neutrophil, this accounts for almost half of **adenosine triphosphate** (ATP) production. Only 2 to 3% of glucose metabolism in the resting neutrophil is metabolized through the hexosemonophosphate shunt, but this is an important pathway, as nicotinamideadenine dinucleotide phosphate (NADP) is reduced to NADPH with superoxide radical formation, which is bactericidal.[389,390]

Function of Neutrophils

Neutrophils are migratory phagocytic cells that leave the systemic circulation exponentially, with an average $T^{\frac{1}{2}}$ of about 7 h, and survive in tissues and secretions for about another 30 h. They migrate to sites of **inflammation** or **necrosis** under the influence of various factors, which are usually soluble. They are the major defense against invading microbes and interact with the **immune response** and the **complement** system. They have Toll-like receptors (TLR), most of which are possibly concerned in the network that exists between them and histiocytes and dendritic cells. Having made contact with microbes, the neutrophil is able to phagocytose and ultimately destroy them using either oxidative or nonoxidative mechanisms. Neutrophils may also have a secretory role, particularly for **transcobalamin** I and III and lysozyme. They also have an unspecified role in antitumor reactions, probably by promoting T-helper cells Th1 and Th2. Specific functions are considered in greater detail.

Migration

The stages in neutrophil migration have been identified as:

Random contact: circulating neutrophils randomly contact endothelium.

Rolling: locally released inflammatory mediators — e.g., histamine, endotoxin, leu-kotrienes, interleukin (IL)-2, tumor necrosis factor (TNF)-α — activate the endo-thelium and increase expression of adhesion molecules, i.e., cell-cell, CD62P, CD62E, and CD34. Neutrophils after random contact "roll" along the activated endothelium. Expression of CD62L by neutrophils is important for rolling to occur.

Adherence: activation of neutrophils by locally released cytokines (e.g., **granulocyte/macrophage colony stimulating factor** [GM-CSF] and IL-8), chemotaxins (e.g., C5a and formyl-methionyl-leukyl-phenylalanine [FMLP]), and endothelial adhe-sion molecules (e.g., CD62E and CD62P) upregulates expression of integrins b1 and b2, resulting in the neutrophil adhering to the endothelium.

Diapedesis: having adhered to endothelium, the neutrophil migrates through the inter-endothelial junction under the influence of IL-8, platelet activating factor, or GM-CSF. Expression of CD11a/CD18 and CD11b/CD18 by neutrophils and CD54 by endothelial cells (perhaps induced by c-interferon) seems to be important in this process.

TABLE 122

Killing Mechanisms of Neutrophils

Oxidative	Nonoxidative
Superoxide (O_2^-)	Lactoferrin
Hydrogen peroxide (H_2O_2)	Lysozyme
Singlet oxygen ($^1O_2$)	Cathepsin G
Halides	Defensins
Chloramines	Bactericidal/permeability-increasing protein
	Azurocidin
	Acid

Chemotaxis: neutrophils move to the site of inflammation under the influence of chemotaxins that are derived from activated granulocytes, e.g., IL-1, IL-8, TNF-α, and leukotriene b4; from bacteria (e.g., FMLP); and from serum (e.g., C5a). Expression of CD18, CD29, or CD44 by the neutrophil is necessary for this process to occur.

Assessment of migration of neutrophils can be tested by:

Rebuck skin window assay: an *in vivo* test in which superficial layers of the skin are removed and a glass slide applied, to which neutrophils migrate and adhere

Boyden chamber assay: an *in vitro* assay of the ability of neutrophils to actively migrate across a chamber through differing concentrations of chemotaxins

Phagocytosis and Bactericidal Mechanisms

Bacteria or foreign particles become coated by **opsonins** (IgG, complement component C3b, defensins). The advancing neutrophil has receptors for these opsonins, enabling binding of the opsonin/bacteria complex. The pseudopodium then flows around the complex and encloses it, forming a phagosome by endocytosis. Neutrophil granules are then rapidly discharged into the phagosome, ultimately killing the bacteria. Neutrophils have a variety of bactericidal mechanisms at their disposal that may or may not require oxygen (see Table 122).

Oxidative Bactericidal Mechanisms

Respiratory burst: neutrophil phagocytosis results within seconds in a sharp increase in oxygen consumption and the production of superoxide (O_2^-) using NADPH as a substrate.

$$2O_2 + NADPH \xrightarrow{\text{oxidase}} 2O_2^- + NADP^+ + H^+$$

The superoxide is rapidly converted to hydrogen peroxide (H_2O_2) and singlet oxygen ($^1O_2$):

$$2O_2^- + 2H^+ \xrightarrow{\text{superoxide dismutase}} H_2O_2 + {}^1O_2$$

Hydrogen peroxide also oxidizes halogens (chloride or bromide), forming halides (e.g., HOCl):

$$H_2O_2 + Cl^- \xrightarrow{\text{myeloperoxide}} HOCl + OH^-$$

or, via the Fenton reaction, producing hydroxyl radicals (OH):

$$H_2O_2 + Fe_2 \rightarrow Fe^{3+} + OH + OH^-$$

Halides react with amines to form chloramines (RNHCl):

$$HOCl + RNH_2 \rightarrow RNHCl + H_2O_2$$

Singlet oxygen, superoxide, hydroxyl radicals, halides, and chloramines are all oxidizing agents (of varying potency) and toxic to bacterial membranes.

The increased consumption of NADPH activates glucose oxidation by the hexosemon-ophosphate shunt, resulting in increased NADP production. Hydrogen peroxide is either destroyed by catalase or reduced:

$$H_2O_2 + NADPH \xrightarrow{\text{glutathione peroxidase}} H_2O + NADP^+ + H^+$$

It has been shown that growth factors, e.g., G-CSF and GM-CSF, increase neutrophil bactericidal activity by enhancing the respiratory burst.

Respiratory burst assessment can be made by:

Nitroblue tetrazolium test (NBT): in normal peripheral blood, there is a small percentage of neutrophils able to reduce the virtually colorless dye NBT to an insoluble blue/black precipitate. The percentage of neutrophils able to do this markedly increases in systemic bacterial infections. The reduction of NBT is due to superoxide production, and therefore a microscopic test using this principle has been developed.

Chemiluminescence: neutrophils destroy bacteria through the production of superoxide, singlet oxygen, and hydroxyl radicals, which are also capable of interacting with organic substrates to form highly excited compounds that can return to their resting state by emitting energy in the form of light. Chemiluminescence (as this method is called) can be performed on stimulated and unstimulated neutrophils using luminol (myeloperoxidase dependent) and lucigenin (myeloperoxidase independent) as organic substrate and a luminometer to detect the emitted light. Various agents — zymosan, phorbol myristate acetate, latex beads, bacteria, and fungi — can be used to stimulate neutrophils.

Nonoxidative Mechanisms

Studies in oxygen-free environments show that neutrophils are capable of bactericidal activity under these conditions. Several granular components capable of such activity have been identified.

Lactoferrin: this iron-binding glycoprotein has been shown to be (a) bacteristatic by chelating iron that is necessary for normal bacterial metabolism and (b) directly bactericidal to both gram-positive and gram-negative organisms when in the apolactoferrin form.

Lysozyme: this cationic enzyme hydrolyzes N-acetylmuramic acid residues present in the cell wall of certain bacteria (e.g., micrococci or bacilli).

Cathepsin G: this serine protease possesses chymotrypsin-like hydrolytic activity and is bactericidal to both gram-positive and gram-negative bacteria, as well as fungi.

Defensins: these cationic proteins (several types have been described) are bactericidal to gram-positive and gram-negative bacteria, fungi, and enveloped viruses (e.g., herpes simplex, types 1 and 2). Their bactericidal action is due to their ability to disrupt cellular membranes. Interestingly, defensins can act as opsonins and are chemotactic for monocytes.

Bactericidal/permeability-increasing protein: this cationic protein is bactericidal to gram-negative organisms due to disruption of cellular membranes.

Azurocidin: this polypeptide is bactericidal to both gram-negative and gram-positive bacteria as well as fungi.

Acidification: the pH of the neutrophil vesicle is mildly alkaline, but upon forming a phagosome, the pH falls to 3 or lower over about 1 h, which is bactericidal to many gram-positive and gram-negative bacteria as well as fungi.

Assessment of killing can be made by various tests that either involve the ability of live bacteria within neutrophils to incorporate tritiated thymidine, or that develop colony growth (after the neutrophils have been lysed). Similar tests looking at the ability of live *Candida* spp. inside neutrophils to exclude the dye methylene blue or to form colonies (after neutrophil lysis) are available. All of the above tests can be performed with or without opsonins present.

Finally, having ingested bacteria, neutrophils are able to release soluble factors, e.g., IL-8, which recruit more neutrophils to the site.

Neonatal Neutrophil Function

Neutrophils of the newborn, while similarly produced, respond differently to sepsis. The reaction is a rapid fall in the total neutrophil mass. Chemotaxis is also reduced in both newborn and immature infants. Other functions may also differ, but these remain unproven.

NEUTROPHIL ALKALINE PHOSPHATASE SCORE

(NAP score) The intensity of the alkaline phosphatase staining in the secondary granules of **neutrophils** scored $0 \rightarrow + 4$. When 100 neutrophils have been scored, a normal range can be ascertained — typically 15 to 130. Alkaline phosphatase activity is found mainly in mature neutrophils, but there is some activity in metamyelocytes. The activity can be measured by the hydrolysis of naphthol AS-B1 phosphate to aryl naphtholamide that has been coupled with a diazonium such as fast red. A fresh heparinized sample should be used, as enzyme activity diminishes in EDTA (ethylenediaminetetraacetate). The test results in a blue granular reaction product, the intensity of the alkaline phosphatase staining in the secondary granules of neutrophils being scored $0 \rightarrow + 4$ (NAP score).

Low NAP scores are seen in:

Chronic myelogenous leukemia (except when there is infection)

Corticosteroid therapy or disease acceleration

Myelodysplasia

Paroxysmal nocturnal hemoglobinuria

Sickle cell disorders

Acute myeloid leukemia

High scores are seen in:

Leukemoid reactions
Myelofibrosis
Polycythemia rubra vera

The NAP score was formerly useful in distinguishing leukemoid reactions from chronic myelogenous leukemia, but currently this distinction is better made by the more-specific *bcr-abl* fluorescence *in situ* hybridization (FISH) assay.

NEUTROPHIL ANOMALIES
See **Leukocyte anomalies**.

NEUTROPHIL FUNCTION DISORDERS
Defects in the function of **neutrophils**. These dysfunctions interfere with their migration into tissues, passage through vascular endothelial cells, locomotion in tissues, phagocytosis, and intracellular killing. The disorders give rise to mucocutaneous sepsis in the mouth and perianal region and to other local infections, which may result in abscess formation in tissues or lymph nodes. Granulomas sometimes develop.

Congenital Disorders
These may present with infection or by delayed separation of the umbilical stump.

Leukocyte-adhesion deficiency. An autosomally recessive deficiency of **cell adhesion molecules** that are necessary for migration of granulocytes from the circulation to the tissues.

Hyper-IgE syndrome. A syndrome where there is impaired neutrophil locomotion and severe eczema with frequent staphylococcal infections or abscesses associated with high plasma levels of IgE.

Schwachman's syndrome. A syndrome where defects in neutrophil migration are associated with pyogenic infections and exocrine pancreatic insufficiency.

Chronic granulomatous disease. A rare inherited heterogeneous group of disorders where there is failure of granulocytes to produce oxidative radicals and thus to kill phagocytosed bacteria.

Chediak-Higashi syndrome. An autosomally recessive disorder with disturbance of phagolysosome function. It is characterized by giant granules in myeloid cells and associated with oculocutaneous albinism.

Lactoferrin deficiency in cytoplasmic granules. This results in lack of neutrophil response to chemotactic signals and diminution of adhesiveness to the surface of particles. Clinically, it causes recurrent pyogenic infections, particularly deep-seated skin abscesses.

Myeloperoxidase deficiency. A benign autosomally recessive trait occurring in 1:4000 people.

Acquired Disorders

Corticosteroid *therapy.* This can cause impairment of leukocyte-endothelial adhesion and reduce neutrophil margination and migration to any infected tissue. There is a consequent neutrophilia.

Influenzal infections. These can cause transient impairment of phagosome-lysosome fusion, giving rise to a high incidence of staphylococcal pneumonia.

Uncontrolled **diabetes mellitus.**

Hypophosphatemia occurring with intravenous feeding.

Inhibitors of endogenous chemotactic factors. These can occur with **Hodgkin disease** and with **alcohol toxicity** as the cause of hepatic cirrhosis.

NEUTROPHILIA

A raised level of circulating **neutrophils** $>7.0 \times 10^9/l$. There are several different mechanisms responsible for an increased circulating blood neutrophil pool (see Figure 93). If neutrophilia is acute in onset, there is usually an associated "left shift" of the nuclei. Toxic granulation and **Döhle body** formation are particularly likely in acute bacterial infections.

Etiology

For a summary of etiology, see Table 123.

Physical/emotional stimuli: release of epinephrine, norepinephrine, or cortisol mobilizes neutrophils from the bone marrow storage pool and reduces margination. The return to the steady state is prompt once the stimulus has been removed.

Infectious disorders: particularly bacterial.

	Mitotic Pool	Maturation Storage Pool	Margination	Circulation
Marrow Storage	N	↓	↑	↑
Demargination	N	N	↓	↑
Decreased Tissue Migration	N	N	N	↑
Increased Proliferation	↑	↑	↑	↑

FIGURE 93
Mechanisms of neutrophilia.

TABLE 123

Causes of Neutrophilia

Physical/emotional stimuli, e.g., exercise, stress, rage, fear, labor
Infectious disorders
    Bacterial
    Acute, e.g., gram-positive and gram-negative organisms, spirochetes
    Chronic, e.g., tuberculosis
    Fungal, e.g., *Candida* spp., *Aspergillus* spp.
    Viral
Inflammation/tissue damage
    Burns
    Trauma (surgery)
    Rheumatoid arthritis
    Gout
    Infarction
    Pancreatitis
Malignant tumors
    Nonhematopoietic, e.g., stomach, breast, renal
    Hematopoietic
    Chronic myelogenous leukemia, polycythemia rubra vera, essential thrombocythemia
    Hodgkin disease
Drugs
    Corticosteroids
    Lithium
    Etiocholanolene
    Colony stimulating factor, e.g., G-CSF, GM-CSF
Bone marrow stimulation
    Rebound
    Hemorrhage
Miscellaneous, e.g., eclampsia, Cushing's disease, smoking

Inflammation/tissue damage from all causes:

- Acute: neutrophils are released from marrow storage sites, with demargination also occurring, again due to epinephrine or cortisol release.

- Chronic: release of **cytokines**, e.g., interleukin (IL)-1, and colony stimulating factors (CSF), e.g., granulocyte (G-CSF) and granulocyte/macrophage (GM-SCF), causes granulocyte hyperplasia, with a consequent increase in circulating neutrophils.

Malignant tumors: virtually any nonhematopoietic tumors can cause reactive neutrophilia, especially in metastatic disease. In some cases, the tumors have been shown to produce colony stimulating factors, but usually tumor necrosis and super-added infections are the cause. **Myeloproliferative disorders** — e.g., **essential thrombocythemia** (ET), **polycythemia rubra vera** (PRV), and **chronic myelogenous leukemia** (CML) — commonly present with neutrophilia and "left shift," which in the case of CML is usually associated with basophilia and eosinophilia. **Hodgkin disease** (and to a lesser extent **non-Hodgkin lymphoma** and **myelomatosis**) can also cause reactive neutrophilia.

Adverse drug reactions: corticosteroids cause neutrophilia (and occasional left shift) by releasing neutrophils from marrow stores and also preventing their migration into tissues. Colony stimulating factors (e.g., G-CSF and GM-CSF) induce granulopoiesis but also release neutrophils from marrow stores.

Bone marrow stimulation: the phenomenon of rebound neutrophilia following **agran-ulocytosis** or **megaloblastosis** due to a temporary failure in the regulation of marrow stores is well described. **Hemorrhage** probably causes neutrophilia due to epinephrine and cortisol release.

NICOTINAMIDE ADENINE DINUCLEOTIDE PHOSPHATE

(NADP) In its reduced state (NADPH), it is the principal high-energy product of aerobic glucose metabolism via the hexose monophosphate shunt of **red blood cells** and of **neutrophils**. The red cell lacks the machinery to utilize NADPH for energy; instead, NADPH serves as a cofactor in the reduction of **glutathione**. NADPH is generated in the pentose phosphate pathway by **glucose-6-phosphate dehydrogenase** (G6PD) and therefore occupies a central position in protection from oxidative injury. In neutrophils, reduction NADP releases superoxide radicals, which are bactericidal.

NIEMANN-PICK DISEASE

A **lipid-storage disorder** inherited as an autosomally recessive trait with a high prevalence among Ashkenazi Jews. Deficiency of sphingomyelinase results in the accumulation of sphingomyelin in body tissues.

Presentation is usually in infancy with **hepatosplenomegaly**, **lymphadenopathy**, and neurological deficit, e.g., mental retardation, fits, etc.

As with **Gaucher's disease**, there is an adult (nonneuropathic) form that usually presents with hepatosplenomegaly, mild cerebellar dysfunction, and a cherry red spot at the macula.

The diagnosis depends on the demonstration of characteristic Niemann Pick cells within tissues, e.g., **bone marrow**. The large mononuclear **histiocytes** (macrophages), of 20- to 80-μm diameter, have plentiful laminated cytoplasm that contains droplets of sphingomyelin.

The prognosis is very poor, with the infantile form resulting in death within two years. **Allogeneic stem cell transplantation** has been of benefit for a few patients.

Splenectomy may be of benefit in the adult form to relieve splenic pain or the effects of **splenomegaly**.

NITRIC OXIDE

(NO) An endogenously synthesized gas with widespread mediator function.[392] Three NO synthases (NOS) have been described: an endothelial isoform (eNOS), a neuronal isoform (nNOS), and a neutrophil/macrophage or inducible isoform (iNOS). These NOSs cleave nitrogen from the amino acid l-arginine and combine it with molecular oxygen to form NO, which binds to **heme**. The ferrous-heme complex activates the enzyme guanylate cyclase that stimulates cGMP.

The actions of NO include:

Inflammation messenger molecule, where it promotes T-lymphocyte**s** and increases **histiocyte** (macrophage) activation with the production of cytokines: **interleukin** (IL-1), **interferon**-γ (γ-INF), **tumor necrosis factor** (TNF)-α, and the migration inhibitory factor

Relaxation of vascular smooth muscle with increase of intracellular cGMP

Inhibition of **platelet** aggregation

Prevention of adhesion by platelets and **neutrophils** to vessel walls

Role in modulating endothelial-progenitor-cell function in vessel repair

Therapeutic applications are under investigation.

NITRITE TOXICITY
See **Methemoglobinemia**.

NITROBLUE TETRAZOLIUM TEST
(NBT) See **Chronic granulomatous disease**.

NITROUS OXIDE TOXICITY
The hematological effects of excess exposure to nitrous oxide, usually during anesthesia. Exposure to nitrous oxide results in acute **megaloblastosis** before the marrow becomes more severely damaged with **bone marrow hypoplasia**. Such changes are seen within 5 to 24 h of exposure. Circulating **cobalamin** is inactivated, and thymidylate synthesis is depressed, thus reducing **DNA** synthesis. Reduction in hepatic 5'-adenosylmethionine is thought to be associated with neurologic change resembling **subacute combined degeneration of the spinal cord**. The concentration of nitrous oxide in the surgical theater atmosphere around anesthetists should not exceed 200 parts per million.

NK-CELL LYMPHOMA
See **Blastic NK-cell lymphoma**.

NODAL MARGINAL-ZONE B-CELL LYMPHOMA
(Rappaport: well-differentiated lymphocytic, poorly-differentiated lymphocytic, mixed lymphocytic-histiocytic; Lukes-Collins: parafollicular B-cell) See **Marginal-zone B-cell lymphoma**.

NODAL POORLY DIFFERENTIATED, MIXED LYMPHOCYTIC-HISTIOCYTIC TYPE LYMPHOMA
See **Follicular lymphoma**; **Non-Hodgkin lymphoma**.

NONACTIVATED PARTIAL THROMBOPLASTIN TIME
(NAPTT) A variant of the **activated partial thromboplastin time** (APTT), in which no activating agent, such as kaolin or silica, is added. The time taken for the plasma to clot represents (a) the concentration of the components of the intrinsic coagulation cascade and (b) the quantity of activated components of the cascade in the plasma sample, such as factors IXa and Xa. It is a measure of thrombotic potential, but has little clinical interpretive use.

NON-HODGKIN LYMPHOMA

(NHL) A heterogeneous group of **lymphoproliferative disorders** with variations of natural history, pathology, immunology, cytogenetics, and responses to therapy. Their precise cause is unknown in the majority of patients in whom a firm pathologic diagnosis has been established. The prognosis depends upon the histologic subtype, stage, clinical characteristics, and laboratory parameters. It is essential to establish, by surgical biopsy of lymph nodes or extranodal tissue, an accurate histologic diagnosis at the outset to ensure that the most appropriate therapeutic interventions are instituted. They are responsive to treatment, but the extent, type, and duration of responses vary significantly.

Classification

A number of classifications have been developed (see **Lymphoproliferative disorders**). Based upon the morphological appearances, **immunophenotyping**, and **cytogenetic analysis**, the World Health Organization (WHO) has recently published the most recent classification — WHO Classification of Tumors of Hematopoietic and Lymphoid Tumors — (see Appendix), which is now the accepted standard classification.

    Precursor B- and T-Cell Neoplasms
        Precursor B-lymphoblastic leukemia/lymphoma
        Precursor T-lymphoblastic leukemia/lymphoma
    Mature B-Cell Neoplasms
        Chronic lymphocytic leukemia/small lymphocytic lymphoma
        B-cell prolymphocytic lymphoma
        Lymphoplasmacytic lymphoma/Waldenström macroglobulinemia
        Splenic marginal-zone lymphoma
        Hairy cell leukemia
        Plasma-cell neoplasms
        Extranodal marginal-zone B-cell lymphoma (MALT lymphoma)
        Nodal marginal-zone B-cell lymphoma
        Follicular lymphoma
        Mantle-cell lymphoma
        Diffuse large B-cell lymphoma
        Mediastinal (thymic) large B-cell lymphoma
        Intravascular large B-cell lymphoma
        Primary effusion lymphoma
        Burkitt lymphoma/leukemia
        Lymphomatoid granulomatosis
    Mature T-Cell and NK-Cell Lymphoma
        T-cell prolymphocytic leukemia
        T-cell large granular lymphocytic leukemia
        Aggressive NK-cell leukemia
        Adult T-cell leukemia/lymphoma
        Extranodal NK/T-cell lymphoma, nasal type

Enteropathy-type T-cell lymphoma

Hepatosplenic T-cell lymphoma

Subcutaneous panniculitis-like T-cell lymphoma

Blastic NK-cell lymphoma

Mycosis fungoides/Sézary syndrome

Primary cutaneous CD30-positive T-cell lymphoproliferative disorders

    Primary cutaneous anaplastic large-cell lymphoma (C-ALCL)

    Lymphomatoid papulosis

    Borderline lesions

Angioimmunoblastic T-cell lymphoma

Peripheral T-cell lymphoma, unspecified

Anaplastic large-cell lymphoma

Immunodeficiency-Associated Lymphoproliferative Disorders

    Lymphoproliferative disorders associated with primary immune disorders

    Human immunodeficiency virus-related lymphomas

    Posttransplant lymphoproliferative disorders

    Methotrexate-associated lymphoproliferative disorders

Grading has been applied according to a marking formulation (Table 124).

TABLE 124

Non-Hodgkin Lymphoma Grading According to the Working Formulation

Low grade

A Small lymphocytic
B Follicular small cleaved cell
C Follicular mixed cell

Intermediate grade

D Follicular large cell
E Diffuse small cleaved cell
F Diffuse mixed cell
G Diffuse large cell

High grade

H Immunoblastic
I Lymphoblastic
J Small noncleaved cell

Miscellaneous

Composite malignant lymphoma
Mycosis fungoides
Extramedullary plasmacytoma
Unclassified
Other

Clinical Features

These are variable, depending upon the type and location. They usually include **lymphad-enopathy** with a cytopenia dependent upon **bone marrow** involvement. Many have pyrexia, weight loss, and night sweats, sometimes referred to as "B" symptoms.

Prognosis

A predictive model, the International Index, based on clinical pretreatment characteristics and the relative risk of death in patients with intermediate histology, has been developed.[332,393] Those that are associated with survival include:

Age (<60 years vs. >60 years)

Lactate dehydrogenase, LDH (normal or >normal)

Performance status (0,1 vs. 2–4)

Stage (I/II vs. III/IV)

Extranodal involvement (1 site vs. >1 site)[393]

Patients were divided into different risk groups based on the number of risk factors, with predicted 5-year survivals of 73% in the low-risk group with zero or one factor; 51% for low–intermediate risk with two factors; 43% for high–intermediate with three factors; and 26% for high risk with four or five factors. Age is an important consideration in the prognosis of intermediate histology non-Hodgkin lymphoma. Patients older than 60 years have complete remission rates that are similar or only slightly lower than those patients less than or equal to age 60, but older patients are much less likely to maintain their complete remission.

Developments in the biological heterogeneity of the intermediate-histology non-Hodgkin lymphomas are providing evidence of other biologically important correlates that impact on survival and may serve as independent prognostic factors. Patients with a proliferative index of greater than 80%, as measured by Ki67, have a 1-year survival of 18%, whereas those with an index of less than 80% have an 82% 1-year survival. CD44 isoform expression predicts outcome. Elevated serum interleukin (IL)-6 levels are associated with a poorer clinical outcome. Patients with bcl-6 expression have a better outcome. The t(14,18) translocation and *bcl-2* overexpression may be linked to a shortened survival in intermediate-histology non-Hodgkin lymphoma, but these may be overcome by **rituximab** therapy. Microarray studies are now identifying new prognostic factors. These and future developments will lead to future prognostic factor schemas that will not only be pathologically but also clinically based.

In low-grade non-Hodgkin lymphomas, the International Index has been demonstrated to be predictive of survival. It has been reported that an LDH above normal and b2-microglobulin above normal predict survival in follicular lymphomas. S-phase kinetics have predicted survival in studies of patients with low-grade lymphoma. Other reported prognostic factors in low-grade non-Hodgkin lymphomas include response to initial therapy, stage, extent of helper T-cell infiltrate, a normal hemoglobin level, and absence of inter-follicular fibrosis. Microarray studies have identified prognostic groups. Despite these observations, a uniformly accepted risk-factor-analysis program is not applied in the low-grade non-Hodgkin lymphomas at present. The REAL classification (see **Lymphoprolif-erative disorders**) and further-defined subsets with pertinent biological correlates should aid in the development of a more applicable risk-factor index in low-grade lymphomas.

Treatment of Non-Hodgkin Lymphoma — General Principles

Cytotoxic agent therapy (chemotherapy) is the primary underlying treatment modality of most types of NHL. Not all non-Hodgkin lymphomas are approached in the same fashion. The discussions that follow about chemotherapy do not necessarily have applications to AIDS-related lymphoproliferative disorders, posttransplant lymphoproliferative disorders, or the extranodal lymphomas, which include primary central nervous system, gastric, thyroid, testicular, bone, pulmonary, female reproductive organs, skin, or primary extradural areas. The indications for radiation therapy and/or surgery are small but include patients presenting with isolated gastric, central nervous system, testicular, bowel, orbital, pulmonary, or cutaneous involvement.

Follicular Lymphomas

Of patients with low-grade NHL, 80 to 90% have stage III or IV disease and are not curable with standard treatment regimens.[394] Observation is the initial treatment of choice in asymptomatic patients who do not have bulky disease. In patients with advanced disease, around 60% require therapy at a median interval of 3 years from the time of diagnosis. Spontaneous regression, either partial or complete, occurs in about 25% of patients. The median survival of patients with follicular center lymphoma is 10 years. For those patients with "B" symptoms (see above), bulky disease, progressive disease, peripheral blood cytopenias, or who wish to be treated, options with the goal of achieving a clinical complete remission include single-agent daily oral **chlorambucil**, intravenous CVP (**cyclophosphamide, vincristine**, and **prednisone** or **prednisolone**), CVP with total lymphoid irradiation, or CHOP (cyclophosphamide, Adriamycin, vincristine, and prednisone or prednisolone). Trials are proceeding in which rituximab is incorporated with a CVP regimen, rituximab as maintenance following a CVP regimen, and rituximab-CHOP with the hope of improved disease-free survival. The median disease-free interval is 17 months. Even with clinically complete remissions, minimal residual disease is still present, as determined by molecular genetic techniques. Patients who relapse respond to re-treatment with similar drugs or different programs. The remission duration is usually shorter with subsequent treatments. Rituximab was the first monoclonal antibody approved in the U.S. for treatment of relapsed follicular lymphoma. Radioisotopes conjugated to monoclonal antibodies are now approved in the U.S. The two agents available are ibritumomab tiuxetan and yttrium 90 and the second tositumomab and iodine 131. Patients have been reported to respond longer to these agents than the previous chemotherapy agents, and approximately one-third have long-term remissions.

Interferon has been shown to have activity in low-grade lymphomas. Patients treated with more-aggressive chemotherapy regimens tend to achieve a remission earlier, but these remissions are no more durable, do not improve disease-free survival, and do not increase overall survival. Trials with **autologous stem cell transplantation** (SCT) are underway for those patients who relapse or who are young. This may lead to extended disease-free survival, but not to overall survival. Studies evaluating the role of autologous SCT for early consolidation in follicular center lymphomas are currently underway.

Other treatment options for patients who relapse or who are refractory to standard treatment include:

Older drugs administered in different fashions

Splenectomy

2'-Chlorodeoxyadenosine

2'-Deoxycoformycin

Fludarabine

Interleukin-4

Gene therapy

The overall response rates (complete remissions and partial remissions) with fludarabine in previously treated patients have been 38 to 55%. The response rates to 2'-deoxycofor-mycin have been 17 to 29%, and the response to 2-chlorodeoxyadenosine ranges from 43 to 75%. The major toxicities of the purine nucleoside analogs are not only myelosuppression, but immunosuppression with complications that include infections secondary to herpes zoster virus, **cytomegalovirus**, and pneumocystis pneumonia and, less commonly, **graft-versus-host-disease**. The CD4 and CD8 lymphocyte counts decrease significantly with these agents.

Diffuse Large B-Cell Lymphomas

In patients with intermediate-histology NHL, anthracycline chemotherapy regimens are the hallmark of therapy, with complete remission rates of 60 to 80% of patients with stage II to IV disease and long-term disease-free survival as predicted by the International Index. No statistically significant difference has been found[395] in 3-year disease-free survival between treatment regimes of CHOP (cyclophosphamide, Adriamycin, vincristine, and prednisone or prednisolone), m-BACOP (methotrexate, bleomycin, doxorubicin, cyclophosphamide, vincristine, and dexamethasone), ProMACE-CytaBOM (cyclophosphamide, doxorubicin, etoposide, prednisone or prednisolone, cytarabine, bleomycin, vincristine, methotrexate, and leucovorin), and MACOP-B (methotrexate, doxorubicin, cyclophosphamide, oncovin, bleomycin, and prednisone or prednisolone).[396] Data from multiple sources report relapse rates of 7% for years 2 to 5 and 30 to 40% of patients cured. Rituximab added to CHOP chemotherapy has improved the disease-free survival time and overall survival.[397]

Trials are now underway comparing standard treatment with the incorporation of autologous SCT as primary treatment, but the treatment results are inconclusive at present. Unselected patients who are in complete remission probably do not benefit from high-dose cytotoxic therapy. It is difficult at the present time to demonstrate a role for peripheral-blood stem cell transplantation (PBSCT) as part of the initial management of diffuse large B-cell lymphoma. CHOP is the gold standard of treatment in the initial management of intermediate non-Hodgkin lymphoma.

Multiple nonrandomized and predominantly single-institution studies have demonstrated that patients who relapse and who are treated with very high doses of therapy and autologous SCT may enter a complete remission and potentially be cured. In patients who relapse after a complete remission with CHOP and who enter a complete remission with re-treatment with standard chemotherapy regimens followed by autologous SCT, the autologous cure rates are 35 to 40%. Autologous SCT is considered the standard treatment of high- and intermediate-grade NHL in sensitive relapse.

High-Grade Lymphomas

Lymphoblastic lymphoma is characterized immunohistologically by an immature T-cell phenotype that is **terminal deoxynucleotidyl transferase** (TdT)-positive. It usually involves young adults, with a predominance of males with mediastinal masses. In contrast to the other non-Hodgkin lymphomas, there is a propensity for relapse in the central nervous system (CNS). There are different management strategies for adult and pediatric patients. Various regimens have been used in the treatment of lymphoblastic lymphoma incorporating intensive chemotherapy with CNS prophylaxis. Some programs have included extended maintenance, but there is no agreement at this time about the best

duration of maintenance therapy.[398] Previous management strategies included CHOP/
L-**asparaginase** regimen plus CNS prophylaxis that includes intrathecally administered
methotrexate and cranial irradiation. Approximately 75% of patients without bone marrow involvement and with normal lactate dehydrogenase (LDH) are cured with these
treatment modalities. Otherwise, regimens under evaluation include hyper CVAD
(cyclophosphamide, vincristine, doxorubicin, dexamethazone) and leukemia regimens
incorporating radiation therapy to the mediastinum and cranial radiation.

Reports of the results of experiences in stem cell/bone marrow transplantation have
included heterogeneous patient populations, which makes these forms of treatment difficult to interpret. Bone marrow transplant studies have shown survival rates in 60 to 75%.
Allogeneic SCT should be considered in high-risk patients.

Satisfactory results have been achieved for treatment of small noncleaved cell lymphoma
treated with an aggressive, high-intensity, brief-duration chemotherapy program that
included cyclophosphamide and etoposide with 7 days of prednisone, followed by bleomycin and vincristine at weekly intervals, with methotrexate intravenously for the first
cycle and doxorubicin added to the second cycle.

New Biological Approaches to Lymphoma Treatment

Rituximab, a chimeric mouse/human anti-CD20 monoclonal antibody, which is a
   human c-1 k antibody with mouse-variable regions. CD20, an antigen important
   in cell-cycle initiation and differentiation, is strongly expressed in over 90% of B-
   cell lymphomas. *In vitro* studies have demonstrated that this antibody binds
   human complement, and lysis of lymphoid B-cell lines has been demonstrated.
   Adverse reactions have included fever, skin rash, nausea, rigors, orthostatic
   hypotension, and bronchospasm. The majority of patients who have been so
   treated have had low-grade lymphomas. Other antibodies are under development
   and clinical evaluation.

Radioisotope conjugated to monoclonal antibodies.

Suicide-gene therapy,[399] in which tumor cells are genetically modified, which under
   appropriate circumstances will kill tumor cells. This has been termed "bystander
   effect" because the dying gene-modified cells are toxic to nearby unmodified
   tumor cells.

Angiogenesis inhibitors.

Bcl-2 antisense molecules.

NONTROPICAL SPRUE

(Adult celiac disease) See **Intestinal disorders**.

NORMOBLAST

(Erythroblasts) Red blood cell precursors normally found in the bone marrow as the later
part of erythropoiesis (see **Hematopoiesis**). Normoblasts are differentiated into three types
— basophilic normoblasts, polychromatic normoblasts, and orthochromatic normoblasts
— based upon the depth of cytoplasmic Romanowsky staining, which is a reflection of
acidity dependent upon hemoglobin content. Nuclear change occurs simultaneously with
a round, deeply chromatic nucleus in the basophilic normoblast, pyknotic nuclei in the
polychromatic normoblast, and nuclear remnants in the orthochromatic normoblasts. The
latter cells are seen in the peripheral blood in disorders associated with **normoblastemia**.

Morphological maturation occurs from the earliest cell, the proerythroblast, which is programmed for **hemoglobin** production as well as for proliferation. Maturation continues through basophilic erythroblasts, polychromatic erythroblasts, and finally orthochromatic erythroblasts to reticulocytes, which become mature erythrocytes or red blood cells:

Proerythroblast → Basophilic erythroblast → Polychromatophilic erythroblast
→ Orthochromatic erythroblast → Reticulocyte

The pronormoblast or proerythroblast is the largest cell (14- to 19-μm diameter) and the most primitive morphologically identifiable erythroid cell. This cell contains little cytoplasm (<20% total volume), with a prominent nucleolus within coarse nuclear chromatin.

The basophilic erythroblast is smaller, with a lower nuclear cytoplasmic ratio and with more deep-blue cytoplasm rich in ribosomal RNA. Nuclear condensation to patchy heterochromatin commences at this stage.

Polychromatic erythroblasts herald the first visible cytoplasmic hemoglobin and are yet smaller (12- to 15-μm diameter), with a decreasing nuclear/cytoplasmic ratio. These cells contain maximal numbers of mitochondria, which can be easily seen when containing pathological quantities of iron in **sideroblastic anemia**. Nuclear chromatin condensation is progressively increasing. The control of nuclear chromatin condensation is unclear, but it is thought to relate to cytoplasmic (and nuclear) hemoglobin concentration, which at an appropriate level causes nuclear processes to cease.

Over a period of 3 to 4 days, from a single polychromatic erythroblast, 8 to 16 fully hemoglobinized orthochromatic erythroblasts are formed. These have a diameter of 8 to 12 μm, with heavily condensed pyknotic chromatin. Nuclear extrusion occurs at this stage to leave the reticulocyte. Nuclear extrusion is an active process mediated by the cytoskeletal proteins and is complete in 5 to 60 min.

Reticulocytes are larger than mature erythrocytes, with polychromatic cytoplasm containing significant quantities of ribosomal RNA, which can be stained as aggregates by supravital dyes to produce the characteristic "reticulum" network. Maturation of reticulocytes results in reducing quantities and less clumping of this reticulum. Electron microscopy identifies a folded, irregular surface that only matures to the biconcave disc at the mature erythrocyte stage. Hemoglobin synthesis ceases upon reticulocyte maturation.

Erythropoiesis can be demonstrated by imaging marrow, liver, and spleen with ^{99m}Tc sulfur colloid or ^{111}Indium, even though these isotopes primarily label the monocyte-macrophage system. Their uptake is similar to ^{59}Fe, and they can be used to demonstrate erythroid tissue, but accurate quantitation of total erythropoiesis is made by measuring the rate of red blood cell production (see **Ferrokinetics**).

NORMOBLASTEMIA

(Erythroblastemia) Nucleated red blood cells (NRBC) in the peripheral blood. (See **Nucleated red blood cell counting**.) These cells can be seen with any severe **anemia**. Large numbers are found in **hemolytic disease of the newborn**. In the adult, they are found in chronic **myeloproliferative disorders**, **leukemias**, and carcinomatosis, usually as part of a **leukoerythroblastic anemia**. NRBCs are also found following **splenectomy** and in sickle cell crises. Their presence is often an agonal event in cyanotic heart failure, respiratory failure, and septicemia.

NORMOCHROMIC ANEMIA

See **Normocytic anemia**.

NORMOCYTES

(Erythrocytes; mature **red blood cells**) Biconcave disks that vary little in size and shape. In well-prepared blood smears, they have smooth contours and diameters of approximately 7 μm (range 6.0 to 8.5 μm), comparable with the diameter of the small mature lymphocyte nucleus, which serves as a useful micrometer. Because the cell is biconcave, the stained normal red cell shows an area of central pallor, which should not exceed one-third of the total surface area. A small number of red cells (usually less than 10%) may be oval in shape and <0.1% fragmented forms may be present.

NORMOCYTIC ANEMIA

Anemia associated with reduced levels of **hemoglobin**, with normal levels of mean cell volume (MCV) and mean cell hemoglobin (MCH) and normocytic normochromic **red blood cells** upon **peripheral-blood film examination**. Differentiation depends upon investigation in a stepwise process (see Figure 94). The initial investigations are a **reticulocyte count** and **bone marrow** aspirate/trephine biopsy. If these are normal, it is probably an **anemia of chronic disorders**, for which a clinical cause should be sought. A dyserythropoietic marrow suggests a form of **myelodysplasia**. The definitive diagnosis may be established from the bone marrow changes seen in **acute leukemias**, **myeloproliferative disorders**, **myelomatosis**, **myelofibrosis**, and by carcinoma metastases with **infiltrative myelophthiasis**.

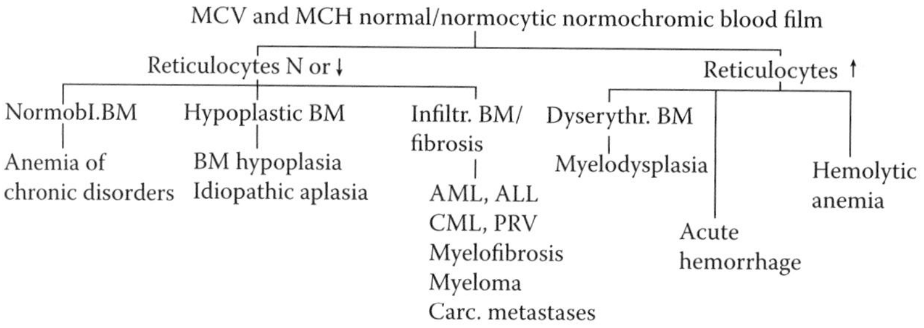

FIGURE 94

Flow diagram of the investigation of normocytic anemia. (Adapted from Bates, I. and Bain, B.J., Approach to the diagnosis and classification of blood disorders, in *Dacie and Lewis Practical Haematology*, 10th ed., Churchill-Livingston Elsevier, 2006, Figure 23.3. With permission.)

NOSE DISORDERS

See **Nasal disorders**.

NUCLEAR CONTOUR INDEX

(NCI) An expression of the degree of nuclear convolution. It is based upon a semiautomated analysis of electron microsopic scanning and is calculated by dividing the nuclear perimeter by the square root of the nuclear diameter. T-lymphocytes normally have higher NCIs than non-T-lymphocytes. The index can therefore be used to differentiate benign from malignant cells, particularly in the diagnosis of **mycosis fungoides/Sézary syndrome**.

NUCLEATED RED BLOOD CELL COUNTING

(NRBC) Measurement of the number of circulating immature erythrocytes (nucleated red blood cells, orthochromatic erythroblasts of **erythropoiesis**). It necessary to produce accurate NRBC counts for no other reason than the generation of correct total and differential leukocyte counts.

Small numbers of NRBCs are found in the cord blood of normal neonates. The neonatologist will always wish to have an NRBC count when there is a history of fetal hypoxia, or asphyxia, in growth-retarded infants and in premature infants. Large numbers of circulating NRBCs, most of which derive from sites of **extramedullary hematopoiesis** in the liver and spleen, are very characteristic of **hemolytic disease of the newborn**. In adults, erythroblastemia (**normoblastemia**) occurs in leukemia, myeloproliferative syndromes, particularly myelosclerosis, and carcinomatosis due to either extramedullary hematopoiesis or perturbation of bone marrow architecture, or both. Here, the number of circulating NRBCs may be comparatively low and associated with immature granulocytes, resulting in the so-called leukoerythroblastic blood picture. NRBCs may be present in the peripheral blood of adults with sickle cell disorder, particularly during painful crises, and also in **thalassemia**.

Until very recently, the only method for NRBC enumeration was by microscopy, expressing the count as the number of NRBC/100 WBC. Not only was this a laborious procedure, it was also imprecise, with reported coefficients of variation (COV) ranging from 30 to 110%. Many automated blood cell counters already display flags indicating the presence of NRBC; however, these are limited by low sensitivity and specificity of flagging performance and do not produce a count. More recently, a number of **flow cytometry** methods using monoclonal antibody labels have been published and indicate less imprecision, with COV consistently below 20%; however, the technique is time-consuming, expensive, requires considerable flow cytometry expertise, and the result is not produced in a clinically meaningful time scale. Commercially available routine hematology analyzers incorporating accurate NRBC counting have recently been introduced using a variety of flow cytometry techniques.

NUCLEOSIDES

The base-ribose moieties of **nucleotides**. They occur in **RNA** as adenosine, guanosine, cytidine, and uridine and, in **DNA**, in their dexy forms, with the exception of uridine, which is here replaced by thymidine as deoxythymidine.

Nucleotide receptors are an emerging family of regulatory molecules in blood cells. They effect cell proliferation, differentiation, chemotaxis, release of cytokines or lysosomal constituents, and generation of reactive oxygen or nitrogen species upon stimulation of cells with extracellular **adenosine triphosphate** (ATP). Stimulation occurs through plasma membrane receptors: perinergic P2 receptors.

NUCLEOTIDE

A molecule comprising either a **purine** or **pyrimidine** base bonded to either a phosphorylated ribose or a deoxyribose moiety, as in DNA. They are phosphorylated **nucleosides** synthesized *de novo* in most cells, although the mature red cell depends on the so-called salvage pathway for its supply of purine nucleotides. The adenine phosphoribosyltransferase (APRT) and hypoxanthine-guanine phosphoribosyltransferase (HGPRT) reactions incorporate adenine or hypoxanthine or guanine into nucleotides, respectively.

Metabolic disorders

- Deficiency of APRT occurs as an autosomally recessive disorder resulting in nephrolithiasis due to the formation of deoxyadenine stones.
- Deficiency of HGPRT is inherited as a sex-linked disorder resulting in hyper-uricemia and a neurologic syndrome characterized by self-mutilation, the **Lesch-Nyhan syndrome**.
- Hyperactivity of **adenosine deaminase**.
- Deficiency of **pyrimidine 5'-nucleotidase**.
- Deficiency of **adenylate kinase** (AK).

NULL CELL

Lymphocytes that carry no CD markers. Proliferation of B-lymphocytes between pre-B-cells and early mature B-cells may be categorized as "null-CLL" (see **Chronic lymphatic leukemia**).

NUTRITIONAL DEFICIENCY DISORDERS

The effects of nutritional deficiency on hematopoietic tissues.

Starvation with failure of protein metabolism leads to impaired hematopoiesis. Loss of fat and parenchyma in bone marrow only occurs with extreme malnutrition. Immunodeficiency may be a consequence.

Deficiency of single nutrients, either alone or as part of general malnutrition:

- **Iron — iron deficiency**
- **Folic acid — megaloblastosis**
- **Cobalamin** — megaloblastosis
- **Ascorbic acid** — scurvy

Infants with malnutrition develop **acanthocytes** of their red blood cells.

O

ONCOGENESIS

See also **Leukemogenesis**.

Malignant transformation of cells. It arises from a series of events inducing **DNA** damage, which ultimately results in a selective growth advantage for that cell and its clonal progeny.

ONCOSTATIN M

(OSM) An **interleukin** of the IL-6 group produced by activated leukocytes, of uncertain biological role, perhaps important in **hematopoiesis**.

ONCOVIN

See **Vinca alkaloids**.

OPHTHALMIC DISORDERS

The changes in the eyes occurring with hematological disorders.

Pallor of conjunctivae, the simplest clinical sign of **anemia**.

Jaundice of sclera may occur with all forms of **hemolytic anemia**.

Hemorrhage into conjunctivae or retinas, a valuable clinical sign of hemostatic disorder, usually **thrombocytopenia**.

Increased vascularity of conjunctivae in **polycythemia rubra vera**.

Uveitis, episcleritis, scleritis, keratitis, and retinal vasculitis with systemic **vasculitis**, **sarcoidosis**, and **Vogt-Koyanaygi-Harada's disease**.[401]

Small infiltrates of conjunctiva, sclera, and cornea with **non-Hodgkin lymphoma**, **amyloidosis, ankylosing spondylitis, Kaposi's sarcoma**, and **Wegener's granulomatosis**.

Sickle cell disorder changes:

- "Comma" sign in conjunctival vessels, which contain densely packed, sickled red cells
- Segmental flow in conjunctival vessels due to sludging of blood
- Widened veins and tortuosity of large retinal vessels, which may result in retinal hemorrhage
- Retinal hemorrhage

Proptosis caused by orbital tumors of any hematological malignancy, but particularly in children with retinal **myeloid sarcoma** (chloroma), ocular adnexial **marginal-zone B-cell lymphoma** (MALT lymphoma), or other type of **non-Hodgkin lymphoma**, the latter types associated with increased tortuosity and microaneurysms of the retinal vessels.

Retinal vasculitis with **systemic lupus erythematosus, giant-cell arteritis**, and Wegener's granulomatosis. Small infiltrates of non-Hodgkin lymphoma are associated with increased tortuosity and microaneurysms of the retinal vessels. **Cytomegalovirus** (CMV) retinitis is associated with **acquired immunodeficiency syndrome** (AIDS). In all of these conditions, retinal vein occlusion can occur.

OPSONINS

Substances that coat foreign particles, e.g., bacteria, and enhance clearance by binding to specific receptors on **neutrophil** or **monocyte** membranes, enabling efficient phagocytosis to occur. **Antibodies** IgG (G1 and G3 subclasses), C3b, fibronectin, and defensins can all act as opsonins. The binding of IgG to the Fc receptor also activates neutrophil-killing mechanisms. C3b is generated via the alternative complement pathway, but is less efficient at stimulating phagocytosis than IgG. The importance of opsonization is not only that pathogens will be taken up by phagocytes and so be destroyed, but also that **immune complexes** will be removed from the circulation, thereby avoiding pathological consequences.

ORAL MUCOSA DISORDERS

Changes in hematological disorders of the mouth, including the lips and tongue.

Color:
- Pallor due to **anemia**
- Hyperemia due to **polycythemia**
- Cyanosis due to **methemoglobinemia**
- Lead deposition of gums

Glossitis; smooth tongue; angular stomatitis with nutritional deficiency of **iron, folic acid, cobalamin**, or riboflavin

Aphthous ulceration and geographic tongue with **reactive arthropathy** (Reiter's syndrome)

"Hairy leukoplakia," particularly of the tongue with **Epstein-Barr virus** infection **(infectious mononucleosis)**

Oral mucositis and ulceration associated with **agranulocytosis**, neutropenia, and **immunodeficiency**, particularly associated with high-dose chemoradiotherapy[400]

Behçet's disease

Gingivitis with gum hyperplasia with **monocytic leukemia, cyclic neutropenia**, and agranulocytosis

Infiltrates of **granulomas**, histiocytomas, macroglobulin (IgA linear deposition), and **reactive lymphoid hyperplasia** (follicular lymphoid hyperplasia of the palate, and angiolymphoid hyperplasia with eosinophilia)

Petechiae, ecchymoses, and purpura due to **thrombocytopenia, platelet-function disorders**, or **vascular purpura**

Hemorrhage occurring with:
- Scurvy due to **ascorbic acid** deficiency, often associated with purulent infection
- Monocytic leukemia associated with gum hyperplasia
- **Von Willebrand Disease** and **hemophilia A and B**

- Hereditary hemorrhagic telangiectasia
- Peutz-Jeghers syndrome polyposes
- **Dental hemorrhage**, persistent following extraction or other surgery

Hemangiomas and lymphangiomas

Neoplasia: **Burkitt lymphoma**, **plasmacytomas** of the jaw and submucosa, and other **lymphoproliferative disorders**; these all may be associated with **acquired immunodeficiency syndrome**

OROYA FEVER

See **Bartonellosis**.

ORTHOCHROMATIC ERYTHROBLAST

(Late normoblast) See **Erythropoiesis**; **Nucleated red blood cell counting**.

ORTHOSTATIC PURPURA

Extravasation of blood that occurs on the legs after prolonged standing. It is due to increased transmural pressure in blood vessels. It is seen most frequently in the elderly, those with varicose veins and valvular incompetence, or chronic use of tight-fitting clothing. Chronically high orthostatic pressure, as in chronic venous stasis associated with varicose veins, is accompanied by recurrent episodes and may lead to the development of persistent purpuric macules or **hemosiderin**-laden yellow brown macules. There is no demonstrable defect in **hemostasis**. High local orthostatic pressure accounts for the increase in purpura seen in the lower limbs in patients with generalized purpura.

OSLER-RENDU-WEBER DISEASE

See **Hereditary hemorrhagic telangiectasia**.

OSMOTIC FRAGILITY TEST

The measurement of the ability of red blood cells to take up a certain amount of water before lysing.[89] The major determinant in this process is the surface-to-volume ratio of the cells. **Spherocytes** (regardless of cause) have an increased surface-to-volume ratio, and their ability to take in water is more limited than normal, i.e., osmotic fragility is increased. Very thin red cells have a decreased surface-to-volume ratio and can take up more water than normal before lysing. Their osmotic fragility is therefore decreased. This is characteristic of the red blood cells in **iron deficiency** and in **thalassemia**. In the enzymopathies, the osmotic fragility is usually normal.

Small volumes of blood are mixed with a large excess of buffered saline solutions of varying concentration. The fraction of red blood cells lysed at each saline concentration is determined colorimetrically. The test is normally carried out at room temperature (15 to 25°C). When blood is incubated at 37°C for 24 h, the red cells continue to metabolize glucose for energy to maintain their normal size and shape. If the cells are defective and have increased energy requirements, this period of incubation will stress the abnormal cells more than those of the control and thus exaggerate the response to testing. This procedure helps when borderline results are obtained with the standard method.

With either method, the position of the osmotic-fragility curve is compared with that of the control, and the mean corpuscular fragility (MCF) is recorded. The fragility curve can shift to the left or to the right. The normal range for MCF at pH 7.4 and 20°C is 4.0 to 4.45 g/l NaCl in fresh samples (4.65 to 5.9 g/l NaCl after incubation at 37°C for 24 h). An abnormal result is indicative of a red blood cell abnormality, but a normal result does not exclude this possibility.

OSTEOCLAST
See **Histiocytes**.

OSTEOGENESIS IMPERFECTA
A rare hereditary connective tissue disorder affecting 1 in every 30,000 of the population. In the autosomally dominant form, lysines of type I **collagen** are overhydroxylated. This results in defective bone matrix and brittle bones, resulting in skeletal deformity and fractures. Affected individuals characteristically have blue sclera. Some patients show easy **bruising** and **hemorrhag**e — epistaxes, hemoptyses, and occasionally intracranial bleeding. Treatment is supportive only.

OSTEOSCLEROSIS
New bone formation associated with fibrosis of the **bone marrow**. It is characteristically seen in **idiopathic myelofibrosis** (30 to 70% of patients), but also occurs in **polycythemia rubra vera**, secondary carcinoma (breast, prostate), Paget's disease, and in end-stage renal disease with secondary hyperparathyroidism.

OSTEOSCLEROTIC MYELOMA
(POEMS syndrome) A rare disorder characterized by polyneuropathy, organomegaly, endocrinopathy, monoclonal gammopathy, and skin hyperpigmentation.[402] Some cases are associated with herpes virus/human herpes virus 8. Secretion of RANKL (receptor activator of nuclear factor κ-ligand) by myeloma cells stimulates osteoclast formation, while overexpression of the *DKKI* gene that causes production of dickkopfl, a Wnt signalling antagonist (see **Cell-signal transduction**), inhibits osteoblasts.[403] Single or multiple osteosclerotic bone lesions are important features. Elevated levels of **interleukin**-6 and **tumor necrosis factor**-α are present. A major feature is sensorimotor polyneuropathy with predominating motor disability. Cranial nerves are not involved, except for papilledema. **Hepatosplenomegaly** and **lymphadenopathy** occur. Hyperpigmentation, hypertrichosis, gynecomastia and testicular atrophy are often present. **Polycythemia** or **thrombocytosis** may occur. Almost 90% of patients have a serum M-protein, and in most cases it is of the l light-chain class. The size of the M-protein is always less than 3 g/dl. **Bence-Jones proteinuria**, renal insufficiency, hypercalcemia, and skeletal fractures are rare. The **bone marrow** aspirate and biopsy usually contain less than 5% **plasma cells**. Radiation of the osteosclerotic lesion is frequently beneficial. **Autologous stem cell transplantation** is beneficial for patients with widespread involvement who cannot be treated with local irradiation.[404]

OVALOCYTE
See **Elliptocytes; Melanesian ovalocytosis**.

OXYGEN AFFINITY TO HEMOGLOBIN

Oxyhemoglobin is the compound where **hemoglobin** is reversibly bound to oxygen. The efficiency of oxygen transport from the lungs to the tissues depends upon the level of oxygen affinity of the hemoglobin.

Oxygen-Dissociation Curve

This is the arithmetic plot of the hemoglobin oxygen saturation (ordinate) against the partial pressure of oxygen (abscissa) (see Figure 65). The oxygen affinity is expressed as $P_{50}O_2$. This designates the oxygen tension at which hemoglobin is half-saturated and is taken from the midpoint of the oxygen-dissociation curve. This partial pressure of oxygen of 26 mm Hg in normal cells compares with that of room air at 100 mm Hg, pulmonary alveoli at 95 mm Hg, and systemic arterial blood at 90 mm Hg.

Delivery of oxygen is determined by pO_2 of the tissues. The steep portion of the oxygen-dissociation curve allows a relatively large amount of oxygen to be unloaded for a small decrement in pO_2. With increasing oxygen affinity, the oxygen-dissociation curve is "shifted to the left." High values for the pO_2 are characteristic of a lower hemoglobin affinity. A right shift in the dissociation facilitates oxygen delivery. The three primary determinants of the pO_2 are temperature, pH, and 2,3-diphosphoglycerate (2,3-DPG) concentration. The **Bohr effect** is an important buffer system of the body. When blood reaches the tissues where the oxygen tension is lower and the hydrogen ion concentration is increased by lactic acid and carbon dioxide, the Bohr shift of the oxygen-dissociation curve makes more oxygen available.

Oxygen-Affinity Disorders

These are hereditary hemoglobinopathies in which there is altered affinity of hemoglobin and oxygen. Some of these hemoglobins with altered affinity are unstable. Oxygen affinity may be either increased (e.g., Hb Koln) or decreased (e.g., Hb Hammersmith). Measurement of the oxygen-dissociation curve in these disorders demonstrates a decreased Hill's constant, which occurs only as a consequence of a structural change of hemoglobin. Electrophoretic techniques to detect hemoglobins with altered oxygen affinity are only helpful when the amino acid substitution changes the charge of the molecule.

High-Affinity Hemoglobinopathies

Increased hemoglobin-oxygen affinity occurs due to either a defect in globin structure or to an abnormality in 2,3-DPG production or binding. The first combination of high-affinity hemoglobin with secondary **erythrocytosis** was described in 1966 (Hb Chesapeake). Many other examples have been found, these being principally due to missense mutations that cause β-chain amino acid substitutions. A small number of α-chain substitutions have been described. These are all autosomally dominant heterozygotes, presumably with homozygosity being lethal. High-oxygen-affinity hemoglobins release oxygen to the tissues less easily than normal. This causes tissue hypoxia, relieved only by an increase in oxygen-carrying capacity achieved by secondary erythropoiesis.

Affected patients have recurrent headache, feelings of facial fullness, and plethora. The hemoglobin concentration, hematocrit, and red cell mass are elevated. Erythropoietin titers are slightly elevated. The $P_{50}O_2$ is decreased. In about 50% of cases, abnormal hemoglobin can be detected using starch gel electrophoresis. In the remainder, agar gel electrophoresis or isoelectric focusing is necessary to detect the abnormal hemoglobin (see **Hemoglobin electrophoresis**). Most patients lead normal lives, and therapeutic intervention with **phlebotomy** or other maneuvers is seldom required.

Low-Affinity Hemoglobinopathies

In these disorders, a low hemoglobin level can provide sufficient tissue oxygenation despite the presence of anemia. Examples are Hb M-Boston, Hb M-Iwate, and Hb Hammersmith. The oxygen-dissociation curve is shifted to the right; the output of erythropoietin is decreased; and mild **anemia** results. Diagnosis depends upon demonstration of an abnormal oxygen-dissociation curve or a normal 2,3-DPG level, or by estimating the oxygen-dissociation curve of hemoglobin that has been stripped of 2,3-DPG by extensive dialysis against bis-tris buffer.

OXYHEMOGLOBIN

See **Oxygen affinity to hemoglobin**.

P

PACKED-CELL VOLUME

(PCV; hematocrit; red cell volume fraction) The relative volume occupied by **red blood cells** in capillary or venous specimens of whole blood. This is a key measurement underpinning much of hematological diagnosis, the calibration of almost all hematology blood cell counters being traceable back to the PCV. Reference ranges for hematocrit and red cell indices depend on the validity of this calibration, as does the assignment of values to calibrators and controls. It is measured by subjecting the blood to sufficient centrifugal force to pack the cells into as small a volume as possible. It is expressed as a decimal fraction, e.g., 0.45 l/l. **Automated blood cell counting** calculates the PCV by multiplying the measurement of the red cell count by the mean cell volume (MCV). Inversely, PCV, when measured, is used in calculations of MCV and mean cell hemoglobin concentration (MCHC) (see **Red blood cell indices**).

Working methods for measuring the PCV are:

Micromethod using disposable glass capillary tubes 7.5 mm long with a uniform bore of 1 mm and centrifugation at 10,000 to 15,000 g for 5 min.

Macromethod using a Wintrobe hematocrit tube of uniform bore of 2.5 to 3.0 mm graduated for 100 mm from its base and centrifugation at 2000 to 2300 g for 30 min.

In both methods, the length of the red cell column after centrifugation relative to the total length of the column gives the PCV reading. The manually spun PCV is affected by several variables, including trapped plasma, white cell and platelet contamination of the red cell layer, indistinct margin between red and white cell layers, nonflat tube seals, red cell dehydration, and the oxygenation state of the red cells. Fortunately, these biases counterbalance each other so the net error in the PCV is small, <1 PCV unit. Even this small error can have wide-reaching consequences when applied globally, although it has little impact on the individual patient. For example, an error of 1 PCV unit can lead to the rejection of up to 3.5% potential blood donors. In this context, PCV errors of this magnitude are unacceptable. There is a clear need for an unbiased reference method for the PCV to validate working methods. Several methods have been proposed, but they are complex, time consuming, and may require radiolabeled reagents that are difficult to obtain and may require special licensing before use. The ICSH (International Council for Standardization in Hematology) Expert Panel on Cytometry recommends a hemoglobin/MCHC-based reference method that eliminates all six errors listed previously.[405]

Reference ranges in health for those living at sea level are:

Adult males	0.47 ± 0.07 l/l
Adult females	0.42 ± 0.05 l/l
Neonates (term, cord)	0.54 ± 0.10 l/l
Children 3 to 6 years	0.40 ± 0.04 l/l
Children 10 to 12 years	0.41 ± 0.04 l/l

There is considerable variation in values between venous and capillary blood and particularly in the spleen, where it can be as high as 0.8. The PCV is higher in infancy and childhood (see **Reference Range Tables**) and declines with age. It is raised at altitudes above 2500 m and can fluctuate with exercise and with changes in posture. Fully oxygenated blood has a value about 2% higher than deoxygenated blood.

Errors in estimation arise from:

Hemoconcentration due to prolonged application of a tourniquet during venepuncture and specimen collection

Excessive EDTA anticoagulation (short draw)

Inadequate mixing of the sample prior to measurement

Incorrect centrifugation time or speed

PANCREAS DISORDERS

The hematological disorders associated with disorders of the pancreas:

Venous thromboembolic disease with acute pancreatitis

Hereditary hemochromatosis affecting the liver, pancreas, and lymph nodes

Folic acid or **cobalamin** deficiency from malabsorption with pancreatic insufficiency due to chronic pancreatitis or carcinoma of the pancreas

Zollinger-Ellison syndrome with multiple gastrinomas of the pancreas, causing inability to release cobalamin from R binders, resulting in cobalamin malabsorption

Arterial thrombosis associated with **diabetes mellitus**

PANCYTOPENIA

Reduction in the concentration of all formed elements in the peripheral blood, i.e., the simultaneous existence of **anemia**, **leukopenia**, and **thrombocytopenia**. It is found in association with:

Bone marrow replacement (**acute leukemia, Hodgkin disease, non-Hodgkin lymphoma, histiocytic sarcoma, infiltrative myelophthiasis** (metastatic malignancy), **idiopathic myelofibrosis, myelomatosis, hairy cell leukemia**)

Aplastic anemia (hereditary and acquired forms, including **chemical toxic disorders** and **ionizing radiation**)

Bone marrow hypoplasia, particularly from **cytotoxic agent** therapy

Dyserythropoiesis, including **megaloblastosis**

Infection with **human immunodeficiency virus, parvovirus, ehrlichiosis,** brucellosis, military **tuberculosis, cytomegalovirus,** HHV-6, or visceral **leishmaniasis**

Splenomegaly

Systemic lupus erythematosus

Autoimmune lymphoproliferative syndrome

Adverse drug reactions, particularly phenacetin, para-amino salicylic acid, sulfonamides, rifampicin, and quinine

PAPPENHEIMER BODIES

Basophilic **red blood cell** inclusions, often in small clusters near the periphery of the cell. They are composed of **ferritin** aggregates, or of **mitochondria** or phagosomes containing aggregated ferritin. They often occur in **reticulocytes**. The associated disorders include:

Splenectomy (post)

Sideroblastic anemia

Lead toxicity

PARAPROTEINEMIA

See **Monoclonal gammopathies**.

PARIETAL CELL ANTIBODIES

See **Gastric disorders**.

PAROXYSMAL COLD HEMOGLOBINURIA

(PCH) A very rare form of **cold autoimmune hemolytic anemia** characterized by acute episodes of massive hemolysis following cold exposure. The disease was frequently diagnosed during the latter half of the 19th century because of its supposed association with congenital or tertiary syphilis. Now PCH occurs as an acute febrile illness associated with viral syndromes, particularly the childhood exanthems. There is usually one self-limited attack of acute **intravascular hemolysis** with **hemoglobinuria**. The prognosis is good. A chronic form of the disorder is characterized by recurrent episodes of hemolysis precipitated by exposure to cold temperature.

The cause of autoantibody production in PCH is unknown. There are no known racial or genetic predispositions. During severe chilling, blood flowing through skin capillaries is exposed to low temperatures. The antibody (see **Donath-Landsteiner test**) is biphasic, and early-acting complement components bind to red blood cells at lowered temperatures. Upon return of the cells to 37°C in the central circulation, the cells are rapidly lysed by activation of the terminal complement sequence through C9. The Donath-Landsteiner antibody dissociates from the red blood cells at body temperature.

Constitutional symptoms are prominent during the paroxysm. After cold exposure, the patient develops aching pains in the back or legs and abdominal cramps. Chills and fever follow. Urine passed after onset of symptoms typically shows hemoglobinuria, which, with the general symptoms, lasts a few hours.

The **hemoglobin** level can drop rapidly in a severe attack. Chronic anemia, raised **reticulocyte count**, hemoglobinemia, and hyperbilirubinemia may be present, depending on the frequency and severity of attacks. **Complement** titers are depressed during an acute episode because of consumption in the hemolytic reaction. **Spherocytes** and **erythrophagocytosis** by monocytes and neutrophils are typically found on the blood film during an attack. **Leukopenia** is seen early in the attack, followed by **neutrophilia**. The urine may be dark red or brown due to the presence of hemoglobin or methemoglobin.

The **direct antiglobulin (Coombs) test** is positive during and following a paroxysm, but negative between attacks. The positive reaction is due to coating of surviving red cells with complement. The Donath-Landsteiner antibody is a nonagglutinating IgG that binds in the cold and readily dissociates at room temperature and above. The antibody is detected *in vitro* by the biphasic Donath-Landsteiner test. In this test, the patient's fresh serum is incubated initially with red cells at 4°C and the mixture warmed to 37°C. Intense hemolysis occurs. Antibody titers rarely exceed 1:16. The Donath-Landsteiner antibody usually has specificity for the **P blood group** antigen. This is a unique IgG complement-activating antibody that hemolyzes P1 and P2 red cells but not p or Pk cells. The anti-P autoantibody usually binds at temperatures below 20°C. If the cells are exposed to anti-P at low temperatures in the presence of complement, C1q attaches to the membrane, and if the suspension is warmed to 25°C or higher, the complement activity concludes with hemolysis.

PCH must be distinguished from chronic **cold-agglutinin disease**, which manifests episodic hemolysis and hemoglobinuria, a distinction made primarily in the laboratory. Patients with PCH lack high titers of **cold agglutinins**. Other disorders with similar clinical presentation are distinguished by history and by appropriate serologic studies.

Acute attacks can be prevented by avoidance of cold. Further treatment is rarely necessary. If for some reason **red blood cell transfusion** is required, P-antigen-positive cells would have to be used due to the rarity of P-antigen-negative donors. Blood should be transfused using an in-line blood warmer at 37°C and keeping the patient warm. Most patients with chronic idiopathic PCH survive for many years despite occasional paroxysms of hemolysis. **Splenectomy** is not likely to be of help, but **plasmapheresis** may temporarily reduce hemolysis.

PAROXYSMAL NOCTURNAL HEMOGLOBINURIA

(PNH) An acquired intrinsic **red blood cell** defect due to hematopoietic stem cell deficiency of glycosylphosphidylinositol (GPI). It is postulated that an environmental toxin induces a mutation of *PIG-A* gene of hematopoietic stem cells. The abnormal clone expands within the bone marrow, probably due to decreased NK-cell activity compared with normal cells, thus replacing the hematopoietic stem cell pool and giving rise to **aplastic anemia**, dyserythropoiesis (**myelodysplasia**), or to PNH, where red blood cells, and to a lesser extent granulocytes and platelets, have a deficiency of surface proteins, leading to **complement-mediated lysis.**[406,407]

Pathophysiology

The characteristic defect of increased sensitivity of red blood cells to complement-mediated lysis, either by the classical or by the alternative pathways. This can be precipitated by various factors:

Lowering of pH (**acidified serum test**)

Reduction in ionic strength (**sucrose lysis test**)

Coating of red blood cells with antibody such as anti-A

Increase in magnesium concentration

Treatment with cobra venom

Previously, it was thought that lowering of blood pH during sleep explained nocturnal hemoglobinuria, but this is disputed.

Several red blood cell membrane proteins attached to GPI are deficient in PNH, e.g., leukocyte alkaline phosphatase, acetyl cholinesterase, urokinase plasminogen activator, and several other proteins that regulate complement function. The hereditary absence of CD59 antigen in these patients is critical in producing clinical PNH with a significant degree of hemolysis. The conversion of N-acetylglucosamine and glucosamine-phospho-inositol to monolipids is defective, so that GPI cannot be manufactured normally.

Absence of "decay accelerating factor" (DAF/CD55) — which acts to accelerate the destruction of erythrocyte-bound C3 convertase, leading to its action on surface-bound C3d being amplified — was thought by some to be the basic pathologic lesion. Granulocytes and platelets, and possibly lymphocytes, also show increased complement-mediated lysis.

Occasionally the abnormal clone disappears completely, but transition to an acute leukemia, though rare, has been well documented.

Clinical Features

Although the classical passage of dark brown urine first thing in the morning occurs in only a minority of patients, all show clinical or laboratory signs of a chronic **intravascular hemolysis**. This can be initiated by:

Episodes of infection

Strenuous exercise

Surgery

Injection of radiological contrast dyes

Transfusion of whole blood

Free plasma hemoglobin absorbs nitric oxide that is essential for smooth-muscle function. Iron lost in the urine as hemosiderin or hemoglobin results in **iron deficiency**, which may be exaggerated by gastrointestinal tract bleeding associated with severe **thrombocytopenia** from bone marrow aplasia or dyserythropoiesis. Treatment with oral iron has been known to exacerbate the hemolysis, mainly because the iron raises erythrocyte output, including PNH cells. This added hemolysis is not clinically significant.

Symptoms during a paroxysm are caused by disturbances of smooth-muscle function. They include abdominal pain of a colicky nature, and there may be abdominal tenderness, dysphagia, erectile failure, and severe lethargy. Renal tract manifestations include hypo-posthenuria, abnormal tubular function, and impaired creatinine clearance, though significant renal impairment is very rare. Severe headaches and pain in the eyes are common. The most serious complication is **venous thromboembolic disease** due to platelet GPI deficiency and activation by complement. There is a predilection for intra-abdominal veins. Hepatic vein thrombosis (Budd-Chiari syndrome) carries a particularly poor prognosis. Microthrombi in the pulmonary vasculature results occasionally in pulmonary hypertension. Thrombosis of major cerebral vessels is rare, but in some the headaches are due to thrombosis of small vessels in the cerebral cortex. Pregnancy with PNH carries a high risk of severe thrombotic complications. Anticoagulation throughout pregnancy may be required.

The clinical course is variable. Median survival is 10 years, the common causes of morbidity and mortality being thrombosis and bone marrow failure.

Laboratory Features

Anemia of variable severity occurs in most, but not all, patients, and is usually associated with a mild-to-moderately raised **reticulocyte count**. This may be associated either with a **macrocytic anemia** or with **myelodysplasia** (MDS), but if there is severe **iron deficiency**, there will be a **microcytic anemia**.

Neutropenia and **thrombocytopenia** are usually present, but seldom to a severe degree. The **bone marrow** usually shows a variable degree of erythroid hyperplasia, but it may be hypoplastic or show dyserythropoietic changes. In iron-deficient patients, stainable iron is not present.

Hemoglobinuria occurs in a minority of patients, but **hemosiderinuria** is usually demonstrable.

Screening tests are the **acidified serum test** and the **sucrose lysis tests**. Confirmation of diagnosis is demonstration of a deficiency of GPI-linked molecules on the surface of hematopoietic cells by **flow cytometry** demonstration of **immunophenotypes** $CD14^+$, $CD16^+$, $CD24^+$, $CD55^+$, and $CD59^+$.

Management

Supportive care with **red blood cell transfusion** is often needed. Theoretically, washed red blood cells should be used to avoid transfusing complement, but unwashed packed red blood cells are probably equally satisfactory.

Iron deficiency should be corrected by attention to the source of red blood cell loss as well as giving oral iron supplements.

Lifelong prophylactic oral anticoagulation with **warfarin** is needed for those patients experiencing recurrent venous thromboses. Treatment of the Budd-Chiari syndrome is by **thrombolytic therapy** followed by oral anticoagulants. Because of the high risk of thrombosis, where possible, elective surgery should be avoided.

Androgens and glucocorticoids have been used with mixed results and are probably only useful as short-term measures in patients with aplastic anemia, where they occasionally result in improvement of peripheral-blood counts. **Erythropoietin** may benefit a minority of patients. Trials of complement inhibitor **eculzimab** are in progress; preliminary results show excellent control of hemolysis. **Allogeneic stem cell transplantation** is potentially curative in younger patients, though overall results have been rather disappointing compared with those obtained for uncomplicated aplastic anemia.

PARVOVIRUS

Direct infection of erythroid precursors by a small, nonenveloped single-stranded DNA virus consisting of 5500 nucleotides. A particular strain, *parvovirus B19*, is pathogenic to humans and causes a variety of different diseases (see Table 125). Their lack of envelope and small DNA content render them extremely stable to heat and lipid solvents, and hence they are readily transmitted by blood transfusion and coagulation factor concentrates (see **Transfusion transmitted infection**). Immunoglobulin (Ig) G and IgM can be detected by enzyme-linked immunosorbent assay and immunofluorescence.

TABLE 125

Disorders Induced by Parvovirus B19 Infection

Fifth disease [a]
Polyarthropathy
Pure red cell aplasia
Hydrops fetalis
Transient erythroblastopenia of childhood

[a] Also called erythema infectiosum or "slapped cheek" disease.

About 50% of the population has serological evidence of past infection. Immunocompetent patients develop acute self-limiting illnesses, or even asymptomatic seroconversion, and no specific therapy is required. Those who have a chronic hemolytic anemia, e.g., hereditary spherocytosis and sickle cell disease, may suffer from infections that cause erythroid **aplastic crisis**. Most patients have a spontaneous remission, but some have a persistent **pure red cell aplasia**. Characteristically, giant pronormoblasts are found in the bone marrow of acutely infected patients, with confirmation provided by detection of specific IgM or IgG in the serum and by a polymerase chain reaction (PCR) in patients unable to mount an immune response.

PATHOGEN-RECOGNITION MOLECULES

(PRM) Receptors on the surface of cells, particularly **histiocytes** (macrophages), that recognize pathogens. These can enter the cell and either become resident in the cytoplasm or are destroyed.

Toll-like receptors (TLRs), upon activation, induce intercellular signaling pathways that activate microcidal responses. Other PRMs are nucleotide-binding oligomerization domain proteins (NODs). Genetic mutations affecting NODs are associated with Blau's syndrome (arthritis, skin rashes, and uveitis), Crohn's disease (see **Intestinal tract disorders** — chronic inflammatory disease), and **sarcoidosis**.

P BLOOD GROUPS

A specific antigen–antibody system arising from the P and globoside blood group systems and the globoside collection of blood group antigens; located on **red blood cells** (RBCs), **lymphocytes**, and **monocytes** (see **Blood groups**).

Biochemistry

Like the **ABO(H)**, **Lewis**, and **I blood group** antigens, the P blood group antigens are carbohydrates. Production of the Pk, P, and P1 antigens is through different biosynthetic pathways: the P1 antigen has been assigned to the P blood group system, the P antigen to the globoside system, and Pk and Luke (LKE) antigens to a separate "collection." Pk and P antigens are also detected on erythroblasts, fibroblasts, and vascular endothelium cells.

Genetics and Phenotypes

Biosynthetic pathways are complex and incompletely understood. The phenotypes are summarized in Table 126. As with ABH antigens, genes code for glycosyl transferases that catalyze the transfer of monosaccharides onto carbohydrate precursors.

TABLE 126

Phenotypes of the P and Associated Blood Group Systems

RBC Phenotype	Frequency in Caucasians
P1 + (P1)	75%
P1 – (P2)	25%
P1k	very rare
P2k	very rare
p	very rare

Antibodies and Their Clinical Significance

Anti-P1 is sometimes found in the plasma of P1 (P2) persons. The antibody is almost always naturally occurring and is of the IgM class. Anti-P1 is sometimes erroneously called anti-P (see below).

The apparent incidence of anti-P1 is dependent upon the pretransfusion testing methods in use, as most examples show activity only at temperatures below 37°C. At temperatures below 20°C, anti-P1 can be demonstrated in a high proportion of patients' sera.

Anti-P1 only very rarely causes hemolytic **blood transfusion complications** if incompatible RBCs are transfused. It can be ignored for the purposes of transfusion if it is inactive at 37°C in direct agglutination or antiglobulin methods. Provision of blood for patients with anti-P1 should not be difficult.

Anti-P1 does not cause **hemolytic disease of the newborn** (HDN) for the same reasons that ABO HDN is rare (see **ABO (H) blood groups**).

Anti-PP1Pk (anti-Tja) is regularly found in the plasma of individuals with the very rare phenotype p, and anti-P is regularly found in P1k and P2k individuals. These antibodies can be either IgM or IgG.

Anti-Tja and anti-P are usually complement binding and hemolytic *in vitro*. They can give rise to severe blood transfusion complications involving **intravascular hemolysis** if incompatible RBCs are transfused. It is therefore essential to select allogeneic blood of the correct phenotype, but the rarity of these phenotypes is such that this will not always be possible. **Autologous blood transfusion** will be the method of choice whenever possible.

There appears to be a significantly increased risk of spontaneous abortion in women who have the p phenotype, in particular if anti-Tja of the IgG3 subclass is present.

Patients with **paroxysmal nocturnal hemoglobinuria** usually have auto anti-P that acts as a biphasic hemolysin.

PEARSON-MARROW-PANCREAS SYNDROME
See **Sideroblastic anemia**.

PEL-EBSTEIN FEVER
See **Hodgkin disease**.

PELGER-HUËT ANOMALY

A nuclear hypolobulation of granulocytes (best seen in **neutrophils**). The condition can be inherited or acquired. The inheritance is autosomally dominant (incidence 1:1000 to 1:10,000), and typically bilobed neutrophils are seen, with occasional mononuclear forms (Figure 95). In the rare homozygous patient, all neutrophils are mononuclear. Pelger-Huët cells appear to be functionally normal, and increased infections are not seen.

The acquired form (often called pseudo-Pelger-Huët cells) classically occurs in **myelodysplasia**, but other conditions, e.g., myxedema, can also give rise to this appearance. The Pelger-Huët cells of myelodysplasia are often hypogranular and functionally defective.

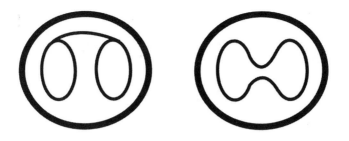

FIGURE 95
Diagrammatic representation of nuclear shapes in Pelger-Huët anomaly.

PELIOSIS

The occurrence of cystic spaces in the **spleen** and **liver** filled with blood. They are usually an incidental finding in the parafollicular areas of lymphoid follicles, particularly in the marginal zones. They can cause **hepatosplenomegaly** and are associated with **thrombocytopenia.**

PENTOSE PHOSPHATE PATHWAY

See **Red blood cell** — metabolism.

PERFORINS

Monomeric proteins present in the granules of NK **lymphocytes** and **cytotoxic T-lymphocytes**. They are probably produced in the **spleen**. They are activated by the cascade of serine proteases, one of which is immunologically homologous with the ninth component of complement, to form pores in the target cell membrane.

Deficiency of perforins is an autosomally recessive disorder of dysregulated immune response: familial **hemophagocytic lymphohistiocytosis**.[262a] The patients have a febrile illness with hepatosplenomegaly, **pancytopenia**, hyperglyceridemia, CSF pleocytosis (50%), and widespread neurological abnormalities. The bone marrow is hypoplastic or aplastic, with **hemophagocytosis** evident.[408]

PERIPHERAL-BLOOD-FILM EXAMINATION

Microscopic examination of a stained blood film. This provides information for:

Diagnosis of blood disorders

Checking the blood count parameters obtained from automated instruments

Blood films are routinely prepared from EDTA specimens up to 3 h following venepuncture. Alternatively, they can be prepared using freshly collected capillary or nonanticoagulated venous blood. Use of nonanticoagulated blood results in platelet clumping. The traditional wedge smear remains the most popular. It should be 2.5 to 3.0 cm long and 2.0 to 2.5 cm wide and have a smooth surface free from holes and ridges with a short feather edge. The film is stained by a Romanowsky method, the principal components of which are Azure B and Eosin Y.

Low-power examination is undertaken first to ensure satisfactory distribution of cells. The feather edge should then be examined to exclude the presence of platelet clumping or leukocyte clumping, which would render platelet and differential leukocyte counting unreliable. A readable area of the film is where red cells are evenly distributed, just touching but without appreciable overlap. Systematic examination of red cells, leukocytes, and platelets is then undertaken,[409] with particular attention to the following characteristics:

Red blood cells

1. Depth of staining (hemoglobin content)
    * Normochromia, **hyperchromia**
    * **Anisochromia, polychromasia**
    * Hypochromia
2. Size of cells
    * **Normocyte** (discocyte), **macrocyte**
    * **Microcyte**, dimorphism
    * Anisocytosis
3. **Poikilocytosis**
    Regular
       * **Codacyte** (target cell)
       * **Dacrocyte** (teardrop cell)
       * **Drepanocyte** (sickle cell)
       * **Elliptocyte** (ovalocyte)
       * **Leptocyt**e (thin cell)
       * Microspherocyte
       * Spherocyte
       * Pincer cell
       * Stomatocyte
    Irregular
       * Crenated cells (artifacts)
       * **Pyknocyte** (irregular contracted cell, bite cell)
       * Spiculated (Burr) cells: **acanthocyte, echinocyte**
       * Fragmented cells: **keratocyte** (horn cell), **schistocyte** (helmet cell)
4. Inclusions
    * **Nucleated red blood cell** (orthochromatic normoblasts)
    * Howell-Jolly bodies
    * **Basophilic stippling** (punctate basophilia)
    * Heinz bodies
    * Pappenheimer bodies
    * **Malaria** *Plasmodium* spp.
    * **Bartonella** parasites
    * **Babesia** parasites

Granulocytes (**neutrophils, eosinophils**, and **basophils**)

1. Nucleus
   - Segmentation — hypersegmented and band-form **neutrophils**
   - Pelger-Huët anomaly
   - Pseudo-Pelger cells
   - May-Hegglin anomaly
   - Pyknocytosis
2. Granules
   - Toxic granulation
   - Alder-Reilly anomaly
3. Vacuoles
4. Inclusions — bacteria, **ehrlichosis** morulae
5. **Dohle bodies**
6. Immature granuloctyes (**myeloblasts, myelocytes, metamyelocytes**)

Mononuclear cells (**monocytes** and **lymphocytes**)

1. "Reactive" lymphocytes — **plasma cells, Mott cells**
2. **Atypical mononuclear cells**
3. **Immunoblasts** (Turk cells)
4. **Lymphoblasts, monoblasts**, lymphoma cells

Platelets

1. **Giant platelets — Bernard-Soulier syndrome**
2. Granular depletion — **Gray platelet syndrome**
3. Platelet clumping
4. **Platelet satellitism**

PERIPHERAL-BLOOD STEM-CELL TRANSPLANTATION

(PBSC transplantation) See **Autologous bone marrow transplantation**.

PERIPHERAL T-CELL LYMPHOMA, UNSPECIFIED

(Rapaport: diffuse poorly differentiated lymphoma, diffuse mixed lymphocytic-histiocytic lymphoma; Lennert: lymphoepithelioid cell lymphoma; Lukes-Collins: T-immunoblastic lymphoma) See also **Lymphoproliferative disorders**; **Non-Hodgkin lymphoma**.

A predominantly nodal heterogeneous group of aggressive T-cell lymphomas. They are characterized by having widespread disordered lymphoid follicles showing a diffuse or occasionally interfollicular cellular proliferation. This ranges from atypical small cells to medium-sized or large cells; most contain a mixed population of small and large atypical cells, and even those with a predominance of medium-sized or large cells often contain a broad spectrum of cell sizes. The neoplastic cells often have irregular nuclei and vary considerably in size and shape, with occasional large, hyperchromatic cells that may resemble **Reed-Sternberg** (RS) cells, but true RS cells are rare or absent. Admixed eosinophils or epithelioid histiocytes may be numerous. There is a T-zone variant and a lymphoepithelioid cell variant. The **immunophenotype** of the tumor cells shows variation in

T-cell-associated antigens (CD3$^{+/-}$, CD2$^{+/-}$, CD5$^{+/-}$, CD7$^{-/+}$) and B-cell-associated antigens. Most nodal cases are CD4$^+$, CD8$^-$, and CD30$^+$ in the large-cell variants. Cytogenetic analysis usually shows rearrangement of *TCR* genes, with Ig genes being germline. The cellular origin is a peripheral T-cell in various stages of transformation.

Clinical Features

These comparatively uncommon tumors comprise less than 15% of lymphomas in Europe and the U.S., but are more common in other parts of the world. Patients are usually adults presenting with nodal involvement but also with hepatosplenomegaly, bone marrow infiltration, and other visceral disturbances. Occasionally, **eosinophilia** or hematophagocytic syndromes are present. The clinical course is usually aggressive and the prognosis poor, with overall survival rates of 30% at 5 years, even with treatment, as relapses are more common than in B-cell lymphomas of similar histologic grades.

Staging
See **Lymphoproliferative disorders**.

Treatment
See also **Non-Hodgkin lymphoma**.
The standard treatment strategy has been CHOP chemotherapy (see **Cytotoxic agents**) at initial presentation and **peripheral stem cell transplantation** (PSCT) for relapsed patients.

PERNICIOUS ANEMIA
See **Addisonian pernicious anemia**.

PETECHIAE
Red or bluish lesions of less than 3 mm in diameter visible in the skin, deep to the epidermis. They fail to disappear upon application of pressure, as they occur due to extravasation of blood. In contrast, lesions with an intact vasculature, vascular lesions, will blanch upon pressure. Petechiae tend to occur in crops and resolve over 3 to 5 days. They are a common manifestation of purpura (see **Hemorrhagic disorders**).

PEUTZ-JEGHER'S SYNDROME
See **Oral mucosa disorders**.

PHAGOCYTOSIS
The internalization of particulate matter by cells into cytoplasmic vesicles. It is a form of **endocytosis** in which large particles (e.g., cell debris) are taken up and engulfed into a cell. The particle is progressively surrounded by cell pseudopodia in which **actin**-binding proteins accumulate. **Lysosomes** fuse with the particle, so that the resulting endocytic vesicle is converted into a **phagosome**.

Phagocytes are **neutrophils**, **monocytes**, and **histiocytes** (macrophages) derived from the same myeloid progenitor stem cell. Activation of phagocytes initiates "respiratory burst," which kills phagocytosed organisms.

PHAGOSOMES
See **Phagocytosis**.

PHARYNGEAL DISORDERS
The changes in the pharynx associated with hematological disorders:

Pharyngitis with sore throat, dysphagia, tonsillar swelling, and ulceration with **agran-ulocytosis**

Massive tonsillar swelling with membrane and exudate in **infectious mononucleosis**

Purpuric hemorrhages in **thrombocytopenia** and **platelet-function disorders**

Epithelial webs in the posterior wall with **iron deficiency** (Plummer-Vinson, Patterson-Kelly syndromes)

Postcricoid carcinoma, complicating epithelial webbing

Lymphoproliferative disorder of the tonsils

Nasopharyngeal carcinoma associated with **Epstein-Barr virus** (EBV) infection; adoptive immunotherapy to EBV-specific T-cells has a potential therapeutic effect

PHENOTYPE
The observable characteristics of a cell or organism resulting from the interaction between its genetic components and the environment. Most commonly used in hematology to define cell-surface antigenic profile, such as the use of **monoclonal antibodies** in **immunophenotyping** for the diagnosis of leukemias and antisera for the characterization of red blood cell antigens (see **Blood groups**).

PHI BODIES
Polyribosomal spindle-shaped hydrogen peroxide-positive parent organelles in the cytoplasm of blast cells from patients with **acute myeloid leukemia**. They are named after their resemblance to the Greek letter f (ϕ). They are also present in leukemic promyelocytes and in the blast cells of **myelodysplasia** (MDS).

PHILADELPHIA (Ph) CHROMOSOME
See **Cytogenetic analysis; Chronic myelogenous leukemia**.

PHLEBOTOMY
The removal of blood for therapy of **erythrocytosis** or **iron overload** or for blood component donation (see **Blood donation; Autologous blood transfusion**). In polycythemia rubra vera and other forms of severe erythrocytosis, therapeutic phlebotomy is performed to reduce the circulating red blood cell mass, to alleviate hyperviscosity, and to induce iron depletion. The removal of one unit of blood (440 to 500 ml) once weekly is usually adequate. In most patients with hemochromatosis or other forms of iron overload who do not have severe anemia, a similar phlebotomy schedule is used. Persons with severe **hyperviscosity** or iron overload sometimes tolerate and benefit from phlebotomy twice

weekly. In women, the elderly, or persons of small body mass, the removal of one-half unit of blood per session is sometimes better tolerated than larger volumes. Patients with hyperviscosity or symptoms of hypovolemia after phlebotomy may benefit from the infusion of crystalloid solutions of the approximate volume of the blood removed. One unit of blood in persons with a normal **hemoglobin** concentration contains approximately 200 mg **iron**, permitting estimation of the quantity of iron removed over a series of phlebotomy treatments ("quantitative phlebotomy").

PHOSPHODIESTERASE INHIBITORS

A group of substances that potentiate drugs acting on **platelets** and that are dependent on cyclic nucleotide generation. They act by reducing the breakdown of cyclic nucleotides, effectively increasing the concentration of cAMP, which favors the movement of Ca^{2+} into the dense bodies, where it is metabolically inert. **Dipyrimadole** inhibits **adenosine** reuptake as well as inhibition of cAMP and cGMP phosphodiesterase, leading to maintained levels of intraplatelet cAMP. This potentiates the action of **prostacyclin** (PGI2). Theophylline, a specific cAMP phosphodiesterase inhibitor, also has antiplatelet effects. Dipyrimadole is also a potent vasodilator and may lead to faintness/headaches. Its actions are synergistic with those of **aspirin**.

PHOSPHOFRUCTOKINASE

See **Phosphohexose kinase**.

6-PHOSPHOGLUCONATE DEHYDROGENASE

An enzyme of the hexose monophosphate shunt of **red blood cell** metabolism, deficiency of which is rare and only gives rise to hemolysis if the patient is taking primaquine.

PHOSPHOGLUCONATE PATHWAY

(Pentose phosphate pathway) See **Red blood cell** — metabolism.

PHOSPHOGLYCERATE KINASE

An enzyme of the Embden-Meyerhof pathway of **red blood cell** metabolism responsible for the conversion of 1,3-diphosphoglycerate (1,3-DPG) to 3-phosphoglycerate, thus generating **adenosine triphosphate** (ATP) from **adenosine diphosphate** (ADP). It is the only enzyme of the anaerobic glycolytic pathway known to be encoded on the X chromosome. Deficiency of this enzyme causes the level of 2,3-diphosphoglycerate (2,3-DPG) to rise and that of ATP to fall, which reduces the Na^+K^+ pump of its preferred source of energy, causing Na^+ accumulation in the cell, with eventual lysis. It is a rare enzyme deficiency, but those cases studied have shown considerable polymorphism. Roughly one-half of affected males show **hemolytic anemia** of variable severity accompanied by neurological and mental abnormalities. Females show mosaicism and may have mild hemolysis. If severe hemolysis exists in phosphoglycerate kinase deficiency, the anemia is ameliorated by **splenectomy**. Neurologic complications, however, rather than hemolysis, tend to dominate the clinical features.

PHOSPHOGLYCEROMUTASE

An enzyme of the Embden-Meyerhof pathway of **red blood cell** metabolism, which reversibly catalyzes the reaction 3-phosphoglycerate to 2-phosphoglycerate. Deficiency of this enzyme has not been reported.

PHOSPHOHEXOSE ISOMERASE

An enzyme of the Embden-Meyerhof pathway of **red blood cell** metabolism that catalyzes the conversion of glucose 6-phosphate to fructose 6-phosphate. Deficiency causes a **hemolytic anemia** that responds to **splenectomy**. Progressive neurological degeneration may be associated.

PHOSPHOHEXOSE KINASE

(Phosphofructokinase) An enzyme of the Embden-Meyerhof pathway of **red blood cell** metabolism, deficiency of which results in mild **hemolytic anemia**.

PHOSPHOTRIOSE DEHYDROGENASE

See **Glyceraldehyde-3-phosphate dehydrogenase**.

PHYSICALLY INDUCED DISORDERS

The hematological disorders induced by physical agents:

Traumatic disorders causing **hemorrhage** and **disseminated intravascular coagulation**

Mechanical damage to red blood cells: **macrovascular hemolytic anemia**, cardiac hemolytic anemia, **march hemoglobinuria**

Burns hemolytic anemia

Ultraviolet light as a possible etiological factor of **lymphoproliferative disorders**

Ionizing radiation affecting bone marrow

Heat stroke causing **neutrophil** botyroid nuclei

PHYTOHEMAGGLUTININ

See **Mitogens**.

PICKWICKIAN SYNDROME

Hypoxemia with secondary **erythrocytosis** resulting from inadequate ventilatory stimulation by the respiratory center. Extreme obesity is usual, with **cyanosis**, somnolence, and hypercapnia. Pulmonary vasoconstriction and eventually irreversible pulmonary hypertension occurs, compounding hypoxemia. Hypoventilation particularly occurs during sleep (apnea), with oxygen desaturation. Loss of weight usually (but not always) results in a return to normal ventilation, and some patients may benefit from sublingual medroxy-progesterone.

PINCER RED BLOOD CELLS

Red blood cells having the shape of pincers, as seen in peripheral-blood smears. The mechanism leading to this morphological abnormality is a partial deficiency of band 3 protein in the red cell membrane. This is an autosomally dominant hereditary defect. Due to the instability of this band 3 protein, the morphology becomes more obvious in aging cells. Its clinical behavior, consequences, and treatment are similar to those of **hereditary spherocytosis**, and it is recognized as one of the red cell membrane-linked protein defects leading to this disorder.

PINOCYTOSIS

The uptake of soluble substances by **granulocytes**. The mechanism is analogous to **phago-cytosis**, in which particulate matter is taken up, the processes being virtually identical, i.e., projections from the cell membrane surround the substance and form a vesicle that is then internalized. The process is not visible by light microscopy. Pinocytosis occurs via specific membrane-bound receptors, e.g., light-density lipoprotein, **granulocyte colony stimulating factor** (G-CSF), **granulocyte/macrophage colony stimulating factor** (GM-CSF), etc., or may be receptor independent.

PITUITARY GLAND DISORDERS

The interrelationship of hematological and pituitary gland disorders. The **neuropeptide** axis provides direct communication. The hormones produced by the pituitary gland in this interaction include prolactin, macrophage-inhibiting factor, and substance P. These hormones antagonize **corticosteroids** and stimulate T and B-lymphocytes and **histiocytes** (macrophages). Any immune disorder may therefore induce a change in the hypothalamic-pituitary-adrenal axis. Specific pituitary disorders include:

Hypopituitarism as a consequence of:
- Hemorrhage or infarction (pituitary apoplexy)
- Infiltration by **sarcoidosis** or **histiocytosis**
- Lymphocytic hypophysitis in a general autoimmune disorder

Anterior pituitary adenoma variable effects:
- **Normocytic anemia** as a result of thyroid and androgen deficiency
- **Erythrocytosis** through the effect of growth hormone on renal release of eryth-ropoietic factor
- **Eosinopenia** and **lymphopenia**

PIVKAS

(Protein-induced vitamin K absence or antagonist) See **Vitamin K**.

PLASMA

The fluid in which the blood cells are suspended. It contains many proteins, some of which are enzymes, e.g., **coagulation factors** and erythropoietin as well as electrolytes and carbohydrates.

PLASMABLAST

Undifferentiated **plasma cells** that have a central large nucleus containing several prominent nucleoli and relatively scant cytoplasm. Some may be binucleate or multinucleate. They are a characteristic feature of plasma cell neoplasms (**myelomatosis**) and **plasmacytomas**, where an increase in plasmablasts may confer a worse prognosis.

PLASMA CELL

See also **Lymphocytes** — B-cells; **Immunoglobulins**.
An end-stage B-lymphocyte with an eccentric round nucleus with "clock-face" **chromatin** pattern. The cytoplasm is strongly basophilic apart from a perinuclear light-staining Golgi body. They are rarely seen in normal peripheral blood but are prominent in chronic inflammatory disorders. Large numbers occur in the peripheral blood and bone marrow in association with **plasma cell neoplasms**.

PLASMA CELL MYELOMA

See **Myelomatosis**.

PLASMA CELL NEOPLASMS

Immunosecretory disorders resulting from the expansion of a single clone of immunoglobulin-secreting, terminally differentiated end-stage **B-lymphocytes**. These monoclonal proliferations of either plasma cells or plasmacytoid lymphocytes are characterized by secretion of a single homogeneous immunoglobulin: M-protein (see **Monoclonal gammopathies**). The neoplasms are classified as:

Plasma cell myelomas (**myelomatosis**)

- Nonsecretory myeloma
- Indolent myeloma
- Smoldering myeloma
- Plasma cell leukemia

Plasmacytoma

- Solitary plasmacytoma of bone
- Extramedullary plasmacytoma

  Immunoglobulin deposition diseases

  Primary **amyloidosis**

  Monoclonal light and heavy chain deposition diseases

  Osteosclerotic myeloma (POEMS syndrome)

  Heavy-chain diseases (HCD)

  γ-HCD

  μ-HCD

  α-HCD

PLASMACYTIC LYMPHOCYTIC LYMPHOMA

See **Lymphoplasmacytic lymphoma/Waldenström macroglobulinemia, Non-Hodgkin lymphoma**.

PLASMACYTOID DENDRITIC CELL
See **Dendritic reticulum cells**.

PLASMACYTOMAS
Clonal proliferations of **plasma cells** that are cytologically and immunophenotypically identical to those of plasma cell myeloma (**myelomatosis,** multiple myeloma), but manifest a localized osseous or extraosseous growth pattern.[410]

Solitary Plasmacytoma of Bone
The diagnosis depends on histologic evidence of a plasma cell tumor. In addition, complete skeletal radiographs must show no other lesions; the bone marrow aspirate must contain no evidence of multiple myeloma; and immunoelectrophoresis or immunofixation of the serum and concentrated urine should show no M-protein. Exceptions to the last criterion occur, but therapy of the solitary lesion usually results in disappearance of the M-protein. Treatment consists of radiation in the range of 40 to 50 Gy. Overt multiple myeloma develops in approximately 55% of patients during 10 years of follow-up. New bone lesions or local recurrence develops in 10% of cases. There is no evidence that adjuvant chemotherapy influences the incidence of conversion to multiple myeloma.

Extramedullary Plasmacytoma
This is a plasma cell tumor that arises outside the bone marrow. It is located in the upper respiratory tract in approximately 80% of cases. However, solitary extramedullary plasmacytomas can occur in virtually any organ. They usually spread locally, but multiple myeloma may develop. There is a predominance of IgA M-proteins. The diagnosis is based on the finding of a plasma cell tumor in an extramedullary location and the absence of multiple myeloma upon bone marrow examination, radiography, and appropriate studies of serum and urine. Treatment consists of tumoricidal **radiotherapy**. Regional recurrence occurs in approximately one-fourth of patients, but development of typical multiple myeloma is uncommon.

PLASMA EXCHANGE
See **Hemapheresis**.

PLASMAPHERESIS
See **Hemapheresis**.

PLASMA THROMBOPLASTIN ANTECEDENT
See **Factor XI**.

PLASMA THROMBOPLASTIN COMPONENT
See **Factor IX**.

PLASMA VOLUME

See also **Blood volume**.
The quantity of liquid plasma in the circulation.

PLASMIDS

Circular **DNA** that occurs naturally in microorganisms and is used for joining DNA fragments in large quantities.

PLASMIN

See **Fibrinogen**.

PLASMINOGEN

See **Fibrinolysis**.

PLATELET

Nonnucleated cytoplasmic fragments derived from bone marrow **megakaryocytes**. They are smooth, biconvex disks, 1 to 4 μm in diameter, with a role in primary hemostasis. Adequate numbers are therefore required for normal hemostasis.

Once released from the marrow, platelets are sequestered in the spleen for 24 to 48 h. The spleen can contain up to 30% of the normal circulating mass of platelets. This proportion can be significantly increased in **splenomegaly**, leading to **thrombocytopenia**. It is not known if these represent newly formed cells. Significant platelet pools may also exist in the lungs.

The reference range for platelets in peripheral blood is 140 to $400 \times 10^9/l$. Platelet counting shows variations in individuals with exercise, stress, and menstrual cycle. Racial differences are observed, with lower reference ranges seen in some Mediterranean races (see **Mediterranean thrombocytopenia**).

Normal platelet life span is 8 to 14 days, dependent upon the method by which platelet survival time is measured. Platelets are removed from the circulation by the reticuloendothelial system on the basis of senescence rather than by random utilization. However, there is a small fixed component to platelet turnover due to random utilization of platelets that maintain vascular integrity.

Platelet Structure

The main structural features are shown diagrammatically in Figure 96. The cell membrane lipid bilayer is partially or completely penetrated by a range of glycoprotein molecules. These function as receptors for a range of different agonists, adhesive proteins, coagulation factors, and other platelets. Specific membrane glycoproteins have been characterized with their associated functions (see Table 127).

The most abundant glycoproteins on the platelet surface are **glycoprotein IIb/IIIa** (GpIIb/IIIa). These two glycoproteins form a heterodimer and carry receptors for **fibrinogen**, **Von Willebrand Factor** (VWF), and fibronectin (adhesive proteins). The GpIIb/IIIa complex is a member of the integrin family of **cellular adhesion molecules**. Glycoprotein Ib is also important, as this contains a receptor for Von Willebrand Factor and thrombin. This receptor plays an essential part in the platelet–vessel wall interaction. A less well-defined group of 7-transmembrane domain glycoproteins can be released in **adenosine**

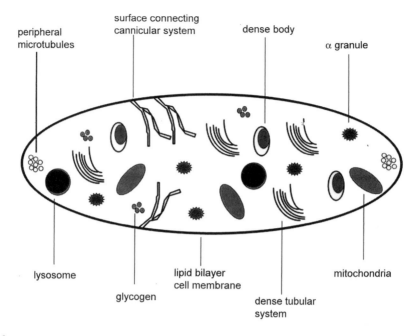

FIGURE 96
Diagrammatic representation of platelet ultrastructure.

TABLE 127

Important Platelet Membrane Glycoproteins

Glycoprotein	10^3 Copies per Platelet	Receptors
Ia	2–4	collagen
IIa	5–10	fibronectin, laminin
Ic	3–6	fibronectin, laminin
Ib/IX	25–30	Von Willebrand Factor, thrombin
IIb/IIIa	40–50	fibrinogen, Von Willebrand Factor, fibronectin, vitronectin
IV	—	collagen, thrombospondin
V	—	thrombin

deaminase (ADA) **adrenaline-** and **thrombin**-mediated aggregation. Deficiencies of platelet membrane glycoproteins lead to disorders of platelet function. In addition, membrane glycoproteins act as specific alloantigenic sites. The cell membrane also contains phospholipid, associated with **prostaglandin** synthesis, calcium mobilization, and localization of coagulant activity to the platelet surface.

Below the cell membrane lies the peripheral band of microtubules, which functions as the cell cytoskeleton. The microtubules maintain the discoid shape in the resting platelet but disassemble upon platelet aggregation and then reappear toward the center of the cell, entrapping granules. This is thought to help the release reaction.

The surface-connected canalicular system is an extensive system of plasma membrane invaginations. It increases the surface area across which membrane transport can occur and through which platelet granules can discharge their contents during the secretory phase of platelet aggregation. Dilated canaliculi are the probable explanation for vacuoles seen in normal platelets.

The dense tubular system is smooth **endoplasmic reticulum**, which is the site of prostaglandin synthesis and sequestration/release of calcium ions.

Platelets also contain many organelles, including **mitochondri**a, glycogen granules, **lysosomes**, peroxisomes, and two types of platelet-specific granules. Platelet-specific granules are either dense osmophilic granules (dense bodies, δ-granules) or α-granules. Dense bodies contain 60% of the platelet storage pool of adenine nucleotides (such as adenosine diphosphate) and serotonin. Dense-body adenine nucleotides do not readily exchange with other adenine nucleotides in the platelet metabolic pool. The α-granules contain a series of different proteins, some of which are platelet specific and some of which are found in the plasma or other cell types, such as coagulation factors. Major contents of α-granules include VWF, platelet factor 4, β-thromboglobulin, thrombospondin, **factor V**, fibrinogen, fibronectin, platelet-derived growth factor, high-molecular-weight kininogen, and tissue plasminogen activator inhibitor-1. Contents of the platelet-specific granules are secreted in response to aggregating stimuli.

Platelets have a number of specific antigens on their surface, many associated with platelet membrane glycoproteins Ia, Ib, Iib, IIIa, and possibly IV and V, such as HPA-1A associated with glycoprotein IIIa. These may be shared with other cells that have the adhesion receptors GP Ia and IIIa, which include vascular endothelium cells, smooth-muscle cells, fibroblasts, and activated T-cells. There are no naturally occurring antibodies, these only arising from reaction to transfused platelets or placental transfer.

Platelets also express **human leukocyte antigen** (HLA) class I antigens and **ABO blood group** antigens. These are of importance in immunological refractoriness to **platelet transfusions**.

Platelet Function

See also **Hemostasis**.

In the presence of vessel wall injury, escaping platelets come into contact with and adhere to **collagen** and subendothelial bound **Von Willebrand Factor** (VWF), through glycoprotein Ib. Glycoprotein IIb/IIIa is then exposed, via VWF binding, and forms a second binding site for VWF. In addition, with exposure of the GpIIb/IIIa site, fibrinogen may be bound to promote platelet aggregation. Within seconds of adhesion to the vessel wall, platelets begin to undergo a shape change, possibly due to ADP released from the damaged cells or other platelets exposed to the subendothelium. Platelets become more spherical and put out pseudopods, which enables platelet-platelet interaction. The peripheral microtubules become centrally apposed, forcing the granules toward the surface and the surface-connected canalicular system. Platelets then undergo a specific release reaction of their granules. The intensity of the release reaction is dependent upon the intensity of the stimulus. With the shape change, there is also further exposure of the GpIIb/IIIa complex, with further binding of fibrinogen. As fibrinogen is a dimer, it can form a direct bridge between platelets or act as a substrate for the lectinlike protein **thrombospondin**. With the enhancement of platelet-platelet interaction, platelet aggregation ensues. Platelet aggregation causes activation, secretion, and release from other platelets, thus leading to a self-sustaining cycle that results in the formation of a **platelet plug.**

The binding of agonist to platelet receptors not only leads to expression of fibrinogen receptors (glycoprotein IIb/IIIa), but also a series of **cell signal transduction** events that mediate the release reaction (see Figure 97).

Agonist receptor interaction leads to activation of guanine nucleotide-binding proteins (G protein) and hydrolysis of plasma membrane phospholipids (phosphotidyl inositides) by phospholipase C (PLC). The inositol triphosphates formed act as ionophores and mobilize calcium ions in the cytosol (cytosol calcium) from the dense tubular system, and lead also to an influx of calcium from outside of the cell. Diacylglycerol, also formed within the G protein/PLC pathway, activates protein kinase C, which in turn phosphorylates a

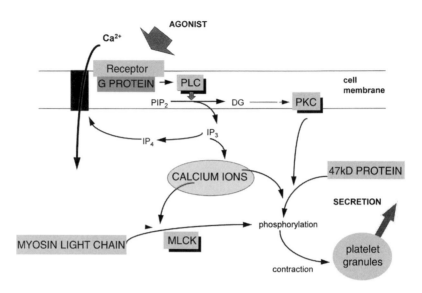

FIGURE 97

Cell signal transduction in the platelet-release reaction. PIP2, phosphotidylinositol; IP3 and IP4, inositol tri- and tetraphosphates; PLC, phospholipase C; PKC, phosphokinase C; DG, diacylglycerol; MLCK, myosin light-chain kinase.

47-kDa contractile protein. Together with the calcium-dependent phosphorylation of myosin light chain, these reactions induce contraction and secretion of granule contents. Cyclic AMP/adenyl cyclase exerts regulatory control over these reactions (high levels of cAMP reduce cytosol calcium concentration) and are in turn regulated by G protein activity. In addition, prostaglandin (cyclic endoperoxides and thromboxane A2) synthesized from membrane phospholipids may bind to specific receptors and further stimulate these processes.

Among the content of platelet α-granules are several coagulation factors, such as factor V, fibrinogen, and **high-molecular-weight kininogen**. Upon secretion from the α-granule, these factors reach high local concentrations. Platelets function to provide a local phospholipid surface for these factors to work upon, particularly factor V. This procoagulant activity of platelets is not seen in resting platelets. It is due to release of granule contents and redistribution of platelet membrane phospholipids upon secretion and activation.

Platelet function is studied by the use of platelet aggregating agents whose effects are measured by a platelet aggregometer.

Platelet Disorders

These are due to either platelet deficiency (**thrombocytopenia**) or **platelet-function disorders**. Their treatment depends entirely upon the cause.

Platelet Antagonists

Inhibition of platelet aggregation can be induced by the action of drugs. These antiplatelet drugs are widely used for prevention of thrombosis (see **Arterial thrombosis; Venous thromboembolic disease**). **Aspirin**, ticlopidine, and clopidogrel are effective for long-term prophylaxis.[82,86]

PLATELET-AGGREGATING AGENTS
See **Platelet-function testing**.

PLATELET AGGREGOMETERS
See **Platelet-function testing**.

PLATELET COUNTING
Estimation of the number of **platelets** circulating in a body fluid, usually the peripheral blood (see **Cytometry**). Platelet counting can be achieved by manual phase contrast microscopy, aperture impedance technology, optical (two-angle light scatter and fluorescence) technologies, and immunological flow cytometry. A reference method for platelet counting has been described by ICSH.[411]

Platelet counts are used for:

Screening for blood disorders

Transfusion management in **thrombocytopenia**

Monitoring **cytotoxic therapy** or recovery from **allogeneic** or **autologous stem cell transplantation**

The clinical value of platelet counts is at its greatest between 10 and $50 \times 10^9/l$; paradoxically, it is within this range where the accuracy of all methods is most doubtful. At low concentrations, different counts can be produced in the impedance and optical channels of the same automated instrument due to different interferences. These differences are algorithmically exploited in some instruments to produce a "correct" count.

The physiological count range is 150 to $400 \times 10^9/l$, showing minor diurnal variation and day-to-day variation. Females have counts about 20% higher than males, falling with menstruation. Around 2% of those living around the Mediterranean have counts less than $90 \times 10^9/l$ (see **Mediterranean thrombocytopenia**). There are no differences with age, apart from infants having levels at the lower end of the normal range. Strenuous exercise raises the level by 30 to 40%. The values for infants and children are given in **Reference Range Table X**.

PLATELET FACTOR 4
A cationic polypeptide synthesized by **megakaryocytes**. In platelets, it exists as polypeptide α-granules that inhibit collagenases of **neutrophils** and fibroblasts.

PLATELET-FUNCTION DISORDERS
The abnormalities of **platelets**, apart from thrombocytopenia, that can give rise to a platelet-induced **hemorrhagic disorder**. A prolonged **bleeding time** in a patient with a normal platelet count suggests the presence of a disorder of platelet function. Platelet-function disorders are either inherited or acquired. Acquired disorders are more common in practice. Bleeding in platelet-function disorders arises because of failure in one or more steps involved in platelet plug formation, such as aggregation, adhesion, secretion, or procoagulant activity. Platelet-function disorders can be broadly grouped as outlined in Table 128.

TABLE 128

Disorders of Platelet Function

Inherited Disorders

Disorders of platelet membranes
Deficiency of membrane glycoproteins
    Bernard-Soulier syndrome
    Glanzmann's thrombasthenia
    Rare membrane protein deficiency
Deficiency of platelet procoagulant activity
Disorders of platelet secretion
    Defects of secretory mechanism
        Weak agonist response defects
        Cyclooxygenase deficiency
        Thromboxane synthetase deficiency
    Deficiency of **platelet granules**
        Storage pool disorders
        α-Granule deficiency (**Gray-platelet syndrome**)
        δ-Granule deficiency
        Combined α- and δ-granule deficiency

Acquired Disorders

Hematological disorders
    Dysproteinemic purpura
    Leukemias
    Myelodysplasia
    Myeloproliferative disorders
Systemic disorders
    Autoimmune disorders
    Burns (severe)
    Cardiopulmonary bypass
    Disseminated intravascular coagulation
    Hemolytic uremic syndrome
    Hypoglycemia (chronic)
    Liver disorders
    Renal tract disorders
    Thrombotic thrombocytopenic purpura
    Transplant rejection
Drugs
    Alcohol
    Aspirin
    Cephalosporins
    Dextran
    Dipyrimadole
    Heparin
    Nonspecific anti-inflammatory drugs (NSAIDS)
    Penicillins
    Selective serotonin-uptake inhibitors
    Radiographic contrast agents
    Sulfinpyrazones
Prostaglandin overproduction

Inherited Disorders

Disorders of Platelet Membranes

These may be due to deficiency of membrane glycoproteins that are important for platelet aggregation. **Bernard-Soulier syndrome** is due to deficiency of glycoprotein Ib/IX complex on the platelet surface. Platelets deficient in glycoprotein Ib are unable to interact with subendothelial **Von Willebrand Factor** (VWF) and adhere via VWF to subendothelial matrix. **Glanzmann's thrombasthenia** is due to deficiency of **glycoprotein IIb/IIIa** on the platelet surface. Due to the deficiency of GpIIb/IIIa complex, Glanzmann's platelets contain an insufficient number of binding sites for fibrinogen and for VWF to provide adequate support for platelet aggregation. Isolated cases of other platelet membrane glycoprotein deficiencies are reported.

Deficiency of Platelet Procoagulant Activity

(**Scott's syndrome**) This is a rare bleeding disorder. The defining characteristic is the absence of Ca^{2+}-stimulated exposure of phosphatidylserine (PS) from the inner leaflet of the plasma membrane bilayer to the cell surface. This normally provides a surface for the assembly of the "Xase" and "prothrombinase" complexes of the coagulation. Consequently the "Xase" and "prothrombinase" complexes cannot assemble on the platelet membrane. The **adenosine triphosphate** (ATP)-binding cassette transporter A1 (ABCA1) has been implicated as a potential cause. Bleeding problems are corrected by **platelet transfusion** of normal platelets.

In these rare cases of bleeding, it is proposed that they are due to insufficient binding sites for **factor X** and **factor V**. Consequently the "tenase" and "prothrombinase" complexes cannot assemble on the platelet membrane. Bleeding problems are again corrected by platelet transfusion of normal platelets.

Disorders of Platelet Secretory Mechanism

These arise from abnormalities of either platelet secretory mechanisms or from deficiencies of one or more types of platelet granules. These disorders lead to a mild-to-moderate bleeding tendency, characterized by features of platelet-based **hemorrhage**, **bruising**, epistaxis, menorrhagia, postoperative bleeding, and bleeding after dental extraction.

Disorders due to failure of platelet secretory mechanisms constitute a heterogeneous group. In most cases, the causes of the failure of the secretory/release mechanism are unknown. Exceptions include a few patients who have definable **cyclooxygenase** or **thromboxane A2** synthetase deficiency. Platelet-aggregation studies show impairment of aggregation to weak agonists such as ADP, adrenaline/epinephrine, or low concentration of **collagen**. Characteristic impairment is seen as first-wave aggregation, but with absence of second wave. Higher concentrations of agonists may lead to second-wave aggregation, albeit slow. Failure to aggregate in response to **arachidonic acid** implies a defect in the prostaglandin pathways. Defects of secretion without a clear cause can be grouped as weak agonist response defects, so-called WARD syndrome. Bleeding in patients with secretory defects can be managed by transfusion of normal platelets. **DDAVP** is often used as a first-line treatment, as it improves hemostatic function in patients with such platelet-function disorders. The mechanism of this improvement is unknown. DDAVP may be sufficient for surgical prophylaxis if patients are known to respond. Local control of bleeding and concomitant use of antifibrinolytic agents such as **tranexamic acid** are important adjunctive measures.

Storage-Pool Disorders of Platelets Acquired Disorders[412]

Myeloproliferative disorders and **myelodysplasia** may be associated with defective platelet function, in addition to the observed **thrombocytosis** or **thrombocytopenia**. No one particular mechanism is responsible for the observed defects of function. Impairment of platelet function is frequently seen in **dysproteinemic purpura**, including impaired platelet aggregation and adhesion, which correlated well with bleeding tendency.

Renal disorders are a very common cause of an acquired platelet-function defect. Bleeding in renal disease is multifactorial, although platelet dysfunction is a major component. Platelets in renal disease show a diverse range of biochemical defects. Function is impaired by urea, guanidinosuccinic acid, and other phenolic metabolites known to accumulate in renal failure. Dialysis corrects, in part, platelet-function defects by removal of interfering compounds. Transfused normal platelets will acquire the same defect as native platelets after a few hours in the circulation.

Liver disorders often cause a platelet-function defect. The mechanisms involved are unknown and are probably multifactorial.

Systemic disorders include autoimmune disorders, **immune thrombocytopenic purpura** (ITP), **thrombotic thrombocytopenic purpura** (TTP), **hemolytic uremic syndrome** (HUS), **disseminated intravascular coagulation** (DIC), cardiopulmonary bypass, transplant rejection, burns, and valvular heart disease. Such diverse conditions can induce an acquired storage pool defect. This can be mediated by damage to circulating platelets, either mechanical or immune, which leads to partial release of platelet granule contents. Continued circulation of such depleted platelets can lead to bleeding through a partial storage-pool-like defect. Chronic hypoglycemia can produce a similar defect by failure of normal-platelet glucose-based metabolism.

Adverse drug reactions[174,412] (see Table 128) comprise the commonest causes of an acquired platelet-function disorder. Of the drugs that impair platelet function, **aspirin** is by far the most important. Aspirin inhibits cyclooxygenase and results in platelets failing to synthesize prostaglandin endoperoxides and thromboxane A2. The acetylation of cyclooxygenase by aspirin is irreversible. Other non-specific anti-inflammatory drugs (NSAIDs) have similar effects but a shorter effect. Most NSAIDs only inhibit platelets for the duration while they or their metabolites are in the circulation. Aspirin and NSAIDs produce a profound platelet release defect/failure of the secretory mechanism. Platelet aggregation studies show only first-wave aggregation with ADP, adrenaline/epinephrine, and collagen. Arachidonic acid-induced aggregation is classically absent. The antiplatelet effects of these drugs can readily be treated by their discontinuance. **Phosphodiesterase inhibitors** such as dipyrimadole are sometimes used prophylactically, being synergistic to aspirin. Prostacyclin can also be used for its synergistic effect with heparin for cardiopulmonary bypass procedures and to support renal dialysis. b-Lactam antibiotics (e.g., penicillins, cephalosporins) are known to impair platelet function. The effects are seen only in patients receiving large doses of parenteral antibiotics. The mechanism of action is unclear, but may be due to the adsorption of the antibiotic onto the platelet membrane, which then blocks multiple receptor-agonist interactions. In addition, they may also have an effect upon calcium influx in response to platelet stimulation. A number of foods are known to have effects upon platelets and platelet function. For example, garlic is known to inhibit both fibrinogen binding to platelets and platelet aggregation, while ethanol has been shown to decrease platelet interaction with endothelial cells. Diets deficient in arachidonic acid but rich in omega-3 fatty acids lead to a reduction in the synthesis of thromboxane A2 (TxA2) and the synthesis of TxA3 — the latter having no platelet aggregatory effect. In contrast, within the vessel wall, synthesis is reduced and the synthesis of PGI3 is increased; the latter is a potent antiaggregating agent.

Prostaglandin overproduction (Bartter's syndrome) is a rare metabolic disorder that may cause an acquired platelet-function disorder. It can be corrected by a high sodium intake.

Acquired platelet disorders vary in the degree to which they might lead to clinically significant bleeding *per se*. Principles of treatment include correcting or treating the underlying disorder as far as possible. Adjunctive treatments may include normal platelet transfusion or use of DDAVP.

PLATELET-FUNCTION TESTING

The range of tests available for the estimation of the activity of a subject's **platelets**.[413] A careful drug history must always be taken prior to platelet-function testing to ensure that drugs that affect platelet function have not been ingested. Ideally, none of these drugs should have been ingested in the preceding 2 weeks.

A global estimation of platelet function can be made using the **bleeding time** test. Reproducible results require the use of a template methodology. Prolongation of bleeding time in the face of a normal platelet count is indicative of a platelet-function disorder.

Methods of assessing platelet adhesion are time consuming and unsuitable for clinical application. Platelet aggregation by a range of agents that cause platelets to aggregate *in vitro* is widely used to measure and assess platelet function using an aggregometer.

Platelet Aggregometers

These instruments are based on the principle that the absorbance of platelet-rich plasma falls as platelets aggregate. The amount and rate of fall are dependent on platelet reactivity to the agonist added if other variables (temperature, mixing speed, and platelet count) are constant. Changes in light absorbance (or transmission) are recorded on a chart recorder, and percentage platelet aggregation to each agonist is determined using platelet-poor plasma as control. Platelet aggregometry is subject to a number of technical variables, including preparation of platelet-rich plasma, time since preparation, pH, and system optics.

Platelet-Aggregating Agents

These include **adenosine diphosphate** (ADP), **adrenaline**/epinephrine, **serotonin**, **thrombin**, **collagen**, **arachidonic acid**, **prostaglandins**, **thromboxane**, and **ristocetin**. All, with the exception of ristocetin, are present in the circulation or endothelium/subendothelium. *In vitro* behavior in response to these agonists is not reflected by *in vivo* behavior in all cases. This is due to the artificial nature of the platelet aggregometer and the multiple factors that determine *in vitro* platelet aggregation. Of these agents, five (ADP, ristocetin, collagen, arachidonic acid, and adrenaline/epinephrine) are commonly used in platelet-function tests.

ADP is used at low concentrations to cause primary reversible "first phase" aggregation. Platelets may disaggregate after the first phase in the absence of continuing stimulus or as a consequence of functional defect. Higher concentrations of ADP cause irreversible "second phase" aggregation, associated with the release of dense δ- and α-granules, and the use of ADP alone may mask subtle functional defects.

Adrenaline/epinephrine response is similar to that seen with ADP. Some normal subjects have impaired responses to low-dose adrenaline/epinephrine.

Collagen aggregation response is preceded by a lag phase of 10 to 60 sec. The lag phase is inversely related to the concentration of collagen used. Platelets then move into a single

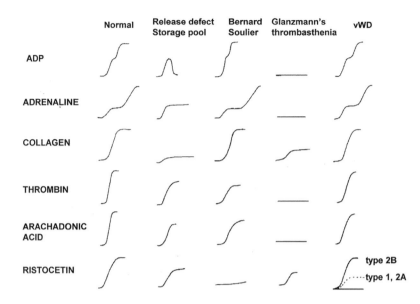

FIGURE 98

Examples of platelet aggregation tracings. Individual tracings run from bottom left (0% aggregation at time 0) to top right (maximum aggregation seen after exposure to agonist).

wave of aggregation due to granule release and activation of arachidonic acid pathways. High-dose collagen leads to aggregation by release reactions alone and bypasses the prostaglandin-mediated pathways. Both high and low collagen responses should be determined.

Arachidonic acid induces thromboxane generation and granule release. Absence of a response to arachidonic acid is indicative of a defect in the arachidonic acid metabolic pathways.

Ristocetin reacts with **Von Willebrand Factor** (VWF) to modulate the Von Willebrand platelet glycoprotein Ib reaction. It does not activate any other mechanism of platelet aggregation or granule release. Poor responses are seen in patients with low levels of VWF. Increased response to low-dose ristocetin is indicative of either type 2B **Von Willebrand Disease** (VWD) or pseudo-VWD.

Different agonists have different patterns of aggregation. The rate and extent of the aggregation of the sample are determined by the aggregometer. Characteristic aggregation patterns are seen in different disorders (see Figure 98).

Defects seen in response to weak agonists should always be repeated for confirmation. Platelets may sometimes aggregate in the absence of an agonist, so-called spontaneous platelet aggregation. This phenomenon is associated with arterial occlusive events.

Reference Center Testing

Nucleotide content of platelets. This determination is particularly important for the diagnosis of storage pool disorders.

Platelet release reaction. This can be studied by use of bioluminescence techniques.

PLATELET INDICES

See also **Automated blood cell counting**.

Calculations of platelet properties derived from platelet measurements.

Plateletcrit. Product of the platelet count and the mean platelet volume (MPV) — a measurement produced by automated blood cell counters. It has no clinical utility.

Mean platelet volume (MPV). The platelet equivalent of the mean cell volume (MCV) produced by multiparameter blood cell counters. The instability of platelet volume in **ethylenediamine tetraacetic acid** (EDTA) precludes its use. With aperture-impedance systems, there is an increase in MPV with increasing time from venous sampling. Conversely, with light-scattering instruments, the MPV decreases upon storage.

Platelet distribution width (PDW). The platelet equivalent to the **red blood cell distribution width** (RDW). The reference range in health is instrument specific. The only clinical utility described for the PDW is in differentiating between reactive **thrombocytosis** (normal PDW) and thrombocytosis associated with **chronic myeloproliferative disorders** (increased PDW).

Reticulated platelet (immature platelet fraction, IPF). Newly released platelets containing **RNA** are the platelet analog of the red blood cell **reticulocyte**. A new automated method to reliably quantitate reticulated platelets, expressed as the immature platelet fraction (IPF), has been developed. The clinical utility of the IPF has been established in the differential diagnosis between **thrombocytopenia** due to bone marrow failure and increased peripheral consumption, particularly **immune thrombocytopenia purpura** (ITP) and **thrombotic thrombocytopenic purpura** (TTP).

Mean platelet component (MPC). An experimental parameter generated by a single commercial instrument that measures platelet granularity using a laser source and light-scattering technology as a putative index of platelet activation.

PLATELETPHERESIS
See **Hemapheresis**.

PLATELET PLUG
The mass of aggregated platelet material formed after initial damage to a blood vessel as a primary response in **hemostasis**. It is an aggregate of platelets that have adhered at the site of vessel injury. It is soft and friable but functions to slow blood loss and allow secondary hemostasis/fibrin clot to develop and seal the vascular defect.

PLATELET-POOR PLASMA
See **Blood components for transfusion** — preparation and storage.

PLATELET-RICH PLASMA
(PRP) See **Blood components for transfusion** — preparation and storage.

PLATELET SATELLITISM
The presence of platelet rosettes around **neutrophils**. It is a cause of an artifactually low **platelet count**, particularly on automated counters. Examination of the blood film will

clearly demonstrate platelet rosetting as the cause of **thrombocytopenia**. It is of no clinical significance.

PLATELET SURVIVAL MEASUREMENT

Determination of the time spent by platelets in the circulation.[25] It is an important test that helps to define the cause of **thrombocytopenia**, but the procedure is time consuming. Older methods, in patients with decreased platelet production, include measuring the decline in platelet count following transfusion of fresh platelets. Commonly used methods of measuring platelet survival nowadays involve radiolabeling platelets. This can be done using either [51]chromium, [111]indium, or [113m]indium compounds. The ^{111}In compounds are less radiotoxic than ^{51}Cr compounds and are preferred for quantitative scintillation camera imaging. (See **Radionuclides**.)

Following injection of radiolabeled platelets, there is a rapid initial drop in the first 10 to 20 min, followed by a much slower fall. This initial rapid decline is due to a combination of splenic pooling and removal of platelets that have been damaged in the radiolabeling process. The survival time is taken as the time after injection at which survival falls to zero. Estimates of survival using ^{111}In-labeled platelets are 9 to 10 days. Platelet survival curves obtained in such studies fit a linear curve, implying that aging determines platelet life span. It is estimated that approximately 30% of platelets are pooled in the spleen, exchanging freely with circulating platelets. Using ^{111}In-labeled platelets, liver and spleen appear to be responsible for more than 80% of the clearance of senescent platelets, via reticuloendothelial phagocytosis. Approximately 50% of platelets are removed from the circulation by the spleen, 35% by the liver, and the remainder by the bone marrow, lymph nodes, and other tissues. Following splenectomy, 75% are removed from the circulation by the liver.

Platelet production can increase two- to eightfold from a baseline of about 40×10^9 platelets/l/day. In situations where platelet survival time is decreased, thrombocytopenia will only ensue when the survival time is so short that it outstrips increased platelet production.

PLATELET TRANSFUSION

The transfusion of platelet components of donated blood, the indications for platelet transfusion, and considerations concerning platelet transfusion refractoriness.[414]

Platelets are prepared by the differential centrifugation of units of whole blood or by apheresis (see **Blood components for transfusion** — preparation and storage; **Hemapheresis**). Units of random donor platelets derived from single units of whole blood are generally pooled prior to administration. Apheresis platelets are usually called units of single-donor platelets. It is now usual to prepare and issue platelets as an adult therapeutic dose. This contains on average 3.0×10^{11} platelets in 200 ml of plasma. For pediatric use, adult therapeutic doses can be split into smaller volumes or a single unit of random donor platelets transfused.

Indications for Transfusion
General guidelines

A decision to transfuse platelets is based on whether or not there is active clinical bleeding, deciding whether the patient is responsive to transfusions from unselected donors, and

TABLE 129

Guidelines for Transfusion of Platelets

Count (10⁹/l)	Decision
<10	Transfusions should be given to stable patients without any evidence of bleeding.
<20	Transfusions indicated where there is additional risk factors for bleeding, including fever, sepsis, and coexistent coagulopathy.
<50	Transfusions indicated where invasive procedures are planned or there is ongoing hemorrhage.
<100	Transfusions should be given if invasive procedures are planned in the eye or brain or if there is hemorrhage at these sites.

monitoring the platelet count (see Table 129). At a **platelet count** of 10 to $100 \times 10^9/l$, the bleeding time shows a linear inverse correlation with the platelet count, suggesting that the count itself is a reliable indicator of platelet-mediated hemostasis. At platelet counts of $<10 \times 10^9/l$, the bleeding time is >30 min.

Platelet transfusion is not normally helpful in **immune thrombocytopenic purpura** (ITP) and is contraindicated in patients with **thrombotic thrombocytopenic purpura** (TTP).

Platelet-Function Disorders

Congenital Disorders

These involve defective membrane receptors and granules, as well as abnormal biochemical pathways, and include **Glanzmann's thrombasthenia, Wiskott-Aldrich syndrome**, **Bernard-Soulier syndrome**, **May-Hegglin anomaly**, and **Chediak-Higashi syndrome**.

Acquired Disorders

These include drugs, commonly **aspirin**, that inhibit cyclooxygenase. Patients with uremia, **myeloproliferative disorders**, and **myelodysplasia** often have platelet dysfunction. Cardiopulmonary bypass procedures cause platelet dysfunction, since platelets become activated and then nonfunctional in the extracorporeal circuit.

Transfusion is indicated in patients who are known to have disordered platelet function, who undergo elective surgical procedures, and in cases where there is active clinical bleeding. Because such patients are often chronically transfused, it may be important to prevent **human leukocyte antigen** (HLA) alloimmunization and subsequent transfusion refractoriness (see **Blood components for transfusion** — complications).

Platelet Refractoriness to Transfusion

After transfusion, the platelet count ideally should be $>50 \times 10^9/l$. Failure to increment by $>10 \times 10^9/l$ at 1 to 20 h posttransfusion on two occasions is a simple practical indicator of refractoriness, which should be investigated. The likely causes include:

Nonimmune refractoriness: a much more common disorder than immune-mediated refractoriness. Accelerated platelet consumption is found in patients who have had a stem cell transplantation (SCT), disseminated intravascular coagulation (DIC), antibiotic or amphotericin-B therapy, **splenomegaly**, or hemorrhage.

Immune-mediated refractoriness: caused by the presence of antibodies. The most important are human leukocyte antigen (HLA) antibodies; others are platelet-specific alloantibodies, platelet autoantibodies, ABO antibodies, and drug-induced platelet antibodies.

   Leukodepletion of blood components for transfusion, both platelet transfusion and **red blood cell transfusion**, reduces alloimmunization and refractoriness but does not impact on refractoriness due to nonimmune causes.

Management of Refractoriness

The patient should be evaluated to detect nonimmune causes of platelet consumption, e.g., infection, splenomegaly, or DIC. A serum sample should be tested for HLA antibodies. Testing for platelet-specific alloantibodies and platelet autoantibodies is performed using the indirect platelet immunofluorescence test, although this also detects HLA antibodies. Therefore, a test for platelet-specific antibodies, such as the monoclonal antibody-specific immobilization of platelet antigen assay, should also be used.

   Immune-mediated platelet refractoriness is managed by providing HLA-matched platelets collected by apheresis. This is straightforward for patients with common HLA phenotypes, but there may be few donors for patients with uncommon phenotypes. If there is a good clinical response to the transfusions of HLA-matched platelets, then they are indicated for subsequent transfusions. If, however, there is no improvement in the corrected posttransfusion count, this might be due to the presence of factors associated with nonimmune platelet consumption or the presence of platelet-specific antibodies. Of patients with HLA antibodies, 70% respond normally to random platelet transfusion, so the presence of antibodies is not a diagnostic test. If matching for ABO, HLA, and HPA antigens is unsuccessful, platelet cross matching may be helpful in identifying compatible donors in some of these cases and may be used as the primary strategy for the selection of compatible platelets. Nonimmune-platelet refractoriness is managed by treating the underlying cause and by increasing the dose of platelets.

PLEIOTROPY

The ability of allelic substitutions at a **gene** locus, or of a cell product, to affect or to be involved in the development of more than one aspect of a **phenotype**. Many protein kinases are pleiotropic.

PLEOMORPHIC SMALL, MEDIUM, AND LARGE T-CELL HTLV-1 + LYMPHOMA
See **Adult T-cell leukemia/lymphoma**; **Non-Hodgkin lymphoma**.

PLETHYSMOGRAPHY
See **Venous thromboembolic disease**.

PLUMBISM
See **Lead toxicity**.

PLUMMER-VINSON SYNDROME
See **Iron deficiency**.

POEMS SYNDROME
Polyneuropathy, Organomegaly, Endocrinopathy, Monoclonal Gammopathy, and Skin changes. (See **Osteosclerotic myeloma**.)

POIKILOCYTOSIS

Any change of **red blood cell** morphology where the cells display an abnormal shape (see **Peripheral-blood film examination**). It occurs in healthy subjects at high altitudes and frequently in any severe anemia. The abnormality in shape may suggest a diagnosis, e.g., **sickle cell disorder**, **red blood cell fragmentation syndrome**, **acanthocytes**, or a group of diagnoses, e.g., **teardrop poikilocytes**.

POINT-OF-CARE TESTING

(POCT; near-patient testing, NPT) The performance of laboratory tests near the patient.[415,416] The location may be at the hospital bedside, in the home, in the physician's or medical practitioner's clinic, or within a hospital at a site outside the central laboratory. This type of practice is fast developing, but at present in hematology it is generally restricted to measurement of **hemoglobin** and INR monitoring for oral anticoagulant therapy (see **Warfarin**). Instruments are available for **automated blood cell counting**, but their use is limited by cost. **Peripheral-blood film examination** is also possible, but limitation of expertise in microscopy restricts its performance.

The requirements for quality and competence are as important as for testing in a central laboratory, but POCT brings its own set of specific problems, such as:

Evaluation of POCT instruments and systems

Purchase and installation of equipment

Maintenance of equipment, consumable supplies, and reagents

Training, certification, and recertification of POCT system operators

Quality control and quality assurance

Once these limitations have been met and overcome, POCT offers distinct advantages in faster decision making by the physician and consequently earlier commencement of treatment. Improved convenience for the patient may increase compliance and give greater patient satisfaction. POCT is, however, a less economic form of testing. **Quality management** is a particular problem because treatment is likely to be initiated before any checks to ensure the reliability of test results can be applied.

POKEWEED MITOGEN

(PWM) A lectin from the root of *Phytolacca americana*, which is mitogenic to both B and T-lymphocytes. PWM binds to carbohydrate molecules on the lymphocyte. *In vitro*, PWM causes activation of B-lymphocytes and terminal differentiation, ultimately resulting in the production of polyclonal **immunoglobulin**. However, it is thought that T-lymphocytes are necessary for B-lymphocyte activation by PWM to occur. T-lymphocyte activation causes increased cellular cytotoxicity.

Various assays using PWM have been developed and are now widely used to investigate immunodeficiency disorders, e.g., **human immunodeficiency virus** (HIV) infection.

POLYAGGLUTINABILITY

A property of **red blood cells** (RBCs) whereby they are usually agglutinated by normal adult human plasma. The usual cause is the transient action, *in vivo* or *in vitro*, of bacterial enzymes, which expose carbohydrate structures on the RBC membrane that are normally hidden. T and Tk polyagglutinability are two such examples. Tn polyagglutinability is different in that the cause is somatic mutation of stem cells, with a resulting deficiency in a galactosyl transferase needed for the normal carbohydrate expression on the RBC membrane. Tn polyagglutinability is usually persistent. Rarely, polyagglutinability may be due to genetic transferase or other enzyme abnormalities, leading to the inheritance of an antigen for which the corresponding antibody is regularly present in plasma, as in HEMPAS (see **Congenital dyserythropoietic anemias**). **Lectins** can be used serologically to determine the nature of the polyagglutinability. Transfusion of plasma to patients (particularly infants), where the RBCs are polyagglutinable, may give rise to blood transfusion complications. In these circumstances, only plasma-depleted blood products (e.g., washed RBCs) should be transfused. Erroneous blood groups in pretransfusion testing can occur if the test RBCs are polyagglutinable, but most monoclonal antibodies do not cause problems.

POLYARTERITIS NODOSA

A multisystem arteritis particularly affecting small- and medium-sized vessels[155] (originally described in 1866 by Kussmaul and Maier). Histologically, all three layers of the blood vessel are affected with an inflammatory cell infiltrate. Early lesions reveal intimal edema, swelling of muscle fibers, and fibrinoid necrosis resulting in luminal narrowing. Later lesions consist of intimal proliferation and fibrosis with lymphocyte and plasma cell infiltration. Ultimately, there is aneurysm formation, thrombosis, ischemia, and infarction. The lesions often start at arterial bifurcations and spread distally.

The cause is unknown, but 25 to 40% of patients are hepatitis-B surface-antigen positive, and immune complex deposition has been suggested, **antineutrophilic cytoplasmic antibodies** being detected. There is also a high incidence among patients with **hairy cell leukemia**.

Clinical Features

Polyarteritis typically affects middle-aged males (M:F ratio 3:1), but all ages can be affected. Any artery or segment of artery can become involved, but commonly, arteries of the kidney, heart, gastrointestinal tract, liver, nerves, muscle, and skin are most commonly affected. Onset is often insidious, with low-grade fever, malaise, anorexia, weight loss, myalgia, and arthralgia.

The kidney is the most frequently affected organ (80%), and renal complications are the commonest cause of death. Uremia, proteinurea, hematuria, and hypertension are usual, with aneurysms and infarction of the main renal artery or arcuate and interlobular arteries commonly seen. Acute necrotizing glomerulonephritis is present in 60%.

Cardiac involvement (70 to 80%) is the second commonest cause of death due to myocardial infarction, acute pericarditis, hypertensive heart disease, and arrhythmias.

Gastrointestinal involvement (60 to 70%) manifests as bleeding, ischemia, or bowel perforation. Involvement of the nervous system may be peripheral (mononeuritis multiplex) due to involvement of the vasa nervorum or central (subarachnoid hemorrhage, cerebral infarction, etc.). Clinically detectable skin nodules are only found in 10 to 15% of patients, but various other skin manifestations, e.g., urticaria, livedo reticularis, and digital infarctions, are more common.

Laboratory Features

Normochromic, **normocytic anemia** with **neutrophilia** is usual, sometimes with **thrombocytosis**. There is polyclonal gammopathy and hypocomplementemia. Antinuclear factor (ANF) is positive in 10 to 30% and rheumatoid factor (RF) in 30%, but usually only at low titers. The **erythrocyte sedimentation rate** (ESR) is usually elevated.

Definitive diagnosis depends on an adequate tissue biopsy, although angiography is useful if biopsy is too hazardous.

Variants

Microscopic angiitis: a disorder predominantly limited to the renal vessels without granuloma formation. P-ANCA is often found.

Churg-Strauss syndrome: a widespread angiitis with granuloma nodules giving rise to purpura and necrotizing arteritis with eosinophilic infiltrate with late renal involvement. Early symptoms include asthma and fever associated with anemia and **eosinophilia.**

Treatment and Prognosis

Untreated, the prognosis is extremely poor, but high-dose **corticosteroids** with additional immunosuppressive agents, e.g., **azathioprine** and **cyclophosphamide**, have markedly improved survival. Renal failure and hypertension should be treated with conventional therapies. The overall five-year survival is now 80%.

POLYCHROMASIA

The bluish-gray appearance of immature **reticulocytes** on Romanowsky-stained blood films due to uptake of eosin by **hemoglobin** and of basic dyes by residual **ribonucleic acid** (RNA). Blue-black staining occurs when there is extramedullary erythropoiesis.

POLYCHROMATIC ERYTHROBLAST

(Intermediate normoblast) See **Erythropoiesis**.

POLYCYTHEMIA

See **Erythrocytosis**; **Polycythemia rubra vera**.

POLYCYTHEMIA RUBRA VERA

(PRV) A chronic clonal hematopoietic stem cell disorder characterized by overproduction of morphologically normal **red blood cells, white blood cells**, and **platelets** in the absence of a definable cause and with suppression of normal polyclonal hematopoiesis.

Pathogenesis

There are no known predisposing factors but a single somatic activating point mutation in the *JAK2* gene has been found in many patients (see **Cell-signal transduction**). There is a 1.2:1 male preponderance, with a median age of onset of 61 years. The condition appears to be more common in Jews and uncommon in blacks, and the incidence in relatives is higher than in the general population.

Erythrocytosis distinguishes this disorder from chronic myeloproliferative disorders, **idiopathic myelofibrosis** and **essential thrombocytosis**. Although erythropoietin-independent *in vitro* erythroid colony formation is a feature of the disease, this abnormality is not specific for the disorder and is not found in the majority of polycythemia vera erythroid progenitor cells. Microvascular thrombosis occurs due to interaction between activated platelets and endothelial cells, and macrovascular thrombosis occurs due to laminar flow disturbance by the elevated red cell mass, activated platelets, and thrombin generation.

Clinical Features

Onset is usually insidious, with nonspecific symptomatology such as malaise, headache, dizziness, distressing pruritus, extremity pain, or some form of **thrombosis**. Cerebrovascular accidents, myocardial infarction, deep venous thrombosis, pulmonary embolism, and hepatic or portal vein thrombosis are well recognized, with the latter events particularly common in women. **Hemorrhage** and **bruising** occur in up to 25% of cases, with gastrointestinal bleeding particularly troublesome, as PRV patients have up to four times the incidence of peptic ulceration.

   Splenomegaly may not be present at onset, but it develops in most patients and can be massive; hepatomegaly is less common. Gout is seen in 10% of cases, although 50 to 70% of patients have **hyperuricemia**.

Laboratory Features

Red blood cell morphology is usually normal, but microcytes (due to **iron deficiency** secondary to hemorrhage) or macrocytes (due to **folic acid** deficiency) can occur. **Hemoglobin F** may be slightly raised. **Granulocytosis** (12 to $30 \times 10^9/l$) that is left-shifted occurs in 25 to 60% of patients. **Basophilia** and **eosinophilia** are also not unusual. The **NAP score** is high in 70% of patients. **Thrombocytosis** (500 to $2000 \times 10^9/l$) is common (80%), and platelet aggregation may be abnormal in response to epinephrine/adrenaline. Platelet morphology may be bizarre, with macrothrombocytes not uncommon.

   Bone marrow examination usually reveals trilineage hematopoietic cell hyperplasia but may be normal. **Myelofibrosis** can be present at onset or develop as part of the natural history of the disease. It has no prognostic significance. **Cytogenetic analysis** shows abnormality in up to 25%, with trisomy 8, trisomy 9, and 20q the most common findings.

Diagnosis[184]

Absolute erythrocytosis as established by determination of red cell mass and plasma volume is mandatory for establishing the diagnosis (see **Blood volume**). The quartet of erythrocytosis, granulocytosis, thrombocytosis, and splenomegaly makes the diagnosis simple, but in 40% of cases, all of these classical features are not seen. A scoring system has therefore been proposed by the Polycythemia Vera Study Group for diagnosing PRV (see Table 130),[417] but many patients initially lack all the necessary diagnostic features, particularly the red cell mass. A new evolving approach to the diagnosis includes the following characteristics: a hemoglobin and hematocrit of greater than 95% confidence limits; polycythemia rubra vera-related features: thrombocytosis, leukocytosis, microcytosis, splenomegaly, pruritus, erythromelalgia, and unusual thromboses; and abnormal bone marrow. The serum erythropoietin level, if elevated or very low, can differentiate PRV from secondary erythrocytosis. Newer tests, such as *PRV-1 mRNA* overexpression or impaired *Mpl* expression, have yet to be clinically validated. There are a number of patients in whom it is impossible to initially definitely make the diagnosis of PRV, and the term "idiopathic erythrocytosis" is appropriate in this situation.

TABLE 130

Diagnostic Scoring System for Polycythemia Rubra Vera

Major Criteria	Minor Criteria

1. Red cell mass
    Male   >36 ml/kg
    Female >32 ml/kg
	1. Thrombocytosis >400 × 10^9/l

2. O_2 saturation >92%	2. Granulocytosis >12 × 10^9/l
3. Splenomegaly	3. LAP >100
	4. Serum B_{12} >900 pg/ml
	5. B_{12} binding capacity >2200 pg/ml

Note: PRV is diagnosed if: (a) all three major criteria are present or (b) the first two of the major criteria plus any two minor criteria are present.

Source: Based on criteria used by the Polycythemia Vera Study Group.[417]

Treatment

There is no known cure for this chronic disease, and therapy is therefore aimed at controlling the consequences of the overactive bone marrow and extramedullary hematopoiesis. Keeping the **packed-cell volume** (PCV) <0.45 l/l in men and <0.42 l/l in women by **phlebotomy** will prevent thrombotic events. Asymptomatic thrombocytosis requires no treatment. Interferon- and hydroxyurea can be used to control leukocytosis or thrombocytosis and extramedullary hematopoiesis.[417] Anagrelide is also useful in controlling thrombocytosis.

    PUVA light therapy, **interferon-α**, or **hydroxyurea** can be used for intractable pruritus if antihistamines or antidepressants are not effective, and **allopurinol** will control hyperuricemia. Epsilon aminocaproic acid or tranexamic acid can control platelet-related hemorrhage, while aspirin can alleviate erythromelalgia. In some patients,[418] splenectomy may be necessary for intractable splenomegaly.

Prognosis

The natural history of polycythemia is variable because the disease is not monolithic. In a few patients, the disease is rapidly progressive, but most can look forward to many decades of productive life. Myelofibrosis is a feature of the natural history of PRV but is not itself a prognostic marker.

POLYMERASE CHAIN REACTION
See **Molecular genetic analysis**.

POLYMORPHIC LYMPHOID PROLIFERATION
See **Posttransplant lymphoproliferative disorders**.

POLYMORPHIC RETICULOSIS
See **Extranodal NK/T-cell lymphoma, nasal type**; **Non-Hodgkin lymphoma**.

POLYMORPHONUCLEAR NEUTROPHIL
See **Neutrophils**.

POLYMYALGIA RHEUMATICA
(Polymyositis; giant-cell arteritis; temporal arteritis) The hematological changes occurring in patients with polymyalgia rheumatica and closely related giant-cell arteritis.[419] These changes include:

Erythrocyte sedimentation rate elevated to a marked degree in most, but in a few it is normal

Plasma **viscosity** raised

C-reactive protein raised

Polyclonal hypergammaglobulinemia

Rouleaux of red blood cells in peripheral-blood smears

Anemia of chronic disorders

POLYMYOSITIS
An inflammatory disorder of striated muscle causing proximal muscle weakness. In some patients there is involvement of the skin — dermatomyositis. Persons with **human leukocyte antigens** HLA-B8/DR 3 are genetically predisposed. The associated hematological changes are:

Mild **anemia of chronic disorders**

Leukocytosis with **eosinophilia**

Erythrocyte sedimentation rate markedly elevated

POLYNEUROPATHY, ORGANOMEGALY, ENDOCRINOPATHY, MONOCLONAL GAMMOPATHY, AND SKIN CHANGES
(POEMS syndrome) See **Osteosclerotic myeloma**.

POORLY DIFFERENTIATED LYMPHOCYTIC LYMPHOMA
See **Non-Hodgkin lymphoma**.

PORCINE FACTOR VIII
See **Coagulation factor concentrates**.

PORPHORINURIA
See **Porphyrias**.

PORPHYRIA CUTANEA TARDA
See also **Porphyrias**.

TABLE 131

Laboratory Features of the Porphyrias

Porphyria	Erythrocytes	Plasma	Urine	Feces
ALA synthase deficiency (X-linked sideroblastic anemia)	ZnPP	—	—	—
ALA dehydratase deficiency porphyria	ZnPP	—	ALA	—
Acute intermittent porphyria	—	—	ALA, PBG	—
Chronic erythropoietic porphyria	Uro I, Copro I	Uro I, Copro I	Uro, 7-Carboxyl	—
Porphyria cutanea tarda	—	Uro, 7-Carboxyl	Uro, 7-Carboxyl	Uro, 7-Carboxyl, Isocopro
Hepatoerythropoietic porphyria	ZnPP	Uro, 7-Carboxyl	Uro, 7-Carboxyl	Uro, 7-Carboxyl, Isocopro
Hepatic coproporphyria	—	Copro	Copro, ALA, PBG	Copro
Variegate porphyria	—	Proto	ALA, PBG	Proto
Erythropoietic porphyria	Proto	Proto	—	Proto

Note: ALA = aminolevulinic acid; 7-Carboxyl = 7-carboxylporphyrin; Copro = coproporphyrin; Isocopro = isocoproporphyrin; PBG = porphobilimogen; Proto = protoporphyrin; Uro = uroporphyrin; ZnPP = zinc protoporphyrin.

Source: Modified from Sassa, S., The hematologic aspects of porphyria, in *Williams Hematology*, 6th ed., Beutler, E., Lichtman, M.A., Coller, B.S., Kipps, T.J., and Seligsohn, U., Eds., McGraw-Hill, New York, 2001.

The clinical disorder resulting from an autosomally dominant partial deficiency of the hepatic uroporphyrinogen decarboxylase; it is the most common of the porphyrias. The characteristic clinical features are: photosensitivity with vesiculo-bullous eruptions, mainly on the hands and face; hyperpigmentation; and excessive facial hair growth. Severe scarring of the skin may suggest scleroderma. There is excessive tissue iron loading with elevation of the serum transferrin saturation and ferritin levels and hepatic siderosis in many patients that sometimes cause hepatic cirrhosis. Many patients have common mutations of the hemochromatosis-associated *HFE* gene on Ch6p (especially *HFE C282Y* or *H63D 104*). The severity of **iron overload** can be estimated by analyses of a liver specimen obtained by biopsy or by quantitative **phlebotomy.** The urinary excretion of uroporphyrinogens I and III and, to a lesser extent, coproporphyrinogen are increased (Table 131).

All patients should abstain from alcohol ingestion or exposure to benzene-related compounds, especially those encountered in industrial settings. In women, estrogen or progesterone therapy can exacerbate porphyria cutanea tarda. Photosensitivity resolves in many patients who achieve iron depletion with therapeutic phlebotomy. Although the mechanism of action is unclear, therapy with chloroquine may be helpful in phlebotomy-resistant patients. Chloroquine is concentrated in the liver and may complex with the porphyrin or stimulate its release from hepatocytes.

PORPHYRIAS

A group of metabolic disorders involving the specific enzymes of the **heme** biosynthetic pathway. These disorders usually have a genetic basis, although the enzyme deficiency in porphyria cutanea tarda may be either inherited or acquired. Porphyrias are characterized biochemically by patterns of accumulation and excretion of intermediates or oxidized intermediates of the heme biosynthetic pathway. The enzyme deficiencies do not profoundly impair total cellular heme synthesis and content. Rates of heme synthesis are usually normal and may even be increased in some patients with porphyria.

TABLE 132

Association of Enzyme Deficiency with Clinical Porphyria

Enzyme	Clinical Disorder [a]
d-Aminolevulinic acid synthase	X-linked **sideroblastic anemia** (E)
d-Aminolevulinic acid dehydratase	ALA-dehydratase deficiency porphyria
Porphobilinogen deaminase	**Acute intermittent porphyria** (H)
Uroporphyrinogen III cosynthase (partial)	**Congenital erythropoietic porphyria** (E)
Uroporphyrinogen III cosynthase (severe)	Hepatoerythropoietic porphyria
Uroporphyrinogen decarboxylase	**Porphyria cutanea tarda** (H)
Coproporphyrinogen oxidase	**Hereditary coproporphyria** (H)
Protoporphyrinogen oxidase	**Variegate porphyria** (H)
Ferrochelatase	**Erythropoietic porphyria** (E)

[a] H = hepatic form; E = erythroid form.

Many symptoms of porphyria are nonspecific and mimic those of more common disorders. The major manifestation in some patients is cutaneous photosensitivity. In some cases there are poorly understood effects on the nervous system that can be life threatening, but these manifestations occur only in those types of porphyria in which porphyrin precursors accumulate. The diagnosis rests on a high level of suspicion and appropriate laboratory testing.

Classification

Porphyrias are divided into erythropoietic and hepatic types, depending on whether the excess production of porphyrin precursors and porphyrins occurs in the **red blood cells** or **liver**, although some porphyrias possess both "erythroid" and "hepatic" features. Porphyrias with neurovisceral features are also called acute porphyrias, and those with cutaneous photosensitivity are called cutaneous porphyrias. Latent forms probably exist in all types. A specific type of porphyria is associated with specific red blood cell enzyme deficiencies (see Table 132). Chromosomal locations for several of the enzymes have been defined.[172]

In acute porphyrias, the effects of drugs are most important (see Table 133). In affected patients, life-threatening attacks of porphyria can occur after exposure to certain commonly prescribed drugs. Many individual drugs have been categorized as unsafe because they have caused acute attacks in humans or are porphyrinogenic in animals or in *in vitro* systems (see *British Natural Formulary*, 51, 499, 2006).

Biochemical Diagnosis[420]

In porphyrias, porphyrins accumulate in normoblasts, skin, and liver. Their presence can be detected by ultraviolet light, birefringency of protoporphyrin crystals under polarized light, fluorescence microscopy, or electron microscopy. Diagnosis is based on patterns of accumulation and excretion of the intermediates of the heme biosynthetic pathway in red blood cells, urine, and feces. Table 131 summarizes the patterns in the different types of porphyria.

Physicochemical Properties of the Porphyrins

Figure 99 shows the structural formula of the porphyrin nucleus and gives a diagrammatic representation of some naturally occurring porphyrins.

TABLE 133

Drug Groups Associated with Acute Porphyria

Amphetamines	Ergot derivatives
Anabolic steroids	Gold salts
Antidepressants	Hormone replacement therapy
Antihistamines	Progestogens
Barbiturates	Statins
Contraceptives, steroids	Sulphonamides
Ergot derivatives	Sulphonylureas

Individual Drugs Associated with Acute Porphyria

Aceclofenac	Diazepam	Mebeverine	Pivmecillinam
Alcohol	Diclofenac	Mefanamic Acid	Porfimer
Amioderone	Doxycycline	Meprobamate	Probenecid
Baclofen	Econazole	Methyldopa	Pyrazinamide
Bromocriptine	Erythromycin	Metoclopramide	Ribabutin
Busulphan	Etamsylate	Metolazone	Rifampicin
Carbergoline	Ethionamide	Metronidazole	Ritonavir
Carbamazepine	Ethosuximide	Metyrapone	Simvastin
Carisoprodol	Etomidate	Miconazole	Spironolactone
Choral Hydrate	Fenfluramine	Mifepristone	Sulfinpyrazone
Chorambucil	Flupentixol	Minoxidil	Sulpiride
Chloramphenicol	Griseofulvin	Nalidixic Acid	Tamoxifen
Chloroform	Halothane	Nifedipine	Temoporfin
Clindamycin	Hydralazine	Nitrofurantoin	Theophylline
Clonidine	Hyoscine	Orphenadrine	Tiagabine
Cocaine	Indapamide	Oxcarbazepine	Tinidazole
Colistin	Indinavir	Oxybutynin	Topiramate
Cyclophosphamide	Isometheptene Mucate	Oxycodone	Triclofos
Cycloserine	Isoniazid	Oxytetracycline	Trimetroprim
Danazol	Ketamine	Pentazocine	Valproate
Dapsone	Ketoconazole	Pentoxifylline (oxypentifylline)	Verapamil
Dexfenfluramine	Ketorolac	Phenoxybenzamine	Xipamide
Dextropropoxyphene	Lidocaine (lignocaine)	Phenytion	Zulopenthixol

Source: *British National Formulary 51*, 9.8.2, 2006, p. 503 and updated twice per year. With permission.

The porphyrin molecule is formed by four pyrrole rings joined by four methane bridges. The tetrapyrrole can have two to eight carboxylic side chains. If each pyrrole ring contains two different acidic side chains, acetyl and propionyl, there are four possible isomers (I, II, III, and IV). The asymmetric type III isomer and the symmetric type I isomer occur most frequently in nature. Uroporphyrin is water soluble and is excreted predominantly by the kidney. In contrast, protoporphyrin is very hydrophobic and is eliminated by hepatobiliary excretion into the intestinal tract. Coproporphyrin has intermediate properties.

The electronic configuration of tetrapyrroles favors the absorption of radiant energy by porphyrins. The violet spectrum of the Soret band (400 to 410 nm) is the region of greatest energy absorbance. A lesser degree of absorbance occurs in the visible band (580 to 650 nm). Absorbing radiant energy results in raising orbital electrons from their ground state to singlet and triplet excited-state energy levels. Single excited molecules are short-lived and emit fluorescence, whereas triplet excited porphyrins are long-lived and emit phosphorescence. Triplet excited porphyrins possess sufficient energy to convert dissolved molecular oxygen to singlet oxygen, the main cytotoxic event in photocytotoxicity and in photodynamic therapy.

When porphyrins are increased in the blood, the patient's skin becomes very sensitive to sunlight, and blisters form readily. Cutaneous lesions include:

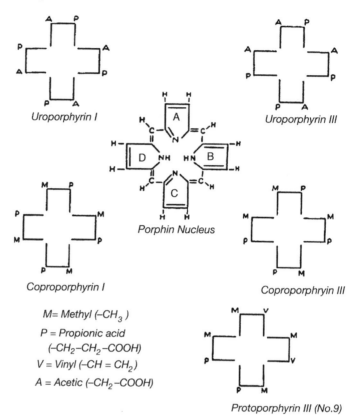

FIGURE 99
Structural formula of the porphyrin nucleus and diagrammatic representation of some naturally occurring porphyrins. (Prepared by G.E. Cartwright; reproduced from Wintrobe, M.M., *Clinical Hematology*, 6th ed., Henry Kimpton, London, 1967. With permission.)

Acute erythematous edematous purpuric syndrome, produced by hydrophobic porphyrins, which occurs in erythropoietic protoporphyria

Subacute bullous and erosive syndrome found in the other forms of porphyrias that are produced by hydrophilic porphyrins

Both types of cutaneous photosensitivity can be caused by lipid peroxidation or by complement activation.

Clinical Features and Treatment

These depend upon the form of the porphyria and must be considered separately for each type:

Congenital erythropoietic porphyria
Hereditary coproporphyria

Erythropoietic protoporphyria
Acute intermittent porphyria
Variegate porphyria
Porphyria cutanea tarda

PORPHYRIA VARIEGATA
See **Variegate porphyria**.

PORTAL CIRCULATION
The vascular system that returns blood from the **spleen**, pancreas, stomach, small intestine, and the greater part of the colon to the **liver**. In the liver, the portal vein divides to form hepatic sinusoids. The system allows the products of digestion to pass through the liver before entering the general circulation. Obstruction to the hepatic vein, intrahepatic sinusoids, portal, or splenic veins causes **portal hypertension**, leading to esophageal varices with chronic **hemorrhag**e and to **splenomegaly**.

PORTAL HYPERTENSION
Increase of venous pressure in the **portal circulation** (see **Liver**). This may be due to:

   Blockage of the portal vein before the liver as a result of portal vein thrombosis, arteriovenous fistula, and **splenomegaly**

   Disruption or change in the liver architecture by cirrhosis, schistosomiasis, **sarcoidosis**, **granulomas** (tuberculosis), alcoholic hepatitis, massive fatty change, and nodular regenerative hyperplasia, sometimes following regression posttreatment for hematological malignancy

   Posthepatic blockage, as in hepatic vein thrombosis (Budd-Chiari syndrome)

The complications include esophageal varices with chronic **hemorrhage,** ascites, and splenomegaly.

PORT WINE NEVI
See **Hemangiomas**.

POSITIVE-EMISSION TOMOGRAPHY
(PET) A tomographic technique with ^{18}F used in staging and therapeutic monitoring of **lymphoproliferative disorders**, particularly **Hodgkin lymphoma** and histologically aggressive **non-Hodgkin lymphoma**.

POSITIVE HEMATOPOIETIC REGULATORS
See **Hematopoietic regulators**.

POSTTRANSFUSION PURPURA

(PTP) An acute and usually severe **thrombocytopenia** occurring 5 to 10 days after **blood component transfusion**. Most patients are women who have been previously pregnant and are alloimmunized against **platelet** antigens, typically HPA-1a. It is postulated that donor platelets present in red blood cell concentrates provoke a robust alloimmune response in the previously alloimmunized recipient. The donor platelets are rapidly destroyed. In addition, the recipient's own antigen-negative platelets that are unreactive with alloantibodies are also destroyed. In the acute phase, antibody is bound to autologous platelets and alloantibodies may be eluted. Either alloantibodies reactive with a pseudospecificity or immune complexes are formed that bind specifically to the patient's platelets. The diagnosis is based on clinical findings and the presence of severe thrombocytopenia in the presence of (usually) anti-HPA-1a. Treatment is with intravenous **immunoglobulin** (IVIG) at a total dose of 2 g/kg over 2 to 5 days and **corticosteroids**; **plasma exchange** has also been used.

POSTTRANSPLANT LYMPHOPROLIFERATIVE DISORDERS

Lymphomas developing in patients following transplantation, particularly **allogeneic stem cell transplantation**. They are predominantly extranodal in presentation and related to **Epstein-Barr virus** (EBV) infection.[182] Risk factors to develop the disease include patients who are EBV negative and cytomegalovirus (CMV) negative pretransplant, especially if the donors are positive. These disorders are broadly defined as:

Early lesions
Polymorphic lesions
Monomorphic lesions
Hodgkin lymphoma

The monomorphic lesions are **diffuse large B-cell lymphoma**; **Burkitt lymphoma**; and **peripheral T-cell, unspecified**.

Treatment

Approaches to the initial management are different from those used for patients who have lymphomas not related to transplantation. Initial withdrawal of immunosuppression leads to a response in about 40% of patients. **Rituximab** as a single agent results in initial responses in 50 to 66% of patients. In patients who have become EBV positive, treatment with EBV-specific cytotoxic T-cells (CTLs) has been used.

PRECIPITIN REACTION

An antigen–antibody reaction in which precipitation occurs.

PRECURSOR B-LYMPHOBLASTIC LEUKEMIA/LYMPHOMA

See also **Acute lymphoblastic leukemia**; **Lymphoproliferative disorders**; **Non-Hodgkin lymphoma**.
(B LBL; Rapaport: diffuse poorly differentiated lymphocytic lymphoma) A high-grade B-lymphoblast non-Hodgkin leukemia/lymphoma. The use of the terms "leukemia" or

"lymphoma" is arbitrary in some patients, since biologically it is the same disease. Characteristically, these cells are slightly larger than small **lymphocytes** but smaller than those of a **diffuse large B-cell lymphoma**. They have round or convoluted nuclei, fine chromatin, inconspicuous nucleoli, and scant, faintly basophilic cytoplasm. Mitoses are frequent. They are morphologically identical to precursor B-lymphoblasts, immunophenotyping being required to distinguish precursor B from precursor T-lymphoblasts. The **immunophenotype** of B-lymphoblasts is TdT$^+$, CD19$^+$, CC79a$^+$, CD22$^+$, CD20$^{-/+}$, CD10$^-$, HLA-Dr$^+$, SIg$^-$, cMu$^{-/+}$, CD34$^{+/-}$. **Cytogenetic analysis** is particularly important in this disease, as it is of prognostic significance. Favorable cytogenetics include t(12,21) and hyperdiploid greater than 50 chromosomes. These are the two most common changes, together representing about 50% of cases. Unfavorable cytogenetics include t(9,22), t(4,11), t(1,19), and hypodiploidy. The cell-line origin of the lymphoma is a bone marrow-derived precursor B-cell.

Clinical Features

It is necessary to differentiate this lymphoma from B-ALL (acute lymphoblastic leukemia). The lymphoma patients present with a cell mass lesion but with only 25% or fewer lymphoblasts in the bone marrow. Children are far more commonly affected than adults, the lymphoma presenting as acute lymphoblastic leukemia in 80% of cases with bone marrow and peripheral-blood involvement. A small proportion of cases present as solid tumors, most often with lesions in skin, bone, and lymph nodes, with or without leukemia. These solid tumors are clinically indistinguishable from precursor T-lymphoblastic lymphomas. The disease is highly aggressive but prognosis is good with treatment.

Staging

See **Lymphoproliferative disorders**.

Treatment

See also **Acute lymphoblastic leukemia**; **Non-Hodgkin lymphoma**.
Children with precursor B-lymphoblastic leukemia/lymphoma have about an 80% cure rate following intensive chemotherapy, but adults have a poor prognosis.

PRECURSOR T-LYMPHOBLASTIC LEUKEMIA/LYMPHOMA

See also **Acute lymphoblastic leukemia**; **Lymphoproliferative disorders**; **Non-Hodgkin lymphoma**.
(T-LBL; Rapaport: poorly differentiated lymphocytic lymphoma) A high-grade T-lymphoblast disorder with biologic unity in the leukemia phase and with non-Hodgkin lymphoma presentation. Characteristically, the disorder has cells that are slightly larger than small **lymphocytes**, nuclei with finely dispersed chromatin, inconspicuous nucleoli, and scant cytoplasm. They are morphologically identical to those of precursor B-lymphoblasts. The **immunophenotype** is usually CD7$^+$, CD3$^+$, other T-cell-associated antigens variable, TCRab or gd$^+$ or no TCR, TdT$^+$, CD1a$^{+/-}$, often CD4,8 double positive or coexpression, Ig$^-$, B-cell-associated antigens$^-$; occasional cases express natural killer (NK) antigens. Immunophenotyping is necessary to distinguish this tumor from precursor B-lymphoblastic neoplasms and, occasionally, from peripheral B- or T-cell neoplasms. **Cytogenetic analysis** shows rearrangement of *TCR* genes, with translocation in the alpha and delta T-cell receptors. Activating mutations of the transmembrane receptor *NOTCH1* are common. In some cases, IgH rearrangement may be seen. The probable origin of the tumor is a precursor T-lymphoblast, either prothymocyte, early thymocyte, or common thymocyte.

Clinical Features

Patients are predominantly adolescent and young adult males, but older adults are occasionally affected. It accounts for 40% of childhood lymphomas and 15% of acute lymphoblastic leukemias (ALL). Patients present with rapidly enlarging mediastinal (thymic) masses and/or peripheral **lymphadenopathy**. The disease is a lymphomatous one if the patient presents with a mass lesion and 25% or fewer lymphoblasts in the marrow. T-cell leukemia typically presents with a high blast count and a mediastinal mass. Untreated, it is rapidly fatal, usually terminating with features of ALL, commonly with central nervous system involvement. The tumor is highly aggressive but potentially treatable.

Staging

See **Lymphoproliferative disorders**.

Treatment

See also **Acute lymphoblastic leukemia; Non-Hodgkin lymphoma**.
The prognosis in childhood T-ALL treated with aggressive therapeutic protocols is comparable with B-ALL.

PREDNISOLONE

See **Corticosteroids**.

PREDNISONE

See **Corticosteroids**.

PREGNANCY

Hematological changes and disorders associated with pregnancy. These are affected by the interaction between the fetus and the mother.

Physiological Changes

Immunological

The fetus expresses paternal allogeneic antigens and so has the potential to be rejected by maternal T-cells. The placenta expresses high levels of **human leukocyte antigen** HLA-G, which helps it to evade effects of natural killer (NK) **lymphocytes**. The placenta also secretes Th2 **cytokines**, which inhibit T-cell responses. During pregnancy, estrogen exerts a systematic reduction in T-cell activity. It has the effect on B-cells of increasing their synthesis of IgG and IgA. The fetus is supplied with some immunological memory. **Major histocompatibility antigens** are absent from the trophoblast, so that allogeneic responses do not occur and fetal stem cells bind maternal **immunoglobulin**. All of these mechanisms may explain the nonrejection of the fetus as an allograft or the nonproduction of maternal **antibodies** to paternal **antigen** present in the fetus.

Blood Volume

Plasma volume (**blood volume**) increases in the second trimester to a peak of around 43% of the nonpregnant level at 34 to 36 weeks, falling during the second week of the puerperium. The rise is even greater in subsequent pregnancies.
  Red blood cell mass shows a linear increase throughout pregnancy by about 17 to 25% due to increased erythropoiesis followed by a slow decline in the puerperium. Due to

plasma dilution, average levels are **hemoglobin** 11 g/dl and **packed-cell volume** (PCV) 0.32 to 0.34 l/l, with normal mean corpuscular hemoglobin concentration (MCHC) (see **Red blood cell indices**).

Leukocytes

Neutrophilia reaches a peak of around $10 \times 10^9/l$ at the 30th week, with a fourfold increase during labor, the level dropping late in the puerperium. Eosinophils, basophils, lymphocytes, and monocytes remain unchanged numerically, but lymphocyte function and cell-mediated immunity are depressed due to an increased concentration of glycoproteins coating the cell surfaces.

Platelets

Platelets fall slightly after the 26th week, with a rise in the mean platelet volume. After parturition, platelet levels temporarily rise, with increased platelet adhesiveness.

Coagulation

Factors VII, VIII, VIIIC, X, and **IX** increase during pregnancy, but **Factor XI** levels fall by 60 to 70% and **Factor XIII** by 50%. Factors II and V remain constant. Plasma **fibrinogen** rises to around 6.0 g/l, with an increase in the high-molecular-weight fibrin/fibrinogen complexes. At parturition, there is release of placental tissue factor (thromboplastin), with a consequent fall in fibrinogen and a rise of fibrin degradation products — physiological disseminated intravascular coagulation (DIC) but with a rapid return to normal within 1 h of delivery. The first 10 days postpartum are associated with a rise in all coagulation factors. Maternal thrombophilia does not cause fetal growth restriction.

Hematological Disorders Complicating Pregnancy

Anemia

Physiological **dilutional anemia** not requiring treatment.

Iron deficiency as a result of limited **iron** reserves before pregnancy, with additional iron requirements from the fetus ending in a negative iron balance for the mother. Each pregnancy results in the loss of 680 mg iron, equivalent to 1300 ml blood. This anemia requires the administration of iron supplement during pregnancy. Puerperal anemia can follow an iron deficiency of pregnancy, particularly if there has been postpartum hemorrhage.

Folic acid deficiency anemia, usually a **macrocytic anemia**, but by the third trimester **megaloblastosis** can occur unless folate supplements are administered. Malabsorption may contribute, but the deficiency is mainly due to increased fetal requirements. Prophylactic folic acid during pregnancy prevents this anemia. Deficiency of folates gives rise to neural-tube defects, which lead to anencephaly and spina bifida. Folic acid taken during the first trimester and preferably from 1 month prior to conception protects the fetus from these disorders. Prevention is legislated in North America by fortifying grain and pasta with folic acid.

Obstetric Hemorrhage

Antepartum hemorrhage due to:
- Placenta previa requiring immediate **red blood cell transfusion** and obstetric management as appropriate.

- Abruptio placentae with afibrinogenemia related to **disseminated intravascular coagulation** (DIC), requiring immediate treatment by transfusion of **fresh-frozen plasma**, **fibrinogen**, red blood cells, and platelets, with delivery of the fetus as soon as possible.

Postpartum hemorrhage, usually caused by retained placental products or obstetric trauma. Blood loss must be immediately replaced by red blood cell transfusion and the obstetric cause corrected. If hemorrhage persists, coagulation factor deficiency must be considered, identified, and treated appropriately.

Thrombocytopenia[286,421]

"Benign" gestational thrombocytopenia (incidental thrombocytopenia of pregnancy). Occurs in about 7% of pregnancies but has no clinical consequence for either mother or fetus.

Immune thrombocytopenic purpura (ITP). This may arise for no obvious cause during pregnancy; management is similar to other causes of ITP.

HELLP (hemolysis, elevated liver enzymes, low platelets, and pregnancy). A syndrome of hemolysis, raised liver enzymes, and thrombocytopenia of no known cause but requiring urgent termination of pregnancy to avoid maternal cerebral hemorrhage and hepatic failure.

Consumption coagulopathy. Due to hypercoagulable states (see below).

Coagulation Disorders

Physiological changes are accentuated by:

Preeclampsia and eclampsia with hypertension, proteinuria, and edema, followed by grand mal seizures and a form of **microangiopathic hemolytic anemia**

Thrombotic thrombocytopenic purpura (TTP), with an 80% fetal mortality due to abruptio placentae (see above)

Postpartum **hemolytic uremic syndrome**, usually following a normal delivery, with acute renal failure requiring renal dialysis

Amniotic fluid embolism, which is an uncommon but usually fatal maternal complication due to **fibrinolysis** causing persistent bleeding from the genital tract and venepuncture sites; cardiopulmonary resuscitation and immediate delivery is necessary, with transfusion of fresh-frozen plasma and platelets followed after delivery by heparin to arrest the cycle of thrombosis, coagulation deficiency, and hemolysis

Retained dead fetus, rarely causing consumption coagulopathy and requiring induction if delivery has not occurred spontaneously

Acquired inhibitors of coagulation factors, usually factor VIII IgG antibodies, which may disappear spontaneously or in response to corticosteroid therapy

Acute hepatic failure occurring with preeclampsia and associated with severe coagulopathy of DIC type, requiring immediate delivery and correction of the coagulation disorders

Antiphospholipid antibody syndrome (APLS) leads to high fetal loss due to Annexin V, a potent anticoagulant produced by villous trophoblasts, binding to phospholipids, and inhibiting coagulation

Postpartum **venous thromboembolic disease** is common, requiring treatment with heparin followed by warfarin

Effects of Pregnancy on Established Hematological Disorders

Sickle cell disorder (SCD). There is a mortality of 20% for the mother and 50% for the fetus due to high oxygen extraction by the placenta, which induces sickling, usually in the third trimester. Premature delivery is common. Major SCD infarctions may follow in the postpartum period.

Thalassemia. Anemia increases with pregnancy. There is an increased incidence of abortion due to hydrops fetalis.

Immune thrombocytopenic purpura (ITP). Spontaneous abortion may occur due to perinatal hemorrhage in the fetus or to intra- or postpartum hemorrhage. Neonatal ITP may follow. Treatment with either corticosteroids or **splenectomy** carries hazards for the fetus.

Inherited and acquired coagulation disorders. **Hemophilia A** or **B** is rare, but **Von Willebrand Disease** is present in 1% of pregnancies. Invasive procedures during pregnancy and delivery should only be performed where **DDAVP** and **coagulation factor concentrates** are available.

Antiphospholipid-antibody syndrome. An increased risk of intrauterine growth retardation, stillbirth, and abortion due to fetal thrombosis occurs, which can be avoided by prophylactic LMW-heparin or aspirin.

Anticoagulant therapy. **Warfarin** is teratogenic for the fetus in early pregnancy and dangerous for the mother at parturition due to uncontrolled bleeding. Subcutaneous low-weight **heparin** is indicated if long-term anticoagulation is essential, but the risks must be carefully assessed.

Polycythemia rubra vera. This is a rare association, which can result in abortion. The PCV level should be held below 0.47 l/l by **phlebotomy**.

Malignancy. There is no known effect of pregnancy on malignancy itself, but some hematological malignancies can affect the course of the pregnancy, depending upon the type, its rate of progression, and treatment. **Acute myeloid leukemia** (AML) has occasionally been transmitted to the fetus.

Regimes of cytotoxic agent therapy. Pregnancy particularly affects those that include **methotrexate**, which is abortifacient, leading to a high incidence of fetal death or malformation. Therapy for low-grade **non-Hodgkin lymphoma** or **chronic myelogenous leukemia** should be withheld until parturition. Abortion or cesarean section is indicated for **Hodgkin disease**. With AML in the first trimester, abortion is recommended, but when diagnosed in the second or third trimesters, treatment is necessary. Future reproductive function is affected by all cytotoxic drugs.

Red blood cell transfusion. Particular care is necessary with regard to **pretransfusion testing**.[422]

PREKALLIKREIN

A precursor plasma protein that is involved in **contact activation** of blood coagulation. It is a 619-amino acid protein that exists in two forms, of molecular weight 85,000 and 88,000. These different forms represent different degrees of glycosylation of the protein. Prekallikrein circulates as a complex with **high-molecular-weight kininogen**. This complex is adsorbed onto negatively charged surfaces. Catalyzed by factor XIIa, prekallikrein is then converted to active **kallikrein** by cleavage of a single peptide bond. It circulates at a concentration of 2.5 mg/ml. Prekallikrein deficiency is of little clinical significance. Rare abnormal forms of prekallikrein have been described; these aberrant forms are not activated by factor XII.

PRELEUKEMIA

See **Myelodysplasia**.

PRELYMPHOMAS

See also **Lymphomatoid granulomatosis**; **Lymphomatoid papulosis**; **Angioimmunoblastic T-cell lymphoma**; **Monoclonal B-cell lymphocytosis**; **Monoclonal gammopathy** of undetermined significance (MGUS).
Lymphoproliferative disorders that are initially benign but with a strong propensity to transform to malignant lymphomas.

PREMATURE INFANT HEMATOLOGY

The hematological status of infants born before the 40th week of pregnancy. The levels of **hemoglobin**, red blood cells, **and hematocrit** are similar to those of the neonate at birth but fall precipitously, hemoglobin by 1 g/dl per week between 2 and 9 weeks of age; the lower the birth weight, the greater is the fall (anemia of prematurity). As the disorder is probably due to deficiency of erythropoietin, management by **red blood cell transfusion** can sometimes be avoided by treatment with recombinant human erythropoietin.[423] The place of **iron** supplementation varies with age of prematurity.[424] **Blood volume** levels are 90 to 105 ml/kg.

The **neutrophil** defense system is underdeveloped, with increased risks of infection. These risks are further increased by low levels of **immunoglobulins**, **complement** components, **fibronectin**, and **cytokines**.

Hemorrhagic disease of the newborn due to **vitamin K** deficiency occurs more frequently than in mature infants, severity being proportional to any associated coagulation factor deficiency, which is related directly to gestational age.

Values for coagulation tests and inhibitors in healthy premature infants are given in the **Reference Range Tables II, IV, XVI,** and **XVIII**.

PRENATAL DIAGNOSIS

See **Hemoglobinopathies**; **Hemolytic disease of the newborn**; **Down syndrome**.

PRETRANSFUSION TESTING

The procedures required prior to **red blood cell transfusion** to provide cells that have an acceptable survival after transfusion.[425] No *in vitro* test can guarantee the detection, in the prospective recipient of a blood transfusion, of all clinically important **alloantibodies** directed against blood groups, and tests designed to assess the extent of interaction between the recipient's **reticuloendothelial system** and donor red cells are too lengthy and complex to be usable routinely.

Most fatalities due to serologic incompatibility are the result of mismatches in the **ABO(H) blood groups**. It is important to prevent the production of alloantibodies to the **Rh ("Rhesus") blood group** D antigen, particularly for premenopausal females. Antibodies outside the ABO and Rh systems are much less likely to give rise to a hemolytic **blood**

transfusion complication, although it is likely that there is an underreporting of mild and delayed reactions (see **Blood groups**).

General Strategy

Serologic testing. Blood component units that are provided by transfusion centers and blood banks will usually have had the full range of mandatory serologic and microbiologic tests performed. Local or national policies need to be followed regarding any further serologic testing of donations and for components sourced elsewhere (e.g., autologous units).

ABO and RhD grouping of the recipient. Forward and reverse ABO group testing must be performed, except in the case of infants, in whose plasma ABO alloantibodies may not yet be present (see **ABO (H) blood groups**). ABO and RhD grouping is best performed using high-quality IgM monoclonal reagents. The ABO and RhD groups obtained should be checked against previous results. Any discrepancy or other anomaly in grouping should be resolved before transfusion. Such anomalies include:

- Polyagglutinability
- The presence of auto- or alloagglutinins, or the presence of weak ABO alloagglutinins (as in very young or elderly individuals, or patients with agammaglobulinemia), leading to an incorrect interpretation of the reverse ABO group
- Autoagglutination, leading to the incorrect interpretation of the forward ABO group
- Inherited weak or variant antigens of the ABO (H) and Rh blood group systems
- Change in blood group following **allogeneic stem cell transplantation**, or apparent change, e.g., following the transfusion of O RhD-negative blood to a person of another blood group
- Incorrect labeling of the sample at **venepuncture** or in the laboratory, or other technical or procedural errors

Red cell antibody screen using the recipient's plasma. The aim of this is to establish the presence of an irregular red cell alloantibody or antibodies of clinical significance, usually by testing against two or three examples of group O red blood cells (RBCs) of known phenotype, selected to cover a wide range of RBC antigens. This alerts the medical staff to any possible delay in the supply of compatible blood and provides the laboratory with time to identify the causative antibody and select suitable units. The **indirect antiglobulin (Coombs) test** (IAT) is the most appropriate method for screening. The use of unpooled red cells having homozygous expressions of RBC antigens is desirable, where possible, to maximize the sensitivity of the screen. When an antibody is detected in the screen, its specificity should be determined using an identification panel, and its clinical significance assessed, even if an antibody has previously been identified, as new specificities may have emerged. The IAT, enzyme techniques for antibody detection, and low ionic polycation techniques can be used in antibody identification.

Final donor RBC selection. This should include a cross-match procedure to exclude incompatibility between donor and recipient. There are two approaches:

- Total reliance is placed on the antibody screen to detect irregular antibodies. Exclusion of an inadvertent ABO mismatch is undertaken either by a direct agglutination ("immediate spin") serologic cross match using patient's serum and donor RBCs, or by computer verification ("electronic cross match") of the ABO groups.

- In addition to the mandatory serologic and microbiologic tests, further serologic evidence of donor RBC compatibility is provided using the IAT. This approach should always be used if the recipient is known to have, or have had, irregular antibodies.

It may be very difficult to find suitable blood for transfusion to a patient whose plasma contains an alloantibody with specificity for a high-frequency antigen and that can cause severe hemolytic blood transfusion complications. Suitable blood may need to be found from frozen blood inventories. Consideration should also be given to **autologous blood transfusion**. Other irregular alloantibodies may not be clinically significant, but may cause delay in the provision of suitable blood.

Quality Assurance

The majority of ABO-incompatible transfusions are the result of procedural errors, often outside the laboratory, such as incorrect labeling of the sample. It is essential that samples and request forms be fully completed at the time of venepuncture and that checks be carried out at every stage. Computer systems can minimize the chance of some errors occurring. Manual or computer systems must be capable of identifying "who did what, and how" at every stage. Blood transfusion complications and "near misses" should be documented and reviewed.

All reagents should meet national standards, be properly controlled, and be used as directed by the manufacturer. The indirect antiglobulin (Coombs) test is the most appropriate method for screening.

PRIMARY ACQUIRED SIDEROBLASTIC ANEMIA

See **Sideroblastic anemia**.

PRIMARY EFFUSION LYMPHOMA

See also **Lymphoproliferative disorders**; **Non-Hodgkin lymphoma**.
(Body cavity-based lymphoma) A rare lymphoma of large B-lymphocytes, typically associated with **HIV infection** and universally associated with human herpes virus 8 (HHV-8)/Kaposi's sarcoma herpes virus (KSHV). The **immunophenotype** of the tumor cells is CD45$^+$ and negative for B-cell-associated antigens.

Clinical Features

Primary effusion lymphoma most characteristically involves the pleura but also the pericardium and peritoneal cavities and gastrointestinal tract. This is an extremely aggressive lymphoma with a median survival of 6 months.

Staging

See **Lymphoproliferative disorders**.

Treatment

See **Non-Hodgkin lymphoma**.

PRIMARY IMMUNODEFICIENCY DISORDERS
See **Immunodeficiency**.

PRIMARY MEDIASTINAL (THYMIC) LARGE B-CELL LYMPHOMA
See also **Lymphoproliferative disorders; Non-Hodgkin lymphoma**.
(Malignant thymoma) An aggressive B-cell non-Hodgkin lymphoma of the **thymus**. The tumor is composed of large B-lymphoid cells with variable nuclear features, resembling **centroblasts**, large **centrocytes**, or multilobed cells, often with pale cytoplasm. **Reed-Sternberg**-like cells may be present. Many tumors have fine compartmentalizing sclerosis. The **immunophenotype** is often Ig$^-$, but it can include B-cell-associated antigens$^+$ (CD20$^+$, CD19$^+$), CD45$^{+/-}$, CD30$^{-/+}$ (weak), CD15$^-$. **Cytogenetic analysis** shows rearrangement of Ig heavy- and light-chain genes but without specific abnormality. Its cellular origin is presumed to be a thymic B-lymphocyte.

Clinical Features
A lymphoma of the mediastinum arising usually in the fourth decade, particularly in women. There is a locally invasive anterior mediastinal mass originating in the thymus, resulting in respiratory airway compression and superior vena caval obstruction. Relapses are extranodal, including liver, gastrointestinal tract, kidneys, ovaries, and central nervous system.

Treatment
See also **Non-Hodgkin lymphoma**.
Patients respond to systemic **cytotoxic agent** therapy (R-CHOP) followed by **radiotherapy** to the mediastinum.

PRIMARY PULMONARY HYPERTENSION
(PPH) An uncommon disorder characterized by increased pulmonary artery pressure and increased pulmonary vascular resistance without an obvious cause.[426] There is an overall female preponderance, with a peak incidence in the third and fourth decades of life. While there are numerous associated disorders, no specific cause has been identified. Some patients show mutations in chromosome 2q 31-33 and in the *BMPR2* (bone morphogenetic protein receptor) gene, a member of the TGF-β family.

Clinical diagnosis is based on the finding of dyspnea with a raised jugular venous pressure. Radiologically, there is an increase in size of the pulmonary artery with clear lung fields. Confirmatory diagnosis depends upon evidence of raised pulmonary artery pressures by cardiac catheterization.

There are three pathological arterial disturbances, clinically indistinguishable:

Plexogenic pulmonary arthropathy. Here there is medial smooth-muscle hypertrophy with concentric laminar fibrosis and plexiform lesions. This occurs in 30 to 60% of cases and affects younger persons.

Thrombotic pulmonary arteriopathy. Medial hypertrophy is here associated with eccentric intimal fibrosis and fibroelastic pads. These are the fibrous remnants of an organized **thrombosis** as part of a chronic thromboembolism. This lesion accounts for 40 to 50% of cases, with an equal sex incidence, and is associated with raised **factor VIII** levels. Some patients have anticardiolipid antibodies.

Pulmonary veno-occlusive disease. A rare condition (<10% of patients) of widespread intimal proliferation and fibrosis of the intrapulmonary veins.

Treatment for cardiac failure is with oxygen therapy, vasodilators, and anticoagulants. **Prostacyclin** (Epoprostenol) has been claimed to improve the quality of life. Endothelin-receptor antagonists are under trial. Heart and lung transplantation are the most effective treatment for younger patients. Surgical removal of a large thrombus is indicated.

PRIMARY THROMBOCYTHEMIA
See **Essential thrombocythemia**.

PRION DISEASE
Disorders resulting from alterations in the structure of proteins termed prions (proteinaceous infectious particles).[427,(16)] This is expressed in its normal form as a 27- to 30-kDa protein, PrPc, both on human cells and as free protein in **plasma**. It is encoded by *PRNP*, a gene on chromosome 20. In prion disease, the amino acid sequence of prion protein does not change, but alterations occur in its conformational structure PrPc of proteases and is associated with neuronal damage and encephalopathy (see Figure 100).

The abbreviated term for protease-resistant prion protein is PrPres. The encephalopathy has a characteristic histological spongiform appearance from which the alterative name for this group of disorders, the transmissible spongiform encephalopathies (TSEs), is derived.

Mutations in *PRNP* have been described in association with inherited encephalopathy, but these are rare. A prion disease in humans, called kuru, was associated with cannibalism, but as the brain of the dead was eaten by relatives, it was initially difficult to differentiate epidemiologically between oral transmission and inherited predisposition.[431] Subsequently, it was shown to be transmissible. Sporadic cases of prion disease, known as Creutzfeldt-Jacob disease (CJD), are also rare. The spread of the prion disease known

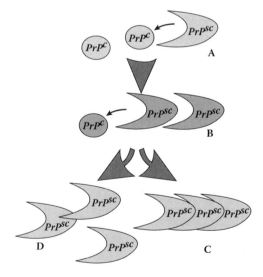

FIGURE 100
Schematic diagram of the prion hypothesis. Molecules of PrPSc interact directly with the PRPC causing the structural conversion of the PrPC to form further PrPSc (A and B). This conversion culminates in the formation of polymeric aggregates of PrPSC (C) or in the formation of further infective material (D). Note that this simplified diagram does not consider the different mechanisms of structural change[428] or include other molecules possibly involved in prion conversion.[429] (From Burthem J. and Roberts, D.J., *Br. J. Haematol.*, 122, 3–9, 2003. With permission of Blackwell Publ.)

as bovine spongiform encephalopathy (BSE) throughout the British Isles and subsequently abroad clearly demonstrated the potential impact of feeding animal remains to animals of the same species and highlighted the transmissible/infectious nature of prion disease. The discovery and rise in incidence of a variant of CJD (vCJD) in humans following the BSE epidemic in cows aroused concerns, confirmed in animal experiments, that the causative agent of BSE can cross species barriers if a large enough dose of abnormal prion is ingested orally.[432] Measures were taken to contain the spread of BSE internationally, but fears of a second epidemic in humans remained. Prion disease as a **transfusion-transmitted infection** was clearly documented in sheep experiments.[433,434] All cases of prion disease, including vCJD in the U.K., are reported to a central surveillance unit.[17] The full chain from donor to recipient is investigated for evidence of transmission if the affected individual has been a donor or recipient of blood components or products. Three cases of possible transmission by **red blood cell transfusion** have been reported.[435,436, 436a] This has further increased the pressure internationally to take measures to reduce the risk of prion transmission by transfusion, as there is no proven method of inactivating the infectious agent. When available, pooled blood components for transfusion should be replaced by recombinant proteins. In countries with a relatively high incidence of vCJD, donors from countries with a lower incidence can be used, if feasible, for products with long shelf lives. In the U.K., where the theoretical risk of encountering an asymptomatic donor with vCJD is highest, further measures have been taken. Fresh-frozen plasma for children is derived from non-U.K. donors, and all blood components are leukocyte depleted. Measures to prevent the possible transmission of prion disease by the use of surgical instruments are also undertaken, with neurosurgical procedures attracting the greatest scrutiny.[437] The relative risks and cost effectiveness of these strategies requires careful assessment.[437a] In the meantime, U.K. policy on blood donation excludes those who have previously received a blood transfusion.

As the differences between normal and abnormal prion protein are subtle, tests suitable for large-scale, cost-effective screening remain elusive. Published methods for testing blood or other body fluids lack sensitivity.[438] The diagnosis of affected cases depends on histopathology and laborious assays for PrPres.

There has been a decline in the number of reported cases in the U.K. since 2000.[439] All clinically affected individuals have been homozygous for methionine at a polymorphic site, codon 129, in the prion protein. This suggests that heterozygotes and cysteine homozygotes may be resistant to infection, but it remains possible that such individuals will present later and contribute to a second peak in incidence, as may recipients of blood from donors who are asymptomatic carriers. There is currently no proven treatment for affected individuals, and prion diseases are universally fatal. Further information on the current published evidence for therapeutic agents is available on-line.[17]

PROACCELERIN
See **Factor V.**

PROCARBAZINE

A methylhydrazine derivative used for **cytotoxic agent** therapy. Metabolic activation generates proximal reactions, causing methylation of **DNA** and subsequent chromosomal damage, including breaks and translocation. It is used in the treatment of **Hodgkin disease. Adverse drug reactions** include myelosuppression and **hypersensitivity** skin rashes. It is a mild monoamine oxidase inhibitor, but dietary restriction is not necessary. Alcohol ingestion may cause a disulfiram-like reaction.

PROCONVERTIN
See **Factor VII**.

PROERYTHROBLAST
See **Erythropoiesis**.

PROLYMPHOCYTIC LEUKEMIA/LYMPHOMA
(B-PLL) See **Chronic lymphatic leukemia; Non-Hodgkin lymphoma**.

PROMYELOCYTE
See **Neutrophils** — maturation and morphology.

PROMYELOCYTIC LEUKEMIA
See **Acute myeloid leukemia**.

PROSTACYCLIN
(Epoprostenol: PGI2) A **prostaglandin** synthesized from **arachidonic acid** via the **cyclooxygenase** pathway. It is derived largely from the vascular endothelium, although a small amount is produced by smooth muscle. It has potent platelet antiaggregatory effects and can disaggregate platelets that have already aggregated. Prostacyclin has the greatest antiplatelet aggregatory effect of all prostaglandins. It is metabolized to 6-keto-PGF1a, which is inactive.

Prostacyclin production is stimulated by thrombin, bradykinin, or adenine nucleotides. These agonists stimulate phospholipase C and arachidonic acid metabolism. Prostacyclin is metabolized in the systemic circulation and has a half-life of approximately 3 min.

The antiaggregatory effects of prostacyclin are mediated through the stimulation of adenyl cyclase and the cAMP second messenger system. Elevated levels of cAMP favor movement of Ca^{2+} into the dense bodies of the platelet, where it is inert, thus effectively decreasing intracellular Ca^{2+} levels. Intracellular Ca^{2+} is required for many functions/enzymes of the platelet. Basal endothelial prostacyclin levels are low, and free prostacyclin is bound to red cell. In contrast to its anti-aggregatory actions, prostacyclin is only mildly antiadhesive to platelets. The effects on aggregation of prostacyclin are only important in pathological platelet aggregation when adenosine diphosphate (ADP), adenosine triphosphate (ATP), and serotonin are released and thrombin generated.

Pharmacological agents such as streptokinase can release prostacyclin. This may be important in the balance of coagulation/fibrinolysis in thrombolytic therapy. Prostacyclin is also a potent vasodilator. It has been used clinically for its antiplatelet effects in cardiopulmonary bypass, renal dialysis, thrombotic thrombocytopenic purpura, and in peripheral vascular disease. It is usually given as a continuous infusion, as it has only a short half-life (2 to 3 min). Its actions are synergistic with heparin. The major side effect is hypotension.

PROSTAGLANDINS
A family of lipid-soluble hormonelike molecules, which can be regarded as derivatives from a hypothetical 20-carbon acid, prostanoic acid. Prostaglandins are produced via the **cyclooxygenase** pathways of **arachidonic acid** metabolism, a 20-carbon unsaturated

fatty acid. All prostaglandins contain a five-carbon cyclopentane ring. There are nine classes that encompass at least 16 different prostaglandins. Different prostaglandins are produced in different tissues in the body and occur in virtually all tissues. All prostaglandins are produced and released in response to a stimulus, and generally have potent biological activity at a local level. They are not stored, nor do they exist free in tissues. The functions of prostaglandins are very diverse, from immunomodulatory roles, through **inflammation**, to modulation of **hemostasis** and **platelet** function via **thromboxane** A2 and **prostacyclin**.

PROTEASOME

An abundant enzyme complex present in the cytoplasm and nucleus of all eukaryotic cells.[440] The primary function of the proteasome is to degrade proteins. Subunits termed ubiquitins are concerned with degradation to peptides that can be presented on **human leukocyte antigens** — Class 1 MHC antigens. As well as removing aberrant proteins from cells, the proteasome degrades and recycles various short-lived proteins. By regulating the turnover of these proteins, the proteasome plays a critical role in the maintenance of cellular homeostasis. Over 80% of all cellular proteins are recycled through the proteasome; substrates include cell-cycle regulators, signaling molecules, tumor suppressors, transcription factors, and antiapoptotic proteins.

Disruption of the tumorigenic regulatory proteins interferes with the systematic activation of signaling pathways required for tumor cell growth and survival and results in inhibited tumor growth and subsequent tumor cell death. Inhibition of the proteasome was therefore identified as a potential therapeutic target for anticancer therapy.[441] **Bortezomib** (PS-341, Velcade), the first proteasome inhibitor evaluated in human clinical trials, has been approved by the U.S. Food and Drug Administration for use in patients with refractory or relapsed **myelomatosis**. Preclinical study results show that bortezomib suppresses tumor cell growth, induces **apoptosis**, overcomes resistance to standard **cytotoxic agents** and **radiotherapy**, and inhibits **angiogenesis**. It is undergoing trial for the treatment of refractory relapsing myeloma and renal carcinoma.

PROTEIN C

A two-chain **vitamin K**-dependent **serine protease** synthesized by the liver, comprising a Gla-rich light chain (MW 22,000) and a heavy chain containing the serine active site (MW 40,000). Together with its cofactor, protein S, protein C serves to inactivate the cofactors of coagulation — factors Va and VIIIa (see Figure 101). It is coded for on chromosome 2q14-q21.

Protein C is homologous to **factors IX, VII**, and **X**. Thrombin generated during activation of the coagulation cascade binds to thrombomodulin (Tm), a protein found on the surface of the vascular endothelium. Thrombin bound to thrombomodulin has no procoagulant activity but is a potent activator of protein C and so functions as an anticoagulant. Tm accelerates the rate of activation of protein C approximately 20,000-fold. Activation of protein C occurs through a specific proteolytic cleavage at Arg 169–Leu 170. As activated protein C (APC) is a serine protease, it is inhibited by antithrombin and by a specific inhibitor — protein C inhibitor (PCI), also known as plasminogen activator inhibitor type 3 (PAI-3) — which both binds to and is activated by heparin and other glycosaminoglycans.

APC inhibits factors Va and VIIIa by specific proteolytic cleavages. This rate of inhibition is accelerated 200,000-fold in the presence of protein S. APC also stimulates fibrinolysis by binding to PAI-1, thus effectively increasing the relative concentration of tissue plasminogen activator (tPA). Activated protein C has been shown to have anti-inflammatory

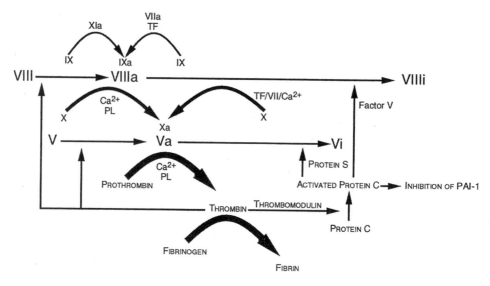

FIGURE 101
Diagrammatic representation of protein C and protein S activity.

properties by retardation of neutrophil migration. This activity can be used to control severe sepsis and **neutropenia** by treatment with rhAPC (recombinant activated protein C).

Assay of Protein C

This can be performed both functionally and immunologically. Clotting-based functional assays may be sensitive to the factor V Leiden mutation. Chromogenic assays are available for determining functional protein C activity. Protein C is activated by a snake venom (PROTAC C™), and the rate of generation of APC is measured by a specific chromogenic substrate. As no **reference material** is available, each laboratory should establish its own normal range, which will usually be 0.6 U/ml or 70 to 140%. Levels for premature and full-term infants are given in **Reference Range Tables XVII** and **XVIII**.

Deficiency of Protein C

This is inherited in both dominant and recessive forms. Its prevalence in the healthy blood donor population is estimated to be as high as 1 in 500, and it is present in 5 to 10% of patients with **venous thromboembolic disease**. Levels of protein C are low at birth and increase over the first six months of life, although they may not reach adult values until late childhood or adolescence. A deficiency of protein C is associated with an increased risk of venous thromboembolic disease. Homozygotes with <5 U/dl functional protein C may present in the neonatal period with neonatal **purpura fulminans**. Individuals with 5 to 10 U/dl may be asymptomatic until the second or third decade of life but are at a significantly increased risk of thromboembolic disease. Deficiencies of protein C may be associated with **warfarin**-induced skin necrosis.

A protein C concentrate is available for treatment of congenital deficiency.

Resistance to Activated Protein C (APCr)

This term describes the poor anticoagulant response observed in some individuals when activated protein C (APC) is added to their plasma. The anticoagulant property of APC lies in its capacity to inhibit the cofactors Va and VIIIa by limited proteolytic cleavage. In 95% of cases of APC resistance, a single-point mutation within the factor V gene (1691 CGAÆ CAA) results in an arginine-to-glutamine substitution in codon 506 (Arg506Gln — known as factor V Leiden) and the abolition of an APC cleavage site. Factor V, but not the factor V Leiden variant, functions as a cofactor for APC in the inactivation of VIIIa. Individuals with the factor V Leiden mutation, therefore, fail to inactivate both Va and VIIIa as rapidly as an unaffected individual. Factor Va can, however, be inactivated via a second APC cleavage site at arginine 306, but this takes place some ten times slower than cleavage at arginine 506. This reduced but measurable inactivation of the factor V Leiden variant may explain why APC resistance is a mild risk factor for thrombosis. APC resistance is associated primarily with venous thromboembolic disease.[77] It does not appear be associated with an increased risk of either cardiovascular disease or cerebrovascular disease.

Clotting-based assays are employed to detect phenotypic APC resistance, and various DNA-based tests are employed to detect the presence of the factor V Leiden gene mutation. The clotting-based tests are affected by a number of variables, including sample preparation, pregnancy, the presence of a lupus anticoagulant, heparin and warfarin, and various clotting factor deficiencies. High levels of factor VIII may also give rise to spurious results. In addition, APC resistance may give rise to false-positive results with prothrombin-based functional assays for protein S and some clotting-based assays for protein C.

APC resistance is found in between 20 and 50% of venous thrombosis patients, depending upon the criteria for patient selection, and between 3 and 10% of normal asymptomatic individuals. Approximately 95% of cases of APC resistance show the Arg506Gln mutation, but in the remaining 5% of cases, the cause of the resistance remains unclear, and in particular, these cases do not demonstrate mutations at any of the other APC cleavage sites in either Va or VIIIa. APC resistance shows marked geographical segregation and is rare in some parts of the world. Acquired APC resistance of a similar magnitude occurs in women taking third-generation contraceptive pills.

PROTEIN S

A **vitamin K**-dependent protein that circulates in plasma as a single-chain glycoprotein of molecular weight 60 kDa. It is synthesized by the **liver**, **vascular endothelium**, and **megakaryocytes** (and is subsequently stored in the α-granules of the **platelets**). It is not a serine protease. Approximately 60% of protein S is complexed to C4b-binding protein, while the remainder is unbound. It is only the unbound or "free" protein S that is physiologically active. Protein S has a $T^{1/2}$ of approximately 60 h. Protein S functions as a cofactor for the inactivation of factors Va and VIIIa by activated **protein C** (see Figure 102). Endothelial cells synthesize protein S, and the endothelial cell surface promotes the inactivation of Va by activated protein C. Both bound and unbound "free" protein S bind to negatively charged phospholipids exposed on the surface of activated platelets. Free Protein S, although having no direct activity, forms a calcium-dependent complex with activated protein C, enhancing the binding of activated protein C to phospholipid and thereby the functional activity of protein C

Assay of Protein S

This can be performed both functionally and immunologically. Total protein S antigen is commonly measured by ELISA **immunoassay**. Free protein S antigen is assayed in plasma following precipitation of the bound component by PEG 8000 or by using specific monoclonal antibodies. Functional protein S assays commonly involve a prothrombin-based assay using protein S-deficient plasma, in which protein C is activated by Protac C™. Such assays are sensitive to the presence of APC resistance, resulting in false positives. As no **reference material** is available, each laboratory should establish its own normal range, which will usually be around 0.9 U/ml. Protein S levels fall in pregnancy. Values for premature and full-term infants are given in **Reference Range Tables XVII** and **XVIII**.

Deficiency of Protein S

This has been estimated to affect between 4 and 5% of patients with thrombotic disease and, by extrapolation, approximately 1 in 33,000 of the general (Dutch) population. Acquired protein S deficiency is seen in **pregnancy**, in patients with **sickle cell disorder**, and with some viral infections, e.g., **human immunodeficiency virus** (HIV) and varicella. Protein S levels are low at birth and increase with age, reaching adult values before 6 months of age. The majority of patients with protein S deficiency appear to be heterozygous for an inherited defect.

Protein S deficiency is conventionally classified into types I to III based upon immunological and functional assays of protein S:

Type I. Parallel decrease in both total and free protein S

Type II. Functional protein S activity is decreased, whereas the levels of total and free protein S are normal

Type III. Normal concentrations of total protein S, but reduced levels of free protein S

It is probable that types I and III deficiencies are phenotypic variants of the same genetic mutation. The clinical manifestations of protein S deficiency are superficial thrombophlebitis and **venous thromboembolic disease**. Protein S deficiency also appears to be associated with an increased risk of premature **arterial thrombosis**. Cases of **warfarin**-induced skin necrosis and neonatal **purpura fulminans** have been reported in association with protein S deficiency.

PROTEIN Z-DEPENDENT PROTEASE INHIBITOR

Protein Z-dependent protease inhibitor (ZPI) is a single-chain glycoprotein. Protein Z is a **vitamin K**-dependent protein that does not have any inherent enzymatic activity. It is a cofactor for ZPI. The mechanism of its interaction with ZPI is unknown. In plasma, ZPI and protein Z circulate as a complex, with ZPI in a molar excess and all circulating protein Z bound to ZPI.

The precise physiological role of ZPI is uncertain, but *in vitro* studies suggest that it functions as an inhibitor of **coagulation of blood**. It potentially acts at two sites, inhibiting the activated factors X (Xa) and XI (XIa).

The inhibition of factor Xa requires calcium and phospholipid and is enhanced approximately 1000-fold in the presence of protein Z. ZPI may act at the platelet phospholipids membrane during coagulation, differing from the other major factor Xa inhibitor, antithrombin, which is assumed to act at the endothelial surface bound to heparan sulfate.

ZPI inhibition of factor XIa is via a different mechanism, which does not require protein Z, calcium ions, or phospholipid. This inhibitory activity is enhanced by **heparin**, suggesting that ZPI may be a heparin-binding serpin.

Preliminary data has suggested that defects in the ZPI and in the protein Z axis may be associated with arterial and venous **thrombosis**.

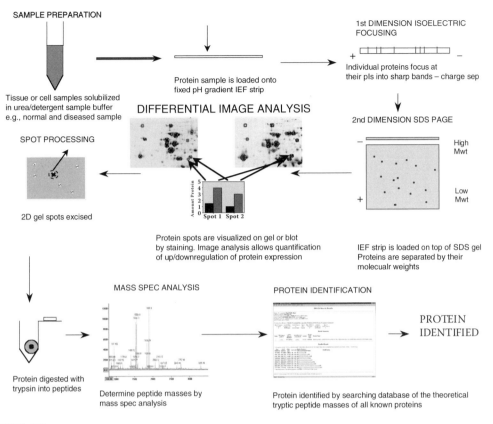

FIGURE 102
The proteomic process.

PROTEOMICS

The global study of protein expression in cell, tissue, or fluid samples. This approach has advantages over study of mRNA expression, at least in part due to the fact that proteins are closer to the final biological processes than RNA. Furthermore, there is often a discordance between mRNA levels and protein levels. However, proteomics is a slow and resource-intensive technology that is still being refined and therefore has to be considered an investigative tool at the current time (see Figure 102).

Proteomics relies on appropriate sample collection into buffers that prevent protein degradation and solubilization. Particular target proteins may be of low abundance and therefore require concentration and removal of interacting agents, e.g., albumin or platelets from blood.

The protein components require separation, usually by one- or two-dimensional SDS PAGE (sodium dodecyl sulfate–polyacrylamide gel electrophoresis). The separated proteins are then identified using tryptic digestion and mass spectrometry (MS) or liquid chromatography coupled to MS (LC/MS). Proteins found in fluids such as urine or blood can also be separated and analyzed using MALDI (matrix-associated laser desorption/ionization) or SELDI (surface-enhanced laser desorption/ionization). These techniques rely on placing the sample on a chip and ionizing the proteins using a laser beam (see Figure 103). The ionized proteins have a variable charge-to-mass ratio and therefore migrate to the analyzer at different rates (time of flight, TOF). On this basis, the proteins can be separated and analyzed.

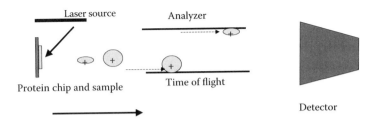

FIGURE 103
Comparison of MALDI and SELDI mass spectroscopy.

PROTHROMBIN

(Factor II) A **vitamin K**-dependent glycoprotein. It is synthesized by the liver, circulating in plasma as a single-chain protein of 71.6 kDa molecular weight, at a concentration of approximately 100 mg/ml (1.5 μM). Prothrombin, in common with other coagulation factors, is synthesized with a hydrophobic signal peptide (residues 43 → −19), which is cleaved by a signal peptidase within the rough **endoplasmic reticulum**. The propeptide of prothrombin (residues 18 →−1) is removed by an intracellular propeptidase following c-carboxylation to leave the mature protein of 579 amino acids. In common with other vitamin K-dependent clotting factors, prothrombin undergoes a series of posttranslational modifications in which the N-terminal ten glutamate residues are c-carboxylated to c-carboxyglutamic acid (the so-called Gla residues). These Gla residues are crucial for the binding of Ca^{2+} ions by prothrombin, and their absence results in a dysfunctional pro-thrombin molecule.

Prothrombin circulates at a concentration of 10.0 to 15.0 mg/dl. Congenital prothrombin deficiency is a rare autosomal, incompletely recessive disorder that leads to easy bruising, menorrhagia, and hemorrhage. Heterozygotes with activity levels around 50% of normal are generally asymptomatic, and only homozygous patients with activity levels <10% of normal experience bleeding problems.

Prothrombin Activation

Enzymatic activation of prothrombin to thrombin — with the release of a large activation peptide (consisting of the Gla domain, helix, loop, and both **kringles**, termed fragment 1+2 [designated F1+2 or F1.2 or F1-2]) — can occur in two ways:

1. An initial cleavage at arginine 271, resulting in the generation of prethrombin 2 and fragment 1+2. These two fragments remain noncovalently associated. A second cleavage in the enzymatically inactive prethrombin 2 at arginine 320 results in the generation of the active α-thrombin molecule and dissociation of fragment 1+2. This second cleavage occurs between the heavy and light chains of thrombin, but does not release them as separate fragments because of a disulfide bond between Cys22 and Cys439.

2. An initial cleavage at arginine 320, resulting in the generation of meizothrombin. Subsequent cleavage of meizothrombin at arginine 284 liberates the active a-thrombin and fragment 1+2. Subsequent autocatalytic cleavage by thrombin of fragment 1+2 at arg155 results in the generation of two separate fragments, fragment 1 and fragment 2.

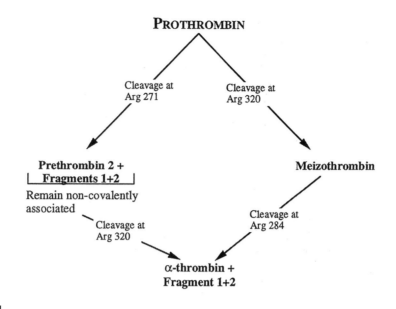

FIGURE 104
Diagrammatic representation of prothrombin activation.

Current evidence suggests that it is the first of these two activation pathways that is important. Efficient activation of prothrombin by factor Xa requires Va, Ca^{2+}, and a suitable negatively charged phospholipid surface, the so-called prothrombinase complex. Kinetic analysis shows that this combination of components is some 300,000 times more efficient than factor Xa alone in the activation of prothrombin. Human prothrombin, fragment 1+2, and α-thrombin are also susceptible to proteolysis by thrombin at residue 155 and 284. However, cleavage of prothrombin by thrombin results in the generation of prethrombin I and the release of fragment 1. The release of fragment 1 results in a failure of prethrombin I to bind to phospholipid membranes, and consequently no acceleration of thrombin activation occurs. Although conversion of prethrombin I to α-thrombin can occur, the failure of prethrombin I to bind to phospholipid membranes results in any thrombin generated rapidly diffusing away.

Prothrombin activation is diagrammatically represented in Figure 104.

PROTHROMBINASE COMPLEX

A complex that assembles on phospholipid membrane surfaces comprising factor Xa, factor Va complex, together with calcium ions. Its function is to convert **prothrombin** to **thrombin**. Assembly of the complex on phospholipid membranes facilitates enzyme-substrate alignment and interaction. Formation of the complex proceeds in three phases: binding of factor Va to membrane surfaces, binding of factor Xa to membrane surfaces, and interaction of bound factor Va and factor Xa in the presence of calcium. Factor Xa can convert prothrombin to thrombin alone. This reaction is 300,000-fold slower than via the prothrombinase complex. Membrane surfaces involved are those of **platelets**, **vascular endothelium**, **lymphocytes**, and **monocytes**.

PROTHROMBIN GENE MUTATION

A variant in the untranslated region of the prothrombin gene (*G20210A*). This variant is associated with higher levels of plasma **prothrombin**. The high levels of prothrombin are believed to enhance thrombin generation and thereby promote thrombosis. This gene mutation is associated with an increased risk of venous thromboembolism (VTE). The mutation is found in heterozygous form in ≈1–2% of Caucasian populations. It is present in ≈10% of those presenting with first VTE. Having the mutation raises the relative risk of thrombosis between two- to sixfold.

PROTHROMBIN TIME

(PT) The time taken for citrated plasma to clot at 37°C after calcium has been added to plasma that has had an optimal amount of **tissue factor** (thromboplastin) previously added.[19] The PT is a measure of the overall efficiency of the extrinsic pathway of blood coagulation. It is dependent upon **factors II, V, VII, X**, and **fibrinogen**. It is most sensitive to changes in factors V, VII, and X and less so on factor II. It is relatively insensitive to changes in fibrinogen and should not be used to detect minor changes in fibrinogen concentration. Times obtained are very dependent upon the origin of the thromboplastin. Human brain extracts have been used historically, but these are no longer used because of the risk of viral and prion infection. The majority of natural thromboplastins used are of rabbit brain or lung origin. The use of recombinant tissue factor is becoming increasingly common. Each preparation has a different sensitivity to clotting factor deficiencies and defects, and particularly those defects induced by oral anticoagulants such as warfarin. Indeed, some thromboplastins are not sensitive at all to an isolated factor VII deficiency. Prolongation of the prothrombin time is seen commonly in **liver disorders**, **warfarin** overdosage, **vitamin K** deficiency, and **disseminated intravascular coagulation** (DIC).

The potential variation in thromboplastin sensitivity is of particular importance in monitoring oral anticoagulant control. A thromboplastin preparation that has been calibrated against the World Health Organization **reference material** for thromboplastin should always be used. Following calibration against such a standard reference material, a thromboplastin reagent is given an **international sensitivity index** (ISI). For patients on oral anticoagulants, the results should be expressed as the **international normalized ratio** (INR). This is a derivative of the prothrombin time and the ISI of the thromboplastin used. As prothrombin times and prothrombin ratios would otherwise vary widely dependent upon assay and reagents used, use of the INR allows standardization of therapeutic ranges. The INR should only be used for patients who are on oral anticoagulants and not used as shorthand for communicating PTs in patients who are not on anticoagulants. Values for PTs at various ages are given in the **Reference Range Tables I, XV**, and **XVI**.

PROTOCOPROPORPHYRINURIA

See **Variegate porphyrias**.

PROTOFIBRILS

Two-stranded polymers of **fibrin** that are formed as intermediates in fibrin assembly following **thrombin** cleavage of **fibrinogen**.

PROTOZOAN INFECTION DISORDERS

See also **Leishmaniasis; Malaria; Trypanosomiasis**.
The hematological changes occurring with protozoan infections. These are listed in Table 134.

TABLE 134

Protozoan Infections Causing Hematological Disorders

Infection	Hematological Change
Amebiasis	**neutrophilia**, **anemia** from intestinal hemorrhage, **liver disorder**
Babesiosis	**hemolytic anemia**
Chaga disease	**lymphocytosis, lymphadenopathy**, hepatosplenomegaly
Leishmaniasis	**pancytopenia**, hepatosplenomegaly, lymphadenopathy, hemolytic anemia, parasitized
(visceral)	**histiocytes** (macrophages) in bone marrow (Leishman-Donovan bodies**)**
Malaria	hemolytic anemia
Toxoplasmosis	
Acquired	hemolytic anemia, lymphadenopathy, **atypical lymphocytes** with parasitemia
Congenital	hepatosplenomegaly, **thrombocytopenia, leukemoid reaction**

PRUSSIAN BLUE (PERL'S) REACTION

The standard staining reaction for nonhemoglobin ferric **iron**. It is achieved by the reaction of ferric iron with potassium ferrocyanide to form an insoluble blue-colored compound, ferriferrocyanide. Prussian blue-reactive iron is often detected in marrow **histiocytes** (macrophages) of normal persons (storage iron as **ferritin** and **hemosiderin**), in a minority of normal marrow erythroblasts (as cytoplasmic ferritin deposits), and in the parenchymal cells and macrophages of the liver or other organs of persons with untreated **hereditary hemochromatosis** or other forms of **iron overload**.

PSEUDOERYTHROCYTOSIS

See **Erythrocytosis** — relative erythrocytosis.

PSEUDO-GAUCHER CELL

Gaucher-like cells with "crinkled" cytoplasm. They occur in the bone marrow of patients with **chronic myelogenous leukemia**, and may also be found in **thalassemia**, **congenital dyserythropoietic anemia** type 1, **Hodgkin disease**, **myelomatosis**, **chronic lymphatic leukemia**, and **human immunodeficiency virus** (HIV) infection. Glucocerebrosidase levels are normal in these conditions, and it is thought that the ability of patients to metabolize globoside released from cellular membranes is exceeded.

PSEUDOPOLYCYTHEMIA

See **Erythrocytosis**.

PSEUDOTHROMBOCYTOPENIA

A common artifactual laboratory abnormality in which artificially low **platelet counts** are recorded. It may arise through **platelet satellitism**, poor **venepuncture** technique and blood collection (leading to *in vitro* thrombin generation), overfilling of sample tubes, or EDTA-induced pseudothrombocytopenia, the commonest cause being due to platelet-reactive antibodies that react with platelets at low calcium concentrations. Pseudothrombocytopenia is therefore commonly seen in blood samples taken into EDTA anticoagulants. Examination of the blood film often shows clumping of platelets. The correct platelet count can be determined by measuring the platelet count in a sample collected in an alternative anticoagulant, such as trisodium citrate.

PSEUDOTUMORS

Space-occupying lesions that result from repeated subperiosteal or intraosseous hemor-rhage occurring in **hemophilia A** and **hemophilia B.** They lead to bone destruction, followed by new bone formation and expansion of bone. They are of two types: adult type, occurring proximally in the pelvis of the femur; and childhood type, occurring distally to elbows and knees. Childhood-type pseudotumors have a better prognosis. Early management is conservative, with immobilization and adequate factor replacement. Aspiration of a pseudotumor is contraindicated. If lesions progress, surgical excision is indicated to avoid potential fracture, pressure symptoms, or fistula formation.

PSEUDO-VON WILLEBRAND DISEASE

(Pseudo-VWD) A primary platelet disorder due to an abnormal platelet receptor that has an increased affinity for normal **Von Willebrand Factor** (VWF). Phenotypically, pseudo-VWD is similar to type 2B VWD, as both arise from an increased platelet–VWF interaction. Both show an increased platelet aggregation response to low-dose ristocetin (0.5 mg/ml). Mild **thrombocytopenia** is often seen. Pseudo-VWD can be distinguished from type 2B VWD by addition of normal human cryoprecipitate to patient platelet-rich plasma. In pseudo-VWD, this will induce platelet aggregation, as the increased reactivity resides in the platelets, but it will not in type 2B, where the increased reactivity is due to abnormal VWF. Pseudo-VWD can be inherited or acquired.

PSEUDOXANTHOMA ELASTICUM

A hereditary disorder of elastic tissue giving rise to **vascular purpura**. Spontaneous joint and muscle bleeds are rare. Four disease types are recognized:

Type I. Dominant inheritance — small yellowish papules that, in the elderly, are soft and lax and hang in folds; flexurally distributed, especially in the groin, axillae, and neck; may be associated with peripheral vascular disease and coronary artery disease; degeneration of Bruch's membrane may occur, leading to early blindness

Type II. Dominant inheritance — small yellowish papules, but flatter and fewer than in type I; retinal and vascular changes are mild; affected individuals have blue sclera

Type III. Recessive and resembles type I, but only mild vascular and retinal degeneration

Type IV. Rare recessive form of the disorder; skin changes are extensive and generalized

PSORIATIC ARTHROPATHY

See **Skin disorders**.

PSYCHOGENIC PURPURA

See **Bruising**.

PTEROYLGLUTAMIC ACID

See **Folic acid**.

PULMONARY EMBOLISM
See **Venous thromboembolism**.

PULMONARY HYPERTENSION
See **Primary pulmonary hypertension**.

PUNCTATE BASOPHILIA
See **Basophilic stippling**.

PURE RED CELL APLASIA
(PRCA) Severe **anemia** with **reticulocyte counts** less than 1% and mature erythroblasts in the **bone marrow** less than 0.5%. **Blackfan-Diamond syndrome** can be regarded as an inherited form of the disorder. Acquired forms present as either an acute **aplastic crisis** or a chronic anemia, either idiopathic or in association with a wide variety of disorders (see Table 135)

TABLE 135

Causes of Pure Red Cell Aplasia

Congenital
  Transient erythroblastopenia of childhood
  Blackfan-Diamond syndrome
Acquired
  Acute aplastic crisis due to infection
  Parvovirus B19 infection
  Human immunodeficiency virus
  Epstein-Barr virus
  Chronic pure red cell aplasia
Idiopathic
  Systemic lupus erythematosus
  Rheumatoid arthritis
  Connective tissue disease
  Thymomas
  T-cell large granular lymphocyte leukemia
  Waldenström macroglobulinemia
  Adverse drug reactions: phenytoin, carbamazepine, chloramphenicol
  Pregnancy
  Protein malnutrition (Kwashiorkor)
  ABO-incompatible transplantation

Pathogenesis

Two mechanisms responsible for chronic PRCA have been proposed: immune mediated and direct cytotoxicity. However, it is often unclear which is the mechanism responsible and, indeed, both may be operational. For example, phenytoin can be directly toxic to erythroid precursor cells, but IgG — capable of suppressing both allogeneic and autologous erythroid colonies — has been found in patients on phenytoin with PRCA. Both humoral inhibitors (usually IgG) and cell-mediated inhibition have been described.

Causes of Pure Red Cell Aplasia

Acute PRCA

This is infective in origin and gives rise to **aplastic crisis**.

Idiopathic Chronic PRCA

This occurs at any age, but is usually found in adults aged 20 to 50, some of whom will have a weakly positive **direct antiglobulin (Coombs) test**.

Thymomas

These are present in about 5% of cases, and an estimated 7% of patients with thymomas have PRCA. Histologically, about 90% are spindle cell tumors, of which 20% are considered malignant. The remaining tumors are mostly lymphomas.

Clinical Features

Symptoms are proportional to the speed of the onset and the degree of normocytic anemia. Serum **iron** is high and **transferrin** usually completely saturated. Erythropoietin levels are elevated.

Thymomas may be asymptomatic at diagnosis of PRCA, but these are usually easily diagnosed with a lateral chest roentgenogram, mediastinal computerized tomography (CT), or magnetic resonance imaging (MRI).

Management (Apart from Aplastic Crisis)

Initial investigations should include a search for underlying causes, and all drugs with potential to cause PRCA should be discontinued.

Corticosteroids are often effective in both idiopathic and secondary PRCA, but prolonged therapy may be necessary. Other immunosuppressive agents have been used, with **ciclosporin** particularly useful. In unresponsive patients, **red blood cell transfusion** allows symptomatic relief, but **iron chelation** will be necessary to avoid **iron overload**. Recombinant erythropoietin has been reported to be of benefit. Thymectomy in patients with thymomas benefits about 50% of cases, but there is a high surgical mortality and relapse rate.

PURINE

A nitrogenous base of nucleic acids, **nucleosides**, and their derivatives. The commonest purines are adenine and guanine.

PURPURA

The appearance of multiple purple patches in the skin or mucous membranes due to subcutaneous or submucosal extravasation of blood. Similar lesions can be seen ophthalmoscopically in the retina. They have a clinical significance similar to **bruising, ecchymoses**, and **petechiae**, being an outward sign of a hemorrhagic disorder arising from **thrombocytopenia, platelet-function disorders**, and **vascular purpura**.

PURPURA FULMINANS

Injury to the microvasculature resulting in massive intravascular thrombosis[442] (see **Disseminated intravascular coagulation** and **skin disorders**). The disorder has been described in association with homozygous **protein C** deficiency, homozygous **protein S** deficiency, acquired protein S deficiency, and as secondary to varicella infections and meningococcal septicemia and **anti-phospholipid antibody syndrome**. Replacement of protein C is used in clinical practice, but is unlikely to be confirmed by a clinical trial due to practical difficulties.

PUVA LIGHT THERAPY

(Psoralen and ultraviolet A radiation therapy) A treatment used in many dermatological conditions, including those of hematological origin. Psoralen is given either orally or topically, followed a few hours later by irradiation with ultraviolet A. The combination of the two therapies (neither works alone) results in the production of phototoxic reactions, which are limited to the skin. Many courses may be required, depending on the severity of the original skin lesion. PUVA has proved useful in treating patients with **mycosis fungoides** and **urticaria pigmentosa**. PUVA has also been beneficial in the treatment of aquagenic pruritus in patients with **polycythemia rubra vera** who have failed supportive treatment. It is, however, cumbersome and carries the risk of late-onset melanoma and nonmelanoma skin cancers. Some patients may respond to ultraviolet (UVB) therapy.

PYKNOCYTE

(Bite cells) Irregularly contracted **red blood cells** with an irregular gap in the membrane due to removal of Heinz bodies by the spleen. They are particularly seen in patients with drug- or chemically induced **hemolytic anemia**.

The term "infantile pyknocytosis" describes a transient hemolytic anemia of unknown cause. Transfused red blood cells acquire the deformity and are destroyed prematurely, but the extrinsic agent has not been identified.

PYRIMIDINE

A nitrogenous base of nucleic acids, **nucleosides**, and their derivatives. The commonest pyrimidines are cytosine, thymine, and uracil.

PYRIMIDINE 5′-NUCLEOTIDASE

An enzyme involved in **nucleotide** metabolism, where it catalyzes the hydrolytic dephosphorylization of nucleoside monophosphate to nucleosides and inorganic phosphate. Deficiency of this enzyme is common, particularly in persons of Mediterranean, African, and Jewish descent. The defect is transmitted as an autosomally recessive trait caused by mutation of the P5′ N-1 gene. Heterozygotes are hematologically and biochemically normal, other than possessing only 50% of enzyme activity. Homozygotes have **hemolytic anemia** of variable severity, with enzyme levels about 5 to 10% of normal. Results of **splenectomy** are unpredictable.

An acquired form of pyrimidine nucleotidase deficiency occurs in **lead toxicity** and other heavy metal toxicity disorders — **arsenic**, **copper**, **mercury toxicity**.

PYRUVATE KINASE

(PK) A major enzyme of the Embden-Meyerhof pathway of **red blood cell** metabolism. A number of pyruvate kinase isoenzymes exist, the dominant ones being present in liver, muscle, leukocytes, platelets, and red blood cells. They are identifiable by distinct bio-chemical differences. PK catalyzes the second adenosine triphosphate (ATP)-generative step in glycolysis. Production is under the control of the *PK-LR* gene. There are over 150 mutations associated with deficiency disorders.

Deficiency[443]

The first enzymopathy deficiency recognized to cause **hereditary nonspherocytic hemolytic anemia**. Next to glucose-6-phosphate dehydrogenase (G-6PD) deficiency, this is the most common inherited red cell enzyme deficiency. The disorder is inherited as an autosomally recessive trait, affected individuals being either homozygous or doubly het-erozygous. Heterozygotes are hematologically normal, whereas homozygotes and double (compound) heterozygotes demonstrate chronic hemolytic anemia of variable severity, from the asymptomatic to severe transfusion dependency. The latter are often seen in the neonatal period. Severe deficiency may also manifest as **hydrops fetalis**. The double heterozygotes contain mixtures of two mutant isoenzyme forms. The affected red cells become rigid because of reduced ATP production. Although worldwide in distribution, most patients are of northern European or Mediterranean descent.

Clinical features include anemia, which may be severe but frequently produces only mild symptoms because of a shift to the right in the oxygen-dissociation curve (due to increase in 2,3-DPG). **Splenomegaly** and jaundice are usual, and gallstones are frequent. Frontal "bossing" may be apparent. Acute exacerbations of the chronic hemolytic process are associated with infections, pregnancy, or surgery. Aplastic crises occur in association with parvovirus and related infections (see **Pure red cell aplasia**).

In addition to anemia and elevated **mean corpuscular hemoglobin** (MCH), there is an increase in the **reticulocyte count** associated with erythroid hyperplasia in the bone mar-row. The peripheral-blood film is characteristic, showing "prickle" cells (**acanthocytes**, **echinocytes**, **spherocytes**, and xerocytes), all resulting from intracellular dehydration due to reduced ATP and Na$^+$, K$^+$-ATPase pump activity. The biochemical features of hemolysis are present. Autohemolysis is increased but is not corrected by glucose. The diagnosis is confirmed by direct enzyme assay. **Splenectomy**, indicated in those with repeated trans-fusion requirement, may alleviate the anemia but does not cure it.

Q

QUALITY MANAGEMENT IN THE HEMATOLOGY LABORATORY

The total quality management requirements for the performance of hematological investigations. Technical quality control of procedures, internal quality control, has been practiced since the introduction of devices for automated testing. External quality assurance of procedures, proficiency testing, was also soon established. More recently, the concept of total quality management to include the management of laboratory personnel and their environment has been introduced. This includes matters concerned with safety in the laboratory. These measures are applicable to all laboratories, but those handling investigation of patients have particular concerns that extend from turnaround time for emergency tests to ethics concerning the handling of specimens. Most countries have developed standards with regard to quality management, some of which are regulatory, and there are now a number of international standards developed by the International Standards Organization (ISO).[444,445]

The matters concerned can be summarized as:

Management system requirements

Organization and management

Quality management system

Document control

Referral of examinations to other laboratories

External services and supplies

Control of nonconformities

Consultative services and resolution of complaints

Preventive actions

Corrective actions

Quality and technical records

Internal audits

Management review

Resources and technical requirements

Personnel

Accommodation and environmental conditions

Laboratory equipment

Preexamination procedures

Examination procedures

Assuring the quality of examination procedures

Postexamination procedures

Reporting results

Alterations and amendments to reports

Laboratory information systems

Ethics in laboratory medicine

Special considerations must be given for laboratories undertaking hematological examinations[444a] and when **point-of-care testing** is in operation.

QUEBEC SYNDROME

An autosomally dominant bleeding disorder. **Platelets** have a severe deficiency of **multimerin**, a high molecular weight multimeric protein that is stored with factor V in platelet alpha granules. Thrombocytopenia is sometimes seen. Platelet aggregation defect is most striking with adrenaline. Quebec platelets appear to have large amounts of urokinase-like protease activity. Hence, it often responds clinically very effectively to fibrinolytic inhibitors.

R

RADIATION

See **Ionizing radiation**.

RADIOACTIVE PHOSPHORUS

(^{32}P) A **radionuclide** producing **ionizing radiation** by β-ray emissions, which is usually formulated as a dibasic sodium salt. ^{32}P localizes to tissues with a high phosphorus content, with rapid uptake by dividing cells, e.g., **bone marrow**. It can be given orally or, more usually, by intravenous injection. The $T^{1/2}$ is about 14 days, but the hematological effects take up to 6 to 8 weeks. ^{32}P has been used in the past in the treatment of **myeloproliferative disorders**, especially **polycythemia rubra vera** and **essential thrombocythemia**, where excellent myelosuppression was achieved in most, but not all patients, particularly those with exuberant **extramedullary hematopoiesis**. A dose of 7 to 12×10^6 MBq (2 to 4 mCi) is usually given, which may be repeated after 3 months if inadequate disease control is achieved. Responses usually last 6 to 12 months, but other myelosuppressive agents, e.g., **hydroxyurea** or **interferon-α** may also be necessary.

The adverse effects of ^{32}P are prolonged **pancytopenia**, which rarely occurs, and a very high risk of leukemia. Indeed, its leukemogenic effect in polycythemia vera has led to the abandonment of ^{32}P as a therapeutic agent in this disorder except in the unusual circumstance where other therapies have failed and life span is limited.

RADIOIMMUNOTHERAPY

(RIT) A form of treatment for malignancy whereby a specific **monoclonal antibody** with a radionuclide attached is used to target and destroy specific malignant cells. Its principal application is in the treatment of **follicular lymphoma** by ^{131}iodine tositumomab and ^{90}yttrium ibritumomab, using a CD20 antibody.[446] It has been shown to be effective in advanced disease with low short-term toxicity.[202b,447–448] Hypothyroidism is a long-term **adverse drug reaction**.

RADIONUCLIDES

(Radioisotopes) Specific atoms with a nucleus of protons and neutrons surrounded by an orbit of electrons.[449] Each isotope occupies the same place in the periodic table, but with different atomic weights and varying stability. The rate of radioactive decay is measured in becquerels, 1 unit being equivalent to 1 disintegration/sec. Clinically, the megabecquerel (MBq) is used, being 10^6 disintegrations/sec. The old mCi unit is now little used (1 Ci = 3.7×10^{10} sec^{-1} = 3.7×10^{10} Bq; 1 millicurie (mCi) = 3.7×10^7 Bq or 37 MBq; 1 microcurie (µCi) = 3.7×10^4 Bq or 0.0037 MBq. The specific activity of a pure nuclide is derived from the half-life measured in seconds.

Diagnostic Uses in Hematology

Cell labeling:

- **Red blood cells**, using ^{51}Cr or ^{99m}Tc for **blood volume, erythrokinetics, red blood cell survival measurement**
- **White blood cells**, using ^{111}In to detect granulocytes in abscesses, particularly in the abdomen
- **Platelets**, using ^{111}In for life span

Protein labeling:

- Albumin, using ^{131}I or ^{99m}Tc for plasma volume
- Fibrinogen using ^{131}I for location of deep-vein thrombosis

Iron labeling: ^{59}Fe for **ferrokinetics** in **autoimmune hemolytic anemia, aplastic anemia, myelofibrosis**, and **sickle cell disorders**

Organ imaging: **Spleen** and **liver**, where the size, shape, position, and integrity can be determined by visualizing uptake of damaged red cells labeled with ^{51}Cr or ^{99m}Tc, or of colloids labeled with an l-emitting radionuclide such as ^{99m}Tc. Counting is performed by a scintillation camera, quantitation requiring calibration of the imaging assembly using appropriate radioactive phantoms.

Absorption and excretion studies:

- Iron — see **Ferrokinetics**
- Cobalamins — see **Schilling test**
- Whole body counting by using ^{59}Fe or ^{58}Co to study absorption and turnover of iron and cobalamin, respectively (now of very limited use due to cost)

Plasma clearance studies: See **Ferrokinetics**

Autoradiography: Photographic nuclear emulsion can be closely applied to histologic or cytologic preparations of **bone marrow** to record disintegration as latent images, which are subsequently developed into grains of silver. It is used to provide information of molecular movements within a single cell by either light or electron microscopy.

Radioisotope dilution assays (radioassays by competitive binding): Here, a known amount of radioactive "hot" analyte is diluted by nonradioactive "cold" analyte" in the test serum, in which the test substance has been released from any serum binders by either heating or by chemical means. A measured volume of the mixed serum is then bound to a specific binding protein, insufficient in amount to bind all of the "hot" analyte. The bound analyte is then separated from the free analyte and its radioactivity counted.

- ^{58}Co for **cobalamin**
- ^{125}I for **folic acid** and folates
- ^{57}Co-B$_{12}$ for intrinsic factor and intrinsic factor antibodies (see **Gastric disorders**)
- ^{59}Fe for **ferritin**
- ^{125}I for immunoradiometric assay of **factor VIII:C Ag; β-thromboglobulin; platelet factor-4**

Therapeutic Uses in Hematology

This is confined to **radioactive phosphorus** for the treatment of **polycythemia rubra vera** and **essential thrombocythemia** when other forms of therapy have failed.

RADIOTHERAPY

The treatment of hematological disorders by external **ionizing radiation**. This damages cells by generating free radicals from intracellular water and molecular oxygen, which interacts with cell membranes and causes irreparable breaks in nuclear double-stranded **DNA**. Conventional radiotherapy is given in daily doses (fractions), with the objective of achieving optimum cumulative dosage for tumor control while minimizing the side effects on neighboring normal tissue. High-energy linear accelerators make it possible to penetrate tissues selectively and to triangulate on the tumor target (conformed radiation). This has provided the ability to vary daily dosage fractionation, increase the efficiency, and decrease the late complications.[225–229] It is usually given in combination with **cytotoxic agents** or alone when chemotherapy has failed.

Indications

Hodgkin disease (HD). Administered to the affected lymph nodes, usually following chemotherapy. In nodular lymphocyte-predominant Hodgkin lymphoma, it is the standard treatment with involved-field radiation therapy.

- Involved field (IF), which includes the site of clinically involved lymph node group and is now the standard radiation therapy when administered in any stage of Hodgkin lymphoma and is administered in limited doses of 35 to 45 Gy
- Extended field (EF) — treatment of multiple involved and uninvolved lymph node groups on one side of the diaphragm (mantle or inverted Y) or on both sides of the diaphragm (total or subtotal lymphoid irradiation)
- Waldeyer field — upper cervical nodes
- Mantle field — lower cervical, supraclavicular, infraclavicular, axillary, hilar, and mediastinal nodes
- Inverted Y — subdiaphragmatic, paraaortic, and ileal
- Endolymphatic — by infusion of ^{131}I-lipiodol pedally for retroperitoneal nodes
- Total nodal — mantle plus inverted Y
- Hemibody irradiation
- Total body (TBI) with compatible allogeneic bone marrow transplantation rescue

Non-Hodgkin lymphoma (NHL). Given to reduce nodes as for HD in the following circumstances:

- Isolated, symptomatic nodes in doses of 35 to 45 Gy
- Reduction of lymphoid mass when situated dangerously to life — tracheal, meningeal, adjacent to major blood vessels
- Failure of chemotherapy — usually TBI

Acute lymphoblastic leukemia (ALL). Cranial irradiation in doses of 16 to 24 Gy to prevent craniospinal relapse

Myelomatosis. Palliative therapy in doses of 20 to 30 Gy for disabling pain from a well-defined focus

Splenic irradiation, in doses of 20 to 30 Gy for:

- Relief of painful **splenomegaly**
- Reduction of splenomegaly in **myelofibrosis**, **autoimmune hemolytic anemia**, and acquired **immune thrombocytopenic purpura** when splenectomy is contraindicated

Complications

Bone marrow hypoplasia with anemia, leukopenia, thrombocytopenia leading to hemorrhage, and overwhelming bacterial or viral infection

Radiation sickness (malaise, nausea, vomiting)

Skin erythema and desquamation, temporary epilation

Acute cerebral edema and parasthesias from craniospinal irradiation

Inhibition of ovulation and spermatogenesis when the gonads have been exposed

Radiation nephritis following kidney exposure — hematuria, proteinuria, hypertension, and uremia

Neoplasia and leukemia (secondary)

Teratogenic effects on the fetus if exposure occurs during the first 6 months of pregnancy

RAPOPORT-LUEBERING SHUNT

Diphosphoglycerate (2,3-DPG) pathway

The production of 2,3-DPG is by diversion of 1,3-diphosphoglycerate from a step in which ATP is generated, to one where a high-energy phosphate is not metabolized.

RAS GENES

Oncogenes initially discovered through their ability to cause rat sarcomas. Mutation in these genes is associated with human **oncogenesis**. The oncoproteins encoded by *ras* genes that bind GTP and GDP act as GTP-activated switches in **cell-signal transduction**.

R-BINDER PROTEINS

A group of immunologically similar proteins, each having a molecular weight in the region of 60 kDa, synthesized by gastric mucosal cells and secreted into gastric juice. They bind to **cobalamins** but differ from intrinsic factor because of their rapid electrophoretic mobility.

There is a very rare congenital disorder in which failure of secretion by the gastric mucosa occurs. Affected individuals do not have megaloblastic change or any other manifestations of cobalamin deficiency, although the serum cobalamin levels are well below the reference range.

REACTIVE ARTHROPATHY

Sterile synovitis with a positive but not obligatory association with **human leukocyte antigens** HLA-B27. An association with intestinal tract infection by *Campylobacter jejuni* has been claimed. The increased susceptibility to inflammation is probably a consequence of:

T-cell receptor selection of bacterial antigen

Mode of presentation of bacteria-derived peptides to T-cells

Molecular mimicry causing autoimmunity against HLA-B27 or other self-antigens

The hematological disorders include:

Normocytic anemia

Erythrocyte sedimentation rate elevated

Leukocytosis

Thrombocytosis

REACTIVE LYMPHOID HYPERPLASIA

Benign increase of lymphoid tissue in **lymph nodes, liver,** and **spleen**. It takes two forms:

Follicular hyperplasia with prominent germal centers occurring with bacterial infections, subacute bacterial endocarditis, and chronic autoimmune disorders such as **systemic lupus erythematosus** and **rheumatoid arthritis**

Generalized without germinal center involvement occurring with **infectious mononucleosis**, other viral infections, and allograft rejection

RECALCIFICATION TIME

The time taken for citrated plasma to clot at 37°C after calcium alone has been added back. It is dependent upon **factors II, V, VIII, IX, X, XI, XII, fibrinogen, prekallikrein,** and **high-molecular-weight kininogen**. It has a very wide reference range due to variable contact activation of plasma samples. It is of little clinical value.

RECEPTOR-MEDIATED ENDOCYTOSIS

Cell membrane changes necessary to engulf particles by invagination with the aid of cell surface receptors such as low-density lipoprotein (LDL) and **transferrin**. The process is followed by resealing of the cell membrane. It is the means by which **malaria** *Plasmodia* spp. enter **red blood cells**.

RECOMBINANT FACTOR VIIA

(rFVIIa) A recombinant factor concentrate used for the treatment of patients with either **hemophilia A** or **hemophilia B** who have inhibitors. It is also used in the treatment of some inheritable **platelet** disorders as an alternative to transfusion with platelet concentrate. Factor VII is manufactured in baby hamster kidney (BHK) cells cultured in media containing no human protein or derivatives. During the chromatographic purification process, the factor VII is autoactivated to produce factor VIIa. Recombinant factor VIIa exerts its hemostatic effect only after interaction with tissue factor (TF) at the site of injury. Recombinant factor VIIa has a short half-life ($T^{1/2}$) ≈ 2 to 3 h, less in children. One possible mechanism proposed for the activity of high-dose rFVIIa is by overcoming the inhibitory effect of tissue factor pathway inhibitor (TFPI) on FVII-based thrombin generation, thus allowing normal thrombin generation in the absence of FVIII or FIX. Generally, the dose administered is in the range 90 to 120 µg/kg initially every 2 h. These dose leads to VII levels in excess of 3000 U/dl. The mode of action of rFVIIa is therefore pharmacological rather than physiological. It has proven to be safe and effective and is increasingly seen as a possible universal hemostatic agent.

Anecdotal reports in the literature have shown infusion of rFVIIa to be effective at arresting severe hemorrhage in the context of a major trauma, unremitting gastrointestinal hemorrhage, and postsurgical bleeding. The mechanism of its hemostatic action in such situations is not entirely clear, but it may be due to a combination of inhibition of TFPI pathways and enhancement of thrombin generation on platelet surfaces. It has proven to be safe and effective and is increasingly seen as a possible universal hemostatic agent, but it is extremely expensive.[449a]

RECOMBINANT FACTOR VIII

(rFVIIIC) See **Coagulation factor concentrates**.

RECTAL DISORDERS

See **Intestinal tract disorders**.

RED BLOOD CELL

(Erythrocytes; red blood corpuscles; RBC) The mature end cells of the **erythron**. They are the most abundant cells circulating in the blood vessels. In healthy adults, **erythropoiesis** occurs exclusively in the bone marrow and depends critically on the availability of adequate supplies of **iron**, **folate**, and **cobalamin** (vitamin B_{12}). Maturation of erythroblasts occurs largely in the marrow; extrusion of erythroblasts into the circulation causes loss of their nuclei, although some **hemoglobin** synthesis continues for a few days once the cells reach the circulation. Because normal circulating red blood cells do not have nuclei, they do not conform to strict definitions of a mammalian cell.

Mature erythrocytes consist of a membrane that encloses a solution of hemoglobin, carbohydrates, enzymes, and electrolytes. Their primary purpose is to transport oxygen from the lungs to tissues. The uptake of oxygen in average persons is about 250 ml/min (see **Oxygen affinity for hemoglobin**). In resting persons, erythrocytes spend approximately 780 msec in pulmonary capillaries; transit time through the lungs decreases with activity. The surface area for exchange of oxygen and other bases is approximately 3000 m^2 and is directly proportional to the size of the red blood cell mass. The biconcave disk shape of mature erythrocytes provides a great surface-to-volume ratio that favors rapid gas transport and enhances the deformability of erythrocytes, all of which must traverse small capillaries.

The reference range of circulating red blood cells is $5.0 \pm 0.5 \times 10^{12}/l$ (men) and $4.3 \pm 0.5 \times 10^{12}/l$ (women) (see **Red blood cell counting**). After 100 to 120 days in the circulation, glycolytic metabolism within the red blood cells declines, and the senescent cells are removed by **histiocytes** (macrophages) in the spleen, liver, and bone marrow.

Red Blood Cell Membrane[450,451]

This forms a boundary between the cell interior, with its highly concentrated solution of hemoglobin, and the plasma surrounding it. It serves as a barrier to maintain a concentration of ions and metabolites inside the cell, which differs markedly from that in the plasma environment. In addition it contains pumps and channels for the movement of sodium, potassium, calcium, and oxidized glutathione. It also facilitates the passage of glucose and other small molecules, and it is responsible for the basic structural integrity of the red blood cell and for maintaining its biconcave shape (see Figure 105).

All lipids in the mature red cell exist in the membrane and are partially responsible for many of the membrane properties, e.g., passive cation permeability and mechanical flexibility. Of these lipids, 95% are phospholipids and nonesterified cholesterol, both being present in almost equal molar amounts, with considerable homeostatic activity between them. Membrane lipids are arranged in a bimolecular layer, this layer being penetrated by the integral membrane proteins.

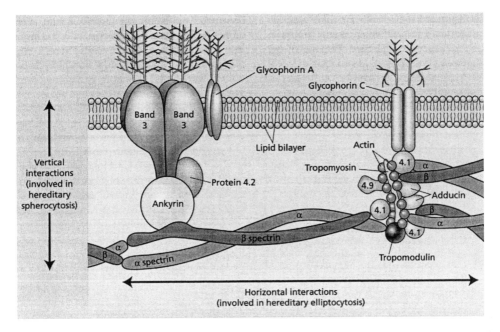

FIGURE 105
Red blood cell membrane cytoskeleton proteins. (Modified from Bain, B.J., *Blood Cells: A Practical Guide*, 3rd ed., J.B. Lippincott, Philadelphia, 2002. With permission.)

The red blood cell contains a network of fibrous proteins that effectively laminate the inner surface of the membrane to form the membrane skeleton. The proteins extracted from red blood cell membranes have been assigned names.[452,453] The designation of the bands was based on their mobility in a sodium dodecyl sulfate (SDS) acrylamide gel system (polyacrylamide gel electrophoresis [PAGE]). The slowest migrating band was band (or protein) 1, the next slowest band was band 2, etc. Subbands were designated with decimals. Later, some bands were renamed (e.g., band 1 to α spectrin; band 2 to β spectrin). Protein 4.1 has never been renamed. These bands have their protein designations, respective genes and chromosome locations (Table 136). Mutations of genes encoding some of these proteins are responsible for hereditary types of anemia associated with alterations in erythrocyte shape. Most underlying mutations of these types of anemia are private.[453a]

Red Blood Cell Blood-Group Antigens and Antibodies
See also **Blood groups**.

RBC Antigens
Most of these **antigens** are synthesized by the red blood cells, with the exception of the **Lewis blood group** system, where they are adsorbed onto the red cell membrane from plasma. Some antigens are only found on red blood cells, but others, particularly those secreted by salivary glands, are distributed throughout the body. The RBC antigens of the blood group systems are expressed on RBC membrane proteins; the amino acid sequences of most are known. Many of these proteins have been shown to be associated with certain functions; for example, **Duffy blood group** antigens are carried on a glycoprotein that is a receptor for several cytokines. The number of antigenic determinants per red cell varies between about 10^3 (Knops antigens) and 10^7 (P antigen) copies.

TABLE 136

Major Red Blood Cell Membrane Proteins

Band	Protein	Gene Symbol	Chromosomal Location	Involvement in Hemolytic Anemias [a]
1	α spectrin	SPTA1	1q22-q23	HE, HS
2	β spectrin	SPTB	14q23-q24.2	HE, HS
2.1	ankyrin	ANK1	8p11.2	HE
2.9	α adducin	ADDA	4p16.3	N
2.9	β adducin	ADDB	2p13-2p14	N
3	anion exchanger-1	EPB33	17q21-qter	HS, SAO, HAc
4.1	protein 4.1	EL11	1p3-p34.2	HE
4.2	pallidin	EB42	15q15-q21	HS
4.9	dematin	EPB49	8p21.1	N
4.9	p55	MPP1	Xq28	N
5	β actin	ACTB	7qter-q22	N
5	tropomodulin	TMOD	9q22	N
6	G-3P-D	GAPD	12p13.31-p13.1	N
7	stomatin	EPB72	9q33-q34	HSt
7	tropomyosin	TPM3	1q31	N
PAS-1	glycophorin A	GYPA	4q28-q31	HE
PAS-2	glycophorin C	GYPC	2q14-q21	HE
PAS-3	glycophorin B	GYPB	4q28-q31	N
	glycophorin D	GYBD	2q14-q21	N
	glycophorin E	GYBE	4q28-q31	N

Note: These results are based on scanning of SDS-PAGE gels of erythrocyte membranes prepared from healthy blood donors.

[a] HS = hereditary spherocytosis; HE = hereditary elliptocytosis; SAO = Southeast Asian ovalocytosis; HAc = hereditary acanthocytosis; HSt = hereditary stomatocytosis; N = no disease state identified.

Source: Table adapted from Gallagher, P.G. and Forget, B.G., The red cell membrane, in *Williams Hematology*, 6th ed., Beutler, E., Lichtman, M.A., Coller, B.S., Kipps, T.J., and Seligsohn, U., Eds., McGraw-Hill, New York, 2001. With permission.

Antibodies to RBC Antigens

Most RBC **antibodies** are **immunoglobulins** of the IgG (usually IgG1 or IgG3 or a mixture of both), IgM, or, rarely, IgA class. IgM antibodies to RBCs are often **agglutinins**, whereas IgG antibodies are usually only detectable serologically using methods such as the **indirect antiglobulin (Coombs) test**.

Clinical Importance of Red Cell Blood Groups

Compatibility for Blood Transfusion

Alloantibodies directed against RBC antigens may be found in persons needing a blood transfusion. It is essential that these RBC alloantibodies be detected by **pretransfusion testing**, because many have the capacity to cause a hemolytic blood transfusion complication. The significance of this blood incompatibility is dependent upon a number of factors.

> *The frequency of the antibody*: The ABO system is the most important in blood transfusion because anti-A and -B are found in almost all people who lack the corresponding antigens, i.e., the antibodies are "naturally occurring." Most other RBC antibodies are not naturally occurring and are produced in an immune response following RBC transfusion (which includes the transfer of fetal blood into the maternal circulation). Anti-D is the most frequently encountered immune RBC

antibody in most populations because of the high immunogenicity of the RhD antigen. After anti-D, other Rh system antibodies (mainly anti-c and anti-E), anti-Fya, anti-K, and anti-Jka are the RBC antibodies most likely to be encountered in pretransfusion testing.

The ability of the antibody to effect the destruction of RBCs: This functional activity will depend upon the Ig class and subclass of the antibody and the ability of the antibody to activate complement. Complement-binding 37°C-active IgM antibodies may give rise to substantial intravascular destruction of RBCs. Anti-A and anti-B are by far the most commonly found of these antibodies, but other much rarer antibodies, such as anti-Vel and anti-PP1Pk, may also cause destruction through this route. A single IgM molecule may initiate, through the classical complement pathway, the sequential binding of serum complement proteins, leading to the formation of the membrane-attack complex responsible for cell lysis. Complement-mediated phagocytosis also leads to extravascular destruction of RBCs, predominantly in the liver. Many IgG antibodies do not bind complement, but are able to effect extravascular RBC destruction, predominantly in the spleen. IgG3 antibodies appear to be more effective at causing RBC lysis, through interaction with the Fc receptors of mononuclear phagocytes, than IgG1 antibodies.

Hemolytic Disease of the Newborn (HDN)

IgG alloantibodies cause **hemolytic disease of the newborn,** as other Ig classes are not transferred by placental Fc receptors. Anti-D is commonly involved, but antibodies to blood groups other than the Rh blood groups can be implicated. Prenatal screening must therefore include testing for RBC alloantibodies and determining the blood group of the mother so that immunoprophylaxis can be given to those who are RhD negative.

Autoimmune Hemolytic Anemia and Drug-Associated Immune Hemolytic Anemia

See also **Immune hemolytic anemia.**

Autoantibodies with specificity for antigens present on autologous RBCs are found in the plasma of patients with autoimmune hemolytic anemia. Many RBC autoantibodies, such as anti-I, react only at temperatures well below 37°C and do not give rise to a significantly reduced survival of autologous RBCs. The exception is in cold autoimmune hemolytic anemia, in which a potent autoantibody active at a temperature above 30°C is generally present. IgG autoantibodies active at 37°C are the causative agents of warm autoimmune hemolytic anemia. Drug-associated immune hemolytic anemia has a variety of causes, and although all involve the accelerated destruction of RBCs, not all of them involve antibodies directed against blood group antigens.

Red Blood Cell Metabolism

Glucose is the normal energy source for the red cell, but it can be replaced by fructose or galactose.[454] Energy is required to maintain:

Hemoglobin iron in the divalent form

Red-cell enzymes and hemoglobin in the active reduced form

The biconcave shape of the cell

The major glucose-supported metabolic pathways maintain stability of the erythrocyte membrane and solubility of intracellular hemoglobin. The pathways are complex and integrated (see Figure 106).

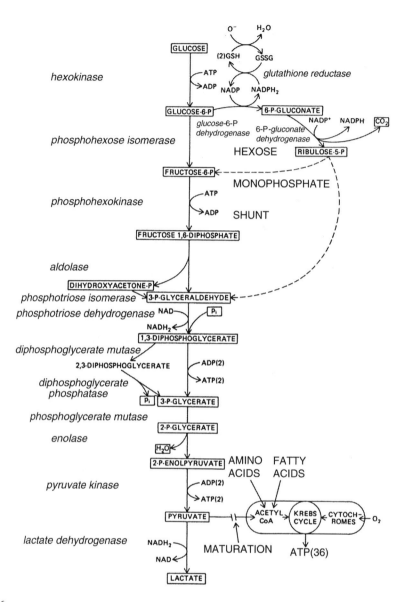

FIGURE 106

Metabolism of glucose by red blood cells. (From Jandl, J.H., *Blood: Pathophysiology*, Blackwell, Oxford, 1991, chap. 4. With permission.)

The metabolic pathways are:

Embden-Meyerhof Pathway of Anaerobic Glycolysis

Glucose is catabolized to pyruvate or lactate with production of ATP. This accounts for about 90% of glucose oxidation. The enzymes involved are:

Aldolase   Hexokinase
Diphosphoglycerate mutase   Lactate dehydrogenase
Diphosphoglycerate phosphatase   Phosphoglycerate mutase
Enolase   Phosphohexokinase

> Phosphohexose isomerase   Phosphotriose isomerase
> Phosphotriose dehydrogenase   Pyruvate kinase

Two important products are formed:

Adenosine triphosphate (ATP), the major high-energy compound in red cells, of which
there is a maximal net gain of two moles (see 2,3-diphosphoglycerate [2,3-DPG]
pathway below)

Reduced nicotinamide-adenine dinucleotide (NADH), a compound essential for the reduction of **methemoglobin**

Hexose-Monophosphate Shunt

(Pentose phosphate pathway) The pentose phosphate pathway of aerobic glycolysis
accounts for about 10% of glucose oxidation. The relative activity of this pathway is
regulated by the concentration of oxidized glutathione (GSSG). When this increases, there
is an increase in the glucose that enters the hexose monophosphate shunt. The end product
is reduced nicotinamide-adenine dinucleotide phosphate (NADPH), which serves as a
cofactor in the reduction of glutathione. The enzymes involved in this pathway are:

Glucose 6-phosphate dehydrogenase

Glutathione reductase

6-Phosphogluconate dehydrogenase

Ribulosephosphate epimerase

Ribosephosphate isomerase

Transketolase

Transaldolase

Methemoglobin Reductase Pathway

The methemoglobin reductase pathway maintains heme in its ferrous state.

Phosphogluconate Pathway

This couples oxidative metabolism with NADP and glutathione reductase, thereby counteracting environmental oxidants and preventing denaturation of globin. The enzymes of
this pathway are glucose 6-phosphate dehydrogenase and glutathione reductase.

2,3-Diphosphoglycerate Pathway

(2,3-DPG; Rapoport-Luebering shunt) The production of 2,3-DPG by diversion of 1,3-
diphosphoglycerate from a step in which ATP is generated to one where a high-energy
phosphate is not metabolized. The proportion of 1,3-DPG following the direct ATP pathway rather than the 2,3-DPG pathway depends on cellular ATP and ADP levels: when
ATP falls and ADP rises, 1,3-DPG passes through the ATP step, and vice versa. The
enzymes involved are diphosphoglyceromutase (DPGM) and diphosphoglycerate phosphatase (DPGP).

2,3-DPG also affects tissue oxygen delivery by reversibly combining with deoxygenated
hemoglobin and decreasing hemoglobin oxygen affinity (see **Hemoglobin**).

The large number of enzymes required by the metabolic pathways are formed largely
in the **normoblasts** of the bone marrow and to a lesser extent in the **reticulocyte**. The
mature red blood cell does not have the capacity to synthesize new enzyme molecules,

so the energy-depleted red blood cell quickly transforms from a biconcave disk to a sphere and, as such, is rapidly removed from the circulation by the **spleen**.

The normal values for red blood cell enzyme activities and glutathione content are given in **Reference Range Table IX**.

Erythropoiesis
Measurement of Erythropoiesis

Normal red blood cell production is extremely effective, and most red blood cells live, or have the potential to live, a normal life span. Under certain conditions, however, a fraction of red blood cell production is ineffective, with destruction of nonviable red blood cells either within the marrow or shortly after the cells reach the blood. Effective erythropoiesis is most simply estimated by determining the reticulocyte count. This count is usually expressed as the percentage of red blood cells that are reticulocytes, but it can also be expressed as the total number of circulating reticulocytes per liter of blood (reticulocyte % × RBC per liter). An elevated reticulocyte count may give an erroneous impression of the actual rate of red cell production because of premature release of reticulocytes into the circulation. To correct for this premature release, some workers calculate a reticulocyte index to compensate (see **Reticulocyte count**).

Ineffective Red Blood Cell Production

Ineffective erythropoiesis is suspected when the reticulocyte count is low or is normal or only slightly increased in the presence of erythroid hyperplasia in the bone marrow. In certain disorders, such as **Addisonian pernicious anemia**, **thalassemia**, and **sideroblastic anemia**, ineffective erythropoiesis is a major component of total erythropoiesis. This can be quantified by **ferrokinetics**. Using ferrokinetic methods, ineffective erythropoiesis is calculated as the difference between total plasma iron turnover and erythrocyte iron turnover plus storage iron turnover.

Total Erythropoiesis

This is the sum of effective and ineffective red cell production and can be estimated from marrow examination by first determining the relative content of fat and hematopoietic tissue. The myeloid/erythroid ratio is then determined. These, taken in conjunction with the red cell count and the reticulocyte count, will usually provide quantitative information on the rate and effectiveness of red blood cell production.

Erythropoiesis can be demonstrated by imaging marrow, liver, and spleen with ^{99m}Tc sulfur colloid or ^{111}Indium, even though these isotopes primarily label the monocyte-macrophage system. Their uptake is similar to ^{59}Fe, and they can be used to demonstrate erythroid tissue, but accurate quantitation of total erythropoiesis is made by measuring the rate of red blood cell production (see **Ferrokinetics**).

Red Blood Cell Removal

Elimination of senescent red cells requires three distinct processes:

Recognition of cells to be removed

Arrest of flowing cells

Phagocytosis

The vascular characteristics of the **spleen**, **liver**, and **bone marrow** provide these requirements. As red cells age, they lose their ability to deform as a result of lipid and protein loss from the cell membrane. There is no evidence of loss of metabolic effectiveness as red blood cells age. Identification of aging cells is the most controversial aspect; IgG-mediated recognition by the monocyte-macrophage system seems most likely. There is autologous binding by IgG to cryptic antigenic sites, the Fc portion of the IgG molecule being preferentially recognized by the phagocytic cells. The splenic structure and circulation represent a unique and sensitive filter system. The spleen has two flow paths, major and minor. The major flow path is high velocity from trabeculae through white pulp and marginal zone directly into venous sinuses. The minor flow path, where senescent cells are most likely removed, enters the red pulp and is retarded by phagocytic cells and narrow sinus apertures in the range of 0.5 to 3.0 μm, the narrowest channels in the entire human vasculature. **Red blood cell survival measurement** can be used to estimate the degree and site of red cell destruction.

Disorders of Red Blood Cells

Deficiency of circulating red blood cells (**anemia**): commonly arises from **hemorrhage**, **hemolysis**, or failure of bone marrow production: **aplastic anemia**, **bone marrow hypoplasia**

Abnormalities in hemoglobin formation: **hemoglobinopathy**

Excess red blood cell production: **erythrocytosis**, **polycythemia rubra vera**

Intrinsic red cell disorders: **red blood cell membrane disorders**, **red blood cell enzyme deficiencies**

RED BLOOD CELL APLASIA

See **Pure red blood cell aplasia**.

RED BLOOD CELL COUNTING

See also **Cytometry**; **Automated blood-cell counting**.

The estimation of the number of red blood cells in a sample of body fluid, usually the peripheral blood.[455] Manual techniques are inaccurate and have been replaced by instrument counting. Where these instruments are unavailable, **packed-cell volume** estimation is a useful surrogate.

The reference ranges,[455a] expressed in units of $10^{12}/l$, are:

Adult male	5.5 ± 1.0
Adult female	4.8 ± 1.0
Neonates (term, cord)	4.7 ± 0.8
Children, 1 year	4.5 ± 0.8
Children, 6 to 12 years	4.6 ± 0.6

More detailed ranges for neonates and children are given in **Reference Range Tables IV** and **V**.

RED BLOOD CELL DISTRIBUTION WIDTH

(RDW) A measurement of **anisocytosis** derived from the **red blood cell volume** distribution produced by **automated blood-cell counting**. The RDW is expressed either as the coefficient of variation (%) or the standard deviation (fl) after excluding the tails of the size distribution curve. The main value of the RDW is in distinguishing microcytosis due to **iron deficiency** (increased RDW) from that due to **thalassemia** (normal RDW). The reference range in health is instrument specific.

RED BLOOD CELL ENZYME DEFICIENCIES

Hematological disorders occurring in association with deficiency of a **red blood cell** enzyme. Although many enzyme deficiencies have now been recognized, only about half are associated with hematological disorder. The molecular basis of many erythroenzymopathies has now been established.[456] **Glucose-6-phosphate dehydrogenase** (G6PD) deficiency and **pyruvate kinase** (PK) deficiency were the first to be described. With the exception of G6PD deficiency, which is very common in certain populations, these are unusual or rare disorders. The modes of inheritance of the enzyme deficiencies vary. Both G6PD deficiency and phosphoglycerate kinase deficiency are sex-linked, the other severe deficiencies of glycolytic enzymes being inherited as autosomally recessive disorders. Autosomally dominant inheritance is characteristic of partial enzyme deficiencies. The clinical and laboratory features vary with each enzyme. Their levels of activity are given in **Reference Range Table VIII**. Deficiencies that give rise to clinical disorders are:

Aldolase

Diphosphoglycerate mutase

Enolase

Glucose-6-phosphate dehydrogenase

Gamma-glutamyl-cysteine synthetase

Glutathione

Glutathione peroxidase

Glutathione reductase

Glutathione synthetase

Glyceraldehyde-3-phosphate dehydrogenase

Hexosekinase

Phosphofructosekinase

6-Phosphogluconate dehydrogenase

Phosphoglycerate kinase

Phosphohexose isomerase

Pyrimidine-5′-nucleotidase

Pyruvate kinase

Triose phosphate isomerase

RED BLOOD CELL FRAGMENTATION SYNDROMES

See **Fragmentation hemolytic anemia**.

RED BLOOD CELL FRAGMENT COUNTING

Estimation of the number of red blood cell fragments in peripheral blood. This is useful for diagnosis and follow-up of several diseases, including **hemolytic uremic syndrome** and transplantation-associated **thrombotic microangiopathy** (TMA). While this quantification still relies on manual counting, automated methods are being developed for both flagging and counting.

RED BLOOD CELL INDICES

Calculations of red blood cell properties derived from red blood cell measurements.

Mean Cell Hemoglobin (MCH)

The ratio of hemoglobin to red cell count. This indicates the amount of hemoglobin in the average cell and is calculated by the following formula expressed in picograms (pg):

$$MCH = \frac{\text{hemoglobin (g/l)}}{\text{red blood cell count} (10^{12}/l)}$$

The **reference range** in health for the MCH is 29.5 ± 2.5 pg. The required operating range for any technique is 15 to 50 pg. False elevation of the instrument-derived MCH occurs in association with the increased turbidity due to nonlysed red cells in certain hemoglobinopathies, hyperbilirubinemia, cryoglobulinemia, and paraproteinemia.

Reference ranges in health (pg) are:

Adults	29.5 ± 2.5
Children, 1 year	27.0 ± 4.0
Children, 3 to 12 years	27.0 ± 3.0

Ranges for neonates and infants are given in **Reference Range Tables IV** and **V.**

Mean Cell Hemoglobin Concentration (MCHC)

The ratio of the hemoglobin to the **packed-cell volume** (PCV) provides the concentration of hemoglobin in the average red cell expressed as grams of hemoglobin per deciliter of packed red blood cells:

$$MCHC = \frac{\text{hemoglobin (g/dl)}}{\text{vol. packed red blood cells (l/l)}}$$

The reference range for the MCHC in health is 32.5 ± 2.5 g/l. Computational errors may give a very low MCHC on certain blood counters when hyperlipidemia or a potent **cold agglutinin** is present. True elevation of the MCHC is so uncommon that when it occurs the possibility of error must be considered.

Reference ranges (expressed in g/l) are:

Adults and children	32.5 ± 2.5
Neonates (term, cord)	30.0 ± 2.7

More-detailed ranges for neonates and infants are given in **Reference Range Tables IV and V**.

Mean Cell Volume (MCV)

The MCV of red blood cells is calculated from the red blood cell count and the volume of packed red blood cells (PCV) by a formula designed to express the result in femtoliters ($\times 10^{-15}$ l):

$$MCV = \frac{PCV\ (10^3 l/l)}{\text{red blood cell count } (10^{12}/l)}$$

Fully automated counters measure the MCV simultaneously with the red blood cell count, either from the mean height of pulses generated during the red blood cell count or from the sum of the pulse heights divided by the number of pulses. Accuracy is limited, however, by shape differences that exist between fresh red blood cells and the calibrator (shape factor). Red blood cell membrane defects (e.g., in **hereditary spherocytosis**) reduce flexibility of the cells and result in greater differences in the measurement of MCV between manual and automated methods. The observed MCV value increases with increasing time from venesection, particularly at ambient temperatures.

The healthy reference range for the MCV is 86 ± 10 fl. The required operating range is 50 to 150 fl. False elevation of the MCV occurs in hyperosmolar states and in association with warm and cold **agglutinins**. False reduction of MCV is seen in hypoosmolar states.

Reference ranges expressed in femtoliters are:

Adults	86 ± 10
Neonates (term)	120 ± 10
Children, 1 year	78 ± 8
Children, 3 to 12 years	84 ± 8

More-detailed ranges for neonates and infants are given in the **Reference Range Tables IV and V**.

RED BLOOD CELL MASS

(RCM) See **Blood volume** — red cell volume.

RED BLOOD CELL MEMBRANE DISORDERS

Quantitative and qualitative abnormalities in **red blood cell** membrane constituents.[451] These result in a large number of changes in red cell morphology, often leading to **hemolytic anemia**. They include:

Hereditary spherocytosis (HS)

Hereditary elliptocytosis (HE)

Common hereditary elliptocytosis (with discoid elliptocytes)

Ovalocytic hereditary elliptocytosis

Stomatocytic hereditary elliptocytosis (Melanesian ovalocytosis, Southeast Asian HE)

Hereditary infantile pyropoikilocytosis

Acanthocytosis, both hereditary and acquired

Paroxysmal nocturnal hemoglobinuria

Rhesus (Rh) null phenotype

RED BLOOD CELL PRESERVATION

See **Blood components for transfusion**.

RED BLOOD CELL SURVIVAL MEASUREMENT

The normal physiological red blood cell life span is 100 to 120 days, but each method of measurement has its own normal or **reference range**. Isotopic methods have now replaced the immunological recognition of transfused compatible red cells.[449]

Cohort Method

This method depends on biosynthetic incorporation of the label into developing cells. The labels used are glycine, containing labeled nitrogen (^{15}N), radioactive carbon (^{14}C), or radioactive iron (either ^{55}Fe or ^{59}Fe). The main disadvantage of cohort labeling is the need for prolonged periods of sampling when the life span is only moderately reduced. In addition, the label from destroyed red blood cells may be reutilized and tag new cohorts of red blood cells, making it difficult to interpret results.

Random Label Methods

Ashby differential agglutination technique, which consists of transfusing compatible but immunologically marked cells. This method is now of historical interest only.

Isotope method, using ^{51}Cr. The chromium ion penetrates the red cell membrane and binds to the β- and γ-chains of globin. Unfortunately, these bonds are not covalent, and there is a continuous elution of the isotope, varying from 0.5 to 2.9% per day.

In the normal person, the red blood cell life span is finite at about 120 days (random destruction 0.06% to 0.4% per day). When ^{51}Cr is used as the label, about 1% of label is eluted per day, and the survival curve becomes exponential, with a half-life of about 30 days. A correction factor for chromium elution should be applied using linear coordinates. If the data lie on a straight line, the destruction is by senescence and the life span can be calculated as twice the half-life. If the data indicate exponential disappearance and it is necessary to use a semilogarithmic paper to depict the data on a straight line, the destruction is random and the life span is 1.44 times the half-life.

Indirect Methods

Study of bilirubin metabolism and carbon monoxide excretion can be used.

RED BLOOD CELL TRANSFUSION

(RBC transfusion) Most frequently, blood transfusions are given to replace blood loss by **hemorrhage** or to correct **anemia**. They must always be preceded by some form of **pretransfusion testing**.

General Indications for RBC Transfusion[195,457]

Medical Patients

Careful patient assessment and scrutiny of laboratory results must be undertaken prior to administering a transfusion. This should establish whether anemia is acute or chronic in nature and whether it has arisen from nutritional/hematinic deficiency or as a result of acute or chronic hemorrhage. The mean cell volume (MCV) is a useful guide here. It is important to establish whether anemia might have resulted from **autoimmune hemolytic anemia**, in which case treatment with **corticosteroids** may be more appropriate. Hematinic therapy should be considered for patients with **iron deficiency** and **cobalamin** or **folic acid** deficiency, even if anemia is profound, since transfusion in this group of often elderly patients may result in heart failure.

Surgical Patients

Detailed studies have shown that the transfusion of blood in patients undergoing routine surgery does not correlate with the degree of preoperative anemia, surgical blood loss, or the postoperative/discharge hematocrit. For coronary bypass surgery, the proportion of patients transfused varies from 28 to 100%, and for patients undergoing hip replacement, 58 to 98%, depending upon the practice of the surgical team. In general, the principal reasons for transfusing patients undergoing elective surgery are the policies regarding blood transfusion within that hospital. It is therefore not possible to give definitive guidelines for the transfusion of patients who undergo surgery. However, it is clearly important to establish a transfusion trigger that will prompt replacement of surgical blood loss. It is generally accepted that during surgery this is likely to be a hemoglobin (Hb) of 8 g/dl rather than 10 g/dl. A large randomized trial in critically ill patients supported the hypothesis that a conservative transfusion trigger of 7 g/dl not only led to less donor red cell transfusion, but may lead to a better survival in younger, fitter patients when compared with a liberal trigger of 9 g/dl.[448] Postoperatively, the decision to transfuse depends not only on hemoglobin or hematocrit levels, but also on clinical symptoms. Patients with uncorrected coronary artery disease and other cardiopulmonary conditions may require a higher Hb concentration.

Specific Indications for RBC Transfusion

Hemorrhagic Shock[458,459]

This follows the loss of >30% of the circulating blood volume within a few hours. If the loss of one blood volume occurs within 24 h, massive blood transfusion is required. Shock is frequently associated with disordered **hemostasis** due to prolonged hypotension and incomplete resuscitation. There may be a preexisting **hemorrhagic disorder**.

Prolonged hypotension and incomplete resuscitation strongly correlate with microvascular bleeding, which occurs as generalized oozing from catheter and venepuncture sites and surgical drains, mucosal bleeding, petechiae, and bruising. It is often thought that coagulopathy and bleeding occur as the result of dilution of coagulation factors. However, a significant dilution does not occur until more than two blood volumes have been lost. In addition, studies from the kinetics of exchange transfusion show that levels of a given blood constituent diminish only gradually, since losses are compensated for by increased *de novo* synthesis or release of constituents from stores. The procedure of transfusion in these circumstances should be:

Establish a good vascular access, usually a minimum of two large-bore cannulae.

Take samples for **hemoglobin**, **packed-cell volume** (PCV), **platelet counting**, and coagulation screen (**prothrombin time** [PT], **activated partial thromboplastin time** [APTT], **fibrinogen**), **pretransfusion testing**, and biochemical profile.

Restore circulating **blood volume**, initially with crystalloid to a volume of 2 l.

Establish monitoring of electrocardiogram, central venous pressure, blood pressure, and urinary output.

Transfuse SAG-M (saline, adenine, glucose, and mannitol) red blood cells (see **Blood components for transfusion**) plus colloid to maintain PCV >30%. Blood should be warmed and passed through a standard macroaggregate (200 mcm) filter.

Review coagulation parameters. For PT or APTT ratio >1.5, give fresh-frozen plasma, 15 ml/kg. For fibrinogen <0.8 g/l, give 10 units (5 ml/kg) of cryoprecipitate.

The platelet count usually falls to subnormal levels when roughly one blood volume has been lost or replaced. Transfuse one adult therapeutic dose (see **Platelet transfusion**) if platelet count $<50 \times 10^9/l$ and bleeding continues, or if platelets $<100 \times 10^9/l$ and there is known or suspected evidence of platelet dysfunction.

Management of massive hemorrhage using protocols such as those described above are considered to have been successful when the underlying cause of bleeding has been treated and further hemorrhage arrested, the hematocrit is 30%, and coagulation parameters are normal, i.e., PT and APTT ratio <1.5.

Sickle Cell Disorders

The goal of transfusion therapy is to improve oxygen-carrying capacity and to prevent vascular occlusion. The hemoglobin in patients with sickle cell disease is usually around 8.0 g/dl, and it is important to avoid indiscriminate red blood cell transfusion, which may increase blood **viscosity** and aggravate sickling. Three transfusion strategies are employed:

Simple additive transfusion. This is indicated for patients with symptomatic anemia, acute splenic, or hepatic sequestration, and in patients with aplastic crisis or accelerated hemolysis. Transfusion may be required to replace acute blood loss for other reasons.

Hypertransfusion. The objective is to maintain the level of HbS (the sickle cell trait) at 20% with an Hb between 10 and 14 g/dl by administering regular additive transfusions every 3 weeks or 6 weeks. It is usual to precede these with an exchange transfusion. The possible indications are:

• Pregnancy

• Three months after prosthetic joint surgery

• Following a stroke for up to 3 years

It is necessary to prevent **iron overload** by administering chelation therapy with desferrioxamine.

Exchange transfusion. This can be used to adjust hematocrit and HbS levels swiftly during severe or life-threatening events. The indications are acute chest or girdle syndromes, acute priapism, cerebrovascular infarction, and preparation for major surgery. Red blood cell exchange is best performed using a cell separator (see **Hemapheresis**).

Transfusions in sickle cell disease are more likely to be complicated by hyperviscosity, iron overload, and viral infections. In addition, at least 60% of multiple transfused patients develop nonhemolytic febrile transfusion reactions (see **Blood transfusion complications**). These may be reduced by the use of **leukodepletion of blood components for transfusion**. Alloimmunization to red blood cell alloantigens occurs in approximately 50% of transfused patients. Extended phenotyping of patients for the Rh, K, MNS, Kidd, and Duffy **blood group** systems simplifies the identification of antibodies that may subsequently form after multiple transfusions. In patients with sickle cell disease, 70% of antibodies are formed against Rh and K antigens. Prospective matching for Rh and K antigens has therefore been advocated in some treatment centers to reduce alloimmunization.

Thalassemia

For β-**thalassemia**, major regular transfusion results in improved quality of life and prevention of growth retardation, which is near normal until age 11, although there may be some stunting of stature thereafter. By suppressing endogenous **erythropoiesis**, the need for **splenectomy** has been lowered, and the age of patients with thalassemia at splenectomy has risen. Transfusions are given regularly at 2- to 4-week intervals to maintain a mean hemoglobin level of 12 g/dl. Splenectomy is undertaken when the annual blood requirement exceeds 1.5 times the standard transfusional needs of splenectomized patients and when subcutaneous infusion of desferrioxamine is required to prevent iron overload.

Paroxysmal Nocturnal Hemoglobinuria (PNH)

Paroxysmal nocturnal hemoglobinuria is an uncommon acquired clonal stem cell disorder that results in red blood cells that have an abnormal sensitivity to complement-mediated lysis. Originally, saline-washed red blood cells were advocated for patients with PNH. More recent experience suggests that unwashed red cells are satisfactory and do not exacerbate hemolysis. They should be group specific and leukodepleted by filtration.

Myelodysplasia (MDS)

Patients with refractory anemia, sideroblastic anemia, and other forms of **myelodysplasia** require frequent red blood cell transfusions. Because a significant proportion of these patients will experience clinically important nonhemolytic febrile transfusion reactions, it is likely that they will need to be either leukoreduced or leukodepleted. Iron overload, even in those receiving regular transfusions, may not be a problem, since the anemia may be caused in part by bleeding due to thrombocytopenia, and the disease may have a relatively short clinical course. In some patients with sideroblastic anemia where storage iron is increased, there is no bleeding due to thrombocytopenia, and the clinical course may be protracted; in this case, chelation therapy should be considered.

Red Blood Cell Substitutes

The main theoretical advantages of these materials include unlimited availability, lack of infection, and no requirement for compatibility testing (see Table 137). They can also be used where there is a cultural objection to the transfusion of human blood.

Perfluorochemicals (PFC)

These are emulsions of highly fluorinated hydrocarbons. The first that was developed for clinical use was Fluosol-DA. This contains perfluoro (PF)-decalin and PF-tripropylamine at a ratio of 7:3. An emulsifier called pluronic-F68 is also incorporated, and the solution

TABLE 137

Ideal Characteristics of Blood Substitutes

Adequate oxygen and carbon dioxide transport
Normal oncotic and osmotic pressures
Viscosity equal to that of normal plasma
Lack of side effects: nontoxic, nonallergenic
Acceptable intravascular half-life
Readily available, sterilizable, and inexpensive

is formulated with glucose, electrolytes, and hydroxyethyl starch. Widespread clinical use of Fluosol and other PFCs has been limited by their inherent disadvantages. These are:

A linear oxygen-dissociation curve, which requires a high concentration of inspired oxygen

A relatively low PFC concentration of 20%

Instability, requiring frozen storage and subsequent reconstitution

Blockade of reticuloendothelial system cells following phagocytosis of PFC particles, which can increase the likelihood of infection

Activation of complement by pluronic-F68, causing anaphylaxis

Newer PFCs such as F-octyl bromide (perflubron), developed as an intravenous contrast agent, have better oxygen affinity, a short tissue-retention time, and do not require pluronic-F68 to form stable emulsions.

PFCs might be used in hemorrhagic shock, correction of preoperative anemia, and carbon monoxide inhalation. Fluosol-DA has been transfused to several hundred human subjects, but it is currently licensed only by the U.S. Food and Drug Administration for distal perfusion in percutaneous transluminal coronary angioplasty. PFCs may have applications in oncology, since they act as radiation sensitizers. Phase I/II studies have been performed in patients with small-cell lung cancer. Controlled trials have yet to be performed.

Hemoglobin Solutions (HS)

These are usually made from outdated blood and are prepared by lysis of washed red cells followed by centrifugation and filtration under pressure. Initially, HS caused renal damage in some patients due to red cell stroma; this must be removed in preparation. Inherent disadvantages are a short half-life and high oxygen affinity, since the hemoglobin molecules have lost 2',3-diphosphoglycerate (2,3-DPG), which allows oxygen dissociation. These problems can be overcome by polymerizing the hemoglobin molecules with glutaraldehyde, adding pyridoxal-5-phosphate to replace lost 2,3-DPG and sodium ascorbate to prevent methemoglobin formation. Recently, human HS have been produced by cross-linking their α-subunits with diaspirin. This product can be heat treated to inactivate virus, and then non-cross-linked hemoglobin and other unwanted proteins are eliminated by precipitation. Subsequently, HS is filtered and frozen for long-term storage. Clinical trials in hemorrhagic shock and stroke are being performed.

Biosynthetic Hemoglobin

The human hemoglobin gene can be expressed in *Escherichia coli*, yeast, or mammalian cells produced in sufficient quantities to be of therapeutic use. A human hemoglobin with a mutant β-globin and fused α-subunits has been produced using *E. coli*. Clinical trials are awaited.

RED BLOOD CELL VOLUME

See also **Blood volume**.

(Red blood cell mass) The quantity of **red blood cells** in the circulation.

RED BLOOD CELL VOLUME FRACTION

See also **Blood volume**.

The relative volume occupied by **red blood cells** in the circulation.

RED URINE

The abnormal color of urine, usually due to hematuria as occurs in any **hemorrhagic disorder**. It may also be found with **intravascular hemolysis** as hemoglobinuria, as myoglobinuria, and in some forms of **porphyria**. Other causes of red urine are due to drugs such as anthracyclines or pyridium and the ingestion of beetroot.

REED-STERNBERG CELL

(RS cells) Large round binuclear cells seen in expanded mantle zones of lymphoid follicles enmeshed by dendritic reticulum cells in **lymph nodes** from patients with **Hodgkin lymphoma**. RS cells are CD30$^+$ and react strongly with **monoclonal antibodies** to transferrin receptor (OKT 9$^+$). They have rearranged immunoglobulin genes, 60% showing B-cell phenotype. They are cytokine-producing cells: **tumor necrosis factor, interleukin** (IL)-1, -5, -6, and -9. The postulated cell of origin is a mature B-cell at the germinal stage of differentiation and, in rare cases, peripheral (postthymic) T-cells.

REFERENCE MATERIAL

Material for laboratory tests that (a) is sufficiently homogeneous and stable with respect to one or more specified properties and (b) has been established to be fit for its intended use in a measurement procedure. Reference materials are used to verify the accuracy of an analytical process (measurement system) used in routine practice. Reference materials must be traceable to a nationally certified reference material or an internationally certified reference preparation. Certified reference materials have been similarly defined as "a reference material characterized by a metrologically valid procedure by one or more specified properties, accompanied by a certificate that states the value of the specified property, its associated uncertainty, and a statement of metrological traceability" (ISO-REMCO).[460]

In hematology, the following reference materials are recognized:

Hemoglobincyanide prepared by the Dutch National Institute for Public Health and Environmental Protection

Bovine, human, and rabbit thromboplastins held by the European Union Measurement and Testing Programme (BCR) and the Central Laboratory of the Netherlands Red Cross Blood Transfusion Service (CLB)

Monosized latex particles suitable for red blood cell and platelet sizing held by BCR

TABLE 138

Available Hematology Reference Materials

General Hematology	Immunohematology
Erythropoietin, human, urinary	
Erythropoietin, rDNA-derived	anti-A blood-typing serum
Ferritin, human	anti-B blood-typing serum
Hemiglobincyanide	
HbA2	
HbF	anti-D (anti-Rh0) incomplete blood-typing serum
Folate, whole blood	anti-D (anti-Rh0) complete blood-typing serum
C-reactive protein (CRP)	anti-E complete blood-typing serum
Coagulation	**Immunology**
Antithrombin III, plasma	human immunoglobulins IgG, IgA, and IgM
Antithrombin III, concentrate	human serum immunoglobulin IgE
	antinuclear factor, homogeneous
Factor VIII: C concentrate, human	FITC-conjugated sheep antihuman IgG, IgM
Factors II, VII, IX, X, concentrate	
Factor VIII and Von Willebrand Factor, plasma	
Factor VIII concentrate	—
Factor IXa concentrate	human serum complement components C1q, C4, C5, factor B, and functional CH50
Heparin, low molecular weight	—
Plasma fibrinogen	—
Plasmin	—
Platelet factor 4	—
Proteins, plasma	—
Protein C, plasma	—
Streptokinase	—
α-Thrombin, human	—
β-Thromboglobulin	—
Tissue plasminogen activator, recombinant	—
Thromboplastin, bovine, combined	—
Thromboplastin, human recombinant	—
Thromboplastin, rabbit, plain	—
Von Willebrand factor, concentrate	—
Urokinase, high molecular weight	—

Source: From *Dacie and Lewis Practical Haematology*, 10th ed., Churchill-Livingstone, Edinburgh, 2006. With permission.

Table 138 lists the reference materials suitable for general use held at the National Institute for Biological Standards and Control, South Mimms, U.K., European Union (BCR) Institute for Reference Materials and Measurement (IRMM) Gest, Belgium and Central Laboratory, a Netherlands Red Cross Blood Transfusion Service (CLB) Amsterdam, Netherlands.

REFERENCE METHOD

A clearly and exactly described technique for a particular determination that, in the opinion of a competent authority, provides sufficiently accurate and precise laboratory data to support its use in assessing the validity of other laboratory methods.[444a] The accuracy of the reference method must be established by use of a **reference material**. Due to the lack of such materials, there are few strictly defined reference methods in hematology. The following have been described by the International Council for Standardization in Haematology (ICSH):

Hemoglobinometry of human blood

Determination of packed-cell volume (PCV)

Enumeration of erythrocytes and leukocytes

Enumeration of platelets

Staining blood and bone marrow films by azure B and eosin Y

One-stage prothrombin time

Erythrocyte sedimentation rate (ESR) test on human blood

Selected or recommended methods and guidelines have been prepared by ICSH for many hematological techniques.[41,44,130,131,183,188,411]

REFERENCE RANGES

(Reference intervals) The numerical measure of a quantity, given as a range of reference values in a representative population based upon samples collected in a standardized manner and analyzed by a standardized method, preferably a **reference method**, if available.[455a] The International Federation of Clinical Chemistry (IFCC) and the International Council for Standardization in Haematology (ICSH) have provided recommendations on the use of reference values.[460,461]

Physiological Factors Affecting Reference Ranges

Gender, particularly affecting red blood cell count levels and erythrocyte sedimentation rate

Age, with great variation from birth up to 12 years

Body mass affecting blood volume

Ethnic origin, particularly affecting levels of leukocytes

Environment, especially altitude and red blood cells, e.g., physiological erythrocytosis

Posture at time of sampling

Diet at time of sampling

Technical Factors Affecting Reference Ranges

Reference-range values are obtained using a specified technique of known precision and accuracy in an adequately described population. The term "reference value" is used for such population results. The values of the reference ranges are affected by:

Venepuncture technique

Analytical method used

The reference population (defined on age, sex, race, health condition, etc.) is composed of "reference individuals" obtained by adequate representative sampling (the "reference sample group"). Reference populations can be constructed in health and disease. From the totality of values, a "reference interval" can be determined that usually contains 95% of the central values, with the lowest and highest 2.5% values falling outside the reference limits. The samples should preferably be collected in the morning before breakfast, the last meal having been eaten not later than 21:00 h (9 P.M.) on the previous evening. Provided that the results follow a normal (Gaussian) distribution, calculation of the mean and standard deviation is simple, and the reference interval is defined as ±2 standard deviations. However, biological data are frequently lognormally distributed, and in this case, the logarithms of the observed values have a Gaussian distribution.

Such ideal reference values have only been determined for red blood cells and their indices. Other parameters are based upon ±2 standard deviations of an undefined "normal" population. These are given in the **Reference Range Tables**.

REFRACTORY ANEMIA

See **Myelodysplasia**; **Sideroblastic anemia**.

REFRACTORY MEGALOBLASTIC ANEMIA

See **Megaloblastosis**; **Myelodysplasia**.

REGRESSING ATYPICAL HISTIOCYTOSIS

See **Primary cutaneous CD30⁺ T-cell lymphoproliferative disorder**; **Anaplastic large-cell lymphoma**; **Non-Hodgkin lymphoma**.

REGULATED ON ACTIVATION, NORMAL T-CELL EXPRESSED AND SECRETED

(RANTES, CCL5) A natural **chemokine** ligand for CCR5, which is a chemokine receptor on **monocytes** and **lymphocytes**. It serves as a coreceptor for **human immunodeficiency virus** (HIV).

RELAPSING POLYCHONDRITIS

A systemic **autoimmune disorder** particularly affecting the cartilage of large joints. There is sometimes an associated inflammatory disorder of the eyes.

RENAL TRACT DISORDERS

The association of renal tract disorders and hematology.

Hematological Disorders Occurring with Renal Tract Disorders

Anemia:

- Impaired erythropoietin production in chronic renal failure, giving rise to normocytic anemia with **echinocytes** ("burr cells"), possibly due to uremic inhibitors
- **Hemolytic anemia** with **spherocytes** due to raised blood urea nitrogen and other nonexcreted toxic metabolites
- **Microangiopathic hemolytic anemia** from **hemolytic uremic syndrome** or **thrombotic thrombocytopenic purpura**
- **Dilutional anemia** from expanded plasma volume
- **Iron deficiency** from hematuria

Thrombocytopenia:
- Bone marrow hypoplasia
- Microangiopathic hemolytic anemia
- Immune complexes in circulation, as occurs in **polyarteritis nodosa** or acute glomerulonephritis
- Renal allograft rejection with increased platelet consumption

Platelet-function disorders due to accumulation of guanidinosuccinic acid, a product of L-arginine metabolism, that acts on platelet factor 3 and ADP-induced aggregation to cause uremic bleeding; an increase of nitric oxide also occurs

Erythrocytosis from increased erythropoietin production by the kidney in patients with:
- Renal tumors (Wilm's tumor, renal carcinoma, renal cysts)
- Renal hypoxia (secondary polycythemia)
- Renal transplantation

Effects of Hematological Disorders on the Renal Tract

Sickle cell disorder causing diminished perfusion of vasa recta of renal medulla, resulting in:
- Renal papillary necrosis
- Renal tubular dysfunction, leading to impaired sodium reabsorption and hyponatremia, hyposthenuria, renal tubular acidosis, and bouts of hyperkalemic hyperchloremic metabolic acidosis
- Hematuria due to areas of infarction
- Nephrotic syndrome due to mesangial proliferation induced by chronic glomerular engorgement and capillary impedance; immune complexes may be involved in this pathogenesis

Myelomatosis:
- Myeloma cast nephropathy (Bence-Jones proteinuria)
- Tubular compression atrophy by large hard casts in response to which serosal epithelial cells undergo uneven hyperplasia, causing light chain proteinuria; these changes can lead to acute renal failure, which frequently responds to restoration of renal perfusion and to "loop" diuretics, which inhibit calcium reabsorption
- Glomerular mesangial lesions due to the cumulative effect of paraprotein deposition over a period of years; proliferation of mesangial cells occurs with increase in the mesangial matrix
- Disseminated tissue deposition of paraproteins, both amyloid and K chains; this results in proteinuria but without K chains, as these are trapped in the kidney
- Deposition of paraproteins — amyloid and nonamyloid — from light chains or their fragments (IgA nephropathy)
- Nephrocalcinosis

Hemophilia-related disorders causing intermittent hematuria, sometimes with clot formation in the renal pelvis or ureters, giving rise to renal colic; epsilon-amino caproic acid (EACA) is contraindicated as a hemostatic agent due to the risk of inducing renal failure by the formation of a permanent plastic cast

Henoch-Schönlein purpura

Polyarteritis nodosa

Lymphadenopathy due to lymphomas of the paraaortic or iliac nodes, causing obstruction at the renal pelvis, ureters, or bladder and leading to pyelitis, cystitis, hematuria, or renal failure

Urate nephropathy from **hyperuricemia**, a hazard of **cytotoxic agent** therapy for leukemia/lymphoma, where massive cell cytolysis can cause marked increases in urate excretion with renal obstruction; this complication of cytotoxic therapy can be avoided by the prophylactic administration of **allopurinol**

Goodpasture's syndrome, where IgG autoantibodies bind with a glycoprotein in the basement membrane of the glomeruli; antibasement membrane antibodies trigger an inflammatory response that causes hematuria and acute renal failure

Immune complex deposition, causing inflammation of the glomeruli; this can take one of two forms, depending upon the size of the immune complexes, the rate at which they are produced, and the duration of their production

- Glomerulonephritis with proteinuria, hematuria, hypertension, and rapid-onset renal failure
- Nephrotic syndrome with proteinuria and gradual-onset renal failure

The origin of the immune complexes are poststreptococcal infection, **adverse drug reactions**, **lupus erythematosus**, Goodpasture's syndrome, and **Wegener's granulomatosis**.

REPTILASE TIME

A thrombin time test in which **thrombin** is replaced by reptilase, a purified enzyme from the snake *Bothrops atrox*.[19] Reptilase splits peptide A from **fibrinogen** and is not neutralized by antithrombin-heparin. Because it is not inhibited by antithrombin-heparin, it provides a useful reagent for identifying **heparin** in plasma samples. If the thrombin time is prolonged, the reptilase time will be normal compared with the thrombin time because of the presence of heparin. However, the reptilase time will remain prolonged in the presence of raised fibrin degradation production or hypofibrinogenemia.

RESPIRATORY TRACT DISORDERS

The association of respiratory tract disorders with hematology. These are commonly caused by **bacterial infection**, **viral infection**, **mycoplasma infection**, or malignancy. Specific associations are:

Iron deficiency from recurrent hemoptysis.

Secondary **erythrocytosis** from chronic hypoxia.

Macrocytes seen in peripheral blood in association with chronic airway obstruction, the pathogenesis of which is unexplained.

Eosinophilia coexisting with pulmonary eosinophilic infiltration (Loeffler's syndrome, pulmonary eosinophilia),[462] arising from the following disorders, which are all self-limiting or respond to **corticosteroid** therapy:

- **Hypersensitivity** to *Aspergillus fumigatis*
- Allergic alveolitis of farmers and bird fanciers
- Metazoan infection by *Ascaris lumbricoides*, *Filaria* spp., *Trichuris trichuria*, *Paragonimus* spp., *Taenia saginata*, or *Taenia solium*
- Drug **hypersensitivity** reactions to naproxen, phenylbutazone, sulidac, tolfenamic acid, glafenine, or fenbrufen
- Atopic asthma

Pulmonary eosinophilic granuloma, a local **histiocytosis** with marked eosinophilic infiltration of a granuloma primarily affecting the lungs. Diagnosis can only be made by lung biopsy. Spontaneous remission usually occurs, sometimes aided by corticosteroid therapy.

Pulmonary infiltrates of **Wegener's granulomatosis, hypereosinophilic syndrome,** and **polyarteritis nodosa**.

Goodpasture's syndrome. IgG autoantibodies bind with glycoprotein in the basement membrane of the lung. Antibasement membrane antibodies trigger an inflammatory response. This damages the basement membrane leading to internal hemorrhage and hemoptysis.

Pulmonary **hemosiderosis**, a rare disorder of marked iron deposition in the lung parenchyma allied to **iron deficiency**. The cause is unknown but has been considered to result from allergy to (unspecified) inhaled substances, ingestion of cow's milk, or to be associated with autoimmune disorders. There may also be focal glomerulonephritis, as occurs in Goodpasture's syndrome. It has been suggested that the distinction between idiopathic pulmonary hemosiderosis and Goodpasture's syndrome may be artificial. It is predominantly a disease of childhood, the most consistent clinical features being cough, hemoptysis, failure to gain weight, and fatigue. These symptoms result from repeated intrapulmonary hemorrhages, with the appearance of iron-laden macrophages in the lung. A considerable amount of iron is therefore lost through the sputum, resulting in iron deficiency. Progression of the disease is accompanied by radiological miliary stippling of the lung, perihilar fibrosis, and hilar lymph node enlargement seen radiologically. The iron deficiency responds to oral iron therapy, but the course is otherwise very variable, most patients appearing to derive some benefit from immunosuppressive therapy. Prolonged remissions have been reported. Death results from heart failure or massive pulmonary hemorrhage.

Alloleukoagglutinins. These can arise as a **blood transfusion complication**. They cause minor pulmonary infiltrates, which rapidly clear. Massive blood transfusion may result in the adult respiratory distress syndrome. This may also arise from aggregated platelets, granulocytes, and amorphous debris. All of these conditions can be prevented by transfusing blood through a nylon-mesh filter.

Anaphylaxis: rhinitis or asthma as a complication of **immunotherapy**.

Sickle cell disorder. An acute chest syndrome can be caused by a combination of lung parenchyma infarction and infection.

Hodgkin and **non-Hodgkin lymphoma**. This is usually an extension from tumors of the hilar or mediastinal lymph nodes. The most common types are **mediastinal (thymic) large B-cell lymphoma; diffuse large B-cell lymphoma; peripheral T-cell lymphoma;** and **extranodal NK/T-cell lymphoma, nasal type**.

Acquired immunodeficiency syndrome. Lymphocytic alveolitis, asthma due to increase in circulating IgE, Kaposi's sarcoma, interstitial pneumonitis, and primary pulmonary hypertension.

Venous thromboembolism. Pulmonary embolism as a common complication of deep-vein thrombosis. Diagnosis can only be confirmed by lung scans using ^{99m}Tc-macroaggregated albumin or by pulmonary angiography. Treatment is by **thrombolytic therapy** or **heparin.**

RESTRICTION ENDONUCLEASES

The class of bacterial enzymes that cut **DNA** at specific sites. In bacteria, their function is to destroy foreign DNA, such as that of bacteriophages. Type I restriction endonucleases occur as a complex with the methylase and a polypeptide that binds to the recognition site on DNA. They are often not very specific and cut at a remote site. Type II restriction endonucleases are the classic experimental tools. They have very specific recognition and cutting sites. The recognition sites are short, four to eight nucleotides, and are usually palindromic sequences. Because both strands have the same sequence running in opposite directions, the enzymes make double-stranded breaks, which, if the site of cleavage is off-center, generates fragments with short single-stranded tails; these can hybridize to the tails of other fragments and are called sticky ends. They are generally named according to the bacterium from which they were isolated (first letter of genus name and the first two letters of the specific name). The bacterial strain is identified next, and multiple enzymes are given Roman numerals. For example, the two enzymes isolated from the R strain of *E. coli* are designated Eco RI and Eco RII.

Restriction enzymes are used widely in experimental techniques for DNA manipulation. (See **Molecular genetic analysis.**)

RESTRICTION-FRAGMENT-LENGTH POLYMORPHISMS

(RFLP) Areas of variability in the size of fragments formed when **DNA** is digested by **restriction endonucleases**. Many bacteria elaborate these enzymes, which reproducibly cleave double-stranded DNA at palindromic sites, i.e., DNA sequences that read in one direction on one strand and in the opposite direction on the other strand. Several hundred restriction endonucleases are commercially available and are useful both for cloning DNA and for analyzing its structure. The size of restriction fragments is most readily identified using the Southern blot technique. One of the most powerful uses of restriction endonucleases is in the detection of genetic variability. These have proved useful, particularly in the prenatal diagnosis of **sickle cell disorders** and **thalassemias**.

RETICULAR CELL

See **Bone marrow** — microenvironment.

RETICULAR DYSGENESIS

See **Neutropenia.**

TABLE 139

Causes of Reticulin Fibrosis of Bone Marrow

Malignancy

Hematopoietic
   Acute leukemia (lymphocytic, myeloid and megakaryocytic)
   Chronic myelogenous leukemia
   Hodgkin lymphoma
   Non-Hodgkin lymphoma
   Hairy cell leukemia
Nonhematopoietic
   Metastatic carcinoma (breast, prostate)

Nonmalignant Conditions

Gray platelet syndrome
Infection
Human immunodeficiency virus, **Tuberculosis**
Renal osteodystrophy
Hyperparathyroidism
Systemic lupus erythematosus
Vitamin D deficiency
Irradiation (thorium dioxide)

RETICULAR FIBROSIS

The term used to designate **bone marrow** fibrosis by a form of **collagen,** usually un-cross-linked, that is argyrophilic, staining black when exposed to silver, but not staining with the classical trichrome stain used to identify collagen. However, except for the degree of cross linking, argyrophilic collagen is not different from nonargyrophilic collagen, since the argyrophilia is really a property of matrix substances, such as hyaluronic acid deposited on the collagen, and such matrix deposition is reduced in highly cross-linked collagen. Increased reticulin is seen in many malignant and nonmalignant conditions and is always a reactive process (see Table 139). A **leukoerythroblastic** reaction due to **extramedullary hematopoiesis** may be present. The diagnosis is made by bone marrow biopsy.

RETICULOCYTE

See also **Erythropoiesis**.

Immature **red blood cells** recognized in vitally stained unfixed preparations. This is because their constituent ribosomes react with certain basic dyes (e.g., new methylene blue, brilliant cresyl blue, azure B) to form a blue-black precipitate of filaments and granules. Various stages of reticulocyte maturation are recognized, depending on the amount of precipitated material present: those with the largest amount of material being the most immature forms; those with only a few dots or strands being the least mature. This maturation has been classified by Heilmeyer into stages I to IV, stage I being the least mature. Assuming that reticulocytes are released normally from the bone marrow and remain in the circulation for the normal period of time, their enumeration reflects activity of erythropoiesis. Maturation of reticulocytes from extrusion of **orthochromatic normo-blast** nucleus to complete loss of ribosomes and ribonucleic acid is thought to take 4 days, of which only the last day is in the peripheral circulation.

RETICULOCYTE COUNTING

Estimation of the number of **reticulocytes** in a sample of body fluid, usually peripheral blood.[463] The procedure as routinely performed using **cytometry** counting chambers or stained **peripheral-blood films** is labor intensive and imprecise. While clinical interest centered only on elevated reticulocyte counts for the recognition of hemorrhage, hemolysis, or response to hematinic therapy, measurement imprecision was tolerable, but this is no longer the case. The era of cytotoxic chemotherapy, of bone marrow transplantation and peripheral stem cell infusion, and of growth factor therapy demands precise and accurate reticulocyte counts at low levels. Automated reticulocyte counters using cytometry techniques promise to remove many of the difficulties inherent in the visual method. Fluorochrome-labeling **flow cytometry** for reticulocyte counting uses acridine orange, auramine O, dimethyloxacarbocyanine, ethidium bromide, pyronin Y, thioflavine T, and thiazole orange. Most require up to 30 min incubation for uptake of the fluorochrome and are thus suitable only as semiautomated methodologies. Ethidium bromide, on the other hand, at relatively high pH enters the cell in only a few minutes, while auramine O requires only a few seconds. A further benefit accruing from flow cytometry reticulocyte counting is that the fluorescence measured is proportional to the amount of RNA present in the cell. As a result, these methods can produce information on maturation of the reticulocyte population.[463]

Reference ranges for the reticulocyte count are:

Proportional count	0.5 to 1.5%
Absolute count	25 to 85 $\times$ 10^9/l

Proportional values for infants and children are given in **Reference Range Tables IV** and **V**.

RETICULOCYTE INDICES

Calculations of **reticulocyte** properties derived from reticulocyte measurements.[464] Reticulocytes are larger than mature **red blood cells**, and their size gradually diminishes during maturation. Diameter reduction correlates with progression through the Heilmeyer classes towards full maturation. Reticulocytes are also less dense than mature red blood cells. There are thus potential parameters based on reticulocyte maturation phenomena. Certain reticulocyte indices have recently emerged as automated measurements. With laser-based technology, simultaneous measurement of cell size and cell content is possible on red blood cells and reticulocytes. Stained reticulocytes are separated from unstained red blood cells and a number of reticulocyte parameters generated, e.g., mean corpuscular volume (MCVr), corpuscular hemoglobin concentration mean (CHCMr), and mean hemoglobin content of reticulocytes (CHr). Hematological indices, which have gained merit in the assessment of functional **iron** status, include CHr and %HYPO (proportion of hypochromic cells). Until recently, both parameters had been restricted to the analyzers from a single manufacturer. However, now a second manufacturer has produced an equivalent parameter, the so-called RET-Y (reticulocyte hemoglobin equivalent).[465] Several studies have established this equivalence. While clinical uses are being identified for these new parameters, their full acceptance will not be achieved until fully standardized nomenclature, definitions, and measurement technologies including reference procedures are produced.

Using **flow cytometry**, it is possible to assess reticulocyte maturation by measuring the intensity of fluorescence or staining: the greater the intensity, the more immature the reticulocyte. Subpopulations are produced by the insertion of one or two electronic thresholds to produce low- and high- or low-, middle-, and high-intensity ratios, respectively. General agreement has recently been obtained among experts that the immature reticulocyte fraction (IRF) concept (i.e., quantitation of the youngest reticulocyte population and expression as a fraction) is a good indicator of erythropoietic activity. The IRF is available on many latest-generation automated blood count analyzers. Clinical uses of the IRF include monitoring of bone marrow regeneration after chemotherapy; determining the timing of peripheral-blood stem cell collection; monitoring the response to **erythropoietin** in chronic renal failure, **acquired immunodeficiency syndrome** (AIDS), myelodysplasia, and autologous blood donation; monitoring neonatal transfusion needs; detection of aplastic crises and immune hemolytic anemia; and monitoring of renal transplant engraftment.

RETICULOENDOTHELIAL SYSTEM

The cell group with endothelial and reticular attributes, thereby showing a common phagocytic behavior to dyestuffs (Ashoff). The term is commonly used to group the phagocytic cells (**histiocytes**/macrophages) of the **bone marrow, spleen, lymph nodes**, and **liver** (Küpffer cells).

RETICULUM CELL

See **Bone marrow** — microenvironment; **Dendritic reticulum cells**.

RETINOIDS

A group of **cytotoxic agents** used for chemotherapy of hematological malignancies. At low concentration, retinoic acid induces differentiation of **promyelocytes** to **granulocytes** *in vitro*. Naturally occurring isomers are all-*trans*-retinoic acid, which promotes terminal differentiation of promyelocytes, and 13-*cis*-retinoic acid, which promotes differentiation of myeloid blast cells.

13-*cis*-Retinoic acid has differentiation activity in **myelodysplasia** and has produced remissions in **acute myeloblastic leukemia** — M3 AML.[466] However, all-*trans*-retinoic acid is more effective in M3 AML and, in daily oral divided dosages of 50 mg/m^2/day given for up to 100 days, produces remission rates of 60 to 95%. Adverse drug reactions are termed the retinoic acid syndrome: fever, dyspnea, acute respiratory distress, pulmonary infiltrates, pleural effusion, hypotension, edema, and hepatic and renal failure. Immediate treatment is necessary. Elevated serum transaminase levels occur in 50% of patients. Other reactions include pancreatitis, cardiac arrhythmias, and benign intracranial hypertension.

REVERSE TRANSCRIPTASE

An enzyme that governs the transcription of **RNA** to **DNA** (see **Human immunodeficiency virus**; **Leukemogenesis**). It can be used to detect expression-specific mRNA sequences (see **Molecular genetic analysis**).

RHEOLOGY OF BLOOD

The study of blood flow and its effects on blood cells.[467] Pulsatile blood flow is converted to steady flow by the elasticity of large artery walls. Branching, tapering, and curving of vessels results in separation of the blood cells into streams with areas of turbulence. Kinetic energy is thereby diverted from linear flow to high-velocity axial streaming. In arteries, the denser **red blood cells** enter the midaxial stream, with **lymphocytes** more peripheral, while the lighter **granulocytes**, **monocytes**, and **platelets** travel in the marginal stream. With reduction in vessel diameter, particularly in capillaries, the red cells become positioned crosswise to the axis of flow and hence have a diameter greater than the vessel itself. To allow their passage, red blood cells are deformed physiologically, the degree of deformability depending upon their viscoelasticity, which declines with their age. Blood flow varies with its viscosity, which is largely dependent upon large plasma proteins and the mass of the circulating red cells. Neither leukocytes nor platelets contribute to whole-blood viscosity. Platelets interact with the vessel wall endothelium, and these reactions are affected by the shear rate of the blood and by the size and deformability of the **red cell mass**. This rheological action of red cells directs platelets toward the endothelial lining of the vessels, thereby enhancing the potential for their adhesion. The **acute phase response** to tissue injury affects blood rheology and is the most frequent cause of pathological disturbances.

When suspended in plasma, the red cells form stacks or **rouleaux**, the adhesion forces responsible being generated by large plasma proteins: **fibrinogen**, α2-macroglobulin, and some **immunoglobulins**, with fibrinogen being the most effective.

Methods for studying blood rheology include **viscosity** of whole blood and of plasma, **aggregate formation of red blood cells**, and the **deformability of red blood cells**.

RHESUS (RL) BLOOD GROUPS

A specific **antigen–antibody** system located on **red blood cells** (RBCs).

Biochemistry

Two well-characterized transmembrane polypeptides (RhD and RhCE) carrying the Rh antigens have been described. RhD-negative individuals lack the RhD polypeptide. Amino acid substitutions on the extracellular loops of the RhD and RhCE polypeptides account for polymorphisms of D and the C/c and E/e polymorphisms.

Genetics and Phenotypes

The Rh antigens are determined by the products of the adjacent genes *RhD* and *RhCE* on chromosome 1. Forty-nine antigens have been described within the Rh system. As the RhD antigen is the most immunogenic of these, it is usually sufficient to classify individuals as RhD positive or RhD negative. Although the CDE terminology was based on an incorrect premise, it is in common use because of its relative simplicity in clinical situations. In CDE terminology, haplotypes at the Rh loci are described by three letters, each of which defines an antigen that is expressed on RBCs. For example, the genotype CDe/cDE results in the expression of C, c, D, E, and e antigens. The absence of the D antigen is denoted by "d," although there is no d antigen. Most RhD-negative individuals have the genotype cde/cde, i.e., they lack the D antigen but express c and e antigens. Altered antigen expressions are sometimes designated by a superscript, e.g., C^w. A single letter is used to represent Rh haplotypes in an alternative abbreviated notation. The genotypes and phenotypes are summarized in Table 140.

TABLE 140

Genotypes and Phenotypes of the Rh Blood Group System

Probable Genotype (CDE)	Phenotype (Abbreviated)	RhD Status	Frequency in Caucasians (%)
CDe/cde	R1r	+	32
CDe/CDe	R1R1	+	17
cde/cde	rr	–	15
cDE/cde	R2r	+	11
CDe/cDE	R1R2	+	11
cDe/cde	Ror	+	2
cDE/cDE	R2R2	+	2
Cde/cde	r'r	–	0.8
cdE/cde	r''r	–	0.9

These phenotype frequencies are very different in different ethnic groups; for example, the incidence of the D antigen is less than 1% in Japanese, Chinese, and American Indians.

Antibodies and Their Clinical Significance

Anti-D is almost always immune in origin. It is usually an IgG immunoglobulin, but there may also be an IgM (agglutinating) component. As it is standard practice to transfuse RhD-negative individuals only with RhD-negative blood, the formation of anti-D following transfusion is unusual. The incidence of anti-D formation during or following pregnancy has fallen dramatically since the introduction of Rh-immune globulin. This should be given to RhD-negative women at 28 weeks gestation, within 72 h of delivery, and whenever an obstetric event occurs (e.g., abortion) that may result in the transfer of fetal RBCs into the maternal circulation.

After anti-D, anti-c and anti-E are the next most common immune Rh antibodies. These antibodies are usually found in RhD-positive people of the probable genotype CDe/CDe (R1R1). The c and E antigens are much less immunogenic than D, but the antibodies are relatively common because it is not usual to match for antigens other than ABO and D. Naturally occurring antibodies, with specificities such as anti-E, anti-C, and anti-Cw, are sometimes encountered. They are often weakly reactive *in vitro* by the antiglobulin (Coombs) test, and they are sometimes found using enzyme techniques for antibody detection or low-ionic-polycation techniques. With the probable exception of anti-c, it is unlikely that these enzyme-only antibodies are of any clinical significance.

Immune Rh antibodies do not usually bind complement, but they may give rise to severe hemolytic blood transfusion complications if incompatible RBCs are transfused. Rh antibodies, in particular anti-D, are the most common cause of **hemolytic disease of the newborn** and cause the most significant morbidity.

Rh Blood Grouping and Selection of Blood for Transfusion

Determination of the RhD group of a patient is an essential requirement of **pretransfusion testing**. Unlike the position with ABO, it is not possible to perform a "reverse group" to confirm the presence or absence of antigen on the individual's RBCs. Properly validated procedures are essential to ensure that RhD-negative individuals, in particular premenopausal females, are identified and RhD-negative blood transfused. In patients who are likely to receive multiple transfusions over a long period of time (e.g., **sickle cell disorder** patients), consideration should be given to the selection of blood that is matched for Rh antigens other than RhD.

The use of group O RhD-negative blood in emergency situations when the recipient's blood group is not known should be restricted to a minimum because:

O RhD-negative blood is usually in short supply.

Antibodies such as anti-c may be present in the recipient's serum and may give rise to blood transfusion complications.

RHESUS (Rh) NULL PHENOTYPE

A rare phenotype of the **Rh ("Rhesus") blood groups**, where Rh antigens are not expressed on **red blood cells** (RBCs). In most cases, the Rh-null phenotype is caused by homozygosity for a variant of *RHAG*, the gene on chromosome 6 coding for the Rh-associated polypeptide that is essential for the expression of Rh antigens. However, the phenotype rarely may be the result of homozygosity for a silent or amorphous gene at the Rh locus on chromosome 1. Most Rh-null individuals have some degree of **hemolytic anemia**, exhibiting **stomato-cytes** and **spherocytes**. Because glycophorin B is also reduced on the membrane of Rh-null RBCs, the U antigen (see **MNS blood groups**) may also be difficult to detect. Rh-null individuals can produce a variety of Rh antibodies, including anti-Rh29, which reacts with all RBCs except those of the Rh-null phenotype, and may cause **blood transfusion complications** and **hemolytic disease of the newborn**. Selection of suitable blood for these people will be extremely difficult, and **autologous blood transfusion** should be considered whenever possible.

RHEUMATOID ARTHRITIS

A chronic inflammatory synovitis of mainly peripheral joints associated with antigens of **human leukocyte antigens** Class II MHC. HLA-DR4 occurs in 50 to 70% of patients. The possession of a specific pentapeptide (QK/RAA) in the third allelic hypervariable region of *HLA-DRβ-1* increases susceptibility. While the triggering antigen has not been identified, it is probably due to a delayed **hypersensitivity** reaction to an unknown stimulus upon CD4$^+$ T-**lymphocytes**, reacting with **monocytes**, **histiocytes**, **plasma cells**, **dendritic reticulum cells**, and fibroblasts that have migrated into the synovia of affected joints. The majority of symptoms are caused by the abundant **cytokines** produced by lymphocytes IL-1 and IL-8, TNF-α, GM-CSF, **chemokines** produced by macrophages (macrophage inflammatory protein and monocyte chemoattractant protein), and IL-6 produced by fibroblasts. The synovial fibroblasts have high levels of the adhesion molecule, vascular cell adhesion molecule (VCAM-1), which supports B-cell survival and differentiation, and decay-accelerating factor (DAF), which prevents complement-induced cell lysis. These molecules may facilitate the formation of ectopic lymphoid tissue in synovium. The T-cell cytokines stimulate B-cells to produce rheumatoid factor: IgM anti-IgG autoantibody. This is not specific to rheumatoid arthritis, as it occurs during many forms of chronic infection, but in rheumatoid arthritis it may result in **immune complex** formation within joints. This inflammatory reaction caused by ongoing T-cell activation may be maintained by the local production of rheumatoid factor and continuous stimulation of macrophages via IgG Fc receptors. A matrix of **metalloproteases** and **osteoclasts** causes irreversible damage as a result of **complement** fixation. Specific anticyclic citrullinated peptides are associated with persistence of arthritis and joint damage, but these may also be found in patients with early disease or even prior to evidence of clinical disorder.[469]

Hematological Changes

The consequent secondary changes[468] are usually proportional to severity of the disease and include:

Anemia of chronic disorders

Iron deficiency due to chronic gastrointestinal hemorrhage following **adverse drug reactions**

Aplastic anemia from sulfasalazine, penicillamine, gold, nonsteroidal analgesics, or cytotoxic drugs

Basophilic stippling of red blood cells

Felty's syndrome (rheumatoid arthritis, splenomegaly, lymphadenopathy, granulo-cytopenia, thrombocytopenia, and usually anemia); the bone marrow is normal or hypercellular; the mechanism of any neutropenia is probably due to immune destruction of neutrophils

Thrombocytosis of moderate degree in active disease

Elevated **erythrocyte sedimentation rate** due to raised immunoglobulins

Treatment

Immunosuppression by corticosteroids, **ciclosporin**, cyclophosphamide, and methotrexate has been used for many years, but recent developments include the use of cytokines.[164,470–472] These include:

B-cell depletion with anti-CD20 **monoclonal antibody**. Combination with intravenous cyclophosphamide has resulted in prolonged remissions.[461]

Receptor fusion protein. Cytotoxic T-lymphocyte-associated antigen4-IgG1 (CTLA4Ig, abatacept). This protein binds to CD80 and CD86 on antigen-presenting cells, blocking their engagement of CD28 on T-cells and thereby preventing T-cell activation.[473]

Antitumor necrosis factor-α agents. **Infliximab**, adalimumab, and **etanercept** have all been reported from trials as resulting in an improvement in function and quality of life over treatment with methotrexate.

Rituximab and anti-IL-6 monoclonal antibodies are undergoing investigation.

All of these immunosuppressants are best administered early before severe joint damage has occurred. They are contraindicated in the presence of active infections, tuberculosis, cardiac failure, multiple sclerosis, malignancy, and pregnancy.

RIBONUCLEIC ACID

(RNA) A nucleic acid that yields ribose upon hydrolysis. It carries out the instructions encoded in **DNA** and is responsible for synthesis of proteins. As most biological activities are carried out by proteins, the accurate synthesis of proteins is critical to the proper functioning of cells and organisms.

Three kinds of RNA molecules perform different but cooperative functions in protein synthesis:

Messenger RNA (mRNA) carries the genetic information copied from DNA in the form of a series of three-base code "words," each of which specifies a particular amino acid. The process is called "translation" because it converts the genetic information of nucleic acid sequences, composed of four distinct **nucleotides**, to the polypeptide sequence composed of 20 distinct amino acids. A unit of the genetic code consisting of three nucleotides (triplet), named a codon, corresponds to one amino acid, except for the final codon stop.

Transfer RNA (tRNA) is the key to deciphering the code words in mRNA. Each type of amino acid has its own type of tRNA, which binds it and carries it to the growing end of a polypeptide chain if the next code word on mRNA calls for it. The correct tRNA with its attached amino acid is selected at each step because each specific tRNA molecule contains a three-base sequence that can base-pair with its complementary code word in the mRNA

Ribosomal RNA (rRNA) associates with a set of proteins to form ribosomes. These complex structures, which physically move along an mRNA molecule, catalyze the assembly of amino acids into protein chains. They also bind tRNAs and various accessory molecules necessary for protein synthesis. Ribosomes are composed of a large and small subunit, each of which contains its own rRNA molecule or molecules.

RNA Processing

After the synthesis of specific mRNA, it is processed in the nucleus. Heteronuclear or pre-mRNA contains both exon and intron sequences. Before RNA can be transported from the nucleus to the cytoplasm, RNA processing is necessary. A 5′ terminal-cap structure containing a monomethylated terminal-G residue (m7GpppG) facilitates the export of all RNA transcripts from the nucleus. The 3′ addition of poly-A RNA is also important. In most cases, the removal of introns by splicing is also necessary for the nuclear export of RNA.

Intrinsic to RNA processing is RNA splicing and alternative splicing. RNA splicing is the process by which intronic sequences are removed to give rise to fully processed mRNA that has the 5′ cap, exon sequences, and the poly-A tail. There will also be exon sequences at the 5′ end that are untranslated (hence the term "5′UTR") and a 3′ untranslated end that often contains AU-rich sequences (3′UTR).

There are three main sequences required for splicing:

5′ donor sequence with a consensus of AG/GURAGU at the exon-intron boundary

3′ recipient sequence AG/G at the exon-intron boundary

Branch-site sequence CACUGAC, usually found in 30 nucleotides 5′ of the 3′ end of the intron

Recognition of the splice sites occurs by a complex of *trans*-acting RNA molecules found in the nucleus — small nuclear RNAs (snRNA) and ribonuclear proteins (snRNPs) — as well as additional individual factors not fully characterized. This is called a **spliceosome**.

Just as splicing of RNA is tightly regulated, alternative splicing that involves the differential use of splicing functions — often in a cell-specific manner — can occur. This can result in using an alternative recipient site for splicing (found either in an intron or a downstream exon) to change the coding sequence. Mutations in acceptor and donor sites will also lead to alternative splicing.

Once RNA is processed, it is exported from the nucleus to the cytoplasm and is chaperoned by ribonuclear proteins that protect it from RNAse degradation. How this process is regulated and what signals control preferential transport of one RNA transcript over another are not yet known.

RNA Stability and Degradation

Once the RNA has reached the cytoplasm, the rate of production of the protein that it encodes can also be regulated. The mRNA may be sequestered away from the ribosome so that it is not translated. The efficiency with which it is translated can vary, and the rate at which the message is degraded can be controlled. It has recently been shown that AU-rich sequences in the 3′UTR of mRNA encoding of certain cytokines (e.g., GM-CSF) and oncogenes (e.g., *c-fos*) are important in stabilizing the RNA from degradation. Addition of AU regions to the normally stable β-globin RNA results in a decrease in the β-globin half-life to only 30 min. Similarly, sequences in the 3'UTR of the transferring receptor have been found to modulate its mRNA stability in response to cellular iron stores (see **Translational regulation**). The AUUA elements and the **iron** response element formed in the 3'UTR of **transferrin** promote degradation of the poly-A' tail, which in turn leads to rapid cleavage of the mRNA.

RIBOSE PHOSPHATE ISOMERASE

An enzyme of the hexose phosphate shunt in **red blood cell** metabolism.

RIBOSOMES

The smallest and most numerous of the cell organelles, and the sites of protein synthesis. They are composed of protein and **RNA**, and are manufactured in the nucleolus of the nucleus. Ribosomes are either found free in the cytoplasm, where they make proteins for the cell's own use, or they are found attached to the rough **endoplasmic reticulum**, where they make proteins for export from the cell. They are often found in groups called polysomes. All eukaryotic ribosomes are of the larger, "80S" type.

RICHTER SYNDROME

(Richter's transformation) A transformation of lymphocytes in cases of **chronic lymphocytic leukemia** to a **diffuse large B-cell lymphoma**. It is a solid tumor, probably arising from B-cell blastic crisis involving the bone marrow and spleen. Treatment is as for diffuse large B-cell lymphoma. The prognosis is poor, with frequent relapses. **Autologous stem cell transplantation** should be considered upon first remission following combination chemotherapy.

RICKETTSIAL INFECTION DISORDERS

The hematological effects induced by rickettsial infections. No significant hematological changes are common, with infection by rickettsiae causing typhus, rickettsial pox, trench fever, or Q fever.

   In Rocky Mountain spotted fever, the capillary walls of all organs and tissues are damaged, causing a local **thrombosis** with areas of **hemorrhage** and necrosis. Here there is an associated **leukocytosis** and **thrombocytopenia.** If the illness is severe, peripheral circulatory failure occurs, with loss of water into tissues and a lowering of the **plasma volume**. In those whose red blood cells are deficient in **glucose 6-phosphate dehydrogenase**, acute **hemolysis** may occur.

RIEDER CELL

Lymphocytes with scant cytoplasm but deeply cloven or "cracked" nuclei. They occur occasionally in **chronic lymphatic leukemia**.

RISTOCETIN

An antibiotic derived from actinomyete species *Nocardia lurida*. It was found to induce **thrombocytopenia**, so it is no longer in clinical use. Ristocetin is used to test **platelet function** because it induces platelet aggregation by acting as a cofactor for **Von Willebrand Factor**–platelet interactions. It supports platelet aggregation by modulating and enhancing the Von Willebrand Factor–platelet GpIb interaction. Ristocetin-induced platelet aggregation and Von Willebrand Factor ristocetin cofactor activity represent the activity of the GpIb binding site of the Von Willebrand Factor protein.

RITUXIMAB

An anti-CD20 **monoclonal antibody** interfering with the activation and differentiation of B-lymphocytes but without effect on hematopoietic **stem cells**, pre-B-lymphocytes, and **plasma cells** and so has no adverse effect on regeneration of B-cells or of immunoglobulin production. It is given by intravenous infusion for treatment of **follicular lymphoma** and **diffuse large B-cell lymphoma**.[172] Severe **cytokine release syndrome** — characterized by fever, rash, angioedema, bronchospasm, and severe dyspnea — is a not infrequent adverse drug reaction, occurring 1 to 2 h following infusion. Patients should be given an analgesic and antihistamine before each infusion to reduce the incidence of these effects. Premedication with corticosteroids should also be considered. **Tumor lysis syndrome** may occur in those patients with bulky lymphomas.

RNA

See **Ribonucleic acid**.

ROCKY MOUNTAIN SPOTTED FEVER

See **Rickettsial infection disorders**.

ROSAI-DORFMANN DISEASE

See **Histiocytosis**.

ROULEAUX FORMATION OF RED BLOOD CELLS

The alignment in **aggregation** of **red blood cells** resembling stacked rolls of coins. The rate of rouleaux settling is the **erythrocyte sedimentation rate**. It is increased with the level of **fibrinogen** or the presence in the plasma of anisometric macromolecules such as dextrans, polyvinylpyrolidine (PVP), or carboxymethyl cellulose, all of which cause a reduction in the gap between red blood cells. **Reticulocytes** may form agglutinates when in increased numbers and form mature red blood cells when coated by **antibody.**

RUSSELL BODIES

Hyaline intracytoplasmic refractive spherules, which are pink or cherry-red upon Romanowsky staining. They are composed of condensed or crystallized proteins synthesized by plasma cells seen with **plasma cell neoplasms**.

RUSSELL VIPER VENOM TIME

(Dilute Russell viper venom time; DRVVT) The time taken in activation of **factor X** by diluted venom from the Russell viper in the presence of phospholipid and calcium ions.[19] The lupus anticoagulant prolongs this time by binding to the phospholipid and preventing the action of the venom (see **Antiphospholipid antibody syndrome**). Dilution of the venom makes this a sensitive test for detecting the presence of lupus anticoagulant. Defects in **contact activation** and deficiency of coagulation **factor VIII** or **IX** do not influence this test. False-positive results may be obtained in patients receiving intravenous **heparin**, and the interpretation will be difficult in those receiving **coumarin** drugs.

S

SALIVARY GLAND DISORDERS

Hematological disorders affecting salivary glands are:

Hemangiomas
Lymphoproliferative disorders

SARCOIDOSIS

A multisystem immunological **granuloma** formation, particularly of lymphoid tissue associated with depression of delayed **hypersensitivity**. The cause remains unknown, but it is associated with genetic mutations of the immunodefense molecule NOD2, which is a Toll-like receptor (TLR) of T-lymphocytes. There is a hyperactive T-helper-1 (Th1)-biased CD4$^+$ T-cell response. It is postulated that this hyperactivity is due to loss of immunoregulation by CD1d-restricted natural killer T-cells (NKT).[474] Mycobacterialike 16 S **RNA** sequences are sometimes found in **lymph nodes**.

The granulomas occur in lymph nodes (particularly mediastinal), lungs, bones, joints, nervous tissue, and skin. The clinical features are fever, malaise, **lymphadenopathy**, and, later, dyspnea. Manifestations of immunological dysfunction include **lymphopenia**, but with an increase of CD4 lymphocytes in bronchoalveolar lavage fluid, increased antigen-expressing capacity, and adhesion molecules on alveolar macrophages. There is disturbed cytokine function of tumor necrosis factor, interleukin-2, and prostaglandin (PGE-2). There is no known effective treatment, but spontaneous resolution usually occurs within 6 months of diagnosis. If this has not occurred, a trial of **corticosteroid** therapy is justified.[475] Epidemiological studies confirm an increased incidence of subsequent **lymphoproliferative disorders** of all types.

SATELLITISM

See **Platelet satellitism**.

SATURNISM

See **Lead toxicity**.

SCHILLING TEST

A radionuclide method for measuring **cobalamin** absorption using the level of urinary excretion.[147a] An oral dose of ^{57}Co- or ^{58}Co-labeled cobalamin is given by mouth to a fasting patient, and at the same time an intramuscular dose of nonradioactive hydroxocobalamin (flushing dose) is administered. A 24-h urine sample is then collected and the radioactivity measured. The normal urinary excretion is >10% in the first 24 h. In patients with **Addisonian pernicious anemia** and in those with cobalamin deficiency associated with intestinal malabsorption, the excretion is <5%. In this last situation, the test is then repeated with simultaneous oral administration of intrinsic factor. In pernicious anemia this causes correction of the radioactive cobalamin excretion, whereas there is no correction when the cause is an intestinal defect. Sources of error in this test are incomplete urine collection, renal disease with impaired glomerular filtration, inadequate flushing of the radioactive cobalamin, and prior administration of other radionuclides.

SCHISTOCYTE

(Helmet cell, fragmented cell) Split **red blood cells** found in **microangiopathic hemolytic anemias**, heart valve prostheses, **burns** causing hemolytic anemia, and **march hemoglobinuria**.

SCHUMM TEST

See **Methemalbumin**.

SCHWACHMAN-DIAMOND SYNDROME

(SDS) A rare autosomally recessive disorder with bone marrow aplasia, dwarfism, and pancreatic exocrine insufficiency. The bone marrow involvement is variable among subjects, severe congenital **neutropenia** being the originally reported feature. An increasing number of patients with late (early adolescence) complete bone marrow aplasia (severe neutropenia, severe **thrombocytopenia**, and moderate **aplastic anemia**) are being reported. Neutropenia is the most important morbidity-conditioning factor, and infections are the most frequent cause of death.

The underlying pathogenesis is unknown. A full-length SDS protein (of otherwise unknown function) is not detected in leukocytes of patients with the disorder, and mutations in the *SDS* gene (located at 7q11) have been reported. The mechanisms linking this unique abnormality with the SDS phenotype have not been found.

Treatment for SDS should be considered from a multidisciplinary point of view. Replacement drug treatment for pancreatic insufficiency and endocrine monitoring are required. There are contradictory reports on whether growth hormone can render any benefit in SDS. The most relevant part of the medical treatment, however, concerns bone marrow aplasia. A common approach is to start the search for a compatible **bone marrow** donor as soon as diagnosis is made and to start patients on antifungal prophylaxis as soon as they are found to be profoundly neutropenic. **Allogeneic stem cell transplantation** (SCT) is the only curative approach. Although no improvement in the endocrinological and pancreatic alterations can be expected, SCT effectively corrects the most important life-limiting abnormality. The choice of the conditioning regime is conflictive, however. Although a common myeloablative regime for aplastic anemia can be administered, a higher incidence of secondary tumors (mainly of gastrointestinal origin) has been reported. Nonmyeloablative regimes with reduced-dose cyclophosphamide, an immunoglobulin (ATG, ALG) or CAMPATH, and full-dose fludarabine are common practice.

SCLERODERMA

(Progressive systemic sclerosis) An **autoimmune disorder** of unknown cause with widespread endothelial damage resulting in release of **cytokines**. These include endothelin-1, which causes vasoconstriction. Intimal damage leads to increasing vascular permeability and activation of adhesion molecules: E selectin, VCAM-1, and ICAM-1. Migrating lymphocytes produce interleukin (IL)-2 and express surface antigens CD3, CD4, and CD5. Other cytokines released include IL-1, IL-4, IL-6, and platelet-derived growth factor (PDGF) with activation of fibroblasts. The vascular damage to small vessels leads to chronic ischemia. The fibroblasts synthesize increased amounts of **collagen** types I and III. About 80% of patients have anti-nuclear antibodies. Other hematological changes include:

Anemia of chronic disorders

Hemorrhagic **telangiectases**

Thrombocytopenia associated with renal failure

Malabsorption of **cobalamin** or **folic acid**

SCOTT'S SYNDROME
See **Platelet-function disorders**.

SEA-BLUE HISTIOCYTOSIS
Accumulation of ceroids in **histiocytes** of the **lymphoreticular system**. It is characterized by tissue infiltration by large mononuclear cells (20 to 60 μm) with characteristic sea-blue or blue-green cytoplasmic granules on Giemsa stain. These are ceroids composed of phospholipid and glycosphingolipids, similar to lipofuscin as a result of oxidation and polymerization of unsaturated lipids, that are engulfed into macrophages. They are seen in the cells of the **bone marrow**, **spleen**, and in **peripheral-blood films**.

The disorder presents as:

Primary sea-blue histiocytosis, where patients can present at any age with **splenomegaly**, **hepatomegaly**, and **thrombocytopenia**, with occasional hepatic cirrhosis and a neurological deficit. Rarely, the condition is an autosomally recessive disorder. The disorder usually follows a benign course, although a few deaths due to hemorrhage or hepatic failure have been reported.

Secondary sea-blue histiocytes associated with **immune thrombocytopenic purpura**, **myeloproliferative disorders**, **thalassemia**, and **sickle cell disorders**.

SELECTINS
See **Cellular adhesion molecules**.

SENILE ANGIOMA
See **Senile vascular lesions**.

SENILE PURPURA
See **Senile vascular lesions**.

SENILE TELANGIECTASIA
See **Senile vascular lesions**.

SENILE VASCULAR LESIONS

A range of vascular anomalies associated with increasing age due to decreased vascular integrity of the vessel wall, thus causing **vascular purpura** (see **Angiodysplasia**).

Senile angioma. Benign angiomatous skin lesions, originally called Campbell de Morgan spots, appear as bright red macules and shiny domed purple papules from midlife onward, becoming larger and more numerous with age. They are part of the physiological aging process and are of cosmetic importance only.

Senile purpura. Large, flat, purplish bruises with a sharp outline, seen predominantly on the forearms and back of the hands. They occur due to weakening of vascular structures and supporting subcutaneous tissues with age. Similar patterns of purpura are seen in patients on corticosteroids and in those with Cushing's syndrome. Bleeding from vascular structures often occurs into the cornified epithelium. Lesions may persist for many months, as such bleeding is inaccessible to macrophages.

Senile telangiectasias. These occur in the fair-skinned and are generally accompanied by other ultraviolet-induced skin changes. They are usually of fine caliber and seen on exposed skin.

Venous lakes. Soft, symptomless dark papules. They are usually found on exposed areas of the face, neck, lips, and ears. They are rarely more than several millimeters in size and are of cosmetic significance only.

SERINE PROTEASE

A group of enzymes that possess a serine residue at the center of the molecule. They catalyze many coagulation proteins, existing in plasma in an inactive precursor form (zymogen) and activated by other enzymes. Examples of serine proteases include all **vitamin K**-dependent coagulation proteins (factor II, **factors VII, IX, X, protein C**), **factors XI, XII, prekallikrein**, plasminogen, and plasminogen activators (see **Fibrinolysis**). The ancestral serine protease is trypsin. The zymogen of trypsin is trypsinogen. During activation, a small preactivation peptide is cleaved off trypsinogen to release the enzymatically active moiety trypsin. Serine proteases have a high degree of homology and contain a number of characteristic domains such as **kringles**, gla (calcium binding) residues, and epidermal growth-factor domains. **Factor V, factor VIII**, and **high-molecular-weight kininogen** are not serine proteases, but nonenzymatic cofactors.

SERINE PROTEASE INHIBITORS

(Serpins) A diverse family of **serine protease** inhibitors found in plants, animals, bacteria, and viruses. It includes many of the key inhibitors of **coagulation**, e.g., antithrombin, heparin cofactor II, protein C inhibitor, plasminogen inactivators, and α2-antiplasmin. They are believed to have evolved from a common ancestral protein, some 500 million years ago. Although many members of this family have retained the presumed inhibitory activity of their ancestor, others have evolved more-specialized functions. However, the function of some serpins is unknown, e.g., ovalbumin, the major protein component of chicken egg white. Members of this family of proteins share a common structure and can be confidently aligned in terms of their primary, secondary, and tertiary structures. They act by irreversibly binding to their target, which is inactivated by conformational changes. Some 13 "class B" serpins are designated as ov-serpins because of their structural resemblance to ovalbumin; these are found within cells, including several in immature or mature leukocytes. The role of these protease inhibitors is not entirely clear, but they obviously protect against intracellular proteases, possibly including cathepsins in myeloid cells and perhaps caspases, suggesting that they may play a role in protecting against apoptosis.

Examples of serine proteases include:

Neutrophil elastase and proteinase 3 (myeloblastin), which are both found in primitive myeloid cells

Monocyte neutrophil elastase inhibitor (MNEI), found in monocytes and neutrophils

Plasminogen activator inhibitor 2 (PAI-2), expressed in monocytes and endothelial cells, upregulated by pro-inflammatory cytokines

Proteinase inhibitor 6 (PI-6), expressed in myeloid cells, endothelial cells, and platelets

Proteinase inhibitor 9 (PI-9, granzyme B inhibitor), widely expressed, particularly in T and NK-cells

Proteinase inhibitor 10 (PI-10, bomapin), which appears to be present in myeloid cells

Peptide hormone precursors (angiotensinogen, hormone carriers)

Thyroxin-binding globulin (TBG)

Cortisol-binding globulin (CBG)

SEROTONIN

(5-hydroxytryptamine, 5HT) A vasoactive mediator that is stored in **mast cells** and **platelets**. It can increase vascular permeability, dilate capillaries, induce platelet aggregation, and produce contraction of smooth muscle — vasoconstriction. Platelet serotonin is taken up from plasma and is stored in platelet dense bodies

SERUM

The liquid remaining after removal of blood cells and fibrin following **coagulation of blood** or plasma. It is extensively used for the study of plasma proteins as well as of **fibrinogen**, enzymes, hormones, and electrolytes.

SERUM PROTHROMBIN CONVERTION ACCELERATOR

See **Factor VII**.

SEVERE COMBINED IMMUNODEFICIENCY — HUMAN

See **Immunodeficiency** — inherited.

SEVERE COMBINED IMMUNE DEFICIENCY — MOUSE

(SCID-hu mouse) An immunologically deficient mouse that has provided a valuable tool for the study of hematopoietic differentiation and hematological malignancy. B-17 SCID/SCID mice are deficient in the major immunoglobulin subtypes with lymphopenia, but they have a normal hematocrit, marrow erythropoiesis, megakaryopoiesis, and myelopoiesis, including macrophages and natural killer (NK) cells. The functional defect is in the V, D, and J regions of the **immunoglobin** gene.

SCID mice have been used for analysis of NK function and the role of **lymphocytes** in the response to infections such as leprosy, pneumocystis, **human immunodeficiency virus**, and **cytomegalovirus**. However, a major use of the SCID mouse is to produce the hybrid state of human and murine hematopoiesis, the SCID-hu mouse.

Initial studies used intraperitoneal injection of human peripheral-blood lymphocytes, which can survive for at least 6 months. The main site of engraftment is intraperitoneal, but interestingly, this model is not associated with **graft-versus-host disease**, but is used in the study of hematolymphoid development and autoimmunity.

Subcutaneous implantation of human fetal liver, fetal thymus, or fetal lymphoid produces sustained engraftment for several months within the local tissue, although this system is limited by the low level of engraftment and the low cell number. The most widely used approach is now the intravenous injection of human bone marrow cells into sublethally irradiated SCID mice. Alternate-day intraperitoneal injections of the growth factors **stem cell factor** (SCF), PIXY 321 (a fusion of granulocyte/macrophage colony stimulating factor [GM-CSF] and interleukin [IL]-3), and **erythropoietin** (EPO) support a high level of engraftment (10 to 50%) of normal human marrow within 30 days. Bone marrow is the preferred site of engraftment.

Many hematological tumors have also been successfully engrafted into SCID mice. **Myelomatosis**, **non-Hodgkin lymphoma**, **Hodgkin disease** lymphoma, **acute lympho-blastic leukemia**, and **chronic lymphocytic leukemias** were initially studied, and more recent attention has been directed at the successful engraftment of acute myeloid leukemias. **Myelodysplasia** of the bone marrow appears to slow engraftment, which is often delayed compared with **acute myeloid leukemia**.

This model has great potential for the study of the pathogenesis and therapy of human leukemia and as a model for minimal residual disease, but it is limited by variable engraftment between animals and by the nonphysiological requirement for maintenance of the tumors by hematopoietic growth factors.

SEX CHROMATIN
See **Neutrophils**.

SÉZARY SYNDROME
See **Mycosis fungoides/Sézary syndrome**.

SICKLE CELL DISORDER
Conditions in which **red blood cells** (RBCs) due to mutant **hemoglobins** adopt a sickle shape at low oxygen tensions, with resultant clinical manifestations.[154,237–239,476–478] The disorders include:

Sickle cell anemia (the homozygous state for the sickle cell gene — hemoglobin S), the prototype of all the sickling disorders

Hemoglobin S/C disease

Hemoglobin S/D disease and sickle cell β-**thalassemia**

Sickle cell trait — heterozygosity for **hemoglobin S** (Hb S) and the most benign form of the disorder, in which the relative abundance of HbA prevents sickling under most physiologic conditions

The frequency of Hb S is greatest in sub-Saharan Africa (20 to 40% of some native populations are heterozygotes). The frequency of sickle cell trait is approximately 8% in African-Americans; lower frequencies occur in the Middle East and in aboriginal Indians. The prevalence of sickle hemoglobin is high in areas where **malaria** is endemic, consistent with a selective advantage afforded by Hb S in persons who contract this disease.

Sickle Cell Anemia

This is due to homozygosity for a mutation that encodes a substitution of valine for glutamic acid at the six position of the β-globin chain. When a RBC sickles and unsickles repeatedly, the RBC membrane becomes damaged and the RBC becomes irreversibly sickled, even when the oxygen pressure is increased. Irreversibly sickled cells have a short life span, and the severity of hemolysis is directly related to the number of circulating cells that are irreversibly sickled. Secondary changes in RBC metabolism (involving potassium and the calcium pump) and in membrane structure occur, although the primary defect is in the hemoglobin. Many inherited and acquired factors influence the pathogenesis of clinical symptoms. Consequently, sickling disorders vary greatly in clinical severity, from the potentially lethal state characteristic of sickle cell anemia to the almost symptom-free sickle cell trait. Both intracellular and extracellular factors influence sickling. There is a positive correlation between the concentration of Hb S in RBCs and the occurrence of sickling. The presence of certain other abnormal hemoglobins such as C, D, and O-Arab with Hb S increases the propensity for sickling. On the other hand, some hemoglobins protect from sickling, e.g., high concentrations of **hemoglobin F** (Hb F).

Deoxygenation for a sufficient interval is the most important determinant of sickling. Sickling occurs at lesser reductions in oxygen tension as the concentration of Hb S increases. For example, sickling may begin to occur in patients with sickle cell anemia at oxygen tension 40 mm Hg, whereas the sickling may require oxygen tension as low as 14 mm Hg in persons with sickle cell trait. A low pH also favors sickling. RBC located in areas of vascular stasis are more vulnerable to sickling that may result in microvascular obstruction and infarction. Low temperatures also tend to precipitate sickling crises due to associated vasoconstriction. Variation in extracellular factors accounts for the natural pattern of the disease, with periods of well-being interspersed with periods when crises abound. Infection is associated with an increase in the frequency of crises for a variety of reasons, including dehydration, acidosis, and hypoxia.

Newborn infants are protected from sickling for the first 8 to 10 weeks of life because of the high level of Hb F. As Hb F levels decline, the manifestations of sickle cell disease appear. Mild anemia exists by age 3 months, there is **splenomegaly** by age 6 months, and the first vaso-occlusive crisis has appeared in 50% of patients by the age of 1 year.

Acute Sickle Cell Crises[479]

These are a consequence of interactive adhesion of sickled red cells to the microvascular endothelium. They may present as general sickle cell crises with recurring attacks of pain involving skeleton, chest, and abdomen, or as crises affecting a specific region:

Abdominal crises: result from infarcts of mesentery and produce severe abdominal pain and peritoneal irritation; bowel sounds are present; symptoms usually persist for 4 to 5 days.

Aplastic crises: due to a temporary arrest of red cell production, usually due to infection (e.g., **parvovirus** B19); result in a profound decline in the hemoglobin level; this self-limited process typically lasts 5 to 10 days.

Bone and joint crises: occur as patient gets older; sludging of blood in larger bones of the extremities, spine, rib cage, and periarticular structures; severe pain results from ischemia; the specific pattern of pain repeats itself in a given patient.

Central nervous system crises: children and young adults are most susceptible to sudden occlusions of cerebral vessels resulting in hemiparesis, seizures, and altered consciousness, from which there may be incomplete resolution.

Hand and foot syndrome: sudden onset of painful dactylitis, which may last for up to 2 weeks; recurrence is seen until age 3 years.

Hematological crises: acute onset of anemia due to aplastic crises, splenic sequestration, and, in some cases, to infarction of significant volumes of bone marrow.

Infectious crises: a major complication frequently requiring hospitalization during the first 5 years. *Streptococcus pneumoniae* is the most common infecting organism, with predilection for blood and spinal fluid; the risk of bacterial meningitis is very high, and the mortality rate is 10 to 35%; in older persons, gram-negative bacteria are the primary offenders (*Salmonella osteomyelitis* is common); loss of splenic function is the principal cause.

Pulmonary crises: often associated with infection.

Splenic sequestration crises: seen in infants and young children with splenomegaly; splenic size increases further; hypovolemic shock and death may supervene in a few hours; may be recurrent until age 5 to 6 years, by which time splenic infarction and fibrosis have occurred.

Vaso-occlusive crises: sudden onset of obstruction of the microcirculation caused by intravascular sickling; usually associated with infection in children, but no obvious relationship in adults.

Chronic Sickle Cell Disease

Bone and joint disease: ischemic necrosis of femoral head; ischemic damage to vertebral body growth plates; "hair on end" pattern on skull X-ray; joint effusion, pain, and, less frequently, synovial hemosiderosis.

Cardiac disorders: high-output cardiac failure due to chronic anemia, recurrent pulmonary infarction, and myocardial hemosiderosis.

Renal tract disorders: hyposthenuria occurs in most patients after the age of 1 year. Later, the kidneys become enlarged, and collecting systems become distorted; papillary necrosis may occur; gross pleomorphic histological changes are seen; hematuria (often unilateral) is an uncommon complication.

Hepatobiliary features: liver is usually enlarged; modest impairment of function; high frequency of pigment gallstones.

Leg ulcers: breakdown of skin over the malleoli and distal part of the legs is common; small-vessel stasis and zinc deficiency probably contribute to the intractable nature of these lesions.

Ophthalmic disorders: conjunctival signs include dark red, comma-shaped vascular fragments; nonproliferative and proliferative retinal changes secondary to vaso-occlusive disease, which may bleed and lead to visual defects.

Pregnancy: a potentially serious problem; maternal mortality may be as high as 20% if medical care is suboptimal; during late pregnancy and the postpartum period, infarcts of lungs, kidneys, and brain may occur; there is an increased incidence of toxemia of pregnancy, cardiac failure, and postpartum endometritis; fetal survival is significantly less than normal. During pregnancy, there can be an increase in the frequency and severity of crises, particularly in the third trimester. The induction of premature labor puts both the mother and infant at risk.

Priapism: due to obstruction of venous return in the corpora cavernosa caused by sickled red cells; this is usually self-limited and of short duration, but can produce impotence if recurrent or prolonged.

Respiratory tract disorders: can result in severe impairment of pulmonary function due to repeated infections and infarcts.

Zinc deficiency accentuates all of these chronic sickle cell disorders.

Laboratory Features
See **Hemoglobinopathies**.

Treatment
General measures involve maintenance of hydration, correction of electrolyte disturbance, early treatment of infection, and adequate control of pain by analgesics, including opiates if necessary (narcotic addiction can be a problem) during crises; **folic acid** supplements should be given in pregnancy, during aplastic crises, and when dietary intake of folate is likely to be suboptimal; zinc deficiency should be corrected.

Red blood cell transfusion exerts beneficial effects by improving tissue oxygenation, temporarily suppressing production of Hb S-containing RBCs, and reducing the tendency to sickling by diluting the recipient's RBC and Hb S concentration. Partial-exchange transfusion is the preferred method, because this limits the net gain of iron. Chronic transfusion introduces increased risks of transfusion **iron overload**, viral hepatitis, **AIDS**, and RBC sensitization. Transfusion therapy should be reserved for dangerously increasing anemia, pain crises protracted beyond 7 days, pregnancy, progressive organ damage, and surgery. **Iron chelation** may be required if transfusion iron overload develops. Transfusion of RBCs sufficient to reduce the concentration of sickle cells below one-third will limit the frequency of painful crises. To pursue this policy effectively, two units of packed RBCs must be transfused every two weeks in adults. Hypertransfusion to switch off Hb S production has also been tried, but should not be contemplated as routine management. RBC transfusion in patients undergoing surgery appears to reduce the risk of perioperative crises.

Induction of hemoglobin F production has been achieved using **hydroxyurea** by mouth daily. This raises Hb F levels from 2 to 20%, reduces the frequency and severity of sickling, and improves patient weight, function, and general well-being. This drug can be given over long periods without toxicity, although significant increase in the **mean cell volume** (MCV) occurs.

Treatment can prevent overwhelming infections by such organisms as *Streptococcus pneumoniae*, *Hemophilus influenzae*, and *Salmonella* spp. As a prophylactic measure, young children should receive polyvalent pneumococcal vaccine and be started on prophylactic oral penicillin. Patients and their relatives must be instructed on procedures to follow if a fever develops.

Priapism can usually be managed by analgesia alone or in combination with nifepidine therapy. This complication, when lasting more than 1 day, can be relieved by performance of a corpora spongiosum shunt.

Leg ulcers require meticulous local care to prevent infection. The application of a zinc oxide bandage encourages healing, although hyperbaric oxygen therapy is necessary in some cases. Rest and elevation of the limb and control of secondary infection are also beneficial.

Many chemical agents have been employed with a goal of reducing sickling via a variety of mechanisms, although no otherwise acceptable drug has been shown to prevent or control vaso-occlusive crises satisfactorily.

Allogeneic stem cell transplantation is an effective treatment, but the risks and costs of transplantation must be set against the natural history of the underlying disease.

Sickle Cell Trait

Hb S trait (heterozygous Hb S) is a common genetic abnormality (7 to 8% of African-Americans) that is not associated with clinical disease or anemia. The frequency of Hb S trait in Africa is even higher than in American blacks and is often compounded by thalassemia. In the uncomplicated case, the blood profile is normal, and sickle cells are not observed in blood films. The hemoglobin electrophoresis pattern is diagnostic. In sickle cell trait, there is always more Hb A present than Hb S. A small number of patients experience transient attacks of hematuria. Severe exertion and dehydration or exposure to very low oxygen tensions may, however, precipitate sickle crisis and sudden death. Rarely, splenic infarction has been reported in flight at high altitudes in unpressurized aircraft. Life expectancy is the same as for persons without Hb S. RBC life span is normal, and painful crises and organ damage do not usually occur.

Hemoglobin S/C Disease

The disorder arising from Hb S/C double heterozygosity. Although less common than Hb S/S, the same manifestations as sickle cell disorder are found, but the frequency and severity of vaso-occlusive crises are lower and splenic infarction is less common. Unlike persons with Hb SS, most patients with Hb S/C disease have an enlarged spleen. Patients frequently present with gross hematuria. The blood smear shows prominent target cells; Hb C crystals may be present, but true sickle cells are not seen.

Hemoglobin S/D Punjab

A relatively mild form of sickle cell disease found mainly in northwestern India due to compound heterozygosity for Hb S and Hb D Punjab. Hemoglobin S/D Punjab is uncommon in persons of sub-Saharan African origin.

SIDEROBLASTIC ANEMIA

Anemias characterized by the presence of "ringed" sideroblasts in the bone marrow. These are nucleated erythroid precursors that have a perinuclear ring or collar of large Prussian blue (iron) granules. Electron microscopy reveals that these granules are iron-filled mitochondria or mitochondrial fragments. There are four major categories of sideroblastic anemia.

X-Linked Sideroblastic Anemia

(Hereditary or familial X-linked sideroblastic anemia) This group of uncommon disorders is due to the occurrence of coding or promoter region mutations of the erythroid-specific ALA synthase gene (*ALAS2*). Many cases occur in hemizygous males, whose anemia is discovered at widely varying ages, depending on severity of anemia and other features.

It is probable that some mutations are lethal in males *in utero* or soon after birth. Hypochromia and microcytosis predominate in some cases, although macrocytosis is also common. The bone marrow reveals erythroid hyperplasia, numerous ringed sideroblasts, and delayed cytoplasmic maturation of erythroblasts. **Transferrin** saturation and serum **ferritin** levels are often elevated; free erythrocyte protoporphyrins are increased. Some women also develop sideroblastic anemia due to *ALAS2* mutations, usually as a function of skewed X-inactivation. Anemia is mild and requires no specific therapy in some cases. Anemia in some patients improves with pyridoxine therapy (150 to 200 mg daily). Severe anemia may require red blood cell transfusion. In some patients, intestinal absorption of iron is increased and **iron overload** may develop, although most patients do not have typical mutations in autosomal hemochromatosis-associated genes. **Phlebotomy** or **iron chelation** therapy is needed to alleviate iron overload and prevent injury to liver and other target organs in such cases.

Myelodysplasia with Ringed Sideroblasts

(Refractory anemia with ring sideroblasts; Primary acquired sideroblastic anemia) This variant of **myelodysplasia** is a clonal marrow disorder characterized by anemia unresponsive to hematinics, reticulocytopenia, erythroid hyperplasia with ringed sideroblasts, and erythrokinetic evidence of ineffective erythropoiesis. This disorder typically appears in the fifth to seventh decades and is more prevalent in men. Acute leukemia develops in 7 to 10% of cases, typically months to years after diagnosis. There are no specific cytogenetic or FISH (fluorescence *in situ* hybridization) markers for this morphologic variant of myelodysplasia.

Some patients present with symptoms of anemia, although many are asymptomatic. Physical examination reveals nonspecific findings. The hemoglobin concentration is usually 8 to 10 g/dl; **mean cell volume** (MCV) is consistently elevated. **Peripheral-blood film examination** shows anisocytosis, poikilocytosis, schistocytes, and macrocytes. **Pappenheimer bodies** and coarse basophilic stippling are sometimes present. Occasional giant platelets may be observed, often with multiple cytoplasmic vacuoles; **thrombocytosis** is common. Granulocytes show various abnormalities, including pseudo-Pelger-Huët morphology. **Bone marrow** shows hyperplastic, megaloblastoid erythropoiesis with ringed sideroblasts, some of which may have cytoplasmic vacuolation. Increased numbers of macrophages are sometimes present. Erythroleukemia shares many of these morphologic abnormalities. The prevalence of common *HFE* mutations is similar to that in the general population. For many patients, the disorder is chronic and indolent, and anemia is best treated with regular **erythropoietin** injections, thalidomide, or amifostine. For patients with unresponsive anemia, periodic transfusion of packed erythrocytes is necessary. In advanced stages in which there is progression to acute leukemia, various chemotherapy agents may beneficial in carefully selected patients. If transfusional siderosis develops, iron chelation therapy may be necessary.

Drug- or Chemical-Associated Sideroblastic Anemias

Anemia and ringed sideroblasts caused by marrow injury due to:

Excess **alcohol** consumption

Heavy metal toxicity: lead, zinc, arsenic

Adverse drug reaction: isoniazid, chloromycetin, lincomycin, penicillamine, busulfan, cycloserine

Cessation of exposure to the offending agent (or successful alleviation of heavy metal intoxication) resolves anemia and induces return of normal iron metabolism by erythroblasts.

Pearson Marrow-Pancreas Syndrome

This disorder, due to deletion and duplication of mitochondrial **DNA**, comprises refractory sideroblastic anemia, vacuolated marrow precursors, and exocrine pancreas insufficiency. Many affected individuals die in early life.

SIDEROCYTE

Red blood cells containing siderotic granules detected by the **Prussian blue (Perl's) reaction**. They do not occur normally in blood, but are present in some persons with various forms of **sideroblastic anemia** and often appear after **splenectomy**. Similar intracytoplasmic granules detected on Romanowsky stains are called **Pappenheimer bodies**.

SIDEROPENIA

See **Iron deficiency**.

SIDEROPHILIN

See **Transferrin**.

SIDEROTIC GRANULES

Water-insoluble complex granules of ferric **iron**, lipid, protein, and carbohydrate. This siderotic material (**hemosiderin**) reacts with acidified potassium ferrocyanide — **Prussian blue (Perl's) reaction** — to form a deep blue precipitate, ferriferrocyanide. This material also stains by Romanowsky dyes (**Pappenheimer bodies**). In the bone marrow, siderotic granules occur in the cytoplasm of normal late erythroblasts and in the cells seen in various forms of **sideroblastic anemia**.

SINUSOIDAL LARGE-CELL LYMPHOMA

See **Anaplastic large-cell (CD30⁺) lymphoma**; **Non-Hodgkin lymphoma**.

SJÖGREN'S SYNDROME

A syndrome of xerophthalmia and xerostomia due to infiltration of lacrimal and salivary glands by **lymphocytes**, **plasma cells**, and **histiocytes** (macrophages).[480] There is an association with **human leukocyte antigen** (HLA B8/DR3). Other hematological changes include:

> **Immune complexes** with paraproteinemia and/or cryoglobulinemia, resulting in **hyperviscosity** syndrome. IgG hypergammaglobulinemia and positivity of **rheumatoid arthritis** (RA) factor, **antinuclear factor** (ANF) antibodies, and antimitochondrial

antibodies (Sm, Ro/SS-A & La/SS-B) may also be present. Organ-specific autoanti-bodies sometimes include red blood cell autoantibodies, shown by a **direct antiglob-ulin (Coombs) test**, resulting in an **autoimmune hemolytic anemia**. The erythrocyte sedimentation rate (ESR) may be increased or decreased.

Normocytic anemia.

Lymphopenia and/or **granulocytopenia.**

Thrombocytopenia.

Lymphadenopathy due to **non-Hodgkin lymphoma**. There is a 44-fold risk of devel-opment. Approximately 85% are MALT lymphomas. See **Extranodal marginal-zone B-cell lymphoma.**

SKELETAL DISORDERS

The disorders of bone cortex arising from disorders of bone.

Bone pain or tenderness. Common with **acute lymphoblastic leukemia** (ALL) and blast crisis of **acute myeloid leukemia** (AML); frequent with **myelomatosis**; uncommon with **non-Hodgkin lymphoma, Hodgkin disease**, and **myelofibrosis**; and rare with **chronic lymphatic leukemia, polycythemia rubra vera**, and **chronic myel-ogenous leukemia.**

Small stature due to retarded growth associated with prolonged and severe **iron defi-ciency** and malabsorption in childhood, **sickle cell disorder** (SCD), and **thalas-semia.**

Pathological fractures, most commonly with myelomatosis and **infiltrative myeloph-thiasis**, especially metastases of carcinoma.

Frontal bossing of the skull with chronic marrow hyperplasia, as occurs in **thalassemia.**

Osteomyelitis as a result of infection, frequently by salmonella in SCD.

Osteoporosis affecting the sternum, ribs, clavicles, scapulae, and pelvis occurring with SCD, ALL, AML, **mastocytosis**, and prolonged **heparin** or **corticosteroid** therapy. Vertebral collapse or cupping of the vertebral bodies often occurs with thalassemia.

Osteomalacia due to increased secretion of fibroblast growth factor with **adult T-cell leukemia/lymphoma** and myelomatosis.

Histiocytosis. Eosinophilic granuloma.

SKIN

The external ectoderm provides a major physical barrier to pathogens. Beneath this layer of cells, the fibroblastic connective tissue contains **dendritic reticulum cells** (Langerhans cells) and epidermal CD8$^+$ T-lymphocytes with γδT-cell receptors to form a cutaneous immune system. These cells are also located in the underlying dermis. The dendritic reticulum cells originate in the skin and pass via the blood and lymph (Veil cells) before migrating into the deeper tissues.

SKIN DISORDERS

The association of skin disorders with hematological disorders.[481,482]

Skin Changes in Hematological Disorders

Color and pigmentation:

- Pallor, particularly of the palms of the hands, an unreliable sign of **anemia** due to wide variations in dermal thickness
- Jaundice from any cause, including **hemolytic anemia**
- Erythema, particularly of the face, associated with all forms of **erythrocytosis**
- **Cyanosis** due to hypoxia, which, when caused by **methemoglobinemia**, has a leaden tint
- Grayish-bronze pigmentation of the whole body with **hereditary hemochromatosis**
- Hyaline plaques in skin folds, particularly inguinal, by **amyloidosis** deposits

Purpura bruising:

- Punctate from capillary hemorrhage or ecchymoses from diffuse bruising due to **vascular purpura** or **thrombocytopenia** (Color Figures 3, 4)
- Bruising and necrosis in disseminated intravascular coagulation
- Purpuric lichenoid dermatitis: occurs in dependent areas of the body, caused by capillary proliferation with inflammation and rupture of skin vessels; chronic extravasation of **red blood cells** results in deposition of **hemosiderin** in the skin with pigmentation

Pruritus, erythema, eczema, psoriasis, and local nodules with **polycythemia rubra vera**, **Hodgkin disease**, or **non-Hodgkin lymphoma**

Papulonodular rashes and urticaria with **hypereosinophilic syndrome**, **graft-versus-host disease**, **sarcoidosis**, and **Vogt-Koyanaygi-Harada's disease**

Ulceration of the legs in chronic **sickle cell disorders**, **thalassemia**, and **hereditary spherocytosis**

Urticaria pigmentosa: a benign local accumulation of **mast cells** in the skin of adults, with **mastocytosis** affecting the bone marrow, spleen, liver, and lymph nodes; it is associated with anemia and **leukopenia** or **monocytosis**, **lymphocytosis**, **eosinophilia**, and **thrombocytopenia**. Prothrombin deficiency occurs with **liver disorders**; mast cell leukemia is an occasional progression

Pyoderma gangrenosum in **chronic myelogenous leukemia** and **myelomatosis**

Herpes zoster as a complication of **immunodeficiency**, particularly with treatment for hematological malignancy (Color Figures 7, 8; also Figure 113)

Blisters and bullae due to photosensitivity in **porphyria cutanea tarda**

Angioedema with deficiency of **complement** C1 inhibitor

Vasculitis of dermal vessels, particularly **leukocytoclastic vasculitis**, **Wegener's granulomatosis**, and **polyarteritis nodosa**

Seborrheic dermatitis, folliculitis, Bartonellosis, fungal infections, and herpes zoster in association with **acquired immunodeficiency syndrome** (AIDS)

Psoriasis with arthropathy due to activated CD4 **lymphocytes**, probably a reaction to infection by group A streptococci; the activated T-cells release **interleukin** (IL)-1, **tumor-necrosis factor** (TNF)-α, and **interferon**. Anti-TNF-α monoclonal antibodies may be effective therapy.

Infiltration of the skin by cutaneous non-Hodgkin lymphoma:

- **Mycosis fungoides** (Sézary syndrome): the specific lymphoma of the skin occurring in stages:

Stage I: erythema or papules only

Stage II: plaque infiltration of small irregular T-lymphocytes

Stage III: tumor formation by T-cells

Stage IV: generalized erythroderma

(The stages are related to prognosis with 10-year relative survival rates of 100% in Stage I, 67% in Stage II, and about 40% in Stages III and IV disease. In advanced disease, usually of some years' duration, there may be infiltration of the bone marrow, liver, spleen, lymph nodes, lungs, kidneys, thyroid, pancreas, and myocardium.)

- T- and B-cell lymphomas, which give rise to pruritus, erythema, eczema, and local nodules, typically with an elevated pink pearl-like appearance. The histological types are:

  T-cell lymphomas: **precursor T-lymphoblastic leukemia/lymphoma; adult T-cell leukemia/lymphoma; anaplastic large-cell lymphoma; peripheral T-cell lymphoma, unspecified**

  B-cell lymphomas: **precursor B-lymphoblastic leukemia/lymphoma, chronic lymphocytic leukemia**/small lymphocytic lymphoma, **extranodal marginal-zone B-cell lymphoma** (MALT lymphoma)

- Kaposi's sarcoma

Hematological Effects of Skin Disorders

Anemia

- **Iron deficiency** caused by:

  Dermatological disorders with associated intestinal tract disorders, which are a source of chronic **hemorrhag**e and malabsorption of **iron**

  Hereditary hemorrhagic telangiectasia

  Angiokeratoma corpus diffusum universale (**Fabry's disease**): an X-linked incompletely recessive disorder of glycolipid metabolism with skin telangiectasia; there is associated elevation of the **erythrocyte sedimentation rate** and **neutrophilia**

  Peutz-Jegher's syndrome: a genetically determined disorder of multiple polyposis of the lips, with lesions in the stomach or colon giving rise to chronic hemorrhage and which may undergo malignant transformation

  Malignant atrophic papulosis: a necrotizing vasculitis of the skin and intestinal tract with chronic hemorrhage

- **Megaloblastosis** from malabsorption:

  "Dermatographic enteropathy": a variety of dermatoses associated with malabsorption of **folic acid**

  Dermatitis herpetiformis: associated with gluten enteropathy and **splenic hypofunction**

- **Folic acid** deficiency arising from increased requirement by tissues in exfoliative dermatitis and psoriasis being treated with methotrexate

- **Anemia of chronic disorders**

Neutrophilia

- Many dermatoses, with particularly high levels in polyarteritis nodosa and Behçet's syndrome
- Acute febrile neutrophilic dermatosis (Sweet's syndrome)

Eosinophilia

- Commonly associated with atopic eczema, **adverse drug reactions, metazoan infections, Behçet's disease**, and **polyarteritis nodosa**

SKIN PUNCTURE

The procedure for obtaining specimens of capillary blood for laboratory analysis when **venepuncture** is difficult, as in infants or small children.[483] Site of puncture is usually the lateral aspect of the thumb or finger except for infants, where the heel is used. After warming and cleaning the site, puncture is made by a sterile disposable needle. A free flow of blood is obtained by light pressure. Blood is collected into capillary tubes with or without anticoagulant absorbed on to the wall of the tube. When filled, each tube is plugged with plastacene before being transported to the laboratory for testing.

Analytical values are less accurate than those obtained following venepuncture, but they may be within acceptable clinical limits for some analytes.

SMALL-CELL CEREBRIFORM LYMPHOMA

See **Mycosis fungoides/Sézary syndrome; Non-Hodgkin lymphoma; Lymphoproliferative disorders**.

SMALL CLEAVED CELL, DIFFUSE OR NODULAR LYMPHOMA

See **Mantle-cell lymphoma; Non-Hodgkin lymphoma; Lymphoproliferative disorders**.

SMALL CLEAVED FOLLICULAR CENTER CELL LYMPHOMA

See **Mantle-cell lymphoma; Non-Hodgkin lymphoma; Lymphoproliferative disorders**.

SMALL CLEAVED, LARGE CLEAVED, OR LARGE NONCLEAVED FOLLICULAR CENTER CELL LYMPHOMA

See **Follicular lymphoma; Non-Hodgkin lymphoma; Lymphoproliferative disorders**.

SMALL LYMPHOCYTIC B-CELL LYMPHOMA

See **Chronic lymphocytic leukemia; Extranodal marginal B-cell lymphoma; Non-Hodgkin lymphoma; Lymphoproliferative disorders**.

SMALL LYMPHOCYTIC, PLASMACYTOID LYMPHOMA

See **Lymphoplasmacytic lymphoma/Waldenström macroglobulinemia; Non-Hodgkin lymphoma; Lymphoproliferative disorders**.

SMALL-, MIXED-, AND LARGE-CELL DIFFUSE LYMPHOCYTIC PLASMACYTOID LYMPHOMA

See **Lymphoplasmacytic lymphoma/Waldenström macroglobulinemia; Non-Hodgkin lymphoma; Lymphoproliferative disorders**.

SMALL NONCLEAVED CELL LYMPHOMA

See **Burkitt lymphoma; Non-Hodgkin lymphoma; Lymphoproliferative disorders**.

SMALL NONCLEAVED CELL, NON-BURKITT'S LYMPHOMA

See **High-grade lymphoma, Burkitt-like; Non-Hodgkin lymphoma; Lymphoproliferative disorders**.

SMEAR CELL

Degenerate lymphocytes seen in **peripheral blood films** and smears of **bone marrow** aspirate. They are a common feature in material from patients with **chronic lymphocytic leukemia**.

SMOKING DISORDERS

See **Tobacco**.

SNAKE VENOM DISORDERS

The hematological effects of snake venoms.[484,485] These vary with the snake species, but all venomous snakes have glands behind each eye connected via ducts to hollow retractile fangs at the front of the mouth that can bite. Cobras and coral snakes cause neuromuscular disturbances but only minor disorders of **hemostasis**. The coagulation changes occurring with each species are shown in Table 141. Bites from these snakes may result in defibrination with **hemorrhage** and **disseminated intravascular coagulation**.

Persistent hypotension, **neutrophilia**, electrocardiographic abnormalities, or limb swelling within 4 h of the bite are indications for treatment with the appropriate antivenin and human tissue plasminogen activator. Less severe effects respond to local removal of the venom with surgical toilet and observation.

TABLE 141

Coagulation Changes Induced by Various Snake Venoms

Snake Species	Coagulation Effect
Saw-scaled viper	**prothrombin** activation
Russell viper	**factor IX** activation
	factor Va enhancement
Malayan pit viper	fibrinopeptide A cleavage
	intravascular hemolysis
Eastern diamondback rattlesnake	fibrinopeptide A cleavage
Western diamondback rattlesnake	**fibrinolysis**
	thrombocytopenia

Russell viper venom is used in the diluted Russell viper venom time test for lupus anticoagulant, and a purified enzyme from the venom of *Bothrops atrox* replaces thrombin in the **reptilase time test**.

SODIUM/POTASSIUM PUMP
See **Cation pump**.

SOLID-PHASE TECHNIQUES
Laboratory methods using plastic 96-well microplates for the detection of **alloantibodies** to **red blood cells** (RBCs). The methods can be used as an alternative to tube- or **column-agglutination** variants of the **indirect antiglobulin (Coombs) test** for antibody screening in **pretransfusion testing**. In one widely available system, red cell antibodies are "captured" by red cell ghosts bound to the solid phase. Following washing, antibodies bound to the solid phase are visualized by the addition of indicator red cells, which in a positive reaction form a monolayer over the bottom surface of the microplate well, but which in a negative reaction are centrifuged to the center of the well.

SOUTH AFRICAN GENETIC PORPHYRIA
See **Variegate porphyria**.

SPECTRIN
See **Cell locomotion**; **Hereditary spherocytosis**; **Red blood cell** — membrane.

SPHEROCYTE
An anomaly of **red blood cell** morphology due to a decrease in surface area:volume ratio, usually because of loss of cell membrane, resulting in red blood cells with a reduced diameter. **Microspherocytes** also have a reduced volume. The causes are:

Inherited
- Hereditary elliptocytosis
- Hereditary infantile pyropoikilocytosis
- Hereditary spherocytosis

Acquired
- **Blood transfusion complication** of ABO incompatibility
- Warm autoimmune hemolytic anemia
- *Clostridium* bacterial infection
- Hemolytic disease of the newborn
- Administration of anti-D to RhD-positive patients (i.e., for **immune thrombocytopenic purpura**)
- Heinz body hemolytic anemia
- Drug-induced immune hemolytic anemia

- **Burns** hemolytic anemia
- **Snake venom disorders**
- Hypophosphatemia
- Acute **paroxysmal cold hemoglobinuria**

SPLEEN

An organ of the **lymphoid system** situated in the left hypochondrium of the abdominal cavity having anatomical relations with the diaphragm, stomach, pancreas, left kidney, left adrenal gland, and the colon. It has a flattened oblong shape of about 125 mm in length, weighing 135 ± 30 g but varying with age from a peak at puberty.

Blood supply is by the splenic artery and venous drainage by the splenic vein into the portal system.

Accessory spleens, splenunculi, are common. They are most often located in the splenic hilum, gastrosplenic ligament, omentum, or tail of the pancreas. Splenic tissue has also been found in the testes. Polysplenia is a rare condition in which the bulk of splenic tissue is divided into two or more nearly equally sized tissue masses. A wandering spleen without attachment to the diaphragm can lead to torsion, with dramatic abdominal pain.

Embryology

Evidence of a spleen occurs at the 5th week of gestation, arising from mesenchymal tissue. Vascularization occurs during the 6th to 7th week, followed by migration of hematopoietic cells with evidence of phagocytosis. At the 15th to 17th week, splenic lobules with central arteries are evident, and by the 18th week there is lymphoid colonization by T-lymphocytes. Development of B-cell areas occurs around the 23rd week, but functional activity is delayed until birth and not complete until 1 year after birth.

Structure

A diagrammatic representation is given in Figure 107.

The spleen is surrounded by a thick fibrous capsule with inward-traversing trabeculae, penetrated at the hilum by blood vessels. The splenic artery arborizes into small central arteries along a branching network of collagen and elastic fibers. These are further subdivided, the fibrous adventitia becoming replaced by tightly packed T-cells to form a cylindrical perivascular sheath held by circumferential layers of reticular meshwork (T-zones, periarteriolar lymphoid sheaths [PALS]). Primary T-cell activation occurs in this area, antigen being presented by interdigitating **dendritic reticulum cells**. This zone becomes progressively reduced in thickness until the vessel branches into slender penicilli invested by only one or two layers of lymphocytes. The sheaths have associated nodular lymphoid follicles (Malpighi) surrounded by a marginal zone of B-cells (**centrocytes**) interspersed with **immunoblasts**. Each follicle has a germinal center of rapidly proliferating activated B-cells (centroblasts). There are also primary follicles of resting B-cells. The sheaths and follicles form the white pulp of the spleen, which is embedded in a red pulp, the marginal zone being populated by both T and B-lymphocytes and into which **monocytes** migrate and become **histiocytes** (macrophages). The marginal-zone B-cells make responses to polysaccharide antigens (TI antigens). The red pulp is an elaborate network of anastomosing sinuses lined by vascular endothelium with splenic cords (Billroth) in which there are mainly **red blood cells** with many monocytes/macrophages as well as lymphocytes.

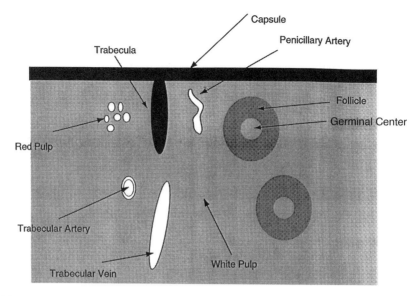

FIGURE 107
Diagrammatic representation of the histological structure of the spleen.

Physiological Function

Filtration. Normally, the spleen contains 40 to 80 ml of blood, which flows through the spleen at a rate of 200 to 300 ml per minute, which is 4 to 5% of the cardiac output, with a transit time of 30 to 60 sec. From the arteriolar capillaries, most blood flows directly into the sinuses of the red pulp, about 2% normally passing via the cords. Blood entering the spleen has its lymphocyte-rich plasma "skimmed off" into the perivascular sheaths and marginal zones, where they spend most of their life span, the B-lymphocytes moving into the follicles. Tuftsin, a tetrapeptide formed in the spleen, binds to receptor sites of neutrophils, monocytes, **natural killer (NK) cells**, and macrophages to stimulate phagocytosis.

The lowering of the oxygen tension, ambient pH, and glucose concentration causes the red cells to be placed under stress. Any depletion in cellular **adenosine triphosphate** (ATP) with instability of the membrane, fragmentation, or inflexibility renders them susceptible to phagocytosis. In addition, there is a "pitting" mechanism of forcible exocytosis, unique to the splenic cords, whereby any abnormalities of the cell membrane such as red blood cells at the end of their life, spherocytes, or sickle cells are trapped, and any inclusions, such as **Howell-Jolly bodies**, **Heinz bodies**, or **siderotic granules**, are removed. **Hemoglobin** is degraded, but no **iron** is stored in the spleen. Normally structured cells are returned to the circulation, but splenic macrophages remove parasitized red blood cells from the circulation.

Particulate matter, including bacteria, in the residual plasma is arrested in the marginal zone and the Billroth cords by the phagocytes, the plasma rapidly flowing through the sinuses into trabecular veins, which drain into the splenic vein. Autonomic nerves, lining the trabeculae, provide the nerve supply to the spleen. Adrenergic stimulation results in release of red blood cells and especially of platelets.

*Reservoir for **platelets***: 20 to 40% of circulating platelets are held in the spleen. Platelet pooling to a level of around 30% occurs physiologically and is proportional to the size of the spleen.

Immune response: splenic macrophages in the Billroth cords or in the mantle or marginal zones present circulating bacteria and antigen to T-lymphocytes, which stimulate B-lymphocytes to produce **antibody,** particularly of the **immunoglobulin** M (IgM) type. They are also activators of the **complement** system. The spleen may be the production site of autoantibodies. Of splenic lymphocytes, 15 to 20% are exchangeable and in a state of constant turnover.

Hematopoiesis: fetal spleens contain hematopoietic cells from the 12th week, but the spleen as a site of active hematopoiesis is doubtful.

Assessment of Function

This is directly related to the amount of splenic tissue and indirectly to its size. This can be determined by radionuclide imaging using ^{99m}Tc- or ^{51}Cr-labeled red cells (see **Radionuclides**), ultrasonic scanning, X-ray, computerized tomography, and nuclear magnetic resonance imaging.[486] "Pitting" of red blood cell membranes can be seen by scanning electron microscopy.

Disorders of the Spleen[487]

See **Splenic hyperfunction**; **Splenic hypofunction**; **Splenomegaly**.

SPLENECTOMY

The indications and the effects of surgical removal of the spleen. The two surgical approaches are abdominal splenectomy and laparascopic splenectomy. With careful observation, blunt splenic injury can be treated conservatively. The presence of **splenomegaly** is never by itself an indication for removal.

Indications

Disorders for which splenectomy has to be considered are:

Rupture, either spontaneously or by accidental trauma

Splenic cysts

Splenic tumors or malignancy of adjacent organs with spread

Hereditary spherocytosis or elliptocytosis

Immune thrombocytopenic purpura (ITP)

Immune hemolytic anemia

Pyruvate kinase deficiency of red blood cells

Hypersplenic effects of **myeloid metaplasia** or **lymphoproliferative disorders**

Staging of lymphoproliferative disorders (now a rare indication)

Effects of Splenectomy

See **Splenic hypofunction**.

SPLENIC HYPERFUNCTION

See also **Hypersplenism**.

The causes and effects of increased splenic activity. It is always associated with splenic hyperplasia and **splenomegaly**, but this may not be clinically evident.

Causes of Splenic Hyperfunction

Response of **lymphocytes/histiocytes** (macrophages) to:

- Acute infections: bacterial septicemia, viral hepatitis, Epstein-Barr virus (**infectious mononucleosis**), cytomegalovirus, salmonelloses, brucellosis, typhus, relapsing fever, toxoplasmosis, tularemia
- Hyperreactive malarial splenomegaly
- Subacute and chronic infections: subacute bacterial endocarditis, **tuberculosis**, brucellosis, syphilis, **babesiosis**, **leishmaniasis**, schistosomiasis, trypanosomiasis, histoplasmosis, and other systemic fungal infections; **human immunodeficiency virus**; **granulomas**
- Noninfectious inflammatory disorders: **rheumatoid arthritis**, **Felty's syndrome**, **systemic lupus erythematosus**, rheumatic fever, serum sickness, **sarcoidosis**, berylliosis

Trapping of **red blood cells** in splenic sinusoids, usually associated with **hemolytic anemia,** due to:

- **Hereditary spherocytosis** or **elliptocytosis**
- Other disorders of the red blood cell membrane or metabolism
- Immune hemolytic anemia
- **Thalassemia** and **sickle cell disorders**, including hemoglobin S/C disease and sickle-thalassemia

Myeloproliferative disorders, particularly **chronic myelogenous leukemia**

Myeloid metaplasia with **lipid storage disorders** and **infiltrative myelophthiases**

Lymphoproliferative disorders, particularly **Hodgkin disease, chronic lymphocytic leukemia, hairy cell leukemia, angioimmunoblastic T-cell lymphoma, angiofollicular lymph node hyperplasia,** and **myelomatosis.**

Acute leukemias: **acute lymphoblastic leukemia, acute myeloid leukemia,** and **juvenile-myelomonocytic leukemia**

Systemic **mastocytosis**

Histiocytosis: systemic, hemophagocytic histiocytosis; **histiocytic sarcoma**

Effects of Splenic Hyperfunction

Pancytopenia

Red blood cell pooling: the spleen becomes a reservoir of red blood cells, such that exchange with the peripheral circulatory cells is reduced, flow taking 30 to 60 min. This reservoir can be as great as 50% of the total **red cell mass**, acting as an arteriovenous shunt. Most of the slow-moving cells are in the cordal compartment. To maintain the **blood volume**, the red cells are replaced by plasma, the plasma volume being proportional to the splenic enlargement, and a **dilutional anemia** ensues.

Platelet pooling: this is proportional to splenic size and, in massive splenomegaly, can be up to 90% of the total platelet mass. It can be mobilized by adrenergic stimulation.

Granulocyte pooling: only occurs with massive splenomegaly, particularly with Felty's syndrome.

SPLENIC HYPOFUNCTION

The causes and the effects of partial or total splenic deficiency.

Causes of Splenic Hypofunction

Splenic agenesis: (asplenia) a rare congenital absence of the spleen, usually associated with severe malformation of the heart, great vessels, and biliary atresia.

Splenic infarction: (autosplenectomy) recurrent infarction results eventually in splenic fibrosis and functional ablation. This occurs in advanced **sickle cell disorders** (SCD), **essential thrombocythemia, thrombotic thrombocytopenic purpura,** systemic vasculides (particularly **polyarteritis nodosa**), **myeloproliferative disorders,** and **lymphoproliferative disorders.**

Functional hyposplenia due to failure of the splenic filtering function of red blood cells occurs in children with SCD at 2 to 3 years, with **intestinal tract disorders** such as gluten enteropathy, Crohn's disease, ulcerative colitis, and with gut-associated dermatitis herpetiformis. It also occurs for unexplained reasons following **allogeneic stem cell transplantation** and in chronic **graft-versus-host disease.** Increased red blood cell "pitting" can be seen in older people, but it has no clinical significance.

Splenectomy.

Ionizing radiation, either accidental or therapeutic.

Immunodeficiency, particularly from **cytotoxic agent** therapy, reduces the ability of white pulp cells to respond to antigens.

Effects of Splenic Hypofunction

Circulation of Abnormal Red Cells Not Removed by the Spleen

These cells or bodies include **acanthocytes, echinocytes, leptocytes** (target cells), **Howell-Jolly bodies, Heinz bodies, siderocytes, nucleated red blood cells,** and **Pappenheimer bodies.**

Infection

Splenectomized or asplenic patients have an increased susceptibility to infection.[488,489] Overwhelming postsplenectomy infection (OPSI) can occur, but it is uncommon (1 to 10%). Septicemia is usually due to infection by *Streptococcus pneumoniae, Hemophilus influenzae* type b, or *Neisseria meningitides.* The mortality rate is up to 60%. Children are particularly at risk. The comparatively high incidence is possibly due to changes in immunoglobulin levels, deficient primary antibody responses to specific spleen-dependent antigens (e.g., polysaccharides), change in neutrophil function, and decreased ability to phagocytose bacteria due to lack of splenic tuftsin and properdin.

Early Postoperative Infection

This is usually due to either a subphrenic abscess or pulmonary and pleural sepsis.

Late Postoperative Infection

This usually occurs within the first 2 years postsplenectomy, but in up to one-third of patients at least 5 years elapse. The rate in adults is only 1 to 2% of cases, the incidence after post-traumatic splenectomy being only slightly higher than similar infections in the rest of the population. The rate is over 10% in children and in those with underlying disorders such as sickle cell disorders, **thalassemia**, **Hodgkin disease**, and immunodeficiency.

Prevention of Infection

This is by immunization with polyvalent pneumococcal vaccine and *H. influenzae* type b vaccine given 2 weeks or more prior to an elective splenectomy. It is less likely to be effective when given postoperatively or in immunodeficient patients. It is best avoided in **pregnancy** and should be delayed for at least 6 months after suppressive chemotherapy or **radiotherapy**. Vaccination with meningococcal vaccine for group A and C disease should be restricted to those at specific risk. Booster doses of pneumococcal vaccine at 5- to 10-year intervals is recommended, but for those whose splenectomy was for immune thrombocytopenic purpura, a relapse of this disorder can occur. Annual influenza immunization may help reduce secondary bacterial infection.

Lifelong oral phenoxymethylpenicillin prophylaxis is advised postsplenectomy by some physicians, but it is not administered uniformly and compliance rates are low. Patients allergic to penicillin should be offered erythromycin. They must be informed of the risks and told that, at a minimum, they must take an antibiotic when they develop the symptoms of a febrile infection, however minor. Asplenic patients traveling to malarious areas should be strongly advised to take malarial prophylactic drugs.

Should infection occur, a sample of blood for blood culture must be obtained prior to immediate treatment with intravenous or intramuscular benzylpenicillin. If the patient has an allergy to penicillin, cefotaxime or ceftriaxone should be given instead. For those who have been taking antibiotics prophylactically or who may have penicillin-resistant organisms, and in children under 5 years of age, a broader-spectrum antibiotic should be given. Patients should be informed of the risk of infection and given the necessary advice should they develop a febrile illness.

SPLENIC MARGINAL-ZONE LYMPHOMA

(Lukes-Collins: small lymphocytic lymphoma; Working Formulation: small lymphocytic lymphoma) See also **Lymphoproliferative disorders**; **Non-Hodgkin lymphoma**.

A rare, low-grade B-cell non-Hodgkin lymphoma limited to the lymphoid follicles of the spleen, splenic hilar lymph nodes, and bone marrow. **Lymphocytes** with polar villi (villous lymphocytes) may be seen in the peripheral blood. The characteristic histological pattern is involvement of both the mantle and marginal zone of the splenic white pulp, usually with a central residual germinal center, which can be either atrophic or hyperplastic. Red pulp involvement may be prominent. The neoplastic cells range from small lymphocytes in the mantle zone to larger cells with irregular nuclei and pale cytoplasm in the marginal zone. The **immunophenotype** is SIg$^+$ (M > G or A), CIg$^+$, B-cell-associated antigens$^+$ (CD20$^+$, CD79a$^+$), CD5$^-$, CD10$^-$, CD23$^-$, CD43$^{-/+}$, and CD11c$^-$. Upon **cytogenetic analysis**, the loss of chromosome 7q21-32 has been described. The origin is a peripheral B-cell with partial differentiation to a splenic marginal-zone cell.

Clinical Features

Association with infection by the hepatitis C virus has been claimed. Patients typically have **splenomegaly** and may have **anemia** and **thrombocytopenia**, usually without peripheral lymphadenopathy. Abnormal lymphocytes may be found in the circulating blood, and the bone marrow is usually involved. Some patients may have mild monoclonal gammopathy. The course is indolent unless treated, and transformation to a **diffuse large B-cell lymphoma** may occur.

Treatment

Splenectomy may result in prolonged remission.

SPLENOMEGALY

The causes and effects of enlargement of the spleen. Evidence by palpation is only detectable when the size has doubled. Earlier enlargement can be detected by **radionuclide** imaging using ^{99m}Tc- or ^{51}Cr-labeled red cells, ultrasonic scanning, X-ray, computerized tomography, and nuclear magnetic resonance imaging.[486]

Causes of Splenomegaly

The pathogenesis of splenomegaly is variable, often with more than a single process operating. The relative incidence and extent of splenomegaly is subject to enormous geographical variation. Massive splenomegaly occurs more frequently in tropical countries (tropical splenomegaly).

Splenic hyperfunction with hyperplasia

Congestion with raised splenic venous pressure:

- Intrahepatic **portal hypertension**: portal cirrhosis, postnecrotic and biliary cirrhosis, hereditary hemochromatosis, Wilson's disease, congenital fibrosis
- Portal vein obstruction: thrombosis, cavernous malformation, porta-hepatis obstruction by lymphadenopathy
- Splenic vein obstruction: thrombosis, angiomatous malformation, aneurysm of splenic artery
- Hepatic vein obstruction by thrombus or tumor (Budd-Chiari syndrome)
- Cardiac: congestive heart failure, constrictive pericarditis

Infiltration:

- Storage disorders

  Lipid storage disorders: Gaucher's disease, Niemann-Pick disease

  Mucopolysaccharidoses: Hurler's syndrome, Hunter's syndrome

  Cholesterol esters in familial high-density lipoprotein deficiency (Tangier disease)
- **Amyloidosis** deposition
- **Sea-blue histiocytosis**
- Vascular tumors:

  Littoral cell angioma from the sinus lining or littoral cells of the red pulp arising from a disorder of both histiocytic and endothelial differentiation

Hemangiopericytoma of spindle-shaped cells around a vascular channel
Angiosarcoma, hemangioendotheliomas, hamartoma
- **Kaposi's sarcoma**
- Fibrous tumors: fibroma, fibrosarcoma
- Carcinoma metastases from colon, pancreas, lung, breast, ovaries, and melanomas
- Cysts
Echinococcus
Pseudocysts from trauma or postinfection
Peliosis
Mesothelial cysts of clear fluid surrounded by a layer of epithelial cells and fibrous tissue arising from an embryonic defect
- Unexplained
Idiopathic nontropical splenomegaly
Megaloblastosis
Iron deficiency
Osteopetrosis
Hereditary hemorrhagic telangiectasia
Hyperthyroidism

Effects of Splenomegaly

Splenic hyperfunction.

Secondary portal hypertension with massive splenomegaly where the spleen is acting as an arteriovenous shunt, the intrasplenic pressure is increased, and the splenic vasculature is expanded so that the total blood flow through the portal vein is increased severalfold. Bleeding from esophageal and gastric varices can then occur without obstruction to the blood flow.

Traumatic rupture: this occurs particularly with splenomegaly due to **infectious mononucleosis** and **malaria**.

SPLENOSIS

Intraabdominal dissemination and implantation of splenic tissue as a consequence of **splenectomy**, especially when performed for splenic trauma. It can result in relapse when splenectomy has been performed for treatment of **immune thrombocytopenic purpura** or **hereditary spherocytosis**.

SPONDYLOARTHRITIDES

See **Ankylosing spondylitis; Reactive arthritis; Psoriatic arthropathy; Intestinal disorders** — Crohn's disease.

SPUR CELL

See **Acanthocytes**.

STABLE FACTOR
See **Factor VII**.

STANOZOL
See **Anabolic steroids**.

STEM CELL
Cells with both self-renewal, proliferative, and differentiative capacity. Self-renewal is defined as the capacity of the cell to generate progeny with identical self-renewal and proliferative capability. Human embryonic stem cells arise as a cluster or cell mass within the lining of the yolk sac. Their development is controlled by stem cell and SOX genes, but they must remain together and be influenced by environmental feeders. Human embryonic stem cells are pluripotent cells (see Figure 108), giving rise by default to ectodermal stem cells from which skin and neurons develop. These can, in turn, differentiate to mesodermal cells by default, or they can differentiate to endodermal stem cells from which the abdominal organs such as the pancreas and liver develop, or they can differentiate to mesenchymal stem cells and hematopoietic stem cells. The latter are capable of repopulating the hematopoietic system of a sublethally irradiated host animal (see **Hematopoiesis**).

Stem cell transplantation after stimulation by **granulocyte colony stimulating factor** (G-CSF) and **granulocyte/macrophage colony stimulating factor** (GM-CSF) is replacing **bone marrow transplantation**, but is limited by economic resources. AMD3100, a reversible bicyclam agonist of CXCR4 (receptor for stem cell-derived factor 1), is under investigation. A long-term complication of stem cell transplantation is induction of osteoporosis. In the future, human embryonic stem cells may be the source for transplantation of specific organs.

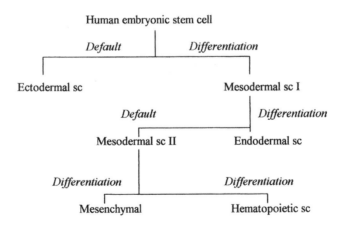

FIGURE 108
Hypothetical stem cell development.

STEM CELL FACTOR
(SCF) A growth factor derived from **bone marrow** stroma essential for **hematopoiesis**. It acts on CD34-positive cells, and its receptor is c-kit CD117.

STEM CELL LEUKEMIA
See **Acute leukemia**.

STERCOBILINOGEN
See **Hemoglobin** — degradation.

STOMACH DISORDERS
See **Gastric disorders**.

STOMATOCYTE
A morphological anomaly of **red blood cells**, where the cells have a slitlike central pallor upon Romanowsky-staining of the peripheral blood. These are seen in small numbers in some normal individuals, but should never exceed 0.5% of red blood cells. They are occasionally an artifact of blood-film preparation when they appear in zones throughout the film rather than diffusely.

Hereditary Stomatocytosis

A group of inherited, autosomally dominant disorders characterized by mild-to-moderate **hemolytic anemia**, the presence of up to 20% of red blood cells in the peripheral blood as stomatocytes, increased **osmotic fragility**, and increased **autohemolysis**, which is reduced by adding glucose. Some cases are asymptomatic, while others have moderate **anemia** and intermittent jaundice. The **hemoglobin** concentration is seldom less than 8 to 10 g/dl. The **reticulocyte count** is elevated, and there may be **splenomegaly**. Most patients have little or no response to **splenectomy**.

Acquired Stomatocytosis

Occurs with acute **alcohol toxicity**, cirrhosis, and other hepatobiliary disorders, neoplasms, and cardiovascular disorders. Large numbers (10 to 20%) suggest the existence of hereditary stomatocytosis (see above).

STORAGE-POOL DISORDERS OF PLATELETS
Deficiency of platelet-storage granules, giving rise to **platelet-function disorders** that can occur because of the lack of active components in such granules. Platelet cytoplasm contains four types of granules:

Dense (δ)-granules containing **adenosine diphosphate** (ADP), **adenosine triphosphate** (ATP), calcium, serotonin, and pyrophosphate

α-Granules containing a range of proteins

Lysosomes containing a variety of acid hydrolases

Microperoxisomes containing peroxidase activity

The term "storage-pool disorder" (SPD) refers to absence of δ-granules or α-granules. Deficiencies are:

Deficiency of δ-granules alone, termed δ-SPD

Combined deficiency of both granules, termed αδ-SPD

Deficiency of α-granules alone, termed α-SPD or gray-platelet syndrome

Deficiency of Dense Granules

(δ-SPD) Deficiency may occur as an isolated hereditary abnormality or as part of a number of distinct syndromes (see **Hermansky-Pudlak syndrome**; **Chediak-Higashi syndrome**; **Wiscott-Aldrich syndrome**; **Thrombocytopenia**; **Ehlers-Danlos syndrome**; **Osteogenesis imperfecta**). The isolated hereditary form is often inherited as an autosomally dominant condition. The pathogenesis is not clear; some patients show a complete deficiency of dense granules, whereas others have a partial reduction in parallel to the degree of the biochemical defect.

Patients with δ-SPD present with a mild-to-moderate bleeding tendency, characteristic of a platelet disorder with predominant mucocutaneous bleeding problems. Platelet counts and platelet morphology are typically normal. The **bleeding time** is usually prolonged. Platelet-function studies classically show only primary/first-wave aggregation in response to weak agonists such as ADP, adrenaline/epinephrine, and low concentrations of **thrombin**. Response to low-level **collagen** may also be reduced, although responses to high levels of collagen and thrombin may be near normal. A proportion of patients may have normal aggregation studies with all agonists. Platelet nucleotide analysis shows a low level of ADP and a low total nucleotide level. As δ-granule ADP is reduced, it does not contribute to the whole-platelet content of ADP and, as a result, δ-granule-deficient platelets have an elevated ATP:ADP ratio. This alteration in ATP:ADP ratio is helpful in diagnosis when characteristic platelet-aggregation abnormalities are absent.

Combined Granule Deficiency

These patients do not have a more severe bleeding problem than those with pure δ-SPD. The distinction is clinically unimportant.

Gray-Platelet Syndrome

A rare, autosomally dominant disorder that takes its name from the appearance of platelets on a stained blood film. Platelets are larger than normal, often reduced in number, and have a homogeneous gray appearance due to an almost total lack of platelet α-granules. All other platelet organelles are present. These platelets are deficient in α-granules and also profoundly low in **platelet factor 4**, **β-thromboglobulin**, **thrombospondin**, **fibrinogen**, **Von Willebrand Factor**, fibronectin, and platelet-derived growth factor.

The syndrome presents as a mild bleeding disorder with characteristic mucocutaneous bleeding. It may be associated with **myelofibrosis**. Platelet-aggregation studies provide variable results among different affected individuals, with a range of defects being described. Abnormal responses to thrombin are one of the more consistent abnormalities seen. Patients may also show increased levels of plasma β-thromboglobulin.

Treatment

Bleeding episodes are treated by platelet transfusion from a healthy donor. **DDAVP** (1-desamino-8-d-arginine vasopressin) may be of some use in mild bleeding problems. Local measures and concomitant use of antifibrinolytic agents such as tranexamic acid are important adjuncts.

STRAWBERRY NEVI
See **Hemangiomas**.

STREPTOKINASE
See **Fibrinolysis**.

STRESS ERYTHROCYTOSIS
See **Erythrocytosis**.

STUART-PROWER FACTOR
See **Factor X**.

SUBACUTE COMBINED DEGENERATION OF THE SPINAL CORD
A degenerative nervous system disorder in which there is demyelination of the white matter of both dorsal and lateral columns of the spinal cord and of peripheral nerves due to failure of the **cobalamin**-dependent enzyme (methylmalonyl-CoA mutase). These complications of **Addisonian pernicious anemia** due to cobalamin deficiency are less common than previously, due to improved diagnosis and treatment. About 30% of such patients experience mild neurological features (paresthesias and minor cerebral symptoms), and less than 10% have symptoms and signs of spinal cord involvement. Because multiple nerve pathways are involved, the term "subacute combined degeneration of the spinal cord" is used. Patients with dorsal column involvement develop clumsiness and the gait becomes uncoordinated and ataxic. The lateral columns (pyramidal tracts) become involved in the later and more severe stages, with muscular weakness; a spastic or "scissors" gait develops. The plantar response becomes extensor. Hyperreflexia and clonus may be present. Disorders of cerebral function such as irritability, drowsiness, emotional instability, and confusion with dementia may occur. Optic neuropathy with amblyopia is a late development.

Although progression of the neurologic manifestations is prevented by cobalamin, the degree to which established lesions reverse varies. The shorter the duration of symptoms and signs, the more complete the recovery. Features of long duration may be lessened by treatment, but some residual dysfunction is to be expected. Neurologic response tends to be slow, and up to 6 months may elapse before a maximum response is achieved.

SUBCUTANEOUS PANNICULITIS-LIKE T-CELL LYMPHOMA
(REAL subcutaneous panniculitic T-cell lymphoma) See also **Non-Hodgkin lymphoma**. A rare lymphoma presenting with multiple subcutaneous nodules. Histology reveals a subcutaneous infiltrate with small cells to larger transformed cells. The cells are $CD8^+$. There is rearrangement of T-cell receptor genes.

Clinically, the patients present with skin nodules, and some patients may present with **pancytopenia**, fever, and **hepatosplenomegaly** (hemophagocytic syndrome). Treatment is by combination **cytotoxic agent** chemotherapy.

SUCROSE LYSIS TEST

(SLT) A screening test sometimes used for the diagnosis of **paroxysmal nocturnal hemo-globinuria** (PNH) based on the observation that red cells from PNH patients are readily lysed by complement components (see **Acidified serum lysis test**) at low ionic concentration, while normal red cells are not lysed.[2] The patient's washed red cells are incubated with an iso-osmotic solution of sucrose together with fresh normal group AB- or ABO-compatible serum. After centrifugation, red cell lysis is compared against a control incubated with serum diluted in isotonic saline. Lysis is typically 10 to 80%. Red blood cells from patients with leukemia or myelosclerosis may show mild lysis, but the degree is always less than 10%. The test is negative for red blood cells from patients with the HEMPAS type of **congenital dyserythropoietic anemia**. Confirmation of the diagnosis of PNH requires the demonstration of GP1-linked molecules within the red blood cell membrane by **flow cytometry**.

SULFHEMOGLOBINEMIA

The presence in blood of poorly characterized **hemoglobin** derivatives with absorption of light at 620 nm, even in the presence of cyanide. Sulfhemoglobin is thought to contain one excess sulfur atom bound to the heme ring. Sulfhemoglobinemia is associated with rapid degradation of hemoglobin, as occurs with **intravascular hemolysis**.

SUPEROXIDE DISMUTASE

A metalloprotein containing **copper** that is present in **red blood cells**. It appears to play a prominent role in the protection of **hemoglobin** and other red cell components against a highly reactive superoxide anion. It is also present in the **neutrophil** cytosol.

SWEET'S SYNDROME

See **Acute febrile neutrophilic dermatosis**.

SYSTEMIC DISORDERS

A group of disorders affecting a multiplicity of tissues and organs, including connective tissue. Apart from **sarcoidosis**, they are associated with the following **autoimmune diseases** and forms of **vasculitis**:

Behçet's disease	**Rheumatoid arthritis**
Churg-Strauss vasculitis	**Sjögren's syndrome**
Giant-cell arteritis	**Systemic lupus erythematosus**
Kawasaki's disease	Takayasu's arteritis
Microscopic polyarteritis	**Vogt-Koyanaygi-Harada's disease**
Polyarteritis nodosa	**Wegener's granuloma**
Relapsing polychondritis	

SYSTEMIC INFLAMMATORY RESPONSE

See **Acute-phase inflammatory response**.

SYSTEMIC LUPUS ERYTHEMATOSUS

(SLE) An **autoimmune disorder** mediated by **immune complexes** (hypersensitivity type III) of unknown cause. At least 95% of patients have **antinuclear factor antibodies**, and of these patients, **lupus erythematosus cells** can be demonstrated in around 75%.[490,490a] Other antibodies, often present before symptoms, include anti-Ro, anti-La, anti-phospholipid, and rheumatoid factor (25%). The disorder is more common in females and the concordance rate for identical twins is about 60%. There is an increased frequency of **human leukocyte antigens** HLA-B8 and DR3 in Caucasians and HLA-DR2 in Japanese. There is an associated inherited deficiency of **complement** C2 and C4. Patients with low levels of C1q, MBL, or DNAse, which normally clear protein debris, are at greater risk of complications. Failure to remove immune complexes leads to **vasculitis**, particularly affecting the renal glomeruli. Failure of endothelial **apoptosis** may lead to early onset of atherosclerosis.[491] This has been associated with a mannose-binding lectin-variant allele.[492]

Other hematological changes include:

Anemia:

- **Anemia of chronic disorders**
- Anemia of **renal tract disorders**
- **Iron deficiency** due to gastrointestinal **hemorrhage** from adverse drug reactions or from **thrombocytopenia**
- **Autoimmune hemolytic anemia**, typically with a positive **direct antibody (Coombs) test** for IgG and the C3 component of complement on the red blood cells
- **Microangiopathic hemolytic anemia** due to deposition of immune complexes, which, by activation of the coagulation pathway, cause fibrin deposition with red blood cell fragmentation, hemolysis, and thrombocytopenia

Basophilic stippling of red blood cells

Leukopenia with **neutropenia** and **lymphopenia**, often related to circulating immune complexes but also due to accelerated apoptosis

Thrombocytopenia related to immune complexes or acute immune thrombocytopenic purpura

Thrombosis due to **antiphospholipid antibody syndrome**

Splenic hypoplasia due to unexplained splenic atrophy

Bone marrow necrosis with hyperplasia

Immunological function disturbances:

- Loss of **T-lymphocyte** suppressor function
- Increased production of **interleukins** IL-1 and IL-2
- Raised levels of **immunoglobulins** IgG and IgM

Raised **erythrocyte sedimentation rate**

Treatment is by immunosuppression of autoantibodies. This is by **corticosteroids** and, if unresponsive, **azathioprine** or **cyclophosphamide**. Recently, therapy with a monoclonal antibody, **rituximab**, and intravenous cyclophosphamide has been reported to give a good outcome.[493]

T

TARGET CELLS
See **Codacytes**.

TARTRATE-RESISTANT ACID PHOSPHATASE
Isoenzyme 5 occurring in lysosomelike bodies, cytoplasmic vesicles, and Golgi apparatus saccules of **B-lymphocytes** in **hairy cell leukemia**. The presence of this enzyme may be of diagnostic value.

TAY-SACHS DISEASE
A **lipid storage disorder** inherited as an autosomally recessive trait. Deficiency of hexosaminidase A results in the accumulation of ganglioside GM2 within neural tissues. As with several similar disorders, it is prevalent in Ashkenazi Jews (incidence 1:2000). Presentation is within 6 months of birth, with rapidly progressive neurological deficit, mental retardation, deafness, blindness (cherry red spot at macula), and fits. Ultimately, a vegetative state ensues, followed shortly by death. Diagnosis is made by tissue biopsy or enzyme estimation in leukocytes or tissue fibroblasts.

T-CELLS
(T-lymphocytes) See **Lymphocytes**.

T-CELL ANOMALIES
See **Immunodeficiency**.

T-CELL CHRONIC LYMPHOCYTIC LEUKEMIA
See **T-cell large granular lymphocyte leukemia**; **T-cell prolymphocytic leukemia**.

T-CELL LARGE GRANULAR LYMPHOCYTE LEUKEMIA
(T-LGL; T-cell chronic lymphocytic leukemia) A heterogeneous **lymphoproliferative disorder** characterized by a persistent (>6 months) increase in the number of peripheral blood large granular **lymphocytes** (LGLs), usually within the range 2 to $20 \times 10^9/l$, without a clearly identified cause. It comprises only 2 to 3% of small lymphocytic leukemias. Apart from the peripheral blood, the bone marrow, liver, and spleen are involved, but very rarely the lymph nodes. The cells have a mature T-cell **immunophenotype**, indicating the postulated cell of origin. The cells express FasL on their surface, which induces Fas/FasL **apoptosis**. Mechanisms of autoantibody destruction similar to those of rheumatoid arthritis (RA) are also active

Most cases have an indolent clinical course. **Neutropenia** is frequent. Moderate **splenomegaly** is the common clinical feature. Associated **rheumatoid arthritis**, circulating **immune complexes**, and hypergammaglobulinemia are common. Transformation to a **peripheral T-cell lymphoma** occurs in a minority of cases. Those requiring symptomatic treatment may benefit from a course of **ciclosporin** A, **methotrexate, cyclophosphamide**, and **corticosteroids. Splenectomy** removes the symptoms of hypersplenomegaly but will not benefit any form of cytopenia.

T-CELL LYMPHOMAS

Lymphomas of T-**lymphocyte** precursor or T-lymphocyte origin include:

Precursor T-cell neoplasms
- **Precursor T-lymphoblastic leukemia/lymphoma**
- **Blastic NK-cell lymphoma**

Mature T-cell and NK-cell neoplasms
- **T-cell prolymphocytic leukemia**
- **T-cell large granular lymphocyte leukemia**
- **Aggressive NK-cell leukemia**
- **Extranodal T-cell lymphoma, nasal type**
- **Enteropathy-type T-cell lymphoma**
- **Hepatosplenic T-cell lymphoma**
- **Subcutaneous panniculitis-like T-cell lymphoma**
- **Mycosis/fungoides/Sézary syndrome**
- Primary cutaneous **anaplastic large-cell lymphoma**
- **Peripheral T-cell lymphoma, unspecified**
- **Angioimmunoblastic T-cell lymphoma**
- Anaplastic large-cell lymphoma

T-cell proliferations of uncertain malignant potential
- **Lymphomatoid papulosis**

T-CELL ONTOGENY

See **Lymphocytes**.

T-CELL PROLYMPHOCYTIC LEUKEMIA

(T-PLL; REAL: T-cell chronic lymphocytic leukemia) See also **Lymphoproliferative disorders; Non-Hodgkin lymphoma**.

An aggressive T-cell leukemia characterized by cells with prominent nucleoli, some nuclear irregularity, and abundant cytoplasm — prolymphocytic leukemia; some tumors have cells that may resemble those of **chronic lymphocytic leukemia** (CLL). The cells are usually nongranular, but have focal granular paranuclear positivity on acid phosphatase and acid nonspecific esterase stains. **Lymph node** involvement is diffuse and paracortical, with sparing of follicles; it differs from CLL in lack of pseudofollicles. Prominent small vessels of the high endothelial venule type may be numerous, and often contain atypical small- to medium-sized lymphoid cells. In the **spleen**, the red pulp may be infiltrated, and in the **liver** the hepatic sinusoids. **Bone marrow** involvement is usually diffuse and may show increased reticulin. The **immunophenotype** is CD2$^+$, CD3$^+$, CD7$^+$, CD4$^+$ (60% of cases), CD4$^+$8$^+$ (25%), and CD4$^-$8$^+$ with CD25$^-$ (rare). **Cytogenetic analysis** shows clonal rearrangement of *TCR* genes: inv 14(q11;q32) in 75%; trisomy 8q. The lymphoma arises from T-cells with a mature immunophenotype.

Clinical Features

T-CLL/PLL comprises about 2% of cases with CLL but up to 20% of PLL. Patients have a **lymphocytosis** ($>100 \times 10^9$/l) with a 20% frequency of cutaneous involvement or mucosal infiltrates. Bone marrow, spleen, liver, and lymph nodes may be involved. It is more aggressive than CLL, with a median survival of only 1 year, even with treatment.

Staging

See **Lymphoproliferative disorders**.

Treatment

See **Non-Hodgkin lymphoma**. Recent introductions have included **cytotoxic agent** combinations of CAMPATH-H, pentostatin, CHOP, and **autologous stem cell transplantation**.

TEARDROP POIKILOCYTE

See **Poikilocytosis**.

TELANGIECTASIA

Dilatation of capillary vessels and arterioles, being a variety of angioma. There are a number of forms:

Hereditary hemorrhagic telangiectasia

Hereditary lymphatic telangiectasia

Senile vascular lesions

Associated with **scleroderma**

TELOMERES

See also **Genes**.

Tandem **DNA** repeat sequences sited at the end of a eukaryotic **chromosome**. They are replicated by a specific enzyme, telomerase, composed of both RNA and protein.

TEMPORAL ARTERITIS

See **Polymyalgia rheumatica**.

TENIPOSIDE
See **Epidophyllotoxins**.

TERMINAL DEOXYNUCLEOTIDYL TRANSFERASE
(TdT) An enzyme found in **thymus** cells, rarely in lymphoid progenitors of normal **bone marrow**. The enzyme TdT inserts nontemplated or N-nucleotides into the junctions between gene segments in T-cell receptor and immunoglobulin V-region genes, being therefore responsible for lymphoid antigen specificity. Neoplasias arising from these cells frequently express this enzyme, and TdT$^+$ cells are seen in large numbers in leukemia, especially **precursor T-lymphoblastic leukemia/lymphoma** (T-ALL) and, less commonly, in **chronic myelogenous leukemia** (CML) blast crisis cells. Weak positivity is seen also in about half of all patients with M1, M2, and M4 **acute myeloid leukemia** (AML). TdT is not specific and can be used only as an adjunct to the immunological diagnosis and classification of leukemia in bone marrow and peripheral blood. It can also be used as a diagnostic tool for the detection of leukemia cells in other biological samples (cerebrospinal fluid [CSF], pleural fluid), although the significance of its presence is debatable, as no standard normal counts have been established.

THALASSEMIA
See also **Hemoglobinopathies**.

Disorders in which there is a reduced rate of synthesis by one or more of the **globin-chain genes**, leading to imbalance of globin-chain synthesis, defective **hemoglobin** production, and damage to **red blood cells** or their precursors from the effects of the globin subunits produced in excess.[236] It is a quantitative rather than a qualitative disorder. The frequency and severity of the various types of thalassemia depend on the racial background. α-Thalassemia is common in Southeast Asia, whereas β-thalassemia is commonly seen in Mediterranean and African populations. Thalassemias are classified on the basis of the specific globin whose production is reduced. They are further defined by the relative production of other globins. Defects have been identified in all globin genes, and more than one type of thalassemic defect can occur in a single patient. At a molecular level, the mechanisms for thalassemia production are:

   Gene deletion

   Reduced or altered transcription

   Splicing mutations

   RNA-modification variants

   Translation-termination variants

   Gene-fusion variants

The following main varieties are recognized:

   α^0-thalassemia (absence of α-chain production)

   α^+-thalassemia (partial deficiency of α-chain production)

   β^0-thalassemia (absence of β-chain production)

   β^+-thalassemia (partial deficiency of β-chain production)

Other forms of thalassemia include δ-β thalassemia, hemoglobin Lepore syndromes, some forms of hereditary persistence of HbF (fetal hemoglobin), and decreased production of δ-chains with reduced HbA2 in heterozygotes and absent HbA2 in homozygotes.

Thalassemia Syndromes

The nature of the genetic mutation determines the clinical presentation of thalassemia and ranges from asymptomatic patients to those with moderate or severe anemia. All, however, demonstrate red cell hypochromia and microcytes, reflecting decreased synthesis of globin chains. Racial background and family history assist with diagnosis. Patients with the most severe forms of the disease are usually from high-prevalence population groups and have parents who are heterozygous for the globin-chain defect.

The severity of anemia in patients with thalassemia is classified by hemoglobin level: thalassemia major (Hb 3 to 6 g/dl), thalassemia intermedia (Hb 6 to 10 g/dl), and thalassemia minor (Hb 10 to 14 g/dl). The phenotypes and genotypes of patients with thalassemia are listed in Table 142.

TABLE 142

Clinical Phenotypes and Genotypes of Thalassemias

	Genotype
Thalassemia Major	
β-thalassemia major	β^0/β^0
β-thalassemia-Mediterranean	$\beta^+Medit/\beta^+Medit$
β-thalassemia/Hb E disease	$\beta^0/\beta E$
Thalassemia Intermedia	
α-thalassemia	$-\alpha/-\alpha, \beta/\beta^0$
β-thalassemia Africa	$\beta^+Africa/\beta^+Africa$
β-thalassemia major (high HbF)	β^0/β^0
Thalassemia Minor	
α-thalassemia-1	$-\alpha/\alpha\alpha$
α-thalassemia-2	$-\alpha/-\alpha$
β-thalassemia trait	β/β^0 or $^+$
δβ-thalassemia	$\beta/\beta^+\delta^+$
Hb constant spring	$\alpha\alpha cs/\alpha\alpha cs$
Lepore trait	$\beta/\beta Lepore$
Hemoglobin H disease	α-thal-2/α-thal-1
	α-thal-1/constant spring
Hemoglobin Bart's	$\gamma 4$

Diagnosis[237–240,494]

Screening tests for hemoglobinopathies and specific differentiation by **hemoglobin electrophoresis** is the basis of thalassemia diagnosis. These data can be supported by measurement of globin-chain synthesis ratios, performed either on bone marrow normoblasts or on peripheral-blood reticulocytes incubated with radiolabeled amino acids for short periods. The globin chains are precipitated and separated by column chromatography, and the radioactivity is measured in each chain species. In normal cells, α- and β-chain synthesis rates are equal, but unequal rates occur in many types of thalassemia. The technique is especially informative in two situations. The first is a rare form of heterozygous β-thalassemia, so-called silent β-thalassemia, in which the HbA2 level is normal and red cell indices are in the respective reference ranges; this variant is found among relatives of patients with thalassemia major. The second situation occurs when heterozygous α-thalassemia is suspected, a pregnancy is contemplated, and there is a possibility that the offspring would develop hemoglobin Bart's **hydrops fetalis**. Genotypic analysis is not essential to a clinical care program, but it can be of practical value. It permits a more accurate prognosis and permits predictions about the genotypes to be expected in other family members. **Allogeneic stem cell transplantation** and future **gene therapy** of thalassemia will require genotyping.

α-Thalassemias

These disorders are due to a subnormal rate of synthesis of hemoglobin α-chain, usually due to a deletion mutation. The α-chains are encoded by two duplicated genes on chromosome 16; thus, deletion of all four genes is required to suppress α-chain synthesis completely. This results in the formation of **hemoglobin Bart's** ($\gamma4$) or **hemoglobin H** ($\beta4$). Varying abnormalities are produced when fewer genes are deleted. The resulting clinical syndromes are listed in Table 143.

TABLE 143

α-Thalassemia Syndromes

No. of Genes Deleted	Genotype	Clinical Phenotype
1	$-\alpha/\alpha\alpha$	Silent α-thalassemia
2	$--/\alpha\alpha$ or $-\alpha/-\alpha$	α-Thalassemia trait
3	$--/-\alpha$	Hb H disease
4	$--/--$	Hydrops fetalis

Silent Carrier State for α-Thalassemia

Red blood cell morphology is normal and there are no clinical manifestations. Hemoglobin characteristics are HbA 98 to 100%, Hb Bart's 0 to 2%.

α-Thalassemia Trait

The clinical and hematological picture is very mild. Hb characteristics are HbA 90 to 95%, Hb Bart's 5 to 10%.

Hemoglobin H Disease

A **hemolytic anemia** with a Hb level that is typically 7 to 9 g/dl. The erythrocytes are severely hypochromic, and Hb H inclusions can be demonstrated by supravital staining. The specific diagnosis is made in the laboratory with the detection of Bart's hemoglobin and Hb H. About 25% of hemoglobin is HbA.

Hydrops Fetalis

Infants affected with **hydrops fetalis** are either stillborn or die shortly after birth. Because no α-chains are synthesized, there is no HbA, HbF, or HbA2. Fetal chains form tetramers ($\gamma4$) — Hb Bart's, which cannot transport oxygen effectively. Hb Bart's constitutes 80 to 90% of the total; the remainder is Hb H.

TABLE 144

Severity of β-Thalassemias

Clinical Designation	Hb (g/dl)
β-thalassemia major	2.5–6.0
β-thalassemia intermedia	6.0–9.5
β-thalassemia minor	9.5–13.5

TABLE 145

Genotypes of β-Thalassemia Syndromes

β-Globin Allele	Heterozygote	Homozygote
β^0	Thalassemia minor	Thalassemia major
β^+Medit	Thalassemia minor	Thalassemia major
β^+African	Thalassemia minor	Thalassemia intermedia

β-Thalassemias

These disorders are due to subnormal synthesis of hemoglobin β-chains. As with other forms of thalassemia, classification of β-thalassemias employs the clinical manifestations and results in three forms (Table 144). The clinical classification corresponds to a series of genotypes, i.e., several genotypes can lead to the same clinical form (Table 145).

Genetic heterogeneity explains the appearance of severely affected offspring of less severely affected parents. Double heterozygotes are common, and the severity of their anemia is typically intermediate between that of the two types of homozygotes.

β-Thalassemia Major

The two most important alleles are β^0 and β^+Medit. In the β^0 form, no β-chains are produced from this allele. Although the defect is genetically heterogeneous, most mutations are due to single base changes, e.g., production of a stop codon leading to premature chain termination, or involvement of an intron-splicing site, thus preventing production of effective β-messenger. β^+ mutations are also heterogeneous. The β^+Medit form most often arises from an intron-splicing mutation, but this is correctly recognized by the processing enzyme part of the time. Thus, a small amount of messenger is produced, although the resulting defect is severe.

Clinical and Laboratory Features

Affected infants are clinically normal at birth because they have large proportions of hemoglobin F. By six months of age, when the production of HbF has ceased, affected infants develop severe anemia, **hepatosplenomegaly**, and failure to thrive. Mild jaundice may also be present. Hypochromic, **microcytic anemia** is present, but iron stores are normal. Nucleated red blood cells are commonly present in the peripheral blood; there is polychromasia and an elevated **reticulocyte count**. The bone marrow shows marked erythroid hyperplasia. As a result of lifelong bone marrow hyperplasia, the bones become trabeculated with cortical thinning on radiographs, giving the typical "hair on end" pattern in the skull. Cholelithiasis is common.

β-Thalassemia Intermedia

The diagnosis is based on the presence of hypochromic, microcytic anemia in the presence of adequate iron stores. HbF levels are greatly elevated. However, thalassemia intermedia is considerably milder than in the major form, and the red cell morphology is less severely disturbed. Moderate splenomegaly is common. Affected patients usually tolerate moderate anemia without transfusion, but may have increased susceptibility to infection. Life span is slightly reduced. Genetic counseling and testing of family members is indicated.

β-Thalassemia Minor

Patients have mild anemia, often detected incidentally or during a family screening program. The red blood cells are hypochromic and microcytic. There is mild poikilocytosis, and few target cells are present. The red blood cell life span is slightly reduced. The spleen is sometimes palpable. Life span is typically normal and no treatment is indicated.

Other β-Thalassemia Syndromes

Patients who are double heterozygotes for a β-thalassemia gene and a β-hemoglobinopathy gene. The most common examples are sickle/β-thalassemia and Hb C/β-thalassemia.

Hemoglobin C/β-Thalassemia

This mixed hemoglobinopathy is characterized by mild hemolysis associated with splenomegaly and frequent target cells in the blood. It is recorded mainly in northern and western Africa.

Hemoglobin E/β-Thalassemia

This mixed hemoglobinopathy is characterized by a severe deficiency in β-chain production and a clinical picture of severe β-thalassemia. The blood film reveals a typical thalassemia pattern: hemoglobins consist of E, F, and A2. There is usually no HbA because the β^0 form of thalassemia is particularly common in areas where Hb E is also common. This disorder occurs worldwide.

Hemoglobin S/β-Thalassemia

The disorders arising from Hb S/β^0-thalassemia double heterozygosity depend upon the type of β-thalassemia mutation. The association of Hb S with β^0-thalassemia or with the more severe forms of β^+ result in a clinical disorder similar to sickle cell anemia. The interaction of the sickle cell gene with the milder forms of β^+-thalassemia results in disorders similar to sickle cell trait. Sickle cell thalassemia occurs in parts of Africa, Greece, and Italy. In all of these interactions, one parent has sickle cell trait and the other β-thalassemia trait.

Genetic Counseling

Once the diagnosis is made, **genetic counseling** is mandatory. Because the parents and two-thirds of apparently normal offspring are heterozygotes, the family must be evaluated hematologically, informed about their status, and educated about their medical condition and its genetic implications. The importance of misdiagnosis as **iron deficiency** must be stressed.

Treatment

With appropriate treatment (**red blood cell transfusion**, **iron chelation**, and treatment of infection), these patients may survive for many years. However, premature death usually occurs due to cardiac failure, intercurrent infection, or complications of chronic iron overload. No drug therapy has proved to be satisfactory. Transplantation using umbilical cord or marrow cells to eliminate the defective globin gene may become a widespread treatment.

THERMOGRAPHY
See **Venous thromboembolic disease**.

THESAUROCYTE
See **Flame cells**.

THIAMIN-RESPONSIVE MEGALOBLASTIC ANEMIA
A very rare, severe form of **megaloblastosis** accompanied by sensorineural deafness and **diabetes mellitus** in infancy. The defect is due to defective high-affinity thiamine transport, leading to reduced synthesis of ribose. It responds to treatment with thiamine, 25 to 100 mg daily.

THICK-BLOOD FILM
The preparation of concentrated blood cells on a glass slide. It is particularly useful in the diagnosis of clinically mild **malaria**, in which the organism may be missed on a thin film. Two to three drops of blood are placed on a slide and stained for 15 min without prior fixation. During staining, the red blood cells are lysed. The stain is then allowed to run off, and the preparation is carefully washed with distilled water and dried at room temperature. Since the cells are lysed, the intracellular location of the parasites is not recognized.

THIN CELL
See **Leptocytes**.

6-THIOGUANINE
See **Antimetabolites**.

THROMBOASTHENIA
See **Glanzmann's thrombasthenia**.

THROMBIN
The enzyme derived from **prothrombin** activation, which splits **fibrinogen** to fibrin and **fibrinopeptides A and B**. It is the most potent physiological activator of **platelets**, causing shape change, the generation of thromboxane A2 (TxA2), **adenosine diphosphate** (ADP) release, and ultimately aggregation. Binding of thrombin to platelets occurs via a specific

thrombin receptor. Thrombin also activates the cofactors of coagulation **factor V** and **factor VIII**, thereby accelerating the coagulation cascade many thousandfold. It activates **factor XIII** to XIIIa, resulting in cross linking and subsequent stabilization of the fibrin clot, and also activates **factor XI** to become factor XIa. Thrombin bound to thrombomodulin loses its procoagulant activity but becomes a potent activator of **protein C**. In addition to its procoagulant and anticoagulant activities, thrombin also has important roles in cellular growth, cellular activation, and the regulation of cellular migration.

THROMBIN-ACTIVATABLE FIBRINOLYSIS INHIBITOR

(TAFI) A carboxypeptidase B-like proenzyme. It has a molecular weight of 55 kDa. The gene is located on the long arm of chromosome 13. It is synthesized in the liver and circulates in the blood at a concentration of about 275 nmol/l. TAFI can be activated by **thrombin** or **plasmin**, whereupon it downregulates **fibrinolysis**. Clot lysis is slowed by the cleavage of the C-terminal lysine and arginine residues from partially degraded fibrin. This prevents the binding of plasminogen and tissue plasminogen activator (t-PA) for fibrinolytic activity. It is believed that there is no naturally occurring physiological inhibitor to TAFI, and it decays spontaneously because it is thermally unstable. It has a half-life of ≈15 min. **Thrombomodulin** enhances thrombin activation of TAFI by more than 1000-fold, suggesting that the thrombin-thrombomodulin complex is the physiological activator of TAFI. Activated **protein C** can upregulate fibrinolysis by limiting the activation of TAFI via the attenuation of thrombin production. In acquired coagulation disorders such as **disseminated intravascular coagulation**, TAFI antigen levels are reduced.

THROMBIN-GENERATION TESTS

Overall functional tests of the coagulation system in either platelet-poor plasma (PPP) or platelet-rich plasma (PRP), where the addition effect of platelets is also measured. **Thrombin** generation was historically a cumbersome technique involving repeated subsampling. The recently developed use of special fluorogenic thrombin substrates allows automated monitoring of thrombin concentration in PPP and PRP. These allow a graph to be produced of thrombin generated against time. The area under the curve is termed the endogenous thrombin potential (ETP), which quantifies the enzymatic "work" that thrombin can do during its lifetime.

THROMBIN TIME

(TT) The time taken for citrated plasma to clot at 37°C after thrombin is added to plasma.[19] Typically bovine thrombin is used. It measures the conversion of **fibrinogen** to fibrin. The thrombin time is dependent upon concentration and reaction of fibrinogen. Consequently, both quantitative (hypo/afibrinogenemia) and qualitative defects (dysfibrinogenemia) of fibrinogen will prolong the thrombin time. The thrombin time may also be prolonged by the presence of inhibitory substances, including fibrinogen/fibrin degradation products, hypoalbuminemia, and heparin. Prolongation of the thrombin time by heparin contamination can be determined by using the reptilase time, an adaptation of the thrombin time. The thrombin time forms the basis of the Clauss fibrinogen assay. The reference range is 15 to 19 sec, with slightly longer times for premature and full-term infants (see **Reference Range Tables XV** and **XVI**).

THROMBOCYTHEMIA

See **Essential thrombocythemia**.

THROMBOCYTOPENIA

A **platelet count** below the lower limit of the **reference range** $150 \times 10^9/l$, but lower in neonates and infancy. Thrombocytopenia occurs when platelets are lost from the circulation faster than they can be replaced. **Pseudothrombocytopenia** must be excluded as a cause of thrombocytopenia before considering other causes by examination of the blood film.

True thrombocytopenia may result from failure of platelet production, an increased rate of destruction of platelets, or loss/removal from the circulation. A combination of mechanisms may account for thrombocytopenia in some situations. For example, in **chronic lymphocytic leukemia**, thrombocytopenia may arise from a combination of defective marrow production due to marrow infiltration, immune-mediated destruction due to circulating autoantibodies, and splenic sequestration secondary to **splenomegaly**. However, as a general rule, when thrombocytopenia is due to defective marrow production, there is a reduction in the number of megakaryocytes in the bone marrow, whereas in thrombocytopenia due to decreased platelet survival, there is an increase in the number of marrow megakaryocytes. Bone marrow examination is an integral part of the investigation of thrombocytopenia, unless the cause is obvious.

Thrombocytopenia rarely leads to bleeding unless the platelet count is below $75 \times 10^9/l$. There is no clear relationship between severity of thrombocytopenia and severity of bleeding symptoms. However, spontaneous hemorrhage may be expected with platelet counts below $10 \times 10^9/l$. The **bleeding time** is prolonged in parallel to the platelet count below $75 \times 10^9/l$.

The causes of thrombocytopenia are listed in Table 146. An abbreviated list of drugs associated with thrombocytopenia is given in Table 147. After defective marrow production of platelets, a large group of immune mechanisms accounts for decreased platelet survival. There is a range of mechanisms involved in immune destruction of platelets, from classical **immune thrombocytopenic purpura** to nonspecific immune mechanisms. Raised levels of platelet-associated **immunoglobulin** may be seen in many of these other immune-destructive mechanisms. These may be due to platelet alloantibody or autoantibody bound to the platelet surface or to passive adsorption of antigen–antibody immune complex or antibody to a nonplatelet antigen that has been adsorbed onto the platelet surface. Any of these mechanisms will lead to removal of platelets by the reticuloendothelial system. Conditions that are associated with raised platelet-associated immunoglobulin and nonspecific immune destruction include **malaria, lymphoproliferative disorders**, **viral infection disorder** (AIDS)-associated thrombocytopenia, **systemic lupus erythematosus**, preeclampsia, and some drug-induced thrombocytopenias (see Table 147).

Treatment of thrombocytopenia is variable, but broadly is directed at the underlying cause, with immunomodulation by corticosteroids and/or immunoglobulin if appropriate. Platelet transfusion is only indicated when the count falls below $20 \times 10^9/l$, and in children a short high-dose course of prednisone/prednisolone is usually sufficient. Splenectomy is only indicated when there has been no improvement after 12 to 24 months.

TABLE 146

Causes of Thrombocytopenia

Failure of Platelet Production

As part of a generalized bone marrow failure
   Hereditary **aplastic anemia**
   Acquired **bone marrow hypoplasia**
   Myelodysplasia
   Ionizing radiation
Megakaryocyte abnormalities
   Hereditary thrombocytopenias
   Congenital **megakaryocytic hypoplasia**
   Virus infections disorders
Bone marrow infiltration
   Acute leukemias (ALL, AML)
   Lymphoproliferative disorders
   Myelomatosis
   Myeloproliferative disorders
   Myelofibrosis
   Osteopetrosis
   Carcinoma metastases
Metabolic disorders
   Megaloblastosis
   Renal disorders — uremia
   Alcohol toxicity
Drug-induced (see Table 147)

Decreased Platelet Survival

Immune mediated
   Platelet alloantibodies
      Neonatal alloimmune thrombocytopenia
      Posttransfusion purpura
   Platelet autoantibodies
      Immune thrombocytopenic purpura
   Other immune mechanisms
      Systemic lupus erythematosus
      Malaria
      Acquired immunodeficiency syndrome
      Viral infections (e.g., infectious mononucleosis)
   Lymphoproliferative disorders
   Drug induced (see Table 147)

Increased Platelet Consumption

   Disseminated intravascular coagulation
   Microangiopathic hemolytic anemias (e.g., **hemolytic
      uremic syndrome; thrombotic thrombocytopenic purpura**)
   Cardiopulmonary bypass

Loss/Sequestration of Platelets

Massive **red blood cell transfusion**
Splenomegaly
Acute venous thromboembolic disease

TABLE 147

Drugs Associated with Thrombocytopenia

Failure of Platelet Production

Predictable
  Cytotoxic agent therapy
Idiosyncratic/occasional
  Chloramphenicol
  Cotrimoxazole
  Penicillamine

Immune Mechanisms (Actual or probable)

Anti-inflammatory drugs
  Aspirin
  Paracetamol
  Gold
Antibiotics
  Penicillins
  Cephalosporins
  Sulfonamides
  Trimethoprim
  Para-aminosalicylate
  Rifampicin
Anticonvulsants
  Valproate
  Diazepam
  Phenytoin
  Carbamazepine
Diuretics
  Thiazides
  Frusemide
Antidiabetic agents
  Sulfonylureas
Others
  Digitoxin/digoxin
  Heparin
  Methyldopa
  Quinine/quinidine
  Cimetidine
  α-Interferon

THROMBOCYTOSIS

An increased level of circulating **platelets** above the **reference range** count of $400 \times 10^9/$ l in which normal platelet morphology and function are maintained.[495,496] It is a response to the action of thrombopoietin regulating the differentiation of megakaryocytes and their proliferation. The causes are wide (see Table 148).

TABLE 148

Causes of Thrombocytosis

Primary

Clonal	**essential thrombocythemia** (ET)
Reactive	endogenously raised thrombopoietin
	Interleukin 6
	Catacholamines with **inflammation**, **malignancy**, stress
Familial	heterogeneous dimer form of ET

Secondary

Transient	acute **hemorrhage**, recovery from thrombocytopenia
	inflammation, exercise, trauma
Sustained	**iron deficiency, hemolytic anemia, hyposplenism**
	malignancy, e.g., carcinoma metastases, **Hodgkin disease**
	chronic inflammatory disorders, e.g., **intestinal tract disorders, tuberculosis**
	connective tissue disorders, polymyositis (temporal arteritis), **rheumatoid arthritis**
	adverse reaction to drugs, e.g., vincristine, all-*trans*-retinoic acid, cytokines, growth factors

THROMBOELASTOGRAM

(TEG) An instrument that enables global assessment of hemostasis from a single blood sample.[497] It records the reaction of platelets with the coagulation cascade from the time of the initial platelet-fibrin interaction to eventual clot lysis. The tracings that are generated can provide rapid information (<20 to 30 min) on clotting-factor activity, platelet function, and any clinically significant fibrinolytic process. The TEG consists of two mechanical parts: a heated cuvette (37°C) and a pin that is suspended freely from a torsion wire. The freshly placed blood is placed in the cuvette; as clot formation begins, the fibrin strands couple the motion of the cup to the pin, and this is transmitted through the pin to give the characteristic TEG trace (see Figure 110).

Figure 110 shows a diagrammatic representation of a thromboelastogram. In this figure, *r* equals the reaction time and represents the time for the sample placement in the cuvette until the TEG tracing reaches 2 mm in amplitude (normal range 6 to 8 min). This value represents the rate of initial fibrin formation and is related to the functional clotting-factor activity and to any circulating inhibitor activity. Prolongation of the *r* time is seen with deficiencies of coagulation factors, anticoagulation (heparin), or in cases of severe hypofibrinogenemia. A small *r* value may be present in some hypercoagulable syndromes. *K* equals the clot formation time (normal range 3 to 6 min) and is measured from the *r* time to the point at which the amplitude of the tracing reaches 20 mm. This value represents the time taken for a fixed degree of viscoelasticity to be achieved by the forming clot as a result of the fibrin formation. It reflects the activity of the intrinsic pathway, fibrinogen, and platelets. The angle *a* (normal range 50 to 60°) equals the angle formed by the slope of the TEG tracing from the *r* value to the *K* value. It denotes the speed at which solid clot forms. Decreased *K* values are seen in hypofibrinogenemia and thrombocytopenia. MA equals the maximum amplitude (normal range 50 to 60 mm) and is the greatest amplitude on the TEG trace; it reflects the strength of the fibrin clot. It is a direct function of the maximum dynamic properties of fibrin and platelets. Platelet abnormalities, whether quantitative or qualitative, substantially affect the MA value. The A60 is the MA – 5 mm and is the amplitude of the TEG tracing 60 min after MA is achieved (normal range MA – 5 mm). It is a measure of clot lysis or retraction. The clot lysis index (CLI: normal range >85%) is derived as A60/MA × 100(%). It measures the amplitude as a function of time and reflects the loss of clot integrity as a result of lysis (see Figure 109).

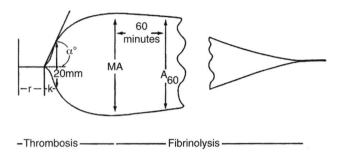

FIGURE 109
Chart of thromboelastogram tracing. (From Mallett, S.V. and Cox, D.J.A., *Br. J. Anaesthesia*, 69, 307, 1992. With permission.)

The TEG is widely used in liver and cardiac surgery, where a complete evaluation of hemostasis and fibrinolysis is required. TEG provides information about fibrinolytic activity and platelet function that is not generally available from routine coagulation tests. As a measure of global hemostasis, the method is recently seeing a return of interest and value.

THROMBOEMBOLISM

See **Arterial thrombosis; Thrombosis; Venous thromboembolic disease**.

β-THROMBOGLOBULIN

An 81-amino-acid-specific protein of **platelets** with a low affinity for **heparin**. It is the degradation product of platelet basic protein. β-Thromboglobulin may have a role as a neutrophil chemotactic agent and possibly in platelet maturation. Its normal level in plasma is <50 ng/ml. High levels of plasma β-thromboglobulin are seen in **venous thromboembolic disease**, myocardial infarction, **disseminated intravascular coagulation**, and cancer. Low levels are seen in **gray-platelet syndrome** and δ-**storage pool disorders of platelets**.

THROMBOLYTIC THERAPY

See **Fibrinolysis**.

THROMBOMODULIN

A **vascular endothelium** cell-surface protein that binds **thrombin** (IIa), rendering it inactive; however, through its marked enhancement of **protein C** activation, it becomes a potent anticoagulant. Thrombomodulin has a molecular weight of 78 kDa. It is present on all endothelial cells, with the exception of brain vessels, venules in lymph nodes, and sinusoidal lining cells of the liver. Each endothelial cell has between 30,000 and 55,000 molecules of thrombomodulin, and this accounts for 50 to 60% of the thrombin-binding capacity of endothelium. Bound thrombin-thrombomodulin complex rapidly converts protein C to activated protein C. Activated protein C acts as a natural anticoagulant by degrading factor Va and factor VIIIa on membrane surfaces in the presence of **protein S**. Once thrombin is bound, the complex is then internalized, thrombin broken down, and thrombomodulin returned to the cell surface. Thrombin bound to thrombomodulin cannot activate platelets or cleave **fibrinogen**. Thrombomodulin expression is decreased in inflammatory states. A deficiency of thrombomodulin has been reported to cause **venous thromboembolic disease**.

THROMBOPHILIA

See also **Thrombosis**.

Hypercoagulable disorders, both inherited and acquired, that are recognized to be associated with an increased risk of **arterial thrombosis** and **venous thromboembolic disease**. High-normal **factor VII** levels are a primary risk factor for ischemic heart disease, possibly through increased thrombin generation. Similarly, high-normal **fibrinogen** levels are a strong risk factor for ischemic heart disease and cerebrovascular disease, possibly by inducing platelet aggregation and increasing plasma/blood viscosity. Elevation in fibrinogen may be one of the mechanisms by which **tobacco** smoking exerts its effects upon the circulation. Raised levels of **homocysteine**, possibly associated with **folic acid** deficiency, may be an unproven factor. In addition, elevated levels of **factors VIII, IX,** or **XI** (whether hereditary or acquired) appear to be independently associated with an increased risk of venous thromboembolism.

Hereditary disorders are increasingly recognized as having an etiological significance.[498] These include:

Deficiency of **protein C** or **protein S**

Deficiency of **antithrombin III**

Resistance of coagulation **factor V** to breakdown by activated protein C (APC resistance) in those with factor V-Leiden

Prothrombin 3′ in translated region gene variant

Women with familial thrombophilia, especially when there is a combined deficiency or antithrombin deficiency, have an increased risk of fetal loss, particularly of stillbirth. Once a family has been identified as having some form of thrombophilia, screening of first-degree relatives has to be considered.[499] If carried out it, must be associated with counseling with regard to risk factors associated with certain lifestyles, such as smoking and the use of the contraceptive pill. A contraindication to such screening is the generation of psychological effects and problems related to insurance. Due to the high risk of thrombosis during or following pregnancy, screening of female relatives who are of childbearing age may be justified.

THROMBOPHLEBITIS

Inflammation of a vein wall associated with an organizing thrombus. The site is usually a superficial vein, but it often occurs with an indwelling catheter associated with local sepsis. The inflammatory changes extend outside the vessel wall, causing firm adhesion of the thrombus. Embolization is extremely rare, so there is no indication for treatment other than local palliative measures to relieve pain. Limb stasis should be avoided to prevent the complication of venous thromboembolic disease.

THROMBOPLASTIN

See **Tissue factor; International sensitivity index**.

THROMBOPOIESIS
See **Hematopoiesis**; **Megakaryocytes**.

THROMBOPOIETIN
See **Megakaryocytes**.

THROMBOSIS
One of the commonest hematological disorders (see **Hematology**) was defined by Virchow in 1856 as the formation of a solid or semisolid mass from the constituents of blood within the vascular system during life. He recognized that abnormalities of blood flow, changes in the structure of the vessel wall, and the coagulability of blood were factors in its pathogenesis.

The structure of a thrombus is based on fibrin, which enmeshes platelets, red blood cells, and leukocytes. Its composition varies with the site of the thrombosis, depending upon the degree of blood turbulence. With rapid flow, as in large arteries, the thrombi consist mainly of aggregated platelets, while with slow circulatory flow, as through veins, they consist mainly of red blood cells. Once formed, thrombi undergo progressive structural change. **Leukocytes** are increasingly attracted by chemotactic factors released from aggregating platelets or proteolytic fragments of plasma proteins that have become incorporated into the thrombus. The aggregated platelets swell and lyse, to be gradually replaced by more fibrin, which is eventually digested by **fibrinolysis**. In the meantime, there is local circulatory obstruction with infarction, which can become more widespread due to embolization of thrombotic material. Arterial thrombi that are nonocclusive may become incorporated into the vessel wall and accelerate any obstruction due to atherosclerosis.

Pathogenesis
This differs with the site of thrombus formation, but is largely determined by the occurrence of local stasis and vessel wall damage in any part of the cardiovascular system.

Venous thrombosis commonly occurs in the deep veins of the legs or pelvis, less frequently in the upper limbs, retina, brain, or mesentery. They are often "silent" until embolization, particularly to the lungs, occurs (see **venous thromboembolic disease**).

Arterial thrombosis occurs in coronary, cerebral, carotid, or peripheral vessels of the lower limbs, usually associated with a site of atheromatous plaque formation.

Intracardiac thromboses originate on damaged or prosthetic valves, over ischemic myocardium, or in dilated or dyskinetic chambers. They are usually asymptomatic until embolization occurs.

Microcirculatory thrombi arise with **disseminated intravascular coagulation**, causing small-vessel ischemic necrosis, often of renal glomeruli, followed by or coincident with consumption of platelets and coagulation factors (consumption coagulopathy), leading to a **hemorrhagic disorder**.

Embolization can extend the ischemic effects of a thrombus and is a common complication: venous to the lungs (pulmonary embolism), intracardiac and carotid to the brain.

Hypercoagulable State
See also **Thrombophilia**.

This is probably necessary for initiation of clinical thrombosis. In this context, thrombosis is an overactivity of **hemostasis** by a series of zymogens (coagulation factors), their respective **serine proteases**, and cofactors, or a deficiency of their inhibitors, including those affecting fibrinolysis. The origin of thrombosis therefore depends upon a change in the physiological balance between thrombogenic and protective factors, the extent varying with the site within the cardiovascular system.

Thrombogenic factors include:

Disturbance of **vascular endothelium**

Exposure of subendothelial **collagen**

Activation of platelets by collagen or circulating platelet agonists

Activation of **coagulation factors**

Inhibition of **fibrinolysis**

Stasis

Protective factors include:

Intact vascular endothelium

Neutralization of activated coagulation factors by:
- Endothelial cell bound components — **heparan sulfate, thrombomodulin**
- Natural serine protease inhibitors — **antithrombin III, heparin cofactor II, protein C** inhibitor, α2-antiplasmin, α2-macroglobulin

Dilution of activated coagulation factors

Clearance of activated coagulation factors by the liver

Natural anticoagulants protein C and **protein S**

Disruption of platelet aggregates by rapid blood flow

Dissociation of fibrin by fibrinolysis

Risk factors for thrombosis management and treatment[500] vary with the location of the thrombosis: arterial thrombosis, disseminated intravascular coagulation, venous thromboembolism.

THROMBOSPONDIN

An adhesive protein (molecular weight 450 kDa) that is produced by both **vascular endothelium** and **platelets**. It differs slightly, dependent upon site of production. It is a trimer of identical subunits. Platelet thrombospondin is released from α-granules of platelets upon thrombin stimulation. Endothelial-derived thrombospondin is found in the subendothelium. Thrombospondin is important in maintenance of platelet adhesion in high-shear vessels. In addition to its role as an adhesive, protein thrombospondin may also modulate smooth-muscle proliferation and endothelial cell movement. Thrombospondin synthesis is increased in **inflammation** and trauma.

THROMBOTEST

A method of measuring overall clotting activity by using a specific commercial reagent — the "Thrombotest reagent" (Nyegaard Co.), which contains bovine plasma, bovine thromboplastin, and cephalin.[25] The Thrombotest can be used on plasma, capillary whole blood, and citrated whole blood. The test is performed by adding the Thrombotest reagent to the blood/plasma sample and measuring the clotting time. The percentage activity is then derived from a standard curve provided by the manufacturer. The therapeutic range lies between 6 to 11%. It is used mainly in Scandinavian countries to monitor **warfarin** dosage.

THROMBOTIC MICROANGIOPATHY

Widespread arteriolar **thrombosis**, seen post**transplantation**.

THROMBOTIC PULMONARY HYPERTENSION

A vascular disorder of the pulmonary arteries in which the pressure exceeds 30/15 mm Hg. It is commonly associated with raised cardiac atrial pressure and with diffuse pulmonary disease.[426] It occurs as chronic thromboembolism, the cause of which is unknown. It is associated with raised **factor VIII** level, and some patients have anticardiolipid antibodies (see **Lupus anticoagulants**). Treatment is surgical removal of the clot.

THROMBOTIC THROMBOCYTOPENIC PURPURA

(TTP) A severe multisystem disorder characterized by fever, **thrombocytopenia, microangiopathic hemolytic anemia** (MAHA), impaired renal function, and fluctuating neurological signs/symptoms, first described by Moschcowitz in 1925. Untreated, the mortality for TTP approaches 90%.

Pathogenesis

TTP and hemolytic uremic syndrome are different clinical manifestations of a common pathophysiological process arising from different causes that results in intravascular platelet aggregation. TTP arises from deficiency of a **Von Willebrand Factor**-cleaving metalloprotease (termed a disintegrin) and a metalloprotease with eight thrombospondin-1 like domains (**ADAMTS-13**). This enzyme is essential to cleave multimers of unusually large Von Willebrand Factor anchored to **vascular endothelium**. The deficiency arises in a number of forms (see Table 149).

TABLE 149

Forms of ADAMTS 13 Deficiency

Form of Deficiency	TTP Presentation
Genetic mutation	familial, relapsing
Autoantibodies	acquired idiopathic
Transient	single episode
Recurrent	intermittent
Thienopyridine-associated	ticlopidine therapy
Pregnancy (with/without antibodies)	associated pregnancy

The characteristic pathological feature of TTP is a hyaline thrombus consisting of platelet aggregates and fibrin that occludes capillaries and arterioles of all tissues, but particularly the kidneys, pancreas, heart, adrenals, and brain. It is these microthrombi that give rise to the characteristic clinical and laboratory features of TTP. Current evidence suggests a defect in endothelial cells leading to a decrease in PGI2 (**prostacyclin**) or the presence of a plasma factor that alters its stability and function. The increased tendency for circulating **platelets** to aggregate is related to the release into the circulation of very high-molecular-weight Von Willebrand Factor (VWF) multimers that bind to platelet membrane receptors, thus promoting platelet adhesion and, subsequently, aggregation.

Clinical Features

The disease presents between the ages of 10 and 60 years. TTP can take several clinical courses, including a single acute episode, relapsing TTP, chronic TTP, and childhood/familial TTP. The more common acute fulminant variety can result in death in a few days. In this form, neurologic features are most prominent, including aphasia, behavioral changes, paresis, and seizures. The state of consciousness may fluctuate or eventually progress to coma. **Hemorrhagic disorders** are next in frequency, usually generalized **purpura**. **Hemolysis** can be severe. Fever occurs in most patients.

In the chronic variety of the disease, symptoms persist or are recurrent over months and sometimes years. **Disseminated intravascular coagulation** (DIC) can present with similar clinical features, but in patients with TTP, the laboratory abnormalities of DIC are absent.

Laboratory Features

Most patients can be diagnosed on the basis of history, examination, blood counts, blood film, and urinalysis[360,361]:

Anemia with increased **reticulocyte count** and **fragmentation of red blood cells**

Thrombocytopenia

Proteinuria, microscopic or gross hematuria

Biochemical changes of hemolysis with raised lactic dehydrogenase (LDH)

Blood urea nitrogen level elevated in at least 50% of patients

Unusually large, high-molecular-weight VWF multimers with ADAMTS-13 deficiency upon assay

Biopsies that demonstrate the characteristic microvascular lesions can be helpful but, are not essential to make the diagnosis

Treatment

The mainstay is plasma exchange (3 to 5 l daily) by **plasmapheresis** with **fresh-frozen plasma** or cryoprecipitate, which should be continued until the platelet count is stable. Plasma infusions should be given only if plasma exchange is not readily available. Platelet transfusion can be hazardous. Familial cases require plasma containing ADAMTS-13 every 3 weeks. Suppression of ADAMTS-13 antibodies can be achieved by high-dose corticosteroids or four to eight weekly doses of **rituximab**. Splenectomy may be of value in cases of refractory or relapsing TTP. Despite the use of optimal therapy, TTP can remain irreversible and fatal in a number of patients, but the majority now recover.

THROMBOXANE A2

A prostaglandin synthesized from **arachidonic acid** via the cyclooxygenase pathway and thromboxane synthetase. It has potent platelet-aggregating effects and is a potent vasoconstrictor. The aggregatory action of thromboxane is mediated by an increase in platelet cytosol calcium and inhibition of adenyl cyclase/cAMP (**cyclic adenosine monophosphate**), thereby directly opposing the action of **prostacyclin**. Thromboxane is one of many platelet-aggregation mediators. It is not essential for platelet aggregation, and it functions primarily to amplify the effect of other biological mediators. It can diffuse across the platelet plasma membrane to activate platelets. Thromboxane is an unstable prostaglandin with a half-life of approximately 30 sec. This short half-life helps to confine the spread of platelet activation to the original site of injury.

THYMOCYTE

See **Lymphocytes**.

THYMUS

An organ of the **lymphoreticular system** situated in the superior mediastinum anterior to the great vessels and the heart. It is derived embryologically from two types of epithelial cells, ectodermal and epidermal, arising from the third bronchial epithelial pouch. **Lymphocyte** stem cells have migrated from the bone marrow to undergo development. The thymus increases in size until puberty, after which it gradually regresses.

Structure

The thymus consists of two lobes, closely joined by a thin capsule of connective tissue, that subdivide into a number of lobules (see Figure 110). It reaches its greatest size at puberty, weighing 30 to 40 g. Each lobule has a cortex of densely packed lymphocytes (thymocytes) with a medulla of loosely distributed lymphocytes, epithelial cells (reticular cells), and **histiocytes** (macrophages). The lymphoid progenitor cells migrate to the subcapsular area, the mature cells migrating to the medulla. Within the medulla are Hassall's corpuscles of flattened epithelial cells that are phenotypically distinct from those in the cortex, although both are derived from pharyngeal epithelium. T-cell development in the

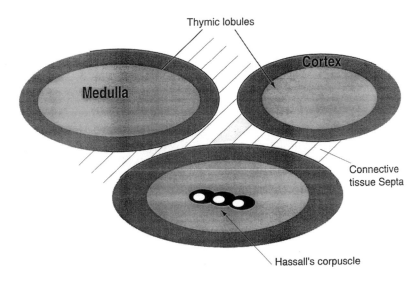

FIGURE 110
Diagrammatic representation of the structure of the thymus.

thymus depends upon the epithelial components of both cortex and medulla. In adult life, the organ normally regresses in size and function.

Function

This is primarily T-cell ontogeny, occurring in the fetus and continuing into early adult life. The colony forming unit-lymphocytes (CFU-L) stem cells either die in the cortex or mature to move deeper into the cortex and eventually into the medulla, where they acquire distinctive surface and cytoplasmic markers before passing into the peripheral circulation. The total time for passage of these cells through the thymus is about 24 h.

Local thymic microenvironmental factors differentiate thymocytes into inducer-helper and suppressor cytotoxic cells whose functions have been genetically programmed. All early cortical thymocytes bear the T10 surface marker, but medullary thymocytes bear either T4 or T8 surface markers.

Disorders of the Thymus Gland

Thymic hypoplasia associated with lymphopenia, all causes

Thymic hyperplasia

Reactive lymphoproliferation

Follicular hyperplasia with **autoimmune disorders**: myasthenia gravis, **systemic lupus erythematosus, rheumatoid arthritis, pure red cell aplasia**

Thymoma

THYMOMAS

A collective term used for tumors of the thymus gland. The histological types are:

Epithelial (80%) derived from branchial epithelium

Lymphocytic: **mediastinal large B-cell lymphoma, precursor T-cell lymphoma, Hodgkin disease**

Mixed

Others: carcinoid, seminoma, teratoma

Thymomas can occur at any age, but peak incidence is in the fifth and sixth decades. Most patients are diagnosed by chance, but large tumors will cause compressive symptoms, e.g., hoarseness, dyspnea, and dysphagia. Systemic complications can occur, including myasthenia gravis (10 to 50%), **pure red cell aplasia** (5%), or hypogammaglobulinemia.

The treatment of choice is surgical removal, with survival determined by the extent of tumor spread. In patients with complete surgical resection, overall survival is similar to that of the general population, but only 12.5% long-term survival is seen in patients with invasive tumors.

THYROID GLAND DISORDERS

The hematological associations with the thyroid gland.

Physiological Control of Thyroid Hormones

Production of renal **erythropoietin**

Level of red blood cell **2,3-diphosphoglycerate** dehydrogenase (2,3-DPG)

Hypothyroidism

Macrocytic anemia with decreased erythropoiesis

Megaloblastosis due to **cobalamin** deficiency associated with intrinsic factor autoantibodies as part of a general autoimmunity disorder

Hyperthyroidism

The thyroid is stimulated by autoantibodies that bind to the thyroid-stimulating hormone receptor (Type II **hypersensitivity** reaction). The autoantibody is related to **human leukocyte antigen** HLA allele DR3. This cross-reacts with an orbital antigen to give rise to exophthalmos. The autoantibody can also cross the placenta, causing transient neonatal hyperthyroidism. Other associations are:

Microcytic anemia with increased erythropoiesis but without increase in blood volume and increased red cell turnover

Increase in red blood cell 2,3-DPG, allowing an increase in **oxygen affinity to hemoglobin** and increased availability of oxygen for the tissues

Neutropenia following antithyroid drug therapy

Immune thrombocytopenic purpura in association with thyroid autoantibodies

Swelling by Infiltration by Cells

See **Lymphoproliferative disorders**.

T-IMMUNOBLASTIC LYMPHOMA

See **Anaplastic large-cell lymphoma**; **Angiocentric lymphoma**; **Hodgkin lymphoma**; **Peripheral T-cell lymphoma, unspecified**.

TISSUE FACTOR

(Thromboplastin) A 263-amino acid integral transmembrane protein (molecular weight 45 kDa), with a short 23-amino acid hydrophobic domain, which spans the membrane. It is coded for by a short gene of 12.4 kb on chromosome 1p 21. Tissue factor is found on the surface of **vascular endothelium** cells, but it is also constitutively expressed by many nonvascular tissues. It can be upregulated by cellular stimulation on monocytes and vascular endothelium. Tissue factor is a receptor for **factor VII** and is required for the initiation of blood coagulation. It acts as a cofactor enhancing the proteolytic activity of factor VIIa toward **factor IX** and **factor X.** It binds factor VII via calcium ions. It is the large extracellular domain that initiates coagulation by binding and activating factor VII. The 21-amino acid cytoplasmic domain is involved in intracellular signaling. This is also stimulated by binding factor VII, but the function of this signaling is unknown. While tissue factor is found on the surface of vascular endothelium cells, it is also constitutively expressed by many nonvascular tissues. It can be upregulated by cellular stimulation on monocytes and vascular endothelium. Lying on the surface of the vascular endothelium, it provides a protective envelope around blood vessels and organs that can initiate coagulation/hemostasis as soon as there is any leakage of blood.

TISSUE FACTOR PATHWAY INHIBITOR

(TFPI) A 32-kDa protein that inhibits **factor Xa, tissue factor–factor VII**, forming a 1:1:1 complex. It has a unique structure with a negatively charged amino terminus followed by three Kunitz-type inhibitory domains and a positively charged carboxy terminus. TFPI forms a quaternary complex with tissue factor, factor VII, and factor X. In this complex, the catalytic site of factor Xa is bound to the second of the three inhibitory Kunitz domains in TFPI, with the catalytic site of factor VIIa bound to the first inhibitory domain. TFPI can bind to preformed factor X–factor VII–tissue factor complex, or to factor X–TFPI complex binding to factor VII–tissue factor. Although TFPI requires factor X for effective inhibition of factor VII–tissue factor, it can directly inhibit factor VII at high concentration.

As much as 50 to 90% of TFPI is associated with and synthesized by **vascular endothelial** cells. Approximately 5 to 10% is associated with platelets. The remainder is bound in plasma to lipoproteins (80% with low density and 20% with high density). Less than 5% of TFPI is uncomplexed. It is released into plasma following infusion of **heparin**, presumably derived from the vascular endothelium. Levels can increase up to tenfold after heparin infusion. Levels are low in the newborn but achieve adult normal levels by age 6 months. It is not known whether deficiency of TFPI is a cause for disease and thrombosis.

TISSUE MACROPHAGES

See **Histiocytes**.

TISSUE PLASMINOGEN ACTIVATORS

(t-PA) See **Fibrinolysis**.

T-LARGE-CELL ANAPLASTIC (KI-1) LYMPHOMA

See **Angioimmunoblastic T-cell lymphoma**; **Non-Hodgkin lymphoma**; **Peripheral T-cell lymphoma, unspecified**.

T-LYMPHOBLASTIC LYMPHOMA

See **T-cell prolymphocytic leukemia/lymphoma**; **Non-Hodgkin lymphoma**.

T-LYMPHOCYTIC LEUKEMIA/LYMPHOMA

See **Chronic lymphocytic leukemia/small-cell lymphoma**.

T-LYMPHOCYTIC, PROLYMPHOCYTIC LEUKEMIA

See **T-cell prolymphocytic leukemia**.

TOBACCO

(Smoking disorders) The hematological effects of excessive tobacco smoking (tobacco abuse).

Arterial thrombosis: this is a well-recognized epidemiological risk, but the mechanism is unknown. The most likely explanation is pulmonary alveolar reaction causing increased **platelet** activation, with adhesion to vascular endothelium increasing the risk of both irreversible atherosclerosis and venous thromboembolism. The increased incidence is related to the length of time of smoking in years.

Macrocytosis, **neutrophilia**, **monocytosis**, and **lymphocytosis** due to bone marrow stimulation from decreased oxygen delivery arising from reduced oxygen perfusion through the lungs and raised levels of carboxyhemoglobin — a low-level secondary **polycythemia**. The degree of these changes is related directly to the degree of smoking.

Erythrocytosis due to reduced **plasma volume** of unknown cause and secondary erythrocytosis due to increased blood levels of carbon monoxide.

Cobalamin conversion to cyanocobalamin, causing a relative B_{12} deficiency. Those with existing cobalamin deficiency develop "tobacco amblyopia" but without megaloblastosis.

Ascorbic acid levels in plasma are reduced but return to normal levels upon cessation of smoking.

Chronic polyclonal B-cell **lymphocytosis** associated with **Epstein-Barr virus** infection.

Pulmonary Langerhans cell **histiocytosis** incidence increases.

Chromosomal abnormalities in fetal epithelial cells associated with smoking during pregnancy.

α-TOCOPHEROL

See **Vitamin E**.

TONGUE DISORDERS

See **Mouth disorders**.

TOPOISOMERASE II INHIBITORS

A group of **cytotoxic agents** that block the intracellular enzyme topoisomerase. They include etoposide, teniposide, razoxane, and the antibiotics doxorubicin, daunorubicin, and mitoxantrone (see **Anthracyclines**). They have been implicated in **leukemogenesis**, particularly **myelodysplasia**-therapy-related.

TOTAL IRON-BINDING CAPACITY

(TIBC) See **Transferrin**.

TOURNIQUET TEST

(Hess's test) A clinical procedure used to demonstrate increased **capillary fragility**. A sphygmomanometer cuff is applied to the upper arm and inflated to 80 mm Hg for 5 min. Two or three minutes after the cuff is deflated, the number of petechiae in a 3-cm-diameter area are counted. More than 20 petechiae are abnormal. Appearance of purpuric spots is taken as a sign of increased capillary fragility or significant **thrombocytopenia**. The test is redundant in the presence of widespread purpura.

TOXIC-SHOCK SYNDROME

Overwhelming peripheral cardiac failure caused by infection of virulent strains of staphylococcal or streptococcal bacteria that secrete peptide toxins. Some of these toxins are superantigens that bind to specific TCR β-chains and directly activate T-lymphocytes. The T-cells release **cytokines**, causing intense hypotension and peripheral cardiac failure.

TRANEXAMIC ACID

See **Fibrinolysis** — antifibrinolytic agents.

TRANSCOBALAMINS

Plasma proteins that transport **cobalamin** (vitamin B_{12}) throughout the body. Transcobalamin I (TC-I) carries most of the circulating cobalamin. Transcobalamin II (TC-II) is the plasma protein that mediates the transport of cobalamin to the tissues. Although it carries only a small proportion of the cobalamin in plasma, it is the protein to which newly acquired cobalamin is first bound. It is a simple protein with a molecular mass of 38 kDa and is synthesized by many cells, including enterocytes, hepatocytes, mononuclear phagocytes, fibroblasts, and hematopoietic precursors in the marrow. TC-II provides only 10 to 25% of the total plasma cobalamin, but it provides the majority of the total unsaturated cobalamin-binding capacity of plasma. Transcobalamin III (TC-III) has been described, but is often undetectable in the plasma. Transcobalamins I and III are sometimes referred to as the plasma R proteins.

TC-I and TC-II are normally present in the plasma in trace quantities (7 and 20 μg/l, respectively). Increased plasma levels of TC-I are found in the chronic **myeloproliferative disorders**, benign **neutrophilia**, **chronic myelogenous leukemia**, in metastatic cancer, and occasionally in association with hepatoma. Increases in TC-II levels are found in the chronic **myeloproliferative disorders**, liver disorders, inflammatory disorders, and in **Gaucher's disease**.

Transcobalamin II Deficiency

An autosomally recessive disorder causing a **megaloblastosis** present in early infancy. The danger is that severe **anemia** develops with a normal cobalamin level. If it is not diagnosed quickly, irreversible nervous system damage ensues. These infants are healthy at birth. In addition to severe anemia, recurrent bacterial infections are also a problem. The diagnosis is made by measuring the TC-II-specific cobalamin-binding capacity of the serum. This assay is available in only a small number of specialized laboratories.

TRANSCRIPTIONAL REGULATION

The process by which **DNA** is transcribed to **RNA**. This process takes place in the nucleus and occurs at a baseline level in a cell to provide housekeeping proteins to maintain cell viability, or in a specialist cell to transcribe RNA for specific cellular functions. Transcription is a tightly ordered process; for this to take place in an orderly fashion, the transcribed DNA must be accessible to a complex of transcription enzymes and coenzymes (Figure 111).

DNA is packaged within the cell as elaborately folded **chromatin**. The most basic level of chromatin organization is the nucleosome, which consists of 146 base pairs of DNA wrapped 1.7 times around a core of eight **histone** proteins. Nucleosomes are subsequently packaged into higher-order structures that combine to form the **chromosome**. These layers of folding regulate access of transcription regulators to the DNA, which in the genome can be found in a variety of states. At one end of the spectrum there are regions that are free of nucleosomes and readily accessible to RNA polymerase and the transcriptional machinery. Other loci may be complexed with histone octamers into arrays of nucleosomes so that transcription is dependent on covalent modifications of histones and the presence of sequence-specific DNA-binding transcription factors. Lastly, nucleosomes may be tightly wound in a solenoidal fashion into a 30-nm fiber in the presence of accessory proteins to form nonaccessible heterochromatin, such as that found at the **centromere**, **telomeres**, and the inactivated X chromosome. To change from one state to the other, chromatin undergoes covalent modifications and physical remodeling

Transcription of most of the structural and regulatory genes in the eukaryotic genome is carried out by RNA polymerase II. The polymerase itself has no specificity in its ability to begin transcription of DNA and must be guided to promoter regions on the DNA by sequence-specific transcription factors. These promoter and enhancer regions are regulatory sequences that govern the temporal, spatial, and homeostatic expression of genes by recruiting transcription factors and, subsequently, the RNA polymerase II enzyme. These factors act in concert to increase or decrease the chance that a particular locus is transcribed.

Transcription factors have a modular design, which includes DNA-binding domains and protein-protein interaction domains. These protein domains facilitate interaction with the cofactors required to activate or repress transcription and are hence called activation or repression domains. Many transcription factors recruit coactivators and corepressors to establish a chromatin modification and remodeling complex at the site of target gene promoters and enhancers. Disruption of the normal function of transcriptional coactivators or corepressors is a common step in malignant transformation.

Transcriptional Activation

In general, transcription factors responsible for activating transcription recruit coactivator proteins to form a transcriptional complex. Coactivators enable more-efficient transcription by the transcription factors by offering them a specific interface for docking onto DNA-binding sites. The p300 and CBP coactivator proteins have at least two known roles in the transcription complex. They act as a bridge between transcription factors and the basal transcriptional machinery. They are also histone acetyl transferases (HATs), which acetylate both histone tails and transcription factors. Histone acetylation facilitates transcriptional activation (see **Histones**).

One example of transcriptional activation occurs in the absence of retinoic acid (see **Retinoids**). The retinoic acid receptor (RARα) and its leukemogenic fusion variant PML-RARα are transcription factors that are constitutively complexed with corepressor molecules. When bound by all-trans retinoic acid (ATRA), the proteins undergo a conformational change that obliterates corepressor-binding sites and exposes new surfaces that can bind coactivator proteins. As a result, transcriptional repression is reversed, and genes essential for cellular maturation and differentiation are reactivated. Hence, treatment of **acute promyelocytic leukemia** (APL) with ATRA is truly a form of transcriptional therapy.

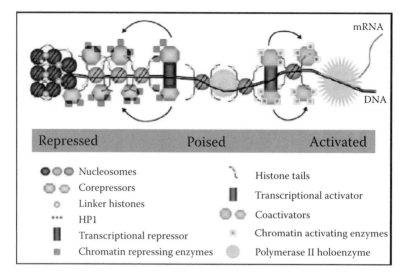

FIGURE 111

Transcriptional regulation. Chromatin is the substrate for transcriptional regulation. A simplified gene locus is shown, where the chromatin is shown in the repressed, poised, or transcriptionally active state. A transcriptional repressor recruits corepressors, which in turn recruit repressing enzymes. These enzymes introduce specific covalent modifications of histone tails that offer binding sites facilitating recruitment of additional corepressors and recessive enzymes. Eventually, binding of linker histones and heterochromatic protein 1 (HPI) proteins results in highly condensed heterochromatin. Conversely, a transcriptional activator recruits coactivators, which then recruit activating enzymes. These introduce activating histone modifications such as acetylation, which makes DNA accessible to the general transcription machinery and trigger the activation of polymerase in the poised position, with chromatin available for transcription or repression. Most likely, a majority of genes exists in a continuum where they are more- or less-accessible transcriptional repressors or activators. (From Melnick, A.M., Adelson, K., and Licht, J.D., The therapeutic basis of transcriptional theory of cancer: can it be put into practice? *J. Clin. Oncol.*, 23, 3957, 2005. With permission.)

Transcriptional Repression

Transcriptional repressor proteins often contain conserved, structured domains that recruit corepressors and chromatin modification machinery. Two of the best-characterized and conserved repression motifs are the BTB/POZ (bric-a-brac, tramtrack, broad complex/pox virus zinc finger) domain and the KRAB domain (Krüppel-associated box). The BTB/POZ transcriptional repressor domain is found in oncogenes such as PLZF-RARα (promyelocytic leukemia zinc finger), which is the fusion protein associated with retinoid refractory t(11;17)-associated APL, and Bcl-6, which is frequently constitutively expressed in **B-cell lymphoma**. BTB domains are found in approximately 60 human transcription factors. Corepressors that bind to transcription-factor-repression domains recruit other components of the transcriptional machinery, including histone deacetylases. Corepressors can, in fact, activate the deacetylases that modify chromatin to effect transcriptional silencing. Transcriptional corepressors such as N-CoR (nuclear receptor corepressor), SMRT (silencing mediator of retinoid and thyroid receptors), Sin3a, KAP1, and ETO (821)/MTG8 were identified both by their ability to interact with sequence-specific DNA-binding transcription factors, and by their presence in transcription complexes purified from cells. These factors match transcriptional repressors to the enzymes required for chromatin silencing. Many corepressors have specific domains that interact with modified forms of chromatin, enabling them to lock in place and maintain the silenced chromatin state. There can be large numbers of repressor, corepressor, and effector (chromatin modifying) molecules in each complex, making it a significant challenge to identify therapeutic targets.

TRANSFERRIN

(Tf; iron-binding protein; siderophilin) An elongated glycoprotein of 79,570 Da that migrates electrophoretically with the beta globulins. Normal human plasma contains 2 to 3 g/l of transferrin, carrying about 1.4 μg Fe/mg transferrin. Normally, approximately one-third of the transferrin iron-binding sites are saturated with **iron** (transferrin or iron saturation); the remainder represents unsaturated iron-binding capacity (UIBC). The total iron-binding capacity (TIBC) in healthy adults is 250 to 390 μg/dl (45 to 70 μmol/l). Levels in infants are lower. Genetic polymorphisms of transferrin are common and may account for race/ethnicity differences in serum iron levels and transferrin saturation. Transferrin also transports a variety of physiologic and nonphysiologic divalent metal cations other than iron.

The major function of transferrin is to transport iron from absorptive cells in the small intestine, storage macrophages, or other sites to marrow erythroblasts for hemoglobin production. Transferrin binds to specific membrane receptors of erythroblasts or **reticulocytes**. Transferrin receptor is the ubiquitous surface glycoprotein of actively growing and dividing cells, composed of disulfide-linked polypeptide chains, each 90 kDa, that react with OKT-9. Diferric transferrin binds to membrane transferrin receptors preferentially. The principal mechanism for cellular iron uptake is **endocytosis** of the transferrin–transferrin receptor complex. The transferrin–transferrin receptor complex forms clusters in pits of the cell membrane, which become internalized in coated vesicles that fuse with endosomes. Transferrin receptor-2 mediates uptake of transferrin-bound iron by the liver after the classical transferrin receptor is downregulated by **iron overload**, and thus may explain the increased susceptibility of the liver to iron loading in **hereditary hemochromatosis**. Mutations of the *TFR2* gene (7q22) cause an autosomally recessive form of hemochromatosis, often severe and of early age of onset.

Iron is released from transferrin by acidification of the endosomes. This is a continuous process that occurs whether or not transferrin has been bound. Transferrin and transferrin receptors are recycled from cytosol to membrane. Regulation of the iron balance of the cell depends largely on stimulation of transferrin receptors and ferritin synthesis. This regulatory mechanism involves the effect of iron on iron-responsive elements in the untranslated regions of transferrin receptor mRNA and ferritin mRNA, respectively. Thus, **iron deficiency** encourages the production of transferrin receptors (and transferrin), and iron repletion favors the production of ferritin, an iron-storage protein. Most nonerythroid cells also bear specific surface receptors for transferrin and thus receive iron in a manner that is similar to that of developing erythroid cells. Transferrin receptor production is upregulated by erythropoietin.

In some conditions, transferrin-mediated delivery of iron is significantly altered. In HFE hemochromatosis and certain other iron overload disorders, transferrin is often fully saturated with iron. Fully saturated transferrin and the presence of non-transferrin-bound iron in the plasma favor iron deposition in hepatocytes and other parenchymal cells. In HFE hemochromatosis, mutant HFE protein may also increase the influx of transferrin-bound iron into erythroblasts, causing an increase in mean corpuscular volume and hemoglobin content. Congenital absence of transferrin (atransferrinemia), a rare disorder, is caused by null Tf alleles (see **Atransferrinemia**). In the absence of functional transferrin, excess iron is absorbed by the intestine and is subsequently deposited in the liver, pancreas, spleen, and other organs. Affected patients have severe hypochromic, microcytic **anemia** due to lack of delivery of sufficient iron to erythroblasts. Treatment consists of regular infusions of plasma or purified transferrin to provide transferrin. Antitransferrin antibodies have been described in a few patients who developed mild iron deficiency anemia otherwise similar to that observed in patients with atransferrinemia.

TRANSFORMING GROWTH FACTOR B

(TGF-β) One of a large family of **cytokines** and growth factors, some of which are important in fetal development. Despite its name, TGF-β is a potent inhibitor of the proliferation of a range of cells, including **lymphocytes** and endothelial cells. Its anti-inflammatory effect is probably due to its growth-inhibitory effects. It also promotes **collagen** production and is involved in wound healing.

TRANSFUSION-RELATED LUNG INJURY

(TRALI) See **Blood components for transfusion** — complications.

TRANSFUSION-TRANSMITTED INFECTIONS

(TTI) The infections transmitted through the transfusion of blood components (see Table 150).

Incidence of Transfusion-Transmitted Infection

This is comparatively low, considering the large number of blood transfusions given each year, but the personal damage to an individual patient can be great. The most frequent and severe infections are those caused by bacteria. Viral transmission also occurs, but the public perception of viral risk may be exaggerated (see Table 151).

TABLE 150

Microorganisms Transmissible by Blood Transfusion

Plasma Borne

Bacterial	diphtheroids
	Pseudomonas aeruginosa
	Salmonella spp.
	Serratia marcescens
	Staphylococcus aureus
	Staphylococcus epidermidis
	Treponema pallidum
Viral	hepatitis A virus (HAV) — rare, no carrier state
	hepatitis B virus (HBV) — variants, carrier state
	hepatitis C virus (HCV) — common, carrier state
	hepatitis D virus (HDV) — delta agent, HBV required as "helper" agent
	human immune deficiency virus (HIV-1, HIV-2) (carrier and latent states)
	parvovirus B19
	West Nile virus
Prion disease	variant Creutzfeldt-Jacob disease (vCJD)
	Creutzfeldt-Jacob disease (CJD)
Protozoa	*Trypanosoma cruzi*

White Blood Cell-Associated

Herpes viruses	**cytomegalovirus** (CMV) (latent state)
	Epstein-Barr virus (EBV) (latent state)
Retroviruses	**human immune deficiency virus** (HIV-1, HIV-2) (latent state)
	human T-lymphotropic virus (HTLV-I, HTLV-II) (latent state)

Red Blood Cell-Associated

Babesiosis
Bartonellosis
Malaria

TABLE 151

Risk of Transmission of Certain Viruses

	U.S.	U.K.
HBV [a]	1:200,000	1:900,000
HCV	1:2,000,000	1:30,000,000
HIV	1:2,000,000	1:8,000,000

[a] In the U.S., testing for anti-HBC and alanine aminotransferase (ALT) impacts on the incidence of transmission.

Exclusion of Transfusion-Transmitted Infections from Donated Blood

All donors must be carefully selected and screened using a questionnaire to exclude those who may transmit infection (see **Blood donation**). This includes:

Those who have traveled to certain countries in the world (malaria, plasmodia, *Trypanosoma cruzi*, West Nile virus)

Those belonging to "high risk" groups, e.g., homosexuals, intravenous drug abusers (HBV, HCV, HIV-1 and -2)

Those with recent or current gastrointestinal disturbance (*Yersinia enterocolica*, *Escherichia coli* 0157:H7)

Those whose siblings, parents, or grandparents have developed Creutzfeldt-Jacob disease, who have been treated with human growth hormone, or who have been resident or had previous transfusions in the U.K. (see **Prion disease**)

The requirements for testing donor blood to exclude TTI vary from country to country. Different testing methodologies are employed, e.g., hemagglutination, radioimmunoassay, and enzyme immunoassay; the latter is most commonly used in Western Europe and the U.S. In the U.K., blood is screened for HBV, HCV, HIV-1 and -2, HTLV-I and -II, and syphilis. In the U.S., testing is also performed for HBC antibody, HTLV-I plus -II, and for alanine aminotransferase (ALT).

Clinical Consequences of Infection

Treponema pallidum

(Syphilis) Recipients usually develop a typically secondary eruption. Screening is largely a public health measure, but also allows for the exclusion of donors who have been infected by at least one venereal disease. The prevalence of syphilis is declining, and spirochetes are unlikely to survive in blood stored for more than 72 h at 4 to 6°C. The need for continued testing is still debated; serologic testing does not prevent all cases of transfusion, since in primary syphilis there may be spirochetemia with negative serologic tests. Moreover, many subjects give a biologically false-positive test.

Malaria Plasmodia

Parasites can survive storage at 2 to 6°C and can be transmitted by virtually all blood components. In the U.S., the most common causes of transfusion-associated malaria are *Plasmodium malariae* followed by *P. falciparum* and *P. vivax*. Recipients may develop unexplained fever, rigors, and shivering, together with headache, myalgia, and malaise.

Trypanosoma cruzi

This is the causative agent of Chagas disease and is most prevalent in rural areas of South and Central America. The reduviid bug that transmits it is associated with poor living conditions. A chagoma develops at the site of a bite, and this is followed after 10 to 14 days by parasitemia associated with fever, **lymphadenopathy**, and **splenomegaly**. Mortality may be as high as 10%. There follows an asymptomatic period that may last for decades. Finally, a chronic disease develops that causes cardiomyopathy, megaesophagus, and megacolon. The risk of transmission by contaminated blood in carriers is 12 to 50%. *Trypanosoma cruzi* survives low temperatures (including freezing), and 20% of acute cases are asymptomatic. Prevention of transmission in high-prevalence areas is by screening of donations for antibody to *T. cruzi* or addition of gentian violet to blood.

Bacterial Infection

(See Table 150 and **Blood component transfusion** — complications.) In bacterial infections other than with *Treponema pallidum*, the donor is usually the source of infection where blood is carefully processed in clean environments using closed-system collection sets. Gram-positive cocci such as *Staphylococcus epidermidis* are frequently found on the skin; they do not grow well at 4°C but proliferate well in platelet concentrates incubated at 22°C. They and other organisms may exhibit log phase growth during the storage of platelet concentrates prior to transfusion, and this can cause severe septicemia and toxemia leading to death in recipients of infected blood products.

Yersinia enterocolica

This is associated with bacteremia in patients who have minor gastrointestinal disturbances, often prior to donation. The organisms grow well at 4°C and have been reported to cause lethal sepsis. Other bacteria, including *Salmonella* spp., have caused fatal post-transfusion sepsis.

Parvovirus B19

This causes fifth disease, a childhood exanthem that manifests as fever, rash, and polyarthropathy. Following transfusion of infected blood, red blood cell precursors in the recipient's bone marrow become infected, leading to maturation arrest and **pure red cell aplasia**. This can be associated with aplastic crises in patients with increased red cell turnover, e.g., patients with **sickle cell disorders** or **hemolytic anemias**.

Human Immunodeficiency Virus-1

(HIV) It has been calculated that as many as 95% of recipients of an HIV-positive blood donation will seroconvert. Patients may develop an acute "flu-like" illness followed by a latent phase. The **acquired immunodeficiency syndrome** (AIDS) develops on average 7 to 10 years after transfusion.

Hepatitis B Virus

(HBV) The incubation is on average 4 months (2 to 6 months). It is clinically manifest as fever, anorexia, weakness, vomiting, and abdominal pain. Less than 50% of infections are icteric. Five to 10% of HBV infections become chronic, with an increased risk of chronic liver disease and hepatoma.

Hepatitis C Virus

(HCV) The incubation period is a mean of 8 weeks (2 to 26 weeks). Most cases are clinically mild, and only a minority develop jaundice. Liver function tests fluctuate after infection;

50% have an elevated ALT (alanine aminotransferase) for >1 year. Of blood donors who are HCV positive, 70% have histologic evidence of chronic hepatitis in the absence of symptoms. Of HCV carriers, 15 to 20% are at risk of developing cirrhosis, and 0.2% will develop hepatoma, usually 20 years or more after infection.

Cytomegalovirus

(CMV) Transfusion-associated CMV is of clinical importance in CMV-negative, low-birth-weight (<1.5 kg) babies, transplant recipients, and patients with HIV infection. Transfusion-associated infection can cause hepatitis, pneumonitis, retinitis, central nervous system disease, gastrointestinal disease, **thrombocytopenia**, and **anemia**. Transmission of infection can be prevented by providing CMV-seronegative or leukodepleted cellular blood components for transfusion.

Abnormal Prion-Related Protein Associated with Creutzfeldt-Jacob Disease

See **Prion disease**.

TRANSIENT ERYTHROBLASTOPENIA OF CHILDHOOD

(TEC) A disorder occurring in previously healthy children of 1 to 3 years of age, characterized by normochromic, **normocytic anemia**; reduced **reticulocyte count**; absence of erythroid precursors in the **bone marrow**; and spontaneous recovery within 4 to 8 weeks. **Thrombocytopenia** and **leukopenia** are infrequent. Some patients exhibit neurological abnormalities, e.g., fits, ataxia, hemiparesis, and papilledema, with spontaneous recovery. During recovery, erythroid precursors can be found in peripheral blood, and this phase can be preceded by bone pain.

   The cause remains unknown. However, in many cases, a history of viral illness (either respiratory or gastrointestinal) often precedes the diagnosis, so **parvovirus** infection has been suggested as a possible origin. Most children (60%) have been found to have circulating immunoglobulin G (IgG) antibodies during the acute phase that inhibit the formation of erythroid colonies *in vitro*. Similarly, inhibition of **erythropoiesis** of cellular origin has been identified in 25% of patients. Thus, an **autoimmune disorder** (triggered or not by a preceding viral infection such as parvovirus B19) directed at early erythroid precursors has been proposed as the most likely pathogenesis. Red blood cell enzyme (**pyrimidine 5′-nucleotidase**) levels have been reported to be low in TEC, but this probably reflects the age of the red cells present.

TRANSLATIONAL REGULATION

The control of the change of proteins from mRNA within a complex of **ribosomes** called the polysome. Translational control of proteins and their role in cellular physiology is increasingly being recognized as a pivotal mechanism or regulation. It affects a wide range of mRNAs, including a large number that encode transcription factors and **cell cycle** regulators. Translation is the synthesis of a protein according to the genetic information. It is the second step of gene expression, following transcription, and is a universal and essential step for life. The process is called translation because it converts the genetic information of nucleic acid sequences, composed of four distinct nucleotides, to the polypeptide sequence composed of 20 distinct amino acids. A unit of the genetic code consisting of three nucleotides (triplet), named a codon, corresponds to one amino acid, except for the final stop codon. The conversion of information is often referred to as "decoding." Being highly accurate (with an error rate of only 5×10^{-4} per amino acid

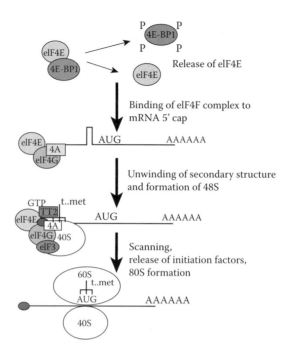

FIGURE 112

Translation initiation in eukaryotes. The 4E-BPs are hyperphosphorylated to release eIF4E so that it can interact with the 5′ cap, and the eIF4E initiation complex is assembled. The interaction of poly(A)binding protein with the initiation complex and circularization of the mRNA is not depicted in the diagram. The secondary structure of the 5′UTR is melted; the 40S ribosome subunit is bound to eIF3; and the ternary complex consisting of eIF2, GTP, and the Met-tRNA are recruited to the mRNA. The ribosome scans the mRNA in a 5′→3′ direction until an AUG start codon is found in the appropriate sequence context. The initiation factors are released, and the large ribosomal subunit is recruited. (From Meric, F. and Hunt, K.K., Translational initiation in cancer: a novel target for therapy, *Molecular Cancer Ther.*, 1, 971, 2002. With permission.)

residue) and fast (15 amino acids incorporated per second at 37°C), translation is one of the most complex biological processes; it involves a number of factors in a coordinate manner and consumes a vast proportion (about 80%) of energy that cells produce.

The translation process is divided into three steps, namely initiation, elongation, and termination. The small subunit of the ribosome is bound to mRNA with the assistance of protein factors named eukaryotic initiation factors (eIFs) of which eIF 2 and 4 are best characterized. Translation starts with the dissociation of an 80S ribosome into 40S and 60S subunits (see Figure 112). The 40S subunit attaches to met-TRNA and phosphorylated eIF2 to form a preinitiation complex. The preinitiation complex then binds at or near the m7GpppG 5′ cap structure with another initiation factor eIF4. The complex then scans the 5′ untranslated region of the mRNA for an AUG start codon and joins the 60S at this AUG start site. Translation can now begin. Scanning of the 5′UTR is therefore an important step in translation. Elements in the 5′UTR can therefore modulate translation and may be one of the sites for the posttranscriptional regulation of gene expression. Although the scanning model is the commonly utilized mechanism for eukaryotic translation, internal ribosomal entry (IRES) can be an alternative means of translation.

Apart from the elongation and initiation factors, translation can also be regulated by *trans*-acting proteins that interact with *cis*-RNA elements. Perhaps the best characterized are the mRNAs whose protein products are involved in **iron** metabolism. The mRNAs encoding **ferritin** heavy and light chains, erythroid 5-**aminolevulinate synthase** (ALA), and transferrin receptor (TfR) contain stem-loop structures that function as iron-responsive elements (IREs) and to which the regulatory protein (IRE-BP) binds when cells are depleted of iron. This results in decreased ferritin translation via the 5'UTR IRE-BP binding and increased mRNA stability of transferrin by the binding of the IRE to the IRE in the 3'UTR. Conversely, if there is an increase in cellular iron, no binding of IRE-BP occurs, and there is increased ferritin translation with concomitant decrease in transferrin stability.

TRANSPLACENTAL HEMORRHAGE
See **Hemolytic disease of the newborn.**

TRANSPLANTATION
The transfer of living tissue from one location or body to another. Autologous transplantation refers to tissue returning to the same individual after a period of storage outside the body, usually in a frozen state. Syngeneic transplantation is between identical twins. Allogeneic transplantation is between nonidentical members of the same species and, in humans, successful transplantation requires **ABO (H) blood group** and **human leukocyte antigen** (HLA-A, -B, -C, and -DR) allele matching. In hematological practice, transplantation is either autologous or allogeneic by cells derived either from the bone marrow or peripheral blood. Allogeneic transplantation carries the risk of rejection or **graft-versus-host disease.**

Solid-Organ Transplantation

Kidney: matching for HLA is beneficial. DR > B > A. Benefit is seen for both graft and patient survival. A cytotoxic cross match is usually carried out prior to transplantation. If subjects are presensitized by pregnancy, transfusion, or a prior graft, selection of a compatible donor may be difficult. Transplantation of cross-match-incompatible kidneys may result in hyperacute rejection.

Pancreas: matching for HLA-DR improves survival. The role of A and B matching is unproven.

Liver: matching for HLA-A and -B antigens improves graft survival, whereas matching for HLA-DR has minimal impact.

Heart: If <2 -B and -DR antigens are mismatched, then graft survival is better than if 2 -B and -DR antigens are mismatched. Graft survival is also impaired if the cytotoxic cross match is positive.

Stem cell/bone marrow transplantation: see **Allogeneic stem cell transplantation** and **Autologous stem cell transplantation.**

The other complications of transplantation include **bacterial, viral,** and **fungal infection disorders** and **posttransplant lymphoproliferative disorders.**

TRAUMATIC DISORDERS
The hematological effects of tissue injury by external violence. This includes major accidents with severe **hemorrhage** and tissue necrosis leading to acute **disseminated intravascular coagulation.** Minor trauma to capillaries induces the responses of **hemostasis.**

TRAUMATIC (MECHANICAL) HEMOLYTIC ANEMIA

See **Macrovascular hemolytic anemia**.

TRIOSE PHOSPHATE ISOMERASE

The most active enzyme in the anaerobic glycolytic pathway of **red blood cell** metabolism. Its metabolic role is to catalyze interconversion of the two trioses, dihydroxyacetone phosphate and glyceraldehyde 3-phosphate, both of which are formed by the action of **aldolase**. A specific enzyme assay system is available.

Deficiency has been found in patients with **hereditary nonspherocytic hemolytic anemia** associated with a severe neuromuscular disorder. This is a rare disorder inherited as an autosomal recessive. Most patients die in the first decade of life, usually from cardiopulmonary failure. Mild compensated hemolysis is usually present, but the **reticulocyte count** is slightly elevated, as is the serum **lactic dehydrogenase** level. The peripheral blood shows changes of **fragmentation hemolytic anemia**, including helmet cells and triangular cells. There may be mild **hemoglobinemia**, reduced **haptoglobin** level, and **hemosiderinuria** with **iron deficiency**.

TROPICAL SPLENOMEGALY

See **Hyperreactive malarial splenomegaly; Splenomegaly**.

TROPICAL SPRUE

See **Cobalamins; Folic acid; Intestinal disorders**.

TRYPANOSOMIASIS

The hematological effects arising from infection with protozoa of the genus *Trypanosoma*. *Trypanosoma cruzi* is the agent of American trypanosomiasis (Chaga disease), common in Latin America and southern U.S. *Trypanosoma brucei* is the agent of African trypanosomiasis (sleeping sickness), which gives rise to no specific hematological changes.

Trypanosoma cruzi resides in mammals, including domestic animals, and is transmitted by the reduviid bug as vector. Here the trypanosome has flagellae (promastigote), which are shed after infection of a mammal (amastigote). When an infected bug bites a human, usually on the face, it defecates and leaves feces infected with the trypanosome. These proliferate and cause a local pruritis. As a consequence of irritation, rubbing the wound leads to the organism entering the bloodstream. From here, the trypanosomes infect myocytes and neurones of the peripheral and central nervous system, where they proliferate and differentiate into trypanomastigotes. The cell eventually ruptures, releasing the organisms into the tissues and bloodstream. An inflammatory response leads to destruction of autonomic ganglia in the heart and gastrointestinal tract. Parasitemia results in **hepatosplenomegaly** and **lymphadenopathy**. The diagnosis is made by the identification of trypanosomes in a **peripheral-blood film** or **buffy coat** film. Enzyme immunoassay (ELISA) tests are available. Treatment is the oral administration of nifurtimox, which results in 70 to 95% parasitologic cure.

TUBERCULOSIS

The hematological changes occurring with mycobacterium **tuberculosis** infection.[101]

The bacterium stimulates **histiocytes** (macrophages) by binding a specific Toll-like receptor (TLR) that recognizes mycobacterial lipoproteins. This stimulates phagocytosis and secretion of IL-12 and nitric oxide. Mycobacterial peptides are presented by macrophages to Th1 cells, which respond by the secretion of **tumor necrosis factor** (TNF) and **interferon**-γ, resulting in a primary **granuloma**. Because the mycobacteria evade killing by macrophages, they become, with the help of T-lymphocytes, sealed off inside a **phagosome**. The macrophages become giant cells and epithelioid cells. The center of the granuloma becomes hypoxic and undergoes caseous necrosis. If the infection is not contained, there is widespread dissemination in the form of miliary tuberculosis. Reactivation of tuberculous infection is common, especially in association with **acquired immunodeficiency syndrome** (AIDS).

Disturbed bone marrow function gives rise to either **neutrophilia**, **monocytosis**, and **thrombocytosis (leukemoid reaction)** or **neutropenia**, **monocytopenia**, **lymphopenia**, and **thrombocytopenia**, possibly progressing to **aplastic anemia**, which is more frequent with miliary tuberculosis. Granulomas with caseation in the bone marrow may occur, which progress to **bone marrow necrosis** without a fibrotic reaction.

Anemia of chronic disorders.

Splenomegaly with the effects of hypersplenism.

Immune thrombocytopenic purpura in miliary tuberculosis.

Antituberculous therapy results in resolution of the hematological disorder, but it may induce a **sideroblastic anemia** due to pyridoxine antagonism.

TUFTSIN

A tetrapeptide, formed in the **spleen**, that promotes **phagocytosis** by **histiocytes** (macrophages).

TUMOR ANTIGENS

Antigens carried by tumor cells that may distinguish them from normal cells.[18] Both B- and T-**lymphocyte**-reactive antigens may be present on tumor cells, but (in human cancer) these are usually self-antigens rather than neoantigens generated by the carcinogenic process.

Early research with murine tumors demonstrated effective and specific **immunity** to tumors induced by certain viruses (e.g., SV40), by ultraviolet (UV) irradiation, and by chemical carcinogens. Immunity to virus-induced tumors is clearly due to the presence of viral antigens. Immunity to UV- and chemical carcinogen-induced tumors is more complex.

Studies of chemically induced tumors indicated that each independently induced tumor bore unique antigens, in that immunity did not cross-react (or very rarely), leading to the idea of tumor-specific antigens (TSAs: TSSAs [TS surface antigens] or TSTAs [TS transplantation antigens]). The main response to these antigens is cellular. They do not induce antibodies, and so their isolation and characterization has, in general, not been possible. The probable explanation for the nature of murine TSAs is that the mutational effects of the carcinogen treatment have produced single amino acid alterations in T-cell epitopes that are insufficient to generate antibody responses. TSAs, with the exceptions mentioned below, have not been demonstrated in the case of human tumors. In any case, the general idea has now been discredited with the realization that the murine tumors are not good models for human cancer. In particular, the strongest immune responses somewhat paradoxically occur with rapidly growing tumors induced with high doses of carcinogen. Tumors considered as better models of human cancer — "spontaneous" tumors, or tumors induced with low concentrations of carcinogen — are weakly immunogenic or nonimmunogenic.

There is now excellent evidence that, in some human cancers, what may be referred to as tumor-associated antigens (TAAs) do exist. These are antigens that are not unique to a particular individual tumor but are expressed by many tumors of the same histological type or are more widespread. Melanomas provide the best example: here, immune responses have been demonstrated to tyrosinase, expressed uniquely by melanocytes, and to products of the MAGE family of genes, expressed more widely by melanomas and other tumors (e.g., breast). In both of these cases, **cytotoxic T-lymphocyte** responses have occurred, and T-**lymphocyte** lines have been used as probes to isolate the target antigen. In the case of colorectal cancers, T-cell responses to mucin have been convincingly demonstrated. For some TAAs, antibody responses also occur (e.g., gp100 in melanoma), but the general rule does appear to be T-cell responses only.

What appears to be the case is that all TAAs demonstrated in human cancers are (a) normal cell products (i.e., nonmutant) that would normally be expressed by the target cell (e.g., tyrosinase in melanoma), (b) normal products abnormally expressed (e.g., MAGE, which appears to be normally expressed only in testis or in regenerating tissue), or (c) normal products expressed in an abnormal manner (e.g., mucin in colorectal cancer). It can therefore be argued that antitumor responses are, in effect, autoimmune.

The only situation where TSAs specific to individual tumors, equivalent to murine TSAs, can be said to exist in human tumors is in the case of idiotypic determinants on the surface of lymphoid malignancies. However, closely analogous to this may be missense mutations in oncogenes such as p53; antibodies to these (and to other mutant oncogenes) have been detected, but their role in cancer is not understood.

The driving force behind research into cancer immunology has been the possibility of development of **immunotherapy** or vaccination. Recently, antibody therapy for breast cancer using monoclonal antibody to the receptor her2/neu/erbb2 (herceptin), overexpressed in many breast carcinomas, or to CD_{20} (**rituximab**), expressed by **B-cell lymphomas**, has been shown to be successful. The production of vaccines to human papilloma virus types 16 and 18 promises to reduce the incidence of cervical cancer. Many other immune therapies and anticancer vaccines are being researched.

TUMOR-LYSIS SYNDROME

The symptoms due to the combined effects of hyperkalemia, hyperphosphatemia, hypocalcemia, and **hyperuricemia** with urate precipitation, uremia, and renal failure, caused by massive lysis of tumor cells. It usually occurs following **cytotoxic agent** therapy of large **acute lymphoblastic leukemia** or **Burkitt lymphoma/leukemia** where there is high tumor burden present. It usually responds to treatment by hydration, diuresis, and **allopurinol** therapy.[501]

TUMOR-NECROSIS FACTOR

(TNF) Receptors for growth factors and **cytokines** released by activated T-lymphocytes involved in processes concerned with **inflammation**, infarction, and **apoptosis**. There are two subtypes: -α, also known as cachectin, and -β, also known as lymphotoxin (LT). They are part of a family of 19 cytokines produced by activated T-lymphocytes (-α) or **monocytes** activated by bacterial lipopolysaccharide (-β). There are two receptors, TNF-R1 and TNF-R2, the former expressed universally and the latter mostly by endothelial cells and **leukocytes**. Depending on the receptors expressed and their cross talk, TNF binding mediates either cell activation (e.g., enhancement of **human leukocyte antigen** [HLA] expression) or cell death via the apoptosis-inducing "death domain" of TNF-R1.

The action of TNF-α includes activation of **vascular endothelium** and liver cells to produce proteins that mediate the **acute phase response**. These reactions include:

Changes in vascular permeability

Induction of matrix **metalloproteinases** and destruction of extracellular matrix

Stimulation of **angiogenesis** by activation of vascular endothelial growth factor (VEGF)

Stimulation of leukocyte migration through **cell-adhesion molecules**: intercellular-adhesion molecule-1, E-selectin, and vascular cell-adhesion molecule-1

In HIV infection, TNF-related apoptosis-inducing factor contributes to T-cell depletion by the sensitivity of uninfected $CD4^+$ cells to this factor.

A number of drugs (e.g., **infliximab**, **etanercept**) have been developed to inhibit the activity of TNF-α. They have been shown to be effective in the treatment of **autoimmune disorders**, **rheumatoid arthritis**, and Crohn's disease but carry the risk of **adverse drug reactions** affecting the central nervous system and of inducing immunosuppression, thereby increasing the risk of opportunistic infection.

TYROSINE KINASE

(TK) An enzyme that catalyzes the transfer of phosphate from **adenosine triphosphate** (ATP) to tyrosine residues in polypeptides. The human genome contains 90TK- and 43TK-like genes, the products of which regulate cellular proliferation, survival, differentiation, function, and motility. **Imatinib** is an inhibitor of *BCR-ABL* TK in **chronic myelogenous leukemia**.[502]

T-ZONE LYMPHOMA

See **Peripheral T-cell lymphoma**; **Non-Hodgkin lymphoma**.

U

UBIQUITINATION

The degradation of cellular proteins, initially by binding them covalently to ubiquitins, which target them for rapid hydrolysis by a **proteasome**. Ubiquitins are bound, and often modulated, by ubiquitin-conjugating enzymes such as ubiquitin kinases, of which Cdc34 is required for **DNA** replication and Ubc9 is needed for the M-phase of the **cell cycle**.

ULTRASONOGRAPHY

See **Venous thromboembolic disease**.

UNSTABLE HEMOGLOBINS

Structural variants of **hemoglobin** resulting from a mutation (deletion or substitution) that changes the amino acid sequence of one of the globin chains. These changes consequently weaken the forces that maintain the structure of hemoglobin, allowing the molecule to become denatured and precipitated as insoluble globulins. These precipitates often attach to the cell membrane and are recognized as **Heinz bodies**, the presence of which in blood does not become prominent until after **splenectomy**.

More than 100 unstable hemoglobins have been described; mutations in the alpha-chain gene account for more than 80%. These disorders are inherited as autosomally dominant traits; affected individuals are heterozygous for the pertinent mutations. **Hemoglobinopathies** associated with unstable hemoglobins are characterized by the presence of Heinz bodies (congenital Heinz body hemolytic anemia). Unstable hemoglobins can coexist with other hemoglobinopathies.

Hemoglobin instability occurs for a variety of reasons:

Replacement of an amino acid, which contacts the **heme** group

Replacement of nonpolar by polar residues at the interior of the molecule

Deletions or insertions of amino acids, especially when critical helical regions are involved

Replacement of intersubunit contacts, particularly those between the αβ1 and β1 chains

Anomalous introduction of proline into a helix

Amino acid substitution in areas of the molecule where the atoms are very tightly packed

Clinical Features

The physical examination findings can include jaundice, **splenomegaly**, and pallor (if **anemia** is severe). Some patients excrete dark urine, presumably due to the presence of dipyrrole pigments. In some patients, **methemoglobin** may develop and cause **cyanosis**.

Any **hemolysis** is usually compensated, particularly when the unstable hemoglobin has an increased **oxygen affinity for hemoglobin** (left-shift of oxygen dissociation curve). Infections or exposure to "oxidant" drugs, especially sulfonamides, may precipitate hemolytic episodes. Some infants may have a very unstable hemoglobin so that chronic hemolysis may become evident during the first year of life.

Laboratory Features

The mean cell hemoglobin (MCH) is usually diminished. The blood film shows **pyknocytes** in variable numbers, with some **basophilic stippling** of red blood cells. The **reticulocyte count** is often elevated out of proportion to the degree of anemia, particularly with high-oxygen-affinity hemoglobins. The isopropanol precipitation test is the most convenient means of demonstrating an unstable hemoglobin. Further investigation includes **hemoglobin electrophoresis**, although a normal pattern does not exclude an unstable hemoglobin. The oxygen affinity is often altered, and determination of the $P_{50}O_2$ is helpful. Definitive diagnosis is achieved by physical separation of the abnormal hemoglobin, globin-chain separation, and peptide analysis, or by sequencing the mutant gene. The diagnosis of unstable hemoglobin should be considered in all patients with **hereditary nonspherocytic hemolytic anemia**, especially those with **hypochromia** and disproportionate elevation of the reticulocyte count.

Treatment

Most patients have a benign course, so treatment is usually not indicated. "Oxidant" drugs must be avoided. Splenectomy is beneficial in some patients with splenomegaly and severe hemolysis, but this should be avoided in the high-oxygen-affinity variants.

UROBILINOGEN

See also **Hemoglobin** — degradation.
Urobilin and its reduced form, urobilinogen, are formed by bacterial action on bile pigments in the intestine. Excretion is increased in **hemolysis**, but this is an unreliable indicator, since its excretion is also increased in liver dysfunction. Schlesinger's zinc method provides a qualitative test.[502a] Iodine is added to urine to convert urobilinogen to urobilin. This is then mixed in a suspension of zinc acetate in ethanol and centrifuged. A green fluorescence appears in the supernatant if urobilin is present.

UROKINASE

See **Fibrinolysis**.

UROLOGICAL DISORDERS

Hematological disorders causing abnormalities in the urinary tract and male genitalia.

Priapism due to:

- Sickling of red blood cells in the corpus cavernosum in **sickle cell disorders**
- **Thrombosis** of corpus cavernosum in **chronic myelogenous leukemia** and **essential thrombocythemia**

Testicular swelling in relapsed **acute lymphoblastic leukemia**

Hypogonadism from **cytotoxic agent** therapy

Ureteric calculi from **hyperuricemia** due to severe tumor lysis of hematological malignancy

Ureteric obstruction due to retroperitoneal fibrosis associated with **lymphoproliferative disorders**

UROPOD

A cellular protrusion of **neutrophils, monocytes**, and natural killer (NK)-**lymphocytes** concerned in **cell locomotion**. The acquisition of a cell polarity is a crucial requirement for migration, activation, and **apoptosis** of **leukocytes**. The polarization of leukocytes involves the formation of two distinct poles: the leading edge: the attachment cell site to the substrate, allowing directional movements of the cell; and on the opposite side, the uropod, mostly involved in (a) the recruitment of bystander leukocytes through LFA-1/ICAM-dependent cell-cell interactions and (b) a variety of leukocyte activities, including activation and apoptosis. A complex system of signal-transduction molecules (see **Cell-signal transduction**), including **tyrosine kinases**, lipid kinases, second messengers, and members of the Rho family of small GTPases (guanosine triphosphatase), is thought to regulate the cytoskeletal rearrangements underlying leukocyte polarization and migration. However, the uropod takes shape in neutrophils, monocytes, and natural killer (NK) cells, and the formation of this cell protrusion seems to exert an important role in immune interactions. In fact, the polarization sites of leukocytes are involved in a complex cross talk between cells and extracellular matrix components, and a number of receptors and counterreceptors crowd into the contact sites to allow efficient cell-to-cell or cell-substrate interaction. The membrane/cytoskeleton interaction plays a crucial role in tuning these activities and in "predisposing" leukocytes to their function through the acquisition of a polarized phenotype.

URTICARIAL PURPURA

Reactions associated with or resulting in urticaria due to **hypersensitivity**, with local release of vasoactive products that leads to increased vascular permeability of capillaries and small venules. This in turn allows some extravasation of blood and development of **purpura**. If urticaria occurs in patients with **thrombocytopenia**, significant purpura may be seen within the urticarial lesions.

URTICARIA PIGMENTOSA

See **Mastocytosis**.

V

VARIEGATE PORPHYRIA
See also **Porphyrias**.
(Porphyria variegata; protocoproporphyria; South African genetic porphyria) The clinical disorder arising from an inherited autosomal-dominant partial deficiency of protoporphyrinogen oxidase. This disorder is widespread geographically, but in South Africa, Finland, and Taiwan, it is more common than **acute intermittent porphyria**.

Clinical Features
There is considerable variation in clinical expression, but presentation before puberty is uncommon. Affected persons may experience abdominal pain, tachycardia, vomiting, constipation, hypertension, neuropathy, back pain, confusion, bulbar paralysis, psychiatric complaints, fever, urinary frequency, or dysuria. Skin manifestations are similar to those of **porphyria cutanea tarda** and **hereditary coproporphyria**. Neurovisceral manifestations are indistinguishable from those that occur in acute intermittent porphyria and hereditary coproporphyria. Skin manifestations usually occur apart from neurovisceral manifestations. As in acute intermittent porphyria, acute attacks of variegate porphyria can be precipitated by exposure to many drugs, of which barbiturates, sulfonamides, antibiotics, anticonvulsants, and alcohol are the most common. Attacks can also be precipitated by the onset of menses, high progesterone-containing contraceptives, pregnancy, stress, or a sudden decrease in calorie intake.

Laboratory Features
See **Porphyrias**.

Treatment
(See also **Acute intermittent porphyria**).
Acute attacks require treatment with glucose. Hematin infusions and other measures employed for the management of attacks of acute intermittent porphyria are usually effective. Measures to protect the skin from sunlight are helpful. Harmful drugs must be avoided. Screening for latent cases in families of affected individuals is recommended.

VASCULAR ENDOTHELIUM
The monolayer of cells lining the inner surface of blood vessel walls that forms the extensive distribution system for blood throughout the body. The total surface area of the vascular endothelium in a 70-kg male is 1000 m². The cells are derived from an endothelial progenitor cell that in turn is derived from the hematopoietic **stem cell**. Endothelial cells are attached to and rest upon the subendothelium. The subendothelium is an extracellular matrix secreted by the endothelial cells. The innermost basal layer is composed of **collagen** IV and V separating the endothelium from connective tissue containing elastin, collagen

II, mucopolysaccharides (including heparan sulfate, dermatan sulfate, chondroitin sulfate), laminin, **fibronectin, Von Willebrand Factor,** vitronectin, **thrombospondin,** and occasionally **fibrin,** all components being synthesized by the endothelial cells. There is also an admixture of cells — pericytes, myocytes, fibroblasts, **mast cells,** and **histiocytes.** At the periphery, the vessel is surrounded by contractile smooth muscle and an outer layer of loosely packed collagen I, the intima, through which capillaries (vasa vasorum) penetrate. The main difference between arteries and veins is the content of elastic fibers and smooth muscle, both being greater in arteries. Capillaries have little of either, being mainly endothelium supported by a thin layer of connective tissue.

The endothelium and subendothelium form a selectively impermeable layer, resistant to the passive transfer of fluid and cellular elements of blood, but permeable to gases. Cells can pass through the endothelium at sites of inflammation by a process of adherence to and then migration between endothelial cells. This is aided by the common lymphatic and endothelial and vascular endothelial receptor-1 (CLEVER-1). Subendothelia can act as a physical barrier in the absence of endothelial cells. An uninterrupted vascular tree is necessary for survival. The ability of the vasculature to maintain a nonleaking system and blood fluid is therefore essential. If a vessel is disrupted and leakage occurs, the coagulation system and platelets close the defect temporarily until cellular repair of the defect takes place. If a vessel is occluded by thrombus, blood flow may be reestablished by lysing the clot or recanalizing the occluded vessel. These properties are among the main functional characteristics of the vascular endothelial cell.

Functions of Vascular Endothelium

Maintenance of blood flow and integrity of the circulation.

Control of vascular permeability, thus preventing local edema.

Maintenance of vascular tone, blood pressure, and blood flow by inducing **vasoconstriction** and vasodilatation. This is achieved by secretion of renin, endothelin, endothelial-derived relaxing factor (EDRF), adenosine, **prostacyclin,** and surface enzymes that can convert or inactivate other vasoactive peptides such as angiotensin and **bradykinin.**

Antiplatelet aggregation: endothelial cells are intrinsically nonthrombogenic. Platelets adhere to subendothelium rather than endothelial cells. This is due to endothelial production of components that inhibit platelet aggregation, such as prostacyclin, EDRF, and adenosine.

Anticoagulation: cell-surface heparan sulfate enhances the effect of **antithrombin** in forming thrombin-antithrombin complexes. Local subendothelial heparan sulfate has a similar effect. Exogenous heparin can bind to endothelial cells and promote antithrombin inactivation of thrombin. Perhaps the major anticoagulant property of endothelium is via enhancement of **protein C** pathway inactivation **factor V**a and **factor VIII**a. This inactivation is markedly enhanced by **thrombomodulin,** an endothelial cell-surface protein that binds thrombin and enhances the ability of thrombin to activate protein C. Endothelium secretes protease nexin 1. This inactivates thrombin by covalent binding to the thrombin-active site. This complex formation is enhanced by heparan sulfate.

Coagulation induction: with trauma to blood vessels, the endothelium functions as an induction component to coagulation pathways. Endothelial cells produce **tissue factor** in response to injury, bind **factors IX, X, V,** and **high-molecular-weight**

kininogen; contain **factor XIII** activity; and produce endothelin to induce vaso-constriction.

Fibrinolysis induction: endothelial cells secrete several components active in fibrin-olysis. These include plasminogen activators and plasminogen activator inhibitor. These components are bound to the endothelial cell surface to enable assembly of active complexes.

Repair of vessel wall: simple minor injuries are repaired by migration of adjacent cells and subsequent endothelial cell proliferation. More severe vessel wall injuries require migration and proliferation of smooth-muscle cells and fibroblasts. Endo-thelium secretes components that are active in the repair process by enhancing smooth-muscle migration and fibroblast function. These include a protein resem-bling platelet-derived growth factor, vascular permeability factor, and fibroblast growth factor. Endothelial cells are also responsive to platelet-derived endothelial growth factor and **transforming growth factor-β**.

Interactive properties: the endothelium interacts with **leukocytes**. This is critical in the migration of leukocytes into areas of inflammation. The endothelial cells are active in transcytosing cytokines to the luminal surface for presentation to leuko-cytes. This interaction is mediated by **cell-adhesion molecules** present on both endothelial cells and leukocytes.

Endothelial binding sites for cytokines include glycosaminoglycans (GAGs) and Duffy antigen/receptor (DARC).

Production of the vascular endothelial growth factor (VEGF), which has the following actions[502b]:

- Induces **angiogenesis**
- Has role in **hematopoiesis**
- Inhibits maturation of **dendritic reticulum cells**
- Increases bone absorption by **osteoclast** activity and osteoclast chemotaxis

Disorders of Vascular Endothelium

Structural anomalies
- Congenital **hemangiomas**
- Hereditary hemorrhagic telangiectasia
- Senile vascular lesions
- Aneurysms giving rise to **disseminated intravascular coagulation**

Trauma: **hemostasis** or **hemorrhage**

Inflammatory
- Vasculitis
- Atherosclerosis leading to **arterial thrombosis**

Neoplasia: hemangiosarcoma

Functional
- Von Willebrand Disease
- Primary pulmonary hypertension

VASCULAR PURPURA

Lesions that represent breakdown in primary **hemostasis** unassociated with primary defects in blood coagulation or platelets. An accurate history and examination of the lesions is important in establishing the cause of purpuric lesions. The family history and childhood history are particularly important. Tests of defective vascular function such as **capillary fragility tests** and capillary **bleeding time** can be performed, but they may be of little interpretive value. A classification of vascular purpuras is given in Table 152.

TABLE 152

Classification of Vascular Purpura

Hereditary Vascular Malformations

Connective tissue disorders
  Ehlers-Danlos syndrome
  Marfan's syndrome
  Osteogenesis imperfecta
  Pseudoxanthoma elasticum
Vascular disorders
  Hereditary hemorrhagic telangiectasia
  Kasabach-Merritt syndrome

Acquired Vascular Purpura

Increased transmural pressure
  Mechanical purpura
  Orthostatic purpura
Decreased mechanical integrity
  Angiodysplasia
  Senile vascular lesions: senile purpura; senile telangiectasia;
    venous lakes; senile angioma; angiodysplasia
  Amyloidosis of vessels
  Ascorbic acid deficiency: scurvy
  Corticosteroid excess
Trauma to vessels
  Factitious purpura: abuse, self abuse
Infectious purpura
  Bacterial infection disorders: meningococcal, streptococcal
  Viral hemorrhagic disorders
  Kaposi's sarcoma
Drug-induced purpura
Embolic purpura
Inflammatory/**urticarial purpura**
Thrombotic
  Disseminated intravascular coagulation; purpura fulminans
Multiple mechanisms
  Dysproteinemic purpura
  Vasculitis
  Henoch-Schönlein purpura

Unknown

Psychogenic purpura
Autoerythrocyte sensitization
Autosensitivity to DNA

VASCULITIS

(Hypersensitivity vasculitis; vasculitides) Immune-mediated **inflammation** of blood vessels as part of a **hypersensitivity** reaction.[503,504] Patients with collagen vascular diseases and systemic vasculitis may have an array of vasculitic purpuric lesions due to the cutaneous vasculitis. The origin of the **immune complex** antigen may be multiple and various, including infectious agents, drugs, and complexes in **systemic lupus erythematosus** and **rheumatoid arthritis**. Deposition of immune complexes upon the endothelium leads to activation of **platelets**, release of vasoactive peptides, and increased vascular permeability. Immune complexes may then penetrate to the subendothelium and deeper elements. Activation of complement at these sites attracts leukocytes to the vessel wall, which then release lysosomal enzymes that damage and degrade vascular wall structures (**leukocytoclastic vasculitis**). This allows extravasation of blood and the development of purpura, particularly affecting the lower legs, buttocks, forearms, and hands. Subsequent **thrombosis** and vascular occlusion may occur.

Syndromes associated with cutaneous vasculitis depend upon the size of the greatest number of vessels involved. The precise cause is unknown, but primary causes include serum sickness and serum sicknesslike illnesses (see Table 153). There is an association with **antineutrophilic cytoplasmic antibodies** (ANCA) in **Wegener's granulomatosis**, **Churg-Strauss granuloma**, and microscopic polyangiitis.[505] Secondary causes are those of **vascular purpura**, particularly drugs. Infection can affect all grades of vessel, so only common examples are given.

Prognosis is excellent in the absence of systemic involvement, when it becomes poor, particularly in the elderly. **Immunosuppression** by **corticosteroids** and **cyclophosphamide** in varying doses has been the mainstay of pharmacological therapy, but intravenous **immunoglobulin** is becoming widely used. **Plasmapheresis** has been used in some patients with advantage. Azathioprine is used for long-term maintenance therapy. **Ritaximab** is under clinical trial.

TABLE 153

Syndromes of Vasculitis

Vessel Size	Primary	Secondary
Capillaries	**Henoch-Schönlein purpura**	Drugs
	Cryoglobulinemia	Hepatitis C
	Leukocytoclastic vasculitis	
Arterioles	**Wegener's granuloma**	**Rheumatoid arthritis**
	Microscopic polyangiitis	
	Churg-Strauss syndrome	**Systemic lupus erythematosus**
	HIV infection	
	Drug induced	
	Malignancy	
Medium arteries	**Polyarteritis nodosa**	Hepatitis B
	Kawasaki disease	**Hairy cell leukemia**
Aorta and large arteries	Giant cell arteritis	Rheumatoid disorders
	Takayasu's arteritis	Syphilis
	Behçet's disease	Tuberculosis
	Cogan's syndrome	
	Isolated CNS angiitis	

VASOCONSTRICTION

Contraction of a blood vessel wall mediated by neurogenic (α2 adrenergic systems) or chemical mechanisms. If an artery is severed, neurogenic spasm leads to reflex vasoconstriction, followed by a more prolonged myogenic contraction/vessel-wall spasm.

VEGANS

Strict vegetarians who avoid all dairy products and eggs. This is found occasionally among Western vegetarians, but the original adherents were Hindu. The associated **cobalamin** deficiency presents with mild **megaloblastosis**, glossitis, and neurologic disturbances. Concomitant **iron deficiency** occurs because of the large quantity of phytic acid ingested as part of the whole-grain bread diet.

VEILED CELL

Nonphagocytic monocytoid cells derived from the pluripotential hematopoietic stem cell and residing in the T-**lymphocyte**-dependent areas of lymphoid follicles. They have a convoluted single nucleus and long "veils" of motile cytoplasm. Veiled cells resemble **Langerhans cells**, and it is thought that they transport antigen to the T-cell-dependent areas from the periphery, e.g., dermis, via the afferent lymphatic system. Having reached the **lymph node**, they become typical interdigitating **dendritic reticulum cells** and present antigen (probably in conjunction with **major histocompatibility complex** [MHC] class II antigen) to T-lymphocytes.

VENA CAVAL FILTERS

A filter placed within the vena cava to prevent emboli in **venous thromboembolic disease**, blocking the venous passage from lower-limb deep venous thrombosis (DVT). Indications for the placement of a filter in the vena cava include patients with proximal DVT who cannot be treated by anticoagulation due to bleeding risks, or those with problematic recurrent pulmonary embolism (PE) despite adequate anticoagulation.[505a] Filters are effective in reducing fatal and nonfatal PE in the short term (3 months), but over the longer term (2 years) the risk of recurrent DVT is increased. If a long-term filter is used, anticoagulants should be considered, if possible, to reduce the risk of recurrent DVT. Currently, vena caval filters are permanent, but removable/temporary filters are undergoing clinical trial.

VENEPUNCTURE

(Venesection) The procedure for obtaining samples of venous blood for laboratory analysis or testing.[483] With the present high levels of accuracy and precision of modern analyzers, it is important that there be similar levels in preanalysis procedures. Specimens can be collected into a syringe by a sterile needle or into an evacuated tube system. Analytical levels can change with stress, so it important that the donor be comfortable and at rest prior to specimen collection. For certain analyses, it is necessary for the donor to be fasting or to have avoided certain foods or drugs, e.g., aspirin or fats before platelet-function testing. When taking specimens from persons at high risk of carrying an infection, the operator should wear protective gloves.

   The appropriate size and type of syringe or evacuated container must be selected, and the veins of the donor inspected prior to venepuncture. The first choice of vein is one in the antecubital fossa of the arm, but those of the wrist, dorsum of the hand, or of the ankle may be equally suitable. Areas of scarring or where an intravenous needle is in place must be avoided. Compression of the limb above the proposed site of venepuncture will help distend the selected vein, but this maneuver must not be used when collecting blood for calcium or lactate assays. If a tourniquet is to be used for compression, it must not be applied for more than 1 min prior to venepuncture, as hemoconcentration will occur after longer periods. The chosen site should be cleansed with 70% v/v isopropylalcohol or a similar skin antiseptic. When the specimen is for culture of microorganisms, chlorhexidine in isopropylalcohol should be used for cleansing. The site should be allowed to dry to avoid hemolysis of the specimen. With compression applied and the vein distended, the needle is inserted into the vein, compression released, and the required volume of blood withdrawn. The needle is then removed and pressure applied via sterile wool or gauze to the venepuncture site and the limb raised. Finally, when skin allergy is suspected, the site of puncture should be covered by an adhesive dressing or gauze bandage.

The specimen from the syringe should be transferred to the appropriate container, filled to the mark if necessary, the container stoppered, and, when anticoagulation is required, inverted several times. All containers must then be correctly labeled with donor identification and, where appropriate, "high risk" labels applied. Upon completion, needles must be deposited in a designated puncture-proof container and syringes disposed of according to local safety arrangements. The phlebotomist must then wash the hands and meet all other local hygiene requirements. Specimens should be transported, preferably in an upright position, at the appropriate temperature, fulfill the local health and safety requirements, and be dispatched to the laboratory without delay, accompanied by a completed request form.

VENESECTION

See **Venepuncture**.

VENOGRAPHY

See **Venous thromboembolic disease**.

VENOUS CEREBRAL THROMBOSIS

A rare form of venous thrombosis affecting the cerebral veins and intracranial venous sinuses. Apart from being a complication of head injury, it mainly occurs in young females associated with either oral contraceptives or obstetric delivery. There is also a genetic risk factor. Treatment with **heparin** is questionable, but recovery without any neurological disturbance occurs in 80% of cases.

VENOUS LAKES

See **Senile vascular lesions**.

VENOUS THROMBOEMBOLIC DISEASE

The pathogenesis, cause, complications, and treatment of **thrombosis** occurring in the venous system.[506] Thrombi usually arise in the deep veins of the lower limbs (popliteal, femoral, iliac), from which embolization occurs to the lungs — pulmonary emboli (PE). The thrombi are in areas of local stasis, being mainly composed of red blood cells enmeshed in fibrin. The pathogenic stimulus is probably provided by increased platelet aggregation secondary to thrombin formation and by increased availability of phospholipid for **coagulation factor** activation.

Risk Factors

These are well defined. Thrombosis often results from several of these factors working together. Important risk factors include increasing age, surgery, trauma, immobilization, paresis, and malignancy. Most hospitalized patients have one or more of these risk factors for venous thromboembolism (VTE), and so deep-vein thrombosis (DVT) is common in hospitalized patients. The risk of developing DVT after hip-replacement surgery has been estimated to be as high as 50% of patients when thromboprophylaxis is not used. The risk of developing DVT in certain patients immobilized with a medical illness is similarly high, with approximately 40 to 50% of patients admitted with stroke or myocardial infarction developing detectable venous thrombosis. There is overlap of both hereditary and acquired risk factors with those of **arterial thrombosis**, but the overall incidence is greater for venous thrombosis, and there are particular risk factors that apply to venous thromboembolic disease. These include:

Hypercoagulable state (thrombophilia)
- Deficiency of **antithrombin III, protein C**, or **protein S**
- Presence of **antiphospholipid antibody**
- Increased levels of **fibrinogen** and/or **factors VII, VIII, IX**, and **X**
- Increased activated protein C resistance (APC resistance) in those with factor V Leiden or taking the third-generation oral contraceptive pill
- Presence of **prothrombin gene mutation**
- Disturbed **plasminogen** activation

Stasis, usually due to immobilization; sometimes associated with increased **viscosity** of whole blood from **erythrocytosis** or **essential thrombocythemia**, or plasma viscosity from raised levels of fibrinogen

Estrogens have been implicated on epidemiological evidence when used as oral contraceptives and for hormonal replacement therapy in menopause

Infection directly induced by central-line catheters and other implements for diagnostic or therapeutic infusion; the most frequent cause in children where renal thrombosis occurs in those under 1 year of age

Malignancy: due to shared common risk factors, commonly with tumors of the stomach and pancreas

Patients can be stratified by risk as follows:

Low risk
- Minor surgery (<30 min) + no risk factors other than age
- Major surgery (>30 min), age <40 years + no other risk factors
- Minor trauma or medical illness

Moderate risk
- Major general, urological, gynecological, cardiothoracic, vascular, or neurological surgery + age >40 years or other risk factor
- Major medical illness or malignancy
- Major trauma or burn
- Minor surgery, trauma or illness in patients with previous DVT, PE, or thrombophilia

High risk
- Fracture or major orthopedic surgery of pelvis, hip, or lower limb
- Major pelvic or abdominal surgery for cancer

- Major surgery, trauma, or illness in patient with previous DVT, PE, or thrombophilia
- Major lower limb amputation

Clinical Features

Deep-vein thrombosis (DVT) of the lower limbs and its complications are a serious and potentially fatal disorder. Although it often complicates the course of other diseases, it is not uncommon in otherwise healthy individuals. DVTs usually originate in the deep veins of the calf due to sluggish blood flow in these areas. More rarely, thromboses may originate in the ileofemoral and popliteal veins, especially following trauma or surgery. Signs of a DVT include swelling of the limbs (with or without edema), warmth, tenderness, dilatation of the superficial veins, and cyanosis or pallor of the limbs. Varicose veins, ulceration, or superficial thrombophlebitis may also be present. Thromboses may also occur at other sites, e.g., axillary and subclavian veins, cerebral venous sinuses, and mesenteric veins.

Diagnostic Procedures[507]

Venography

Ascending venography outlines the deep venous system of the lower limbs, including the external and common iliac veins. It is a difficult technique that requires experience to perform and to interpret correctly. Inadequate venograms may occur in 4 to 12% of examinations. Adverse reactions are primarily hypersensitivity reactions to the contrast media, but damage to the vascular endothelium leading to superficial thrombophlebitis or even deep venous thromboses has been reported and may occur in 1 to 2% of cases.

Noninvasive Procedures

These are favored for the diagnosis of venous thromboses because of their relative ease and lack of adverse reactions:

Impedance plethysmography (IPG): a noninvasive method that detects volume changes in the leg. Impedance plethysmography is the most widely performed and relies upon changes in blood volume in the calf produced by inflation and deflation of a pneumatic thigh cuff that results in changes in electrical resistance (impedance). These changes are decreased in patients with obstructed venous drainage, e.g., thromboses of the popliteal or more proximal veins. IPG detects approximately 92% of proximal thrombi (popliteal/femoral/iliac veins), but only 20% of distal thromboses (calf). A negative IPG does not, therefore, exclude a calf thrombus or a nonocclusive proximal vein thrombosis. Recent studies have cast doubt on the reliability of IPGs for the detection of venous thromboses; these studies are based on cases where venography has clearly demonstrated a proximal vein thrombosis that was missed by IPG. There is also a significant recurrent thrombosis rate in patients with negative IPGs.

Real-time B-mode ultrasonography with compression: an accurate and sensitive technique that detects approximately 97% of proximal venous thromboses. The frequency of inconclusive examinations is between 0 and 5%. As with IPG, ultrasonic (U/S) examination should be repeated over 7 days to detect extending thrombus. Compression U/S is more accurate than IPG for the detection of proximal vein thromboses in asymptomatic patients.

Doppler ultrasonography using a Doppler U/S flow-velocity detector: a beam of U/S, which is normally reflected by the erythrocytes, is directed percutaneously at the underlying vein. If blood flow is stationary, then no sound is heard because the incident and reflected beams are at the same frequency. However, if the blood is moving, then the beam is reflected at a different frequency, thus showing a Doppler shift. Doppler U/S is very sensitive to the presence of proximal thrombi, but the interpretation is very subjective and the test requires considerable skill and experience. However, the test is relatively inexpensive and rapid to perform.

^{125}I-*fibrinogen leg scans*: these are probably of most benefit in patients with clinically suspected recurrent venous thromboses and where venography or noninvasive tests have been nondiagnostic.

Thermography: a simple noninvasive technique that utilizes an infrared camera to detect changes in temperature between limbs. It cannot detect thromboses in the pelvic veins or the inferior vena cava and has a significant false-positive rate.

D-dimer assay: a reliable procedure, provided that the technique is subjected to careful quality assurance.[166] A positive result, however, does not confirm the presence of VTE, but a negative test can be used as strong evidence that clotting has not occurred. When the D-dimer result is used in conjunction with pretest probability scoring systems for DVT and PE, a negative (i.e., below some preordained cutoff value) D-dimer test can reduce the number of imaging investigations required for the diagnosis of VTE. However, if there is a high clinical suspicion of VTE, diagnostic tests should proceed in spite of a normal D dimer. It should be noted that a D-dimer assay is of no use in those with a high clinical probability of VTE. In elderly or inpatients, D dimer retains a high negative predictive value, but it is normal in less than 20% of patients and, hence, not very useful. Clinicians must be aware of the characteristics of the test used in their hospital.

Complications

DVT can cause death from pulmonary embolism. It can also result in recurrence or extension of a preexisting thrombus or the development of venous insufficiency leading to venous ulceration. Emboli that originate in the proximal veins (popliteal, ileal, and femoral) are usually larger than emboli originating in more distal veins and, as such, carry a much greater risk of pulmonary embolism.

Prevention[508]

Thromboprophylaxis is highly effective clinically, and it is cost effective. PE is the most common preventable cause of hospital death. Clinically VTE is often silent, but symptomatic DVT and PE are both common in patients who do not receive prophylaxis. The in-hospital case-fatality rate of VTE is estimated at 12%. Approximately 10% of hospital deaths may be due to PE. Investigating and treating symptomatic patients is costly. Patients are at increased risk of future VTE and may develop postthrombotic syndrome. Postthrombotic syndrome is chronic pain, swelling, and occasional ulceration of the skin of the leg, and it occurs in approximately 30% of patients who have a DVT, although only a quarter of these are severely affected. The U.K. Agency for Healthcare Research and Quality has made a systematic review that ranked 79 patient-safety interventions based on the strength of the evidence supporting more widespread implementation of these procedures.[509] The highest ranked safety practice was the "appropriate use of prophylaxis to prevent VTE in patients at risk."

A number of therapeutic options exist for the prevention of venous thromboembolism, both physical and pharmacological. Patients at low risk are best treated with elastic stockings rather than antithrombotic drugs. Those at high risk, particularly as a consequence of surgery, are most effectively treated with drugs.[510–512]

Physical Methods

These include intermittent pneumatic compression and graded-compression elastic stockings or the insertion of a **vena caval filter**.

Pharmacological Methods

These include **heparin** (unfractionated or low-molecular-weight heparin [LMWH]), the vitamin K antagonists, e.g., **warfarin**, and various platelet inhibitors. Multiple clinical trials and metanalysis studies have evaluated pharmacological methods of thromboprophylaxis utilizing low-dose unfractionated heparin or LMWH-based regimes. All have shown reduced frequency of asymptomatic DVT, symptomatic disease, and fatal PE for general surgical patients. In addition, medical patients will also benefit. Risk reduction is on the order of 60 to 70%. Comprehensive guidance is available on use of thromboprophylaxis.

Treatment

Venous thromboembolic disease, when confirmed, requires rapid anticoagulation with heparin for typically 3 to 5 days, followed by a period of anticoagulation with an orally active anticoagulant such as warfarin, the INR (international normalized ratio) being maintained between 2.0 and 3.0. The duration of oral anticoagulation must be tailored to the individual patient, taking into account any risk factors, the severity of the thrombosis and its complications, and whether this was the first event or one of many.[511] It should rarely be prolonged beyond 6 to 12 months. For those with recurrent venous thrombosis long term, low-intensity warfarin therapy (INR 1.5 to 2.0) is effective.

Use of anticoagulants in the context of surgery leads to concern about excess bleeding risk. Despite the perceived risk of bleeding, abundant data from metanalyses and placebo-controlled, blinded, randomized clinical trials have demonstrated little or no increase in the rates of clinically important bleeding with low-dose heparin regimes.

Fixed-dose subcutaneous LMWH may be used for treating uncomplicated DVTs. Treatment with thrombolytic agents is not generally recommended for the initial treatment of venous thromboses, but it may be indicated for patients with specific clinical situations, such as heparin sensitivity, threatened limb survival, and severe pulmonary embolism. Wearing of compression stockings within 2 to 3 weeks of diagnosis for at least 2 years may reduce the incidence of the postthrombotic syndrome: leg edema and ulceration.

Newer oral anticoagulants are under trial, but all have adverse drug reactions. These include **Fondaparinux** 500 and **Ximelagatran**.

Where anticoagulation is contraindicated, **vena caval filters** are of use.

VESSEL WALL

See **Vascular endothelium**.

VILLOUS LYMPHOCYTE

Characteristic **lymphocytes** with polar villi seen in the peripheral blood of patients with **splenic marginal-zone lymphoma**.

VINBLASTINE
See **Vinca alkaloids**.

VINCA ALKALOIDS

Cell-phase-specific **cytotoxic agents** — vincristine and vinblastine — extracted from the periwinkle plant. They have their major antitumor effect by binding to cellular microtubular protein, thus causing mitotic arrest.

Vincristine (Oncovin)

Used intravenously (IV) for the treatment of **acute lymphoblastic leukemia** (ALL) and in cytotoxic regimes such as CHOP for **non-Hodgkin lymphoma**. It is metabolized in the liver and excreted via the kidneys. Its most important side effect is neurotoxicity, most patients losing tendon reflexes and many suffering troublesome peripheral sensory loss. Motor weakness is less common, and neurotoxic side effects usually resolve slowly when the drug is discontinued. Autonomic neuropathy causes constipation and abdominal pain.

Vinblastine

A drug often used as a substitute for vincristine when vincristine neurotoxicity is a limiting factor in disease treatment. Its main side effect is **bone marrow hypoplasia**, but as with vincristine, it is also neurotoxic with similar, but less troublesome, side effects. Other toxic effects include gastrointestinal tract upsets, alopecia, and tiredness.

VINCRISTINE
See **Vinca alkaloids**.

VIRAL HEMORRHAGIC FEVERS

Severe pyrexial illnesses with **hemorrhagic disorders**. They are caused by **viral infection disorders** with arenavirus (lassa fever) and filoviruses (Ebola virus, Marburg disease, West Nile virus). The initial outbreak was in the Crimea in 1944 and in the Congo in 1956. The occurrence is now particularly associated with the Congo and Nile valleys. The illness presents with fever, headache, vomiting, diarrhea, and **lymphadenopathy**. Hemorrhage occurs from the nasal and oral mucosa and from the gastrointestinal, tract with purpura and bruising to follow. Death occurs within 6 days. There is no known treatment.

It has been postulated[513] that similar viruses were the cause of the hemorrhagic plagues of the 14th through 17th centuries. A recurrence of hemorrhagic plague therefore remains a possibility.

VIRAL INFECTION DISORDERS

See also **Cytomegalovirus; Epstein-Barr virus; Human immunodeficiency virus; Human lymphotropic virus; Parvovirus**.

The consequences of virus infection on hematological tissue and substances. These can be general or specific to the virus group. Many are blood borne and are responsible for **transfusion-transmitted infection**.

General Effects

Most cells infected by virus are killed by **cytotoxic T-lymphocytes** upon the recognition of viral peptides bound to **human leukocyte antigens**/major histocompatibility complex (HLA/MHC). The consequent reactions give rise to a variety of effects.

Neutropenia of mild degree within the first 48 h of infection and lasting for 4 to 5 days. **Döhle bodies** occur in the neutrophil cytoplasm. Myelosuppression occurs in **HIV** and influenzal virus infections. **Neutrophilia** occurs when there is significant tissue damage. These changes are a consequence of fluctuation in the release of neutrophils from the storage pool.

Anemia of chronic disorders.

Hemolytic anemia in those with **glucose 6-phosphate dehydrogenase** deficiency of red blood cells, particularly with infection by viral hepatitis, **cytomegalovirus** (CMV), Coxsackie, influenza A, herpes simplex, rubella, and varicella.

Paroxysmal cold hemoglobinuria with measles and other nonspecific viral infections.

Lymphocytosis with infection by rubella and CMV viruses.

Impaired **platelet** function due to viral binding of platelet receptors occurs with infection by measles, CMV, Epstein-Barr virus (EBV), **dengue**, influenza, and rubella. This results in **endocytosis**, aggregation release reaction, and sometimes platelet lysis.

Immune thrombocytopenic purpura in rubella, morbilli, varicella, EBV, HIV, and many nonspecific viral infections.

Hemophagocytosis: this is most frequently associated with CMV, EBV, human herpes virus 6, HIV, and parvovirus. **Pancytopenia** is due to hemophagocytosis in the bone marrow and lymph nodes associated with a histiocytic hyperplasia. There are also the changes associated with **liver disorders**.

Disseminated intravascular coagulation (DIC): this occurs with chicken pox and measles, particularly in the immunosuppressed. A severe form occurs in viral hemorrhagic fevers, where it is triggered by complement-mediated activation of the coagulation cascade.

There are suggestions that viruses of the herpes group may have causal associations with **lymphoproliferative disorders**.

Specific Virus Effects

Retroviruses

Human immunodeficiency viruses (HIV): **T-lymphocyte** helper cells are receptors for this virus, causing the host to develop immunodeficiency with resulting opportunistic infections, particularly by CMV, herpes simplex, and herpes zoster — **acquired immunodeficiency syndrome**. Some patients progress to develop **non-Hodgkin lymphoma** (usually **B-cell lymphomas**).

Human T-cell leukemia lymphoma virus (HTLV-1) is associated with **adult T-cell leukemia/lymphoma**. The infection is mostly found in Southeast Asia and the Caribbean countries.

Herpes Viruses

These inhibit MHC expression, preventing recognition by cytotoxic T-cells, immunity being by the **natural killer** (NK) lymphocytes.

Epstein-Barr virus

(EBV) These viruses secrete a protein homologous to Bcl-2 molecule, which prevents the cell from undergoing **apoptosis**. B-cells are invaded with consequent activation and transformation, the cells expressing BLAST-1 and BLAST-2 antigens. T-cells become transformed to **atypical lymphocytes**, seen in **infectious mononucleosis**. Another response is the formation of **cold agglutinins** with antibodies to **I blood group** antigens. Virus-associated hemophagocytic syndrome (VAHS) associated with fatal infectious mononucleosis is probably related. The virus has been isolated from cell lines derived from **Burkitt lymphoma** and is associated with **Hodgkin lymphoma**.

Cytomegalovirus

Cytomegalovirus infects epithelial cells and monocytes, producing enlarged cells with nuclear and perinuclear inclusions (owl-eye cells). A viral protein is formed that mimics MHC class I antigens and stimulates expression of HLA-E, effectively inhibiting NK-cells. The outcome is an infectious mononucleosislike illness with **pancytopenia** in those immunosuppressed. In infants, there is massive **splenomegaly**.

Human Herpes Virus-8

This is the **Kaposi's sarcoma** virus (KSHV) that induces endothelial growth by increasing heme oxygenase (HO-1) activity, leading to **angiogenesis**. Hypoxia is a fertile environment for this infection. It may also be the causal agent for multicentric **angiofollicular lymph node hyperplasia** (Castleman), **primary effusion lymphoma**, **osteoclastic myeloma**, and pulmonary hypertension.

Parvovirus B19

Viremia here results in the virus entering erythroid precursors and causing transient **pure red cell aplasia**, particularly when associated with marrow hyperplasia of chronic hemolysis — **sickle cell disorder**, **hereditary spherocytosis**, **pyruvate kinase** deficiency. A similar disturbance is seen with infection by dengue.

Measles Virus

These viruses invades B and T-lymphocytes and monocytes, leading to impaired function and the secretion of **interferon**. Persistence of the infection within the cells may be of importance in the pathogenesis of subacute sclerosing panencephalitis. It is associated with paroxysmal cold hemoglobinuria.

Respiratory Syncytial Virus

These invade mononuclear cells, particularly T-lymphocytes, resulting in a lower T4/T8 lymphocyte ratio.

Hepatitis Group Viruses

Hepatitis B Virus

These replicate in hepatocytes but are not cytopathic. Antibodies to the hepatitis B surface protein (HbsAg) can protect from infection of hepatocytes. In its absence, there is a Th1-type response, and hepatitis-B-specific CD4⁺ and CD8⁺ T-cells migrate to the liver. These cells secrete interferon-γ, which inhibits viral replication but gives rise to inflammation and chronic active hepatitis, leading to cirrhosis and hepatoma.

Hepatitis C Virus

This is a single-stranded RNA flavivirus with six genotypes distributed variously through-out the world. It is the cause of aggressive hepatitis in **hemophilia**. It also causes **amega-karyocytic thrombocytopenia**, which usually progresses to **aplastic anemia**. It has been associated with **splenic marginal-zone lymphoma**. The infection may respond to treat-ment with interferon.

Flaviviruses

These replicate in the macrophages of lymph nodes, causing generalized **lymphadenop-athy** and the production an IgM followed by an IgG, the syndrome of dengue.

Arenaviruses

These infect **histiocytes** (macrophages) and possibly cause the release of mediators of cellular and vascular damage, giving rise to **viral hemorrhagic fever** — lassa fever.[513]

Filoviruses

These suppress the **innate immune** response by blocking the action of α-interferon. Infected monocytes express tissue factor and initiate disseminated intravascular coagula-tion. The viruses (Ebola fever, Marburg disease) also damage vascular endothelium.[514]

Adenoviruses

An emerging pathogen as a cause of **immunodeficiency**. Prevention and treatment are particularly important for patients who have received an allogeneic or autologous stem cell transplantation.[515]

VIRUS-ASSOCIATED HEMANGIOPHAGOCYTIC SYNDROME

(VAHS) See **Histiocytosis** — hemophagocytic histiocytosis.

VISCOSITY OF WHOLE BLOOD AND PLASMA

The viscosity of blood as the cause of its intrinsic resistance to flow, which arises from the internal friction between circulating cells and plasma molecules as they flow through vessels.[467] Viscosity is calculated as the ratio of the pressure gradient to flow rate multiplied by a pressure constant (K1), which is determined from the pressure-flow relationships of standard calibration fluids whose viscosity is known. Under steady flow conditions, fluids undergo shearing (concentric cylindrical layers of fluid sliding telescopically over each other). The outermost layer is stationary, while the innermost (axial) layers travel at peak velocity. The coefficient of dynamic viscosity of the fluid (n) is the ratio of shear stress to shear rate measured in millipascal seconds (mPa·sec). The shear stress (t) is the force per unit area applied to a fluid layer, which produces movement relative to an adjacent layer. The shear rate (g) is the velocity gradient between two adjacent fluid layers.

Whole-blood viscosity is the resistance to shearing motion of blood flow, mainly due to circulating red blood cells. Measurement gives an overall estimate of the rheological properties of blood. It should be measured at a standard temperature (37 or 25°C) and at both high ($\approx$2000 sec^{-1}) and low ($\approx$1 sec^{-1}) shear rates. The main determinants are the **packed-cell volume** (PCV), with which there is a logarithmic relationship, plasma viscos-ity, **aggregate formation of red blood cells**, and **deformability of red blood cells**.

TABLE 154

Ranges for Normal and Abnormal Levels of Viscosity

	Absolute Viscosity		Relative Viscosity
	37°C	25°C	
Distilled water	0.69	0.89	1.00
Plasma viscosity			
Reference range	1.16–1.35	1.50–1.72	1.67–1.94
Population range	1.14–1.50	1.46–1.94	1.65–2.17
Acute reactions	1.35–1.95	1.72–2.51	1.95–2.83
Chronic reactions	1.35–1.55	1.72–2.00	1.95–2.25
?Paraproteinemia	>2.0	>2.5	>2.9
Serum viscosity			
Reference range	1.09–1.23	1.40–1.60	1.57–1.80
Population range	1.08–1.36	1.39–1.60	1.57–1.97
?Paraproteinemia	>1.85	>2.4	>2.7

Note: The table shows reference and population ranges (mean ± 2 standard deviations) for plasma and serum viscosity at 37 and 25°C, and relative to water. The usual zones for plasma viscosity in acute- and chronic-phase protein reactions are also indicated, as are cutoff points above which paraproteins are very likely in plasma and serum.

Source: Lowe, G.D.O., personal communication.

Plasma viscosity is the resistance to shearing motion of plasma, mainly due to proteins of large molecular size, especially those with proximal axial asymmetry — **fibrinogen** and some **immunoglobulins**. Values do not vary with the rate of shear, except when paraproteins and **cryoglobulins** form shear-dependent reversible aggregates.

The instruments used for measurement are rotational viscometers set at a constant shear rate or a constant shear stress. With capillary viscometers, the value is determined from the relationship between pressure and flow rate. With rolling-ball viscometers, a metal ball in a cylindrical tube filled with test plasma is rolled downward, and the motion of the ball is timed over a fixed distance. The temperature must be kept constant at a quoted level. The results are related to the viscosity of distilled water and are therefore relative values. Test results are used to assess the **acute-phase response**. The normal average is 1.24 mPa·sec at 37°C and 1.60 mPa·sec at 25°C. There is no difference between levels in males and females. The ranges for normal and abnormal levels are given in Table 154.

VITAMIN A DEFICIENCY ANEMIA

A chronic deficiency **anemia** morphologically indistinguishable from that caused by **iron deficiency**. Mean cell volume (MCV) and mean cell hemoglobin (MCH) (see **Red blood cell indices**) are reduced, and serum **iron** levels may be low; however, liver and marrow iron stores are increased. The administration of iron therapy does not improve the anemia. Nutritional surveys in developing countries demonstrate a strong relationship between serum levels of vitamin A and blood hemoglobin concentration. Vitamin A deficiency anemia may be as common as iron deficiency in developing countries.

VITAMIN B GROUP DEFICIENCIES

Hematological disorders arising from deficiencies of the vitamin B group.

Vitamin B_6: deficiency is caused by isoniazide, the antituberculosis agent, disturbing vitamin B_6 metabolism. A **microcytic anemia** can develop that can be corrected by large doses of pyridoxine.

Riboflavin: deficiency is nutritional and results in a decrease in red blood cell **glutathione** reductase activity. This is not associated with hemolytic anemia or oxidant-induced injury. Human volunteers deprived of riboflavin experimentally and treated with a riboflavin antagonist developed **pure red cell aplasia**.

Pantothenic acid: when artificially induced in humans, deficiency is not associated with anemia, although it is in swine.

Niacin: deficiency in dogs results in the appearance of a macrocytic anemia. The anemia that occurs in human pellagra cannot be ascribed to niacin deficiency.

VITAMIN B₁₂

See **Cobalamins**.

VITAMIN C

See **Ascorbic acid**.

VITAMIN D

(Calciferol) Two secosteroids, vitamin D_2 (ergocalciferol) and vitamin D_3 (cholecalciferol), produced by photolysis in the skin from naturally occurring sterol precursors. Their main physiological function is regulation of bone mineral homeostasis. The liver and kidney metabolize both forms of vitamin D to become active 1,25(OH)2D3, which can be recognized by tissue receptors, i.e., vitamin D receptors (VDRs). In addition to bone and other tissues, these VDRs are found on the cells of hematopoietic and lymphoid systems. The 1,25(OH)2D3 regulates **cytokine** and **interleukin** (IL)-2 release from **lymphocytes** and **tumor-necrosis factor** (TNF) from **monocytes**. It also enhances differentiation of hematopoietic tissue, osteoblasts, and osteoclasts.

VITAMIN E

(α-tocopherol) A fat-soluble vitamin that is a mixture of tocopherols occurring in vegetables (particularly asparagus), seed oils (soybean, sunflower, nuts), and whole grain. Its action as an antioxidant has been claimed to affect:

Oxidative modification of low-density lipoproteins with modulation of **platelet** function

Physiological nervous function

Prevention of some genetic disorders

Prevention of red blood cell **hemolysis** due to an abnormality in **glutathione** synthesis or to deficiency of **glucose 6-phosphate dehydrogenase** in neonates and those with fat malabsorption

Reduction of irreversible sickling in **sickle cell disorders**

Immune response reactions

It has been used pharmacologically for the reduction of hemolysis in patients with **hereditary nonspherocytic hemolytic anemia**, where it was thought to bolster antioxidative activity, and for the prevention of acanthocytes in patients with **abetalipoproteinemia**. Deficiency can cause hemolysis, particularly in neonates when iron-supplemented milk is used for feeding.[516,517]

VITAMIN K

An essential fat-soluble vitamin that serves as a cofactor for a carboxylase enzyme that is required for the c-carboxylation of the glutamic acid residues in coagulation **factors** II, **VII, IX, X, proteins C**, and **protein S**. This carboxylation allows these proteins to bind Ca^{2+} and subsequently to negatively charged phospholipid membranes.

Three forms of vitamin K exist:

Vitamin K_1: phytonadione, a major form of vitamin K found in green plants and vegetables

Vitamin K_2: menaquinone, vitamin K synthesized by the bacterial flora of the gut

Vitamin K_3: menadione, a synthetic water-soluble form of vitamin K

The primary source of vitamin K (vitamin K_1) is dietary, in leafy green plants and vegetables, but a small amount of vitamin K is also synthesized by the gut flora (vitamin K_2). Vitamin K in the diet is absorbed in the ileum and requires bile salts for its absorption. In the absence of an adequate dietary intake, the stores of vitamin K can be depleted within 1 week, but vitamin K synthesized by the gut flora can partially compensate, and thus dietary deficiencies of vitamin K rarely lead to any clinical sequelae. The requirements for vitamin K are approximately 100 to 200 µg/day. Vitamin K_1 obtained from food sources is reduced to vitamin KH_2 by a vitamin reductase. Vitamin KH_2 is then oxidized to vitamin K epoxide (vitamin KO) in a reaction that is coupled to carboxylation of glutamic acid residues on coagulation factors.

Causes of Deficiency

Hemorrhagic disease of the newborn

Dietary deficiency (rare) in postsurgical patients on a restricted fat-free diet, especially when receiving bowel-sterilizing oral antibiotics or complicated by infection

Intestinal tract disorders of malabsorption due to **nontropical sprue**

Chronic biliary obstruction

Therapy with **coumarin** drugs

Deficiency from any cause leads to disturbed synthesis of clotting factors (II, VII, IX, X, and proteins C and S) that are not c-carboxylated and, therefore, are incapable of binding Ca^{2+}. Such dysfunctional proteins are termed **PIVKAS** (proteins induced by the presence of a vitamin K antagonist).

Deficiencies are readily treated with a water-soluble preparation, menadiol sodium phosphate. In severe cases, rapid correction with fresh-frozen plasma or factor VII and IX concentrates may be required. Allergic reactions to vitamin K given intravenously have been reported. In neonates, where the gut is sterile and thus where there is no synthesis by *Escherichia coli*, intramuscular injection of phytomenadione is required.

All newborn infants have some degree of vitamin K deficiency, as shown by their low plasma concentration and low levels of factors II, VII, IX, and X (see **Neonatal hematology**). To counteract this and thus prevent hemorrhagic disease of the newborn, 1.0 mg vitamin K can be given by intramuscular injection as a prophylactic measure.[518]

VOGT-KOYANAYGI-HARADA'S DISEASE

A systemic inflammatory disease due to **cell-mediated immunity** affecting melanocytes; commonly presents with an accompanying uveitis.

VON WILLEBRAND DISEASE

(VWD) An autosomally dominant or autosomally recessive hereditary deficiency of **Von Willebrand Factor** (VWF) due to either a quantitative or qualitative deficiency of VWF. It is the commonest of the inherited bleeding disorders. Up to 1% of a population may have VWD, as defined by reduced levels of Von Willebrand Factor, but only about 125 per million have a clinically significant bleeding disorder. Quantitative defects of VWF are thought to account for approximately 75 to 80% of all patients with VWD. However, with improved understanding of molecular pathology of VWD, this figure is changing, and more cases of qualitative VWD are being recognized. Qualitative defects of VWF leading to VWD are characterized by the presence of an abnormal and dysfunctional protein. VWD is subclassified on the basis of the type of defect into type 1, 2, and 3 VWD (see Table 155). Type 2 VWD is further subclassified, dependent upon the type of functional defect present (see below). Subclassification is important in determining therapy.

TABLE 155

Classification of Von Willebrand Disease

Subclassification	Type of Defect	VWF Protein	Proportion of Cases (%)
1	Quantitative partial deficiency	Normal	75
2	Qualitative functional deficiency	Abnormal	20
3	Quantitative complete deficiency	Absent	5

Type 1 and type 2 VWD are traditionally described as an autosomally dominant disorder, with type 2N and type 3 VWD being compound heterozygous or homozygous states. The genetics of VWD are paradoxically becoming more complex with increasing knowledge about the range of mutations that appear to be associated with VWD. Genes other than the VWF locus may also have a major influence on VWF levels in putative VWD, the most important being the blood group. In addition to reduced levels of VWF, reduced levels of **factor VIII** are also seen in VWD. This arises because VWF is the physiological carrier protein for factor VIII. In the absence of VWF, factor VIII is prematurely degraded.

Clinical Features

VWD usually presents as a mild to moderate bleeding disorder with easy bruising, epistaxis, menorrhagia, and prolonged bleeding after surgery, dental extractions, and trauma. The lower the VWF levels, the more profound is the constellation of symptoms. A characteristic of bleeding in VWD is the prominence of mucosal "platelet type" bleeding, due to impaired VWF-platelet-endothelium interactions in small vessels with high shear stresses. Type 3 VWD presents a more severe picture and, in association with very low levels of factor VIII activity, may be associated with hemarthrosis and muscle bleeds, as seen in **hemophilia A**. Mild VWD may not present until well into adulthood, and then only after significant hemostatic challenge.

Laboratory Features

The diagnosis is classically suggested by a prolonged **bleeding time** and isolated prolongation of the **activated partial thromboplastin time** (APTT), and it is confirmed by finding low levels of VWF antigen and activity with reduced levels of factor VIII upon specific assay. Diagnosis in mildly affected patients may be more difficult. Mildly affected patients may have a normal bleeding time and APTT. The diagnosis can, therefore, only be excluded by finding normal levels of VWF antigen and activity upon specific assay. In addition, VWF levels may vary in patients in response to stress; thus in suspected mild VWD it is important to repeat testing on different occasions to confirm results.

VWF ristocetin cofactor activity (VWF:RCo) is the most popular measure of VWF activity. It is a functional measure of the ability of VWF to bind platelet glycoprotein Ib (GpIb) in the presence of ristocetin (a glycopeptide synthesized by *Nocardia lurida*). This activity assay is performed by measuring the agglutination of normal fixed platelets in dilutions of test plasma containing an excess of ristocetin. The result depends on the presence of high-molecular-weight (HMW) multimers and an intact GpIb binding site. The assay has high interassay and interlaboratory variability.

There is increasing interest in measuring VWF activity as the capacity of VWF to bind collagen (VWF:CB). The use of VWF:CB as a measure of VWF activity is becoming widespread and complements the VWF:RCo assay. The current method is ELISA (enzyme-linked immunosorbent assay) based. VWF:CB is very dependent on the presence of HMW multimers and an intact collagen-binding site. The type of collagen used is an important variable in the performance of the assay. It has not replaced the VWF:RCo assay.

In addition to reduced levels of VWF antigen and activity (VWF:RCo), patients with VWD may also demonstrate abnormal platelet aggregation responses to ristocetin (see **Platelet-function testing**). Ristocetin induces platelet-VWF interaction and, in turn, platelet aggregation. In the presence of reduced levels of VWF, ristocetin-induced platelet aggregation is impaired. However, in type 2B VWD (see below), ristocetin-induced platelet aggregation is enhanced rather than impaired, with aggregation occurring at low levels of added ristocetin (0.5 mg/ml).

Subclassification of VWD is based upon analysis of VWF levels and VWF multimers. Decreased plasma VWF concentration with a normal multimeric distribution is consistent with a quantitative deficiency of VWF (type 1). Structural abnormalities, such as loss of high-molecular-weight and intermediate-molecular-weight multimers, are consistent with qualitative defects of VWF (type 2). Due to lack of VWF, no multimers are seen in type 3 VWD. This forms the basis of the primary classification of VWD. The secondary classification of type 2 VWD is defined upon absence of HMW VWF multimers, definition of platelet-dependent function of VWF (ristocetin-induced platelet aggregation, increased or decreased), and definition of factor VIII binding (see Table 156).

Previous classifications of type 2 VWD involved many different subtypes. All qualitative defects associated with absent HMW multimers and reduced platelet-dependent functions are now grouped as type 2A VWD. A range of differing candidate mutations have been described for type 2A VWD.

Type 2B VWD is due to an increased affinity of VWF for platelet GpIb, leading to an increased platelet-VWF interaction *in vivo*. This increased affinity, due to mutations within the VWF GpIb binding site, leads to *in vivo* platelet aggregation and consumption of HMW VWF multimers. As a consequence of this, continuing increase in platelet-VWF interaction

TABLE 156

Classification of Von Willebrand Disease Type 2

Subtype	High-Molecular-Weight Multimers	Ristocetin-Induced Platelet Aggregation	Factor VIII Binding Capacity
2A	Absent	Decreased	Normal
2B	Reduced/absent	Increased	Normal
2M	Normal	Decreased	Normal
2N	Normal	Normal	Markedly reduced

for type 2B VWD is often associated with a mild, consumptive **thrombocytopenia**. Type 2B VWD is phenotypically similar to **pseudo-Von Willebrand Disease**.

Type 2N VWD is due to a defect/mutation in the factor VIII binding site of VWF. Other than markedly reduced binding of factor VIII, type 2N VWF protein is normal in all other functions and quantity. Type 2N is phenotypically similar to mild **hemophilia A** and should be excluded in all cases of mild hemophilia A and in women who appear to be hemophilia A carriers but with no previous family history.

It requires diagnosis by use of a specific factor VIII VWF binding assay. Type 2M VWD is defined as a qualitative variant of VWF with decreased platelet-dependent function that is not caused by the absence of HMW multimers. In essence, it is similar to type 2A, but multimer analysis is normal. Candidate mutations have been identified around the specific GpIb binding site that mediates the platelet-dependent function of VWF. Type 2M is becoming increasingly recognized, and a number of type 1 cases may often be reclassified as type 2M upon closer examination.

Treatment

This is dependent upon the severity of VWD and bleeding problem. **DDAVP** will raise levels of WF and factor VIII by two- to fourfold and is the mainstay of treatment for mild VWD. Responses to DDAVP are tachyphylactic, a factor that must be considered for prolonged courses of treatment. **Tranexamic acid**, used in conjunction with DDAVP or alone, is a useful adjunctive therapy. Use of DDAVP is controversial in type 2B VWD, where it could release further abnormal VWF, which aggravates the preexisting thrombocytopenia and *in vivo* platelet aggregation, potentially leading to **disseminated intravascular coagulation** and **thrombosis.** Although there is data supporting DDAVP in type 2B, it is the opinion of many clinicians that DDAVP is contraindicated in patients with type 2B VWD. DDAVP is of limited value in type 2A and type 2M VWD, as release of further dysfunctional protein will fail to correct the bleeding diathesis. For severe type 3 VWD, type 2 variants, or moderate type 1 VWD, where DDAVP alone is likely to be insufficient in terms of rise required or duration of proposed treatment, VWF replacement therapy is required. Virus-inactivated, intermediate-purity factor VIII **coagulation factor concentrate**, which contains well-preserved VWF, or high-purity VWF concentrates are now the treatments of choice. Although **cryoprecipitate** contains a normal distribution of VWF multimers and is an effective treatment for VWD, it is not virus-inactivated and thus is to be used only if concentrate therapy fails. Intermediate-purity factor VIII concentrates that contain well-preserved VWF do not contain a normal complement of HMW VWF multimers, but they do contain a sufficient amount to be effective in the majority of situations. Platelet concentrates may also be useful in difficult bleeding problems in VWD. Rarely, patients with severe VWD may develop alloantibodies to transfused VWF (inhibitors). These patients are often extremely difficult to manage.

VON WILLEBRAND FACTOR

(VWF) A large, complex, multimeric glycoprotein (Gp) crucial to primary **hemostasis.** VWF has a large complex gene of 178 kb on the short arm of chromosome 12. VWF is synthesized in **vascular endothelial** cells and bone marrow **megakaryocytes**. Following translation, it undergoes extensive intracellular processing. It has a basic subunit of 400 kDa. The 2813-amino acid primary translation product undergoes processing in the **endoplasmic reticulum** to remove a signal peptide, thereby forming pro-VWF dimers. Subsequently, during passage through the **Golgi apparatus**, the VWF propeptide mediates the assembly of the 500-kDa pro-VWF dimers into a range of multimeric structures with molecular weights of up to 20,000 kDa. Extensive posttranslational modification of VWF occurs, including glycosylation, sulfation, and cleavage of the propeptide. The mature VWF protein is secreted directly into plasma or the subendothelial matrix, or else it is stored in endothelial cell Weibel-Palade bodies and platelet alpha granules.

The largest multimers, recognized to be the most hemostatically effective, are present in endothelial cell Weibel-Palade bodies and platelet alpha granules, but these are detected only transiently in normal blood following their acute release in response to vascular injury or stimulation. High-molecular-weight multimers released into plasma from endothelial cells are broken down by the action of thrombospondin and a specific protease called **ADAMTS-13**.

VWF has an essential function in primary hemostasis to facilitate platelet adhesion and aggregation in the presence of high fluid-shear stress forces, such as occur in the microvasculature. High-molecular-weight VWF multimers are of particular importance for the mediation of interactions between subendothelium exposed at sites of vascular injury and the platelet-membrane glycoprotein Ib (GPIb) and **glycoprotein IIb/IIIa** (GPIIb/IIIa). The exact nature of the interaction of VWF with exposed subendothelium is uncertain; however, a binding site for fibrillar collagens is located within the VWF A3 domain. The interaction of VWF with subendothelium and the effect of shear forces is thought to induce a conformational change in VWF, exposing the platelet GpIb binding site within the VWF A1 domain. The initial reaction of VWF with platelet GPIb is rapid but of low affinity. The resultant slowing of platelet travel along the vessel wall enables the less rapid but higher affinity binding of VWF to GPIIb/IIIa on the surface of activated platelets to take place via the RGD sequence located within the VWF C1 domain. This latter binding interaction induces platelet plug formation and is particularly important under conditions of high-shear stress.

VWF also has an essential role in stabilizing **factor VIII** in the circulation. The VWF binding site for factor VIII is located in the D'/D3 domains, within the amino-terminal 272 amino acid residues of the mature VWF subunit. In the presence of VWF, the half-life of plasma factor VIII is approximately 8 to 12 h; in the absence of VWF, the half-life has been estimated to be less than 1 h. VWF itself has a half-life of $\approx$ 24 h.

Enzyme-linked immunosorbent assay (ELISA) gives normal adult concentrations of 0.5 to 2.0 IU/ml. Levels for premature and full-term infants are given in **Reference Range Tables XV** and **XVI**.

Deficiency arises from hereditary **Von Willebrand Disease** or, occasionally, from acquired Von Willebrand Disease.

W

WALDENSTRÖM MACROGLOBULINEMIA

A malignant lymphoplasma-cell-proliferative B-cell disorder that produces a large IgM paraprotein.

Clinical Features

The median age at diagnosis is approximately 65 years, and there is a slight male predominance. Weakness, fatigue, oronasal bleeding, weight loss, and visual or neurologic disturbances are the most common presenting symptoms. Bone pain is rare. Fever, night sweats, and weight loss may be seen. Pallor, **hepatosplenomegaly**, and peripheral **lymphadenopathy** are the most frequent physical findings. Retinal lesions, including hemorrhages, exudates, and venous congestion with vascular segmentation ("sausage" formation) may be impressive. Peripheral neuropathy may occur.

Laboratory Features

Almost all patients have a normochromic **normocytic anemia**. The frequently increased plasma volume spuriously reduces the **hemoglobin** and **hematocrit** levels. A tall, narrow peak or dense band of g-mobility constituting the IgM M-protein is seen on the electrophoretic pattern. Seventy-five percent of patients have a k-light chain. About 10% of macroglobulins are cryoprecipitable. A monoclonal light chain is found in the urine in almost 80% of patients. The **bone marrow** aspirate is often hypocellular, but biopsy specimens are hypercellular and extensively infiltrated with **lymphocytes** and **plasma cells**. The number of **mast cells** is often increased.

Treatment

Therapy should be withheld until the patient is symptomatic. The most frequently used agent is **chlorambucil** administered on a daily oral schedule or intermittently at 4- to 6-week intervals. The use of cyclophosphamide or a combination of **alkylating agents** such as vincristine, BCNU, melphalan, cyclophosphamide, and prednisone or prednisolone is beneficial. Fludarabine or 2-chlorodeoxyadenosine has produced impressive results. Interferon-α2 may also be beneficial. Patients with symptomatic hyperviscosity syndrome should be treated with plasmapheresis. Extended dose **ritaximab** in combination with other pharmaceutical agents has been recommended.[518a]

WARFARIN

(Warfarin sodium; 4-hydroxy coumarin; Coumadin; Marevan) A coumarin drug used as an orally administered anticoagulant.[519] Warfarin inhibits the formation of **vitamin K** by inhibiting the enzyme vitamin K epoxide reductase. This in turn inhibits a vitamin K-dependent carboxylase enzyme (by removing substrate) that is necessary for the carboxylation of the vitamin K-dependent coagulant proteins. Warfarin therefore leads to impaired posttranslational modification of factor II and **factors VII**, **IX**, and **X** and the natural anticoagulants **proteins C** and **S**. The factors therefore are dysfunctional, and warfarin produces an anticoagulant/antithrombotic effect. Warfarin is rapidly absorbed from the gut, and in the circulation, approximately 90 to 99% of warfarin is bound to albumin and other plasma proteins. Warfarin acts as an anticoagulant by reducing the functional levels of all the vitamin K-dependent clotting factors. The antithrombotic effects of warfarin are more associated with the reduction of factors II and X and the swifter reduction in factors VII and IX. Therapeutic anticoagulation is only achieved when the levels of factors II and X drop, which is generally after about 4 days.[57] Warfarin has generally been given as a loading dose of 10 mg on day 1, 10 mg on day 2, and 5 mg on day 3; more recently, it has been given as a standard dose of 5 mg daily, monitoring the efficacy of anticoagulation on the third day by measuring the **international normalized ratio** (INR). In elderly patients or those with abnormal liver function, the dosage of warfarin will need to be reduced, and induction should be less than 5 mg daily. Warfarin is frequently given at the same time as **heparin**, and only when the INR is in the therapeutic range for 2 days or more is heparin discontinued. Both heparin and warfarin can be safely monitored by the **activated partial thromboplastin time** (APTT) and INR, respectively, when both are coadministered.

The degree of anticoagulation is dependent upon the precise indication for warfarin. Thus, patients with multiple recurrent pulmonary emboli or metallic heart valves will require to have their INR maintained at between 3.0 to 4.5, while a patient with atrial fibrillation may achieve significant benefit with an INR of between 2.0 and 3.0. A target INR of 2.0 to 3.0 is also generally satisfactory for treatment of deep-vein thrombosis, pulmonary embolism, and recurrent venous thrombosis if the recurrence occurred when the patient was off anticoagulants. The duration of therapy depends on the indication, but for a first event of significance, it is usually approximately 6 months. Long-term treatment depends on the clinical circumstances and risk of recurrence. This is currently an area of great controversy.

Monitoring of therapy has traditionally been performed in hospital, but the introduction of methods for near-patient testing of prothrombin time measurement, aided by computerized dosage schedules, allows it to be performed at primary-care institutions.[516] In these circumstances, it is essential that the same levels of quality assurance, training, and audit be maintained.[196b] Self-management of oral anticoagulation has been assessed and detailed recommendations made.[520–522]

Adverse Drug Reactions

These are commonly a consequence of drug interaction, which may be related to *CYP* genes (see *British National Formulary*, 51, Appendix I.)

Hemorrhage

Bleeding is the major adverse reaction of warfarin. Significant bleeding and, indeed, fatal bleeding events may occur at therapeutic INRs, but the risks of bleeding increase dramatically with increase of the INR and with increasing age of the patients. Although actual estimates vary, the overall risk of major bleeding in response to administration of oral anticoagulants is generally accepted to be around 2 events per 100 treatment-years. Bleeding risk significantly increases with an INR of >5.0. The management of patients with a high INR is dependent upon whether or not there is evidence of bleeding and whether this is major or minor. Options for treatment include simply stopping warfarin, **vitamin K** administration, or transfusion of **fresh-frozen plasma** (FFP) or **coagulation-factor concentrates** (see Table 157). Some thought must also be paid to the consequences of reversal,

TABLE 157

Management of Hemorrhage Due to Excess Warfarin

INR and Symptoms	Action
3.0–6.0 (target INR 2.5)	Reduce warfarin dose or stop
4.0–6.0 (target INR 3.5)	Restart warfarin when INR <5.0
6.0–8.0	Stop warfarin
No bleeding or minor bleeding	Restart warfarin when INR <5.0
>8.0	Stop warfarin
No bleeding or minor bleeding	Restart warfarin when INR <5.0
	If other risk factors for bleeding, give 0.5–2.5 mg vitamin K
Major bleeding	Stop warfarin
	Give prothrombin complex concentrate 50 U/kg or FFP 15 ml/kg
	Give 5 mg vitamin K

notably risk of **thrombosis**, although the severity of the precipitating bleeding event will assume priority in the main.

Once warfarin is stopped, the INR will decline, but this may take several days, depending on the height of the INR. Vitamin K given orally or intravenously effectively reverses the effects of warfarin. A dose as low as 0.5 mg will reduce an INR >5.0 to therapeutic levels within 24 h. Doses of 2.5 mg or greater will completely reverse warfarin, whereas large doses (5 mg or greater) will lead to some degree of warfarin resistance, making reanticoagulation after the acute event has subsided more difficult.

Coagulation-factor concentrates are now recommended as the treatment of choice for the rapid reversal of oral anticoagulants, in the face of bleeding. Concentrates used for anticoagulant reversal are prothrombin complex concentrates that contain factors II, IX, and X and variable amounts of factor VII. These products rapidly and effectively reverse overanticoagulated patients. The correction of factor IX is more effective with these types of products than FFP. Following infusion, the INR will have corrected almost immediately.

Teratogenicity

For this reason, warfarin cannot be given in the first trimester of pregnancy, when major embryogenesis is taking place. Many centers, therefore, manage such women by administering heparin during the first and third trimesters and warfarin during the second trimester, although warfarin is not entirely safe at any stage of pregnancy.

WARM AUTOIMMUNE HEMOLYTIC ANEMIA

(AIHA) The cause, pathogenesis, clinical and laboratory features, treatment, course, and prognosis of **hemolytic anemias** due to **antibodies** reacting at 37°C.

Pathogenesis

In the majority of patients, the cause of AIHA is unknown.[91] Secondary causes are **lymphoproliferative disorders (chronic lymphatic leukemia [CLL], Hodgkin disease, non-Hodgkin lymphoma, Waldenström macroglobulinemia)** and **autoimmune disorders (systemic lupus erythematosus, rheumatoid arthritis** scleroderma, ulcerative colitis). Other associated disorders are ovarian dermoid cysts and teratomas, **Kaposi's sarcoma, immunodeficiency,** and **viral infection disorders,** particularly in childhood. In some patients, the **autoantibodies** are directed against a single target tissue, whereas in others there appears to be a more global disturbance of the immune system, with autoantibodies being directed against an array of target tissues. In patients with autoantibodies directed against an identifiable **red blood cell** antigen, prolonged survival of red cells lacking that **antigen** is seen, in contrast to the rapid clearance of cells possessing the antigen. There is, in general, an inverse relationship between the quantity of red blood cell-bound antibody and red blood cell survival. Decreased survival is due to trapping of the cells by macrophages in the Billroth cords of the spleen and to a lesser extent by the **Küpffer cells** in the liver. The process leads to sphering, fragmentation, and ingestion of the trapped cells. In warm antibody AIHA, the red cells are coated with **immunoglobulin** G (IgG) autoantibodies, with or without **complement**. The **histiocyte** (macrophage) plasma membrane has receptors for the Fc region of IgG (particularly IgG1 and IgG3) as well as receptors for opsonic fragments of C3 (C3b and C3bi) and C4. When they are present together on the red blood cell surface, IgG and C3 act together as opsonins to enhance trapping and **phagocytosis**. Most commonly, partial phagocytosis of the coated red blood cell results in only partial loss of membrane, with the noningested remainder of the cell assuming the lowest ratio of surface area to volume, which is a sphere. Spherocytes are less deformable than normal red cells and are thus more likely to undergo destruction following further passage through the microvasculature, particularly in the spleen. The degree of spherocytosis correlates well with the severity of the disease in any patient.

Clinical Features

AIHA is not a rare disease. The majority of patients are over 40 years of age at the time of diagnosis. Warm-antibody AIHA is very variable in clinical characteristics and course. Symptoms of **anemia** usually draw attention to the disease, although occasionally jaundice may be the presenting feature. Insidious progression is the rule, but the occasional patient may suddenly, over the course of a few days, develop severe anemia and jaundice. Moderate **splenomegaly** is suggestive of primary AIHA; the appearance of massive splenomegaly suggests an underlying lymphoproliferative disorder. Infection, surgery, physical trauma, and **pregnancy** are all recognized to aggravate the process.

Laboratory Features

The degree of anemia is variable, depending on the rate of **hemolysis** and the compensatory erythropoiesis. Macrocytosis with reticulocytosis is usual, together with moderate **leukocytosis**. Concurrent **neutropenia** and **thrombocytopenia** may occur (**Evan's syndrome**). The peripheral-blood smear shows **polychromasia** with **microspherocytes** (microspherocytosis). In chronic disorders, the plasma will show a raised level of **bilirubin**, whereas in acute disorders, there will a fall in **haptoglobin** and maybe **hemoglobinemia**. Mild degrees of hemolysis increase urinary **urobilinogen**, and severe cases show **hemoglobinuria** with **hemosiderinuria**.

The diagnosis of AIHA depends on the demonstration of immunoglobulin or **complement** bound to the patient's red blood cells. To screen for this, a broad-spectrum antiglobulin reagent is used (see **Direct antiglobulin [Coombs] test**). This test contains antibodies directed against human immunoglobulin and complement components (mainly C3). If agglutination occurs, antisera reacting selectively with immunoglobulin or with complement components are used to define the specific pattern of red cell sensitization. Using this method, three patterns are recognized: IgG coating (20 to 66%), IgG/complement coating (24 to 63%), and complement coating alone (7 to 14%). All patterns are associated with accelerated red blood cell destruction, but it is not possible to distinguish those giving rise to hemolysis from those that are clinically harmless. In addition to bound antibody, some patients exhibit "free" or plasma antibody. This is detected by the **indirect antiglobulin (Coombs) test**. Patients with AIHA with a positive indirect antiglobulin test should also have a positive direct antiglobulin test. The presence of a positive indirect test with a negative direct test suggests that an alloantibody stimulated by pregnancy or transfusion exists. Of these antibodies, 50 to 70% show specificity for the **Rh blood group**, and some of the remainder are positive for another **blood group** antigen.

Treatment

Immunosuppression

Corticosteroids and dexamethazone downregulate Fcγ receptor and thus decrease phagocytosis. About 20% of patients with warm antibody AIHA so treated obtain complete remission, whereas 10% show no or only minimal response. The best responses are seen in primary types. Initial response rates of 80% occur. Usually, treatment is given orally, commencing with 60 mg **prednisolone/prednisone** daily for 10 to 15 days, tapering the dose as response develops. Acutely ill patients benefit from intravenous administration during the first few days. It may be possible to tail off steroids completely, or it may be found that a small maintenance dose is required. Careful follow-up is necessary, often for several years. Relapse indicates consideration of further courses of glucocorticoid therapy, **splenectomy**, or other immunosuppressive drugs.

There is no direct proof that other immunosuppressive drugs inhibit antibody synthesis, but encouraging responses have occurred in some patients. This form of therapy in warm antibody AIHA should be reserved only for patients who have failed to respond to glucocorticoids and splenectomy or who are poor surgical risks. The drugs that may help are **cyclophosphamide** (60 mg/m^2) or **azathioprine** (80 mg/m^2) given orally daily. This therapy can be continued for 6 months while awaiting a response, provided that the drug is tolerated. Blood counts must be monitored regularly.

Immunoglobulin given intravenously is the treatment of choice for hemolytic anemia associated with **systemic lupus erythematosus**. Anti-D IgG given to Rh group (D)-positive patients may be of value by coating red blood cells and blocking Fc receptors. **Danazol** (a nonvirilizing anabolic steroid), **ciclosporin**, and **rituximab** intravenously may hold some promise, but efficacy has not been established in controlled studies.

To support **bone marrow** compensation, **folic acid** supplementation is recommended.

Red Blood Cell Transfusion

Red blood cell transfusion is only necessary in the severely anemic individual and is not desirable if it can be avoided. To support bone marrow compensation, **folic acid** supplementation is recommended. The main problems regarding red blood cell transfusion are:

Virtual impossibility of finding totally compatible cells, thus requiring initiation of a policy of transfusing the least incompatible cells

Rapid destruction of the transfused cells, possibly as fast as that of the patient's own cells

Further induction of red blood cell autoantibodies

Splenectomy

Approximately one-third of patients require maintenance therapy with glucocorticoids indefinitely in a dose greater than 15 mg daily. These patients are candidates for splenectomy because, with Ig antibodies, it removes the primary site of red cell destruction. It may also remove the site of antibody production. Results of splenectomy vary widely; however, at least two-thirds of those undergoing surgery respond completely. The continuation of hemolysis following splenectomy is partly related to continuing high levels of autoantibody, with red cell destruction in the Küpffer cells of the liver.

Course and Prognosis

The actuarial survival at 10 years is reported as 73%. Primary warm-antibody AIHA, however, generally pursues an unpredictable course characterized by remissions and relapses. There are no good predictors of outcome. Thromboembolic episodes (deep-vein thromboses and splenic infarcts) are common during the course of the disease. Pulmonary emboli, infection, and severe uncontrollable anemia are the usual causes of death. Prognosis in secondary AIHA is largely dependent on the course of the underlying disease.

WEGENER'S GRANULOMATOSIS

A rare condition characterized by necrotizing granulomatous **vasculitis** (both arteries and veins) affecting the renal and respiratory tracts.[523] It is an **autoantibody** disorder against neutrophil proteinase 3: classical **antineutrophilic cytoplasmic antibodies** (cANCA). When **neutrophils** are activated, proteinase 3 is expressed on the cell surface and becomes accessible to cANCA. Binding inhibits migration and stimulates oxidative burst and enzyme release. This damages vessel walls and induces **inflammation**. A second set of cANCA autoantibodies reacts with neutrophil myeloperoxidase in the glomeruli. Both reactions are **hypersensitivity** type III, with the immune complexes forming **granulomas** similar to those seen in **tuberculosis**, with infiltration and prominent necrosis of **lymphocytes**, **plasma cells**, **macrophages**, and giant (multinucleated) cells. The cause remains unknown, but inhalation of a stimulating agent is the most likely possibility.

Clinical Features

The condition is usually acute in onset, with males and females equally affected. The usual age of onset is 20 to 50 years. Both the upper and lower respiratory tracts are affected. Sinusitis (90%) and otitis media (40%) are common, with destruction of the nasal cartilage leading to saddle-nose deformity characteristic. The chest X-ray is abnormal in virtually all patients, with nodules (single or multiple) present, some of which have cavitated. Transient pulmonary infiltrates are seen, but hilar lymphadenopathy is unusual.

The classical renal lesion is acute focal glomerulonephritis with fibrinoid necrosis, which rapidly progresses to acute renal failure. Hypertension is unusual. Skin disorders include **purpura** due to leukocytoclasia, papules from necrotizing vasculitis, livedo reticularis, and pyoderma gangrenosa. Other organs, e.g., eyes, joints, heart, and nervous system, may be affected in a minority of patients, but rarely cause life-threatening complications.

Laboratory Features

Raised **erythrocyte sedimentation rate**

Raised **C-reactive protein**

Normocytic anemia due to **microangiopathic hemolytic anemia**

Granulocytosis

Thrombocytosis

Polyclonal gammopathy, or more usually, immunoglobulin A (IgA) alone is raised antinuclear factor is typically negative, but low-titer rheumatoid factor may be found

Microscopic hematuria with or without proteinuria indicates renal involvement

Antineutrophilic cytoplasmic antibody level acts as a marker of disease activity

The definitive diagnosis depends on tissue renal biopsy, but access to tissue and interpretation may be difficult.

Treatment and Prognosis

Untreated, the prognosis is very poor, with over 80% of patients dying within one year. Death is usually due to renal or respiratory failure, the latter often complicated by a superinfection.

Corticosteroids alone are usually ineffective, but the addition of **cyclophosphamide** (2mg/kg/day) results in over 90% of patients achieving remission. After 1 year, the dose can be tapered or replaced by **methotrexat**e or daily **azathioprine** if there is little renal involvement. Relapses can occur if therapy is stopped, but remission can often be reinduced. With renal failure, **plasmapheresis** is indicated, and successful renal transplantation has been performed. Treatment with **infliximab** is under clinical trial.

WELL-DIFFERENTIATED LYMPHOCYTIC, PLASMACYTOID, DIFFUSE MIXED LYMPHOCYTIC AND HISTIOCYTIC LYMPHOMA

See **Chronic lymphocytic leukemia/lymphoma; Lymphoplasmacytic lymphoma/Waldenström macroglobulinemia; Lymphoproliferative disorders; Marginal-zone B-cell lymphoma; Non-Hodgkin lymphoma.**

WEST NILE VIRUS

(WNV) A mosquito-borne flavivirus. Humans are normally incidental hosts, and birds are the main reservoir of infection transmission. Human infection has been documented for decades in parts of Africa and Europe, but it was first recognized in the U.S. in 1999. Subsequently, it has become the predominant circulating arthropod in the U.S. with >15,000 human cases and >600 fatalities since 1999.[19] Peak incidence in the U.S. occurs between July and September. It classically causes meningoencephalitis and is associated with fever, headache, muscle pain and weakness, bone and joint pain, skin rash, abdominal pain, vomiting, diarrhea, **lymphadenopathy**, painful eyes, and seizures. In 2002, 4200 cases were reported to the Centers for Disease Control and Prevention (CDC). In August 2002, the CDC was notified of the first case of suspected **transfusion-transmitted infection** of WNV.[524] Further cases soon followed in which blood components and solid-organ transplants were implicated in transmission of WNV.[525]

Subsequent analysis of reported cases suggested that WNV could be transmitted by platelets, leukodepleted red blood cells, and fresh-frozen plasma. No donors had IgM WNV antibodies, and circulating levels of WNV were sometimes near the limits of detection of nucleic acid amplification assays available.[525]

The incubation period following transmission by a mosquito bite in healthy individuals is 2 to 14 days, but it may be longer in immunosuppressed individuals. Patients with WNV-associated illness following transfusion had incubation periods of 2 to 21 days (median 10 days). The median incubation period in four transplant recipients was 13.5 days.[526]

Screening for WNV in blood donors is aimed at detecting viral RNA.[527,528] Screening for nucleic acid in a blood donor population is made more cost effective by creating "minipools" of several donations on which the assay is performed. If found positive, each individual donation is then screened. Failure of detection by minipool screening resulting in WNV transmission was described in 2004.[525] Of 540 positive donations detected by the

American Red Cross (ARC) in 2003–2004, 27% were detectable only by testing individual donations. Minipool screening on plasma from 16 donations was implemented by the ARC on June 23, 2003. Individual donation testing is undertaken in areas of high incidence (>1:1000 samples), as data from the 2002 epidemic suggested that, at the peak of the epidemic, one viremic donor may go undetected for every four detected.[529] The future significance of this infection remains to be seen.[530,531]

No specific treatment is available. In severe cases, treatment consists of supportive care that often involves hospitalization, intravenous fluids, respiratory support, and prevention of secondary infections. Several clinical trials are ongoing. Those that meet specific criteria are listed by CDC.[19]

WHITE BLOOD CELL

See **Leukocyte**.

WHITE BLOOD CELL COUNTING

The estimation of the total number of white blood cells (**leukocytes**) in a sample of body fluid, usually peripheral blood.[455] The blood sample is diluted with a solution that lyses red blood cells. This is the method whether the white cells are being counted manually in a cytometry counting chamber or by an automated method. A carefully performed visual white cell count is perfectly acceptable, although it is time consuming. Instrument-rated white cell counts have a COV of 1 to 3%, whereas that for that for the manual method approaches 10%. Counts performed on electronic instruments can be falsely elevated in the presence of cryoglobulins or cryofibrinogen, clumped platelets, fibrin strands, EDTA-induced platelet aggregation, nucleated red blood cells, or nonlysed red cells in certain congenital hemolytic anemias.

The reference range for adults is 4.0 to $10.0 \times 10^9/l$; ranges for infants and children are given in **Reference Range Table X**.

WISCOTT-ALDRICH SYNDROME

(WAS) An X-linked single-gene **immunodeficiency** disorder.[532,533] It is characterized by **thrombocytopenia**, severe eczema, and **lymphoproliferative disorders**. The genetic disorder gives rise to a deficiency of a 66-kDa protein (WASp), which acts via Cdc42, a small GTPase (guanosine triphosphatase) of the Ras-related group. Ddc42 controls the formation of cell-surface projections in **cell locomotion** and also has involvement in cell proliferation. This is the probable cause of disturbed T- and B-**lymphocyte** interactions and also lack of platelet integrity. Treatment of the disorder is by **splenectomy** and/or **allogeneic stem cell transplantation**.[534]

X

XEROCYTE

A dehydrated variant of a **stomatocyte**. It is seen in patients with hereditary xerocytosis, a rare **hereditary nonspherocytic hemolytic anemia**, where the peripheral-blood film shows codacytes and stomatocytes.

XIMELAGATRAN

An anticoagulant at present undergoing investigation for possible clinical use. It is a dipeptide mimetic of the region of **fibrinopeptide A** that interacts with the active site on **thrombin** and hence is a direct thrombin inhibitor. This is unlike heparin or warfarin, which are indirect inhibitors acting through antithrombin (**heparin**) or an inhibitor of coagulation protein production (**warfarin**). As a direct thrombin inhibitor, it can inhibit both free and clot-bound thrombin. Theoretically, this produces more-effective anticoagulation by inhibition of clot-bound thrombin, believed to be responsible for thrombus propagation. Ximelagatran is a prodrug of the active metabolite melagatran and can be administered orally. After absorption, it is hydrolyzed to melagatran. Ximelagatran's oral bioavailability is approximately 18 to 24%, resulting in low interindividual variability in resultant melagatran plasma levels. It has been demonstrated to have a relatively wide therapeutic window in terms of bleeding and antithrombotic effect compared with warfarin. As a result, it does not require monitoring and can be administered in a fixed dosing regime. There is an apparent lack of drug-drug and drug-food interactions with ximelagatran, which would offer a major clinical and practical advantage over warfarin. Clinical studies have demonstrated ximelagatran to be comparable in efficacy to warfarin and low-molecular-weight heparins (LMWH) for prophylaxis of venous thromboembolism; comparable to warfarin for stroke prevention in the setting of atrial fibrillation; and, when combined with aspirin, possibly more effective than aspirin alone at preventing major adverse cardiovascular events in patients with a recent myocardial infarction. Adverse effects with ximelagatran primarily involve bleeding complications, which appear comparable to those occurring with standard anticoagulant treatment (i.e., warfarin and LMWH). Ximelagatran has also been demonstrated to cause transient increases in liver enzymes, the significance of which is unclear. Ximelagatran currently remains an investigational drug.

Z

ZAHN LINES

Laminations of pale gray platelets/fibrin and darker red coagulated blood seen in **arterial thrombosis**. These laminations result from irregular deposition, in fast moving streams, of layers of platelet masses and fibrin alternating with layers of red blood cells that have become enmeshed in precipitated fibrin.

ZIEVE'S SYNDROME

See **Acanthocytes; Liver disorders**.

ZINC

An electrolyte of human **red blood cells** (RBCs), normally present in a concentration of 0.153 μmol/ml RBC (0.463 μmol/g hemoglobin). It appears to be an important constituent of the red blood cell, as deficiency is associated with some hematological disorders. The level of zinc protoporphyrin can be measured.[236]

Zinc Deficiency

Certain clinical features are common to some patients with **sickle cell disorders** and those who are zinc deficient. These include delayed onset of puberty and hypogonadism in females, characterized by decreased facial, pubic, and axillary hair, short stature and low body weight, rough skin, and poor appetite. A careful trial of zinc supplementation to individuals with sickle cell anemia has resulted in significant improvement in secondary sexual characteristics and increased serum testosterone in male adults, an increment in longitudinal growth and body weight in adolescents who were retarded in growth, reversal of the dark-adaptation abnormality, and correction of anergy in adult subjects with sickle cell disease. Zinc supplementation to patients with chronic leg ulcers has proved beneficial in limited trials. Zinc behaves as an antisickling agent, possessing membrane-stabilizing properties, binding to hemoglobin, and increasing oxygen affinity *in vitro*. In pharmacological doses, it has decreased the number of irreversibly sickled cells. It is not yet clear if it has a role in the prevention of crises.

Zinc deficiency has also been associated with acrodermatitis enteropathica, which in turn has been associated with combined **immunodeficiency**.

ZOLLINGER-ELLISON SYNDROME

See **Gastric disorders**.

Appendix I

WHO Classification of Tumors of Hematopoietic and Lymphoid Tissues

CHRONIC MYELOPROLIFERATIVE DISEASES

Chronic myelogenous leukemia
Chronic neutrophilic leukemia
Chronic eosinophilic leukemia/hypereosinophilic leukemia
Polycythemia vera
Chronic idiopathic myelofibrosis
Essential thrombocythemia
Chronic myeloproliferative disease, unclassifiable

MYELODYSPLASTIC/MYELOPROLIFERATIVE DISEASES

Chronic myelomonocytic leukemia
Atypical chronic myeloid leukemia
Juvenile myelomonocytic leukemia
Myelodysplastic/myeloproliferative diseases, unclassifiable

MYELODYSPLASTIC SYNDROMES

Refractory anemia
Refractory anemia with ring sideroblasts
Refractory cytopenia with multilineage dysplasia
Refractory anemia with excess blasts
Myelodysplastic syndrome associated with isolated del(5q) chromosome abnormality
Myelodysplastic syndrome, unclassifiable

ACUTE MYELOID LEUKEMIAS (AML)
Acute Myeloid Leukemias with Recurrent Cytogenetic Abnormalities

AML with t(8;21)(q22;q22), (AML1/ETO)
AML with inv(16)(p13q22) of t(16;16)(p13'q22), (*CBFβ/MYH11*)
Acute promyelocytic leukemia (t(15;17)(q22;q12), (*PML/RARα*) and variants)
AML with 11q23 (*MLL*) abnormalities

Acute Myeloid Leukemia with Multilineage Dysplasia

With prior myelodysplastic syndrome
Without prior myelodysplastic syndrome

Acute Myeloid Leukemia and Myelodysplastic Syndrome, Therapy Related

Alkylating agent related
Topoisomerase II inhibitor related

Acute Myeloid Leukemia Not Otherwise Categorized

Acute myeloid leukemia, minimally differentiated
Acute myeloid leukemia without maturation
Acute myeloid leukemia with maturation
Acute myelomonocytic leukemia
Acute monoblastic and monocytic leukemia
Acute erythroid leukemia
Acute megakaryoblastic leukemia
Acute basophilic leukemia
Acute panmyelosis with myelofibrosis
Myeloid sarcoma

Acute Leukemia of Ambiguous Lineage

B-CELL NEOPLASMS
Precursor B-Cell Neoplasm

Precursor lymphoblastic leukemia/lymphoma

Mature B-Cell Neoplasms

Chronic lymphocytic leukemia/small lymphocytic lymphoma
B-cell prolymphocytic leukemia
Lymphoplasmacytic lymphoma
Splenic marginal zone lymphoma
Hairy cell lymphoma
Plasma cell lymphoma
Solitary plasmacytoma of bone
Extraosseous plasmacytoma

Extranodal marginal zone B-cell lymphoma of mucosa-associated lymphoid tissue (MALT-lymphoma)

Nodal marginal zone B-cell lymphoma

Follicular lymphoma

Mantle cell lymphoma

Diffuse large B-cell lymphoma

Mediastinal (thymic) large B-cell lymphoma

Intravascular large B-cell lymphoma

Primary effusion lymphoma

Burkitt lymphoma/leukemia

B-Cell Proliferations of Uncertain Malignant Potential

Lymphomatoid granulomatosis

Post-transplant lymphoproliferative disorder, polymorphic

T-CELL AND NK-CELL NEOPLASMS
Precursor T-Cell Neoplasms

Precursor T-lymphoblastic leukemia/lymphoma

Blastic NK-cell lymphoma

Mature T-Cell And NK-Cell Neoplasms

T-cell prolymphocytic leukemia

T-cell large granular lymphocytic leukemia

Aggressive NK-cell leukemia

Extranodal NK/T-cell lymphoma, nasal type

Enteropathy-type T-cell lymphoma

Hepatosplenic T-cell lymphoma

Subcutaneous panniculitis-like T-cell lymphoma

Mycosis fungoides

Sézary syndrome

Primary cutaneous anaplastic large-cell lymphoma

Peripheral T-cell lymphoma, unspecified

Angioimmunoblastic T-cell lymphoma

Anaplastic large-cell lymphoma

T-Cell Proliferations of Uncertain Malignant Potential

Lymphomatoid papulosis

HODGKIN LYMPHOMA

Nodular lymphocyte predominant Hodgkin lymphoma
Classical Hodgkin lymphoma
Nodular sclerosis classical Hodgkin lymphoma
Lymphocyte-rich classical Hodgkin lymphoma
Mixed cellularity classical Hodgkin lymphoma
Lymphocyte-depleted classical Hodgkin lymphoma

HISTIOCYTIC AND DENDRITIC CELL NEOPLASMS
Macrophage/Histiocytic Neoplasm

Histiocytic sarcoma

Dendritic Cell Neoplasms

Langerhans cell histiocytosis
Langerhans cell sarcoma
Interdigitating dendritic cell sarcoma/tumor
Follicular dendritic cell sarcoma/tumor
Dendritic cell sarcoma, not otherwise specified

MASTOCYTOSIS

Cutaneous mastocytosis
Indolent systemic mastocytosis
Systemic mastocytosis with associated clonal, hematological nonmast cell lineage disease
Aggressive systemic mastocytosis
Mast cell leukemia
Mast cell sarcoma
Extracutaneous mastocytoma

Appendix II

FAB Classification of Leukemia/Myelodysplasia

The French-American-British (FAB) classification was generally accepted as a useful and reproducible morphological categorization of AML and ALL (acute lymphoblastic leukemia). The morphological diagnosis of ALL from AML was one of exclusion, with cell cytoplasm that lacks granularity and does not show myeloperoxidase activity.

ACUTE MYELOBLASTIC LEUKEMIA

AML M0

Proposed relatively recently by the FAB group, its recognition relies mainly on immunophenotyping. MO blasts display no distinctive morphological characteristics, with peroxidase and Sudan black stains being essentially negative. Immunologically, blasts express the myeloid associated markers CD13 or CD33 in the presence of negative lymphoid markers.

AML M1

Marrow aspirates contain 90% or more of medium-to-large blasts with few, if any, cytoplasmic granules, Auer rods, or vacuoles. At least 3% of blasts will be peroxidase positive.

AML M2 (acute promyelocytic leukemia [APL])

More than 10% of the nucleated cells have matured to the promyelocyte stage or beyond. Many patients show frank dysplastic changes, including nuclear-cytoplasmic asynchrony, pseudo-Pelger-Hüet appearance, and abnormalities of granulation, especially hypogranulation and giant granules. Auer rods may be found at any stage from myeloblast to mature granulocyte. The 8;21 chromosomal translocation is found in 18% of these patients. (Translocation t (15;17) fused to retinoic acid receptor α(RARα) gene to PML gene on chromosome 15 gives rise to fusion protein PML-RARα, suggesting that the disruption of RARα is the cause of APL.)

AML M3

A common hypergranular type and a less usual microgranular variant are recognized. The hypergranular type has a bone marrow population of abnormal promyelocytes packed with coarse red-purple granules, which often obscure the nucleus. The so-called faggot cells may contain sheaves of Auer rods. The nucleus is lobulated, reniform, or bilobed and is best appreciated in the microgranular variant, in which the cytoplasm contains only a fine red dusting of granules. The peroxidase stain is strongly positive, even when few granules are present. Virtually all cases are associated with the 15;17 chromosomal translocation.

AML M4

Differentiation is seen along both myeloid and monocytic lineages. Bone marrow blasts represent >30% of nucleated cells, many showing monocytic features. The monocytic component in the peripheral blood is $>5 \times 10^9$ cells/l, including monoblasts, promonocytes, and monocytes. These cells show nonspecific esterase (NSE) positivity and, with a combined esterase procedure, cells containing both NSE and the granulocyte enzyme chloroacetate esterase can be demonstrated. M4E is a distinct subtype characterized by an increased number of abnormal eosinophilic precursors containing prominent basophilic granules and associated abnormalities (especially inversion) of chromosome 16.

AML M5

In M5a, large monoblasts with ample basophilic cytoplasm constitute 80% or more of the bone marrow monocytic component. These blasts may contain vacuoles and/or small azurophilic granules. The nucleus is round or convoluted, with a large nucleolus and the cell membrane usually irregular with pseudopodia. In M5b there are mostly promonocytes and abnormal monocytes in bone marrow and peripheral blood. The nuclei of most cells are folded, with inconspicuous nucleoli. For both subtypes, peroxidase stain is negative and the NSE reaction is strongly positive. Serum and urinary lysozyme levels are significantly increased.

AML M6

Erythroid precursors are >50% and blasts >30% of nucleated erythroid cells, showing moderate to marked dysplastic features, e.g., nuclear lobulation, multinuclearity, karyorrhexis, and cytoplasmic vacuoles. Intense periodic acid-Schiff (PAS) block positivity may be seen in erythroid cells and increased iron incorporation with or without ringed sideroblasts. Care is needed in differentiating M6 from MDS (myelodysplasia), where erythroid blasts are nearly always <30%.

AML M7

Diagnosis depends on either (1) demonstration of platelet peroxidase by ultracytochemistry or (2) immunophenotyping to identify CD41+ glycoprotein or the CD61+ glycoprotein on the blast surface. Blast morphology is polymorphic, with some cells resembling L1 lymphoblasts (see below) with scant cytoplasm and dense chromatin, while others mimic L2 (or MO) blasts. Sudan black and peroxidase stains are negative, with variable PAS and esterase positivity.

ACUTE LYMPHOBLASTIC LEUKEMIA
FAB L1

The cells are relatively homogeneous small blasts up to twice the size of small lymphocytes. Cytoplasm is scanty and the nucleoli are absent or poorly visualized. Occasionally, vacuoles with a few azurophilic granules are seen.

FAB L2

The cells are heterogeneous and the nuclei irregularly shaped, often folded or indented with prominent nucleoli. The cytoplasm varies in quantity and may contain occasional azurophilic granules.

FAB L3

The cells are large, homogeneous blasts with deeply basophilic cytoplasm containing prominent, well-defined vacuoles. The nucleus is regular with prominent nucleoli.

MYELODYSPLASIA

Refractory Anemia (RA)

Less than 1% blasts in the peripheral blood, with marrow dysplasia in one or more cell lines and less than 5% blasts.

RA with Ring Sideroblasts (RAS)

Less than 1% blasts in peripheral blood, with marrow dysplasia in one or more cell lines and less than 5% blasts plus ring sideroblasts in more than 15% erythroblasts.

RA with Excess Blasts (RAEB)

Less than 5% blasts in peripheral blood marrow dysplasia in one or more cell lines plus 5 to 20% blasts.

RAEB in Transformation (RAEB-t)

Greater than 5% blasts (or less than 5% with Auer rods) in the peripheral blood, with marrow dysplasia in one or more cell lines plus 20 to 30% blasts or 5 to 20% with Auer rods.

CHRONIC MYELOMONOCYTIC LEUKEMIA (CMML)

Any form of myelodysplasia plus $>1 \times 10^9/l$ monocytes.

Reference Range Tables

TABLE I

Reference Ranges of Hematological Values for Adults

Blood Cells

Red Blood Cells (RBCs)

Hemoglobin A	
Males	13.0–17.0 g/dl
Females	12.0–15.0 g/dl
Hemoglobin A_2	2.2%–3.5%
Hemoglobin F	<1.0%
Mean cell hemoglobin (MCH)	27–32 pg
Mean cell hemoglobin concentration (MCHC)	31.5–34.5 g/dl
Mean cell volume (MCV)	83–101 fl
Packed cell volume (PCV)/hematocrit (Hct)	
Males	0.40–0.50 l/l
Females	0.36–0.46 l/l
Red blood cell count	
Males	$4.5–5.5 \times 10^{12}/l$
Females	$3.8–4.8 \times 10^{12}/l$
Red cell diameter: dry films	6.7–7.7 μm
Red cell distribution width (RDW)	
As coefficient of variation (CV)	12.8% ± 1.2%
As standard deviation (SD)	42.5% ± 3.5 fl
Reticulocyte count	0.5%–2.5% RBC
	$50–100 \times 10^9/l$

White Blood Cells (WBCs)

Total WBC count	$4.0–10.0 \times 10^9/l$
Differential WBC count	
Neutrophils	$2.0–7.0 \times 10^9/l$ (40–80%)
Eosinophils	$0.02–0.5 \times 10^9/l$ (1–6%)
Basophils	$0.02–0.1 \times 10^9/l$ (<1–2%)
Lymphocytes	$1.0–3.0 \times 10^9/l$ (20–40%)
Monocytes	$0.2–1.0 \times 10^9/l$ (2–10%)

Blood Volume

Red cell volume	
Males	25–35 ml/kg
Females	20–30 ml/kg
Plasma volume	40–50 ml/kg
Total blood volume	
Males	65–85 ml/kg
Females	60–80 ml/kg

Hemostasis Values

Antithrombin III	0.75–1.25 U/ml
Bleeding time (template method)	2.5–9.5 min
Euglobulin clot lysis time	90–240 min

TABLE I (continued)

Reference Ranges of Hematological Values for Adults

Fibrinogen (plasma)	2.0–4.0 g/l
Heparin cofactor II	55–145%
Activated partial thromboplastin time (APTT)	30–40 sec
Plasminogen	0.75–1.60 U/ml
Platelets	150–400 × 10⁹/l
Platelet factor 4	<10 ng/ml
Protein C	
Function	0.70–1.40 U/ml
Antigen	0.61–1.32 U/ml
Protein S	
Function	0.78–1.37 U/ml
Free	0.68–1.52 U/ml
Prothrombin time (PT)	
Recombinant	11–16 sec
Thromboplastin	10–12 sec
Thrombin time (TT)	15–19 sec
β-Thromboglobulin	<50 ng/ml

Plasma Values

Hemoglobin	10–14 mg/l
Methemoglobin	<2.0%
Haptoglobin	
By radial immunodiffusion	0.8–2.7 g/l
By hemoglobin-binding capacity	0.3–2.0 g/l
Iron (serum)	10–30 μmol/l (70–180 μg/dl)
Iron binding capacity — total (TIBC)	45–70 μmol/l (250–400 μg/dl)
Ferritin	
Males	15–200, median 100 ng/ml
Females	15–200, median 40 ng/ml
Transferrin	2.0–3.0 g/l
Cobalamin serum	180–640 ng/l
Folate	
Serum	3–20 μg/l
Red cell	160–640 μg/l

Note: Neonatal and childhood values are given in separate reference range tables below.

Source: Adapted from Lewis, S.M. in *Dacie and Lewis Practical Haematology*, 10th ed., Churchill-Livingstone, Elsevier, Edinburgh, 2006. With permission.

TABLE II

Hemoglobin Concentrations (g/dl) for Iron-Sufficient Preterm Infants

Age	Birth Weight (g)	
	1000–1500	1501–2000
2 weeks	11.7–18.4	11.8–19.6
1 month	8.7–15.2	8.2–15.0
2 months	8.8–11.5	9.4–11.4
3 months	9.8–11.2	9.3–11.8
4 months	9.1–13.1	9.1–13.1
5 months	10.2–14.3	10.4–13.0
6 months	9.4–13.8	10.7–12.6

Source: Derived from Lundstrom. U., Siimes, M.A., and Dallman, P.R., *J. Pediatr.*, 91, 878, 1977. With permission.

TABLE III

Hemoglobin F and Hemoglobin A_2 in the First
Year of Life (Measured by Electrophoresis)

Age	Hemoglobin F (%)	Hemoglobin A_2
1–7 days	61.9–81.8	—
2 weeks	65.4–81.3	—
1 month	46.4–70.3	0–0.8
2 months	31.8–60.9	0.4–2.2
3 months	13.5–54.9	0.4–2.3
4 months	9.7–25.9	0.8–2.1
5 months	1.8–18.4	0.6–2.5
6 months	2.9–13.8	1.4–2.5
8 months	2.2–12.1	1.4–2.6
10 months	1.9–3.3	1.9–2.8
12 months	1.5–4.9	2.0–2.7

Source: Derived from Schröter, W. and Nafz, C.,
Helvetica Paediatrica Acta, 36, 519, 1981. With permission.

TABLE IV

Red Blood Cell Reference Ranges on First Postnatal Day during the Last 16 Weeks of Gestation

	Gestational Age (weeks)							
	24–25	26–27	28–29	30–31	32–33	34–35	36–37	Term
Weight (g)	540–910	799–1187	1046–1302	1218–1702	1624–2008	1666–2248	2032–2458	—
RBC × 10^{12}/l	4.12–5.08	4.28–5.18	3.87–5.37	4.05–5.53	4.24–5.76	5.04–5.14	4.59–5.95	5.07–5.21
Hb (g/dl)	17.9–20.9	16.5–21.5	17.5–21.1	16.9–21.3	16.5–20.5	17.5–21.7	17.5–20.9	17.9–21.5
PCV/Hct (l/l)	59–67	54–70	53–67	52–68	52–68	54–68	57–71	53.6–68.4
MCV (fl)	135–135	118–146	118–144	114–140	107–139	112–132	108–134	110–128
Retic. (%)	5.5–6.5	6.4–9.8	5.0–10.0	3.8–7.8	3.1–6.9	2.3–5.5	2.4–6.0	1.8–4.6

Note: RBC, red blood cells; Hb, hemoglobin; PCV/Hct, packed cell volume/hematocrit; MCV, mean cell volume; Retic., reticulocyte count.

Source: Derived from Zaizov, R. and Matoth, Y., *Am. J. Hematol.*, 1, 276, 1976. With permission.

TABLE V

Red Blood Cell Reference Ranges from Birth to 18 Years

	Hemoglobin (g/dl)	PCV/Hct (l/l)	RBC (10^{12}/l)	MCV (fl)	MCH (pg)	MCHC (g/dl)	Retics (%)
Birth (cord blood)	13.5–19.5	42–62	3.9–5.5	98–118	31–37	30–36	1.8–4.6
1–3 days (capillary)	14.5–22.5	45–67	4.0–6.6	95–121	31–37	29–37	1.5–4.5
1 week	13.5–17.5	42–66	3.9–6.3	88–126	28–40	28–38	0.1–0.9
2 weeks	12.5–16.5	39–63	3.6–6.2	86–124	28–40	28–38	0.2–0.8
1 month	10.0–14.0	31–55	3.0–5.4	85–123	28–40	29–37	0.4 –1.2
2 months	9.0–4.0	28–42	2.7–4.9	77–115	26–34	29–37	0.9–2.3
3–6 months	9.5–14.0	29–41	3.1–4.5	74–108	25–35	30–36	0.4–1.0
0.5–2 years	10.5–13.5	33–39	3.7–5.3	70–86	23–31	30–36	0.2–1.8
2–6 years	12.5–13.5	34–40	3.9–5.3	75–87	24–30	31–39	0.2–1.8
6–12 years	11.5–13.5	35–45	4.0–5.2	77–95	25–31	31–37	0.2–1.8
12–18 years							
Male	13.0–16.0	37–50	4.5–5.3	78–98	25–35	31–37	0.2–1.8
Female	12.0–14.0	36–41	4.1–5.1	78–102	25–35	31–37	0.2–1.8

Note: PCV/Hct, packed cell volume/hematocrit; RBC, red blood cells; MCV, mean cell volume; MCH, mean cell hemoglobin; MCHC, mean cell hemoglobin concentration; Retics, reticulocyte count.

Source: Derived from Dallman, P.R., *Pediatrics*, 16th ed., Appleton-Century-Crofts, Norwalk, CT, 1977. With permission.

TABLE VI

Mean Serum Iron and Iron Saturation Percentage

Age (years)	Serum Iron (μg/dl)	Saturation (%)
0.5–2	16–120	6–38
2–6	20–124	7–43
6–12	23–123	7–43
18+	48–136	18–46

Source: From Koerper, M.A. and Dallman, P.R., *Pediatr. Res.*, 11, 473, 1977. With permission.

TABLE VII

Values of Serum Iron, Total Iron-Binding Capacity, and Transferrin Saturation from Infants during the First Year of Life

	Age (months)						
	0.5	1	2	4	6	9	12
SI							
Median (μmol/l)	22	22	16	15	14	15	14
95% range (μmol/l)	11–36	10–31	3–29	3–29	5–24	6–24	6–28
Median	120	125	87	84	77	84	78
95% range (μg/dl)	63–201	58–172	15–159	18–164	28–135	34–135	35–155
TIBC							
Mean ± SD (μmol/l)	34 ± 8	36 ± 8	44 ± 10	54 ± 7	58 ± 9	61 ± 7	64 ± 7
Mean ± SD (μg/dl)	191 ± 43	199 ± 43	246 ± 55	300 ± 51	321 ± 51	341 ± 42	358 ± 38
S%							
Median	68	63	34	27	23	25	23
95% range	30–99	35–94	21–63	7–53	10–43	10–39	10–47

Note: SI = serum iron; TIBC = total iron-binding capacity; S%, iron saturation (%)/transferrin saturation (%).

Source: From Saarinen, U.M. and Siimes, M.A., *J. Pediatr.*, 91, 875–877, 1977. With permission.

TABLE VIII

Red Blood Cell Enzyme Activities

Enzyme	Activity in Normal Adult RBC (IU/g Hb) [a]	Mean Activity in Newborn RBC as Percentage of Mean (100%) Activity in Normal Adult RBC
Aldolase	3.19 ± 0.86	140
Enolase	5.39 ± 0.83	250
Glucose phosphate isomerase	60.80 ± 1.00	162
Glucose 6-phosphate dehydrogenase	8.34 ± 1.59	174
WHO method	12.1 ± 2.09	—
Glutathione peroxidase	30.82 ± 4.65	56
Glyceraldehyde phosphate dehydrogenase	226 ± 41.9	170
Hexokinase	1.78 ± 0.38	239
Lactate dehydrogenase	200 ± 26.5	132
NADH-methemoglobin reductase (at 30°C)	19.2 ± 3.85	Increased
Phosphofructokinase	11.01 ± 2.33	97
Phosphoglycerate kinase	320 ± 36.1	165
Pyruvate kinase	15.0 ± 1.99	160
6-Phosphogluconate dehydrogenase	8.78 ± 0.78	150
Triose phosphate isomerase	211 ± 39.7	101

[a] Mean ± 1 SD at 37°C.

Source: From Hinchcliffe, R.F. and Lilleyman, J.S., Eds., *Practical Paediatric Haematology: A Laboratory Worker's Guide to Blood Disorders in Children*, John Wiley and Sons, New York, 1987. With permission.

TABLE IX

Peripheral Blood Leukocytes Reference Ranges ($\times 10^9$/l)

Age	Total Leukocytes	Neutrophils [a]	Lymphocytes	Monocytes (mean)	Eosinophils (mean)
Birth	9.0–30.0	6.0–26.0	2.0–11.0	1.1	0.4
12 hours	3.0–38.0	6.0–28.0	2.0–11.0	1.2	0.5
24 hours	9.4–34.0	5.0–21.0	2.0–11.5	1.1	0.5
1 week	5.0–21.0	1.5–10.0	2.0–17.0	1.1	0.5
2 weeks	5.0–20.0	1.0–9.5	2.0–17.0	1.0	0.4
1 month	5.0–19.5	1.0–9.0	2.5–16.5	0.7	0.3
6 months	6.0–17.5	1.0–8.5	4.0–13.5	0.6	0.3
1 year	6.0–17.5	1.5–8.5	4.0–10.5	0.6	0.3
2 years	6.0–17.0	1.5–8.5	3.0–9.5	0.5	0.3
4 years	5.5–15.5	1.5–8.5	2.0–8.0	0.5	0.3
6 years	5.0–14.5	1.5–8.0	1.5–7.0	0.4	0.2
8 years	4.5–13.5	1.5–8.0	1.5–6.8	0.4	0.2
10 years	4.5–13.5	1.8–8.0	1.5–6.5	0.4	0.2
16 years	4.5–13.0	1.8–8.0	1.2–5.2	0.4	0.2
21 years	4.5–11.0	1.8–7.7	1.0–4.8	0.3	0.2

[a] Includes band forms at all ages and a small number of metamyelocytes and myelocytes in the first few days of life.

Source: Derived from Dallman, P.R., in *Pediatrics*, 16th ed., Rudolph, A.M., Ed., Appleton-Century-Crofts, Norwalk, CT, 1997, p. 1178. With permission.

TABLE X

Reference Ranges of Platelets in Peripheral Blood

Age	Platelets ($\times 10^9$/l)
Preterm, 27–31 weeks	215–313
Preterm, 32–36 weeks	220–360
Term infant	232–378
Children	250–350

Source: Derived from Oski, F.A. and Naiman, J.L., in *Hematologic Problems in the Newborn*, 3rd ed., W.B. Saunders, Philadelphia, 1982. With permission.

TABLE XI

Reference Ranges for Folic Acid (µg/l)

Age	Range
Serum Folate	
Normal premature infants	
1–4 days	7.17–52.00
2–3 weeks	4.12–15.62
1–2 months	2.81–11.25
2–3 months	3.56–11.82
3–5 months	3.85–16.50
5–7 months	6.00–12.25
Normal children	
1 year	3.00–35.00
1–6 years	4.12–21.15
1–10 years	6.6–16.5
Red Cell Folate	
Infants <1 year	74–995
Children 1–11 years	96–364

Source: Derived from Shojania, A. and Gross, S., *J. Pediatr.*, 64, 323, 1964. With permission.

TABLE XII

Bone Marrow Cell Populations of Normal Infants in Tibial Bone Marrow (mean % ± SD)

Cell Type	Month										
	0	1	2	3	4	5	6	9	12	15	18
Lymphocyte	15.54 ± 6.67	48.94 ± 9.30	45.00 ± 7.95	45.74 ± 13.47	48.64 ± 9.88	48.96 ± 10.82	49.80 ± 9.04	50.62 ± 9.50	49.37 ± 13.22	44.41 ± 9.76	45.48 ± 9.56
Plasma cells	0.00 ± 0.02	0.02 ± 0.06	0.02 ± 0.05	0.00 ± 0.02	0.01 ± 0.03	0.05 ± 0.11	0.03 ± 0.07	0.01 ± 0.03	0.03 ± 0.07	0.07 ± 0.12	0.06 ± 0.08
Proerythroblasts	0.02 ± 0.06	0.01 ± 0.14	0.13 ± 0.19	0.10 ± 0.13	0.05 ± 0.10	0.07 ± 0.10	0.09 ± 0.12	0.07 ± 0.09	0.02 ± 0.04	0.07 ± 0.12	0.08 ± 0.13
Basophilic erythroblasts	0.24 ± 0.25	0.34 ± 0.33	0.57 ± 0.41	0.40 ± 0.33	0.24 ± 0.24	0.47 ± 0.33	0.32 ± 0.24	0.31 ± 0.24	0.30 ± 0.25	0.38 ± 0.37	0.50 ± 0.34
Polychromatic erythroblasts	13.06 ± 6.78	6.90 ± 4.45	13.06 ± 3.48	10.51 ± 3.39	6.84 ± 2.58	7.55 ± 2.35	7.30 ± 3.60	7.73 ± 3.39	6.83 ± 3.75	6.04 ± 1.56	6.97 ± 3.56
Orthochromatic erythroblasts	0.69 ± 0.73	0.54 ± 1.88	0.66 ± 0.82	0.70 ± 0.87	0.34 ± 0.30	0.46 ± 0.51	0.38 ± 0.56	0.39 ± 0.48	0.37 ± 0.51	0.50 ± 0.65	0.44 ± 0.49
Neutrophils											
Promyelocytes	0.79 ± 0.91	0.76 ± 0.65	0.78 ± 0.68	0.76 ± 0.80	0.59 ± 0.51	0.87 ± 0.80	0.67 ± 0.66	0.41 ± 0.34	0.69 ± 0.71	0.67 ± 0.58	0.64 ± 0.59
Myelocytes	3.95 ± 2.93	2.50 ± 1.48	2.03 ± 1.14	2.24 ± 1.70	2.32 ± 1.59	2.73 ± 1.82	2.22 ± 1.25	2.07 ± 1.20	2.32 ± 1.14	2.48 ± 0.94	2.49 ± 1.39
Metamyelocytes	19.37 ± 4.84	11.34 ± 3.59	11.27 ± 3.38	11.93 ± 13.09	6.04 ± 3.63	11.89 ± 3.24	11.02 ± 3.12	11.80 ± 3.90	11.10 ± 3.82	12.48 ± 7.45	12.42 ± 4.15
Bands	28.89 ± 7.56	14.10 ± 4.63	13.15 ± 4.71	14.60 ± 7.5	13.93 ± 6.13	14.07 ± 5.48	14.00 ± 4.58	14.08 ± 4.53	14.02 ± 4.88	15.17 ± 4.20	14.20 ± 5.23
Mature neutrophils	7.37 ± 4.64	3.64 ± 2.97	3.07 ± 2.45	3.48 ± 1.62	4.27 ± 2.69	3.77 ± 2.44	4.85 ± 2.69	3.97 ± 2.29	5.65 ± 3.92	6.94 ± 3.88	6.31 ± 3.91
Total eosinophils	2.70 ± 1.27	2.61 ± 1.40	2.50 ± 1.22	2.54 ± 1.22	2.37 ± 4.13	1.98 ± 0.86	2.08 ± 1.16	1.74 ± 1.08	1.92 ± 1.09	3.39 ± 1.93	2.70 ± 2.16
Total basophils	0.12 ± 0.20	0.07 ± 0.16	0.08 ± 0.10	0.09 ± 0.09	0.11 ± 0.14	0.09 ± 0.13	0.10 ± 0.13	0.11 ± 0.13	0.13 ± 0.15	0.27 ± 0.37	0.10 ± 0.12
Monocytes	0.88 ± 0.85	1.01 ± 0.89	0.91 ± 0.83	0.68 ± 0.56	0.75 ± 0.75	1.29 ± 1.06	1.21 ± 1.01	1.17 ± 0.97	1.46 ± 1.52	1.68 ± 1.09	2.12 ± 1.59
Megakaryocytes	0.06 ± 0.15	0.05 ± 0.09	0.10 ± 0.13	0.06 ± 0.09	0.06 ± 0.06	0.08 ± 0.09	0.04 ± 0.07	0.09 ± 0.12	0.05 ± 0.08	0.00 ± 0.00	0.07 ± 0.12
Unknown blasts	0.31 ± 0.31	0.62 ± 0.50	0.58 ± 0.50	0.63 ± 0.60	0.56 ± 0.53	0.50 ± 0.37	0.56 ± 0.48	0.42 ± 0.50	0.37 ± 0.33	0.46 ± 0.32	0.43 ± 0.45
Unknown cells	0.22 ± 0.34	0.21 ± 0.25	0.16 ± 0.24	0.19 ± 0.21	0.23 ± 0.25	0.23 ± 0.25	0.10 ± 0.15	0.14 ± 0.17	0.11 ± 0.14	0.13 ± 0.18	0.20 ± 0.23
Damaged cells	5.79 ± 2.78	5.50 ± 2.46	5.09 ± 1.78	4.75 ± 2.30	4.80 ± 2.29	4.86 ± 1.25	5.04 ± 1.08	4.89 ± 1.60	5.34 ± 2.19	4.99 ± 1.96	5.05 ± 2.15

Source: Derived from Rosse, C., Kraemer, M.J. et al., *J. Lab. Clin. Med.*, 89, 1225–1240, 1977. With permission.

TABLE XIII

References Ranges for Coagulation Tests in Healthy Full-Term Infants during First 6 Months of Life

	Day 1	Day 5	Day 30	Day 90	Day 180
Prothrombin time (sec)	10.1–15.9	10.0–15.3	10.0–14.3	10.0–14.2	10.7–13.9
Activated partial thromboplastin time (sec)	31.3–54.5	25.4–59.8	32.0–55.2	29.0–50.1	28.1–42.9
Thrombin clotting time (sec)	19.0–28.3	18.0–29.2	19.4–29.2	20.5–29.7	18.8–31.2
Fibrinogen (g/l)	1.67–3.99	1.62–4.62	1.62–3.78	1.50–3.79	1.50–3.87
Factor II (U/ml)	0.26–0.70	0.33–0.93	0.34–1.02	0.45–1.05	0.60–1.16
Factor V (U/ml)	0.34–1.08	0.45–1.45	0.62–1.34	0.48–1.32	0.55–1.27
Factor VII (U/ml)	0.28–1.04	0.35–1.43	0.42–1.38	0.39–1.43	0.47–1.27
Factor VIII C (U/ml)	0.50–1.78	0.50–1.54	0.50–1.57	0.50–1.25	0.50–1.09
Von Willebrand Factor (U/ml)	0.50–2.87	0.50–2.54	0.50–2.46	0.50–2.06	0.50–1.97
Factor IX (U/ml)	0.15–0.91	0.15–0.91	0.21–0.81	0.21–1.13	0.36–1.36
Factor X (U/ml)	0.12–0.68	0.19–0.79	0.31–0.87	0.35–1.07	0.38–1.18
Factor XI (U/ml)	0.10–0.66	0.23–0.87	0.27–0.79	0.41–0.97	0.49–1.34
Factor XII (U/ml)	0.13–0.93	0.11–0.83	0.17–0.81	0.25–1.09	0.39–1.15
Factor XIIIa (U/ml)	0.27–1.31	0.44–1.44	0.39–1.47	0.36–1.72	0.46–1.62
Factor XIIIb (U/ml)	0.30–1.22	0.32–1.80	0.39–1.73	0.48–1.84	0.50–1.70

Source: Derived from Andrew, M., Paes, B., Milner, R. et al., *Blood*, 70, 165, 1987. With permission.

TABLE XIV

Reference Ranges for Coagulation Tests in Healthy Premature Infants (30–36 Weeks Gestation) during the First 6 Months of Life

	Day 1	Day 5	Day 30	Day 90	Day 180
Prothrombin time (sec)	10.6–16.2	10.0–15.3	10.0–13.6	10.0–14.6	10.0–15.0
Activated partial thromboplastin time (sec)	27.5–79.4	26.9–74.1	26.9–62.5	28.3–50.7	21.7–53.3
Thrombin clotting time (sec)	19.2–30.4	18.8–29.4	18.8–29.9	19.4–30.8	18.9–31.5
Fibrinogen (g/l)	1.50–3.73	1.60–4.18	1.50–4.14	1.50–3.52	1.50–3.60
Factor II (U/ml)	0.20–0.77	0.29–0.85	0.36–0.95	0.30–1.06	0.51–1.23
Factor V (U/ml)	0.41–1.44	0.46–1.54	0.48–1.56	0.59–1.39	0.58–1.46
Factor VII (U/ml)	0.21–1.13	0.30–1.38	0.21–1.45	0.31–1.43	0.47–1.51
Factor VIII C (U/ml)	0.50–2.13	0.53–2.05	0.50–1.99	0.58–1.88	0.50–1.87
Von Willebrand Factor (U/ml)	0.78–2.10	0.72–2.19	0.66–2.16	0.75–1.84	0.54–1.58
Factor IX (U/ml)	0.19–0.65	0.14–0.74	0.13–0–80	0.25–0.93	0.50–1.20
Factor X (U/ml)	0.11–0.71	0.19–0.83	0.20–0.92	0.35–0.99	0.35–1.19
Factor XI (U/ml)	0.08–0.52	0.13–0.69	0.15–0.71	0.25–0.93	0.46–1.10
Factor XII (U/ml)	0.10–0.66	0.09–0.69	0.11–0.75	0.15–1.07	0.22–1.42
Factor XIIIa (U/ml)	0.32–1.08	0.57–1.45	0.51–1.47	0.71–1.55	0.65–1.61
Factor XIIIb (U/ml)	0.35–1.27	0.68–1.58	0.57–1.57	0.75–1.67	0.67–1.63

Source: Derived from Andrew, M., Paes, B., Milner, R. et al., *Blood*, 72, 1651–1657, 1988. With permission.

TABLE XV

Reference Ranges for Inhibitors of Coagulation in the Healthy Full-Term Infant during the First 6 Months of Life (mean ± 1 SD)

Inhibitors	Day 1	Day 5	Day 30	Day 90	Day 180
Antithrombin III (U/ml)	0.63 ± 0.12	0.67 ± 0.13	0.78 ± 0.15	0.97 ± 0.12	1.04 ± 1.10
α2-Antiplasmin (U/ml)	0.85 ± 0.15	1.00 ± 0.15	1.00 ± 0.12	1.08 ± 0.16	1.11 ± 0.14
Complement C1 inhibitor (U/ml)	0.72 ± 0.18	0.90 ± 0.15	0.89 ± 0.21	1.15 ± 0.22	1.41 ± 0.26
Heparin cofactor II (U/ml)	0.43 ± 0.25	0.48 ± 0.24	0.47 ± 0.20	0.72 ± 0.37	1.20 ± 0.35
Protein C (U/ml)	0.35 ± 0.09	0.42 ± 0.11	0.43 ± 0.11	0.54 ± 0.13	0.59 ± 0.11
Protein S (U/ml)	0.36 ± 0.12	0.50 ± 0.14	0.63 ± 0.15	0.86 ± 0.16	0.87 ± 0.16

Source: Derived from Andrew, M., Paes, B., Milner, R. et al., *Blood*, 70, 165, 1987. With permission.

TABLE XVI

Reference Ranges for Inhibitors of Coagulation in Healthy Premature Infants during the First 6 Months of Life

Inhibitor	Day 1	Day 5	Day 30	Day 90	Day 180
Antithrombin III (U/ml)	0.14–0.62	0.30–0.82	0.37–0.81	0.45–1.21	0.52–1.28
α2-Antiplasmin (U/ml)	0.40–1.16	0.49–1.13	0.55–1.23	0.64–1.48	0.77–1.53
Complement C1 inhibitor (U/ml)	0.31–0.99	0.45–1.21	0.40–1.24	0.60–1.68	0.96–2.04
Heparin cofactor II (U/ml)	0.00–0.80	0.00–0.69	0.15–0.71	0.20–1.11	0.45–1.40
Protein C (U/ml)	0.12–0.44	0.11–0.51	0.15–0.59	0.23–0.67	0.31–0.83
Protein S (U/ml)	0.14–0.38	0.13–0.61	0.22–0.90	0.40–1.12	0.44–1.20

Source: Derived from Andrew, M., Paes, B., Milner, R. et al., *Blood*, 72, 1651, 1988. With permission.

References

1. Henry, S. and Samuelsson, B., ABO polymorphisms and their putative biological relationships with disease, in *Human Blood Cells: Consequences of Genetic Polymorphism and Variations*, King, M.-J., Ed., Imperial College Press, London, 2000, pp. 1–103.
2. Regan, F., Newlands, M., and Bain, B.J., Acquired haemolytic anaemias, in *Dacie and Lewis Practical Haematology*, 10th ed., Lewis, S.M., Bain, B.J., and Bates, I., Eds., Churchill Livingstone, Elsevier, Philadelphia, 2006, pp. 239–270.
3. Young, N.S. and Maciejewski, J., The pathophysiology of acquired aplastic anemia, *New Engl. J. Med.*, 336, 1365–1372, 1997.
4. Brodsky, R.A. and Jones, R.J., Aplastic anemia, *Lancet*, 365, 1647–1656, 2005.
5. Young, N.S., The etiology of acquired aplastic anemia, *Rev. Clin. Exp. Haematol.*, 4, 236–259, 2000.
6. Feibbe, W.E., Telomerase mutations in aplastic anemia, *New Engl. J. Med.*, 352, 1481–1483, 2005.
7. British Committee for Standards in Haematology, Guidelines on the diagnosis and management of acquired aplastic anemia, *Br. J. Haematol.*, 123, 782–801, 2003.
8. Killick, S.B. and Marsh, J.C.W., Aplastic anemia: management, *Blood Rev.*, 28, 39–42, 2000.
9. Fauci, A.S., Multifactorial nature of human immunodeficiency virus infection, *Science*, 262, 1011–1018, 1993.
10. Jolles, S., de Loes, S.K., Johnson, M., and Janossy, G., Primary HIV-1 infection: a new medical emergency? *Br. Med. J.*, 312, 1243–1244, 1996.
11. Scarletti, G., Pediatric HIV infection, *Lancet*, 348, 863–867, 1996.
12. Yeni, P.G., Hammer, S.M., and Carpenter, C.C. et al., Antiviral treatment for adult HIV infection in 2002: updated recommendations of the International AIDS Society: USA panel, *JAMA*, 88, 222–235, 2002.
13. Dybul, M., Fauci, A.S., Bartlett, J.G., Kaplan, J.E., and Pay, A.K., Panel on clinical practices for treatment of HIV: guidelines for using retroviral agents among HIV-infected adults and adolescents, *Ann. Int. Med.*, 17, 381–433, 2002.
14. Lehrman, G., Hogue, I.B., Palmer, S. et al., Depletion of latent HIV-1 infection *in vivo*: a proof-of-concept study, *Lancet*, 366, 549–555, 2005.
15. Sepkowitz, K.A. and Armstrong, D., Treatment of opportunistic infections in AIDS, *Lancet*, 346, 588–589, 1995.
16. Bridges, S.H. and Sarver, N., Gene therapy and immune restoration for HIV disease, *Lancet*, 345, 427–432, 1995.
17. Loveday, C. and Hill, A., Prediction of progression to AIDS with serum HIV-1 RNA and CD4 count, *Lancet*, 345, 790–791, 1995.
18. Rinder, M.R., Richard, R.E., and Rinder, H.M., Acquired Von Willebrand Disease: a concise review, *Am. J. Hematol.*, 54, 139–145, 1997.
19. Laffan, M. and Manning, R., Investigation of haemostasis, in *Dacie and Lewis Practical Haematology*, 9th ed., Lewis, S.M., Bain, B.J., and Bates, I., Eds., Churchill Livingstone, Elsevier, Philadelphia, 2001, pp. 379–440.
20. Iperen, C.E., Wiel van de, C.E., and Mark, J.J.M., Acute event-related anaemia, *Br. J. Haematol.*, 115, 739–743, 2002.
21. Greaves, M.F., Aetiology of acute leukaemia, *Lancet*, 349, 344–349, 1997.
22. Russell, N.H., Biology of acute leukemia, *Lancet*, 349, 118–121, 1997.
23. Sawyers, C.L., Molecular genetics of acute leukemia, *Lancet*, 349, 196–200, 1997.
24. Burnett, A.K. and Eden, O.B., The treatment of acute leukaemia, *Lancet*, 349, 270–275, 1997.
25. Greaves, M., Childhood leukemia, *Br. Med. J.*, 324, 283–287, 2002.

26. British Committee for Standards in Haematology, The role of cytology, cytochemistry, immunophenotyping and cytogenetic analysis in the diagnosis of haematological malignancies, *Clin. Lab. Haematol.* 18, 231–236, 1996.

27. British Committee for Standards in Haematology, Revised guidelines for immunophenotyping of acute leukemia and chronic lymphoproliferative disorders, *Clin. Lab. Haematol.*, 24, 1–13, 2002.

28. Pui, C.H., Scrappe, M., Masera, G. et al., Ponte di Legno Working Group statement on the right of children to have full access to essential treatment and report of the Sixth International Childhood Acute Lymphoblastic Leukemia Workshop, *Leukemia*, 18, 1043–1053, 2004.

29. Piu, C.H., Relling, M.V., and Downing, D.R., Acute lymphoblastic leukaemia, *New. Engl. J. Med.*, 350, 1535–1548, 2004.

30. Winick, N.J., Carroll, W.L., and Hunger, S.P., Childhood leukemia: new advances and challenges, *New Engl. J. Med.*, 351, 601–603, 2004.

31. Jeha, S., Kantarjian, H., Irwin, D., Shen, V., Shenoy, S., Blaney, S., Camitta, B., and Pui, C.H., Efficacy and safety of rasburicase, a recombinant urate oxidase (Elitek), in the management of malignancy-associated hyperuricemia in pediatric and adult patients: final results in a multicenter compassionate use trial, *Leukemia*, 19, 34–38, 2005.

32. Grimwade, D. and Haferlich, T., Gene expression profiling in acute myeloid leukemia, *New Engl. J. Med.*, 350, 1535–1548, 2004.

33. Coco, F.L., Diverio, D., Falini, B., Biondi, A., Nervi, C., and Pelicci, P.G., Genetic diagnosis and molecular monitoring in the management of acute promyelocytic leukemia, *Blood*, 94, 12–22, 1999.

34. Tagi, T., Morimoto, M., Eguchi, S. et al., Identification of a gene expression syndrome associated with pediatric acute myeloid leukemia prognosis, *Blood*, 102, 1849–1856, 2003.

35. Redner, R.L., Rush, E.A., Faas, S., Rudert, W.A., and Corey, S.J., The t(5;17) variant of acute promyelocytic leukemia expresses a nucleo plasmin-retinoic acid receptor fusion, *Blood*, 87, 882–886, 1996.

36. Grisendi, S. and Pandolfi, P.P., NPM mutations in acute myelogenous leukemia, *New Engl. J. Med.*, 352, 291–292, 2005.

37. Schiffer, C.A., Hematopoietic growth factors as adjuncts to the treatment of acute myeloid leukemia, *Blood*, 88, 3675–3685, 1996.

38. Coco, F.L., Cimino, G., Breccia, M. et al., Gezutab (Myelotag) as a single agent for molecular relapsed acute promyelocytic leukemia, *Blood*, 104, 1995–1999, 2004.

39. Burnett, A.K., Goldstone, A.H., Stevens, R.M.F. et al., Randomised comparison of additional bone-marrow transplantation to intensive chemotherapy for acute myeloid leukemia in first remission: results of MRC AML 10 trial, *Lancet*, 350, 700–708, 1998.

40. Barbei, T., Finazzi, G., and Falened, A., The impact of All-*trans*-retinoic acid in coagulopathy of acute promyelocytic leukemia, *Blood*, 93, 399–425, 1999.

41. International Council for Standardization in Haematology, Guidelines on the selection of laboratory tests for monitoring the acute phase response, *J. Clin. Pathol.*, 41, 1203–1212, 1988.

42. Weiss, M., Moldawer, L.L., and Schneider, E.M., Granulocyte colony stimulating factor to prevent progression of systemic non-responsiveness in systemic inflammatory response syndrome, *Blood*, 93, 425–439, 1999.

43. Pirmohamed, M. and Park, P.K., Genetic susceptibility to adverse drug reactions: review, *Trends Pharmacol. Sci.*, 22, 298–205, 2001.

43a. Weinshilbourn, R., Inheritance and drug response, *New Engl. J. Med.*, 348, 529–537, 2003.

44. International Council for Standardization in Haematology (Expert Panel on Blood Rheology), Guidelines for measurement of blood viscosity and erythrocyte deformability, *Clin. Hemorrheol.*, 6, 439–453, 1986.

45. Pisciotta, A.V., Drug induced agranulocytosis peripheral destruction of polymorphonuclear leukocytes and their marrow precursors, *Blood Rev.*, 4, 226–237, 1990.

45a. Li, K., Li, C.K., Fok, T.F. et al. Neonatal blood: a source of haematopoetic stem cells for transplantation? *Lancet*, 351, 647, 1998.

46. Slavin, S., Nagler, A., Naparstek, E. et al., Non-myeloablative stem cell transplantation and cell therapy as an alternative to conventional bone marrow transplantation with lethal cytoreduction for the treatment of malignant and non-malignant hematologic diseases, *Blood*, 91, 756–763, 1998.

47. Jacobsohn, D.A., Duerst, R., Tse, W., and Kletzel, M., Reduced intensity haemopoietic stem-cell transplantation for treatment of non-malignant diseases in children, *Lancet*, 364, 156–162, 2004.

48. Hudson, B.G., Tryggvasen, M.D., Sundarmoorthy, M. et al., Alport's syndrome, Goodpasture's syndrome and Type IV collagen, *New Engl. J. Med.*, 348, 2543–2556, 2003.

49. Merlin, G. and Belloti, V., Molecular mechanisms of amyloidosis, *New Engl. J. Med.*, 349, 583–596, 2003.

50. Kyle, R.A. and Gertz, M.A., Primary systemic amyloidosis: clinical and laboratory features in 474 cases, *Semin. Hematol.*, 32, 45–59, 1995.

51. British Committee for Standards in Haematology, Guidelines on the diagnosis and management of amyloidosis (AL), *Br. J. Haem.*, 125, 681–700, 2004.

52. Dispenzieri, A., Kyle, R.A., Lacy, M.Q., Therneau, T.M., Larson, D.R., Plevak, M.F., Rajkumar, S.V., Fonseca, R., Greipp, P.R., Witzig, T.E., Lust, J.A., Zeldenrust, S.R., Snow, D.S., Hayman, S.R., Litzow, M.R., Gastineau, D.A., Tefferi, A., Inwards, D.J., Micallef, I.N., Ansell, S.M., Porrata, L.F., Elliott, M.A., and Gertz, M.A., Superior survival in primary systemic amyloidosis patients undergoing peripheral blood stem cell transplantation: a case-control study, *Blood*, 103, 3960–3963, 2004.

53. Weiss, G. and Goodnough, L.T., Anemia of chronic disorders, *New Engl. J. Med.*, 352, 1011–1023, 2005.

54. Cicardi, M. and Agostoni, A., Hereditary angioedema, *New Engl. J. Med.*, 334, 1666–1667, 1996.

55. Lachman, P., Peters, K., and Walport, M.J., Eds., *Clinical Aspects of Immunology*, Blackwell, Oxford, 1993, p. 306.

56. Robb-Smith, A.H.T. and Taylor, C.R., *Lymph Node Biopsy*, Miller Hayden, London, 1981, p. 100.

57. Macarty, M.J., Vukelja, S.J., Banks, P.M., and Weiss, R.B., Angiofollicular hyperplasia (Castleman's disease), *Cancer Treat. Rev.*, 21, 291–310, 1995.

58. Swales, C. and Bowness, P., Anti-tumor necrosis factor in seronegative spondyloarthritis, *Clin. Med.*, 5, 219–222, 2005.

59. Lewis, S.M., Miscellaneous tests, in *Dacie and Lewis Practical Haematology*, 9th ed., Lewis, S.M., Bain, B.J., and Bates, I., Eds., Churchill Livingstone, Edinburgh, 2001, pp. 595–607.

60. Levine, J.S., Branch, D.W., and Raunch, J., The antiphospholipid antibody syndrome, *New Engl. J. Med.*, 346, 752–763, 2002.

61. British Committee for Standards in Haematology, Guidelines for the investigation and management of the antiphospholipid syndrome, *Br. J. Haematol.*, 109, 704–715, 2000.

62. Roubey, R.A.S. and Hoffman, M., From antiphospholipid syndrome to antibody-mediated thrombosis, *Lancet*, 350, 1491–1493, 1997.

63. Paydas, S., Koçak, R., Zorludemir, S., and Baslamisli, F., Bone marrow necrosis in antiphospholipid syndrome, *J. Clin. Pathol.*, 50, 261–262, 1997.

64. Lockshin, M.D. and Erkan, D., Treatment of antiphospholipid syndrome, *New Engl. J. Med.*, 349, 1177–1179, 2003.

65. Kutteh, W.H., Antiphospholipid antibody associated with recurrent pregnancy loss: treatment with heparin and low dose aspirin is superior to low dose aspirin alone, *Am. J. Obstet. Gynecol.*, 174, 1584–1589, 1996.

66. Triplett, D.A. and Asherson, R.A., Pathophysiology of catastrophic antiphospholipid syndrome, *Am. J. Haematol.*, 65, 154–159, 2000.

67. Renehan, A.G., Booth, C., and Potten, C.S., What is apoptosis, and why is it important? *Br. Med. J.*, 322, 1536–1538, 2001.

68. Berliner, J., Navab, M., Fogelman, A.M. et al., Atherosclerosis: basic mechanisms, *Circulation*, 91, 2488–2496, 1995.

69. Libby, P., Atheroma: more than a mush, *Lancet*, 348 (Suppl. 1), 4–7, 1996.

70. Maseri, A., Biasucci, L.M., and Liuzzo, G., Inflammation in ischaemic heart disease, *Br. Med. J.*, 312, 1049–1050, 1996.

71. Berliner, J.A. and Watson, A.D., A role for oxidized phospholipids in atherosclerosis, *New Engl. J. Med.*, 353, 9–11, 2005.

72. Hansson, G.K., Inflammation, atherosclerosis and coronary artery disease, *New Engl. J. Med.*, 352, 1685–1695, 2005.

73. Wolf, P.A. and Cupples, A.L., Epidemiology of stroke, in *Oxford Textbook of Geriatrics*, Evans, G. and Williams, T.F., Eds., Oxford University Press, Oxford, 1992, pp. 304–312.

74. Robetorye, R. and Rodgers, G.M., Update on selected inherited thrombotic disorders, *Am. J. Hematol.*, 68, 256–268, 2001.

75. Bick, R.L., Cancer-associated thrombosis, *New Engl. J. Med.*, 349, 109–111, 2003.

76. Meade, T.W., Fibrinogen and cardiovascular disease, *J. Clin. Pathol.*, 50, 13–15, 1997.

77. Vandenbrouke, J.P. and Rosendaal, F.R., End of the line for "third-generation-pill" controversy? *Lancet*, 349, 1113–1114, 1997.

78. Shinton, R.A., Lifelong exposures and the potential for stroke prevention: the contribution of cigarette smoking, exercise, and body fat, *J. Epidemiol. Comm. Health*, 51, 138–143, 1997.

79. Kannel, W.B., Prevalence, incidence and mortality of coronary heart disease, in *Atherosclerosis and Coronary Artery Disease*, Vol. 1, Fuster, V., Ross, R., and Topol, E.J., Eds., Lippincott-Raven, Philadelphia, 1996, chap. 2.

80. Collins, R., Peto, R., Baigent, C., and Sleigt, P., Aspirin, heparin and fibrinolytic therapy in suspected acute myocardial infarction, *New Engl. J. Med.*, 336, 847–860, 1997.

81. Lange, R.A. and Hillis, L.D., Concurrent antiplatelet and fibrinolytic therapy, *New Engl. J. Med.*, 352, 1248–1250, 2005.

82. COMMIT (clopidogrel and metaprolol in myocardial infarction trail) Cooperation Group, The addition of clopidogrel to aspirin in 45,852 patients with acute myocardial infarction: randomized placebo-controlled trial, *Lancet*, 366, 1607–1621, 2005.

83. Rothberg, M.B., Celestin, C., Fiore, L.D., Lawler, E., and Cook, J.R., Warfarin plus aspirin after myocardial infarction or the acute coronary syndrome: meta-analysis with estimates of rank and benefit, *Ann. Int. Med.*, 143, 241–250, 2005.

84. Vaughan, C.J., Murphy, M., and Buckley, B.M., Statins do more than just lower cholesterol, *Lancet*, 348, 1079–1082, 1996.

85. Heart Protection Study Collaborative Group, MRC/BHF heart protection study of cholesterol lowering with simvastin in 20,536 high-risk individuals: a randomised placebo-controlled trial, *Lancet*, 360, 7–22, 2002.

86. CAPRIE Steering Committee, A randomized, blinded, trial of clopdogrel versus aspirin in patients at risk of ischemic effects (CAPRIE), *Lancet*, 348, 1329–1337, 1996.

87. The Medical Research Council's General Practice Research Framework, Thrombosis prevention trial: randomised trial of low-intensity oral anticoagulation with warfarin and low-dosage aspirin in the primary prevention of ischaemic heart disease in men at increased risk, *Lancet*, 351, 233–241, 1998.

88. Jandl, J.H., Physiology of red cells, in *Blood: Textbook of Hematology*, 2nd ed., Little, Brown, Boston, 1996, chap. 3.

89. Roper, D., Layton, M., and Lewis, S.M., Investigation of the hereditary haemolytic anaemias: membrane and enzyme abnormalities, in *Dacie and Lewis Practical Haematology*, 9th ed., Lewis, S.M., Bain, B.J., and Bates, I., Eds., Churchill Livingstone, Edinburgh, 2001, pp. 205–237.

90. Mackay, I.R., Tolerance and autoimmunity, *Brit. Med. J.*, 321, 93–96, 2000.

91. Dacie, J.V., The Haemolytic Anaemias, Vol. 3, *The Autoimmune Haemolytic Anaemias*, 3rd ed., Churchill Livingstone, London, 1995.

92. Provan, D., Autoimmune haematological disorders, *Clin. Med. JRCPL*, 1, 447–451, 2001.

93. British Committee for Standards in Haematology, Guidelines for the collection, processing and storage of human bone marrow and peripheral blood stem cells for transplantation, *Trans. Med.*, 4, 165–172, 1994.

94. British Committee for Standards in Haematology, Blood Transfusion Task Force, Guidelines for autologous transfusion: I, pre-operative autologous donation, *Transf. Med.*, 3, 307–316, 1993.

95. British Committee for Standards in Haematology, Blood Transfusion Task Force, Guidelines for autologous transfusion: II, perioperative haemodilution and cell salvage, *Br. J. Anaesth.*, 78, 768–771, 1997.

96. Muller, U., Exadaktylos, A., Roeder, C., Pisan, M., Eggli, S., and Juni, P., Effect of a flow chart on use of blood transfusions in primary total hip and knee replacement: prospective before and after study, *Br. Med. J.*, 328, 934–938, 2004.

97. Cushner, F.D., Hawes, T., Kessler, D., Hill, K., and Scuderi, G.R., Orthopaedic-induced anemia: the fallacy of autologous donation programs, *Clin. Orthopaed. Related Res.*, 431, 145–149, 2005.

98. Dixon, S., James, V., Hind, D., and Currie, C.J., Economic analysis of the implementation of autologous transfusion technologies throughout England, *Int. J. Tech. Assessment Health Care*, 21, 234–239, 2005.

99. Nath, A. and Pogrel, M.A., Preoperative autologous blood donation for oral and maxillofacial surgery: an analysis of 913 patients, *J. Oral Maxillofacial Surg.*, 63, 347–349, 2005.

100. Catling, S. and Joels, L., Cell salvage in obstetrics: the time has come, *Br. J. Obstet. Gynaecol.*, 112, 131–132, 2005.

101. Bannister, B.A., Haematological changes in viral infections and bacterial infection, in *Infection and Haematology*, Jenkins, G.J. and Williams, J.D., Eds., Butterworth & Heinemann, Oxford, 1994, p. 295–314.

102. Denburg, J.A., Basophil and mast cell lineages *in vitro* and *in vivo*, *Blood*, 79, 846–857, 1992.

103. Falcone, F.H., Hass, H., and Gibbs, B.F., The human basophil: a new appreciation of its role in immune responses, *Blood*, 96, 4028–4038, 2000.

104. British Committee for Standardisation in Haematology, Blood Transfusion Task Force, Guidelines on gamma irradiation of blood components for the prevention of transfusion-associated graft-versus-host disease, *Transf. Med.*, 6, 261–271, 1996.

105. British Committee for Standards in Haematology, Guidelines for the administration of blood and blood components and the management of transfused patients, *Transf. Med.*, 9, 227–239, 1999.

106. British Committee for Standards in Haematology, Guidelines for the administration of blood products: transfusion of infants and neonates, *Transf. Med.*, 4, 63–69, 1994.

107. British Committee for Standards in Haematology, Guidelines on the clinical use of leucocyte depleted blood components, *Transf. Med.*, 8, 59–71, 1998.

108. Wyncoll, D.L.A. and Evans, T.W., Acute respiratory distress syndrome, *Lancet*, 354, 497–502, 1999.

109. Ware, L.B. and Matthay, M.A., The acute respiratory distress syndrome, *New Engl. J. Med.*, 342, 1301–1308, 1334–1349, 2000.

110. Kleinman, S., Caulfield, T., Chan, P., Davenport, R., McFarland, J., McPhedran, S., Meade, M., Morrison, D., Pinsent, T., Robillard, P., and Slinger, P., Toward an understanding of transfusion-related acute lung injury: statement of a consensus panel, *Transfusion*, 44, 1774–1789, 2004.

111. Kopko, P.M., Paglieroni, T.G., Popovsky, M.A., Muto, K.N., MacKenzie, M.R., and Holland, P.V., TRALI: correlation of antigen–antibody and monocyte activation in donor-recipient pairs, *Transfusion*, 43, 177–184, 2003.

112. Silliman, C.C., Paterson, A.J., Dickey, W.O., Stroneck, D.F., Popovsky, M.A., Caldwell, S.A., and Ambruso, D.R., The association of biologically active lipids with the development of transfusion-related acute lung injury: a retrospective study, *Transfusion*, 7, 719–726, 1997.

112. Kopko, P.M., Paglieroni, T.G., Popovsky, M.A., Muto, K.N., MacKenzie, M.R., and Holland, P.V., TRALI: correlation of antigen–antibody and monocyte activation in donor-recipient pairs, *Transfusion*, 43, 177–184, 2003.

113. Popovsky, M.A. and Moore, S.B., Diagnostic and pathogenetic considerations in transfusion-related acute lung injury, *Transfusion*, 25, 573–577, 1985.

114. Palfi, M., Soren, B., Ernerudh, J., and Berlin, G.A., A randomized controlled trial of transfusion-related acute lung injury: is plasma from multiparous blood donors dangerous? *Transfusion*, 41, 317–322, 2001.

115. Wallis, J.P., Lubenko, A., Wells, A.W., and Chapman, C.E., Single hospital experience of TRALI, *Transfusion*, 43, 1053–1059, 2003.

116. Silliman, C.C., Boshkov, L.K., Mehdizadehkashi, Z., Elzi, D.J., Dickey, W.O., Podlosky, L., Clarke, G., and Ambruso, D.R., Transfusion-related acute lung injury: epidemiology and a prospective analysis of etiologic factors, *Blood*, 101, 454–462, 2003.

117. Holness, L., Knippen, M.A., Simmons, L., and Lachenbruch, P.A., Fatalities caused by TRALI, *Transfusion Med. Rev.*, 18, 184–188, 2004.

118. Win, N., Ranasinghe, E., and Lucas, G., Transfusion-related acute lung injury: a 5-year look-back study, *Transf. Med.*, 12, 387–389, 2002.

119. Daniels, G., *Human Blood Groups*, Blackwell, Oxford, 1995.

120. Reid, M.E. and Lomas-Francis, C., *The Blood Group Antigen Facts Book*, 2nd ed., Elsevier, London, 2004.

120a. Henry, S. and Samuelsson, B., ABO polymorphisms and their putative biological relationships with disease, in *Human Blood Cells: Consequences of Genetic Polymorphisms and Variations*, King, M.-J., Ed., Imperial College Press, London, 2000, pp. 1–103.

121. Hoffbrand, A.V. and Pettit, J.E., *Essential Haematology*, 3rd ed., Blackwell, Oxford, 1993.

122. Osgood, E.E. and Seaman, A.J., The cellular composition of normal bone marrow as obtained by sternal puncture, *Physiolog. Rev.*, 24, 46–49, 1944.

123. Vaughan, S.L. and Brockmyre, F., Normal bone marrow as obtained by sternal puncture, in *Blood: Morphologic Hematol.*, 1, 54, 1947.

124. Bain, B.J., Clark, D.M., Lampert, I.A., and Wilkins, B.S., *Bone Marrow Pathology*, 3rd ed., Blackwell, Oxford, 2001.

125. Gartner, S. and Kaplan, H.S., Long-term culture of human bone marrow cells, *Proc. Natl. Acad. Sci. U.S.A.*, 77, 4756–4759, 1980.

126. Johnson, A. and Dorshkind, K., Stromal cells in myeloid and lymphoid long-term bone marrow cultures can support hemopoietic lineages and modulate their production of hemopoietic growth factors, *Blood*, 68, 1348–1354, 1986.

127. Smith, R.R.L. and Spival, J.L., Marrow necrosis in anorexia nervosa and in involuntary starvation, *Br. J. Haematol.*, 60, 525–530, 1985.

128. Richardson, P.G., Sonneveld, P., Schuster, M.W. et al., Bortezomib or high-dose dexamethazone for relapsed multiple myeloma, *New Engl. J. Med.*, 352, 2487–2498, 2005.

129. Dispenzieri, A., Bortezomib for myeloma: much ado about something, *New Engl. J. Med.*, 352, 2546–2548, 2005.

130. International Council for Standardization in Haematology, Rules and Operating Procedures, ICSH, Glasgow, 1991.

131. International Council for Standardization in Haematology, The assignment of values to fresh blood used for calibrating automated blood cell counters, *Clin. Lab. Haematol.*, 10, 203–212, 1988.

131a. Bevan, D.H., Cardiac by-pass haemostasis, *Br. J. Haematol.*, 104, 208–219, 1999.

132. Andrian, U.H. and Engelhardt, B., α4 Integrins as therapeutic targets in autoimmune disease, *New Engl. J. Med.*, 348, 68–72, 2003.

133. British Committee for Standards in Haematology, Guidelines on the insertion and management of central venous lines, *Br. J. Haematol.*, 98, 1041–1047, 1997.

134. Irons, R.D., Ed., *Toxicology of the Blood and Bone Marrow*, Raven Press, New York, 1985.

135. Luster, A.D., Chemokines: chemotactic chemokines that mediate inflammation, *New Engl. J. Med.*, 338, 436–445, 1998.

136. Klien, A.D., Noel, P., Akin, C. et al., Elevated serum tryptase kinase levels identify a subset of patients with a myeloproliferative variant of idiopathic hypereosinophilic syndrome associated with fibrosis, poor prognosis and imatinib responsiveness, *Blood*, 101, 4660–4666, 2003.

137. Chiorazzi, N., Rai, K.R., and Ferrarini, M., Mechanisms of disease: chronic lymphatic leukemia, *New Engl. J. Med.*, 352, 804–813, 2005.

138. British Committee for Standards in Haematology, Guidelines on the diagnosis and management of chronic lymphatic leukemia, *Br. J. Haematol.*, 125, 294–317, 2004.

139. Orchard, J.A., Ibbotson, R.E., Davis, Z. et al., ZAP-70 expression and prognosis in chronic lymphocytic leukemia, *Lancet*, 363, 105–111, 2004.

140. Byrd, J.C., Rai, K., and Bercedis, L., Addition of rituximab to fludarabine may prolong progression: free survival and overall survival in patients with previously untreated chronic lymphatic leukemia: an updated retrospective comparative analysis of CALGB 9712 and CALGB 9011, *Blood*, 105, 49–53, 2005.

141. Deininger, M.W.M., Goldman, J., and Melo, J.V., The molecular biology of chronic myeloid leukemia, *Blood*, 96, 3343–3356, 2000.

142. British Committee for Standards in Haematology, Position paper on the therapeutic use of imatinib mesylate in chronic myeloid leukemia, *Br. J. Haematol.*, 119, 268–272, 2002.

142a. O'Brien, S.G., Guilhot, F., Larsden, R.A. et al., Imatinib compared with interferon and low dose cytarabine for newly diagnosed chronic-phase chronic myeloid leukemia, *New Engl. J. Med.*, 348, 994–1004, 2003.

143. van den Besselaar, A.M.H.P. and Bertina, R.M., Standardization and quality control in blood coagulation assays, in *Quality Assurance in Haematology*, Lewis, S.M. and Verwilghen, R.L., Eds., Baillière Tindall, London, 1988.

144. Barbara, J. and Flanagan, P., Blood transfusion risk: protecting against the unknown, *Br. Med. J.*, 316, 717–718, 1998.

145. Shinton, N.K., Vitamin B$_{12}$ and folate metabolism, *Br. Med. J.*, 1, 556–559, 1972.

146. Carmell, R., Current concepts in cobalamin deficiency, *Annu. Rev. Med.*, 51, 357–375, 2000.

147. British Committee for Standards in Haematology, Guidelines on the investigation and diagnosis of cobalamin and folate deficiencies, *Clin. Lab. Haematol.*, 16, 101–115, 1994.

148. Hamilton, M. and Blackmore, S., Investigation of megaloblastic anaemia, In *Dacie and Lewis Practical Haematology*, 10th ed., Lewis, S.M., Bain, B.J., and Bates, I., Eds., Churchill Livingstone, Elsevier, Philadelphia, 2006, pp. 161–185.

149. Regan, F., Newlands, M., and Bain, B.J., Acquired haemolytic anaemias, in *Dacie and Lewis Practical Haematology*, 9th ed., Lewis, S.M., Bain, B.J., and Bates, I., Eds., Churchill Livingstone, Edinburgh, 2001, pp. 213–214.

150. Grocott, M.P.W. and Hamilton, M.A., Resuscitation fluids, *Vox Sang.*, 82, 1–8, 2001.

151. The SAFE study investigators, A comparison of albumin and saline for fluid resuscitation in the intensive care unit, *New Engl. J. Med.*, 350, 2247–2256, 2004.

152. British Committee for Standards in Haematology, Guidelines on the use of colony stimulating factors in haematological malignancy, *Br. J. Haematol.*, 123, 22–33, 2003.

153. Marks, P.W. and Mitus, A.J., Congenital dyserythropoietic anemias, *Am. J. Hematol.*, 51, 55–63, 1996.

154. Dacie, J.V., The Haemolytic Anaemias, Vols. 1 and 2, *The Hereditary Haemolytic Anaemias*, 3rd ed., Churchill Livingstone, London, 1995.

155. Hughes, G.R.V., *Connective Tissue Diseases*, 3rd ed., Blackwell, Oxford, 1994.

156. Poller, L., Ed., *Oral Anticoagulation*, Edward Arnold, London, 1996.

157. World Health Organisation Expert Committee on Biological Standardization: human C-reactive protein, *WHO Tech. Rep. Series*, 760, 21–22, 1987.

158. Tall, A.R., C-reactive protein reassessed, *New Engl. J. Med.*, 350, 1540–1551, 2004.

159. Ridker, P.M., Cannon, C.P., and Morrow, D., C-reactive protein levels and outcomes of statin therapy, *New Engl. J. Med.*, 352, 20–28, 2005.

160. British Committee for Standards in Haematology, Guidelines for the use of fresh frozen plasma, cryoprecipitate and cryosupernatant, *Br. J. Haematol.*, 126, 11–28, 2004.

161. British Committee for Standards in Haematology, Evaluation of biodepleted plasmas, *Thrombosis and Haemostasis*, 70, 433–437, 1993.

162. Staudt, L.M., Molecular diagnosis of hematological cancers, *New Engl. J. Med.*, 348, 1777–1785, 2003.

163. Callard, R. and Gearing, A., Eds., *The Cytokine Factsbook*, Academic Press, London, 1994.

164. Feldmann, M., Brennan, F.M., and Main, R.N., Role of cytokines in rheumatoid arthritis, *Annu. Rev. Immunol.*, 14, 397–440, 1996.

165. Griffiths, P.D., Cytomegalovirus and Epstein-Barr virus infections, in *Topley and Wilson's Principles of Bacteriology, Virology and Immunity*, Vol. 4, 8th ed., Parker, M.T. and Collier, L.H., Eds., Edward Arnold, London, 1990, pp. 442–449.

165a. Bates, I. and Mendelow, B., Haematology in under-resourced laboratories, in *Dacie and Lewis Practical Haematology*, 9th ed., Lewis, S.M., Bain, B.J., and Bates, I., Eds., Churchill Livingstone, Edinburgh, 2001, pp. 673–698.

166. British Committee for Standards in Haematology, Diagnosis of deep vein thrombosis in symptomatic outpatients and the potential for clinical assessment and D-Dimer assays to reduce the need for diagnostic imaging, *Br. J. Haematol.*, 124, 15–25, 2004.

167. Reid, C.D.L., The dendritic cell lineage in haemopoiesis, *Br. J. Haematol.*, 96, 217–223, 1997.

168. Hart, D.N.J., Dendritic cell biology evolves into clinical application, *Lancet*, 365, 102–104, 2005.

169. Fonseca, R., Yamakawa, M., Nakamura, S., et al., Follicular dendritic cell sarcoma and interdigitating dendritic cell sarcoma, *Am. J. Hematol.*, 59, 161–167, 1998.

170. Gibbons, R.V., Dengue: an escalating problem, *Br. Med. J.*, 324, 1563–1566, 2002.

171. National Committee for Clinical Laboratory Standards, document H20-A, Reference Differential Count (Proportional) and Evaluation of Instrumental Methods, NCCLS, Wayne, PA, 1992.

172. British Committee for Standards in Haematology, Position paper on the therapeutic use of Rituximab in CD20-positive diffuse large B-cell non-Hodgkin's lymphoma, *Br. J. Haematol*, 121, 44–48, 2003.

173. Knowles, S.M. and Regan, F., Blood cell antigens and antibodies:erythrocytes, platelets and granulocytes, in *Dacie and Lewis Practical Haematology*, 9th ed., Lewis, S.M., Bain, B.J., and Bates, I., Eds., Churchill, Livingstone, Edinburgh, 2001, pp. 481–522.

174. George, J.N. and Shattil, S.J., Acquired disorders of platelet function, in *Haematology, Basic Practice and Principles*, 2nd ed., Hoffman, R., Benz, E.J., Shattil, S.J. et al., Eds., Churchill Livingstone, New York, 1995, chap. 130.

175. Brouqui, Ph., Dumler, J.S., Lienhard, R., Brossal, M., and Raoult, D., Human granulocytosis ehrlichiosis in Europe, *Lancet*, 346, 782–783, 1995.

176. Parsonnet, J. and Isaacson, P.G., Bacterial infection and MALT lymphoma, *New Engl. J. Med.*, 350, 213–215, 2004.

177. American Association of Blood Banks, *Technical Manual*, 14th ed., Brecher, M.E., Ed., AABB, Bethesda, MD, 2002.

178. Brito-Babapulle, F., The eosinophilias, including the idiopathic hypereosinophilic syndrome, *Br. J. Haematol.*, 121, 203, 2003.

179. Burton, G.H., Rash and pulmonary eosinophilia associated with brufen, *Br. Med. J.*, 300, 82–83, 1990.

180. Samter, M., Frank, M.M., Austen, K.F., and Clamen, H.N., *Immunological Diseases*, Little, Brown, Boston, 1988.

181. Rowe, M., Epstein-Barr virus and lymphoid malignancy, in *Recent Advances in Haematology*, Vol. 6, Hoffbrand, A.V. and Brenner, M.K., Eds., Churchill Livingstone, London, 1992, chap. 11.

182. Haque, T., Wilkie, E.M., and Taylor, C., Treatment of Epstein-Barr virus: positive post-transplantation lymphoproliferative disease with partly HLA-matched allogeneic cytotoxic T-cells, *Lancet*, 360, 436–442, 2002.

183. International Council for Standardization in Haematology (Expert Panel on Blood Rheology), Recommendations for measurement of erythrocyte sedimentation rate, *J. Clin. Pathol.*, 46, 198–203, 1993.

184. British Committee for Standards in Haematology, Guidelines on the diagnosis and management of polycythemia/erythrocytosis, *Br. J. Haematol.*, 130, 174–195, 2005.

185. Drenth, J.P.H. and Michiels, J.J., Three types of erythromyelalgia, *Br. Med. J.*, 301, 454–455, 1990.

186. Chatterjee, P.K., Pleiotropic renal actions of erythropoietin, *Lancet*, 365, 1890–1892, 2005.

187. Wasserman, L.R., Balcerak, S.P., Berk, P.D. et al., Influence of therapy in polycythemia rubra vera, *Trans. Am. Assoc. Physicians*, 94, 30–38, 1981.

188. International Council for Standardization of Haematology, Recommendations for ethylenediaminetetraacetic acid for blood cell counting and sizing, *Am. J. Clin. Pathol.*, 100, 371–372, 1993.

189. Vandenbrouke, J.P., Koster, T., Briet, E. et al., Increased risk of venous thrombosis in oral contraceptive users who are carriers of factor V Leiden mutation, *Lancet*, 344, 1454–1457, 1994.

189a. Hay, C.R.M., Why do inhibitors arise in patients with Haemophilia A? *Br. J. Haematol.*, 105, 584–90, 1999.

190. Hoffman, H.M., Rosengren, S., Boyle, D.L. et al., Prevention of cold-associated acute inflammation in familial cold autoinflammatory syndrome by interleukin-1 receptor antagonist, *Lancet*, 364, 1779–1783, 2004.

191. Deistt, F.L., Emile, J.-F., Rieeux-Laucat, F. et al., Clinical, immunological and pathological consequences of Fas-deficient conditions, *Lancet*, 348, 719–723, 1996.

192. British Committee for Standards in Haematology, Antenatal serology testing in pregnancy, *Br. J. Obstet. Gynaecol.* 103, 195–196, 1996.

193. Lo, Y.-M.D., Non-invasive prenatal diagnosis using fetal cells in maternal blood, *J. Clin. Pathol.*, 47, 1060–1065, 1994.

194. British Committee for Standards in Haematology, Guidelines for the fetal diagnosis of globin gene disorders, *J. Clin. Pathol.*, 47, 199–204, 1994.

195. British Committee for Standards in Haematology, Transfusion guidelines for neonates and older children, *Br. J. Haematol.*, 124, 433–453, 2004.
196. British Committee for Standards in Haematology, Guidelines on fibrinogen assays, *Br. J. Haematol.*, 121, 396–404, 2003.
196a. Cesarman-Maus, G. and Hajjar, K.A., Molecular mechanisms of fibrinolysis, *Br. J. Haematol.*, 129, 307–321, 2005.
196b. Chalmers, R.A. and Gibson, B.E.S., Thrombolytic therapy in the management of thromboembolic disease, *Br. J. Haematol.*, 104, 14–21, 1999.
197. Groner, W. and Simson, E., *Practical Guide to Modern Hematology Analysers*, John Wiley & Sons, New York, 1995.
198. Ormerod, M.G., *Flow Cytometry: A Practical Approach*, 2nd ed., IRL Press, Oxford University Press, Oxford, 1994.
199. British Committee for Standards in Haematology, Guidelines for the flow cytometric enumeration of CD34+ haematopoietic stem cells, *Clin. Lab. Haematol.*, 21, 301–308, 1999.
200. Wilken, D.E.L., MTHFR 677c Æ mutation, folate intake, neural tube defect and risk of cardiovascular disease, *Lancet*, 350, 603–604, 1997.
201. Schwartz, R.S., Autoimmune folate deficiency and the rise and fall of "Horror Autoxicus," *New Engl. J. Med.*, 352, 1948–1950, 2005.
202. Mills, J.L., von Kohorn, I., Conley, M.R. et al., Low vitamin B12 concentrations in patients without anemia: the consequences of fortification of grain, *Am. J. Clin. Nutr.*, 77, 1474–1477, 2003.
202a. Kaminski, M.S., Tuck, M., Estes, J. et al., ^{131}I-Tositumomab therapy for follicular lymphoma, *New Engl. J. Med.*, 352, 441–448, 2005.
202b. Connors, J.M., Radioimmunotherapy: hot new treatment for lymphoma, *New Engl. J. Med.*, 352, 496–467, 2005.
203. Shohan, S. and Levitz, S.M., The immune responses to fungal infections, *Br. J. Haematol.*, 129, 569–582, 2005.
204. Prentice, H., Grant, H.G., Kibble, C.C., and Prentice, A.G., Towards a targeted risk based antifungal strategy in neutropenic patients, *Br. J. Haematol.*, 110, 273–284, 2000.
205. Lovat, L.B., Age related changes in gut physiology and nutritional status, *Gut*, 38, 306–309, 1996.
206. Tsai, H.H., Helicobacter pylori for the general physician, *J.R. Coll. Physicians Lond.*, 31, 478–482, 1997.
207. Jmoudiak, M. and Futerman, A.H., Gaucher disease: pathological mechanisms and modern management, *Br. J. Haematol.*, 129, 178–188, 2005.
208. Mistry, P.K., Wraight, E.P., and Cox, T.M., Therapeutic delivery of proteins to macrophages: implications for treatment of Gaucher's disease, *Lancet*, 348, 1555–1559, 1996.
209. Nogushii, P., Risks and benefits of gene therapy, *New Engl. J. Med.*, 348, 193–194, 2003.
210. Ramaswarmy, S., Translating cancer genomics into clinical oncology, *New Engl. J. Med.*, 350, 1814–1816, 2004.
211. Swirsky, D. and Bain, B.J., Erythrocyte and leucocyte cytochemistry: leukemia classification, in *Dacie and Lewis Practical Haematology*, 9th ed., Lewis, S.M., Bain, B.J., and Bates, I., Eds., Churchill Livingstone, Edinburgh, 2001, pp. 311–333.
212. Wahner-Roedler, D.L., Witzig, T.E., Loehrer, L.L., and Kyle, R.A., Gamma-heavy chain disease: review of 23 cases, *Medicine*, 82, 236–250, 2003.
213. Wahner-Roedler, D.L., Heavy chain disease, in *Myeloma: Biology and Management*, Malpas, J.S., Bergsagel, D.E., Kyle, R.A., and Anderson, K.C., Eds., Oxford University Press, New York, 1995; reprint, Saunders, Philadelphia, 2004.
214. Brouet, J.-C., Clauvel, J.-P., Danon, F. et al., Biologic and clinical significance of cryoglobulin, *Am. J. Med.*, 57, 775–788, 1974.
215. British Committee for Standards in Haematology, Guidelines for the clinical use of a blood cell separator, *Clin. Lab. Haematol.*, 12, 141–158, 1990.
216. Adams, P.F. and Benson, V., Current estimates from the National Heart Interview Survey, U.S., 1989, *Vital and Health Statistics*, DHHS Pub. (PHS) 90, 504, U.S. Government Printing Office, Washington, DC, 1990.

217. Wolf, P.A. and Cupples, A.L., Epidemiology of stroke, in *Oxford Textbook of Geriatrics*, Evans, G. and Williams, T.F., Eds., Oxford University Press, Oxford, 1992, pp. 304–312.

218. Anderson, F.A., Jr., Wheeler, H.B., Goldberg, R.J. et al., A population-based perspective of the hospital incidence and case-fatality rates of deep vein thrombosis and pulmonary embolism: the Worcester DVT study, *Arch. Int. Med.*, 151, 933–938, 1991.

219. Jandl, J.H., Hemoglobinopathies, in *Blood: Textbook of Hematology*, 2nd ed., Little, Brown, Boston, 1996, chap. 13.

220. Jandl, J.H., Hypochromic anemias and disorders of iron metabolism, in *Blood: Textbook of Hematology*, 2nd ed., Little, Brown, Boston, 1996, chap. 6.

221. Parkin, D.M., Muir, C.S., and Whelan, S.L., Cancer incidence in five continents, in *IARC Scientific Publications*, Vol. 7, No. 143, Lyon, France, 1997.

222. Linet, M.S. and Devesa, S.S., Epidemiology of leukaemia, overview and patterns of occurrence, in *Leukemia*, 7th ed., Henderson, E.S., List, T.A., and Greaves, M.F., Eds., Saunders, Philadelphia, 2002.

223. Jandl, J.H., Hemopoietic malignancies, in *Blood: Textbook of Hematology*, 2nd ed., Little, Brown, Boston, 1996, chap. 20.

224. Department of Health Office of Population Censuses and Surveys, Hospital In-patient Enquiry: In-patient and Day Case Trends, Her Majesty's Stationery Office, London, 1989.

225. Quaglino, D. and Hayhoe, F.G.J., Chronic myeloproliferative disorders, in *Haematological Oncology: Clinical Practice*, Churchill Livingstone, London, 1992, chap. 4.

226. Quaglino, D. and Hayhoe, F.G.J., Chronic lymphatic leukaemia and related disorders, in *Haematological Oncology: Clinical Practice*, Churchill Livingstone, London, 1992, chap. 5.

227. Quaglino, D. and Hayhoe, F.G.J., Multiples myeloma and other differentiated B-cell malignancies, in *Haematological Oncology: Clinical Practice*, Churchill Livingstone, London, 1992, chap. 9.

228. Quaglino, D. and Hayhoe, F.G.J., Hodgkin's disease, in *Haematological Oncology: Clinical Practice*, Churchill Livingstone, London, 1992, chap. 7.

229. Quaglino, D. and Hayhoe, F.G.J., Non-Hodgkin's lymphoma, in *Haematological Oncology: Clinical Practice*, Churchill Livingstone, London, 1992, chap. 8.

230. Jandl, J.H., Hemophilias, in *Blood: Textbook of Hematology*, 2nd ed., Little, Brown, Boston, 1996, chap. 32.

231. Taichman, R.S., Blood and bone, two tissues whose fates are intertwined to create the hematopoietic stem cell niche, *Blood*, 105, 2631–2639, 2004.

232. Lehmann, H. and Huntsman, R.G., *Man's Haemoglobins*, Elsevier, Amsterdam, 1966.

233. Dickerson, R.E., X-ray analysis and protein structure, in *The Proteins*, 2nd ed., Neurath, H., Ed., Academic Press, New York, 1964, p. 634.

234. Hoffbrand, A.V. and Lewis, S.M., *Tutorials in Postgraduate Medicine: Haematology*, William Heinemann, London, 1966.

235. Lewis, S.M., Stott, G.J., and Wynn, K.J., An inexpensive and reliable new haemoglobin colour scale for assessing anaemia, *J. Clin. Pathol.*, 51, 21–24, 1998.

236. Wild, B. and Bain, B.J., Investigations of abnormal haemoglobins and thalassaemia, in *Dacie and Lewis Practical Haematology*, 9th ed., Lewis, S.M., Bain, B.J., and Bates, I., Eds., Churchill Livingstone, Edinburgh, 2001, pp. 271–310.

237. British Committee for Standards in Haematology, Guidelines on laboratory diagnosis of haemoglobinopathies, *Br. J. Haematol.*, 101, 783–792, 1998.

238. British Committee for Standards in Haematology, Guidelines for haemoglobinopathy screening, *Clin. Lab. Haematol.*, 10, 87–94, 1988.

239. British Committee for Standards in Haematology, Guidelines for the fetal diagnosis of globin gene disorders, *J. Clin. Pathol.*, 47, 199–204, 1994.

240. Chudwin, D.S. and Rucknagel, D.L., Immunological quantification of hemoglobin F and A2, *Clin. Chem. Acta*, 50, 413–418, 1974.

241. Makler, M.T. and Pesce, A.J., ELISA assay for measurement of hemoglobin A and hemoglobin F, *Am. J. Clin. Pathol.*, 74, 673–676, 1980.

242. Betke, K., Marti, H.R., and Schlicht, I., Estimation of small percentage of fetal hemoglobin, *Nature*, 184, 1877, 1959.

243. Jonxis, J.H.P. and Visser, H.K.A., Determination of low percentages of fetal hemoglobin in blood of normal children, *Am. J. Dis. Ch.*, 92, 588–591, 1956.

244. Wild, B.J. and Stephens, A.D., The use of automated HPLC to detect and quantitate hemoglobins, *Clin. Lab. Haematol.*, 19, 171–176, 1997.

245. Dacie, J.V., The Haemolytic Anaemias, Vol. 4, *Secondary and Symptomatic Haemolytic Anaemias*, 3rd ed., Churchill Livingstone, London, 1995.

246. Mollison, P., *Blood Transfusion in Clinical Medicine*, 7th ed., Mollison, P., Ed., Blackwell, Oxford, 1997.

247. Contreras, M., The prevention of Rh haemolytic disease of the fetus and newborn: general background, *Br. J. Obstet. Gynaecol.*, 105 (Suppl. 18), 7–10, 1998.

248. Robson, S.C., Lee, D., and Urbaniak, S., Anti-D immunoglobulin in Rh D prophylaxis, *Br. J. Obstet. Gynaecol.*, 105, 129–133, 1998.

249. British Committee for Standards in Haematology, Guidelines for the estimation of fetal-maternal haemorrhage, *Transf. Med.*, 9, 87–92, 1999.

250. MacKenzie, I.Z., Bichler, J., Mason, G.C. et al., Efficacy and safety of a new, chromatographically purified rhesus (D) immunoglobulin, *Eur. J. Obstet. Gynaecol., Reprod. Biol.*, 117, 154–161, 2004.

251. National Blood Transfusion Service Immunoglobulin Working Party, Recommendations for the use of anti-D immunoglobulin, *Prescriber's J.*, 31, 137–145, 1991.

252. Fitzpatrick, M., Haemolytic uraemic syndrome and *E. coli 0157*, *Br. Med. J.*, 318, 684–685, 1999.

252a. British Committee for Standards in Haematology, Guidelines on the diagnosis and management of microangiopathic haemolytic anaemias, *Br. J. Haematol.*, 120, 556–573, 2003.

253. Giangrande, P.L.F., Hepatitis in haemophilia, *Br. J. Haematol.*, 103, 1–9, 1998.

254. Draper, G.J. and McNinch, A.W., Vitamin K for neonates: the controversy, *Br. Med. J.*, 308, 867–868, 1994.

255. Niedermaier, G. and Briner, V., Henoch-Schönlein syndrome induced by carbidopa/levodopa, *Lancet*, 349, 1071–1072, 1997.

255a. Donadio, J.V. and Grande, J.P., IgA nephropathy, *New Engl. J. Med.*, 347, 738–748, 2002.

256. Ganz, T., Hepcidin, a key regulator of iron metabolism and mediator of anemia of inflammation, *Blood*, 102, 783–788, 2003.

257. Beutler, E., Genetic iron bound hemochromatosis: clinical effects of HLA-H mutations, *Lancet*, 349, 296–297, 1997.

258. Roberts, A.G., Whatley, S.D., Morgan, R.R. et al., Increased frequency of the haemochromatosis Cys282Tyr mutation in sporadic porphyria cutanea tarda, *Lancet*, 349, 321–323, 1997.

259. Adams, P.C., Screening for haemochromatosis: producing or preventing illness? *Lancet*, 366, 269–271, 2005.

260. British Committee for Standards in Haematology, Guidelines on the diagnosis and management of hereditary spherocytosis, *Br. J. Haematol.*,126, 455–474, 2004.

261. Jandl, J.H., Lymphocytes and plasma cells, in *Blood: Textbook of Hematology*, 2nd ed., Little, Brown, Boston, 1996, chap. 8.

262. Seljelid, J. and Eskeland, T., The biology of macrophages, general principles and properties, *Eur. J. Haematol.*, 151, 267–275, 1993.

263. Diepstra, A., Niens, M., Vellenga, E. et al., Association with HLA Class I in Epstein-Barr-virus-positive and with HLA Class III in Epstein-Barr-virus-negative Hodgkin's lymphoma, *Lancet*, 365, 2216–2224, 2005.

264. Cannellos, G.P., Anderson, J.R., Propert, K.J. et al., Chemotherapy of advanced Hodgkin's disease with MOPP alternating with ABVD, *New Engl. J. Med.*, 327, 1478–1484, 1992.

265. van Leeuwen, F.E., Klokman, W.J., Hagenbeck, A. et al., Second cancer risk following Hodgkin's disease: a 20-year follow-up study, *J. Clin. Oncol.*, 12, 312–325, 1994.

266. Linch, D.C., Winfield, D., Goldstone, A.H. et al., Dose intensification with autologous bone marrow transplantation in relapsed and resistant Hodgkin's disease: results of a BNLI randomized trial, *Lancet*, 341, 1051–1054, 1993.

267. Horning, S.J., in *William's Hematology*, 6th ed., Beutler, E., Lichtman, M.A., Copller, B.S., Kipps, T.J., and Seligsohn, U., Eds., McGraw-Hill, New York, 2001, p. 1224.

268. Bonadona, G., Valgussa, P., and Santoro, A., Alternating non-cross resistant combination therapy or MOPP in stage IV Hodgkin's disease: a report of 8-year results, *Ann. Int. Med.*, 104, 739, 1986.

269. Ambinder, R., Infection and lymphoma, *New Engl. J. Med.*, 349, 1309–1312, 2003.

270. Peggs, K.S., Hunter, A., and Chopra, R., Clinical evidence of a graft-versus-Hodgkin's-lymphoma effect after reduced-intensity allogeneic transplantation, *Lancet*, 365, 1934–1941, 2005.

271. Still, R.A. and McDowell, I.F.W., Clinical implications of plasma homocysteine measurement in cardiovascular disease, *J. Clin. Pathol.*, 51, 183–188, 1998.

272. Raisz, L.G., Homocysteine and osteoporotic fractures: culprit or bystander? *New Engl. J. Med.*, 350, 2089–2090, 2004.

273. Lewis, S.J., Ebrahim, S., and Smith, G.D., Meta-analysis of MTHFR 677C→T polymorphism and coronary heart disease: does totality of evidence support causal role of homocysteine and preventive potential of folate? *Br. Med. J.*, 331, 1053–1056, 2005.

274. Mortimer, P.P. and Loveday, C., The virus and the tests, in *ABC of AIDS*, 5th ed., Adler, M.W., Ed., British Medical Journal Publishing Group, London, 2001, pp. 6–11.

275. Duncan, S.R., Scott, S., and Duncan, C.J., Reappraisal of the historical selective pressures for the CCR5-δ32 mutation, *J. Med. Genetics*, 42, 205–208, 2005.

275a. UNAids, Report on Aids Epidemic Update 2005: Joint United Nations Programme on HIV/ Aids (UNAids), UNAIDS, Geneva.

275b. Report on the Global AIDS Epidemic: Executive Summary, UNAIDS, Geneva.

276. Morris, A., Hewitt, C., and Young, S., The major histocompatibility complex: its genes and their roles in antigen presentation, *Molecular Aspects Med.*, 15, 377–403, 1994.

277. Staba, S.L., Escolar, M.L., Poe, M. et al., Cord-blood transplants from unrelated donors in patients with Hurler's syndrome, *New Engl. J. Med.*, 350, 1960–1969, 2004.

278. Liesveld, J.L. and Abbound, C.N., The hypereosinophilic syndrome, *Blood Rev.*, 5, 29–37, 1991.

278a. Schwartz, R., The eosinophilic syndrome and the biology of cancer, *New Engl. J. Med.*, 348, 1199–1200, 2003.

278b. Klien, A.D., Noel, P., Akin, C. et al., Elevated serum tryptase kinase levels identify a subset of patients with a myeloproliferative variant of idiopathic hypereosinophilic syndrome associated with fibrosis, poor prognosis and imatinib responsiveness, *Blood*, 101, 4660–4666, 2003.

279. Yu, L.C., Twu, Y.-C., Chou, M.L., Reid, M.E. et al., The molecular genetics of the human *I* locus and molecular background to explain the partial association of the *i* phenotype with congenital cataracts, *Blood*, 101, 2081–2088, 2003.

280. Petz, L.D. and Garratty, G., *Immune Hemolytic Anemias*, 2nd ed., Churchill Livingstone, London, 2004.

281. Warner, M. and Kelton, J.K., Laboratory investigation of immune thrombocytopenia, *J. Clin. Pathol.*, 50, 5–12, 1997.

282. George, J.N., Woolf, S.H., Raskob, G.E. et al., Idiopathic thrombocytopenic purpura: a practical guideline developed by explicit methods for the American Society of Hematology, *Blood*, 88, 3–40, 1996.

283. Cines, D.B. and Blanchiffe, V.S., Immune thrombocytopenic purpura, *New Engl. J. Med.*, 346, 995–1008, 2002.

284. George, J.N. and Veseley, S.K., Immune thrombocytopenic purpura: let the treatment fit the patient, *New Engl. J. Med.*, 349, 903–905, 2003.

285. American Society of Hematology ITP Practice Guideline Panel, Diagnosis and treatment of thrombocytopenic purpura: recommendations of the American Society of Hematology, *Ann. Int. Med.*, 126, 319–326, 1997.

286. British Committee for Standards in Haematology, Guidelines for the investigation and management of idiopathic thrombocytopenic purpura in adults, children and in pregnancy, *Br. J. Haematol.*, 120, 574–596, 2003.

287. Roberts, I.A.G. and Murray, N.A., Management of thrombocytopenia in neonates, *Br. J. Haematol.*, 105, 864–870, 1999.

288. Lilleyman, J.S., Management of childhood immune thrombocytopenic purpura, *Br. J. Haematol.*, 105, 871–875, 1999.

289. Wilken, D.E.L., Treatment dilemma in childhood idiopathic thrombocytopenic purpura, *Lancet*, 350, 602–603, 1997.

290. Roitt, I.M., Brostoff, J., and Male, D.I.K., *Immunology*, 3rd ed., Mosby, St. Louis, 1993.

291. Kaczmarski, R.S., Mufti, G.J., Moxham, J. et al., CD4+ lymphocytopenia due to common variable immunodeficiency mimicking AIDS, *J. Clin. Pathol.*, 47, 364–366, 1994.

292. Gardulf, A., Anderson, V., Bjorkander, J. et al., Subcutaneous immunoglobulin replacement in patients with primary antibody deficiencies: safety and costs, *Lancet*, 345, 365–369, 1995.

293. Wengler, G.S., Lanfranchi, A., Frusca, T. et al., *In utero* bone marrow transplantation of parental CD34 hemopoietic cells to a patient with X-linked severe immunodeficiency (SCID XI), *Lancet*, 348, 1484–1487, 1996.

294. Cavazzana-Calvo, M. and Fischer, A., Efficacy of gene therapy for SCID is being confirmed, *Lancet*, 364, 2155–2156, 2004.

295. Pileri, S.A. and Sabattini, E., A rational approach to immunohistochemical analysis of malignant lymphoma on paraffin wax sections, *J. Clin. Pathol.*, 50, 2–4, 1997.

296. British Committee for Standards in Haematology, Immunophenotyping in the diagnosis of chronic lymphoproliferative diseases, *J. Clin. Pathol.*, 47, 871–875, 1994.

297. British Committee for Standards in Haematology, Revised guidelines for immunophenotyping of acute leukemia and chronic lymphoproliferative disorders, *Clin. Lab. Haematol.*, 24, 1–13, 2002.

298. British Committee for Standards in Haematology, Guidelines for enumeration of CD4+ T-lymphocytes in immunosuppressed individuals, *Clin. Lab. Haematol.*, 19, 231–242, 1997.

299. Hotchkiss, R.S. and Karl, I., The pathophysiology and treatment of sepsis, *New Engl. J. Med.*, 348, 138–150, 2003.

300. Annane, D., Bellisnat, E., and Cavaillon, J.-M., Septic shock, *Lancet*, 365, 63–78, 2005.

301. Morris, A. and Zvetkova, I., Cytokine research: the interferon paradigm, *J. Clin. Pathol.*, 50, 635–639, 1997.

302. Dorman, S.E., Picard, C., Lammas, D. et al., Clinical features of dominant and recessive interferon-γ receptor 1 deficiencies, *Lancet*, 364, 2113–2121, 2004.

303. Finter, N.B., The naming of cats — and alpha-interferons, *Lancet*, 348, 348–349, 1996.

304. Plötz, S.G., Simon, H.-U., Darsano, U. et al., Use of an anti-interleukin-5 antibody in the hypereosinophilic syndrome with eosinophilic dermatitis, *New Engl. J. Med.*, 349, 2334–2339, 2003.

305. Papanicolau, D.A., Wilder, R.L., and Manolagus, C., The pathophysiologic roles of interleukin-6 in human disease, *Ann. Int. Med.*, 128, 127–137, 1998.

306. Mannon, P.J., Fuss, I.J., Mayer, L. et al., Antiinterleukin 12 antibody for active Crohn's disease, *New Engl. J. Med.*, 351, 2069–2079, 2004.

307. Maki, M. and Collin, P., Coeliac disease, *Lancet*, 349, 1755–1759, 1997.

308. McManus, R. and Kellener, D., Celiac disease: the villain unmasked? *New Engl. J. Med.*, 348, 2573–2574, 2003.

309. Watson, R.G.P., Diagnosis of celiac disease, *Br. Med. J.*, 330, 739–740, 2005.

310. Podolsky, D.K., Inflammatory bowel disease, *New Engl. J. Med.*, 347, 417–429, 2002.

311. Meinzer, U. and Hugot, J.-P., Nod2 and Crohn's disease: many connected highways, *Lancet*, 365, 1752–1754, 2005.

312. Eksteen, B., Miles, A.E., Grant, A.J. et al., Lymphocyte homing in the pathogenesis of extraintestinal manifestations of inflammatory bowel disease, *Clin. Med.*, 4, 173–180, 2004.

313. Glasstone, S., *Sourcebook on Atomic Energy*, D. Van Nostrand, New York, 1950.

314. Pollard, E.C., The biological action of ionizing radiation, *Am. Sci.*, 57, 206, 1969.

315. Morgan, K.Z. and Turner, J.E., Eds., *Principles of Radiation Protection*, Vol. 1, John Wiley & Sons, New York, 1967.

316. National Council on Radiation Protection and Measurements, Medical X-ray and Gamma-ray Protection for Energies up to 20 MeV, NCRD Report 33, National Council on Radiation Protection and Measurements, Bethesda, MD, 1969.

317. Mettler, F.A. and Upton, A.C., *Medical Aspects of Ionizing Radiation*, W.B. Saunders, Philadelphia, 1995.

318. Cardis, E., Vrijheid, M., and Blettner, E., Risk of cancer after low doses of ionising radiation: retrospective cohort study in 15 countries, *Br. Med. J.*, 331, 77–80, 2005.
319. Worwood, M., Iron-deficiency anaemia and iron overload, in *Dacie and Lewis Practical Haematology*, 9th ed., Lewis, S.M., Bain, B.J., and Bates, I., Eds., Churchill Livingstone, Edinburgh, 2001, pp. 115–128.
320. Olivieri, N.F. and Brittenham, G.M., Iron-chelating therapy and the treatment of thalassemia, *Blood*, 89, 739–761, 1997.
321. Pippard, M.J., Detection of iron overload, *Lancet*, 349, 73, 1997.
322. Vogt, B. and Frey, F.J., Inhibition of angiogenesis in Kaposi's sarcoma by captopril, *Lancet*, 349, 1148, 1997.
323. Krown, S.E., Kaposi's sarcoma: what's human chorionic gonadotropin got to do with it? *New Engl. J. Med.*, 335, 1309–1310, 1996.
324. Newberger, J.W., Treatment of Kawasaki disease, *Lancet*, 347, 1128, 1996.
325. Takagi, N., Kihara, M., Yamaguchi, S. et al., Plasma exchange in Kawasaki disease, *Lancet*, 346, 1307, 1995.
326. Waldron, H.A. and Scott, A., Metals — lead, in *Hunter's Diseases of Occupations*, 8th ed., Raffle, P.A.B., Adams, P.H., Baxter, P.J., and Lee, W.R., Eds., Edward Arnold, London, 1994, pp. 92–100.
327. Saeed, M., Khalil, A.G., Elhassen, A.M.A. et al., Serum erythropoietin concentration in anaemia of visceral leishmaniasis (kala-azar) before and during antimonial therapy, *Br. J. Haematol.*, 100, 720–724, 1998.
328. Nyhan, W.L. and Wong, D.F., New approaches to understanding Lesch-Nyhan disease, *New Engl. J. Med.*, 334, 1602–1604, 1996.
329. Pedersen-Bjergaard, J., Insights into leukemogenesis from therapy-related leukaemia, *New Engl. J. Med.*, 352, 1591–1594, 2005.
330. Lachman, P., Peters, K., and Walport, M.J., Eds., *Clinical Aspects of Immunology*, Blackwell, Oxford, 1993, chap. 70.
331. Mortimer, P.S. and Levick, J.R., Chronic peripheral oedema: the critical role of the lymphatic system, *Clin. Med.*, 4, 448–453, 2004.
332. National Cancer Institute, Study of classification of non-Hodgkin's lymphoma: summary and description of a working formulation for clinical usage, *Cancer*, 49, 2112–2135, 1982.
333. Lennert, K., Mohri, N., Stein, H. et al., The histopathology of malignant lymphoma, *Br. J. Haematol*, 31 (Suppl.), 193–203, 1975.
334. Lukes, R. and Collins, R., Immunologic characterization of human malignant lymphomas, *Cancer*, 34, 1488, 1974.
335. Harris, N.L., Jaffe, E.S., Stein, H., Banks, P.M., Chan, J.K.C., Cleary, M.L., Delsol, G., de Wolf-Peeters, C., Falini, B., Gatter, K.C., Grogan, T.M., Isaacson, P.G., Knowles, D.M., Mason, D.Y., Muller-Hermelinj, H.-K., Pileri, S.A., Piris, M.A., Ralfkiaer, E., and Warnke, R.A., A revised European-American classification of lymphoid neoplasms: a proposal from the International Lymphoma Study Group, *Blood*, 84, 1361–1392, 1994.
336. Carbone, P.P., Kaplan, H.S., and Musshoff, K., Report of the Committee on Hodgkin's Disease Staging Classification, *Cancer Res.*, 31, 1860–1861, 1971.
337. Lister, T.A. and Crowther, D., Staging for Hodgkin's disease, *Semin. Oncol.*, 17, 696–703, 1990.
338. Crocker, J., Ed., *Cell Proliferation in Lymphomas*, Blackwell, Oxford, 1993.
339. Vellodi, A., Lysozyme storage disease, *Br. J Haematol.*, 128, 413–431, 2005.
340. World Health Organization, 19th Expert Committee on Malaria, WHO report, Geneva, 1993.
341. Facer, C.A., Haematological aspects of malaria, in *Infection and Haematology*, Jenkins, G.J. and Williams, J.D., Eds., Butterworth & Heinemann, Oxford, 1994.
342. British Committee for Standards in Haematology, Malaria Working Party of the General Haematology Task Force, The laboratory diagnosis of malaria, *Clin. Lab. Haematol.*, 19, 165–170, 1997.
343. Seesod, N., Nopparat, P., Hendrum, A. et al., An integrated system using immunomagnetic separation polymerase-chain-reaction and colorimetric detection for diagnosis of *Plasmodium falciparum*, *Am. J. Trop. Med. Hygiene*, 56, 322–328, 1997.
344. British Medical Association and Royal Pharmaceutical Society, *British National Formulary*, 50, Antimalarials, London, Sept. 2005, pp. 328–330.

345. White, N.J., The treatment of malaria, *New Engl. J. Med.*, 335, 800–805, 1997.
346. Kremsner, P.G. and Krishna, S., Antimalarial combinations, *Lancet*, 364, 285–294, 2004.
347. Maitland, K.M., Nadel, S., Pollard, A.J. et al., Management of severe malaria in children: proposed guidelines for the United Kingdom, *Br. Med. J.*, 331, 337–343, 2005.
348. Passvol, G., Malaria and resistance genes: they work in wondrous ways, *Lancet*, 348, 1532–1533, 1996.
349. Olliaro, P., Nevill, C., LeBas, J. et al., Systematic review of amodiaquine treatment in uncomplicated malaria, *Lancet*, 348, 1196–1201, 1996.
350. National Center for Infectious Diseases Travelers Health; *The Yellow Book: Health Information for International Travel 2003–2004*, Centers for Disease Control, Atlanta, 2004.
351. World Health Organization, International Travel and Health: Vaccination Requirements and Health Advice, WHO, Geneva, 2004.
352. Mufti, G.J., Flandrin, G., Schaefer, H.-E., Sandberg, A.A., and Kanfer, E.J., *An Atlas of Malignant Haematology*, Martin Dunitz, London, 1996.
353. Hart, I.R. and Fidler, I.J., Cancer invasion and metastases, *Q. Rev. Biol.*, 55, 121–142, 1980.
354. Weisenburger, D.D. and Armitage, J.O., Mantle cell lymphoma: an entity comes of age, *Blood*, 87, 4483–4494, 1996.
355. Bain, B.J., Systemic mastocytosis and other mast cell neoplasms, *Br. J. Haematol.*, 106, 9–17, 1999.
356. Pardanini, A., Brockman, S.R., Paternoster, S.F. et al., The FIP1L1-PDGFRA fusion; prevalence and clinicopathologic correlates in 89 consecutive patients with moderate or severe eosinophilia, *Blood*, 104, 3038–3045, 2004.
357. Basser, R.L., Rasko, J.E.J., Clarke, K. et al., Thrombopoietic effects of pegylated recombinant human megakaryocyte growth and development factor (PEG-rHuMGDF) in patients with advanced cancer, *Lancet*, 348, 1279–1281, 1996.
358. Percy, M.J., Gillespie, J.S., Savage, G. et al., Familial idiopathic methemoglobinemia revisited: original cases reveal 2 novel mutations in NADH-cytochrome b5 reductase, *Blood*, 100, 3447–3449, 2002.
359. Santini, V., Kantarjian, H.M., and Issa, J.P., Changes in DNA methylation in neoplasia: pathophysiology and therapeutic implications, *Ann. Int. Med.*, 134, 573–586, 2001.
360. British Committee for Standards in Haematology, Guidelines on the diagnosis and management of microangiopathic haemolytic anaemias, *Br. J. Haematol.*, 120, 556–573, 2000.
361. Moake, J.L., Mechanisms of disease: thrombotic microangiopathy, *New Engl. J. Med.*, 347, 589–600, 2002.
362. British Committee for Standards in Haematology, The use and evaluation of leucocyte monoclonal antibodies in diagnostic laboratories, *Clin. Lab. Haematol.*, 18, 1–5, 1996.
363. Martin, E.E., Rawstron, A.C., Ghia, P. et al., Diagnostic criteria for monoclonal B-cell lymphocytosis, *Br. J. Haematol.*, 130, 325–332, 2005.
364. Kyle, R.A. and Rajkumar, S.V., Monoclonal gammopathies of undetermined significance: a review, *Immunological Rev.*, 194, 112–139, 2003.
365. Kyle, R.A., Therneau, T.M., Rajkumar, S.V., Larson, D.R., Plevak, M.F., and Melton, L.J., III, Long-term follow-up of 241 patients with monoclonal gammopathy of undetermined significance: the original Mayo Clinic series 25 years later, *Mayo Clin. Proc.*, 79, 859–866, 2004.
366. Kyle, R.A., Therneau, T.M., Rajkumar, S.V. et al., A long-term study of prognosis in monoclonal gammopathy of undetermined significance, *New Engl. J. Med.*, 346, 564–569, 2002.
367. Lorincz, A.L., Cutaneous T-cell lymphoma (mycosis fungoides), *Lancet*, 347, 871–876, 1996.
368. Diamandidou, E., Cohen, P.R., and Kurzrock, R., Mycosis fungoides and Sézary syndrome, *Blood*, 88, 2385–2409, 1996.
369. Dippel, E., Schrag, H., Goerdt, S. et al., Extracorporeal photopheresis and interferon-α in advanced cutaneous T-cell lymphoma, *Lancet*, 350, 32–33, 1997.
370. British Committee for Standards in Haematology, Guidelines for the diagnosis and therapy of adult myelodysplastic syndromes, *Br. J. Haematol.*, 120, 187–200, 2003.
371. Cazzola, M. and Malcovati, L., Myelodysplastic syndromes: coping with ineffective hematopoiesis, *New Engl. J. Med.*, 352, 536–538, 2005.
372. Greenberg, P., Cox, C., Le Beau, M. et al., International scoring system for evaluating prognosis in myelodysplastic syndromes, *Blood*, 89, 2079–2088, 1997.

373. Sutton, L., Chastang, C., Ribaud, P. et al., Factors influencing outcome *de novo* myelodysplastic syndromes treated by allogeneic bone marrow transplantation: a long term study of 71 patients, *Blood*, 88, 358–365, 1996.

374. Neiman, R.S., Barcos, M., Berard, C. et al., Granulocytic sarcoma: a clinicopathologic study of 61 biopsied cases, *Cancer*, 48, 1426–1437, 1981.

375. Malpas, J.S., Bergsagel, D.E., Kyle, R.A., and Anderson, K.C., *Myeloma: Biology and Management*, Oxford University Press, New York, 1995; reprint, Saunders, Philadelphia, 2004.

376. Kyle, R.A., Therneau, T.M., Rajkumar, S.V., Larson, D.R., Plevak, M.F., and Melton, L.J., Incidence of multiple myeloma in Olmsted County, Minnesota: trend over 6 decades, *Cancer*, 101, 2667–2674, 2004.

377. Kyle, R.A., Gertz, M.A., Witzig, T.E., Lust, J.A., Lacy, M.Q., Dispenzieri, A., Fonseca, R., Rajkumar, S.V., Offord, J.R., Larson, D.R., Plevak, M.E., Therneau, T.M., and Greipp, P.R.., Review of 1027 patients with newly diagnosed multiple myeloma, *Mayo Clin. Proc.*, 78, 21–33, 2003.

378. British Committee for Standards in Haematology, Guidelines in the diagnosis and management of multiple myeloma, *Br. J. Haematol.*, 115, 522–540, 2001.

379. Kyle, R.A. and Rajkumar, S.V., Multiple myeloma: drug therapy, *New Engl. J. Med.*, 351, 1860–1873, 2004.

380. British Committee for Standards in Haematology, Thalidomide in multiple myeloma: current status and future prospects, *Br. J. Haematol.*, 120, 18–26, 2003.

381. Child, J.A., Morgan, G.J., Davies, F.E. et al., High-dose chemotherapy with hematopoietic stem-cell rescue for multiple myeloma, *New Engl. J. Med.*, 348, 1875–1883, 2003.

382. Spivak, J.L., The myeloproliferative disorders, in *Blood: Principles and Practice of Hematology*, Handin, R.I., Lux, S.E., and Stossel, T.P., Eds., Lippincott, Williams and Wilkins, Philadelphia, 2003, pp. 379–432.

383. Baxter, E.J., Scott, L.M., Campbell, P.J. et al., Acquired mutation of the tyrosine kinase JAK2 in human myeloproliferative disease, *Lancet*, 365, 1054–1061, 2005.

383a. Wei, W. and Sham, J.S.T., Nasopharyngeal carcinoma, *Lancet*, 365, 2041–2054, 2005.

384. Miller, D.H., Khan, O.A., Sheremata, W.A. et al., A controlled trial of natalizumab for relapsing multiple sclerosis, *New Engl. J. Med.*, 348, 15–23, 2003.

385. Ghosh, S., Goldin, E., Gordon, F.H. et al., Natalizumab for active Crohn's disease, *New Engl. J. Med.*, 348, 24–32, 2003.

386. Reinhardt, W., Danoff, S., King, R. et al., Rheology of fetal and maternal blood, *Pediatr. Res.*, 19, 147–153, 1985.

387. British Committee for Standards in Haematology, The investigation and management of neonatal haemostasis and thrombosis, *Br. J. Haematol.*, 119, 295–309, 2002.

388. Hughes, R.A.C., Britton, T., and Richards, M., Effects of lymphoma on the peripheral nervous system, *J. R. Soc. Med.*, 87, 526–530, 1994.

389. Hampton, M.B., Kettle, A.J., and Winterbourn, C.C., Inside the neutrophil phagosome: oxidants, myeloperoxidase, and bacterial killing, *Blood*, 92, 3007, 1998.

390. Babier, B.M., NADPH oxidase: an update, *Blood*, 93, 1464, 1999.

391. Neth, O.W., Bajaj-Elliott, M., Turner, M.W. et al., Susceptibility of infection with neutropenia: the role of the innate immune system, *Br. J. Haematol.*, 129, 713–722, 2005.

392. Bhager, K. and Vallance, P., Nitric oxide 9 years on, *J. R. Soc. Med.*, 89, 667–673, 1996.

393. Shipp, M.A. and Harrington, D.P., A predictive model for aggressive non-Hodgkin's lymphoma: the International Non-Hodgkin's Lymphoma Prognostic Factor Project, *New Engl. J. Med.*, 329, 987–994, 1993.

394. Dana, B.W., Dahlberg, S., Nathwani, B.N. et al., Long-term follow-up of patients with low-grade malignant lymphomas treated with doxorubicin-based chemotherapy or chemoimmunotherapy, *J. Clin. Oncol.*, 11, 644–651, 1993.

395. Fisher, R.I., Gaynor, E.R., Dahlerg, S. et al., Comparison of CHOP versus m-BACOP versus Pro-MACE-cytaBOM versus MACOP-B in patients with intermediate or high-grade non-Hodgkin's lymphoma, *New Engl. J. Med.*, 328, 1002–1006, 1993.

396. Santini, G., Salvagio, L., Leoni, P. et al., VACOP-B versus VACOP-B plus autologous bone marrow transplantation for advanced diffuse non-Hodgkin's lymphoma: results of a prospective randomised trial by the non-Hodgkin's lymphoma co-operative study group, *J. Clin. Oncol.*, 16, 2796, 1998.

397. Foon, K.A. and Fisher, R.I., in *William's Hematology*, 6th ed., Beutler, E., Lichtman, M.A., Copller, B.S., Kipps, T.J., and Seligsohn, U., Eds., McGraw-Hill, New York, 2001, p. 1250.

398. Coleman, C.N., Picozzi, V.J., Jr., Cox, R.S. et al., Treatment of lymphoblastic lymphomas in adults, *J. Clin. Oncol.*, 4, 1628–1637, 1986.

399. Freeman, S.M., Ramesh, R., and Marrogi, A.J., Immune system in suicide-gene therapy, *Lancet*, 349, 2–3, 1997.

400. Garfunkel, A.A., Oral mucositis: the search for a solution, *New Engl. J. Med.*, 351, 2649–2651, 2004.

401. McCluskey, P. and Powell, R.J., The eye in systemic inflammatory disease, *Lancet*, 364, 2125–2133, 2004.

402. Dispenzieri, A., Kyle, R.A., Lacy, M.Q., Rajkumar, S.V., Therneau, T.M., Larson, D.R., Greipp, P.R., Witzig, T.E., Basu, R., Suarez, G.A., Fonseca, R., Lust, J.A., and Gertz, M.A., POEMS syndrome: definitions and long-term outcome, *Blood*, 101, 2496–2506, 2003.

403. Glass, D.A., Patel, M.S., and Karsenty, G., A new insight into formation of osteosclerotic lesions in multiple myeloma, *New Engl. J. Med.*, 349, 2479–2480, 2003.

404. Dispenzieri, A., Moreno-Aspitia, A., Suarez, G.A., Lacy, M.Q., Colon-Otero, G., Tefferi, A., Litzow, M.R., Roy, V., Hogan, W.J., Kyle, R.A., and Gertz, M.A., Peripheral blood stem cell transplantation in 16 patients with POEMS syndrome, and a review of the literature, *Blood*, 104, 3400–3407, 2004.

405. International Council for Standardization in Haematology, Recommendations for a reference method for the packed cell volume, *Clin. Lab. Hematol.*, 7, 148–170, 2001.

406. Parker, C.J., Historical aspects of paroxysmal nocturnal haemoglobinuria, *Br. J. Haematol.*, 117, 3–22, 2002.

407. Schwartz, R.S., Black morning, yellow sunsets: a day with PNH, *New Engl. J. Med.*, 350, 537–538, 2004.

408. Katano, H. and Cohen, I.I., Perforin and lymphohistiocytic proliferative disorder, *Br. J. Haematol.*, 128, 739–750, 2005.

409. Bain, B.J., *Blood Cells: A Practical Guide*, 3rd ed., J.B. Lippincott, Philadelphia, 2002.

410. British Committee for Standards in Haematology, Guidelines on the diagnosis and management of solitary plasmacytoma of bone and solitary extramedullary plasmacytoma, *Br. J. Haematol.*, 124, 717–726, 2004.

411. International Council for Standardization in Haematology, Platelet counting by the RBC/platelet method: a reference method, *Am. J. Clin. Pathol.*, 115, 460–464, 2001.

412. Rao, A.K. and Carvallo, A.C.A., Acquired qualitative platelet defects, in *Hemostasis and Thrombosis*, Colman, R.W., Hirsh, J., Marder, V.J., and Salzman, E.W., Eds., J.B. Lippincott, Philadelphia, 1993, pp. 685–704.

413. British Committee for Standards in Haematology, Guidelines for platelet function testing, *J. Clin. Pathol.*, 41, 1322–1330, 1988.

414. British Committee for Standards in Haematology, Guidelines on the use of platelet transfusions, *Br. J. Haematol.*, 122, 10–23, 2003.

415. Price, C.P. and Hicks, J.M., *Point-of-Care-Testing*, AACC Press, Washington, DC, 1999.

416. Medical Devices Agency, Management and Use of IVD Point-of-Care-Test Devices, MDA DB2002(03), London, 2002.

417. Kaplan, M.E., Mack, K., Goldberg, J.D. et al., Long term management of polycythaemia rubra vera with hydroxyurea: progress report, *Sem. Haematol.*, 23, 167–171, 1986.

418. Spivak, J., Daily aspirin: only half the answer, *New Engl. J. Med.*, 350, 99–100, 2004.

419. Salvarani, C., Cantini, F., Bioardi, L., and Hunder, G.G., Polymyalgia and giant cell arteritis, *New Engl. J. Med.*, 347, 261–271, 2002.

420. Sassa, S., The hematologic aspects of porphyria, in *William's Hematology*, 6th ed., Beutler, E., Lichtman, M.A., Coller, B.S., Kipps, T.J., and Seligsohn, U., Eds., McGraw-Hill, New York, 2001.

421. Walker, I.D., Walker, J.J., Colvin, B.T. et al., Investigation and management of haemorrhagic disorders in pregnancy, *J. Clin. Pathol.*, 47, 100–108, 1994.

422. British Committee for Standards in Haematology, Addendum for guidelines for blood grouping, and red cell antibody testing during pregnancy, *Transf. Med.*, 9, 99, 1999.

423. Dallman, P.R., Anemia of prematurity: the prospects of avoiding blood transfusions by treatment with recombinant human erythropoietin, *Advanced Pediatr.*, 40, 385–403, 1993.

424. Lundstrom, U., Siimes, M.A., and Dallman, P.R., At what age does iron supplementation become necessary in low-birth-weight infants? *J. Pediatr.*, 91, 878, 1977.

425. British Committee for Standardisation in Haematology, Guidelines for compatibility procedures in blood transfusion laboratories, *Transf. Med.*, 14, 59–73, 2004.

426. Lang, I.M., Chronic thromboembolic pulmonary hypertension: not so rare after all, *New Engl. J. Med.*, 350, 2236–2238, 2004.

427. Prusiner, S.B., Shattuck lecture: neurodegenerative diseases and prions, *New Engl. J. Med.*, 344, 1516–1526, 2001.

428. Caughey, B., Interactions between prion protein isoforms: the kiss of death, *Trends Biochem. Sci.*, 26, 235–242, 2001.

429. Prusiner, S.B., Scott, M.R., DeArmond, S.J., and Cohen, F.E., Prion protein biology, *Cell*, 93, 337–348, 1998.

430. Burthem, J. and Roberts, D.J., *Br. J. Haematol.*, 122, 3–9, 2003.

431. Gajdusek, D.C. and Zigas, V., Degenerative disease of the central nervous system in New Guinea: the endemic occurrence of "kuru" in the native population, *New Engl. J. Med.*, 14, 974–978, 1957.

432. Lasmézas, C.I., Comoy, E., Hawkins, S. et al., Risk of oral infection with bovine spongiform encephalopathy agent in primates, *Lancet*, 365, 781–783, 2005.

433. Houston, F., Foster, J.D., Chong, A., Hunter, A., and Bostock, C.J., Transmission of BSE by blood transfusion in sheep, *Lancet*, 356, 999–1000, 2000.

434. Hunter, N., Foster, J., Chong, A. et al., Transmission of prion diseases by blood transfusion, *J. Gen. Virol.*, 83, 2897–2905, 2002.

435. Pincock, S., Patient's death from vCJD may be linked to blood transfusion, *Lancet*, 363, 43, 2004.

436. Pedern, A.H., Head, M.W., Ritchie, D.L., Bell, J.E., and Ironside, J.W., Preclinical vCJD after blood transfusion in a PRNP codon 129 heterozygous patient, *Lancet*, 264, 527–529, 2004.

436a. Wroe, S.J., Pal, S., Siddique, D. et al., Clinical presentation and pre-mortem diagnosis of variant Creutzfeldt-Jacob disease associated with blood transfusion: a case report, *Lancet*, 368, 2061–2067, 2006.

437. Fichet, G., Comoy, E., Duval, C. et al., Novel methods for disinfection of prion contaminated medical devices, *Lancet*, 364, 521–526, 2004.

437a. Wilson, K. and Ricketts, M.N., A third episode of transfusion-derived vCJD, *Lancet*, 368, 2037–2038, 2006.

438. Wadsworth, J.D.F., Joiner, S., Hill, A.F. et al., Tissue distribution of protease resistant prion protein in variant Creutzfeldt-Jakob disease using a highly sensitive immunoblotting assay, *Lancet*, 358, 171–180, 2001.

439. Head, M.W. and Ironside, J.W., Mad cows and monkey business: the end of vCJD? *Lancet*, 365, 730–731, 2005.

440. Tansey, W.P.T., Death, destruction and proteasome, *New Engl. J. Med.*, 351, 393–394, 2004.

441. Mitchell, B.S., The proteosome: an emerging therapeutic target in cancer, *New Engl. J. Med.*, 348, 2597–2598, 2003.

442. Smith, O.P. and White, B., Infectious purpura fulminans: diagnosis and treatment, *Br. J. Haematol.*, 104, 202–207, 1999.

443. Zanella, A., Fermo, E., Bianchi, P. et al., Red cell pyruvate kinase deficiency: molecular and clinical aspects, *Br. J. Haematol.*, 130, 11–25, 2005.

444. International Organization for Standards 15189, Medical Laboratories: Particular Requirements for Quality and Competence, ISO, Geneva, 2003.

444a. Lewis, S.M., Quality assurance, in *Dacie and Lewis Practical Haematology*, 9th ed., Lewis, S.M., Bain, B.J., and Bates, I., Eds., Churchill Livingstone, Edinburgh, 2001, pp. 657–671.

445. International Organization for Standards 15190, Medical Laboratories: Requirements for Safety, ISO, Geneva, 2003.

446. Illidge, T.M. and Johnson, P.W.M., The emerging role of RIT in haematological malignancy, *Br. J. Haematol.*, 108, 679, 2000.

447. Gordon, L.I., Molina, A., Witzig, T. et al., Durable responses after ibritumab tiuxetan radioimmunotherapy for CD20+ B-cell lymphoma: long-term follow-up of a phase ½ study, *Blood*, 103, 4429–4431, 2004.

448. Cheson, B.D., Radioimmunotherapy of non-Hodgkin lymphoma, *Blood*, 101, 391–398, 2003.

449. Dokal, I. and Lewis, S.M., Diagnostic radioisotopes in haematology, in *Dacie and Lewis Practical Haematology*, 9th ed., Lewis, S.M., Bain, B.J., and Bates, I., Eds., Churchill Livingstone, Edinburgh, 2001, pp. 357–378.

449a. Mittal, S. and Watson, H.G., A critical appraisal of the use of recombinant factor VIIa in acquired bleeding conditions, *Br. J. Haematol*, 133, 355–363, 2006.

450. Gallagher, P.G. and Forget, B.G., The red cell membrane, in *William's Hematology*, 6th ed., Beutler, E., Lichtman, M.A., Coller, B.S., Kipps, T.J., and Seligsohn, U., Eds., McGraw-Hill, New York, 2001.

451. Tse, W.T. and Lux, S.E., Red blood cell membrane disorders, *Br. J. Haematol.*, 104, 2–13, 1999.

452. Fairbanks, G., Steck, T.L., and Wallach, D.F., Electrophoretic analysis of the major polypeptides of the human erythrocyte membrane, *Biochemistry*, 10, 2606–2617, 1971.

453. Steck, T.L., Fairbanks, G., and Wallach, D.F., Disposition of the major proteins in the isolated erythrocyte membrane: proteolytic dissection, *Biochemistry*, 10, 2617–2624, 1971.

453a. McMullin, M., The molecular basis of disorders of the red cell membrane, *J. Clin. Pathol.*, 52, 245–248, 1999.

454. Jandl, J.H., *Blood: Pathophysiology*, Blackwell, Oxford, 1999, chap. 4.

455. Bain, B.J., Lewis, S.M., and Bates, I., Basic haematological techniques, in *Dacie and Lewis Practical Haematology*, 9th ed., Lewis, S.M., Bain, B.J., and Bates, I., Eds., Churchill Livingstone, Edinburgh, 2001, pp. 25–57.

455a. Lewis, S.M., Reference ranges and normal values, in *Dacie and Lewis Practical Haematology*, 9th ed., Lewis, S.M., Bain, B.J., and Bates, I., Eds., Churchill Livingstone, Edinburgh, 2001, pp. 11–24.

456. Miwa, S. and Fujii, H., Molecular basis of erythroenzymopathies associated with hereditary hemolytic anemia, *Am. J. Hematol.*, 51, 122, 1996.

457. British Committee for Standards in Haematology, The clinical use of red cell transfusion, *Br. J. Haematol.*, 113, 24–31, 2001.

458. Herbert, P.C., Wells, G., Blajchman, M.A. et al., A multicenter, randomized, controlled clinical trial of transfusion requirements in critical care: transfusion requirements in critical care investigators: Canadian Critical Care Trials Group, *New Engl. J. Med.*, 340, 409–417, 1999.

459. British Committee for Standards in Haematology, Guidelines for transfusion of massive blood loss, *Clin. Lab. Haematol.*, 10, 265–273, 1988.

460. International Standards Organization, Terms and Definitions in Connection with Reference Materials, ISO Guide 30, Geneva, 1981.

461. Verwilghen, R.L., The use of reference values, in *Quality Assurance in Haematology*, Lewis, S.M. and Verwilghen, R.L., Eds., Baillière Tindall, London, 1988, pp. 33–41.

462. Tanoue, L.T., The eosinophilic pneumonias, in *Fishman's Manual of Pulmonary Diseases and Disorders*, 3rd ed., Fishman, A.R., Elias, J.A., Fishman, J.A., Grippi, M.A., Kaiser, L.R., and Senior, R.M., McGraw-Hill, New York, 2002, chap. 33.

463. d'Onofrio, G., Zini., G., and Rowan, R.M., Reticulocyte counting: methods and clinical applications, in *Advanced Laboratory Methods in Haematology*, Rowan, R.M., Van Assendelft, O.W., and Preston, F.E., Eds., Arnold, London, 2002, pp. 78–126.

464. Briggs, C., Rogers, R., Thompson, B. et al., New red cell parameters as potential markers of functional iron deficiency, *Infus. Ther. Transfus. Med.*, 28, 256–262, 2001.

465. Franck, S., Linssen, J., Messinger, M. et al., Clinical utility of the RET-Y in the diagnosis of iron-restricted erythropoiesis, *Clin. Chem.*, 50, 1240, 2004.

466. Diverio, D., Falini, B., Biondi, A., Nervi, C., and Pelicci, P.G., Genetic diagnosis and molecular monitoring in the management of acute promyelocytic leukemia, *Blood*, 94, 12, 1999.

467. Lowe, G.D.O., Ed., Blood rheology and hyperviscosity syndromes, *Balliere's Clin. Hematol.*, 1(3), 597–867, 1987.

468. Ball, G.V. and Koopman, W.J., *Clinical Rheumatology*, W.B. Saunders, Philadelphia, 1986.

469. Visser, H., le Cessie, S., Vos, K., Breedveld, F.C., and Hazes, J.M.W., How to diagnose rheumatoid arthritis early: a prediction model for persistent (erosive) arthritis, *Arthritis Rheum.*, 46, 357–365, 2002.

470. Olsen, N.J. and Stein, M., New drugs for rheumatoid arthritis, *New Engl. J. Med.*, 350, 2167–2179, 2004

471. Östör, A.J., Beyond methotrexate biologic therapy in rheumatoid arthritis, *Clin. Med.*, 5, 222–226, 2005.

472. Tsokos, G.C., B-cells be gone: B-cell depletion in the treatment of rheumatoid arthritis, *New Engl. J. Med.*, 350, 2546–2548, 2004.

473. Kremer, J.M., Westhovens, R., Leon, M. et al., Rheumatoid arthritis: treatment hopeful with fusion protein-cytotoxic T-lymphocyte-associated antigen4-IgG1 (CTLA4Ig): binds to CD80 and CD86 on antigen presenting cells, blocking their engagement of CD28 on T-cells and preventing T-cell activation, *New Engl. J. Med.*, 349, 1907–1915, 2003.

474. Ho, L.-P., Urban, D.R., Thickett, D.R. et al., Deficiency of a subset of T-cells with immunoregulatory properties in sarcoidosis, *Lancet*, 365, 1062–1071, 2005.

475. Newman, L.S., Rose, C.S., and Maier, L.A., Marked progress in sarcoidosis, *New Engl. J. Med.*, 336, 1224–1234, 1997.

476. Sergeant, G., *Sickle Cell Disease*, 2nd ed., Oxford University Press, Oxford, 1992.

477. Sergeant, G.R., Sickle-cell disease, *Lancet*, 350, 725–729, 1997.

478. Bunn, H.F., Pathogenesis and treatment of sickle cell disease, *New Engl. J. Med.*, 337, 762–769, 1997.

479. British Committee for Standards in Haematology, Management of acute painful crisis in sickle cell disease, *Br. J. Haematol.*, 120, 744–752, 2003.

480. Fox, R.I., Sjögren's syndrome, *Lancet*, 366, 321–331, 2005.

481. Arnold, H.L., Odum, R.B., and James, W.D., *Andrew's Diseases of the Skin*, W.B. Saunders, Philadelphia, 1990.

482. Moschella, S.L. and Hurley, H.J., *Dermatology*, W.B. Saunders, Philadelphia, 1992.

483. National Committee for Clinical Laboratory Standards, SC2-L, *Specimen Collection*, NCCLS, Wayne, 1997.

484. Minton, S., Clinical hemostatic disorders caused by venoms, in *Disorders of Hemostasis*, 3rd ed., Ratnoff, O.D., and Forbes, C.D., Eds., W.B. Saunders, Philadelphia, 1996, chap. 18.

485. Gold, B.S., Dart, R.C., and Barish, R.A., Bites of venomous snakes, *New Engl. J. Med.*, 347, 356, 2002.

486. Meiman, R.S. and Orazi, A., *Disorders of the Spleen*, 2nd ed., Saunders, Philadelphia, 2002.

487. Royal, H.D., Brown, M.L., Drum, D.E. et al., Procedure guideline for hepatic and splenic imaging, *J. Nuclear Med.*, 39, 1114–1116, 1998.

488. British Committee for Standards in Haematology, Guidelines for the prevention and treatment of infections in patients with an absent or dysfunctional spleen, *Clin. Med.*, 2, 440–443, 2002.

489. Newland, A., Provan, D., and Myint, S., Preventing severe infection after splenectomy, *Br. Med. J.*, 331, 417–418, 2005.

490. Schmerling, R.H., Antibodies in systemic lupus erythematosus: there before you know it, *New Engl. J. Med.*, 349, 1499–1420, 2003.

490a. Leandro, M.J., Edwards, J.C., Cambridge, D. et al., An open study of B-lymphocyte depletion in systemic lupus erythematosus, *Arthritis Rheum.*, 46, 2673–2677, 2002.

491. Hahn, B.H.H., Systemic lupus erythematosus and accelerated atherosclerosis, *New Engl. J. Med.*, 349, 2379–2380, 2003.

492. Øhlenschlaeger, T., Garred, P., Madsen, H.O. et al., Mannose-binding lectin variant alleles and the risk of arterial thrombosis in systemic lupus erythematosus, *New Engl. J. Med.*, 351, 260–267, 2004.

493. D'Cruz, D.P. and Hughes, G.R.V., Treatment of lupus nephritis, *Br. Med. J.*, 330, 377–378, 2005.

494. British Committee for Standards in Haematology, Guidelines for the investigation of alpha and beta thalassaemia traits, *J. Clin. Pathol.*, 47, 289–295, 1994.

495. Schafer, A.I., Thrombocytosis, *New Engl. J. Med.*, 350, 1211–1219, 2004.

496. Dame, C. and Sutor, A.H., Primary and secondary thrombocytosis in childhood, *Br. J. Haematol.*, 129, 165–177, 2005.

497. Mallett, S.V. and Cox, D.J.A., Thromboelastography, *Br. J. Anaesth.*, 69, 307–313, 1992.

498. Florell, S.R. and Rodgers, G.M., Inherited thrombotic disorders: an update, *Am. J. Hematol.*, 54, 53–60, 1997.

499. British Committee for Standards in Haematology, Investigation and management of hereditary thrombophilia, *Br. J. Haematol.*, 114, 512–528, 2001.

500. Shapiro, S., Treating thrombosis in the 21st century, *New Engl. J. Med.*, 349, 1762–1764, 2003.

501. Mitchell, S., Cairo, M.S., and Bishop, M., Tumour lysis syndrome: new therapeutic strategies and classification, *Br. J. Haematol.*, 127, 3–11, 2004.

502. Kraus, D.S. and van Etten, R.A., Tyrosine kinases as targets for cancer therapy, *New Engl. J. Med.*, 352, 172–187, 2005.

502a. Lewis, S.M. and Roper, D., Laboratory methods in the investigation of haemolytic anaemias, in *Dacie and Lewis Practical Haematology*, 10th ed., Churchill-Livingstone, Elsevier, 2006, p. 187.

502b. Podar, K. and Anderson, K.C., The pathophysiologic role of VEGF in hematologic malignancies: therapeutic implications, *Blood*, 105, 1383–1395, 2005.

503. Savage, C.O.S., Harper, L., and Adu, D., Primary systemic vasculitis, *Lancet*, 349, 553–557, 1997.

504. Scott, D.G.I. and Watts, R.A., Classification and epidemiology of systemic vasculitis, *Br. J. Rheumat.*, 33, 897–900, 1994.

505. Langford, C.A., Treatment of ANCA-associated vasculitis, *New Engl. J. Med.*, 349, 3–4, 2003.

505a. Baglin, T.P., Brush, J., and Streiff, M., on behalf of British Committee for Standards in Haematology, Guidelines on the use of vena caval filters, *Br. J. Haematol.*, 134, 590–595, 2006.

506. Bockenstadt, P., Venous thromboembolism, *New Engl. J. Med.*, 349, 1203–1204, 2003.

507. Hirsh, J. and Hoak, J., Management of deep vein thrombosis and pulmonary embolism: a statement for health professionals, *Circulation*, 93, 2212–2245, 1996.

508. Geerts, W.H., Pinco, E.F., Heit, J.A. et al., Prevention of venous thromboembolism: the Seventh ACCP Conference on Antithrombotic and Thrombolytic Therapy, *Chest*, 126 (Suppl. 3), 338S–400S, 2004.

509. Agency for Healthcare Research and Quality, Making health care safer: a critical analysis of patient safety practices, *Evid. Rep. Technol. Assess.*, 43, i–x, 1–668, 2001.

510. Ridker, P.M., Goldhaber, S.Z., Danielson, M.I.A., Rosenberg, Y. et al., Long term, low intensity, warfarin therapy for the prevention of recurrent venous thromboembolism, *New Engl. J. Med.*, 348, 1425–1434, 2003.

511. Büller, H.R. and Prims, M.H., Secondary prophylaxis for venous thromboembolism, *New Engl. J. Med.*, 349, 702–703, 2003.

512. Duguid, D.L., Oral anticoagulant therapy for venous thromboembolism, *New Engl. J. Med.*, 336, 433–434, 1997.

513. Scott, S. and Duncan, C., *Return of the Black Death*, Wiley, Chichester, 2005.

514. Peters, C.J., Marburg and Ebola: arming ourselves against the deadly filoviruses, *New Engl. J. Med.*, 352, 2571–2573, 2005.

515. Ljungman, P., Prevention and treatment of viral infections in stem cell transplant recipients, *Br. J. Haematol.*, 118, 44–57, 2002.

516. Swann, I.L. and Dendra, J.R., Anemia, vitamin E deficiency and failure to thrive in an infant, *Clin. Lab. Haematol.*, 20, 61–63, 1998.

517. von Kries, R., Neonatal vitamin K prophylaxis: the Gordian knot still awaits untying, *Br. Med. J.*, 316, 161–162, 1998.

518. Zipursky, A., Prevention of vitamin K deficiency bleeding in newborns, *Br. J. Haematol.*, 104, 430–437 1999.

518a. Johnson, S.A., Birchall, J., Luckie, C. et al., Guidelines on the management of Waldenström macroglobulinaemia, *Br. J. Haematol.*, 132, 683–697, 2006.

519. British Committee for Standards in Haematology, Guidelines on oral anticoagulation, 3rd ed., *Br. J. Haematol.*, 101, 374–387, 1998.

520. Baglin, T.P., Keaney, D.M., and Watson, H.G., for British Committee for Standards in Haematology, Guidelines on oral anticoagulation (warfarin): third edition–2005 update, *Br. J. Haematol.*, 132, 277–285, 2006.

521. British Committee for Standards in Haematology, Recommendations for patient self-management of oral anticoagulation, *Br. Med. J.*, 323, 985–998, 2001.

522. Fitzmaurice, D.A., Murray, E.T., McCahon, D. et al., Self management of oral anticoagulation; randomised trial, *Br. Med. J.*, 331, 1057–1059, 2005.

523. Bacon, P.A., The spectrum of Wegener's granulomatosis and disease relapse, *New Engl. J. Med.*, 252, 330–332, 2005.

524. Harington, T., Kuehnert, M.J., Kamel, H. et al., West Nile virus infection transmitted by blood transfusion, *Transfusion*, 43, 1018–1022, 2003.

525. Imawato, M., Jernigan, D.B., Guasch, A. et al., Transmission of West Nile virus from an organ donor to four transplant recipients, *New Engl. J. Med.*, 348, 2196–2203, 2003.

526. Pealer, L.N., Marfin, A.A., Petersen, L.R. et al., Transmission of West Nile virus through blood transfusion in the United States in 2002, *New Engl. J. Med.*, 349, 1236–1245, 2003.

527. Saldanha, J., Shead, S., Heath, A. et al., Collaborative study to evaluate a working reagent for West Nile virus RNA detection by nucleic acid testing, *Transfusion*, 44, 97–102, 2004.

528. Busch, M.P., Caglioti, M.T., Robertson, E.F. et al., Screening for West Nile virus RNA by nucleic acid amplification testing, *New Engl. J. Med.*, 353, 460–467, 2005.

529. de Oliviera, A.M., Beecham, B.D., Montgomery, S.P. et al., West Nile virus blood transfusion related infection despite nucleic acid testing, *Transfusion*, 44, 1695–1699, 2004.

530. Stramer, S.L., Fang, C.T., Foster, G.A. et al., West Nile virus among blood donors in the United States, 2003 and 2004, *New Engl. J. Med.*, 353, 451–459, 2005.

531. Petersen, L.R. and Epstein, J.S., Problem solved? West Nile virus and transfusion safety, *New Engl. J. Med.*, 353, 516–517, 2005.

532. Featherstone, C., How does one gene cause Wiskott-Aldrich syndrome? *Lancet*, 348, 950, 1996.

533. Remold-O'Donnell, E., Rosa, F.S., and Kenney, D.M., Defects in Wiskott-Aldrich syndrome, *Blood*, 87, 2621–2631, 1996.

Tables Reference Range

Andrew, M., Paes, B., Milner, R. et al., Development of the human coagulation system in the full-term infant, *Blood*, 70, 165–172, 1987.

Andrew, M., Paes, B., Milner, R. et al., Development of the human coagulation system in the premature infant, *Blood*, 72, 1651–1657, 1988.

Dallman, P.R., Blood and blood-forming tissues, in *Pediatrics*, 16th ed., Rudolph, A., Ed., Appleton-Century-Crofts, Norwalk, CT, 1977, p. 111.

Dallman, P.R., Blood and blood-forming tissues, in *Pediatrics*, 16th ed., Rudolph, A., Ed., Appleton-Century-Crofts, Norwalk, CT, 1977, p. 1178.

Hinchcliffe, R.F. and Lilleyman, J.S., Eds., *Practical Paediatric Haematology: a Laboratory Worker's Guide to Blood Disorders in Children*, John Wiley & Sons, New York, 1987.

Kobayashi, R., Ariga, T., Nonoyama, S. et al., Outcome in patients with Wiscott-Aldrich syndrome following stem cell transplantation: an analysis of 57 patients in Japan, *Br. J. Haematol.*, 135, 362–366, 2006.

Koerper, M.A. and Dallman, P.R., Serum iron concentrations and transferrin saturation are lower in normal children than in adults, *Pediatr. Res.*, 11, 473, 1977.

Lanzkowsky, P., *Manual of Pediatric Hematology*, 2nd ed., Churchill Livingstone, New York, 1995.

Oski, F.A. and Naiman, J.L., Normal blood values in the newborn period, in *Hematologic Problems in the Newborn*, 3rd ed., W.B. Saunders, Philadelphia, 1982.

Price, D.C. and Ries, C., Hematology, in *Nuclear Medicine in Clinical Pediatrics*, Handmaker, H. and Lowenstein, J.M., Eds., Society of Nuclear Medicine, New York, 1975, p. 279.

Rosse, C., Kraemar, M.J., Dillon, T.L. et al., Bone marrow cell populations of normal infants: the predominance of lymphocytes, *J. Lab. Clin. Med.*, 89, 1225–1240, 1977.

Saarinen, U.M. and Siimes, M.A., Developmental changes in serum iron, total-iron-binding-capacity, and transferrin in infancy, *J. Pediatr.*, 91, 875–877, 1977.

Schröter W. and Naffz, C., Diagnostic significance of hemoglobin F and A2 levels in homo- and heterozygous b-thalassemia during infancy, *Helvetica Paediatrica Acta*, 36, 519, 1981.

Shojania, A. and Gross, S., Folic acid deficiency and prematurity, *J. Pediatr.*, 64, 323, 1964.

Zaizov, R. and Matoth, Y., Red cell values on the first postnatal day during the last 16 weeks of gestation, *Am. J. Hematol.*, 1, 276, 1976.

Internet References

(1) www.shotuk.org

(2) http://www.fda.gov/cber/ltr/trali101901.htm

(3) https://secure.blood.co.uk/c11_cant.asp

(4) http://www.redcross.org/services/biomed/0,1082,0_557_,00.html

(5) http://www.researchd.com/rdicdabs/cdindex.htm (dated 7/3: listings up to CD247).

(6) http://pathologyoutlines.com/cdmarkers.html (dated 5/4: listings up to CD266).

(7) http://www.immunologylink.com/CDantigen.htm (dated 2000: listings up to CD166).

(8) http://flowsite.hitchcock.org/AboutFlow/cd_table.html (dated 9/2:listings up to CD166).

(9) http://www.cochrane.org/reviews/en/ab001208.html

(10) http://www.copewithcytokines.de/cope.cgi (last updated July 2003).

(11) http://www.rndsystems.com/asp/b_index.asp ("Cytokine Bulletins" and "Cytokine Mini-Reviews," R&D Systems Inc.)

(12) http://www.rndsystems.com/asp/g_sitebuilder.asp?BodyId=2 ("Cytokine Bulletins" and "Cytokine Mini-Reviews," R&D Systems Inc.)

(13) http://www.nejm.org

(14) http://www.anthonynolan.org.uk/HIG/nomen/reports/nomen_reports.html

(14a) htpp//www.unaids.org/epi2005/index.html

(15) www.cdc.gov/travel/yb/toc.htm/

(16) www.who.int/ith/

(17) http://www.cjd.ed.ac.uk/

(18) http://www.cancerimmunity.org/statics/databases.htm

(19) www.cdc.gov/ncidod/dvbid/westnile

(20) www.cdc.gov/mmwr/preview/mmwrhtml/rr5514al.htm

Bibliography

Adler, M.W., Ed., *ABC of AIDS*, 5th ed., British Medical Journal Publishing Group, London, 2001.

American Association of Blood Banks, *Technical Manual*, 14th ed., Brecher, M.E., Ed., AABB, Bethesda, MD, 2002.

Bain, B.J., *Blood Cells: A Practical Guide*, J.B. Lippincott, Philadelphia, 1989.

Bain, B.J., Classification of acute leukemia: the need to incorporate cytogenetic and molecular genetic information, *J. Clin. Pathol.*, 51, 420–423, 1998.

Beutler, E., Lichtman, M.A., Coller, B.S., Kipps, T.J., and Seligsohn, U., Eds., *William's Hematology*, McGraw-Hill, New York, 2001.

Bick, R.L., Bennett, J.M., Brynes, R.K. et al., Eds., *Hematology, Clinical and Laboratory Practice*, Mosby, St. Louis, 1993.

Bloom, W. and Fawcett, D.W., Eds., *A Textbook of Histology*, Chapman and Hall, New York, 1994.

British National Formulary, 51, British Medical Association and Royal Pharmaceutical Society, London, 2005.

Chanarin, I., *Megaloblastic Anaemias*, 1st ed., Blackwell, Oxford, 1969.

Colman, R.W., Hirsh, J., Marder, V.J., and Salzman, E.W., *Hemostasis and Thrombosis*, J.B. Lippincott, Philadelphia, 1993.

Dacie, J.V., The Haemolytic Anaemias, Vol. 1, *The Hereditary Haemolytic Anaemias*, 3rd ed., Churchill Livingstone, London, 1995.

Dacie, J.V., The Haemolytic Anaemias, Vol. 2, *The Hereditary Haemolytic Anaemias*, 3rd ed., Churchill Livingstone, London, 1995.

Dacie, J.V., The Haemolytic Anaemias, Vol. 3, *The Autoimmune Haemolytic Anaemias*, 3rd ed., Churchill Livingstone, London, 1995.

Dacie, J.V., The Haemolytic Anaemias, Vol. 4, *Secondary or Symptomatic Haemolytic Anaemias*, Churchill Livingstone, London, 1995.

Daniels, G., *Human Blood Groups*, Blackwell, Oxford, 1995.

Delamore, I.W. and Lui Yin, J.A., *Haematological Aspects of Systemic Disease*, Balliere Tindall, London, 1990.

DeVita, V.T., Jr., Hellman, S., and Rosenberg, S.A., Eds., *Cancer: Principles and Practice of Oncology*, J.B. Lippincott, Philadelphia, 1995.

DeVita, V.T., Jr., Hellman, S., and Rosenberg, S.A., Eds., *AIDS Etiology: Diagnosis, Treatment and Prevention*, 4th ed., Lippincott-Raven, Philadelphia, 1997.

Garraty, G., Ed., *Immunobiology of Transfusion Medicine*, Marcel Dekker, New York, 1994.

Greer, J.P., Foerster, J., Lukens, J.N., Rodgers, G.M. et al., *Wintrobe's Clinical Hematology*, Lippincott, Williams and Wilkens, Philadelphia, 2004.

Hardisty, R.M. and Weatherall, D.J., *Blood and Its Disorders*, 2nd ed., Blackwell, Oxford, 1982.

Hinchcliffe, R.F. and Lilleyman, J.S., Eds., *Practical Paediatric Haematology: A Laboratory Worker's Guide to Blood Disorders in Children*, John Wiley & Sons, New York, 1987.

Hoffbrand, A.V. and Pettit, J.E., *Essential Haematology*, 4th ed., Blackwell, Oxford, 2001.

Jandl, J.H., *Blood: Textbook of Hematology*, 2nd ed., Little, Brown, Boston, 1996.

Jenkins, G.J. and Williams, J.D., Eds., *Infection and Haematology*, Butterworth & Heinemann, Oxford, 1994.

Kuby, J., *Immunology*, 2nd ed., W.H. Freeman, New York, 1994.

Lachman, P., Peters, K., and Walport, M.J., Eds., *Clinical Aspects of Immunology*, Blackwell, Oxford, 1993.

Lewis, S.M., Bain, B.J., and Bates, I., *Dacie and Lewis Practical Haematology*, 10th ed., Churchill-Livingstone, Elsevier, Philadelphia, 2006.

Malpas, J.S., Bergsagel, D.E., Kyle, R.A., and Anderson, K.C., *Myeloma: Biology and Management*, Oxford University Press, New York, 1995; reprint, Saunders, Philadelphia, 2004.

Mollison, P.L., Engelfriet, C.P., and Contreras, M., Eds., *Blood Transfusion in Clinical Medicine*, 10th ed., Blackwell, London, 1997.

Mufti, G.J., Flandrin, G., Schaefer, H.-E., Sandberg, A.A., and Kanfer, E.J., *An Atlas of Malignant Hematology*, Martin Dunitz, London, 1996.

Pamphilon, D.H., Ed., *Modern Transfusion Medicine*, CRC Press, Boca Raton, FL, 1995.

Ratnoff, O.D. and Forbes, C.D., *Disorders of Hemostasis*, W.B. Saunders, Philadelphia, 1996.

Rédel, G.E., *Encyclopedic Dictionary of Genetics, Genomics and Proteomics*, Wiley-Liss, New York, 2003.

Roitt, I.M., Brostoff, J., and Male, D.I.K., *Immunology*, 3rd. ed., Mosby, St. Louis, 1993.

Samter, M., Frank, M.M., Austen, K.F., and Clamen, H.N., *Immunological Diseases*, Little, Brown, Boston, 1988.

Sande, M.A. and Velberding, P.A., Eds., *The Medical Management of AIDS*, 2nd ed., W.B. Saunders, Philadelphia, 1990.

Sergeant, G., *Sickle Cell Disease*, 2nd ed., Oxford University Press, Oxford, 1992.

United States Pharmacopoeia, Rockville, MD, 2006.

World Health Organization. *Classification of Tumours of Haematopoietic and Lymphoid Tissues*, Jaffe, E.S., Harris, N.L., Stein, H., and Vardiman, J.W. IARC Press, Lyon, France, 2001.

Index

A Library